Universe

UNIVERSE

SEVENTH EDITION

Roger A. Freedman
University of California, Santa Barbara

William J. Kaufmann III
San Diego State University

W. H. FREEMAN AND COMPANY
NEW YORK

ACQUISITIONS EDITORS:	Patrick Farace, Valerie Raymond
DEVELOPMENT EDITOR:	Randi Rossignol
NEW MEDIA AND SUPPLEMENTS EDITORS:	Brian Donnellan, Victoria Anderson
PROJECT EDITOR:	Bradley Umbaugh
MARKETING MANAGER:	Mark Santee
COVER AND TEXT DESIGNER:	Victoria Tomaselli
ILLUSTRATIONS:	Fine Line Illustrations
ILLUSTRATION COORDINATOR:	Cecilia Varas
PHOTO EDITOR AND RESEARCHER:	Vikii Wong
PRODUCTION MANAGER:	Julia DeRosa
COPY EDITOR AND INDEXER:	Louise B. Ketz
COMPOSITION:	Sheridan Sellers – W. H. Freeman Electronic Publishing Center
MANUFACTURING:	RR Donnelley & Sons Company

Library of Congress Cataloging-in-Publication Control Number: 2004103077
ISBN: 0-7167-9884-0   (EAN: 9780716798842)

W. H. Freeman and Company
41 Madison Avenue
New York, NY 10010
Houndmills, Basingstoke RG21 6XS, England

www.whfreeman.com

DEDICATION

Dedicated to Lee Johnson Kaufmann
and Caroline Robillard-Freedman,
strong survivors

and to the memory of
S/Sgt. Ann Kazmierczak Freedman, WAC

Roger A. Freedman is on the faculty of the Department of Physics at the University of California, Santa Barbara. He grew up in San Diego, California and was an undergraduate at the University of California campuses in San Diego and Los Angeles. He did his doctoral research in nuclear theory and its astrophysical applications at Stanford University under the direction of Professor J. Dirk Walecka. Dr. Freedman came to UCSB in 1981 after three years of teaching and doing research at the University of Washington.

(Regina Rivera)

Dr. Freedman holds a commercial pilot's license, and when not teaching or writing he can frequently be found flying with his wife, Caroline. He has flown across the United States and Canada.

William J. Kaufmann III was the author of the first four editions of *Universe*. Born in New York City on December 27, 1942, he often visited the magnificent Hayden Planetarium as he was growing up. Dr. Kaufmann earned his bachelor's degree *magna cum laude* in physics from Adelphi University in 1963, a master's degree in physics from Rutgers in 1965, and a Ph.D. in astrophysics from Indiana University in 1968. At 27 he became the youngest director of any major observatory in the United States when he took the helm of the Griffith Observatory in Los Angeles. During his career he also held positions at San Diego State University, UCLA, Caltech, and the University of Illinois. Throughout his professional life as a scientist and educator, Dr. Kaufmann worked to bridge the gap between the scientific community and the general public to help the public share in the advances of astronomy. A prolific author, his many books include *Black Holes and Warped Spacetime, Relativity and Cosmology, The Cosmic Frontiers of General Relativity, Exploration of the Solar System, Planets and Moons, Stars and Nebulas, Galaxies and Quasars,* and *Supercomputing and the Transformation of Science.* Dr. Kaufmann died in 1994.

Contents Overview

Contents

PART I INTRODUCING ASTRONOMY

1 | Astronomy and the Universe 1

2 | Knowing the Heavens 20

3 | Eclipses and the Motion of the Moon 42

4 | Gravitation and the Waltz of the Planets 61

PART II PLANETS AND MOONS

(NASA: J. Bell, Cornell University and M. Wolff, SSI)

PART III STARS AND STELLAR EVOLUTION

(Hubble Heritage Team, AURA/STScI/NASA)

PART IV GALAXIES AND COSMOLOGY

(Simulations performed at the National Center for Supercomputer Applications by Andrey Kravtsov/University of Chicago and Anatoly Klypin/NMSU)

Preface

From its first edition, Bill Kaufmann's *Universe* has been a tremendous help to my introductory astronomy students. They have benefited from Bill's compelling prose, his clear exposition, and his emphasis on *how* astronomers learn about the universe. I've benefited as well. The in-depth presentation in *Universe* — never "watered down," yet always accessible to the beginning student — has made it possible for me to teach a more satisfying course that gives students a real insight into the nature of science. And so it's been my special pleasure to carry on Bill's work by developing *Universe* into its new Seventh Edition.

New features make the Seventh Edition a more powerful learning tool

When I took over the stewardship of *Universe* in 1996, my goals were to continue its emphasis on the process of scientific discovery; to make it easier for students to grasp the concepts of astronomy; to help students develop the skills needed to solve quantitative problems; and to make the book as modern, lively, and interesting as I possibly could. Above all, I've worked to enhance the strengths that have made *Universe* a standard in astronomy education, including its particular focus on *how scientists work and think*. With this goal in mind, I've added a number of new features to this edition.

Increased emphasis on ▶ comparative planetology

Two new chapters — Chapter 7, *Comparative Planetology I: Our Solar System*, and Chapter 8, *Comparative Planetology II: The Origin of Our Solar System* — provide a comparative overview of the planets. Chapter 7 discusses what we learn from cratering of terrestrial worlds and from studies of planetary magnetic fields. Chapter 8 includes recent ideas about planetary origins, including both the core accretion and disk instability models of Jovian planet formation. A new Chapter 14, *Jupiter and Saturn: Lords of the Planets*, is built around comparisons between the two largest planets.

7 Comparative Planetology I: Our Solar System

Fifty years ago, astronomers knew precious little about the solar system. Even the best telescopes provided images of the planets that were frustratingly hazy and indistinct. Of asteroids, comets, and the satellites of the planets, we knew even less.

Today, our knowledge of the solar system has grown many thousandfold, due almost entirely to robotic spacecraft. (The illustration shows one such robotic explorer, the joint NASA/European Space Agency spacecraft *Cassini*, as it approached Saturn in 2004.) Spacecraft have flown at close range past all the planets except Pluto, revealing details that astronomers of an earlier generation could only dream about. We have landed spacecraft on the Moon, Venus, and Mars and dropped a probe into the immense atmosphere of Jupiter. This is truly the golden age of solar system exploration.

In this chapter we paint in broad outline our present understanding of the solar system. We will see that the planets come in a variety of sizes and chemical compositions. There is also rich variety among the moons of the planets and among smaller bodies we call asteroids, comets, and Kuiper belt objects. We will investigate the nature of craters on the Moon and other worlds of the solar system. And by exploring the magnetic fields of planets, we will be able to peer inside other worlds and learn about their interior compositions.

In Chapter 8 we will put together these observations to build a picture of how the solar system formed some

Questions to Guide Your Reading

As you read the sections of this chapter, look for the answers to the following questions:

7-1 Are all the other planets similar to Earth, or are they very different?

7-2 Do other planets have moons like Earth's Moon?

7-3 How do astronomers know what the other planets are made of?

7-4 Are all the planets made of basically the same material?

8 Comparative Planetology II: The Origin of Our Solar System

What did our solar system look like before the planets were fully formed? The answer may lie in this remarkable image from the Hubble Space Telescope. At the center of the image is an immense disk of gas and dust centered on the young, Sunlike star HD 141569A. (The light from the star itself was blocked out within the telescope to make the rather faint disk more visible.) Astronomers strongly suspect that in its infancy our own Sun was surrounded by such a disk and that the planets coalesced from it.

In this chapter we will examine the evidence that led astronomers to this picture of the origin of the planets. We will see how the abundances of different chemical elements in the solar system indicate that the Sun and planets formed from a thin cloud of interstellar matter some 4.56 billion years ago, an age determined by measuring the radioactivity of meteorites. We will learn how the nature of planetary orbits gives important clues to what happened as this cloud contracted and how evidence from meteorites reveals the chaotic conditions that existed within the cloud. And we will see how this cloud eventually evolved into the solar system that we see today.

In the past few years astronomers have been able to test this picture of planetary formation by examining disks around young stars (like the one in the accompanying image). Most remarkably of all, they have discovered planets in orbit around dozens of other stars. These recent observations provide valuable information about how our own system of planets came to be.

Planets are thought to form within the disks surrounding young stars. (NASA, M. Clampin (STScI), H. Ford (JHU), G. Illingworth (UCO/Lick), J. Krist (STScI), D. Ardila (JHU), D. Golimowski (JHU), the ACS Science Team and ESA)
R I **V** U X G

Questions to Guide Your Reading

As you read the sections of this chapter, look for the answers to the following questions:

8-1 What must be included in a viable theory of the origin of the solar system?

8-2 Why are some elements (like gold) quite rare, while others (like carbon) are more common?

8-3 How do we know the age of the solar system?

8-4 How do astronomers think the solar system formed?

8-5 Did all of the planets form in the same way?

8-6 Are there planets orbiting other stars? How do astronomers search for other planets?

◀ **Enhanced focus on the process of astronomy**

It's essential that students learn not only *what* we know about our fascinating physical universe but also *how* we know it. The scientific way of knowing, based on observation and continual questioning, is at the heart of both the organization and the features of *Universe*. To help underscore the importance of the scientific method, I've added new **Science in Process** statements. These highlight sections that have a special focus on the process of astronomical discovery.

A more intuitive art program ▶

Students often need help interpreting complex figures. To help them, I've added balloon captions to many of the figures. The balloons contain explanatory text that clarifies the most important features of each figure.

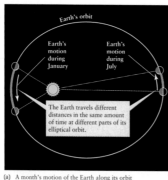

(a) A month's motion of the Earth along its orbit

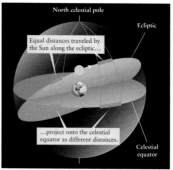

(b) A day's motion of the Sun along the ecliptic

figure 2-20

Why the Sun Is a Poor Timekeeper There are two main reasons that the Sun is a poor timekeeper. (a) The Earth's speed along its orbit varies during the year. It moves fastest when closest to the Sun in January and slowest when farthest from the Sun in July. Hence, the apparent speed of the Sun along the ecliptic is not constant. (b) Because of the tilt of the Earth's rotation axis, the ecliptic is inclined with respect to the celestial equator. Therefore, the projection of the Sun's daily progress along the ecliptic onto the celestial equator (shown in blue) varies during the year. This causes further variations in the length of the apparent solar day.

◀ **S. T. A. R. — a new problem-solving rubric**

Many students find it easier to solve quantitative problems when they know what steps to follow. I've rewritten all the worked examples in *Universe* (found in boxes called **Tools of the Astronomer's Trade**) to follow a logical and consistent sequence of steps called **S.T.A.R.**: assess the Situation, select the Tools, find the Answer, and Review the answer and explore its significance. Students who follow these steps in their own work will more rapidly become proficient problem solvers.

Observing Projects ▶

In addition to the *Starry Night Backyard*™ planetarium software, the CD-ROM that comes with this book includes a special version of *Deep Space Explorer*™ from Space.com. *Deep Space Explorer*™ allows the user to observe a tremendous number of astronomical objects from any perspective and to "fly" through space while doing so. Every chapter of *Universe* includes one or more **Observing Projects** that use *Deep Space Explorer*™.

Observing Projects

43. Use the *Deep Space Explorer*™ program to examine the terrestrial planets Mercury, Venus, Earth, and Mars and the asteroid Ceres. In the left-hand part of the window, under the heading **Solar System** select **Explore.** Then click on the name of each planet or asteroid to view it in detail. You can zoom in and zoom out using the buttons at the upper left of the window (an upward-pointing triangle and a downward-pointing triangle). You can also rotate the planet or asteroid by putting the mouse cursor over the image, holding down the mouse button, and moving the mouse. (On a two-button mouse, hold down the left mouse button.) Describe each planet or asteroid's appearance. Which planet or planets have clouds? Which planet or asteroid shows the heaviest cratering? Which show evidence of liquid water?

Collaborative Exercises

47. A scientific theory is fundamentally different than the everyday use of the word "theory." List and describe any three scientific theories of your choice and creatively imagine an additional three hypothetical theories that are not scientific. Briefly describe what is scientific and what is nonscientific about each of these theories.

48. Angular measure describes how far apart two objects appear to an observer. From where you are currently sitting, estimate the angular distance between the floor and the ceiling at the front of the room you are sitting in, the angular distance between the two people sitting closest to you, and the angular size of a clock or an exit sign on the wall. Draw sketches to illustrate each answer and describe how each of your answers would change if you were standing in the very center of the room.

49. Astronomers use powers of ten to describe the distances to objects. List an object or place that is located at very roughly each of the following distances from you: 10^{-2} m, 10^0 m, 10^1 m, 10^3 m, 10^7 m, 10^{10} m, and 10^{20} m.

◀ Collaborative Exercises for group problem-solving

New **Collaborative Exercises** are designed for the ever-increasing number of instructors who have their students work in groups to solve problems. By seeing how others think and learn, students can further develop their individual problem-solving skills.

Up-to-date information that shows ▶ science in the making

I've brought *Universe* up to date with the latest discoveries, ideas, and images from all aspects of the subject we love. These include:

- New material on Kuiper belt objects, Pluto's status as a planet, and relic magnetism on Mars (Chapter 7)

- Updated discussion of the greenhouse effect, including the latest data on CO_2 concentrations and global temperatures (Chapter 9)

- New information on Martian water, the Martian climate, and results from *Mars Odyssey*, *Mars Express*, and the *Mars Exploration Rovers* (Chapters 13 and 30)

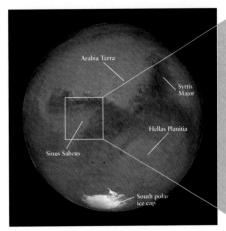

(a) Mars from the Hubble Space Telescope

(b) Closeup of Sinus Sabeus region

Figure 13-4   R I **V** U X G

Martian Craters (a) This image, made during the favorable opposition of 2003, shows cratered regions on the side of Mars opposite to that in the image that opens this chapter. Arabia Terra is dotted with numerous flat-bottom craters. Syrtis Major was first identified by Christiaan Huygens in 1659. A single titanic impact carved out Hellas Planitia, which is 5 times the size of Texas. (b) This mosaic of images from the *Viking Orbiter 1* and *2* spacecraft shows an extensively cratered region located south of the Martian equator. (a: NASA; J. Bell, Cornell University; and M. Wolff, SSI; b: USGS)

- Recent findings about Jupiter's belts and zones and Jupiter's magnetosphere (Chapter 14)

- New ideas about the possible origins of Uranus and Neptune (Chapter 16)

- An expanded and updated discussion of the solar neutrino problem, including results from SNO (Chapter 18)

Figure 14-12 R **I** V U X G

A Radio View of Jupiter Instruments on board the *Cassini* spacecraft recorded this false-color map of decimetric emission from Jupiter at a wavelength of 2.2 cm. The emission comes from a region elongated parallel to the planet's magnetic equator, which is tilted relative to its geographic equator. A visible-light image of Jupiter is superimposed on the radio image. (NASA/JPL)

- Updated information on the supermassive black hole at the center of the Milky Way and the Milky Way's closest satellite galaxy, the Canis Major Dwarf (Chapters 25 and 26)

- New data relating to the accelerating universe from WMAP, supernova studies, and cluster observations (Chapter 28)

- New material on primordial sound waves and polarization of the cosmic background radiation (Chapters 28 and 29)

WEB LINK 25.1   VIDEO 25.1   **Figure 25-23** R **I** V U X G

Stars Orbiting Sagittarius A* The diagram shows the results of ten years' observations of S2, one of several stars that orbit an unseen object at the position of the Sagittarius A* radio source. (The date of each observation is given in fractions of a year: for example, 2001.50 is July 1, 2001.) Note how rapidly S2 moved during 2002, when it was closest to Sagittarius A*. The actual plane of the orbit is not face-on to us, so S2 did not come as close to Sagittarius A* as the diagram might suggest. (ESO)

A number of popular features remain from previous editions

Questions to Guide Your Reading

As you read the sections of this chapter, look for the answers to the following questions:

7-1 Are all the other planets similar to Earth, or are they very different?

7-2 Do other planets have moons like Earth's Moon?

7-3 How do astronomers know what the other planets are made of?

7-4 Are all the planets made of basically the same material?

7-5 What is the difference between an asteroid and a comet?

7-6 Why are craters common on the Moon but rare on the Earth?

7-7 Why do interplanetary spacecraft carry devices for measuring magnetic fields?

7-8 Do all the planets have a common origin?

◀ **Questions to Guide Your Reading**

A brief overview and question set guide the reader into each chapter's topics. The answer to each question can be found within the corresponding section of the chapter.

Wavelength tabs ▶

In addition to using ordinary telescopes, astronomers rely heavily on special telescopes that are sensitive to nonvisible forms of light. To help students appreciate what astronomers learn from these different kinds of observations, wavelength tabs appear with all of the images in *Universe*. The highlighted letter on each tab shows whether the image was made with Radio waves, Infrared radiation, Visible light, Ultraviolet light, X rays, or Gamma rays. The images of the Martian south polar ice cap shown here are visible light, V, and infrared, I, images.

(a) Visible-light image   (b) Infrared image of carbon dioxide ice   (c) Infrared image of water ice

R I **V** U X G   R **I** V U X G   R **I** V U X G

Figure 30-4
Water at the Martian South Pole (a) This visible-light image shows the south polar cap of Mars, but does not indicate its chemical composition. But by using a camera tuned to different wavelengths of infrared light, the *Mars Express* spacecraft was able to identify the distinctive reflections of (b) an upper layer of carbon dioxide ice and (c) a lower layer of water ice. Other observations have shown that there is also water ice at the Martian north pole. (ESA-OMEGA)

> **CAUTION** ⚠
> Note that while Figure 7-1 shows the orbits of the planets, it does not show the planets themselves. The reason is simple: If Jupiter, the largest of the planets, were to be drawn to the same scale as the rest of this figure, it would be a dot just 0.0002 cm across—about $^1/_{300}$ of the width of a human hair and far too small to see without a microscope. The planets themselves are *very* small compared to the distances between the planets. Indeed, while an airliner traveling at 1000 km/h (620 mi/h) can fly around the Earth in less than two days, at this speed it would take 17 *years* to fly from the Earth to the Sun. The solar system is a very large and very empty place!

◀ **Confronting misconceptions**

Many people think that the Earth is closer to the Sun in summer than in winter and that the phases of the Moon are caused by the Earth's shadow falling on the Moon. But in fact these "commonsense" ideas are incorrect! (Chapters 2 and 3 explain why.)

Throughout *Universe*, paragraphs signaled by the **Caution** icon ⚠ alert the reader to conceptual pitfalls like these.

Bringing astronomy down to Earth ▶

To learn astronomical ideas, it can be helpful to relate them to everyday experiences on Earth. Throughout *Universe* you'll find **analogy** paragraphs that relate, for example, the motions of the planets to children on a merry-go-round and the bending of light in a telescope lens to a car driving on sand.

The Heavens on the Earth boxes also show how the principles that astronomers use can explain everyday phenomena, from why diet soft drink cans float to the color of the daytime sky.

> **ANALOGY**
> To see one simple answer to this question, notice that a large turkey or roast taken from the oven will stay warm inside for hours, but a single meatball will cool off much more rapidly. The reason is that the meatball has more surface area relative to its volume, and so it can more easily lose heat to its surroundings. A planet or satellite also tends to cool down as it emits electromagnetic radiation into space (see Section 5-3); the smaller the planet or satellite, the greater its surface area relative to its volume, and the more readily it can radiate away heat. Both the Earth and Moon were probably completely molten when they first formed, but because the Moon (diameter 3476 km) is so much smaller than the Earth (diameter 12,756 km), it has lost much of its internal heat and has a much more solid interior.

Key Words		
Terms preceded by an asterisk (*) are discussed in the Boxes.	*escape speed, p. 156	magnetometer, p. 161
asteroid, p. 157	ices, p. 153	meteoroid, p. 158
asteroid belt, p. 157	impact crater, p. 158	minor planet, p. 157
average density, p. 150	Jovian planet, p. 149	molecule, p. 152
chemical composition, p. 151	*kinetic energy, p. 155	spectroscopy, p. 152
comet, p. 157	Kuiper belt, p. 157	terrestrial planet, p. 149
dynamo, p. 161	Kuiper belt object, p. 157	
	liquid metallic hydrogen, p. 162	

◀ **Key Words**

A list of key words appears at the end of each chapter, along with the number of the page where each word is introduced.

◀ Key Ideas

Students can get the most out of these brief chapter summaries by using them in conjunction with the notes they take while reading.

Questions and Projects

Items from these sections are designed to be assigned as homework. The questions help students review and apply concepts from the text. Some ask students to analyze images in the text or to evaluate how the mass media portray concepts in astronomy. *Web/CD-ROM Questions* challenge students to work with animations and physical concepts on the *Universe* CD-ROM or Web site, or to research topics on the World Wide Web. *Problem-Solving Tips and Tools* provide guidance for the more advanced questions. *Observing Tips and Tools* accompany many *Observing Projects*, which make frequent use of the *Starry Night Backyard*™ planetarium software and *Deep Space Explorer*™ software provided on the CD-ROM. Near the end of the book is a section of **Answers to Selected Questions.**

SETH SHOSTAK

Searching for Extraterrestrial Life

Dr. Seth Shostak is Senior Astronomer at the SETI Institute in California, and participates in the world's most comprehensive search for extraterrestrial intelligence. He has a physics degree from Princeton University, and a Ph.D. in astronomy from the California Institute of Technology. Dr. Shostak studied galaxies at observatories and universities worldwide before joining the SETI Institute in 1990. In addition to his research, he keeps busy promoting science to the general public with lectures, writings, and a weekly radio show.

For more than four decades, a small number of researchers have been training their telescopes on the skies, looking for signals that would prove that someone out there is at least as smart as we are. It's a project known as SETI, the Search for Extraterrestrial Intelligence, and it's something I've been doing for more than a dozen years.

Of course, looking for aliens with aptitude is only a very small part of the hunt for extraterrestrial life in general. Our spacecraft continue to scout the rusty surface of Mars, hoping to find conclusive evidence that this planet once had (or maybe still has) some simple biology. We've also mounted robotic expeditions to a few of the Jovian moons, where cryptic life might possibly eke out a living. This continuing effort to uncover living things nearby is quite exciting for me. Even though such life would probably be less impressive than bathtub mold, the discovery of even a bit of biology on another world of our solar system would strongly suggest that cosmic life is as common as crabgrass.

world might be perceived as dangerous, and expect that the news would be covered up. My work in SETI has convinced me that this isn't true. There's no policy of secrecy, and no mechanism for enforcing it. Within a few days of finding an interesting signal, astronomers around the world (and their friends and relatives) would all know about it, and so would you. Every telescope on Earth would be pointed in the direction of the broadcast.

A few facts about the signal would quickly emerge. On the basis of the slow change in frequency, we could deduce some parameters of the aliens' world—for example, the length of their day and year. We could speedily identify their sun, allowing us to estimate the temperature of their home planet.

But what about the message? No self-respecting extraterrestrial would transmit an empty signal; it will surely contain information. If they are deliberately beaming their broadcasts our way—if they are trying to get in touch—then they may fashion the content in a way that just about anyone,

◀ Guest Essays

Several chapters in the textbook end with essays by scientists at the cutting edge of astronomy. Robert P. Kirshner of Harvard University describes his research on the accelerating universe in the Chapter 28 essay, John N. Bahcall of the Institute for Advanced Study wrote on the physics of the Sun in an essay that follows Chapter 18, and Seth Shostak of the SETI Institute, contributed the essay "Searching for Extraterrestrial Life" at the end of Chapter 30. Their essays give a sense of their excitement about astronomy. For a full list of essays, see the Contents Overview on pages vii and viii.

Media and Supplements Package

FOR STUDENTS

The Student CD-ROM

Packaged with this book are two unique software products:

- *Starry Night Backyard*™ is award-winning planetarium software that is the ideal electronic companion for an astronomy course. It features more than 1,000,000 stars, an **Atlas Mode** that provides a full view of the sky and the horizon, and the capability to view the sky from any point in the solar system.

- NEW! *Deep Space Explorer*™ reports the position, type, angular velocity, size, and shape of up to 28,000 of the nearest galaxies. Some 30,000 nearby stars are yours to explore, including those known to have planets. The program allows you to observe objects from any angle.

The Free Student Companion Web Site

www.whfreeman.com/universe7e
serves as a study guide, featuring:

- **Learning Objectives** for each chapter to help students formulate study strategies.

- **Self-Quizzes** offer randomized questions and answers with instant feedback referring to specific sections in the text, to help students study, review, and prepare for exams. Instructors can access results through an online database or they can have them e-mailed directly to their account.

- **Active Integrated Media Modules** ▶ take students deeper into key topics from the text.

Topics include:

— Determining Distances to the Stars and Beyond
— Doppler Effect
— Nearest Stars
— Relativistic Redshift

The Doppler Effect

Introduction | Animation | Calculator | Quiz

Observer sees redshift

Observer sees blueshift

Source velocity = 0.5 times the speed of light

Start/Stop | Velocity - | Velocity + | Reset

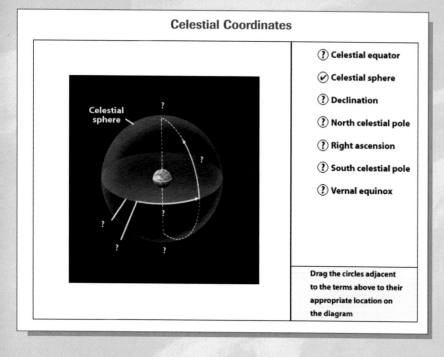

Celestial Coordinates

Celestial sphere

(?) Celestial equator

(✓) Celestial sphere

(?) Declination

(?) North celestial pole

(?) Right ascension

(?) South celestial pole

(?) Vernal equinox

Drag the circles adjacent to the terms above to their appropriate location on the diagram

◄ • **Interactive Drag & Drop Exercises** based on text illustrations help students grasp the vocabulary in context.

• **Flashcard exercises** offer help with vocabulary and definitions.

• *Starry Night Backyard*™ and *Deep Space Explorer*™ **Observing Exercises**—In-depth lab projects that make use of the student editions of *Starry Night Backyard*™ and *Deep Space Explorer*™ on the student CD-ROM.

• **Current Events in Astronomy** keep students up to date with Web links, photos, and information from a variety of resources. Updated every month.

• **Animations and Videos**, both original and NASA-created, are keyed to specific chapters. Updated regularly.

• **History of Astronomy Links.**

• **Print Links** that point users to external sites to find printed materials.

• **Web Links** to Internet sites that contain supplemental material about key topics found in the textbook.

• **Looking Deeper** articles in PDF format that extend topics discussed in the text. These include the search for gravitational waves, how to interpret the shapes of spectral lines, and discoveries about the inner core of the Earth.

FOR INSTRUCTORS

Interactive Web Modules for Any Astronomy Course

Astronomy Online, by Timothy F. Slater, University of Arizona

Booklet and Web site Activation Code, 0-7167-9669-4

Available in HTML and WebCT formats, this site can serve as either the online component for a lecture-based course or a complete online source for a distance learning course. The site combines content from Freeman's astronomy textbooks with powerful features such as online quizzing, animations and simulations, drag-and-drop exercises, active integrated media modules, Web links, and much more. The course can be used as is or tailored to meet the instructor's individual needs.

The HTML version can be accessed via:
www.whfreeman.com/aol

The WebCT version can be ordered via the WebCT content showcase at:
www.webct.com

On-Line Course Materials (WebCT, Blackboard)

As a service for adopters, we will provide content files in the appropriate online course format, including the instructor and student resources for this text. The files can be used as is, or can be customized to fit specific needs. Course outlines, pre-built quizzes, links, activities, and a whole array of materials are included.

Test Bank CD-ROM

(Windows, Macintosh, WebCT and Blackboard formats on one disc),
William J. F. Wilson and T. Alan Clark, University of Calgary, 0-7167-8771-7

More than 3000 multiple-choice questions, section-referenced, designated as being either "quantitative" or "qualitative." The easy-to-use CD version includes Windows, Mac, WebCT and Blackboard versions on a single disc, in a format that lets you add, edit, resequence, and print questions to suit your needs. The CD is also the access point for...

Diploma Online Testing, from the Brownstone Research Group
(Accessible via the test bank CD-ROM)
With Diploma, instructors can easily create and administer secure exams over a network and over the Internet, with questions that incorporate multimedia and interactive exercises. The program lets you restrict tests to specific computers or time blocks, and includes an impressive suite of gradebook and result-analysis features.

Online Quizzing, *powered by Questionmark*

(Accessible via the Web site, **www.whfreeman.com/universe7e**)

With Questionmark's Perception, instructors can easily and securely quiz students online using pre-written, multiple-choice questions for each text chapter (not from the test bank). Students receive instant feedback and can take the quizzes multiple times. Randomized questioning creates a different chapter question set each time a student takes a quiz, with answers reshuffled. Instructors can go into a protected Web site to view results by quiz, student, or question, or can get weekly results via e-mail.

PowerPoint® Lecture Presentation

A set of online lecture presentations created in PowerPoint allows instructors to tailor their lectures to suit their own needs using images and notes from the textbook.

Instructor's CD-ROM, 0-7167-8701-6

To help instructors create lecture presentations, Web sites, and other resources, this CD-ROM allows instructors to **search** and **export** all the resources contained below by key-term:

- **All text images in JPEG, GIF, and PowerPoint format**

- **Animations, Videos, and AIMMs** and more found on the CD-ROM/Web site.

- Instructor's Manual and Resource Guide, Mark Hollabaugh, Normandale College

    The printable, editable electronic *Instructor's Manual* includes worked-out solutions to all end-of-chapter problems, detailed chapter outlines, classroom-tested teaching hints and strategies. The extensive resource guide covers student and instructor reading materials, audiovisual material, and discussion/paper topics.

Overhead Transparencies, 0-7167-9629-5

100 full-color transparencies of key illustrations, photos, and tables from the text.

The Once and Future Cosmos

Scientific American Special Edition, 2002, 0-7167-5708-7

This new *Scientific American* Special Edition takes your students to the frontiers of cosmological inquiry, clarifying the new developments and what they mean.

New Light on the Solar System

Scientific American Special Edition, 2003, 0-7167-8630-6

This special edition of *Scientific American* provides the latest developments about our corner of the cosmos, in articles written by the experts who are leading the investigations.

◀ **Observing Projects using *Starry Night Backyard*™, William Wilson, T.A. Clark, and Marcel Bergman, University of Calgary,** 0-7167-5621-8

Available for packaging with the text, and compatible with both Mac and PC, this book contains seventeen comprehensive lab activities for *Starry Night Backyard*™.

Acknowledgements

I would like to thank my colleagues who carefully scrutinized the manuscript of this edition. This is a stronger and better textbook because of their conscientious efforts:

Robert R. J. Antonucci, *University of California, Santa Barbara*
Richard Bowman, *Bridgewater College*
Richard Christie, *Okanagan University College*
Steven Desch, *Arizona State University*
Juhan Frank, *Louisiana State University*
Joshua Gundersen, *University of Miami*
Melinda Hutson, *Portland Community College*
Darell Johnson, *Missouri Western State University*
Michael Joner, *Brigham Young University*
Lauren Jones, *Denison University*
Steve Kawaler, *Iowa State University*
William Keel, *University of Alabama*
Franck Marchis, *University of California, Berkeley*
Glen Miller, *Pasadena City College*
Edward M. Murphy, *University of Virginia*
Gerald Newsom, *Ohio State University*
Ronald P. Olowin, *Saint Mary's College*
Michael J. O'Shea, *Kansas State University*
J. Douglas Patterson, *Johnson County Community College*
Charles J. Peterson, *University of Missouri*
Richard J. Rand, *University of New Mexico*
Frederick Ringwald, *Cal State University, Fresno*
Douglas Scott, *University of British Columbia*
John Sepikas, *Pasadena City College*
Caroline Simpson, *Florida International University*
Larry K. Smith, *Snow College*

(JPL/NASA)

I would also like to thank the many people whose advice on previous editions has had an ongoing influence:

Robert Allen, *University of Wisconsin, La Crosse*; Robert R. J. Antonucci, *University of California, Santa Barbara*; Alice L. Argon, *Harvard-Smithsonian Center for Astrophysics*; Grant Bazan, *Las Positas Community College*; Omer Blaes, *University of California, Santa Barbara*; David Van Blerkom, *University of Massachusetts*; David Bruning, *University of Wisconsin, Parkside*; Spencer L. Buckner, *Austin Peay State University*; John M. Burns, *Mt. San Antonio College*; Bel Campbell, *University of New Mexico*; George A. Carlson, *Citrus College*; Bruce W. Carney, *University of North Carolina*; Bradley W. Carroll, *Weber State University*; George L. Cassiday, *University of Utah*; Karen G. Castle, *Diablo Valley*

College; Kim Coble, *University of California, Santa Barbara;* Malcolm Coe, *The University of Southampton;* John J. Cowan, *University of Oklahoma;* John E. Crawford, *McGill University;* Roger B. Culver, *Colorado State University;* David Dahl, *St. Olaf College;* Steve Danford, *University of North Carolina, Greensboro;* Robert Dick, *Carleton University;* James N. Douglas, *University of Texas at Austin;* Robert J. Dukes, Jr., *College of Charleston;* Robert A. Egler, *North Carolina State University;* Debra Meloy Elmegreen, *Vassar College;* David S. Evans, *University of Texas, Austin;* George W. Ficken, Jr., *Cleveland State University;* Andrew Fraknoi, *Astronomical Society of the Pacific;* Juhan Frank, *Louisiana State University;* Richard French, *Wellesley College;* Mary V. Frohne, *Western Illinois University;* Pamela L. Gay, *University of Texas, Austin;* Robert M. Geller, *University of California, Santa Barbara;* Steven Giddings, *University of California, Santa Barbara;* Donna Hurlbut Gifford, *Pima Community College;* Owen Gingerich, *Harvard University;* Paul F. Goldsmith, *University of Massachusetts;* J. Richard Gott III, *Princeton University;* Donald Gudehus, *Georgia State University;* Austin F. Gulliver, *Brandon University;* Buford M. Guy, *Cleveland State Community College;* Carl Gwinn, *University of California, Santa Barbara;* Bruce Hanna, *Old Dominion University;* Eric Harpell, *Las Positas Community College;* Bernadette Londak Harris, *B.C. Open University;* Charles L. Hartley, *Hartwick College;* Paul A. Heckert, *Western Carolina University;* Bill Herbst, *Wesleyan University;* Paul Hintzen, *California State University, Long Beach;* Paul Hodge, *University of Washington;* Douglas P. Hube, *University of Alberta;* Icko Iben, Jr., *Pennsylvania State University;* Scott B. Johnson, *Idaho State University;* Dave Kary, *Citrus College;* Michael Kaufman, *San Jose State University;* B. Alexander King III, *Austin Peay State University;* Arthur Kosowsky, *Rutgers University;* David Kriegler, *University of Nebraska, Omaha;* John K. Lawrence, *California State University, Northridge;* Andrew Lazarewicz, *Boston College;* Ntungwa Maasha, *Coastal Georgia Community College;* Marie E. Machacek, *Northeastern University;* Paul Marquard, *Casper College;* Laurence A. Marschall, *Gettysburg College;* Donald H. Martins, *University of Alaska, Anchorage;* Margaret Mazzolini, *Swinburne University;* Raymond C. McNeil, *Northern Kentucky University;* Dimitri Mihalas, *University of Illinois;* J. Ward Moody, *Brigham Young University;* Gerald H. Newsom, *The Ohio State University;* Robert M. O'Connell, *College of the Redwoods;* Erin O'Connor, *Allan Hancock College;* C. Robert O'Dell, *Rice University;* L. D. Opplinger, *Western Michigan University;* Fritz Osell, *Leeward Community College;* William Parke, *George Washington University;* Stanton J. Peale, *University of California, Santa Barbara;* John R. Percy, *University of Toronto;* Eric R. Peterson, *De Anza College;* Robert L. Pompi, *State University of New York, Binghamton;* Harrison B. Prosper, *Florida State University;* Carlton Pryor, *Rutgers University;* James L. Regas, *California State University, Chico;* Terry Retting, *University of Notre Dame;* Tina Riedinger, *University of Tennessee;* James A. Roberts, *University of North Texas;* Charles W. Rogers, *Southwestern Oklahoma State University;* Roger Romani, *Stanford University;* Michael Ruiz, *University of North Carolina, Asheville;* Kenneth S. Rumstay, *Valdosta State College;* Richard Saenz, *California Polytechnic State University;* Thomas F. Scanlon, *Grossmont College;* Richard L. Sears, *University of Michigan;* James Shea, *University of Wisconsin, Parkside;* Isaac Shlosman, *University of Kentucky;* Alan F.

Sill, *Texas Tech University;* Caroline Simpson, *Florida International University;* Michael L. Sitko, *University of Cincinnati;* David B. Slavsky, *Loyola University of Chicago;* Jim F. Smeltzer, *Northwest Missouri State University;* Larry K. Smith, *Snow College;* Darryl Stanford, *City College of San Francisco;* Ron Stoner, *Bowling Green State University;* Carol Strong, *University of Alabama, Huntsville;* Joseph S. Tenn, *Sonoma State University;* Donald Terndrup, *The Ohio State University;* Colin Terry, *Ventura College;* Gordon B. Thomson, *Rutgers University;* Charles R. Tolbert, *University of Virginia;* Virginia Trimble, *University of California, Irvine;* Thomas Tsung, *Grossmont College;* George F. Tucker, *The Sage Colleges;* Bruce A. Twarog, *University of Kansas;* Stephen Walton, *California State University, Northridge;* George Wegner, *Dartmouth College;* David Weinberg, *The Ohio State University;* Donat G. Wentzel, *University of Maryland;* Joseph C. Wesney, *Greenwich High School;* Nicholas Wheeler, *Reed College;* Raymond E. White, *University of Arizona;* A. B. Whiting, *U.S. Naval Academy;* Lynda Williams, *San Francisco State University;* Louis Winkler, *The University of Pennsylvania;* and Robert L. Zimmerman, *University of Oregon.*

I'm particularly grateful to Mark Hollabaugh, Bruce Jakosky, Bob Kirshner, and Seth Shostak, whose new essays for this edition greatly enhance its coverage, and to John Bahcall, Geoff Marcy, and John Ruhl, who were kind enough to update their essays in light of recent developments.

Many others have participated in the preparation of this book and I thank them for their efforts. I am particularly grateful to Patrick Farace, my acquisitions editor, and Randi Rossignol, my development editor. Victoria Anderson, my supplements editor, supervised the superb Web site and CD-ROMs. Special thanks go to Bradley Umbaugh, my project editor, who skillfully guided the manuscript through production and to Sheridan Sellers, W. H. Freeman Electronic Publishing Center, who composed the book. Vicki Tomaselli is credited with the book's beautiful design and, for the artwork, I thank Fine Line Illustrations and Cecilia Varas, who coordinated the illustrations. Vikii Wong, photo editor and researcher, was responsible for obtaining the many beautiful images found in the book. My thanks go to Louise B. Ketz, who copy-edited and indexed the book; to Michele Kornegay and Barbara Hults, who proofread various stages of the book; and to Julia de Rosa, production manager.

On a personal note, I would like to thank my father, Richard Freedman, for first cultivating my interest in space many years ago, and for his personal contributions to the exploration of the universe as an engineer for the Atlas and Centaur launch vehicle programs. Most of all, I thank my charming wife, Caroline, for putting up with my long nights slaving over the computer!

Although I have made a concerted effort to make this edition error-free, some mistakes may have crept in unbidden. I would appreciate hearing form anyone who finds an error or wishes to comment on the text. You may e-mail or write me.

(EUROPEAN SOUTHERN OBSERVATORY)

Roger A. Freedman
Department of Physics
University of California, Santa Barbara
Santa Barbara, CA 93106
airboy@physics.ucsb.edu

To the Student

HOW TO GET THE MOST FROM *UNIVERSE*

If you're like most students just opening this textbook, you're enrolled in one of the few science courses you'll take in college. As you study astronomy, you'll probably do relatively little reading compared to a literature or history course—at least in terms of the number of pages. But your readings will be packed with information, much of it new to you and (I hope) exciting. You can't read this textbook like a novel and expect to learn much from it. Don't worry, though. I wrote this book with you in mind. In this section I'll suggest how *Universe* can help you succeed in astronomy, and take you on a guided tour of the book and media.

Apply these techniques to studying astronomy

- **Read before a lecture** You'll get the most out of your astronomy course if you read each chapter *before* hearing a lecture about its subject matter. That way, many of the topics will already be clear in your mind, and you'll understand the lecture better. You'll be able to spend more of your listening and note-taking time on the more challenging ideas presented in the lecture.

- **Take notes as you read and make use of office hours** Keep a notebook handy as you read, and write down the key points of each section so that you can review them later. If any parts of the section don't seem clear on first reading, make a note of them, too, including the page numbers. Once you've gone through the chapter, re-read it with special emphasis on the ideas that gave you trouble the first time. If you're still unsure after the lecture, consult your instructor, either during office hours or after class. Bring your notes with you so your instructor can see which concepts are giving you trouble. Once your instructor has helped clarify things for you, revise your notes so you'll remember your new-found insights. You'll end up with a chapter summary in your own words. This will be a tremendous help when studying for exams!

- **Make use of your fellow students** Many students find it useful to form study groups for astronomy.

You can hash out challenging topics with each other and have a good time while you're doing it. But make sure that you write up your homework by yourself, because the penalties for copying or plagiarizing other students' work can be severe in the extreme. Some students find individual assistance useful too. If you think a tutor will be helpful, link up with one early. Getting a tutor late in the course, in the belief that you'll be able to catch up with what you missed earlier on, is almost always a pointless exercise.

- **Take advantage of the Web site and CD-ROM** Take some time to explore the *Universe* Web site (**www.whfreeman.com/universe7e**) and the CD-ROM that comes packaged with this book. On both of these you'll find review materials, animations, videos, interactive exercises, flashcards, and many other features keyed to chapters in *Universe*. All these features are designed to help you learn and enjoy astronomy, so make sure to take full advantage of them.

- **Try astronomy for yourself with your star charts,** *Starry Night Backyard*™, **and** *Deep Space Explorer*™ At the back of this book you'll find a set of star charts for the each month of the year in the Northern Hemisphere. (For a set of Southern Hemisphere star charts, see the *Universe* Web site.) Star charts can get you started with your own observations of the universe. Hold the chart overhead in the same orientation as the compass points, with the southern horizon toward the south and the western horizon toward the west. (To save strain on your arms, you may want to cut these pages out of the book.) The CD-ROM packaged with this textbook includes the easy-to-use *Starry Night Backyard*™ planetarium program, which you can use to view the sky on any date and time as seen from anywhere on Earth. It also includes the *Deep Space Explorer*™ program, which allows you to "fly" through the universe and make close-up observations of planets, moons, and galaxies.

Before you study Universe, take this quiz

Universe has many features designed to help you succeed in your study of astronomy. To get the most from it, understanding these features and knowing how to use them are essential. To make sure that you do, first read through the Preface on the preceding pages. Then take this brief quiz.

If you can answer all the questions, you're ready to begin studying astronomy! (You can check your answers on page Q-1 at the back of the book.)

1. Which specially labeled paragraphs alert you to common misconceptions and conceptual pitfalls?

2. Which specially labeled paragraphs draw analogies between ideas in astronomy and aspects of everyday life?

3. In many chapters, in addition to the numbered sections, you will also find material set off in Boxes. Which type of Box provides extra help with solving mathematical problems? Which relates astronomical principles to phenomena here on Earth?

4. Throughout the book you will encounter icons labeled "Web Link," "Animation," "Video," "AIMM," or "Looking Deeper." Where should you look to find the information to which these icons refer?

5. What are the *Starry Night Backyard*™ and *Deep Space Explorer*™ programs? Where can you find copies to install on your own computer? How much do they cost?

6. Many of the figures in this book are accompanied by the letters R I V U X G, with one of the letters highlighted. For instance, Figure 6-34*a* on page 141 has the letters R I **V** U X G, while Figure 6-34*d* has R I V U **X** G. What is the significance of the highlighted letter?

7. Where can you find self-tests and review material for each chapter of *Universe*?

8. Refer to the Appendices at the back of this book. On which page(s) of *Universe* would you look to find the following? (a) the value of the Stefan-Boltzmann constant; (b) the average orbital speed of Mars; (c) the average distance from the center of the planet Pluto to its moon Charon; (d) the distance in light-years to the star Proxima Centauri.

9. Refer to the Index at the back of this book. On which page(s) of *Universe* would you find the following terms described? (a) spicule; (b) refraction; (c) tidal force; (d) aphelion.

10. Refer to the Answers to Selected Questions at the back of this book. What is the answer to Question 37 of Chapter 5?
(**NOTE:** Your instructor may assign as homework some of the questions whose answers can be found in the Answers to Selected Questions. If so, your instructor will expect you to write out and explain your calculations to show how this answer is obtained.)

11. Where in this book can you find star charts for each month of the year?

Here's the most important advice of all

I haven't mentioned the most important thing you should do when studying astronomy: *Have fun!* Of all the different kinds of scientists, astronomers are among the most excited about what they do and what they study. Let some of that excitement about the universe rub off on you, and you'll have a great time with this course and with this textbook.

In preparing this edition of *Universe*, I tried very hard to make it the kind of textbook that a student like you would find useful. I'm very interested in your comments and opinions! Please feel free to send me e-mail or write me, and I will respond personally.

Best wishes for success in your studies!

Roger A. Freedman
Department of Physics
University of California, Santa Barbara
Santa Barbara CA 93106
airboy@physics.ucsb.edu

1 Astronomy and the Universe

Imagine yourself in the desert on a clear, dark, moonless night, far from the glare of city lights. As you gaze upward, you see a panorama that no poet's words can truly describe and that no artist's brush could truly capture. Literally thousands of stars are scattered from horizon to horizon, some of them strikingly bright (such as Sirius, the brightest star in the sky, near the top of this photograph). As you watch, the entire spectacle swings slowly overhead from east to west as the night progresses. If you are very lucky indeed, you may even see a shower of meteors like those shown here.

For thousands of years people have looked up at the heavens and contemplated the universe. Like our ancestors, we find our thoughts turning to profound questions as we gaze at the stars. How was the universe created? Where did the Earth, Moon, and Sun come from? What are the planets and stars made of? And how do we fit in? What is our place in the cosmic scope of space and time?

To wonder about the universe is a particularly human endeavor. Our curiosity, our desire to explore and discover, and, most important, our ability to reason about what we have discovered are qualities that distinguish us from other animals. The study of the stars transcends all boundaries of culture, geography, and politics. In a literal sense, astronomy is a universal subject—its subject is the entire universe.

The Leonid meteors trail across the sky in the constellation Canis Major.
(Courtesy of Wally Pacholka) R I **V** U X G

Questions to Guide Your Reading

As you read the sections of this chapter, look for the answers to the following questions:

1-1 What methods do scientists use to expand our understanding of the universe?

1-2 What makes up our solar system?

1-3 What are the stars? Do they last forever?

1-4 What are galaxies? What do astronomers learn by studying them?

1-5 How does measuring angles help astronomers learn about objects in the sky?

1-6 What is powers-of-ten notation, and why is it useful in astronomy?

1-7 Why do astronomers measure distances in astronomical units, light-years, and parsecs?

1-8 How does studying the cosmos help us on Earth?

1-1 | To understand the universe, astronomers use the laws of physics to construct testable theories and models

Astronomy has a rich heritage that dates back to the myths and legends of antiquity. Centuries ago, the heavens were thought to be populated with demons and heroes, gods and goddesses. Astronomical phenomena were explained as the result of supernatural forces and divine intervention.

The course of civilization was greatly affected by a profound realization: *The universe is comprehensible.* This awareness is one of the great gifts to come to us from ancient Greece. Greek astronomers discovered that by observing the heavens and carefully reasoning about what they saw, they could learn something about how the universe operates. For example, as we shall see in Chapter 3, they measured the size of the Earth and understood and predicted eclipses. Modern science is a direct descendant of the intellectual endeavors of these ancient Greek pioneers.

Like art, music, or any other human creative activity, science makes use of intuition and experience. But the approach used by scientists to explore physical reality differs from other forms of intellectual endeavor in that it is based fundamentally on *observation, logic,* and *skepticism.* This approach, called the **scientific method,** requires that our ideas about the world around us be consistent with what we actually observe.

The scientific method goes something like this: A scientist trying to understand some phenomenon proposes a **hypothesis,** which is a collection of ideas that seems to explain the phenomenon. It is in developing hypotheses that scientists are at their most creative, imaginative, and intuitive. But their hypotheses must always agree with existing observations and experiments, because a discrepancy with what is observed implies that the hypothesis is wrong. (The exception is if the scientist thinks that the existing results are wrong and can give compelling evidence to show that they are wrong.) The scientist then uses logic to work out the implications of the hypothesis and to make predictions that can be tested. A hypothesis is on firm ground only after it has accurately forecast the results of new experiments or observations. (In practice, scientists typically go through these steps in a less linear fashion than we have described.)

Scientists describe reality in terms of **models,** which are hypotheses that have withstood observational or experimental tests. A model tells us about the properties and behavior of some object or phenomenon. A familiar example is a model of the atom, which scientists picture as electrons orbiting a central nucleus. Another example, which we will encounter in Chapter 18, is a model that tells us about physical conditions (for example, temperature, pressure, and density) in the interior of the Sun. A well-developed model uses mathematics—one of the most powerful tools for logical thinking—to make detailed predictions. For example, a successful model of the Sun's interior should describe in detail how the temperature changes as you look deeper into the Sun. For this reason, mathematics is one of the most important tools used by scientists.

A body of related hypotheses can be pieced together into a self-consistent description of nature called a **theory.** An example from Chapter 4 is the theory that the planets are held in their orbits around the Sun by the Sun's gravitational force. Without models

and theories there is no understanding and no science, only collections of facts.

 In everyday language the word "theory" is often used to mean an idea that looks good on paper, but has little to do with reality. In science, however, a good theory is one that explains reality very well and that can be applied to new phenomena. An excellent example is the theory of gravitation (Chapter 4), which was devised by the English scientist Isaac Newton in the seventeenth century to explain the orbits of the six planets known at that time. When astronomers of later centuries discovered the planets Uranus, Neptune, and Pluto, they found that these planets also moved in accordance with Newton's theory. The same theory describes the motions of satellites around the Earth as well as the orbits of planets around other stars (see Chapter 8).

Skepticism is an essential part of the scientific method. New hypotheses must be able to withstand the close scrutiny of other scientists. The more radical the hypothesis, the more skepticism and critical evaluation it will receive from the scientific community, because the general rule in science is that extraordinary claims require extraordinary evidence. That is why scientists as a rule do not accept claims that people have been abducted by aliens and taken aboard UFOs. The evidence presented for these claims is flimsy, secondhand, and unverifiable.

At the same time, scientists must be open-minded. They must be willing to discard long-held ideas if these ideas fail to agree with new observations and experiments, provided the new data have survived evaluation. (If an alien spacecraft really did land on Earth, scientists would be the first to accept that aliens existed—provided they could take a careful look at the spacecraft and its occupants.) That is why scientific knowledge is always provisional. As you go through this book, you will encounter many instances where new observations have transformed our understanding of Earth, the planets, the Sun and stars, and indeed the very structure of the universe.

Theories that accurately describe the workings of physical reality have a significant effect on civilization. For example, basing his conclusions in part on observations of how the planets orbit the Sun, Isaac Newton deduced a set of fundamental principles that describe how *all* objects move. These theoretical principles, which we will encounter in Chapter 4, work equally well on Earth as in the most distant corner of the universe. They represent our first complete, coherent description of the behavior of the physical universe. **Newtonian mechanics** had an immediate practical application in the construction of machines, buildings, and bridges. It is no coincidence that the Industrial Revolution followed hard on the heels of these theoretical and mathematical advances inspired by astronomy.

Newtonian mechanics and other physical theories have stood the test of time and been shown to have great and general validity. Proven theories of this kind are collectively referred to as the **laws of physics.** Astronomers use these laws to interpret and understand their observations of the universe. The laws governing light and its relationship to matter are of particular importance, because the only information we can gather about distant stars and galaxies is in the light that we receive from them. Using the physical laws that describe how objects absorb and emit light, astronomers have measured the temperature of the Sun and even learned what the Sun is

Figure 1-1 R I **V** U X G

A Telescope in Space Because it orbits outside the Earth's atmosphere in the vacuum of space, the Hubble Space Telescope (HST) can detect not only visible light but also ultraviolet and near-infrared light coming from distant stars and galaxies. These forms of nonvisible light are absorbed by our atmosphere and hence are difficult or impossible to detect with a telescope on the Earth's surface. This photo of HST was taken by the crew of the space shuttle *Columbia* after a servicing mission in 2002. (NASA)

made of. By analyzing starlight in the same way, they have discovered that our own Sun is a rather ordinary star and that the observable universe may contain billions of stars just like the Sun.

An important part of science is the development of new tools for research and new techniques of observation. As an example, until fairly recently everything we knew about the distant universe was based on visible light. Astronomers would peer through telescopes to observe and analyze visible starlight. By the end of the nineteenth century, however, scientists had begun discovering forms of light invisible to the human eye: X rays, gamma rays, radio waves, microwaves, and ultraviolet and infrared radiation.

As we will see in Chapter 6, in recent years astronomers have constructed telescopes that can detect such nonvisible forms of light (**Figure 1-1**). These instruments give us views of the universe

vastly different from anything our eyes can see. These new views have allowed us to see through the atmospheres of distant planets, to study the thin but incredibly violent gas that surrounds our Sun, and even to observe new solar systems being formed around distant stars. Aided by high-technology telescopes, today's astronomers carry on the program of careful observation and logical analysis begun thousands of years ago by their ancient Greek predecessors.

1-2 | By exploring the planets, astronomers uncover clues about the formation of the solar system

SCIENCE IN PROCESS *How exploring the Moon and planets has revolutionized our understanding of Earth*

The science of astronomy allows our intellects to voyage across the cosmos. We can think of three stages in this voyage: from the Earth to the solar system, from the solar system to the stars, and from stars to galaxies and the grand scheme of the universe.

The star we call the Sun and all the celestial bodies that orbit the Sun—including Earth, the other eight planets, all their various moons, and smaller bodies such as asteroids and comets—make up the **solar system.** Since the 1960s a series of unmanned spacecraft has visited and explored all the planets except Pluto (**Figure 1-2**). Using the remote "eyes" of such spacecraft, we have flown over Mercury's cratered surface, peered beneath Venus's poisonous cloud cover, and discovered enormous canyons and extinct volcanoes on Mars. We have found active volcanoes on a moon of Jupiter, seen the rings of Saturn and Uranus up close, and looked down on the active atmosphere of Neptune.

Along with rocks brought back by the Apollo astronauts from the Moon (the only world beyond Earth yet visited by humans), this new information has revolutionized our understanding of the origin and evolution of the solar system. We have come to realize that many of the planets and their satellites were shaped by collisions with other objects. Craters on the Moon and on many other worlds are the relics of innumerable impacts by bits of interplanetary rock. The Moon may itself be the result of a catastrophic collision between the Earth and a planet-sized object shortly after the solar system was formed. Such a collision could have torn sufficient material from the primordial Earth to create the Moon.

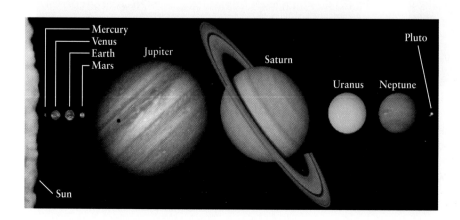

Figure 1-2 R I **V** U X G

The Sun and Planets to Scale This montage of images from various spacecraft and ground-based telescopes shows the relative sizes of the nine planets and the Sun. The Sun is so large compared to the planets that only a portion of it fits into this illustration. The distances from the Sun to each planet are *not* shown to scale; the actual distance from the Sun to the Earth, for instance, is 12,000 times greater than the Earth's diameter. (Calvin J. Hamilton and NASA/JPL)

The discoveries that we have made in our journeys across the solar system are directly relevant to the quality of human life on our own planet. Until recently, our understanding of geology, weather, and climate was based solely on data from the Earth. Since the advent of space exploration, however, we have been able to compare and contrast other worlds with our own. This new knowledge gives us valuable insight into our origins, the nature of our planetary home, and the limits of our natural resources.

1-3 | By studying stars and nebulae, astronomers discover how stars are born, grow old, and die

The nearest of all stars to Earth is the Sun. Although humans have used the Sun's warmth since the dawn of our species, it was only in the 1920s and 1930s that physicists figured out how the Sun shines. At the center of the Sun, thermonuclear reactions—so called because they require extremely high temperatures—convert hydrogen (the Sun's primary constituent) into helium. This violent process releases a vast amount of energy, which eventually makes its way to the Sun's surface and escapes as light (**Figure 1-3**). By 1950 physicists could reproduce such thermonuclear reactions here on Earth in the form of a hydrogen bomb (**Figure 1-4**). While such thermonuclear weapons are capable of destroying life on our planet, peaceful applications of this same process may provide a clean source of energy sometime in the next several decades.

All the stars you can see in the nighttime sky also shine by thermonuclear reactions. These reactions consume the material of which stars are made, which means that stars cannot last forever. Rather, they must form, evolve, and eventually die.

Much of what we know about the life stories of stars comes from the study of huge clouds of interstellar gas, called **nebulae** (singular **nebula**), that are found scattered across the sky. Within some nebulae, such as the Orion Nebula shown in **Figure 1-5**, stars are born from the material of the nebula itself. Other nebulae reveal what happens when thermonuclear reactions stop and a star dies after millions or billions of years. Some stars that are far more massive than the Sun end their lives with a spectacular detonation called a **supernova** (plural **supernovae**) that blows the star apart. The Crab Nebula (**Figure 1-6**) is a striking example of a remnant left behind by a supernova.

Dying stars can produce some of the strangest objects in the sky. Some dead stars become **pulsars,** which spin dizzily at rates of tens or hundreds of rotations per second. And some stars end their lives as almost inconceivably dense objects called **black holes,** which are surrounded by gravity so powerful that nothing—not even light—can escape. Even though a black hole itself emits essentially no radiation, a number of black holes have been discovered by Earth-orbiting telescopes. This is done by detecting the X rays emitted by gases falling toward a black hole.

 CAUTION Astronomers often use biological terms such as "birth" and "death" to describe stages in the evolution of inanimate objects like stars. Keep in mind that such terms are

Figure 1-3   R I **V** U X G
Our Star, the Sun   The Sun is a typical star. Its diameter is about 1.39 million kilometers (roughly a million miles), and its surface temperature is about 5500°C (10,000°F). The Sun draws its energy from thermonuclear reactions occurring at its center, where the temperature is about 15 million degrees Celsius. (Celestron International)

Figure 1-4   R I **V** U X G
A Thermonuclear Explosion   A hydrogen bomb uses the same physical principle as the thermonuclear reactions at the Sun's center: the conversion of matter into energy by nuclear reactions. This thermonuclear detonation on October 31, 1952, had an energy output equivalent to 10.4 million tons of TNT. This is a mere ten-billionth of the amount of energy released by the Sun in one second. (Defense Nuclear Agency)

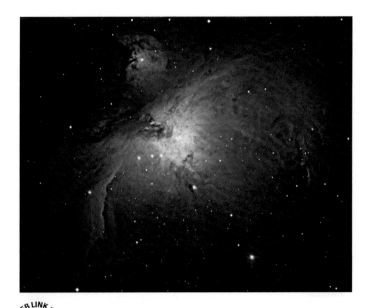

Figure 1-5 R I **V** U X G
The Orion Nebula—Birthplace of Stars This beautiful
nebula is a stellar "nursery" where stars are formed out of the nebula's gas.
Intense ultraviolet light from newborn stars excites the surrounding gas and
causes it to glow. Many of the stars embedded in this nebula are less than a
million years old, a brief interval in the lifetime of a typical star. The Orion
Nebula is some 1500 light-years from Earth and is about 30 light-years
across. (David Malin Images/Anglo-Australian Observatory)

Figure 1-6 R I **V** U X G
The Crab Nebula—Wreckage of an Exploded Star
When a dying star exploded in a supernova, it left behind this elegant
funeral shroud of glowing gases blasted violently into space. A thousand
years after the explosion these gases are still moving outward at about
1800 km/s (roughly 4 million miles per hour). The Crab Nebula is 6500
light-years from Earth and about 10 light-years across. (The FORS Team,
VLT, European Southern Observatory)

used only as *analogies*, which help us visualize these stages. They
are not to be taken literally!

During their death throes, stars return the gas of which they are
made to interstellar space. (Figure 1-6 shows these expelled gases
expanding away from the site of a supernova explosion.) This gas
contains heavy elements—that is, elements heavier than hydrogen
and helium—that were created during the star's lifetime by thermonuclear reactions in its interior. Interstellar space thus becomes
enriched with newly manufactured atoms and molecules. The Sun
and its planets were formed from interstellar material that was
enriched in this way. This means that the atoms of iron and nickel
that make up the body of the Earth, as well as the carbon in our
bodies and the oxygen we breathe, were created deep inside ancient
stars. By studying stars and their evolution, we are really studying
our own origins.

1-4 | By observing galaxies, astronomers learn about the origin and fate of the universe

Stars are not spread uniformly across the universe but are grouped
together in huge assemblages called **galaxies.** Galaxies come in a
wide range of shapes and sizes. A typical galaxy, like the Milky
Way, of which our Sun is part, contains several hundred billion
stars. Some galaxies are much smaller, containing only a few million stars. Others are monstrosities that devour neighboring galaxies in a process called "galactic cannibalism."

Our Milky Way Galaxy has arching spiral arms like those of the
galaxy shown in Figure 1-7. These arms are particularly active sites
of star formation. In recent years, astronomers have discovered a
mysterious object at the center of the Milky Way with a mass millions of times greater than that of our Sun. It now seems certain
that this curious object is an enormous black hole.

Some of the most intriguing galaxies appear to be in the throes
of violent convulsions and are rapidly expelling matter. The centers of these strange galaxies, which may harbor even more massive black holes, are often powerful sources of X rays and radio
waves.

Even more awesome sources of energy are found still deeper in
space. At distances so great that their light takes billions of years to
reach Earth, we find the mysterious **quasars.** Although quasars look
like nearby stars (Figure 1-8), they are among the most distant and
most luminous objects in the sky. A typical quasar shines with the
brilliance of a hundred galaxies. Detailed observations of quasars
imply that they draw their energy from material falling into enormous black holes.

The motions of distant clusters of galaxies reveal that we live in
an expanding universe. Extrapolating into the past, we learn that
the universe must have been born from an incredibly dense state
(perhaps infinitely dense) more than 13 billion years ago. A variety
of evidence indicates that at that moment—the beginning of time—
the universe began with a cosmic explosion, known as the **Big Bang,**

Figure 1-7 R I **V** U X G

A Galaxy This spectacular galaxy, called M63, contains about a hundred billion stars. M63 has a diameter of about 60,000 light-years and is located about 35 million light-years from Earth. Along this galaxy's spiral arms you can see a number of glowing clumps. Like the Orion Nebula in our own Milky Way Galaxy (Figure 1-5), these are sites of active star formation. (Subaru Telescope, National Astronomical Observatory of Japan)

which occurred throughout all space. Events shortly after the Big Bang dictated the present nature of the universe.

Thanks to the combined efforts of astronomers and physicists, we are making steady advances in understanding these cosmic events. This understanding may reveal the origin of some of the most basic properties of physical reality. Studying the most remote galaxies may also answer questions about the ultimate fate of the universe. Will it continue expanding forever, or will it someday stop and collapse back in on itself?

The work of unraveling the deepest mysteries of the universe requires specialized tools, including telescopes, spacecraft, and computers. But for many purposes the most useful device for studying the universe is the human brain itself. Our goal in this book is to help you use *your* brain to share in the excitement of scientific discovery.

In the remainder of this chapter we introduce some of the key concepts and mathematics that we will use in subsequent chapters. Study these carefully, for you will use them over and over again throughout your own study of astronomy.

Figure 1-8 R I **V** U X G

A Quasar The two bright starlike objects in this image look almost identical, but they are dramatically different. The object on the left is indeed a star that lies a few hundred light-years from Earth. But the "star" on the right is actually a quasar about 9 billion light-years away. To appear so bright even though they are so distant, quasars like this one must be some of the most luminous objects in the universe. The other objects in this image are galaxies like that in Figure 1-7. (Charles Steidel, California Institute of Technology; and NASA)

1-5 | Astronomers use angles to denote the positions and apparent sizes of objects in the sky

Whether they study planets, stars, galaxies, or the very origins of the universe, astronomers must know where to point their telescopes. For this reason, an important part of astronomy is keeping track of the positions of objects in the sky. *Angles* and a system of *angular measure* are essential tools for this aspect of astronomy (**Figure 1-9**).

An **angle** is the opening between two lines that meet at a point. **Angular measure** describes the size of an angle exactly. The basic unit of angular measure is the **degree**, designated by the symbol °. A full circle is divided into 360°, and a right angle measures 90° (Figure 1-9*a*). As Figure 1-9*b* shows, if you draw lines from your eye to each of the two "pointer stars" in the Big Dipper, the angle between these lines—that is, the **angular distance** between these two stars—is about 5°. (These two stars "point" to Polaris, or the North Star, as described in Chapter 2.) The angular distance between the

(a)

The angular distance between the two pointer stars at the front of the Big Dipper is about 5°, roughly 10 times the angular diameter of the Moon.

(b)

The angular distance between the stars at the top and bottom of the Southern Cross is about 6°.

(c)

figure 1-9

Angles and Angular Measure (a) Angles are measured in degrees (°). There are 360° in a complete circle and 90° in a right angle. For example, the angle between the vertical direction (directly above you) and the horizontal direction (toward the horizon) is 90°. The angular diameter of the full moon in the sky is about ½°. (b) The seven bright stars that make up

the Big Dipper can be seen from anywhere in the northern hemisphere. The angular distance between the two "pointer stars" at the front of the Big Dipper is about 5°. (c) The four bright stars that make up the Southern Cross can be seen from anywhere in the southern hemisphere. The angular distance between the stars at the top and bottom of the cross is about 6°.

stars that make up the top and bottom of the Southern Cross, which is visible from south of the equator, is about 6° (Figure 1-9c).

Astronomers also use angular measure to describe the apparent size of a celestial object—that is, what fraction of the sky that object seems to cover. For example, the angle covered by the diameter of the full moon is about ½° (Figure 1-9a). We therefore say that the **angular diameter** (or **angular size**) of the Moon is ½°. Alternatively, astronomers say that the Moon **subtends** an angle of ½°. Ten full moons could fit side by side between the two pointer stars in the Big Dipper.

The adult human hand held at arm's length provides a means of estimating angles, as **Figure 1-10** shows. For example, your fist covers an angle of 10°, whereas the tip of your finger is about 1° wide. You can use various segments of your index finger extended to arm's length to estimate angles a few degrees across.

To talk about smaller angles, we subdivide the degree into 60 **arcminutes** (also called minutes of arc), which is commonly abbreviated as 60 arcmin or 60′. An arcminute is further subdivided into 60 **arcseconds** (or seconds of arc), usually written as 60 arcsec or 60″. Thus,

$$1° = 60 \text{ arcmin} = 60'$$
$$1' = 60 \text{ arcsec} = 60''$$

For example, on January 1, 2004, the planet Saturn had an angular diameter of 20.6 arcsec as viewed from Earth. That is a convenient, precise statement of how big the planet appeared in Earth's sky on that date. (Because this angular diameter is so small, to the naked eye Saturn appears simply as a point of light. To see any detail on Saturn, such as the planet's rings, requires a telescope.)

If we know the angular size of an object as well as the distance to that object, we can determine the actual linear size of the object (measured in kilometers or miles, for example). **Box 1-1** describes how this is done.

figure 1-10

Estimating Angles with Your Hand Various parts of the adult human hand extended to arm's length can be used to estimate angular distances and angular sizes in the sky.

box 1-1 TOOLS OF THE ASTRONOMER'S TRADE

The Small-Angle Formula

You can estimate the angular sizes of objects in the sky with your hand and fingers (Figure 1-10). Using rather more sophisticated equipment, astronomers can measure angular sizes to a fraction of an arcsecond. Keep in mind, however, that *angular* size is not the same as *actual* size. As an example, if you extend your arm while looking at a full moon, you can completely cover the Moon with your thumb. That's because from your perspective, your thumb has a larger angular size (that is, it subtends a larger angle) than the Moon. But the actual size of your thumb (about 2 centimeters) is much less than the actual diameter of the Moon (more than 3000 kilometers).

The accompanying figure shows how the angular size of an object is related to its linear size. Part **a** of the figure shows that for a given angular size, the more distant the object, the larger its actual size. For example, your thumb held at arm's length just covers the full moon; the angular size of your thumb and the Moon are about the same, but the Moon is much farther away and is far larger in linear size. Part **b** shows that for a given linear size, the angular size decreases the farther away the object. This is why a car looks smaller and smaller as it drives away from you.

We can put these relationships together into a single mathematical expression called the **small-angle formula.** Suppose that an object subtends an angle α (the Greek letter alpha) and is at a distance *d* from the observer, as in part **c** of the figure. If the angle α is small, as is almost always the case for objects in the sky, the linear size (*D*) of the object is given by the following expression:

The small-angle formula

$$D = \frac{\alpha d}{206{,}265}$$

D = linear size of an object
α = angular size of the object, in arcsec
d = distance to the object

The number 206,265 is required in the formula. Mathematically, it is equal to the number of arcseconds in a

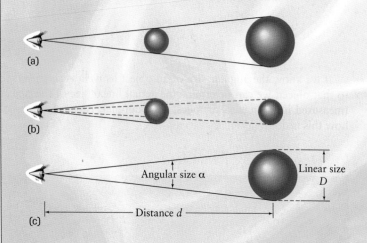

(a)

(b)

Angular size α — Linear size D

Distance d

(c)

(a) Two objects that have the same angular size may have different linear sizes if they are at different distances from the observer. (b) For an object of a given linear size, the angular size is larger the closer the object is to the observer. (c) The small-angle formula relates the linear size D of an object to its angular size α and its distance d from the observer.

1-6 | Powers-of-ten notation is a useful shorthand system for writing numbers

Astronomy is a subject of extremes. Astronomers investigate the largest structures in the universe, including galaxies and clusters of galaxies. But they must also study atoms and atomic nuclei, among the smallest objects in the universe, in order to explain how and why stars shine. They also study conditions ranging from the incredibly hot and dense centers of stars to the frigid near-vacuum of interstellar space. To describe such a wide range of phenomena, we need an equally wide range of both large and small numbers.

Astronomers avoid such confusing terms as "a million billion billion" by using a standard shorthand system called **powers-of-ten notation.** All the cumbersome zeros that accompany a large number are consolidated into one term consisting of 10 followed by an

exponent, which is written as a superscript. The exponent indicates how many zeros you would need to write out the long form of the number. Thus,

$$
\begin{aligned}
10^0 &= 1 \text{ (one)} \\
10^1 &= 10 \text{ (ten)} \\
10^2 &= 100 \text{ (one hundred)} \\
10^3 &= 1000 \text{ (one thousand)} \\
10^4 &= 10{,}000 \text{ (ten thousand)} \\
10^6 &= 1{,}000{,}000 \text{ (one million)} \\
10^9 &= 1{,}000{,}000{,}000 \text{ (one billion)} \\
10^{12} &= 1{,}000{,}000{,}000{,}000 \text{ (one trillion)}
\end{aligned}
$$

and so forth. The exponent also tells you how many tens must be multiplied together to give the desired number, which is why the exponent is also called the **power of ten.** For example, ten thou-

TOOLS OF THE ASTRONOMER'S TRADE continued

complete circle (that is, 360°) divided by the number 2π (the ratio of the circumference of a circle to that circle's radius).

The following examples show two different ways to use the small-angle formula. In both examples we follow a four-step process: Evaluate the *situation* given in the example, decide which *tools* are needed to solve the problem, use those tools to find the *answer* to the problem, and *review* the result to see what it tells you. Throughout this book, we'll use these same four steps in *all* examples that require the use of formulas. We encourage you to follow this four-step process when solving problems for homework or exams. You can remember these steps by their acronym: *S.T.A.R.*

EXAMPLE: On July 26, 2003, Jupiter was 943 million kilometers from Earth and had an angular diameter of 31.2 arcseconds. From this information, calculate the actual diameter of Jupiter in kilometers.

Situation: The astronomical object in this example is Jupiter, and we are given its distance d and its angular size α (the same as angular diameter). Our goal is to find Jupiter's diameter D.

Tools: The equation to use is the small-angle formula, which relates the quantities d, α, and D. Note that when using this formula, the angular size α *must* be expressed in arcseconds.

Answer: The small-angle formula as given is an equation for D. Plugging in the given values $\alpha = 31.2$ arcsec and $d = 943$ million km,

$$D = \frac{31.2 \times 943{,}000{,}000 \text{ km}}{206{,}265} = 143{,}000 \text{ km}$$

Because the distance d to Jupiter is given in kilometers, the diameter D is also in kilometers.

Review: Does our answer make sense? From Appendix 2 at the back of this book, the equatorial diameter of Jupiter measured by spacecraft flybys is 142,984 km, so our calculated answer is very close.

EXAMPLE: Under excellent conditions, a telescope on Earth can see details with an angular size as small as 1 arcsec. What is the greatest distance at which you could see details as small as 1.7 m (the height of a typical person) under these conditions?

Situation: Now the object in question is a person, whose linear size D we know. Our goal is to find the distance d at which the person has an angular size α equal to 1 arcsec.

Tools: Again we use the small-angle formula to relate d, α, and D.

Answer: We first rewrite the formula to solve for the distance d, then plug in the given values $D = 1.7$ m and $\alpha = 1$ arcsec:

$$d = \frac{206{,}265\, D}{\alpha} = \frac{206{,}265 \times 1.7 \text{ m}}{1} = 350{,}000 \text{ m} = 350 \text{ km}$$

Review: This is much less than the distance to the Moon, which is 384,000 km. Thus, even the best telescope on Earth could not be used to see an astronaut walking on the surface of the Moon.

---

sand can be written as 10^4 ("ten to the fourth" or "ten to the fourth power") because $10^4 = 10 \times 10 \times 10 \times 10 = 10{,}000$.

In powers-of-ten notation, numbers are written as a figure between one and ten multiplied by the appropriate power of ten. The approximate distance between Earth and the Sun, for example, can be written as 1.5×10^8 kilometers (or 1.5×10^8 km for short). Once you get used to it, this is more convenient than writing "150,000,000 kilometers" or "one hundred and fifty million kilometers." (The same number could also be written as 15×10^7 or 0.15×10^9, but the preferred form is *always* to have the first figure be between 1 and 10.)

Note that most electronic calculators use a shorthand for powers-of-ten notation. To enter the number 1.5×10^8, you first enter 1.5, then press a key labeled "EXP" or "EE," then enter the exponent 8. (The EXP or EE key takes care of the "$\times 10$" part of the expression.) The number will then appear on your calculator's display as

"1.5 E 8," "1.5 8," or some variation of this; typically the "$\times 10$" is not displayed as such. There are some variations from one kind of calculator to another, so you should spend a few minutes reading over your calculator's instruction manual to make sure you know the correct procedure for working with numbers in powers-of-ten notation. You will be using this notation continually in your study of astronomy, so this is time well spent.

CAUTION! Confusion can result from the way that calculators display powers-of-ten notation. Since 1.5×10^8 is displayed as "1.5 8" or "1.5 E 8," it is not uncommon to think that 1.5×10^8 is the same as 1.5^8. That is not correct, however; 1.5^8 is equal to 1.5 multiplied by itself eight times, or 25.63, which is not even close to $150{,}000{,}000 = 1.5 \times 10^8$. Another, not uncommon, mistake is to write 1.5×10^8 as 15^8. If you are inclined to do this, perhaps you are thinking that you can multiply 1.5 by 10, then

10^{-15}	10^{-10}	10^{-5}	1	10^{5}	10^{10}	10^{15}	10^{20}	10^{25}

Size of a proton — Size of an atom — Size of a virus — Size of a human — Diameter of the Earth — Diameter of the Sun — Distance from Earth to Sun — Distance to the nearest star beyond the Sun — Diameter of the Galaxy — Size of the observable universe

figure 1-11

Examples of Powers-of-Ten Notation The scale gives the sizes of objects in meters, ranging from subatomic particles at the left to the entire observable universe on the right. The photograph at the left shows tungsten atoms, 10^{-10} meter in diameter. Second from left is the crystalline skeleton of a diatom (a single-celled organism), 10^{-4} meter (0.1 millimeter) in size. At the center is the Taj Mahal, about 60 meters tall and within reach of our unaided senses. On the right, looking across the Indian Ocean toward the south pole, we see the curvature of the Earth, about 10^{7} meters in diameter. At the far right is a galaxy, 10^{21} meters (100,000 light-years) in diameter. (Courtesy of Scientific American Books; NASA; and photograph by David Malin from the Anglo-Australian Observatory)

tack on the exponent later. This also does not work; 15^8 is equal to 15 multiplied by itself eight times, or 2,562,890,625, which again is nowhere near 1.5×10^8. Reading over the manual for your calculator will help you to avoid these common errors.

You can use powers-of-ten notation for numbers that are less than one by using a minus sign in front of the exponent. A negative exponent tells you to *divide* by the appropriate number of tens. For example, 10^{-2} ("ten to the minus two") means to divide by 10 twice, so $10^{-2} = 1/10 \times 1/10 = 1/100 = 0.01$. This same idea tells us how to interpret other negative powers of ten:

10^0 = 1 (one)

10^{-1} = 1/10 = 0.1 (one tenth)

10^{-2} = $1/10 \times 1/10 = 1/10^2 = 0.01$ (one hundredth)

10^{-3} = $1/10 \times 1/10 \times 1/10 = 1/10^3 = 0.001$ (one thousandth)

10^{-4} = $1/10 \times 1/10 \times 1/10 \times 1/10 = 1/10^4 = 0.0001$ (one ten-thousandth)

10^{-6} = $1/10 \times 1/10 \times 1/10 \times 1/10 \times 1/10 \times 1/10 = 1/10^6 = 0.000001$ (one millionth)

10^{-12} = $1/10 \times 1/10 \times 1/10 \times 1/10 \times 1/10 \times 1/10 \times 1/10 \times 1/10 \times 1/10 \times 1/10 \times 1/10 \times 1/10 = 1/10^{12} = 0.000000000001$ (one trillionth)

and so forth.

As these examples show, negative exponents tell you how many tenths must be multiplied together to give the desired number. For example, one ten-thousandth, or 0.0001, can be written as 10^{-4}

("ten to the minus four") because $10^{-4} = 1/10 \times 1/10 \times 1/10 \times 1/10 = 0.0001$.

A useful shortcut in converting a decimal to powers-of-ten notation is to notice where the decimal point is. For example, the decimal point in 0.0001 is four places to the left of the "1," so the exponent is –4, that is, $0.0001 = 10^{-4}$.

You can also use powers-of-ten notation to express a number like 0.00245, which is not a multiple of 1/10. For example, $0.00245 = 2.45 \times 0.001 = 2.45 \times 10^{-3}$. (Again, the standard for powers-of-ten notation is that the first figure is a number between one and ten.) This notation is particularly useful when dealing with very small numbers. A good example is the diameter of a hydrogen atom, which is much more convenient to state in powers-of-ten notation (1.1×10^{-10} meter, or 1.1×10^{-10} m) than as a decimal (0.00000000011 m) or a fraction (110 trillionths of a meter.)

Because it bypasses all the awkward zeros, powers-of-ten notation is ideal for describing the size of objects as small as atoms or as big as galaxies (Figure 1-11). Box 1-2 explains how powers-of-ten notation also makes it easy to multiply and divide numbers that are very large or very small.

1-7 | Astronomical distances are often measured in astronomical units, parsecs, or light-years

Astronomers use many of the same units of measurement as do other scientists. They often measure lengths in meters (abbreviated m), masses in kilograms (kg), and time in seconds (s). (You can read more about these units of measurement, as well as techniques

box 1-2 TOOLS OF THE ASTRONOMER'S TRADE

Arithmetic with Powers-of-Ten Notation

Using powers-of-ten notation makes it easy to multiply numbers. For example, suppose you want to multiply 100 by 1000. If you use ordinary notation, you have to write a lot of zeros:

$$100 \times 1000 = 100,000 \text{ (one hundred thousand)}$$

By converting these numbers to powers-of-ten notation, we can write this same multiplication more compactly as

$$10^2 \times 10^3 = 10^5$$

Because 2 + 3 = 5, we are led to the following general rule for *multiplying* numbers expressed in terms of powers of ten: Simply *add* the exponents.

EXAMPLE: $10^4 \times 10^3 = 10^{4+3} = 10^7$

To *divide* numbers expressed in terms of powers of ten, remember that $10^{-1} = 1/10$, $10^{-2} = 1/100$, and so on. The general rule for any exponent n is

$$10^{-n} = \frac{1}{10^n}$$

In other words, dividing by 10^n is the same as multiplying by 10^{-n}. To carry out a division, you first transform it into

multiplication by changing the sign of the exponent, and then carry out the multiplication by adding the exponents.

EXAMPLE: $\dfrac{10^4}{10^6} = 10^4 \times 10^{-6} = 10^{4+(-6)} = 10^{4-6} = 10^{-2}$

Usually a computation involves numbers like 3.0×10^{10}, that is, an ordinary number multiplied by a factor of 10 with an exponent. In such cases, to perform multiplication or division, you can treat the numbers separately from the factors of 10^n.

EXAMPLE: We can redo the first numerical example from Box 1-1 in a straightforward manner by using exponents:

$$D = \frac{31.2 \times 943,000,000 \text{ km}}{206,265}$$

$$= \frac{31.2 \times 10 \times 9.43 \times 10^8}{2.06265 \times 10^5} \text{ km}$$

$$= \frac{3.1 \times 9.43 \times 10^{1+8-5}}{2.06265} \text{ km} = 14.3 \times 10^4 \text{ km}$$

$$= 1.43 \times 10 \times 10^4 \text{ km} = 1.43 \times 10^5 \text{ km}$$

for converting between different sets of units, in **Box 1-3**.) Like other scientists, astronomers often find it useful to combine these units with powers of ten and create new units using prefixes. As an example, the number 1000 (= 10^3) is represented by the prefix "kilo," and so a distance of 1000 meters is the same as 1 kilometer (1 km). Here are some of the most common prefixes, with examples of how they are used:

$$
\begin{aligned}
\text{one-billionth meter} &= 10^{-9} \text{ m} = 1 \text{ nanometer} \\
\text{one-millionth second} &= 10^{-6} \text{ s} = 1 \text{ microsecond} \\
\text{one-thousandth arcsecond} &= 10^{-3} \text{ arcsec} = 1 \text{ milliarcsecond} \\
\text{one-hundredth meter} &= 10^{-2} \text{ m} = 1 \text{ centimeter} \\
\text{one thousand meters} &= 10^3 \text{ m} = 1 \text{ kilometer} \\
\text{one million tons} &= 10^6 \text{ tons} = 1 \text{ megaton}
\end{aligned}
$$

In principle, we could express all sizes and distances in astronomy using units based on the meter. Indeed, we will use kilometers to give the diameters of the Earth and Moon, as well as the Earth-Moon distance. But, while a kilometer (roughly equal to three-fifths of a mile) is an easy distance for humans to visualize, a megameter (10^6 m) is not. For this reason, astronomers have devised units of

measure that are more appropriate for the tremendous distances between the planets and the far greater distances between the stars.

When discussing distances across the solar system, astronomers use a unit of length called the **astronomical unit** (abbreviated AU). This is the average distance between Earth and the Sun:

$$1 \text{ AU} = 1.496 \times 10^8 \text{ km} = 92.96 \text{ million miles}$$

Thus, the average distance between the Sun and Jupiter can be conveniently stated as 5.2 AU.

 To talk about distances to the stars, astronomers use two different units of length. The **light-year** (abbreviated ly) is the distance that light travels in one year. This is a useful concept because the speed of light in empty space always has the same value, 3.00×10^5 km/s (kilometers per second) or 1.86×10^5 mi/s (miles per second). In terms of kilometers or astronomical units, one light-year is given by

$$1 \text{ ly} = 9.46 \times 10^{12} \text{ km} = 63,240 \text{ AU}$$

This distance is roughly equal to 6 trillion miles.

CAUTION Keep in mind that despite its name, the light-year is a unit of distance and *not* a unit of time. As an example, Proxima Centauri, the nearest star other than the Sun, is a distance of 4.2 light-years from Earth. This means that light takes 4.2 years to travel to us from Proxima Centauri.

Physicists often measure interstellar distances in light-years because the speed of light is one of nature's most important numbers. But many astronomers prefer to use another unit of length, the *parsec*, because its definition is closely related to a method of measuring distances to the stars.

Imagine taking a journey far into space, beyond the orbits of the outer planets. As you look back toward the Sun, Earth's orbit subtends a smaller angle in the sky the farther you are from the Sun. As **Figure 1-12** shows, the distance at which 1 AU subtends an angle of 1 arcsec is defined as **1 parsec** (abbreviated pc):

$$1 \text{ pc} = 3.09 \times 10^{13} \text{ km} = 3.26 \text{ ly}$$

The distance from Earth to Proxima Centauri can be stated as 1.3 pc as well as 4.2 ly. Whether you choose to use parsecs or light-years is a matter of personal taste.

For even greater distances, astronomers commonly use **kiloparsecs** and **megaparsecs** (abbreviated kpc and Mpc). As we saw

before, these prefixes simply mean "thousand" and "million," respectively:

$$1 \text{ kiloparsec} = 1 \text{ kpc} = 1000 \text{ pc} = 10^3 \text{ pc}$$
$$1 \text{ megaparsec} = 1 \text{ Mpc} = 1{,}000{,}000 \text{ pc} = 10^6 \text{ pc}$$

For example, the distance from Earth to the center of our Milky Way Galaxy is about 8 kpc, and the galaxy shown in Figure 1-7 is about 11 Mpc away.

Some astronomers prefer to talk about thousands or millions of light-years rather than kiloparsecs and megaparsecs. Once again, the choice is a matter of personal taste. As a general rule, astronomers use whatever yardsticks seem best suited for the issue at hand and do not restrict themselves to one system of measurement. For example, an astronomer might say that the supergiant star Antares has a diameter of 860 million kilometers and is located at a distance of 185 parsecs from Earth.

1-8 | Astronomy is an adventure of the human mind

An underlying theme of this book is that the universe is rational. It is not a hodgepodge of unrelated things behaving in unpredictable ways. Rather, we find strong evidence that fundamental laws of physics govern the nature of the universe and the behavior of everything in it. These unifying concepts enable us to explore realms far removed from our earthly experience. Thus, a scientist can do experiments in a laboratory to determine the properties of light or the behavior of atoms and then use this knowledge to investigate the structure of the universe.

The discovery of fundamental laws of nature has had a profound influence on humanity. These laws have led to an immense number of practical applications that have fundamentally transformed commerce, medicine, entertainment, transportation, and other aspects of our lives. In particular, space technology has given us instant contact with any point on the globe through communication satellites. Space technology has also made possible accurate weather forecasts from meteorological satellites, precise navigation to any point on Earth using signals from the satellites of the Global Positioning System (GPS), and long-term monitoring of Earth's climate and environment from orbit (**Figure 1-13**).

As important as the applications of science are, the pursuit of scientific knowledge for its own sake is no less important. We are fortunate to live in an age in which this pursuit is in full flower. Just as explorers such as Columbus and Magellan discovered the true size of our planet in the fifteenth and sixteenth centuries, astronomers of the twenty-first century are exploring the universe to an extent that is unparalleled in human history. Indeed, even the voyages into space imagined by such great science fiction writers as Jules Verne and H. G. Wells pale in comparison to today's reality. Over a few short decades, humans have walked on the Moon, sent robot spacecraft to dig into the Martian soil and explore the satellites of Jupiter, and used the most powerful telescopes ever built to probe the limits of the observable universe. Never before has so much been revealed in so short a time.

As you proceed through this book, you will learn about the tools that scientists use to explore the natural world, as well as what they

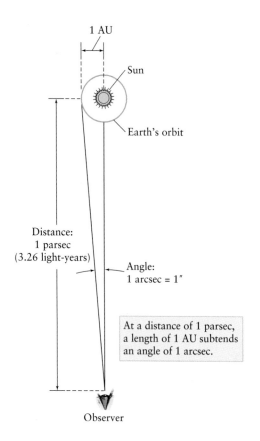

figure 1-12

A Parsec The parsec, a unit of length commonly used by astronomers, is equal to 3.26 light-years. The parsec is defined as the distance at which 1 AU perpendicular to the observer's line of sight subtends an angle of 1 arcsec.

figure 1-13

The Earth's Oceans from Space The *Topex/Poseidon* satellite (a joint project of the United States and France) measures variations in the height of the oceans. Areas of warm water bulge upward by several centimeters, so these measurements track changes in ocean temperature. In this map made from *Topex/Poseidon* data, elevated areas of warm water appear in white and red while depressed areas of cool water appear in blue and purple. These slight temperature variations have a profound effect on Earth's global climate. By monitoring these variations, scientists can give early warning of the potential for floods, droughts, and other severe weather. The resulting savings in lives and property more than pay for the cost of the satellite. (NASA/JPL)

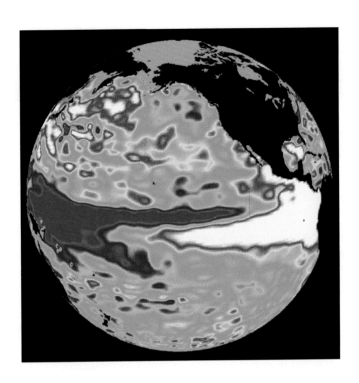

observe with these tools. But, most important, you will see how astronomers build from their observations an understanding of the universe in which we live. It is this search for understanding that makes science more than merely a collection of data and elevates it to one of the great adventures of the human mind. It is an adventure that will continue as long as there are mysteries in the universe—an adventure we hope you will come to appreciate and share.

---

box 1-3 TOOLS OF THE ASTRONOMER'S TRADE

Units of Length, Time, and Mass

To understand and appreciate the universe, we need to describe phenomena not only on the large scales of galaxies but also on the submicroscopic scale of the atom. Astronomers generally use units that are best suited to the topic at hand. For example, interstellar distances are conveniently expressed in either light-years or parsecs, whereas the diameters of the planets are more comfortably presented in kilometers.

Most scientists prefer to use a version of the metric system called the International System of Units, abbreviated **SI** (after the French name Système International). In SI units, length is measured in meters (m), time is measured in seconds (s), and mass (a measure of the amount of material in an object) is measured in kilograms (kg). How are these basic units related to other measures?

When discussing objects on a human scale, sizes and distances are usually expressed in millimeters (mm), centimeters (cm), and kilometers (km). These units of length are related to the meter as follows:

$$1 \text{ millimeter} = 0.001 \text{ m} = 10^{-3} \text{ m}$$
$$1 \text{ centimeter} = 0.01 \text{ m} = 10^{-2} \text{ m}$$
$$1 \text{ kilometer} = 1000 \text{ m} = 10^{3} \text{ m}$$

Although the English system of inches (in.), feet (ft), and miles (mi) is much older than SI, today the English system is actually based on the SI system: The inch is defined to be exactly 2.54 cm. A useful set of conversions is

$$1 \text{ in.} = 2.54 \text{ cm}$$
$$1 \text{ ft} = 0.3048 \text{ m}$$
$$1 \text{ mi} = 1.609 \text{ km}$$

Each of these equalities can also be written as a fraction equal to 1. For example, you can write

$$\frac{0.3048 \text{ m}}{1 \text{ ft}} = 1$$

Fractions like this are useful for converting a quantity from one set of units to another. For example, the *Saturn V* rocket used to send astronauts to the Moon stands about 363 feet tall. How can we convert this height to meters? The trick is to remember that a quantity does not change if you multiply it by 1. Expressing the number 1 by the fraction (0.3048 m)/(1 ft), we can write the height of the rocket as

$$363 \text{ ft} \times 1 = 363 \text{ ft} \times \frac{0.3048 \text{ m}}{1 \text{ ft}} = 111 \frac{\text{ft} \times \text{m}}{\text{ft}} = 111 \text{ m}$$

Continued on following page

EXAMPLE: The diameter of Mars is 6794 km. Let's try expressing this in miles.

CAUTION ! You can get into trouble if you are careless in applying the trick of taking the number whose units are to be converted and multiplying it by 1. For example, if we multiply the diameter by 1 expressed as (1.609 km)/(1 mi), we get

$$6794 \text{ km} \times 1 = 6794 \text{ km} \times \frac{1.609 \text{ km}}{1 \text{ mi}} = 10,930 \frac{\text{km}^2}{\text{mi}}$$

The unwanted units of km did not cancel, so this cannot be right. Furthermore, a mile is larger than a kilometer, so the diameter expressed in miles should be a smaller number than when expressed in kilometers.

The correct approach is to write the number 1 so that the unwanted units *will* cancel. The number we are starting with is in kilometers, so we must write the number 1 with kilometers in the denominator ("downstairs" in the fraction). Thus, we express 1 as (1 mi)/(1.609 km):

$$6794 \text{ km} \times 1 = 6794 \text{ km} \times \frac{1 \text{ mi}}{1.609 \text{ km}}$$

$$= 4222 \text{ km} \times \frac{\text{mi}}{\text{km}} = 4222 \text{ mi}$$

Now the units of km cancel as they should, and the distance in miles is a smaller number than in kilometers (as it must be).

When discussing very small distances such as the size of an atom, astronomers often use the micrometer (μm) or the nanometer (nm). These are related to the meter as follows:

$$1 \text{ micrometer} = 1 \text{ μm} = 10^{-6} \text{ m}$$
$$1 \text{ nanometer} = 1 \text{ nm} = 10^{-9} \text{ m}$$

Thus, 1 μm = 10^3 nm. (Note that the micrometer is often called the micron.)

The basic unit of time is the second (s). It is related to other units of time as follows:

$$1 \text{ minute (min)} = 60 \text{ s}$$
$$1 \text{ hour (h)} = 3600 \text{ s}$$
$$1 \text{ day (d)} = 86,400 \text{ s}$$
$$1 \text{ year (y)} = 3.156 \times 10^7 \text{ s}$$

In the SI system, speed is properly measured in meters per second (m/s). Quite commonly, however, speed is also expressed in km/s and mi/h:

$$1 \text{ km/s} = 10^3 \text{ m/s}$$
$$1 \text{ km/s} = 2237 \text{ mi/h}$$
$$1 \text{ mi/h} = 0.447 \text{ m/s}$$
$$1 \text{ mi/h} = 1.47 \text{ ft/s}$$

In addition to using kilograms, astronomers sometimes express mass in grams (g) and in solar masses ($M_\odot$), where the subscript $\odot$ is the symbol denoting the Sun. It is especially convenient to use solar masses when discussing the masses of stars and galaxies. These units are related to each other as follows:

$$1 \text{ kg} = 1000 \text{ g}$$
$$1 \text{ M}_\odot = 1.99 \times 10^{30} \text{ kg}$$

CAUTION ! You may be wondering why we have not given a conversion between kilograms and pounds. The reason is that these units do not refer to the same physical quantity! A kilogram is a unit of *mass*, which is a measure of the amount of material in an object. By contrast, a pound is a unit of *weight*, which tells you how strongly gravity pulls on that object's material. Consider a person who weighs 110 pounds on Earth, corresponding to a mass of 50 kg. Gravity is only about one-sixth as strong on the Moon as it is on Earth, so on the Moon this person would weigh only one-sixth of 110 pounds, or about 18 pounds. But that person's mass of 50 kg is the same on the Moon; wherever you go in the universe, you take all of your material along with you. We will explore the relationship between mass and weight in Chapter 4.

Key Words

Terms preceded by an asterisk (*) are discussed in the Boxes.

angle, p. 6
angular diameter (angular size), p. 7
angular distance, p. 6
angular measure, p. 6
arcminute (´, minute of arc), p. 7

arcsecond (˝, second of arc), p. 7
astronomical unit (AU), p. 11
Big Bang, p. 5
black hole, p. 4
degree (°), p. 6
exponent, p. 8
galaxy, p. 5

hypothesis, p. 2
kiloparsec (kpc), p. 12
laws of physics, p. 2
light-year (ly), p. 11
megaparsec (Mpc), p. 12
model, p. 2
nebula (*plural* nebulae), p. 4

Key Ideas

Astronomy, Science, and the Nature of the Universe: The universe is comprehensible. The scientific method is a procedure for formulating hypotheses about the universe. These are tested by observation or experimentation in order to build consistent models or theories that accurately describe phenomena in nature.

• Observations of the heavens have helped scientists discover some of the fundamental laws of physics. The laws of physics are in turn used by astronomers to interpret their observations.

The Solar System: Exploration of the planets provides information about the origin and evolution of the solar system, as well as the history and resources of Earth.

Stars and Nebulae: Studying the stars and nebulae helps us learn about the origin and history of the Sun and the solar system.

Galaxies: Observations of galaxies tell us about the origin and history of the universe.

Angular Measure: Astronomers use angles to denote the positions and sizes of objects in the sky. The size of an angle is measured in degrees, arcminutes, and arcseconds.

Powers-of-Ten Notation is a convenient shorthand system for writing numbers. It allows very large and very small numbers to be expressed in a compact form.

Units of Distance: A variety of distance units are used by astronomers. These include the astronomical unit (the average distance from Earth to the Sun), the light-year (the distance that light travels in one year), and the parsec.

Review Questions

1. What is the difference between a theory and a law of physics?

2. Describe the role that skepticism plays in science.

3. Describe one reason why it is useful to have telescopes in space.

4. What caused the craters on the Moon?

5. What role do nebulae like the Orion Nebula play in the life stories of stars?

6. What is the difference between a solar system and a galaxy?

7. What are degrees, arcminutes, and arcseconds used for? What are the relationships among these units of measure?

8. How many arcseconds equal 1°?

9. With the aid of a diagram, explain what it means to say that the Moon subtends an angle of ½°.

10. What is an exponent? How are exponents used in powers-of-ten notation?

11. What are the advantages of using powers-of-ten notation?

12. Write the following numbers using powers-of-ten notation: (**a**) ten million, (**b**) sixty thousand, (**c**) four one-thousandths, (**d**) thirty-eight billion, (**e**) your age in months.

13. How is an astronomical unit (AU) defined? Give an example of a situation in which this unit of measure would be convenient to use.

14. What is the advantage to the astronomer of using the light-year as a unit of distance?

15. What is a parsec? How is it related to a kiloparsec and to a megaparsec?

16. Give the word or phrase that corresponds to the following standard abbreviations: (**a**) km, (**b**) cm, (**c**) s, (**d**) km/s, (**e**) mi/h, (**f**) m, (**g**) m/s, (**h**) h, (**i**) y, (**j**) g, (**k**) kg. Which of these are units of speed? (*Hint:* You may have to refer to a dictionary. All of these abbreviations should be part of your working vocabulary.)

17. In the original (1977) *Star Wars* movie, Han Solo praises the speed of his spaceship by saying, "It's the ship that made the Kessel run in less than 12 parsecs!" Explain why this statement is obvious misinformation.

18. A reporter once described a light-year as "the time it takes light to reach us traveling at the speed of light." How would you correct this statement?

Advanced Questions

Questions preceded by an asterisk (*) involve topics discussed in the Boxes.

Problem-solving tips and tools

The small-angle formula, given in Box 1-1, relates the size of an astronomical object to the angle it subtends. Box 1-3 illustrates how to convert from one unit of measure to another. An object traveling at speed v for a time t covers a distance d given by $d = vt$; for example, a car traveling at 90 km/h (v) for 3 hours (t) covers a distance $d = (90 \text{ km/h})(3 \text{ h}) = 270$ km. Similarly, the time t required to cover a given distance d at speed v is $t = d/v$; for example, if $d = 270$ km and $v = 90$ km/h, then $t = (270 \text{ km})/(90 \text{ km/h}) = 3$ hours.

19. What is the meaning of the letters R I V U X G that appear under some of the figures in this chapter? Why in each case is one of the letters highlighted? (*Hint:* See the section "How to Use This Textbook" that precedes Chapter 1.)

20. The diameter of the Sun is 1.4×10^{11} cm, and the distance to the nearest star, Proxima Centauri, is 4.2 ly. Suppose you want to build an exact scale model of the Sun and Proxima Centauri, and you are using a ball 30 cm in diameter to represent the Sun. In your scale model, how far away would Proxima Centauri be from the Sun? Give your answer in kilometers, using powers-of-ten notation.

21. How many Suns would it take, laid side by side, to reach the nearest star? Use powers-of-ten notation. (*Hint:* See the preceding question.)

22. A hydrogen atom has a radius of about 5×10^{-9} cm. The radius of the observable universe is about 14 billion light-years. How many times larger than a hydrogen atom is the observable universe? Use powers-of-ten notation.

23. The Sun's mass is 1.99×10^{30} kg, three-quarters of which is hydrogen. The mass of a hydrogen atom is 1.67×10^{-27} kg. How many hydrogen atoms does the Sun contain? Use powers-of-ten notation.

24. The average distance from the Earth to the Sun is 1.496×10^8 km. Express this distance (**a**) in light-years and (**b**) in parsecs. Use powers-of-ten notation. (**c**) Are light-years or parsecs useful units for describing distances of this size? Explain.

25. The speed of light is 3.00×10^8 m/s. How long does it take light to travel from the Sun to Earth? Give your answer in seconds, using powers-of-ten notation. (*Hint:* See the preceding question.)

26. When the *Voyager 2* spacecraft sent back pictures of Neptune during its flyby of that planet in 1989, the spacecraft's radio signals traveled for 4 hours at the speed of light to reach Earth. How far away was the spacecraft? Give your answer in kilometers, using powers-of-ten notation. (*Hint:* See the preceding question.)

27. The star Procyon is 3.50 pc from Earth. (**a**) What is the distance to Procyon in kilometers? Use powers-of-ten notation. (**b**) How long does it take for light emanating from Procyon to reach Earth? Give your answer in years. (*Hint:* You do not need to know the value of the speed of light.)

28. The age of the universe is about 14 billion years. What is this age in seconds? Use powers-of-ten notation.

*29. Explain where the number 206,265 in the small-angle formula comes from.

*30. At what distance would a person have to hold a European 2-euro coin (which has a diameter of about 2.6 cm) in order for the coin to subtend an angle of (a) 1°? (b) 1 arcmin? (c) 1 arcsec? Give your answers in meters.

*31. A person with good vision can see details that subtend an angle of as small as 1 arcminute. If two dark lines on an eye chart are 2 millimeters apart, how far can such a person be from the chart and still be able to tell that there are two distinct lines? Give your answer in meters.

*32. The average distance to the Moon is 384,000 km, and the Moon subtends an angle of ½°. Use this information to calculate the diameter of the Moon in kilometers.

*33. Suppose your telescope can give you a clear view of objects and features that subtend angles of at least 2 arcsec. What is the diameter of the smallest crater you can see on the Moon? (*Hint:* See the preceding question.)

*34. On January 11, 2003, the planet Venus was a distance of 0.719 AU from Earth. The diameter of Venus is 12,104 km. What was the angular size of Venus as seen from Earth on January 11, 2003? Give your answer in arcminutes.

Discussion Questions

35. Scientists assume that "reality is rational." Discuss what this means and the thinking behind it.

36. All scientific knowledge is inherently provisional. Discuss whether this is a weakness or a strength of the scientific method.

37. How do astronomical observations differ from those of other sciences?

Web/CD-ROM Questions

38. Use the links given in the *Universe* Web site, Chapter 1, to learn about the Orion Nebula (Figure 1-5). Can the nebula be seen with the naked eye? Does the nebula stand alone, or is it part of a larger cloud of interstellar material? What has been learned by examining the Orion Nebula with telescopes sensitive to infrared light?

39. Use the links given in the *Universe* Web site, Chapter 1, to learn more about the Crab Nebula (Figure 1-6). When did observers on Earth see the supernova that created this nebula? Does the nebula emit any radiation other than visible light? What kind of object is at the center of the nebula?

40. Search the World Wide Web to learn more about *Topex/Poseidon* (see Figure 1-13) and other space missions that study the Earth. What are an "El Niño" and a "La Niña"? How can they be detected from space? How can the speed of winds over the oceans be measured by an orbiting satellite?

41. Access the AIMM (Active Integrated Media Module) called "Small-Angle Toolbox" in Chapter 1 of the *Universe* CD-ROM or Web site. Use this to determine the diameters in kilometers of the Sun, Saturn, and Pluto given the following distances and angular sizes:

Object	Distance (km)	Angular size (″)
Sun	1.5×10^8	1800
Saturn	1.5×10^9	16.5
Pluto	6.3×10^9	0.06

Observing Projects

42. On a dark, clear, moonless night, can you see the Milky Way from where you live? If so, briefly describe its appearance. If not, what seems to be interfering with your ability to see the Milky Way?

43. Look up at the sky on a clear, cloud-free night. Is the Moon in the sky? If so, does it interfere with your ability to see the fainter stars? Why do you suppose astronomers prefer to schedule their observations on nights when the Moon is not in the sky?

44. Look up at the sky on a clear, cloud-free night and note the positions of a few prominent stars relative to such reference markers as rooftops, telephone poles, and treetops. Also note the location from where you make your observations. A few hours later, return to that location and again note the positions of the same bright stars that you observed earlier. How have their positions changed? From these changes, can you deduce the general direction in which the stars appear to be moving?

45. Use the CD-ROM that accompanies this book to install the *Starry Night Backyard*™ planetarium software on your computer. Use *Starry Night Backyard*™ to determine when the Moon is visible today during the day and when it is visible tonight. Determine which, if any, of the following planets are visible tonight: Mercury, Venus, Mars, Jupiter, and Saturn. *Hints:* (1) Feel free to experiment with *Starry Night Backyard*™. You can always return to your starting screen by selecting **Go Home** from the **View** menu. To see the sky from your actual location on Earth, select **Viewing Location...** in the **View** menu, then scroll through the list of locations to find your city or town. (2) To change your viewing direction, move the mouse until the cursor changes into a little hand. Then, hold down the mouse button (on a Windows computer, the left button) as you move the mouse and you will move the sky. (3) Use the Control Panel at the top of the main window to change the time and date that is displayed as well as how rapidly time appears to change. (4) Use the **Find...** command in the **Edit** menu to locate specific planets or stars by name. (5) To learn about any object in the sky, point the cursor at the object and double-click the mouse.

You can find even more information about the program in the *Starry Night Backyard*™ manual (in *Starry Night Backyard*™, select **User's Guide** or **Manual** in the **Help** menu).

46. Use the CD-ROM that accompanies this book to install the *Deep Space Explorer*™ software on your computer. Then use *Deep Space Explorer*™ to investigate the Milky Way Galaxy. When the program starts up, you will see the Milky Way as it would appear from a great distance. A box near the bottom of the screen shows you your distance from the Sun, which lies within the Milky Way. (Moving the cursor over any object on the screen shows you the name of the object, the object's distance from your position in space, and the object's apparent magnitude—a measure of how bright it appears from your position.) (**a**) You can rotate the Milky Way (as though you were viewing it from different angles) by putting the mouse cursor over the image, holding down the mouse button, and moving the mouse. (On a two-button mouse, hold down the left mouse button.) How would you describe the shape of the Milky Way? (**b**) You can zoom in and zoom out using the buttons at the upper left of the window (an upward-pointing triangle and a downward-pointing triangle). Zoom in until you can see the planets in their orbits around the Sun, then zoom back out until you can again see the entire Milky Way Galaxy. Are the Sun and planets located at the center of the Milky Way? How would you describe their location?

Collaborative Exercises

47. A scientific theory is fundamentally different than the everyday use of the word "theory." List and describe any three scientific theories of your choice and creatively imagine an additional three hypothetical theories that are not scientific. Briefly describe what is scientific and what is nonscientific about each of these theories.

48. Angular measure describes how far apart two objects appear to an observer. From where you are currently sitting, estimate the angular distance between the floor and the ceiling at the front of the room you are sitting in, the angular distance between the two people sitting closest to you, and the angular size of a clock or an exit sign on the wall. Draw sketches to illustrate each answer and describe how each of your answers would change if you were standing in the very center of the room.

49. Astronomers use powers of ten to describe the distances to objects. List an object or place that is located at very roughly each of the following distances from you: 10^{-2} m, 10^{0} m, 10^{1} m, 10^{3} m, 10^{7} m, 10^{10} m, and 10^{20} m.

SANDRA M. FABER

Why Astronomy?

Sandra M. Faber, professor of astronomy at the University of California, Santa Cruz, and astronomer at Lick Observatory, was intrigued by the origins of the universe when she entered Swarthmore College. She completed her doctorate in astronomy at Harvard University but did her dissertation at the Carnegie Institution's Department of Terrestrial Magnetism, where she was influenced by the distinguished astronomer Vera Rubin. Dr. Faber chaired the now legendary group of astronomers called the Seven Samurai, who surveyed the nearest 400 elliptical galaxies and discovered a new mass concentration, the Great Attractor. She is the recipient of many honors and awards for her research, including the Bok Prize from Harvard University in 1978. She was elected to the National Academy of Sciences in 1985 and in 1986 won the coveted Dannie Heineman Prize from the American Astronomical Society, given in recognition of a sustained body of especially influential astronomical research. She is also on the Board of Trustees of the Carnegie Institution.

As you study astronomy, it is appropriate for you to ask, "Why am I studying this subject? What good is it for human beings in general and for me in particular?" Astronomy, it must be admitted, does not offer the same practical benefits as other sciences. So how can astronomy be important to your life?

On the most basic level, I like to think of astronomy as providing the introductory chapters for the ultimate textbook on human history. Ordinary texts start with recorded history, going back some 3000 years. For events before that, we consult archeologists and anthropologists, who tell us about the early history of our species. For knowledge of the time before that, we consult paleontologists, biologists, and geologists about the origin and evolution of life and the evolution of our planet, altogether going back some five billion years. Astronomy tells us about the vast stretch of time before the origin of the Earth, the ten billion years or so that saw the formation of the Sun and solar system, the Milky Way Galaxy, and the origin of the universe in the Big Bang. Knowledge of astronomy is essential for a well-educated person's view of history.

Astronomy challenges our belief system and impels us to put our "philosophical house" in order. Take the origin of the world, for example. According to Genesis in the Bible, the world and all in it were created in six days by the hand of God. However, the ancient Egyptians believed that the Earth arose spontaneously from the infinite waters of the eternal universe, Nun, and Alaskan legends taught that the world was created by the conscious imaginings of a creator deity named Father Raven.

The modern astronomical story of the creation of the Earth differs from all these in that it claims to be buttressed by physics and by observational fact. The Sun, it is asserted, formed via gravitational collapse from a dense cloud of interstellar dust and gas about five billion years ago. The planets coalesced at the same time as condensates within the swirling solar nebula. This process took not weeks or days but several hundred thousand years. Confidence in this theory stems from our looking out into the Galaxy and seeing young stars actually form in this way.

At issue here, really, is the deep question of how we are to gain information about the nature of the physical world—whether by revelation and intuition or by logic and observation. Where science stops and faith begins is a thorny issue for all human beings, but particularly so for astronomers—and for astronomy students.

Astronomy cultivates our notions about cosmic time and cosmic evolution. It is all too easy, given the short span of human life, to overlook the fact that the universe is a dynamic place. When I first entered astronomy, I found myself handicapped by a conservative mind-set that assumed, subtly, that celestial objects were unchanging. This might seem strange for someone whose avowed interest was in learning how galaxies formed. Such are the vagaries of the human mind! Fortunately, I lost this bias as I grew older, mainly, I think, by observing evolution all around me. Many things that had seemed immutable to me as a child—social structures, customs, even the physical environment—turned out not to be so. With that realization, the veil fell from my eyes, and I was able to accept cosmic evolution as a core concept in my thinking about the universe.

Following closely on the concept of evolution is the concept of fragility—if something can change, it might even actually disappear someday. The most obvious example is the limited lifetime of our Sun. In another five billion years or so, the Sun will enter its death throes, during which it will swell up and brighten to 1000 times its present luminosity and, in the process, incinerate the Earth.

Five billion years is far enough in the future that neither you nor I feel any personal responsibility for preparing the human race to meet this challenge. However, there are many other cosmic catastrophes that will befall us in the meantime. The Earth will be hit by sizable pieces of space junk; eroded impact craters show this happens all the time—every few million years or so. Debris thrown up into the atmosphere from these events could be so extensive as to totally alter the climate for an extended period of time, possibly leading to mass extinctions. There will be enormous volcanic eruptions that will make Mount St. Helens look like a small firecracker. Another Ice Age, which will totally disrupt modern agriculture, is virtually certain within the next 20,000 years, unless we cook the Earth ourselves first by burning too much fossil fuel.

Closer to the human sphere, such common notions as the inevitability of human progress, the desirability of endless economic growth, the ability of the Earth to support its growing human population, are all based on limited experience and may not—and probably *will* not—prove viable in the long run. Consequently, we must rethink who we are as a species and what is our proper activity on Earth. Contemplating cosmic time puts us into the appropriate frame of mind for grappling with these issues. These are problems that are very long range and involve the whole human race yet are totally under human control and are vital to our long-term survival and well-being.

Astronomy is essential to developing a perspective on human existence and its relation to the cosmos. What question could be grander and at the same time more relevant to us? We can hardly contemplate human existence in general without taking at least a brief look at our own lives.

Earth is a mote in the vast cosmic sea, and our lives, in the evolutionary scheme of the universe, would seem at face value to be insignificant. Should that thought terrify us, depriving our lives of any meaning? Should we seek to provide meaning by introducing an intelligent creator? Or should we be fully comforted merely to know that this great universe has given birth to us in an entirely natural, perhaps even inevitable, chain of events?

Not long ago I visited the Southern Hemisphere for the first time to observe the sky from Chile. At that latitude, the center of the Milky Way passes overhead, where it makes a grand show, not the miserable, fuzzy patch that is visible, low on the horizon, from North America. Stepping out on the observatory catwalk one morning before dawn, I saw for the first time the galactic center in all its glory soaring overhead. At that instant, my perceptions underwent a profound alteration. One moment I was standing in an earthbound "room" with the Galaxy painted on the "ceiling." The next instant, the walls of this earthly room fell away, and I was floating freely in outer space, viewing the center of the Galaxy from an outpost on its periphery. The sense of depth was breathtaking: Sparkling star clusters and dense dust clouds were etched against the blazing inner bulge of the Galaxy many thousands of light-years distant. I saw the Galaxy as an immense, three-dimensional object in space for the first time. It became "real." At that moment, my sense of place in the universe underwent a fundamental and irreversible change. I remained a citizen of the Earth, but I also became a citizen of the Galaxy.

Many astronomers believe that the ultimate, proper concept of "home" for the human race is our universe. It seems more and more likely that there are a very large number of other universes. Virtually none of them would appear to resemble ours; they may have different numbers of space and time dimensions—perhaps even no space or time—and very different physical laws. The vast majority probably are incapable of harboring intelligent life as we know it.

The parallel with Earth is striking. Among the solar system planets, only Earth can support human life. Among the great number of planets that likely exist in our Galaxy, only a small fraction are apt to be such that we could call them home. The fraction of hospitable universes is likely to be smaller still. Our universe would seem, then, to be the ultimate instance of "home": a sanctuary in a vast sea of inhospitable universes.

I began this essay by talking about history and ended with questions that border on the ethical and religious. Astronomy is like that: It offers a modern-day version of Genesis—and of Apocalypse, too. Astronomy stimulates and challenges us on our deepest, most human levels. I hope that during this course you will be able to take time out to contemplate the broader implications of what you are studying. This is one of the rare opportunities in life to think about who you are and where you and the human race are going. Don't miss it.

2 Knowing the Heavens

It is a clear night atop Mauna Kea, a dormant volcano on the island of Hawaii. As you gaze toward the north, as in this time-exposure photograph, you find that the stars are not motionless. Rather, they move in counterclockwise circles around a fixed point above the northern horizon. Stars close to this point never dip below the horizon, while stars farther from the fixed point rise in the east and set in the west. These motions fade from view when the Sun rises in the east and illuminates the sky. The Sun, too, arcs across the sky in the same manner as the stars. At day's end, when the Sun sets in the west, the panorama of stars is revealed for yet another night.

These observations are at the heart of *naked-eye astronomy*—the sort that requires no equipment but human vision. Naked-eye astronomy cannot tell us what the Sun is made of or how far away the stars are. For such purposes we need tools such as the Canada-France-Hawaii Telescope, housed within the dome shown in the photograph. (The red trails were made by vehicles on the road to the telescope.) But by studying naked-eye astronomy, you will learn the answers to equally profound questions such as why there are seasons, why the night sky is different at different times of year, and why the night sky looks different in Australia than in North America. In discovering the answers to these questions, you will learn how the Earth moves through space and will begin to understand our true place in the cosmos.

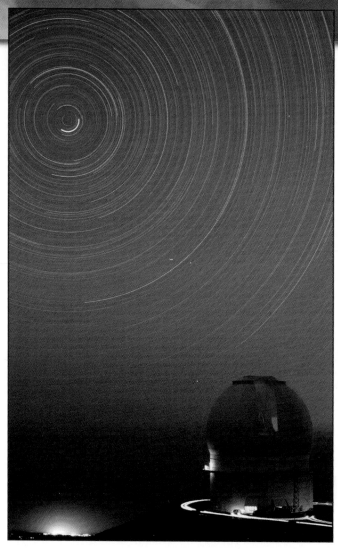

The Earth's rotation makes stars appear to trace out circles in the sky. (Richard J. Wainscoat, University of Hawaii) R I **V** U X G

Questions to Guide Your Reading

As you read the sections of this chapter, look for the answers to the following questions:

2-1 What role did astronomy play in ancient civilizations?

2-2 Are the stars that make up a constellation actually close to one another?

2-3 Are the same stars visible every night of the year? What is so special about the North Star?

2-4 Are the same stars visible from any location on Earth?

2-5 What causes the seasons? Why are they opposite in the northern and southern hemispheres?

2-6 Has the same star always been the North Star?

2-7 Can we use the rising and setting of the Sun as the basis of our system of keeping time?

2-8 Why are there leap years?

2-1 | Naked-eye astronomy had an important place in ancient civilizations

 Positional astronomy—the study of the positions of objects in the sky and how these positions change—has roots that extend far back in time. Four to five thousand years ago, the inhabitants of the British Isles erected stone structures, such as Stonehenge, that suggest a preoccupation with the motions of the sky. Alignments of these stones appear to show where the Sun rose and set at key times during the year. A similar structure in the New World is the Medicine Wheel, a circular ring of stones constructed by the Plains Indians atop a windswept plateau in Wyoming. Certain stones in this ring mark the rising points of the Sun and certain bright stars on the first day of summer.

Aztec, Mayan, and Incan architects in Central and South America also designed buildings with astronomical orientations. At the ruined city of Tiahuanaco in Bolivia, the walls of the Temple of the Sun were aligned north-south and east-west with an accuracy of better than one degree. The great Egyptian pyramids, built around 3000 B.C., are likewise oriented north-south and east-west with remarkable precision.

In addition to temples and tombs, some ancient buildings appear to have been dedicated expressly to astronomy. One of the best examples was built a thousand years ago in the Mayan city of Chichén Itzá, on the Yucatán Peninsula (**Figure 2-1**). The Caracol's cylindrical tower contains windows aligned with the northernmost and southernmost rising and setting points of both the Sun and the planet Venus. A similar four-story adobe building, probably constructed during the fourteenth century, is located at the Casa Grande site in Arizona.

These structures bear witness to an awareness of naked-eye astronomy by the peoples of many cultures. Many of the concepts of modern positional astronomy come to us from these ancients, including the idea of dividing the sky into constellations.

2-2 | Eighty-eight constellations cover the entire sky

 Looking at the sky on a clear, dark night, you might think that you can see millions of stars. Actually, the unaided human eye can detect only about 6000 stars. Because half of the sky is below the horizon at any one time, you can see at most about 3000 stars. When ancient peoples looked at these thousands of stars, they imagined that groupings of stars traced out pictures in the sky. Astronomers still refer to these groupings, called **constellations** (from the Latin for "group of stars").

You may already be familiar with some of these pictures or patterns in the sky, such as the Big Dipper, which is actually part of the large constellation Ursa Major (the Great Bear). Many constellations, such as Orion in **Figure 2-2**, have names derived from the myths and legends of antiquity. Although some star groupings vaguely resemble the figures they are supposed to represent (see Figure 2-2c), most do not.

The term "constellation" has a broader definition in present-day astronomy. On modern star charts, the entire sky is divided into

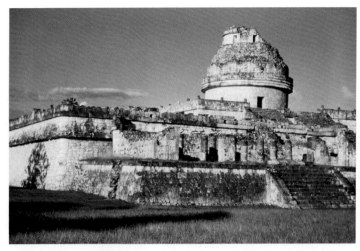

Figure 2-1 R I **V** U X G

The Caracol at Chichén Itzá The architecture of this Mayan observatory in the Yucatán, built about A.D. 1000, is based on alignments with important celestial events. Mayan astronomers measured the motions of celestial bodies with great precision and developed a very accurate calendar based on the motions of the planet Venus. (Courtesy of E. C. Krupp)

88 regions, each of which is called a constellation. For example, the constellation Orion is now defined to be an irregular patch of sky whose borders are shown in Figure 2-2b. When astronomers refer to the "Great Nebula" M42 in Orion, they mean that as seen from Earth this nebula appears to be within Orion's patch of sky. Some constellations cover large areas of the sky (Ursa Major being one of the biggest) and others very small areas (Crux, the Southern Cross, being the smallest). But because the modern constellations cover the entire sky, every star lies in one constellation or another.

CAUTION! When you look at a constellation's star pattern, it is tempting to conclude that you are seeing a group of stars that are all relatively close together. In fact, most of these stars are nowhere near one another. As an example, Figure 2-2b shows the distances in light-years to four stars in Orion. Although Bellatrix (Arabic for "the Amazon") and Mintaka ("the belt") appear to be close to each other, Mintaka is actually more than 600 light-years farther away from us. The two stars only *appear* to be close because they are in nearly the same direction as seen from Earth. The same illusion often appears when you see an airliner's lights at night. It is very difficult to tell how far away a single bright light is, which is why you can mistake an airliner a few kilometers away for a star trillions of times more distant.

The star names shown in Figure 2-2b are from the Arabic. For example, Betelgeuse means "armpit," which makes sense when you look at the star atlas drawing in Figure 2-2c. Other types of names are also used for stars. For example, Betelgeuse is also known as α Orionis (α is the Greek letter alpha) and as HD 39801.

(a)

(b)

(c)

figure 2-2 R I **V** U X G

Three Views of Orion From December through March, the constellation Orion is easily seen high above the southern horizon from North America or Europe. From Australia or South America, it appears high above the northern horizon. **(a)** This photograph of Orion shows many more stars than can be seen with the naked eye. **(b)** A portion of a modern star atlas shows the distances in light-years (ly) to some of the stars in Orion. The yellow lines show the borders between Orion and its neighboring constellations (labeled in capitals). **(c)** This fanciful drawing from a star atlas published in 1835 shows Orion the Hunter as well as other celestial creatures. (a: Luke Dodd/Science Photo Library/Photo Researchers; c: Courtesy of Janus Publications)

CAUTION ! A number of unscrupulous commercial firms offer to name a star for you for a fee. The money that they charge you for this "service" is real, but the star names are not; none of these names are recognized by astronomers. If you want to use astronomy to commemorate your name or the name of a friend or relative, consider making a donation to your local planetarium or science museum. The money will be put to much better use!

2-3 | The appearance of the sky changes during the course of the night and from one night to the next

Go outdoors soon after dark, find a spot away from bright lights, and note the patterns of stars in the sky. Do the same a few hours later. You will find that the entire pattern of stars (as well as the Moon, if it is visible) has shifted its position. New constellations will have risen above the eastern horizon, and some will have disappeared below the western horizon. If you look again before dawn, you will see that the stars that were just rising in the east when the night began are now low in the western sky. This daily motion, or **diurnal motion,** of the stars is apparent in time-exposure photographs (see the photograph that opens this chapter).

If you repeat your observations on the following night, you will find that the motions of the sky are almost but not quite the same. The same constellations rise in the east and set in the west, but a few minutes earlier than on the previous night. If you look again after a month, the constellations visible at a given time of night (say, midnight) will be noticeably different, and after six months you will see an almost totally different set of constellations. Only after a year has passed will the night sky have the same appearance as when you began.

Why does the sky go through diurnal motion? Why do the constellations slowly shift from one night to the next? As we will see, the answer to the first question is that the Earth *rotates* once a day around an axis from the north pole to the south pole, while the answer to the second question is that the Earth also *revolves* once a year around the Sun.

To understand diurnal motion, note that at any given moment it is daytime on the half of the Earth illuminated by the Sun and nighttime on the other half (**Figure 2-3**). The Earth rotates from west to east, making one complete rotation every 24 hours, which is why there is a daily cycle of day and night. Because of this rotation, stars appear to us to rise in the east and set in the west, as do the Sun and Moon.

Figure 2-4 helps to further explain diurnal motion. It shows two views of the Earth as seen from a point above the north pole.

figure 2-3 R I **V** U X G

Day and Night on the Earth At any moment, half of the Earth is illuminated by the Sun. As the Earth rotates from west to east, your location moves from the dark (night) hemisphere into the illuminated (day) hemisphere and back again. This image was recorded in 1992 by the *Galileo* spacecraft as it was en route to Jupiter. (JPL/NASA)

At the instant shown in Figure 2-4*a*, it is day in Asia but night in most of North America and Europe. Figure 2-4*b* shows the Earth 4 hours later. Four hours is one-sixth of a complete 24-hour day, so the Earth has made one-sixth of a rotation between Figures 2-4*a* and 2-4*b*. Europe is now in the illuminated half of the Earth (the Sun has risen in Europe), while Alaska has moved from the illuminated to the dark half of the Earth (the Sun has set in Alaska). For a person in California, in Figure 2-4*a* the time is 8:00 P.M. and the constellation Cygnus (the Swan) is directly overhead. Four hours later, the constellation over California is Andromeda (named for a mythological princess). Because the Earth rotates from west to east, it appears to us on Earth that the entire sky rotates around us in the opposite direction, from east to west.

We described earlier that in addition to the diurnal motion of the sky, the constellations visible in the night sky also change slowly over the course of a year. This happens because the Earth orbits, or revolves around, the Sun (**Figure 2-5**). Over the course of a year, the Earth makes one complete orbit, and the darkened, nighttime side of the Earth gradually turns toward different parts of the heavens. For example, as seen from the northern hemisphere, at midnight in late July the constellation Cygnus is close to overhead; at midnight in late September the constellation Andromeda is close to overhead; and at midnight in late November the constellation Perseus (commemorating a mythological hero) is close to overhead. If you follow a particular star on successive evenings, you will find that it rises approximately 4 minutes earlier each night, or 2 hours earlier each month.

Constellations can help you find your way around the sky. For example, if you live in the northern hemisphere, you can use the Big Dipper in Ursa Major to find the north direction by drawing a straight line through the two stars at the front of the Big Dipper's

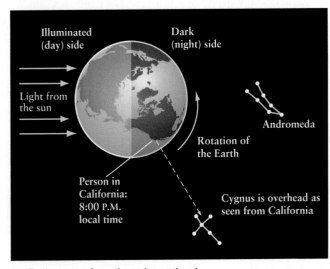

(a) Earth as seen from above the north pole

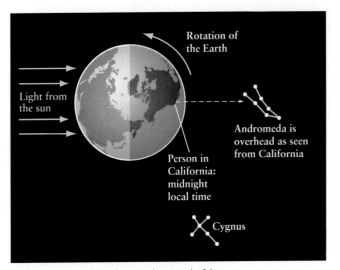

(b) 4 hours (one-sixth of a complete rotation) later

figure 2-4

Why Diurnal Motion Happens The diurnal (daily) motion of the stars, the Sun, and the Moon is a consequence of the Earth's rotation. (a) This drawing shows the Earth from a vantage point above the north pole. In this drawing, for a person in California the local time is 8:00 P.M. and the constellation Cygnus is

directly overhead. (b) Four hours later, the Earth has made one-sixth of a complete rotation to the east. As seen from Earth, the entire sky appears to have rotated to the west by one-sixth of a complete rotation. It is now midnight in California, and the constellation directly over California is Andromeda.

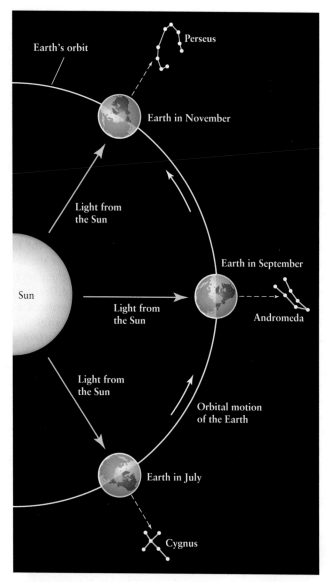

figure 2-5

Why the Night Sky Changes During the Year As the Earth orbits around the Sun, the nighttime side of the Earth gradually turns toward different parts of the sky. Hence the particular stars that you see in the night sky are different at different times of the year. This figure shows which constellation is overhead at midnight local time—when the Sun is on the opposite side of the Earth from your location—during different months for observers at midnorthern latitudes (including the United States). If you want to view the constellation Andromeda, the best time of the year to do it is in late September, when Andromeda is nearly overhead at midnight.

bowl (Figure 2-6). The first moderately bright star you come to is Polaris, also called the North Star because it is located almost directly over the Earth's north pole. If you draw a line from Polaris straight down to the horizon, you will find the north direction.

As Figure 2-6 shows, by following the handle of the Big Dipper you can locate the bright reddish star Arcturus in Boötes (the Shepherd) and the prominent bluish star Spica in Virgo (the Virgin). The

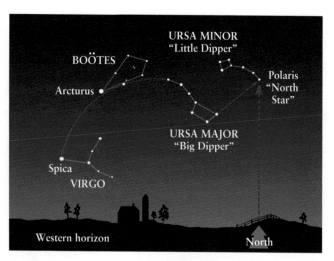

figure 2-6

The Big Dipper as a Guide The North Star can be seen from anywhere in the northern hemisphere on any night of the year. This star chart shows how the Big Dipper can be used to point out the North Star as well as the brightest stars in two other constellations. The chart shows the sky at around 11 P.M. (daylight savings time) on August 1. Due to the Earth's orbital motion around the Sun, you will see this same view at 1 A.M. on July 1 and at 9 P.M. on September 1. The angular distance from Polaris to Spica is 102°.

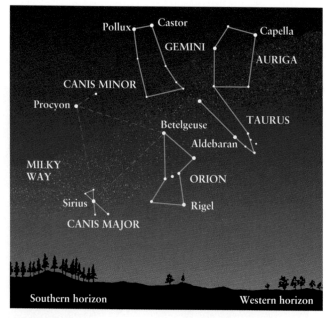

figure 2-7

The "Winter Triangle" This star chart shows the view toward the southwest on a winter evening in the northern hemisphere (around midnight on January 1, 10 P.M. on February 1, or 8 P.M. on March 1). Three of the brightest stars in the sky make up the "winter triangle," which is about 26° on a side. In addition to the constellations involved in the triangle, the chart shows the prominent constellations Gemini (the Twins), Auriga (the Charioteer), and Taurus (the Bull).

saying "Follow the arc to Arcturus and speed to Spica" may help you remember these stars, which are conspicuous in the evening sky during the spring and summer.

During winter in the northern hemisphere, you can see some of the brightest stars in the sky. Many of them are in the vicinity of the "winter triangle" (Figure 2-7), which connects bright stars in the constellations of Orion (the Hunter), Canis Major (the Large Dog), and Canis Minor (the Small Dog).

A similar feature, the "summer triangle," graces the summer sky in the northern hemisphere. This triangle connects the brightest stars in Lyra (the Harp), Cygnus (the Swan), and Aquila (the Eagle) (Figure 2-8). A conspicuous portion of the Milky Way forms a beautiful background for these constellations, which are nearly overhead during the middle of summer at midnight.

A wonderful tool to help you find your way around the night sky is the planetarium program *Starry Night Backyard*™, which you will find on the CD-ROM that accompanies this book. In addition, at the end of this book you will find a set of selected star charts for the evening hours of all 12 months of the year. You may find stargazing an enjoyable experience, and *Starry Night Backyard*™ and the star charts will help you identify many well-known constellations.

Note that all the star charts in this section and at the end of this book are drawn for an observer in the northern hemisphere. If you live in the southern hemisphere, you can see constellations that are not visible from the northern hemisphere, and vice versa. In the next section we will see why this is so.

2-4 | It is convenient to imagine that the stars are located on a celestial sphere

Many ancient societies believed that all the stars are the same distance from the Earth. They imagined the stars to be bits of fire imbedded into the inner surface of an immense hollow sphere, called the **celestial sphere**, with the Earth at its center. In this picture of the universe, the Earth was fixed and did not rotate. Instead, the entire celestial sphere rotated once a day around the Earth from east to west, thereby causing the diurnal motion of the sky. The picture of a rotating celestial sphere fit well with naked-eye observations, and for its time was a useful model of how the universe works. (We discussed the role of models in science in Section 1-1).

We now know that this simple model of the universe is not correct. Diurnal motion is due to the rotation of the Earth, not the rest of the universe. Furthermore, as we learned when discussing the constellations in Section 2-2, it is impossible to tell with the naked eye how far away the stars are. Indeed, the stars that you can see with the naked eye range from 4.2 to more than 1000 light-years away, and telescopes allow us to see objects at distances of billions of light-years.

Thus, astronomers now recognize that the celestial sphere is an *imaginary* object that has no basis in physical reality. Nonetheless, the celestial sphere model remains a useful tool of positional astronomy. If we imagine, as did the ancients, that the Earth is stationary and that the celestial sphere rotates around us, it is relatively easy to specify the directions to different objects in the sky and to visualize the motions of these objects.

Figure 2-9 depicts the celestial sphere, with the Earth at its center. (A truly proportional drawing would show the celestial sphere as being millions of times larger than the Earth.) We picture the stars as points of light that are fixed on the inner surface of the celestial sphere. If we project the Earth's equator out into space, we obtain the **celestial equator**. The celestial equator divides the sky into northern and southern hemispheres, just as the Earth's equator divides the Earth into two hemispheres.

If we project the Earth's north and south poles into space, we obtain the **north celestial pole** and the **south celestial pole**. The two celestial poles are where the Earth's axis of rotation (extended out into space) intersects the celestial sphere (see Figure 2-9). The star Polaris is less than 1° away from the north celestial pole, which is why it is called the North Star or the Pole Star.

The point in the sky directly overhead an observer anywhere on Earth is called that observer's **zenith**. The zenith and celestial sphere are shown in Figure 2-10 for an observer located at 35° north latitude (that is, at a location on the Earth's surface 35° north of the equator). The zenith is shown at the top of Figure 2-10, so the Earth and the celestial sphere appear "tipped" compared to Figure 2-9. At any time, an observer can see only half of the celestial sphere; the other half is below the horizon, hidden by the body of the Earth. The hidden half of the celestial sphere is darkly shaded in Figure 2-10.

For an observer anywhere in the northern hemisphere, including the observer in Figure 2-10, the north celestial pole is always above

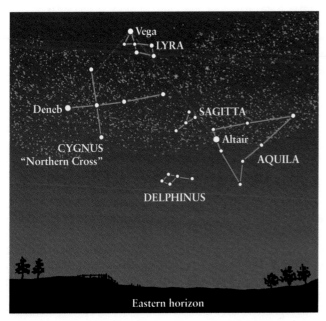

figure 2-8

The "Summer Triangle" This star chart shows the eastern sky as it appears in the evening during spring and summer in the northern hemisphere (around 1 A.M. daylight savings time on June 1, around 11 P.M. on July 1, and around 9 P.M. on August 1). The angular distance from Deneb to Altair is about 38°. The constellations Sagitta (the Arrow) and Delphinus (the Dolphin) are much fainter than the three constellations that make up the triangle.

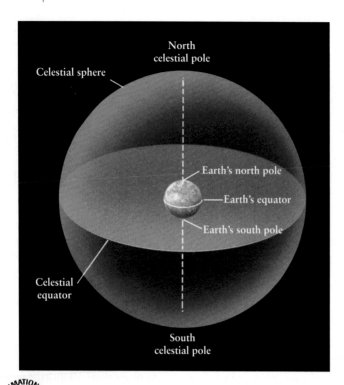

figure 2-9

The Celestial Sphere The celestial sphere is the apparent sphere of the sky. The view in this figure is from the outside of this (wholly imaginary) sphere. The Earth is at the center of the celestial sphere, so our view is always of the *inside* of the sphere. The celestial equator and poles are the projections of the Earth's equator and axis of rotation out into space. The celestial poles are therefore located directly over the Earth's poles.

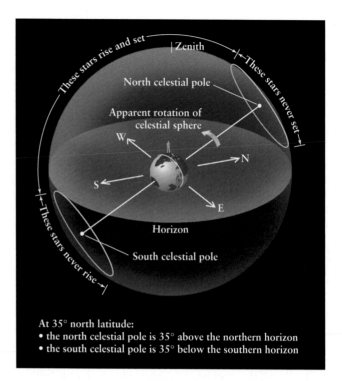

At 35° north latitude:
• the north celestial pole is 35° above the northern horizon
• the south celestial pole is 35° below the southern horizon

figure 2-10

The View from 35° North Latitude To an observer at 35° north latitude (roughly the latitude of Los Angeles, Atlanta, Tel Aviv, and Tokyo), the north celestial pole is always 35° above the horizon. Stars within 35° of the *north* celestial pole are circumpolar; they trace out circles around the north celestial pole during the course of the night, and are always above the horizon on any night of the year. Stars within 35° of the *south* celestial pole are always below the horizon and can never be seen from this latitude. Stars that lie between these two extremes rise in the east and set in the west.

the horizon. As the Earth turns from west to east—or, equivalently, as the celestial sphere appears to turn from east to west—stars sufficiently near the north celestial pole revolve around the pole, never rising or setting. Such stars are called **circumpolar**. For example, as seen from North America or Europe, Polaris is a circumpolar star and can be seen at any time of night on any night of the year. The photograph that opens this chapter shows the circular trails of stars around the north celestial pole as seen from Hawaii (at 20° north latitude). Stars near the south celestial pole revolve around that pole but always remain below the horizon of an observer in the northern hemisphere. Hence, these stars can never be seen by the observer in Figure 2-10. Stars between those two limits rise in the east and set in the west.

For an observer at 35° south latitude (the latitude of Sydney, Australia), the roles of the north and south celestial poles are reversed: Objects close to the *south* celestial pole are circumpolar, that is, they revolve around that pole and never rise or set. For an observer in the southern hemisphere, stars close to the *north* celestial pole are always below the horizon and can never be seen. Hence, Australian astronomers never see the North Star but are able to see other stars that are forever hidden from North American or European observers.

For observers at most locations on Earth, stars rise and set at an angle to the horizon (**Figure 2-11a**). To see why this is so, notice that the rotation of the celestial sphere carries stars across the sky in paths that are parallel to the celestial equator. If you stand at the north pole, the north celestial pole is directly above you at the zenith (Figure 2-9) and the celestial equator lies all around you at the horizon. Hence, as the celestial sphere rotates, the stars appear to move parallel to the horizon (Figure 2-11b). If instead you stand on the equator, the celestial equator passes from the eastern horizon through the zenith to the western horizon. The north and south celestial poles are 90° away from the celestial equator, so they lie on the northern and southern horizons respectively. As the celestial sphere rotates around an axis from pole to pole, the stars rise and set straight up and down—that is, in a direction perpendicular to the horizon (Figure 2-11c). At any location on Earth between the equator and either pole, the rising and setting motions of the stars are at an angle intermediate between Figures 2-11b and 2-11c. The particular angle depends on the latitude.

Using the celestial equator and poles, we can define a coordinate system to specify the position of any star on the celestial sphere. As **Box 2-1** describes, the most commonly used coordinate system uses two angles, *right ascension* and *declination*, that are

(a) At middle northern latitudes

(b) At the north pole

(c) At the equator

figure 2-11 R I **V** U X G
The Apparent Motion of Stars at Different Latitudes
As the Earth rotates, stars appear to rotate around us along paths that are parallel to the celestial equator. **(a)** As shown in this long time exposure, at most locations on Earth the rising and setting motions are at an angle to the horizon that depends on the latitude. **(b)** At the north pole (latitude 90° north) the stars appear to move parallel to the horizon. **(c)** At the equator (latitude 0°) the stars rise and set along vertical paths. (a: David Miller/DMI)

box 2-1 TOOLS OF THE ASTRONOMER'S TRADE

Celestial Coordinates

To denote the positions of objects in the sky, astronomers use a system based on *right ascension* and *declination*. Declination corresponds to latitude. As shown in the illustration, the **declination** of an object is its angular distance north or south of the celestial equator, measured along a circle passing through both celestial poles. It is measured in degrees, arcminutes, and arcseconds (see Section 1-5).

Right ascension corresponds to longitude. It is measured from a point on the celestial equator called the vernal equinox. This point is one of two locations where the Sun crosses the celestial equator during its apparent annual motion, as we discuss in Section 2-5. In the Earth's northern hemisphere, spring officially begins when the Sun reaches the vernal equinox in late March. The **right ascension** of an object is the angular distance from the vernal equinox eastward along the celestial equator to the circle used in measuring its declination (see illustration). Astronomers measure this angular distance in time units (hours, minutes, and seconds), corresponding to the time required for the celestial sphere to rotate through this angle. For example, suppose there is a star at your zenith right now with right ascension $6^h\ 0^m\ 0^s$. Two hours and 30 minutes from now, there will be a different object at your zenith with right ascension $8^h\ 30^m\ 0^s$.

 The coordinates of the bright star Rigel for the year 2000 are R.A. = $5^h\ 14^m\ 32.2^s$, Decl. = $-8°\ 12'\ 06''$. (R.A. and Decl. are abbreviations for right ascension and declination.) A minus sign on the declination indicates that the star is south of the celestial equator; a plus sign (or no sign at all) indicates that an object is north of the celestial equator. As we discuss in Box 2-2, right ascension helps determine the best time to observe a particular object.

It is important to state the year for which a star's right ascension and declination are valid. This is so because of precession, which we discuss in Section 2-6.

(Continued on following page)

TOOLS OF THE ASTRONOMER'S TRADE continued

EXAMPLE: What are the coordinates of a star that lies exactly halfway between the vernal equinox and the south celestial pole?

Situation: Our goal is to find the right ascension and declination of the star in question.

Tools: We use the definitions depicted in the figure.

Answer: Since the circle used to measure this star's declination passes through the vernal equinox, this star's right ascension is R.A. = 0^h 0^m 0^s. The angle between the celestial equator and south celestial pole is 90° 0′ 0″, so the declination of this star is Decl. = –45° 0′ 0″.

Review: The declination in this example is negative because the star is in the southern half of the celestial sphere.

EXAMPLE: At midnight local time you see a star with R.A. = 2^h 30^m 0^s at your zenith. When will you see a star at your zenith with R.A. = 21^h 0^m 0^s?

Situation: If you held your finger stationary over a globe of the Earth, the longitude of the point directly under your finger would change as you rotated the globe. In the same way, the right ascension of the point directly over your head (the zenith) changes as the celestial sphere rotates. We use this concept to determine the time in question.

Tools: We use the idea that a change in right ascension of 24^h corresponds to an elapsed time of 24 hours and a complete rotation of the celestial sphere.

Answer: The time required for the sky to rotate through the angle between the stars is the difference in their right ascensions: 21^h 0^m 0^s – 2^h 30^m 0^s = 18^h 30^m 0^s. So the second star will be at your zenith 18½ hours after the first one, or at 6:30 P.M. the following evening.

Review: Our answer was based on the idea that the celestial sphere makes *exactly* one complete rotation in 24 hours. If this were so, from one night to the next each star would be in exactly the same position at a given time. But because of the way that we customarily measure time, the celestial sphere makes slightly more than one complete rotation in 24 hours. (We explore the reasons for this in Box 2-2.) As a result, our answer is in error by about 3 minutes. For our purposes, this is a small enough error that we can ignore it.

---

quite like longitude and latitude on Earth. These coordinates tell us in what direction we should look to see the star. To locate the star's true position in three-dimensional space, we must also know the distance to the star.

2-5 | The seasons are caused by the tilt of Earth's axis of rotation

In addition to rotating on its axis every 24 hours, the Earth revolves around the Sun—that is, it *orbits* the Sun—in about 365¼ days. As we travel with the Earth around its orbit, we experience the annual cycle of seasons. But why *are* there seasons? Furthermore, the seasons are opposite in the northern and southern hemispheres. For example, February is midwinter in North America but midsummer in Australia. Why should this be?

The reason why we have seasons, and why they are different in different hemispheres, is that the Earth's axis of rotation is not perpendicular to the plane of the Earth's orbit. Instead, as Figure 2-12 shows, it is tilted about 23½° away from the perpendicular. The Earth maintains this tilt as it orbits the Sun, with the Earth's north pole pointing toward the north celestial pole. (This stability is a hallmark of all rotating objects. A top will not fall over as long as it is spinning, and the rotating wheels of a motorcycle help to keep the rider upright.)

During part of the year, when the Earth is in the part of its orbit shown on the left side of Figure 2-12, the northern hemisphere is tilted toward the Sun. As the Earth spins on its axis, a point in the northern hemisphere spends more than 12 hours in the sunlight. Thus, the days there are long and the nights are short, and it is summer in the northern hemisphere. The summer is hot not only because of the extended daylight hours but also because the Sun is high in the northern hemisphere's sky. As a result, sunlight strikes the ground at a nearly perpendicular angle that heats the ground efficiently (Figure 2-13a). During this same time of year in the southern hemisphere, the days are short and the nights are long, because a point in this hemisphere spends fewer than 12 hours a day in the sunlight. The Sun is low in the sky, so sunlight strikes the surface at a grazing angle that causes little heating (Figure 2-13b), and it is winter in the southern hemisphere.

Half a year later, the Earth is in the part of its orbit shown on the right side of Figure 2-12. Now the situation is reversed, with winter in the northern hemisphere (which is now tilted away from the Sun) and summer in the southern hemisphere. During spring and autumn, the two hemispheres receive roughly equal amounts of illumination from the Sun, and daytime and nighttime are of equal length everywhere on Earth.

 CAUTION A common misconception is that the seasons are caused by variations in the distance from the Earth to the Sun. According to this idea, the Earth is closer to the Sun in sum-

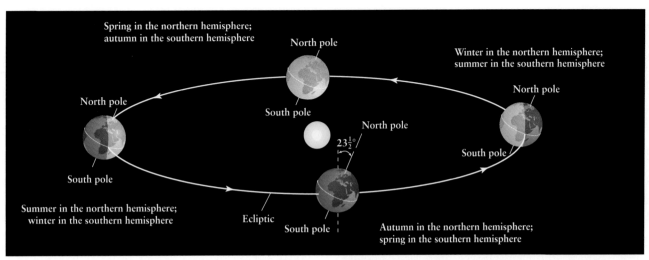

figure 2-12

The Seasons The Earth's axis of rotation is inclined 23½° away from the perpendicular to the plane of the Earth's orbit. The north pole is aimed at the north celestial pole, near the star Polaris. The Earth maintains this orientation as it orbits the Sun. Consequently, the amount of solar illumination and the number of daylight hours at any location on Earth vary in a regular pattern throughout the year. This is the origin of the seasons.

mer and farther away in winter. But in fact, the Earth's orbit around the Sun is very nearly circular, and the Earth-Sun distance varies only about 3% over the course of a year. (The Earth's orbit only *looks* elongated in Figure 2-12 because this illustration shows an oblique view.) We are slightly closer to the Sun in January than in July, but this small variation has little influence on the cycle of the seasons. Also, if the seasons were really caused by variations in the Earth-Sun distance, the seasons would be the same in both hemispheres!

As the Earth orbits around the Sun, the Sun's position (as seen from the Earth) gradually shifts with respect to the background stars. As **Figure 2-14** shows, the Sun therefore appears to trace out a circular path called the **ecliptic** on the celestial sphere. (This word suggests that the path traced out by the Sun has something

to do with eclipses. We will discuss the connection in Chapter 3.) Because there are 365¼ days in a year and 360° in a circle, the Sun appears to move along the ecliptic at a rate of about 1° per day. This motion is from west to east, that is, in the direction opposite to the apparent motion of the celestial sphere.

ANALOGY It may help to envision the celestial sphere as a merry-go-round rotating clockwise, and the Sun as a restless child who is walking slowly around the merry-go-round's rim in the counterclockwise direction. During the time it takes the child to make a round trip, the merry-go-round rotates 365¼ times.

The ecliptic can also be thought of as the projection onto the celestial sphere of the plane of the Earth's orbit. This is *not* the

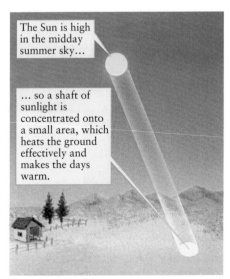

The Sun is high in the midday summer sky…

… so a shaft of sunlight is concentrated onto a small area, which heats the ground effectively and makes the days warm.

(a) The Sun in summer

The Sun is low in the midday winter sky…

… so the same shaft of sunlight is spread out over a larger area and less heating of the ground takes place.

(b) The Sun in winter

figure 2-13

Solar Energy in Summer and Winter At different times of the year, sunlight strikes the ground at different angles. (a) In summer, sunlight is concentrated and the days are also longer, which further increases the heating. (b) In winter the sunlight is less concentrated, the days are short, and little heating of the ground takes place. This accounts for the low temperatures in winter.

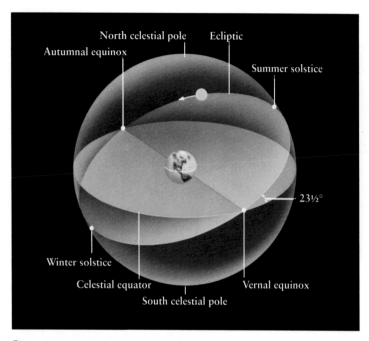

figure 2-14

The Ecliptic, Equinoxes, and Solstices The ecliptic is the apparent annual path of the Sun as projected onto the celestial sphere. Inclined to the celestial equator by 23½° because of the tilt of the Earth's axis of rotation, the ecliptic intersects the celestial equator at two points, called equinoxes. The northernmost point on the ecliptic is the summer solstice, and the southernmost point is the winter solstice. The Sun is shown in its approximate position for August 1.

stice that the Sun stops moving northward on the celestial sphere. At this point, the Sun is as far north of the celestial equator as it can get. It marks the location of the Sun at the moment summer begins in the northern hemisphere (about June 21). At the beginning of the northern hemisphere's winter (about December 21), the Sun is farthest south of the celestial equator at a point called the **winter solstice.**

Because the Sun's position on the celestial sphere varies slowly over the course of a year, its daily path across the sky (due to the Earth's rotation) also varies with the seasons (**Figure 2-15**). On the first day of spring or the first day of fall, when the Sun is at one of the equinoxes, the Sun rises directly in the east and sets directly in the west.

When the northern hemisphere is tilted away from the Sun and it is winter in the northern hemisphere, the Sun rises in the southeast. Daylight lasts for fewer than 12 hours as the Sun skims low over the southern horizon and sets in the southwest. Northern hemisphere nights are longest when the Sun is at the winter solstice.

The closer you get to the north pole, the shorter the winter days and the longer the winter nights. In fact, anywhere within 23½° of the north pole (that is, north of latitude 90° − 23½° = 66½° N) the Sun is below the horizon for 24 continuous hours at least one day of the year. The circle around the Earth at 66½° north latitude is called the **Arctic Circle** (**Figure 2-16**). The corresponding region around the south pole is bounded by the **Antarctic Circle** at 66½° south latitude. At the time of the winter solstice, explorers south of the Antarctic Circle enjoy "the midnight sun," or 24 hours of continuous daylight (Figure 2-16a).

During summer in the northern hemisphere, when the northern hemisphere is tilted toward the Sun, the Sun rises in the northeast and sets in the northwest. The Sun is at its northernmost position at

same as the plane of the Earth's equator, thanks to the 23½° tilt of the Earth's rotation axis shown in Figure 2-12. Instead, the ecliptic and the celestial equator are inclined to each other by that same 23½° angle.

The ecliptic and the celestial equator intersect at only two points, which are exactly opposite each other on the celestial sphere. Each point is called an **equinox** (from the Latin for "equal night"), because when the Sun appears at either of these points, day and night are each about 12 hours long at all locations on Earth. The term "equinox" is also used to refer to the date on which the Sun passes through one of these special points on the ecliptic.

On about March 21 of each year, the Sun crosses northward across the celestial equator at the **vernal equinox.** This marks the beginning of spring in the northern hemisphere ("vernal" is from the Latin for "spring"). On about September 22 the Sun moves southward across the celestial equator at the **autumnal equinox,** marking the moment when fall begins in the northern hemisphere. Since the seasons are opposite in the northern and southern hemispheres, for Australians and South Africans the vernal equinox actually marks the beginning of autumn. The names of the equinoxes come from astronomers of the past who lived north of the equator.

Between the vernal and autumnal equinoxes lie two other significant locations along the ecliptic. The point on the ecliptic farthest north of the celestial equator is called the **summer solstice.** "Solstice" is from the Latin for "solar standstill," and it is at the summer sol-

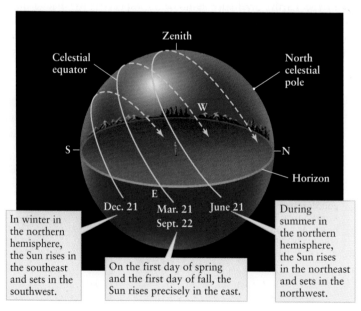

figure 2-15

The Sun's Daily Path Across the Sky This drawing shows the apparent path of the Sun during the course of a day on four different dates. Like Figure 2-10, this drawing is for an observer at 35° north latitude.

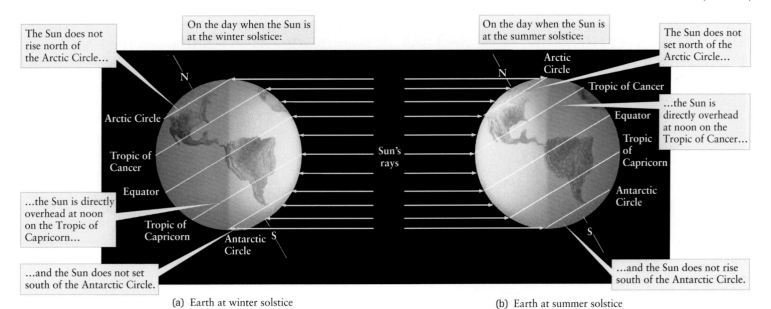

The Sun does not rise north of the Arctic Circle…

On the day when the Sun is at the winter solstice:

The Sun does not set north of the Arctic Circle…

Arctic Circle

Tropic of Cancer

Equator

Tropic of Cancer

…the Sun is directly overhead at noon on the Tropic of Cancer…

Tropic of Capricorn

Equator

Sun's rays

Tropic of Capricorn

Antarctic Circle

…the Sun is directly overhead at noon on the Tropic of Capricorn…

On the day when the Sun is at the summer solstice:

Arctic Circle

…and the Sun does not set south of the Antarctic Circle.

Antarctic Circle

…and the Sun does not rise south of the Antarctic Circle.

(a) Earth at winter solstice

(b) Earth at summer solstice

Figure 2-16

Tropics and Circles Four important latitudes on Earth are the Arctic Circle (66½° north latitude), Tropic of Cancer (23½° north latitude), Tropic of Capricorn (23½° south latitude), and Antarctic Circle (66½° south latitude). These drawings show the significance of these latitudes when the Sun is **(a)** at the winter solstice and **(b)** at the summer solstice.

the summer solstice, giving the northern hemisphere the greatest number of daylight hours. At the summer solstice the Sun does not set at all north of the Arctic Circle and does not rise at all south of the Antarctic Circle (Figure 2-16*b*).

The variations of the seasons are much less pronounced close to the equator. Between the **Tropic of Capricorn** at 23½° south latitude and the **Tropic of Cancer** at 23½° north latitude, the Sun is directly overhead—that is, at the zenith—at high noon at least one day a year. Outside of the tropics, the Sun is never directly overhead, but is always either south of the zenith (as seen from locations north of the Tropic of Cancer) or north of the zenith (as seen from south of the Tropic of Capricorn).

2-6 | The Moon helps to cause precession, a slow, conical motion of Earth's axis of rotation

The Moon is by far the brightest and most obvious naked-eye object in the nighttime sky. Like the Sun, the Moon slowly changes its position relative to the background stars; unlike the Sun, the Moon makes a complete trip around the celestial sphere in only about four weeks, or about a month. (The word "month" comes from the same Old English root as the word "moon.") Ancient astronomers realized that this motion occurs because the Moon orbits the Earth in roughly four weeks. In one hour the Moon moves on the celestial sphere by about ½°, or roughly its own angular size.

The Moon's path on the celestial sphere is never far from the Sun's path (that is, the ecliptic). This is because the plane of the Moon's orbit around the Earth is inclined only slightly from the plane of the Earth's orbit around the Sun. The Moon's path varies somewhat from one month to the next, but always remains within

a band called the **zodiac** that extends about 8° on either side of the ecliptic. Twelve famous constellations—Aries, Taurus, Gemini, Cancer, Leo, Virgo, Libra, Scorpius, Sagittarius, Capricornus, Aquarius, and Pisces—lie along the zodiac. The Moon is generally found in one of these 12 constellations. (Thanks to a redrawing of constellation boundaries in the mid-twentieth century, the zodiac actually passes through a thirteenth constellation—Ophiuchus, the Serpent Bearer—between Scorpius and Sagittarius.) As it moves along its orbit, the Moon appears north of the celestial equator for about two weeks and then south of the celestial equator for about the next two weeks. We will learn more about the Moon's motion, as well as why the Moon goes through phases, in Chapter 3.

The Moon not only moves around the Earth but, in concert with the Sun, also causes a slow change in the Earth's rotation. This is because both the Sun and the Moon exert a gravitational pull on the Earth. We will learn much more about gravity in Chapter 4; for now, all we need is the idea that gravity is a universal attraction of matter for other matter.

The gravitational pull of the Sun and the Moon affects the Earth's rotation because the Earth is slightly fatter across the equator than it is from pole to pole: Its equatorial diameter is 43 kilometers (27 miles) larger than the diameter measured from pole to pole. The Earth is therefore said to have an "equatorial bulge." Because of the gravitational pull of the Moon and the Sun on this bulge, the orientation of the Earth's axis of rotation gradually changes.

The Earth behaves somewhat like a spinning top, as illustrated in Figure 2-17. If the top were not spinning, gravity would pull the top over on its side. When the top is spinning, gravity causes the top's axis of rotation to trace out a circle, producing a motion called **precession.**

As the Sun and Moon move along the zodiac, each spends half its time north of the Earth's equatorial bulge and half its time south of it.

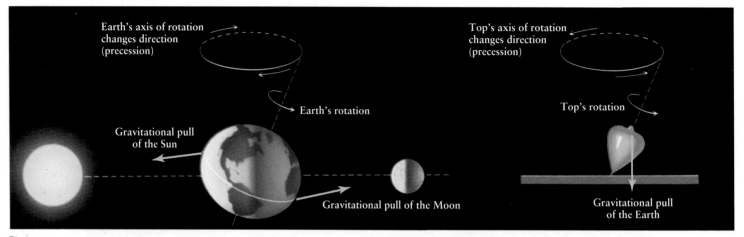

figure 2-17

Precession The gravitational pull of the Moon and the Sun on the Earth's equatorial bulge causes the Earth to precess. As the Earth precesses, its axis of rotation slowly traces out a circle in the sky, like the shifting axis of a spinning top.

The gravitational pull of the Sun and Moon tugging on the equatorial bulge tries to twist the Earth's axis of rotation to be perpendicular to the plane of the ecliptic. But because the Earth is spinning, the combined actions of gravity and rotation cause the Earth's axis to trace out a circle in the sky, much like what happens to the toy top. As the axis precesses, it remains tilted about 23½° to the perpendicular.

As the Earth's axis of rotation slowly changes its orientation, the north and south celestial poles—which are the projections of that axis onto the celestial sphere—change their positions relative to the stars. At present, the north celestial pole lies within 1° of the star Polaris, which is why Polaris is the North Star. But 5000 years ago, the north celestial pole was closest to the star Thuban in the constellation of Draco (the Dragon). Thus, that star and not Polaris was the North Star. And 12,000 years from now, the North Star will be the bright star Vega in Lyra (the Harp). It takes 26,000 years for the north celestial pole to complete one full precessional circle around the sky (**Figure 2-18**). The south celestial pole executes a similar circle in the southern sky.

Precession also causes the Earth's equatorial plane to change its orientation. Because this plane defines the location of the celestial equator in the sky, the celestial equator precesses as well. The intersections of the celestial equator and the ecliptic define the equinoxes (see Figure 2-14), so these key locations in the sky also shift slowly from year to year. For this reason, the precession of the Earth is also called the **precession of the equinoxes**. The first person to detect the precession of the equinoxes, in the second century B.C., was the Greek astronomer Hipparchus, who compared his own observations with those of Babylonian astronomers three centuries earlier. Today, the vernal equinox is located in the constellation Pisces (the Fishes). Two thousand years ago, it was in Aries (the Ram). Around the year A.D. 2600, the vernal equinox will move into Aquarius (the Water Bearer).

CAUTION Astrological terms like the "Age of Aquarius" involve boundaries in the sky that are not recognized by astronomers and are generally not even related to the positions of the constellations. For example, most astrologers would call

a person born on March 21, 1985, an "Aries" because the Sun was supposedly in the direction of that constellation on March 21. But due to precession, the Sun was actually in the constellation Pisces on that date! Indeed, astrology is *not* a science at all, but merely a col-

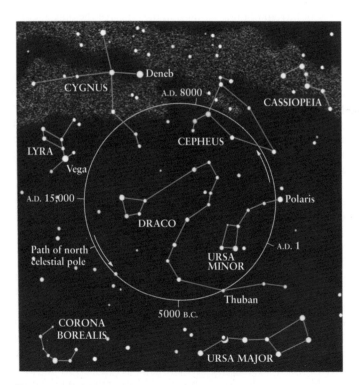

figure 2-18

Precession and the Path of the North Celestial Pole As the Earth precesses, the north celestial pole slowly traces out a circle among the northern constellations. At present, the north celestial pole is near the moderately bright star Polaris, which serves as the North Star. Twelve thousand years from now the bright star Vega will be the North Star.

lection of superstitions and hokum. Its practitioners use some of the terminology of astronomy but reject the logical thinking that is at the heart of science. James Randi has more to say about astrology and other pseudosciences in his essay "Why Astrology Is Not Science" at the end of this chapter.

The astronomer's system of locating heavenly bodies by their right ascension and declination, discussed in Box 2-1, is tied to the positions of the celestial equator and the vernal equinox. Because of precession, these positions are changing, and thus the coordinates of stars in the sky are also constantly changing. These changes are very small and gradual, but they add up over the years. To cope with this difficulty, astronomers always make note of the date (called the **epoch**) for which a particular set of coordinates is precisely correct. Consequently, star catalogs and star charts are periodically updated. Most current catalogs and star charts are prepared for the epoch 2000. The coordinates in these reference books, which are precise for January 1, 2000, will require very little correction over the next few decades.

2-7 | Positional astronomy plays an important role in keeping track of time

 SCIENCE IN PROCESS *How ancient scholars developed a system of timekeeping based on the Sun*

Astronomers have traditionally been responsible for telling time. This is because we want the system of timekeeping used in everyday life to reflect the position of the Sun in the sky. Thousands of years ago, the sundial was invented to keep track of **apparent solar time.** To obtain more accurate measurements astronomers use the **meridian.** As Figure 2-19 shows, this is a north-south circle on the celestial sphere that passes through the zenith (the point directly overhead) and both celestial poles. *Local noon* is defined to be when the Sun crosses the **upper meridian,** which is the half of the meridian above the horizon. At *local midnight,* the Sun crosses the **lower meridian,** the half of the meridian below the horizon; this crossing cannot be observed directly.

The crossing of the meridian by any object in the sky is called a **meridian transit** of that object. If the crossing occurs above the horizon, it is an *upper* meridian transit. An **apparent solar day** is the interval between two successive upper meridian transits of the Sun as observed from any fixed spot on the Earth. Stated less formally, an apparent solar day is the time from one local noon to the next local noon, or from when the Sun is highest in the sky to when it is again highest in the sky.

Unfortunately, the Sun is not a good timekeeper (Figure 2-20). The length of an apparent solar day (as measured by a device such as an hourglass) varies from one time of year to another. There are two main reasons why this is so, both having to do with the way in which the Earth orbits the Sun.

The first reason is that the Earth's orbit is not a perfect circle; rather, it is an ellipse, as Figure 2-20a shows in exaggerated form. As we will learn in Chapter 4, the Earth moves more rapidly along its orbit when it is near the Sun than when it is farther away. Hence, the Sun appears to us to move more than 1° per day along the eclip-

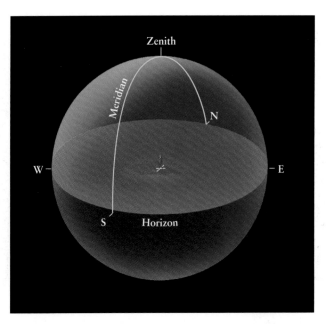

figure 2-19

The Meridian The meridian is a circle on the celestial sphere that passes through the observer's zenith (the point directly overhead) and the north and south points on the observer's horizon. The passing of celestial objects across the meridian can be used to measure time. The upper meridian is the part above the horizon, and the lower meridian (not shown) is the part below the horizon.

tic in January, when the Earth is nearest the Sun, and less than 1° per day in July, when the Earth is farthest from the Sun. By itself, this effect would cause the apparent solar day to be longer in January than in July.

The second reason why the Sun is not a good timekeeper is the 23½° angle between the ecliptic and the celestial equator (see Figure 2-14). As Figure 2-20b shows, this causes a significant part of the Sun's apparent motion when near the equinoxes to be in a north-south direction. The net daily eastward progress in the sky is then somewhat foreshortened. At the summer and winter solstices, by contrast, the Sun's motion is parallel to the celestial equator. Thus, there is no comparable foreshortening around the beginning of summer or winter. This effect by itself would make the apparent solar day shorter in March and September than in June or December. Combining these effects with those due to the Earth's noncircular orbit, we find that the length of the apparent solar day varies in a complicated fashion over the course of a year.

To avoid these difficulties, astronomers invented an imaginary object called the **mean sun** that moves along the celestial equator at a uniform rate. (In science and mathematics, "mean" is a synonym for "average.") The mean sun is sometimes slightly ahead of the real Sun in the sky, sometimes behind. As a result, mean solar time and apparent solar time can differ by as much as a quarter of an hour at certain times of the year.

Because the mean sun moves at a constant rate, it serves as a fine timekeeper. A **mean solar day** is the interval between successive upper meridian transits of the mean sun. It is exactly 24 hours long,

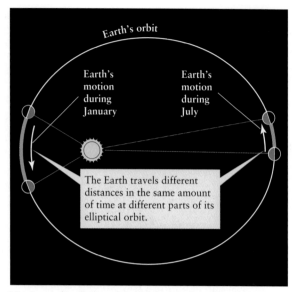

(a) A month's motion of the Earth along its orbit

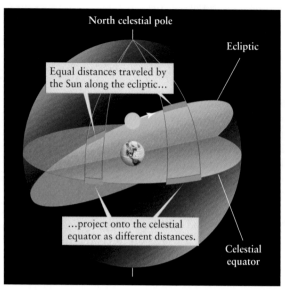

(b) A day's motion of the Sun along the ecliptic

figure 2-20

Why the Sun Is a Poor Timekeeper There are two main reasons that the Sun is a poor timekeeper. **(a)** The Earth's speed along its orbit varies during the year. It moves fastest when closest to the Sun in January and slowest when farthest from the Sun in July. Hence, the apparent speed of the Sun along the ecliptic is not constant. **(b)** Because of the tilt of the

Earth's rotation axis, the ecliptic is inclined with respect to the celestial equator. Therefore, the projection of the Sun's daily progress along the ecliptic onto the celestial equator (shown in blue) varies during the year. This causes further variations in the length of the apparent solar day.

the average length of an apparent solar day. One 24-hour day as measured by your alarm clock or wristwatch is a mean solar day.

 Time zones were invented for convenience in commerce, transportation, and communication. In a time zone, all clocks and watches are set to the mean solar time for a meridian of longitude that runs approximately through the center of the zone. Time zones around the world are generally centered on meridians of longitude at 15° intervals. In most cases, going from one time zone to the next requires you to change the time on your wristwatch by exactly 1 hour. The time zones for most of North America are shown in Figure 2-21.

In order to coordinate their observations with colleagues elsewhere around the globe, astronomers often keep track of time using Coordinated Universal Time, somewhat confusingly abbreviated UTC or UT. This is the time in a zone that includes Greenwich, England, a seaport just outside of London where the first internationally accepted time standard was kept. In former times UTC was known as Greenwich Mean Time. In North America, Eastern Standard Time (EST) is 5 hours different from UTC; 9:00 A.M. EST is 14:00 UTC. Coordinated Universal Time is also used by aviators and sailors, who regularly travel from one time zone to another.

Although it is natural to want our clocks and method of timekeeping to be related to the Sun, astronomers often use a system that is based on the apparent motion of the stars. This system, called **sidereal time,** is useful when aiming a telescope. Most observatories are therefore equipped with a clock that measures sidereal time, as discussed in Box 2-2.

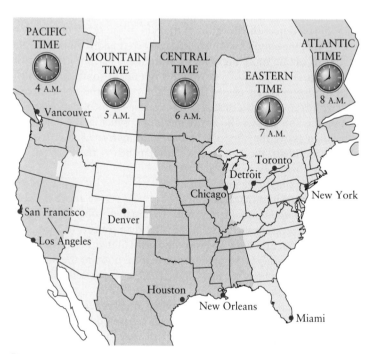

figure 2-21

Time Zones in North America For convenience, the Earth is divided into 24 time zones, generally centered on 15° intervals of longitude around the globe. There are four time zones across the continental United States, making for a 3-hour time difference between New York and California.

box 2-2   TOOLS OF THE ASTRONOMER'S TRADE

Sidereal Time

If you want to observe a particular object in the heavens, the ideal time to do so is when the object is high in the sky, on or close to the upper meridian. This minimizes the distorting effects of the Earth's atmosphere, which increase as you view closer to the horizon. For astronomers who study the Sun, this means making observations at local noon, which is not too different from noon as determined using mean solar time. For astronomers who observe planets, stars, or galaxies, however, the optimum time to observe depends on the particular object to be studied. The problem is this: Given the location of a given object on the celestial sphere, when will that object be on the upper meridian?

To answer this question, astronomers use *sidereal time* rather than solar time. It is different from the time on your wristwatch. In fact, a sidereal clock and an ordinary clock even tick at different rates, because they are based on different astronomical objects. Ordinary clocks are related to the position of the Sun, while sidereal clocks are based on the position of the vernal equinox, the location from which right ascension is measured. (See Box 2-1 for a discussion of right ascension.)

Regardless of where the Sun is, midnight sidereal time at your location is defined to be when the vernal equinox crosses your upper meridian. (Like solar time, sidereal time depends on where you are on Earth.) A **sidereal day** is the time between two successive upper meridian passages of the vernal equinox. By contrast, an apparent solar day is the time between two successive upper meridian crossings of the Sun. The illustration shows why these two kinds of day are not equal. Because the Earth orbits the Sun, the Earth must make one complete rotation plus about 1° to get from one local solar noon to the next. This extra 1° of rotation corresponds to 4 minutes of time, which is the amount by which a solar day exceeds a sidereal day. To be precise:

$$1 \text{ sidereal day} = 23^h\ 56^m\ 4.091^s$$

where the hours, minutes, and seconds are in mean solar time.

One day according to your wristwatch is one mean solar day, which is exactly 24 hours of solar time long. A **sidereal clock** measures sidereal time in terms of sidereal hours, minutes, and seconds, where one sidereal day is divided into 24 sidereal hours. This explains why a sidereal clock ticks at a slightly different rate than your wristwatch. As a result, at some times of the year a sidereal clock will show a very different time than an ordinary clock. (On local noon on March 21, when the Sun is at the vernal equinox, a sidereal clock will say that it is midnight. Do you see why?)

We can now answer the question in the opening paragraph. The vernal equinox, whose celestial coordinates are R.A. = $0^h\ 0^m\ 0^s$, Decl. = 0° 0′ 0″, crosses the upper meridian at midnight sidereal time (0:00). The autumnal equinox, which is on the

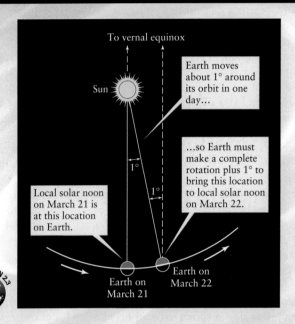

opposite side of the celestial sphere at R.A. = $12^h\ 0^m\ 0^s$, Decl. = 0° 0′ 0″, crosses the upper meridian 12 sidereal hours later at noon sidereal time (12:00). As these examples illustrate, *any* object crosses the upper meridian when the sidereal time is equal to the object's right ascension. That is why astronomers measure right ascension in units of time rather than degrees, and why right ascension is always given in sidereal hours, minutes, and seconds.

EXAMPLE: Suppose you want to observe the bright star Spica (Figure 2-6), which has epoch 2000 coordinates R.A. = $13^h\ 25^m\ 11.6^s$, Decl. = −11° 9′ 41″. What is the best time to do this?

Situation: Our goal is to find the time when Spica passes through your upper meridian, where it can best be observed.

Tools: We use the idea that a celestial object is on your upper meridian when the sidereal time equals the object's right ascension.

Answer: Based on the given right ascension of Spica, it will be best placed for observation when the sidereal time at your location is about 13:25. (Note that sidereal time is measured using a 24-hour clock.)

Review: By itself, our answer doesn't tell you what time on your wristwatch (which measures mean solar time) is best for observing Spica. That's why most observatories are equipped with a sidereal clock.

While sidereal time is extremely useful in astronomy, mean solar time is still the best method of timekeeping for most earthbound purposes. All time measurements in this book are expressed in mean solar time unless otherwise stated.

2-8 | Astronomical observations led to the development of the modern calendar

Just as the day is a natural unit of time based on the Earth's rotation, the year is a natural unit of time based on the Earth's revolution about the Sun. However, nature has not arranged things for our convenience. The year does not divide into exactly 365 whole days. Ancient astronomers realized that the length of a year is approximately 365¼ days, so the Roman emperor Julius Caesar established the system of "leap years" to account for this extra quarter of a day. By adding an extra day to the calendar every four years, he hoped to ensure that seasonal astronomical events, such as the beginning of spring, would occur on the same date year after year.

Caesar's system would have been perfect if the year were exactly 365¼ days long and if there were no precession. Unfortunately, this is not the case. To be more accurate, astronomers now use several different types of years. For example, the **sidereal year** is defined to be the time required for the Sun to return to the same position with respect to the stars. It is equal to 365.2564 mean solar days, or $365^d \, 6^h \, 9^m \, 10^s$.

The sidereal year is the orbital period of the Earth around the Sun, but it is *not* the year on which we base our calendar. Like Caesar, most people want annual events—in particular, the first days of the seasons—to fall on the same date each year. For example, we want the first day of spring to occur on March 21. But spring begins when the Sun is at the vernal equinox, and the vernal equinox moves slowly against the background stars because of precession. Therefore, to set up a calendar we use the **tropical year,** which is equal to the time needed for the Sun to return to the vernal equinox. This period is equal to 365.2422 mean solar days, or $365^d \, 5^h \, 48^m \, 46^s$. Because of precession, the tropical year is 20 minutes and 24 seconds shorter than the sidereal year.

Caesar's assumption that the tropical year equals 365¼ days was off by 11 minutes and 14 seconds. This tiny error adds up to about three days every four centuries. Although Caesar's astronomical advisers were aware of the discrepancy, they felt that it was too small to matter. However, by the sixteenth century the first day of spring was occurring on March 11.

The Roman Catholic Church became concerned because Easter kept shifting to progressively earlier dates. To straighten things out, Pope Gregory XIII instituted a calendar reform in 1582. He began by dropping ten days (October 4, 1582, was followed by October 15, 1582), which brought the first day of spring back to March 21. Next, he modified Caesar's system of leap years.

Caesar had added February 29 to every calendar year that is evenly divisible by four. Thus, for example, 2004, 2008, and 2012 are all leap years with 366 days. But we have seen that this system produces an error of about three days every four centuries. To solve the problem, Pope Gregory decreed that only the century years evenly divisible by 400 should be leap years. For example, the years 1700, 1800, and 1900 (which would have been leap years according to Caesar) were not leap years in the improved Gregorian system, but the year 2000, which can be divided evenly by 400, *was* a leap year.

We use the Gregorian system today. It assumes that the year is 365.2425 mean solar days long, which is very close to the true length of the tropical year. In fact, the error is only one day in every 3300 years. That won't cause any problems for a long time.

Key Words

Terms preceded by an asterisk (*) are discussed in the Boxes.

Antarctic Circle, p. 30
apparent solar day, p. 33
apparent solar time, p. 33
Arctic Circle, p. 30
autumnal equinox, p. 30
celestial equator, p. 25
celestial sphere, p. 25
circumpolar, p. 26
constellation, p. 21
*declination, p. 27
diurnal motion, p. 22
ecliptic, p. 29

epoch, p. 33
equinox, p. 30
lower meridian, p. 33
mean solar day, p. 33
mean sun, p. 33
meridian, p. 33
meridian transit, p. 33
north celestial pole, p. 25
positional astronomy, p. 21
precession, p. 31
precession of the equinoxes, p. 32
*right ascension, p. 27
*sidereal clock, p. 35
*sidereal day, p. 35

sidereal time, p. 34
sidereal year, p. 36
south celestial pole, p. 25
summer solstice, p. 30
time zone, p. 34
tropical year, p. 36
Tropic of Cancer, p. 31
Tropic of Capricorn, p. 31
upper meridian, p. 33
vernal equinox, p. 30
winter solstice, p. 30
zenith, p. 25
zodiac, p. 31

Key Ideas

Ideas preceded by an asterisk (*) are discussed in the Boxes.

Constellations and the Celestial Sphere: It is convenient to imagine the stars fixed to the celestial sphere with the Earth at its center.

• The surface of the celestial sphere is divided into 88 regions called constellations.

Diurnal (Daily) Motion of the Celestial Sphere: The celestial sphere appears to rotate around the Earth once in each 24-hour period. In fact, it is actually the Earth that is rotating.

• The poles and equator of the celestial sphere are determined by extending the axis of rotation and the equatorial plane of the Earth out to the celestial sphere.

• *The positions of objects on the celestial sphere are described by specifying their right ascension (in time units) and declination (in angular measure).

Seasons and the Tilt of the Earth's Axis: The Earth's axis of rotation is tilted at an angle of about 23½° from the perpendicular to the plane of the Earth's orbit.

• The seasons are caused by the tilt of the Earth's axis.

• Over the course of a year, the Sun appears to move around the celestial sphere along a path called the ecliptic. The ecliptic is inclined to the celestial equator by about 23½°.

• The ecliptic crosses the celestial equator at two points in the sky, the vernal and autumnal equinoxes. The northernmost point that the Sun reaches on the celestial sphere is the summer solstice, and the southernmost point is the winter solstice.

Precession: The orientation of the Earth's axis of rotation changes slowly, a phenomenon called precession.

• Precession is caused by the gravitational pull of the Sun and Moon on the Earth's equatorial bulge.

• Precession of the Earth's axis causes the positions of the equinoxes and celestial poles to shift slowly.

• *Because the system of right ascension and declination is tied to the position of the vernal equinox, the date (or epoch) of observation must be specified when giving the position of an object in the sky.

Timekeeping: Astronomers use several different means of keeping time.

• Apparent solar time is based on the apparent motion of the Sun across the celestial sphere, which varies over the course of the year.

• Mean solar time is based on the motion of an imaginary mean sun along the celestial equator, which produces a uniform mean solar day of 24 hours. Ordinary watches and clocks measure mean solar time.

• *Sidereal time is based on the apparent motion of the celestial sphere.

The Calendar: The tropical year is the period between two passages of the Sun across the vernal equinox. Leap year corrections are needed because the tropical year is not exactly 365 days. The sidereal year is the actual orbital period of the Earth.

Review Questions

1. How are constellations useful to astronomers? How many stars are not part of any constellation?

2. A fellow student tells you that only the stars in Figure 2-2b that are connected by blue lines are part of the constellation Orion. How would you respond?

3. What is the celestial sphere? Why is this ancient concept still useful today?

4. Imagine that someone suggests sending a spacecraft to land on the surface of the celestial sphere. How would you respond to such a suggestion?

5. What is the celestial equator? How is it related to the Earth's equator? How are the north and south celestial poles related to the Earth's axis of rotation? Where on Earth would you have to be for the celestial equator to pass through your zenith?

6. How many degrees is the angle from the horizon to the zenith? Does your answer depend on what point on the horizon you choose?

7. Why can't a person in Antarctica use the Big Dipper to find the north direction?

8. Is there any place on Earth where you could see the north celestial pole on the northern horizon? If so, where? Is there any place on Earth where you could see the north celestial pole on the western horizon? If so, where? Explain your answers.

9. How do the stars appear to move over the course of the night as seen from the North Pole? As seen from the equator? Why are these two motions different?

10. Using a diagram, explain why the tilt of the Earth's axis relative to the Earth's orbit causes the seasons as we orbit the Sun.

11. Give two reasons why it's warmer in summer than in winter.

12. What is the ecliptic? Why is it tilted with respect to the celestial equator? Does the Sun appear to move along the ecliptic, the celestial equator, or neither? By about how many degrees does the Sun appear to move on the celestial sphere each day?

13. Where on Earth do you have to be in order to see the north celestial pole directly overhead? What is the maximum possible elevation of the Sun above the horizon at that location? On what date can this maximum elevation be observed?

14. What are the vernal and the autumnal equinoxes? What are the summer and winter solstices? How are these four points related to the ecliptic and the celestial equator?

15. At what point on the horizon does the vernal equinox rise? Where on the horizon does it set? (*Hint:* See Figure 2-15.)

16. How does the daily path of the Sun across the sky change with the seasons? Why does it change?

17. Where on Earth do you have to be in order to see the Sun at the zenith? Will it be at the zenith every day? Explain.

18. What causes precession of the equinoxes? How long does it take for the vernal equinox to move 1° along the ecliptic?

19. What is the (fictitious) mean sun? What path does it follow on the celestial sphere? Why is it a better timekeeper than the actual Sun in the sky?

20. Why is it convenient to divide the Earth into time zones?

21. Why is the time given by a sundial not necessarily the same as the time on your wristwatch?

22. What is the difference between the sidereal year and the tropical year? Why are these two kinds of year slightly different in length? Why are calendars based on the tropical year?

23. When is the next leap year? Was 2000 a leap year? Will 2100 be a leap year?

Advanced Questions

Questions preceded by an asterisk (*) involve topics discussed in the Boxes.

Problem-solving tips and tools

 To help you visualize the heavens, it is worth taking the time to become familiar with various types of star charts. These include the simple star charts at the end of this book, the monthly star charts published in such magazines as *Sky & Telescope* and *Astronomy,* and the more detailed maps of the heavens found in star atlases.

 One of the best ways to understand the sky and its motions is to use the *Starry Night Backyard*™ computer program on the CD-ROM that accompanies this book. This easy-to-use program allows you to view the sky on any date and at any time, as seen from any point on Earth, and to animate the sky to visualize its diurnal and annual motions.

You may also find it useful to examine a planisphere, a device consisting of two rotatable disks. The bottom disk shows all the stars in the sky (for a particular latitude), and the top one is opaque with a transparent oval window through which only some of the stars can be seen. By rotating the top disk, you can immediately see which constellations are above the horizon at any time of the year. A planisphere is a convenient tool to carry with you when you are out observing the night sky.

---

24. On November 1 at 8:30 P.M. you look toward the eastern horizon and see the bright star Bellatrix (shown in Figure 2-2*b*) rising. At approximately what time will Bellatrix rise one week later, on November 8?

25. Figure 2-4 shows the situation on September 21, when Cygnus is highest in the sky at 8:00 P.M. local time and Andromeda is highest in the sky at midnight. But as Figure 2-5 shows, on July 21 Cygnus is highest in the sky at midnight. On July 21, at approximately what local time is Andromeda highest in the sky? Explain your reasoning.

26. Figure 2-5 shows which constellations are high in the sky (for observers in the northern hemisphere) in the months of July, September, and November. From this figure, would you be able to see Perseus at midnight on May 15? Draw a picture to justify your answer.

27. Figure 2-6 shows the appearance of Polaris, the Little Dipper, and the Big Dipper at 11 P.M. (daylight savings time) on August 1. Sketch how these objects would appear on this same date at (**a**) 8 P.M. and (**b**) 2 A.M. Include the horizon in your sketches, and indicate the north direction.

28. Figure 2-6 shows the appearance of the sky near the North Star at 11 P.M. (daylight savings time) on August 1. Explain why the sky has this same appearance at 1 A.M. on July 1 and at 11 P.M. on September 1.

29. The time-exposure photograph that opens this chapter shows the trails made by individual stars as the celestial sphere appears to rotate around the Earth. (**a**) For approximately what length of time was the camera shutter left open to take this photograph? (**b**) The stars in this photograph (taken in Hawaii, at 20° north latitude) appear to rotate around one of the celestial poles. Which celestial pole is it? As seen from this location, do the stars move clockwise or counterclockwise around this celestial pole? (**c**) If you were at 20° south latitude, which celestial pole could you see? In which direction would you look to see it? As seen from this location, do the stars move clockwise or counterclockwise around this celestial pole?

30. (**a**) Redraw Figure 2-10 for an observer at the north pole. (*Hint:* The north celestial pole is directly above this observer.) (**b**) Redraw Figure 2-10 for an observer at the equator. (*Hint:* The celestial equator passes through this observer's zenith.) (**c**) Using Figure 2-10 and your drawings from (a) and (b), justify the following rule, long used by navigators: The latitude of an observer in the northern hemisphere is equal to the angle in the sky between that observer's horizon and the north celestial pole. (**d**) State the rule that corresponds to (c) for an observer in the southern hemisphere.

31. The photograph that opens this chapter was taken next to the Canada-France-Hawaii Telescope (CFHT) atop Mauna Kea in Hawaii. The telescope is at longitude 155° 28′ 18″ west and latitude 19° 49′ 36″ north. (**a**) By making measurements on the photograph, find the approximate angular width and angular height of the photo. (**b**) How far (in degrees, arcminutes, and arcseconds) from the south celestial pole can a star be and still be circumpolar as seen from CFHT?

32. Is there any place on Earth where all the visible stars are circumpolar? If so, where? Is there any place on Earth where none of the visible stars is circumpolar? If so, where? Explain your answers.

33. The yellow plane in Figure 2-14 is called the *plane of the ecliptic.* Explain why this is the same as the plane in which the Earth orbits around the Sun.

34. Figure 2-15 shows the daily path of the Sun across the sky on March 21, June 21, September 22, and December 21 for an observer at 35° north latitude. Sketch a drawing of this kind for (**a**) an observer at 35° south latitude; (**b**) an observer at the equator; and (**c**) an observer at the north pole.

35. Suppose that you live at a latitude of 40° N. What is the elevation of the Sun above the southern horizon at noon at the time of the winter solstice? Explain your reasoning. Include a drawing as part of your explanation.

36. In the northern hemisphere, houses are designed to have "southern exposure," that is, with the largest windows on the southern side of the house. But in the southern hemisphere houses are designed to have "northern exposure." Why are houses designed this way, and why is there a difference between the hemispheres?

37. The city of Mumbai (formerly Bombay) in India is 19° north of the equator. On how many days of the year, if any, is the Sun at the zenith at midday as seen from Mumbai? Explain your answer.

38. Ancient records show that 2000 years ago, the stars of the Southern Cross were visible in the southern sky from Greece. Today, however, these stars cannot be seen from Greece. What accounts for this change?

39. The Great Pyramid at Giza has a tunnel that points toward the north celestial pole. At the time the pyramid was built, around 2600 B.C., toward which star did it point? Toward which star does this same tunnel point today? (See Figure 2-18.)

40. Unlike western Europe, Imperial Russia did not use the revised calendar instituted by Pope Gregory XIII. Explain why the Russian Revolution, which started on November 7, 1917, according to the modern calendar is called "the October revolution" in Russia. What was this date according to the Russian calendar at the time? Explain.

*41. What is the right ascension of a star that is on the meridian at midnight at the time of the autumnal equinox? Explain.

*42. The coordinates on the celestial sphere of the summer solstice are R.A. = 6^h 0^m 0^s, Decl. = +23° 27′. What are the right ascension and declination of the winter solstice? Explain your answer.

*43. Because 24 hours of right ascension takes you all the way around the celestial equator, $24^h = 360°$. What is the angle in the sky (measured in degrees) between a star with R.A. = 8^h 0^m 0^s, Decl. = 0° 0′ 0″ and a second star with R.A. = 11^h 20^m 0^s, Decl. = 0° 0′ 0″? Explain your answer.

*44. On a certain night, the first star in Advanced Question 43 passes through the zenith at 12:30 A.M. local time. At what time will the second star pass through the zenith? Explain your answer.

*45. At local noon on March 21, when the Sun is at the vernal equinox, a sidereal clock will say that it is midnight. Explain why.

*46. (a) What is the sidereal time when the vernal equinox rises? (b) On what date is the sidereal time nearly equal to the solar time? Explain.

*47. How would the sidereal and solar days change (a) if the Earth's rate of rotation increased, (b) if the Earth's rate of rotation decreased, and (c) if the Earth's rotation were retrograde (that is, if the Earth rotated about its axis opposite to the direction in which it revolves about the Sun)?

48. The image of the Earth shown top right was made by the *Galileo* spacecraft while en route to Jupiter. South America is at the center of the image and Antarctica is at the bottom of the image. (a) In which month of the year was this image made? Explain your reasoning. (b) When this image was made, was the Earth relatively close to the Sun or relatively distant from the Sun? Explain your reasoning.

(NASA/JPL)

R I **V** U X G

Discussion Questions

49. Examine a list of the 88 constellations. Are there any constellations whose names obviously date from modern times? Where are these constellations located? Why do you suppose they do not have archaic names?

50. Why is it useful to astronomers to have telescopes in both the southern hemisphere and the northern hemisphere? (See Figure 2-10.)

51. Describe how the seasons would be different if the Earth's axis of rotation, rather than having its present 23½° tilt, were tilted (a) by 0° or (b) by 90°.

52. In William Shakespeare's *Julius Caesar* (act 3, scene 1), Caesar says:

> *But I am constant as the northern star,*
> *Of whose true-fix'd and resting quality*
> *There is no fellow in the firmament.*

Translate Caesar's statement about the "northern star" into modern astronomical language. Is the northern star truly "constant"? Was the northern star the same in Shakespeare's time (1564–1616) as it is today?

Web/CD-ROM Questions

53. Search the World Wide Web for information about the national flags of Australia, New Zealand, and Brazil and the state flag of Alaska. Which stars are depicted on these flags? Explain any similarities or differences among these flags.

54. Some people say that on the date that the Sun is at the vernal equinox, and only on this date, you can stand a raw egg on end. Others say that there is nothing special about the vernal equinox, and that with patience you can stand a raw egg on end on any day of the year. Search the World Wide Web for information about this story and for hints about how to stand an egg on end. Use these hints to try the experiment yourself on a day when the Sun is *not* at the vernal equinox. What do you conclude about the connection between eggs and equinoxes?

55. Use the U.S. Naval Observatory Web site to find the times of sunset and sunrise on (**a**) your next birthday and (**b**) the date this assignment is due. (**c**) Are the times the same for the two dates? Explain why or why not.

Observing Projects

Observing tips and tools

Moonlight is so bright that it interferes with seeing the stars. For the best view of the constellations, do your observing when the Moon is below the horizon. You can find the times of moonrise and moonset in your local newspaper or on the World Wide Web. Each monthly issue of the magazines *Sky & Telescope* and *Astronomy* includes much additional observing information.

56. On a clear, cloud-free night, use the star charts at the end of this book to see how many constellations of the zodiac you can identify. Which ones were easy to find? Which were difficult? Are the zodiacal constellations the most prominent ones in the sky?

57. Examine the star charts that are published monthly in such popular astronomy magazines as *Sky & Telescope* and *Astronomy*. How do they differ from the star charts at the back of this book? On a clear, cloud-free night, use one of these star charts to locate the celestial equator and the ecliptic. Note the inclination of the Milky Way to the ecliptic and celestial equator. The Milky Way traces out the plane of our galaxy. What do your observations tell you about the orientation of the Earth and its orbit relative to the galaxy's plane?

58. Suppose you wake up before dawn and want to see which constellations are in the sky. Explain how the star charts at the end of this book can be quite useful, even though chart times are given only for the evening hours. Which chart most closely depicts the sky at 4:00 A.M. tomorrow morning? Set your alarm clock for 4:00 A.M. to see if you are correct.

59. Use the *Starry Night Backyard*™ program to observe the diurnal motion of the sky. (**a**) First set *Starry Night Backyard*™ to display the sky as seen from where you live. Select **Viewing Location...** in the **View** menu and click on the **List** button to find your city or town. Then, using the hand cursor, center your field of view on the northern horizon (if you live in the northern hemisphere) or the southern horizon (if you live in the southern hemisphere). In the Control Panel at the top of the main window, click on the control for the rate of time flow (immediately to the right of the date and time display) and set the discrete time step to 1 minute. Then click on the "Forward" button (a triangle that points to the right). To see the stars during the daytime, turn off daylight (select **Hide Daylight** in the **View** menu). Do the stars appear to rotate clockwise or counterclockwise? Explain in terms of the Earth's rotation. Are any of the

stars circumpolar? (**b**) Now center your field of view on the southern horizon (if you live in the northern hemisphere) or the northern horizon (if you live in the southern hemisphere). Describe what you see. Are any of these stars circumpolar?

60. Use the *Starry Night Backyard*™ program to observe the Sun's motion on the celestial sphere. To see the entire celestial sphere, including the part below the horizon, select **Guides > Atlas** in the **Go** menu. Then center on the Sun by using the **Find...** command in the **Edit** menu. (**a**) In the Control Panel at the top of the main window, click on the control for the rate of time flow (immediately to the right of the date and time display) and set the discrete time step to 1 day. Then click on the "Forward" button (a triangle that points to the right). Observe the Sun for a full year of simulated time. How does the Sun appear to move against the constellations? What path does it follow? Does it ever change direction? (**b**) Stop the motion using the "Stop" button (a black square). Then select **Constellations > Identify** in the **Go** menu to display the constellations and their boundaries. In the Control Panel at the top of the main window, click on the downward-pointing arrow immediately to the right of the date and time display and select **Now**. (Alternatively, you can type today's date into the date display.) Again center on the Sun using the **Find...** command in the **Edit** menu. In which constellation is the Sun located today? Is this the same as the astrological sign for today's date? Explain in terms of precession. (**c**) Again set the discrete time step to 1 day and click on the "Forward" button. Through which constellations does the Sun appear to pass over the course of a year? (You can use the "Stop" button—a black square—to stop the motion at any point.)

61. Use the *Deep Space Explorer*™ program to investigate the orbits of the Earth and some of the other planets around the Sun. In the left-hand part of the window under the heading **Solar System**, click on the triangle next to the word **Explore** and then click on **Inner Solar System**. You can zoom in and zoom out using the buttons at the upper left of the window (an upward-pointing triangle and a downward-pointing triangle). You can also rotate the solar system by putting the mouse cursor over the image, holding down the mouse button, and moving the mouse. (On a two-button mouse, hold down the left mouse button.) (**a**) We learned in Section 2-7 that the Earth's orbit is not a perfect circle. Rotate the solar system until you are looking at the orbits face-on. Can you tell that the Earth's orbit is not circular? Which of the four inner planets (Mercury, Venus, Earth, and Mars) have orbits that are clearly noncircular? (You may want to use a ruler to measure the distance from the Sun to various points on each planet's orbit. If the distance is the same at all points, the orbit is circular; otherwise, it's noncircular.) (**b**) Rotate the solar system until you are looking at the orbits edge-on. Do the other planets' orbits lie in the same plane as the plane of the Earth's orbit (the plane of the ecliptic)? Which planet's orbit appears to be tilted the most from the Earth's orbit? (You may need to use the zoom controls to find out.) As seen from the other planets of the solar system, would the Sun appear to follow exactly the same path across the celestial sphere as seen from Earth?

JAMES RANDI

Why Astrology Is Not Science

James ("the Amazing") Randi works tirelessly to expose trickery so that others can relish the greater wonder of science. As a magician, he has had his own television show and an enormous public following. As a lecturer, he addresses teachers, students, and others worldwide. His newsletter and column for *The Skeptic* are key resources for educators. His many books include *Flim-Flam!, The Faith Healers,* and *The Mask of Nostradamus,* about a legendary con man with secrets of his own.

Mr. Randi helped found the Committee for the Scientific Investigation of Claims of the Paranormal, and his $10,000 prize for "the performance of any paranormal event … under proper observing conditions" has gone unclaimed for more than 25 years. An amateur archeologist and astronomer as well, he lives in Florida with several untalented parrots and the occasional visiting magus.

I'm involved in the strange business of telling folks what they should already know. I meet audiences who believe in all sorts of impossible things, often despite their education and intelligence. My job is to explain how science differs from the unproven, illogical assumptions of pseudoscience—and why it matters. Perhaps my best example is the difference between astronomy and astrology.

Both astrology and astronomy arose from the wonders of the night sky, from the stars to comets, planets, the Sun, and the Moon. Surely, humans have long reasoned, there must be some meaning in their motions. Surely the Moon's effect on tides hints at hidden "causes" for strange events. *Judiciary* (literally "judging") astrology therefore attempted to foretell the future—our earthly future. To serve it, *horary* (literally "hourly") astrology carefully tracked the heavens.

It is the latter that has become astronomy. Thanks to its process of careful measurement and testing, we now understand more about the true nature of the starry universe than astrologers could ever have imagined. With the birth of a new science, astronomers had a logical framework based on physical causes and systematic observations.

Astrology remains a popular delusion. Far too many believe today that patterns in the sky govern our lives. They accept the vague tendencies and portents of seers who cast horoscopes. They shouldn't. Just a glance at the tenets of astrology provides ample evidence of its absurdity.

An individual is said to be born under a sign. To the astrologer, the Sun was located "in" that sign at the moment of birth. (Stars are not seen in the daytime, but no matter—a calculation tells where the Sun is.) Each sign takes its name from a constellation, a totally imaginary figure invented for our convenience in referring to stars. Different cultures have different mythical figures up there, and so different schools of astrology assign different meanings to the signs they use.

In the spirit of equal-opportunity swindling, astrologers divide up the year fairly, ignoring variations in the size of constellations. Since Libra is tiny, while Virgo is huge, they chop some of the sky off Virgo and add it—along with bits of Scorpio—to bring Libra up to size. The Sun could well be declared "in" Libra when it is actually outside that constellation.

It gets worse. Science constantly challenges itself and changes. The rules of astrology could not, although they were made up thousands of years ago, and since then the "fixed" stars have moved. In particular, precession of the equinoxes has shifted objects in the sky relative to our calendar. The constellations have changed but astrology has not. If you were born August 7, you are said to be a Leo, but the Sun that day was really in the same part of the sky as the constellation Cancer.

With a theory like this to back it up, we should not be surprised at the bottom line: *A pseudoscience does not work.* Test after test has checked its predictions, and the result is always the same. One such investigator is Shawn Carlson of the University of California, San Diego. As he put it in *Nature* magazine, astrology is "a hopeless cause." Johannes Kepler, the pioneering astronomer, himself cast horoscopes, but they are little remembered today. Owen Gingerich, a historian of science at Harvard, puts it well: Kepler was the astrologer who destroyed astrology.

Astronomy works, and it works very well indeed. That isn't easy. Because we humans tend to find what we want in any body of data, it takes science's careful process of observation, creative insight, and critical thinking to understand and predict changes in nature. As I write, a transit of Ganymede is due next Thursday at 21:47:20. At exactly that time, the satellite of Jupiter will cross in front of its planet as seen from Earth, and yet most of us will never know it. Still other moons of Jupiter may hold fresh clues to the formation of our entire solar system and the conditions for life elsewhere.

For most people, astronomy has too little fantasy or money in it, and they will never experience the beauty in its predictions. The dedicated labors of generations of scientists have enabled us to perform a genuine wonder.

3 Eclipses and the Motion of the Moon

On August 11, 1999, millions of humans stood along a narrow corridor of land that extended from England through Europe and the Middle East to India. For a few brief minutes, they saw the rare cosmic spectacle shown in this photograph taken in Turkey: a total solar eclipse. As the Moon covered the disk of the Sun, the stars and planets became visible in the darkening sky, and the solar corona—the Sun's thin outer atmosphere, which glows with an unearthly pearlescent light—was revealed.

Not everyone will ever see the Moon cover the Sun in this way. But anyone can find the Moon in the sky and observe how its appearance changes from night to night, from new moon to full moon and back again. The times when the Moon rises and sets also differ noticeably from one night to the next.

In this chapter our subject is how the Moon moves as seen from the Earth. We will explore why the Moon goes through a regular cycle of phases, and how the Moon's orbit around the Earth leads to solar eclipses as well as lunar eclipses. We will also see how ancient astronomers used their observations of the Moon to determine the size and shape of the Earth, as well as other features of the solar system. Thus, the Moon—which has always loomed large in the minds of poets, lovers, and dreamers—has also played a key role in the development of our modern picture of the universe.

 WEB LINK 3.1

The Sun in total eclipse, August 11, 1999. (Fred Espenak)
R I **V** U X G

Questions to Guide Your Reading

As you read the sections of this chapter, look for the answers to the following questions:

3-1 Why does the Moon go through phases?

3-2 Is there such a thing as the "dark side of the Moon"?

3-3 What is the difference between a lunar eclipse and a solar eclipse?

3-4 How often do lunar eclipses happen? When one is taking place, where do you have to be to see it?

3-5 How often do solar eclipses happen? Why are they visible only from certain special locations on Earth?

3-6 How did ancient astronomers deduce the sizes of the Earth, the Moon, and the Sun?

3-1 | The phases of the Moon are caused by its orbital motion

As seen from Earth, both the Sun and the Moon appear to move from west to east on the celestial sphere—that is, relative to the background of stars. They move at very different rates, however. The Sun takes one year to make a complete trip around the imaginary celestial sphere along the path we call the *ecliptic* (Section 2-5). By comparison, the Moon takes only about four weeks. In the past, these similar motions led people to believe that both the Sun and the Moon orbit around the Earth. We now know that only the Moon orbits the Earth, while the Earth-Moon system as a whole (**Figure 3-1**) orbits the Sun. (In Chapter 4 we will learn how this was discovered.)

One key difference between the Sun and the Moon is the nature of the light that we receive from them. The Sun emits its own light. So do the stars, which are objects like the Sun but much farther away, and so does an ordinary light bulb. By contrast, the light that we see from the Moon is reflected light. This is sunlight that has struck the Moon's surface, bounced off, and ended up in our eyes here on Earth.

CAUTION! You probably associate *reflection* with shiny objects like a mirror or the surface of a still lake. In science, however, the term refers to light bouncing off any object. You see most objects around you by reflected light. When you look at your hand, for example, you are seeing light from the Sun (or from a light fixture) that has been reflected from the skin of your hand and into your eye. In the same way, moonlight is really sunlight that has been reflected by the Moon's surface.

Figure 3-1 shows both the Moon and the Earth as seen from a spacecraft. When this image was recorded, the Sun was far off to the right. Hence, only the right-hand hemispheres of both worlds were illuminated by the Sun; the left-hand hemispheres were in darkness and are not visible in the picture. In the same way, when we view the Moon from the Earth, we see only the half of the Moon that faces the Sun and is illuminated. However, not all of the illuminated half of the Moon is necessarily facing us. As the Moon moves around the Earth, from one night to the next we see different amounts of the illuminated half of the Moon. These different appearances of the Moon are called **lunar phases.**

INTERACTIVE EXERCISE 3.1 Figure 3-2 shows the relationship between the lunar phase visible from Earth and the position of the Moon in its orbit. For example, when the Moon is at position A, we see it in roughly the same direction in the sky as the Sun. Hence, the dark hemisphere of the Moon faces the Earth. This phase, in which the Moon is barely visible, is called **new moon.**

As the Moon continues around its orbit from position A in Figure 3-2, more of its illuminated half becomes exposed to our view. The result, shown at position B, is a phase called **waxing crescent moon** ("waxing" is a synonym for "increasing"). About a week after new moon, the Moon is at position C; we then see half of the

Figure 3-1 R I **V** U X G

The Earth and the Moon This picture of the Earth and the Moon was taken in 1992 by the *Galileo* spacecraft on its way toward Jupiter. The Sun, which provides the illumination for both the Earth and the Moon, was far to the right and out of the camera's field of view when this photograph was taken. (NASA/JPL)

Moon's illuminated hemisphere and half of the dark hemisphere. This phase is called **first quarter moon.**

CAUTION! Despite the name, a first quarter moon appears to be *half* illuminated, not one-quarter illuminated! The name means that this phase is one-quarter of the way through the complete cycle of lunar phases.

About four days later, the Moon reaches position D in Figure 3-2. Still more of the illuminated hemisphere can now be seen from Earth, giving us the phase called **waxing gibbous moon** ("gibbous" is another word for "swollen"). When you look at the Moon in this phase, as in the waxing crescent and first quarter phases, the illuminated part of the Moon is toward the west. Two weeks after new moon, when the Moon stands opposite the Sun in the sky (position E), we see the fully illuminated hemisphere. This phase is called **full moon.**

Over the following two weeks, we see less and less of the Moon's illuminated hemisphere as it continues along its orbit, and the Moon is said to be *waning* ("decreasing"). While the Moon is waning, its illuminated side is toward the east. The phases are called **waning**

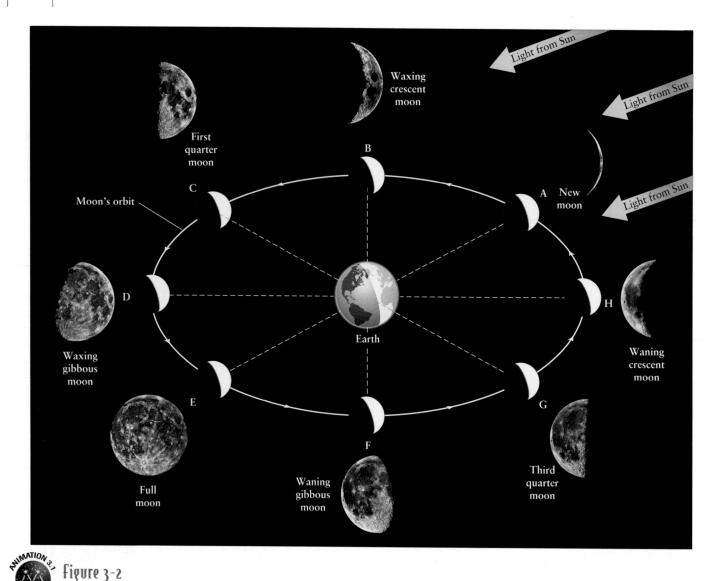

Figure 3-2

Why the Moon Goes Through Phases This figure shows the Moon at eight positions on its orbit, along with photographs of what the Moon looks like at each position as seen from Earth. The changes in phase occur because light from the Sun illuminates one half of the Moon, and as the Moon orbits the Earth we see varying amounts of the Moon's illuminated half. It takes about 29½ days for the Moon to go through a complete cycle of phases. (Photographs from Lick Observatory)

gibbous moon (position F), **third quarter moon** (position G, also called *last quarter moon),* and **waning crescent moon** (position H). The Moon takes about four weeks to complete one orbit around the Earth, so it likewise takes about four weeks for a complete cycle of phases from new moon to full moon and back to new moon.

Figure 3-2 also explains why the Moon is often visible in the daytime, as shown in **Figure 3-3.** From any location on Earth, about half of the Moon's orbit is visible at any time. For example, if it is midnight at your location, you are in the middle of the dark side of the Earth that faces away from the Sun. At that time you can easily see the Moon at positions C, D, E, F, or G. If it is midday at your location, you are in the middle of the Earth's illuminated side, and the Moon will be easily visible if it is at positions A, B, C, G, or H. (The Moon is so bright that it can be seen even against the bright blue sky.) You can see that the Moon is prominent in the midnight sky for about half of its orbit, and prominent in the midday sky for the other half.

A very common misconception about lunar phases is that they are caused by the shadow of the *Earth* falling on the Moon. As Figure 3-2 shows, this is not the case at all. Instead, phases are simply the result of our seeing the illuminated half of the Moon at different angles as the Moon moves around its orbit. To help you better visualize how this works, **Box 3-1** describes how you can simulate the cycle shown in Figure 3-2 using ordinary objects on Earth. (As we will learn in Section 3-3, the Earth's shadow does indeed fall on the Moon on rare occasions. When this happens, we see a lunar eclipse.)

figure 3-3 R I **V** U X G

The Moon During the Day The Moon can be seen during the daytime as well as at night. The time of day or night when it is visible depends on its phase. When the Moon is full, as shown in this photograph, it is directly opposite the Sun in the sky (see position E in Figure 3-2): it rises in the east when the Sun is setting in the west, and sets in the west when the Sun is rising in the east. (PhotoDisc)

3-2 | The Moon's rotation always keeps the same face toward the Earth

The phase of the Moon is always changing. But one constant aspect of the Moon is that it always keeps essentially the same hemisphere, or face, toward the Earth. Thus, you will always see the same craters and mountains on the Moon, no matter when you look at it; the only difference will be the angle at which these surface features are illuminated by the Sun. (You can verify this by carefully examining the photographs of the Moon in Figure 3-2.)

CAUTION ! Why is it that we only ever see one face of the Moon? You might think that it is because the Moon does not rotate (unlike the Earth, which rotates around an axis that passes from its north pole to its south pole). To see that this *cannot* be the case, consider **Figure 3-4**. This figure shows the Earth and the orbiting Moon from a vantage point far above the Earth's north pole. In this figure two craters on the lunar surface have been colored, one in red and one in blue. If the Moon did not rotate on its axis, as in Figure 3-4*a*, at some times the red crater would be visible from Earth, while at other times the blue crater would be visible. Thus, we would see different parts of the lunar surface over time, which does not happen in reality.

If the Moon did not rotate, we could see all sides of the Moon

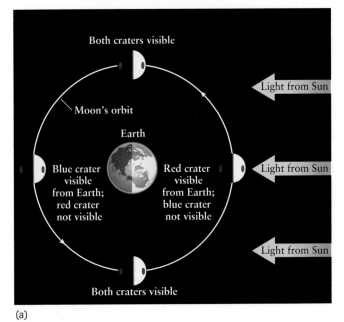

(a)

In fact the Moon does rotate, and we see only one face of the Moon

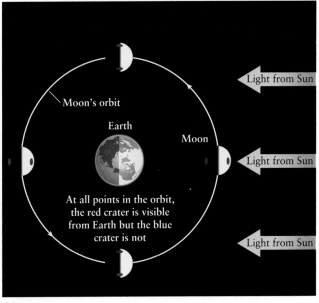

(b)

ANIMATION 3.2 figure 3-4
The Moon's Rotation These diagrams show the Moon at four points in its orbit as viewed from high above the Earth's north pole. (a) If the Moon did not rotate, then at various times the red crater would be visible from Earth while at other times the blue crater would be visible. Over a complete orbit, the entire surface of the Moon would be visible. (b) In reality, the Moon rotates on its north-south axis. Because the Moon makes one rotation in exactly the same time that it makes one orbit around the Earth, we only see one face of the Moon.

box 3-1 THE HEAVENS ON THE EARTH

Phases and Shadows

Figure 3-2 shows how the relative positions of the Earth, Moon, and Sun explain the phases of the Moon. You can visualize lunar phases more clearly by doing a simple experiment here on Earth. All you need are a small round object, such as an orange or a baseball, and a bright source of light, such as a street lamp or the Sun.

In this experiment, you play the role of an observer on the Earth looking at the Moon, and the round object plays the role of the Moon. The light source plays the role of the Sun. Hold the object in your right hand with your right arm stretched straight out in front of you, with the object directly between you and the light source (position A in the accompanying illustration). In this orientation the illuminated half of the object faces away from you, like the Moon when it is in its new phase (position A in Figure 3-2).

Now, slowly turn your body to the left so that the object in your hand "orbits" around you (toward positions C, E, and G in the illustration). As you turn, more and more of the illuminated side of the "moon" in your hand becomes visible, and it goes through the same cycle of phases—waxing crescent, first quarter, and waxing gibbous—as does the real Moon. When you have rotated through half a turn so that the light source is directly behind you, you will be looking face on at the illuminated side of the object in your hand. This corresponds to a full moon (position E in Figure 3-2). Make sure your body does not cast a shadow on the "moon" in your hand—that would correspond to a lunar eclipse!

As you continue turning to the left, more of the unilluminated half of the object becomes visible as its phase moves through waning gibbous, third quarter, and waning crescent. When your body has rotated back to the same orientation that you were in originally, the unilluminated half of your hand-held "moon" is again facing toward you, and its phase is again new. If you continue to rotate, the object in your hand repeats the cycle of "phases," just as the Moon does as it orbits around the Earth.

The experiment works best when there is just one light source around. If there are several light sources, such as in a room with several lamps turned on, the different sources will create multiple shadows and it will be difficult to see the phases of your hand-held "moon." If you do the experiment outdoors using sunlight, you may find that it is best to perform it in the early morning or late afternoon when shadows are most pronounced and the Sun's rays are nearly horizontal.

In fact, the Moon always keeps the same face toward us because it *is* rotating, but in a very special way: It takes exactly as long to rotate on its axis as it does to make one orbit around the Earth. This situation is called **synchronous rotation.** As Figure 3-4*b* shows, this keeps the crater shown in red always facing the Earth, so that we always see the same face of the Moon. In Chap-

ter 4 we will learn why the Moon's rotation and orbital motion are in step with each other.

An astronaut standing at the spot shown in red in Figure 3-4*b* would spend two weeks (half of a lunar orbit) in darkness, or lunar nighttime, and the next two weeks in sunlight, or lunar daytime. Thus, as seen from the Moon, the Sun rises and sets, and no

To distant constellation

Sun

Earth's orbit

2. As seen from Earth, when the Moon is at point 1 it appears to be in the same direction as a certain constellation.

3. After one *sidereal* month, as seen from Earth the Moon again appears to be in the direction of the same constellation.

1. At new moon, the Moon lies between the Sun and the Earth.

New moon

New moon

4. After one *synodic* month, the Moon is again new and lies between the Sun and the Earth.

figure 3-5
The Sidereal and Synodic Months The sidereal month is the time the Moon takes to complete one full revolution around the Earth with respect to the background stars. However, because the Earth is constantly moving along its orbit about the Sun, the Moon must travel through slightly more than 360° of its orbit to get from one new moon to the next. Thus, the synodic month—the time from one new moon to the next—is longer than the sidereal month.

part of the Moon is perpetually in darkness. This means that there really is no "dark side of the Moon." The side of the Moon that constantly faces away from the Earth is properly called the *far* side. The Sun rises and sets on the far side just as on the side toward the Earth. Hence, the blue crater on the far side of the Moon in Figure 3-4*b* is in sunlight for half of each lunar orbit.

The time for a complete lunar "day"—the same as the time that it takes the Moon to rotate once on its axis—is about four weeks. (Remember that it takes the same time for one complete lunar orbit.) It also takes about four weeks for the Moon to complete one cycle of its phases as seen from Earth. This regular cycle of phases inspired our ancestors to invent the concept of a month. For historical reasons, none of which has much to do with the heavens, the calendar we use today has months of differing lengths. Astronomers find it useful to define two other types of months, depending on whether the Moon's motion is measured relative to the stars or to the Sun. Neither corresponds exactly to the familiar months of the calendar.

The **sidereal month** is the time it takes the Moon to complete one full orbit of the Earth, as measured with respect to the stars. This true orbital period is equal to about 27.32 days. The **synodic month**, or *lunar month,* is the time it takes the Moon to complete one cycle of phases (that is, from new moon to new moon or from full moon to full moon) and thus is measured with respect to the Sun rather than the stars. The length of the "day" on the Moon is a synodic month, not a sidereal month.

The synodic month is longer than the sidereal month because the Earth is orbiting the Sun while the Moon goes through its phases. As **Figure 3-5** shows, the Moon must travel *more* than 360° along its orbit to complete a cycle of phases (for example, from one new moon to the next). Because of this extra distance,

the synodic month is equal to about 29.53 days, about two days longer than the sidereal month.

Both the sidereal month and synodic month vary somewhat from one orbit to another, the latter by as much as half a day. The reason is that the Sun's gravity sometimes causes the Moon to speed up or slow down slightly in its orbit, depending on the relative positions of the Sun, Moon, and Earth. Furthermore, the Moon's orbit changes slightly from one month to the next.

3-3 | Eclipses occur only when the Sun and Moon are both on the line of nodes

From time to time the Sun, Earth, and Moon all happen to lie along a straight line. When this occurs, the shadow of the Earth can fall on the Moon or the shadow of the Moon can fall on the Earth. Such phenomena are called **eclipses**. They are perhaps the most dramatic astronomical events that can be seen with the naked eye.

A **lunar eclipse** occurs when the Moon passes through the Earth's shadow. This occurs when the Sun, Earth, and Moon are in a straight line, with the Earth between the Sun and Moon so that the Moon is at full phase (position E in Figure 3-2). At this point in the Moon's orbit, the face of the Moon seen from Earth would normally be fully illuminated by the Sun. Instead, it appears quite dim because the Earth casts a shadow on the Moon.

A **solar eclipse** occurs when the Earth passes through the Moon's shadow. As seen from Earth, the Moon moves in front of the Sun.

Once again, this can happen only when the Sun, Moon, and Earth are in a straight line. However, for a solar eclipse to occur, the Moon must be between the Earth and the Sun. Therefore, a solar eclipse can occur only at new moon (position A in Figure 3-2).

CAUTION! Both new moon and full moon occur at intervals of 29½ days. Hence, you might expect that there would be a solar eclipse every 29½ days, followed by a lunar eclipse about two weeks (half a lunar orbit) later. But in fact, there are only a few solar eclipses and lunar eclipses per year. Solar and lunar eclipses are so infrequent because the plane of the Moon's orbit and the plane of the Earth's orbit are not exactly aligned, as Figure 3-6 shows. The angle between the plane of the Earth's orbit and the plane of the Moon's orbit is about 5°. Because of this tilt, new moon and full moon usually occur when the Moon is either above or below the plane of the Earth's orbit. When the Moon is not in the plane of the Earth's orbit, the Sun, Moon, and Earth cannot align perfectly, and an eclipse cannot occur.

In order for the Sun, Earth, and Moon to be lined up for an eclipse, the Moon must lie in the same plane as the Earth's orbit around the Sun. This is called the **plane of the ecliptic** because its

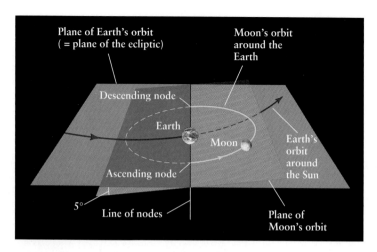

Figure 3-6

The Inclination of the Moon's Orbit This drawing shows the Moon's orbit around the Earth (in yellow) and part of the Earth's orbit around the Sun (in red). The plane of the Moon's orbit (shown in brown) is tilted by about 5° with respect to the plane of the Earth's orbit, also called the plane of the ecliptic (shown in blue). These two planes intersect along a line called the line of nodes.

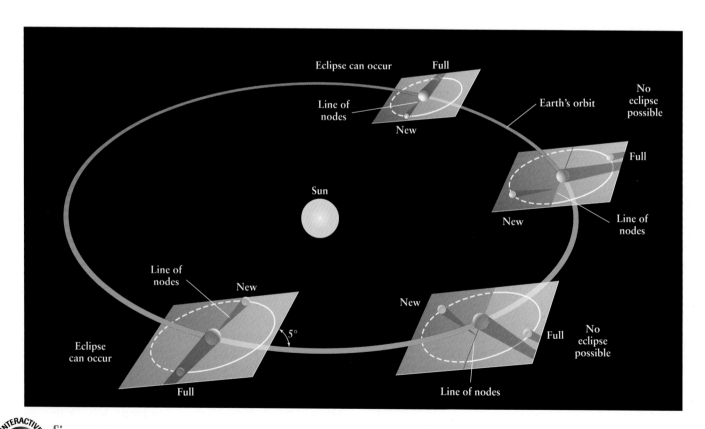

Figure 3-7

Conditions for Eclipses Eclipses can take place only if the Sun and Moon are both very near to or on the line of nodes. Only then can the Sun, Earth, and Moon all lie along a straight line. A solar eclipse occurs only if the Moon is very near the line of nodes at new moon; a lunar eclipse occurs only if the Moon is very near the line of nodes at full moon. If the Sun and Moon are not near the line of nodes, the Moon's shadow cannot fall on the Earth and the Earth's shadow cannot fall on the Moon.

projection onto the celestial sphere is the same as the path that the Sun appears to us to follow among the stars—that is, the ecliptic (see Figure 2-14). Thus, whenever an eclipse occurs, the Moon appears from Earth to be on the ecliptic (which is how the ecliptic gets its name).

The planes of the Earth's orbit and the Moon's orbit intersect along a line called the **line of nodes,** shown in Figure 3-6. The line of nodes passes through the Earth and is pointed in a particular direction in space. Eclipses can occur only if the line of nodes is pointed toward the Sun—that is, if the Sun lies on or near the line of nodes—and if, at the same time, the Moon lies on or very near the line of nodes. Only then do the Sun, Earth, and Moon lie in a line straight enough for an eclipse to occur (**Figure 3-7**).

 Anyone who wants to predict eclipses must know the orientation of the line of nodes. But the line of nodes is gradually shifting because of the gravitational pull of the Sun on the Moon. As a result, the line of nodes rotates slowly westward. Astronomers calculate such details to fix the dates and times of upcoming eclipses.

There are at least two—but never more than five—solar eclipses each year. The last year in which five solar eclipses occurred was 1935. The least number of eclipses possible (two solar, zero lunar)

happened in 1969. Lunar eclipses occur just about as frequently as solar eclipses, but the maximum possible number of eclipses (lunar and solar combined) in a single year is seven.

3-4 | Lunar eclipses can be either total, partial, or penumbral, depending on the alignment of the Sun, Earth, and Moon

The character of a lunar eclipse depends on exactly how the Moon travels through the Earth's shadow. As **Figure 3-8** shows, the shadow of the Earth has two distinct parts. In the **umbra,** the darkest part of the shadow, no portion of the Sun's surface can be seen. A portion of the Sun's surface is visible in the **penumbra,** which therefore is not quite as dark. Most people notice a lunar eclipse only if the Moon passes into the Earth's umbra. As this umbral phase of the eclipse begins, a bite seems to be taken out of the Moon.

The inset in Figure 3-8 shows the different ways in which the Moon can pass into the Earth's shadow. When the Moon passes through only the Earth's penumbra (Path 1), we see a **penumbral eclipse.** During a penumbral eclipse, the Earth blocks only part of

 Fiġure 3-8
Three Types of Lunar Eclipse People on the nighttime side of the Earth see a lunar eclipse when the Moon moves through the Earth's shadow. In the umbra, the darkest part of the shadow, the Sun is completely covered by the Earth. The penumbra is less dark because only

part of the Sun is covered by the Earth. The three paths show the motion of the Moon if the lunar eclipse is penumbral (Path 1), total (Path 2), or partial (Path 3). The inset shows these same paths, along with the umbra and penumbra, as viewed from the Earth.

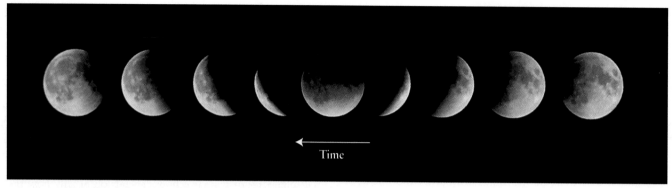

figure 3-9 R I **V** U X G

A Total Lunar Eclipse This sequence of nine photographs was taken over a 3-hour period during the lunar eclipse of January 20, 2000. The sequence, which runs from right to left, shows the Moon moving through the Earth's umbra. During the total phase of the eclipse (shown in the center), the Moon has a distinct reddish color. (Fred Espenak, NASA/ Goddard Space Flight Center; ©2000 Fred Espenak, MrEclipse.com)

the Sun's light and so none of the lunar surface is completely shaded. Because the Moon still looks full but only a little dimmer than usual, penumbral eclipses are easy to miss. If the Moon travels completely into the umbra (Path 2), a **total lunar eclipse** occurs. If only part of the Moon passes through the umbra (Path 3), we see a **partial lunar eclipse.**

If you were on the Moon during a total lunar eclipse, the Sun would be hidden behind the Earth. But some sunlight would be visible through the thin ring of atmosphere around the Earth, just as you can see sunlight through a person's hair if they stand with their head between your eyes and the Sun. As a result, a small amount of light reaches the Moon during a total lunar eclipse, and so the Moon does not completely disappear from the sky as seen from Earth. Most of the light that passes through the Earth's atmosphere is red, and thus the eclipsed Moon glows faintly in reddish hues (**Figure 3-9**).

Lunar eclipses occur at full moon, when the Moon is directly opposite the Sun in the sky. Hence, a lunar eclipse can be seen at any place on Earth where the Sun is below the horizon (that is, where it is nighttime). A lunar eclipse has the maximum possible duration if the Moon travels directly through the center of the umbra. The Moon's speed through the Earth's shadow is roughly 1 kilometer per second (3600 kilometers per hour, or 2280 miles per hour), which means that **totality**—the period when the Moon is completely within the Earth's umbra—can last for as long as 1 hour and 42 minutes.

 On average, two or three lunar eclipses occur in a year. **Table 3-1** lists all ten lunar eclipses from 2004 to 2008. Of all lunar eclipses, roughly one-third are total, one-third are partial, and one-third are penumbral.

table 3-1	**Lunar Eclipses, 2004–2008**		
Date	Type	Where visible	Duration of totality (h = hours, m = minutes)
2004 May 4	Total	South America, Europe, Africa, Asia, Australia	1h 16m
2004 October 28	Total	Americas, Europe, Africa, central Asia	1h 21m
2005 April 24	Penumbral	Eastern Asia, Australia, Pacific, Americas	—
2005 October 17	Partial	Asia, Australia, Pacific, North America	—
2006 March 14	Penumbral	Americas, Europe, Africa, Asia	—
2006 September 7	Partial	Europe, Africa, Asia, Australia	—
2007 March 3	Total	Americas, Europe, Africa, Asia	1h 14m
2007 August 28	Total	Eastern Asia, Australia, Pacific, Americas	1h 31m
2008 February 21	Total	Central Pacific, Americas, Europe, Africa	51m
2008 August 16	Partial	South America, Europe, Africa, Asia, Australia	—

Eclipse predictions by Fred Espenak, NASA/Goddard Space Flight Center. All dates are given in standard astronomical format: year, month, day.

3-5 | Solar eclipses can be either total, partial, or annular, depending on the alignment of the Sun, Earth, and Moon

As seen from Earth, the angular diameter of the Moon is almost exactly the same as the angular diameter of the far larger but more distant Sun—about 0.5°. Thanks to this coincidence of nature, the Moon just "fits" over the Sun during a **total solar eclipse.**

A total solar eclipse is a dramatic event. The sky begins to darken, the air temperature falls, and winds increase as the Moon gradually covers more and more of the Sun's disk (Figure 3-10). All nature responds: Birds go to roost, flowers close their petals, and crickets begin to chirp as if evening had arrived. As the last few rays of sunlight peek out from behind the edge of the Moon and the eclipse becomes total, the landscape around you is bathed in an eerie gray or, less frequently, in shimmering bands of light and dark. Finally, for a few minutes the Moon completely blocks out the dazzling solar disk and not much else. The **solar corona**—the Sun's thin, hot outer atmosphere, which is normally too dim to be seen—blazes forth in the darkened daytime sky (see the inset photograph in Figure 3-10). It is an awe-inspiring sight.

CAUTION If you are fortunate enough to see a solar eclipse, keep in mind that the only time when it is safe to look at the Sun is during **totality,** when the solar disk is blocked by the Moon and only the solar corona is visible. Viewing this magnificent spectacle cannot harm you in any way. But you must *never*

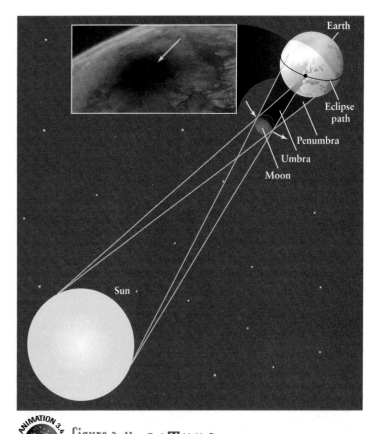

ANIMATION 3.4 **Figure 3-11** R I **V** U X G
The Geometry of a Total Solar Eclipse During a total solar eclipse, the tip of the Moon's umbra reaches the Earth's surface. As the Earth and Moon move along their orbits, this tip traces an eclipse path across the Earth's surface. People within the eclipse path see a total solar eclipse as the tip moves over them. Anyone within the penumbra sees only a partial eclipse. The inset photograph was taken from the *Mir* space station during the August 11, 1999, total solar eclipse (the same eclipse shown in the photo that opens this chapter). The tip of the umbra appears as a black spot on the Earth's surface. At the time the photograph was taken, this spot was 105 km (65 mi) wide and was crossing the English Channel at 3000 km/h (1900 mi/h). (Photograph by Jean-Pierre Haigneré, Centre National d'Etudes Spatiales, France/GSFS)

look directly at the Sun when even a portion of its intensely brilliant disk is exposed. *If you look directly at the Sun at any time without a special filter approved for solar viewing, you will suffer permanent eye damage or blindness.*

To see the remarkable spectacle of a total solar eclipse, you must be inside the darkest part of the Moon's shadow, also called the umbra, where the Moon completely blocks the Sun. Because the Sun and the Moon have nearly the same angular diameter as seen from Earth, only the tip of the Moon's umbra reaches the Earth's surface (Figure 3-11). As the Earth rotates, the tip of the umbra traces an **eclipse path** across the Earth's surface. Only those locations within the eclipse path are treated to the spectacle of a total solar eclipse. The inset in Figure 3-11 shows the dark spot on the Earth's surface produced by the Moon's umbra.

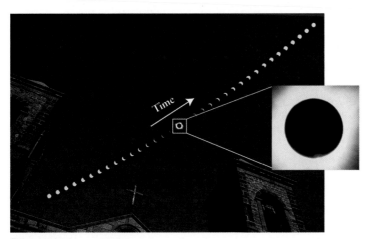

Figure 3-10 R I **V** U X G
A Total Solar Eclipse This multiple-exposure photograph shows the stages of the total solar eclipse of July 11, 1991, as seen from La Paz, Baja California. The individual exposures were taken at 5-minute intervals as the Sun and Moon moved across the sky from left to right. Notice how the Moon progressively covers then uncovers the Sun. The exposure enclosed by a white box shows the Moon completely blocking the light of the Sun's disk, making the faint solar corona visible. The inset photograph is an enlarged view of the eclipsed Sun surrounded by the corona. (Akira Fujii, Hiroyuki Tomioka, and Yonematsu Shiono; inset: Fred Espenak)

Immediately surrounding the Moon's umbra is the region of partial shadow called the penumbra. As seen from this area, the Sun's surface appears only partially covered by the Moon. During a solar eclipse, the Moon's penumbra covers a large portion of the Earth's surface, and anyone standing inside the penumbra sees a **partial solar eclipse.** Such eclipses are much less interesting events than total solar eclipses, which is why astronomy enthusiasts strive to be inside the eclipse path. If you are within the eclipse path, you will see a partial eclipse before and after the brief period of totality (see Figure 3-10).

The width of the eclipse path depends primarily on the Earth-Moon distance during totality. The eclipse path is widest if the Moon happens to be at **perigee,** the point in its orbit nearest the Earth. In this case the width of the eclipse path can be as great as 270 kilometers (170 miles). In most eclipses, however, the path is much narrower.

In some eclipses the Moon's umbra does not reach all the way to the Earth's surface. This can happen if the Moon is at or near **apogee,** its farthest position from Earth. In this case, the Moon appears too small to cover the Sun completely. The result is a third type of solar eclipse, called an **annular eclipse.** During an annular eclipse, a thin ring of the Sun is seen around the edge of the Moon (**Figure 3-12**). The length of the Moon's umbra is nearly 5000 kilometers (3100 miles) less than the average distance between the Moon and the Earth's surface. Thus, the Moon's shadow often fails to reach the Earth even when the Sun, Moon, and Earth are properly aligned for an eclipse. Hence, annular eclipses are slightly more common—as well as far less dramatic—than total eclipses.

Even during a total eclipse, most people along the eclipse path observe totality for only a few moments. The Earth's rotation, coupled with the orbital motion of the Moon, causes the umbra to race eastward along the eclipse path at speeds in excess of 1700 kilometers per hour (1060 miles per hour). Because of the umbra's high speed, totality never lasts for more than 7½ minutes. In a typical total solar eclipse, the Sun-Moon-Earth alignment and the Earth-Moon distance are such that totality lasts much less than this maximum.

 The details of solar eclipses are calculated well in advance. They are published in such reference books as the *Astronomical Almanac* and are available on the World Wide Web. **Figure 3-13** shows the eclipse paths for all total solar eclipses from 1997 to 2020. **Table 3-2** lists all the total, annular, and partial eclipses from 2004 to 2008, including the maximum duration of totality for total eclipses.

Ancient astronomers achieved a limited ability to predict eclipses. In those times, religious and political leaders who were able to predict such awe-inspiring events as eclipses must have made a tremendous impression on their followers. One of three priceless manuscripts to survive the devastating Spanish Conquest shows that the Mayan astronomers of Mexico and Guatemala had a fairly reliable method for predicting eclipses. The great Greek astronomer Thales of Miletus is said to have predicted the famous eclipse of 585 B.C., which occurred during the middle of a war. The sight was so unnerving that the soldiers put down their arms and declared peace.

In retrospect, it seems that what ancient astronomers actually produced were eclipse "warnings" of various degrees of reliability

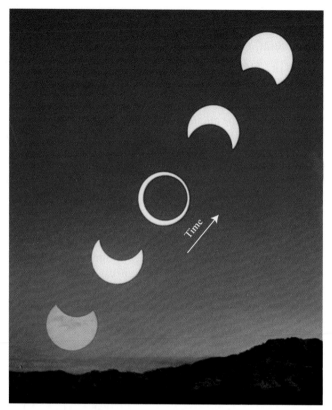

Figure 3-12 R I **V** U X G

An Annular Solar Eclipse This composite of six photographs taken at sunrise in Costa Rica shows the progress of an annular eclipse of the Sun on December 24, 1973. (Five photographs were made of the Sun, plus one of the hills and sky.) Note that at mideclipse the limb, or outer edge, of the Sun is visible around the Moon. (Courtesy of Dennis di Cicco)

rather than true predictions. Working with historical records, these astronomers generally sought to discover cycles and regularities from which future eclipses could be anticipated. **Box 3-2** describes how you might produce eclipse warnings yourself.

3-6 | Ancient astronomers measured the size of the Earth and attempted to determine distances to the Sun and Moon

SCIENCE IN PROCESS *The ideas of geometry made it possible for Greek scholars to measure cosmic distances*

The prediction of eclipses was not the only problem attacked by ancient astronomers. More than 2000 years ago, centuries before sailors of Columbus's era crossed the oceans, Greek astronomers were fully aware that the Earth is not flat. They had come to this conclusion using a combination of observation and logical deduction, much like modern scientists. The Greeks noted that during

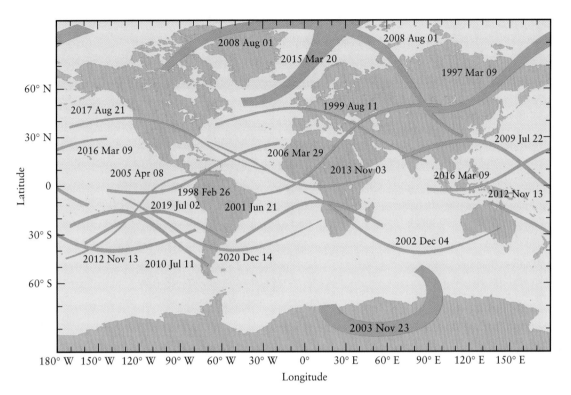

figure 3-13

Eclipse Paths for Total Eclipses, 1997–2020 This map shows the eclipse paths for all 18 total solar eclipses occurring from 1997 through 2020. In each eclipse, the Moon's shadow travels along the eclipse path in a generally eastward direction across the Earth's surface. (Courtesy of Fred Espenak, NASA/Goddard Space Flight Center)

lunar eclipses, when the Moon passes through the Earth's shadow, the edge of the shadow is always circular. Because a sphere is the only shape that always casts a circular shadow from any angle, they concluded that the Earth is spherical.

Around 200 B.C., the Greek astronomer Eratosthenes devised a way to measure the circumference of the spherical Earth. It was known that on the date of the summer solstice (the first day of sum-

mer; see Section 2-5) in the town of Syene in Egypt, near present-day Aswan, the Sun shone directly down the vertical shafts of water wells. Hence, at local noon on that day, the Sun was at the zenith (see Section 2-4) as seen from Syene. Eratosthenes knew that the Sun never appeared at the zenith at his home in the Egyptian city of Alexandria, which is on the Mediterranean Sea almost due north of Syene. Rather, on the summer solstice in Alexandria, the position

table 3-2	Solar Eclipses, 2004–2008			
Date	**Type**	**Where visible**		**Notes**
2004 April 19	Partial	Antarctica, southern Africa		74% eclipsed
2004 October 14	Partial	Northeast Asia, Hawaii, Alaska		93% eclipsed
2005 April 8	Annular and Total	New Zealand, North and South America		Annular along part of path; maximum duration of totality 0m 42s
2005 October 3	Annular	Europe, Africa, southern Asia		—
2006 March 29	Total	Africa, Europe, western Asia		Maximum duration of totality 4m 7s
2006 September 22	Annular	South America, western Africa, Antarctica		—
2007 March 19	Partial	Asia, Alaska		87% eclipsed
2007 September 11	Partial	South America, Antarctica		75% eclipsed
2008 February 7	Annular	Antarctica, eastern Australia, New Zealand		—
2008 August 1	Total	Northeast North America, Europe, Asia		Maximum duration of totality 2m 27s

Eclipse predictions by Fred Espenak, NASA/Goddard Space Flight Center. All dates are given in standard astronomical format: year, month, day.

box 3-2 TOOLS OF THE ASTRONOMER'S TRADE

Predicting Solar Eclipses

Suppose that you observe a solar eclipse in your hometown and want to figure out when you and your neighbors might see another eclipse. How would you begin?

First, remember that a solar eclipse can occur only if the line of nodes points toward the Sun at the same time that there is a new moon (see Figure 3-7). Second, you must know that it takes 29.53 days (one synodic month) to go from one new moon to the next. Because solar eclipses occur only during new moon, you must wait several whole lunar months for the proper alignment to occur again.

However, there is a complication: The line of nodes gradually shifts its position with respect to the background stars. It takes 346.6 days to move from one alignment of the line of nodes pointing toward the Sun to the next identical alignment. This period is called the **eclipse year.**

Therefore, to predict when you will see another solar eclipse, you need to know how many whole lunar months equal some whole number of eclipse years. This will tell you how long you will have to wait for the next virtually identical alignment of the Sun, the Moon, and the line of nodes. By trial and error, you find that 223 lunar months is the same length of time as 19 eclipse years, because

$$223 \times 29.53 \text{ days} = 19 \times 346.6 \text{ days} = 6585 \text{ days}$$

This calculation is accurate to within a few hours. A more accurate calculation gives an interval, called the **saros,** that is about one-third day longer, or 6585.3 days (18 years, 11.3 days). Eclipses separated by the saros interval are said to form an *eclipse series.*

You might think that you and your neighbors would simply have to wait one full saros interval to go from one solar eclipse to the next. However, because of the extra one-third day, the Earth

will have rotated by an extra 120° (one-third of a complete rotation) when the next solar eclipse of a particular series occurs. The eclipse path will thus be one-third of the way around the world from you. Therefore, you must wait three full saros intervals (54 years, 34 days) before the eclipse path comes back around to your part of the Earth. The illustration shows a series of six solar eclipse paths, each separated from the next by one saros interval.

There is evidence that ancient Babylonian astronomers knew about the saros interval. However, the discovery of the saros is more likely to have come from lunar eclipses than solar eclipses. If you are far from the eclipse path, there is a good chance that you could fail to notice a solar eclipse. Even if half the Sun is covered by the Moon, the remaining solar surface provides enough sunlight for the outdoor illumination not to be greatly diminished. By contrast, anyone on the nighttime side of the Earth can see an eclipse of the Moon unless clouds block the view.

---

of the Sun at local noon was about 7° south of the zenith (**Figure 3-14**). This angle is about one-fiftieth of a complete circle, so he concluded that the distance from Alexandria to Syene must be about one-fiftieth of the Earth's circumference.

In Eratosthenes's day, the distance from Alexandria to Syene was said to be 5000 stades. Therefore, Eratosthenes found the Earth's circumference to be

$$50 \times 5000 \text{ stades} = 250,000 \text{ stades}$$

Unfortunately, no one today is sure of the exact length of the Greek unit called the stade. One guess is that the stade was about one-sixth of a kilometer, which would mean that Eratosthenes obtained a circumference for the Earth of about 42,000 kilometers. This is remarkably close to the modern value of 40,000 kilometers.

Eratosthenes was only one of several brilliant astronomers to emerge from the so-called Alexandrian school, which by his time had

a distinguished tradition. One of the first Alexandrian astronomers, Aristarchus of Samos, had proposed a method of determining the relative distances to the Sun and Moon, perhaps as long ago as 280 B.C.

Aristarchus knew that the Sun, Moon, and Earth form a right triangle at the moment of first or third quarter moon, with the right angle at the location of the Moon (**Figure 3-15**). He estimated that, as seen from Earth, the angle between the Moon and the Sun at first and third quarters is 87°, or 3° less than a right angle. Using the rules of geometry, Aristarchus concluded that the Sun is about 20 times farther from us than is the Moon. We now know that Aristarchus erred in measuring angles and that the average distance to the Sun is about 390 times larger than the average distance to the Moon. It is nevertheless impressive that people were trying to measure distances across the solar system more than 2000 years ago.

Aristarchus also made an equally bold attempt to determine the relative sizes of the Earth, Moon, and Sun. From his observations of how long the Moon takes to move through the Earth's shadow dur-

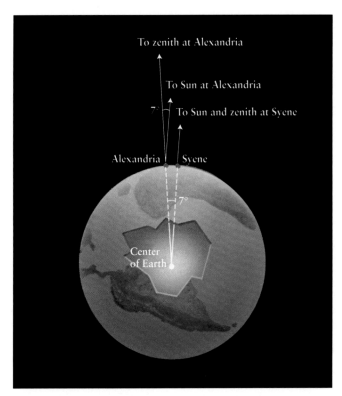

figure 3-14

Eratosthenes's Method of Determining the Diameter of the Earth
Around 200 B.C., Eratosthenes noticed that the Sun is about 7° south of the zenith at Alexandria when it is directly overhead at Syene. This angle is about one-fiftieth of a circle, so the distance between Alexandria and Syene must be about one-fiftieth of the Earth's circumference.

table 3-3	Comparison of Ancient and Modern Astronomical Measurements	
	Ancient measure (km)	Modern measure (km)
Earth's diameter	13,000	12,756
Moon's diameter	4,300	3,476
Sun's diameter	9×10^4	1.39×10^6
Earth-Moon distance	4×10^5	3.84×10^5
Earth-Sun distance	10^7	1.50×10^8

ing a lunar eclipse, Aristarchus estimated the diameter of the Earth to be about 3 times larger than the diameter of the Moon. To determine the diameter of the Sun, Aristarchus simply pointed out that the Sun and the Moon have the same angular size in the sky. Therefore, their diameters must be in the same proportion as their distances (see part *a* of the figure in Box 1-1). In other words, because Aristarchus thought the Sun to be 20 times farther from the Earth than the Moon, he concluded that the Sun must be 20 times larger than the Moon. Once Eratosthenes had measured the Earth's circumference, astronomers of the Alexandrian school could estimate the diameters of the Sun and Moon as well as their distances from Earth.

Table 3-3 summarizes some ancient and modern measurements of the sizes of Earth, the Moon, and the Sun and the distances between them. Some of these ancient measurements are far from the modern values. Yet the achievements of our ancestors still stand as impressive applications of observation and reasoning and important steps toward the development of the scientific method.

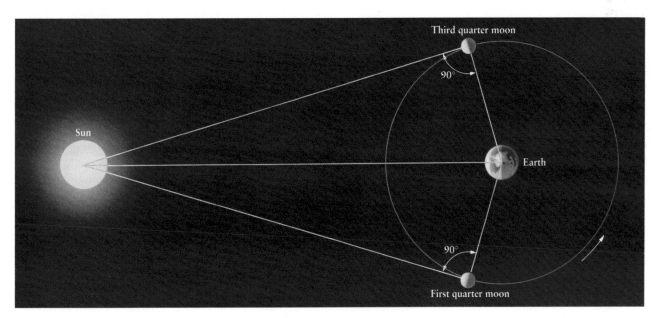

figure 3-15

Aristarchus's Method of Determining Distances to the Sun and Moon Aristarchus knew that the Sun, Moon, and Earth form a right triangle at first and third quarter phases. Using geometrical arguments, he calculated the relative lengths of the sides of these triangles, thereby obtaining the distances to the Sun and Moon.

Key Words

Terms preceded by an asterisk (*) are discussed in the Boxes.

annular eclipse, p. 52
apogee, p. 52
eclipse, p. 47
eclipse path, p. 51
*eclipse year, p. 54
first quarter moon, p. 43
full moon, p. 43
line of nodes, p. 49
lunar eclipse, p. 47
lunar phases, p. 43

new moon, p. 43
partial lunar eclipse, p. 50
partial solar eclipse, p. 52
penumbra (*plural* penumbrae), p. 49
penumbral eclipse, p. 49
perigee, p. 52
plane of the ecliptic, p. 48
*saros, p. 54
sidereal month, p. 47
solar corona, p. 51
solar eclipse, p. 47
synchronous rotation, p. 46

synodic month, p. 47
third quarter moon, p. 44
totality (lunar eclipse), p. 50
totality (solar eclipse), p. 51
total lunar eclipse, p. 50
total solar eclipse, p. 51
umbra (*plural* umbrae), p. 49
waning crescent moon, p. 44
waning gibbous moon, pp. 43–44
waxing crescent moon, p. 43
waxing gibbous moon, p. 43

Key Ideas

Lunar Phases: The phases of the Moon occur because light from the Moon is actually reflected sunlight. As the relative positions of the Earth, the Moon, and the Sun change, we see more or less of the illuminated half of the Moon.

Length of the Month: Two types of months are used in describing the motion of the Moon.

• With respect to the stars, the Moon completes one orbit around the Earth in a sidereal month, averaging 27.32 days.

• The Moon completes one cycle of phases (one orbit around the Earth with respect to the Sun) in a synodic month, averaging 29.53 days.

The Moon's Orbit: The plane of the Moon's orbit is tilted by about 5° from the plane of the Earth's orbit, or ecliptic.

• The line of nodes is the line where the planes of the Moon's orbit and the Earth's orbit intersect. The gravitational pull of the Sun gradually shifts the orientation of the line of nodes with respect to the stars.

Conditions for Eclipses: During a lunar eclipse, the Moon moves through the Earth's shadow. During a solar eclipse, the Earth passes through the Moon's shadow.

• Lunar eclipses occur at full moon, while solar eclipses occur at new moon.

• Either type of eclipse can occur only when the Sun and Moon are both on or very near the line of nodes. If this condition is not met, the Earth's shadow cannot fall on the Moon and the Moon's shadow cannot fall on the Earth.

Umbra and Penumbra: The shadow of an object has two parts: the umbra, within which the light source is completely blocked, and the penumbra, where the light source is only partially blocked.

Lunar Eclipses: Depending on the relative positions of the Sun, Moon, and Earth, lunar eclipses may be total (the Moon passes completely into the Earth's umbra), partial (only part of the Moon passes into the Earth's umbra), or penumbral (the Moon passes only into the Earth's penumbra).

Solar Eclipses: Solar eclipses may be total, partial, or annular.

• During a total solar eclipse, the Moon's umbra traces out an eclipse path over the Earth's surface as the Earth rotates. Observers outside the eclipse path but within the penumbra see only a partial solar eclipse.

• During an annular eclipse, the umbra falls short of the Earth, and the outer edge of the Sun's disk is visible around the Moon at mideclipse.

The Moon and Ancient Astronomers: Ancient astronomers such as Aristarchus and Eratosthenes made great progress in determining the sizes and relative distances of the Earth, the Moon, and the Sun.

Review Questions

Questions preceded by an asterisk (*) involve topics discussed in the Boxes.

1. (a) Explain why the Moon exhibits phases. (b) A common misconception about the Moon's phases is that they are caused by the Earth's shadow. Use Figure 3-2 to explain why this not correct.

2. How would the sequence and timing of lunar phases be affected if the Moon moved around its orbit (a) in the same direction, but at twice the speed; (b) at the same speed, but in the opposite direction? Explain your answers.

3. Is the far side of the Moon (the side that can never be seen from Earth) the same as the dark side of the Moon? Explain.

4. Astronomers sometimes refer to lunar phases in terms of the *age* of the Moon. This is the time that has elapsed since new moon phase. Thus, the age of a full moon is half of a 29½ -day synodic period, or approximately 15 days. Find the approximate age of (a) a waxing crescent moon; (**b**) a third quarter moon; (**c**) a waning gibbous moon.

5. (**a**) If you lived on the Moon, would you see the Sun rise and set, or would it always be in the same place in the sky? Explain. (**b**) Would you see the Earth rise and set, or would it always be in the same place in the sky? Explain using Figure 3-4.

6. If you lived on the Moon, would you see the Earth go through phases? If so, would the sequence of phases be the same as those of the Moon as seen from Earth, or would the sequence be reversed? Explain using Figure 3-2.

7. What is the difference between a sidereal month and a synodic month? Which is longer? Why?

8. On a certain date the Moon is in the direction of the constellation Gemini as seen from Earth. When will the Moon next be in the direction of Gemini: one sidereal month later, or one synodic month later? Explain.

9. What is the difference between the umbra and the penumbra of a shadow?

10. Why doesn't a lunar eclipse occur at every full moon and a solar eclipse at every new moon?

11. What is the line of nodes? Why is it important to the subject of eclipses?

12. What is a penumbral eclipse of the Moon? Why do you suppose that it is easy to overlook such an eclipse?

13. The maximum duration of totality of a lunar eclipse is 1 hour, 42 minutes. But none of the lunar eclipses listed in Table 3-1 lasts this long. Why is this?

14. Can one ever observe an annular eclipse of the Moon? Why or why not?

15. If you were looking at the Earth from the side of the Moon that faces the Earth, what would you see during (**a**) a total lunar eclipse? (**b**) a total solar eclipse? Explain your answers.

16. If there is a total eclipse of the Sun in April, can there be a lunar eclipse three months later in July? Why or why not?

17. Which type of eclipse—lunar or solar—do you think most people on Earth have seen? Why?

18. How is an annular eclipse of the Sun different from a total eclipse of the Sun? What causes this difference?

***19.** What is the saros? How did ancient astronomers use it to predict eclipses?

20. How did Eratosthenes measure the size of the Earth?

21. How did Aristarchus try to estimate the distance from the Earth to the Sun and Moon?

Advanced Questions

Problem-solving tips and tools

To estimate the average angular speed of the Moon along its orbit (that is, how many degrees around its orbit the Moon travels per day), divide 360° by the length of a sidereal month. It is helpful to know that the saros interval of 6585.3 days equals 18 years and 11⅓ days if the interval includes four leap years, but is 18 years and 10⅓ days if it includes five leap years.

22. The dividing line between the illuminated and unilluminated halves of the Moon is called the *terminator*. The terminator appears curved when there is a crescent or gibbous moon, but appears straight when there is a first quarter or third quarter moon (see Figure 3-2). Describe how you could use these facts to explain to a friend why lunar phases cannot be caused by the Earth's shadow falling on the Moon.

23. What is the phase of the Moon if it is highest in the sky at (**a**) midnight; (**b**) sunrise; (**c**) noon; (**d**) sunset? (*Hint:* See Figure 3-2.)

24. What is the phase of the Moon if it rises at (**a**) midnight; (**b**) sunrise; (**c**) noon; (**d**) sunset? (*Hint:* Use your answers to Question 23. How much time elapses from when the Moon rises to when it is highest in the sky?)

25. Suppose it is the first day of autumn in the northern hemisphere. What is the phase of the Moon if the Moon is located at (**a**) the vernal equinox? (**b**) the summer solstice? (**c**) the autumnal equinox? (**d**) the winter solstice? (*Hint:* Make a drawing showing the relative positions of the Sun, Earth, and Moon. Compare with Figure 3-2.)

26. This photograph of the Earth was taken by the crew of the *Apollo 8* spacecraft as they orbited the Moon. A portion of the lunar surface is visible at the right-hand side of the photo. In this photo, the Earth is oriented with its north pole approximately at the top. When this photo was taken, was the Moon waxing or waning as seen from Earth? Explain your answer with a diagram.

(NASA/JSC) R I **V** U X G

27. (a) The Moon moves noticeably on the celestial sphere over the space of a single night. To show this, calculate how long it takes the Moon to move through an angle equal to its own angular diameter ($\frac{1}{2}°$) against the background of stars. Give your answer in hours. (b) Through what angle (in degrees) does the Moon move during a 12-hour night? Can you notice an angle of this size? (*Hint:* See Figure 1-10.)

28. During an occultation, or "covering up," of Jupiter by the Moon, an astronomer notices that it takes the Moon's edge 90 seconds to cover Jupiter's disk completely. If the Moon's motion is assumed to be uniform and the occultation was "central" (that is, center over center), find the angular diameter of Jupiter. (*Hint:* Assume that Jupiter does not appear to move against the background of stars during this brief 90-second interval. You will need to convert the Moon's angular speed from degrees per day to arcseconds per second.)

29. How many more sidereal months than synodic months are there in a year? Explain.

30. Suppose the Earth moved a little faster around the Sun, so that it took a bit less than one year to make a complete orbit. If the speed of the Moon's orbit around the Earth were unchanged, would the length of the sidereal month be the same, longer, or shorter than it is now? What about the synodic month? Explain your answers.

31. If the Moon revolved about the Earth in the same orbit but in the opposite direction, would the synodic month be longer or shorter than the sidereal month? Explain your reasoning.

32. One definition of a "blue moon" is the second full moon within the same calendar month. There is usually only one full moon within a calendar month, so the phrase "once in a blue moon" means "hardly ever." Why are blue moons so rare? Are there any months of the year in which it would be impossible to have two full moons? Explain your answer.

33. You are watching a lunar eclipse from some place on the Earth's night side. Will you see the Moon enter the Earth's shadow from the east or from the west? Explain your reasoning.

34. The total lunar eclipse of July 16, 2000, was visible from Australia. The duration of totality was 1 hour, 47 minutes. Was this total eclipse also visible from Europe, on the opposite side of the Earth? Explain your reasoning.

35. During a total solar eclipse, the Moon's umbra moves in a generally eastward direction across the Earth's surface. Use a drawing like Figure 3-11 to explain why the motion is eastward, not westward.

36. A total solar eclipse was visible from Baja California on July 11, 1991 (see Figure 3-10). Draw what the eclipse would have looked like as seen from Arizona, to the north of the path of totality. Explain the reasoning behind your drawing.

37. Figures 3-11 and 3-13 show that the path of a total eclipse is quite narrow. Use this to explain why a glow is visible all around the horizon when you are viewing a solar eclipse during totality (see the photograph that opens this chapter).

38. (a) Suppose the diameter of the Moon were doubled, but the orbit of the Moon remained the same. Would total solar eclipses be more common, less common, or just as common as they are now? Explain. (b) Suppose the diameter of the Moon were halved, but the orbit of the Moon remained the same. Explain why there would be *no* total solar eclipses.

39. Just as the distance from the Earth to the Moon varies somewhat as the Moon orbits the Earth, the distance from the Sun to the Earth changes as the Earth orbits the Sun. The Earth is closest to the Sun at its *perihelion;* it is farthest from the Sun at its *aphelion.* In order for a total solar eclipse to have the maximum duration of totality, should the Earth be at perihelion or aphelion? Assume that the Earth-Moon distance is the same in both situations. As part of your explanation, draw two pictures like Figure 3-11, one with the Earth relatively close to the Sun and one with the Earth relatively far from the Sun.

*40. On December 4, 2002, residents of southern Africa were treated to a total solar eclipse. (a) On what date and over what part of the world will the next total eclipse of that series occur? Explain. (b) On what date might you next expect a total eclipse of that series to be visible from southern Africa? Explain.

Discussion Questions

41. Describe the cycle of lunar phases that would be observed if the Moon moved around the Earth in an orbit perpendicular to the plane of the Earth's orbit. Would it be possible for both solar and lunar eclipses to occur under these circumstances? Explain your reasoning.

42. How would a lunar eclipse look if the Earth had no atmosphere? Explain your reasoning.

43. In his 1885 novel *King Solomon's Mines,* H. Rider Haggard described a total solar eclipse that was seen in both South Africa and in the British Isles. Is such an eclipse possible? Why or why not?

44. Why do you suppose that total solar eclipse paths fall more frequently on oceans than on land? (You may find it useful to look at Figure 3-13.)

45. Examine Figure 3-13, which shows all of the total solar eclipses from 1997 to 2020. What are the chances that you might be able to travel to one of the eclipse paths? Do you think you might go through your entire life without ever seeing a total eclipse of the Sun?

 ## Web/CD-ROM Questions

 46. Access the animation "The Moon's Phases" in Chapter 3 of the *Universe* Web site or CD-ROM. This shows the Earth-Moon system as seen from a vantage point looking down onto the north pole. (a) Describe where you would be on the diagram if you are on the equator and the time is 6:00 P.M. (b) If it is 6:00 P.M. and you are standing on Earth's equa-

tor, would a third-quarter moon be visible? Why or why not? If it would be visible, describe its appearance.

47. Search the World Wide Web for information about the next total lunar eclipse. Will the total phase of the eclipse be visible from your location? If not, will the penumbral phase be visible? Draw a picture showing the Sun, Earth, and Moon when the totality is at its maximum duration, and indicate your location on the drawing of the Earth.

48. Search the World Wide Web for information about the next total solar eclipse. Through which major cities, if any, does the path of totality pass? What is the maximum duration of totality? At what location is this maximum duration observed? Will this eclipse be visible (even as a partial eclipse) from your location? Draw a picture showing the Sun, Earth, and Moon when the totality is at its maximum duration, and indicate your location on the drawing of the Earth.

 49. Access the animation "A Solar Eclipse Viewed from the Moon" in Chapter 3 of the *Universe* Web site or CD-ROM. This shows the solar eclipse of August 11, 1999, as viewed from the Moon. Using a diagram, explain why the stars and the Moon's shadow move in the directions shown in this animation.

Observing Projects

50. Observe the Moon on each clear night over the course of a month. On each night, note the Moon's location among the constellations and record that location on a star chart that also shows the ecliptic. After a few weeks, your observations will begin to trace the Moon's orbit. Identify the orientation of the line of nodes by marking the points where the Moon's orbit and the ecliptic intersect. On what dates is the Sun near the nodes marked on your star chart? Compare these dates with the dates of the next solar and lunar eclipses.

51. It is quite possible that a lunar eclipse will occur while you are taking this course. Look up the date of the next lunar eclipse in Table 3-1, on the World Wide Web, or in the current issue of a reference such as the *Astronomical Almanac* or *Astronomical Phenomena*. Then make arrangements to observe this lunar eclipse. You can observe the eclipse with the naked eye, but binoculars or a small telescope will enhance your viewing experience. If the eclipse is partial or total, note the times at which the Moon enters and exits the Earth's umbra. If the eclipse is penumbral, can you see any changes in the Moon's brightness as the eclipse progresses?

 52. Use the *Starry Night Backyard*™ program to observe the motion of the Moon. (**a**) Display the entire celestial sphere, including the part below the horizon (select **Guides > Atlas** in the **Go** menu). Center on the Moon by using the **Find...** command in the **Edit** menu. In the Control Panel at the top of the main window, click on the control for the rate of time flow (immediately to the right of the date and time display) and set the discrete time step to 1 day. Then click on the "Forward" button (a triangle that points to the right). How does the Moon appear to move against the background of stars? Does it ever change direction? (**b**) Using the "Step time forward" button (just to the right of the "Forward" button), determine how many days elapse between successive times when the Moon is on the ecliptic. (If you don't see a green line representing the ecliptic, select **Show View Options Panel** in the **View** menu, then make sure the box "The Ecliptic" is checked.) Then move forward in time to a date when the Moon is on the ecliptic *and* either full or new. What type of eclipse will occur on that date? Confirm your answer by comparing with Tables 3-1 and 3-2 or with lists of eclipses on the World Wide Web.

 53. Use the *Deep Space Explorer*™ program to examine the Moon as seen from space. In the left-hand part of the window under the heading **Solar System**, click on the triangle next to the word **Explore** and then click on **Moon**. You can zoom in and zoom out using the buttons at the upper left of the window (an upward-pointing triangle and a downward-pointing triangle). You can also rotate the Moon by putting the mouse cursor over the image, holding down the mouse button, and moving the mouse. (On a two-button mouse, hold down the left mouse button.) (**a**) Rotate the Moon around to see it from different perspectives, as though you were flying a spaceship around the Moon. How does the phase of the Moon change as you rotate it around? (*Hint:* Compare with Box 3-1.) (**b**) Rotate the Moon until you can also see the Sun. Explain how your observations show that the phases of the Moon cannot be caused by the Earth's shadow falling on the Moon.

Collaborative Exercise

54. Using a bright light source at the center of a darkened room, or a flashlight, use your fist held at arm's length to demonstrate the difference between a full moon and a lunar eclipse. (Use yourself or a classmate as the Earth.) How must your fist "orbit" the Earth so that lunar eclipses do not happen at every full moon? Create a simple sketch to illustrate your answers.

MARK HOLLABAUGH

Archaeoastronomy and Ethnoastronomy

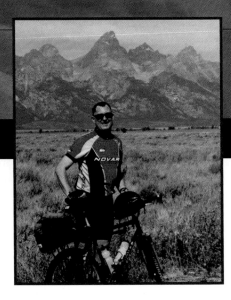

Mark Hollabaugh teaches physics and astronomy at Normandale Community College, in Bloomington, Minnesota. A graduate of St. Olaf College, he earned his Ph.D. in science education at the University of Minnesota. He also taught at St. Olaf College, Augsburg College, and the U.S. Air Force Academy. As a boy he watched the dance of the northern lights and the flash of meteors. Meeting *Apollo 13* commander Jim Lovell and going for a ride in Roger Freedman's airplane are as close as he came to his dream of being an astronaut.

Many years ago I read astronomer John Eddy's *National Geographic* article about the Wyoming Medicine Wheel. A few years later I visited this archaeological site in the Bighorn Mountains and started on the major preoccupation of my career.

Archaeoastronomy combines astronomy and archaeology. You may be familiar with sites such as Stonehenge or the Mayan ruins of the Yucatán (see Figure 2-1). You may not know that there are many archaeological sites in the United States that demonstrate the remarkable understanding a people, usually known as the Anasazi, had about celestial motions. Chaco Canyon, Hovenweep, and Chimney Rock in the Four Corners area of the Southwest preserve ruins from this ancient Pueblo culture.

Moonrise at Chimney Rock in southern Colorado provides a good example of the Anasazi's knowledge of the lunar cycles. If you watched the rising of the Moon for many, many years, you would discover that the Moon's northernmost rising point undergoes an 18.6-year cycle. Dr. McKim Malville of the University of Colorado discovered that the Anasazi who lived there knew of the lunar standstill cycle and watched the northernmost rising of the Moon between the twin rock pillars of Chimney Rock.

My own specialty is the ethnoastronomy of the Lakota, or Teton Sioux, who flourished on the Great Plains of what are now Nebraska, the Dakotas, and Wyoming. Ethnoastronomy combines ethnography with astronomy. As an ethnoastronomer, I am less concerned with physical evidence in the form of ruins and more interested in myths, legends, religious belief, and current practices. In my quest to understand the astronomical thinking and customs of the nineteenth-century Lakota, I have traveled to museums, archives, and libraries in Nebraska, Colorado, Wyoming, South Dakota, and North Dakota. I frequently visit the Pine Ridge and Rosebud Reservations in South Dakota. The Lakota, and other Plains Indians, had a rich tradition of understanding celestial motions and developed an even richer explanation of why things appear the way they do in the sky.

My first professional contribution to ethnoastronomy was in 1996 at the Fifth Oxford International Conference on Astronomy in Culture held in Santa Fe, New Mexico. I had noticed that images of eclipses often appeared in Lakota winter counts—their method of making an historic record of events. The great Leonid meteor shower of 1833 appears in almost every Plains Indian winter count. As I looked at the hides or in ledger books recording these winter counts, I wondered why the Sun, the Moon and the stars are so common among the Lakota of 150 years ago. My curiosity led me to look deeper: What did the Lakota think about eclipses? Why does their central ritual, the Sun Dance, focus on the Sun? Why do so many legends involve the stars?

The Lakota observed lunar and solar eclipses. Perhaps they felt they had the power to restore the eclipsed Sun or Moon. In August 1869, an Indian agency physician in South Dakota told the Lakota there would be an eclipse. When the Sun disappeared from view, the Lakota began firing their guns in the air. In their minds, they were more powerful than the white doctor because the result of their action was to restore the Sun.

The Lakota used a lunar calendar. Their names for the months came from the world around them. October was the moon of falling leaves. In some years, there actually are 13 new moons, and they often called this extra month the "Lost Moon." The lunar calendar often dictated the timing of their sacred rites. Although the Lakota were never dogmatic about it, they preferred to hold their most important ceremony, the Sun Dance, at the time of the full moon in June, which is when the summer solstice occurs.

Why did the Lakota pay attention to the night sky? Lakota elder Ringing Shield's statement about Polaris, recorded in the late nineteenth century, provides a clue: "One star never moves and it is *wakan*. Other stars move in a circle about it. They are dancing in the dance circle." For the Lakota, a driving force in their culture was a quest to understand the nature of the sacred, or *wakan*. Anything hard to understand or different from the ordinary was *wakan*.

Reaching for the stars, as far away as they are, was a means for the Lakota to bring the incomprehensible universe a bit closer to the earth. Their goal was the same as what Dr. Sandra M. Faber says in her essay "Why Astronomy?" at the end of Chapter 1, "a perspective on human existence and its relation to the cosmos."

4 Gravitation and the Waltz of the Planets

In July 1969 humans first journeyed from the Earth to the surface of the Moon. By studying the rocks that the astronauts collected on the Moon during this and later missions, geologists learned how the surface of our nearest celestial neighbor has evolved over the past several billion years. These discoveries revolutionized our understanding of the very nature of the Moon.

But several other revolutions in human understanding had to take place before these discoveries could be made. One revolution overthrew the ancient idea that the Earth is an immovable object at the center of the universe, around which moved the Sun, the Moon, and the planets. In this chapter we will learn how Nicolaus Copernicus, Tycho Brahe, Johannes Kepler, and Galileo Galilei helped us understand that the Earth is itself one of several planets orbiting the Sun.

We will learn, too, about Isaac Newton's revolutionary discovery of why the planets move in the way that they do. This was just one aspect of Newton's immense body of work, which included formulating the fundamental laws of physics and developing a precise mathematical description of the force of gravity—the force that holds the planets in their orbits.

Newton's laws apply on Earth as well as in the heavens. They also govern the flight of spacecraft, which made missions to the Moon possible. That is why astronaut William Anders radioed the following while en route from the Moon to the Earth: "I think Isaac Newton is doing most of the driving right now."

 WEB LINK 4.1
July 1969: With the Earth in the distance, the Apollo 11 lander *Eagle* orbits over the Moon. (Michael Collins, *Apollo 11*, NASA)   R I **V** U X G

Questions to Guide Your Reading

As you read the sections of this chapter, look for the answers to the following questions:

4-1 How did ancient astronomers explain the motions of the planets?

4-2 Why did Copernicus think that the Earth and the other planets go around the Sun?

4-3 How did Tycho Brahe attempt to test the ideas of Copernicus?

4-4 What paths do the planets follow as they move around the Sun?

4-5 What did Galileo see in his telescope that confirmed that the planets orbit the Sun?

4-6 What fundamental laws of nature explain the motions of objects on Earth as well as the motions of the planets?

4-7 Why don't the planets fall into the Sun?

4-8 What keeps the same face of the Moon always pointed toward the Earth?

4-1 | Ancient astronomers invented geocentric models to explain planetary motions

Since the dawn of civilization, scholars have attempted to explain the nature of the universe. The ancient Greeks were the first to use the principle that still guides scientists today: *The universe can be described and understood logically.* For example, more than 2500 years ago Pythagoras and his followers put forth the idea that nature can be described with mathematics. About 200 years later, Aristotle asserted that the universe is governed by physical laws.

As they attempted to create a model (see Section 1-1) of how the universe works, most Greek scholars assumed that the Sun, the Moon, the stars, and the planets revolve about a stationary Earth. A model of this kind, in which the Earth is at the center of the universe, is called a **geocentric model**. Similar ideas were held by the scholars of ancient China.

Today we recognize that the stars are not merely points of light on an immense celestial sphere. But in fact this is how the ancient Greeks regarded the stars in their geocentric model of the universe. To explain the diurnal motions of the stars, they assumed that the celestial sphere was *real,* and that it rotated around the stationary Earth once a day. The Sun and Moon both participated in this daily rotation of the sky, which explained their rising and setting motions. To explain why the Sun and Moon both move slowly with respect to the stars, the ancient Greeks imagined that both of these objects orbit around the Earth.

ANALOGY Imagine a merry-go-round that rotates clockwise as seen from above, as in **Figure 4-1a**. As it rotates, two children walk slowly counterclockwise at different speeds around the merry-go-round's platform. Thus, the children rotate along with the merry-go-round and also change their positions with respect to the merry-go-round's wooden horses. This scene is analogous to the way the ancient Greeks pictured the motions of the stars, Sun, and Moon. In their model, the celestial sphere rotated to the west around a stationary Earth (Figure 4-1b). The stars rotate along with the celestial sphere just as the wooden horses rotate along with the merry-go-round in Figure 4-1a. The Sun and Moon are analogous to the two children; they both turn westward with the celestial sphere, making one complete turn each day, and also move slowly eastward at different speeds with respect to the stars.

The geocentric model of the heavens also had to explain the motions of the planets. The ancient Greeks and other cultures of that time knew of five planets: Mercury, Venus, Mars, Jupiter, and Saturn, each of which is a bright object in the night sky. For example, when Venus is at its maximum brilliancy, it is 16 times brighter than the brightest star. (By contrast, Uranus, Neptune, and Pluto are quite dim and were not discovered until after the invention of the telescope.)

Like the Sun and Moon, all of the planets rise in the east and set in the west once a day. And like the Sun and Moon, from night to night the planets slowly move on the celestial sphere, that is, with respect to the background of stars. However, the character of this motion on the celestial sphere is quite different for the planets. Both the Sun and the Moon always move from west to east on the celestial sphere, that is, opposite the direction in which the celestial sphere appears to rotate. Furthermore, the Sun and the Moon each move at relatively constant speeds around the celestial sphere. (The Moon's speed is faster than that of the Sun: It travels all the way around the celestial sphere in about a month while the Sun takes an entire year.) By contrast, each of the planets appears to wander back and forth on the celestial sphere, moving from west to east at most times but at other times from east to west. (The name *planet* is well deserved; it comes from a Greek word meaning "wanderer.") As a planet wanders, its speed on the celestial sphere can vary substantially.

You can see this distinctive behavior in **Figure 4-2**, which shows the motion of Mars with respect to the background of stars dur-

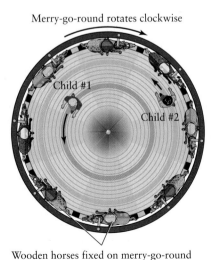

Merry-go-round rotates clockwise

Child #1

Child #2

Wooden horses fixed on merry-go-round

(a) A rotating merry-go-round

Celestial sphere rotates to the west

Moon

Sun

Earth

Stars fixed on celestial sphere

(b) The Greek geocentric model

figure 4-1

A Merry-Go-Round Analogy (a) Two children walk at different speeds around a rotating merry-go-round with its wooden horses. (b) In an analogous way, the ancient Greeks imagined that the Sun and Moon move around the rotating celestial sphere with its fixed stars. Thus, the Sun and Moon move from east to west across the sky every day and also move slowly eastward from one night to the next relative to the background of stars.

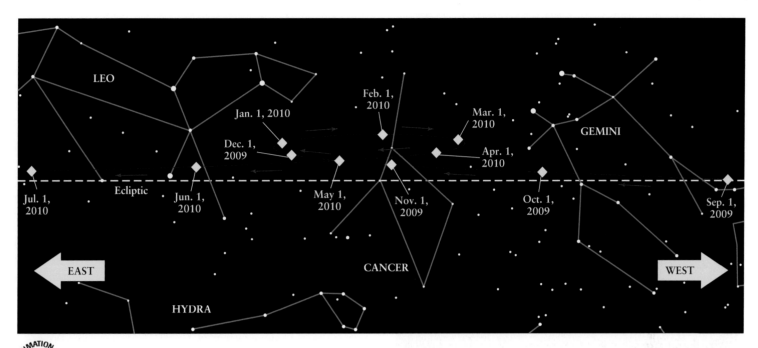

figure 4-2

The Path of Mars in 2009–2010 From September 2009 through July 2010, Mars will move across the constellations Gemini, Cancer, and Leo. Mars's motion will be direct (from west to east, or from right to left in this figure) most of the time but will be retrograde (from east to west, or from left to right in this figure) during January and February 2010. Notice that the speed of Mars relative to the stars is not constant: The planet travels farther across the sky from November 1 to December 1 than it does from December 1 to January 1.

ing 2009 and 2010. In the language of the merry-go-round analogy in Figure 4-1, the Greeks imagined the planets as children walking around the rotating merry-go-round but who keep changing their minds about which direction to walk!

As seen from Earth, the planets wander primarily across the 12 constellations of the zodiac. (Figure 4-2 shows that during 2009–2010, Mars moves through Gemini, Cancer, and Leo.) As we learned in Section 2-5, these constellations encircle the sky in a continuous band centered on the ecliptic, which is the apparent path of the Sun across the celestial sphere.

If you observe a planet as it travels across the zodiac from night to night, you will find that the planet usually moves slowly eastward against the background stars. This eastward progress is called **direct motion**. Occasionally, however, the planet will seem to stop and then back up for several weeks or months. This occasional westward movement is called **retrograde motion**. Both direct and retrograde motions are much slower than the apparent daily rotation of the sky caused by the Earth's rotation. Hence, they are best detected by mapping the position of a planet against the background stars from night to night over a long period. Figure 4-2 is a map of just this sort. Mars undergoes retrograde motion about every 22½ months; all the other planets go through retrograde motion, but at different intervals.

CAUTION On a map of the Earth with north at the top, west is to the left and east is to the right. Why, then, is *east* on the left and *west* on the right in Figure 4-2? The answer is that a map of the Earth is a view looking downward at the ground from above, while a star map like Figure 4-2 is a view looking upward at the sky. If the constellation Cancer in Figure 4-2 were directly overhead, Leo would be toward the eastern horizon and Gemini would be toward the western horizon.

Explaining the nonuniform motions of the five planets was one of the main challenges facing the astronomers of antiquity. The Greeks developed many theories to account for retrograde motion and the loops that the planets trace out against the background stars. One of the most successful and enduring models was originated by Apollonius of Perga and by Hipparchus in the second century B.C. and expanded upon by Ptolemy, the last of the great Greek astronomers, during the second century A.D. **Figure 4-3a** sketches the basic concept, usually called the **Ptolemaic system.** Each planet is assumed to move in a small circle called an **epicycle**, whose center in turn moves in a larger circle, called a **deferent**, which is centered approximately on the Earth. Both the epicycle and deferent rotate in the same direction, shown as counterclockwise in Figure 4-3a.

As viewed from Earth, the epicycle moves eastward along the deferent. Most of the time the eastward motion of the planet on its epicycle adds to the eastward motion of the epicycle on the deferent (Figure 4-3b). Then the planet is seen to be in direct (eastward) motion against the background stars. However, when the planet is on the part of its epicycle nearest Earth, the motion of the planet along the epicycle is opposite to the motion of the epicycle along the deferent. The planet therefore appears to slow down and halt its usual eastward movement among the constellations, and actually

(a)

(b)

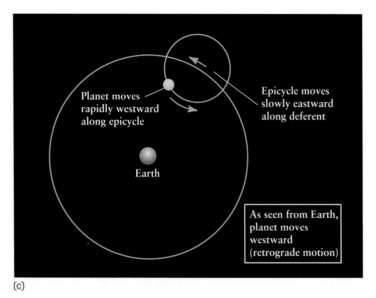

(c)

figure 4-3

A Geocentric Explanation of Retrograde Motion

(a) The ancient Greeks imagined that each planet moves along an epicycle, which in turn moves along a deferent centered approximately on the Earth. The planet moves along the epicycle more rapidly than the epicycle moves along the deferent. (b) At most times the eastward motion of the planet on the epicycle adds to the eastward motion of the epicycle on the deferent. Then the planet moves eastward in direct motion as seen from Earth. (c) When the planet is on the inside of the deferent, its motion along the epicycle is westward. Because this motion is faster than the eastward motion of the epicycle on the deferent, the planet appears from Earth to be moving westward in retrograde motion.

goes backward in retrograde (westward) motion for a few weeks or months (Figure 4-3c). Thus, the concept of epicycles and deferents enabled Greek astronomers to explain the retrograde loops of the planets.

Using the wealth of astronomical data in the library at Alexandria, including records of planetary positions for hundreds of years, Ptolemy deduced the sizes and rotation rates of the epicycles and deferents needed to reproduce the recorded paths of the planets. After years of tedious work, Ptolemy assembled his calculations into 13 volumes, collectively called the *Almagest*. His work was used to predict the positions and paths of the Sun, Moon, and planets with unprecedented accuracy. In fact, the *Almagest* was so successful that it became the astronomer's bible, and for more than 1000 years, the Ptolemaic system endured as a useful description of the workings of the heavens.

A major problem posed by the Ptolemaic system was a philosophical one: It treated each planet independent of the others. There

was no rule in the *Almagest* that related the size and rotation speed of one planet's epicycle and deferent to the corresponding sizes and speeds for other planets. This problem made the Ptolemaic system very unsatisfying to many Islamic and European astronomers of the Middle Ages. They felt that a correct model of the universe should be based on a simple set of underlying principles that applied to all of the planets.

The idea that simple, straightforward explanations of phenomena are most likely to be correct is called **Occam's razor,** after William of Occam (or Ockham), the fourteenth-century English philosopher who first expressed it. (The "razor" refers to shaving extraneous details from an argument or explanation.) Although Occam's razor has no proof or verification, it appeals to the scientist's sense of beauty and elegance, and it has helped lead to the simple and powerful laws of nature that scientists use today. In the centuries that followed,

Occam's razor would help motivate a new and revolutionary view of the universe.

4-2 | Nicolaus Copernicus devised the first comprehensive heliocentric model

During the first half of the sixteenth century, a Polish lawyer, physician, canon of the church, and gifted mathematician named Nicolaus Copernicus (Figure 4-4) began to construct a new model of the universe. His model, which placed the Sun at the center, explained the motions of the planets in a more natural way than the Ptolemaic system. As we will see, it also helped lay the foundations of modern physical science. But Copernicus was not the first to conceive a Sun-centered model of planetary motion.

Imagine riding on a fast racehorse. As you pass a slowly walking pedestrian, he appears to move backward, even though he is traveling in the same direction as you and your horse. In the third century B.C., this sort of simple observation inspired the Greek astronomer Aristarchus to suggest a straightforward explanation of retrograde motion.

In Aristarchus's **heliocentric** (Sun-centered) **model**, all the planets, including Earth, revolve about the Sun. Different planets take different lengths of time to complete an orbit, so from time to time one planet will overtake another, just as a fast-moving horse overtakes a person on foot. When the Earth overtakes Mars, for example, Mars appears to move backward in retrograde motion, as Figure 4-5 shows. Thus, in the heliocentric picture, the occasional retrograde motion of a planet is merely the result of the Earth's motion.

As we saw in Section 3-6, Aristarchus demonstrated that the Sun is bigger than the Earth (see Table 3-3). This made it sensible to imagine the Earth orbiting the larger Sun. He also imagined that the Earth rotated on its axis once a day, which explained the daily rising and setting of the Sun, Moon, and planets and the diurnal motions of the stars. To explain why the apparent motions of the planets never take them far from the ecliptic, Aristarchus proposed that the orbits of the Earth and all the planets must lie in nearly the same plane. (Recall from Section 2-5 that the ecliptic is the projection onto the celestial sphere of the plane of the Earth's orbit.)

Figure 4-4
Nicolaus Copernicus (1473–1543) Copernicus was the first person to work out the details of a heliocentric system in which the planets, including the Earth, orbit the Sun. (E. Lessing/Magnum)

This heliocentric model is conceptually much simpler than an Earth-centered system, such as that of Ptolemy, with all its "circles upon circles." In Aristarchus's day, however, the idea of an orbiting, rotating Earth seemed inconceivable, given the Earth's apparent stillness and immobility. Nearly 2000 years would pass before a heliocentric model found broad acceptance.

In the years after 1500, Copernicus came to realize that a heliocentric model has several advantages beyond providing a natural explanation of retrograde motion. In the Ptolemaic system, the arrangement of the planets—that is, which are close to the Earth

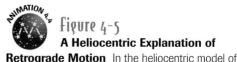

Figure 4-5
A Heliocentric Explanation of Retrograde Motion In the heliocentric model of Aristarchus, the Earth and the other planets orbit the Sun. The Earth travels around the Sun more rapidly than Mars. Consequently, as the Earth overtakes and passes this slower-moving planet, Mars appears for a few months (from points 4 through 6) to fall behind and move backward with respect to the background of stars.

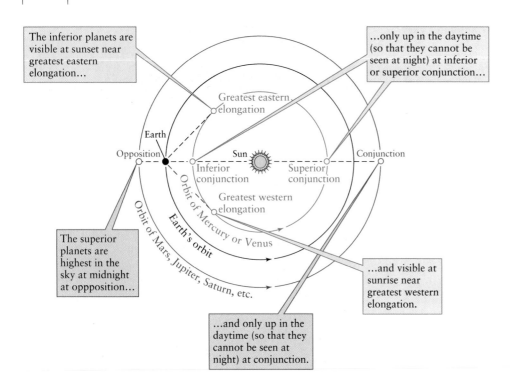

The inferior planets are visible at sunset near greatest eastern elongation...

...only up in the daytime (so that they cannot be seen at night) at inferior or superior conjunction...

The superior planets are highest in the sky at midnight at oppposition...

...and visible at sunrise near greatest western elongation.

...and only up in the daytime (so that they cannot be seen at night) at conjunction.

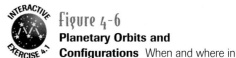

Figure 4-6

Planetary Orbits and Configurations When and where in the sky a planet can be seen from Earth depends on the size of its orbit and its location on that orbit. Inferior planets have orbits smaller than the Earth's, while superior planets have orbits larger than the Earth's. (Note that in this figure you are looking down onto the solar system from a point far above the Earth's northern hemisphere.)

and which are far away—was chosen in large part by guesswork. But using a heliocentric model, Copernicus could determine the arrangement of the planets without ambiguity.

Copernicus realized that because Mercury and Venus are always observed fairly near the Sun in the sky, their orbits must be smaller than the Earth's. Planets in such orbits are called **inferior planets** (Figure 4-6). The other visible planets—Mars, Jupiter, and Saturn—are sometimes seen on the side of the celestial sphere opposite the Sun, so these planets appear high above the horizon at midnight (when the Sun is far below the horizon). When this happens, the Earth must lie between the Sun and these planets. Copernicus therefore concluded that the orbits of Mars, Jupiter, and Saturn must be larger than the Earth's orbit. Hence, these planets are called **superior planets.**

Three additional planets, Uranus, Neptune, and Pluto, were discovered after the telescope was invented (and after the death of Copernicus). All three can be seen at times in the midnight sky, so these are also superior planets with orbits larger than the Earth's.

The heliocentric model also explains why planets appear in different parts of the sky on different dates. Both inferior planets (Mercury and Venus) go through cycles: The planet is seen in the west after sunset for several weeks or months, then for several weeks or months in the east before sunrise, and then in the west after sunset again.

Figure 4-6 shows the reason for this cycle. When Mercury or Venus is visible after sunset, it is near **greatest eastern elongation.** (The angle between the Sun and a planet as viewed from Earth is called the planet's **elongation.**) The planet's position in the sky is as far east of the Sun as possible, so it appears above the western horizon after sunset (that is, to the east of the Sun) and is often called an "evening star." At **greatest western elongation,** Mercury or Venus is as far west of the Sun as it can possibly be. It then rises before the Sun, gracing the predawn sky as a "morning star" in the

east. When Mercury or Venus is at **inferior conjunction,** it is between us and the Sun, and it is moving from the evening sky into the morning sky. At **superior conjunction,** when the planet is on the opposite side of the Sun, it is moving back into the evening sky.

A superior planet such as Mars, whose orbit is larger than the Earth's, is best seen in the night sky when it is at **opposition.** At this point the planet is in the part of the sky opposite the Sun and is highest in the sky at midnight. This is also when the planet appears brightest, because it is closest to us. But when a superior planet like Mars is located behind the Sun at **conjunction,** it is above the horizon during the daytime and thus is not well placed for nighttime viewing.

The Ptolemaic system has no simple rules relating the motion of one planet to another. But Copernicus showed that there *are* such rules in a heliocentric model. In particular, he found a correspondence between the time a planet takes to complete one orbit—that is, its **period**—and the size of the orbit.

Determining the period of a planet takes some care, because the Earth, from which we must make the observations, is also moving. Realizing this, Copernicus was careful to distinguish between two different periods of each planet. The **synodic period** is the time that elapses between two successive identical configurations as seen from Earth—from one opposition to the next, for example, or from one conjunction to the next. The **sidereal period** is the true orbital period of a planet, the time it takes the planet to complete one full orbit of the Sun relative to the stars.

The synodic period of a planet can be determined by observing the sky, but the sidereal period has to be found by calculation. Copernicus figured out how to do this (Box 4-1). Table 4-1 shows the results for all of the planets.

To find a relationship between the sidereal period of a planet and the size of its orbit, Copernicus still had to determine the relative distances of the planets from the Sun. He devised a straightforward

box 4-1 TOOLS OF THE ASTRONOMER'S TRADE

Relating Synodic and Sidereal Periods

We can derive a mathematical formula that relates a planet's sidereal period (the time required for the planet to complete one orbit) to its synodic period (the time between two successive identical configurations). To start with, let's consider an inferior planet (Mercury or Venus) orbiting the Sun as shown in the accompanying figure. Let P be the planet's sidereal period, S the planet's synodic period, and E the Earth's sidereal period or sidereal year (see Section 2-8), which Copernicus knew to be nearly 365¼ days.

The rate at which the Earth moves around its orbit is the number of degrees around the orbit divided by the time to complete the orbit, or $360°/E$ (equal to a little less than 1° per day). Similarly, the rate at which the inferior planet moves along its orbit is $360°/P$.

During a given time interval, the angular distance that the Earth moves around its orbit is its rate, $(360°/E)$, multiplied by the length of the time interval. Thus, during a time S, or one synodic period of the inferior planet, the Earth covers an angular distance of $(360°/E)S$ around its orbit. In that same time, the inferior planet covers an angular distance of $(360°/P)S$. Note, however, that the inferior planet has gained one full lap on the Earth, and hence has covered 360° more than the Earth has (see the figure). Thus, $(360°/P)S = (360°/E)S + 360°$. Dividing each term of this equation by $360°S$ gives

For an inferior planet:

$$\frac{1}{P} = \frac{1}{E} + \frac{1}{S}$$

P = inferior planet's sidereal period
E = Earth's sidereal period = 1 year
S = inferior planet's synodic period

A similar analysis for a superior planet (for example, Mars, Jupiter, or Saturn) yields

For a superior planet:

$$\frac{1}{P} = \frac{1}{E} - \frac{1}{S}$$

P = superior planet's sidereal period
E = Earth's sidereal period = 1 year
S = superior planet's synodic period

Using these formulas, we can calculate a planet's sidereal period P from its synodic period S. Often astronomers express P,

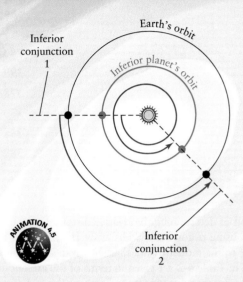

E, and S in terms of years, by which they mean Earth years of approximately 365.26 days.

EXAMPLE: Jupiter has an observed synodic period of 398.9 days, or 1.092 years. What is its sidereal period?

Situation: Our goal is to find the sidereal period of Jupiter, a superior planet.

Tools: Since Jupiter is a superior planet, we use the second of the two equations given above to determine the sidereal period P.

Answer: We are given the Earth's sidereal period $E = 1$ year and Jupiter's synodic period $S = 1.092$ years. Using the equation $1/P = 1/E - 1/S$,

$$\frac{1}{P} = \frac{1}{1} - \frac{1}{1.092} = 0.08425,$$

so

$$P = \frac{1}{0.08425} = 11.87 \text{ years}$$

Review: Our answer means that it takes 11.87 years for Jupiter to complete one full orbit of the Sun. This is greater than the Earth's 1-year sidereal period because Jupiter's orbit is larger than the Earth's orbit. Jupiter's synodic period of 1.092 years is the time from one opposition to the next, or the time that elapses from when the Earth overtakes Jupiter to when it next overtakes Jupiter. This is so much shorter than the sidereal period because Jupiter moves quite slowly around its orbit. The Earth overtakes it a little less often than once per Earth orbit, that is, at intervals of a little bit more than a year.

table 4-1	Synodic and Sidereal Periods of the Planets	
Planet	Synodic period	Sidereal period
Mercury	116 days	88 days
Venus	584 days	225 days
Earth	—	1.0 year
Mars	780 days	1.9 years
Jupiter	399 days	11.9 years
Saturn	378 days	29.5 years
Uranus	370 days	84.1 years
Neptune	368 days	164.9 years
Pluto	367 days	248.6 years

geometric method of determining the relative distances of the planets from the Sun using trigonometry. His answers turned out to be remarkably close to the modern values, as shown in Table 4-2.

The distances in Table 4-2 are given in terms of the astronomical unit, which is the average distance from Earth to the Sun (Section 1-7). Copernicus did not know the precise value of this distance, so he could only determine the *relative* sizes of the orbits of the planets. One method used by modern astronomers to determine the astronomical unit is to measure the Earth-Venus distance very accurately using radar. At the same time, they measure the angle in the sky between Venus and the Sun, and then calculate the Earth-Sun distance using trigonometry. In this way, the astronomical unit is found to be 1.496×10^8 km (92.96 million miles). Once the astronomical unit is known, the average distances from the Sun to each of the planets in kilometers can be determined from Table 4-2. A table of these distances is given in Appendix 2.

By comparing Tables 4-1 and 4-2, you can see the unifying relationship between planetary orbits in the Copernican model: The far-

ther a planet is from the Sun, the longer it takes to travel around its orbit (that is, the longer its sidereal period). That is so for *two* reasons: (1) the larger the orbit, the farther a planet must travel to complete an orbit; and (2) the larger the orbit, the slower a planet moves. For example, Mercury, with its small orbit, moves at an average speed of 47.9 km/s (107,000 mi/h). Saturn travels around its large orbit much more slowly, at an average speed of 9.64 km/s (21,600 mi/h). The older Ptolemaic model offers no such simple relations between the motions of different planets.

At first, Copernicus assumed that Earth travels around the Sun along a circular path. He found that perfectly circular orbits could not accurately describe the paths of the other planets, so he had to add an epicycle to each planet. (This was *not* to explain retrograde motion, which Copernicus realized was because of the differences in orbital speeds of different planets, as shown in Figure 4-5. Rather, the small epicycles helped Copernicus account for slight variations in each planet's speed along its orbit.) Even though he clung to the old notion that orbits must be made up of circles, Copernicus had shown

table 4-2	Average Distances of the Planets from the Sun	
Planet	Copernican value (AU*)	Modern value (AU)
Mercury	0.38	0.39
Venus	0.72	0.72
Earth	1.00	1.00
Mars	1.52	1.52
Jupiter	5.22	5.20
Saturn	9.07	9.55
Uranus	—	19.19
Neptune	—	30.07
Pluto	—	39.54
*1 AU = 1 astronomical unit = average distance from the Earth to the Sun.		

that a heliocentric model could explain the motions of the planets. He compiled his ideas and calculations into a book entitled *De revolutionibus orbium coelestium* (On the Revolutions of the Celestial Spheres), which was published in 1543, the year of his death.

For several decades after Copernicus, most astronomers saw little reason to change their allegiance from the older geocentric model of Ptolemy. The predictions that the Copernican model makes for the apparent positions of the planets are, on average, no better or worse than those of the Ptolemaic model. The test of Occam's razor does not really favor either model, because both use a combination of circles to describe each planet's motion.

More concrete evidence was needed to convince scholars to abandon the old, comfortable idea of a stationary Earth at the center of the universe. The story of how this evidence was accumulated begins nearly 30 years after the death of Copernicus, when a young Danish astronomer pondered the nature of a new star in the heavens.

4-3 | Tycho Brahe's astronomical observations disproved ancient ideas about the heavens

On November 11, 1572, a bright star suddenly appeared in the constellation of Cassiopeia. At first, it was even brighter than Venus, but then it began to grow dim. After 18 months, it faded from view. Modern astronomers recognize this event as a supernova explosion, the violent death of a massive star.

In the sixteenth century, however, the vast majority of scholars held with the ancient teachings of Aristotle and Plato, who had argued that the heavens are permanent and unalterable. Consequently, the "new star" of 1572 could not really be a star at all, because the heavens do not change; it must instead be some sort of bright object quite near the Earth, perhaps not much farther away than the clouds overhead.

The 25-year-old Danish astronomer Tycho Brahe (1546–1601) realized that straightforward observations might reveal the distance to the new star. It is common experience that when you walk from one place to another, nearby objects appear to change position against the background of more distant objects. This phenomenon, whereby the apparent position of an object changes because of the motion of the observer, is called **parallax.** If the new star was nearby, then its position should shift against the background stars over the course of a single night because Earth's rotation changes our viewpoint. Figure 4-7 shows this predicted shift. (Actually, Tycho believed that the heavens rotate about the Earth, as in the Ptolemaic model, but the net effect is the same.)

Tycho's careful observations failed to disclose any parallax. The farther away an object is, the less it appears to shift against the background as we change our viewpoint, and the smaller the parallax. Hence, the new star had to be quite far away, farther from Earth than anyone had imagined. This was the first evidence that the "unchanging" stars were in fact changeable.

Tycho also attempted to measure the parallax of a bright comet that appeared in 1577 and, again, found it too small to measure. Thus, the comet also had to be far beyond the Earth. Furthermore, because the comet's position relative to the stars changed from night to night, its motion was more like a planet than a star. But most

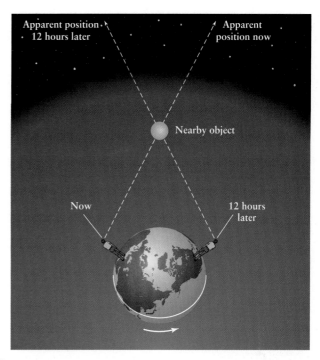

figure 4-7

A Nearby Object Shows a Parallax Shift Tycho Brahe argued that if an object is near the Earth, an observer would have to look in different directions to see that object over the course of a night and its position relative to the background stars would change. Tycho failed to measure such changes for a supernova in 1572 and a comet in 1577. He therefore concluded that these objects were far from the Earth.

scholars of Tycho's time taught that the motions of the planets had existed in unchanging form since the beginning of the universe. If new objects such as comets could appear and disappear within the realm of the planets, the conventional notions of planetary motions needed to be revised.

Tycho's observations showed that the heavens are by no means pristine and unchanging. This discovery flew in the face of nearly 2000 years of astronomical thought. In support of these revolutionary observations, the king of Denmark financed the construction of two magnificent observatories for Tycho's use on the island of Hven, just off the Danish coast. The bequest for these two observatories— Uraniborg ("heavenly castle") and Stjerneborg ("star castle")— allowed Tycho to design and have built a set of astronomical instruments vastly superior in quality to any earlier instruments (Figure 4-8). With this state-of-the-art equipment, Tycho proceeded to measure the positions of stars and planets with unprecedented accuracy. In addition, he and his assistants were careful to make several observations of the same star or planet with different instruments, in order to identify any errors that might be caused by the instruments themselves. This painstaking approach to mapping the heavens revolutionized the practice of astronomy and is used by astronomers today.

A key goal of Tycho's observations during this period was to test the ideas Copernicus had proposed decades earlier about the Earth going around the Sun. Tycho argued that if the Earth was in motion,

Figure 4-8

Tycho Brahe (1546–1601) Observing This contemporary illustration shows Tycho Brahe with some of the state-of-the-art measuring apparatus at Uraniborg, one of the two observatories that he built under the patronage of Frederik II of Denmark. (This magnificent observatory lacked a telescope, which had not yet been invented.) The data that Tycho collected were crucial to the development of astronomy in the years after his death. (Photo Researchers, Inc.)

Figure 4-9

Johannes Kepler (1571–1630) By analyzing Tycho Brahe's detailed records of planetary positions, Kepler developed three general principles, called Kepler's laws, that describe how the planets move about the Sun. Kepler was the first to realize that the orbits of the planets are ellipses and not circles. (E. Lessing/Magnum)

then nearby stars should appear to shift their positions with respect to background stars as we orbit the Sun. Tycho failed to detect any such parallax, and he concluded that the Earth was at rest and the Copernican system was wrong.

On this point Tycho was in error, for nearby stars do in fact shift their positions as he had suggested. But even the nearest stars are so far away that the shifts in their positions are less than an arcsecond, too small to be seen with the naked eye. Tycho would have needed a telescope to detect the parallax that he was looking for, but the telescope was not invented until after his death in 1601. Indeed, the first accurate determination of stellar parallax was not made until 1838.

Although he remained convinced that the Earth was at the center of the universe, Tycho nonetheless made a tremendous contribution toward putting the heliocentric model on a solid foundation. From 1576 to 1597, he used his instruments to make comprehensive measurements of the positions of the planets with an accuracy of 1 arcminute. This is as well as can be done with the naked eye and was far superior to any earlier measurements. Within the reams of data that Tycho compiled lay the truth about the motions of the planets. The person who would extract this truth was a German mathematician who became Tycho's assistant in 1600, a year before the great astronomer's death. His name was Johannes Kepler (Figure 4-9).

4-4 | Johannes Kepler proposed elliptical paths for the planets about the Sun

The task that Johannes Kepler took on at the beginning of the seventeenth century was to find a model of planetary motion that agreed completely with Tycho's extensive and very accurate observations of planetary positions. To do this, Kepler found that he had to break with an ancient prejudice about planetary motions.

Astronomers had long assumed that heavenly objects move in circles, which were considered the most perfect and harmonious of all geometric shapes. They believed that if a perfect God resided in heaven along with the stars and planets, then the motions of these bodies must be perfect too. Against this context, Kepler dared to try to explain planetary motions with noncircular curves. In particular, he found that he had the best success with a particular kind of curve called an **ellipse**.

You can draw an ellipse by using a loop of string, two thumbtacks, and a pencil, as shown in Figure 4-10a. Each thumbtack in the figure is at a **focus** (plural **foci**) of the ellipse; an ellipse has two foci. The longest diameter of an ellipse, called the **major axis**, passes through both foci. Half of that distance is called the **semimajor axis** and is usually designated by the letter *a*. A circle is a special case of an ellipse in which the two foci are at the same

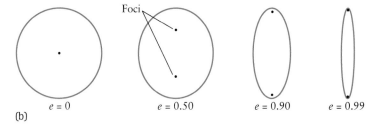

(a)

(b)

$e = 0$ $e = 0.50$ $e = 0.90$ $e = 0.99$

figure 4-10

Ellipses (a) To draw an ellipse, use two thumbtacks to secure the ends of a piece of string, then use a pencil to pull the string taut. If you move the pencil while keeping the string taut, the pencil traces out an ellipse. The thumbtacks are located at the two foci of the ellipse. The major axis is the greatest distance across the ellipse; the semimajor axis is half of this distance. (b) A series of ellipses with the same major axis but different eccentricities. An ellipse can have any eccentricity from $e = 0$ (a circle) to just under $e = 1$ (virtually a straight line).

point (this corresponds to using only a single thumbtack in Figure 4-10*a*). The semimajor axis of a circle is equal to its radius.

By assuming that planetary orbits were ellipses, Kepler found, to his delight, that he could make his theoretical calculations match precisely to Tycho's observations. This important discovery, first published in 1609, is now called **Kepler's first law:**

The orbit of a planet about the Sun is an ellipse with the Sun at one focus.

The semimajor axis *a* of a planet's orbit is the average distance between the planet and the Sun.

CAUTION! The Sun is at one focus of a planet's elliptical orbit, but there is *nothing* at the other focus. This "empty focus" has geometrical significance, because it helps to define the shape of the ellipse, but plays no other role.

Ellipses come in different shapes, depending on the elongation of the ellipse. The shape of an ellipse is described by its **eccentricity,** designated by the letter *e*. The value of e can range from 0 (a circle) to just under 1 (nearly a straight line). The greater the eccentricity, the more elongated the ellipse. Figure 4-10*b* shows a few examples of ellipses with different eccentricities. Because a circle is a special case of an ellipse, it is possible to have a perfectly circular orbit. But all of the planets have orbits that are at least slightly elliptical, with eccentricities that range from 0.007 for Venus (the most nearly circular of any planetary orbit) to 0.250 for Pluto (the most elongated orbit).

Once he knew the shape of a planet's orbit, Kepler was ready to describe exactly *how* it moves on that orbit. As a planet travels in an elliptical orbit, its distance from the Sun varies. Kepler realized that the speed of a planet also varies along its orbit. A planet moves most rapidly when it is nearest the Sun, at a point on its orbit called **perihelion.** Conversely, a planet moves most slowly when it is farthest from the Sun, at a point called **aphelion** (**Figure 4-11**).

figure 4-11
Kepler's First and Second Laws

According to Kepler's first law, a planet travels around the Sun along an elliptical orbit with the Sun at one focus. According to his second law, a planet moves fastest when closest to the Sun (at perihelion) and slowest when farthest from the Sun (at aphelion). As the planet moves, an imaginary line joining the planet and the Sun sweeps out equal areas in equal intervals of time (from *A* to *B* or from *C* to *D*). By using these laws in his calculations, Kepler found a perfect fit to the apparent motions of the planets.

After much trial and error, Kepler found a way to describe just how a planet's speed varies as it moves along its orbit. Figure 4-11 illustrates this discovery, also published in 1609. Suppose that it takes 30 days for a planet to go from point A to point B. During that time, an imaginary line joining the Sun and the planet sweeps out a nearly triangular area. Kepler discovered that a line joining the Sun and the planet also sweeps out exactly the same area during any other 30-day interval. In other words, if the planet also takes 30 days to go from point C to point D, then the two shaded segments in Figure 4-11 are equal in area. **Kepler's second law** can be stated thus:

A line joining a planet and the Sun sweeps out equal areas in equal intervals of time.

This relationship is also called the **law of equal areas.** In the idealized case of a circular orbit, a planet would have to move at a constant speed around the orbit in order to satisfy Kepler's second law.

ANALOGY A good analogy for Kepler's second law is a twirling ice skater. If the skater pulls her arms straight in to her body, she spins faster; if she lets her arms extend away from her body, her rate of spin decreases. In the same way, a planet in an elliptical orbit travels at a higher speed when it moves closer to the Sun (toward perihelion) and travels at a lower speed when it moves away from the Sun (toward aphelion).

Kepler's second law describes how the speed of a given planet changes as it orbits the Sun. Kepler also deduced from Tycho's data a relationship that can be used to compare the motions of *different* planets. Published in 1618 and now called **Kepler's third law,** it states a relationship between the size of a planet's orbit and the time the planet takes to go once around the Sun:

The square of the sidereal period of a planet is directly proportional to the cube of the semimajor axis of the orbit.

Kepler's third law says that the larger a planet's orbit—that is, the larger the semimajor axis, or average distance from the planet to the Sun—the longer the sidereal period, which is the time it takes the planet to complete an orbit. From Kepler's third law one can show that the larger the semimajor axis, the slower the average speed at which the planet moves around its orbit. (By contrast, Kepler's *second* law describes how the speed of a given planet is sometimes faster and sometimes slower than its average speed.) This qualitative relationship between orbital size and orbital speed is just what Aristarchus and Copernicus used to explain retrograde motion, as we saw in Section 4-2. Kepler's great contribution was to make this relationship a quantitative one.

It is useful to restate Kepler's third law as an equation. If a planet's sidereal period P is measured in years and the length of its semimajor axis a is measured in astronomical units (AU), where 1 AU is the average distance from the Earth to the Sun (see Section 1-7), then Kepler's third law is

Kepler's third law

$$P^2 = a^3$$

P = planet's sidereal period, in years
a = planet's semimajor axis, in AU

If you know either the sidereal period of a planet or the semimajor axis of its orbit, you can find the other quantity using this equation. Box 4-2 gives some examples of how this is done.

We can verify Kepler's third law for all of the planets, including those that were discovered after Kepler's death, using data from Tables 4-1 and 4-2. If Kepler's third law is correct, for each planet the numerical values of P^2 and a^3 should be equal. This is indeed true to very high accuracy, as Table 4-3 shows.

Kepler's laws are a landmark in the history of astronomy. They made it possible to calculate the motions of the planets with better

table 4-3	A Demonstration of Kepler's Third Law ($P^2 = a^3$)			
Planet	**Sidereal period** P (years)	**Semimajor axis** a (AU)	P^2	a^3
Mercury	0.24	0.39	0.06	0.06
Venus	0.61	0.72	0.37	0.37
Earth	1.00	1.00	1.00	1.00
Mars	1.88	1.52	3.53	3.51
Jupiter	11.86	5.20	140.7	140.6
Saturn	29.46	9.55	867.9	871.0
Uranus	84.10	19.19	7,072	7,067
Neptune	164.86	30.07	27,180	27,190
Pluto	248.60	39.54	61,800	61,820

box 4-2 — TOOLS OF THE ASTRONOMER'S TRADE

Using Kepler's Third Law

Kepler's third law relates the sidereal period P of an object orbiting the Sun to the semimajor axis a of its orbit:

$$P^2 = a^3$$

You must keep two essential points in mind when working with this equation:

1. The period P *must* be measured in years, and the semimajor axis a *must* be measured in astronomical units (AU). Otherwise you will get nonsensical results.

2. This equation applies *only* to the special case of an object, like a planet, that orbits the Sun. If you want to analyze the orbit of the Moon around the Earth, of a spacecraft around Mars, or of a planet around a distant star, you must use a different, generalized form of Kepler's third law. We discuss this alternative equation in Section 4-7 and Box 4-4.

EXAMPLE: The average distance from Venus to the Sun is 0.72 AU. Use this to determine the sidereal period of Venus.

Situation: The average distance from the Venus to the Sun is the semimajor axis a of the planet's orbit. Our goal is to calculate the planet's sidereal period P.

Tools: To relate a and P we use Kepler's third law, $P^2 = a^3$.

Answer: We first cube the semimajor axis (multiply it by itself twice):

$$a^3 = (0.72)^3 = 0.72 \times 0.72 \times 0.72 = 0.373$$

According to Kepler's third law this is also equal to P^2, the square of the sidereal period. So, to find P, we have to "undo" the square, that is, take the square root. Using a calculator, we find

$$P = \sqrt{P^2} = \sqrt{0.373} = 0.61$$

Review: The sidereal period of Venus is 0.61 years, or a bit more than seven Earth months. This makes sense: A planet with a smaller orbit than the Earth's (an inferior planet) must have a shorter sidereal period than the Earth.

EXAMPLE: A certain small asteroid (a rocky body a few tens of kilometers across) takes eight years to complete one orbit around the Sun. Find the semimajor axis of the asteroid's orbit.

Situation: We are given the sidereal period $P = 8$ years, and are to determine the semimajor axis a.

Tools: As in the preceding example, we relate a and P using Kepler's third law, $P^2 = a^3$.

Answer: We first square the period:

$$P^2 = 8^2 = 8 \times 8 = 64$$

From Kepler's third law, this is also equal to a^3. To determine a, we must take the *cube root* of a^3, that is, find the number whose cube is 64. If your calculator has a cube root function, denoted by the symbol $\sqrt[3]{}$, you can use it to find that the cube root of 64 is 4: $\sqrt[3]{64} = 4$. Otherwise, you can determine by trial and error that the cube of 4 is 64:

$$4^3 = 4 \times 4 \times 4 = 64$$

Because the cube of 4 is 64, it follows that the cube root of 64 is 4 (taking the cube root "undoes" the cube).

With either technique you find that the orbit of this asteroid has semimajor axis $a = 4$ AU.

Review: The period is greater than 1 year, so the semimajor axis is greater than 1 AU. Note that $a = 4$ AU is intermediate between the orbits of Mars and Jupiter (see Table 4-3). Many asteroids are known with semimajor axes in this range, forming a region in the solar system called the asteroid belt.

---

accuracy than any geocentric model ever had, and they helped to justify the idea of a heliocentric model. Kepler's laws also pass the test of Occam's razor, for they are simpler in every way than the schemes of Ptolemy or Copernicus, both of which used a complicated combination of circles.

But the significance of Kepler's laws goes beyond understanding planetary orbits. These same laws are also obeyed by spacecraft orbiting the Earth, by two stars revolving about each other in a binary star system, and even by galaxies in their orbits about each other. Throughout this book, we shall use Kepler's laws in a wide range of situations.

As impressive as Kepler's accomplishments were, he did not prove that the planets orbit the Sun, nor was he able to explain why planets move in accordance with his three laws. These advances were made by two other figures who loom large in the history of astronomy: Galileo Galilei and Isaac Newton.

4-5 Galileo's discoveries with a telescope strongly supported a heliocentric model

 SCIENCE IN PROCESS *Galileo used the latest technology of his time to radically transform our picture of the universe*

When Dutch opticians invented the telescope during the first decade of the seventeenth century, astronomy was changed forever. The scholar who used this new tool to amass convincing evidence that the planets orbit the Sun, not the Earth, was the Italian mathematician and physical scientist Galileo Galilei (Figure 4-12).

While Galileo did not invent the telescope, he was the first to point one of these new devices toward the sky and to publish his observations. Beginning in 1610, he saw sights of which no one had ever dreamed. He discovered mountains on the Moon, sunspots on the Sun, and the rings of Saturn, and he was the first to see that the Milky Way is not a featureless band of light but rather "a mass of innumerable stars."

One of Galileo's most important discoveries with the telescope was that Venus exhibits phases like those of the Moon (Figure 4-13). Galileo also noticed that the apparent size of Venus as seen through his telescope was related to the planet's phase. Venus appears small at gibbous phase and largest at crescent phase. There

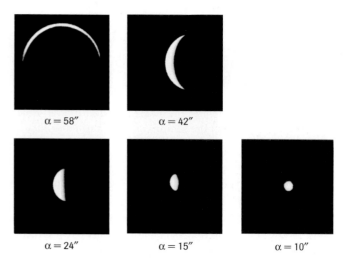

$\alpha = 58''$ $\alpha = 42''$

$\alpha = 24''$ $\alpha = 15''$ $\alpha = 10''$

Figure 4-13   R I **V** U X G

The Phases of Venus This series of photographs shows how the appearance of Venus changes as it moves along its orbit. The number below each view is the angular diameter α of the planet in arcseconds. Venus has the largest angular diameter when it is a crescent, and the smallest angular diameter when it is gibbous (nearly full). (New Mexico State University Observatory)

is also a correlation between the phases of Venus and the planet's angular distance from the Sun.

Figure 4-14 shows that these relationships are entirely compatible with a heliocentric model in which the Earth and Venus both go around the Sun. They are also completely *incompatible* with the Ptolemaic system, in which the Sun and Venus both orbit the Earth. To explain why Venus is never seen very far from the Sun, the Ptolemaic model had to assume that the deferents of Venus and of the Sun move together in lockstep, with the epicycle of Venus centered on a straight line between the Earth and the Sun (Figure 4-15). In this model, Venus was never on the opposite side of the Sun from the Earth, and so it could never have shown the gibbous phases that Galileo observed.

Galileo also found more unexpected evidence for the ideas of Copernicus. In 1610 Galileo discovered four moons, now called the Galilean satellites, orbiting Jupiter (Figure 4-16). He realized that they were orbiting Jupiter because they appeared to move back and forth from one side of the planet to the other. Figure 4-17 shows confirming observations made by Jesuit observers in 1620. Astronomers soon realized that the larger the orbit of one of the moons around Jupiter, the slower that moon moves and the longer it takes that moon to travel around its orbit. These are the same relationships that Copernicus deduced for the motions of the planets around the Sun. Thus, the moons of Jupiter behave like a Copernican system in miniature.

Galileo's telescopic observations constituted the first fundamentally new astronomical data in almost 2000 years. Contradicting prevailing opinion and religious belief, his discoveries strongly suggested a heliocentric structure of the universe. Although cautioned by the Roman Catholic Church not to promote such a heliocentric model, Galileo nonetheless persisted and was condemned to spend

Figure 4-12

Galileo Galilei (1564–1642) Galileo was one of the first people to use a telescope to observe the heavens. He discovered craters on the Moon, sunspots on the Sun, the phases of Venus, and four moons orbiting Jupiter. His observations strongly suggested that the Earth orbits the Sun, not vice versa. (Art Resource)

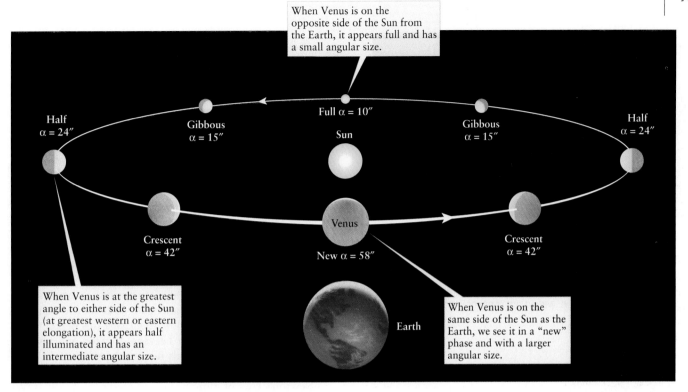

When Venus is on the opposite side of the Sun from the Earth, it appears full and has a small angular size.

Half
α = 24″

Gibbous
α = 15″

Full α = 10″

Sun

Gibbous
α = 15″

Half
α = 24″

Crescent
α = 42″

Venus

Crescent
α = 42″

New α = 58″

When Venus is at the greatest angle to either side of the Sun (at greatest western or eastern elongation), it appears half illuminated and has an intermediate angular size.

Earth

When Venus is on the same side of the Sun as the Earth, we see it in a "new" phase and with a larger angular size.

Figure 4-14

The Changing Appearance of Venus Explained in a Heliocentric Model A heliocentric model, in which the Earth and Venus both orbit the Sun, provides a natural explanation for the changing appearance of Venus shown in Figure 4-13.

his latter years under house arrest "for vehement suspicion of heresy." Nevertheless, there was no turning back.

While Galileo's observations showed convincingly that the Ptolemaic model was entirely wrong and that a heliocentric model is the more nearly correct one, he was unable to provide a complete explanation of why the Earth should orbit the Sun and not vice versa. The first person who was able to provide such an explanation was the Englishman Isaac Newton, born on Christmas Day of 1642, a dozen years after the death of Kepler and the same year that Galileo died. While Kepler and Galileo revolutionized our understanding of planetary motions, Newton's contribution was far greater: He deduced the basic laws that govern all motions on Earth as well as in the heavens.

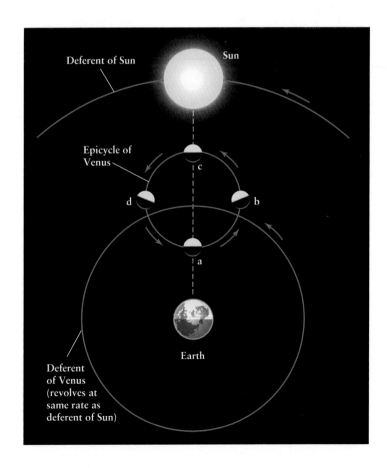

Deferent of Sun

Sun

Epicycle of Venus

c

d

b

a

Earth

Deferent of Venus (revolves at same rate as deferent of Sun)

Figure 4-15

The Appearance of Venus in the Ptolemaic Model In the geocentric Ptolemaic model the deferents of Venus and the Sun rotate together, with the epicycle of Venus centered on a line (shown dashed) that connects the Sun and the Earth. In this model an Earth observer would never see Venus as more than half illuminated. (At positions *a* and *c*, Venus appears in a "new" phase; at positions *b* and *d*, it appears as a crescent. Compare with Figure 3-2, which shows the phases of the Moon.) Because Galileo saw Venus in nearly fully illuminated phases, he concluded that the Ptolemaic model must be incorrect.

figure 4-16 · R I V U X G

Jupiter and Its Largest Moons This photograph, taken by an amateur astronomer with a small telescope, shows the four Galilean satellites alongside an overexposed image of Jupiter. Each satellite is bright enough to be seen with the unaided eye, were it not overwhelmed by the glare of Jupiter. (Courtesy of C. Holmes)

figure 4-17

Early Observations of Jupiter's Moons In 1610 Galileo discovered four "stars" that move back and forth across Jupiter from one night to the next. He concluded that these are four moons that orbit Jupiter, much as our Moon orbits the Earth. This drawing shows notations made by Jesuit observers on successive nights in 1620. The circle represents Jupiter and the stars its moons. Compare the drawing numbered 13 with the photograph in Figure 4-16. (Yerkes Observatory)

4-6 | Isaac Newton formulated three laws that describe fundamental properties of physical reality

Until the mid-seventeenth century, virtually all attempts to describe the motions of the heavens were *empirical,* or based directly on data and observations. From Ptolemy to Kepler, astronomers would adjust their ideas and calculations by trial and error until they ended up with answers that agreed with observation.

Isaac Newton (**Figure 4-18**) introduced a new approach. He began with three quite general statements, now called **Newton's laws of motion.** These laws, deduced from experimental observation, apply to all forces and all bodies. Newton then showed that Kepler's three laws follow logically from these laws of motion and from a formula for the force of gravity that he derived from observation.

In other words, Kepler's laws are not just an empirical description of the motions of the planets, but a direct consequence of the fundamental laws of physical matter. Using this deeper insight into the nature of motions in the heavens, Newton and his successors were able to describe accurately not just the orbits of the planets but also the orbits of the Moon and comets.

Newton's laws of motion describe objects on Earth as well as in the heavens. Thus, we can understand each of these laws by considering the motions of objects around us. We begin with **Newton's first law of motion,** or **law of inertia:**

A body remains at rest, or moves in a straight line at a constant speed, unless acted upon by a net outside force.

By **force** we mean any push or pull that acts on the body. An *outside* force is one that is exerted on the body by something other than the body itself. The net, or total, outside force is the combined effect of all of the individual outside forces that act on the body.

Right now, you are demonstrating the first part of Newton's first law. As you sit in your chair reading this, there are two outside forces acting on you: The force of gravity pulls you downward, and the chair pushes up on you. These two forces are of equal strength but of opposite direction, so their effects cancel—there is no *net* outside force. Hence, your body remains at rest as stated in Newton's first law. If you try to lift yourself out of your chair by grabbing your knees and pulling up, you will remain at rest because this force is not an outside force.

 The second part of Newton's first law, about objects in motion, may seem to go against common sense. If you want to make this book move across the floor in a straight line at a constant speed, you must continually push on it. You might therefore think that there *is* a net outside force, the force of your push. But another force also acts on the book—the force of friction as the book rubs across the floor. As you push the book across the floor, the force of your push exactly balances the force of friction, so again there is no net outside force. The effect is to make the book move in a straight line at constant speed, just as Newton's first law says. If you stop pushing, there will be nothing to balance the effects of friction. Then there will be a net outside force and the book will slow to a stop.

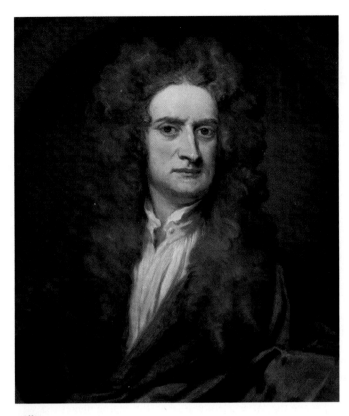

figure 4-18

Isaac Newton (1642–1727) Using mathematical techniques that he devised, Isaac Newton formulated the law of universal gravitation and demonstrated that the planets orbit the Sun according to simple mechanical rules. (National Portrait Gallery, London)

Newton's first law tells us that a force must be acting on the planets. A planet moving through empty space encounters no friction, so it would tend to fly off into space along a straight line if there were no other outside force acting on it. Because this does not happen, Newton concluded that there must be a force that acts continuously on the planets to keep them in their elliptical orbits.

Newton's second law describes how the motion of an object changes if there is a net outside force acting on it. To appreciate Newton's second law, we must first understand three quantities that describe motion—speed, velocity, and acceleration.

Speed is a measure of how fast an object is moving. Speed and direction of motion together constitute an object's **velocity**. Compared with a car driving north at 100 km/h (62 mi/h), a car driving east at 100 km/h has the same speed but a different velocity.

Acceleration is the rate at which velocity changes. Because velocity involves both speed and direction, acceleration can result from changes in either. Contrary to popular use of the term, acceleration does not simply mean speeding up. A car is accelerating if it is speeding up, and it is also accelerating if it is slowing down or turning (that is, changing the direction in which it is moving).

You can verify these statements about acceleration if you think about the sensations of riding in a car. If the car is moving with a constant velocity (in a straight line at a constant speed), you feel the same as if the car were not moving at all. But you can feel it when the car accelerates in any way: You feel thrown back in your seat if the car speeds up, thrown forward if the car slows down, and thrown sideways if the car changes direction in a tight turn. In **Box 4-3** we discuss the reasons for these sensations, along with other applications of Newton's laws to everyday life.

An apple falling from a tree is a good example of acceleration that involves only an increase in speed. Initially, at the moment the stem breaks, the apple's speed is zero. After 1 second, its downward speed is 9.8 meters per second, or 9.8 m/s (32 feet per second, or 32 ft/s). After 2 seconds, the apple's speed is twice this, or 19.6 m/s. After 3 seconds, the speed is 29.4 m/s. Because the apple's speed increases by 9.8 m/s for each second of free fall, the rate of acceleration is 9.8 meters per second per second, or 9.8 m/s^2 (32 ft/s^2). Thus, the Earth's gravity gives the apple a constant acceleration of 9.8 m/s^2 downward, toward the center of the Earth.

A planet revolving about the Sun along a perfectly circular orbit is an example of acceleration that involves change of direction only. As the planet moves along its orbit, its speed remains constant. Nevertheless, the planet is continuously being accelerated because its direction of motion is continuously changing.

Newton's second law of motion says that the acceleration of an object is proportional to the net outside force acting on the object. In other words, the harder you push on an object, the greater the resulting acceleration. This law can be succinctly stated as an equation. If a net outside force F acts on an object of mass m, the object will experience an acceleration a such that

Newton's second law

$$F = ma$$

F = net outside force on an object
m = mass of object
a = acceleration of object

The **mass** of an object is a measure of the total amount of material in the object. It is usually expressed in kilograms (kg) or grams (g). For example, the mass of the Sun is 2×10^{30} kg, the mass of a hydrogen atom is 1.7×10^{-27} kg, and the mass of an average adult is 75 kg. The Sun, a hydrogen atom, and a person have these masses regardless of where they happen to be in the universe.

It is important not to confuse the concepts of mass and weight. **Weight** is the force of gravity that acts on a body and, like any force, is usually expressed in pounds or newtons (1 newton = 0.225 pound).

We can use Newton's second law to relate mass and weight. We have seen that the acceleration caused by the Earth's gravity is 9.8 m/s^2. When a 50-kg swimmer falls from a diving board, the only outside force acting on her as she falls is her weight. Thus, from Newton's second law ($F = ma$), her weight is equal to her mass multiplied by the acceleration due to gravity:

$$50 \text{ kg} \times 9.8 \text{ m/s}^2 = 490 \text{ newtons} = 110 \text{ pounds}$$

Note that this answer is correct only when the swimmer is on Earth. She would weigh less on the Moon, where the pull of gravity

box 4-3 THE HEAVENS ON THE EARTH

Newton's Laws in Everyday Life

In our study of astronomy, we use Newton's three laws of motion to help us understand the motions of objects in the heavens. But you can see applications of Newton's laws every day in the world around you. By considering these everyday applications, we can gain insight into how Newton's laws apply to celestial events that are far removed from ordinary human experience.

Newton's *first* law, or principle of inertia, says that an object at rest naturally tends to remain at rest and that an object in motion naturally tends to remain in motion. This explains the sensations that you feel when riding in an automobile. When you are waiting at a red light, your car and your body are both at rest. When the light turns green and you press on the gas pedal, the car accelerates forward but your body attempts to stay where it was. Hence, the seat of the accelerating car pushes forward into your body, and it feels as though you are being pushed back in your seat.

Once the car is up to cruising speed, your body wants to keep moving in a straight line at this cruising speed. If the car makes a sharp turn to the left, the right side of the car will move toward you. Thus, you will feel as though you are being thrown to the car's right side (the side on the outside of the turn). If you bring the car to a sudden stop by pressing on the brakes, your body will continue moving forward until the seat belt stops you. In this case, it feels as though you are being thrown toward the front of the car.

Newton's *second* law states that the net outside force on an object equals the product of the object's mass and its acceleration. You can accelerate a crumpled-up piece of paper to a pretty good speed by throwing it with a moderate force. But if you try to throw a heavy rock by using the same force, the acceleration will be much less because the rock has much

more mass than the crumpled paper. Because of the smaller acceleration, the rock will leave your hand moving at only a slow speed.

Automobile airbags are based on the relationship between force and acceleration. It takes a large force to bring a fast-moving object suddenly to rest because this requires a large acceleration. In a collision, the driver of a car not equipped with airbags is jerked to a sudden stop and the large forces that act can cause major injuries. But if the car has airbags that deploy in an accident, the driver's body will slow down more gradually as it contacts the airbag, and the driver's acceleration will be less. (Remember that *acceleration* can refer to slowing down as well as to speeding up.) Hence, the force on the driver and the chance of injury will both be greatly reduced.

Newton's *third* law, the principle of action and reaction, explains how a car can accelerate at all. It is not correct to say that the engine pushes the car forward, because Newton's second law tells us that it takes a force acting from outside the car to make the car accelerate. Rather, the engine makes the wheels and tires turn, and the tires push backward on the ground. (You can see this backward force in action when a car drives through wet ground and sprays mud backward from the tires.) From Newton's third law, the ground must exert an equally large forward force on the car, and this is the force that pushes the car forward.

You use the same principles when you walk: You push backward on the ground with your foot, and the ground pushes forward on you. Icy pavement or a freshly waxed floor have greatly reduced friction. In these situations, your feet and the surface under you can exert only weak forces on each other, and it is much harder to walk.

---

is weaker, and more on Jupiter, where the gravitational pull is stronger. Floating deep in space, she would have no weight at all; she would be "weightless." Nevertheless, in all these circumstances, she would always have exactly the same mass, because mass is an inherent property of matter unaffected by details of the environment. Whenever we describe the properties of planets, stars, or galaxies, we speak of their masses, never of their weights.

We have seen that a planet is continually accelerating as it orbits the Sun. From Newton's second law, this means that there must be a net outside force that acts continually on each of the planets. As we will see in the next section, this force is the gravitational attraction of the Sun.

The last of Newton's general laws of motion, called **Newton's third law of motion,** is the famous statement about action and reaction:

Whenever one body exerts a force on a second body, the second body exerts an equal and opposite force on the first body.

For example, if you weigh 110 pounds, when you are standing up you are pressing down on the floor with a force of 110 pounds. Newton's third law tells us that the floor is also pushing up against your feet with an equal force of 110 pounds.

Newton realized that because the Sun is exerting a force on each planet to keep it in orbit, each planet must also be exerting an equal and opposite force on the Sun. However, the planets are much less massive than the Sun (for example, the Earth has only $1/{300,000}$ of the Sun's mass). Therefore, although the Sun's force on a planet is the same as the planet's force on the Sun, the planet's much smaller mass gives it a much larger acceleration, according to Newton's second law. This is why the planets circle the Sun instead of vice versa. Thus, Newton's laws reveal the reason for our heliocentric solar system.

4-7 | Newton's description of gravity accounts for Kepler's laws and explains the motions of the planets

Tie a ball to one end of a piece of string, hold the other end of the string in your hand, and whirl the ball around in a circle. As the ball "orbits" your hand, it is continuously accelerating because its velocity is changing. (Even if its speed is constant, its direction of motion is changing.) In accordance with Newton's second law, this can happen only if the ball is continuously acted on by an outside force—the pull of the string. The pull is directed along the string toward your hand. In the same way, Newton saw, the force that keeps a planet in orbit around the Sun is a pull that always acts toward the Sun. That pull is **gravity,** or **gravitational force.**

Newton's discovery about the forces that act on planets led him to suspect that the force of gravity pulling a falling apple straight down to the ground is fundamentally the same as the force on a planet that is always directed straight at the Sun. In other words, gravity is the force that shapes the orbits of the planets. What is more, he was able to determine how the force of gravity depends on the distance between the Sun and the planet. His result was a law of gravitation that could apply to the motion of distant planets as well as to the flight of a football on Earth. Using this law, Newton

achieved the remarkable goal of deducing Kepler's laws from fundamental principles of nature.

To see how Newton reasoned, think again about a ball attached to a string. If you use a short string, so that the ball orbits in a small circle, and whirl the ball around your hand at a high speed, you will find that you have to pull fairly hard on the string (Figure 4-19a). But if you use a longer string, so that the ball moves in a larger orbit, and if you make the ball orbit your hand at a slow speed, you only have to exert a light tug on the string (Figure 4-19b). The orbits of the planets behave in the same way; the larger the size of the orbit, the slower the planet's speed (Figure 4-19c, d). By analogy to the force of the string on the orbiting ball, Newton concluded that the force that attracts a planet toward the Sun must decrease with increasing distance between the Sun and the planet.

Using his own three laws and Kepler's three laws, Newton succeeded in formulating a general statement that describes the nature of the gravitational force. Newton's **law of universal gravitation** is as follows:

Two bodies attract each other with a force that is directly proportional to the mass of each body and inversely proportional to the square of the distance between them.

This law states that *any* two objects exert gravitational pulls on each other. Normally, you notice only the gravitational force that the Earth exerts on you, otherwise known as your weight. In fact, you feel gravitational attractions to *all* the objects around you. For

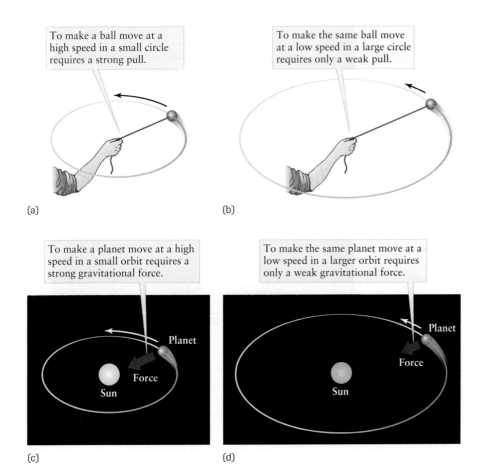

To make a ball move at a high speed in a small circle requires a strong pull.

(a)

To make the same ball move at a low speed in a large circle requires only a weak pull.

(b)

To make a planet move at a high speed in a small orbit requires a strong gravitational force.

Planet
Force
Sun

(c)

To make the same planet move at a low speed in a larger orbit requires only a weak gravitational force.

Planet
Force
Sun

(d)

figure 4-19

An Orbit Analogy (a) To make a ball on a string move at high speed around a small circle, you have to exert a substantial pull on the string. (b) If you lengthen the string and make the same ball move at low speed around a large circle, much less pull is required. (c), (d) Similarly, a planet that orbits close to the Sun moves at high speed and requires a substantial gravitational force from the Sun, while a planet in a large orbit moves at low speed and requires less gravitational force to stay in orbit.

example, this book is exerting a gravitational force on you as you read it. But because the force exerted on you by this book is proportional to the book's mass, which is very small compared to the Earth's mass, the force is too small to notice. (It can actually be measured with sensitive equipment.)

Consider two 1-kg objects separated by a distance of 1 meter. Newton's law of universal gravitation says that the force is directly proportional to the mass, so if we double the mass of one object to 2 kg, the force between the objects will double. If we double both masses so that we have two 2-kg objects separated by 1 meter, the force will be $2 \times 2 = 4$ times what it was originally (the force is directly proportional to the mass of *each* object). If we go back to two 1-kg masses, but double their separation to 2 meters, the force will be only one-quarter its original value. This is because the force is inversely proportional to the square of the distance: If we double the distance, the force is multiplied by a factor of

$$\frac{1}{2^2} = \frac{1}{4}$$

Newton's law of universal gravitation can be stated more succinctly as an equation. If two objects have masses m_1 and m_2 and are separated by a distance r, then the gravitational force F between these two objects is given by the following equation:

Newton's law of universal gravitation

$$F = G\left(\frac{m_1 m_2}{r^2}\right)$$

F = gravitational force between two objects
m_1 = mass of first object
m_2 = mass of second object
r = distance between objects
G = universal constant of gravitation

If the masses are measured in kilograms and the distance between them in meters, then the force is measured in newtons. In this formula, G is a number called the **universal constant of gravitation**. Laboratory experiments have yielded a value for G of

$$G = 6.67 \times 10^{-11} \text{ newton} \cdot \text{m}^2/\text{kg}^2$$

We can use Newton's law of universal gravitation to calculate the force with which any two bodies attract each other. For example, to compute the gravitational force that the Sun exerts on the Earth, we substitute values for the Earth's mass ($m_1 = 5.98 \times 10^{24}$ kg), the Sun's mass ($m_2 = 1.99 \times 10^{30}$ kg), the distance between them ($r = 1$ AU $= 1.5 \times 10^{11}$ m), and the value of G into Newton's equation. We get

$$F_{\text{Sun-Earth}} = 6.67 \times 10^{-11} \left[\frac{(5.98 \times 10^{24}) \times (1.99 \times 10^{30})}{(1.50 \times 10^{11})^2}\right]$$

$$= 3.53 \times 10^{22} \text{ newtons}$$

If we calculate the force that the Earth exerts on the Sun, we get exactly the same result. (Mathematically, we just let m_1 be the Sun's

mass and m_2 be the Earth's mass instead of the other way around. The product of the two numbers is the same, so the force is the same.) This is in accordance with Newton's third law: Any two objects exert *equal* gravitational forces on each other.

Because there is a gravitational force between any two objects, Newton concluded that gravity is also the force that keeps the Moon in orbit around the Earth. It is also the force that keeps artificial satellites in orbit. But if the force of gravity attracts two objects to each other, why don't satellites immediately fall to Earth? Why doesn't the Moon fall into the Earth? And, for that matter, why don't the planets fall into the Sun?

To see the answer, imagine (as Newton did) dropping a ball from a great height above the Earth's surface, as in Figure 4-20. After you drop the ball, it, of course, falls straight down (path A in Figure 4-20). But if you *throw* the ball horizontally, it travels some distance across the Earth's surface before hitting the ground (path B). If you throw the ball harder, it travels a greater distance (path C). If you could throw at just the right speed, the curvature of the ball's path will exactly match the curvature of the Earth's surface (path E). Although the Earth's gravity is making the ball fall, the Earth's surface is falling away under the ball at the same rate. Hence, the ball does not get any closer to the surface, and the ball

 Figure 4-20
An Explanation of Orbits If a ball is dropped from a great height above the Earth's surface, it falls straight down (A). If the ball is thrown with some horizontal speed, it follows a curved path before hitting the ground (B, C). If thrown with just the right speed (E), the ball goes into circular orbit; the ball's path curves but it never gets any closer to the Earth's surface. If the ball is thrown with a speed that is slightly less (D) or slightly more (F) than the speed for a circular orbit, the ball's orbit is an ellipse.

is in circular orbit. So the ball in path E is in fact falling, but it is falling *around* the Earth rather than *toward* the Earth.

A spacecraft is launched into orbit in just this way—by throwing it fast enough. The thrust of a rocket is used to give the spacecraft the necessary orbital speed. Once the spacecraft is in orbit, the rocket turns off and the spacecraft falls continually around the Earth.

CAUTION An astronaut on board an orbiting spacecraft feels "weightless," but this is *not* because she is "beyond the pull of gravity." The astronaut is herself an independent satellite of the Earth, and the Earth's gravitational pull is what holds her in orbit. She feels "weightless" because she and her spacecraft are falling *together* around the Earth, so there is nothing pushing her against any of the spacecraft walls. You feel the same "weightless" sensation whenever you are falling, such as when you jump off a diving board or ride the free-fall ride at an amusement park.

If the ball in Figure 4-20 is thrown with a slightly slower speed than that required for a circular orbit, its orbit will be an ellipse (path D). An elliptical orbit also results if instead the ball is thrown a bit too fast (path F). In this way, spacecraft can be placed into any desired orbit around the Earth by adjusting the thrust of the rockets.

Just as the ball in Figure 4-20 will not fall to Earth if given enough speed, the Moon does not fall to Earth and the planets do not fall into the Sun. The planets acquired their initial speeds around the Sun when the solar system first formed 4.56 billion years ago. Figure 4-20 shows that a circular orbit is a very special case, so it is no surprise that the orbits of the planets are not precisely circular.

CAUTION Orbiting satellites do sometimes fall out of orbit and crash back to Earth. When this happens, however, the real culprit is not gravity but air resistance. A satellite in a relatively low orbit is actually flying through the tenuous outer wisps of the Earth's atmosphere. The resistance of the atmosphere slows the satellite and changes a circular orbit like E in Figure 4-20 to an elliptical one like D. As the satellite sinks to lower altitude, it encounters more air resistance and sinks even lower. Eventually, it either strikes the Earth or burns up in flight due to air friction. By contrast, the Moon and planets orbit in the near-vacuum of interplanetary space. Hence, they are unaffected by this kind of air resistance, and their orbits are much more long-lasting.

Using his three laws of motion and his law of gravity, Newton found that he could prove Kepler's three laws mathematically. Kepler's first law, concerning the elliptical shape of planetary orbits, proved to be a direct consequence of the $1/r^2$ factor in the law of universal gravitation. (Had the nature of gravity in our universe been different, so that this factor was given by a different function such as $1/r$ or $1/r^3$, elliptical orbits would not have been possible.) The law of equal areas, or Kepler's second law, turns out to be a consequence of the Sun's gravitational force on a planet being directed straight toward the Sun.

Newton also demonstrated that Kepler's third law follows logically from his law of gravity. Specifically, he proved that if two objects with masses m_1 and m_2 orbit each other, the period P of their orbit and the semimajor axis a of their orbit (that is, the average distance between the two objects) are related by the equation

$$P^2 = \left[\frac{4\pi^2}{G(m_1 + m_2)} \right] a^3$$

This equation is called **Newton's form of Kepler's third law.** It is valid whenever two objects orbit each other because of their mutual gravitational attraction. It is invaluable in the study of binary star systems, in which two stars orbit each other. If the orbital period P and semimajor axis a of the two stars in a binary system are known, astronomers can use this formula to calculate the sum m_1+m_2 of the masses of the two stars. Within our own solar system, Newton's form of Kepler's third law makes it possible to learn about the masses of planets. By measuring the period and semimajor axis for a satellite, astronomers can determine the sum of the masses of the planet and the satellite. (The satellite can be a moon of the planet or a spacecraft that we place in orbit around the planet. Newton's laws apply in either case.) Box 4-4 gives an example of using Newton's form of Kepler's third law.

Newton also discovered new features of orbits around the Sun. For example, his equations soon led him to conclude that the orbit of an object around the Sun need not be an ellipse. It could be any one of a family of curves called conic sections.

A **conic section** is any curve that you get by cutting a cone with a plane, as shown in Figure 4-21. You can get circles and ellipses by slicing all the way through the cone. You can also get two types

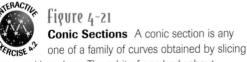

figure 4-21
Conic Sections A conic section is any one of a family of curves obtained by slicing a cone with a plane. The orbit of one body about another can be any one of these curves: a circle, an ellipse, a parabola, or a hyperbola.

box 4-4   TOOLS OF THE ASTRONOMER'S TRADE

Newton's Form of Kepler's Third Law

Kepler's original statement of his third law, $P^2 = a^3$, is valid only for objects that orbit the Sun. (Box 4-2 shows how to use this equation.) But Newton's form of Kepler's third law is much more general: It can be used in *any* situation where two bodies of masses m_1 and m_2 orbit each other. For example, Newton's form is the equation to use for a moon orbiting a planet or a satellite orbiting the Earth. This equation is

Newton's form of Kepler's third law

$$P^2 = \left[\frac{4\pi^2}{G(m_1 + m_2)} \right] a^3$$

P = sidereal period of orbit, in seconds
a = semimajor axis of orbit, in meters
m_1 = mass of first object, in kilograms
m_2 = mass of second object, in kilograms
G = universal constant of gravitation = 6.67×10^{-11}

Notice that P, a, m_1, and m_2 *must* be expressed in these particular units. If you fail to use the correct units, your answer will be incorrect.

EXAMPLE: Io is one of the four large moons of Jupiter discovered by Galileo (see Figure 4-16). It orbits at a distance of 421,600 km from the center of Jupiter and has an orbital period of 1.77 days. Determine the combined mass of Jupiter and Io.

Situation: We are given Io's orbital period P and semimajor axis a (the distance from Io to the center of its circular orbit, which is at the center of Jupiter). Our goal is to find the sum of the masses of Jupiter (m_1) and Io (m_2).

Tools: Because this is not an orbit around the Sun, we must use Newton's form of Kepler's third law to relate P and a. This relationship also involves $m_1 + m_2$, whose sum ($m_1 + m_2$) we are asked to find.

Answer: To solve for $m_1 + m_2$, we rewrite the equation in the form

$$m_1 + m_2 = \frac{4\pi^2 a^3}{GP^2}$$

To use this equation, we have to convert the distance a from kilometers to meters and convert the period P from days to seconds. There are 1000 meters in 1 kilometer and 86,400 seconds in 1 day, so

$$a = (421,600 \text{ km}) \times \frac{1000 \text{ m}}{1 \text{ km}} = 4.216 \times 10^8 \text{ m}$$

$$P = (1.77 \text{ days}) \times \frac{86,400 \text{ s}}{1 \text{ day}} = 1.529 \times 10^5 \text{ s}$$

We can now put these values and the value of G into the above equation:

$$m_1 + m_2 = \frac{4\pi^2(4.216 \times 10^8)^3}{(6.67 \times 10^{-11})(1.529 \times 10^5)^2} = 1.90 \times 10^{27} \text{ kg}$$

Review: Io is very much smaller than Jupiter, so its mass is only a small fraction of the mass of Jupiter. Thus, $m_1 + m_2$ is very nearly the mass of Jupiter alone. We conclude that Jupiter has a mass of 1.90×10^{27} kg, or about 300 times the mass of Earth. This technique can be used to determine the mass of any object that has a second, much smaller object orbiting it. Astronomers use this technique to find the masses of stars, black holes, and entire galaxies of stars.

---

of open curves called **parabolas** and **hyperbolas**. If you were to throw the ball in Figure 4-20 with a fast enough speed, it would follow a parabolic or hyperbolic orbit and would fly off into space, never to return. Comets hurtling toward the Sun from the depths of space sometimes follow hyperbolic orbits.

Newton's ideas turned out to be applicable to an incredibly wide range of situations. Using his laws of motion, Newton himself proved that the Earth's axis of rotation must precess because of the gravitational pull of the Moon and the Sun on the Earth's equatorial bulge (see Figure 2-17). In fact, all the details of the orbits of the planets and their satellites could be explained mathematically with a body of knowledge built on Newton's work that is today called **Newtonian mechanics.**

Not only could Newtonian mechanics explain a variety of known phenomena in detail, but it could also predict new phenomena. For example, one of Newton's friends, Edmund Halley, was intrigued by three similar historical records of a comet that had been sighted at intervals of 76 years. Assuming these records to be accounts of the same comet, Halley used Newton's methods to work out the details of the comet's orbit and predicted its return in 1758. It was first sighted on Christmas night of 1757, a fitting memorial to Newton's birthday. To this day the comet bears Halley's name (Figure 4-22).

Another dramatic success of Newton's ideas was their role in the discovery of the eighth planet from the Sun. The seventh planet, Uranus, was discovered accidentally by William Herschel in 1781 during a telescopic survey of the sky. Fifty years later, however, it

figure 4-22   R I **V** U X G

Comet Halley This most famous of all comets orbits the Sun with an average period of about 76 years. During the twentieth century, the comet passed near the Sun in 1910 and again in 1986. It will next be prominent in the sky in 2061. This photograph shows how the comet looked in 1986. (David Malin, Anglo-Australian Observatory/Royal Observatory, Edinburgh)

was clear that Uranus was not following its predicted orbit. John Couch Adams in England and Urbain Le Verrier in France independently calculated that the gravitational pull of a yet unknown, more distant planet could explain the deviations of Uranus from its orbit. Le Verrier predicted that the planet would be found at a certain location in the constellation of Aquarius. A brief telescopic search on September 23, 1846, revealed the planet Neptune within 1° of the calculated position. Before it was sighted with a telescope, Neptune was actually predicted with pencil and paper.

Because it has been so successful in explaining and predicting many important phenomena, Newtonian mechanics has become the cornerstone of modern physical science. Even today, as we send astronauts into Earth orbit and spacecraft to the outer planets, Newton's equations are used to calculate the orbits and trajectories of these spacecraft.

In the twentieth century, scientists found that Newton's laws do not apply in all situations. A new theory called *quantum mechanics* had to be developed to explain the behavior of matter on the very smallest of scales, such as within the atom and within the atomic nucleus. Albert Einstein developed the *theory of relativity* to explain what happens at very high speeds approaching the speed of light and in places where gravitational forces are very strong. For many purposes in astronomy, however, Newton's laws are as useful today as when Newton formulated them more than three centuries ago.

4-8 | Gravitational forces between the Earth and Moon produce tides

We have seen how Newtonian mechanics explains why the Moon stays in orbit around the Earth. It also explains why there are ocean tides, as well as why the Moon always keeps the same face

toward the Earth. Both of these are consequences of *tidal forces*—an aspect of gravity that deforms planets and reshapes galaxies.

Tidal forces are differences in the gravitational pull at different points in an object. As an illustration, imagine that three billiard balls are lined up in space at some distance from a planet, as in Figure 4-23a. According to Newton's law of universal gravitation, the force of attraction between two objects is greater the closer the two objects are to each other. Thus, the planet exerts more force on the 3-ball (in red) than on the 2-ball (in blue), and exerts more force on the 2-ball than on the 1-ball (in yellow). Now, imagine that the three balls are released and allowed to fall toward the planet. Figure 4-23b shows the situation a short time later. Because of the differences in gravitational pull, a short time later the 3-ball will have moved farther than the 2-ball, which will in turn have moved farther than the 1-ball. But now imagine that same motion from the perspective of the 2-ball. From this perspective, it appears as though the 3-ball is pulled toward the planet while the 1-ball is pushed away (Figure 4-23c). These apparent pushes and pulls are called tidal forces.

figure 4-23

The Origin of Tidal Forces (a) Imagine three identical billiard balls placed some distance from a planet and released. (b) The closer a ball is to the planet, the more gravitational force the planet exerts on it. Thus, a short time after the balls are released, the blue 2-ball has moved farther toward the planet than the yellow 1-ball, and the red 3-ball has moved farther still. (c) From the perspective of the 2-ball in the center, it appears that forces have pushed the 1-ball away from the planet and pulled the 3-ball toward the planet. These forces are called tidal forces.

The Moon has a similar effect on the Earth as the planet in Figure 4-23 has on the three billiard balls. The arrows in **Figure 4-24a** indicate the strength and direction of the gravitational force of the Moon at several locations on the Earth. The side of the Earth closest to the Moon feels a greater gravitational pull than the Earth's center does, and the side of the Earth that faces away from the Moon feels less gravitational pull than does the Earth's center. This means that just as for the billiard balls in Figure 4-23, there are tidal forces acting on the Earth (Figure 4-24b). These tidal forces exerted by the Moon try to elongate the Earth along a line connecting the centers of the Earth and the Moon and try to squeeze the Earth inward in the direction perpendicular to that line.

Because the body of the Earth is largely rigid, it cannot deform very much in response to the tidal forces of the Moon. But the water in the oceans can and does deform into a football shape, as **Figure 4-25a** shows. As the Earth rotates, a point on its surface goes from where the water is shallow to where the water is deep and back again. This is the origin of low and high ocean tides. (In this simplified description we have assumed that the Earth is completely covered with water. The full story of the tides is much more complex, because the shapes of the continents and the effects of winds must also be taken into account.)

The Sun also exerts tidal forces on the Earth's oceans. (The tidal effects of the Sun are about half as great as those of the Moon.)

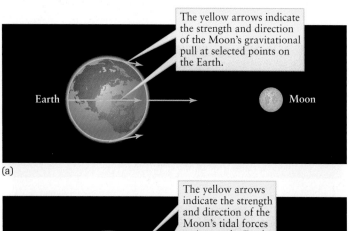

The yellow arrows indicate the strength and direction of the Moon's gravitational pull at selected points on the Earth.

Earth

Moon

(a)

The yellow arrows indicate the strength and direction of the Moon's tidal forces acting on the Earth.

Moon

From the perspective of the center of the Earth

(b)

Figure 4-24

Tidal Forces on the Earth (a) The Moon exerts different gravitational pulls at different locations on the Earth. (b) At any location, the tidal force equals the Moon's gravitational pull at that location minus the gravitational pull of the Moon at the center of the Earth. These tidal forces tend to deform the Earth into a nonspherical shape.

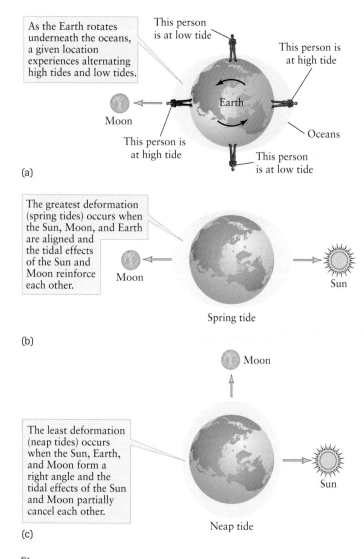

As the Earth rotates underneath the oceans, a given location experiences alternating high tides and low tides.

This person is at low tide

This person is at high tide

Moon

Earth

Oceans

This person is at high tide

This person is at low tide

(a)

The greatest deformation (spring tides) occurs when the Sun, Moon, and Earth are aligned and the tidal effects of the Sun and Moon reinforce each other.

Moon

Sun

Spring tide

(b)

Moon

The least deformation (neap tides) occurs when the Sun, Earth, and Moon form a right angle and the tidal effects of the Sun and Moon partially cancel each other.

Sun

Neap tide

(c)

Figure 4-25

High and Low Tides (a) The gravitational forces of the Moon and Sun deform the Earth's oceans, giving rise to low and high tides. (b), (c) The strength of the tides depends on the relative positions of the Sun, Moon, and Earth.

When the Sun, Moon, and the Earth are aligned, which happens at either new moon or full moon, the tidal effects of the Sun and Moon reinforce each other and the tidal distortion of the oceans is greatest. This produces large shifts in water level called **spring tides** (Figure 4-25b). At first quarter and last quarter, when the Sun and Moon form a right angle with the Earth, the tidal effects of the Sun and Moon partially cancel each other. Hence, the tidal distortion of the oceans is the least pronounced, producing smaller tidal shifts called **neap tides** (Figure 4-25c).

 Note that spring tides have nothing to do with the season of the year called spring. Instead, the name refers to the way that the ocean level "springs up" to a greater than

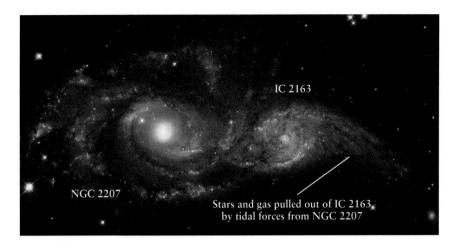

Figure 4-26   R I **V** U X G

Tidal Forces on a Galaxy For millions of years the galaxies NGC 2207 and IC 2163 have been moving ponderously past each other. The larger galaxy's tremendous tidal forces have drawn a streamer of material a hundred thousand light-years long out of IC 2163. If you lived on a planet orbiting a star within this streamer, you would have a magnificent view of both galaxies. NGC 2207 and IC 2163 are respectively 143,000 light-years and 101,000 light-years in diameter. Both galaxies are 114 million light-years away in the constellation Canis Major. (NASA and the Hubble Heritage Team, AURA/STScI)

normal height. Spring tides occur whenever there is a new moon or full moon, no matter what the season of the year.

Just as the Moon exerts tidal forces on the Earth, the Earth exerts tidal forces on the Moon. Soon after the Moon formed some 4.56 billion years ago, it was molten throughout its volume. The Earth's tidal forces deformed the molten Moon into a slightly elongated shape, with the long axis of the Moon pointed toward the Earth. The Moon retained this shape and orientation when it cooled and solidified. To keep its long axis pointed toward the Earth, the Moon spins once on its axis as it makes one orbit around the Earth—that is, it is in synchronous rotation (Section 3-2). Hence, the same side of the Moon always faces the Earth, and this is the side that we see. For the same reason, most of the satellites in the solar system are in synchronous rotation, and thus always keep the same side facing their planet.

Tidal forces are also important on scales much larger than the solar system. Figure 4-26 shows two spiral galaxies, like the one in Figure 1-7, undergoing a near-collision. During the millions of years that this close encounter has been taking place, the tidal forces of the larger galaxy have pulled an immense streamer of stars and interstellar gas out of the smaller galaxy.

Many galaxies, including our own Milky Way Galaxy, show signs of having been disturbed at some time by tidal interactions with other galaxies. By their effect on the interstellar gas from which stars are formed, tidal interactions can actually trigger the birth of new stars. Our own Sun and solar system may have been formed as a result of tidal interactions of this kind. Hence, we may owe our very existence to tidal forces. In this and many other ways, the laws of motion and of universal gravitation shape our universe and our destinies.

Key Words

acceleration, p. 77
aphelion, p. 71
conic section, p. 81
conjunction, p. 66
deferent, p. 63
direct motion, p. 63
eccentricity, p. 71
ellipse, p. 70
elongation, p. 66
epicycle, p. 63
focus (of an ellipse; *plural* foci), p. 70
force, p. 76
geocentric model, p. 62
gravitational force, p. 79
gravity, p. 79
greatest eastern elongation, p. 66
greatest western elongation, p. 66
heliocentric model, p. 65

hyperbola, p. 82
inferior conjunction, p. 66
inferior planet, p. 66
Kepler's first law, p. 71
Kepler's second law, p. 72
Kepler's third law, p. 72
law of equal areas, p. 72
law of inertia, p. 76
law of universal gravitation, p. 79
major axis (of an ellipse), p. 70
mass, p. 77
neap tides, p. 84
Newtonian mechanics, p. 82
Newton's first law of motion, p. 76
Newton's form of Kepler's third law, p. 81
Newton's second law of motion, p. 77
Newton's third law of motion, p. 78
Occam's razor, p. 64

opposition, p. 66
parabola, p. 82
parallax, p. 69
perihelion, p. 71
period (of a planet), p. 66
Ptolemaic system, p. 63
retrograde motion, p. 63
semimajor axis (of an ellipse), p. 70
sidereal period, p. 66
speed, p. 77
spring tides, p. 84
superior conjunction, p. 66
superior planet, p. 66
synodic period, p. 66
tidal forces, p. 83
universal constant of gravitation, p. 80
velocity, p. 77
weight, p. 77

Key Ideas

Apparent Motions of the Planets: Like the Sun and Moon, the planets move on the celestial sphere with respect to the background of stars. Most of the time a planet moves eastward in direct motion, in the same direction as the Sun and the Moon, but from time to time it moves westward in retrograde motion.

Geocentric Model: Ancient astronomers believed the Earth to be at the center of the universe. They invented a complex system of epicycles and deferents to explain the direct and retrograde motions of the planets on the celestial sphere.

Heliocentric Model: Copernicus's heliocentric (Sun-centered) theory simplified the general explanation of planetary motions.

• In a heliocentric system, the Earth is one of the planets orbiting the Sun.

• A planet undergoes retrograde motion as seen from Earth when the Earth and the planet pass each other.

• The sidereal period of a planet, its true orbital period, is measured with respect to the stars. Its synodic period is measured with respect to the Earth and the Sun (for example, from one opposition to the next).

Elliptical Orbits and Kepler's Laws: Copernicus thought that the orbits of the planets were combinations of circles. Using data collected by Tycho Brahe, Kepler deduced three laws of planetary motion: (1) the orbits are in fact ellipses; (2) a planet's speed varies as it moves around its elliptical orbit; and (3) the orbital period of a planet is related to the size of its orbit.

Evidence for the Heliocentric Model: The invention of the telescope led Galileo to new discoveries that supported a heliocentric model. These included his observations of the phases of Venus and of the motions of four moons around Jupiter.

Newton's Laws of Motion: Isaac Newton developed three principles, called the laws of motion, that apply to the motions of objects on Earth as well as in space. These are (1) the law of inertia, (2) the relationship between the net outside force on an object and the object's acceleration, and (3) the principle of action and reaction. These laws and Newton's law of universal gravitation can be used to deduce Kepler's laws. They lead to extremely accurate descriptions of planetary motions.

• The mass of an object is a measure of the amount of matter in the object. Its weight is a measure of the force with which the gravity of some other object pulls on it.

• In general, the path of one object about another, such as that of a planet or comet about the Sun, is one of the curves called conic sections: circle, ellipse, parabola, or hyperbola.

Tidal Forces: Tidal forces are caused by differences in the gravitational pull that one object exerts on different parts of a second object.

• The tidal forces of the Moon and Sun produce tides in the Earth's oceans.

• The tidal forces of the Earth have locked the Moon into synchronous rotation.

Review Questions

1. In what direction does a planet move relative to the stars when it is in direct motion? When it is in retrograde motion? How do these compare with the direction in which we see the Sun move relative to the stars?

2. (a) In what direction does a planet move relative to the horizon over the course of one night? (b) The answer to (a) is the same whether the planet is in direct motion or retrograde motion. What does this tell you about the speed at which planets move on the celestial sphere?

3. What is an epicycle? How is it important in Ptolemy's explanation of the retrograde motions of the planets?

4. What is the significance of Occam's razor as a tool for analyzing theories?

5. How did the models of Aristarchus and Copernicus explain the retrograde motion of the planets?

6. How did Copernicus determine that the orbits of Mercury and Venus must be smaller than the Earth's orbit?

7. At what configuration (for example, superior conjunction, greatest eastern elongation, and so on) would it be best to observe Mercury or Venus with an Earth-based telescope? At what configuration would it be best to observe Mars, Jupiter, or Saturn? Explain your answers.

8. Which planets can never be seen at opposition? Which planets can never be seen at inferior conjunction? Explain your answers.

9. What is the difference between the synodic period and the sidereal period of a planet?

10. What is parallax? What did Tycho Brahe conclude from his attempt to measure the parallax of a supernova and a comet?

11. What observations did Tycho Brahe make in an attempt to test the heliocentric model? What were his results? Explain why modern astronomers get different results.

12. What are the foci of an ellipse? If the Sun is at one focus of a planet's orbit, what is at the other focus?

13. What are Kepler's three laws? Why are they important?

14. At what point in a planet's elliptical orbit does it move fastest? At what point does it move slowest? At what point does it sweep out an area at the fastest rate?

15. A line joining the Sun and an asteroid is found to sweep out an area of 6.3 AU^2 during 2007. How much area is swept out during 2008? Over a period of five years?

16. The orbit of a spacecraft about the Sun has a perihelion distance of 0.1 AU and an aphelion distance of 0.4 AU. What is the semimajor axis of the spacecraft's orbit? What is its orbital period?

17. A comet with a period of 125 years moves in a highly elongated orbit about the Sun. At perihelion, the comet comes very close to the Sun's surface. What is the comet's average distance from the Sun? What is the farthest it can get from the Sun?

18. What observations did Galileo make that reinforced the heliocentric model? Why did these observations contradict the older model of Ptolemy? Why could these observations not have been made before Galileo's time?

19. What are Newton's three laws? Give an everyday example of each law.

20. How much force do you have to exert on a 3-kg brick to give it an acceleration of 2 m/s^2? If you double this force, what is the brick's acceleration? Explain.

21. What is the difference between weight and mass?

22. What is your weight in pounds and in newtons? What is your mass in kilograms?

23. Suppose that the Earth were moved to a distance of 3.0 AU from the Sun. How much stronger or weaker would the Sun's gravitational pull be on the Earth? Explain.

24. How far would you have to go from Earth to be completely beyond the pull of its gravity? Explain.

25. What are conic sections? In what way are they related to the orbits of planets in the solar system?

26. Why was the discovery of Neptune an important confirmation of Newton's law of universal gravitation?

27. What is a tidal force? How do tidal forces produce tides in the Earth's oceans?

28. What is the difference between spring tides and neap tides?

Advanced Questions

Questions preceded by an asterisk (*) involve topics discussed in the Boxes.

Problem-solving tips and tools

Box 4-1 explains sidereal and synodic periods in detail. The semimajor axis of an ellipse is half the length of the long, or major, axis of the ellipse. For data about the planets and their satellites, see Appendices 1, 2, and 3 at the back of this book. If you want to calculate the gravitational force that you feel on the surface of a planet, the distance r to use is the planet's radius (the distance between you and the center of the planet). Boxes 4-2 and 4-4 show how to use Kepler's third law in its original form and in Newton's form.

29. Figure 4-2 shows the retrograde motion of Mars as seen from Earth. Sketch a similar figure that shows how Earth would appear to move against the background of stars during this same time period as seen by an observer on Mars.

*30. The synodic period of Mercury (an inferior planet) is 115.88 days. Calculate its sidereal period in days.

*31. Table 4-1 shows that the synodic period is *greater* than the sidereal period for Mercury, but the synodic period is *less* than the sidereal period for Jupiter. Draw diagrams like the one in Box 4-1 to explain why this is so.

*32. A general rule for superior planets is that the greater the average distance from the planet to the Sun, the more frequently that planet will be at opposition. Explain how this rule comes about.

33. In 2003, Mercury was at greatest western elongation on February 4, June 3, and September 26. It was at greatest eastern elongation on April 16, August 14, and December 9. Does Mercury take longer to go from eastern to western elongation, or vice versa? Why do you suppose this is the case?

34. Explain why the semimajor axis of a planet's orbit is equal to the average of the distance from the Sun to the planet at perihelion (the *perihelion distance*) and the distance from the Sun to the planet at aphelion (the *aphelion distance*).

35. A certain asteroid is 2 AU from the Sun at perihelion and 6 AU from the Sun at aphelion. (a) Find the semimajor axis of the asteroid's orbit. (b) Find the sidereal period of the orbit.

36. A comet orbits the Sun with a sidereal period of 64.0 years. (a) Find the semimajor axis of the orbit. (b) At aphelion, the comet is 31.5 AU from the Sun. How far is it from the Sun at perihelion?

37. One trajectory that can be used to send spacecraft from the Earth to Mars is an elliptical orbit that has the Sun at one focus, its perihelion at the Earth, and its aphelion at Mars. The spacecraft is launched from Earth and coasts along this ellipse until it reaches Mars, when a rocket is fired to either put the spacecraft into orbit around Mars or cause it to land on Mars. (a) Find the semimajor axis of the ellipse. (*Hint:* Draw a picture showing the Sun and the orbits of the Earth, Mars, and the spacecraft. Treat the orbits of the Earth and Mars as circles.) (b) Calculate how long (in days) such a one-way trip to Mars would take.

38. The mass of the Moon is 7.35×10^{22} kg, while that of the Earth is 5.98×10^{24} kg. The average distance from the center of the Moon to the center of the Earth is 384,400 km. What is the size of the gravitational force that the Earth exerts on the Moon? What is the size of the gravitational force that the Moon exerts on the Earth? How do your answers compare with the force between the Sun and the Earth calculated in the text?

39. Suppose that you traveled to a planet with 4 times the mass and 4 times the diameter of the Earth. Would you weigh more or less on that planet than on Earth? By what factor?

40. The mass of Saturn is approximately 100 times that of Earth, and the semimajor axis of Saturn's orbit is approximately 10 AU. To this approximation, how does the gravitational force that the

Sun exerts on Saturn compare to the gravitational force that the Sun exerts on the Earth? How do the accelerations of Saturn and the Earth compare?

41. On Earth, a 50-kg astronaut weighs 490 newtons. What would she weigh if she landed on Jupiter's moon Callisto? What fraction is this of her weight on Earth? See Appendix 3 for relevant data about Callisto.

42. A satellite is said to be in a "geosynchronous" orbit if it appears always to remain over the exact same spot on Earth. (**a**) What is the period of this orbit? (**b**) At what distance from the center of the Earth must such a satellite be placed into orbit? (**c**) Explain why the orbit must be in the plane of the Earth's equator.

43. Imagine a planet like the Earth orbiting a star with 4 times the mass of the Sun. If the semimajor axis of the planet's orbit is 1 AU, what would be the planet's sidereal period? (*Hint:* Use Newton's form of Kepler's third law. Compared with the case of the Earth orbiting the Sun, by what factor has the quantity $m_1 + m_2$ changed? Has a changed? By what factor must P^2 change?)

44. The photograph that opens this chapter shows the lunar module *Eagle* in orbit around the Moon after completing the first successful lunar landing in July 1969. (The photograph was taken from the command module *Columbia*, in which the astronauts returned to Earth.) The spacecraft orbited 111 km above the surface of the Moon. Calculate the period of the spacecraft's orbit. See Appendix 3 for relevant data about the Moon.

*45. In Box 4-4 we analyzed the orbit of Jupiter's moon Io. Look up information about the orbits of Jupiter's three other large moons (Europa, Ganymede, and Callisto) in the appendices. Demonstrate that these data are in agreement with Newton's form of Kepler's third law.

*46. Suppose a newly discovered asteroid is in a circular orbit with synodic period 1.25 years. The asteroid lies between the orbits of Mars and Jupiter. (**a**) Find the sidereal period of the orbit. (**b**) Find the distance from the asteroid to the Sun.

47. The average distance from the Moon to the center of the Earth is 384,400 km, and the diameter of the Earth is 12,756 km. Calculate the gravitational force that the Moon exerts (**a**) on a 1-kg rock at the point on the Earth's surface closest to the Moon and (**b**) on a 1-kg rock at the point on the Earth's surface farthest from the Moon. (**c**) Find the difference between the two forces you calculated in parts (**a**) and (**b**). This difference is the tidal force pulling these two rocks away from each other, like the 1-ball and 3-ball in Figure 4-23. Explain why tidal forces cause only a very small deformation of the Earth.

Discussion Questions

48. Which planet would you expect to exhibit the greatest variation in apparent brightness as seen from Earth? Which planet would you expect to exhibit the greatest variation in angular diameter? Explain your answers.

49. Use two thumbtacks, a loop of string, and a pencil to draw several ellipses. Describe how the shape of an ellipse varies as the distance between the thumbtacks changes.

Web/CD-ROM Questions

50. (**a**) Search the World Wide Web for information about Kepler. Before he realized that the planets move on elliptical paths, what other models of planetary motion did he consider? What was Kepler's idea of "the music of the spheres"? (**b**) Search the World Wide Web for information about Galileo. What were his contributions to physics? Which of Galileo's new ideas were later used by Newton to construct his laws of motion? (**c**) Search the World Wide Web for information about Newton. What were some of the contributions that he made to physics other than developing his laws of motion? What contributions did he make to mathematics?

 51. Monitoring the Retrograde Motion of Mars. Watching Mars night after night reveals that it changes its position with respect to the background stars. To track its motion, access and view the animation "The Path of Mars in 2009–2010" in Chapter 4 of the *Universe* Web site or CD-ROM. (**a**) Through which two constellations does Mars move? (**b**) On approximately what date does Mars stop its direct (west-to-east) motion and begin its retrograde motion? (*Hint:* Use the "Stop" function on your animation controls.) (**c**) Over how many days does Mars move retrograde?

Observing Projects

52. It is quite probable that within a few weeks of your reading this chapter one of the planets will be near opposition or greatest eastern elongation, making it readily visible in the evening sky. Select a planet that is at or near such a configuration by searching the World Wide Web or by consulting a reference book, such as the current issue of the *Astronomical Almanac* or the pamphlet entitled *Astronomical Phenomena* (both published by the U.S. government). At that configuration, would you expect the planet to be moving rapidly or slowly from night to night against the background stars? Verify your expectations by observing the planet once a week for a month, recording your observations on a star chart.

53. If Jupiter happens to be visible in the evening sky, observe the planet with a small telescope on five consecutive clear nights. Record the positions of the four Galilean satellites by making nightly drawings, just as the Jesuit priests did in 1620 (see Figure 4-17). From your drawings, can you tell which moon orbits closest to Jupiter and which orbits farthest? Was there a night when you could see only three of the moons? What do you suppose happened to the fourth moon on that night?

54. If Venus happens to be visible in the evening sky, observe the planet with a small telescope once a week for a month. On each night, make a drawing of the crescent that you see. From your drawings, can you determine if the planet is nearer or farther from the Earth than the Sun is? Do your drawings show any changes in the shape of the crescent from one week to the next? If so, can you deduce if Venus is coming toward us or moving away from us?

 55. Use the *Starry Night Backyard*™ program to observe the moons of Jupiter. Display the entire celestial sphere (select **Guides > Atlas** in the **Go** menu) and center on the planet Jupiter (use **Find...** in the **Edit** menu). Then use the zoom controls at the right-hand end of the Control Panel (at the top of the main window) to adjust your view so that you can see the planet and its four Galilean satellites. (Click on the binoculars to zoom in and on the person to zoom out.) (**a**) Click on the control for the rate of time flow (immediately to the right of the date and time display) and set the discrete time step to 1 hour. Using the "Step time forward" button (just to the right of the "Forward" button), observe and draw the positions of the moons at 1-hour intervals. (**b**) From your drawings, can you tell which moon orbits closest to Jupiter and which orbits farthest away? Explain your reasoning. (**c**) Are there times when only three of the satellites are visible? What happens to the fourth moon at those times?

 56. Use the *Starry Night Backyard*™ program to observe the changing appearance of Mercury. Display the entire celestial sphere (select **Guides > Atlas** in the **Go** menu) and center on Mercury (use **Find...** in the **Edit** menu), then use the zoom controls at the right-hand end of the Control Panel (at the top of the main window) to adjust your view so that you can clearly see details on the planet's surface. (Click on the binoculars to zoom in and on the person to zoom out.) (**a**) Click on the control for the rate of time flow (immediately to the right of the date and time display) and set the discrete time step to 1 day. Using the "Step time forward" button, observe and record the changes in Mercury's phase and apparent size from one day to the next. (**b**) Explain why the phase and apparent size change in the way that you observe.

 57. Use the *Starry Night Backyard*™ program to observe retrograde motion. Display the entire celestial sphere (select **Guides > Atlas** in the **Go** menu) and center on Mars (use **Find...** in the **Edit** menu). Set the date (in the time display at the upper left of the main window) to January 1, 2005. Then click on the control for the rate of time flow (immediately to the right of the date and time display) and set the discrete time step to 1 day. Then click on the "Forward" button (a triangle that points to the right) and watch the motion of Mars for two years of simulated time. (**a**) For most of the two-year period, does Mars move generally to the left (eastward) or to the right (westward) on the celestial sphere? On what date during this period does this *direct* motion end, so that Mars appears to come to a momentary halt, and *retrograde* motion begin? On what date does retrograde motion end and forward motion begin again? You may want to use the "Stop" button (a black square) and the "Backward" button (a triangle that points to the left) to help you pin down the

exact dates of these events. (**b**) You have been observing the motion of Mars as seen from Earth. To observe the motion of Earth as seen from Mars, select **Viewing Location...** in the **View** menu. In the box that appears on the screen, change the pull-down menus next to the words "View from:" so that they read "the surface of" and "Mars." Then click on the "Set Location" button. Now center the field of view on the Earth (use **Find...** in the **Edit** menu), again set the date to January 1, 2005, set the discrete time step to 1 day, and click on the "Forward" button. As before, watch the motion for two years of simulated time. In which direction does the Earth appear to move most of the time? On what date does its motion change from direct to retrograde? On what date does its motion change from retrograde back to direct? Are these roughly the same dates as you found in part (a)? (**c**) To understand the motions of Mars as seen from Earth and vice versa, observe the motion of the planets from a point above the solar system. Again select **Viewing Location...** in the **View** menu. In the box that appears on the screen, select "stationary location" from the pull-down menu immediately to the right of the words "View from." Then type **0 au** in the X: box, **0 au** in the Y: box, and **5 au** in the Z: box, and click on the "Set Location" button. This will put you at a position 5 AU above the plane of the solar system. Then center the field of view on the Sun (use **Find...** in the **Edit** menu). Once again set the date to January 1, 2005, set the discrete time step to 1 day, and click on the "Forward" button. Watch the motions of the planets for two years of simulated time. On what date during this two-year period is Earth closest to Mars? How does this date compare to the two dates you recorded in part (a) and the two dates you recorded in part (b)? Explain the significance of this. (*Hint:* see Figure 4-5.)

 58. Use the *Deep Space Explorer*™ program to examine the orbits of the planets. In the left-hand part of the window under the heading **Solar System,** click on the triangle next to the word **Explore** and then click on **Inner Solar System.** You can zoom in and zoom out using the buttons at the upper left of the window (an upward-pointing triangle and a downward-pointing triangle). You can also rotate the solar system by putting the mouse cursor over the image, holding down the mouse button, and moving the mouse. (On a two-button mouse, hold down the left mouse button.) (**a**) Rotate the solar system so that you are viewing the orbit of Earth (in green) face-on. Do all of the orbits of the inner planets Mercury, Venus, Earth, and Mars appear to be circles centered on the Sun? Which of these orbits are the most noncircular? (A good way to tell is to hold a ruler next to the screen and measure the distance from the Sun to various points on each planet's orbit. If the orbit is circular, all points on the orbit will be the same distance from the Sun. You can zoom in as needed to get a better view of the orbits.) (**b**) Rotate the solar system so that you are viewing the orbit of Earth (in green) edge-on. Do the other planets all orbit in exactly the same plane as the Earth? If not, are most of the other orbital planes tilted at a relatively small angle or a relatively large angle? Which planet's orbit is tilted at the greatest angle to the Earth's orbital plane? (You may want to zoom out until you can see the orbits of all the planets.)

5 The Nature of Light

In the early 1800s, the French philosopher Auguste Comte argued that because the stars are so far away, humanity would never know their nature and composition. But the means to learn about the stars was already there for anyone to see—starlight. Just a few years after Comte's bold pronouncement, scientists began analyzing starlight to learn the very things that he had deemed unknowable.

We now know that atoms of each chemical element emit and absorb light at a unique set of wavelengths characteristic of that element alone. Some of these atomic "fingerprints" are shown in the accompanying photo, along with the spectra of light from the Sun and from ordinary lamps. By looking for these wavelengths in the light from celestial objects, astronomers have learned that the same chemical elements found on Earth are also found in nearby planets, distant stars, and remote galaxies.

In this chapter we learn about the basic properties of light. Light is a form of energy with the properties of both waves and particles. The light emitted by an object depends upon the object's temperature; we can use this to determine the surface temperatures of stars. By studying the structure of atoms, we will learn why each element emits and absorbs light only at specific wavelengths and will see how astronomers determine what the atmo-

spheres of planets and stars are made of. The motion of a light source also affects wavelengths, permitting us to deduce how fast stars and other objects are approaching or receding. These are but a few of the reasons why understanding light is a prerequisite to understanding the universe.

 The light spectra of different substances and different light sources. (Courtesy of Bausch and Lomb)

Questions to Guide Your Reading

As you read the sections of this chapter, look for the answers to the following questions:

5-1 How fast does light travel? How can this speed be measured?

5-2 Why do we think light is a wave? What kind of wave is it?

5-3 How is the light from an ordinary lightbulb different from the light emitted by a neon sign?

5-4 How can astronomers measure the temperatures of the Sun and stars?

5-5 What is a photon? How does an understanding of photons help explain why ultraviolet light causes sunburns?

5-6 How can astronomers tell what distant celestial objects are made of?

5-7 What are atoms made of?

5-8 How does the structure of atoms explain what kind of light those atoms can emit or absorb?

5-9 How can we tell if a star is approaching us or receding from us?

5-1 | Light travels through empty space at a speed of 300,000 km/s

Galileo Galilei and Isaac Newton were among the first to ask basic questions about light. Does light travel instantaneously from one place to another, or does it move with a measurable speed? Whatever the nature of light, it does seem to travel swiftly from a source to our eyes. We see a distant event before we hear the accompanying sound. (For example, we see a flash of lightning before we hear the thunderclap.)

In the early 1600s, Galileo tried to measure the speed of light. He and an assistant stood at night on two hilltops a known distance apart, each holding a shuttered lantern. First, Galileo opened the shutter of his lantern; as soon as his assistant saw the flash of light, he opened his own. Galileo used his pulse as a timer to try to measure the time between opening his lantern and seeing the light from his assistant's lantern. From the distance and time, he hoped to compute the speed at which the light had traveled to the distant hilltop and back.

Galileo found that the measured time failed to increase noticeably, no matter how distant the assistant was stationed. Galileo therefore concluded that the speed of light is too high to be measured by slow human reactions. Thus, he was unable to tell whether or not light travels instantaneously.

The first evidence that light does *not* travel instantaneously was presented in 1676 by Olaus Rømer, a Danish astronomer. Rømer had been studying the orbits of the moons of Jupiter by carefully timing the moments when they passed into or out of Jupiter's shadow. To Rømer's surprise, the timing of these eclipses of Jupiter's moons seemed to depend on the relative positions of Jupiter and the Earth. When the Earth was far from Jupiter (that is, near conjunction; see Figure 4-6), the eclipses occurred several minutes later than when the Earth was close to Jupiter (near opposition).

Rømer realized that this puzzling effect could be explained if light needs time to travel from Jupiter to the Earth. When the Earth is closest to Jupiter, the image of a satellite disappearing into Jupiter's shadow arrives at our telescopes a little sooner than it does when Jupiter and the Earth are farther apart (Figure 5-1). The range of variation in the times at which such eclipses are observed is about 16.6 minutes, which Rømer interpreted as the length of time required for light to travel across the diameter of the Earth's orbit (a distance of 2 AU). The size of the Earth's orbit was not accurately known in Rømer's day, and he never actually calculated the speed of light. Today, using the modern value of 150 million kilometers for the astronomical unit, Rømer's method yields a value for the speed of light equal to roughly 300,000 km/s (186,000 mi/s).

Almost two centuries after Rømer, the speed of light was measured very precisely in an experiment carried out on Earth. In 1850, the French physicists Armand-Hippolyte Fizeau and Jean Foucault built the apparatus sketched in Figure 5-2. Light from a light source reflects from a rotating mirror toward a stationary mirror 20 meters away. The rotating mirror moves slightly while the light is making the round trip, so the returning light ray is deflected away from the source by a small angle. By measuring this angle and knowing the dimensions of their apparatus, Fizeau and Foucault could deduce the speed of light. Once again, the answer was very nearly 300,000 km/s.

The speed of light in a vacuum is usually designated by the letter c (from the Latin *celeritas*, meaning "speed"). The modern value is $c = 299,792.458$ km/s (186,282.397 mi/s). In most calculations you can use

$$c = 3.00 \times 10^5 \text{ km/s} = 3.00 \times 10^8 \text{ m/s}$$

The most convenient set of units to use for c is different in different situations. The value in kilometers per second (km/s) is often

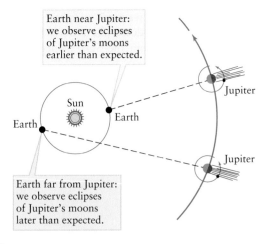

figure 5-1

Rømer's Proof That Light Does Not Travel Instantaneously The timing of eclipses of Jupiter's moons as seen from Earth depends on the Earth-Jupiter distance. Rømer correctly attributed this effect to variations in the time required for light to travel from Jupiter to the Earth.

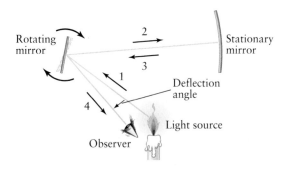

figure 5-2

The Fizeau-Foucault Method of Measuring the Speed of Light Light from a light source (1) is reflected off a rotating mirror to a stationary mirror (2) and from there back to the rotating mirror (3). The ray that reaches the observer (4) is deflected away from the path of the initial beam because the rotating mirror has moved slightly while the light was making the round trip. The speed of light is calculated from the deflection angle and the dimensions of the apparatus.

most useful when comparing c to the speeds of objects in space, while the value in meters per second (m/s) is preferred when doing calculations involving the wave nature of light (which we will discuss in Section 5-2).

Note that the quantity c is the speed of light *in a vacuum*. Light travels more slowly through air, water, glass, or any other transparent substance than it does in a vacuum. In our study of astronomy, however, we will almost always consider light traveling through the vacuum (or near-vacuum) of space.

The speed of light in empty space is one of the most important numbers in modern physical science. This value appears in many equations that describe atoms, gravity, electricity, and magnetism. According to Einstein's special theory of relativity, nothing can travel faster than the speed of light.

5-2 | Light is electromagnetic radiation and is characterized by its wavelength

 SCIENCE IN PROCESS *How scientists discovered the nature of color and the wave nature of light*

Light is energy. This fact is apparent to anyone who has felt the warmth of the sunshine on a summer's day. But what exactly is light? How is it produced? What is it made of? How does it move through space? Scholars have struggled with these questions throughout history.

The first major breakthrough in understanding light came from a simple experiment performed by Isaac Newton around 1670. Newton was familiar with what he called the "celebrated Phenomenon of Colours," in which a beam of sunlight passing through a glass prism spreads out into the colors of the rainbow (Figure 5-3). This rainbow is called a **spectrum** (plural **spectra**).

Until Newton's time, it was thought that a prism somehow added colors to white light. To test this idea, Newton placed a second prism so that just one color of the spectrum passed through it (Figure 5-4). According to the old theory,

this should have caused a further change in the color of the light. But Newton found that each color of the spectrum was unchanged by the second prism; red remained red, blue remained blue, and so on. He concluded that a prism merely separates colors and does not add color. Hence, the spectrum produced by the first prism shows that sunlight is a mixture of all the colors of the rainbow.

Newton suggested that light is composed of particles too small to detect individually. In 1678, however, the Dutch physicist and astronomer Christiaan Huygens proposed a rival explanation. He suggested that light travels in the form of waves rather than particles.

Around 1801, Thomas Young in England carried out an experiment that convincingly demonstrated the wavelike aspect of light. He passed a beam of light through two thin, parallel slits in an opaque screen, as shown in Figure 5-5a. On a white surface some distance beyond the slits, the light formed a pattern of alternating bright and dark bands. Young reasoned that if a beam of light was a stream of particles (as Newton had suggested), the two beams of light from the slits should simply form bright images of the slits on the white surface. The pattern of bright and dark bands he observed is just what would be expected, however, if light had wavelike properties. An analogy with water waves demonstrates why.

ANALOGY Imagine ocean waves pounding against a reef or breakwater that has two openings (Figure 5-5b). A pattern of ripples is formed on the other side of the barrier as the waves come through the two openings and interfere with each other. At certain points, wave crests arrive simultaneously from the two openings. These reinforce each other and produce high waves. At other points, a crest from one opening meets a trough from the other opening. These cancel each other out, leaving areas of still water. This process of combining two waves also takes place in Young's double-slit experiment: The bright bands are regions where waves from the two slits reinforce each other, while the dark bands appear where waves from the two slits cancel each other.

The discovery of the wave nature of light posed some obvious questions. What exactly is "waving" in light? That is, what is it about light that goes up and down like water waves on the ocean? Because we can see light from the Sun, planets, and stars, light

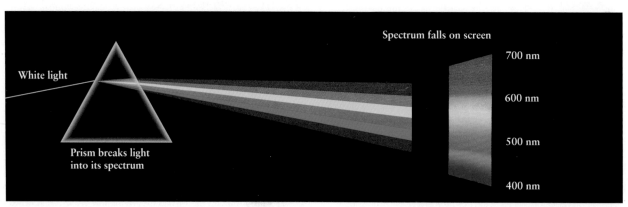

Figure 5-3

A Prism and a Spectrum When a beam of sunlight passes through a glass prism, the light is broken into a rainbow-colored band called a spectrum. The numbers on the right side of the spectrum indicate the wavelengths of different colors of light.

White light

First prism breaks
light into its spectrum

Screen lets only one
color pass through

Second prism bends light,
but does not change its color

figure 5-4

Newton's Experiment on the Nature of Light In a crucial experiment, Newton took sunlight that had passed through a prism and sent it through a second prism. Between the two prisms was a screen with a hole in it that allowed only one color of the spectrum to pass through. This same color emerged from the second prism. Newton's experiment proved that prisms do not add color to light but merely bend different colors through different angles. It also proved that white light, such as sunlight, is actually a combination of all the colors that appear in its spectrum.

waves must be able to travel across empty space. Hence, whatever is "waving" cannot be any material substance. What, then, is it?

The answer came from a seemingly unlikely source—a comprehensive theory that described electricity and magnetism. Numerous experiments during the first half of the nineteenth century demonstrated an intimate connection between electric and magnetic forces. A central idea to emerge from these experiments is the concept of a *field*, an immaterial yet measurable disturbance of any region of space in which electric or magnetic forces are felt. Thus, an electric charge is surrounded by an electric field, and a magnet is surrounded by a magnetic field. Experiments in the early 1800s demonstrated that moving an electric charge produces a magnetic field; conversely, moving a magnet gives rise to an electric field.

In the 1860s, the Scottish mathematician and physicist James Clerk Maxwell succeeded in describing all the basic properties of electricity and magnetism in four equations. This mathematical achievement demonstrated that electric and magnetic forces are really two aspects of the same phenomenon, which we now call **electromagnetism.**

By combining his four equations, Maxwell showed that electric and magnetic fields should travel through space in the form of waves at a speed of 3.0×10^5 km/s—a value exactly equal to the best

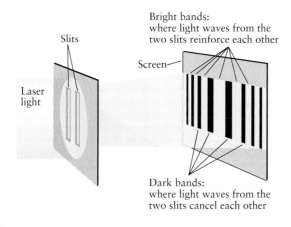

Slits

Laser
light

Screen

Bright bands:
where light waves from the
two slits reinforce each other

Dark bands:
where light waves from the
two slits cancel each other

(a)

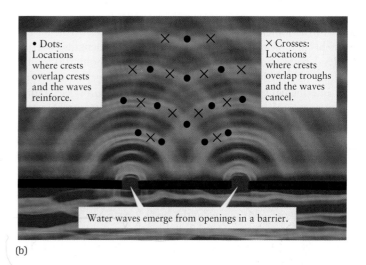

• Dots:
Locations
where crests
overlap crests
and the waves
reinforce.

✕ Crosses:
Locations
where crests
overlap troughs
and the waves
cancel.

Water waves emerge from openings in a barrier.

(b)

figure 5-5

Young's Double-Slit Experiment (a) Thomas Young's classic double-slit experiment can easily be repeated in the modern laboratory by shining light from a laser onto two closely spaced parallel slits. Alternating dark and bright bands appear on a screen beyond the slits. (b) The intensity of light on the screen in (a) is analogous to the height of water waves that pass through a barrier with two openings. (The photograph shows this experiment with water waves in a small tank.) In certain locations, wave crests from both openings reinforce each other to produce extra high waves. At other locations a crest from one opening meets a trough from the other. The crest and trough cancel each other, producing still water. (Eric Schrempp/Photo Researchers)

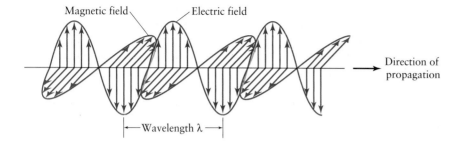

Magnetic field   Electric field

Direction of propagation

←— Wavelength λ —→

figure 5-6

Electromagnetic Radiation All forms of light consist of oscillating electric and magnetic fields that move through space at a speed of 3.00×10^5 km/s $= 3.00 \times 10^8$ m/s. This figure shows a "snapshot" of these fields at one instant. The distance between two successive crests, called the wavelength of the light, is usually designated by the Greek letter λ (lambda).

available value for the speed of light. Maxwell's suggestion that these waves do exist and are observed as light was soon confirmed by experiments. Because of its electric and magnetic properties, light is also called **electromagnetic radiation.**

⚠ **CAUTION** You may associate the term *radiation* with radioactive materials like uranium, but this term refers to anything that radiates, or spreads away, from its source. For example, scientists sometimes refer to sound waves as "acoustic radiation." Radiation does not have to be related to radioactivity!

Electromagnetic radiation consists of oscillating electric and magnetic fields, as shown in Figure 5-6. The distance between two successive wave crests is called the **wavelength** of the light, usually designated by the Greek letter λ (lambda). No matter what the wavelength, electromagnetic radiation always travels at the same speed $c = 3.0 \times 10^5$ km/s $= 3.0 \times 10^8$ m/s in a vacuum.

More than a century elapsed between Newton's experiments with a prism and the confirmation of the wave nature of light. One reason for this delay is that **visible light,** the light to which the human eye is sensitive, has extremely short wavelengths—less than a thousandth of a millimeter—that are not easily detectable. To express such tiny distances conveniently, scientists use a unit of length called the **nanometer** (abbreviated nm), where 1 nm $= 10^{-9}$ m. Experiments demonstrated that visible light has wavelengths covering the range from about 400 nm for violet light to about 700 nm for red light. Intermediate colors of the rainbow like yellow (550 nm) have intermediate wavelengths, as shown in Figure 5-7. (Some astronomers prefer to measure wavelengths in *angstroms*. One angstrom, abbreviated Å, is one-tenth of a nanometer: 1 Å $= 0.1$ nm $= 10^{-10}$ m. In these units, the wavelengths of visible light extend from about 4000 Å to about 7000 Å. We will not use these units in this book, however.)

Maxwell's equations place no restrictions on the wavelength of electromagnetic radiation. Hence, electromagnetic waves could and should exist with wavelengths both longer and shorter than the 400–700 nm range of visible light. Consequently, researchers began to look for *invisible* forms of light. These are forms of electromagnetic radiation to which the cells of the human retina do not respond.

The first kind of invisible radiation to be discovered actually preceded Maxwell's work by more than a half century. Around 1800 the British astronomer William Herschel passed sunlight through a prism and held a thermometer just beyond the red end of the visible spectrum. The thermometer registered a temperature increase, indi-

cating that it was being exposed to an invisible form of energy. This invisible energy, now called **infrared radiation,** was later realized to be electromagnetic radiation with wavelengths somewhat longer than those of visible light.

In experiments with electric sparks in 1888, the German physicist Heinrich Hertz succeeded in producing electromagnetic radia-

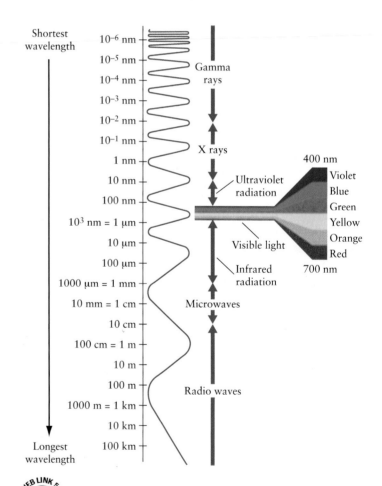

figure 5-7

The Electromagnetic Spectrum The full array of all types of electromagnetic radiation is called the electromagnetic spectrum. It extends from the longest-wavelength radio waves to the shortest-wavelength gamma rays. Visible light occupies only a tiny portion of the full electromagnetic spectrum.

tion with even longer wavelengths of a few centimeters or more. These are now known as **radio waves.** In 1895 another German physicist, Wilhelm Röntgen, invented a machine that produces electromagnetic radiation with wavelengths shorter than 10 nm, now known as **X rays.** The X-ray machines in modern medical and dental offices are direct descendants of Röntgen's invention. Over the years radiation has been discovered with many other wavelengths.

Thus, visible light occupies only a tiny fraction of the full range of possible wavelengths, collectively called the **electromagnetic spectrum.** As Figure 5-7 shows, the electromagnetic spectrum stretches from the longest-wavelength radio waves to the shortest-wavelength gamma rays.

On the long-wavelength side of the visible spectrum, infrared radiation covers the range from about 700 nm to 1 mm. Astronomers interested in infrared radiation often express wavelength in *micrometers* or *microns*, abbreviated μm, where 1 μm = 10^{-3} mm = 10^{-6} m. **Microwaves** have wavelengths from roughly 1 mm to 10 cm, while radio waves have even longer wavelengths.

At wavelengths shorter than those of visible light, **ultraviolet radiation** extends from about 400 nm down to 10 nm. Next are X rays, which have wavelengths between about 10 and 0.01 nm, and beyond them at even shorter wavelengths are **gamma rays.** Note that these rough boundaries are simply arbitrary divisions in the electromagnetic spectrum.

Astronomers who work with radio telescopes often prefer to speak of *frequency* rather than wavelength. The **frequency** of a wave is the number of wave crests that pass a given point in one second. Equivalently, it is the number of complete *cycles* of the wave that pass per second (a complete cycle is from one crest to the next). Frequency is usually denoted by the Greek letter ν (nu). The unit of frequency is the cycle per second, also called the hertz (abbreviated Hz) in honor of Heinrich Hertz, the physicist who first produced radio waves. For example, if 500 crests of a wave pass you in one second, the frequency of the wave is 500 cycles per second or 500 Hz.

In working with frequencies, it is often convenient to use the prefix *mega-* (meaning "million," or 10^6, and abbreviated M) or *kilo-* (meaning "thousand," or 10^3, and abbreviated k). For example, AM radio stations broadcast at frequencies between 535 and 1605 kHz (kilohertz), while FM radio stations broadcast at frequencies in the range from 88 to 108 MHz (megahertz).

The relationship between the frequency and wavelength of an electromagnetic wave is a simple one. Because light moves at a constant speed $c = 3 \times 10^8$ m/s, if the wavelength (distance from one crest to the next) is made shorter, the frequency must increase (more of those closely spaced crests pass you each second). Mathematically, the frequency ν of light is related to its wavelength λ by

Frequency and wavelength of an electromagnetic wave

$$\nu = \frac{c}{\lambda}$$

ν = frequency of an electromagnetic wave (in Hz)

c = speed of light = 3×10^8 m/s

λ = wavelength of the wave (in meters)

For example, hydrogen atoms in space emit radio waves with a wavelength of 21.12 cm. To calculate the frequency of this radiation, we must first express the wavelength in meters rather than centimeters: λ = 0.2112 m. Then we can use the above formula to find the frequency ν:

$$\nu = \frac{c}{\lambda} = \frac{3 \times 10^8 \text{ m/s}}{0.2112 \text{ m}} = 1.42 \times 10^9 \text{ Hz} = 1420 \text{ MHz}$$

Visible light has a much shorter wavelength and higher frequency than radio waves. You can use the above formula to show that for yellow-orange light of wavelength 600 nm, the frequency is 5×10^{14} Hz or 500 *million* megahertz!

While Young's experiment (Figure 5-5) showed convincingly that light has wavelike aspects, it was discovered in the early 1900s that light *also* has some of the characteristics of a stream of particles and waves. We will explore light's dual nature in Section 5-5.

5-3 | An opaque object emits electromagnetic radiation according to its temperature

To learn about objects in the heavens, astronomers study the character of the electromagnetic radiation coming from those objects. Such studies can be very revealing because different kinds of electromagnetic radiation are typically produced in different ways. As an example, on Earth the most common way to generate radio waves is to make an electric current oscillate back and forth (as is done in the broadcast antenna of a radio station). By contrast, X rays for medical and dental purposes are usually produced by bombarding atoms with fast-moving particles extracted from within other atoms. Our own Sun emits radio waves from near its glowing surface and X rays from its corona (see the photo that opens Chapter 3). Hence, these observations indicate the presence of electric currents near the Sun's surface and of fast-moving particles in the Sun's outermost regions. (We will discuss the Sun at length in Chapter 18.)

The simplest and most common way to produce electromagnetic radiation, either on or off the Earth, is to heat an object. The hot filament of wire inside an ordinary lightbulb emits white light, and a neon sign has a characteristic red glow because neon gas within the tube is heated by an electric current. In like fashion, almost all the visible light that we receive from space comes from hot objects like the Sun and the stars. The kind and amount of light emitted by a hot object tell us not only how hot it is but also about other properties of the object.

We can tell whether the hot object is made of relatively dense or relatively thin material. Consider the difference between a lightbulb and a neon sign. The dense, solid filament of a lightbulb makes white light, which is a mixture of all different visible wavelengths, while the thin, transparent neon gas produces light of a rather definite red color and, hence, a rather definite wavelength. For now we will concentrate our attention on the light produced by dense, opaque objects. (We will return to the light produced by gases in Section 5-6.) Even though the Sun and stars are gaseous, not solid, it turns out that they emit light with many of the same properties as light emitted by a hot, glowing, solid object.

Hot:
glows deep red

(a)

Hotter:
glows reddish-orange

(b)

Even hotter:
glows yellowish-white

(c)

figure 5-8

Heating a Bar of Iron This sequence of drawings shows how the appearance of a heated bar of iron changes with temperature. As the temperature increases, the bar glows more brightly because it radiates more energy. The color of the bar also changes because the dominant wavelength of light emitted by the bar decreases as the temperature goes up.

Imagine a welder or blacksmith heating a bar of iron. As the bar becomes hot, it begins to glow deep red, as shown in Figure 5-8a. (You can see this same glow from the coils of a toaster or from an electric range turned on "high.") As the temperature rises further, the bar begins to give off a brighter, reddish-orange light (Figure 5-8b). At still higher temperatures, it shines with a brilliant yellowish-white light (Figure 5-8c). If the bar could be prevented from melting and vaporizing, at extremely high temperatures it would emit a dazzling blue-white light.

As this example shows, the amount of energy emitted by the hot, dense object and the dominant wavelength of the emitted radiation both depend on the temperature of the object. The hotter the object, the more energy it emits and the shorter the wavelength at which most of the energy is emitted. Colder objects emit relatively little energy, and this emission is primarily at long wavelengths.

These observations explain why you can't see in the dark. The temperatures of people, animals, and furniture are rather less than even that of the iron bar in Figure 5-8a. So, while these objects emit radiation even in a darkened room, most of this emission is at wavelengths greater than those of red light, in the infrared part of the

spectrum (see Figure 5-7). Your eye is not sensitive to infrared, and you thus cannot see ordinary objects in a darkened room. But you can detect this radiation by using a camera that is sensitive to infrared light (Figure 5-9).

To better understand the relationship between the temperature of a dense object and the radiation it emits, it is helpful to know just what "temperature" means. The temperature of a substance is directly related to the average speed of the tiny **atoms**—the building blocks that come in distinct forms for each distinct chemical element—that make up the substance. (Typical atoms are about 10^{-10} m = 0.1 nm in diameter, or about 1/5000 as large as a typical wavelength of visible light.)

If something is hot, its atoms are moving at high speeds; if it is cold, its atoms are moving slowly. Scientists usually prefer to use the Kelvin temperature scale, on which temperature is measured in **kelvins** (K) upward from **absolute zero.** This is the coldest possible temperature, at which atoms move as slowly as possible (they can never quite stop completely). On the more familiar Celsius and Fahrenheit temperature scales, absolute zero (0 K) is –273°C and –460°F. Ordinary room temperature is 293 K, 20°C, or 68°F. Box 5-1 discusses the relationships among the Kelvin, Celsius, and Fahrenheit temperature scales.

figure 5-9 R ▮ V U X G

An Infrared Portrait In this image made with a camera sensitive to infrared radiation, the different colors represent regions of different temperature. Red areas (like the man's face) are the warmest and emit the most infrared light, while blue-green areas (including the man's hands and hair) are at the lowest temperatures and emit the least radiation. (Dr. Arthur Tucker/Photo Researchers)

box 5-1 TOOLS OF THE ASTRONOMER'S TRADE

Temperatures and Temperature Scales

Three temperature scales are in common use. Throughout most of the world, temperatures are expressed in **degrees Celsius** (°C). The Celsius temperature scale is based on the behavior of water, which freezes at 0°C and boils at 100°C at sea level on Earth. This scale is named after the Swedish astronomer Anders Celsius, who proposed it in 1742.

Astronomers usually prefer the Kelvin temperature scale. This is named after the nineteenth-century British physicist Lord Kelvin, who made many important contributions to our understanding of heat and temperature. Absolute zero, the temperature at which atomic motion is at the absolute minimum, is –273°C in the Celsius scale but 0 K in the Kelvin scale. Atomic motion cannot be any less than the minimum, so nothing can be colder than 0 K; hence, there are no negative temperatures on the Kelvin scale. Note that we do *not* use degree (°) with the Kelvin temperature scale.

A temperature expressed in kelvins is always equal to the temperature in degrees Celsius plus 273. On the Kelvin scale, water freezes at 273 K and boils at 373 K. Water must be heated through a change of 100 K or 100°C to go from its freezing point to its boiling point. Thus, the "size" of a kelvin is the same as the "size" of a Celsius degree. When considering temperature changes, measurements in kelvins and Celsius degrees are the same.

The now-archaic Fahrenheit scale, which expresses temperature in **degrees Fahrenheit** (°F), is used only in the United States. When the German physicist Gabriel Fahrenheit introduced this scale in the early 1700s, he intended 100°F to represent the temperature of a healthy human body. On the Fahrenheit scale, water freezes at 32°F and boils at 212°F. There are 180 Fahrenheit degrees between the freezing and boiling points of water, so a degree Fahrenheit is only 100/180 = 5/9 as large as either a Celsius degree or a kelvin.

Two simple equations allow you to convert a temperature from the Celsius scale to the Fahrenheit scale and from Fahrenheit to Celsius:

$$T_F = \frac{9}{5}(T_C + 32)$$

$$T_C = \frac{5}{9}(T_F - 32)$$

T_F = temperature in degrees Fahrenheit

T_C = temperature in degrees Celsius

EXAMPLE: A typical room temperature is 68°F. We can convert this to the Celsius scale using the second equation:

$$T_C = \frac{5}{9}(68 - 32) = 20°C$$

To convert this to the Kelvin scale, we simply add 273 to the Celsius temperature. Thus,

$$68°F = 20°C = 293 K$$

The diagram displays the relationships among these three temperature scales.

Figure 5-10 depicts quantitatively how the radiation from a dense object depends on its Kelvin temperature. Each curve in this figure shows the intensity of light emitted at each wavelength by a dense object at a given temperature. In other words, the curves show the spectrum of light emitted by such an object. At any temperature, a hot, dense object emits at all wavelengths, so its spectrum is a smooth, continuous curve with no gaps in it.

The shape of the spectrum depends on temperature, however. An object at relatively low temperature (say, 3000 K) has a low curve,

indicating a low intensity of radiation. The **wavelength of maximum emission**, at which the curve has its peak and the emission of energy is strongest, is at a long wavelength. The higher the temperature, the higher the curve (indicating greater intensity) and the shorter the wavelength of maximum emission.

 Figure 5-10 shows that for a dense object at a temperature of 3000 K, the wavelength of maximum emission is around 1000 nm (1 μm). Because this is an infrared wavelength

The higher the temperature of a blackbody, the shorter the wavelength of maximum emission (the wavelength at which the curve peaks).

Visible light

The higher the temperature of a blackbody, the more light is emitted at *all* wavelengths.

12,000 K

6000 K

3000 K

AIMM 5.1

figure 5-10

Blackbody Curves Each of these curves shows the intensity of light at every wavelength that is emitted by a blackbody (an idealized case of a dense object) at a particular temperature. The rainbow-colored band shows the range of visible wavelengths. The vertical scale has been compressed so that all three curves can be seen; the peak intensity for the 12,000-K curve is actually about 1000 times greater than the peak intensity for the 3000-K curve.

well outside the visible range, you might think that you cannot see the radiation from an object at this temperature. In fact, the glow from such an object *is* visible; the curve shows that this object emits plenty of light within the visible range, as well as at even shorter wavelengths.

The 3000-K curve is quite a bit higher at the red end of the visible spectrum than at the violet end, so a dense object at this temperature will appear red in color. Similarly, the 12,000-K curve has its wavelength of maximum emission in the ultraviolet part of the spectrum, at a wavelength shorter than visible light. But such a hot, dense object also emits copious amounts of visible light (much more than at 6000 K or 3000 K, for which the curves are lower) and thus will have a very visible glow. The curve for this temperature is higher for blue light than for red light, and so the color of a dense object at 12,000 K is a brilliant blue or blue-white. These conclusions agree with the color changes of a heated rod shown in Figure 5-8. The

same principles apply to stars: A star that looks blue has a high surface temperature, while a red star has a relatively cool surface.

These observations lead to a general rule:

The higher an object's temperature, the more intensely the object emits electromagnetic radiation and the shorter the wavelength at which it emits most strongly.

We will make frequent use of this general rule to analyze the temperatures of celestial objects such as planets and stars.

The curves in Figure 5-10 are drawn for an idealized type of dense object called a **blackbody**. A perfect blackbody does not reflect any light at all; instead, it absorbs all radiation falling on it. Because it reflects no electromagnetic radiation, the radiation that it does emit is entirely the result of its temperature. Ordinary objects, like tables, textbooks, and people, are not perfect blackbodies; they reflect light, which is why they are visible. A star such as the Sun, however, behaves very much like a perfect blackbody, because it absorbs almost completely any radiation falling on it from outside. The light emitted by a blackbody is called **blackbody radiation,** and the curves in Figure 5-10 are often called **blackbody curves.**

CAUTION Despite its name, a blackbody does not necessarily *look* black. The Sun, for instance, does not look black because its temperature is high (around 5800 K), and so it glows brightly. But a room-temperature (around 300 K) blackbody would appear very black indeed. Even if it were as large as the Sun, it would emit only about 1/100,000 as much energy. (Its blackbody curve is far too low to graph in Figure 5-10.) Furthermore, most of this radiation would be at wavelengths that are too long for our eyes to perceive.

Figure 5-11 shows the blackbody curve for a temperature of 5800 K. It also shows the intensity curve for light from the Sun, as measured from above the Earth's atmosphere. (This is necessary because the Earth's atmosphere absorbs certain wavelengths.) The peak of both curves is at a wavelength of about 500 nm, near the middle of the visible spectrum. Note how closely the observed intensity curve for the Sun matches the blackbody curve. This is a strong indication that the temperature of the Sun's glowing surface is about 5800 K—a temperature that we can measure across a distance of 150 million kilometers! The close correlation between blackbody curves and the observed intensity curves for most stars is a key reason why astronomers are interested in the physics of blackbody radiation.

Blackbody radiation depends *only* on the temperature of the object emitting the radiation, not on the chemical composition of the object. The light emitted by molten gold at 2000 K is very nearly the same as that emitted by molten lead at 2000 K. Therefore, it might seem that analyzing the light from the Sun or from a star can tell astronomers the object's temperature but not what the star is made of. As Figure 5-11 shows, however, the intensity curve for the Sun (a typical star) is not precisely that of a blackbody. We will see later in this chapter that the *differences* between a star's spectrum and that of a blackbody allow us to determine the chemical composition of the star.

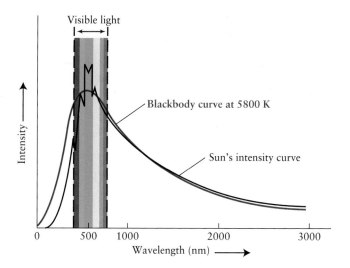

Figure 5-11

The Sun as a Blackbody This graph shows that the intensity of sunlight over a wide range of wavelengths (black curve) is a remarkably close match to the intensity of radiation coming from a blackbody at a temperature of 5800 K (red curve). The measurements of the Sun's intensity were made above the Earth's atmosphere (which absorbs certain wavelengths of sunlight). It's not surprising that the Sun's spectrum peaks at visible wavelengths; the human eye evolved to take advantage of the most plentiful light available.

5-4 | Wien's law and the Stefan-Boltzmann law are useful tools for analyzing glowing objects like stars

The mathematical formula that describes the blackbody curves in Figure 5-10 is a rather complicated one. But there are two simpler formulas for blackbody radiation that prove to be very useful in many branches of astronomy. They are used by astronomers who investigate the stars as well as by those who study the planets (which are dense, relatively cool objects that emit infrared radiation). One of these formulas relates the temperature of a blackbody to its wavelength of maximum emission, and the other relates the temperature to the amount of energy that the blackbody emits.

Figure 5-10 shows that the higher the temperature (T) of a blackbody, the shorter its wavelength of maximum emission (λ_{max}). In 1893 the German physicist Wilhelm Wien used ideas about both heat and electromagnetism to make this relationship quantitative. The formula that he derived, which today is called **Wien's law**, is

Wien's law for a blackbody

$$\lambda_{max} = \frac{0.0029 \text{ K m}}{T}$$

λ_{max} = wavelength of maximum emission of the object (in meters)

T = temperature of the object (in kelvins)

According to Wien's law, the wavelength of maximum emission of a blackbody is inversely proportional to its temperature in kelvins. In other words, if the temperature of the blackbody doubles, its wavelength of maximum emission is halved, and vice versa. For example, Figure 5-10 shows blackbody curves for temperatures of 3000 K, 6000 K, and 12,000 K. From Wien's law, a blackbody with a temperature of 6000 K has a wavelength of maximum emission λ_{max} = (0.0029 K m)/(6000 K) = 4.8×10^{-7} m = 480 nm, in the visible part of the electromagnetic spectrum. At 12,000 K, or twice the temperature, the blackbody has a wavelength of maximum emission half as great, or λ_{max} = 240 nm; this is in the ultraviolet. At 3000 K, just half our original temperature, the value of λ_{max} is twice the original value—960 nm, which is an infrared wavelength. You can see that these wavelengths agree with the peaks of the curves in Figure 5-10.

Remember that Wien's law involves the wavelength of maximum emission in *meters*. If you want to convert the wavelength to nanometers, you must multiply the wavelength in meters by $(10^9 \text{ nm})/(1 \text{ m})$.

Wien's law is very useful for determining the surface temperatures of stars. It is not necessary to know how far away the star is, how large it is, or how much energy it radiates into space. All we need to know is the dominant wavelength of the star's electromagnetic radiation.

The other useful formula for the radiation from a blackbody involves the total amount of energy the blackbody radiates at all wavelengths. (By contrast, the curves in Figure 5-10 show how much energy a blackbody radiates at each individual wavelength.)

Energy is usually measured in **joules** (J), named after the nineteenth-century English physicist James Joule. A joule is the amount of energy contained in the motion of a 2-kilogram mass moving at a speed of 1 meter per second. The joule is a convenient unit of energy because it is closely related to the familiar **watt** (W): 1 watt is 1 joule per second, or 1 W = 1 J/s = 1 J s^{-1}. (The superscript −1 means you are dividing by that quantity.) For example, a 100-watt lightbulb uses energy at a rate of 100 joules per second, or 100 J/s. The energy content of food is also often measured in joules; in most of the world, diet soft drinks are labeled as "low joule" rather than "low calorie."

The amount of energy emitted by a blackbody depends both on its temperature and on its surface area. This makes sense: A large burning log radiates much more heat than a burning match, even though the temperatures are the same. To consider the effects of temperature alone, it is convenient to look at the amount of energy emitted from each square meter of an object's surface in a second. This quantity is called the **energy flux** (F). Flux means "rate of flow," and thus F is a measure of how rapidly energy is flowing out of the object. It is measured in joules per square meter per second, usually written as J/m²/s or J m^{-2} s^{-1}. Alternatively, because 1 watt equals 1 joule per second, we can express flux in watts per square meter (W/m², or W m^{-2}).

In 1879 the Austrian physicist Josef Stefan reported experimental evidence that the flux from a blackbody is proportional to the fourth power of the object's temperature (measured in kelvins). Five years after Stefan announced his law, another Austrian physicist,

box 5-2 TOOLS OF THE ASTRONOMER'S TRADE

Using the Laws of Blackbody Radiation

The Sun and stars behave like nearly perfect blackbodies. Wien's law and the Stefan-Boltzmann law can therefore be used to relate the surface temperature of the Sun or a distant star to the energy flux and wavelength of maximum emission of its radiation. The following examples show how to do this.

EXAMPLE: The maximum intensity of sunlight is at a wavelength of roughly 500 nm = 5.0×10^{-7} m. Use this information to determine the surface temperature of the Sun.

Situation: We are given the Sun's wavelength of maximum emission λ_{max}, and our goal is to find the Sun's surface temperature, denoted by $T_\odot$. (The symbol $\odot$ is the standard astronomical symbol for the Sun.)

Tools: We use Wien's law to relate the values of λ_{max} and $T_\odot$.

Answer: As written, Wien's law tells how to find λ_{max} if we know the surface temperature. To find the surface temperature from λ_{max}, we first rearrange the formula, then substitute the value of λ_{max}:

$$T_\odot = \frac{0.0029 \text{ K m}}{\lambda_{max}} = \frac{0.0029 \text{ K m}}{5.0 \times 10^{-7} \text{ m}} = 5800 \text{ K}$$

Review: This is a very high temperature by Earth standards, about the same as an iron welding arc.

EXAMPLE: Using detectors above the Earth's atmosphere, astronomers have measured the average flux of solar energy arriving at Earth. This value, called the **solar constant,** is equal to 1370 W m^{-2}. Use this information to calculate the Sun's surface temperature. (This provides a check on our result from the preceding example.)

Situation: The solar constant is the flux of sunlight as measured at the Earth. We want to use the value of the solar constant to calculate $T_\odot$.

Tools: It may seem that all we need is the Stefan-Boltzmann law, which relates flux to surface temperature. However, the quantity F in this law refers to the flux measured at the Sun's surface, *not* at the Earth. Hence we will first need to calculate F from the given information.

Answer: To determine the value of F, we first imagine a huge sphere of radius 1 AU with the Sun at its center, as shown in the figure. Each square meter of that sphere receives 1370 watts of power from the Sun, so the total energy radiated by the Sun per second is equal to the solar constant multiplied by the sphere's surface area. The result, called the **luminosity** of the Sun and denoted by the symbol $L_\odot$, is

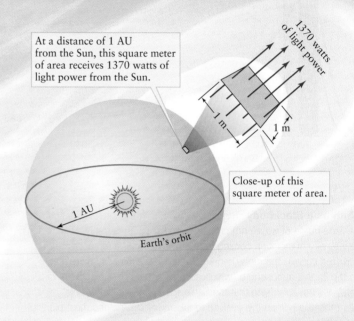

At a distance of 1 AU from the Sun, this square meter of area receives 1370 watts of light power from the Sun.

1370 watts of light power

1 m

1 m

Close-up of this square meter of area.

1 AU

Earth's orbit

$L_\odot = 3.90 \times 10^{26}$ W. That is, in 1 second the Sun radiates 3.90×10^{26} joules of energy into space. Because we know the size of the Sun, we can compute the energy flux (energy emitted per square meter per second) at its surface. The radius of the Sun is $R_\odot = 6.96 \times 10^8$ m, and the Sun's surface area is $4\pi R_\odot^2$. Therefore, its energy flux $F_\odot$ is the Sun's luminosity (total energy emitted by the Sun per second) divided by the Sun's surface area (the number of square meters of surface):

$$F_\odot = \frac{L_\odot}{4\pi R_\odot^2} = \frac{3.90 \times 10^{26} \text{ W}}{4\pi (6.96 \times 10^8 \text{ m})^2} = 6.41 \times 10^7 \text{ W m}^{-2}$$

Once we have the Sun's energy flux $F_\odot$, we can use the Stefan-Boltzmann law to find the Sun's surface temperature $T_\odot$:

$$T_\odot^4 = \frac{F_\odot}{\sigma} = 1.13 \times 10^{15} \text{ K}^4$$

Taking the fourth root (the square root of the square root) of this value, we find the surface temperature of the Sun to be $T_\odot = 5800$ K.

Review: Our result for $T_\odot$ agrees with the value we computed in the previous example using Wien's law. Notice that the solar constant of 1370 W m^{-2} is very much less than $F_\odot$, the flux at the Sun's surface. By the time the Sun's radiation reaches Earth, it is spread over a greatly increased area.

TOOLS OF THE ASTRONOMER'S TRADE continued

EXAMPLE: Sirius, the brightest star in the night sky, has a surface temperature of about 10,000 K. Find the wavelength at which Sirius emits most intensely.

Situation: Our goal is to calculate the wavelength of maximum emission of Sirius (λ_{max}) from its surface temperature T.

Tools: We use Wien's law to relate the values of λ_{max} and T.

Answer: Using Wien's law,

$$\lambda_{max} = \frac{0.0029 \text{ K m}}{T} = \frac{0.0029 \text{ K m}}{10,000 \text{ K}} = 2.9 \times 10^{-7} \text{ m} = 290 \text{ nm}$$

Review: Our result shows that Sirius emits light most intensely in the ultraviolet. In the visible part of the spectrum, it emits more blue light than red light (like the curve for 12,000 K in Figure 5-10), so Sirius has a distinct blue color.

EXAMPLE: How does the energy flux from Sirius compare to the Sun's energy flux?

Situation: To compare the energy fluxes from the two stars, we want to find the *ratio* of the flux from Sirius to the flux from the Sun.

Tools: We use the Stefan-Boltzmann law to find the flux from Sirius and from the Sun, which from the preceding examples have surface temperatures 10,000 K and 5800 K, respectively.

Answer: For the Sun, the Stefan-Boltzmann law is $F_\odot = \sigma T_\odot{}^4$, and for Sirius we can likewise write $F_* = \sigma T_*{}^4$, where the subscripts $\odot$ and $*$ refer to the Sun and Sirius, respectively. If we divide one equation by the other to find the ratio of fluxes, the Stefan-Boltzmann constants cancel out and we get

$$\frac{F_*}{F_\odot} = \frac{T_*{}^4}{T_\odot{}^4} = \frac{(10,000 \text{ K})^4}{(5800 \text{ K})^4} = \left(\frac{10,000}{5800}\right)^4 = 8.8$$

Review: Because Sirius has such a high surface temperature, each square meter of its surface emits 8.8 times more energy per second than a square meter of the Sun's relatively cool surface.

Ludwig Boltzmann, showed how it could be derived mathematically from basic assumptions about atoms and molecules. For this reason, Stefan's law is commonly known as the **Stefan-Boltzmann law.** Written as an equation, the Stefan-Boltzmann law is

Stefan-Boltzmann law for a blackbody

$$F = \sigma T^4$$

F = energy flux, in joules per square meter of surface per second

σ = a constant = 5.67×10^{-8} W m^{-2} K^{-4}

T = object's temperature, in kelvins

The value of the constant σ (the Greek letter sigma) is known from laboratory experiments.

The Stefan-Boltzmann law says that if you double the temperature of an object (for example, from 300 K to 600 K), then the energy emitted from the object's surface each second increases by a factor of $2^4 = 16$. If you increase the temperature by a factor of 10 (for example, from 300 to 3000 K), the rate of energy emission increases by a factor of $10^4 = 10,000$. Thus, a chunk of iron at room temperature (around 300 K) emits very little electromagnetic radiation (and essentially no visible light), but an iron bar heated to 3000 K glows quite intensely.

Box 5-2 gives several examples of applying Wien's law and the Stefan-Boltzmann law to typical astronomical problems.

5-5 | Light has properties of both waves and particles

At the end of the nineteenth century, physicists mounted a valiant effort to explain all the characteristics of blackbody radiation. To this end they constructed theories based on Maxwell's description of light as electromagnetic waves. But all such theories failed to explain the characteristic shapes of blackbody curves shown in Figure 5-10.

In 1900, however, the German physicist Max Planck discovered that he could derive a formula that correctly described blackbody curves if he made two radical assumptions—that electromagnetic energy is emitted in discrete, particlelike packets, and that the energy of each such packet—today called a **photon**—is related to the wavelength of light: The greater the wavelength, the lower the energy of a photon associated with that wavelength. Thus, a photon of red light (wavelength λ = 700 nm) has less energy than a photon of violet light (λ = 400 nm). In this picture, light has a dual personality; it behaves as a stream of particlelike photons, but each photon has wavelike properties. In this sense, the best answer to the question "Is light a wave or a stream of particles?" is "Yes!"

It was soon realized that Planck's photon hypothesis explains more than just the detailed shape of blackbody curves. For example, it explains why only ultraviolet light causes suntans and sunburns. The reason is that tanning or burning involves a chemical reaction in the skin. High-energy, short-wavelength ultraviolet photons can trigger these reactions, but the lower-energy, longer-wavelength photons of visible light cannot. Similarly,

box 5-3 THE HEAVENS ON THE EARTH

Photons at the Supermarket

A beam of light can be regarded as a stream of tiny packets of energy called photons. The Planck relationships $E = hc/\lambda$ and $E = h\nu$ can be used to relate the energy E carried by a photon to its wavelength λ and frequency ν.

As an example, the laser bar-code scanners used at stores and supermarkets emit orange-red light of wavelength 633 nm. To calculate the energy of a single photon of this light, we must first express the wavelength in meters. A nanometer (nm) is equal to 10^{-9} m, so the wavelength is

$$\lambda = (633 \text{ nm})\left(\frac{10^{-9} \text{ m}}{1 \text{ nm}}\right)$$

$$= 633 \times 10^{-9} \text{ m}$$

$$= 6.33 \times 10^{-7} \text{ m}$$

Then, using the Planck formula $E = hc/\lambda$, we find that the energy of a single photon is

$$E = \frac{hc}{\lambda} = \frac{(6.625 \times 10^{-34} \text{ J s})(3 \times 10^8 \text{ m s}^{-1})}{6.33 \times 10^{-7} \text{ m}} = 3.14 \times 10^{-19} \text{ J}$$

This is a very small amount of energy. The laser in a typical bar-code scanner emits 10^{-3} joule of light energy per second, so the number of photons emitted per second is

$$\frac{10^{-3} \text{ joule per second}}{3.14 \times 10^{-19} \text{ joule per photon}} = 3.2 \times 10^{15} \text{ photons per second}$$

This is such a large number that the laser beam seems like a continuous flow of energy rather than a stream of little energy packets.

---

normal photographic film is sensitive to visible light but not to in-frared light; a long-wavelength infrared photon does not have enough energy to cause the chemical change that occurs when film is exposed to the higher-energy photons of visible light.

 Another phenomenon explained by the photon hypothesis is the **photoelectric effect.** In this effect, a metal plate is illuminated by a light beam. If ultraviolet light is used, tiny negatively charged particles called **electrons** are emitted from the metal plate. (We will see in Section 5-7 that the electron is one of the basic particles of the atom.) But if visible light is used, no matter how bright, no electrons are emitted.

In 1905 the great German-born physicist Albert Einstein explained this behavior by noting that a certain minimum amount of energy is required to remove an electron from the metal plate. The energy of a short-wavelength ultraviolet photon is greater than this minimum value, so an electron that absorbs a photon of ultraviolet light will have enough energy to escape from the plate. But an electron that absorbs a photon of visible light, with its longer wavelength and lower energy, does not gain enough energy to escape and so remains within the metal. Einstein and Planck both won Nobel prizes for their contributions to understanding the nature of light.

The relationship between the energy E of a single photon and the wavelength of the electromagnetic radiation can be expressed in a simple equation:

Energy of a photon (in terms of wavelength)

$$E = \frac{hc}{\lambda}$$

E = energy of a photon

h = Planck's constant

c = speed of light

λ = wavelength of light

The value of the constant h in this equation, now called *Planck's constant,* has been shown in laboratory experiments to be

$$h = 6.625 \times 10^{-34} \text{ J s}$$

The units of h are joules multiplied by seconds, called "joule-seconds" and abbreviated J s.

Because the value of h is so tiny, a single photon carries a very small amount of energy. For example, a photon of red light with wavelength 633 nm has an energy of only 3.14×10^{-19} J (**Box 5-3**). This is why we ordinarily do not notice that light comes in the form of photons; even a dim light source emits so many photons per second that it seems to be radiating a continuous stream of energy.

The energies of photons are sometimes expressed in terms of a small unit of energy called the **electron volt** (eV). One electron volt is equal to 1.602×10^{-19} J, so a 633-nm photon has an energy of 1.96 eV. If energy is expressed in electron volts, Planck's constant is best expressed in electron volts multiplied by seconds, abbreviated eV s:

$$h = 4.135 \times 10^{-15} \text{ eV s}$$

Because the frequency ν of light is related to the wavelength λ by $\nu = c/\lambda$, we can rewrite the equation for the energy of a photon as

Energy of a photon (in terms of frequency)

$$E = h\nu$$

E = energy of a photon

h = Planck's constant

ν = frequency of light

The equations $E = hc/\lambda$ and $E = h\nu$ are together called **Planck's law.** Both equations express a relationship between a particlelike property of light (the energy E of a photon) and a wavelike property (the wavelength λ or frequency ν).

The photon picture of light is essential for understanding the detailed shapes of blackbody curves. As we will see, it also helps to explain how and why the spectra of the Sun and stars differ from those of perfect blackbodies.

5-6 | Each chemical element produces its own unique set of spectral lines

In 1814 the German master optician Joseph von Fraunhofer repeated Newton's classic experiment of shining a beam of sunlight through a prism (see Figure 5-3). But this time Fraunhofer subjected the resulting rainbow-colored spectrum to intense magnification. To his surprise, he discovered that the solar spectrum contains hundreds of fine, dark lines, now called **spectral lines.** By

contrast, if the light from a perfect blackbody were sent through a prism, it would produce a smooth, continuous spectrum with no dark lines. Fraunhofer counted more than 600 dark lines in the Sun's spectrum; today we know of more than 30,000. The photograph of the Sun's spectrum in **Figure 5-12** shows hundreds of these spectral lines.

Half a century later, chemists discovered that they could produce spectral lines in the laboratory and use these spectral lines to analyze what kinds of atoms different substances are made of. Chemists had long known that many substances emit distinctive colors when sprinkled into a flame. To facilitate study of these colors, around 1857 the German chemist Robert Bunsen invented a gas burner (today called a Bunsen burner) that produces a clean flame with no color of its own. Bunsen's colleague, the Prussian-born physicist Gustav Kirchhoff, suggested that the colored light produced by substances in a flame might best be studied by passing the light through a prism (**Figure 5-13**). The two scientists promptly discovered that the spectrum from a flame consists of a pattern of thin, bright spectral lines against a dark background. The same kind of spectrum is produced by heated gases such as neon or argon.

Kirchhoff and Bunsen then found that each chemical element produces its own unique pattern of spectral lines. Thus was born in 1859 the technique of **spectral analysis,** the identification of chemical substances by their unique patterns of spectral lines.

A chemical **element** is a fundamental substance that cannot be broken down into more basic chemicals. Some examples are hydrogen, oxygen, carbon, iron, gold, and silver. After Kirchhoff and Bunsen had recorded the prominent spectral lines of all the then-known elements, they soon began to discover other spectral lines in the spectra of vaporized mineral samples. In this way they discovered

Figure 5-12 R I **V** U X G

The Sun's Spectrum Numerous dark spectral lines are seen in this image of the Sun's spectrum. The spectrum is spread out so much that it had to be cut into segments to fit on this page. (N. A. Sharp, NOAO/NSO/ Kitt Peak FTS/AURA/NSF)

1. Add a chemical substance to a flame

2. Send light from the flame through a narrow slit, then through a prism

3. Bright lines in the spectrum show that the substance emits light at specific wavelengths only

Figure 5-13

The Kirchhoff-Bunsen Experiment In the mid-1850s, Gustav Kirchhoff and Robert Bunsen discovered that when a chemical substance is heated and vaporized, the spectrum of the emitted light exhibits a series of bright spectral lines. They also found that each chemical element produces its own characteristic pattern of spectral lines. (In an actual laboratory experiment, lenses would be needed to focus the image of the slit onto the screen.)

elements whose presence had never before been suspected. In 1860 Kirchhoff and Bunsen found a new line in the blue portion of the spectrum of a sample of mineral water. After isolating the previously unknown element responsible for making the line, they named it cesium (from the Latin *caesium*, "gray-blue"). The next year, a new line in the red portion of the spectrum of a mineral sample led them to discover the element rubidium (Latin *rubidium*, "red").

Spectral analysis even allowed the discovery of new elements outside Earth. During the solar eclipse of 1868, astronomers found a new spectral line in light coming from the hot gases at the upper surface of the Sun while the main body of the Sun was hidden by the Moon. This line was attributed to a new element that was named helium (from the Greek *helios*, "sun"). Helium was not discovered on Earth until 1895, when it was found in gases obtained from a uranium mineral.

The spectrum of the Sun, with its dark spectral lines superimposed on a bright background (see Figure 5-12), may seem to be unrelated to the spectra of bright lines against a dark background produced by substances in a flame (see Figure 5-13). But by the early 1860s Kirchhoff's experiments had revealed a direct connection between these two types of spectra. His conclusions are summarized in three important statements about spectra that are today called **Kirchhoff's laws**. These laws, which are illustrated in Figure 5-14, are as follows:

Law 1 A hot opaque body, such as a perfect blackbody, or a hot, dense gas produces a **continuous spectrum**—a complete rainbow of colors without any spectral lines.

Law 2 A hot, transparent gas produces an **emission line spectrum**—a series of bright spectral lines against a dark background.

Law 3 A cool, transparent gas in front of a source of a continuous spectrum produces an **absorption line spectrum**—a series of dark spectral lines among the colors of the continuous spectrum. Furthermore, the dark lines in the absorption spectrum of a particular gas occur at exactly the *same* wavelengths as the bright lines in the emission spectrum of that same gas.

Kirchhoff's laws imply that if a beam of white light is passed through a gas, the atoms of the gas somehow extract light of very specific wavelengths from the white light. Hence, an observer who looks straight through the gas at the white-light source (the blackbody in Figure 5-14) will receive light whose spectrum has dark absorption lines superimposed on the continuous spectrum of the white light. The gas atoms then radiate light of precisely these same wavelengths in all directions. An observer at an oblique angle (that is, one who is not sighting directly through the cloud toward the blackbody) will receive only this light radiated by the gas cloud; the spectrum of this light is bright emission lines on a dark background.

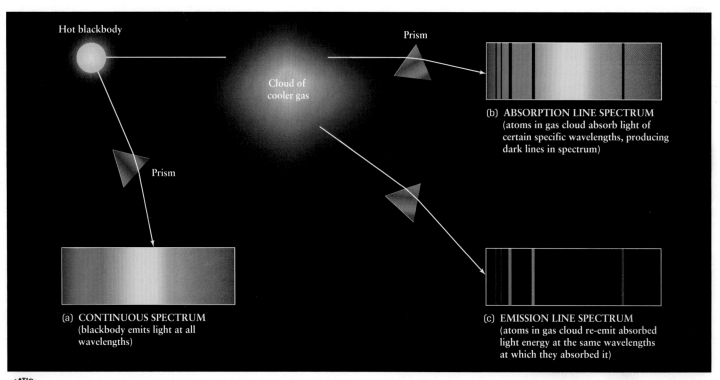

(a) **CONTINUOUS SPECTRUM** (blackbody emits light at all wavelengths)

(b) **ABSORPTION LINE SPECTRUM** (atoms in gas cloud absorb light of certain specific wavelengths, producing dark lines in spectrum)

(c) **EMISSION LINE SPECTRUM** (atoms in gas cloud re-emit absorbed light energy at the same wavelengths at which they absorbed it)

 Figure 5-14
Continuous, Absorption Line, and Emission Line Spectra A hot, opaque body (like a blackbody) emits a continuous spectrum of light (spectrum *a*). If this light is passed through a cloud of a cooler gas, the cloud absorbs light of certain specific wavelengths, and the spectrum of light that passes directly through the cloud has dark absorption lines (spectrum *b*). The cloud does not retain all the light energy that it absorbs but radiates it outward in all directions. The spectrum of this reradiated light contains bright emission lines (spectrum *c*) with exactly the same wavelengths as the dark absorption lines in spectrum *b*. The specific wavelengths observed depend on the chemical composition of the cloud.

Absorption spectrum of the Sun ⟶

Emission spectrum of iron
(in the laboratory on Earth) ⟶

For each emission line of iron, there is a corresponding absorption
line in the solar spectrum; hence there must be iron in the Sun's atmosphere

Figure 5-15 R I **V** U X G

Iron in the Sun The upper part of this figure is a portion of the Sun's spectrum at violet wavelengths, showing numerous dark absorption lines. The lower part of the figure is a corresponding portion of the emission line spectrum of vaporized iron. The iron lines coincide with some of the solar lines, which proves that there is some iron (albeit a relatively small amount) in the Sun's atmosphere. (Carnegie Observatories)

CAUTION! Figure 5-14 shows that light can either pass through a cloud of gas or be absorbed by the gas. But there is also a third possibility: The light can simply bounce off the atoms or molecules that make up the gas, a phenomenon called **light scattering**. In other words, photons passing through a gas cloud can miss the gas atoms altogether, be swallowed whole by the atoms (absorption), or bounce off the atoms like billiard balls colliding (scattering). **Box 5-4** describes how light scattering explains the blue color of the sky and the red color of sunsets.

INTERACTIVE EXERCISE 5.1 Whether an emission line spectrum or an absorption line spectrum is observed from a gas cloud depends on the relative temperatures of the gas cloud and its background. Absorption lines are seen if the background is hotter than the gas, and emission lines are seen if the background is cooler.

For example, if sodium is placed in the flame of a Bunsen burner in a darkened room, the flame will emit a characteristic orange-yellow glow. (This same glow is produced if we use ordinary table salt, which is a compound of sodium and chlorine.) If we pass the light from the flame through a prism, it displays an emission line spectrum with two closely spaced spectral lines at wavelengths of 588.99 and 589.59 nm, in the orange-yellow part of the spectrum. We now turn on a lightbulb whose filament is hotter than the flame and shine the bulb's white light through the flame. The spectrum of this light after it passes through the flame's sodium vapor is the continuous spectrum from the lightbulb, but with two closely spaced *dark* lines at 588.99 and 589.59 nm. Thus, the chemical composition of the gas is revealed by either bright emission lines or dark absorption lines.

Spectroscopy is the systematic study of spectra and spectral lines. Spectral lines are tremendously important in astronomy, because they provide reliable evidence about the chemical composition of distant objects. As an example, the spectrum of the Sun shown in Figure 5-12 is an absorption line spectrum. The continuous spectrum comes from the hot surface of the Sun, which acts like a blackbody. The dark absorption lines are caused by this light passing through a cooler gas; this gas is the atmosphere that surrounds the Sun. Therefore, by identifying the spectral lines present in the solar spectrum, we can determine the chemical composition of the Sun's atmosphere.

Figure 5-15 shows both a portion of the Sun's absorption line spectrum and the emission line spectrum of iron vapor over the same wavelength range. This pattern of bright spectral lines in the lower spectrum is iron's own distinctive "fingerprint," which no other substance can imitate. Because some absorption lines in the Sun's spectrum coincide with the iron lines, some vaporized iron must exist in the Sun's atmosphere.

Spectroscopy can also help us analyze gas clouds in space, such as the nebula NGC 2363 shown in **Figure 5-16**. Such glowing clouds have emission line spectra, because we see them against the

WEB LINK 5.7

Figure 5-16 R I **V** U X G

The Nebula NGC 2363 The glowing gas cloud in this Hubble Space Telescope image lies within a galaxy some 10 million light-years away in the constellation Camelopardalis (the Giraffe). Hot stars within the nebula emit high-energy, ultraviolet photons, which are absorbed by the surrounding gas and heat the gas to high temperature. This heated gas produces light with an emission line spectrum. The particular wavelength of red light emitted by the nebula is 656 nm, characteristic of hydrogen gas. (L. Drissen, J.-R. Roy, and C. Robert/Département de Physique and Observatoire du Mont Mégantic, Université Laval; and NASA)

box 5-4   THE HEAVENS ON THE EARTH

Light Scattering

Light scattering is the process whereby photons bounce off particles in their path. These particles can be atoms, molecules, or clumps of molecules. You are reading these words using photons from the Sun or a lamp that bounced off the page—that is, were scattered by the particles that make up the page.

An important fact about light scattering is that very small particles—ones that are smaller than a wavelength of visible light—are quite effective at scattering short-wavelength photons of blue light, but less effective at scattering long-wavelength photons of red light. This fact explains a number of phenomena that you can see here on Earth.

The light that comes from the daytime sky is sunlight that has been scattered by the molecules that make up our atmosphere (see part *a* of the accompanying figure). Air molecules are less than 1 nm across, far smaller than the wavelength of visible light, so they scatter blue light more than red light—which is why the sky looks blue. Smoke particles are also quite small, which explains why the smoke from a cigarette or a fire has a bluish color.

Distant mountains often appear blue thanks to sunlight being scattered from the atmosphere between the mountains and your eyes. (The Blue Ridge Mountains, which extend from Pennsylvania to Georgia, and Australia's Blue Mountains derive their names from this effect.) Sunglasses often have a red or orange tint, which blocks out blue light. This cuts down on the amount of scattered light from the sky reaching your eyes and allows you to see distant objects more clearly.

Light scattering also explains why sunsets are red. The light from the Sun contains photons of all visible wavelengths, but as this light passes through our atmosphere the blue photons are scattered away from the straight-line path from the Sun to your eye. Red photons undergo relatively little scattering, so the Sun always looks a bit redder than it really is. When you look toward the setting sun, the sunlight that reaches your eye has had to pass through a relatively thick layer of atmosphere (part *b* of the accompanying figure). Hence, a large fraction of the blue light from the Sun has been scattered, and the Sun appears quite red.

The same effect also applies to sunrises, but sunrises seldom look as red as sunsets do. The reason is that dust is lifted into the atmosphere during the day by the wind (which is typically stronger in the daytime than at night), and dust particles in the atmosphere help to scatter even more blue light.

If the small particles that scatter light are sufficiently concentrated, there will be almost as much scattering of red light as of blue light, and the scattered light will appear white. This explains the white color of clouds, fog, and haze, in which the scattering particles are ice crystals or water droplets. Whole milk looks white because of light scattering from tiny fat globules; nonfat milk has only a very few of these globules and so has a slight bluish cast.

Light scattering has many applications to astronomy. For example, it explains why very distant stars in our Galaxy appear surprisingly red. The reason is that there are tiny dust particles in the space between the stars, and this dust scatters blue photons. By studying how much scattering takes place, astronomers have learned about the tenuous material that fills interstellar space.

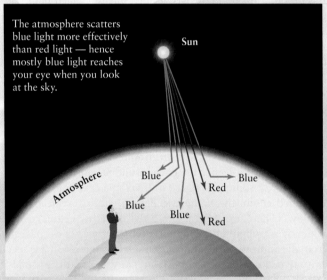

(a) Why the sky looks blue

(b) Why the setting Sun looks red

black background of space. The dominant red color of NGC 2363 is due to an emission line at a wavelength near 656 nm. This is one of the characteristic wavelengths emitted by hydrogen gas, so we can conclude that this nebula contains hydrogen. More detailed analyses of this kind show that hydrogen is the most common element in gaseous nebulae, and indeed in the universe as a whole.

What is truly remarkable about spectroscopy is that it can determine chemical composition at any distance. The 656-nm red light produced by a sample of heated hydrogen gas on Earth is the same as that observed coming from NGC 2363 in Figure 5-16, located about 10 million light-years away. By using the basic principles outlined by Kirchhoff, astronomers have the tools to make chemical assays of objects that are almost inconceivably distant. Throughout this book we will see many examples of how astronomers use Kirchhoff's laws to determine the nature of celestial objects.

To make full use of Kirchhoff's laws, it is helpful to understand why they work. Why does an atom absorb light of only particular wavelengths? And why does it then emit light of only these same wavelengths? Maxwell's theory of electromagnetism (see Section 5-2) could not answer these questions. The answers did not come until early in the twentieth century, when scientists began to discover the structure and properties of atoms.

5-7 | An atom consists of a small, dense nucleus surrounded by electrons

The first important clue about the internal structure of atoms came from an experiment conducted in 1910 by Ernest Rutherford, a gifted physicist from New Zealand. Rutherford and his colleagues at the University of Manchester in England had been investigating the recently discovered phenomenon of radioactivity. Certain radioactive elements, such as uranium and radium, were known to emit particles of various types. One type, the alpha particle, has about the same mass as a helium atom and is emitted from some radioactive substances with considerable speed.

In one series of experiments, Rutherford and his colleagues were using alpha particles as projectiles to probe the structure of solid matter. They directed a beam of these particles at a thin sheet of metal (Figure 5-17). Almost all the alpha particles passed through the metal sheet with little or no deflection from their straight-line paths. To the surprise of the experimenters, however, an occasional alpha particle bounced back from the metal sheet as though it had struck something quite dense. Rutherford later remarked, "It was almost as incredible as if you fired a fifteen-inch shell at a piece of tissue paper and it came back and hit you."

Rutherford concluded from this experiment that most of the mass of an atom is concentrated in a compact, massive lump of matter that occupies only a small part of the atom's volume. Most of the alpha particles pass freely through the nearly empty space that makes up most of the atom, but a few particles happen to strike the dense mass at the center of the atom and bounce back.

Rutherford proposed a new model for the structure of an atom, shown in Figure 5-18. According to this model, a massive, positively charged **nucleus** at the center of the atom is orbited by tiny, negatively charged electrons. Rutherford concluded that at least 99.98% of the mass of an atom must be concentrated in its nucleus, whose diameter is only about 10^{-14} m. (The diameter of a typical atom is far larger, about 10^{-10} m.)

 To appreciate just how tiny the nucleus is, imagine expanding an atom by a factor of 10^{12} to a diameter of 100 meters, about the size of a sports stadium. To this

Radioactive substance emits alpha particles.

Most alpha particles pass through the foil with very little deflection.

A

B

Gold foil (seen edge-on)

Occasionally an alpha particle rebounds (like *A* or *B*), indicating that it has collided with the massive nucleus of a gold atom.

Figure 5-17

Rutherford's Experiment Alpha particles from a radioactive source are directed at a thin metal foil. This experiment provided the first evidence that the nuclei of atoms are relatively massive and compact.

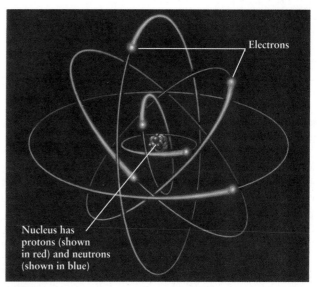

Electrons

Nucleus has protons (shown in red) and neutrons (shown in blue)

Figure 5-18

Rutherford's Model of the Atom Electrons orbit the atom's nucleus, which contains most of the atom's mass. The nucleus contains two types of particles, protons and neutrons.

box 5-5 · TOOLS OF THE ASTRONOMER'S TRADE

Atoms, the Periodic Table, and Isotopes

Each different chemical element is made of a specific type of atom. Each specific atom has a characteristic number of protons in its nucleus. For example, a hydrogen atom has 1 proton in its nucleus, an oxygen atom has 8 protons in its nucleus, and so on.

The number of protons in an atom's nucleus is the **atomic number** for that particular element. The chemical elements are most conveniently listed in the form of a **periodic table** (shown in the figure). Elements are arranged in the periodic table in order of increasing atomic number. With only a few exceptions, this sequence also corresponds to increasing average mass of the atoms of the elements. Thus, hydrogen (symbol H), with atomic number 1, is the lightest element. Iron (symbol Fe) has atomic number 26 and is a relatively heavy element.

All the elements listed in a single vertical column of the periodic table have similar chemical properties. For example, the elements in the far right column are all gases under the conditions of temperature and pressure found at the Earth's surface, and they are all very reluctant to react chemically with other elements.

In addition to nearly 100 naturally occurring elements, the periodic table includes a number of artificially produced elements. Most of these elements are heavier than uranium (symbol U) and are highly radioactive, which means that they decay into lighter elements within a short time of being created in laboratory experiments. Scientists have succeeded in creating only a few atoms of elements 104 and above.

The number of protons in the nucleus of an atom determines which element that atom is. Nevertheless, the same element may have different numbers of neutrons in its nucleus. For example, oxygen (O) has atomic number 8, so every oxygen nucleus has exactly 8 protons. But oxygen nuclei can have 8, 9, or 10 neutrons. These three slightly different kinds of oxygen are called **isotopes.** The isotope with 8 neutrons is by far the most abundant variety. It is written as ^{16}O, or oxygen-16. The rarer isotopes with 9 and 10 neutrons are designated as ^{17}O and ^{18}O, respectively.

The superscript that precedes the chemical symbol for an element equals the total number of protons and neutrons in a nucleus of that particular isotope. For example, a nucleus of the most common isotope of iron, ^{56}Fe or iron-56, contains a total of 56 protons and neutrons. From the periodic table, the atomic number of iron is 26, so every iron atom has 26 protons in its nucleus. Therefore, the number of neutrons in an iron-56 nucleus is $56 - 26 = 30$. (Most nuclei have more neutrons than protons, especially in the case of the heaviest elements.)

It is extremely difficult to distinguish chemically between the various isotopes of a particular element. Ordinary chemical reactions involve only the electrons that orbit the atom, never the neutrons buried in its nucleus. But there are small differences in the wavelengths of the spectral lines for different isotopes of the same element. For example, the spectral line wavelengths of the hydrogen isotope ^{2}H are about 0.03% greater than the wavelengths for the most common hydrogen isotope, ^{1}H. Thus, different isotopes can be distinguished by careful spectroscopic analysis.

Isotopes are important in astronomy for a variety of reasons. By measuring the relative amounts of different

---

scale, the nucleus would be just a centimeter across—no larger than your thumbnail.

We know today that the nucleus of an atom contains two types of particles, **protons** and **neutrons.** A proton has a positive electric charge, equal and opposite to that of an electron. As its name suggests, a neutron has no electric charge—it is electrically neutral. As an example, an alpha particle (such as those Rutherford's team used) is actually a nucleus of the helium atom, with two protons and two neutrons. Protons and neutrons are held together in a nucleus by the so-called strong nuclear force, whose great strength overcomes the electric repulsion between the positively charged protons. A proton and a neutron have almost the same mass, 1.7×10^{-27} kg, and each has about 2000 times as much mass as an electron (9.1×10^{-31} kg). In an ordinary atom there are as many positive protons as there are negative electrons, so the atom

has no net electric charge. Because the mass of the electron is so small, the mass of an atom is not much greater than the mass of its nucleus. That is why an alpha particle has nearly the same mass as an atom of helium.

While the solar system is held together by gravitational forces, atoms are held together by electrical forces. The electric forces attracting the positively charged protons and the negatively charged electrons keep the atom from coming apart. Box 5-5 describes more about the connection between the structure of atoms and the chemical and physical properties of substances made of those atoms.

Rutherford's experiments clarified the structure of the atom, but they did not explain how these tiny particles within the atom give rise to spectral lines. The task of reconciling Rutherford's atomic model with Kirchhoff's laws of spectral analysis was undertaken by the young Danish physicist Niels Bohr, who joined Rutherford's group at Manchester in 1912.

TOOLS OF THE ASTRONOMER'S TRADE continued

isotopes of a given element in a Moon rock or meteorite, the age of that sample can be determined. The mixture of isotopes left behind when a star explodes into a supernova (see Section 1-3) tells astronomers about the processes that led to the explosion.

And knowing the properties of different isotopes of hydrogen and helium is crucial to understanding the nuclear reactions that make the Sun shine. Look for these and other applications of the idea of isotopes in later chapters.

Periodic Table of the Elements

5-8 | Spectral lines are produced when an electron jumps from one energy level to another within an atom

SCIENCE IN PROCESS

Niels Bohr explained spectral lines with a radical new model of the atom

 Niels Bohr began his study of the connection between atomic spectra and atomic structure by trying to understand the structure of hydrogen, the simplest and lightest of the elements. (As we discussed in Section 5-6, hydrogen is also the most common element in the universe.) When Bohr was done, he had not only found a way to explain this atom's spectrum but had also found a justification for Kirchhoff's laws in terms of atomic physics.

The most common type of hydrogen atom consists of a single electron and a single proton. Hydrogen has a simple visible-light spectrum consisting of a pattern of lines that begins at a wavelength of 656.3 nm and ends at 364.6 nm. The first spectral line is called H_α (H-alpha), the second spectral line is called H_β (H-beta), the third is H_γ (H-gamma), and so forth. The closer you get to the short-wavelength end of the spectrum at 364.6 nm, the more spectral lines you see.

The regularity in this spectral pattern was described mathematically in 1885 by Johann Jakob Balmer, a Swiss schoolteacher. The spectral lines of hydrogen at visible wavelengths are today called **Balmer lines,** and the entire pattern from H_α onward is called the **Balmer series.** More than two dozen Balmer lines are seen in the spectrum of the star shown in **Figure 5-19.** Stars in general, including the Sun, have Balmer absorption lines in their spectra, which shows they have atmospheres which contain hydrogen.

figure 5-19 R I V U X G

Balmer Lines in the Spectrum of a Star This portion of the spectrum of the star HD 193182 shows nearly two dozen Balmer lines, from H_{13} through H_{40}. (Numbers are used beyond the first few Balmer lines, so you do not have to memorize the entire Greek alphabet.) The series converges at 364.6 nm, just to the left of H_{40}. This star's spectrum also contains the first 12 Balmer lines (H_α through H_{12}), but they are located beyond the right edge of this photograph. (Carnegie Observatories)

Using trial and error, Balmer discovered a formula from which the wavelengths (λ) of hydrogen's spectral lines can be calculated. Balmer's formula is usually written

$$\frac{1}{\lambda} = R\left(\frac{1}{4} - \frac{1}{n^2}\right)$$

where n can be any integer (whole number) greater than 2. Here R is the *Rydberg constant* ($R = 1.097 \times 10^7$ m^{-1}), named in honor of the Swedish spectroscopist Johannes Rydberg. To get the wavelength λ_α of the spectral line H_α, you first put $n = 3$ into Balmer's formula:

$$\frac{1}{\lambda_\alpha} = (1.097 \times 10^7 \text{ m}^{-1})\left(\frac{1}{4} - \frac{1}{3^2}\right) = 1.524 \times 10^6 \text{ m}^{-1}$$

Then take the reciprocal:

$$\lambda_\alpha = \frac{1}{1.524 \times 10^6 \text{ m}^{-1}} = 6.563 \times 10^{-7} \text{ m} = 656.3 \text{ nm}$$

To get the wavelength of H_β, use $n = 4$, and to get the wavelength of H_γ, use $n = 5$. If you use $n = \infty$ (the symbol ∞ stands for infinity), you get the short-wavelength end of the hydrogen spectrum at 364.6 nm. (Note that 1 divided by infinity equals zero.)

Bohr realized that to fully understand the structure of the hydrogen atom, he had to be able to derive Balmer's formula using the laws of physics. He first made the rather wild assumption that the electron in a hydrogen atom can orbit the nucleus only in certain specific orbits. (This was a significant break with the ideas of Newton, in whose mechanics any orbit should be possible.) Figure 5-20 shows the four smallest of these **Bohr orbits,** labeled by the numbers $n = 1$, $n = 2$, $n = 3$, and so on.

Although confined to one of these allowed orbits while circling the nucleus, an electron can jump from one Bohr orbit to another. For an electron to do this, the hydrogen atom must gain or lose a specific amount of energy. The atom must absorb energy for the electron to go from an inner to an outer orbit; the atom must release energy for the electron to go from an outer to an inner orbit. As an example, Figure 5-21 shows an electron jumping between the $n = 2$ and $n = 3$ orbits of a hydrogen atom as the atom absorbs or emits an H_α photon.

When the electron jumps from one orbit to another, the energy of the photon that is emitted or absorbed equals the difference in energy between these two orbits. This energy difference, and hence the photon energy, is the same whether the jump is from a low orbit to a high orbit (Figure 5-21*a*) or from the high orbit back to the low one (Figure 5-21*b*). According to Planck and Einstein, if two photons have the same energy E, the relationship $E = hc/\lambda$ tells us that they must also have the same wavelength λ. It follows that if an atom can emit photons of a given energy and wavelength, it can also absorb photons of precisely the same energy and wavelength. Thus, Bohr's picture explains Kirchhoff's observation that atoms emit and absorb the same wavelengths of light.

The Bohr picture also helps us visualize what happens to produce an emission line spectrum. When a gas is heated, its atoms move around rapidly and can collide forcefully with each other. These energetic collisions excite the atoms' electrons into high orbits. The electrons then cascade back down to the innermost possible orbit, emitting photons whose energies are equal to the energy differences between different Bohr orbits. In this fashion, a hot gas produces an emission line spectrum with a variety of different wavelengths.

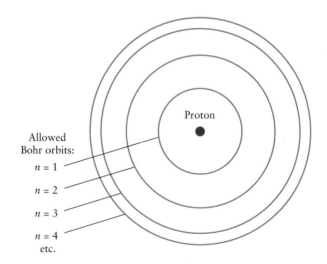

figure 5-20

The Bohr Model of the Hydrogen Atom In this model, an electron circles the hydrogen nucleus (a proton) only in allowed orbits $n = 1, 2, 3$, and so forth. The first four Bohr orbits are shown here. This figure is not drawn to scale; in the Bohr model, the $n = 2, 3$, and 4 orbits are respectively 4, 9, and 16 times larger than the $n = 1$ orbit.

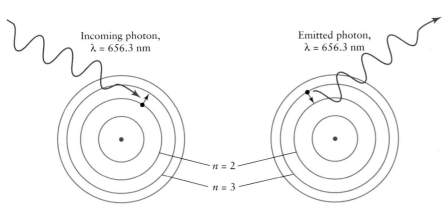

(a) Atom absorbs a 656.3-nm photon; absorbed energy causes electron to jump from the $n = 2$ orbit up to the $n = 3$ orbit

(b) Electron falls from the $n = 3$ orbit to the $n = 2$ orbit; energy lost by atom goes into emitting a 656.3-nm photon

 fiqure 5-21

The Absorption and Emission of an H$_\alpha$ Photon This schematic diagram, drawn according to the Bohr model, shows what happens when a hydrogen atom **(a)** absorbs or **(b)** emits a photon whose wavelength is 656.3 nm.

To produce an absorption line spectrum, begin with a relatively cool gas, so that the electrons in most of the atoms are in inner, low-energy orbits. If a beam of light with a continuous spectrum is shone through the gas, most wavelengths will pass through undisturbed. Only those photons will be absorbed whose energies are just right to excite an electron to an allowed outer orbit. Hence, only certain wavelengths will be absorbed, and dark lines will appear in the spectrum at those wavelengths.

Using his picture of allowed orbits and the formula $E = hc/\lambda$, Bohr was able to prove mathematically that the wavelength λ of the photon emitted or absorbed as an electron jumps between an inner orbit N and an outer orbit n is

Bohr formula for hydrogen wavelengths

$$\frac{1}{\lambda} = R\left(\frac{1}{N^2} - \frac{1}{n^2}\right)$$

N = number of inner orbit

n = number of outer orbit

R = Rydberg constant = $1.097 \times 10^7 \text{ m}^{-1}$

λ = wavelength (in meters) of emitted or absorbed photon

If Bohr let $N = 2$ in this formula, he got back the formula that Balmer discovered by trial and error. Hence, Bohr deduced the meaning of the Balmer series: All the Balmer lines are produced by electrons jumping between the second Bohr orbit ($N = 2$) and higher orbits ($n = 3, 4, 5$, and so on).

Bohr's formula also correctly predicts the wavelengths of other series of spectral lines that occur at nonvisible wavelengths (**Figure 5-22**). Using $N = 1$ gives the **Lyman series**, which is entirely in the ultraviolet. All the spectral lines in this series involve electron tran-

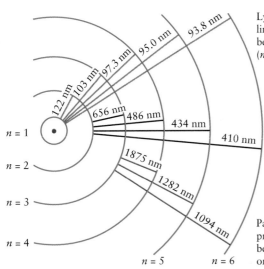

Lyman series (ultraviolet) of spectral lines: produced by electron transitions between the $n = 1$ orbit and higher orbits ($n = 2, 3, 4, \ldots$)

Balmer series (visible and ultraviolet) of spectral lines: produced by electron transitions between the $n = 2$ orbit and higher orbits ($n = 3, 4, 5, \ldots$)

Paschen series (infared) of spectral lines: produced by electron transitions between the $n = 3$ orbit and higher orbits ($n = 4, 5, 6, \ldots$)

fiqure 5-22

Electron Transitions in the Hydrogen Atom This diagram shows the photon wavelengths associated with different electron transitions in hydrogen. In each case, the same wavelength occurs whether a photon is emitted (when the electron drops from a high orbit to a low one) or absorbed (when the electron jumps from a low orbit to a high one). The orbits are not shown to scale.

sitions between the lowest Bohr orbit and all higher orbits (n = 2, 3, 4, and so on). This pattern of spectral lines begins with L$_\alpha$ (Lyman alpha) at 122 nm and converges on L$_\infty$ at 91.2 nm. Using N = 3 gives a series of infrared wavelengths called the **Paschen series.** This series, which involves transitions between the third Bohr orbit and all higher orbits, begins with P$_\alpha$ (Paschen alpha) at 1875 nm and converges on P$_\infty$ at 822 nm. Additional series exist at still longer wavelengths.

Today's view of the atom owes much to the Bohr model, but is different in certain ways. The modern picture is based on **quantum mechanics,** a branch of physics dealing with photons and subatomic particles that was developed during the 1920s. As a result of this work, physicists no longer picture electrons as moving in specific orbits about the nucleus. Instead, electrons are now known to have both particle and wave properties and are said to occupy only certain **energy levels** in the atom.

An extremely useful way of displaying the structure of an atom is with an **energy-level diagram.** Figure 5-23 shows such a diagram for hydrogen. The lowest energy level, called the **ground state,** corresponds to the n = 1 Bohr orbit. Higher energy levels, called **excited states,** correspond to successively larger Bohr orbits.

An electron can jump from the ground state up to the n = 2 level if the atom absorbs a Lyman-alpha photon with a wavelength of 122 nm. Such a photon has energy $E = hc/\lambda$ = 10.2 eV (electron volts; see Section 5-5). That's why the energy level of n = 2 is shown in Figure 5-23 as having an energy 10.2 eV above that of the ground

state (which is usually assigned a value of 0 eV). Similarly, the n = 3 level is 12.1 eV above the ground state, and so forth. Electrons can make transitions to higher energy levels by absorbing a photon or in a collision between atoms; they can make transitions to lower energy levels by emitting a photon.

On the energy-level diagram for hydrogen, the n = ∞ level has an energy of 13.6 eV. (This corresponds to an infinitely large orbit in the Bohr model.) If the electron is initially in the ground state and the atom absorbs a photon of any energy greater than 13.6 eV, the electron will be removed completely from the atom. This process is called **ionization.** A 13.6-eV photon has a wavelength of 91.2 nm, equal to the shortest wavelength in the ultraviolet Lyman series (L$_\infty$). So any photon with a wavelength of 91.2 nm or less can ionize hydrogen. (The Planck formula $E = hc/\lambda$ tells us that the higher the photon energy, the shorter the wavelength.)

As an example, the gaseous nebula NGC 2363 shown in Figure 5-16 has hot stars in its neighborhood that produce copious amounts of ultraviolet photons with wavelengths less than 91.2 nm. Hydrogen atoms in the nebula that absorb these photons become ionized and lose their electrons. When the electrons recombine with the nuclei, they cascade down the energy levels to the ground state and emit visible light in the process. This is what makes the nebula glow.

 The same basic principles that explain the hydrogen spectrum also apply to the atoms of other elements. Electrons in each kind of atom can be only in certain energy levels, so only photons of certain wavelengths can be emitted or absorbed. Because each kind of atom has its own unique arrangement of electron levels, the pattern of spectral lines is likewise unique to that particular type of atom (see the photograph that opens this chapter). These patterns are in general much more complicated than for the hydrogen atom. Hence, there is no simple relationship analogous to the Bohr formula that applies to the spectra of all atoms.

The idea of energy levels explains the emission line spectra and absorption line spectra of gases. But what about the continuous spectra produced by dense objects like the filament of a lightbulb or the coils of a toaster? These objects are made of atoms, so why don't they emit light with an emission line spectrum characteristic of the particular atoms of which they are made?

The reason is directly related to the difference between a gas on the one hand and a liquid or solid on the other. In a gas, atoms are widely separated and can emit photons without interference from other atoms. But in a liquid or a solid, atoms are so close that they almost touch, and thus these atoms interact strongly with each other. These interactions interfere with the process of emitting photons. As a result, the pattern of distinctive bright spectral lines that the atoms would emit in isolation becomes "smeared out" into a continuous spectrum.

 Think of atoms as being like tuning forks. If you strike a single tuning fork, it produces a sound wave with a single clear frequency and wavelength, just as an isolated atom emits light of definite wavelengths. But if you shake a box packed full of tuning forks, you will hear a clanging noise that is a mixture of sounds of all different frequencies and wavelengths. This is directly analogous to the continuous spectrum of light emitted by a dense object with closely packed atoms.

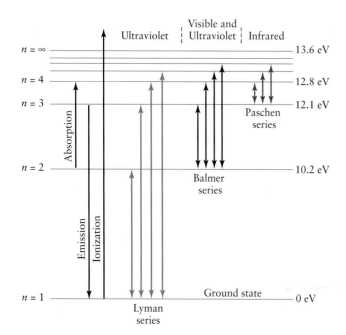

figure 5-23

Energy-Level Diagram of Hydrogen A convenient way to display the structure of the hydrogen atom is in a diagram like this, which shows the allowed energy levels. The diagram shows a number of possible electron jumps, or transitions, between energy levels. An upward transition occurs when the atom absorbs a photon; a downward transition occurs when the atom emits a photon. (Compare with Figure 5-22.)

With the work of such people as Planck, Einstein, Rutherford, and Bohr, the interchange between astronomy and physics came full circle. Modern physics was born when Newton set out to understand the motions of the planets. Two and a half centuries later, physicists in their laboratories probed the properties of light and the structure of atoms. Their labors had immediate applications in astronomy. Armed with this new understanding of light and matter, astronomers were able to probe in detail the chemical and physical properties of planets, stars, and galaxies.

5-9 | The wavelength of a spectral line is affected by the relative motion between the source and the observer

In addition to telling us about temperature and chemical composition, the spectrum of a planet, star, or galaxy can also reveal something about that object's motion through space. This idea dates from 1842, when Christian Doppler, a professor of mathematics in Prague, pointed out that the observed wavelength of light must be affected by motion. **Figure 5-24** shows why. In this figure, a light source is moving from right to left, and the circles represent the crests of waves emitted from the moving source at various positions. Each successive wave crest is emitted from a position slightly closer to the observer on the left, so she sees a shorter wavelength—the distance from one crest to the next—than she would if the source were stationary. All the lines in the spectrum of an approaching source are shifted toward the short-wavelength (blue) end of the spectrum. This phenomenon is called a **blueshift**.

Wave crest 1: emitted when light source was at S₁

Wave crest 2: emitted when light source was at S₂

Wave crests 3 and 4: emitted when light source was at S₃ and S₄, respectively

This observer sees **blueshift**

This observer sees **redshift**

Figure 5-24

The Doppler Effect The wavelength of light is affected by motion between the light source and an observer. The light source shown here is moving, so wave crests 1, 2, etc., emitted when the source was at points S₁, S₂, etc., are crowded together in front of the source but are spread out behind it. Consequently, wavelengths are shortened (blueshifted) if the source is moving toward the observer and lengthened (redshifted) if the source is moving away from the observer. Motion perpendicular to an observer's line of sight does not affect wavelength.

The source is receding from the observer on the right in Figure 5-24. The wave crests that reach him are stretched apart, so that he sees a longer wavelength than he would if the source were stationary. All the lines in the spectrum of a receding source are shifted toward the longer-wavelength (red) end of the spectrum, producing a **redshift**. In general, the effect of relative motion on wavelength is called the **Doppler effect**.

ANALOGY You have probably noticed a similar Doppler effect for sound waves. When a police car is approaching, the sound waves from its siren have a shorter wavelength and higher frequency than if the siren were at rest, and hence you hear a higher pitch. After the police car passes you and is moving away, you hear a lower pitch from the siren because the sound waves have a longer wavelength and a lower frequency.

Suppose that λ_0 is the wavelength of a particular spectral line from a light source that is not moving. It is the wavelength that you might look up in a reference book or determine in a laboratory experiment for this spectral line. If the source is moving, this particular spectral line is shifted to a different wavelength λ. The size of the wavelength shift is usually written as $\Delta\lambda$, where $\Delta\lambda = \lambda - \lambda_0$. Thus, $\Delta\lambda$ is the difference between the wavelength listed in reference books and the wavelength that you actually observe in the spectrum of a star or galaxy.

Doppler proved that the wavelength shift ($\Delta\lambda$) is governed by the following simple equation:

Doppler shift equation

$$\frac{\Delta\lambda}{\lambda_0} = \frac{v}{c}$$

$\Delta\lambda$ = wavelength shift

λ_0 = wavelength if source is not moving

v = velocity of the source measured along the line of sight

c = speed of light = 3.0×10^5 km/s

CAUTION The capital Greek letter Δ, or "delta," is commonly used to denote a change in the value of a quantity. Thus, $\Delta\lambda$ is the change in the wavelength λ due to the Doppler effect. It is *not* equal to a quantity Δ multiplied by a second quantity λ!

The velocity determined from the Doppler effect is called **radial velocity**, because v is the component of the star's motion parallel to our line of sight, or along the "radius" drawn from Earth to the star. Of course, a sizable fraction of a star's motion may be perpendicular to our line of sight. The speed of this transverse movement across the sky does not affect wavelengths if the speed is small compared with c. Box 5-6 includes two examples of calculations with radial velocity using the Doppler formula.

CAUTION The redshifts and blueshifts of stars visible to the naked eye, or even through a small telescope, are only a small fraction of a nanometer. These tiny wavelength changes are far too small to detect visually. (Astronomers were able to detect the tiny Doppler shifts of starlight only after they had developed highly

box 5-6 | TOOLS OF THE ASTRONOMER'S TRADE

Applications of the Doppler Effect

Doppler's formula relates the radial velocity of an astronomical object to the wavelength shift of its spectral lines. Here are two examples that show how to use this remarkably powerful formula.

EXAMPLE: As measured in the laboratory, the prominent H_α spectral line of hydrogen has a wavelength $\lambda_0 = 656.285$ nm. But in the spectrum of the star Vega, the brightest star in the constellation Lyra (the Harp), this line has a wavelength $\lambda = 656.255$ nm. What can we conclude about the motion of Vega?

Situation: Our goal is to use the ideas of the Doppler effect to find the velocity of Vega toward or away from Earth.

Tools: We use the Doppler shift formula, $\Delta\lambda/\lambda_0 = v/c$, to determine Vega's velocity v.

Answer: The wavelength shift is

$$\Delta\lambda = \lambda - \lambda_0 = 656.255 \text{ nm} - 656.285 \text{ nm} = -0.030 \text{ nm}$$

The negative value means that we see the light from Vega shifted to shorter wavelengths—that is, there is a blueshift. (Note that the shift is very tiny and can be measured only using specialized equipment.) From the Doppler shift formula, the star's radial velocity is:

$$v = c\frac{\Delta\lambda}{\lambda_0} = (3.0 \times 10^5 \text{ km/s})\left(\frac{-0.030 \text{ nm}}{656.285 \text{ nm}}\right) = -14 \text{ km/s}$$

Review: The minus sign indicates that Vega is coming toward us at 14 km/s. The star may also be moving perpendicular to the line from Earth to Vega, but such motion produces no Doppler shift.

By plotting the motions of stars such as Vega toward and away from us, astronomers have been able to learn how the Milky Way Galaxy (of which our Sun is a part) is rotating.

From this knowledge, and aided by Newton's universal law of gravitation (see Section 4-7), they have made the surprising discovery that the Milky Way contains roughly 10 times more matter than had once been thought! The nature of this unseen *dark matter* is still a subject of debate.

EXAMPLE: In the radio region of the electromagnetic spectrum, hydrogen atoms emit and absorb photons with a wavelength of 21.12 cm, giving rise to a spectral feature commonly called the *21-centimeter line*. The galaxy NGC 3840 in the constellation Leo (the Lion) is receding from us at a speed of 7370 km/s, or about 2.5% of the speed of light. At what wavelength do we expect to detect the 21-cm line from this galaxy?

Situation: Given the velocity of NGC 3840 away from us, our goal is to find the wavelength as measured on Earth of the 21-centimeter line from this galaxy.

Tools: We use the Doppler shift formula to calculate the wavelength shift $\Delta\lambda$, then use this to find the wavelength λ measured on Earth.

Answer: The wavelength shift is

$$\Delta\lambda = \lambda_0\left(\frac{v}{c}\right) = (21.12 \text{ cm})\left(\frac{7370 \text{ km/s}}{3.0 \times 10^5 \text{ km/s}}\right) = 0.52 \text{ cm}$$

Therefore, we will detect the 21-cm line of hydrogen from this galaxy at a wavelength of

$$\lambda = \lambda_0 + \Delta\lambda = 21.12 \text{ cm} + 0.52 \text{ cm} = 21.64 \text{ cm}$$

Review: The 21-cm line has been redshifted to a longer wavelength because the galaxy is receding from us. In fact, most galaxies are receding from us. This observation is one of the key pieces of evidence that the universe is expanding outward from the Big Bang that took place some 14 billion years ago.

---

sensitive equipment for measuring wavelengths. This was done around 1890, a half-century after Doppler's original proposal.) So if you see a star with a red color, it means that the star really is red; it does *not* mean that it is moving rapidly away from us.

The Doppler effect is an important tool in astronomy because it uncovers basic information about the motions of planets, stars, and galaxies. For example, the rotation of the planet Venus was deduced from the Doppler shift of radar waves reflected from its surface. Small Doppler shifts in the spectrum of sunlight have shown that the entire Sun is vibrating like an immense gong. The back-and-forth Doppler shifting of the spectral lines of certain stars reveals that these

stars are being orbited by unseen companions; from this astronomers have discovered planets around other stars and massive objects that may be black holes. Astronomers also use the Doppler effect along with Kepler's third law to measure the masses of galaxies. These are but a few examples of how Doppler's discovery has empowered astronomers in their quest to understand the universe.

In this chapter we have glimpsed how much can be learned by analyzing light from the heavens. To analyze this light, however, it is first necessary to collect as much of it as possible, because most light sources in space are very dim. Collecting the faint light from distant objects is a key purpose of telescopes. In the next chapter we will describe both how telescopes work and how they are used.

Key Words

Terms preceded by an asterisk (*) are discussed in the Boxes.

absolute zero, p. 96

absorption line spectrum, p. 104

atom, p. 96

*atomic number, p. 108

Balmer line, p. 109

Balmer series, p. 109

blackbody, p. 98

blackbody curve, p. 98

blackbody radiation, p. 98

blueshift, p. 113

Bohr orbits, p. 110

continuous spectrum, p. 104

*degrees Celsius, p. 97

*degrees Fahrenheit, p. 97

Doppler effect, p. 113

electromagnetic radiation, p. 94

electromagnetic spectrum, p. 95

electromagnetism, p. 93

electron, p. 102

electron volt, p. 102

element, p. 103

emission line spectrum, p. 104

energy flux, p. 99

energy level, p. 112

energy-level diagram, p. 112

excited state, p. 112

frequency, p. 95

gamma rays, p. 95

ground state, p. 112

infrared radiation, p. 94

ionization, p. 112

*isotope, p. 108

joule, p. 99

kelvin, p. 96

Kirchhoff's laws, p. 104

light scattering, p. 105

*luminosity, p. 100

Lyman series, p. 111

microwaves, p. 95

nanometer, p. 94

neutron, p. 108

nucleus, p. 107

Paschen series, p. 112

*periodic table, p. 108

photoelectric effect, p. 102

photon, p. 101

Planck's law, p. 103

proton, p. 108

quantum mechanics, p. 112

radial velocity, p. 113

radio waves, p. 95

redshift, p. 113

*solar constant, p. 100

spectral analysis, p. 103

spectral line, p. 103

spectroscopy, p. 105

spectrum (*plural* spectra), p. 92

Stefan-Boltzmann law, p. 101

ultraviolet radiation, p. 95

visible light, p. 94

watt, p. 99

wavelength, p. 94

wavelength of maximum emission, p. 97

Wien's law, p. 99

X rays, p. 95

Key Ideas

The Nature of Light: Light is electromagnetic radiation. It has wavelike properties described by its wavelength λ and frequency ν, and travels through empty space at the constant speed $c = 3.0 \times 10^8$ m/s $= 3.0 \times 10^5$ km/s.

Blackbody Radiation: A blackbody is a hypothetical object that is a perfect absorber of electromagnetic radiation at all wavelengths. Stars closely approximate the behavior of blackbodies, as do other hot, dense objects.

• The intensities of radiation emitted at various wavelengths by a blackbody at a given temperature are shown by a blackbody curve.

• Wien's law states that the dominant wavelength at which a blackbody emits electromagnetic radiation is inversely proportional to the Kelvin temperature of the object: λ_{max} (in meters) = (0.0029 K m)/T.

• The Stefan-Boltzmann law states that a blackbody radiates electromagnetic waves with a total energy flux F directly proportional to the fourth power of the Kelvin temperature T of the object: $F = \sigma T^4$.

Photons: An explanation of blackbody curves led to the discovery that light has particlelike properties. The particles of light are called photons.

• Planck's law relates the energy E of a photon to its frequency ν or wavelength λ: $E = h\nu = hc/\lambda$, where h is Planck's constant.

Kirchhoff's Laws: Kirchhoff's three laws of spectral analysis describe conditions under which different kinds of spectra are produced.

• A hot, dense object such as a blackbody emits a continuous spectrum covering all wavelengths.

• A hot, transparent gas produces a spectrum that contains bright (emission) lines.

• A cool, transparent gas in front of a light source that itself has a continuous spectrum produces dark (absorption) lines in the continuous spectrum.

Atomic Structure: An atom has a small dense nucleus composed of protons and neutrons. The nucleus is surrounded by electrons that occupy only certain orbits or energy levels.

• When an electron jumps from one energy level to another, it emits or absorbs a photon of appropriate energy (and hence of a specific wavelength).

• The spectral lines of a particular element correspond to the various electron transitions between energy levels in atoms of that element.

• Bohr's model of the atom correctly predicts the wavelengths of hydrogen's spectral lines.

The Doppler Shift: The Doppler shift enables us to determine the radial velocity of a light source from the displacement of its spectral lines.

• The spectral lines of an approaching light source are shifted toward short wavelengths (a blueshift); the spectral lines of a receding light source are shifted toward long wavelengths (a redshift).

• The size of a wavelength shift is proportional to the radial velocity of the light source relative to the observer.

Review Questions

1. Approximately how many times around the Earth could a beam of light travel in one second?

2. How long does it take light to travel from the Sun to the Earth, a distance of 1.50×10^8 km?

3. How did Newton show that a prism breaks white light into its component colors, but does not add any color to the light?

4. (a) Describe an experiment in which light behaves like a wave. (b) Describe an experiment in which light behaves like a particle.

5. What is meant by the frequency of light? How is frequency related to wavelength?

6. A cellular phone is actually a radio transmitter and receiver. You receive an incoming call in the form of a radio wave of frequency 880.65 MHz. What is the wavelength (in meters) of this wave?

7. A light source emits infrared radiation at a wavelength of 1150 nm. What is the frequency of this radiation?

8. What is a blackbody? In what way is a blackbody black? If a blackbody is black, how can it emit light? If you were to shine a flashlight beam on a perfect blackbody, what would happen to the light?

9. Using Wien's law and the Stefan-Boltzmann law, explain the color and intensity changes that are observed as the temperature of a hot, glowing object increases.

10. Explain why astronomers are interested in blackbody radiation.

11. Why do astronomers find it convenient to use the Kelvin temperature scale in their work rather than the Celsius or Fahrenheit scale?

12. If you double the Kelvin temperature of a hot piece of steel, how much more energy will it radiate per second?

13. The bright star Bellatrix in the constellation Orion has a surface temperature of 21,500 K. What is its wavelength of maximum emission in nanometers? What color is this star?

14. The bright star Antares in the constellation Scorpius (the Scorpion) emits the greatest intensity of radiation at a wavelength of 853 nm. What is the surface temperature of Antares? What color is this star?

15. How is the energy of a photon related to its wavelength? What kind of photons carry the most energy? What kind of photons carry the least energy?

16. To emit the same amount of light energy per second, which must emit more photons per second: a source of red light, or a source of blue light? Explain.

17. Explain how we know that atoms have massive, compact nuclei.

18. Describe the spectrum of hydrogen at visible wavelengths and explain how Bohr's model of the atom accounts for the Balmer lines.

19. Why do different elements display different patterns of lines in their spectra?

20. What is the Doppler effect? Why is it important to astronomers?

21. If you see a blue star, what does its color tell you about how the star is moving through space? Explain your answer.

Advanced Questions

Questions preceded by an asterisk (*) involve topics discussed in the Boxes.

> **Problem-solving tips and tools**
>
> You can find formulas in Box 5-1 for converting between temperature scales. Box 5-2 discusses how a star's radius, luminosity, and surface temperature are related. Box 5-3 shows how to use Planck's law to calculate the energy of a photon. To learn how to do calculations using the Doppler effect, see Box 5-6.

22. Your normal body temperature is 98.6°F. What kind of radiation do you predominantly emit? At what wavelength (in nm) do you emit the most radiation?

23. What is the temperature of the Sun's surface in degrees Fahrenheit?

24. What wavelength of electromagnetic radiation is emitted with greatest intensity by this book? To what region of the electromagnetic spectrum does this wavelength correspond?

25. Black holes are objects whose gravity is so strong that not even an object moving at the speed of light can escape from their surface. Hence, black holes do not themselves emit light. But it is possible to detect radiation from material falling *toward* a black hole. Calculations suggest that as this matter falls, it is compressed and heated to temperatures around 10^6 K. Calculate the wavelength of maximum emission for this temperature. In what part of the electromagnetic spectrum does this wavelength lie?

*26. Use the value of the solar constant given in Box 5-2 and the distance from the Earth to the Sun to calculate the luminosity of the Sun.

*27. The star Alpha Lupi (the brightest in the constellation Lupus, the Wolf) has a surface temperature of 21,600 K. How much more energy is emitted each second from each square meter of the surface of Alpha Lupi than from each square meter of the Sun's surface?

*28. Jupiter's moon Io has an active volcano named Pele whose temperature can be as high as 320°C. (a) What is the wavelength of maximum emission for the volcano at this temperature? In what part of the electromagnetic spectrum is this? (b) The average temperature of Io's surface is −150°C. Compared with a square meter of surface at this temperature, how much more energy is emitted per second from each square meter of Pele's surface?

*29. The bright star Sirius in the constellation of Canis Major (the Large Dog) has a radius of 1.67 $R_\odot$ and a luminosity of 25 $L_\odot$. (a) What is the energy flux at the surface of Sirius? (b) What is the star's surface temperature?

30. In Figure 5-12 you can see two distinct dark lines at the boundary between the orange and yellow parts of the Sun's spectrum (in the center of the third colored band from the top of the figure). The wavelengths of these dark lines are 588.99 and 589.59 nm. What do you conclude from this about the chemical composition of the Sun's atmosphere? (*Hint:* See Section 5-6.)

31. Since the 1970s, instruments on board balloons and spacecraft have detected 511-keV photons coming from the direction of the center of our Galaxy. (The prefix k means *kilo*, or thousand, so 1 keV = 10^3 eV.) What is the wavelength of these photons? To what part of the electromagnetic spectrum do these photons belong?

32. (a) Calculate the wavelength of P_δ, the fourth wavelength in the Paschen series. (b) Draw a schematic diagram of the hydrogen atom and indicate the electron transition that gives rise to this spectral line. (c) In what part of the electromagnetic spectrum does this wavelength lie?

33. (a) Calculate the wavelength of H_{13}, the spectral line for an electron transition between the $n = 15$ and $n = 2$ orbits of hydrogen. (b) In what part of the electromagnetic spectrum does this wavelength lie? Use this to explain why Figure 5-19 is labeled R I V U̲ X G.

34. Certain interstellar clouds contain a very cold, very thin gas of hydrogen atoms. Ultraviolet radiation with any wavelength shorter than 91.2 nm cannot pass through this gas; instead, it is absorbed. Explain why.

35. (a) Can a hydrogen atom in the ground state absorb an H-alpha (H_α) photon? Explain why or why not. (b) Can a hydrogen atom in the $n = 2$ state absorb a Lyman-alpha (L_α) photon? Explain why or why not.

36. An imaginary atom has just 3 energy levels: 0 eV, 1 eV, and 3 eV. Draw an energy-level diagram for this atom. Show all possible transitions between these energy levels. For each transition, determine the photon energy and the photon wavelength. Which transitions involve the emission or absorption of visible light?

37. The nebula NGC 2363 shown in Figure 5-16 is located within the galaxy NGC 2366 in the constellation Camelopardalis (the Giraffe). This galaxy and the nebula within it are moving away from us at 252 km/s. At what wavelength does the red H_α line of hydrogen (which causes the color of the nebula) appear in the nebula's spectrum?

38. The wavelength of H_β in the spectrum of the star Megrez in the Big Dipper (part of the constellation Ursa Major, the Great Bear) is 486.112 nm. Laboratory measurements demonstrate that the normal wavelength of this spectral line is 486.133 nm. Is the star coming toward us or moving away from us? At what speed?

39. You are given a traffic ticket for going through a red light (wavelength 700 nm). You tell the police officer that because you were approaching the light, the Doppler effect caused a blueshift that made the light appear green (wavelength 500 nm). How fast would you have had to be going for this to be true? Would the speeding ticket be justified? Explain.

Discussion Questions

40. The equation that relates the frequency, wavelength, and speed of a light wave, $\nu = c/\lambda$, can be rewritten as $c = \nu\lambda$. A friend who has studied mathematics but not much astronomy or physics might look at this equation and say: "This equation tells me that the higher the frequency ν, the greater the wave speed c. Since visible light has a higher frequency than radio waves, this means that visible light goes faster than radio waves." How would you respond to your friend?

41. (a) If you could see ultraviolet radiation, how might the night sky appear different? Would ordinary objects appear different in the daytime? (b) What differences might there be in the appearance of the night sky and in the appearance of ordinary objects in the daytime if you could see infrared radiation?

42. The accompanying visible-light image shows the star cluster NGC 3293 in the constellation Carina (the Ship's Keel). What can

(David Malin/Anglo-Australian Observatory)   R I **V** U X G

you say about the surface temperatures of most of the bright stars in this cluster? In what part of the electromagnetic spectrum do these stars emit most intensely? Are your eyes sensitive to this type of radiation? If not, how is it possible to see these stars at all? There is at least one bright star in this cluster with a distinctly different color from the others; what can you conclude about its surface temperature?

43. The human eye is most sensitive over the same wavelength range at which the Sun emits the greatest intensity of radiation. Suppose creatures were to evolve on a planet orbiting a star somewhat hotter than the Sun. To what wavelengths would their vision most likely be sensitive?

44. Why do you suppose that ultraviolet light can cause skin cancer but ordinary visible light does not?

Web/CD-ROM Questions

45. Search the World Wide Web for information about rainbows. Why do rainbows form? Why do they appear as circular arcs? Why can you see different colors?

46. Measuring Stellar Temperatures. Access the Active Integrated Media Module "Blackbody Curves" in Chapter 5 of the *Universe* Web site or CD-ROM. (**a**) Use the module to determine the range of temperatures over which a star's peak wavelength is in the visible spectrum. (**b**) Determine if any of the following stars have a peak wavelength in the visible spectrum: Rigel, $T = 11,000$ K; Deneb, $T = 8400$ K; Arcturus, $T = 4290$ K; Vega, $T = 9500$ K; Mira, $T = 2000$ K.

Observing Projects

47. Turn on an electric stove or toaster oven and carefully observe the heating elements as they warm up. Relate your observations to Wien's law and the Stefan-Boltzmann law.

48. Obtain a glass prism (or a diffraction grating, which is probably more readily available and is discussed in the next chapter) and look through it at various light sources, such as an ordinary incandescent light, a neon sign, and a mercury vapor street lamp. **Do not look at the sun! Looking directly at the Sun causes permanent eye damage or blindness.** Do you have any trouble seeing spectra? What do you have to do to see a spectrum? Describe the differences in the spectra of the various light sources you observed.

49. Use the *Starry Night Backyard*™ program to examine some distant celestial objects. First display the entire celestial sphere (select **Guides > Atlas** in the **Go** menu), then search for objects (i), (ii), and (iii) listed below (select **Find...** in the **Edit** menu). For each object, use the zoom controls at the right-hand end of the Control Panel (at the top of the main window) to adjust your view until you can see the object in detail. For each object, state whether it has a continuous spectrum, an absorption line spectrum, or an emission line spectrum, and explain your reasoning. (i) The Lagoon Nebula in Sagittarius. (*Hint:* See Figure 5-16.) (ii) M31, the great galaxy in the constellation Andromeda. (*Hint:* The light coming from this galaxy is the combined light of hundreds of billions of individual stars.) (iii) The Moon. (*Hint:* Recall from Section 3-1 that moonlight is simply reflected sunlight.)

50. Use the *Starry Night Backyard*™ program to examine the temperatures of several relatively nearby stars. First display the entire celestial sphere (select **Guides > Atlas** in the **Go** menu), then search for each of the stars listed below (select **Find...** in the **Edit** menu). Information for each star can be found by clicking on the **Info** tab at the far left of the *Starry Night Backyard*™ window. For each star, record the surface temperature (listed in the **Info** section of the window under **Other Data**). Then answer the following questions. (**a**) Which of the stars have a longer wavelength of maximum emission λ_{max} than the Sun? Which of the stars have a shorter λ_{max} than the Sun? (**b**) Which of the stars has a reddish color? (i) Altair; (ii) Procyon; (iii) Epsilon Indi; (iv) Tau Ceti; (v) Epsilon Eridani; (vi) Lalande 21185.

51. Use the *Deep Space Explorer*™ program to explore the colors of galaxies. In the left-hand part of the window under the heading **Local Universe,** click on the triangle next to the word **Explore** and then click on **Inside Virgo Cluster.** The Virgo cluster is a grouping of more than 2000 galaxies and about 9 million light-years across, located about 50 million light-years from Earth. You can zoom in and zoom out on the cluster using the buttons at the upper left of the window (an upward-pointing triangle and a downward-pointing triangle). You can also rotate the cluster by putting the mouse cursor over the image, holding down the mouse button, and moving the mouse. (On a two-button mouse, hold down the left mouse button.) (**a**) Zoom in and rotate the image to examine several different galaxies. Are all galaxies the same color? Do all galaxies contain stars of the same surface temperature? (Recall from Section 1-4 that galaxies are assemblages of stars.) (**b**) Examine several spiral galaxies like the one shown in Figure 1-7. What are the dominant colors of the inner regions and outer regions of these galaxies? What can you conclude about the surface temperatures of the brightest stars in the inner regions of spiral galaxies compared to the surface temperatures of those in the outer regions?

Collaborative Exercise

52. The Doppler effect describes how relative motion impacts wavelength. With a classmate, stand up and demonstrate each of the following: (**a**) a blueshifted source for a stationary observer; (**b**) a stationary source and an observer detecting a redshift; and (**c**) a source and an observer both moving in the same direction, but the observer is detecting a redshift. Create simple sketches to illustrate what you and your classmate did.

6 Optics and Telescopes

Many people envision astronomers spending their nights looking through the eyepiece of a telescope, as in the circa-1900 photograph of the astronomer Percival Lowell shown on the left. But most modern telescopes used in astronomical research have no eyepiece at all; instead, they record their images electronically, much like a digital camera. Included among these are all telescopes that operate in space, such as the Hubble Space Telescope shown on the right.

Whether a telescope has an eyepiece or not, and whether it is located on Earth or in space, its fundamental purpose is the same: to gather more light than can the naked eye. In many cases telescopes are used to produce images far brighter and sharper than the eye alone could ever record. Telescopes also produce finely detailed spectra of objects in space. These spectra reveal the chemical composition of nearby planets as well as of distant galaxies, and help astronomers understand the nature and evolution of astronomical objects of all kinds.

Telescopes can also extend our view of the universe by exploring the electromagnetic spectrum at wavelengths that the human eye cannot detect. Radio telescopes have mapped out the structure of our Milky Way Galaxy; with infrared telescopes,

astronomers have peered at stars in the process of formation; and gamma-ray telescopes have detected the most powerful explosions in the universe. The telescope, in all its variations, is by far the most useful tool that astronomers have for collecting data about the universe.

 WEB LINK 6.1 A classic telescope on Earth, and the Hubble Space Telescope in orbit. (left: Lowell Observatory; right: NASA) R I **V** U X G

Questions to Guide Your Reading

As you read the sections of this chapter, look for the answers to the following questions:

6-1 Why is it important that telescopes be large?

6-2 Why do most modern telescopes use a large mirror rather than a large lens?

6-3 Why are observatories in such remote locations?

6-4 Do astronomers use ordinary photographic film to take pictures of the sky? Do they actually look through large telescopes?

6-5 How do astronomers use telescopes to measure the spectra of distant objects?

6-6 Why do astronomers need telescopes that detect radio waves and other nonvisible forms of light?

6-7 Why is it useful to put telescopes in orbit?

6-1 | A refracting telescope uses a lens to concentrate incoming light at a focus

The **optical telescope**—that is, a telescope designed for use with visible light—was invented in the Netherlands in the early seventeenth century. Soon after, Galileo used one of these new inventions for his groundbreaking astronomical observations (see Section 4-5). These first telescopes used *lenses* to make distant objects appear larger and brighter. Telescopes of this same basic design are used today by many amateur astronomers. To understand telescopes of this kind, we need to understand how lenses work.

All lenses, including those used in telescopes, make use of the same physical principle: *Light travels at a slower speed in a dense substance.* Thus, although the speed of light in a vacuum is 3.00×10^8 m/s, its speed in glass is less than 2×10^8 m/s. Just as a woman's walking pace slows suddenly when she walks from a boardwalk onto a sandy beach, so light slows abruptly as it enters a piece of glass. The same woman stepping back onto a boardwalk easily resumes her original pace when she steps back onto the boardwalk; in the same way, light resumes its original speed upon exiting the glass.

A material through which light travels is called a **medium** (*plural media*). As a beam of light passes from one transparent medium into another—say, from air into glass, or from glass back into air—the direction of the light can change. This phenomenon, called **refraction**, is caused by the change in the speed of light.

ANALOGY Imagine driving a car from a smooth pavement onto a sandy beach (**Figure 6-1a**). If the car approaches the beach head-on, it slows down when it enters the sand but keeps moving straight ahead. If the car approaches the beach at an angle, however, one of the front wheels will be slowed by the sand before the other is, and the car will veer from its original direction. In the same way, a beam of light changes direction when it enters a piece of glass at an angle (Figure 6-1b).

Figure 6-2a shows the refraction of a beam of light passing through a piece of flat glass. As the beam enters the upper surface of the glass, refraction takes place, and the beam is bent to a direction more nearly perpendicular to the surface of the glass. As the beam exits from the glass back into the surrounding air, a second refraction takes place, and the beam bends in the opposite sense. (The amount of bending depends on the speed of light in the glass, so different kinds of glass produce slightly different amounts of refraction.) Because the two surfaces of the glass are parallel, the beam emerges from the glass traveling in the same direction in which it entered.

Something more useful happens if the glass is curved into a convex shape (one that is fatter in the middle than at the edges), like the lens in Figure 6-2b. When a beam of light rays passes through the lens, refraction causes all the rays to converge at a point called the **focus**. If the light rays entering the lens are all parallel, the focus occurs at a special point called the **focal point**. The distance from the lens to the focal point is called the **focal length** of the lens.

The case of parallel light rays, shown in Figure 6-2b, is not merely a theoretical ideal. The stars are so far away that light rays

(a) How cars behave

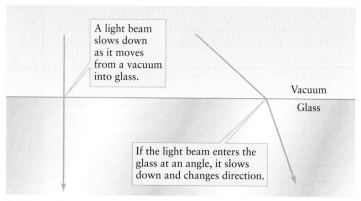

(b) How light beams behave

Figure 6-1

Refraction (a) When a car drives from smooth pavement into soft sand, it slows down. If it enters the sand at an angle, the front wheel on one side feels the drag of the sand before the other wheel, causing the car to veer to the side and change direction. (b) Similarly, light slows down when it passes from a vacuum into glass and changes direction if it enters the glass at an angle.

from them are essentially parallel, as **Figure 6-3** shows. Consequently, a lens always focuses light from an astronomical object to the focal point. If the object has a very small angular size, like a distant star, all the light entering the lens from that object converges onto the focal point. The resulting image is just a single bright dot.

But if the object is *extended*—that is, has a relatively large angular size, like the Moon or a planet—then light coming from each point on the object is brought to a focus at its own individual point. The result is an extended image that lies in the **focal plane** of the lens (**Figure 6-4**), which is a plane that includes the focal point. You can use an ordinary magnifying glass in this way to make an image of the Sun on the ground.

To use a lens to make a photograph of an astronomical object, you would place a piece of film or an electronic detector in the

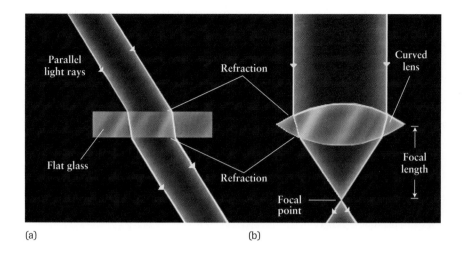

figure 6-2

Refraction and Lenses (a) Refraction is the change in direction of a light ray when it passes into or out of a transparent medium such as glass. When light rays pass through a flat piece of glass, the two refractions bend the rays in opposite directions. There is no overall change in the direction in which the light travels. (b) If the glass is in the shape of a convex lens, parallel light rays converge to a focus at a special point called the focal point. The distance from the lens to the focal point is called the focal length of the lens.

(a)
(b)

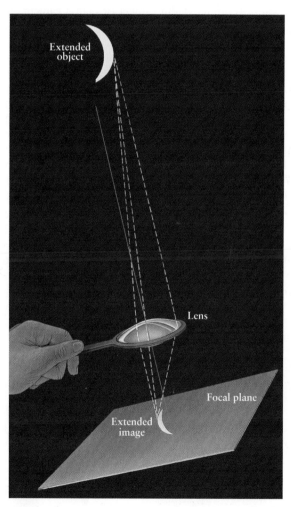

figure 6-3

Light Rays from Distant Objects Are Parallel Light rays travel away in all directions from an ordinary light source. If a lens is located very far from the light source, only a few of the light rays will enter the lens, and these rays will be essentially parallel. This is why we drew parallel rays entering the lens in Figure 6-2b.

figure 6-4

A Lens Creates an Extended Image of an Extended Object Light coming from each point on an extended object passes through a lens and produces an image of that point. All of these tiny images put together make an extended image of the entire object. The image of a very distant object is formed in a plane called the focal plane. The distance from the lens to the focal plane is called the focal length.

Figure 6-5

A Refracting Telescope A refracting telescope consists of a large-diameter objective lens with a long focal length and a small eyepiece lens of short focal length. The eyepiece lens magnifies the image formed by the objective lens in its focal plane (shown as a dashed line). To take a photograph, the eyepiece is removed and the film is placed in the focal plane.

focal plane. An ordinary film camera or digital camera works in the same way for taking photographs of relatively close objects here on Earth. Most observations for astronomical research are done in this way to produce a permanent record of the observation. Furthermore, modern imaging-recording technology is far more sensitive than the human eye, as we will see in Section 6-4. But many amateur astronomers want to view the image visually, and so they add a second lens to magnify the image formed in the focal plane. Such an arrangement of two lenses is called a **refracting telescope**, or **refractor** (Figure 6-5). The large-diameter, long-focal-length lens at the front of the telescope, called the **objective lens**, forms the image; the smaller, shorter-focal-length lens at the rear of the telescope, called the **eyepiece lens**, magnifies the image for the observer.

In addition to the focal length, the other important dimension of the objective lens of a refractor is the diameter. Compared with a small-diameter lens, a large-diameter lens captures more light, produces brighter images, and allows astronomers to detect fainter objects. (For the same reason, the iris of your eye opens when you go into a darkened room to allow you to see dimly lit objects.)

The **light-gathering power** of a telescope is directly proportional to the area of the objective lens, which in turn is proportional to the square of the lens diameter (Figure 6-6). Thus, if you double the diameter of the lens, the light-gathering power increases by a factor of $2^2 = 2 \times 2 = 4$. Box 6-1 describes how to compare the light-gathering power of different telescopes.

Because light-gathering power is so important for seeing faint objects, the lens diameter is almost always given in describing a tele-

Small-diameter objective lens: dimmer image, less detail

Large-diameter objective lens: brighter image, more detail

Figure 6-6 R I **V** U X G

Light-Gathering Power These two photographs of the galaxy M31 in Andromeda were taken using the same exposure time and at the same magnification, but with two different telescopes with objective lenses of different diameters. The right-hand photograph is brighter and shows more detail because it was made using the large-diameter lens, which intercepts more starlight than a small-diameter lens. This same principle applies to telescopes that use curved mirrors rather than lenses to collect light (see Section 6-2). (Association of Universities for Research in Astronomy)

box 6-1   TOOLS OF THE ASTRONOMER'S TRADE

Magnification and Light-Gathering Power

The magnification of a telescope is equal to the focal length of the objective divided by the focal length of the eyepiece. Telescopic eyepieces are usually interchangeable, so the magnification of a telescope can be changed by using eyepieces of different focal lengths.

EXAMPLE: A small refracting telescope has an objective of focal length 120 cm. If the eyepiece has a focal length of 4.0 cm, what is the magnification of the telescope?

Situation: We are given the focal lengths of the telescope's objective and eyepiece lenses. Our goal is to calculate the magnification provided by this combination of lenses.

Tools: We use the relationship that the magnification equals the focal length of the objective (120 cm) divided by the focal length of the eyepiece (4.0 cm).

Answer: Using this relationship,

$$\text{Magnification} = \frac{120 \text{ cm}}{4.0 \text{ cm}} = 30 \text{ (usually written as } 30\times)$$

Review: A magnification of 30× means that as viewed through this telescope, a large lunar crater that subtends an angle of 1 arcminute to the naked eye will appear to subtend an angle 30 times greater, or 30 arcminutes (one-half of a degree). This makes the details of the crater much easier to see.

If a 2.0-cm-focal-length eyepiece is used instead, the magnification will be (120 cm)/(2.0 cm) = 60×. The shorter the focal length of the eyepiece, the greater the magnification.

The light-gathering power of a telescope depends on the diameter of the objective lens; it does not depend on the focal length. The light-gathering power is proportional to the square of the diameter. As an example, a fully dark adapted human eye has a pupil diameter of about 5 mm. By comparison, a small telescope whose objective lens is 5 cm in diameter has 10 times the diameter and $10^2 = 100$ times the light-gathering power of the eye. (Recall that there are 10 mm in 1 cm.)

Hence, this telescope allows you to see objects 100 times fainter than you can see without a telescope.

EXAMPLE: The same relationships apply to reflecting telescopes, discussed in Section 6-2. Each of the two Keck telescopes on Mauna Kea in Hawaii (discussed in Section 6-3; see Figure 6-16) uses a concave mirror 10 m in diameter to bring starlight to a focus. How many times greater is the light-gathering power of either Keck telescope compared to that of the human eye?

Situation: We are given the diameters of the pupil of the human eye (5 mm) and of the mirror of either Keck telescope (10 m). Our goal is to compare the light-gathering powers of these two optical instruments.

Tools: We use the relationship that light-gathering power is proportional to the square of the diameter of the area that gathers light. Hence, the *ratio* of the light-gathering powers is equal to the square of the ratio of the diameters.

Answer: We first calculate the ratio of the diameter of the Keck mirror to the diameter of the pupil. To do this, we must first express both diameters in the same units. Because there are 1000 mm in 1 meter, the diameter of the Keck mirror can be expressed as

$$10 \text{ m} \times \frac{1000 \text{ mm}}{1 \text{ m}} = 10{,}000 \text{ mm}$$

Thus, the light-gathering power of either of the Keck telescopes is greater than that of the human eye by a factor of

$$\frac{(10{,}000 \text{ mm})^2}{(5 \text{ mm})^2} = (2000)^2 = 4 \times 10^6, \text{ or } 4{,}000{,}000$$

Review: Either Keck telescope can gather *4 million* times as much light as a human eye. When it comes to light-gathering power, the bigger the telescope, the better!

---

scope. For example, the Lick telescope on Mount Hamilton in California is a 90-cm refractor, which means that it is a refracting telescope whose objective lens is 90 cm in diameter. By comparison, Galileo's telescope of 1610 was a 3-cm refractor. The Lick telescope has an objective lens 30 times larger in diameter, and so has 30 × 30 = 900 times the light-gathering power of Galileo's instrument.

In addition to their light-gathering power, telescopes are useful because they magnify distant objects. As an example, the angular

diameter of the Moon as viewed with the naked eye is about 0.5°. But when Galileo viewed the Moon through his telescope, its apparent angular diameter was 10°, large enough so that he could identify craters and mountain ranges. The **magnification,** or **magnifying power,** of a telescope is the ratio of an object's angular diameter seen through the telescope to its naked-eye angular diameter. Thus, the magnification of Galileo's telescope was 10°/0.5° = 20 times, usually written as 20×.

The magnification of a refracting telescope depends on the focal lengths of both of its lenses:

$$\text{Magnification} = \frac{\text{focal length of objective lens}}{\text{focal length of eyepiece lens}}$$

This formula shows that using a long-focal-length objective lens with a short-focal-length eyepiece gives a large magnification. Box 6-1 illustrates how this formula is used.

CAUTION   Many people think that the primary purpose of a telescope is to magnify images. But in fact magnification is *not* the most important aspect of a telescope. The reason is that there is a limit to how sharp any astronomical image can be, due either to the blurring caused by the Earth's atmosphere or to fundamental limitations imposed by the nature of light itself. (We will describe these effects in more detail in Section 6-3.) Magnifying a blurred image may make it look bigger but will not make it any clearer. Thus, beyond a certain point, there is nothing to be gained by further magnification. Astronomers put much more store in the light-gathering power of a telescope than in its magnification. Greater light-gathering power means brighter images, which makes it easier to see faint details.

If you were to build a telescope like that in Figure 6-5 using only the instructions given so far, you would probably be disappointed with the results. The problem is that a lens bends different colors of light through different angles, just as a prism does (recall Figure 5-3). As a result, different colors do not focus at the same point, and stars viewed through a telescope that uses a simple lens are surrounded by fuzzy, rainbow-colored halos. **Figure 6-7a** shows this optical defect, called **chromatic aberration.**

One way to correct for chromatic aberration is to use an objective lens that is not just a single piece of glass. Different types of glass can be manufactured by adding small amounts of chemicals to the glass when it is molten. Because of these chemicals, the speed of light varies slightly from one kind of glass to another, and the refractive properties vary as well. If a thin lens is mounted just behind the main objective lens of a telescope, as shown in Figure 6-7b, and if the telescope designer carefully chooses two different kinds of glass for these two lenses, different colors of light can be brought to a focus at the same point.

Chromatic aberration is only the most severe of a host of optical problems that must be solved in designing a high-quality refracting telescope. Master opticians of the nineteenth century devoted their careers to solving these problems, and several magnificent refractors were constructed in the late 1800s (**Figure 6-8**).

Unfortunately, there are several negative aspects of refractors that even the finest optician cannot overcome. First, because faint light must readily pass through the objective lens, the glass from which the lens is made must be totally free of defects, such as the bubbles that frequently form when molten glass is poured into a mold. Such defect-free glass is extremely expensive. Second, glass is opaque to certain kinds of light. Ultraviolet light is absorbed almost completely, and even visible light is dimmed substantially as it passes through the thick slab of glass that makes up the objective lens. Third, it is impossible to produce a large lens that is entirely free of chromatic aberration. Fourth, because the lens can be supported only around its edges, it tends to sag and distort under its own weight. This has adverse effects on the image clarity.

For these reasons and more, few major refractors have been built since the beginning of the twentieth century. Instead, astronomers have avoided all of the limitations of refractors by building telescopes that use a mirror instead of a lens to form an image.

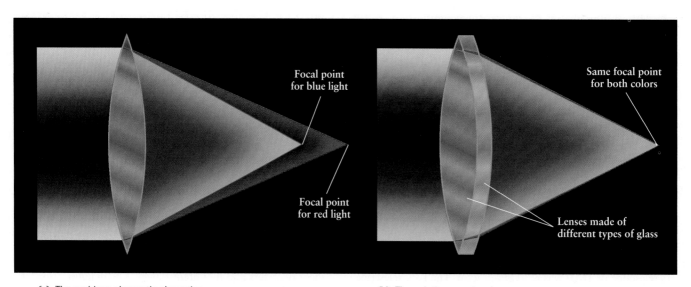

(a) The problem: chromatic aberration   (b) The solution: use two lenses

figure 6-7

Chromatic Aberration (a) A single lens suffers from a defect called chromatic aberration, in which different colors of light are brought to a focus at different distances from the lens. (b) This problem can be corrected by adding a second lens made from a different kind of glass.

(a)

(b)

Figure 6-8 R I **V** U X G

A Large Refracting Telescope (a) This giant refractor, built in 1897, is housed at Yerkes Observatory near Chicago. The objective lens is at the upper end of the telescope tube, and the eyepiece is at the lower end. The telescope tube is 19.5 m (64 ft) long; it has to be this long because the focal length of the objective is just under 19.5 m (see Figure 6-5). The

entire observatory floor can be raised or lowered to keep the eyepiece within reach. (b) This historical photograph shows the astronomer George van Biesbrock with the objective lens of the Yerkes refractor. This lens, the largest ever made, is 102 cm (40 in.) in diameter. (Yerkes Observatory)

6-2 | A reflecting telescope uses a mirror to concentrate incoming light at a focus

Almost all modern telescopes form an image using the principle of **reflection**. To understand reflection, imagine drawing a dashed line perpendicular to the surface of a flat mirror at the point where a light ray strikes the mirror (**Figure 6-9**). The angle i between the

incident (arriving) light ray and the perpendicular is always equal to the angle r between the *reflected* ray and the perpendicular.

In 1663, the Scottish mathematician James Gregory first proposed a telescope using reflection from a concave mirror—one that is fatter at the edges than at the middle. Such a mirror makes parallel light rays converge to a focus (**Figure 6-10**). The distance between the reflecting surface and the focus is the focal length of the mirror. A telescope that uses a curved mirror to make an image

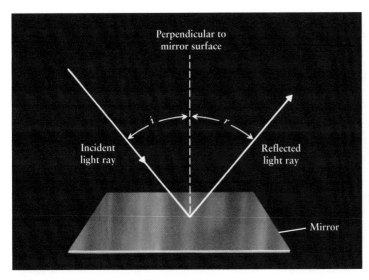

Figure 6-9

Reflection by a Flat Mirror The angle at which a beam of light approaches a mirror, called the angle of incidence (i), is always equal to the angle at which the beam is reflected from the mirror, called the angle of reflection (r). That is, $i = r$.

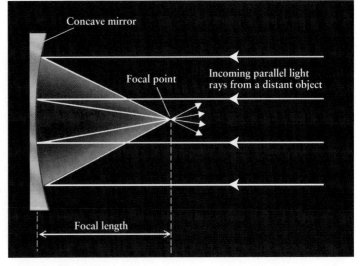

Figure 6-10

Reflection by a Concave Mirror A concave mirror causes parallel light rays to converge to a focus at the focal point. The distance between the mirror and the focus is the focal length of the mirror.

of a distant object is called a **reflecting telescope, or reflector.** Using terminology similar to that used for refractors, the mirror that forms the image is called the **objective mirror** or **primary mirror.**

To make a reflector, an optician grinds and polishes a large slab of glass into the appropriate concave shape. The glass is then coated with silver, aluminum, or a similar highly reflective substance. Because light reflects off the surface of the glass rather than passing through it, defects within the glass—which would have very negative consequences for the objective lens of a refracting telescope—have no effect on the optical quality of a reflecting telescope.

Another advantage of reflectors is that they do not suffer from the chromatic aberration that plagues refractors. This is because reflection is not affected by the wavelength of the incoming light, so all wavelengths are reflected to the same focus. (A small amount of chromatic aberration may arise if the image is viewed using an eyepiece lens.) Furthermore, the mirror can be fully supported by braces on its back, so that a large, heavy mirror can be mounted without much danger of breakage or surface distortion.

Although a reflecting telescope has many advantages over a refractor, the arrangement shown in Figure 6-10 is not ideal. One problem is that the focal point is in front of the objective mirror. If you try to view the image formed at the focal point, your head will block part or all of the light from reaching the mirror.

To get around this problem, in 1668 Isaac Newton simply placed a small, flat mirror at a 45° angle in front of the focal point, as sketched in **Figure 6-11a.** This secondary mirror deflects the light rays to one side, where Newton placed an eyepiece lens to magnify the image. A reflecting telescope with this optical design is appropriately called a **Newtonian reflector.** The magnifying power of a Newtonian reflector is calculated in the same way as for a refractor: The focal length of the objective mirror is divided by the focal length of the eyepiece (see Box 6-1).

Later astronomers modified Newton's original design. The objective mirrors of some modern reflectors are so large that an astronomer could actually sit in an "observing cage" at the undeflected focal point directly in front of the objective mirror. (In practice, riding in this cage on a winter's night is a remarkably cold and uncomfortable experience.) This arrangement is called a **prime focus** (Figure 6-11b). It usually provides the highest-quality image, because there is no need for a secondary mirror (which might have imperfections).

Another popular optical design, called a **Cassegrain focus** after the French contemporary of Newton who first proposed it, also has a convenient, accessible focal point. A hole is drilled directly through the center of the primary mirror, and a convex secondary mirror placed in front of the original focal point reflects the light rays back through the hole (Figure 6-11c).

A fourth design is useful when there is optical equipment too heavy or bulky to mount directly on the telescope. Instead, a series of mirrors channels the light rays away from the telescope to a remote focal point where the equipment is located. This design is called a **coudé focus,** from a French word meaning "bent like an elbow" (Figure 6-11d).

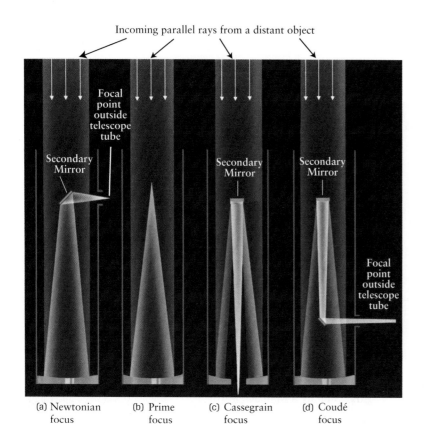

(a) Newtonian focus (b) Prime focus (c) Cassegrain focus (d) Coudé focus

ANIMATION 6.2

figure 6-11

Designs for Reflecting Telescopes Four of the most popular optical designs for reflecting telescopes are shown here. **(a)** The Newtonian focus is used today on some small reflecting telescopes for amateur astronomers. **(b)** Prime focus is used only on some large telescopes; an observer or instrument is placed directly at the focal point, within the barrel of the telescope. **(c)** The Cassegrain focus is used on reflecting telescopes of all sizes, from 90-mm (3.5-in.) reflectors used by amateur astronomers to giant research telescopes on Mauna Kea (see Figure 6-14). **(d)** The coudé focus is useful when large and heavy optical apparatus is to be used at the focal point. Light reflects off the objective mirror to a secondary mirror, then back down to an angled tertiary mirror. With this arrangement, the heavy apparatus at the focal point does not have to move when the telescope is repositioned.

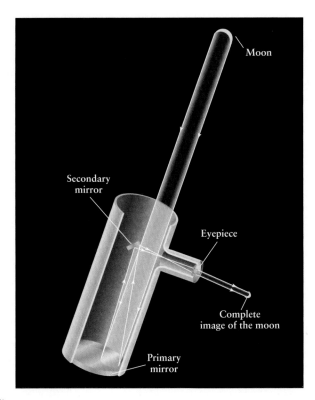

figure 6-12

The Secondary Mirror Does Not Cause a Hole in the Image This illustration shows how even a small portion of the primary (objective) mirror of a reflecting telescope can make a complete image of the Moon. Thus, the secondary mirror does not cause a black spot or hole in the image. (It does, however, make the image a bit dimmer by reducing the total amount of light that reaches the primary mirror.)

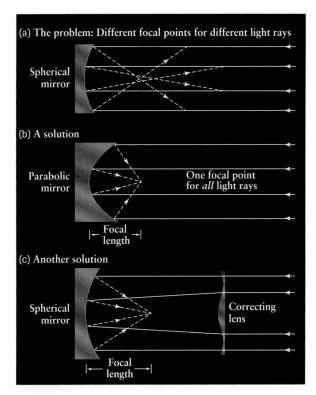

figure 6-13

Spherical Aberration (a) Different parts of a spherically concave mirror reflect light to slightly different focal points. This effect, called spherical aberration, causes image blurring. This difficulty can be corrected by either (b) using a parabolic mirror or (c) using a correcting lens in front of the mirror.

You might think that the secondary mirror in the Newtonian, Cassegrain, and coudé designs shown in Figure 6-11 would cause a black spot or hole in the center of the telescope image. But this does not happen. The reason is that light from every part of the object lands on every part of the primary, objective mirror. Hence, any portion of the mirror can itself produce an image of the distant object, as Figure 6-12 shows. The only effect of the secondary mirror is that it prevents part of the light from reaching the objective mirror, which reduces somewhat the light-gathering power of the telescope.

A reflecting telescope must be designed to minimize a defect called **spherical aberration** (Figure 6-13). At issue is the precise shape of a mirror's concave surface. A spherical surface is easy to grind and polish, but different parts of a spherical mirror have slightly different focal lengths (Figure 6-13a). This results in a fuzzy image.

One common way to eliminate spherical aberration is to polish the mirror's surface to a parabolic shape, because a parabola reflects parallel light rays to a common focus (Figure 6-13b). Unfortunately, the astronomer then no longer has a wide-angle view. Furthermore, unlike spherical mirrors, parabolic mirrors suffer from a defect

called **coma,** wherein star images far from the center of the field of view are elongated to look like tiny teardrops. A different approach is to use a spherical mirror, thus minimizing coma, and to place a thin correcting lens at the front of the telescope to eliminate spherical aberration (Figure 6-13c). This approach is only used on relatively small reflecting telescopes for amateur astronomers.

As of 2004, there were 13 optical reflectors in operation or under construction with primary mirrors more than 8 meters (26.2 feet) in diameter. These are listed in Table 6-1. Figure 6-14 shows the objective and secondary mirrors of one of these, the Gemini North telescope (also called Gillett) atop the dormant volcano Mauna Kea in Hawaii. A near-twin of this telescope, called Gemini South, is in Cerro Pachón, Chile. These twins will allow astronomers to observe both the northern and southern parts of the celestial sphere with essentially the same state-of-the-art instrument. Two other "twins" are the side-by-side 8.4-m objective mirrors of the Large Binocular Telescope in Arizona. Combining the light from these two mirrors gives double the light-gathering power, equivalent to a single 11.8-m mirror. In the following section we will learn about the special characteristics of the other telescopes listed in Table 6-1.

table 6-1 | The World's Largest Optical Telescopes

Telescope	Location	Year of completion	Mirror diameter (m)
Gran Telescopio Canarias	La Palma, Canary Islands, Spain	2004	10.4
Keck II	Mauna Kea, Hawaii	1996	10.0
Keck I	Mauna Kea, Hawaii	1993	10.0
Hobby-Eberly Telescope	McDonald Observatory, Texas	1998	11.0*
South African Large Telescope	Sutherland, South Africa	2004	9.2
Large Binocular Telescope	Mount Graham, Arizona	2004–05	Two 8.4
Subaru	Mauna Kea, Hawaii	1999	8.3
VLT UT 1–Antu	Cerro Paranal, Chile	1998	8.2
VLT UT 2–Kueyen	Cerro Paranal, Chile	1999	8.2
VLT UT 3–Melipal	Cerro Paranal, Chile	2000	8.2
VLT UT 4–Yepun	Cerro Paranal, Chile	2000	8.2
Gemini North (Gillett)	Mauna Kea, Hawaii	1999	8.1
Gemini South	Cerro Pachón, Chile	2000	8.1

*The objective mirror of the Hobby-Eberly Telescope is 11.0 m in diameter, but in operation only an area of 9.2 m in diameter is used to collect light.

figure 6-14 R I **V** U X G

The Gemini North Telescope on Mauna Kea This photograph is a view down the Gemini North telescope, looking toward the 8.1-meter objective mirror (1). Light incident on this mirror is reflected toward the 1.0-meter secondary mirror (2), then through the hole in the objective mirror (3) to the Cassegrain focus (see Figure 6-11c). (NOAO/AURA/NSF)

In addition to the "giants" listed in Table 6-1, several other reflectors around the world have objective mirrors between 3 and 6 meters in diameter, and dozens of smaller but still powerful telescopes have mirrors in the range of 1 to 3 meters. There are thousands of professional astronomers, each of whom has several ongoing research projects, and thus the demand for all of these telescopes is high. On any night of the year, nearly every research telescope in the world is being used to explore the universe.

6-3 | Telescope images are degraded by the blurring effects of the atmosphere and by light pollution

In addition to providing a brighter image, a large telescope also helps achieve a second major goal: It produces star images that are sharp and crisp. A quantity called **angular resolution** gauges how well fine details can be seen. Poor angular resolution causes star images to be fuzzy and blurred together.

To determine the angular resolution of a telescope, pick out two adjacent stars whose separate images are just barely discernible (Figure 6-15). The angle θ (the Greek letter theta) between these stars is the telescope's angular resolution; the *smaller* that angle, the finer the details that can be seen and the sharper the image.

When you are asked to read the letters on an eye chart, what's being measured is the angular resolution of your eye. If you have 20/20 vision, the angular resolution θ of your eye is about 1 arcminute, or 60 arcseconds. (You may want to review the definitions of these angular measures in Section 1-5.) Hence, with the naked eye it is impossible to distinguish two stars less than 1 arcminute apart or to see details on the Moon with an angular size smaller than

Two light sources with angular separation greater than angular resolution of telescope: Two sources easily distinguished

(a)

Light sources moved closer so that angular separation equals angular resolution of telescope: Just barely possible to tell that there are two sources

(b)

figure 6-15 R I **V** U X G

Angular Resolution The angular resolution of a telescope indicates the sharpness of the telescope's images. (a) This telescope view shows two sources of light whose angular separation is greater than the angular resolution. (b) The light sources have been moved together so that their angular separation is equal to the angular resolution. If the sources were moved any closer together, the telescope image would show them as a single source.

this. All the planets have angular sizes (as seen from Earth) of 1 arcminute or less, which is why they appear as featureless points of light to the naked eye.

One factor limiting angular resolution is **diffraction,** which is the tendency of light waves to spread out when they are confined to a small area like the lens or mirror of a telescope. (A rough analogy is the way water exiting a garden hose sprays out in a wider angle when you cover part of the end of the hose with your thumb.) As a result of diffraction, a narrow beam of light tends to spread out within a telescope's optics, thus blurring the image. If diffraction were the only limit, the angular resolution of a telescope would be given by the formula

Diffraction-limited angular resolution

$$\theta = 2.5 \times 10^5 \frac{\lambda}{D}$$

θ = diffraction-limited angular resolution of a telescope, in arcseconds

λ = wavelength of light, in meters

D = diameter of telescope objective, in meters

For a given wavelength of light, using a telescope with an objective of *larger* diameter D *reduces* the amount of diffraction and makes the angular resolution θ *smaller* (and hence better). For example, with visible light with wavelength 600 nm, or 6×10^{-7} m, the diffraction-limited resolution of a 20-cm (8-in.) telescope of the sort used by amateur astronomers would be $\theta = (2.5 \times 10^5)(6 \times 10^{-7}$ m$)/(0.20$ m$) = 0.75$ arcsec. With either of the 10-meter Keck telescopes (see Table 6-1), the angular resolution would be 0.015 arcsec, or 40 times better.

In practice, however, ordinary optical telescopes cannot achieve such fine angular resolution. The problem is that turbulence in the air causes star images to jiggle around and twinkle. Even through the largest telescopes, a star still looks like a tiny blob rather than a pinpoint of light. A measure of the limit that atmospheric turbulence places on a telescope's resolution is called the **seeing disk**. This disk is the angular diameter of a star's image broadened by turbulence. The size of the seeing disk varies from one observatory site to another and from one night to another. At the observatories on Kitt Peak in Arizona and Cerro Tololo in Chile, the seeing disk is typically around 1 arcsec. Some of the very best conditions in the world can be found at the observatories atop Mauna Kea in Hawaii, where the seeing disk is often as small as 0.5 arcsec. This is one reason why so many telescopes have been built there (**Figure 6-16**).

In many cases the angular resolution of a telescope is even worse than the limit imposed by the seeing disk. This occurs if the objective mirror deforms even slightly due to variations in air temperature or flexing of the telescope mount. To combat this, most of the large telescopes listed in Table 6-1 are equipped with an **active optics** system. Such a system adjusts the mirror shape every few seconds to help keep the telescope in optimum focus and properly aimed at its target.

Changing the mirror shape is also at the heart of a more refined technique called **adaptive optics**. The goal of this technique is to compensate for atmospheric turbulence, so that the angular resolution can be smaller than the size of the seeing disk and can even approach the theoretical limit set by diffraction. Turbulence causes the image of a star to "dance" around erratically. Optical sensors monitor this dancing motion 10 to 100 times per second, and a powerful computer rapidly calculates the mirror shape needed to compensate. Fast-acting mechanical devices called *actuators* then deform the mirror accordingly, at a much faster rate than in an active optics system. (In some adaptive optics systems, the actuators deform a small secondary mirror rather than the large objective mirror.)

The views of the planet Neptune in **Figure 6-17** show the dramatic improvement in angular resolution possible with adaptive optics. Images made with adaptive optics are nearly as sharp as if the telescope were in the vacuum of space, where there is no atmospheric distortion whatsoever and the only limit on angular resolution is diffraction. Several of the telescopes listed in Table 6-1 are designed to make use of adaptive optics. At present, however, adaptive optics can be used only with rather bright objects.

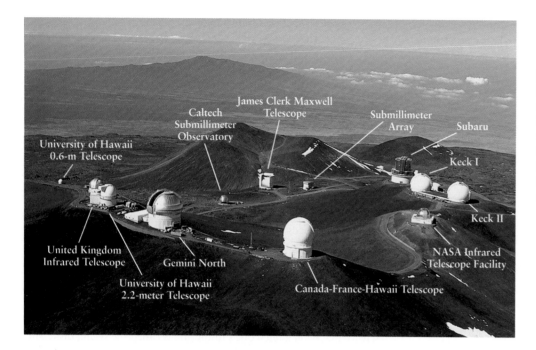

figure 6-16   R I **V** U X G
The Telescopes of Mauna Kea The summit of Mauna Kea—an extinct Hawaiian volcano that reaches more than 4100 m (13,400 ft) above the waters of the Pacific—has nighttime skies that are unusually clear, still, and dark. To take advantage of these superb viewing conditions, Mauna Kea has become the home of many powerful telescopes. (Richard J. Wainscoat, University of Hawaii)

CAUTION

The images in Figure 6-17 are **false color** images: They do not represent the true color of the planet Neptune. False color is often used when the image is made using wavelengths that the eye cannot detect, as with the infrared images in Figure 6-17. A different use of false color is to indicate the relative brightness of different parts of the image, as in the infrared image of a person in Figure 5-9. Throughout this book, we'll always point out when false color is used in an image.

Perhaps the finest angular resolution will be obtained with the Very Large Telescope (VLT), a project of the multinational European Southern Observatory. The VLT is actually four reflecting

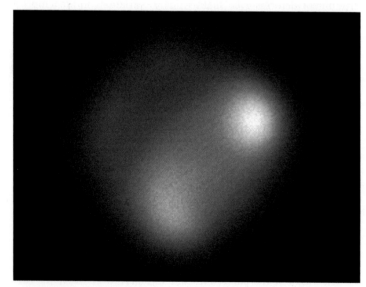

(a) Neptune viewed without adaptive optics

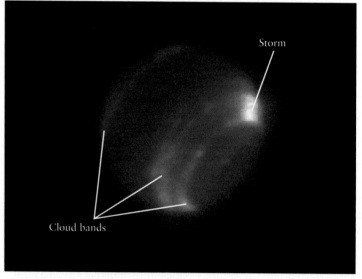

(b) Neptune viewed with adaptive optics

figure 6-17   R **I** V U X G

Using Adaptive Optics to "Unblur" Telescope Images Both of these false-color infrared images of the planet Neptune were made with the 10.0-m Keck II telescope on Mauna Kea. (a) Without adaptive optics, the planet's image is blurred across a seeing disk of 0.4 arcsecond by turbulence in the atmosphere. (b) With an adaptive optics system turned on, the angular resolution improves to 0.05 arcsecond, revealing cloud bands and a giant storm in Neptune's atmosphere. (Courtesy of the W. M. Keck Observatory) telescopes located in adjacent buildings at Cerro Paranal, Chile (Figure 6-18). Each telescope has a relatively flexible 8.2-meter objective mirror equipped with 214 active optics actuators for controlling the mirror's shape. The four reflectors can be used individually to look at different objects, but it will soon be possible to observe the same object simultaneously with all four telescopes and to combine the light received by all four. This will give the same light-gathering power as a single 16.4-m (53.8-ft) mirror. What is more, because the reflectors are spread out over a distance of 200 m (670 ft, or more than two football fields), the angular resolution will be that of a 200-m telescope. In this way it will be possible to achieve the amazing resolution of 0.001 arcsec, which corresponds to being able to distinguish the two headlights on a car located on the Moon! The Keck I and II telescopes on Mauna Kea (see Figure 6-16) will be used together in a similar way.

Light from city street lamps and from buildings also degrades telescope images. This **light pollution** illuminates the sky, making it more difficult to see the stars. You can appreciate the problem if you have ever looked at the night sky from a major city. Only a few of the very brightest stars can be seen, as against the thousands that can be seen with the naked eye in the desert or the mountains. To avoid light pollution, observatories are built in remote locations far from any city lights.

Unfortunately, the expansion of cities has brought light pollution to observatories that in former times had none. As an example, the growth of Tucson, Arizona, has had deleterious effects on observations at the nearby Kitt Peak National Observatory. Efforts have been made to have cities adopt light fixtures that provide safe illumination for their citizens but produce little light pollution. These efforts have met with only mixed success.

One factor over which astronomers have absolutely no control is the weather. Optical telescopes cannot see through clouds, so it is important to build observatories where the weather is usually good. One advantage of mountaintop observatories such as Mauna Kea or Cerro Paranal is that most clouds form at altitudes below the observatory, giving astronomers a better chance of having clear skies.

In many ways the best location for a telescope is in orbit around Earth, where it is unaffected by weather, light pollution, or atmospheric turbulence. We will discuss orbiting telescopes in Section 6-7.

6-4 | An electronic device is commonly used to record the image at a telescope's focus

Telescopes provide astronomers with detailed pictures of distant objects. The task of recording these pictures is called **imaging.**

Astronomical imaging really began in the nineteenth century with the invention of photography. It was soon realized that this new invention was a boon to astronomy. By taking long exposures with a camera mounted at the focus of a telescope, an astronomer can record features too faint to be seen by simply looking through the telescope. Such long exposures can reveal details in galaxies, star clusters, and nebulae that would not be visible to an astronomer looking through a telescope. Indeed, most large, modern telescopes do not have eyepieces at all.

Unfortunately, photographic film is not a very efficient light detector. Only about 1 out of every 50 photons striking photographic film triggers the chemical reaction needed to produce an image. Thus, roughly 98% of the light falling onto photographic film is wasted.

The most sensitive light detector currently available to astronomers is the **charge-coupled device (CCD).** At the heart of a CCD is a semiconductor wafer divided into an array of small light-sensitive squares called picture elements or, more commonly, **pixels** (Figure 6-19). For example, some state-of-the-art CCDs for astronomy have more than 16 million pixels arranged in 4096 rows by 4096 columns. They have about a thousand times more pixels per square centimeter than on a typical computer screen, which means that a CCD of this type can record very fine image details. CCDs with smaller numbers of pixels are used in digital cameras, scanners, and fax machines.

When an image from a telescope is focused on the CCD, an electric charge builds up in each pixel in proportion to the number of photons falling on that pixel. When the exposure is finished, the amount of charge on each pixel is read into a computer, where

Figure 6-18 R I ■V U X G

The Very Large Telescope The four domes of the European Southern Observatory's Very Large Telescope (VLT) are atop a 2640-m (8660-ft) mountain in the Atacama desert of Chile, where the skies are very dark and cloudless and the seeing is excellent. The four 8.2-m telescopes bear the names of celestial objects in the local Mapuche language: Antu (Sun), Kueyen (Moon), Melipal (Southern Cross), and Yepun (Venus, the evening star). (European Southern Observatory)

Figure 6-19   R I ☑ U X G

A Charge-Coupled Device (CCD) This tiny silicon square records images in its 16,777,216 light-sensitive pixels. At the end of each exposure, the recorded data is transferred to a computer. Somewhat less elaborate CCDs are used in consumer electronics. (Institute for Astronomy, University of Hawaii, and Roger Ressmeyer ©1993 Corbis)

the resulting image can be stored in digital form and either viewed on a monitor or printed out. Compared with photographic film, CCDs are some 35 times more sensitive to light (they commonly respond to 70% of the light falling on them, versus 2% for film), can record much finer details, and respond more uniformly to light of different colors. Figure 6-20 shows the dramatic difference between photographic and CCD images. The great sensitivity of CCDs also makes them useful for **photometry,** which is the measurement of the brightnesses of stars and other astronomical objects.

In the modern world of CCD astronomy, astronomers need no longer spend the night in the unheated dome of a telescope. Instead, they operate the telescope electronically from a separate control room, where the electronic CCD images can be viewed on a computer monitor. The control room need not even be adjacent to the telescope. Although the Keck I and II telescopes (see Figure 6-16) are at an altitude of 4100 m (13,500 feet), astronomers can now make observations from a facility elsewhere on the island of Hawaii that is much closer to sea level. This saves the laborious drive to the summit of Mauna Kea and eliminates the need for astronomers to acclimate to the high altitude.

Most of the images that you will see in this book were made with CCDs. Because of their extraordinary sensitivity and their ability to be used in conjunction with computers, CCDs have attained a role of central importance in astronomy.

(a) Using photographic film

(b) Using a CCD

(c) Combined CCD image

Figure 6-20   R I ☑ U X G

Imaging: Photographic Film Versus CCD (a) This negative print (black stars and white sky) shows a portion of the sky as imaged with a 4-meter telescope and photographic film. (b) This negative image shows the same region of the sky as seen with the same telescope, but with the photographic film replaced by a CCD. You can see dozens of stars and galaxies that did not appear at all in the photograph. (c) To produce this color positive view of the same region, a series of CCD images were made using different color filters. These were then combined using a computer image-processing program. (Patrick Seitzer, National Optical Astronomy Observatories)

6-5 | Spectrographs record the spectra of astronomical objects

We saw in Section 5-6 how the spectrum of an astronomical object provides a tremendous amount of information about that object, including its chemical composition and temperature. This is why measuring spectra, or **spectroscopy**, is one of the most important uses of telescopes. Indeed, some telescopes (such as the Hobby-Eberly Telescope listed in Table 6-1) are designed solely for measuring the spectra of distant, faint objects; they are never used for imaging.

An essential tool of spectroscopy is the **spectrograph**, a device that records spectra on a CCD or photographic film. This optical device is mounted at the focus of a telescope. Figure 6-21 shows an older design for a spectrograph, in which a prism is used to form the spectrum of a planet, star, or galaxy. A slit, two lenses, and the prism are arranged to focus the spectrum onto a small piece of photographic film. Next, the exposed portion of the photographic plate is covered, and light from a hot gas (usually an element like iron or argon) is focused on the spectrograph slit. This result is a *comparison spectrum* above and below the spectrum of the planet, star, or galaxy (Figure 6-22). The wavelengths of the bright spectral lines of the comparison spectrum are known from laboratory experiments and can therefore serve as reference markers.

There are drawbacks to this old-fashioned spectrograph. A prism does not disperse the colors of the rainbow evenly: Blue and violet portions of the spectrum are spread out more than the red portion. In addition, because the blue and violet wavelengths must pass through more of the prism's glass than do the red wavelengths (examine Figure 5-3), light is absorbed unevenly across the spectrum. Indeed, a glass prism is opaque to near-ultraviolet light.

A better device for breaking starlight into the colors of the rainbow is a **diffraction grating**, or **grating** for short. This is a piece of glass on which thousands of closely spaced parallel lines have been cut. Some of the finest diffraction gratings have more than 10,000 lines per centimeter, which are usually cut by drawing a diamond back and forth across the glass. The spacing of the lines must be very regular, and the best results are obtained when the grooves are beveled. When light is shone on a diffraction grating, a spectrum is produced by the way in which light waves leaving different parts of the grating interfere with each other. (This same effect produces the

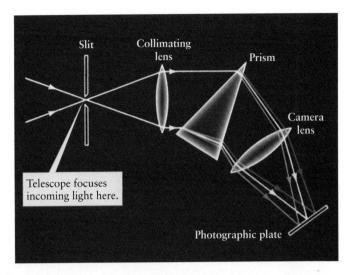

figure 6-21

A Prism Spectrograph This optical device uses a prism to break up the light from a source into a spectrum. The collimating lens directs incoming light rays so that they enter the prism parallel to one another. A second lens, called a camera lens, then focuses the spectrum onto a photographic plate.

rainbow of colors you see reflected from a compact disc or DVD. Information is stored on the disc in a series of closely spaced pits, which act as a diffraction grating.) Figure 6-23 shows the design of a modern grating spectrograph.

In Figure 6-23 a CCD, instead of a photographic plate, is placed at the focus of the spectrograph. When the exposure is finished, electronic equipment measures the charge that has accumulated in each pixel. These data are used to graph light intensity against wavelength. Dark absorption lines in the spectrum appear as depressions or valleys on the graph, while bright emission lines appear as peaks. Figure 6-24 compares two ways of exhibiting a spectrum in which several absorption lines appear. Later in this book we shall see spectra presented in both these ways.

Spectrum of star

Comparison spectrum of iron (at the observatory on Earth)

figure 6-22 R I **V** U X G

A Spectrogram Running across the middle of this photograph is the spectrum of a star. Note the many dark absorption lines, which are characteristic of the light from stars (see Section 5-6). Above and below the star's spectrum are emission lines produced by an iron arc at the observatory. These emission lines serve as a comparison spectrum. (Figure 5-15 shows another example of a spectrogram.) (Palomar Observatory)

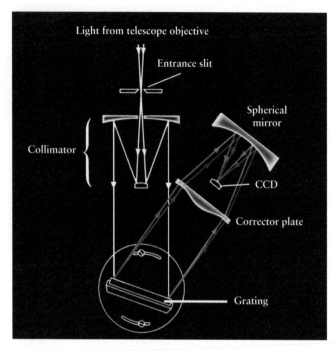

Figure 6-23

A Grating Spectrograph This optical device uses a diffraction grating to break up the light from a source into a spectrum. The collimator ensures that light rays striking the grating are parallel. A corrector lens and mirror then focus the spectrum onto a CCD, which is far more sensitive than a photographic plate.

Figure 6-24

Two Representations of a Spectrum This drawing compares two ways of displaying a spectrum. When a photographic plate is placed at the focus of a spectrograph, a rainbow-colored spectrum is obtained. The dark absorption lines in this example are the Balmer lines of hydrogen (see Section 5-8). When a CCD is placed at the focus of a spectrograph, a computer program can be used to convert the CCD's data into a graph of intensity versus wavelength. Note that absorption lines appear as dips on such a graph.

6-6 | A radio telescope uses a large concave dish to reflect radio waves to a focus

 SCIENCE IN PROCESS *Astronomers use nonvisible wavelengths to reveal hidden aspects of the universe*

For thousands of years, all the information that astronomers gathered about the universe was based on ordinary visible light. In the twentieth century, however, astronomers first began to explore the nonvisible electromagnetic radiation coming from astronomical objects. In this way they have discovered aspects of the cosmos that are forever hidden to optical telescopes.

Astronomers have used ultraviolet light to map the outer regions of the Sun and the clouds of Venus, and used infrared radiation to see new stars and perhaps new planetary systems in the process of formation. By detecting radio waves from Jupiter and Saturn, they have mapped the intense magnetic fields that surround those giant planets; by detecting curious bursts of X rays from space, they have learned about the utterly alien conditions in the vicinity of a black hole. It is no exaggeration to say that today's astronomers learn as much about the universe using telescopes for nonvisible wavelengths as they do using visible light.

Radio waves were the first part of the electromagnetic spectrum beyond the visible to be exploited for astronomy. This happened as a result of a research project seemingly unrelated to astronomy. In the early 1930s, Karl Jansky, a young electrical engineer at Bell Telephone Laboratories, was trying to locate what was causing interference with the then-new transatlantic radio link. By 1932 he realized that one kind of radio noise is strongest when the constellation Sagittarius is high in the sky. The center of our Galaxy is located in the direction of Sagittarius, and Jansky concluded that he was detecting radio waves from an astronomical source.

At first only Grote Reber, a radio engineer living in Illinois, took up Jansky's research. In 1936 Reber built in his backyard the first **radio telescope**, a radio-wave detector dedicated to astronomy. He modeled his design after an ordinary reflecting telescope, with a parabolic metal "dish" (reflecting antenna) measuring 31 ft (10 m) in diameter and a radio receiver at the focal point of the dish.

Reber spent the years from 1938 to 1944 mapping radio emissions from the sky at wavelengths of 1.9 m and 0.63 m. He found radio waves coming from the entire Milky Way, with the greatest emission from the center of the Galaxy. These results, together with the development of improved radio technology during World War II, encouraged the growth of radio astronomy and the construction of new radio telescopes around the world. Radio observatories are as common today as major optical observatories.

Like Reber's prototype, a typical modern radio telescope has a large parabolic dish (**Figure 6-25**). An antenna tuned to the desired frequency is located at the focus (like the prime focus design for optical reflecting telescopes shown in Fig. 6-11*b*). The incoming signal is relayed from the antenna to amplifiers and recording instruments, typically located in a room at the base of the telescope's pier.

 CAUTION The radio telescope in Figure 6-25 looks like a radar dish but is used in a different way. In radar, the dish is used to send out a narrow beam of radio waves. If this beam en-

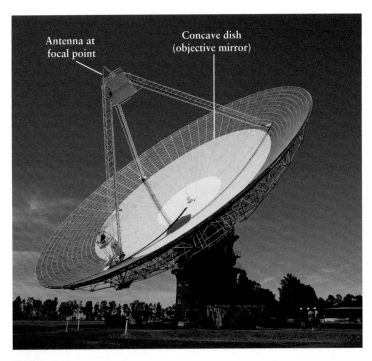

Figure 6-25 R I **V** U X G

A Radio Telescope The dish of the Parkes radio telescope in New South Wales, Australia, is 64 m (210 ft) in diameter. Radio waves reflected from the dish are brought to a focus and collected by an antenna at the focal point. (David Nunuk/Photo Researchers)

angular resolution than a 1-m optical telescope. Because radio radiation has very long wavelengths, astronomers thought that radio telescopes could produce only blurry, indistinct images.

 A very large radio telescope can produce a somewhat sharper radio image, because as the diameter of the telescope increases, the angular resolution decreases. In other words, the bigger the dish, the better the resolution. For this reason, most modern radio telescopes have dishes more than 30 m (100 ft) in diameter. This is also useful for increasing light-gathering power, because radio signals from astronomical objects are typically very weak in comparison with the intensity of visible light. But even the largest single radio dish in existence, the 305-m (1000-ft) Arecibo radio telescope in Puerto Rico, cannot come close to the resolution of the best optical instruments.

A very clever technique makes it possible to produce radio images with excellent resolution. Unlike ordinary light, radio signals can be carried over electrical wires. Consequently, two radio telescopes observing the same astronomical object can be hooked together, even if they are separated by many kilometers. This technique is called **interferometry,** because the incoming radio signals are made to "interfere," or blend together. This makes the combined signal sharp and clear. The effective angular resolution of two such radio telescopes is equivalent to that of one gigantic dish with a diameter equal to the **baseline,** or distance between the two telescopes.

One of the largest arrangements of radio telescopes for interferometry is the Very Large Array (VLA), located in the desert near Socorro, New Mexico (**Figure 6-26**). The VLA consists of

counters an object like an airplane, some of the radio waves will be reflected back to the radar dish and detected by a receiver at the focus of the dish. Thus, a radar dish looks for radio waves *reflected* by distant objects. By contrast, a radio telescope is designed to receive radio waves *emitted* by objects in space.

Many radio telescope dishes, like the one in Figure 6-25, have visible gaps in them like a wire mesh. This does not affect their reflecting power because the holes are much smaller than the wavelengths of the radio waves. The same idea is used in the design of microwave ovens. The glass window in the oven door would allow the microwaves to leak out, so the window is covered by a metal screen with small holes. These holes are much smaller than the 12.2-cm (4.8-in.) wavelength of the microwaves, so the screen reflects the microwaves back into the oven.

One great drawback of early radio telescopes was their very poor angular resolution. Recall from Section 6-3 that angular resolution is the smallest angular separation between two stars that can just barely be distinguished as separate objects. For radio telescopes, the limitation on angular resolution is not atmospheric turbulence, as it is for optical telescopes. Rather, the problem is that angular resolution is directly proportional to the wavelength being observed (see the formula for angular resolution θ in Section 6-3). The longer the wavelength, the larger (and hence worse) the angular resolution and the fuzzier the picture. As an example, a 1-m radio telescope detecting radio waves of 5-cm wavelength has 100,000 times poorer

Figure 6-26 R I **V** U X G

The Very Large Array (VLA) The 27 radio telescopes of the VLA in central New Mexico are arranged along the arms of a Y. The north arm of the array is 19 km long; the southwest and southeast arms are each 21 km long. By spreading the telescopes out along the legs and combining the signals received, the VLA can give the same angular resolution as a single dish many kilometers in radius. (Courtesy of NRAO/AUI)

27 parabolic dishes, each 25 m (82 ft) in diameter. These 27 telescopes are arranged along the arms of a gigantic Y that covers an area 27 km (17 mi) in diameter. By pointing all 27 telescopes at the same object and combining the 27 radio signals, this system can produce radio views of the sky with an angular resolution comparable to that of the very best optical telescopes. (The Very Large Telescope, shown in Figure 6-18, will combine the images from its four optical reflectors to do interferometry with *visible* light. In this way the VLT may achieve even better angular resolution with visible light than the VLA does using radio waves.)

WEB LINK 6.9 Dramatically better angular resolution can be obtained by combining the signals from radio telescopes at different observatories thousands of kilometers apart. This technique is called **very-long-baseline interferometry (VLBI)**. VLBI is used by a system called the Very Long Baseline Array (VLBA), which consists of ten 25-meter dishes at different locations between Hawaii and the Caribbean. Although the ten dishes are not physically connected, they are all used to observe the same object at the same time. The data from each telescope are recorded electronically and processed later. By carefully synchronizing the ten recorded signals, they can be combined just as if the telescopes had been linked together during the observation. With VLBA, features smaller than 0.001 arcsec can be distinguished at radio wavelengths. This angular resolution is 100 times better than a large optical telescope with adaptive optics.

Even better angular resolution can be obtained by adding radio telescopes in space; the baseline is then the distance from the VLBA to the orbiting telescope. The first such space radio telescope, the Japanese HALCA spacecraft, went into operation in 1997. Orbiting at a maximum distance of 21,400 km (13,300 mi) above the Earth's surface, HALCA gives a baseline three times longer—and thus an angular resolution three times better—than can be obtained with Earthbound telescopes alone.

Figure 6-27 shows how optical and radio images of the same object can give different and complementary kinds of information. The visible-light image of Saturn (Figure 6-27a) shows clouds in the planet's atmosphere and the structure of the rings. Like the visible light from the Moon, the light used to make this image is just reflected sunlight. By contrast, the false-color radio image of Saturn (Figure 6-27b) is a record of waves emitted by electrically charged particles outside the planet. These particles emit because of their motion in Saturn's magnetic field, so this radio image provides clues about the conditions in the interior of Saturn where the magnetic field is produced. This information could never be obtained from a visible-light image such as Figure 6-27a.

6-7 | Telescopes in orbit around the Earth detect radiation that does not penetrate the atmosphere

The many successes of radio astronomy show the value of observations at nonvisible wavelengths. But the Earth's atmosphere is opaque to many wavelengths. Other than visible light and radio waves, very little radiation from space manages to penetrate the air we breathe. To overcome this, astronomers have placed a variety of telescopes in orbit.

Figure 6-28 shows the transparency of the Earth's atmosphere to different wavelengths of electromagnetic radiation. The atmo-

(a) R I **V** U X G

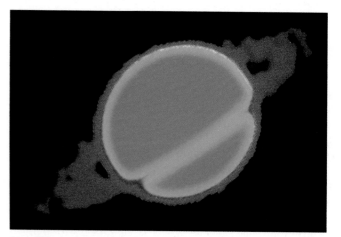

(b) **R** I V U X G

Figure 6-27

Optical and Radio Views of Saturn (a) This picture was taken by a spacecraft 18 million kilometers from Saturn. The view was produced by sunlight reflecting from the planet's cloudtops and rings. (b) This VLA image shows radio emission from Saturn at a wavelength of 2 cm. This emission is caused by electrically charged particles moving within Saturn's strong magnetic field and acting as tiny radio transmitters. In this false-color image, the most intense radio emission is shown in red, the least intense in blue. Intermediate colors of the rainbow represent intermediate levels of radio intensity. Black indicates no detectable radio emission. Note the blue radio "shadow" caused by Saturn's rings where they lie in front of the planet. (a: NASA; b: National Radio Astronomy Observatory)

figure 6-28

The Transparency of the Earth's Atmosphere This graph shows the percentage of radiation that can penetrate the Earth's atmosphere at different wavelengths. Regions in which the curve is high are called "windows" because the atmosphere is relatively transparent at those wavelengths. There are also three wavelength ranges in which the atmosphere is opaque and the curve is near zero: at wavelengths less than about 290 nm, which are absorbed by atmospheric oxygen and nitrogen; between the optical and radio windows, due to absorption by water vapor and carbon dioxide; and at wavelengths longer than about 20 m, which are reflected back into space by ionized gases in the upper atmosphere.

sphere is most transparent in two wavelength regions, the **optical window** (which includes the entire visible spectrum) and the **radio window** (which includes part, but not all, of the radio spectrum). There are also several relatively transparent regions at infrared wavelengths between 1 and 40 μm. Infrared radiation within these wavelength intervals can penetrate the Earth's atmosphere somewhat and can be detected with ground-based telescopes. This wavelength range is called the *near-infrared,* because it lies just beyond the red end of the visible spectrum.

Water vapor is the main absorber of infrared radiation from space, which is why infrared observatories are located at sites with exceptionally low humidity. The site must also be at high altitude to get above as much of the atmosphere's water vapor as possible. One site that meets both criteria is the summit of Mauna Kea in Hawaii, shown in Figure 6-16. (The complete lack of vegetation on the summit attests to its extreme dryness.) Some of the telescopes on Mauna Kea are designed exclusively for detecting infrared radiation. Others, such as the two Keck telescopes, are used for both visible and near-infrared observations (see Figure 6-17).

Even at the elevation of Mauna Kea, water vapor in the atmosphere restricts the kinds of infrared observations that astronomers can make. This situation can be improved by carrying telescopes on board high-altitude balloons or aircraft. But the ultimate solution is to place a telescope in Earth's orbit and radio its data back to astronomers on the ground. The first such orbiting infrared observatory, the Infrared Astronomical Satellite (IRAS), was launched in 1983. During its nine-month mission, IRAS used its 57-cm (22-in.) telescope to map almost the entire sky at wavelengths from 12 to 100 μm.

The IRAS data revealed the presence of dust disks around nearby stars. Planets are thought to coalesce from disks of this kind, so this was the first concrete (if indirect) evidence that there might be planets orbiting other stars. The dust that IRAS detected is warm enough to emit infrared radiation but too cold to emit much visible light, so it remained undetected by ordinary optical telescopes. IRAS also discovered distant, ultraluminous galaxies that emit almost all their radiation at infrared wavelengths.

In 1995 the Infrared Space Observatory (ISO), a more advanced 60-cm reflector with better light detectors, was launched into orbit by the European Space Agency. During its 2½-year mission, ISO made a number of groundbreaking observations of very distant galaxies and of the thin, cold material between the stars of our own Galaxy. Like IRAS, ISO had to be cooled by liquid helium to temperatures just a few degrees above absolute zero. Had this not been done, the infrared blackbody radiation from the telescope itself would have outshone the infrared radiation from astronomical objects. The ISO mission came to an end when the last of the helium evaporated into space.

The next great orbiting infrared observatory is the Spitzer Space Telescope, an 85-cm infrared telescope designed to survey the infrared sky with unprecedented resolution (**Figure 6-29**). This telescope will be used to study brown dwarfs (exotic objects larger than planets but smaller than stars), search for more direct evidence of planets forming around other stars, and probe the interiors of galaxies billions of light-years away. The Spitzer Space Telescope was launched in 2003 on a mission lasting from 2½ to 5 years.

Astronomers are also very interested in observing at ultraviolet wavelengths. These observations can reveal a great deal about hot stars, ionized clouds of gas between the stars, and the Sun's high-temperature corona (see Section 3-5), all of which emit copious amounts of ultraviolet light. The spectrum of ultraviolet sunlight reflected from a planet can also reveal the composition of the planet's atmosphere. However, Earth's atmosphere is opaque to ultraviolet light except for the narrow *near-ultraviolet* range, which extends from about 400 nm (the violet end of the visible spectrum) down to 300 nm.

To see shorter-wavelength *far-ultraviolet* light, astronomers must again make their observations from space. The first ultraviolet telescope was placed in orbit in 1962, and several others

figure 6-29

The Spitzer Space Telescope Launched in 2003, this spacecraft is the largest infrared telescope ever placed in space. Its 85-cm objective mirror and three science instruments, kept cold by 360 liters of liquid helium, enable the Spitzer Space Telescope to observe the universe at wavelengths from 3 to 180 µm. (NASA/JPL-Caltech)

have since followed it into space. Small rockets have also lifted ultraviolet cameras briefly above the Earth's atmosphere. Figure 6-30 shows an ultraviolet view of the constellation Orion, along with infrared and visible views.

The Far Ultraviolet Spectroscopic Explorer (FUSE), which went into orbit in 1999, specializes in measuring spectra at wavelengths from 90 to 120 nm. Highly ionized oxygen atoms, which can exist only in an extremely high-temperature gas, have a characteristic spectral line in this range. By looking for this spectral line in various parts of the sky, FUSE has confirmed that our Milky Way Galaxy (Section 1-4) is surrounded by an immense "halo" of gas at temperatures in excess of 200,000 K. Only an ultraviolet telescope could have detected this "halo," which is thought to have been produced by exploding stars called supernovae (Section 1-3).

Infrared and ultraviolet satellites give excellent views of the heavens at selected wavelengths. But since the 1940s astronomers had dreamed of having one large telescope that could be operated at any

wavelength from the near-infrared through the visible range and out into the ultraviolet. This is the mission of the Hubble Space Telescope (HST), which was placed in a 600-km-high orbit by the space shuttle *Discovery* in 1990 (see the photograph on the right that opens this chapter). HST has a 2.4-meter (7.9-ft) objective mirror and was designed to observe at wavelengths from 115 nm to 1 µm. Like most ground-based telescopes, HST uses a CCD to record images. (In fact, the development of HST helped drive advances in CCD technology.) The images are then radioed back to Earth in digital form.

The great promise of HST was that from its vantage point high above the atmosphere, its angular resolution would be limited only by diffraction. But soon after HST was placed in orbit, astronomers discovered that a manufacturing error had caused the telescope's 2.4-m primary mirror to suffer from spherical aberration. The mirror should have been able to concentrate 70% of a star's light into an image with an angular diameter of 0.1 arcsec. Instead, only 20% of the light was focused into this small area. The remainder was smeared out over an area about 1 arcsec wide, giving images little better than those achieved at major ground-based observatories.

On an interim basis, astronomers used only the 20% of incoming starlight that was properly focused and, with computer processing, discarded the remaining poorly focused 80%. This was practical only for brighter objects on which astronomers could afford to waste light. But many of the observing projects scheduled for HST involved extremely dim galaxies and nebulae.

These problems were resolved by a second space shuttle mission in 1993. Astronauts installed a set of small secondary mirrors whose curvature exactly compensated for the error in curvature of the primary mirror. Once these were in place, HST was able to make truly sharp images of extremely faint objects. Astronomers have used the repaired HST to make discoveries about the nature of planets, the evolution of stars, the inner workings of galaxies, and the expansion of the universe. You will see many HST images in later chapters.

The success of HST has inspired plans for its larger successor, the James Webb Space Telescope or JWST (Figure 6-31). Planned for a 2011 launch, JWST will observe at visible and infrared wavelengths from 600 nm to 28 µm. With its 6.5-m objective mirror—2.5 times the diameter of the HST objective mirror, with six times the light-gathering power—JWST will study faint objects such as planetary systems forming around other stars and galaxies near the limit of the observable universe. Unlike HST, which is in a relatively low-altitude orbit around the Earth, JWST will orbit the Sun some 1.5 million km beyond the Earth. In this orbit the telescope's view will not be blocked by the Earth. Furthermore, by remaining far from the radiant heat of the Earth it will be easier to keep JWST at the very cold temperatures required by its infrared detectors.

Space telescopes have also made it possible to explore objects whose temperatures reach the almost inconceivable values of 10^6 to 10^8 K. Atoms in such a high-temperature gas move so fast that when they collide, they emit X-ray photons of very high energy and very short wavelengths less than 10 nm. X-ray telescopes designed to detect these photons must be placed in orbit, since Earth's atmosphere is totally opaque at these wavelengths.

CAUTION X-ray telescopes work on a very different principle from the X-ray devices used in medicine and dentistry. If you have your foot "X-rayed" to check for a broken bone, a

(a) R I V **U** X G

(b) R **I** V U X G

(c) R I **V** U X G

(d)

figure 6-30

Orion Seen at Ultraviolet, Infrared, and Visible Wavelengths (a) An ultraviolet view of the constellation of Orion was obtained during a brief rocket flight in 1975. This 100-s exposure covers the wavelength range 125–200 nm. (b) The false-color view from the Infrared Astronomical Satellite displays emission at different wavelengths in different colors: red for 100-μm radiation, green for 60-μm radiation, and blue for 12-μm radiation. Compare these images with (c) an ordinary optical photograph and (d) a star chart of Orion. (a: G. R. Carruthers, Naval Research Laboratory; b: NASA; c: R. C. Mitchell, Central Washington University)

piece of photographic film (or an electronic detector) sensitive to X rays is placed under your foot and an X-ray beam is directed at your foot from above. The radiation penetrates through soft tissue but not through bone, so the bones cast an "X-ray shadow" on the film. A fracture will show as a break in the shadow. X-ray telescopes, by contrast, do *not* send beams of X rays toward astronomical objects in an attempt to see inside them. Rather, these telescopes detect X rays that the objects emit on their own.

Astronomers got their first quick look at the X-ray sky from brief rocket flights during the late 1940s. These observations confirmed that the Sun's corona (see Section 3-5) is a source of X rays, and must therefore be at a temperature of millions of kelvins. In 1962 a rocket experiment revealed that objects beyond the solar system also emit X rays.

Since 1970, a series of increasingly sensitive and sophisticated X-ray observatories have been placed in orbit, including NASA's Einstein Observatory, the European Space Agency's Exosat, and the German-British-American ROSAT. These telescopes have shown that other stars also have high-temperature coronae, and have found hot, X-ray-emitting gas clouds so immense that hundreds of galaxies fit inside them. They also discovered unusual stars that emit X rays in erratic bursts. These bursts are now thought to be coming from heated gas swirling around a small but massive object—possibly a black hole.

X-ray astronomy took a quantum leap forward in 1999 with the launch of NASA's Chandra X-ray Observatory and the European Space Agency's XMM-Newton. Named for the Indian-American

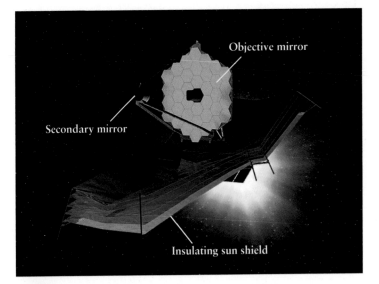

Objective mirror

Secondary mirror

Insulating sun shield

figure 6-31

The James Webb Space Telescope (JWST) The successor to the Hubble Space Telescope, JWST will have an objective mirror 6.5 m (21 ft) in diameter. Like many Earthbound telescopes, JWST will be of Cassegrain design (see Figures 6-11c and 6-14). Rather than using liquid helium to keep the telescope and instruments at the low temperatures needed to observe at infrared wavelengths, JWST will keep cool using a multilayer sunshield the size of two tennis courts. (Courtesy of Northrop Grumman Space Technology)

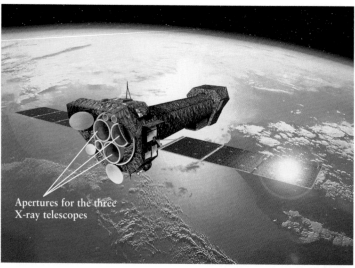

(a) Chandra X-ray Observatory

(b) XMM-Newton

 Figure 6-32

Two Orbiting X-Ray Observatories (a) X rays are absorbed by ordinary mirrors like those used in optical reflectors, but they can be reflected if they graze the mirror surface at a very shallow angle. In the Chandra X-ray Observatory, X rays are focused n this way onto a focal

plane 10 m (33 ft) behind the mirror. (b) XMM-Newton is about the same size as Chandra, and its three X-ray telescopes form images in the same way. (a: NASA/Chandra X-ray Observatory Center/Smithsonian Astrophysical Observatory; b: D. Ducros/European Space Agency)

astrophysicist Subrahmanyan Chandrasekhar, Chandra can view the X-ray sky with an angular resolution of 0.5 arcsec (Figure 6-32*a*). This is comparable to the best ground-based optical telescopes, and more than a thousand times better than the resolution of the first orbiting X-ray telescope. Chandra can also measure X-ray spectra 100 times more precisely than any previous spacecraft and can detect variations in X-ray emissions on time scales as short as 16 microseconds. This latter capability is essential for understanding how X-ray bursts are produced around black holes.

XMM-Newton (for *X*-ray *M*ulti-*m*irror *M*ission) is actually three X-ray telescopes that all point in the same direction (Figure 6-32*b*). Their combined light-gathering power is five times greater than that of Chandra, which makes XMM-Newton able to observe fainter objects. (For reasons of economy, the mirrors were not ground as precisely as those on Chandra, so the angular resolution of XMM-Newton is only about 6 arcseconds.) It also carries a small but highly capable telescope for ultraviolet and visible observations. Hot X-ray sources are usually accompanied by cooler material that radiates at these longer wavelengths, so XMM-Newton can observe these hot and cool regions simultaneously.

Gamma rays, the shortest-wavelength photons of all, help us to understand phenomena even more energetic than those that produce X rays. As an example, when a massive star explodes into a supernova, it produces radioactive atomic nuclei that are strewn across interstellar space. Observing the gamma rays emitted by these nuclei helps astronomers understand the nature of supernova explosions.

Like X rays, gamma rays do not penetrate Earth's atmosphere, so space telescopes are required. The most powerful gamma-ray telescope placed in orbit to date is the Compton Gamma Ray Observa-

 Figure 6-33 R I **V** U X G

The Compton Gamma Ray Observatory (CGRO) This photograph shows the Compton Observatory being deployed from the space shuttle *Atlantis* in 1991. Named in honor of Arthur Holly Compton, an American scientist who made important discoveries about gamma rays, CGRO carried four different gamma-ray detectors. When its mission ended in 2000, CGRO disintegrated as it re-entered our atmosphere. (NASA)

(a)

R **V** U X G

(b)

R I V U X G

(c)

R **I** V U X G

(d)

R I V U **X** G

(e)

R I V U X **G**

tory (CGRO), shown in **Figure 6-33**. One particularly important task for CGRO was the study of gamma-ray bursts, which are brief, unpredictable, and very intense flashes of gamma rays that occur unpredictably from all parts of the sky. By analyzing data from CGRO and other orbiting observatories, astronomers have shown that the sources of these gamma-ray bursts are billions of light-years away. For these bursts to be visible across such great distances, their sources must be among the most energetic objects in the universe. By combining these gamma-ray observations with images made by optical telescopes, astronomers have found that at least some of the gamma-ray bursts emanate from stars that explode catastrophically.

The advantages and benefits of Earth-orbiting observatories cannot be overemphasized. We are no longer limited to the narrow ranges of whatever wavelengths manage to leak through our shimmering, hazy atmosphere (**Figure 6-34**). For the first time, we are really *seeing* the universe.

WEB LINK 6.15

figure 6-34

The Entire Sky at Five Wavelength Ranges These five views show the entire sky at visible, radio, infrared, X-ray, and gamma-ray wavelengths. The entire celestial sphere is mapped onto an oval, with the Milky Way stretching horizontally across the center. The black crescents in the infrared and X-ray images are where data are missing. (GSFC/NASA)

(a) In the visible view the constellation Orion is at the right, Sagittarius in the middle, and Cygnus toward the left. Many of the dark areas along the Milky Way are locations where interstellar dust is sufficiently thick to block visible light.

(b) The radio view shows the sky at a wavelength of 21 cm. This wavelength is emitted by hydrogen atoms in interstellar space. The brightest regions (shown in red) are in the plane of the Milky Way, where the hydrogen is most concentrated.

(c) The infrared view from IRAS shows emission at 100 μm, 60 μm, and 12 μm. Most of the emission is from dust particles in the plane of the Milky Way that have been warmed by starlight.

(d) The X-ray view from ROSAT shows wavelengths of 0.8 nm (blue), 1.7 nm (green), and 5.0 nm (red), corresponding to photon energies of 1500, 750, and 250 eV. Extremely high temperature gas emits these X rays. The white regions, which emit strongly at all X-ray wavelengths, are remnants of supernovae.

(e) The gamma-ray view from the Compton Gamma Ray Observatory includes all wavelengths less than about 1.2×10^{-5} nm (photon energies greater than 10^8 eV). The diffuse radiation from the Milky Way is emitted when fast-moving subatomic particles collide with the nuclei of atoms in interstellar gas clouds. The bright spots above and below the Milky Way are distant, extremely energetic galaxies.

Key Words

active optics, p. 129
adaptive optics, p. 129
angular resolution, p. 128
baseline, p. 135
Cassegrain focus, p. 126
charge-coupled device (CCD), p. 131
chromatic aberration, p. 124
coma, p. 127
coudé focus, p. 126
diffraction, p. 129
diffraction grating, p. 133
eyepiece lens, p. 122
false color, p. 130
focal length, p. 120
focal plane, p. 120
focal point, p. 120

focus (of a lens or mirror), p. 120
grating, p. 133
imaging, p. 131
interferometry, p. 135
light-gathering power, p. 122
light pollution, p. 131
magnification (magnifying power), p. 123
medium (*plural* media), p. 120
Newtonian reflector, p. 126
objective lens, p. 122
objective mirror (primary mirror), p. 126
optical telescope, p. 120
optical window (in the Earth's atmosphere), p. 137

photometry, p. 132
pixel, p. 131
prime focus, p. 126
radio telescope, p. 134
radio window (in the Earth's atmosphere), p. 137
reflecting telescope (reflector), p. 126
reflection, p. 125
refracting telescope (refractor), pp. 122
refraction, p. 120
seeing disk, p. 129
spectrograph, p. 133
spectroscopy, p. 133
spherical aberration, p. 127
very-long-baseline interferometry (VLBI), p. 136

Key Ideas

Refracting Telescopes: Refracting telescopes, or refractors, produce images by bending light rays as they pass through glass lenses.

• Chromatic aberration is an optical defect whereby light of different wavelengths is bent in different amounts by a lens.

• Glass impurities, chromatic aberration, opacity to certain wavelengths, and structural difficulties make it inadvisable to build extremely large refractors.

Reflecting Telescopes: Reflecting telescopes, or reflectors, produce images by reflecting light rays to a focus point from curved mirrors.

• Reflectors are not subject to most of the problems that limit the useful size of refractors.

Angular Resolution: A telescope's angular resolution, which indicates ability to see fine details, is limited by two key factors.

• Diffraction is an intrinsic property of light waves. Its effects can be minimized by using a larger objective lens or mirror.

• The blurring effects of atmospheric turbulence can be minimized by placing the telescope atop a tall mountain with very smooth air. They can be dramatically reduced by the use of adaptive optics and can be eliminated entirely by placing the telescope in orbit.

Charge-Coupled Devices: Sensitive light detectors called charge-coupled devices (CCDs) are often used at a telescope's focus to record faint images.

Spectrographs: A spectrograph uses a diffraction grating and lenses to form the spectrum of an astronomical object.

Radio Telescopes: Radio telescopes use large reflecting dishes to focus radio waves onto a detector.

• Very large dishes provide reasonably sharp radio images. Higher resolution is achieved with interferometry techniques that link smaller dishes together.

Transparency of the Earth's Atmosphere: The Earth's atmosphere absorbs much of the radiation that arrives from space.

• The atmosphere is transparent chiefly in two wavelength ranges known as the optical window and the radio window. A few wavelengths in the near-infrared also reach the ground.

Telescopes in Space: For observations at wavelengths to which the Earth's atmosphere is opaque, astronomers depend on telescopes carried above the atmosphere by rockets or spacecraft.

• Satellite-based observatories provide new information about the universe and permit coordinated observation of the sky at all wavelengths.

Review Questions

1. Describe refraction and reflection. Explain how these processes enable astronomers to build telescopes.

2. Explain why a flat piece of glass does not bring light to a focus while a curved piece of glass can.

3. With the aid of a diagram, describe a refracting telescope. Which dimensions of the telescope determine its light-gathering power? Which dimensions determine the magnification?

4. What is the purpose of a telescope eyepiece? What aspect of the eyepiece determines the magnification of the image? In what circumstances would the eyepiece not be used?

5. Do most professional astronomers actually look through their telescopes? Why or why not?

6. Quite often advertisements appear for telescopes that extol their magnifying power. Is this a good criterion for evaluating telescopes? Explain your answer.

7. What is chromatic aberration? For what kinds of telescopes does it occur? How can it be corrected?

8. With the aid of a diagram, describe a reflecting telescope. Describe four different ways in which an astronomer can access the focal plane.

9. Explain some of the disadvantages of refracting telescopes compared to reflecting telescopes.

10. What kind of telescope would you use if you wanted to take a color photograph entirely free of chromatic aberration? Explain your answer.

11. Explain why a Cassegrain reflector can be substantially shorter than a refractor of the same focal length.

12. No major observatory has a Newtonian reflector as its primary instrument, whereas Newtonian reflectors are extremely popular among amateur astronomers. Explain why this is so.

13. What is spherical aberration? How can it be corrected?

14. What is diffraction? Why does it limit the angular resolution of a telescope? What other physical phenomenon is often a more important restriction on angular resolution?

15. What is active optics? What is adaptive optics? Why are they useful? Would either of these be a good feature to include on a telescope to be placed in orbit?

16. Explain why combining the light from two or more optical telescopes can give dramatically improved angular resolution.

17. What is light pollution? What effects does it have on the operation of telescopes? What can be done to minimize these effects?

18. What is a charge-coupled device (CCD)? Why have CCDs replaced photographic film for recording astronomical images?

19. What is a spectrograph? Why do many astronomers regard it as the most important device that can be attached to a telescope?

20. What are the advantages of a grating spectrograph over a prism spectrograph?

21. Compare an optical reflecting telescope and a radio telescope. What do they have in common? How are they different?

22. Why can radio astronomers make observations at any time during the day, whereas optical astronomers are mostly limited to observing at night? (*Hint:* Does your radio work any better or worse in the daytime than at night?)

23. Why are radio telescopes so large? Why does a single radio telescope have poorer angular resolution than a large optical telescope? How can the resolution be improved by making simultaneous observations with several radio telescopes?

24. What are the optical window and the radio window? Why isn't there an X-ray window or an ultraviolet window?

25. Why is it necessary to keep an infrared telescope at a very low temperature?

26. How are the images made by an X-ray telescope different from those made by a medical X-ray machine?

27. Why must astronomers use satellites and Earth-orbiting observatories to study the heavens at X-ray and gamma-ray wavelengths?

Advanced Questions

Problem-solving tips and tools

You may find it useful to review the small-angle formula discussed in Box 1-1. The area of a circle is proportional to the square of its diameter. Data on the planets can be found in the appendices at the end of this book. Section 5-2 discusses the relationship between frequency and wavelength. Box 6-1 gives examples of how to calculate magnifying power and light-gathering power.

28. Show by means of a diagram why the image formed by a simple refracting telescope is upside down.

29. Ordinary photographs made with a telephoto lens make distant objects appear close. How does the focal length of a telephoto lens compare with that of a normal lens? Explain your reasoning.

30. The observing cage in which an astronomer sits at the prime focus of the 5-m telescope on Palomar Mountain is about 1 m in diameter. Calculate what fraction of the incoming starlight is blocked by the cage.

31. (a) Compare the light-gathering power of the Subaru 8.3-m telescope with that of the Hubble Space Telescope (HST), which has a 2.4-m objective mirror. (b) What advantages does Subaru have over HST? What advantages does HST have over Subaru?

32. Suppose your Newtonian reflector has an objective mirror 20 cm (8 in.) in diameter with a focal length of 2 m. What magnification do you get with eyepieces whose focal lengths are (a) 9 mm, (b) 20 mm, and (c) 55 mm? (d) What is the telescope's diffraction-limited angular resolution when used with orange light of wavelength 600 nm? (e) Would it be possible to achieve this angular resolution if you took the telescope to the summit of Mauna Kea?

33. The Hobby-Eberly Telescope (HET) has a spherical mirror, which is the least expensive shape to grind. Consequently, the telescope has spherical aberration. Explain why this doesn't affect the usefulness of HET for spectroscopy.

34. The four largest moons of Jupiter are roughly the same size as our Moon and are about 628 million (6.28×10^8) kilometers from Earth at opposition. What is the size in kilometers of the smallest surface features that the Hubble Space Telescope (resolution of 0.1 arcsec) can detect? How does this compare with the smallest

features that can be seen on the Moon with the unaided human eye (resolution of 1 arcmin)?

35. The Hubble Space Telescope (HST) has been used to observe the galaxy M100, some 70 million light-years from Earth. (**a**) If the angular resolution of the HST image is 0.1 arcsec, what is the diameter in light-years of the smallest detail that can be discerned in the image? (**b**) At what distance would a U.S. dime (diameter 1.8 cm) have an angular size of 0.1 arcsec? Give your answer in kilometers.

36. Can the Hubble Space Telescope distinguish any features on Pluto? Justify your answer using calculations. (*Hint:* See Appendix 1.)

37. A NASA project named ARISE (Advanced Radio Interferometry between Space and Earth) proposes to place a radio telescope into an even higher orbit than the HALCA telescope. Using this telescope in concert with the VLBA, baselines as long as 50,000 km may be obtainable. Astronomers want to use this combination to study radio emission at a frequency of 1665 MHz from distant objects called quasars. (**a**) What is the wavelength of this emission? (**b**) Taking the baseline to be the effective diameter of this radio-telescope array, what angular resolution can be achieved?

38. The mission of the Submillimeter Wave Astronomy Satellite (SWAS), launched in 1998, is to investigate interstellar clouds within which stars form. One of the frequencies at which it observes these clouds is 557 GHz (1 GHz = 1 gigahertz = 10^9 Hz), characteristic of the emission from interstellar water molecules. (**a**) What is the wavelength (in meters) of this emission? In what part of the electromagnetic spectrum is this? (**b**) Why is it necessary to use a satellite for these observations? (**c**) SWAS has an angular resolution of 4 arcminutes. What is the diameter of its primary mirror?

39. To search for ionized oxygen gas surrounding our Milky Way Galaxy, astronomers aimed the ultraviolet telescope of the FUSE spacecraft at a distant galaxy far beyond the Milky Way. They then looked for an ultraviolet spectral line of ionized oxygen in that galaxy's spectrum. Were they looking for an emission line or an absorption line? Explain.

40. A sufficiently thick interstellar cloud of cool gas can absorb low-energy X rays but is transparent to high-energy X rays and gamma rays. Explain why both part *b* and part *d* of Figure 6-34 reveal the presence of cool gas in the Milky Way. Could you infer the presence of this gas from the visible-light image in Figure 6-34*a*? Explain.

Discussion Questions

41. If you were in charge of selecting a site for a new observatory, what factors would you consider important?

42. Discuss the advantages and disadvantages of using a small telescope in Earth's orbit versus a large telescope on a mountaintop.

Web/CD-ROM Questions

43. Several telescope manufacturers build telescopes with a design called a "Schmidt-Cassegrain." These use a correcting lens in an arrangement like that shown in Figure 6-13*c*. Consult advertisements on the World Wide Web to see the appearance of these telescopes and find out their cost. Why do you suppose they are very popular among amateur astronomers?

44. The Large Zenith Telescope (LZT) in British Columbia, Canada, uses a 5.0-m *liquid* mirror made of mercury. Use the World Wide Web to investigate this technology. How can a liquid metal be formed into the necessary shape for a telescope mirror? What are the advantages of a liquid mirror? What are the disadvantages?

45. Three of the telescopes shown in Figure 6-16—the James Clerk Maxwell Telescope (JCMT), the Caltech Submillimeter Observatory (CSO), and the Submillimeter Array (SMA)—are designed to detect radiation with wavelengths close to 1 mm. Search for current information about JCMT, CSO, and SMA on the World Wide Web. What kinds of celestial objects emit radiation at these wavelengths? What can astronomers see using JCMT, CSO, and SMA that cannot be observed at other wavelengths? Why is it important that they be at high altitude? How large are the primary mirrors used in JCMT, CSO, and SMA? What are the differences among the three telescopes? Which can be used in the daytime? What recent discoveries have been made using JCMT, CSO, or SMA?

46. In 2003 an ultraviolet telescope called GALEX (Galaxy Evolution Explorer) was placed into orbit. Use the World Wide Web to learn about GALEX and its mission. What aspects of galaxies was GALEX designed to investigate? Why is it important to make these observations using ultraviolet wavelengths?

47. At the time of this writing, NASA planned to end the mission of the Hubble Space Telescope by 2011. Consult the Space Telescope Science Institute Web site to learn about plans for HST's final years of operation. Are future space shuttle missions planned to service HST? If so, what changes will be made to HST on such missions? What will become of HST at the end of its mission lifetime?

48. NASA and the European Space Agency (ESA) have plans to launch a number of advanced space telescopes. These include Constellation-X, Herschel, the Gamma-ray Large Area Space Telescope (GLAST), the Space Interferometry Mission (SIM), and the X-ray Evolving Universe Spectroscopy (XEUS) mission. Search the World Wide Web for information about at least two of these. What are the scientific goals of these projects? What is unique about each telescope? What advantages would they have over existing ground-based or orbiting telescopes? What kind of orbit will each telescope be in? When will each be launched and placed in operation?

49. Telescope Magnification. Access the Active Integrated Media Module "Telescope Magnification" in Chapter 6 of the *Universe* Web site or CD-ROM. A common telescope found in department stores is a 3-inch (76-mm) diameter refractor that boasts a magnification of 300 times. Use the magnification calculator to determine the magnifications that are achieved by using each of the following commonly found eyepieces on that telescope: Eyepiece A with focal length 40 mm; Eyepiece B with focal length 25 mm; Eyepiece C with focal length 12 mm; and Eyepiece D with focal length 2.5 mm.

Observing Projects

50. Obtain a telescope during the daytime along with several eyepieces of various focal lengths. If you can determine the telescope's focal length, calculate the magnifying powers of the eyepieces. Focus the telescope on some familiar object, such as a distant lamppost or tree. **DO NOT FOCUS ON THE SUN! Looking directly at the Sun can cause blindness.** Describe the image you see through the telescope. Is it upside down? How does the image move as you slowly and gently shift the telescope left and right or up and down? Examine the eyepieces, noting their focal lengths. By changing the eyepieces, examine the distant object under different magnifications. How do the field of view and the quality of the image change as you go from low power to high power?

51. On a clear night, view the Moon, a planet, and a star through a telescope using eyepieces of various focal lengths and known magnifying powers. (To determine the locations in the sky of the Moon and planets, you may want to use the *Starry Night Backyard*™ program on the CD-ROM that comes with this book. You may also want to consult such magazines as *Sky & Telescope* and *Astronomy* or their Web sites.) In what way does the image seem to degrade as you view with increasingly higher magnification? Do you see any chromatic aberration? If so, with which object and which eyepiece is it most noticeable?

52. Many towns and cities have amateur astronomy clubs. If you are so inclined, attend a "star party" hosted by your local club. People who bring their telescopes to such gatherings are delighted to show you their instruments and take you on a telescopic tour of the heavens. Such an experience can lead to a very enjoyable, lifelong hobby.

53. The field of view of a typical small telescope for amateur astronomers is about 30 arcminutes, or 30′. (Many research telescopes have much smaller fields of view. For instance, the widest field of view available to the Hub-

ble Space Telescope is only 3.4 arcminutes.) Use the *Starry Night Backyard*™ program to simulate the view that such a telescope provides of various celestial objects. First display the entire celestial sphere (select **Guides > Atlas** in the **Go** menu). Then search for objects (i), (ii), (iii), and (iv) listed below (select **Find...** in the **Edit** menu). For each object, use the sliding zoom controls at the right-hand end of the Control Panel (at the top of the main window) to adjust the field of view to approximately 30 arcminutes. Describe how much detail you can see on each object: (i) the Moon; (ii) Jupiter; (iii) Saturn; (iv) M31, the great galaxy in the constellation Andromeda.

54. Use the *Deep Space Explorer*™ program to explore the concept of angular resolution. In the left-hand part of the window under the heading **Earth**, click on the triangle next to the word **Explore** and then click on **Moon**. You can zoom in and zoom out on the cluster using the buttons at the upper left of the window (an upward-pointing triangle and a downward-pointing triangle). You can also rotate the Moon by putting the mouse cursor over the image, holding down the mouse button, and moving the mouse. (On a two-button mouse, hold down the left mouse button.) (**a**) What is the size of the smallest detail that you can see? (You will have to make measurements on the screen using a ruler and compare it to the diameter of the Moon, which is 3476 km.) (**b**) The angular resolution of the Hubble Space Telescope (HST) is 0.1 arcsec. How far away from the Moon could HST be and still be able to resolve details as small as you determined in part (**a**)? Give your answer in kilometers and in astronomical units (AU).

Collaborative Exercises

55. Stand up and have everyone in your group join hands, making as large a circle as possible. If a telescope mirror were built as big as your circle, what would be its diameter? What would be your telescope's diffraction-limited angular resolution for blue light? Would atmospheric turbulence have a noticeable effect on the angular resolution?

56. Are there enough students in your class to stand and join hands and make two large circles that recreate the sizes of the two Keck telescopes? Explain how you determined your answer.

7 Comparative Planetology I: Our Solar System

Fifty years ago, astronomers knew precious little about the solar system. Even the best telescopes provided images of the planets that were frustratingly hazy and indistinct. Of asteroids, comets, and the satellites of the planets, we knew even less.

Today, our knowledge of the solar system has grown many thousandfold, due almost entirely to robotic space-craft. (The illustration shows one such robotic explorer, the joint NASA/European Space Agency spacecraft *Cassini*, as it approached Saturn in 2004.) Spacecraft have flown at close range past all the planets except Pluto, revealing details that astronomers of an earlier generation could only dream about. We have landed space-craft on the Moon, Venus, and Mars and dropped a probe into the immense atmosphere of Jupiter. This is truly the golden age of solar system exploration.

In this chapter we paint in broad outline our present understanding of the solar system. We will see that the planets come in a variety of sizes and chemical compositions. There is also rich variety among the moons of the planets and among smaller bodies we call asteroids, comets, and Kuiper belt objects. We will investigate the nature of craters on the Moon and other worlds of the solar system. And by exploring the magnetic fields of planets, we will be able to peer inside other worlds and learn about their interior compositions.

In Chapter 8 we will put together these observations to build a picture of how the solar system formed some 4.56×10^9 years ago, and how it has evolved since then. But for now, we invite you to join us on a guided tour of the worlds that orbit our Sun.

July 1, 2004: The *Cassini* spacecraft arrives at Saturn (artist's impression). (JPL/NASA)

Questions to Guide Your Reading

As you read the sections of this chapter, look for the answers to the following questions:

7-1 Are all the other planets similar to Earth, or are they very different?

7-2 Do other planets have moons like Earth's Moon?

7-3 How do astronomers know what the other planets are made of?

7-4 Are all the planets made of basically the same material?

7-5 What is the difference between an asteroid and a comet?

7-6 Why are craters common on the Moon but rare on the Earth?

7-7 Why do interplanetary spacecraft carry devices for measuring magnetic fields?

7-8 Do all the planets have a common origin?

7-1 | There are two broad categories of planets: Earthlike and Jupiterlike

Each of the nine planets that orbit the Sun is unique. Only Earth has liquid water and an atmosphere breathable by humans; only Venus has a perpetual cloud layer made of sulfuric acid droplets; only Jupiter has immense storm systems that persist for centuries; and only Pluto always keeps the same face toward its moon. But there are also striking similarities among planets. Volcanoes exist not only on Earth but also on Venus and Mars; there are rings around Jupiter, Saturn, Uranus, and Neptune; and there are impact craters on Mercury, Venus, Earth, and Mars, showing that all of these planets have been bombarded by interplanetary debris.

How can we make sense of the many similarities and differences among the planets? An important step is to organize our knowledge of the planets in a systematic way. There are two useful ways to do this. First, we can contrast the orbits of different planets around the Sun; and second, we can compare the planets' physical properties such as diameter, mass, average density, and chemical composition.

The first eight planets fall naturally into two classes according to the sizes of their orbits. As **Figure 7-1** shows, the orbits of the four

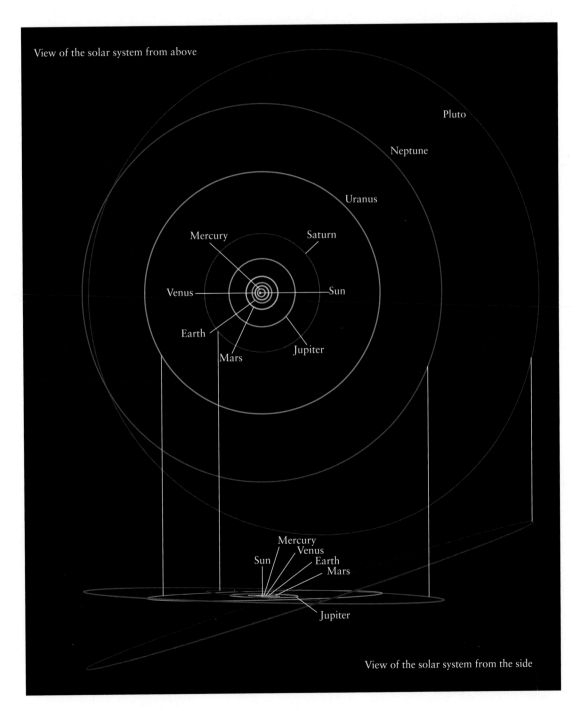

View of the solar system from above

Pluto

Neptune

Uranus

Mercury   Saturn

Venus   Sun

Earth

Mars   Jupiter

Mercury
Venus
Sun   Earth
          Mars

Jupiter

View of the solar system from the side

ANIMATION 7.1

Figure 7-1

The Solar System to Scale This scale drawing shows the orbits of the planets around the Sun. The four inner planets are crowded in close to the Sun, while the five outer planets orbit the Sun at much greater distances. Many smaller objects called asteroids orbit within a belt between the orbits of Mars and Jupiter. On the scale of this drawing, the planets themselves would be much smaller than the diameter of a human hair and too small to see.

table 7-1 | Characteristics of the Planets

	The Inner Planets			
	Mercury	Venus	Earth	Mars
Average distance from Sun (10⁶ km)	57.9	108.2	149.6	227.9
Average distance from Sun (AU)	0.387	0.723	1.000	1.524
Orbital period (years)	0.241	0.615	1.000	1.88
Orbital eccentricity	0.206	0.007	0.017	0.093
Inclination of orbit to the ecliptic	7.00°	3.39°	0.00°	1.85°
Equatorial diameter (km)	4880	12,104	12,756	6794
Equatorial diameter (Earth = 1)	0.383	0.949	1.000	0.533
Mass (kg)	3.302×10^{23}	4.868×10^{24}	5.974×10^{24}	6.418×10^{23}
Mass (Earth = 1)	0.0553	0.8150	1.0000	0.1074
Average density (kg/m³)	5430	5243	5515	3934

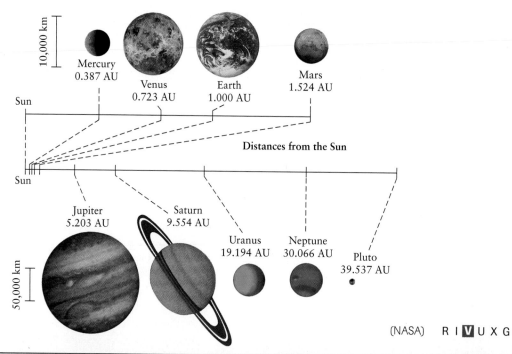

Mercury 0.387 AU
Venus 0.723 AU
Earth 1.000 AU
Mars 1.524 AU
Sun
10,000 km

Distances from the Sun

Sun
Jupiter 5.203 AU
Saturn 9.554 AU
Uranus 19.194 AU
Neptune 30.066 AU
Pluto 39.537 AU
50,000 km

(NASA)   R I **V** U X G

	The Outer Planets				
	Jupiter	Saturn	Uranus	Neptune	Pluto
Average distance from Sun (10⁶ km)	778.3	1429	2871	4498	5915
Average distance from Sun (AU)	5.203	9.554	19.194	30.066	39.537
Orbital period (years)	11.86	29.46	84.10	164.86	248.60
Orbital eccentricity	0.048	0.053	0.043	0.010	0.250
Inclination of orbit to the ecliptic	1.30°	2.48°	0.77°	1.77°	17.15°
Equatorial diameter (km)	142,984	120,536	51,118	49,528	2300
Equatorial diameter (Earth = 1)	11.209	9.449	4.007	3.883	0.180
Mass (kg)	1.899×10^{27}	5.685×10^{26}	8.682×10^{25}	1.024×10^{26}	1.3×10^{22}
Mass (Earth = 1)	317.8	95.16	14.53	17.15	0.0021
Average density (kg/m³)	1326	687	1318	1638	2000

inner planets (Mercury, Venus, Earth, and Mars) are crowded in close to the Sun. In contrast, the orbits of the next four planets (Jupiter, Saturn, Uranus, and Neptune) are widely spaced at great distances from the Sun. The ninth planet, Pluto, is in an even larger orbit than Neptune. As we will see, Pluto is really in a category of its own. Table 7-1 lists the orbital characteristics of all the planets.

CAUTION ! Note that while Figure 7-1 shows the orbits of the planets, it does not show the planets themselves. The reason is simple: If Jupiter, the largest of the planets, were to be drawn to the same scale as the rest of this figure, it would be a dot just 0.0002 cm across—about ¹⁄₃₀₀ of the width of a human hair and far too small to see without a microscope. The planets themselves are *very* small compared to the distances between the planets. Indeed, while an airliner traveling at 1000 km/h (620 mi/h) can fly around the Earth in less than two days, at this speed it would take 17 *years* to fly from the Earth to the Sun. The solar system is a very large and very empty place!

Most of the planets have orbits that are nearly circular. As we learned in Section 4-4, Kepler discovered in the seventeenth century that these orbits are actually ellipses. Astronomers denote the elongation of an ellipse by its *eccentricity* (see Figure 4-10*b*). The eccentricity of a circle is zero, and indeed most planets have orbital eccentricities that are very close to zero. The exceptions are Mercury and Pluto. In fact, Pluto's noncircular orbit sometimes takes it nearer the Sun than its neighbor, Neptune. (Happily, the orbits of Pluto and Neptune are such that the two planets will never collide.)

If you could observe the solar system from a point several astronomical units (AU) above the Earth's north pole, you would see that all the planets orbit the Sun in the same counterclockwise direction. Furthermore, the orbits of the planets all lie in nearly the same plane. In other words, the orbits of the planets are inclined at only slight angles to the plane of the ecliptic, which is the plane of the Earth's orbit around the Sun. (Again, however, Pluto is an exception; the plane of its orbit is tilted at about 17° to the ecliptic, as Figure 7-1 shows.) Furthermore, the plane of the Sun's equator is very closely aligned with the orbital planes of the planets. As we will see

in Chapter 8, these near-alignments are not a coincidence. They provide important clues about the origin of the planets.

When we compare the physical properties of the planets, we again find that they fall naturally into two classes—four small inner planets and four large outer ones—with Pluto as an exception. The four small inner planets are called **terrestrial planets** because they resemble the Earth (in Latin, *terra*). They all have hard, rocky surfaces with mountains, craters, valleys, and volcanoes. You could stand on the surface of any one of them, although you would need a protective spacesuit on Mercury, Venus, or Mars. The four large outer planets are called **Jovian planets** because they resemble Jupiter. (Jove was another name for the Roman god Jupiter.) An attempt to land a spacecraft on the surface of any of the Jovian planets would be futile, because the materials of which these planets are made are mostly gaseous or liquid. The visible "surface" features of a Jovian planet are actually cloud formations in the planet's atmosphere. The photographs in Table 7-1 show the distinctive appearances of the two classes of planets.

The most apparent difference between the terrestrial and Jovian planets is their *diameters*. You can compute the diameter of a planet if you know its angular diameter as seen from Earth and its distance from Earth. For example, on December 30, 2003, Venus was 1.97×10^8 km from Earth and had an angular diameter of 12.7 arcsec. Using the small-angle formula from Box 1-1, we can calculate the diameter of Venus to be 12,100 km (7520 mi). Similar calculations demonstrate that the Earth, with its diameter of about 12,756 km (7926 mi), is the largest of the four inner, terrestrial planets. In sharp contrast, the four outer, Jovian planets are much larger than the terrestrial planets. First place goes to Jupiter, whose equatorial diameter is more than 11 times that of the Earth. Pluto, always the exception, is even smaller than the inner planets, despite being the outermost planet. Its diameter is less than one-fifth that of the Earth. Figure 7-2 shows the Sun and the planets drawn to the same scale. The diameters of the planets are given in Table 7-1.

The *masses* of the terrestrial and Jovian planets are also dramatically different. If a planet has a satellite, you can calculate the planet's mass from the satellite's period and semimajor axis by using Newton's form of Kepler's third law (see Section 4-7 and

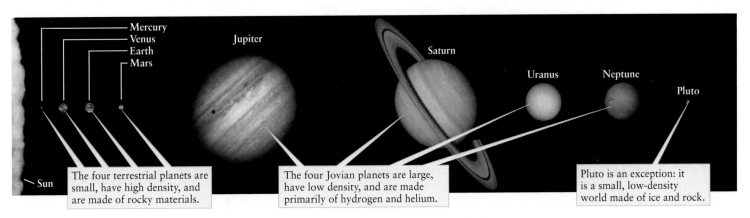

figure 7-2 R I **V** U X G

The Planets to Scale This figure shows the nine planets to the same scale. The four terrestrial planets have orbits nearest the Sun, and the Jovian planets are the next four planets from the Sun. Pluto is an exceptional case. (Calvin J. Hamilton and NASA/JPL)

box 7-1 THE HEAVENS ON THE EARTH

Average Density

Average density—the mass of an object divided by that object's volume—is a useful quantity for describing the differences between planets in our solar system. This same quantity has many applications here on Earth.

A rock tossed into a lake sinks to the bottom, while an air bubble produced at the bottom of a lake (for example, by the air tanks of a scuba diver) rises to the top. These are examples of a general principle: An object sinks in a fluid if its average density is greater than that of the fluid, but rises if its average density is less than that of the fluid. The average density of water is 1000 kg/m^3, which is why a typical rock (with an average density of about 3000 kg/m^3) sinks, while an air bubble (average density of about 1.2 kg/m^3) rises.

At many summer barbecues, cans of soft drinks are kept cold by putting them in a container full of ice. When the ice melts, the cans of diet soda always rise to the top, while the cans of regular soda sink to the bottom. Why is this? The average density of a can of diet soda—which includes water, flavoring, artificial sweetener, and the trapped gas that makes the drink fizzy—is slightly less than the density of water, and so

the can floats. A can of regular soda contains sugar instead of artificial sweetener, and the sugar is a bit heavier than the sweetener. The extra weight is just enough to make the average density of a can of regular soda slightly more than that of water, making the can sink. (You can test these statements for yourself by putting unopened cans of diet soda and regular soda in a sink or bathtub full of water.)

The concept of average density provides geologists with important clues about the early history of the Earth. The average density of surface rocks on Earth, about 3000 kg/m^3, is less than the Earth's average density of 5515 kg/m^3. The simplest explanation is that in the ancient past, the Earth was completely molten throughout its volume, so that low-density materials rose to the surface and high-density materials sank deep into the Earth's interior in a process called *chemical differentiation*. This series of events also suggests that the Earth's core must be made of relatively dense materials, such as iron and nickel. A tremendous amount of other geological evidence has convinced scientists that this picture is correct.

Box 4-4). Astronomers have also measured the mass of each planet except Pluto by sending a spacecraft to pass near the planet. The planet's gravitational pull (which is proportional to its mass) deflects the spacecraft's path, and the amount of deflection tells us the planet's mass. Using these techniques, astronomers have found that the four Jovian planets have masses that are tens or hundreds of times greater than the mass of any of the terrestrial planets. Again, first place goes to Jupiter, whose mass is 318 times greater than the Earth's.

Once we know the diameter and mass of a planet, we can learn something about what that planet is made of. The trick is to calculate the planet's **average density**, or mass divided by volume, measured in kilograms per cubic meter (kg/m^3). The average density of any substance depends in part on that substance's composition. For example, air near sea level on Earth has an average density of 1.2 kg/m^3, water's average density is 1000 kg/m^3, and a piece of concrete has an average density of 2000 kg/m^3. **Box 7-1** describes some applications of the idea of average density to everyday phenomena on Earth.

The four inner, terrestrial planets have very high average densities (see Table 7-1); the average density of the Earth, for example, is 5515 kg/m^3. By contrast, a typical rock found on the Earth's surface has a lower average density, about 3000 kg/m^3. Thus, the Earth must contain a large amount of material that is denser than rock. This information provides our first clue that terrestrial planets have dense iron cores.

In sharp contrast, the outer, Jovian planets have quite low densities. Saturn has an average density less than that of water. This in-

formation strongly suggests that the giant outer planets are composed primarily of light elements such as hydrogen and helium. All four Jovian planets probably have large cores of mixed rock and highly compressed water buried beneath low-density outer layers tens of thousands of kilometers thick.

We can conclude that the following general rule applies to the planets:

The terrestrial planets are made of rocky materials and have dense iron cores, which gives these planets high average densities. The Jovian planets are composed primarily of light elements such as hydrogen and helium, which gives these planets low average densities.

Pluto is again an oddity. Although it is even smaller than the dense inner planets, its average density is closer to that of the giant outer planets. Because its average density of about 2000 kg/m^3 is intermediate between the densities of water ice (that is, frozen water) and of rock, Pluto is probably composed of a mixture of these two substances.

7-2 | Seven large satellites are almost as big as the terrestrial planets

All the planets except Mercury and Venus have satellites. More than 130 satellites are known (Jupiter alone has more than 60), and dozens of other small ones probably remain to be discovered. Like the terrestrial planets, all of these satellites have solid surfaces.

Of the known satellites, seven are larger than Pluto and roughly as big as the planet Mercury. Table 7-2 lists these satellites and shows them to the same scale. Note that the Earth's Moon and Jupiter's satellites Io and Europa have relatively high densities, indicating that these satellites are made primarily of rocky materials. By contrast, the densities of Ganymede, Callisto, Titan, and Triton are all relatively low. Planetary scientists conclude that the interiors of these four satellites also contain substantial amounts of water ice, which is less dense than rock. (In Section 7-4 we will learn about types of frozen "ice" made of substances other than water.)

CAUTION Water ice may seem like a poor material for building a satellite, since the ice you find in your freezer can easily be cracked or crushed. But under high pressure, such as is found in the interior of a large satellite, water ice becomes as rigid as rock. (It also becomes denser than the ice found in ice cubes, although not as dense as rock.) Note that water ice is an important constituent only for satellites in the outer solar system, where the Sun is far away and temperatures are very low. For example, the surface temperature of Titan is a frigid 95 K ($-178°C = -288°F$). In Section 7-4 we will learn more about the importance of temperature in determining the composition of a planet or satellite.

Most of the known satellites are much smaller than those listed in Table 7-2. All of these small satellites have diameters less than 2000 km, and many are just a few kilometers across.

Interplanetary spacecraft have made many surprising and fascinating discoveries about the satellites of the solar system. We now know that Jupiter's satellite Io is the most geologically active world in the solar system, with numerous geyserlike volcanoes that continually belch forth sulfur-rich compounds. The fractured surface of Europa, another of Jupiter's large satellites, suggests that a worldwide ocean of liquid water may lie beneath its icy surface. Saturn's largest satellite, Titan, has an atmosphere nearly twice as dense as the Earth's atmosphere, with a perpetual haze layer that gives it a featureless appearance (see the photographs in Table 7-2).

7-3 | Spectroscopy reveals the chemical composition of the planets

As we have seen, the average densities of the planets and satellites give us a crude measure of their **chemical compositions**—that is, what substances they are made of. For example, the low average density of the Moon (3340 kg/m^3) compared with the Earth (5515 kg/m^3) tells us that the Moon contains relatively little iron or other dense metals. But to truly understand the nature of the planets and satellites, we need to know their chemical compositions in much greater detail than we can learn from average density alone.

The most accurate way to determine chemical composition is by directly analyzing samples taken from a planet's atmosphere and

table 7-2	**The Seven Giant Satellites**						WEB LINK 7.3
	Moon	**Io**	**Europa**	**Ganymede**	**Callisto**	**Titan**	**Triton**
Parent planet	Earth	Jupiter	Jupiter	Jupiter	Jupiter	Saturn	Neptune
Diameter (km)	3476	3642	3130	5268	4806	5150	2706
Mass (kg)	7.35×10^{22}	8.93×10^{22}	4.80×10^{22}	1.48×10^{23}	1.08×10^{23}	1.34×10^{23}	2.15×10^{22}
Average density (kg/m^3)	3340	3530	2970	1940	1850	1880	2050
Substantial atmosphere?	No	No	No	No	No	Yes	No

Moon · Io · Europa · Ganymede · Callisto · Titan · Triton

(JPL/NASA) R I **V** U X G

soil. Unfortunately, of all the planets and satellites, we have such direct information only for the Earth and the three worlds on which spacecraft have landed—Venus, the Moon, and Mars. In all other cases, astronomers must analyze sunlight reflected from the distant planets and their satellites. To do that, astronomers bring to bear one of their most powerful tools, **spectroscopy,** the systematic study of spectra and spectral lines. (We discussed spectroscopy in Sections 5-6 and 6-5.)

Spectroscopy is a sensitive probe of the composition of a planet's *atmosphere*. If a planet has an atmosphere, then sunlight reflected from that planet must have passed through its atmosphere. During this passage, some of the wavelengths of sunlight will have been absorbed. Hence, the spectrum of this reflected sunlight will have dark absorption lines. Astronomers look at the particular wavelengths absorbed and the amount of light absorbed at those wavelengths. Both of these depend on the kinds of chemicals present in the planet's atmosphere and the abundance of those chemicals.

For example, astronomers have used spectroscopy to analyze the atmosphere of Saturn's largest satellite, Titan (see Table 7-2 for a photograph of Titan). The graph in **Figure 7-3a** shows the spectrum of visible sunlight reflected from Titan. (We first saw this method of displaying spectra in Figure 6-24.) The dips in this curve of intensity versus wavelength represent absorption lines. However,

not all of these absorption lines are produced in the atmosphere of Titan (Figure 7-3b). Before reaching Titan, light from the Sun's glowing surface must pass through the Sun's hydrogen-rich atmosphere. This produces the hydrogen absorption line in Figure 7-3a at a wavelength of 656 nm. After being reflected from Titan, the light must pass through the Earth's atmosphere before reaching the telescope; this is where the oxygen absorption line in Figure 7-3a is produced. Only the two dips near 620 nm and 730 nm are caused by gases in Titan's atmosphere.

These two absorption lines are caused not by individual atoms in the atmosphere of Titan but by atoms combined to form **molecules.** For example, two hydrogen atoms (symbol H) can combine with an oxygen atom (O) to form a molecule of water (H_2O). Molecules, like atoms, also produce unique patterns of lines in the spectra of astronomical objects. The absorption lines in Figure 7-3a indicate the presence in Titan's atmosphere of molecules of methane (CH_4, a molecule made of one carbon atom and four hydrogen atoms). This shows that Titan is a curious place indeed, because on Earth, methane is a rather rare substance that is the primary ingredient in natural gas! When we examine other planets and satellites with atmospheres, we find that all of their spectra have absorption lines of molecules of various types.

In addition to visible-light measurements such as those in Figure 7-3a, it is very useful to study the *infrared* and *ultraviolet* spectra of

(a) The spectrum of sunlight reflected from Titan

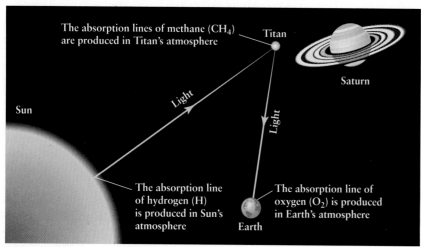

(b) Interpreting Titan's spectrum

figure 7-3

Analyzing the Spectrum of a Satellite (a) This graph shows the spectrum of sunlight reflected from Titan, the only satellite in the solar system with a substantial atmosphere. The dips in the curve are due to absorption by hydrogen atoms (H), oxygen molecules (O_2), and methane molecules (CH_4). Of these, only methane is actually present in Titan's atmosphere. (b) The spectrum in (a) was made by light that was emitted from the Sun's glowing surface, passed through the Sun's atmosphere, was reflected by Titan's atmosphere, and passed through the Earth's atmosphere before reaching a telescope. Only the methane lines are caused by absorption in Titan's atmosphere.

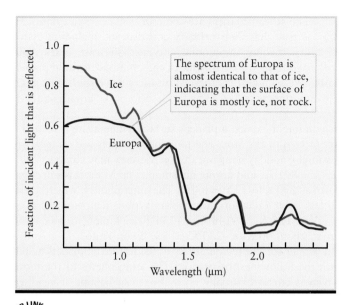

WEB LINK 7.4

figure 7-4
The Spectrum of Jupiter's Moon Europa Infrared light from the Sun that is reflected from the surface of Europa, one of Jupiter's moons, has almost exactly the same spectrum as sunlight reflected from ordinary water ice.

In the graph:
> The spectrum of Europa is almost identical to that of ice, indicating that the surface of Europa is mostly ice, not rock.

7-4 | Hydrogen and helium are abundant on the Jovian planets, whereas the terrestrial planets are composed mostly of heavy elements

Spectroscopic observations from Earth and spacecraft show that the outer layers of the Jovian planets are composed primarily of the lightest gases, hydrogen and helium. In contrast, chemical analysis of soil samples from Venus, Earth, and Mars demonstrate that the terrestrial planets are made mostly of heavy elements, such as iron, oxygen, silicon, magnesium, nickel, and sulfur. Spacecraft images such as Figure 7-5 and Figure 7-6 only hint at these striking differences in chemical composition, which are summarized in Table 7-3.

Temperature plays a major role in determining whether the materials of which planets are made exist as solids, liquids, or gases. Hydrogen and helium are gaseous except at extremely low temperatures and extraordinarily high pressures. By contrast, rock-forming compounds such as iron and silicon are solids except at temperatures well above 1000 K. (You may want to review the discussion of temperature scales in Box 5-1.) Between these two extremes are substances such as water (H_2O), carbon dioxide (CO_2), methane (CH_4), and ammonia (NH_3), which solidify at low temperatures (typically below 100 to 300 K) into solids called **ices**. (In

planetary atmospheres. Many molecules have much stronger spectral lines in these nonvisible wavelength bands than in the visible. As an example, the ultraviolet spectrum of Titan shows that nitrogen molecules (N_2) are the dominant constituent of Titan's atmosphere. Furthermore, Titan's infrared spectrum includes spectral lines of a variety of molecules that contain carbon and hydrogen, indicating that Titan's atmosphere has a very complex chemistry. None of these molecules could have been detected by visible light alone.

Spectroscopy can also provide useful information about the *solid surfaces* of planets and satellites without atmospheres. When light shines on a solid surface, some wavelengths are absorbed while others are reflected. (For example, a plant leaf absorbs red and violet light but reflects green light—which is why leaves look green.) Unlike a gas, a solid illuminated by sunlight does not produce sharp, definite spectral lines. Instead, only broad absorption features appear in the spectrum. By comparing such a spectrum with the spectra of samples of different substances on Earth, astronomers can infer the chemical composition of the surface of a planet or satellite.

As an example, Figure 7-4 shows the infrared spectrum of light reflected from the surface of Jupiter's satellite Europa (Table 7-2 includes a photograph of Europa). Because this spectrum is so close to that of water ice—that is, frozen water—astronomers conclude that water ice is the dominant constituent of Europa's surface. (We saw in Section 7-2 that water ice cannot be the dominant constituent of Europa's *interior*, because the satellite's density is too high.)

Unfortunately, spectroscopy tells us little about what the material is like just below the surface of a satellite or planet. For this purpose, there is simply no substitute for sending a spacecraft to a planet and examining its surface directly.

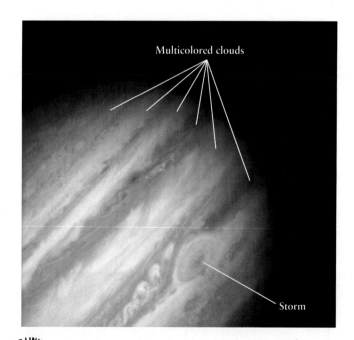

WEB LINK 7.5

figure 7-5 R I **V** U X G
A Jovian Planet This Hubble Space Telescope image gives a detailed view of Jupiter's cloudtops. Jupiter is composed mostly of the lightest elements, hydrogen and helium, which are colorless; the colors in the atmosphere are caused by trace amounts of other substances. The giant storm at lower right, called the Great Red Spot, has been raging for more than 300 years. (Space Telescope Science Institute/JPL/NASA)

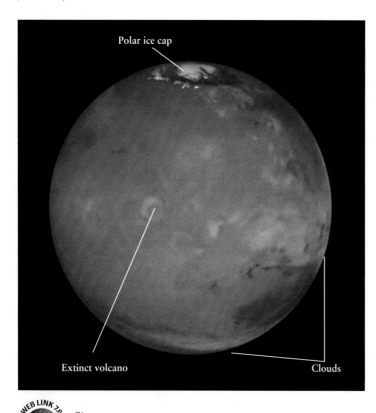

Polar ice cap

Extinct volcano

Clouds

WEB LINK 7.6

Figure 7-6 R I **V** U X G

A Terrestrial Planet Mars is composed mostly of heavy elements such as iron, oxygen, silicon, magnesium, nickel, and sulfur. The planet's red surface can be seen clearly in this Hubble Space Telescope image because the Martian atmosphere is thin and nearly cloudless. Olympus Mons, the extinct volcano to the left of center, is nearly 3 times the height of Mount Everest. (Space Telescope Science Institute/JPL/NASA)

astronomy, frozen water is just one kind of "ice.") At somewhat higher temperatures, they can exist as liquids or gases. For example, clouds of ammonia ice crystals are found in the cold upper atmosphere of Jupiter, but within Jupiter's warmer interior, ammonia exists primarily as a liquid.

CAUTION! The Jovian planets are sometimes called "gas giants." It is true that their primary constituents, including hydrogen, helium, ammonia, and methane, are gases under normal conditions on Earth. But in the interiors of these planets, pressures are so high that these substances are *liquids,* not gases. The Jovian planets might be better described as "liquid giants"!

As you might expect, a planet's surface temperature is related to its distance from the Sun. The four inner planets are quite warm. For example, midday temperatures on Mercury may climb to 700 K (= 427°C = 801°F), and during midsummer on Mars, it is sometimes as warm as 290 K (= 17°C = 63°F). The outer planets, which receive much less solar radiation, are cooler. Typical temperatures range from about 125 K (= –148°C = –234°F) in Jupiter's upper atmosphere to about 40 K (= –233°C = –387°F) on Pluto.

The higher surface temperatures of the terrestrial planets help to explain the following observation: The atmospheres of the terrestrial planets contain virtually no hydrogen or helium. Instead, the atmospheres of Venus, Earth, and Mars are composed of heavier molecules such as nitrogen (N_2, 14 times more massive than a hydrogen molecule), oxygen (O_2, 16 times more massive), and carbon dioxide (22 times more massive). To understand the connection between surface temperature and the absence of hydrogen and helium, we need to know a few basic facts about gases.

The temperature of a gas is directly related to the speeds at which the atoms or molecules of the gas move: The higher the gas temperature, the greater the speed of its atoms or molecules. Furthermore, for a given temperature, lightweight atoms and molecules move more rapidly than heavy ones. On the four inner, terrestrial planets, where atmospheric temperatures are high, low-mass hydrogen molecules and helium atoms move so swiftly that they can escape from the relatively weak gravity of these planets. Hence, the atmospheres that surround the terrestrial planets are composed primarily of more massive, slower-moving molecules such as CO_2, N_2, O_2, and water vapor (H_2O). On the four Jovian planets, low temperatures and relatively strong gravity prevent even lightweight hydrogen and helium gases from escaping into space, and so their atmospheres are much more extensive. The combined mass of Jupiter's atmosphere, for example, is about a million (10^6) times greater than that of the Earth's atmosphere. This is comparable to the mass of the entire Earth! Box 7-2 describes more about the ability of a planet's gravity to retain gases.

table 7-3	**Comparing Terrestrial and Jovian Planets**	INTERACTIVE EXERCISE 7.1
	Terrestrial Planets	Jovian Planets
Distance from the Sun	Less than 2 AU	More than 5 AU
Size	Small	Large
Composition	Mostly rocky materials containing iron, oxygen, silicon, magnesium, nickel, and sulfur	Mostly hydrogen and helium
Density	High	Low

box 7-2 TOOLS OF THE ASTRONOMER'S TRADE

Kinetic Energy, Temperature, and Whether Planets Have Atmospheres

A moving object possesses energy as a result of its motion. The faster it moves, the more energy it has. Energy of this type is called **kinetic energy.** If an object of mass m is moving with a speed v, its kinetic energy E_k is given by

Kinetic energy

$$E_k = \frac{1}{2} mv^2$$

E_k = kinetic energy of an object

m = mass of object

v = speed of object

This expression for kinetic energy is valid for all objects, both big and small, from atoms and molecules to planets and stars, as long as their speeds are slow in relation to the speed of light. If the mass is expressed in kilograms and the speed in meters per second, the kinetic energy is expressed in joules (J).

EXAMPLE: An automobile of mass 1000 kg driving at a typical freeway speed of 30 m/s (= 108 km/h = 67 mi/h) has a kinetic energy of

$$E_k = \frac{1}{2}(1000 \times 30^2) = 450,000 \text{ J} = 4.5 \times 10^5 \text{ J}$$

Consider a gas, such as the atmosphere of a star or planet. Some of the gas atoms or molecules will be moving slowly, with little kinetic energy, while others will be moving faster and have more kinetic energy. The temperature of the gas is a direct measure of the *average* amount of kinetic energy per atom or molecule. The hotter the gas, the faster atoms or molecules move, on average, and the greater the average kinetic energy of an atom or molecule.

If the gas temperature is sufficiently high, typically several thousand kelvins, molecules move so fast that when they collide with one another, the energy of the collision can break the molecules apart into their constituent atoms. Thus, the Sun's atmosphere, where the temperature is 5800 K, consists primarily of individual hydrogen atoms rather than hydrogen molecules. By contrast, the hydrogen atoms in the Earth's atmosphere (temperature 290 K) are combined with oxygen atoms into molecules of water vapor (H_2O).

The physics of gases tells us that in a gas of temperature T (in kelvins), the average kinetic energy of an atom or molecule is

Kinetic energy of a gas atom or molecule

$$E_k = \frac{3}{2} kT$$

E_k = average kinetic energy of a gas atom or molecule, in joules

$k = 1.38 \times 10^{-23}$ J/K

T = temperature of gas, in kelvins

The quantity k is called the Boltzmann constant. Note that the higher the gas temperature, the greater the average kinetic energy of an atom or molecule of the gas. This average kinetic energy becomes zero at absolute zero, or $T = 0$, the temperature at which molecular motion is at a minimum.

At a given temperature, all kinds of atoms and molecules will have the same average kinetic energy. But the average *speed* of a given kind of atom or molecule depends on its mass. To see this, note that the average kinetic energy of a gas atom or molecule can be written in two equivalent ways:

$$E_k = \frac{1}{2} mv^2 = \frac{3}{2} kT$$

where v represents the average speed of an atom or molecule in a gas with temperature T. Rearranging this equation, we obtain

Average speed of a gas atom or molecule

$$v = \sqrt{\frac{3kT}{m}}$$

v = average speed of a gas atom or molecule, in m/s

$k = 1.38 \times 10^{-23}$ J/K

T = temperature of gas, in kelvins

m = mass of the atom or molecule, in kilograms

For a given gas temperature, the greater the mass of a given type of gas atom or molecule, the slower its average speed. (The value of v given by this equation is actually slightly higher than the average speed of the atoms or molecules in the gas, but it is close enough for our purposes here. If you are studying physics, you may know that v is actually the root-mean-square speed.)

(continued)

TOOLS OF THE ASTRONOMER'S TRADE continued

EXAMPLE: What is the average speed of the oxygen molecules that you breathe at a room temperature of 20°C (= 68°F)?

Situation: We are given the temperature of a gas and are asked to find the average speed of the gas molecules.

Tools: We use the relationship $v = \sqrt{3kT/m}$, where T is the gas temperature in kelvins and m is the mass of a single oxygen molecule in kilograms.

Answer: To use the equation to calculate the average speed v, we must express the temperature T in kelvins (K) rather than degrees Celsius (°C). As we learned in Box 5-1, we do this by adding 273 to the Celsius temperature, so 20°C becomes $(20 + 273) = 293$ K. The mass m of an oxygen molecule is not given, but from a reference book you can find that the mass of an oxygen *atom* is 2.66×10^{-26} kg. The mass of an oxygen molecule (O_2) is twice the mass of an oxygen atom, or $2(2.66 \times 10^{-26}$ kg$) = 5.32 \times 10^{-26}$ kg. Thus, the average speed of an oxygen molecule in 20°C air is

$$v = \sqrt{\frac{3(1.38 \times 10^{-23})(293)}{5.32 \times 10^{-26}}} = 478 \text{ m/s} = 0.478 \text{ km/s}$$

Review: This is about 1700 kilometers per hour, or about 1100 miles per hour. Hence, atoms and molecules move rapidly in even a moderate-temperature gas.

In some situations, atoms and molecules in a gas may be moving so fast that they can overcome the attractive force of a planet's gravity and escape into interplanetary space. The minimum speed that an object at a planet's surface must have in order to permanently leave the planet is called the planet's **escape speed**. The escape speed for a planet of mass M and radius R is given by

$$v_{escape} = \sqrt{\frac{2GM}{R}}$$

where $G = 6.67 \times 10^{-11}$ N m²/kg² is the universal constant of gravitation.

The accompanying table gives the escape speed for the Sun, the planets, and the Moon. For example, to get to another planet, a spacecraft must leave Earth with a speed greater than 11.2 km/s (25,100 mi/h).

A good rule of thumb is that a planet can retain a gas if the escape speed is at least 6 times greater than the average speed of the molecules in the gas. (Some molecules are moving slower than average, and others are moving faster, but very few are moving more than 6 times faster than average.) In such a case, very few molecules will be moving fast enough to escape from the planet's gravity.

EXAMPLE: Consider the Earth's atmosphere. We saw that the average speed of oxygen molecules is 0.478 km/s at room temperature. The escape speed from the Earth (11.2 km/s) is much more than 6 times the average speed of the oxygen molecules, so the Earth has no trouble keeping oxygen in its atmosphere.

A similar calculation for hydrogen molecules (H_2) gives a different result, however. At 293 K, the average speed of a hydrogen molecule is 1.9 km/s. Six times this speed is 11.4 km/s, which is slightly higher than the escape speed from the Earth. Thus, the Earth does not retain hydrogen in its atmosphere. Any hydrogen released into the air slowly leaks away into space. On Jupiter, by contrast, the escape speed is so high that even the lightest gases such as hydrogen are retained in its atmosphere. But on Mercury the escape speed is low and the temperature high (so that gas molecules move faster), so Mercury cannot retain any significant atmosphere at all.

Object	Escape speed (km/s)
Sun	618
Mercury	4.3
Venus	10.4
Earth	11.2
Moon	2.4
Mars	5.0
Jupiter	59.5
Saturn	35.5
Uranus	21.3
Neptune	23.5
Pluto	1.3

7-5 | Small chunks of rock and ice also orbit the Sun

In addition to the nine planets, many smaller objects orbit the Sun. Between the orbits of Mars and Jupiter are thousands of rocky objects called **asteroids**. There is no sharp dividing line between planets and asteroids, which is why asteroids are also called **minor planets**. The largest asteroid, Ceres, has a diameter of about 900 km. The next largest, Pallas and Vesta, are each about 500 km in diameter. Still smaller ones, like the asteroid shown in close-up in Figure 7-7, are increasingly numerous. There are thousands of kilometer-sized asteroids and millions of asteroids that are boulder-sized or smaller.

Most (although not all) asteroids orbit the Sun at distances of 2 to 3.5 AU. This region of the solar system between the orbits of Mars and Jupiter is called the **asteroid belt.**

CAUTION One common misconception about asteroids is that they are the remnants of an ancient planet that somehow broke apart or exploded, like the fictional planet Krypton in the comic book adventures of Superman. In fact, the asteroids were probably never part of any planet-sized body. The early solar system is thought to have been filled with asteroidlike objects, most of which were incorporated into the planets. The "leftover" objects that missed out on this process make up our present-day population of asteroids.

Quite far from the Sun, beyond the orbit of Neptune, are chunks of very dirty ice called **comets**. Many comets have highly elongated orbits that occasionally bring them close to the Sun. When this happens, the Sun's radiation vaporizes some of the comet's ices, pro-ducing long flowing tails of gas and dust particles (Figure 7-8). Astronomers deduce the composition of comets by studying the spectra of these tails.

CAUTION Science-fiction movies and television programs sometimes show comets tearing across the night sky like a rocket. That would be a pretty impressive sight—but that is not what comets look like. Like the planets, comets orbit the Sun. And like the planets, comets move hardly at all against the background of stars over the course of a single night (see Section 4-1). If you are lucky enough to see a bright comet, it will not zoom dramatically from horizon to horizon. Instead, it will seem to hang majestically among the stars, so you can admire it at your leisure.

Both asteroids and comets are thought to be debris left over from the formation of the solar system. In the inner regions of the solar system, rocky fragments have been able to endure continuous exposure to the Sun's heat, but any ice originally present would have evaporated. Far from the Sun, chunks of ice have survived for billions of years. Thus, debris in the solar system naturally divides into two families (asteroids and comets), which can be arranged according to distance from the Sun, just like the two categories of planets (terrestrial and Jovian).

Many comets are thought to come from the **Kuiper belt,** a region of the solar system that extends from around the orbit of Neptune to about 500 AU from the Sun. Since the first discovery in 1992, about 1,000 objects made of ice and rock have been identified in the Kuiper belt. Several of these Kuiper belt objects are larger than Ceres, the largest asteroid (934 km in diameter). In fact, the largest known **Kuiper belt object** has more mass

Figure 7-7 R I **V** U X G
An Asteroid The asteroid shown in this image, 433 Eros, is only 33 km (21 mi) long and 13 km (8 mi) wide. Because Eros is so small, its gravity is too weak to have pulled it into a spherical shape. This image was taken in March 2000 by *NEAR Shoemaker,* the first spacecraft to orbit around and land on an asteroid. (*NEAR* Project, Johns Hopkins University, Applied Physics Laboratory, and NASA)

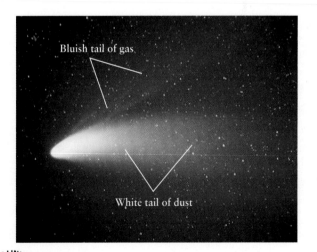

Figure 7-8 R I **V** U X G
A Comet This photograph shows Comet Hale-Bopp as it appeared in April 1997. The solid part of a comet like this is a chunk of dirty ice a few tens of kilometers in diameter. When a comet passes near the Sun, solar radiation vaporizes some of the icy material, forming a bluish tail of gas and a white tail of dust. Both tails can extend for tens of millions of kilometers. (Richard J. Wainscoat, University of Hawaii)

than all known asteroids combined! The total mass of all the objects in the Kuiper belt is thought to be about 100 times the mass of the asteroid belt.

The spectra of Kuiper belt objects are similar to that of Pluto, and like Pluto these objects move in eccentric orbits that are noticeably inclined to the plane of the ecliptic (see Figure 7-1). Indeed, many planetary scientists contend that Pluto should not be thought of as a planet at all, but simply as the largest and best known member of the Kuiper belt. Many researchers consider it likely that with advances in telescope technology, worlds of ice and rock even larger than Pluto will be discovered in the dim, distant recesses of the Kuiper belt.

7-6 | Cratering on planets and satellites is the result of impacts from interplanetary debris

Scientists study craters on a planet or satellite to reveal the age of the surface and the nature of the interior

One of the great challenges in studying planets and satellites is how to determine their internal structures. Are they solid or liquid inside? If there is liquid in the interiors, is the liquid calm or in agitated motion? Because planets are opaque, we cannot see directly into their interiors to answer these questions. But we can gather important clues about the interiors of terrestrial planets and satellites by studying the extent to which their surfaces are covered with craters (**Figure 7-9**). To see how this is done, we first need to understand where craters come from.

The nine planets orbit the Sun in roughly circular orbits. But many of the smaller objects in the solar system, including a number of asteroids and comets, are in more elongated orbits. Such an elongated orbit can put the object on a collision course with a planet or satellite. If the object collides with a Jovian planet, it is swallowed up by the planet's thick atmosphere. (Astronomers actually saw an event of this kind in 1994, when a comet crashed into Jupiter.) But if the object collides with the solid surface of a terrestrial planet or a satellite, the result is an **impact crater** (see Figure 7-9). Such impact craters, found throughout the solar system, offer stark evidence of these violent collisions.

The easiest way to view impact craters is to examine the Moon through a telescope or binoculars. Some 30,000 lunar craters are visible, with diameters ranging from 1 km to several hundred km. Close-up photographs from lunar orbit have revealed millions of craters too small to be seen from Earth (Figure 7-9a). These smaller craters are thought to have been caused by impacts of relatively small objects called **meteoroids,** which range in size from a few hundred meters across to the size of a pebble or smaller. Meteoroids are the result of collisions between asteroids, whose orbits sometimes cross. The chunks of rock that result from these collisions go into independent orbits around the Sun, which can lead them to collide with the Moon or another world.

When German astronomer Franz Gruithuisen proposed in 1824 that lunar craters were the result of impacts, a major sticking point was the observation that nearly all craters are circular. If craters were merely gouged out by high-speed rocks, a rock striking the Moon in any direction except straight downward would have created a noncircular crater. A century after Gruithuisen, it was realized that a meteoroid colliding with the Moon generates a shock wave in the lunar surface that spreads out from the point of impact. Such a shock

(a) A crater on the Moon

(b) A crater on Earth

(c) A crater on Mars

Figure 7-9 R I **V** U X G

Impact Craters These images, all taken from spacecraft, show impact craters on three different worlds. (a) The Moon's surface has craters of all sizes. The large crater near the middle of this image is about 80 km (50 miles) in diameter, equal to the length of San Francisco Bay. (b) Manicouagan Reservoir in Quebec is the relic of a crater formed by an impact over 200 million years ago. The crater was eroded over the ages by the advance and retreat of glaciers, leaving a ring lake 100 km (60 miles) across. (c) Lowell Crater in the southern highlands of Mars is 201 km (125 miles) across. Like the image of the Moon in part (a), there are craters on top of craters. Note the light-colored frost formed by condensation of carbon dioxide from the Martian atmosphere. (a: NASA; b: JSC/NASA; c: NASA/JPL/MSSS) wave produces a circular crater no matter what direction the meteoroid was moving. (In a similar way, the craters made by artillery shells are almost always circular.) Many of the larger lunar craters also have a central peak, which is characteristic of a high-speed impact (see Figure 7-9a). Craters made by other processes, such as volcanic action, would not have central peaks of this sort.

Not all planets and satellites show the same amount of cratering. The Moon is heavily cratered over its entire surface, with craters on top of craters as shown in Figure 7-9a. On Earth, by contrast, craters are very rare: geologists have identified fewer than 200 impact craters on our planet (Figure 7-9b). Our understanding is that both the Earth and the Moon formed at nearly the same time and have been bombarded at comparable rates over their history. Why, then, are craters so much rarer on the Earth than on the Moon?

The answer is that the Earth is a *geologically active* planet: the continents slowly change their positions over eons, new material flows onto the surface from the interior (as occurs in a volcanic eruption), and old surface material is pushed back into the interior (as occurs off the coast of Chile, where the ocean bottom is slowly being pushed beneath the South American continent). These processes, coupled with erosion from wind and water, cause craters on Earth to be erased over time. The few craters found on Earth today must be relatively recent, since there has not yet been time to erase them.

The Moon, by contrast, is geologically *inactive*. There are no volcanoes and no motion of continents (and, indeed, no continents). Furthermore, the Moon has neither oceans nor an atmosphere, so there is no erosion as we know it on Earth. With none of the processes that tend to erase craters on Earth, the Moon's surface remains pockmarked with the scars of billions of years of impacts.

In order for a planet to be geologically active, its interior must be at least partially molten. This is necessary so that continents can slide around on the underlying molten material and so that molten lava can come to the surface, as in a volcanic eruption. Hence, we would expect geologically inactive (and hence heavily cratered) worlds like the Moon to have less molten material in their interiors than does the Earth. Investigations of these inactive worlds bear this out. But *why* is the Moon's interior less molten than the Earth's?

ANALOGY To see one simple answer to this question, notice that a large turkey or roast taken from the oven will stay warm inside for hours, but a single meatball will cool off much more rapidly. The reason is that the meatball has more surface area relative to its volume, and so it can more easily lose heat to its surroundings. A planet or satellite also tends to cool down as it emits electromagnetic radiation into space (see Section 5-3); the smaller the planet or satellite, the greater its surface area relative to its volume, and the more readily it can radiate away heat. Both the Earth and Moon were probably completely molten when they first formed, but because the Moon (diameter 3476 km) is so much smaller than the Earth (diameter 12,756 km), it has lost much of its internal heat and has a much more solid interior.

By considering these differences between the Earth and the Moon, we have uncovered a general rule for worlds with solid surfaces:

The smaller the terrestrial world, the less internal heat it is likely to have retained, and, thus, the less geologic activity it will dis-play on its surface. The less geologically active the world, the older and hence more heavily cratered its surface.

This rule means that we can use the amount of cratering visible on a planet or satellite to estimate the age of its surface and how geologically active it is. As an example, Mercury has a heavily cratered surface, which means that the surface is very old. This accords with Mercury being the smallest of the terrestrial planets (see Table 7-1): Due to its small size, it has lost the internal heat required to sustain geologic activity. On Venus, by comparison, there are only about a thousand craters larger than a few kilometers in diameter, many more than have been found on Earth but only a small fraction of the number on the Moon or Mercury. Venus is only slightly smaller than the Earth, and it has enough internal heat to power the geologic activity required to erase most of its impact craters.

Mars is an unusual case, in that extensive cratering is found only in the higher terrain (Figure 7-9c); the lowlands of Mars are remarkably smooth and free of craters. Thus, it follows that the Martian highlands are quite old, while the lowlands have a younger surface from which most craters have been erased. Considering the planet as a whole, the amount of cratering on Mars is intermediate between that on Mercury and the Earth. This agrees with our general rule, because Mars is intermediate in size between Mercury and Earth. The interior of Mars was once hotter and more molten than it is now, so that geologic processes were able to erase some of the impact craters. A key piece of evidence that supports this picture is that Mars has a number of immense volcanoes (Figure 7-10). These

figure 7-10 R I ▣ U X G
A Martian Volcano Olympus Mons is the largest of the inactive volcanoes of Mars and the largest volcano in the solar system. The caldera, or volcanic crater, at the summit is approximately 70 km across, large enough to contain the state of Rhode Island. This view (actually a mosaic of images taken from orbit) shows the volcano wreathed in mid-morning clouds. (NASA)

were active when Mars was young, but as this relatively small planet cooled down and its interior solidified, the supply of molten material to the volcanoes from the Martian interior was cut off. As a result, all of the volcanoes of Mars are now inactive.

As for all rules, there are limitations and exceptions to the rule relating a world's size to its geologic activity. One limitation is that the four terrestrial planets all have slightly different compositions, which affects the types and extent of geologic activity that can take place on their surfaces. This also complicates the relationship between the number of craters and the age of the surface. An important exception to our rule is Jupiter's satellite Io, which, despite its small size, is the most volcanic world in the solar system (see Section 7-2). Something must be supplying Io with energy to keep its interior hot; this turns out to be Jupiter, which exerts powerful tides on Io as it moves in a relatively small orbit around its planet. These tides cause Io to flex like a ball of clay being kneaded between your fingers, and this flexing heats up the satellite's interior. But despite these limitations and exceptions, the relationships between a world's size, internal heat, geologic activity, and amount of cratering are powerful tools for understanding the terrestrial planets and satellites.

7-7 | A planet with a magnetic field indicates a fluid interior in motion

The amount of impact cratering on a terrestrial planet or satellite provides indirect evidence about whether the planet or satellite has a molten interior. But another, more direct tool for probing the interior of *any* planet or satellite is an ordinary compass, which senses the magnetic field outside the planet or satellite. Magnetic field measurements prove to be an extremely powerful way to investigate the internal structure of a world without having to actually dig into its interior. To illustrate how this works, consider the behavior of a compass on Earth.

The needle of a compass on Earth points north because it aligns with the Earth's *magnetic field*. Such fields arise whenever electrically charged particles are in motion. For example, a loop of wire carrying an electric current generates a magnetic field in the space around it. The magnetic field that surrounds an ordinary bar magnet (**Figure 7-11a**) is created by the motions of negatively charged electrons within the iron atoms of which the magnet is made. The Earth's magnetic field is similar to that of a bar magnet, as Figure

(a)

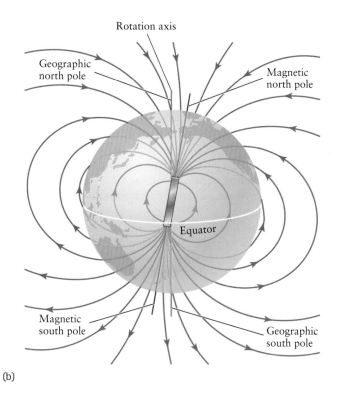

Rotation axis
Geographic north pole
Magnetic north pole
Equator
Magnetic south pole
Geographic south pole

(b)

figure 7-11   R I **V** U X G

The Magnetic Fields of a Bar Magnet and of the Earth (a) This picture was made by placing a piece of paper on top of a bar magnet, then spreading iron filings on the paper. The pattern of the filings show the magnetic field lines, which appear to stream from one of the magnet's poles to the other. (b) The Earth's magnetic field lines have a similar pattern. Although the Earth's field is produced in a different way—by electric

currents in the liquid portion of our planet's interior—the field is much the same as if there were a giant bar magnet inside the Earth. This "bar magnet" is not exactly aligned with the Earth's rotation axis, which is why the magnetic north and south poles are not at the same locations as the true, or geographic, poles. A compass needle points toward the north magnetic pole, not the true north pole. (a: Jules Bucher/Photo Researchers)

7-11*b* shows. The consensus among geologists is that this magnetic field is caused by the motion of the liquid portions of the Earth's interior. Because this molten material (mostly iron) conducts electricity, these motions give rise to electric currents, which in turn produce the Earth's magnetic field. Our planet's rotation helps to sustain these motions and hence the magnetic field. This process for producing a magnetic field is called a **dynamo**.

CAUTION ! While the Earth's magnetic field is similar to that of a giant bar magnet, you should not take this picture too literally. The Earth is *not* simply a magnetized ball of iron. In an iron bar magnet, the electrons of different atoms orbit their nuclei in the same general direction, so that the magnetic fields generated by individual atoms add together to form a single, strong field. But at temperatures above 770°C (1418°F = 1043 K), the orientations of the electron orbits become randomized. The fields of individual atoms tend to cancel each other out, and the iron loses its magnetism. Geological evidence shows that almost all of the Earth's interior is hotter than 770°C, so the iron there cannot be extensively magnetized. The correct picture is that the Earth acts as a dynamo: The liquid iron carries electric currents, and these currents create the Earth's magnetic field.

If a planet or satellite has a mostly solid interior, then the dynamo mechanism cannot work: Material in the interior cannot flow, there are no electric currents, and the planet or satellite does not generate a magnetic field. One example of this is the Moon. As we saw in Section 7-6, the extensive cratering of the lunar surface indicates that the Moon has no geologic activity and must therefore have a mostly solid interior. Measurements made during the *Apollo* missions, in which 12 humans visited the lunar surface between 1969 and 1972, showed that the present-day Moon indeed has no global magnetic field. However, careful magnetic measurements of lunar rocks returned by the *Apollo* astronauts indicate that the Moon *did* have a weak magnetic field when the rocks solidified. These rocks, like the rest of the lunar surface, are very old. Hence, in the distant past the Moon may have had a small amount of molten iron in its interior that acted as a dynamo. This material presumably solidified at least partially as the Moon cooled, so that the lunar magnetic field disappeared.

We have now identified another general rule about planets and satellites:

A planet or satellite with a global magnetic field has liquid material in its interior that conducts electricity and is in motion, generating the magnetic field.

Thus, by studying the magnetic field of a planet or satellite, we can learn about that world's interior. This explains why many spacecraft carry devices called **magnetometers** to measure magnetic fields. In the illustration of the *Cassini* spacecraft that opens this chapter, the magnetometer is located on the long boom that extends down and to the right from the body of the spacecraft. (The glow at the end of the boom is the reflection of the Sun.) Magnetometers are often placed on booms of this sort to isolate them from the magnetic fields produced by electric currents in the spacecraft's own circuitry.

Measurements made with magnetometers on spacecraft have led to a number of striking discoveries. For example, it has been found that Mercury has a planetwide magnetic field like the Earth's, although it is only about 1% as strong as the Earth's field. Mercury has a heavily cratered surface and hence little or no geologic activity, which by itself would suggest that the planet's interior is mostly solid. The magnetic field measurements show that *some* of Mercury's interior must be in the liquid state to act as a dynamo. By contrast, Venus has no measurable planetwide magnetic field, even though the paucity of craters on its surface indicates the presence of geologic activity and hence a hot interior of the planet. One possible reason for the lack of a magnetic field on Venus is that the planet turns on its axis very slowly, taking 243 days for a complete rotation. Because of this slow rotation, the fluid material within the planet is hardly agitated at all, and so may not move in the fashion that generates a magnetic field.

Like Venus, Mars has no planetwide magnetic field. But the magnetometer aboard the *Mars Global Surveyor* spacecraft, which went into orbit around Mars in 1997, found magnetized regions in the cratered highlands of the Martian southern hemisphere (**Figure 7-12**). As we saw in Section 7-6, these portions of the Martian surface are very old, and so would have formed when the planet's interior was still hot and molten. Electric currents in the flowing molten material could then produce a planetwide magnetic field. As surface material cooled and solidified, it became magnetized by the planetwide field, and this material has retained its magnetization over the eons. *Mars Global Surveyor* has not found magnetized areas in the younger terrain of the lowlands. Hence, the

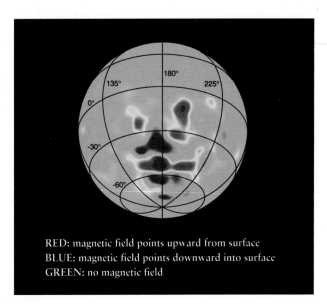

RED: magnetic field points upward from surface
BLUE: magnetic field points downward into surface
GREEN: no magnetic field

figure 7-12

Relic Magnetism on Mars Portions of the highlands of Mars became magnetized during the period early in Martian history when the planet had an extensive magnetic field. The planetwide field has long since disappeared, but the surface regions remained magnetized. The false colors in this illustration indicate the direction and strength of the field at different locations on Mars. (Connerney et al., *Geophys. Res. Lett.*, 28, 4015–4018, 2001) planetwide Martian magnetic field must have shut off by the time the lowland terrain formed.

The most intense planetary magnetic field in the solar system is that of Jupiter: at Jupiter's cloudtops, the magnetic field is about 14 times stronger than the field at the Earth's surface. It is thought that Jupiter's field, like the Earth's, is produced by a dynamo acting deep within the planet's interior. Unlike the Earth, however, Jupiter is composed primarily of hydrogen and helium, not substances like iron that conduct electricity. How, then, can Jupiter have a dynamo that generates such strong magnetic fields?

To answer this question, recall from Section 5-8 that a hydrogen atom consists of a single proton orbited by a single electron. Deep inside Jupiter, the pressure is so great and hydrogen atoms are squeezed so close together that electrons can hop from one atom to another. This hopping motion creates an electric current, just as the ordered movement of electrons in the copper wires of a flashlight constitutes an electric current. In other words, the highly compressed hydrogen deep inside Jupiter behaves like an electrically conducting metal; thus, it is called **liquid metallic hydrogen.**

Laboratory experiments show that hydrogen becomes a liquid metal when the pressure is more than about 1.4 million times ordinary atmospheric pressure on Earth. Recent calculations suggest that this transition occurs about 7000 km below Jupiter's cloudtops. Most of the planet's enormous bulk lies below this level, so there is a tremendous amount of liquid metallic hydrogen within Jupiter. Since Jupiter rotates rapidly—a "day" on Jupiter is just less than 10 hours long—this liquid metal moves rapidly, generating the planet's powerful magnetic field. Saturn also has a magnetic field produced by dynamo action in liquid metallic hydrogen. (The field is weaker than Jupiter's because Saturn is a smaller planet with less internal pressure, so there is less of the liquid metal available.)

Uranus and Neptune also have magnetic fields, but they cannot be produced in the same way: Because these planets are relatively small, the internal pressure is not great enough to turn liquid hydrogen into a metal. Instead, it is thought that both Uranus and Neptune have large amounts of liquid water in their interiors and that this water has molecules of ammonia and other substances dissolved in it. (The fluid used for washing windows has a similar chemical composition.) Under the pressures found in this interior water, the dissolved molecules lose one or more electrons and become electrically charged (that is, they become ionized; see Section 5-8). Water is a good conductor of electricity when it has such electrically charged molecules dissolved in it, and electric currents in this fluid are probably the source of the magnetic fields of Uranus and Neptune.

7-8 | The diversity of the solar system is a result of its origin and evolution

Our brief tour of the solar system has revealed its almost dizzying variety. No two planets are alike, satellites come in all sizes, the extent of cratering varies from one terrestrial planet to another, and the magnetic fields of different planets vary dramatically in their strength and in how they are produced. All of this variety leads us to a simple yet profound question: *Why* are the planets and satellites of the solar system so different from each other?

Among humans, the differences from one individual to another result from heredity (the genetic traits passed on from an individual's parents) and environment (the circumstances under which the individual matures to an adult). As we will find in the following chapter, much the same is true for the worlds of the solar system.

In Chapter 8 we will see evidence that the entire solar system shares a common "heredity," in that the planets, satellites, comets, asteroids, and the Sun itself formed from the same cloud of interstellar gas and dust. The composition of this cloud was shaped by cosmic processes, including nuclear reactions that took place within stars that died long before our solar system was formed. We will see how different planets formed in different environments depending on their distance from the Sun and will discover how these environmental variations gave rise to the planets and satellites of our present-day solar system. And we will see how we can test these ideas of solar system origin and evolution by studying planetary systems orbiting other stars.

Our journey through the solar system is just beginning. In this chapter we have explored space to examine the variety of the present-day solar system; in Chapter 8 we will journey through time to see how our solar system came to be.

| Key Words |

Terms preceded by an asterisk (*) are discussed in the Boxes.

asteroid, p. 157
asteroid belt, p. 157
average density, p. 150
chemical composition, p. 151
comet, p. 157
dynamo, p. 161

*escape speed, p. 156
ices, p. 153
impact crater, p. 158
Jovian planet, p. 149
*kinetic energy, p. 155
Kuiper belt, p. 157
Kuiper belt object, p. 157
liquid metallic hydrogen, p. 162

magnetometer, p. 161
meteoroid, p. 158
minor planet, p. 157
molecule, p. 152
spectroscopy, p. 152
terrestrial planet, p. 149

Key Ideas

Properties of the Planets: All of the planets orbit the Sun in the same direction and in almost the same plane. Most of the planets have nearly circular orbits.

• The four inner planets are called terrestrial planets. They are relatively small (with diameters of 5000 to 13,000 km), have high average densities (4000 to 5500 kg/m³), and are composed primarily of rocky materials.

• The four giant outer planets are called Jovian planets. They have large diameters (50,000 to 143,000 km) and low average densities (700 to 1700 kg/m³) and are composed primarily of hydrogen and helium.

• Pluto is a special case. It is smaller than any of the terrestrial planets and has an intermediate average density of about 1900 kg/m³, suggesting that it is composed of a mixture of ice and rock.

Satellites and Small Bodies in the Solar System: Besides the planets, the solar system includes satellites of the planets, asteroids, comets, and Kuiper belt objects.

• Seven large planetary satellites (one of which is the Moon) are comparable in size to the planet Mercury. The remaining satellites of the solar system are much smaller.

• Asteroids are small, rocky objects, while comets and Kuiper belt objects are made of dirty ice. All are remnants left over from the formation of the planets.

• The Kuiper belt extends far beyond the orbit of Pluto. Pluto can be thought of as the largest member of the Kuiper belt.

Spectroscopy and the Composition of the Planets: Spectroscopy, the study of spectra, provides information about the chemical composition of objects in the solar system.

• The spectrum of a planet or satellite with an atmosphere reveals the atmosphere's composition. If there is no atmosphere, the spectrum indicates the composition of the surface.

• The substances that make up the planets can be classified as gases, ices, or rock, depending on the temperatures at which they solidify. The terrestrial planets are composed primarily of rocky materials, whereas the Jovian planets are composed largely of gas.

Impact Craters: When an asteroid, comet, or meteoroid collides with the surface of a terrestrial planet or satellite, the result is an impact crater.

• Geologic activity renews the surface and erases craters, so a terrestrial world with extensive cratering has an old surface and little or no geologic activity.

• Because geologic activity is powered by internal heat, and smaller worlds lose heat more rapidly, as a general rule smaller terrestrial worlds are more extensively cratered.

Magnetic Fields and Planetary Interiors: Planetary magnetic fields are produced by the motion of electrically conducting liquids inside the planet. This mechanism is called a dynamo. If a planet has no magnetic field, that is evidence that there is little such liquid material in the planet's interior or that the liquid is not in a state of motion.

• The magnetic fields of terrestrial planets are produced by metals such as iron in the liquid state. The stronger fields of the Jovian planets are generated by liquid metallic hydrogen or by water with ionized molecules dissolved in it.

Review Questions

1. Do all the planets orbit the Sun in the same direction? Are all of the orbits circular?

2. What are the characteristics of a terrestrial planet?

3. What are the characteristics of a Jovian planet?

4. In what ways does Pluto not fit the usual classification of either terrestrial or Jovian planets?

5. What is meant by the average density of a planet? What does the average density of a planet tell us?

6. In what ways are the largest satellites similar to the terrestrial planets? In what ways are they different? Which satellites are the largest?

7. The absorption lines in the spectrum of a planet or satellite do not necessarily indicate the composition of the planet or satellite's atmosphere. Why not?

8. What are the differences in chemical composition between the terrestrial and Jovian planets?

9. Why are hydrogen and helium abundant in the atmospheres of the Jovian planets but present in only small amounts in the Earth's atmosphere?

10. What is an asteroid? What is a comet? In what ways are these minor members of the solar system like or unlike the planets?

11. What are the asteroid belt and the Kuiper belt? Where are they located? How do the objects found in these two regions compare?

12. What is one piece of evidence that impact craters are actually caused by impacts?

13. What is the relationship between the extent to which a planet or satellite is cratered and the amount of geologic activity on that planet or satellite?

14. How do we know that the surface of Venus is older than the Earth's surface but younger than the Moon's surface?

15. Why do smaller worlds retain less of their internal heat?

16. How does the size of a terrestrial planet influence the amount of cratering on the planet's surface?

17. How is the magnetic field of a planet different from that of a bar magnet? Why is a large planet more likely to have a magnetic field than a small planet?

18. Could you use a compass to find your way around Venus? Why or why not?

19. If Mars has no planetwide magnetic field, why does it have magnetized regions on its surface?

20. What is liquid metallic hydrogen? Why is it found only in the interiors of certain planets?

Advanced Questions

Questions preceded by an asterisk (*) involve topics discussed in the Boxes.

Problem-solving tips and tools

The volume of a sphere of radius r is $4\pi r^3/3$, and the surface area of a sphere of radius r is $4\pi r^2$. The surface area of a circle of radius r is πr^2. The average density of an object is its mass divided by its volume. To calculate escape speeds, you will need to review Box 7-2. Be sure to use the same system of units (meters, seconds, kilograms) in all your calculations involving escape speeds, orbital speeds, and masses. Appendix 6 gives conversion factors between different sets of units, and Box 5-1 has formulas relating various temperature scales.

---

21. Mars has two small satellites, Phobos and Deimos. Phobos circles Mars once every 0.31891 day at an average altitude of 5980 km above the planet's surface. The diameter of Mars is 6794 km. Using this information, calculate the mass and average density of Mars.

22. Figure 7-3 shows the spectrum of Saturn's largest satellite, Titan. Can you think of a way that astronomers can tell which absorption lines are due to Titan's atmosphere and which are due to the atmospheres of the Sun and Earth? Explain.

*23. (a) Find the mass of a hypothetical spherical asteroid 2 km in diameter and composed of rock with average density 2500 kg/m³. (b) Find the speed required to escape from the surface of this asteroid. (c) A typical jogging speed is 3 m/s. What would happen to an astronaut who decided to go for a jog on this asteroid?

*24. The hypothetical asteroid described in Question 23 strikes the Earth with a speed of 25 km/s. (a) What is the kinetic energy of the asteroid at the moment of impact? (b) How does this energy compare with that released by a 20-kiloton nuclear weapon, like the device that destroyed Hiroshima, Japan, on August 6, 1945? (*Hint:* 1 kiloton of TNT releases 4.2×10^{12} joules of energy.)

*25. Suppose a spacecraft landed on Jupiter's moon Europa (see Table 7-2), which moves around Jupiter in an orbit of radius 670,900 km. After collecting samples from the satellite's surface, the spacecraft prepares to return to Earth. (a) Calculate the escape speed from Europa. (b) Calculate the escape speed from Jupiter at the distance of Europa's orbit. (c) In order to begin its homeward

journey, the spacecraft must leave Europa with a speed greater than either your answer to (a) or your answer to (b). Explain why.

*26. A hydrogen atom has a mass of 1.673×10^{-27} kg, and the temperature of the Sun's surface is 5800 K. What is the average speed of hydrogen atoms at the Sun's surface?

*27. The Sun's mass is 1.989×10^{30} kg, and its radius is 6.96×10^8 m. (a) Calculate the escape speed from the Sun's surface. (b) Using your answer to Question 26, explain why the Sun has lost very little hydrogen over its entire 4.56-billion-year history.

*28. Saturn's satellite Titan has an appreciable atmosphere, yet Jupiter's satellite Ganymede—which is about the same size and mass as Titan—has no atmosphere. Explain why there is a difference.

29. The distance from the asteroid 433 Eros (Figure 7-7) to the Sun varies between 1.13 and 1.78 AU. (a) Find the period of Eros's orbit. (b) Does Eros lie in the asteroid belt? How can you tell?

30. Imagine a planet with roughly the same mass as Earth but located 50 AU from the Sun. (a) What do you think this planet would be made of? Explain your reasoning. (b) On the basis of this speculation, assume a reasonable density for this planet and calculate its diameter. How many times bigger or smaller than Earth would it be?

31. Consider a hypothetical Kuiper belt object located 100 AU from the Sun. (a) What would be the orbital period (in years) of this object? (b) There are 360 degrees in a circle, and 60 arcminutes in a degree. How long would it take this object to move 1 arcminute across the sky? (c) Kuiper belt objects are discovered by looking for "stars" that move on the celestial sphere. Use your answer from part (b) to explain why these discoveries require patience. (d) Discovering Kuiper belt objects also requires large telescopes equipped with sensitive detectors. Explain why.

32. The surfaces of Mercury, the Moon, and Mars are riddled with craters formed by the impact of space debris. Many of these craters are billions of years old. By contrast, there are only a few conspicuous craters on the Earth's surface, and these are generally less than 500 million years old. What do you suppose explains the difference?

33. During the period of most intense bombardment by space debris, a new 1-km-radius crater formed somewhere on the Moon about once per century. During this same period, what was the probability that such a crater would be created within 1 km of a certain location on the Moon during a 100-year period? During a 10^6-year period? (*Hint:* If you drop a coin onto a checkerboard, the probability that the coin will land on any particular one of the board's 64 squares is $^1/_{64}$.)

34. When an impact crater is formed, material (called *ejecta*) is sprayed outward from the impact. (The accompanying photograph of the Moon shows light-colored ejecta extending outward from the crater Copernicus.) While ejecta are found surrounding the craters on Mercury, they do not extend as far from the craters as do ejecta on the Moon. Explain why, using the difference in surface gravity

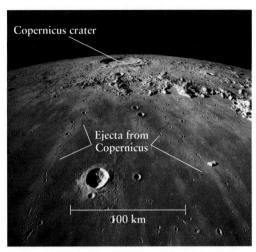

Copernicus crater

Ejecta from Copernicus

100 km

(Courtesy of USRA) R I **V** U X G

between the Moon (surface gravity = 0.17 that on Earth) and Mercury (surface gravity = 0.38 that on Earth).

35. Mercury rotates once on its axis every 58.646 days, compared to 1 day for the Earth. Use this information to argue why Mercury's magnetic field should be much smaller than the Earth's.

36. Suppose *Mars Global Surveyor* had discovered magnetized regions in the lowlands of Mars. How would this discovery have affected our understanding of the evolution of the Martian interior?

Discussion Questions

*37. There are no asteroids with an atmosphere. Discuss why not.

38. The *Galileo* spacecraft in orbit around Jupiter discovered that Ganymede (Table 7-2) has a magnetic field twice as strong as that of Mercury. Does this discovery surprise you? Why or why not?

Web/CD-ROM Questions

39. Search the World Wide Web for information about impact craters on Earth. Where is the largest crater located? How old is it estimated to be? Which crater is closest to where you live?

40. Determining Terrestrial Planet Orbital Periods. Access the animation "Planetary Orbits" in Chapter 7 of the *Universe* Web site or CD-ROM. Focus on the motions of the inner planets at the last half of the animation. Using the stop and start buttons, determine how many days it takes Mars, Venus, and Mercury to orbit the Sun once if Earth takes approximately 365 days.

41. Use a telescope or binoculars to observe craters on the Moon. Make a drawing of the Moon, indicating the smallest and largest craters that you can see. Can you estimate their sizes? For comparison, the Moon as a whole has a diameter of 3476 km. *Hint:* You can see craters most distinctly when the Moon is near first quarter or third quarter (see Figure 3-2). At these phases, the Sun casts long shadows across the portion of the Moon in the center of your field of view, making the variations in elevation between the rims and centers of craters easy to identify. You can determine the phase of the Moon by looking at a calendar or the weather page of the newspaper, by using the *Starry Night Backyard*™ program, or on the World Wide Web.

42. Use the *Starry Night Backyard*™ program to observe magnified images of at least four of the planets. First display the entire celestial sphere (select **Guides > Atlas** in the **Go** menu), then use **Find...** in the **Edit** menu to find the planet you wish to view. Use the sliding zoom controls at the right-hand end of the Control Panel (at the top of the main window) to zoom in until you can see a detailed view of the planet. Describe each planet's appearance. From what you observe, is there any way of knowing whether you are looking at a planet's surface or simply a cloud cover?

Observing Projects

43. Use the *Deep Space Explorer*™ program to examine the terrestrial planets Mercury, Venus, Earth, and Mars and the asteroid Ceres. In the left-hand part of the window, under the heading **Solar System** select **Explore**. Then click on the name of each planet or asteroid to view it in detail. You can zoom in and zoom out using the buttons at the upper left of the window (an upward-pointing triangle and a downward-pointing triangle). You can also rotate the planet or asteroid by putting the mouse cursor over the image, holding down the mouse button, and moving the mouse. (On a two-button mouse, hold down the left mouse button.) Describe each planet or asteroid's appearance. Which planet or planets have clouds? Which planet or asteroid shows the heaviest cratering? Which show evidence of liquid water?

44. Use the *Deep Space Explorer*™ program to examine the Jovian planets Jupiter, Saturn, Uranus, and Neptune. In the left-hand part of the window, under the heading **Solar System** select **Explore**. Then click on the name of each planet to view it in detail. You can zoom in and zoom out using the buttons at the upper left of the window (an upward-pointing triangle and a downward-pointing triangle). You can also rotate the planet by putting the mouse cursor over the image of the planet or asteroid, holding down the mouse button, and moving the mouse. (On a two-button mouse, hold down the left mouse button.) Describe each planet's appearance. Which has the greatest color contrast in its cloudtops? Which has the least color contrast? What can you say about the thickness of Saturn's rings compared to their diameter?

8 Comparative Planetology II: The Origin of Our Solar System

What did our solar system look like before the planets were fully formed? The answer may lie in this remarkable image from the Hubble Space Telescope. At the center of the image is an immense disk of gas and dust centered on the young, Sunlike star HD 141569A. (The light from the star itself was blocked out within the telescope to make the rather faint disk more visible.) Astronomers strongly suspect that in its infancy our own Sun was surrounded by such a disk and that the planets coalesced from it.

In this chapter we will examine the evidence that led astronomers to this picture of the origin of the planets. We will see how the abundances of different chemical elements in the solar system indicate that the Sun and planets formed from a thin cloud of interstellar matter some 4.56 billion years ago, an age determined by measuring the radioactivity of meteorites. We will learn how the nature of planetary orbits gives important clues to what happened as this cloud contracted and how evidence from meteorites reveals the chaotic conditions that existed within the cloud. And we will see how this cloud eventually evolved into the solar system that we see today.

In the past few years astronomers have been able to test this picture of planetary formation by examining disks around young stars (like the one in the accompanying image). Most remarkably of all, they have discovered planets in orbit around dozens of other stars. These recent observations provide valuable information about how our own system of planets came to be.

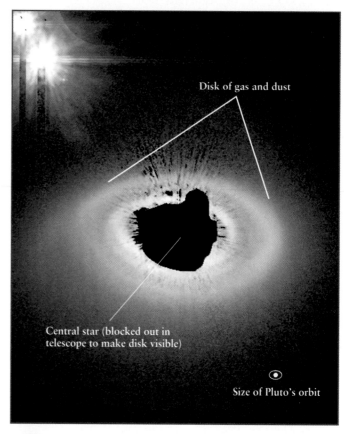

Disk of gas and dust

Central star (blocked out in telescope to make disk visible)

Size of Pluto's orbit

 Planets are thought to form within the disks surrounding young stars. (NASA, M. Clampin (STScI), H. Ford (JHU), G. Illingworth (UCO/Lick), J. Krist (STScI), D. Ardila (JHU), D. Golimowski (JHU), the ACS Science Team and ESA)
R I **V** U X G

Questions to Guide Your Reading

As you read the sections of this chapter, look for the answers to the following questions:

8-1 What must be included in a viable theory of the origin of the solar system?

8-2 Why are some elements (like gold) quite rare, while others (like carbon) are more common?

8-3 How do we know the age of the solar system?

8-4 How do astronomers think the solar system formed?

8-5 Did all of the planets form in the same way?

8-6 Are there planets orbiting other stars? How do astronomers search for other planets?

8-1 | Any model of solar system origins must explain the present-day Sun and planets

How did the Sun and planets form? In other words, where did the solar system come from? This question has tantalized astronomers for centuries. Our goal in this chapter is to examine our current understanding of how the solar system came to be—that is, our current best *theory* of the origin of the solar system.

Recall from Section 1-1 that a theory is not merely a set of wild speculations, but a self-consistent collection of ideas that must pass the test of providing an accurate description of the real world. Since no humans were present to witness the formation of the planets, scientists must base their theories of solar system origins on their observations of the present-day solar system. (In an analogous way, paleontologists base their understanding of the lives of dinosaurs on the evidence provided by fossils that have survived to the present day.) In so doing, they are following the steps of the scientific method that we described in Section 1-1.

What key attributes of the solar system should guide us in building a theory of solar system origins? Among the many properties of the planets that we discussed in Chapter 7, three of the most important are these:

Item 1: The terrestrial planets, which are composed primarily of rocky substances, are relatively small, while the Jovian planets, which are composed primarily of hydrogen and helium, are relatively large (see Sections 7-1 and 7-4).

Item 2: All of the planets orbit the Sun in the same direction, and all of their orbits are in nearly the same plane (see Section 7-1).

Item 3: The terrestrial planets orbit close to the Sun, while the Jovian planets orbit far from the Sun (see Section 7-1).

Any theory that attempts to describe the origin of the solar system must be able to explain how these attributes came to be. We begin by considering what Item 1 tells us; we will return to Items 2 and 3 and the orbits of the planets later in this chapter.

8-2 | The abundances of the chemical elements are the result of cosmic processes

The small sizes of the terrestrial planets compared to the Jovian planets (Item 1 in Section 8-1) suggests that some chemical elements are quite common in our solar system, while others are quite rare. The tremendous masses of the Jovian planets—Jupiter alone has more mass than all of the other planets combined—means that the elements of which they are made, primarily hydrogen and helium, are very abundant. The Sun, too, is made almost entirely of hydrogen and helium: Its average density of 1410 kg/m^3 is in the same range as the densities of the Jovian planets (see Table 7-1), and its absorption spectrum (see Figure 5-12) shows the dominance of hydrogen and helium in the Sun's atmosphere. Hydrogen, the most abundant element, makes up nearly three-quarters of the combined mass of the Sun and planets. Helium is the second most abundant element. Together, hydrogen and helium account for about 98% of the mass of all the material in the solar system. All of the other chemical elements are relatively rare; combined, they make up the remaining 2%.

The dominance of hydrogen and helium is not merely a characteristic of our local part of the universe. By analyzing the spectra of stars and galaxies, astronomers have found essentially the same pattern of chemical abundances out to the farthest distance attainable by the most powerful telescopes. Hence, the vast majority of the atoms in the universe are hydrogen and helium atoms. The elements that make up the bulk of the Earth—mostly iron, oxygen, and silicon—are relatively rare in the universe as a whole, as are the elements of which living organisms are made—carbon, oxygen, nitrogen, and phosphorus, among others. (You may find it useful to review the periodic table of the elements, described in Box 5-5.)

WEB LINK 8.2 There is a good reason for this overwhelming abundance of hydrogen and helium. A wealth of evidence has led astronomers to conclude that the universe began some 13.7 billion years ago with a violent event called the Big Bang (see Section 1-4). Only the lightest elements—hydrogen and helium, as well as tiny amounts of lithium and perhaps beryllium—emerged from the enormously high temperatures following this cosmic event. All the heavier elements were manufactured by stars later, either by thermonuclear fusion reactions deep in their interiors or by the violent explosions that mark the end of massive stars. Were it not for these processes that take place only in stars, there would be no heavy elements in the universe, no planet like our Earth, and no humans to contemplate the nature of the cosmos.

Because our solar system contains heavy elements, it must be that at least some of its material was once inside other stars. But how did this material become available to help build our solar system? The answer is that near the ends of their lives, stars cast much of their matter back out into space. For most stars this process is a comparatively gentle one, in which a star's outer layers are gradually expelled. Figure 8-1 shows a star losing material in this fashion. This ejected material appears as the cloudy region, or **nebulosity** (from *nubes,* Latin for "cloud"), that surrounds the star and is illuminated by it. A few stars eject matter much more dramatically at the very end of their lives, in a spectacular detonation called a *supernova*, which blows the star apart (see Figure 1-6).

No matter how it escapes, the ejected material contains heavy elements dredged up from the star's interior, where they were formed. This material becomes part of the **interstellar medium,** a tenuous collection of gas and dust that pervades the spaces between the stars. As different stars die, they increasingly enrich the interstellar medium with heavy elements. Observations show that new stars form as condensations in the interstellar medium (Figure 8-2). Thus, these new stars have an adequate supply of heavy elements from which to develop a system of planets, satellites, comets, and asteroids. Our own solar system must have formed from enriched material in just this way. Thus, our solar system contains "recycled" material that was produced long ago inside a now-dead star. This "recycled" material includes all of the carbon in your body, all of the oxygen that you breathe, and all of the iron and silicon in the soil beneath your feet.

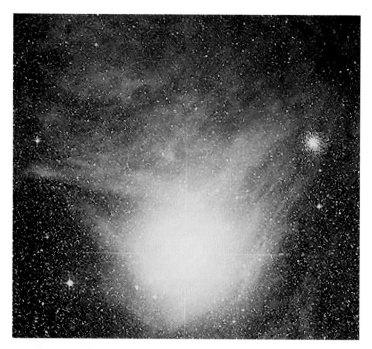

Figure 8-1 R I **V** U X G

A Mass-Loss Star The star Antares is shedding material from its outer layers, forming a cloud around the star. Although this cloud is very thin, we can see it because some of the ejected material has condensed into tiny grains of dust that reflect the star's light. (Dust particles in the air around you reflect light in the same way, which is why you can see them within a shaft of sunlight in a darkened room). Antares lies some 600 light-years from Earth in the constellation Scorpio. (David Malin/Anglo-Australian Observatory)

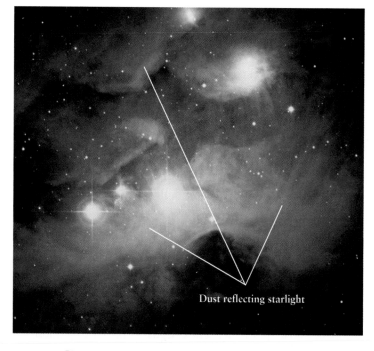

Dust reflecting starlight

Figure 8-2 R I **V** U X G

A Dusty Region of Star Formation Unlike Figure 8-1, which depicts an old star that is ejecting material into space, this image shows young stars in the constellation Orion (the Hunter) that have only recently formed from a cloud of gas and dust. The bluish, wispy appearance of the cloud (called NGC 1973-1975-1977) is caused by starlight reflecting off interstellar dust grains within the cloud (see Box 5-4). The grains are made of heavy elements produced by earlier generations of stars. (David Malin/Anglo-Australian Observatory)

Stars create different heavy elements in different amounts. For example, carbon (as well as oxygen, silicon, and iron) is readily produced in the interiors of massive stars, whereas gold (as well as silver, platinum, and uranium) is created only under special circumstances. Consequently, gold is rare in our solar system and in the universe as a whole, while carbon is relatively abundant (although still much less abundant than hydrogen or helium).

A convenient way to express the relative abundances of the various elements is to say how many atoms of a particular element are found for every trillion (10^{12}) hydrogen atoms. For example, for every 10^{12} hydrogen atoms in space, there are about 100 billion (10^{11}) helium atoms. From spectral analysis of stars and chemical analysis of Earth rocks, Moon rocks, and meteorites, scientists have determined the relative abundances of the elements in our part of the Milky Way Galaxy today. **Figure 8-3** shows the relative abundances of the 30 lightest elements, arranged in order of their **atomic number**. An element's atomic number is the number of protons in the nucleus of an atom of that element. It is also equal to the number of electrons orbiting the nucleus (see Box 5-5). In general, the greater the atomic number of an atom, the greater its mass.

The inset in Figure 8-3 lists the 10 most abundant elements. Note that even carbon (chemical symbol C), the third most abundant element, is quite rare relative to hydrogen (H) and helium (He): There are only 4×10^8 carbon atoms for each 10^{12} hydrogen atoms and each 10^{11} helium atoms. Expressed another way, for each carbon atom in our region of the Milky Way Galaxy, there are about 2500 hydrogen atoms and 250 helium atoms.

In addition to the 10 most abundant elements listed in Figure 8-3, five elements are moderately abundant: sodium (Na), aluminum (Al), argon (Ar), calcium (Ca), and nickel (Ni). These elements have abundances in the range of 10^6 to 10^7 relative to the standard 10^{12} hydrogen atoms. Most of the other elements are much rarer. For example, for every 10^{12} hydrogen atoms in the solar system, there are only six atoms of gold.

The small cosmic abundances of elements other than hydrogen and helium help to explain why the terrestrial planets are so small (Item 1 in Section 8-1). Because the heavier elements required to make a terrestrial planet are rare, only relatively small planets can form out of them. By contrast, hydrogen and helium are so abundant that it was possible for these elements to form large Jovian planets.

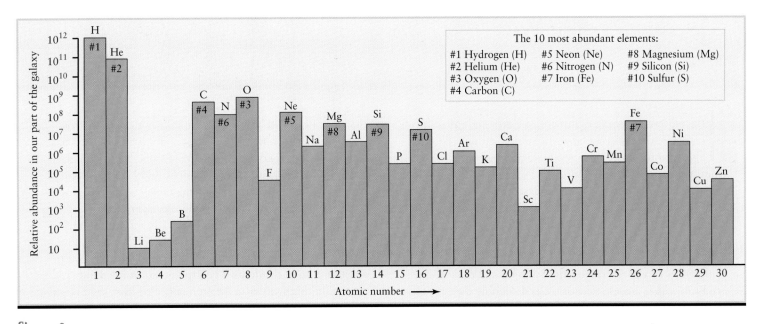

Figure 8-3

Abundances of the Lighter Elements This graph shows the abundances of the 30 lightest elements (listed in order of increasing atomic number) compared to a value of 10^{12} for hydrogen. The inset lists the 10 most abundant of these elements, which are also indicated in the graph.

Notice that the vertical scale is not a linear one; each division on the scale corresponds to a tenfold increase in abundance. All elements heavier than zinc (Zn) have abundances of fewer than 1000 atoms per 10^{12} atoms of hydrogen.

8-3 | The abundances of radioactive elements reveal the solar system's age

 Scientists use radioactive age-dating to measure the age of the solar system

The heavy elements can tell us even more about the solar system: They also help us determine its age. The particular heavy elements that provide us with this information are several which are *radioactive*. Their atomic nuclei are unstable because they contain too many protons or too many neutrons. A radioactive nucleus therefore ejects particles until it becomes stable. In doing so, a nucleus may change from one element to another. Physicists refer to this transmutation as **radioactive decay.** For example, a radioactive form of the element rubidium (atomic number 37) decays into the element strontium (atomic number 38) when one of the neutrons in the rubidium nucleus decays into a proton and an electron (which is ejected from the nucleus). Each type of radioactive nucleus decays at its own characteristic rate, which can be measured in the laboratory. This is the key to a technique called **radioactive age-dating,** which is used to determine the ages of rocks. If a rock contained a certain amount of radioactive rubidium when it first formed, over time more and more of the atoms of rubidium within the rock will decay into strontium atoms. The ratio of the number of strontium atoms the rock contains to the number of rubidium atoms it contains then gives a measure of the age of the rock. Box 8-1 describes radioactive age-dating in more detail.

Scientists have applied techniques of radioactive age-dating to rocks taken from all over the Earth. The results show that most rocks are tens or hundreds of millions of years old, but that some rocks are as much as 4 billion (4×10^9) years old. These results confirm that geologic processes have produced new surface material over the Earth's history, as we concluded from the small number of impact craters found on Earth (see Section 7-6). They also show that the Earth is at least 4×10^9 years old. Radioactive age-dating has also been applied to rock samples brought back from the Moon by the *Apollo* astronauts. The oldest *Apollo* specimen, collected from one of the most heavily cratered and hence most ancient regions of the Moon, is 4.5×10^9 years old. But the oldest rocks found anywhere in the solar system are **meteorites,** the bits of meteoroids that survive passing through the Earth's atmosphere and land on our planet's surface. Radioactive age-dating of meteorites reveals that they are all nearly the *same* age, about 4.56 billion years old. The absence of any younger or older meteorites indicates that these are all remnants of objects that formed at the same time in the early solar system. We conclude that the age of the oldest meteorites, about 4.56×10^9 years, is the age of the solar system itself.

Thus, by studying the abundances of the elements, we are led to a remarkable insight: Some 4.56 billion years ago, a collection of hydrogen, helium, and heavy elements came together to form the Sun and all of the objects that orbit around it. All of those heavy elements, including the carbon atoms in your body and the oxygen atoms that you breathe, were created and cast off by stars that lived and died long before our solar system formed, during the first 8 to 9 billion years of the universe's existence. We are literally made of star dust.

box 8-1 TOOLS OF THE ASTRONOMER'S TRADE

Radioactive Age-Dating

How old are the rocks found on the Earth and other planets? Are rocks found at different locations the same age or of different ages? How old are meteorites? Questions like these are important to scientists who wish to reconstruct the history of our solar system. But simply looking at a rock cannot tell us whether it was formed a thousand years, a million years, or a billion years ago. Fortunately, most rocks contain trace amounts of radioactive elements such as uranium. By measuring the relative abundances of various radioactive isotopes and their decay products within a rock, scientists can determine the rock's age.

As we saw in Box 5-5, every atom of a particular element has the same number of protons in its nucleus. However, different isotopes of the same element have different numbers of neutrons in their nuclei. For example, the common isotopes of uranium are ^{235}U and ^{238}U. Each isotope of uranium has 92 protons in its nucleus (correspondingly, uranium is element 92 in the periodic table; see Box 5-5). However, a ^{235}U nucleus contains 143 neutrons, whereas a ^{238}U nucleus has 146 neutrons.

A radioactive nucleus with too many protons or too many neutrons is unstable; to become stable, it *decays* by ejecting particles until it becomes stable. If the number of protons (the atomic number) changes in this process, the nucleus changes from one element to another.

Some radioactive isotopes decay rapidly, while others decay slowly. Physicists find it convenient to talk about the decay rate in terms of an isotope's **half-life.** The half-life of an isotope is the time interval in which one-half of the nuclei decay. For example, the half-life of ^{238}U is 4.5 billion (4.5×10^9) years. This means that if you start out with 1 kg of ^{238}U, after 4.5 billion years you will have only ½ kg of ^{238}U remaining;

the other ½ kg will have turned into other elements. If you want another half-life, so that a total of 9.0 billion years has elapsed, only ¼ kg of ^{238}U—one-half of one-half of the original amount—will remain. Several isotopes useful for age-dating rocks are listed in the accompanying table.

To see how geologists date rocks, consider the slow conversion of radioactive rubidium (^{87}Rb) into strontium (^{87}Sr). (The periodic table in Box 5-5 shows that the atomic numbers for these elements are 37 for rubidium and 38 for strontium, so in the decay a neutron is transformed into a proton. In this process an electron is ejected from the nucleus.) Over the years, the amount of ^{87}Rb in a rock decreases, while the amount of ^{87}Sr increases. Because the ^{87}Sr appears in the rock due to radioactive decay, this isotope is called *radiogenic.* Dating the rock is not simply a matter of measuring its ratio of rubidium to strontium, however, because the rock already had some strontium in it when it was formed. Geologists must therefore determine how much fresh strontium came from the decay of rubidium after the rock's formation.

To do this, geologists use as a reference another isotope of strontium whose concentration has remained constant. In this case, they use ^{86}Sr, which is stable and is not created by radioactive decay; it is said to be *nonradiogenic.* Dating a rock thus entails comparing the ratio of radiogenic and nonradiogenic strontium ($^{87}Sr/^{86}Sr$) in the rock to the ratio of radioactive rubidium to nonradiogenic strontium ($^{87}Rb/^{86}Sr$). Because the half-life for converting ^{87}Rb into ^{87}Sr is known, the rock's age can then be calculated from these ratios (see the table).

Radioactive isotopes decay with the same half-life no matter where in the universe they are found. Hence, scientists have used the same techniques to determine the ages of rocks from the Moon and of meteorites.

Original radioactive isotope	Final stable isotope	Half-life (years)	Range of ages that can be determined (years)
Rubidium (^{87}Rb)	Strontium (^{87}Sr)	47.0 billion	10 million–4.56 billion
Uranium (^{238}U)	Lead (^{206}Pb)	4.5 billion	10 million–4.56 billion
Potassium (^{40}K)	Argon (^{40}Ar)	1.3 billion	50,000–4.56 billion
Carbon (^{14}C)	Nitrogen (^{14}N)	5730	100–70,000

8-4 | The Sun and planets formed from a solar nebula

We have seen how processes in the Big Bang and within ancient stars produced the raw ingredients of our solar system. But given these ingredients, how did they combine to make the Sun and planets? Astronomers have developed a variety of models for the origin of the solar system. The test of these models is whether they explain the properties of the present-day system of Sun and planets.

Any model of the origin of the solar system must explain why all the planets orbit the Sun in the same direction and in nearly the same plane (Item 2 in Section 8-1). One model that was devised explicitly to address this issue was the *tidal hypothesis,* proposed in the early 1900s. As we saw in Section 4-8, two nearby planets, stars, or galaxies exert tidal forces on each other that cause the objects to elongate. In the tidal hypothesis, another star happened to pass close by the Sun, and the star's tidal forces drew a long filament out of the Sun. The filament material would then go into orbit around the Sun, and all of it would naturally orbit in the same direction and in the same plane. From this filament the planets would condense. However, it was shown in the 1930s that tidal forces strong enough to pull a filament out of the Sun would cause the filament to disperse before it could condense into planets. Hence the tidal hypothesis cannot be correct.

An entirely different model is now thought to describe the most likely series of events that led to our present solar system (Figure 8-4). The central idea of this model dates to the late 1700s, when the German philosopher Immanuel Kant and the French scientist Pierre-Simon de Laplace turned their attention to the manner in which the planets orbit the Sun. Both concluded that the arrangement of the orbits—all in the same direction and in nearly the same plane—could not be mere coincidence. To explain the orbits, Kant and Laplace independently proposed that our entire solar system, including the Sun as well as all of its planets and satellites, formed from a vast, rotating cloud of gas and dust called the **solar nebula** (Figure 8-4*a*). This model is called the **nebular hypothesis.**

The consensus among today's astronomers is that Kant and Laplace were exactly right. In the modern version of the nebular hypothesis, at the outset the solar nebula was similar in character to the nebulosity shown in Figure 8-2 and had a mass somewhat greater than that of our present-day Sun.

Each part of the nebula exerted a gravitational attraction on the other parts, and these mutual gravitational pulls tended to make the nebula contract. As it contracted, the greatest concentration of matter occurred at the center of the nebula, forming a relatively dense region called the **protosun.** As its name suggests, this part of the solar nebula eventually developed into the Sun. The planets formed from the much sparser material in the outer regions of the solar nebula. Indeed, the mass of all the planets together is only 0.1% of the Sun's mass.

When you drop a ball, the gravitational attraction of the Earth makes the ball fall faster and faster as it falls; in the same way, material falling inward toward the protosun would have gained speed. As this fast-moving material ran into the protosun, the energy of the collision was converted into thermal energy, causing the temperature deep inside the solar nebula to climb. This process, in which the gravitational energy of a contracting gas cloud is converted into

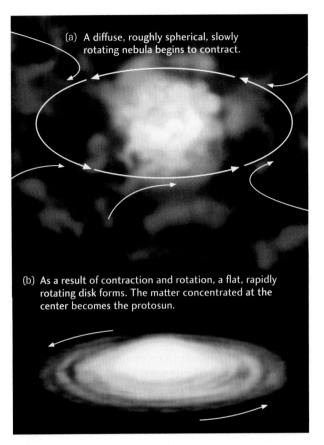

(a) A diffuse, roughly spherical, slowly rotating nebula begins to contract.

(b) As a result of contraction and rotation, a flat, rapidly rotating disk forms. The matter concentrated at the center becomes the protosun.

Figure 8-4

The Birth of the Solar System (a) A cloud of interstellar gas and dust begins to contract because of its own gravity. (b) As the cloud flattens and spins more rapidly around its rotation axis, a central condensation develops that evolves into a glowing protosun. The planets will form out of the surrounding disk of gas and dust.

thermal energy, is called **Kelvin-Helmholtz contraction,** after the nineteenth-century physicists who first described it.

As the newly created protosun continued to contract and become denser, its temperature continued to climb as well. After about 10^5 (100,000) years, the protosun's surface temperature stabilized at about 6000 K, but the temperature in its interior kept increasing to ever higher values as the central regions of the protosun became denser and denser. Eventually, after perhaps 10^7 (10 million) years had passed since the solar nebula first began to contract, the gas at the center of the protosun reached a density of about 10^5 kg/m^3 (a hundred times denser than water) and a temperature of a few million kelvins (that is, a few times 10^6 K). Under these extreme conditions, nuclear reactions that convert hydrogen into helium began in the protosun's interior. When this happened, the energy released by these reactions stopped the contraction and a true star was born. Nuclear reactions continue to the present day in the interior of the Sun and are the source of all the energy that the Sun radiates into space.

If the solar nebula had not been rotating at all, everything would have fallen directly into the protosun, leaving nothing behind to form the planets. Instead, the solar nebula must have had an overall slight rotation, which caused its evolution to follow a different path. As the slowly rotating nebula collapsed inward, it would naturally have tended to rotate faster. This relationship between the size of an object and its rotation speed is an example of a general principle called the **conservation of angular momentum.**

ANALOGY Figure skaters make use of the conservation of angular momentum. When a spinning skater pulls her arms and legs in close to her body, the rate at which she spins automatically increases. (If you are not a figure skater, you can demonstrate this yourself by sitting on an office chair. Sit with your arms outstretched and hold a weight, like a brick or a full water bottle, in either hand. Now use your feet to start yourself rotating, lift your feet off the ground, and then pull your arms inward. Your rotation will speed up quite noticeably.)

As the solar nebula began to rotate more rapidly, it also tended to flatten out (Figure 8-4b). From the perspective of a particle rotating along with the nebula, it felt as though there were a force pushing the particle away from the nebula's axis of rotation. (Likewise, passengers on a spinning carnival ride seem to feel a force pushing them outward and away from the ride's axis of rotation.) This apparent force was directed opposite to the inward pull of gravity, and so it tended to slow the contraction of material toward the nebula's rotation axis. But there was no such effect opposing contraction in a direction parallel to the rotation axis. Some 10^5 (100,000) years after the solar nebula first began to contract, it had developed the structure shown in Figure 8-4b, with a rotating, flattened disk surrounding what will become the protosun. The planets formed from the material in this disk, which explains why their orbits all lie in essentially the same plane and why they all orbit the Sun in the same direction.

There were no humans to observe these processes taking place during the formation of the solar system. But Earth astronomers have seen disks of material surrounding other stars that formed only recently. These are called **protoplanetary disks,** or **proplyds,** because it is thought that planets can eventually form from these disks around other stars. Thus, proplyds are planetary systems that are still "under construction." By studying these proplyds, astronomers are able to examine what our solar nebula may have been like some 4.56×10^9 years ago.

WEB LINK 8.5 Figure 8-5 shows a number of protoplanetary disks in the Orion Nebula, a region of active star formation. A star is visible at the center of each proplyd, which reinforces the idea that our Sun began to shine before the planets were fully formed. (The image that opens this chapter shows an even more detailed view of a disk surrounding a young star.) A study

(a)

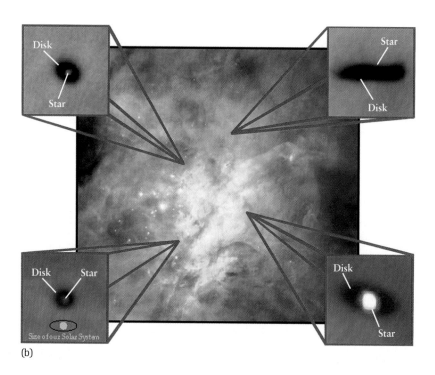

(b)

VIDEO 8.1 **figure 8-5** R I **V** U X G

Protoplanetary Disks (a) The Orion Nebula is a star-forming region located some 1500 light-years from Earth. It is the middle "star" in Orion's "sword" (see Figure 2-2). The smaller, bluish nebula is the object shown in Figure 8-2. (b) This view of the center of the Orion Nebula is a mosaic of Hubble Space Telescope images. The four insets are false-color close-ups of four proplyds, or protoplanetary disks, that lie within the nebula. A young, recently formed star is at the center of each proplyd. (The proplyd at upper right is seen nearly edge-on.) The inset at the lower left shows the size of our own solar system for comparison. (a: Anglo-Australian Observatory image by David Malin; b: C. R. O'Dell and S. K. Wong, Rice University; NASA) of 110 stars in the Orion Nebula detected proplyds around 56 of them, which suggests that systems of planets may form around a substantial fraction of stars. Later in this chapter we will see direct evidence for planets that have formed around stars other than the Sun.

8-5 | The planets formed by the accretion of planetesimals and the accumulation of gases in the solar nebula

We have seen how the solar nebula would have contracted to form a young Sun with a protoplanetary disk, or proplyd, rotating around it. But how did the material in this disk form into planets? And why are the terrestrial planets in the inner solar system, while the giant Jovian planets are in the outer solar system (Item 3 in Section 8-1)?

To understand how the planets, asteroids, and comets formed, we must look at the conditions that prevailed within the solar nebula. The density of material in the part of the nebula outside the protosun was rather low, as was the pressure of the nebula's gas. If the pressure is sufficiently low, a substance cannot remain in the liquid state, but must end up as either a solid or a gas. For a given pressure, what determines whether a certain substance is a solid or a gas is its **condensation temperature.** If the temperature of a substance is above its condensation temperature, the substance is a gas; if the temperature is below the condensation temperature, the substance solidifies into tiny specks of dust or snowflakes. You can often see similar behavior on a cold morning. The air temperature can be above the condensation temperature of water, while the cold windows of parked cars may have temperatures below the condensation temperature. Thus, water molecules in the air remain as a gas (water vapor) but form solid ice particles (frost) on the car windows.

Substances such as water (H_2O), methane (CH_4), and ammonia (NH_3) have low condensation temperatures, ranging from 100 to 300 K. Rock-forming substances have much higher condensation temperatures, in the range from 1300 to 1600 K. The gas cloud from which the solar system formed had an initial temperature near 50 K, so all of these substances could have existed in solid form. Thus, the solar nebula would have been populated by an abundance of small ice particles and solid dust grains like the one shown in Figure 8-6. (Recall from Section 7-4 that "ice" can refer to frozen CO_2, CH_4, or NH_3, as well as frozen water.) But hydrogen and helium, the most abundant elements in the solar nebula, have condensation temperatures so near absolute zero that these substances always existed as gases during the creation of the solar system. You can best visualize the initial state of the solar nebula as a thin gas of hydrogen and helium strewn with tiny dust particles.

This state of affairs changed as the central part of the solar nebula underwent Kelvin-Helmholtz contraction to form the protosun. During this phase, the protosun was actually quite a bit more luminous than the present-day Sun, and it heated the innermost part of the solar nebula to temperatures above 2000 K. Meanwhile, temperatures in the outermost regions of the solar nebula remained below 50 K.

Figure 8-7 shows the probable temperature distribution in the solar nebula at this stage in the formation of our solar system. In the inner regions of the solar nebula, water, methane, and ammonia were vaporized by the high temperatures. Only materials with high condensation temperatures could have remained solid. Of these materials, iron, silicon, magnesium, and sulfur were particularly abundant, followed closely by aluminum, calcium, and nickel. (Most of these elements were present in the form of oxides, which are chemical compounds containing oxygen. These compounds also have high condensation temperatures.) In contrast, ice particles and ice-coated dust grains were able to survive in the cooler, outer portions of the solar nebula.

Small chunks would have formed in the inner part of the solar nebula from collisions between neighboring dust grains. Initially, electric forces (chemical bonds) held them together. Over a few million years, these chunks coalesced into roughly 10^9 asteroidlike objects called **planetesimals,** with diameters of a kilometer or so.

10 μm = 0.01 mm

Figure 8-6

WEB LINK 8.6

A Grain of Cosmic Dust This highly magnified image shows a microscopic dust grain that came from interplanetary space. It entered Earth's upper atmosphere and was collected by a high-flying aircraft. Dust grains of this sort are abundant in star-forming regions like that shown in Figure 8-2. These tiny grains were also abundant in the solar nebula and served as the building blocks of the planets. (Donald Brownlee, University of Washington)

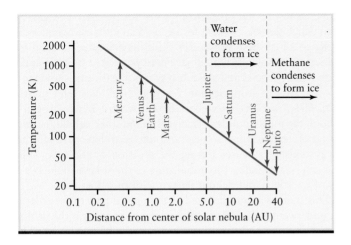

figure 8-7

Temperature Distribution in the Solar Nebula This graph shows how temperatures probably varied across the solar nebula as the planets were forming. Note the general decline in temperature with increasing distance from the center of the nebula. Beyond 5 AU from the center of the nebula, temperatures were low enough for water to condense and form ice; beyond 30 AU, methane (CH_4) could also condense into ice.

ber of chondrules suggests that such shock waves occurred throughout the inner solar nebula over a period of more than 10^6 years. The words "cloud" and "nebula" may suggest material moving in gentle orbits around the protosun; the existence of chondrules suggests that conditions in the inner solar nebula were at times quite violent.

Astronomers use computer simulations to learn how the inner planets could have formed from planetesimals. A computer is programmed to simulate a large number of particles circling a newborn sun along orbits dictated by Newtonian mechanics. As the simulation proceeds, the particles coalesce to form larger objects, which in turn collide to form planets. By performing a variety of simulations, each beginning with somewhat different numbers of planetesimals in different orbits, it is possible to see what kinds of planetary systems would have been created under different initial conditions. Such studies demonstrate that a wide range of initial conditions ultimately lead to basically the same result in the inner solar system: Accretion continues for roughly 10^8 (100 million) years and typically forms four or five terrestrial planets with orbits between 0.3 and 1.6 AU from the Sun.

Figure 8-9 shows one particular computer simulation. The calculations began with 100 planetesimals, each having a mass of 1.2×10^{23} kg. This choice ensures that the total mass (1.2×10^{25} kg) equals the combined mass of the four terrestrial planets (Mercury through Mars) plus their satellites. The initial orbits of planetesi-

These were large enough to be held together by the gravitational attraction of the different parts of the planetesimal for each other. During the next stage, gravitational attraction between the planetesimals caused them to collide and accumulate into still-larger objects called **protoplanets,** which were roughly the size and mass of our Moon. This accumulation of material to form larger and larger objects is called **accretion.** During the final stage, these Moon-sized protoplanets collided to form the terrestrial planets. This final episode must have involved some truly spectacular, world-shattering collisions.

Important clues to this stage in the evolution of the inner solar system come from studies of meteorites, which, as we saw in Section 8-3, are the oldest solid objects known in the solar system. Many of these are fragments of planetesimals that were never incorporated into the planets. Most meteorites contain not only dust grains but also **chondrules,** which are small, glassy, roughly spherical blobs (Figure 8-8). (The "ch" in "chondrule" is pronounced as in "chord.") Liquids tend to form spherical drops (like drops of water), so the shape of chondrules shows that they were once molten.

Attempts to produce chondrules in the laboratory show that the blobs must have melted suddenly, then solidified over the space of only an hour or so. This means they could not have been melted by the temperature of the inner solar nebula, which would have remained high for hundreds of thousands of years after the formation of the protosun. Instead, chondrules must have been produced by sudden, high-energy events that quickly melted material and then permitted it to cool. These events were probably shock waves propagating through the gas of the solar nebula, like the sonic booms produced by an airplane traveling faster than sound. The sheer num-

WEB LINK 8.7 figure 8-8 R I **V** U X G

Chondrules This is not a collection of pebbles, but a cross-section of the interior of a meteorite. The "pebbles" are individual chondrules ranging in size from a few millimeters to a few tenths of a millimeter. Many of the chondrules are coated with dust grains like the one shown in Figure 8-6. (Courtesy of John A. Wood and the Smithsonian Astrophysical Observatory)

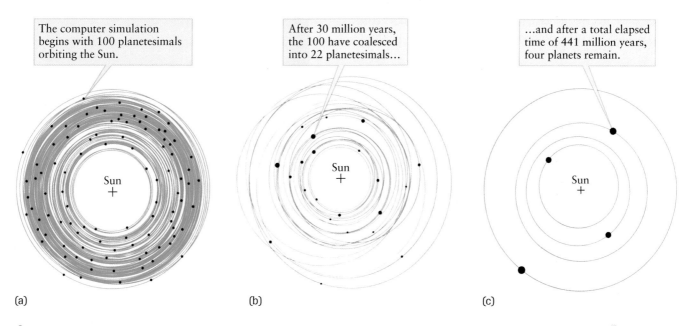

The computer simulation begins with 100 planetesimals orbiting the Sun.

Sun

(a)

After 30 million years, the 100 have coalesced into 22 planetesimals...

Sun

(b)

...and after a total elapsed time of 441 million years, four planets remain.

Sun

(c)

figure 8-9

Accretion of the Terrestrial Planets These three drawings show the results of a computer simulation of the formation of the inner planets. In this simulation, the inner planets were essentially formed after 150 million years. (Adapted from George W. Wetherill)

mals are inclined to each other by angles of less than 5° to simulate a thin layer of particles orbiting the protosun.

After an elapsed time simulating 30 million (3×10^7) years, the 100 original planetesimals have merged into 22 protoplanets (Figure 8-9b). After 79 million (7.9×10^7) years, these have combined into 11 larger protoplanets. Nearly another 100 million (10^8) years elapse before further mergers leave just six growing protoplanets. Finally, Figure 8-9c shows four planets following nearly circular orbits after a total elapsed time of 441 million (4.41×10^8) years. In this particular simulation, the fourth planet from the Sun ends up being the most massive; in our own solar system, the third planet (Earth) is the most massive of the terrestrial planets. Note that due to collisions, the four planets in the simulation end up in orbits that are nearly circular, just like the orbits of most of the planets in our solar system.

As we have seen, the material from which the inner protoplanets accreted was rich in substances with high condensation temperatures. This material remained largely in solid form despite the high temperatures in the inner solar nebula. But as the protoplanets grew, they were heated by the violent impacts of planetesimals as well as the decay of radioactive elements such as uranium, and this heat caused melting. Thus, the terrestrial planets began their existence as spheres of at least partially molten rocky materials. Material was free to move within these molten spheres, so the denser, iron-rich minerals sank to the centers of the planets while the less dense silicon-rich minerals floated to their surfaces. This process is called **chemical differentiation** (see Box 7-1). In this way the terrestrial planets developed their dense iron cores. We saw in Section 7-7 that

this electrically conductive iron gives rise to the magnetic fields of the terrestrial planets.

Like the inner planets, the outer planets may have begun to form by the accretion of planetesimals. The key difference is that ices as well as rocky grains were able to survive in the colder outer regions of the solar nebula. The elements of which ices are made are much more abundant than those that form rocky grains. Thus, much more solid material would have been available to form planetesimals in the outer solar nebula than in the inner part. As a result, solid objects several times larger than any of the terrestrial planets could have formed in the outer solar nebula. Each such object, made up of a mixture of ices and rocky material, could have become the core of a Jovian planet and served as a "seed" around which the rest of the planet eventually grew.

Thanks to the lower temperatures in the outer solar system, gas atoms (principally hydrogen and helium) were moving relatively slowly and so could more easily be captured by the strong gravity of these massive cores (see Box 7-2). Thus, the core of a Jovian protoplanet could have captured an envelope of gas as it continued to grow by accretion. This picture is called the **core accretion model** for the origin of the Jovian planets.

Calculations based on the core accretion model suggest that both rocky materials and gas slowly accumulated for several million years, until the masses of the core and the envelope became equal. From that critical moment on, the envelope pulled in all the gas it could get, dramatically increasing the protoplanet's mass and size. This runaway growth of the protoplanet would have continued until all the available gas was used up. The result was a huge planet with an enormously thick, hydrogen-rich envelope surrounding a rocky core with 5 to 10 times the mass of the Earth. This scenario could have occurred at four different distances from the Sun, thus creating the four Jovian planets.

In the core accretion model, Uranus and Neptune probably did not form at their present locations, respectively about 19 and 30 AU from the Sun. The solar nebula was too sparse at those distances to allow these planets to have grown to the present-day sizes. Instead, it is thought that Uranus and Neptune formed between 4 and 10 AU from the Sun, but were flung into larger orbits by gravitational interactions with Jupiter.

An alternative model for the origin of the Jovian planets, the **disk instability model,** suggests that they formed directly from the gas of the solar nebula. If the gas of the outer solar nebula was not smooth but clumpy, a sufficiently large and massive clump of gas could begin to collapse inward all on its own, like the protosun but on a smaller scale. Such a large clump would attract other gas, forming a very large planet of hydrogen and helium within just a few hundreds or thousands of years. In this model the rocky core of a Jovian planet is not the original nucleus around which the planet grew; rather, it is the result of planetesimals and icy dust grains that were drawn gravitationally into the planet and then settled at the planet's center. Another difference is that in the disk instability model it is possible for Uranus and Neptune to have formed in their present-day orbits, since they would have grown much more rapidly than would be possible in the core accretion model.

Planetary scientists are actively researching whether the core accretion model or the disk instability model is the more correct description of the formation of the Jovian planets. In either case, we can account for the different locations of the terrestrial planets (in the inner solar system) and the Jovian planets (in the outer solar system). Figure 8-10 summarizes our overall picture of the formation of the terrestrial and Jovian planets. Like the terrestrial planets, the Jovian planets were quite a bit hotter during their formation than they are today. A heated gas expands, so these gas-rich planets must also have been much larger than at present. As each planet cooled and contracted, it would have formed a disk like a solar nebula in miniature (see Figure 8-4). Many of the satellites of the Jovian planets are thought to have formed from ice particles and dust grains within these disks.

The Jovian planets may also explain why there are asteroids, Kuiper belt objects, and comets (see Section 7-5). Jupiter, the most massive of the planets, was almost certainly the first to form. Jupiter's tremendous mass would have exerted strong gravitational forces on any nearby planetesimals. Some would have been sent crashing into the Sun, while others would have been sent crashing into the terrestrial planets to form impact craters (see Section 7-6). Still others would have been ejected completely from the solar system. Only a relatively few rocky planetesimals survived to produce the present-day population of asteroids. As a result, asteroids make up only a tiny fraction of the total mass of the present-day solar system: The mass of all known asteroids combined is less than the mass of the Moon.

By contrast, the objects that make up the Kuiper belt (including Pluto) began as icy planetesimals beyond the orbit of Jupiter. The strong gravitational forces from all of the Jovian planets pushed these into orbits even further from the Sun. (As we saw in Section

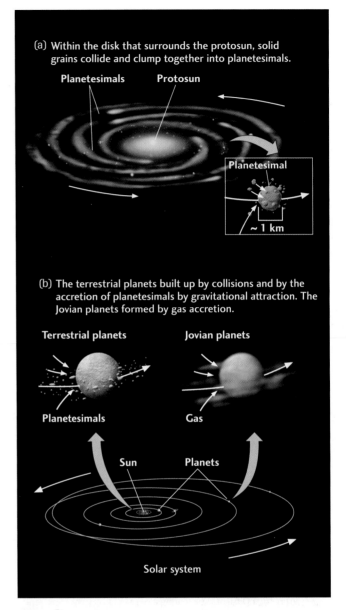

Figure 8-10

Formation of the Planets (a) Planetesimals about 1 km in size formed in the solar nebula from small dust grains sticking together. (b) Planetesimals in the inner solar system grouped together to form the terrestrial planets, as in Figure 8-9. In the outer solar system, the Jovian planets may have begun as terrestrial-like planets that accumulated massive envelopes of hydrogen and helium. Alternatively, the Jovian planets may have formed directly from the gas of the solar nebula.

7-5, the combined mass of the Kuiper belt is much larger than that of the asteroid belt, in part because the materials that form ices are much more abundant than those that form rock.) Some of the smaller icy objects are thought to have been pushed as far as 50,000 AU from the Sun, forming a spherical "halo" around the solar system called the **Oort cloud.** From time to time one of the smaller chunks of ice and rock from either the Kuiper belt or the

Oort cloud is deflected into the inner solar system. If one of these chunks comes close enough to the Sun, it begins to evaporate, producing a visible tail and appearing as a comet (see Figure 7-8).

While the planets, satellites, asteroids, and comets were forming, the protosun was also evolving into a full-fledged star with nuclear reactions occurring in its core (see Section 1-3). The time required for this to occur was about 10^8 years, roughly the same as that required for the formation of the terrestrial planets. Before the onset of nuclear reactions, however, the young Sun probably expelled a substantial portion of its mass into space (Figure 8-11). Magnetic fields within the solar nebula would have funneled a portion of the nebula's mass into oppositely directed **jets** along the rotation axis of the nebula. Figure 8-11*a* shows one such jet emanating from a young star.

In addition, instabilities within the young Sun would have caused it to eject its tenuous outermost layers into space. This brief but intense burst of mass loss, observed in many young stars across the sky, is called a **T Tauri wind,** after the star in the constellation Taurus (the Bull) where it was first identified (see Figure 8-11*b*). Each of the proplyds in Figure 8-5 has a T Tauri star at its center. (The present-day Sun also loses mass from its outer layers in the form of high-speed electrons and protons, a flow called the solar wind. But this is minuscule compared with a T Tauri wind, which causes a star to lose mass 10^6 to 10^7 times faster than in the solar wind.)

With the passage of time, the combined effects of jets, the T Tauri wind, and accretion onto the planets would have swept the solar system nearly clean of gas and dust. With no more interplanetary material to gather up, the planets would have stabilized at roughly their present-day sizes, and the formation of the solar system would have been complete.

8-6 | Astronomers have discovered planets orbiting other stars

If planets formed around our Sun, have they formed around other stars? That is, are there **extrasolar planets** orbiting stars other than the Sun? Our model for the formation of the planets would seem to suggest so. This model is based on the laws of physics and chemistry, which to the best of our understanding are the same throughout the universe. The discovery of a set of planets orbiting another star, with terrestrial planets in orbit close to the star and

Star

0.1 light-year

(a) A jet from a young star

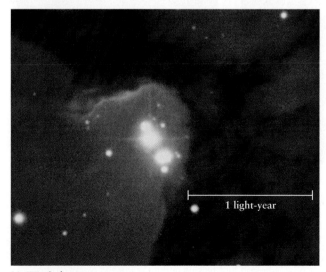

1 light-year

(b) Winds from young stars

WEB LINK 8.8

Figure 8-11 R I **V** U X G

Jets and Winds from Young Stars (a) This young star in the southern constellation Vela (the Sails) is ejecting matter at speeds of hundreds of kilometers per second, forming an immense jet. The jet is about 0.2 light-year (12,500 AU) in length. (b) An outpouring of particles and radiation from the surfaces of these young stars has carved out a cavity in the surrounding dusty material. The stars lie within the Trifid Nebula in the constellation Sagittarius (the Archer). (a: C. Burrows, STScI; J. Hester, Arizona State University; J. Morse, STScI; and NASA; b: David Malin/Anglo-Australian Observatory)

Jovian planets orbiting farther away, would be a tremendous vindication of our theory of solar system formation. It would also tell us that our planetary system is not unique in the universe. Because at least one planet in our solar system—the Earth—has the ability to support life, perhaps other planetary systems could also harbor living organisms.

Since 1995 evidence has accumulated for the existence of more than 100 planets orbiting stars other than the Sun. However, none of these is a terrestrial planet. Most of them are more massive than Jupiter, and some are in eccentric, noncircular orbits quite unlike planetary orbits in our solar system. To appreciate how remarkable these discoveries are, we must look at the process that astronomers go through to search for extrasolar planets.

It is very difficult to make direct observations of planets orbiting other stars. The problem is that planets are small and dim compared with stars; at visible wavelengths, the Sun is 10^9 times brighter than Jupiter and 10^{10} times brighter than the Earth. A hypothetical planet orbiting a distant star, even a planet 10 times larger than Jupiter, would be lost in the star's glare as seen through even the largest telescope on Earth.

Instead, indirect methods must be used to search for extrasolar planets (**Figure 8-12**). One very powerful method is to search for stars that appear to "wobble." If a star has a planet, it is not quite correct to say that the planet orbits the star. Rather, both the planet and the star move in elliptical orbits around a point called the **center of mass**. Imagine the planet and the star as sitting at opposite ends of a very long seesaw; the center of mass is the point where you would have to place the fulcrum in order to make the seesaw balance. Because of the star's much greater mass, the center of mass is much closer to the star than to the planet. Thus, while the planet may move in an orbit that is hundreds of millions of kilometers across, the star will move in a much smaller orbit (Figure 8-12*a*).

As an example, the Sun and Jupiter both orbit their common center of mass with an orbital period of 11.86 years. (Jupiter has more mass than the other eight planets put together, so it is a reasonable approximation to consider the Sun's wobble as being due to Jupiter alone.) Jupiter's orbit has a semimajor axis of 7.78×10^8 km, while the Sun's orbit has a much smaller semimajor axis of 742,000 km. The Sun's radius is 696,000 km, so the Sun slowly wobbles around a point not far outside its surface. If astronomers elsewhere in the Galaxy could detect the Sun's wobbling motion, they could tell that there was a large planet (Jupiter) orbiting the Sun. They could even determine the planet's mass and the size of its orbit, even though the planet itself was unseen.

Detecting the wobble of other stars is not an easy task. One approach to the problem, called the **astrometric method**, involves making very precise measurements of a star's position in the sky relative to other stars. The goal is to find stars whose positions change in a cyclic way (Figure 8-12*b*). The measurements must be made with very high accuracy (0.001 arcsec or better) and, ideally, over a long enough time to span an entire orbital period of the star's motion.

A different approach to the problem is the **radial velocity method** (Figure 8-12*c*). This is based on the Doppler effect, which we described in Section 5-9. A wobbling star will alternately move away from and toward the Earth. This will cause the dark absorption lines in the star's spectrum (see Figures 5-12 and 5-19) to change their wavelengths in a periodic fashion. When the star is moving away from us, its spectrum will undergo a redshift to longer wavelengths. When the star is approaching, there will be a blueshift of the spectrum to shorter wavelengths. These wavelength shifts are very small because the star's motion around its orbit is quite slow. As an example, the Sun moves around its small orbit at only 12.5 m/s (45 km/h, or 28 mi/h). If the Sun was moving directly toward an observer at this speed, the hydrogen ab-

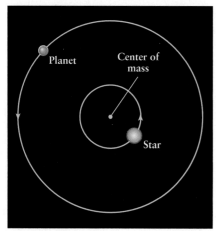

(a) A star and its planet

(b) The astrometric method

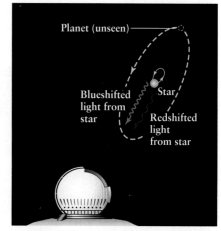

(c) The radial velocity method

Figure 8-12

Detecting Planets Orbiting Other Stars (a) A planet and its star both orbit around their common center of mass, always staying on opposite sides of this point. Even if the planet cannot be seen, its presence can be inferred if the star's motion can be detected. (b) The astrometric method of detecting the unseen planet involves making direct measurements of the star's orbital motion. (c) In the radial velocity method, astronomers measure the Doppler shift of the star's spectrum as it moves alternately toward and away from the Earth. The amount of Doppler shift determines the size of the star's orbit, which in turn tells us about the unseen planet's orbit.

sorption line at a wavelength of 656 nm in the Sun's spectrum would be shifted by only 2.6×10^{-5} nm, or about 1 part in 25 million. Detecting these tiny shifts requires extraordinarily careful measurements and painstaking data analysis.

In 1995, Michel Mayor and Didier Queloz of the Geneva Observatory in Switzerland announced that by using the radial velocity method, they had discovered a planet orbiting the star 51 Pegasi, which is 48 light-years from Earth in the constellation Pegasus (the Winged Horse). Their results were soon confirmed by the team of Geoff Marcy of San Francisco State University and Paul Butler of the University of California, Berkeley, using observations made at the University of California's Lick Observatory. For the first time, solid evidence had been found for a planet orbiting a star like our own Sun. Marcy and Butler, along with other astronomers, have since discovered more than a hundred planets orbiting other stars by means of the radial velocity method. **Figure 8-13** depicts a selection of these discoveries.

The extrasolar planets discovered by the radial velocity method all have masses comparable with or larger than that of Saturn, and thus are presumably Jovian planets made primarily of hydrogen and helium. (This is difficult to confirm directly, because the spectra of the planets are very faint.) According to the picture we presented in Section 8-5, such Jovian planets would be expected to orbit relatively far from their stars, where temperatures were low enough to allow the buildup of a massive envelope of hydrogen and helium gas. But as Figure 8-13 shows, many extrasolar planets are in fact found orbiting very *close* to their stars. For example, the

planet orbiting 51 Pegasi has a mass at least 0.46 times that of Jupiter but orbits only 0.05 AU from its star with an orbital period of just 4.23 days. In our own solar system, this orbit would lie well inside the orbit of Mercury!

Another surprising result is that many of the extrasolar planets found so far have orbits with very large eccentricities (see Figure 4-10*b*). As an example, the planet around 16 Cygni B has an orbital eccentricity of 0.67; its distance from the star varies between 0.6 AU and 2.8 AU. This is quite unlike planetary orbits in our own solar system, where no planet has an orbital eccentricity greater than 0.25.

Do these observations mean that our picture of how planets form is incorrect? If Jupiterlike extrasolar planets such as that orbiting 51 Pegasi formed close to their stars, the mechanism of their formation must have been very different from that which operated in our solar system.

But another possibility is that extrasolar planets actually formed at large distances from their stars and have migrated inward since their formation. If enough gas and dust remain in a disk around a star after its planets form, interactions between the disk material and an orbiting planet will cause the planet to lose energy and to spiral inward toward the star around which it orbits. The planet's inward migration can eventually stop because of subtle gravitational effects from the disk or from the star. Gravitational interactions between the planet and the disk, or between planets in the same planetary system, could also have forced an extrasolar planet into a highly eccentric, noncircular orbit.

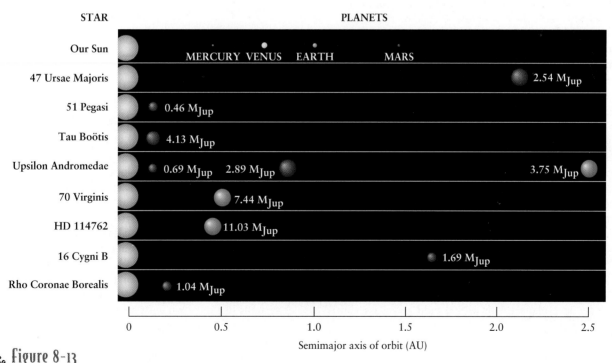

Figure 8-13

A Selection of Extrasolar Planets This figure summarizes what we know about planets orbiting eight other stars. The star name is given at the left of each line. Each planet is shown at its average distance from its star (equal to the semimajor axis of its orbit). Comparison with our own solar system (at the top of the figure) shows how closely these extrasolar planets orbit their stars. The planets are *not* shown to the same scale as their stars, and the relative sizes of the extrasolar planets are estimates only. The mass of each planet—actually a lower limit—is given as a multiple of Jupiter's mass (M_{Jup}). (Adapted from G. Marcy and R. P. Butler)

One piece of evidence that young planets may spiral inward toward their parent stars is the spectrum of the star HD 82943, which has at least two planets orbiting it. The spectrum shows that this star's atmosphere contains a rare form of lithium that is found in planets, but which is destroyed in stars by nuclear reactions within 30 million years. The presence of this exotic form of lithium, known as ^{6}Li, means that at least one planet spiraled so close to the star that it was vaporized and swallowed whole.

Yet another possibility is that some of these extrasolar planets are not planets at all! The problem is that the radial velocity method cannot give precise values for the masses of planets, only lower limits. (An exact determination of the mass would require knowing how the plane of the planet's orbit is inclined to our line of sight. Unfortunately, this angle is not known because the planets themselves are unseen.) Hence, the actual masses of some of the objects shown in Figure 8-13 may be much larger than the values shown in that figure. It is thought that planets more massive than about 13 times the mass of Jupiter are actually **brown dwarfs,** objects that are like the Sun but are insufficiently massive to sustain nuclear reactions in their cores for extended periods (**Figure 8-14**). Like stars, but unlike true planets, brown dwarfs are thought to form directly from gas by gravitational contraction rather than by accretion of planetesimals.

A technique that makes it possible to determine the actual mass of an object orbiting a star is the **transit method.** This method looks for the rare situation in which the object comes between us and its parent star, an event called a **transit.** As in a partial solar eclipse (Section 3-5), this causes a small but measurable dimming of the star's light. If a transit is seen, the orbit must be edge-on to our line of sight. Knowing the orientation of the orbit, the information obtained by radial velocity measurements of the star tells us the true mass (not just a lower limit) of the orbiting object. An added benefit is that the amount by which the star is dimmed during the transit tells us the diameter of the object. A handful of extrasolar planets have been observed transiting their stars, and all of these planets appear to have masses close to that of Jupiter's. These results reinforce the idea that other stars can end up with planets much like those in our solar system.

Observations with the radial velocity method have found some stars with planetary *systems,* that is, multiple planets orbiting the same star. One example is the system of two planets orbiting 47 Ursae Majoris, a star very similar to the Sun. The two planets are in nearly circular orbits with orbital radii of 2.09 and 3.73 AU, intermediate between the orbits of Mars and Jupiter. (The outer planet is beyond the range of distances shown in Figure 8-13.) The inner planet has a mass at least 2.5 times that of Jupiter; the outer planet has a mass at least 0.76 times that of Jupiter. Some simulations of planet formation produce planets with orbital radii and masses in this range. There is also evidence for a system of three planets orbiting Upsilon Andromedae (see Figure 8-13).

Are all stars equally likely to have planets? The answer appears to be no: Only about 5% of the stars surveyed with the radial velocity method show evidence of planets. However, the percentage is 10 to 20% for stars in which heavy elements are at least as abundant as in the Sun. The conclusion is that

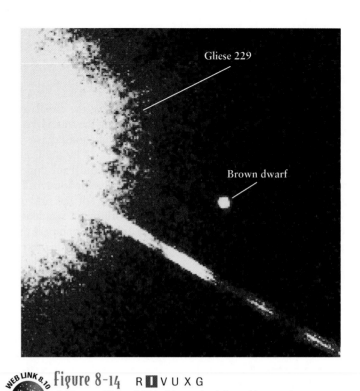

figure 8-14 R ■ V U X G

A Brown Dwarf This false-color infrared image shows a brown dwarf that orbits the star Gliese 229 at an average distance of 32 AU. The brown dwarf is estimated to have a mass of 30 to 55 times the mass of Jupiter (between 0.030 and 0.055 of the Sun's mass) but about the same radius as Jupiter. Its surface temperature is about 1000 K. The spike running diagonally is an illusion caused by the telescope optics. Gliese 229 and its brown dwarf are 19 ly from Earth in the constellation Lepus (the Hare). (S. Kulkarni, Caltech; D. Golimowski, Johns Hopkins University; and NASA)

heavy elements appear to play an important role in the evolution of planets from an interstellar cloud. This agrees with the idea discussed in Section 8-5 that planetesimals (which are composed of heavy elements) are the building blocks of planets.

The study of extrasolar planets is still in its infancy. We know little about how they formed or how their formation may have differed from that of our own solar system. We do not yet know what extrasolar planets look like, and we know little about what they are made of. No Earth-sized extrasolar planets have yet been discovered, because they are too small to be detected by the radial velocity method, the transit method, or the astrometric method using current telescope technology. Hence, we do not yet know whether solar systems like our own exist around other stars.

In the near future, however, advances in adaptive optics (see Section 6-3) and interferometry (Section 6-6) may make it possible to record images of extrasolar planets the size of Jupiter. Orbiting telescopes are now being planned by both NASA and the European Space Agency to search for terrestrial planets around other stars. The next few decades may tell whether our own solar system is an exception or just one of a host of similar systems throughout the Galaxy.

Key Words

Terms preceded by an asterisk (*) are discussed in Box 8-1.

accretion, p. 174

astrometric method (for detecting extrasolar planets), p. 178

atomic number, p. 168

brown dwarf, p. 180

center of mass, p. 178

chemical differentiation, p. 175

chondrule, p. 174

condensation temperature, p. 173

conservation of angular momentum, p. 172

core accretion model, p. 175

disk instability model, p. 176

extrasolar planet, p. 177

*half-life, p. 170

interstellar medium, p. 167

jets, p. 177

Kelvin-Helmholtz contraction, p. 171

meteorite, p. 169

nebulosity, p. 167

nebular hypothesis, p. 171

Oort cloud, p. 176

planetesimal, p. 173

protoplanet, p. 174

protoplanetary disk (proplyd), p. 172

protosun, p. 171

radial velocity method (for detecting extrasolar planets), p. 178

radioactive age-dating, p. 169

radioactive decay, p. 169

solar nebula, p. 171

solar wind, p. 177

T Tauri wind, p. 177

transit, p. 180

transit method (for detecting extrasolar planets), p. 180

Key Ideas

Formation of the Solar System: The most successful model of the origin of the solar system is called the nebular hypothesis. According to this hypothesis, the solar system formed from a cloud of interstellar material called the solar nebula. This occurred 4.56 billion years ago (as determined by radioactive age-dating).

• The chemical composition of the solar nebula, by mass, was 98% hydrogen and helium (elements that formed shortly after the beginning of the universe) and 2% heavier elements (produced much later in the centers of stars, and cast into space when the stars died). The heavier elements were in the form of ice and dust particles.

• The nebula flattened into a disk in which all the material orbited the center in the same direction, just as do the present-day planets.

• The four terrestrial planets formed through the accretion of dust particles into planetesimals, then into larger protoplanets.

• In the core accretion model, the four Jovian planets began as rocky protoplanetary cores, similar in character to the terrestrial planets. Gas then accreted onto these cores in a runaway fashion.

• In the alternative disk instability model, the Jovian planets formed directly from the gases of the solar nebula. In this model the cores formed from planetesimals falling into the planets.

• The Sun formed by gravitational contraction of the center of the nebula. After about 10^8 years, temperatures at the protosun's center became high enough to ignite nuclear reactions that convert hydrogen into helium, thus forming a true star.

Extrasolar Planets: Astronomers have discovered planets orbiting other stars. The planets themselves are not visible; their presence is detected by the "wobble" of the stars around which they orbit.

• Most of the extrasolar planets discovered to date are quite massive and have orbits that are very different from planets in our solar system.

Review Questions

1. The graphite in your pencil is a form of carbon. Where were these carbon atoms formed?

2. What is the interstellar medium? How does it become enriched over time with heavy elements?

3. If hydrogen and helium account for 98% of the mass of all the atoms in the universe, why aren't the Earth and Moon composed primarily of these two gases?

4. How do radioactive elements make it possible to determine the age of the solar system? What are the oldest objects that have been found in the solar system?

5. What is the tidal hypothesis? What aspect of the solar system was it designed to explain? Why was this hypothesis rejected?

6. What is meant by a substance's condensation temperature? What role did condensation temperatures play in the formation of the planets?

7. What is a planetesimal? How did planetesimals give rise to the terrestrial planets?

8. What is meant by accretion?

9. What is a chondrule? How do we know they were not formed by the ambient heat of the solar nebula?

10. Why did the terrestrial planets form close to the Sun while the Jovian planets formed far from the Sun? What are the competing models of how the Jovian planets formed?

11. Why is the combined mass of all the asteroids so small?

12. Explain how our current understanding of the formation of the solar system can account for the following characteristics of the solar system: (**a**) All planetary orbits lie in nearly the same plane. (**b**) All planetary orbits are nearly circular. (**c**) The planets orbit the Sun in the same direction in which the Sun itself rotates.

13. Explain why most of the satellites of Jupiter orbit that planet in the same direction that Jupiter rotates.

14. What are proplyds? What do they tell us about the plausibility of our model of the solar system's origin?

15. What techniques are used to detect planets orbiting other stars? Why is it difficult to use these techniques to detect planets like Earth?

16. Summarize the differences between the planets of our solar system and those found orbiting other stars.

17. Is there evidence that planets have fallen into their parent stars? Explain.

18. What does it mean for a planet to transit a star? What can we learn from such events?

19. A 1999 news story about the discovery of three planets orbiting the star Upsilon Andromedae (see Figure 8-13) stated that "the newly discovered galaxy, with three large planets orbiting a star known as Upsilon Andromedae, is 44 light-years away from Earth." What is wrong with this statement?

Advanced Questions

Questions preceded by an asterisk (*) involve topics discussed in Box 8-1.

Problem-solving tips and tools

The volume of a disk of radius r and thickness t is $\pi r^2 t$. Box 1-1 explains the relationship between the angular size of an object and its actual size. An object moving at speed v for a time t travels a distance $d = vt$; Appendix 6 includes conversion factors between different units of length and time. To calculate the mass of 70 Virginis, review Box 4-4.

---

20. Figure 8-3 shows that carbon, nitrogen, and oxygen are the most abundant elements (after hydrogen and helium). In our solar system, the atoms of these elements are found primarily in the molecules CH_4 (methane), NH_3 (ammonia), and H_2O (water). Why do you suppose this is?

21. (a) If the Earth had retained hydrogen and helium in the same proportion to the heavier elements that exist elsewhere in the universe, what would its mass be? Give your answer as a multiple of the Earth's actual mass, given in Table 7-1. Explain your reasoning. (b) How does your answer to (a) compare with the masses of the Jovian planets? What does your answer imply about how large the cores of the Jovian planets must be?

*22. If you start with 0.80 kg of radioactive potassium (^{40}K), how much will remain after 1.3 billion years? After 2.6 billion years? After 3.9 billion years? How long would you have to wait until there was *no* ^{40}K remaining?

*23. Three-quarters of the radioactive potassium (^{40}K) originally contained in a certain volcanic rock has decayed into argon (^{40}Ar). How long ago did this rock form?

24. Suppose you were to use the Hubble Space Telescope to monitor one of the protoplanetary disks shown in Figure 8-5b. Over the course of 10 years, would you expect to see planets forming within the disk? Why or why not?

25. The protoplanetary disk, or proplyd, at the upper right of Figure 8-5b is seen edge-on. The diameter of the disk is about 700 AU. (a) Make measurements on this image to determine the thickness of the disk in AU. (b) Explain why the disk will continue to flatten as time goes by.

26. The accompanying infrared image shows IRAS 04302+2247, a young star that is still surrounded by a disk of gas and dust. The scale bar at the lower left of the image shows that at the distance of IRAS 04302+2247, an angular size of 2 arcseconds corresponds to a linear size of 280 AU. Use this information to find the distance to IRAS 04302+2247.

(Courtesy of D. Padgett and W. Brandner, IPAC/Caltech; K. Stapelfeldt, JPL; and NASA)

27. The image accompanying Question 26 shows a dark, opaque disk of material surrounding the young star IRAS 04302+2247. The disk is edge-on to our line of sight, so it appears as a dark band running vertically across this image. The material to the left and right of this band is still falling onto the disk. (a) Make measurements on this image to determine the diameter of the disk in AU. Use the scale bar at the lower left of this image. (b) If the thickness of the disk is 50 AU, find its volume in cubic meters. (c) The total mass of the disk is perhaps 2×10^{28} kg (0.01 of the mass of the Sun). How many atoms are in the disk? Assume that the disk is all hydrogen. A single hydrogen atom has a mass of 1.673×10^{-27} kg. (d) Find the number of atoms per cubic meter in the disk. By comparison, the air that you breathe contains about 5.4×10^{25} atoms per cubic meter. By Earth standards, is the disk material thick or thin?

28. If the material in the jet shown in Figure 8-11*a* is moving at 200 km/s, how long ago was the material at the far right-hand end of the jet ejected from the star? Give your answer in years. (You will need to make measurements on the image.)

29. Discuss why the "planet" that orbits the star HD 114762 (see Figure 8-13) could possibly be a brown dwarf, while the planet that orbits 51 Pegasi probably is not.

*30. The planet discovered orbiting the star 70 Virginis, 59 light-years from Earth, moves in an orbit with semimajor axis 0.48 AU and eccentricity 0.40. The period of the orbit is 116.7 days. Find the mass of 70 Virginis. Compare your answer with the mass of the Sun. (*Hint:* The planet has far less mass than the star.)

*31. Because of the presence of Jupiter, the Sun moves in a small orbit of radius 742,000 km with a period of 11.86 years. (**a**) Calculate the Sun's orbital speed in meters per second. (**b**) An astronomer on a hypothetical planet orbiting the star Vega, 25 light-years from the Sun, wants to use the astrometric method to search for planets orbiting the Sun. What would be the angular diameter of the Sun's orbit as seen by this alien astronomer? Would the Sun's motion be discernible if the alien astronomer could measure positions to an accuracy of 0.001 arcsec? (**c**) Repeat part (**b**), but now let the astronomer be located on a hypothetical planet in the Pleiades star cluster, 360 light-years from the Sun. Would the Sun's motion be discernible to this astronomer?

Discussion Questions

32. Propose an explanation why the Jovian planets are orbited by terrestrial-like satellites.

33. Suppose that a planetary system is now forming around some protostar in the sky. In what ways might this planetary system turn out to be similar to or different from our own solar system? Explain your reasoning.

34. Suppose astronomers discovered a planetary system in which the planets orbit a star along randomly inclined orbits. How might a theory for the formation of that planetary system differ from that for our own?

 ## Web/CD-ROM Questions

35. Search the World Wide Web for information about recent observations of protoplanetary disks. What insights have astronomers gained from these observations? Is there any evidence that planets have formed within these disks?

36. In 2000, extrasolar planets with masses comparable to that of Saturn were first detected around the stars HD 16141 (also called 79 Ceti) and HD 46375. Search the World Wide Web for information about these "lightweight" planets. Do these planets move around their stars in the same kind of orbit as Saturn follows around the Sun? Why do you suppose this is? How does the discovery of these planets reinforce the model of planet formation described in Section 7-8?

37. Search the World Wide Web for information about extrasolar planets that transit in front of their parent stars. Has it been possible to detect the atmospheres of these planets? One particularly interesting planet discovered using the transit method is called OGLE-TR-56b; what makes this planet so unusual?

Observing Projects

 38. Use the *Starry Night Backyard*™ program to investigate stars that have planets orbiting them. First display the entire celestial sphere (select **Guides > Atlas** in the **Go** menu), then use **Find...** in the **Edit** menu to find each of the stars listed below. Then click on the **Info** tab on the left-hand side of the *Starry Night Backyard*™ window for full information about the star. For each star, record the luminosity of the star (a measure of the star's total light output). How far from Earth is each star? Which stars are more luminous than the Sun? Which are less luminous? How do you think these differences would have affected temperatures in the nebula in which each star's planets formed (see Figure 8-7)? (i) 47 Ursae Majoris; (ii) 51 Pegasi; (iii) 70 Virginis; (iv) Rho Coronae Borealis.

 39. Use the *Deep Space Explorer*™ program to examine stars that have planets. In the left-hand part of the window, under the heading **Milky Way** select **Explore,** then click on **Extrasolar Planets.** In the star map that appears, each circled star has one or more planets. (You can zoom in and zoom out using the buttons at the upper left of the window (an upward-pointing triangle and a downward-pointing triangle). You can also rotate the star map by putting the mouse cursor over the image, holding down the mouse button, and moving the mouse. On a two-button mouse, hold down the left mouse button.) Click on a circled star to learn more about its properties. Note that the information given for each star includes the *apparent magnitude,* which is a measure of how bright each star appears as seen from Earth. Apparent magnitude uses a "backwards" scale: The greater the apparent magnitude, the *dimmer* the star. Most of the brighter stars you can see with the naked eye from the Earth have apparent magnitudes between 0 and 1, while the dimmest star you can see from a dark location has apparent magnitude 6. Are most of the circled stars visible to the naked eye? List at least two stars that are visible, and include their apparent magnitudes.

GEOFF MARCY

Alien Planets

When **Geoff Marcy** sat down to write this essay, more than 100 planets were known to orbit other stars. He leads the team that has found more than 70 of them. Dr. Marcy became interested in astronomy at the age of 14, when his parents bought him a small reflecting telescope. Since receiving his Ph.D. from the University of California, Santa Cruz, he has studied stars similar to our Sun. He helped show that magnetic regions on their surfaces cause "star spots" and stellar flares. He also showed that brown dwarfs—stars too small to fuse hydrogen—rarely orbit other stars. His work may soon reveal whether a planetary system like ours is common or a quirk of the cosmos. Dr. Marcy is an astronomer at the University of California, Berkeley.

Astronomers have now surveyed 1000 nearby stars in the search for orbiting planets. My team, composed of scientists from UC Berkeley, Carnegie Institute of Washington, UC Santa Cruz, and San Francisco State University, is using the 10-meter Keck telescope in Hawaii to measure the Doppler effect in stars that wobble because of planets orbiting around them. So far, we and other teams have found more than 100 planets, some more massive than Jupiter and some probably less massive than Saturn.

We have now found 10 stars that have multiple planets orbiting them—full planetary systems. Often the planets orbit synchronously, with one planet orbiting the star 2 or 3 times for every orbit of another planet. The synchronization of the planets allows gravity to shepherd their motion, keeping them synchronized forever. We hope to compare the architecture of these planetary systems to that of our own solar system. Is our solar system a common garden-variety type? Do all planetary systems have giant planets orbiting far away from the host star, with small rocky planets orbiting close in?

Remarkably, the planetary systems observed so far seem very different from our own solar system. In some, a giant planet orbits very close to the star. More surprising is that almost all alien planets that orbit beyond 0.1 AU from their star reside in highly elliptical orbits that resemble eccentric comet orbits. Why are these planets in such elliptical orbits so different from the nearly circular orbits of planets in our own solar system? Astronomers are baffled by this difference.

One possibility is that planets usually form as close-knit families around a star. Over time, the planets may gravitationally pull on one another, yanking themselves out of their original circular orbits. These perturbed planets may then venture near other planets, yanking on them, too. Many of the planets may gravitationally slingshot each other completely out of the planetary system. These far-flung planets are destined to roam the darkness of galactic space, cooling to frigid temperatures that are inhospitable to life.

The extrasolar planets cast a mystery back on the Earth. Why is our solar system immune to this chaotic episode that scatters planets into wacky orbits? Imagine the disaster if our Earth were suddenly thrust into a highly elliptical orbit. During half the year, we would be roasting too close to the Sun, with oceans vaporized into an enormous steam bath. The other half of the year our orbit would carry us too far from the Sun, causing all the oceans to freeze over. Life on Earth would be severely challenged to survive in such an elongated orbit.

Why are we humans so lucky as to live in a stable planetary system in which circular orbits ensure that we receive the same warming light intensity from our Sun, all year round? Perhaps the question to ask is: Would intelligent life have evolved here if the Earth suffered from wild temperature swings that would make liquid water rare? We humans, and intelligent species in general, may flourish mostly on worlds that maintain nearly constant temperature. If so, our solar system and the Earth might be a rare galactic quirk that just happens to be suitable for life. We wouldn't be here otherwise. If this picture is correct, our Earth is indeed a precious oasis in our Milky Way Galaxy.

In the next few years, we plan to use the Doppler technique to discover planets having masses only 10 or 20 times that of our Earth. These would likely be rocky planets, allowing liquid water to puddle in ponds and lakes. Who knows what biology might emerge in such extraterrestrial chemistry labs!

In the next decade, we will discover more planetary systems by using the Doppler method. We are working night and day to find Jovian planets that orbit 5 AU from their star, similar to the orbital radius of our Jupiter. Their orbits, circular or elliptical, will shed light on a key question: How common are configurations like our solar system, and how common is advanced life in the universe?

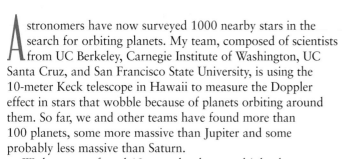

9 The Living Earth

When astronauts made the first journeys from the Earth to the Moon between 1968 and 1972, they often reported that our planet is the most beautiful sight visible from space. Indeed, no other world in the solar system has the Earth's contrast between blue oceans and white clouds. No other world but Earth has the distinctive green color of vegetation. Compared with the Moon, whose surface has been ravaged by billions of years of impacts by interplanetary debris, the Earth seems an inviting and tranquil place.

Yet the appearance of tranquility is deceiving. The seemingly solid surface of the planet is in a state of slow but constant motion, driven by the flow of molten rock within the Earth's interior. Sunlight provides the Earth's atmosphere with the energy that drives our planet's weather, including all the cloud patterns visible in the images shown here.

Unseen in these images are immense clouds of subatomic particles that wander around the outside of the planet in two giant belts, held in thrall by the Earth's magnetic field. And equally unseen are trace gases in the Earth's atmosphere whose abundance may determine the future of our climate, and on which may depend the survival of entire species.

In this chapter our goal is to learn about the Earth's components—its dynamic oceans and atmosphere, its ever-changing surface, and its hot, active interior—and how they interact to make up our planetary home. In later chapters we will use this knowledge as a point of reference for studying the Moon and the other planets.

 WEB LINK 9.1

The two hemispheres of Earth. (NASA)   R I **V** U X G

Questions to Guide Your Reading

As you read the sections of this chapter, look for the answers to the following questions:

9-1   What is the greenhouse effect? How does it affect the average temperature of the Earth?

9-2   Is the Earth completely solid inside? How can scientists tell?

9-3   How is it possible for entire continents to move across the face of the Earth?

9-4   How does our planet's magnetic field protect life on Earth?

9-5   Why is Earth the only planet with an oxygen-rich atmosphere?

9-6   Why are prevailing winds generally from the west over most of North America but generally from the east in Hawaii?

9-7   What are global warming and the "ozone hole"? Why should they concern us?

9-1 | The Earth's atmosphere, oceans, and surface are extraordinarily active

VIDEO 9.1 The crew of an alien spacecraft exploring our solar system might overlook the Earth altogether. Although the largest of the terrestrial planets, with a mass greater than that of Mercury, Venus, and Mars put together, Earth is far smaller than any of the giant Jovian planets (see Table 7-1 and Table 9-1). But a more careful inspection would reveal the Earth to be unique among all the planets that orbit the Sun.

Unlike the arid surfaces of Venus and Mars, the Earth's surface is very wet. Indeed, nearly 71% of the Earth's surface is covered with water (Figure 9-1). Water is also locked into the chemical structure of many Earth rocks, including those found in the driest deserts. Furthermore, the Earth's liquid water is in constant motion. The oceans ebb and flow with the tide, streams and rivers flow down to the sea, and storms whip lake waters into a frenzy.

The Earth's atmosphere is also in a state of perpetual activity. Winds blow at all altitudes, with speeds and directions that change from one hour to the next. The atmosphere is at its most dynamic when water evaporates to form clouds, then returns to the surface as rain or snow (Figure 9-2). While other worlds of the solar system have atmospheres, only the Earth's contains the oxygen that animals (including humans) need to sustain life.

The combined effects of water and wind cause erosion of mountains and beaches. But these are by no means the only forces reshaping the face of our planet. All across the Earth we find evidence that the surface has been twisted, deformed, and folded (Figure 9-3). What is more, new material is continually being added to the Earth's surface as lava pours forth from volcanoes and from immense cracks in the ocean floors. These processes and others work together to renew and refresh our planet's exterior. Thus, although the Earth is some 4.56 billion (4.56×10^9) years old, much of its surface is less than 100 million (10^8) years old. (We saw in Section 8-2 how scientists use radioactive age-dating to determine the ages of rocks and of the Earth as a whole.)

table 9-1	Earth Data
Average distance from the Sun:	1.000 AU = 1.496×10^8 km
Maximum distance from the Sun:	1.017 AU = 1.521×10^8 km
Minimum distance from the Sun:	0.983 AU = 1.471×10^8 km
Eccentricity of orbit:	0.017
Average orbital speed:	29.79 km/s
Orbital period:	365.256 days
Rotation period:	23.9345 hours
Inclination of equator to orbit:	23.45°
Diameter (equatorial):	12,756 km
Mass:	5.974×10^{24} kg
Average density:	5515 kg/m³
Escape speed:	11.2 km/s
Albedo:	0.39
Surface temperature range:	Maximum: 60°C = 140°F = 333 K
	Mean: 14°C = 57°F = 287 K
	Minimum: −90°C = −130°F = 183 K
Atmospheric composition (by number of molecules):	78.08% nitrogen (N_2)
	20.95% oxygen (O_2)
	0.035% carbon dioxide (CO_2)
	about 1% water vapor

figure 9-1 R I V U X G

The Earth's Dynamic Oceans Nearly three-quarters of the Earth's surface is covered with water. In contrast, there is no liquid water at all on Mercury, Venus, Mars, or the Moon. (Farley Lewis/Photo Researchers)

What powers all this activity in the Earth's oceans, atmosphere, and surface? There are three energy sources: radiation from the Sun, the tidal effects of the Moon, and the Earth's own internal heat (Table 9-2).

The Sun is the principal source of energy for the atmosphere. The Earth's surface is warmed by sunlight, which in turn warms the air next to the surface. Hot air is less dense than cool air and so tends to rise (see the discussion of density in Box 7-1). As the air rises, it transfers heat to its surroundings. As a result, the rising air cools and becomes denser. It then sinks downward to be heated again, and the process starts over. This up-and-down motion is called **convection**, and the overall pattern of circulation is called a **convection current**.

figure 9-2 R I V U X G

The Earth's Dynamic Atmosphere This space shuttle image shows thunderstorm clouds over the African nation of Zaire. The tops of thunderstorm clouds frequently reach altitudes of 10,000 m (33,000 ft) or higher. At any given time, nearly 2000 thunderstorms are in progress over the Earth's surface. (JSC/NASA)

figure 9-3 R I V U X G

The Earth's Dynamic Surface These tan-colored ridges, called hogbacks, were once layers of sediment at the bottom of an ancient body of water. Forces within the Earth folded this terrain and rotated the layers into a vertical orientation. The layers were revealed when wind and rain eroded away the surrounding material. These hogback ridges are in the Rocky Mountains of Colorado. (Tom Till/DRK)

table 9-2	The Earth's Energy Sources	
Activity	**Energy Sources**	
Motion of water in oceans, lakes, rivers	Solar energy, tidal forces	
Motion of the atmosphere	Solar energy	
Reshaping of surface	Earth's internal heat	
Life	Solar energy (a few species that live on the ocean floor make use of the Earth's internal heat)	

(You can see convection currents in action by heating water on a stove, as Figure 9-4 shows.) In Section 9-5 we will see how the Earth's rotation modifies the up-and-down motion of the atmosphere to produce a pattern of global circulation.

Solar energy also powers atmospheric activity by evaporating water from the surface. The energy in the water vapor is released when it condenses to form water droplets, like those that make up clouds. In a typical thunderstorm (see Figure 9-2), some 5×10^8 kg of water vapor is lifted to great heights. The amount of energy released when this water condenses is as much as a city of 100,000 people uses in a month!

One complication with solar energy is that the Earth does not absorb all the radiation that it receives from the Sun. About 39% of it is reflected back into space by clouds, snow, ice, and sand. The fraction of incoming sunlight that a planet reflects is called its **albedo** (from the Latin for "whiteness"); thus the Earth's albedo is about 0.39.

Another important effect is that the Earth *emits* radiation into space because of its temperature, in accordance with the laws that describe heated dense objects (see Section 5-3). Wien's law tells us that the wavelength at which such an object emits most strongly (λ_{max}) is inversely proportional to its temperature (T) on the Kelvin scale (see Section 5-4 and Box 5-2). For example, the Sun's surface temperature is about 5800 K, and sunlight has its greatest intensity at a wavelength λ_{max} of 500 nm, in the middle of the visible spectrum. The Earth's average surface temperature of 287 K is far lower than the Sun's, so the Earth radiates most strongly at longer wavelengths in the infrared portion of the electromagnetic spectrum. Temperature also determines the amount of radiation that the Earth emits: The higher the temperature, the more energy it radiates.

The Earth's average surface temperature is nearly constant, which means that on the whole it is neither gaining nor losing energy. Thus, the rate at which the Earth emits energy must equal the rate at which it receives energy from the Sun, not including the solar energy that is reflected back into space. Since the rate of emission depends on temperature, we can use this condition to calculate what the Earth's average surface temperature should be. The result is a very chilly 246 K (−27°C = −17°F). Yet the Earth's actual aver-

age surface temperature is 287 K (14°C = 57°F)! What is wrong with our model? Why is the Earth warmer than we would expect?

The explanation for this discrepancy is called the **greenhouse effect**: The Earth's atmosphere prevents some of the Earth's surface radiation from escaping into space. Certain gases in our atmosphere called **greenhouse gases**, among them water vapor and carbon dioxide, are transparent to visible light but not to infrared radiation. Consequently, visible sunlight has no trouble entering our atmosphere and warming the surface. But the infrared radiation coming from the heated surface is partially trapped by the atmosphere, thus raising the temperatures of both the atmosphere and the surface. As the surface and atmosphere become hotter, they both emit more infrared radiation, part of which is able to escape into space. The temperature levels off when the amount of infrared energy that escapes just balances the amount of solar energy reaching the surface (Figure 9-5). The result is that our planet's surface is some 41°C (74°F) warmer than it would be without the greenhouse effect.

Solar energy also helps to power the oceans. Warm water from near the equator moves toward the poles, while cold polar water returns toward the equator. As we saw in Section 4-8, however, there is back-and-forth motion of the oceans due to the tidal forces of the Moon and Sun. Sometimes these two influences can reinforce each other, as when a storm (caused by solar energy) reaches a coastline at high tide (caused by tidal forces), producing waves strong enough to seriously erode beaches and sea cliffs.

Neither solar energy nor tidal forces can explain the reshaping of the Earth's surface suggested by Figure 9-3. Rather, all this geologic activity is powered by heat flowing from the interior of the Earth itself. The Earth formed by collisions among planetesimals (see Section 8-4), and these collisions heated the body of the Earth. The interior remains hot to this day. An additional source of heat is the

figure 9-4

Convection in the Kitchen Heat supplied to the bottom of a pot warms the water there, making the water expand and lowering its average density. This low-density water rises and transfers heat to its cooler surroundings. The water that began at the bottom thus cools down, becomes denser, and sinks back to the bottom to repeat the process. (Adapted from F. Press, R. Siever, T. Grotzinger, and T. H. Jordan, *Understanding Earth* 4th Ed., W. H. Freeman, 2004)

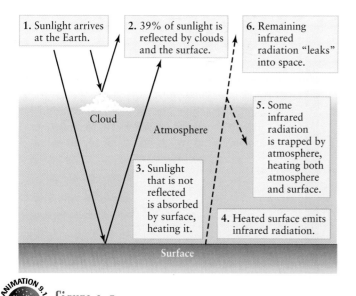

1. Sunlight arrives at the Earth.

2. 39% of sunlight is reflected by clouds and the surface.

6. Remaining infrared radiation "leaks" into space.

Cloud

Atmosphere

5. Some infrared radiation is trapped by atmosphere, heating both atmosphere and surface.

3. Sunlight that is not reflected is absorbed by surface, heating it.

4. Heated surface emits infrared radiation.

Surface

figure 9-5

ANIMATION 9.1

The Greenhouse Effect Sunlight warms the Earth's surface, which due to its temperature emits infrared radiation. Much of this radiation is absorbed by atmospheric water vapor and carbon dioxide, helping to raise the average temperature of the surface. Some infrared radiation does penetrate the atmosphere and leaks into space. In a state of equilibrium, the rate at which the Earth loses energy to space in this way is equal to the rate at which it absorbs energy from the Sun.

decay of radioactive elements such as uranium and thorium deep inside the Earth.

The heat flow from the Earth's interior to its surface is minuscule—just $1/6000$ as great as the flow of energy we receive from the Sun—but it has a profound effect on the face of our planet. To understand the connection between geologic activity on the Earth's surface and heat coming from its interior, we must first learn something about the interior structure of our dynamic Earth.

9-2 | Studies of earthquakes reveal the Earth's layered interior structure

The kinds of rocks found on and near the Earth's surface provide an important clue about our planet's interior. The densities of typical surface rocks are around 3000 kg/m³, but the average density of the Earth as a whole (that is, its mass divided by its volume) is 5515 kg/m³. The interior of the Earth must therefore be composed of substances much denser than those found near the surface. But what is this substance? How did it come to be that the Earth's interior is more dense than its crust? And is the Earth's interior solid like rock or molten like lava?

Iron (chemical symbol Fe) is a good candidate for the substance that makes up most of the Earth's interior. This is so for two reasons. First, iron atoms are quite massive (a typical iron atom has 56 times the mass of a hydrogen atom), and second, iron is relatively abundant. (Figure 8-3 shows that it is the seventh most abundant element in our part of the Milky Way Galaxy.) Other elements such as lead and uranium have more massive atoms, but these elements are quite rare. Hence, the solar nebula could not have had enough of these massive atoms to create the Earth's dense interior. Furthermore, iron is common in meteoroids that strike the Earth, which suggests that it was abundant in the planetesimals from which the Earth formed.

The Earth was almost certainly molten throughout its volume soon after its formation, about 4.56×10^9 years ago. Energy released by the violent impacts of numerous meteoroids and asteroids and by the decay of radioactive isotopes likely melted the solid material collected from the earlier planetesimals. Gravity caused abundant, dense iron to sink toward the Earth's center, forcing less dense material to the surface. **Figure 9-6a** shows this process of *chemical differentiation*. (We discussed chemical differentiation in Section

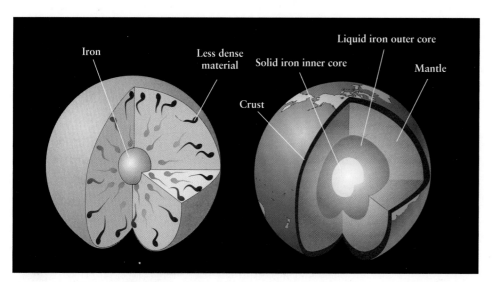

(a) During differentiation, iron sank to the center and less dense material floated upward

(b) As a result of differentiation, the Earth has the layered structure that we see today

figure 9-6

Chemical Differentiation and the Earth's Internal Structure (a) The newly formed Earth was molten throughout its volume. Dense materials such as iron (shown in orange) sank toward the center, while low-density materials (shown in blue) rose toward the surface. (b) The present-day Earth is no longer completely molten inside. A dense, solid iron core is surrounded by a less dense liquid core and an even less dense mantle. The crust, which includes the continents and ocean floors, is the least dense of all; it floats atop the mantle like the skin that forms on the surface of a cooling cup of cocoa.

P waves are longitudinal waves

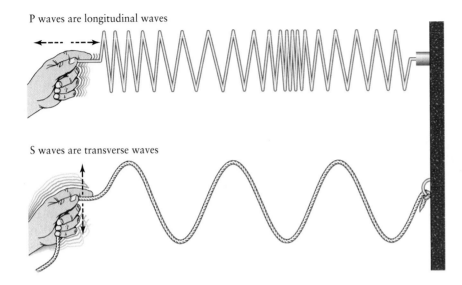

S waves are transverse waves

figure 9-7

Seismic Waves Earthquakes produce two kinds of waves that travel through the body of our planet. One kind, called P waves, are longitudinal waves. They are analogous to those produced by pushing a spring in and out. The other kind, S waves, are transverse waves analogous to the waves produced by shaking a rope up and down.

8-4.) The result was a planet with the layered structure shown in Figure 9-6*b*—a central **core** composed of almost pure iron, surrounded by a **mantle** of dense, iron-rich minerals. The mantle, in turn, is surrounded by a thin **crust** of relatively light silicon-rich minerals. We live on the surface of this crust.

How do we know that this layered structure is correct? The challenge in testing this model is that the Earth's interior is as inaccessible as the most distant galaxies in space. The deepest wells go down only a few kilometers, barely penetrating the surface of our planet. Despite these difficulties, geologists have learned basic properties of the Earth's interior by studying earthquakes and the seismic waves that they produce.

Over the centuries, stresses build up in the Earth's crust. Occasionally, these stresses are relieved with a sudden motion called an **earthquake.** Most earthquakes occur deep within the Earth's crust. The point on the Earth's surface directly over an earthquake's location is called the **epicenter.**

Earthquakes produce three different kinds of **seismic waves,** which travel around or through the Earth in different ways and at different speeds. Geologists use sensitive instruments called **seismographs** to detect and record these vibratory motions. The first type of wave, which is analogous to water waves on the ocean, causes the

rolling motion that people feel around an epicenter. These are called **surface waves** because they travel only over the Earth's surface. The two other kinds of waves, called **P waves** (for "primary") and **S waves** (for "secondary"), travel through the interior of the Earth. P waves are called *longitudinal* waves because their oscillations are parallel to the direction of wave motion, like a spring that is alternately pushed and pulled. In contrast, S waves are called *transverse* waves because their vibrations are perpendicular to the direction in which the waves move. S waves are analogous to waves produced by a person shaking a rope up and down (**Figure 9-7**).

What makes seismic waves useful for learning about the Earth's interior is that they do not travel in straight lines. Instead, the paths that they follow through the body of the Earth are bent because of the varying density and composition of the Earth's interior. We saw in Section 6-1 that light waves behave in a very similar way. Just as light waves bend, or refract, when they pass from air into glass or vice versa (see Figure 6-2), seismic waves refract as they pass through different parts of the Earth's interior. By studying how the paths of these waves bend, geologists can map out the general interior structure of the Earth.

One key observation about seismic waves and how they bend has to do with the differences between S and P waves. When an earthquake occurs, seismographs relatively close to the epicenter record

table 9-3	The Earth's Internal Structure		
Region	Depth below surface (km)	Distance from center (km)	Average density (kg/m^3)
Crust (solid)	0–5 (under oceans) 0–35 (under continents)	6343–6378	3500
Mantle (solid)	from bottom of crust to 2900	3500–6343	3500–5500
Outer core (liquid)	2900–5100	1300–3500	10,000–12,000
Inner core (solid)	5100–6400	0–1300	13,000

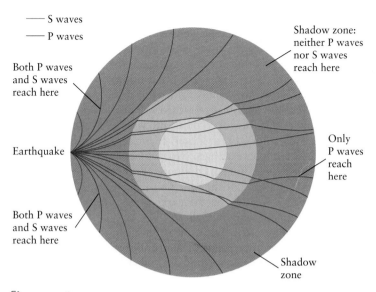

figure 9-8

The Earth's Internal Structure and the Paths of Seismic Waves
Seismic waves follow curved paths because of differences in the density and composition of the material in the Earth's interior. The paths curve gradually where there are gradual changes in density and composition. Sharp bends occur only where there is an abrupt change from one kind of material to another, such as at the boundary between the outer core and the mantle. Only P waves can pass through the Earth's liquid outer core.

both S and P waves, but seismometers on the opposite side of Earth record only P waves. The absence of S waves was first explained in 1906 by British geologist Richard Dixon Oldham, who noted that transverse vibrations such as S waves cannot travel far through liquids. Oldham therefore concluded that our planet has a molten core. Furthermore, there is a region in which neither S waves nor P waves from an earthquake can be detected (**Figure 9-8**). This "shadow zone" results from the specific way in which P waves are refracted at the boundary between the solid mantle and the molten core. By measuring the size of the shadow zone, geologists have concluded that the radius of the molten core is about 3500 km (2200 mi), about 55% of our planet's overall radius but about double the radius of the Moon (1738 km = 1080 mi).

As the quality and sensitivity of seismographs improved, geologists discovered faint traces of P waves in an earthquake's shadow zone. In 1936, the Danish seismologist Inge Lehmann explained that some of the P waves passing through the Earth are deflected into the shadow zone by a small, solid **inner core** at the center of our planet. The radius of this inner core is about 1300 km (800 mi).

 The interior of our planet therefore has a curious structure—a liquid **outer core** sandwiched between a solid inner core and a solid mantle. **Table 9-3** summarizes this structure. To understand this arrangement, we must look at how temperature and pressure inside the Earth affect the melting point of rock.

Both temperature and pressure increase with increasing depth below the Earth's surface. The temperature of the Earth's interior

rises steadily from about 14°C on the surface to nearly 5000°C at our planet's center (**Figure 9-9**).

The Earth's outermost layer, the crust, is only about 5 to 35 km thick. It is composed of rocks for which the **melting point,** or temperature at which the rock changes from solid to liquid, is far higher than the temperatures actually found in the crust. Thus, the crust is solid.

The Earth's mantle, which extends to a depth of about 2900 km (1800 mi), is largely composed of substances rich in iron and magnesium. On the Earth's surface, specimens of these substances have melting points slightly over 1000°C. However, the melting point of a substance depends on the pressure to which it is subjected—the higher the pressure, the higher the melting point. As Figure 9-9 shows, the actual temperatures throughout the mantle are less than the melting point of all the substances of which the mantle is made. Hence, the mantle is primarily solid. The upper levels of the mantle are, however, able to flow slowly and are therefore referred to as being **plastic.**

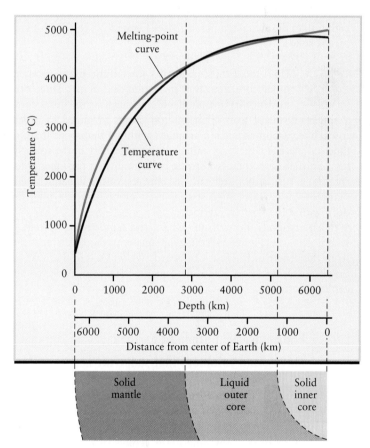

figure 9-9

Temperature and Melting Point of Rock Inside the Earth The temperature (red curve) rises steadily from the Earth's surface to its center. The melting point of the Earth's material (blue curve) is also shown on this graph. Where the temperature is below the melting point, as in the mantle and inner core, the material is solid; where the temperature is above the melting point, as in the outer core, the material is liquid. (Adapted from F. Press, R. Siever, T. Grotzinger, and T. H. Jordan, *Understanding Earth* 4th Ed., W. H. Freeman, 2004)

It may seem strange to think of a plastic material as one that is able to flow. Normally we think of plastic objects as being hard and solid, like the plastic out of which compact discs and video cassettes are made. But to make these objects, the plastic material is heated so that it can flow into a mold, then cooled so that it solidifies. Strictly, the material is only "plastic" when it is able to flow. Maple syrup and tree sap are two other substances that flow at warm temperatures but become solid when cooled.

At the boundary between the mantle and the outer core, there is an abrupt change in chemical composition, from iron-rich materials to almost-pure iron with a small admixture of nickel. Because this nearly pure iron has a lower melting point than the iron-rich materials, the melting-point curve in Figure 9-9 dips below the temperature curve as it crosses from the mantle to the outer core and remains below it down to a depth of about 5100 km (3200 mi). Hence, from depths of about 2900 to 5100 km, the core is liquid.

At depths greater than about 5100 km, the pressure is more than 10^{11} newtons per square meter. This is about 10^6 times ordinary atmospheric pressure, or about 10^4 tons per square inch. Because the melting point of the iron-nickel mixture under this pressure is higher than the actual temperature (see Figure 9-9), the Earth's inner core is solid.

 Until the 1980s, geologists knew little more about the inner core than that it is solid and dense. Since then, evidence has accumulated that suggests the inner core is a single **crystal** of iron—that is, iron atoms arranged in orderly rows, like carbon atoms are arranged within a diamond. If so, the inner core is the largest crystalline object anywhere in the solar system. Furthermore, there are strong indications that the inner core is rotating at a slightly faster rate than the rest of the Earth! These remarkable discoveries suggest that even more surprises may lurk deep within the Earth's interior.

Heat naturally flows from where the temperature is high to where it is low. Figure 9-9 thus explains why there is a heat flow from the center of the Earth outward. We will see in the next section how this heat flow acts as the "engine" that powers our planet's geologic activity.

9-3 | Plate movement produces earthquakes, mountain ranges, and volcanoes that shape the Earth's surface

 SCIENCE IN PROCESS *Evidence from the bottom of the ocean confirmed a theory of moving continents*

One of the most important geological discoveries of the twentieth century was the realization that the Earth's crust is constantly changing. We have learned that the crust is divided into huge **plates** whose motions produce earthquakes, volcanoes, mountain ranges, and oceanic trenches. This picture of the Earth's crust has come to be the central unifying theory of geology, much as the theory of evolution has become the centerpiece of modern biology.

Anyone who carefully examines a map of the Earth might come up with the idea of moving continents. South America, for example, would fit snugly against Africa were it not for the Atlantic Ocean. As **Figure 9-10** shows, the fit between land masses on either side of the Atlantic Ocean is quite remarkable. This observation inspired the German meteorologist Alfred Wegener to advocate "continental drift"—the idea that the continents on either side of the Atlantic Ocean have simply drifted apart. After much research, in 1915 Wegener published the theory that there had originally been a single gigantic supercontinent, which he called Pangaea (meaning "all lands"), that began to break up and drift apart roughly 200 million years ago. (No humans were present to witness this; the earliest humans did not appear on Earth until about 1.9 million years ago. At the time that Pangaea broke up, during what geologists refer to as the early Jurassic period, dinosaurs were the dominant land animals.)

Other geologists refined this theory, arguing that Pangaea must have first split into two smaller supercontinents, which they called Laurasia and Gondwana, separated by what they called the Tethys

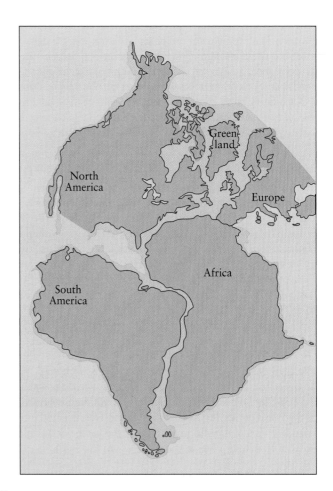

Figure 9-10

Fitting the Continents Together Africa, Europe, Greenland, North America, and South America fit together remarkably well. The fit is especially convincing if the edges of the continental shelves (shown in yellow) are used, rather than today's shorelines. This strongly suggests that these continents were in fact joined together at some point in the past. (Adapted from P. M. Hurley)

Ocean. Gondwana later split into Africa and South America, with Laurasia dividing to become North America and Eurasia. According to this theory, the Mediterranean Sea is a surviving remnant of the ancient Tethys Ocean (**Figure 9-11**).

(a) 237 million years ago: the supercontinent Pangaea

(b) 152 million years ago: the breakup of Pangaea

(c) The continents today

figure 9-11

The Breakup of the Supercontinent Pangaea (a) The shapes of the continents led Alfred Wegener to conclude that over 200 million (2×10^8) years ago, the continents were merged into a single supercontinent, which he called Pangaea. (b) Pangaea first split into two smaller land masses, Laurasia and Gondwana. (c) Over millions of years, the continents moved to their present-day locations. Among the evidence confirming this picture are nearly identical rock formations 200 million years in age that today are thousands of kilometers apart but would have been side by side on Pangaea. (Adapted from F. Press, R. Siever, T. Grotzinger, and T. H. Jordan, *Understanding Earth* 4th Ed., W. H. Freeman, 2004)

Initially, most geologists ridiculed Wegener's ideas. Although it was generally accepted that the continents do "float" on the denser, somewhat plastic mantle beneath them, few geologists could accept the idea that entire continents could move around the Earth at speeds as great as several centimeters per year. The "continental drifters" could not explain what forces could be shoving the massive continents around.

 Beginning in the mid-1950s, however, geologists found evidence that material is being forced upward to the crust from deep within the Earth. Bruce C. Heezen of Columbia University and his colleagues began discovering long mountain ranges on the ocean floors, such as the Mid-Atlantic Ridge, which stretches all the way from Iceland to Antarctica (**Figure 9-12**). During the 1960s, Harry Hess of Princeton University and Robert Dietz of the University of California again carefully examined the floor of the Atlantic Ocean. They concluded that rock from the Earth's mantle is being melted and then forced upward along

figure 9-12

The Mid-Atlantic Ridge This artist's rendition shows the Mid-Atlantic Ridge, an immense mountain ridge that rises up from the floor of the North Atlantic Ocean. It is caused by lava seeping up from the Earth's interior along a rift that extends from Iceland to Antarctica. (Courtesy of M. Tharp and B. C. Heezen) the Mid-Atlantic Ridge, which is in essence a long chain of underwater volcanoes.

The upwelling of new material from the mantle to the crust forces the existing crusts apart, causing **seafloor spreading.** For example, the floor of the Atlantic Ocean to the east of the Mid-Atlantic Ridge is moving eastward and the floor to the west is moving westward. By explaining what fills in the gap between continents as they move apart, seafloor spreading helps to fill out the theories of continental drift. Because of the seafloor spreading from the Mid-Atlantic Ridge, South America and Africa are moving apart at a speed of roughly 3 cm per year. Working backward, these two continents would have been next to each other some 200 million years ago—just as Wegener suggested (see Figure 9-10).

In the early 1960s, geologists began to find additional evidence supporting the existence of large, moving plates. Thus was born the modern theory of crustal motion, which came to be known as **plate tectonics** (from the Greek *tekton,* meaning "builder").

Geologists today realize that earthquakes tend to occur at the boundaries of the Earth's crustal plates, where the plates are colliding, separating, or rubbing against each other. The boundaries of the plates therefore stand out clearly when the epicenters of earthquakes are plotted on a map as in **Figure 9-13.**

 The vast majority of volcanoes also occur at plate boundaries. **Figure 9-14** shows a volcanic eruption at the boundary between the Pacific and Eurasian plates.

 What makes the plates move? The answer is twofold. First, heat flows outward from the Earth's hot core to its cool crust, and second, this heat flows through the mantle by convection, the same process that takes place in the Earth's atmosphere or in a pot of boiling water.

Figure 9-4 shows that convection occurs when a fluid is heated from below. The heat that drives convection in the Earth's outer layers actually comes from very far below, at the boundary between the outer and inner cores. As material deep in the liquid core cools and solidifies to join the solid portion of the core, it releases the energy needed to heat the overlying mantle and cause convection. (In Section 9-1 we described a similar process in Earth's atmosphere: When water vapor in the air cools and forms liquid drops, it releases the energy that powers thunderstorms.)

The upper levels of the mantle that are hot and soft enough to permit an oozing, plastic flow are called the **asthenosphere** (from the Greek *asthenia,* meaning "weakness"). Atop the asthenosphere is a rigid layer, called the **lithosphere** (from the Greek word for "rock.") The lithosphere is divided into plates that ride along the convection currents of the asthenosphere. The crust is simply the uppermost layer of the lithosphere, with a somewhat different chemical composition than the lithosphere's lower regions.

Figure 9-15 shows how convection causes plate movement. Molten subsurface rock seeps upward along **oceanic rifts,** where plates are separating. The Mid-Atlantic Ridge, shown in Figure 9-12, is an oceanic rift. Where plates collide, cool crustal material

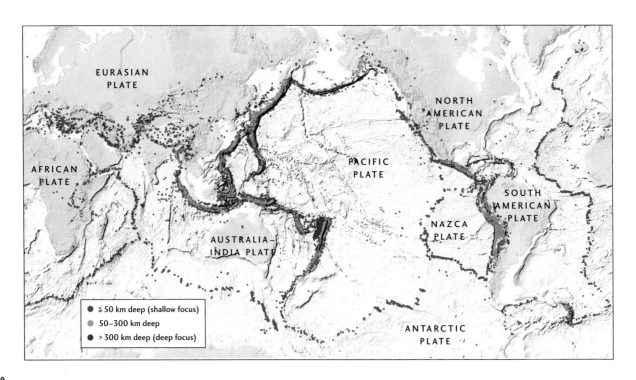

figure 9-13

The Earth's Major Plates The boundaries of the Earth's major plates are the scenes of violent seismic and geologic activity. Most earthquakes occur where plates separate, collide, or rub together. Plate boundaries are therefore easily identified simply by plotting earthquake epicenters (shown here as dots) on a map. The colors of the dots indicate the depths at which the earthquakes originate. (Data from Harvard CMT catalog; plot by M. Boettcher and T. Jordan.)

Figure 9-14   R I $\boxed{\text{V}}$ U X G

An Erupting Volcano On September 30, 1994, a volcano erupted at the boundary of the Pacific and Eurasian plates on Russia's Kamchatka Peninsula. This space shuttle photograph shows an immense cloud of ash belching forth from this volcano. The cloud reached an altitude of 18 km (60,000 ft) above sea level and was carried by the winds for at least 1000 km from the volcano. (*STS-68*, NASA)

from one of the plates sinks back down into the mantle along a **subduction zone.** One such subduction zone is found along the west coast of South America, where the oceanic Nazca plate is being subducted into the mantle under the continental South American plate at a relatively speedy 10 centimeters per year. As the material from the subducted plate sinks, it pulls the rest of its plate along with it, thus helping to keep the plates in motion. New material is added to the crust from the mantle at the oceanic rifts

and is "recycled" back into the mantle at the subduction zones. In this way the total amount of crust remains essentially the same.

  The boundaries between plates are the sites of some of the most impressive geological activity on our planet. Great mountain ranges, such as the Sierras and Cascades along the western coast of North America and the Andes along South America's west coast, are thrust up by

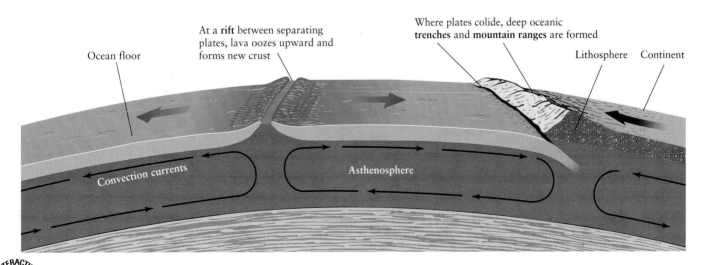

At a **rift** between separating plates, lava oozes upward and forms new crust

Where plates colide, deep oceanic **trenches** and **mountain ranges** are formed

Ocean floor

Lithosphere   Continent

Convection currents

Asthenosphere

Figure 9-15

The Mechanism of Plate Tectonics Convection currents in the asthenosphere, the soft upper layer of the mantle, are responsible for pushing around rigid, low-density crustal plates.

New crust forms in oceanic rifts, where lava oozes upward between separating plates. Mountain ranges and deep oceanic trenches are formed where plates collide.

ongoing collisions between continental plates and the plates of the ocean floor. Subduction zones, where old crust is drawn back down into the mantle, are typically the locations of deep oceanic trenches, such as the Peru-Chile Trench off the west coast of South America. Figure 9-16 and Figure 9-17 show two well-known geographic features that resulted from tectonic activity at plate boundaries.

Plate tectonics also helps to explain geology on the scale of individual rocks and minerals. These are composed of chemical elements (see the periodic table in Box 5-5), but individual elements are rarely found in a pure state. The exceptions include diamonds, which are crystals of carbon, as well as nuggets of gold, silver, and copper. Somewhat more common are solids composed of a particular chemical combination of elements called a *compound*. An example is feldspar, a crystalline material made up of potassium, aluminum, silicon, and oxygen atoms. Diamond and feldspar are both **minerals**—naturally occurring solids composed of a single element or compound. Most common of all are **rocks,** which are solid parts of the Earth's crust that are composed of one or more minerals. For example, granite contains both feldspar and quartz (a mineral containing silicon and oxygen).

Geologic processes create three major categories of rocks (Figure 9-18). **Igneous rocks** result when minerals cool from a molten state. (Molten rock is called **magma** when it is buried below the surface and **lava** when it flows out upon the surface, as in a volcanic eruption.) The ocean floor is made predominantly of a type of igneous rock called basalt (Figure 9-18a). This is just what we would expect if the ocean floor is produced by material welling up from the mantle, as predicted by the theory of seafloor spreading.

Sedimentary rocks are produced by the action of wind, water, or ice. For example, winds can pile up layer upon layer of sand grains. Other minerals present amid the sand can gradually cement the grains together to produce sandstone (Figure 9-18b). Minerals that precipitate out of the oceans can cover the ocean floor with layers of a sedimentary rock called limestone. The motion of tectonic plates can move such rocks to places far from where they were formed. This explains why sedimentary rock from the floor of an ancient ocean can now be found in central Alaska and as hogback ridges in Colorado (see Figure 9-3).

Sometimes igneous or sedimentary rocks become buried far beneath the surface, where they are subjected to enormous pressure and high temperatures. These severe conditions change the structure of the rocks, producing **metamorphic rock** (Figure 9-18c). The presence of metamorphic rocks at the Earth's surface tells us that tectonic activity sometimes lifts up material from deep within the crust. This can happen when two plates collide, as shown in Figure 9-17. The boundaries of the colliding plates are thrust upward to form a new chain of mountains, and buried metamorphic rocks are brought to the surface.

Plate tectonic theory offers insight into geology on the largest of scales, that of an entire supercontinent. In recent years, geologists have uncovered evidence that points to a whole succession of supercontinents that once broke apart and then reassembled. Pangaea is only the most recent supercontinent in this cycle, which repeats about every 500 million (5×10^8) years. As a result, intense episodes of mountain building have occurred at roughly 500-million-year intervals.

Apparently, a supercontinent sows the seeds of its own destruction because it blocks the flow of heat from the Earth's interior. As

Figure 9-16 R I **V** U X G

The Separation of Two Plates The plates that carry Africa and Arabia are moving apart, leaving a great rift that has been flooded to form the Red Sea. This view from orbit shows the northern Red Sea, which splits into the Gulf of Suez and the Gulf of Aqaba. (*Gemini 12,* NASA)

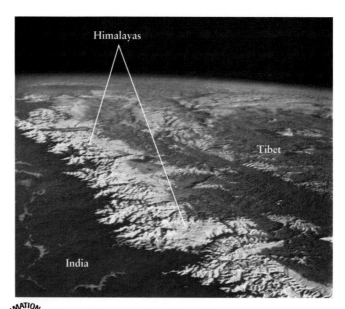

Figure 9-17 R I **V** U X G

The Collision of Two Plates The plates that carry India and China are colliding. Instead of one plate being subducted beneath the other, both plates are pushed upward, forming the Himalayas. Mount Everest is one of the snow-covered peaks near the center of this photograph taken from orbit. (*Apollo 7,* NASA)

(a) Igneous rock (basalt)

(b) Sedimentary rock (sandstone)

(c) Metamorphic rocks (marble, schist)

Figure 9-18 R I 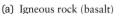 U X G

Igneous, Sedimentary, and Metamorphic Rocks (a) Igneous rocks such as basalt are created when molten materials solidify. This example contains iron-rich minerals that give it its dark color. (b) Sedimentary rocks such as sandstone are typically formed when loose particles of soil or sand are fused into rock by the presence of other minerals, which act as a cement.

(c) Metamorphic rocks are produced when igneous or sedimentary rocks are subjected to high temperatures and pressures deep within the Earth's crust. Marble (left) is formed from sedimentary limestone; schist (right) is formed from igneous rock. (W. J. Kaufmann III; specimens courtesy Mineral Museum, California Division of Mines)

soon as a supercontinent forms, temperatures beneath it rise, much as they do under a book lying on an electric blanket. As heat accumulates, the lithosphere domes upward and cracks. Molten rock from the overheated asthenosphere wells up to fill the resulting fractures, which continue to widen as pieces of the fragmenting supercontinent move apart.

It can take a very long time for the heat trapped under a supercontinent to escape. Although Pangaea broke apart some 200 million years ago (see Figure 9-11), the mantle under its former location is still hot and is still trying to rise upward. As a result, Africa—which lies close to the center of this mass of rising material—sits several tens of meters higher than the other continents.

The changes wrought by plate tectonics are very slow on the scale of a human lifetime, but they are very rapid in comparison with the age of the Earth. For example, the period over which Pangaea broke into Laurasia and Gondwana was only about 0.4% of the Earth's age of 4.56×10^9 years. (To put this in perspective, 0.4% of a human lifetime is about 4 months.) The lesson of plate tectonics is that the seemingly permanent face of the Earth is in fact dynamic and ever-changing.

9-4 | The Earth's magnetic field produces a magnetosphere that traps particles from the solar wind

We saw in Section 7-7 that Earth's magnetic field provides evidence that our planet's interior is partially molten. Our understanding is that the magnetic field is produced by the motion of iron-rich material in the liquid outer core. Because this material is a good conductor of electricity, its motions generate a magnetic field in a mechanism called a *dynamo*.

The same energy source that powers plate tectonics is also responsible for the Earth's dynamo. As we saw in Section 9-3, energy is released when material from our planet's liquid core cools and solidifies onto the solid core. This released energy stirs up the motions of the liquid core. While there is also fluid motion in the mantle, it does not produce an appreciable magnetic field: The silicate rocks that comprise the mantle are poor conductors of electricity, and their motions are a million times slower than those in the outer core.

Studies of ancient rocks reinforce the idea that our planet's magnetism is due to fluid material in motion. When iron-bearing lava cools and solidifies to form igneous rock, it becomes magnetized in the direction of the Earth's magnetic field. By analyzing samples of igneous rock of different ages from around the world, geologists have found that the Earth's magnetic field actually flips over and reverses direction on an irregular schedule ranging from tens of thousands to hundreds of thousands of years. As an example, lava that solidified 30,000 years ago is magnetized in the opposite direction to lava that has solidified recently. Therefore, 30,000 years ago a compass needle would have pointed south, not north! If the Earth were a permanent magnet, like the small magnets used to attach notes to refrigerators, it would be hard to imagine how its magnetic field could spontaneously reverse direction. But computer simulations show that fields produced by moving fluids in the Earth's outer core can indeed change direction from time to time.

The Earth's magnetic field interacts dramatically with charged particles from the Sun. This *solar wind* is a flow of mostly protons and electrons that streams constantly outward from the Sun's upper atmosphere. Near the Earth, the particles in the solar wind move at speeds of roughly 450 km/s, or about a million miles per hour. Because this is considerably faster than sound waves can travel in the very thin gas between the planets, the solar wind is said to be *supersonic*. (Because the gas between the planets is so

INTERACTIVE
EXERCISE 9.2

figure 9-19

The Earth's Magnetosphere The Earth's magnetic field carves out a cavity in the solar wind, shown here in cross section. A shock wave marks the boundary where the supersonic solar wind is abruptly slowed to subsonic speeds. Most of the particles of the solar wind are deflected around the Earth in a turbulent region (colored purple in this drawing). The Earth's magnetic field also traps some charged particles in two huge, doughnut-shaped rings called the Van Allen belts (in blue). This figure shows only a slice through the Van Allen belts.

thin, interplanetary sound waves carry too little energy to be heard by astronauts.)

If the Earth had no magnetic field, we would be continually bombarded by the solar wind. But our planet does have a magnetic field, and the forces that this field can exert on charged particles are strong enough to deflect them away from us. The region of space around a planet in which the motion of charged particles is dominated by the planet's magnetic field is called the planet's **magnetosphere.** Figure 9-19 is a scale drawing of the Earth's magnetosphere, which was discovered in the late 1950s by the first satellites placed in orbit.

When the supersonic particles in the solar wind first encounter the Earth's magnetic field, they abruptly slow to subsonic speeds. The boundary where this sudden decrease in velocity occurs is called a **shock wave.** Still closer to the Earth lies another boundary, called the **magnetopause,** where the outward magnetic pressure of the Earth's field is exactly counterbalanced by the impinging pressure of the solar wind. Most of the particles of the solar wind are deflected around the magnetopause, just as water is deflected to either side of the bow of a ship.

Some charged particles of the solar wind manage to leak through the magnetopause. When they do, they are trapped by the Earth's magnetic field in two huge, doughnut-shaped rings around Earth called the **Van Allen belts.** These belts were discovered in 1958 during the flight of the first successful U.S. Earth-orbiting satellite. They are named after the physicist James Van Allen, who insisted that the satellite carry a Geiger counter to detect charged particles. The inner Van Allen belt, which extends over altitudes of

about 2000 to 5000 km, contains mostly protons. The outer Van Allen belt, about 6000 km thick, is centered at an altitude of about 16,000 km above the Earth's surface and contains mostly electrons.

Sometimes the magnetosphere becomes overloaded with particles. The particles then leak through the magnetic fields at their weakest points and cascade down into the Earth's upper atmosphere, usually in a ring-shaped pattern (Figure 9-20a). As these high-speed charged particles collide with atoms in the upper atmosphere, they excite the atoms to high energy levels. The atoms then emit visible light as they drop down to their ground states, like the excited gas atoms in a neon light (see Section 5-8). The result is a beautiful, shimmering display called an **aurora** (*plural* **aurorae**). These are also called the **northern lights (aurora borealis)** or **southern lights (aurora australis),** depending on the hemisphere in which the phenomenon is observed. Figures 9-20b and 9-20c show the aurorae as seen from orbit and from the Earth's surface.

Occasionally, a violent event on the Sun's surface called a **coronal mass ejection** sends a burst of protons and electrons toward Earth. The resulting auroral display can be exceptionally bright and can often be seen over a wide range of latitudes. Such events also disturb radio transmissions and can damage communication satellites and transmission lines.

It is remarkable that the Earth's magnetosphere, including its vast belts of charged particles, was entirely unknown until a few decades ago. Such discoveries remind us of how little we truly understand and how much remains to be learned even about our own planet.

(a) R I V **U** X G

(b) R I **V** U X G

(c) R I **V** U X G

VIDEO 9.2

figure 9-20

The Aurora An increased flow of charged particles from the Sun can overload the Van Allen belts and cascade toward the Earth, producing aurorae. **(a)** This false-color ultraviolet image from the *Dynamics Explorer 1* spacecraft shows a glowing oval of auroral emission about 4500 km in diameter centered on the north magnetic pole. **(b)** This space shuttle photograph shows the aurora australis over Antarctica. **(c)** This view from Alaska shows aurorae at their typical altitudes of 100 to 400 km. (a: Courtesy of L. A. Frank and J. D. Craven, University of Iowa; b: JSC/NASA; c: Courtesy of S.-I. Akasofu, Geophysical Institute, University of Alaska)

9-5 | The Earth's atmosphere has changed substantially over our planet's history

Both plate tectonics and the Earth's magnetic field paint a picture of a planet with a dynamic, evolving surface and interior. Our atmosphere, too, has changed and evolved substantially over the Earth's 4.56-billion-year history. As we will see, this evolution explains why we have an atmosphere totally unlike that of any other world in the solar system.

When the Earth first formed by accretion of planetesimals (see Section 8-4), gases were probably trapped within the Earth's interior in the same proportions that they were present in the solar nebula. But since the early Earth was hot enough to be molten throughout its volume, most of these trapped gases were released. As we discussed in Section 7-4 and Box 7-2, the Earth's gravity was too weak to prevent hydrogen and helium—the most two common kinds of atoms in the universe, but also the least massive—from leaking away into space. The atmosphere that remained still contained substantial amounts of hydrogen, but in the form of relatively massive molecules of water vapor (H_2O) made by combining two atoms of hydrogen with one atom of oxygen, the third most common element in our part of the Milky Way Galaxy (see Figure 8-3). In fact, water vapor was probably the dominant constituent of the early atmosphere.

Water vapor is one of the greenhouse gases that we discussed in Section 9-1. It traps infrared radiation from the Earth's surface, and its presence in the early atmosphere helped to sustain the high temperature of the surface. But if water vapor had been the *only* constituent of the atmosphere, you would not be reading these words! The reason is that as the molten Earth inevitably cooled, water vapor in the atmosphere condensed into raindrops and fell to Earth to form the oceans. As water vapor was lost from the atmosphere, its contribution to the greenhouse effect weakened, which would have made the surface temperature drop even lower. A further cooling factor is that the young Sun was only about 70% as bright as it is today. The net result is that the entire Earth should have been frozen over! Ice is so reflective that the Earth would have remained frozen even as the Sun aged and became more luminous. Liquid water is absolutely essential to all living creatures, so life as we know it would probably never have evolved on such an icy Earth.

In fact, the first living organisms appeared on Earth within 400 million years after the planet first formed. Hence, the Earth could not have been frozen over for very long, if at all. What saved our planet from a perpetual deep freeze was the presence of carbon, the fourth most common of the elements. When combined with oxygen, carbon forms carbon dioxide (CO_2), a greenhouse gas that remains gaseous at low temperatures. Carbon dioxide would have been released into the atmosphere by volcanic activity, a process called **outgassing** (see Figure 9-14). It would also have been added to the atmosphere by carbon-rich meteors striking the planet's surface. Once CO_2 was in the atmosphere, its greenhouse effect raised the planet's temperature and melted the ice. Some of the water evaporated into the atmosphere, further enhancing the Earth's greenhouse effect.

Fewer than one in 2500 of the molecules in today's atmosphere is carbon dioxide. But CO_2 must have been thousands of times more abundant in the early Earth's atmosphere in order to melt the global ice sheath. Unlike hydrogen or helium, this excess CO_2 was not lost into space; instead, it has been trapped in rocks. Carbon dioxide dissolves in rainwater and falls into the oceans, where it combines with other substances to form a class of minerals called *carbonates*. (Limestone and marble are examples of carbonate-bearing rock.) These form sediments on the ocean floor, which are eventually recycled into the crust by subduction. As an example, marble (Figure 9-18c) is a metamorphic rock formed deep within the crust from limestone, a carbonate-rich sedimentary rock.

This process removed most of the CO_2 from the atmosphere within the first billion (10^9) years after the formation of the Earth. Although this weakened the greenhouse effect, temperatures remained warm as the Sun's brightness increased. The small amounts of atmospheric CO_2 that remain today are the result of a balance between volcanic activity (which releases CO_2 into the atmosphere) and the formation of carbonates (which removes CO_2). (As we will see in Section 9-7, human activity is upsetting this balance.)

The appearance of life on Earth set into motion a radical transformation of the atmosphere. Early single-cell organisms converted energy from sunlight into chemical energy using **photosynthesis,** a chemical process that consumes CO_2 and water and releases oxygen (O_2). Oxygen molecules are very reactive, so originally most of the O_2 produced by photosynthesis combined with other substances to form minerals called oxides. (Evidence for this can be found in rock formations of various ages. The oldest rocks have very low oxide content, while oxides are prevalent in rocks that formed after the appearance of organisms that used photosynthesis.) But as life proliferated, the amount of photosynthesis increased dramatically. Eventually so much oxygen was being produced that it could not all be absorbed to form oxides, and O_2 began to accumulate in the atmosphere. **Figure 9-21** shows how the amount of O_2 in the atmosphere has increased over the history of the Earth.

About 2 billion (2×10^9) years ago, a new type of life evolved to take advantage of the newly abundant oxygen. These new organisms produce energy by consuming oxygen and releasing carbon dioxide—a process called **respiration** that is used by all modern animals, including humans. Such organisms thrived because photosynthetic plants continued to add even more oxygen to the atmosphere.

Several hundred million years ago, the number of oxygen molecules in the atmosphere stabilized at 21% of the total, the same as the present-day value. This value represents a balance between the release of oxygen from plants by photosynthesis and the intake of oxygen by respiration. Thus, the abundance of oxygen is regulated almost exclusively by the presence of life—a situation that has no parallel anywhere else in the solar system.

The most numerous molecules in our atmosphere are nitrogen (N_2), which make up 78% of the total. These, too, are a consequence of the presence of life on Earth. Certain bacteria extract

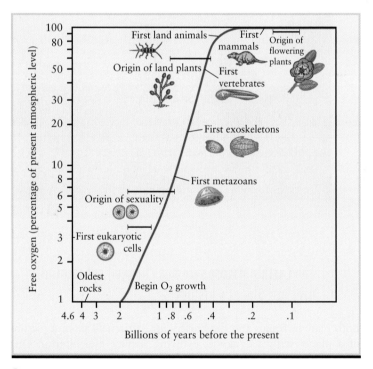

figure 9-21

The Increase in Atmospheric Oxygen This graph shows how the amount of oxygen in the atmosphere (expressed as a percentage of its present-day value) has evolved with time. Note that the atmosphere contained essentially no oxygen until about 2 billion years ago. (Adapted from Preston Cloud, "The Biosphere," *Scientific American,* September 1983, p. 176)

table 9-4	**Chemical Compositions of Three Planetary Atmospheres**			
	Venus	**Earth**	**Mars**	
Nitrogen (N_2)	3.5%	78.08%	2.7%	
Oxygen (O_2)	almost zero	20.95%	almost zero	
Carbon dioxide (CO_2)	96.5%	0.035%	95.3%	
Water vapor (H_2O)	0.003%	about 1%	0.03%	
Other gases	almost zero	almost zero	2%	

oxygen from minerals called nitrates, and in the process liberate nitrogen into the atmosphere. The amount of atmospheric nitrogen is kept in check by lightning. The energy in a lightning flash causes atmospheric nitrogen and oxygen to combine into nitrogen oxides, which dissolve in rainwater, fall into the oceans, and form nitrates.

Table 9-4 shows the dramatic differences between the Earth's atmosphere and those of Venus and Mars. (Mercury, the other terrestrial planet, is too small and has too little gravity to hold an appreciable atmosphere.) The greater intensity of sunlight on Venus caused higher temperatures, which boiled any liquid water and made it impossible for CO_2 to be taken out of the atmosphere and put back into rocks. The atmosphere thus became far denser than our own and rich in greenhouse gases. The result was a very strong greenhouse effect that raised temperatures on Venus to their present value of about 460°C (733 K = 855°F).

Just the opposite happened on Mars, where sunlight is less than half as intense as on Earth. The lower temperatures drove CO_2 from the atmosphere into Martian rocks, and froze all of the water beneath the planet's surface. The atmosphere that remains on Mars has a similar composition to that of Venus but is less than $^1/_{10,000}$ as dense. On neither Venus nor Mars was life able to blossom and transform the atmosphere as it did here on Earth.

We have thus uncovered a general rule about the terrestrial planets:

The closer a terrestrial planet is to the Sun, the more important the greenhouse effect in that planet's atmosphere. The stronger the greenhouse effect, the higher the planet's surface temperature.

This rule suggests that if the Earth had formed a bit closer to or farther from the Sun, temperatures on our planet might have been too low or too high for life ever to evolve. Thus, our very existence is a result of the Earth's special position in the solar system.

9-6 | Circulation in our atmosphere results from convection and the Earth's rotation

While life has shaped our atmosphere's chemical composition, the structure of the atmosphere is controlled by the influence of sunlight. Scientists describe the structure of any atmosphere in terms of two properties, temperature and atmospheric pressure, which vary with altitude.

Atmospheric pressure at any height in the atmosphere is caused by the weight of all the air above that height. The average atmospheric pressure at sea level is defined to be one **atmosphere** (1 atm), equal to 1.01×10^5 N/m^2 or 14.7 pounds per square inch. As you go up in the atmosphere, there is less air above you to weigh down on you. Hence, atmospheric pressure decreases smoothly with increasing altitude.

Unlike atmospheric pressure, temperature varies with altitude in a complex way. Figure 9-22 shows that temperature decreases with increasing altitude in some layers of the atmosphere, but in other layers actually *increases* with increasing altitude. These differences result from the individual ways in which each layer is heated.

Figure 9-22

Temperature Profile of the Earth's Atmosphere This graph shows how the temperature in the Earth's atmosphere varies with altitude. In the troposphere and mesosphere, temperature decreases with increasing altitude; in the stratosphere and thermosphere, temperature actually increases with increasing altitude.

figure 9-23

Circulation Patterns in the Earth's Atmosphere
The dominant circulation in our atmosphere consists of six convection cells, three in the northern hemisphere and three in the southern hemisphere. In the northern temperate region (including the continental United States), the prevailing winds at the surface are from the southwest toward the northeast. Farther south, within the northern tropical region (for example, Hawaii), the prevailing surface winds are from northeast to southwest.

The lowest layer, called the **troposphere**, extends from the surface to an average altitude of 12 km (roughly 7.5 miles, or 39,000 ft). It is heated only indirectly by the Sun. Sunlight warms the Earth's surface, which heats the lower part of the troposphere. By contrast, the upper part of the troposphere remains at cooler temperatures. This vertical temperature variation causes convection currents that move up and down through the troposphere (see Section 8-1). All of the Earth's weather is a consequence of this convection.

Convection on a grand scale is caused by the temperature difference between the Earth's equator and its poles. If the Earth did not rotate, heated air near the equator would rise upward and flow at high altitude toward the poles. There it would cool and sink to lower altitudes, at which it would flow back to the equator. However, the Earth's rotation breaks up this simple convection pattern into a series of **convection cells.** In these cells, air flows east and west as well as vertically and in a north-south direction. The structure of these cells explains why the prevailing winds blow in different directions at different latitudes (**Figure 9-23**).

Almost all the oxygen in the troposphere is in the form of O_2, a molecule made of two oxygen atoms. But in the **stratosphere**, which extends from about 12 to 50 km (about 7.5 to 31 mi) above the surface, an appreciable amount of oxygen is in the form of **ozone,** a molecule made of *three* oxygen atoms (O_3). Unlike O_2, ozone is very efficient at absorbing ultraviolet radiation from the Sun, which means that the stratosphere can directly absorb solar energy. The result is that the temperature actually increases as you move upward in the stratosphere. Convection requires that the temperature must

decrease, not increase, with increasing altitude, so there are essentially no convection currents in the stratosphere.

Above the stratosphere lies the **mesosphere.** Very little ozone is found here, so solar ultraviolet radiation is not absorbed within the mesosphere, and atmospheric temperature again declines with increasing altitude.

 The temperature of the mesosphere reaches a minimum of about −75°C (= −103°F = 198 K) at an altitude of about 80 km (50 mi). This minimum marks the bottom of the atmosphere's thinnest and uppermost layer, the **thermosphere,** in which temperature once again rises with increasing altitude. This is not due to the presence of ozone, because in this very low-density region oxygen and nitrogen are found as individual atoms rather than in molecules. Instead, the thermosphere is heated because these isolated atoms absorb very-short-wavelength solar ultraviolet radiation (which oxygen and nitrogen molecules cannot absorb).

 At altitudes near 300 km the temperature of the thermosphere is about 1000°C (1800°F). This is near the altitude at which the space shuttle and satellites orbit the Earth. Nonetheless, a satellite in orbit does *not* risk being burned up as it moves through the thermosphere. The reason is that the thermosphere is far less dense than the atmosphere at sea level. The high temperature simply means that an average atom in the thermosphere is moving very fast (see Box 7-2). But because the thermosphere is so thin (only about 10^{-11} as dense as the air at sea level), these fast-moving atoms are few and far between. Hence, the thermosphere contains very little energy. Nearly all the heat that an orbiting satellite receives is from sunlight, not the thermosphere.

9-7 | A burgeoning human population is profoundly altering the Earth's biosphere

One of the extraordinary characteristics of the Earth is that it is covered with life. Living organisms thrive in a wide range of environments, from ocean floors to mountaintops and from frigid polar caps to blistering deserts.

Living organisms subsist in a relatively thin layer enveloping the Earth called the **biosphere,** which includes the oceans, the lowest few kilometers of the troposphere, and the crust to a depth of almost 3 kilometers. **Figure 9-24** is a portrait of the Earth's biosphere based on NASA satellite data. The biosphere, which has taken billions of years to evolve to its present state, is a delicate, highly complex system in which plants and animals depend on each other for their mutual survival.

 The state of the biosphere depends crucially on the temperature of the oceans and atmosphere. Even small temperature changes can have dramatic consequences. An example that recurs every three to seven years is the El Niño phenomenon, in which temperatures at the surface of the equatorial Pacific Ocean rise by 2 to 3°C. (Figure 1-13 shows ocean temperatures during the 1997–1998 El Niño.) Ordinarily, water from the cold depths of the ocean is able to well upward, bringing with it nutrients that are used by microscopic marine organisms called phytoplankton that live near the surface (see Figure 9-24). But during an El Niño, the warm surface water suppresses this upwelling, and the phytoplankton starve. This wreaks havoc on organisms such as mollusks that feed on phytoplankton, on the fish that feed on the mollusks, and on the birds and mammals who eat the fish. During the 1982–1983 El Niño, a quarter of the adult sea lions off the Peruvian coast starved, along with all of their pups.

Many different factors can change the surface temperature of our planet. One is that the amount of energy radiated by the Sun can vary up or down by a few tenths of a percent. (Reduced solar brightness may explain the period from 1450 to 1850 when European winters were substantially colder than they are today.)

Other factors are the gravitational influences of the Moon and the other planets. Thanks to these influences, the eccentricity of the Earth's orbit varies with a period of 90,000 to 100,000 years, the tilt of its rotation axis varies between 22.1° and 24.5° with a 40,000-year period, and the orientation of its rotation axis changes due to precession (see Section 2-6) with a 26,000-year period. These variations can affect climate by altering the amount of solar energy that heats the Earth during different parts of the year. They help explain why the Earth periodically undergoes an extended period of low temperatures called an *ice age,* the last of which ended about 11,000 years ago.

As we saw in Sections 9-1 and 9-5, one of the most important factors affecting global temperatures is the abundance of greenhouse gases such as CO_2. Geologic processes can alter this abundance, either by removing CO_2 from the atmosphere (as happens when fresh rock is uplifted and exposed to the air, where it can absorb atmospheric CO_2 to form carbonates) or by supplying new CO_2 (as in volcanic eruptions). But since the 1970s it has become clear that human activity can also affect climate by adding greenhouse gases to the atmosphere.

Our species is having an increasing effect on the biosphere because our population is skyrocketing. **Figure 9-25** shows the sharp rise in the human population that began in the late 1700s with the Industrial Revolution and the spread of modern ideas about hygiene. This rise accelerated in the twentieth century thanks to medical and technological advances ranging from antibiotics to high-yield grains. In 1960 there were 3 billion people on Earth, and in 1975, 4 billion. The 6-billion mark was passed in 1999, and, according to the United Nations Population Fund, there will be 8 billion people on Earth by the year 2025 and 9.4 billion by 2050.

Every human being has basic requirements: food, clothing, and housing. We all need fuel for cooking and heating. To meet these demands, we cut down forests, cultivate grasslands, and build sprawling cities. A striking example of this activity is occurring in the Amazon rain forest of Brazil. Tropical rain forests are vital to our planet's ecology because they absorb significant amounts of

Figure 9-24
The Earth's Biosphere This image, based on data from the *Nimbus 7* spacecraft, shows the distribution of plant life over the Earth's surface. Land colors designate vegetation: dark green for the rain forests, light green and gold for savannas and farmland, and yellow for the deserts. Ocean colors show where freefloating microscopic plants called phytoplankton are found. Phytoplankton are most abundant in the red and orange areas and least abundant in the dark blue areas. (NASA/GSFC)

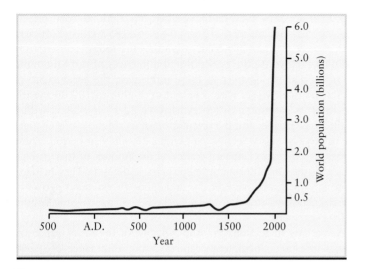

figure 9-25

The Human Population Data and estimates from the U.S. Bureau of the Census, the Population Reference Bureau, and the United Nations Population Fund were combined to produce this graph showing the human population from 500 B.C. to 2000 A.D. Note how the population began to rise in the eighteenth century and the astonishing population increase in the twentieth century.

figure 9-26 R I V U X G

The Deforestation of Amazonia The Amazon, the world's largest rain forest, is being destroyed at a rate of 20,000 square kilometers per year in order to provide land for grazing and farming and as a source for lumber. About 80 percent of the logging is being carried out illegally. (Martin Wendler/Okapia/Photo Researchers)

CO_2 and release O_2 through photosynthesis. Although rain forests occupy only 7% of the world's land areas, they are home to at least 50% of all plant and animal species on Earth. Nevertheless, to make way for farms and grazing land, people simply cut down the trees and set them on fire—a process called slash-and-burn. Such deforestation, along with extensive lumbering operations, is occurring in Malaysia, Indonesia, and Papua New Guinea. The rain forests that once thrived in Central America, India, and the western coast of Africa are almost gone (Figure 9-26).

All this activity, along with the use of petroleum products in automobiles and of coal in power plants, is adding carbon dioxide to the Earth's atmosphere faster than plants and geological processes can extract it. Each year the amount of atmospheric CO_2 increases by about 6 billion tons! Figure 9-27 shows how the carbon dioxide content of the atmosphere has increased since 1958. What is more, industrial and agricultural activities release other greenhouse gases, principally methane (CH_4) and nitrous oxide (N_2O), which are far more effective at increasing the greenhouse effect than CO_2.

Humans did not create the greenhouse effect, but we are responsible for strengthening it by adding greenhouse gases to the atmosphere. Figure 9-28 shows how the global air temperature has changed since the middle of the nineteenth century. The 10 warmest years of the twentieth century all occurred between 1985 and 2000. The UN Intergovernmental Panel on Climate Change predicts that temperatures will continue to rise by an additional 1.4 to 5.8°C during the twenty-first century, a phenomenon called **global warming**. They further predict that rainfall will increase dramatically in some parts of the Earth and decrease in others, ice in the Earth's polar regions will melt, the sea level will rise, and low-lying terrain will flood. Some of these effects are already visible (Figure 9-28*b*).

Human activity is also having potentially disastrous effects on the upper atmosphere. Certain chemicals released into the air—in particular chlorofluorocarbons (CFCs), which have been used in refrigeration and electronics, and methyl bromide, which is used in

figure 9-27

The Carbon Dioxide Content of the Atmosphere This graph shows measurements of atmospheric carbon dioxide in parts per million (ppm). The sawtooth pattern results from plants absorbing more carbon dioxide during the spring and summer. The CO_2 concentration in the atmosphere has increased by 18% since continuous observations started in 1958. By comparison, measurements of Antarctic ice that froze around 1850 shows that the CO_2 concentration then was 280 to 290 ppm. (NOAA/Scripps Institution of Oceanography)

(a) Changes in the Earth's average temperature

(b) Breakup of the Larsen B ice shelf, Antarctica, 2002

figure 9-28   R I **V** U X G
Changes in Global Temperature and the Consequences
(a) Land and sea surface temperatures were combined to produce this graph, which shows a warming trend over the past 140 years. The vertical scale is measured relative to the 1961–1990 average temperature. (b) One sign of this warming trend is the breakup of a large section of the Antarctic ice shelf. This section is likely to have existed for the past 12,000 years.
(a: Hadley Centre for Climate Prediction and Research, U.K. Meteorological Office; b: NASA/GSFC/LaRC/JPL, MIST Team)

fumigation—are destroying the ozone in the stratosphere. We saw in Section 9-6 that ozone molecules absorb ultraviolet radiation from the Sun. Without this high-altitude **ozone layer,** solar ultraviolet radiation would beat down on the Earth's surface with greatly increased intensity. Such radiation breaks apart most of the delicate molecules that form living tissue. Hence, a complete loss of the ozone layer would lead to a catastrophic ecological disaster.

In 1985 scientists discovered an **ozone hole,** a region with an abnormally low concentration of ozone, over Antarctica. Since then this hole has generally expanded from one year to the next (Figure 9-29). Smaller but still serious effects have been observed in the stratosphere above other parts of the Earth. As a result, there has been a worldwide increase in the number of deaths due to skin cancer caused by solar ultraviolet radiation. By international agreement, CFCs are being replaced by compounds that do not deplete stratospheric ozone, and sunlight naturally produces more ozone in the stratosphere. But even in the best of circumstances, the damage to the ozone layer is not expected to heal for many decades.

As we continue to populate our planet, it is clear that we face many challenges. Age-old modes of behavior that helped ensure the survival of our ancestors may not be the wisest courses of action in the future. Contemplating our behavior and our relationship to the Earth brings to mind a Chinese proverb: If we don't change the direction in which we are going, we are likely to end up where we are headed.

October 1979

September 2003

figure 9-29
The Antarctic Ozone Hole These two images compare the distribution of stratospheric ozone over Antarctica in October 1979 and September 2003. False colors in the order of the rainbow show the ozone concentration, with red and orange for high concentration and purple for low. The data show that there has been a net decrease of 50% in ozone over the Antarctic. The amount of ozone at midlatitudes, where most of the human population lives, decreased by 10 to 20% over the same period. (GSFC/NASA)

Key Words

Key Ideas

The Earth's Energy Sources: All activity in the Earth's atmosphere, oceans, and surface is powered by three sources of energy.

• Solar energy is the energy source for the atmosphere. In the greenhouse effect, some of this energy is trapped by infrared-absorbing gases in the atmosphere, raising the Earth's surface temperature.

• Tidal forces from the Moon and Sun help to power the motion of the oceans.

• The internal heat of the Earth is the energy source for geologic activity.

The Earth's Interior: Studies of seismic waves (vibrations produced by earthquakes) show that the Earth has a small, solid inner core surrounded by a liquid outer core. The outer core is surrounded by the dense mantle, which in turn is surrounded by the thin low-density crust.

• Seismologists deduce the Earth's interior structure by studying how longitudinal P waves and transverse S waves travel through the Earth's interior.

• The Earth's inner and outer cores are composed of almost pure iron with some nickel mixed in. The mantle is composed of iron-rich minerals.

• Both temperature and pressure steadily increase with depth inside the Earth.

Plate Tectonics: The Earth's crust and a small part of its upper mantle form a rigid layer called the lithosphere. The lithosphere is divided into huge plates that move about over the plastic layer called the asthenosphere in the upper mantle.

• Plate tectonics, or movement of the plates, is driven by convection within the asthenosphere. Molten material wells up at oceanic rifts, producing seafloor spreading, and is returned to the asthenosphere in subduction zones. As one end of a plate is subducted back into the asthenosphere, it helps to pull the rest of the plate along.

• Plate tectonics is responsible for most of the major features of the Earth's surface, including mountain ranges, volcanoes, and the shapes of the continents and oceans.

• Plate tectonics is involved in the formation of the three major categories of rocks: igneous rocks (cooled from molten material), sedimentary rocks (formed by the action of wind, water, and ice), and metamorphic rocks (altered in the solid state by extreme heat and pressure).

The Earth's Magnetic Field and Magnetosphere: Electric currents in the liquid outer core generate a magnetic field. This magnetic field produces a magnetosphere that surrounds the Earth and blocks the solar wind from hitting the atmosphere.

• A bow-shaped shock wave, where the supersonic solar wind is abruptly slowed to subsonic speeds, marks the outer boundary of the magnetosphere.

• Most of the particles of the solar wind are deflected around the Earth by the magnetosphere.

• Some charged particles from the solar wind are trapped in two huge, doughnut-shaped rings called the Van Allen belts. An excess of these particles can initiate an auroral display.

The Earth's Atmosphere: The Earth's atmosphere differs from those of the other terrestrial planets in its chemical composition, circulation pattern, and temperature profile.

• The Earth's atmosphere evolved from being mostly water vapor to being rich in carbon dioxide. A strong greenhouse effect kept the Earth warm enough for water to remain liquid and to permit the evolution of life.

• The appearance of photosynthetic living organisms led to our present atmospheric composition, about four-fifths nitrogen and one-fifth oxygen.

• The Earth's atmosphere is divided into layers called the troposphere, stratosphere, mesosphere, and thermosphere. Ozone molecules in the stratosphere absorb ultraviolet light.

• Because of the Earth's rapid rotation, the circulation in its atmosphere is complex, with three circulation cells in each hemisphere.

The Biosphere: Human activity is changing the Earth's biosphere, on which all living organisms depend.

• Deforestation and the burning of fossil fuels are increasing the greenhouse effect in our atmosphere and warming the planet.

• Industrial chemicals released into the atmosphere have damaged the ozone layer in the stratosphere.

Review Questions

1. Describe the various ways in which the Earth is unique among the planets of our solar system.

2. Why are typical rocks found on the Earth's surface much younger than the Earth itself?

3. Describe how energy is transferred from the Earth's surface to the atmosphere by both convection and radiation.

4. How does the greenhouse effect influence the temperature of the atmosphere? How does this effect differ from what actually happens in a greenhouse?

5. How does solar energy help power the motion of water in the Earth's oceans?

6. How do we know that the Earth was once entirely molten?

7. Describe the interior structure of the Earth.

8. The deepest wells and mines go down only a few kilometers. What, then, is the evidence that iron is abundant in the Earth's core? That the Earth's outer core is molten but the inner core is solid?

9. The inner core of the Earth is at a higher temperature than the outer core. Why, then, is the inner core solid and the outer core molten instead of the other way around?

10. Describe the process of plate tectonics. Give specific examples of geographic features created by plate tectonics.

11. Explain how convection in the Earth's interior drives the process of plate tectonics.

12. Why do you suppose that active volcanoes, such as Mount St. Helens in Washington State, are usually located in mountain ranges that border on subduction zones?

13. What is the difference between a rock and a mineral?

14. What are the differences among igneous, sedimentary, and metamorphic rocks? What do these rocks tell us about the sites at which they are found?

15. Why do geologists suspect that Pangaea was the most recent in a succession of supercontinents?

16. Why do geologists think that the Earth's magnetic field is produced in the liquid outer core rather than in the mantle?

17. Describe the Earth's magnetosphere. If the Earth did not have a magnetic field, do you think aurorae would be more common or less common than they are today?

18. What are the Van Allen belts?

19. What phenomena tended to make the young Earth freeze over? What other phenomena prevented this from happening?

20. Summarize the history of the Earth's atmosphere. What role has biological activity played in this evolution?

21. Describe the structure of the Earth's atmosphere. Explain how heat from the Sun and the Earth's rotation affect the circulation of the Earth's atmosphere.

22. Carbon dioxide and ozone each make up only a fraction of a percent of our atmosphere. Why, then, should we be concerned about small increases or decreases in the atmospheric abundance of these gases?

Advanced Questions

Problem-solving tips and tools

You can find most of the Earth data that you need in Tables 9-1 and 9-3. You'll need to know that a sphere has surface area $4\pi r^2$ and volume $4\pi r^3/3$, where r is the sphere's radius. You may have to consult an atlas to examine the geography of the South Atlantic. Also, remember that the average density of an object is its mass divided by its volume. Sections 5-3 and 5-4 and Box 5-2 describe how to solve problems involving blackbody radiation, Section 2-6 discusses precession, and Section 4-4 describes the properties of elliptical orbits.

23. The total power in sunlight that reaches the top of our atmosphere is 1.75×10^{17} W. (a) How many watts of power are reflected back into space due to the Earth's albedo? (b) If the Earth had no

atmosphere, all of the solar power that was not reflected would be absorbed by the Earth's surface. In equilibrium, the heated surface would act as a blackbody that radiates as much power as it absorbs from the Sun. How much power would the entire Earth radiate? (c) How much power would one square meter of the surface radiate? (d) What would be the average temperature of the surface? Give your answer in both the Kelvin and Celsius scales. (e) Why is the Earth's actual average temperature higher than the value you calculated in (d)?

24. If the Sun were to stop shining, the only source of heat for the Earth's surface would be the planet's internal heat. This delivers 0.06 watts of power to each square meter of surface. In equilibrium, the surface would act as a blackbody that radiates as much power as it receives from the Earth's interior. What would be the average temperature of the surface if the Sun stopped shining? Express your answer in Kelvins and degrees Celsius.

25. On average, the temperature beneath the Earth's crust increases at a rate of 20°C per kilometer. At what depth would water boil? (Assume the surface temperature is 20°C and ignore the effect of the pressure of overlying rock on the boiling point of water.)

26. What fractions of Earth's total volume are occupied by the core, the mantle, and the crust?

27. What fraction of the total mass of the Earth lies within the inner core?

28. (a) Using the mass and diameter of the Earth listed in Table 9-1, verify that the average density of the Earth is 5500 kg/m³. (b) Assuming that the average density of material in the Earth's mantle is about 3500 kg/m³, what must the average density of the core be? Is your answer consistent with the values given in Table 9-3?

29. Identical fossils of the reptile *Mesosaurus*, which lived 300 million years ago, are found in eastern South America and western Africa and nowhere else in the world. Explain how these fossils provide evidence for the theory of plate tectonics.

30. Africa and South America are separating at a rate of about 3 centimeters per year, as explained in the text. Assuming that this rate has been constant, calculate when these two continents must have been in contact. Today the two continents are 6600 km apart.

31. When the crust under an ocean is pushed toward a continent, the oceanic crust is pushed under the continental crust and undergoes subduction (see the right-hand side of Figure 9-15). What does this tell us about the density of oceanic crust compared to continental crust?

32. The surfaces of Mercury, the Moon, and Mars are riddled with craters formed by the impact of space debris. Many of these are billions of years old. By contrast, there are only a few conspicuous craters on the Earth's surface, and these are generally less than 500 million years old. What do you suppose explains the difference?

33. It was stated in Section 7-7 that iron loses its magnetism at temperatures above 770°C. Use this fact and Figure 9-9 to explain why the Earth's magnetic field cannot be due to a magnetized solid core, but must instead be caused by the motion of electrically conducting material in the liquid core.

34. Fast-moving charged particles can damage living organisms in a number of ways. For example, they can cause mutations if they collide with a DNA molecule within a cell. Explain how the fact that the Earth's interior is molten helps minimize the effect of the solar wind on the biosphere.

35. Most auroral displays have a green color dominated by emission from oxygen atoms at a wavelength of 557.7 nm. (a) What minimum energy (in electron volts) must be imparted to an oxygen atom to make it emit this wavelength? (b) Why is your answer in (a) a *minimum* value?

36. Describe how the present-day atmosphere and surface temperature of the Earth might be different (a) if carbon dioxide had never been released into the atmosphere; (b) if carbon dioxide had been released, but life had never evolved on Earth.

37. The Earth's primordial atmosphere probably contained an abundance of methane (CH_4) and ammonia (NH_3), whose molecules were broken apart by ultraviolet light from the Sun and particles in the solar wind. What happened to the atoms of carbon, hydrogen, and nitrogen that were liberated by this dissociation?

38. The photograph below shows the soil at Daspoort Tunnel near Pretoria, South Africa. The whitish layer that extends from lower left to upper right is 2.2 billion years old. Its color is due to a lack of iron oxide. More recent soils typically contain iron oxide and have a darker color. Explain what this tells us about the history of the Earth's atmosphere.

(Courtesy of H. D. Holland)   R I V U X G

39. Earth's atmospheric pressure decreases by a factor of one-half for every 5.5-km increase in altitude above sea level. Construct a plot of pressure versus altitude, assuming the pressure at sea level is one atmosphere (1 atm). Discuss the characteristics of your graph. At what altitude is the atmospheric pressure equal to 0.001 atm?

40. The Earth is at perihelion on January 3 and at aphelion on July 4. Because of precession, in 15,000 A.D. the amount of sunlight in summer will be more than at present in the northern hemisphere but less than at present in the southern hemisphere. Explain why.

Discussion Questions

41. The human population on Earth is currently doubling about every 30 years. Describe the various pressures placed on the Earth

by uncontrolled human population growth. Can such growth continue indefinitely? If not, what natural and human controls might arise to curb this growth? It has been suggested that overpopulation problems could be solved by colonizing the Moon or Mars. Do you think this is a reasonable solution? Explain your answer.

42. Several scientific studies suggest that the continued burning of fossil and organic fuels by humans is releasing enough CO_2 to stimulate the greenhouse effect and eventually melt the polar ice-caps. Antarctica has an area of 13 million square kilometers and is covered by an icecap that varies in thickness from 300 meters near the coast to 1800 meters in the interior. Estimate the volume of this icecap. Assuming that water and ice have roughly the same density, estimate the amount by which the water level of the world's oceans would rise if Antarctica's ice were to melt completely (see Figure 9-28*b*). What portions of the Earth's surface would be inundated by such a deluge?

Web/CD-ROM Questions

43. The periodic formation and breakup of supercontinents imply that profound environmental changes occur about every 500 million years. What sort of global changes might accompany the formation and breakup of a supercontinent? How might these cycles affect the evolution of life? Use the World Wide Web to research the life-forms that dominated the Earth when Pangaea existed 250 million years ago and when fragments of the preceding supercontinent were as dispersed as today's continents.

44. Search the World Wide Web for information about Rodinia, the supercontinent that predated Pangaea. When did this supercontinent form? When did it break apart? Do we have any evidence of supercontinents that predated Rodinia? Why or why not?

45. Because of plate tectonics, the arrangement of continents in the future will be different from today. Search the World Wide Web for information about "Pangea Ultima," a supercontinent that may form in the distant future. When is it predicted to form? How will it compare to the supercontinent that existed 200 million years ago (see Figure 9-11*a*)?

46. Use the World Wide Web to research how global warming may affect the Earth's future surface temperature. What are some predictions for the surface temperature in 2050? In 2100? What effects may the increased temperature have on human health?

47. Use the World Wide Web to investigate the current status of the Antarctic ozone hole. Is the situation getting better or worse? Is there a comparable hole over the North Pole? Why do most scientists blame the chemicals called CFCs for the existence of the ozone hole?

48. In 1989 representatives of many nations signed a global treaty called the Montreal Protocol to protect the ozone layer. Use the World Wide Web to investigate the current status of this treaty. How many nations are signatories to this treaty? Has the treaty been amended since it was first signed? When are various substances that destroy the ozone layer scheduled to be phased out?

  49. Observing Mountain Range Formation. Access the two animations "The Collision of Two Plates: South America" and "The Collision of Two Plates: The Himalayas" in Chapter 9 of the *Universe* Web site or CD-ROM. (**a**) In which case are the mountains formed by tectonic uplift? In which case are the mountains formed by volcanoes from rising lava? (**b**) For each animation, describe which plate is moving in which direction.

Observing Projects

50. When was the last earthquake near your hometown? How far is your hometown from a plate boundary? What kinds of topography (for example, mountains, plains, seashore) dominate the geography of your hometown area? Do the topography and the frequency of earthquakes seem to be consistent with your hometown's proximity to a plate boundary?

 51. Use the *Starry Night Backyard*™ program to view the Earth from space. First select **Viewing Location...** in the **View** menu and set the viewing location to your city or town. (Use the **List** of cities provided, or use the mouse to click on your approximate position on the **Map.** Then click the **Set Location** button.) In the date and time display in the Control Panel at the top of the main window, click on the downward-pointing arrow and select **Solar noon,** the time at which Sun is highest in the sky on today's date. Move the scrollbar at the right-hand side of the window so that your field of view is straight down. Now use the elevation buttons in the Control Panel (the ones that have arrows pointing up and down) at the top of the window to raise yourself off the surface until you can see the entire Earth. Finally, use the sliding zoom controls at the right-hand end of the Control Panel to show more detail around your city or town. The amount of detail is comparable to the view from a spacecraft a few million kilometers away. (**a**) Can you see any evidence of plate tectonics? Of the presence of life? (**b**) What do you conclude about the importance of sending spacecraft to explore planets at close range?

 52. Use the *Deep Space Explorer*™ program to examine the Earth as seen from space. In the left-hand part of the window under the heading **Solar System,** click on the triangle next to the word **Explore** and then click on **Earth.** You can zoom in and zoom out using the buttons at the upper left of the window (an upward-pointing triangle and a downward-pointing triangle). You can also rotate the Earth by putting the mouse cursor over the image, holding down the mouse button, and moving the mouse. (On a two-button mouse, hold down the left mouse button.) (**a**) What evidence do you see that the Earth has an atmosphere? Explain. (**b**) Eastern Brazil, at the eastern tip of South America, is near the *terminator* (the dividing line between the day and night sides of the Earth). As seen from eastern Brazil, is the sun rising or setting? Explain.

10 Our Barren Moon

Other than the Earth itself, the most familiar world in the solar system is the Moon. It is so near to us that some of its surface features are visible to the naked eye. But the Moon is also a strange and alien place, with dramatic differences from our own Earth. It has no substantial atmosphere, no global magnetic field, and no liquid water of any kind. On the airless Moon, the sky is black even at midday. And unlike the Earth's surface, the surface of the Moon has remained essentially unaltered for billions of years. The boulder in the photograph—shown being examined by *Apollo 17* astronaut Harrison Schmitt, one of only a dozen humans to have visited another world— is actually one of the most recent features on the lunar surface: It tumbled down a mountainside to its present location a mere 800,000 years ago.

Today the Moon is a geologically dead world. But its early history was violent and chaotic. In this chapter, we will learn how the Moon may have formed some 4.56 billion years ago as the result of a titanic collision between the young Earth and a rogue protoplanet the size of Mars. We will see evidence that the young Moon was pelted incessantly for hundreds of millions of years by large chunks of interplanetary debris, producing a cratered landscape. We will discover that the largest of these impacts formed immense circular basins that then flooded with molten lava. You can see the solidified lava as dark patches that cover much of the face of the Moon. In later chapters we will find other evidence that catastrophic collisions have played an important role in shaping the solar system.

10-1 | The Moon's airless, dry surface is covered with plains and craters

Compared with the distances between planets, which are measured in tens or hundreds of millions of kilometers, the average distance from the Earth to the Moon is just 384,400 km

 Walking on the lunar surface, December 1972. (Eugene Cernan, *Apollo 17,* NASA) R I **V** U X G

Questions to Guide Your Reading

As you read the sections of this chapter, look for the answers to the following questions:

10-1 Is the Moon completely covered with craters?

10-2 Has there been any exploration of the Moon since the *Apollo* program in the 1970s?

10-3 Does the Moon's interior have a similar structure to the interior of the Earth?

10-4 How do Moon rocks compare to rocks found on the Earth?

10-5 How did the Moon form?

(238,900 mi). Despite its being so close, the Moon has an angular diameter of just ½°. This tells us that it is a rather small world. The diameter of the Moon is 3476 km (2160 mi), just 27% of the Earth's diameter (Figure 10-1). Table 10-1 lists some basic data about the Moon.

Although the Moon has a small mass (just 1.23% of the Earth's mass), Newton's third law tells us that the Moon exerts just as much gravitational force on the Earth as the Earth does on the Moon (see Section 4-6). Hence, it is not strictly correct to say that the Moon orbits the Earth; rather, they both orbit around a point between their centers called the **center of mass** of the Earth-Moon system. (Because the Earth is so much more massive than the Moon, the center of mass is close to the Earth's center: Indeed, it actually lies within the Earth's interior.) The center of mass then follows an elliptical orbit around the Sun (Figure 10-2a). In a similar way, a spinning wrench sliding along a table actually rotates around its center of mass (Figure 10-2b).

Even without binoculars or a telescope, you can easily see that dark gray and light gray areas cover vast expanses of the Moon's rocky surface. If you observe the Moon over many nights, you will see the Moon go through its phases as the dividing line between the illuminated and darkened portion of the Moon (the **terminator**) shifts. But you will always see the same pattern of dark and light areas, because the Moon keeps the same side turned toward the Earth; you never see the **far side** (or back side) of the Moon. We saw in Section 3-2 that this is the result of the Moon's **synchronous rotation**: It takes precisely the same amount of time for the Moon to rotate on its axis (one lunar "day") as it does to complete one orbit around the Earth.

figure 10-1 R I **V** U X G

Earth and Moon to Scale The *Galileo* spacecraft recorded this image of the Earth (at lower right) and the Moon (at upper left) on December 16, 1992, while en route to Jupiter. Both the Earth and the Moon were about 6.2 million km (3.9 million mi) from the spacecraft, so this image shows them to scale. The Earth has blue oceans, an atmosphere streaked with white clouds, and continents continually being reshaped by plate tectonics. By contrast, the Moon has no oceans, no atmosphere, and no evidence of plate tectonics. (NASA)

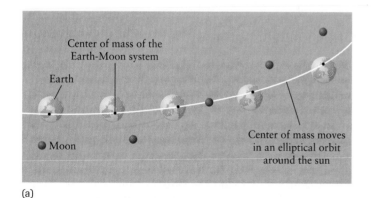

(a)

figure 10-2 R I **V** U X G

Motion of the Earth-Moon System (a) The Earth and Moon both orbit around their center of mass, which in turn follows an elliptical orbit around the Sun. (b) This time-lapse image shows a spinning wrench sliding across a table. The wrench's center of mass (shown by a red cross) moves in a straight line, while the wrench as a whole rotates around the center of mass. (b: Berenice Abbott/Photo Researchers)

(b)

table 10-1	Moon Data
Distance from Earth (center to center):	Average: 384,400 km = 238,900 mi Maximum (apogee): 405,500 km Minimum (perigee): 363,300 km
Eccentricity of orbit:	0.0549
Average orbital speed:	3680 km/h
Sidereal period (relative to fixed stars):	27.322 days
Synodic period (new moon to new moon):	29.531 days
Inclination of lunar equator to orbit:	6.68°
Inclination of orbit to ecliptic:	5.15°
Diameter (equatorial):	3476 km = 2160 mi = 0.272 Earth diameter
Mass:	7.349×10^{22} kg = 0.0123 Earth mass
Average density:	3344 kg/m³
Escape speed:	2.4 km/s
Surface gravity (Earth = 1):	0.17
Albedo:	0.11
Average surface temperatures:	Day: 130°C = 266°F = 403 K Night: −180°C = −292°F = 93 K
Atmosphere:	Essentially none

(Lick Observatory, University of California)   R I **V** U X G

With patience, however, you can see slightly more than half the lunar surface. This is because the Moon appears to wobble slightly over the course of an orbit: It seems to rock back and forth around its north-south axis and to nod up and down in a north-south direction. This apparent periodic wobbling, called **libration**, permits us to view 59% of our satellite's surface.

We stress that the Moon only *appears* to be wobbling. In fact, its rotation axis keeps nearly the same orientation in space as the Moon goes around its orbit. The reason why the Moon seems to rock back and forth is that its orbit around the Earth is slightly elliptical. As a result, the Moon's orbital motion does not keep pace with its rotation at all points around the orbit. Furthermore, because the Moon's rotation axis is not exactly perpendicular to the plane of its orbit, it appears to us to nod up and down.

Clouds always cover some portion of the Earth's surface, but clouds are never seen on the Moon (see Figure 10-1). This is because the Moon is too small a world to have an atmosphere. Its surface gravity is low, only about one-sixth as great as that of the Earth, and thus any gas molecules can easily escape into space (see Box 7-2). Because there is no atmosphere to scatter sunlight, the daytime sky on the Moon appears as black as the nighttime sky on the Earth (see the photograph that opens this chapter).

The absence of an atmosphere means that the Moon can have no liquid water on its surface. Here on Earth, water molecules are kept in the liquid state by the pressure of the atmosphere pushing down on them. To see what happens to liquid water on the airless Moon, consider what happens when the pressure of the atmosphere is reduced. The air pressure in Denver, Colorado (elevation 1650 m, or 5400 ft), is only 83% as great as at sea level. Therefore, in Denver, less energy has to be added to the molecules in a pot of water to make them evaporate, and water boils at only 95°C (= 368 K = 203°F), as compared with 100°C (= 373 K = 212°F) at sea level. (This is why some foods require longer cooking times at high elevations, where the boiling water used for cooking is not as hot as at sea level.) On the Moon, the atmospheric pressure is essentially zero and the boiling temperature is lower than the temperature of the lunar surface. Under these conditions, water can exist as a solid (ice) or a vapor but not as a liquid.

CAUTION! If astronauts were to venture out onto the lunar surface without wearing spacesuits, the lack of atmosphere would *not* cause their blood to boil. That's because our skin is strong enough to contain bodily fluids. In the same way, if we place an orange in a sealed container and remove all the air, the orange retains all of its fluid and does not explode. (It even tastes the same afterward.) The reasons why astronauts wore pressurized spacesuits on the Moon (see the photograph that opens this chapter) were to provide oxygen for breathing, to regulate body temperature, and to protect against ultraviolet radiation from the Sun. Since the Moon has no atmosphere, and thus no ozone layer like that of the Earth (Section 9-7), an unprotected astronaut would suffer a sunburn after just 10 seconds of exposure!

Since the Moon has no atmosphere to obscure its surface, you can get a clear view of several different types of lunar features with a small telescope or binoculars. These include *craters*, the dark-colored *maria*, and the light-colored lunar highlands or *terrae*

(Figure 10-3). Taken together, these different features show the striking differences between the surfaces of the Earth and the Moon.

With an Earth-based telescope, some 30,000 **craters** are visible, with diameters ranging from 1 km to several hundred km. Close-up photographs from lunar orbit have revealed millions of craters too small to be seen from Earth. Following a tradition established in the seventeenth century, the most prominent craters are named after famous philosophers, mathematicians, and scientists, such as Plato, Aristotle, Pythagoras, Copernicus, and Kepler. Virtually all lunar craters, both large and small, are the result of the Moon's having been bombarded by meteoritic material. For this reason, they are also called **impact craters** (see Section 7-6). One piece of evidence that lunar craters are caused by impacts is that many large craters have a central peak, with a shape that could only result from high-speed impacts (see Figure 7-9a).

WEB LINK 10-3 A meteoroid striking the lunar surface generates a shock wave over the surface outward from the point of impact. Such shock waves can spread over an area much larger than the size of the meteoroid that generated the wave. Very large craters more than 100 km across, such as the crater Clavius shown in Figure 10-4, were probably created by the impact of a fast-moving piece of rock only a few kilometers in radius.

The large dark areas on the lunar surface shown in Figure 10-3 are called **maria** (pronounced "MAR-ee-uh"). They form a pattern that looks crudely like a human face, sometimes called "the man in the Moon." The singular form of maria is **mare** (pronounced

figure 10-3 R I **V** U X G

The Moon as Seen from Earth The Moon's diameter (3476 km = 2160 mi) is slightly less than the distance from New York to San Francisco. This image is a composite of photographs made at first quarter and third quarter, when long shadows enhance the surface features. (Lick Observatory, University of California)

figure 10-4 R I **V** U X G

The Crater Clavius This photograph from the 5-meter (200-in.) Palomar telescope shows one of the largest craters on the Moon. Clavius has a diameter of 232 km (144 mi) and a depth of 4.9 km (16,000 ft), measured from the crater's floor to the top of the surrounding rim. The entire New York–Philadelphia metropolitan area would fit nicely within Clavius. (Palomar Observatory)

"MAR-ay"), which means "sea" in Latin. These terms date from the seventeenth century, when observers using early telescopes thought these dark features were large bodies of water. They gave the maria fanciful and romantic names such as Mare Tranquillitatis (Sea of Tranquillity), Mare Nubium (Sea of Clouds), Mare Nectaris (Sea of Nectar), Mare Serenitatis (Sea of Serenity), and Mare Imbrium (Sea of Showers). While modern astronomers still use these names, they now understand that the maria are not bodies of water (which, as we have seen, could not exist on the airless Moon) but rather the remains of huge lava flows. The maria get their distinctive appearance from the dark color of the solidified lava.

On the whole, the maria have far fewer craters than the surrounding terrain. Since craters are caused by meteoritic bombardment, this means that the maria have not been exposed to that bombardment for as long as the surrounding terrain. Hence, the maria must be relatively young, and the lava flows that formed them must have occurred at a later stage in the Moon's geologic history (Figure 10-5). The craters that are found in the maria were caused by the relatively few impacts that have occurred since the maria solidified (Figure 10-6).

Although they are much larger than craters, maria also tend to be circular in shape (see Figure 10-5). This suggests that, like craters, these depressions in the lunar surface were also created by impacts. It is thought that very large meteoroids or asteroids some tens of kilometers across struck the Moon, forming basins. These depressions subsequently flooded with lava that flowed out from the Moon's interior through cracks in the lunar crust. The solidified lava formed the maria that we see today.

An impact large enough to create a mare basin must have thrust upward tremendous amounts of material around the impact site. This explains the mountain ranges that curve in circular arcs around mare basins. (You can see one such mountain range around the edges of Mare Imbrium in Figure 10-5.) Analysis of lunar rocks as well as observations made from lunar orbit have validated this picture of the origin of lunar mountains.

The older, light-colored terrain that surrounds the maria is called **terrae,** from the Latin for "lands." (Seventeenth-century astronomers thought that this terrain was the dry land that surrounded lunar oceans.) Detailed measurements by astronauts in lunar orbit demonstrated that the maria on the Moon's Earth-facing side are 2 to 5 km below the average lunar elevation, while the terrae extend to several kilometers above the average elevation. Terrae are therefore often called **lunar highlands.** The flat, low-lying, dark gray maria cover only 15% of the lunar surface. The remaining 85% is light gray, heavily cratered lunar highlands.

 Remarkably, when the Moon's far side was photographed for the first time by the Soviet *Luna 3* spacecraft in 1959, scientists were surprised to find almost no maria there. Thus, lunar highlands predominate on the side of the

figure 10-5 R I **V** U X G

Mare Imbrium from Earth Mare Imbrium is the largest of 14 dark plains that dominate the Earth-facing side of the Moon. Its diameter of 1100 km (700 mi) is about the same as the distance from Chicago to Philadelphia or from London to Rome. The maria formed after the surrounding light-colored terrain, so they have not been exposed to meteoritic bombardment for as long and hence have fewer craters. (Carnegie Observatories)

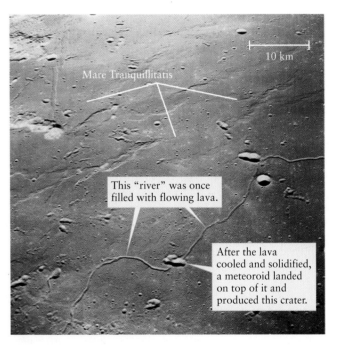

figure 10-6 R I **V** U X G

Details of Mare Tranquillitatis This photograph from lunar orbit reveals numerous tiny craters and cracks in the surface of a typical mare. This color photograph also shows that the Moon's surface is not a uniform gray color. (Figures 10-4 and 10-5 are black-and-white photographs.) The *Apollo 11* landing site is near the center of the top edge of this image. (*Apollo 10,* NASA)

Figure 10-7 R I V U X G

The Near and Far Sides of the Moon This composite image from the *Galileo* spacecraft shows the near (Earth-facing) side of the Moon on the right, with its large, dark, flat maria. On the left is the lunar far side, which is composed almost entirely of light-colored lunar highlands. Mare Orientale lies on the boundary between the near and far sides. The broad dark region at lower left is the South Pole–Aitken Basin, an impact scar some 2100 km (1300 mi) across and 12 km (7 mi) deep. (NASA)

Moon that faces away from the Earth (Figure 10-7). Presumably, the crust is thicker on the far side, so even the most massive impacts there were unable to crack through the crust and let lava flood onto the surface.

Something is entirely missing from our description of the Moon's surface—any mention of plate tectonics. Indeed, no evidence for plate tectonics can be found on the Moon. Recall from Section 9-3 that the Earth's major mountain ranges were created by collisions between lithospheric plates. In contrast, it appears that the Moon is a one-plate world: Its entire lithosphere is a single, solid plate. There are no rifts where plates are moving apart, no mountain ranges of the sort formed by collisions between plates, and no subduction zones. Thus, while lunar mountain ranges bear the names of terrestrial ranges, such as the Alps, Carpathians, and Apennines, they are entirely different in character. Geologically, the Moon is a dead world.

In Section 7-6 we discussed the general rule that a world with abundant craters has an old, geologically inactive surface; the Moon is a prime example of this rule. Without an atmosphere to cause erosion and without plate tectonics, the surface of the Moon changes only very slowly. That is why the Moon's craters, many of which are now known to be billions of years old, remain visible today. On the Earth, by contrast, the continents and seafloor are continually being reshaped by seafloor spreading, mountain building, and subduction.

Thus, any craters that were formed by impacts on the ancient Earth's surface have long since been erased. There are fewer than 200 impact craters on the Earth today; these were formed within the last few hundred million years and have not yet been erased.

10-2 | Manned exploration of the lunar surface was one of the greatest adventures in human history

For thousands of years, storytellers have invented fabulous tales of voyages to the Moon. The seventeenth-century French satirist Cyrano de Bergerac wrote of a spacecraft whose motive power was provided by spheres of morning dew. (After all, reasoned Cyrano, dew rises with the morning sun, so a spacecraft would rise along with it.) In his imaginative tale *Somnium*, Cyrano's contemporary Johannes Kepler (the same Kepler who deduced the laws of planetary motion) used the astronomical knowledge of his time to envision what it would be like to walk on the Moon's surface. In 1865 French author Jules Verne published *From the Earth to the Moon*, a story of a spacecraft sent to the Moon from a launching site in Florida.

Almost exactly a hundred years later, Florida was indeed the starting point of the first voyages made by humans to other worlds. Between 1969 and 1972, 12 astronauts walked on the Moon, and fantasy became magnificent reality.

As in Verne's story, politics had much to do with the motivation for going to the Moon. But in fact, there were excellent scientific reasons to do so. Because the Earth is a geologically active planet, typical terrestrial surface rocks are only a few hundred million years old, which is only a small fraction of the Earth's age. Thus, all traces of the Earth's origins have been erased. By contrast, rocks on the geologically inactive surface of the Moon have been essentially undisturbed for billions of years. Samples of lunar rocks have helped answer many questions about the Moon and have shed light on the origin of the Earth and the birth of the entire solar system.

Lunar missions began in 1959, when the Soviet Union sent three small spacecraft—*Lunas 1, 2,* and *3*—toward the Moon. American attempts to reach the Moon began in the early 1960s with Project Ranger. Equipped with six television cameras, the *Ranger* spacecraft transmitted close-up views of the lunar surface taken in the final few minutes before it crashed onto the Moon. Figure 10-8 shows a view from *Ranger 9* just before it hit the floor of the crater Alphonsus. The *Ranger* images showed far more detail than even the best Earth-based telescope.

A complete high-resolution photographic survey of the lunar surface was the mission of the *Lunar Orbiter* program, under which five spacecraft were launched between August 1966 and August 1967. In all, the five *Lunar Orbiters* sent back a total of 1950 high-resolution pictures covering 99.5% of the Moon. Some of these photographs show features as small as 1 meter across. Because these pictures revealed the lunar surface in close detail, they were essential in helping NASA scientists choose landing sites for the astronauts.

Even after the *Ranger* missions, it was unclear whether the lunar surface was a safe place to land. One leading theory was that billions of years of impacts by meteoroids had ground the Moon's

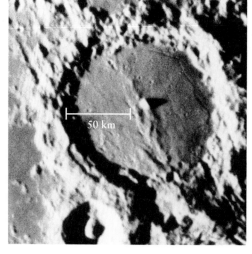

(a) View from *Ranger 9*

(b) Same view from Earth

figure 10-8 R I **V** U X G
The Moon from *Ranger 9* and from Earth (a) Seconds before it crashed onto the Moon on March 24, 1965, *Ranger 9* sent back this picture of the crater Alphonsus. (The white circle shows where the spacecraft hit.) The resolution is far better than in the Earth-based telescopic view shown in (b). Alphonsus is about 129 km (80 mi) in diameter. (NASA)

surface to a fine powder, much like that used by dental hygienists to polish teeth. Perhaps a spacecraft attempting to land would simply sink into a quicksand-like sea of dust! Thus, before a manned landing on the Moon could be attempted, it was necessary to try a soft landing of an unmanned spacecraft. This was the purpose of the *Surveyor* program. Between June 1966 and January 1968, five *Surveyors* successfully touched down on the Moon, sending back pictures and data directly from the lunar surface. These missions demonstrated convincingly that the Moon's surface was as solid as that of the Earth. Figure 10-9 shows an astronaut visiting *Surveyor 3*, 2½ years after it had landed on the Moon.

The first of six manned lunar landings took place on July 20, 1969, when the *Apollo 11* lunar module *Eagle* set down in Mare Tranquillitatis (see Figure 10-6). Astronauts Neil Armstrong and Edwin "Buzz" Aldrin were the first humans to set foot on the Moon. *Apollo 12* also landed in a mare, Oceanus Procellarum (Ocean of Storms). The *Apollo 13* mission experienced a nearly catastrophic inflight explosion that made it impossible for it to land on the Moon. Fortunately, titanic efforts by the astronauts and ground personnel brought it safely home. The remaining four missions, *Apollo 14* through *Apollo 17*, were made in progressively more challenging terrain, chosen to permit the exploration of a wide variety of geologically interesting features. The first era of human exploration of the Moon culminated in December 1972, when *Apollo 17* landed in rugged mountains east of Mare Serenitatis (see the photo that opens this chapter). Figure 10-10 shows one of the *Apollo* bases.

In a curious postscript to the *Apollo* missions, in the 1990s an "urban legend" arose that the Moon landings had never taken place and were merely an elaborate hoax by NASA. While amusing, this legend does not hold up under even the slightest scientific scrutiny. For example, believers in the "Moon hoax" cannot explain why the rocks brought back from the Moon are substantially different in their chemistry from Earth

rocks, as we will see later in this chapter. It makes as much sense to ask "Do you believe that humans actually went to the Moon?" as it does to ask "Do you believe there was a First World War?" or "Do you believe there is such a place as Antarctica?"

figure 10-9 R I **V** U X G

An Astronaut Visits *Surveyor 3* *Surveyor 3* made a soft landing on the Moon in April 1967. It was visited 2½ years later by *Apollo 12* astronaut Alan Bean, who took pieces of the spacecraft back to the Earth for analysis. (The *Apollo 12* lunar module is visible in the distance.) The retrieved pieces showed evidence of damage by micrometeoroid impacts, demonstrating that small bits of space debris are still colliding with the Moon. (Pete Conrad, *Apollo 12*, NASA)

Figure 10-10 R I **V** U X G

The *Apollo 15* Base This photograph shows astronaut James Irwin and the *Apollo 15* landing site at the foot of the Apennine Mountains near the eastern edge of Mare Imbrium. The hill in the background is about 5 km away and rises about 3 km high. At the right is a Lunar Rover, an electric-powered vehicle used by astronauts to explore a greater radius around the landing site. The last three *Apollo* missions carried Lunar Rovers. (Dave Scott, *Apollo 15*, NASA)

Between 1966 and 1976, a series of unmanned Soviet spacecraft also orbited the Moon and soft-landed on its surface. The first soft landing, by *Luna 9*, came four months before the U.S. *Surveyor 1* arrived. The first lunar satellite, *Luna 10*, orbited the Moon four months before *Lunar Orbiter 1*. In the 1970s, robotic spacecraft named *Luna 17* and *Luna 21* landed vehicles that roamed the lunar surface, and *Lunas 16, 20,* and *24* brought samples of rock back to the Earth.

Unmanned spacecraft continued the exploration of the Moon in the 1990s. The *Galileo* spacecraft was launched in 1989 on a mission to Jupiter. On the way, it followed a convoluted trajectory that took it past the Earth-Moon system in December 1990 and December 1992. During these two passes, *Galileo* recorded color images of the Moon with cameras far more advanced than those of the *Lunar Orbiters* (see Figure 10-7).

In 1994 a small spacecraft named *Clementine* spent more than two months observing the Moon from lunar orbit. Although originally intended to test lightweight imaging systems for future defense satellites, *Clementine* proved to be a remarkable tool for lunar exploration. It carried an array of very high-resolution cameras sensitive to ultraviolet and infrared light as well as the visible spectrum. Different types of materials on the lunar surface reflect different wavelengths, so analysis of *Clementine* images made in different parts of the electromagnetic spectrum has revealed the composition of large areas of the Moon's surface (**Figure 10-11**).

One remarkable result of the *Clementine* mission was evidence for a patch of ice tens or hundreds of kilometers across at the Moon's south pole. Deep within the South Pole–Aitken Basin (see Figure 10-7) are places where sunlight never reaches and where temperatures are low enough to prevent ice from evaporating. Radio waves beamed from *Clementine* toward these areas were detected by radio telescopes on the Earth and showed the characteristic signature of reflection from ice.

 Data from the *Lunar Prospector* spacecraft, which went into orbit around the Moon in January 1998, reinforced the *Clementine* findings and indicated that even more ice might lie in deep craters at the Moon's north pole. Perhaps this ice at the lunar poles could help to meet the water requirements of future lunar colonists. The ice may have been deposited by comets (see Section 7-5) that crashed into the lunar surface.

One complication is that the *Lunar Prospector* observations show only that there is an excess of hydrogen atoms at the lunar poles. These could be trapped within water molecules (H_2O) or within other, more exotic minerals. At the end of the *Lunar Prospector* mission in July 1999, the spacecraft was crashed into a crater at the lunar south pole in the hope that a plume of water vapor might rise from the impact site. Although more than a dozen large Earth-based telescopes watched the Moon during

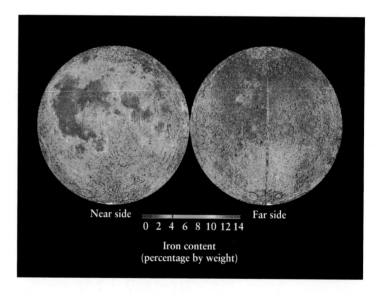

Figure 10-11 R I **V** U X G

Iron on the Moon Images at different wavelengths from the *Clementine* spacecraft were used to make this map of the concentration of iron on the lunar surface. The areas of highest concentration (red) coincide closely with the maria on the near side (compare Figure 10-3), confirming that the maria formed from iron-rich lavas. The lowest iron concentration (blue) is found in the lunar highlands. (No data were collected in the gray areas.) The large green region of intermediate iron concentration on the far side is the South Pole–Aitken Basin (see Figure 10-7). The impact that created this basin may have excavated all the way to the Moon's mantle, in which iron is more abundant than in the crust. (JSC/NASA) the impact, no such plume was observed, nor was any debris of any kind. Radar observations made from Earth in 2003 suggest that if there is ice at the lunar poles, it is in sheets no more than a few centimeters thick. A new generation of European and Japanese unmanned spacecraft scheduled to be launched toward the Moon may help provide a definitive answer about ice at the Moon's poles.

10-3 | The Moon has no global magnetic field but has a small core beneath a thick mantle

We saw in Section 7-6 and Section 7-7 that a small planet or satellite will have little internal heat, and hence is likely to have a mostly solid interior. Lacking substantial amounts of moving molten material in its interior, a small world is unlikely to generate a global magnetic field. As we will see, these general rules describe the Moon very well.

The Moon landings gave scientists an unprecedented opportunity to explore an alien world. The *Apollo* spacecraft were therefore packed with an assortment of apparatus that the astronauts deployed around the landing sites (Figure 10-12). For example, all the missions carried magnetometers, which confirmed that the

Figure 10-12 R I **V** U X G

Seismic Equipment on the Moon This view of the *Apollo 11* base shows astronaut Buzz Aldrin standing alongside a seismometer that detected moonquakes and transmitted data back to the Earth. By combining data from seismometers left at different locations on the Moon, scientists have been able to deduce the structure of the Moon's interior. The *Apollo 11* lunar module and some additional equipment are in the background. (Neil Armstrong, *Apollo 11*, NASA)

present-day Moon has no global magnetic field. However, careful magnetic measurements of lunar rocks returned by the *Apollo* astronauts indicated that the Moon did have a weak magnetic field when the rocks solidified billions of years ago. Hence, the Moon must originally have had a small, molten, iron-rich core. (We saw in Section 7-7 that the Earth's magnetic field is generated by fluid motions in its interior.) The core presumably solidified at least partially as the Moon cooled, so that the lunar magnetic field disappeared.

 More recently, scientists used *Lunar Prospector* to probe the character of the Moon's core. The spacecraft made extensive measurements of lunar gravity and of how the Moon responds when it passes through the Earth's magnetosphere (Section 9-4). When combined with detailed observations of how the Earth's tidal forces affect the motions of the Moon, the data indicate that the present-day Moon has a partially liquid, iron-rich core about 700 km (435 mi) in diameter. While 32% of the Earth's mass is in its core, the core of the Moon contains only 2 to 3% of the lunar mass.

Scientists have used seismic waves to investigate the Moon's interior, just as is done for the Earth (Section 9-2). The *Apollo* astronauts set up seismometers on the Moon's surface. It was found that the Moon exhibits far less seismic activity than the Earth. Roughly 3000 **moonquakes** were detected per year, whereas a similar seismometer on the Earth would record hundreds of thousands of earthquakes per year. Furthermore, typical moonquakes are very weak. A major terrestrial earthquake is in the range from 6 to 8 on the Richter scale, while a major moonquake measures 0.5 to 1.5 on that scale and would go unnoticed by a person standing near the epicenter.

Analysis of the feeble moonquakes reveals that most originate 600 to 800 km below the surface, deeper than most earthquakes. On Earth, the deepest earthquakes mark the boundary between the solid lithosphere and the plastic asthenosphere. This is because the lithosphere is brittle enough to fracture and produce seismic waves, whereas the deeper rock of the asthenosphere oozes and flows rather than cracks. We may conclude that the Moon's lithosphere is about 800 km thick (Figure 10-13). By contrast, the Earth's lithosphere is only 50 to 100 km thick.

The lunar asthenosphere extends down to the Moon's relatively small iron-rich core, which is more than 1400 km below the lunar surface. The asthenosphere and the lower part of the lithosphere are presumably composed of relatively dense iron-rich rocks. The uppermost part of the lithosphere is a lower-density crust, with an average thickness of about 60 km on the Earth-facing side and up to 100 km on the far side. For comparison, the thickness of the Earth's crust ranges from 5 km under the oceans to about 30 km under major mountain ranges on the continents (see Table 9-3). This slightly lopsided distribution of mass between the near and far sides of the Moon helps to keep the Moon in synchronous rotation, so that we always see the same side of the Moon (see Section 4-8).

On the whole, the Moon's interior is much more rigid and less fluid than the interior of the Earth. Plate tectonics requires that there be fluid motion just below a planet's surface, so it is not surprising that there is no evidence for plate tectonics on the lunar surface.

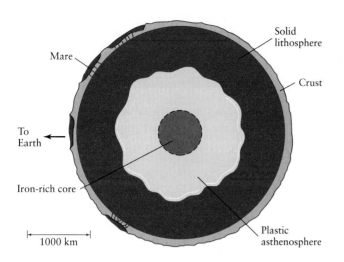

figure 10-13

The Internal Structure of the Moon Like the Earth, the Moon has a crust, a mantle, and a core. The lunar crust has an average thickness of about 60 km on the Earth-facing side but about 100 km on the far side. The crust and solid upper mantle form a lithosphere about 800 km thick. The plastic (nonrigid) asthenosphere extends all the way to the base of the mantle. The iron-rich core is roughly 700 km in diameter. (Compare Figure 9-8.)

Earthquakes tend to occur at the boundaries between tectonic plates, where plates move past each other, collide, or undergo subduction (see Figure 9-13). But there are no plate motions on the Moon. What, then, causes moonquakes? The answer turns out to be the Earth's tidal forces. Just as the tidal forces of the Sun and Moon deform the Earth's oceans and give rise to ocean tides (see Figure 4-25), the Earth's tidal forces deform the solid body of the Moon. The amount of the deformation changes as the Moon moves around its elliptical orbit and the Earth-Moon distance varies. As a result, the body of the Moon flexes slightly, triggering moonquakes. The tidal stresses are greatest when the Moon is nearest the Earth—that is, at perigee (see Section 3-5)—which is just when the *Apollo* seismometers reported the most moonquakes. **Box 10-1** has more information about how the tidal stresses on the Moon depend on its distance from Earth.

Without tectonics, and without the erosion caused by an atmosphere or oceans, the only changes occurring today on the lunar surface are those due to meteoroid impacts. These impacts were also monitored by the seismometers left on the Moon by the *Apollo* astronauts. These devices, which remained in operation for several years, were sensitive enough to detect a hit by a grapefruit-sized meteoroid anywhere on the lunar surface. The data show that every year the Moon is struck by 80 to 150 meteoroids having masses between 0.1 and 1000 kg (roughly ¼ lb to 1 ton).

box 10-1   TOOLS OF THE ASTRONOMER'S TRADE

Calculating Tidal Forces

The gravitational force between any two objects decreases with increasing distance between the objects. This principle leads to a simple formula for estimating the *tidal* force that the Earth exerts on parts of the Moon.

Consider two small objects, each of mass m, on opposite sides of the Moon. Because the two objects are at different distances from the Earth's center, the Earth exerts different forces F_{near} and F_{far} on them (see the accompanying illustration). The consequence of this difference is that the two objects tend to pull away from each other and away from the center of the Moon (see Section 4-8, especially Figure 4-23). The *tidal force* on these two objects, which is the force that tends to pull them apart, is the difference between the forces on the individual objects:

$$F_{tidal} = F_{near} - F_{far}$$

Both F_{near} and F_{far} are both inversely proportional to the square of the distance from the Earth (see Section 4-7). If this distance were doubled, both forces would decrease to $\frac{1}{2}^2 = \frac{1}{4}$ of their original values. But the tidal force F_{tidal}, which is the *difference* between these two forces, decreases even more rapidly with distance: It is inversely proportional to the *cube* of the distance from the Earth to the Moon. That is, doubling the Earth-Moon distance decreases the tidal force to $\frac{1}{2}^3 = \frac{1}{8}$ of its original value. If we use the symbols M_{Earth} for the Earth's mass $(5.974 \times 10^{24}$ kg), d for the Moon's diameter, r for the center-to-center distance from the Earth to the Moon, and G for the

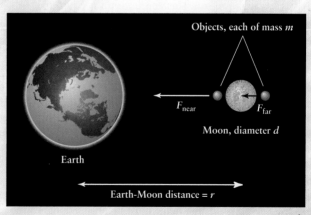

(continued)

TOOLS OF THE ASTRONOMER'S TRADE continued

universal constant of gravitation (6.67×10^{-11} newton m^2/kg^2), the tidal force is given approximately by the formula

$$F_{\text{tidal}} = \frac{2GM_{\text{Earth}}md}{r^3}$$

EXAMPLE: Calculate the tidal force for two 1-kg lunar rocks at the locations shown in the figure that accompanies this box.

Situation: We are asked to calculate a tidal force. We know the masses of the Earth (M_{Earth}) and of each of the two rocks (m). The Earth-Moon distance (r) and the diameter of the Moon (d, equal to the separation between the two rocks) are given in Table 10-1.

Tools: We use the equation given above for the tidal force F_{tidal}.

Answer: To use the given values in the equation, we must first convert the values for the average Earth-Moon distance r and the Moon's diameter d from kilometers to meters:

$$d = 3476 \text{ km} \times \frac{10^3 \text{ m}}{1 \text{ km}} = 3.476 \times 10^6 \text{ m}$$

$$r = 384{,}400 \text{ km} \times \frac{10^3 \text{ m}}{1 \text{ km}} = 3.844 \times 10^8 \text{ m}$$

The tidal force is then

$$F_{\text{tidal}} = \frac{2(6.67 \times 10^{-11})(5.974 \times 10^{24})(1)(3.476 \times 10^6)}{(3.844 \times 10^8)^3}$$

$$= 4.88 \times 10^{-5} \text{ newton}$$

Review: To put our result into perspective, let's compare it to the *weight* of a 1-kg rock on the Moon. The weight of an object is equal to its mass multiplied by the acceleration due to gravity (see Section 4-6). This acceleration is equal to 9.8 m/s^2 on Earth, but is only 0.17 as great on the Moon. So, the weight of a 1-kg rock on the Moon is

$$0.17 \times 1 \text{ kg} \times 9.8 \text{ m/s}^2 = 1.7 \text{ newtons}$$

Thus, the tidal force on lunar rocks is *much* less than their weight. This is a good thing: If the tidal force (which tries to pull a rock away from the Moon's center) were greater than the weight (which pulls it toward the Moon's center), loose objects would levitate off the Moon's surface and fly into space!

The tidal force of the Earth on the Moon deforms the solid body of the Moon, so that it acquires *tidal bulges* on its near and far sides. The mass of each bulge must be proportional to the tidal force that lifts lunar material away from the Moon's center.

So, the above formula tells us that the mass of each bulge is inversely proportional to the cube of the Earth-Moon distance r:

$$m_{\text{bulge}} = \frac{A}{r^3}$$

(Determining the value of the constant A requires knowing how easily the Moon deforms, which is a complicated problem beyond our scope.) We can then use our formula for the tidal force F_{tidal} to tell us the *net* tidal force on the Moon. To do this, we think of the objects of mass m on opposite sides of the Moon as being the tidal bulges. So, we substitute m_{bulge} for m in the tidal force formula:

$$F_{\text{tidal-net}} = \frac{2GM_{\text{Earth}}m_{\text{bulge}}d}{r^3} = \frac{2GM_{\text{Earth}}}{r^3}\frac{A}{r^3}d$$

$$= \frac{2GM_{\text{Earth}}Ad}{r^6}$$

Thus, the net tidal force on the Moon is inversely proportional to the *sixth* power of the Earth-Moon distance. Even a small decrease in this distance can cause a substantial increase in the net tidal force.

EXAMPLE: Compare the net tidal force on the Moon at perigee (when it is closest to the Earth) and at apogee (when it is farthest away).

Situation: We are asked to *compare* the tidal forces on the Moon at two different distances from the Earth. To do this, we will find the *ratio* of the tidal force at perigee to the tidal force at apogee.

Tools: We use the equation given above for the net tidal force on the Moon.

Answer: The Earth-Moon distances at perigee and apogee are given in Table 10-1. If we take the ratio of the tidal forces at these two distances, the constants (including the unknown quantity A) cancel out:

$$\frac{F_{\text{tidal-net-perigee}}}{F_{\text{tidal-net-apogee}}} = \frac{2GM_{\text{Earth}}Ad/r^6{}_{\text{perigee}}}{2GM_{\text{Earth}}Ad/r^6{}_{\text{apogee}}} = \left(\frac{r_{\text{apogee}}}{r_{\text{perigee}}}\right)^6$$

$$= \left(\frac{405{,}500 \text{ km}}{363{,}300 \text{ km}}\right)^6 = (1.116)^6 = 1.93$$

Review: Our result tells us that the net tidal force at perigee is 1.93 times as great (that is, almost double) as at apogee. Tidal forces on the Moon help create the stresses that generate moonquakes, so it is not surprising that these quakes are more frequent at perigee than at apogee.

10-4 | Lunar rocks reveal a geologic history quite unlike that of Earth

One of the most important scientific goals of the *Apollo* missions was to learn as much as possible about the lunar surface. Although only one of the 12 astronauts who visited the Moon was a professional geologist, all of the astronauts received intensive training in field geology, and they were in constant communication with planetary scientists on Earth as they explored the lunar surface.

The Moon has no atmosphere or oceans to cause "weathering" of the surface. But the *Apollo* astronauts found that the lunar surface has undergone a kind of erosion due to billions of years of relentless meteoritic bombardment. This bombardment has pulverized the surface rocks into a layer of fine powder and rock fragments called the **regolith,** from the Greek for "blanket of stone" (**Figure 10-14**). This layer varies in depth from about 2 to 20 m. Just as the rough-surfaced acoustic tile used in ceilings absorbs sound waves, the regolith absorbs most of the sunlight that falls on it. This helps to account for the Moon's very low albedo (see Table 10-1).

In addition to their observations of six different landing sites, the *Apollo* astronauts brought back 2415 individual samples of lunar material totaling 382 kg (842 lb). In addition, the unmanned Soviet spacecraft *Luna 16, 20,* and *24* returned 300 g ($^2/_3$ lb) from three other sites on the Moon. This collection of lunar material has provided important information about the early history of the Moon that could have been obtained in no other way.

All of the lunar samples are igneous rocks; there are no metamorphic or sedimentary rocks (see Section 9-3, especially Figure 9-18). This suggests that most or all of the Moon's surface was once molten. Indeed, Moon rocks are composed mostly of the same minerals that are found in terrestrial volcanic rocks.

Astronauts who visited the maria discovered that these dark regions of the Moon are covered with basaltic rock quite similar to the dark-colored rocks formed by lava from volcanoes on Hawaii and Iceland. The rock of these low-lying lunar plains is called **mare basalt** (**Figure 10-15**).

In contrast to the dark maria, the lunar highlands are composed of a light-colored rock called **anorthosite** (**Figure 10-16**). On the Earth, anorthositic rock is found only in such very old mountain ranges as the Adirondacks in the eastern United States. In comparison to the mare basalts, which have more of the heavier elements like iron, manganese, and titanium, anorthosite is rich in silicon, calcium, and aluminum. Anorthosite is therefore less dense than basalt. During the period when the Moon's surface was molten, the less dense anorthosite rose to the top, forming the terrae that make up the majority of the present-day lunar surface.

Although anorthosite is the dominant rock type in the lunar highlands, most of the rock samples collected there were not pure anorthosite but rather **impact breccias.** These are composites of different rock types that were broken apart, mixed, and fused together by a series of meteoritic impacts (**Figure 10-17**).

Every rock found on Earth, even those from the driest desert, has some water locked into its mineral structure. But lunar rocks are totally dry. With the exception of ice at the lunar poles (see Section

figure 10-14   R I **V** U X G

The Regolith Billions of years of bombardment by space debris have pulverized the uppermost layer of the Moon's surface into powdered rock. This layer, called the regolith, is utterly bone-dry. It nonetheless sticks together like wet sand, as shown by the sharp outline of an astronaut's bootprint. (*Apollo 11,* NASA)

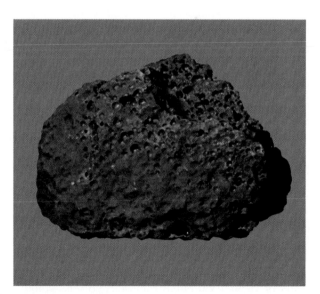

figure 10-15   R I **V** U X G

Mare Basalt This 1.531-kg (3.376-lb) specimen of mare basalt was brought back by the crew of *Apollo 15.* Gas must have dissolved under pressure in the lava from which this rock solidified. When the lava reached the airless lunar surface, the pressure on the lava dropped and the gas formed bubbles. Some of the bubbles were frozen in place as the rock cooled, forming the holes on its surface. (Figure 9-18*a* shows a sample of basalt from Earth.) (NASA)

figure 10-16   R I **V** U X G

Anorthosite The light-colored lunar terrae (highlands) are composed of this ancient type of rock, which is thought to be the material of the original lunar crust. Lunar anorthosites vary in color from dark gray to white; this sample from the *Apollo 16* mission is a medium gray. It measures 18 × 16 × 7 cm. (NASA)

figure 10-17   R I **V** U X G

Impact Breccias These rocks are evidence of the Moon's long history of bombardment from space. Parts of the Moon's original crust were shattered and strewn across the surface by meteoritic impacts. Later impacts compressed and heated this debris, welding it into the type of rock shown here. Such impact breccias are rare on Earth but abundant on the Moon. (NASA)

10-2), there is no evidence that water has ever existed on the Moon. Because the Moon lacks both an atmosphere and liquid water, it is not surprising that no traces of life have ever been found there.

By carefully measuring the abundances of trace amounts of radioactive elements in lunar samples and applying the principles of radioactive age-dating (see Section 8-2), geologists have determined that lunar rocks formed more than 3 billion years ago. Of these ancient rocks, however, anorthosite is much older than the mare basalts. Typical anorthositic specimens from the highlands are between 4.0 and 4.3 billion years old. All these extremely ancient specimens are probably samples of the Moon's original crust. In sharp contrast, all the mare basalts are between 3.1 and 3.8 billion years old. With no plate tectonics to bring fresh material to the surface, no new anorthosites or basalts have formed on the Moon for more than 3 billion years.

By correlating the ages of Moon rocks with the density of craters at the sites where the rocks were collected, geologists have found that the rate of impacts on the Moon has changed over the ages. The ancient, heavily cratered lunar highlands are evidence of an intense bombardment that dominated the Moon's early history. For nearly a billion years, rocky debris left over from the formation of the planets rained down upon the Moon's young surface. As Figure 10-18 shows, this barrage extended from 4.56 billion years ago, when the Moon's surface solidified, until about 3.8 billion years ago. At its peak, this bombardment from space would have produced a new crater of about 1 km radius somewhere on the Moon about once per century. (If this seems like a long time interval, remember that we are talking about a bombardment that lasted hundreds of millions of years and produced millions of craters over that time.)

The frequency of impacts gradually tapered off, however, as meteoroids and planetesimals were swept up by the newly formed planets. This crater-making era ended about the time when the impact of several large planetesimals gouged out the mare basins. A series of lava flows that occurred between 3.8 and 3.1 billion years ago flooded the mare basins with iron-rich magma that oozed up

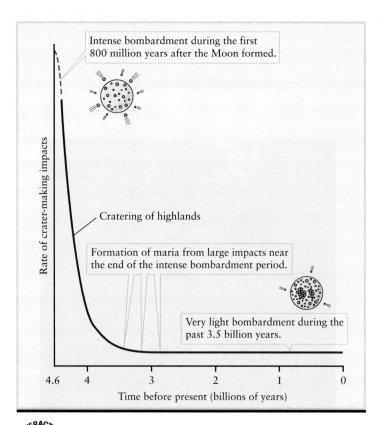

figure 10-18

INTERACTIVE EXERCISE 10.1

The Rate of Crater Formation on the Moon This graph shows the rate at which impact craters formed over the Moon's history. The very early history of cratering is not well understood, because the most ancient craters have been covered up or erased by subsequent craters. To indicate this uncertainty, the leftmost part of the curve is shown as a dashed line. The impact rate for the past 3.5 billion years has been only about 1/10,000 as great as during the most intense bombardment.

from the Moon's still-molten mantle. The relative absence of craters in the maria (see Figure 10-5) tells us that the impact rate over the past 3 billion years has been quite low.

10-5 | The Moon probably formed from debris cast into space when a huge planetesimal struck the proto-Earth

Scientists have refined their theories of the Moon's origin to account for the properties of lunar rocks

The lunar rocks collected by the *Apollo* and *Luna* missions have provided important clues about the history of the Moon. Additional information has come from a simple device that the *Apollo 11, 14,* and *15* missions left on the Moon—a set of reflectors, similar to the orange and red ones found in automobile taillights. Since 1969, scientists on Earth have been firing pulses of laser light at these reflectors and measuring very accurately the length of time it takes each pulse to return to Earth. Knowing the speed of light, they can use this data to calculate the distance to the Moon to an accuracy of just 3 centimeters. Remarkably, they have found that the Moon is moving away from the Earth at a very gradual rate of 3.8 cm per year.

This recession was predicted 70 years before *Apollo* by the British astronomer George Darwin, second son of the famous evolutionist Charles Darwin. George Darwin began with the notion that the Moon's tidal forces elongate the oceans into a football shape (see Section 4-8, especially Figure 4-25). However, the long axis of this "football" does not point precisely at the Moon. The Earth spins on its axis more rapidly than the Moon revolves around the Earth, and this rapid rotation carries the tidal bulge about 10° ahead of a line connecting the Earth and Moon (**Figure 10-19**). This misaligned bulge produces a small but steady gravitational force that tugs the Moon forward and lifts it into an ever larger orbit around the Earth. In other words, Darwin concluded that the Moon must be spiraling away from the Earth—which is just what the laser experiments confirmed.

As the Moon moves away from the Earth, its sidereal period is getting longer in accordance with Kepler's third law (see Section 4-7). At the same time, friction between the oceans and the body of the Earth is gradually slowing the Earth's rotation. The length of the Earth's day is therefore increasing, by approximately 0.002 second per century.

These observations mean that in the distant past, the Earth was spinning faster than it is now and the Moon was much closer. Darwin theorized that the early Earth may have been spinning so fast that a large fraction of its mass tore away, and that this fraction coalesced to form the Moon. This is called the **fission theory** of the Moon's origin. Prior to the *Apollo* program, two other theories were in competition with the fission theory. The **capture theory** posits that the Moon was formed elsewhere in the solar system and then drawn into orbit about the Earth by gravitational forces. By contrast, the **co-creation theory** proposes that the Earth and the Moon were formed at the same time but separately.

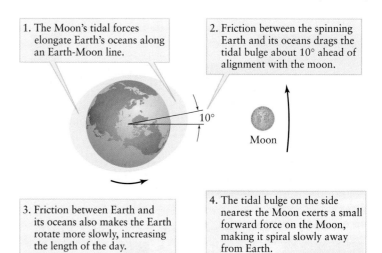

1. The Moon's tidal forces elongate Earth's oceans along an Earth-Moon line.

2. Friction between the spinning Earth and its oceans drags the tidal bulge about 10° ahead of alignment with the moon.

10°

Moon

3. Friction between Earth and its oceans also makes the Earth rotate more slowly, increasing the length of the day.

4. The tidal bulge on the side nearest the Moon exerts a small forward force on the Moon, making it spiral slowly away from Earth.

Figure 10-19

The Moon's Tidal Recession The Earth's rapid rotation drags the tidal bulge of the oceans about 10° ahead of a direct alignment with the Moon. The bulge on the side nearest the Moon exerts more gravitational force than the other, more distant bulge. The net effect is a small forward force on the Moon that makes it spiral slowly away from the Earth.

One fact used to support the fission theory was that the Moon's average density (3344 kg/m³) is comparable to that of the Earth's outer layers, as would be expected if the Moon had been flung out of a rapidly rotating proto-Earth. However, the fission theory predicts that lunar and terrestrial rocks should be similar in composition. In fact, samples returned from the Moon show that they are significantly different.

Volatile elements (such as potassium and sodium) boil at relatively low temperatures under 900°C, whereas **refractory elements** (such as titanium, calcium, and aluminum) boil at 1400°C or higher. Compared with terrestrial rocks, the lunar rocks have slightly greater proportions of the refractory elements and slightly lesser proportions of the volatile elements. This suggests that the Moon formed from material that was heated to a higher temperature than that from which the Earth formed. Some of the volatile elements boiled away, leaving the young moon relatively enriched in the refractory elements.

Proponents of the capture theory used these differences to argue that the Moon formed elsewhere and was later captured by the Earth. They also noted that the plane of the Moon's orbit is close to the plane of the ecliptic, as would be expected if the Moon had originally been in orbit about the Sun but was later captured by the Earth.

There are difficulties with the capture theory, however. If the Earth did capture the Moon intact, then a host of conditions must have been met entirely by chance. The Moon must have coasted to within 50,000 km of the Earth at exactly the right speed to leave a solar orbit and must gone into orbit around the Earth without actually hitting our planet.

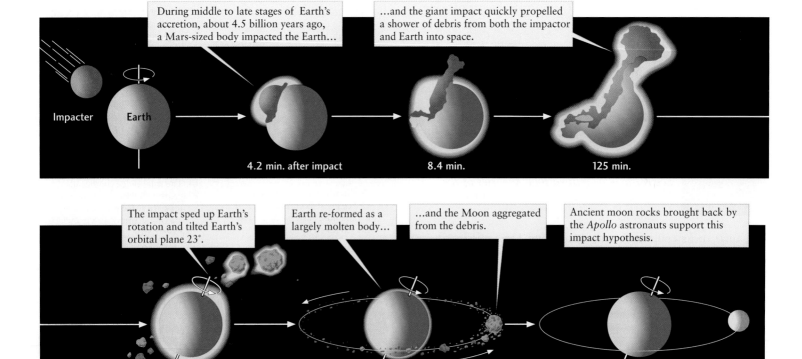

During middle to late stages of Earth's accretion, about 4.5 billion years ago, a Mars-sized body impacted the Earth...

...and the giant impact quickly propelled a shower of debris from both the impactor and Earth into space.

Impacter   Earth

4.2 min. after impact   8.4 min.   125 min.

The impact sped up Earth's rotation and tilted Earth's orbital plane 23°.

Earth re-formed as a largely molten body...

...and the Moon aggregated from the debris.

Ancient moon rocks brought back by the *Apollo* astronauts support this impact hypothesis.

ANIMATION 10.2

Figure 10-20

The Formation of the Moon This figure shows how the Moon could have formed in the aftermath of a collision between a Mars-sized protoplanet and the proto-Earth. (Adapted from F. Press, R. Siever, T. Grotzinger, and T. H. Jordan, *Understanding Earth* 4th Ed., W. H. Freeman, 2004)

Proponents of the co-creation theory argued that it is easier for a planet to capture swarms of tiny rocks. In this theory the Moon formed from just such rocky debris. Great numbers of rock fragments orbited the newborn Sun in the plane of the ecliptic along with the protoplanets. Heat generated in collisions could have baked the volatile elements out of these smaller rock fragments. Then, just as planetesimals accreted to form the proto-Earth in orbit about the Sun, the fragments in orbit about the Earth accreted to form the Moon.

The present scientific consensus is that *none* of these three theories—fission, capture, and co-creation—is correct. Instead, the evidence points toward an idea proposed in 1975 by William Hartmann and Donald Davis and independently by Alastair Cameron and William Ward. In this **collisional ejection theory,** the proto-Earth was struck off-center by a Mars-sized object, and this collision ejected debris from which the Moon formed.

The collisional ejection theory agrees with what we know about the early history of the solar system. We saw in Section 8-4 that smaller objects collided and fused together to form the inner planets. Some of these collisions must have been quite spectacular, especially near the end of planet formation, when most of the mass of the inner solar system was contained in a few dozen large protoplanets. The collisional ejection theory proposes that one such protoplanet struck the proto-Earth 4.56 billion years ago. **Figure 10-20** shows the results of a supercomputer simulation of this cataclysm. In the simulation, energy released during the collision produces a huge plume of vaporized rock that squirts out from the point of impact. As this ejected material cools, it coalesces to form the Moon.

The collisional ejection theory also agrees with many properties of lunar rocks and of the Moon as a whole. For example, rock vaporized by the impact would have been depleted of volatile elements and water, leaving the moon rocks we now know. If the off-center collision took place after chemical differentiation had occurred on the Earth, so that our planet's iron had sunk to its center, then relatively little iron would have been ejected to become part of the Moon. This explains the Moon's low density and the small size of its iron core. It also explains a curious property of the South Pole–Aitken Basin (see Figure 10-7), where an ancient impact excavated the surface down to a depth of 12 km. Data from the *Clementine* spacecraft show that iron concentration in this basin is only 10% (see Figure 10-11), far less than the 20–30% concentration at a corresponding depth below the Earth's surface.

We can now summarize the entire geologic history of the Moon. The surface of the newborn Moon was probably molten for many years, both from heat released during the impact of rock fragments falling into the young satellite and from the decay of short-lived radioactive isotopes. As the Moon gradually cooled, low-density lava floating on the Moon's surface began to solidify into the anorthositic crust that exists today. The barrage of rock fragments that ended about 3.8 billion years ago produced the craters that cover the lunar highlands.

At the end of this crater-making era, more than a dozen asteroid-size objects, each measuring at least 100 km across, struck the Moon and blasted out the vast mare basins. Meanwhile, heat from the decay of long-lived radioactive elements like uranium and thorium began to melt the inside of the Moon. From 3.8 to 3.1 billion years ago, great floods of molten rock gushed up from the lunar interior, filling the impact basins and creating the basaltic maria we see today.

Very little has happened on the Moon since those ancient times. A few relatively fresh craters have been formed, but the astronauts visited a world that has remained largely unchanged for more than 3 billion years. During that same period on the Earth, by contrast, the continents have been totally transformed time and time again (see Section 9-3).

Many questions and mysteries still remain. The six American and three Soviet lunar landings have brought back samples from only nine locations, barely scratching the lunar surface. We still know very little about the Moon's far side and poles. Are there really vast ice deposits at the poles? How much of the Moon's interior is molten? Just how large is its iron core? How old are the youngest lunar rocks? Did lava flows occur over western Oceanus Procellarum only 2 billion years ago, as crater densities there suggest? Could there still be active volcanoes on the Moon? Are there mineral deposits on the lunar surface from which oxygen for life support and rocket fuel could be extracted, making the Moon a springboard for human exploration of the solar system (**Figure 10-21**)? Such questions can be answered only by returning to the Moon.

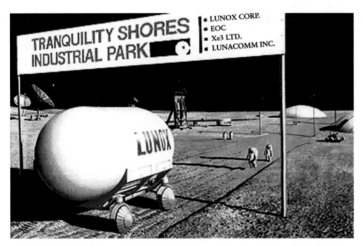

Figure 10-21

The Moon—Site of Future Industry? Volcanic material on the lunar surface contains trapped oxygen, which could be released by heating lunar material along with hydrogen gas. Oxygen and water could also be extracted from ice deposits at the lunar poles. These essentials could support a permanent lunar colony and long-term exploration of the Moon. The Moon's weaker gravity and lack of an atmosphere also make it a useful launching pad for sending payloads farther into the solar system. The rocket fuel for this enterprise would use oxygen "mined" from the Moon. (NASA)

Key Words

anorthosite, p. 221

capture theory (of Moon's formation), p. 223

center of mass, p. 211

co-creation theory (of Moon's formation), p. 223

collisional ejection theory (of Moon's formation), p. 224

crater, p. 213

far side (of the Moon), p. 211

fission theory (of Moon's formation), p. 223

impact breccia, p. 221

impact crater, p. 213

libration, p. 212

lunar highlands, p. 214

mare (*plural* maria), p. 213

mare basalt, p. 221

moonquake, p. 218

refractory element, p. 223

regolith, p. 221

synchronous rotation, p. 211

terminator, p. 211

terrae, p. 214

volatile element, p. 223

Key Ideas

Appearance of the Moon: The Earth-facing side of the Moon displays light-colored, heavily cratered highlands and dark-colored, smooth-surfaced maria. The Moon's far side has almost no maria.

• Virtually all lunar craters were caused by space debris striking the surface. There is no evidence of plate tectonic activity on the Moon.

Internal Structure of the Moon: Much of our knowledge about the Moon has come from human exploration in the 1960s and early 1970s and from more recent observations by unmanned spacecraft.

• Analysis of seismic waves and other data indicates that the Moon has a crust thicker than that of the Earth (and thickest on the far side of the Moon), a mantle with a thickness equal to about 80% of the Moon's radius, and a small iron core.

• The Moon's lithosphere is far thicker than that of the Earth. The lunar asthenosphere probably extends from the base of the lithosphere to the core.

• The Moon has no global magnetic field today, although it had a weak magnetic field billions of years ago.

Geologic History of the Moon: The anorthositic crust exposed in the highlands was formed between 4.0 and 4.3 billion years ago, whereas the mare basalts solidified between 3.1 and 3.8 billion years ago.

• The Moon's surface has undergone very little change over the past 3 billion years.

• Meteoroid impacts have been the only significant "weathering" agent on the Moon. The Moon's regolith, or surface layer of powdered and fractured rock, was formed by meteoritic action.

• All of the lunar rock samples are igneous rocks formed largely of minerals found in terrestrial rocks. The lunar rocks contain no water and also differ from terrestrial rocks in being relatively enriched in the refractory elements and depleted in the volatile elements.

Origin of the Moon: The collisional ejection theory of the Moon's origin holds that the proto-Earth was struck by a Mars-sized protoplanet and that debris from this collision coalesced to form the Moon. This theory successfully explains most properties of the Moon.

• The Moon was molten in its early stages, and the anorthositic crust solidified from low-density magma that floated to the lunar surface. The mare basins were created later by the impact of planetesimals and filled with lava from the lunar interior.

• Tidal interactions between the Earth and Moon are slowing the Earth's rotation and pushing the Moon away from the Earth.

Review Questions

1. Is it correct to say that the Moon orbits the Earth? If not, what is a more correct description?

2. If the Moon always keeps the same face toward the Earth, how is it possible for Earth observers to see more than half of the Moon's surface?

3. Explain why liquid water cannot exist on the surface of the Moon.

4. What kind of features can you see on the Moon with a small telescope?

5. Are impact craters on the Moon the same size as the meteoroids that made the impact? Explain.

6. Describe the differences between the maria and the lunar highlands. Which kind of terrain is more heavily cratered? Which kind of terrain was formed later in the Moon's history? How do we know?

7. What does it mean to say the Moon is a "one-plate world"? What is the evidence for this statement?

8. Why was it necessary to send unmanned spacecraft to land on the Moon before sending humans there?

9. What is the evidence that ice exists at the lunar poles? Is this evidence definitive?

10. Why was it useful for the *Apollo* astronauts to bring magnetometers and seismometers to the Moon?

11. Could you use a magnetic compass to navigate on the Moon? Why or why not?

12. Describe the evidence that (**a**) the Moon has a more solid interior than the Earth and (**b**) the Moon's interior is not completely solid.

13. Explain why moonquakes occur more frequently when the Moon is at perigee than at other locations along its orbit.

14. Why is the Earth geologically active while the Moon is not?

15. What is the regolith? What causes its powdery character?

16. On the basis of moon rocks brought back by the astronauts, explain why the maria are dark-colored but the lunar highlands are light-colored.

17. Briefly describe the main differences and similarities between Moon rocks and Earth rocks.

18. Rocks found on the Moon are between 3.1 and 4.56 billion years old. By contrast, the majority of the Earth's surface is made of oceanic crust that is less than 200 million years old, and the very oldest Earth rocks are about 3 billion years old. If the Earth and Moon are essentially the same age, why is there such a disparity in the ages of rocks on the two worlds?

19. Why do most scientists favor the collisional ejection theory of the Moon's formation?

20. Some people who supported the fission theory proposed that the Pacific Ocean basin is the scar left when the Moon pulled away from the Earth. Explain why this idea is probably wrong.

Advanced Questions

Questions preceded by an asterisk (*) involve topics discussed in Box 10-1.

Problem-solving tips and tools

Recall that the average density of an object is its mass divided by its volume. The volume of a sphere is $4\pi r^3/3$, where r is the sphere's radius. The surface area of a sphere of radius r is $4\pi r^2$, while the surface area of a circle of radius r is πr^2. Recall also that the acceleration of gravity on the Earth's surface is 9.8 m/s². You may find it useful to know that a 1-pound (1-lb) weight presses down on the Earth's surface with a force of 4.448 newtons. You might want to review Newton's universal law of gravitation in Section 4-7. The time to travel a certain distance is equal to the distance traveled divided by the speed of motion. Consult Table 10-1 and the Appendices for any additional data.

---

21. Suppose two worlds (say, a planet and its satellite) have masses m_1 and m_2, and the center-to-center distance between the worlds is r. The distance d_{cm} from the center of world 1 to the center of mass of the system of two worlds is given by the formula

$$d_{cm} = \frac{m_2 r}{m_1 + m_2}$$

(**a**) Suppose world 1 is the Earth and world 2 is the Moon. If the Earth and Moon are at their average center-to-center distance, find the distance from the center of the Earth to the center of mass of the Earth-Moon system. (**b**) Is the Earth-Moon system's center of mass within the Earth? How far below the Earth's surface is it located? (**c**) Find the distance from the center of the Sun (mass 1.989×10^{30} kg) to the center of mass of the Sun-Earth system. How does this compare to the diameter of the Sun? Is it a good approximation to say that the Earth orbits around the center of the Sun?

22. In a whimsical moment during the *Apollo 14* mission, astronaut Alan Shepard hit two golf balls over the lunar surface. Give two reasons why they traveled much farther than golf balls do on Earth.

23. If you view the Moon through a telescope, you will find that details of its craters and mountains are more visible when the Moon is near first quarter phase or third quarter phase than when it is at full phase. Explain why.

24. Temperature variations between day and night are much more severe on the Moon than on Earth. Explain why.

25. Using the diameter and mass of the Moon given in Table 10-1, verify that the Moon's average density is about 3344 kg/m^3. Explain why this average density implies that the Moon's interior contains much less iron than the interior of the Earth.

26. How much would an 80-kg person weigh on the Moon? How much does that person weigh on the Earth?

*27. In Box 10-1 we calculated the tidal force that the Earth exerts on two 1-kg rocks located on the near and far sides of the Moon. We assumed that the Earth-Moon distance was equal to its average value. Repeat this calculation (**a**) for the Moon at perigee and (**b**) for the Moon at apogee. (**c**) What is the ratio of the tidal force on the rocks at perigee to the tidal force at apogee?

28. The last manned spacecraft to land on the Moon was the *Apollo 17* lunar module *Challenger* in December 1972. After lifting off from the lunar surface and returning to the command module *America*, in which they would return to the Earth, the *Apollo 17* astronauts sent the unoccupied *Challenger* to crash into the lunar surface. The seismic waves from this impact were detected by seismometers left by the crews of *Apollo 12, 14, 15, 16*, and *17*. Why was it useful to do this? Why was it not enough to detect seismic waves from naturally occurring moonquakes?

29. When examined back on Earth, pieces of the *Surveyor 3* spacecraft (see Figure 10-9) that were exposed to the lunar environment for 2½ years showed many microscopic craters caused by the impact of tiny micrometeoroids. But a sample of similar material left outside on Earth for 2½ years shows no such impact cratering. Why is there a difference?

30. The youngest lunar anorthosites are 4.0 billion years old, and the youngest mare basalts are 3.1 billion years old. Would you expect to find any impact breccias on the Moon that formed less than 3.1 billion years ago? Explain your answer.

31. In the maria, the lunar regolith is about 2 to 8 meters deep. In the lunar highlands, by contrast, it may be more than 15 meters deep. Explain how the different ages of the maria and highlands can account for these differences.

32. The mare basalts are volcanic rock. Is it likely that active volcanoes exist anywhere on the Moon today? Explain.

33. Calculate the round-trip travel time for a pulse of laser light that is fired from a point on the Earth nearest the Moon, hits a reflector at the point on the Moon nearest the Earth, and returns to its point of origin. Assume that the Earth and Moon are at their average separation from each other.

34. If the tidal recession of the Moon were to continue indefinitely, eventually the length of the day and the lunar month would both be equal to 47 of our present days. At that point the Earth would always keep the same face toward the Moon. Determine the average distance between the Earth and the Moon in this situation. (In fact, the Sun will have come to the end of its lifetime long before this happens.)

35. Before the *Apollo* missions to the Moon, there were two diametrically opposite schools of thought about the history of lunar geology. The "cold moon" theory held that all lunar surface features were the result of impacts. The most violent impacts melted the surface rock, which then solidified to form the maria. The opposite "hot moon" theory held that all lunar features, including maria, mountains, and craters, were the result of volcanic activity. Explain how lunar rock samples show that neither of these theories is entirely correct.

*36. When the Moon originally coalesced, it may have been only one-tenth as far from the Earth as it is now. (**a**) When the Moon first coalesced, was the Earth's tidal force strong enough to lift rocks off the lunar surface? Explain. (**b**) Compared with the net tidal force that the Earth exerts on the Moon today, how many times larger was the net tidal force on the newly coalesced Moon? (This strong tidal force kept the one axis of the Moon oriented toward the Earth, and the Moon kept that orientation after it solidified.)

37. If the Earth and the Moon had formed independently in the same region of the solar system, we would expect that iron would be present in each world in about the same percentage. Explain how the different relative sizes of the Earth's core and the Moon's core suggest that the Moon formed instead by collisional ejection.

Discussion Questions

38. Comment on the idea that without the presence of the Moon in our sky, astronomy would have developed far more slowly.

39. No *Apollo* mission landed on the far side of the Moon. Why do you suppose this was? What would have been the scientific benefits of a mission to the far side?

40. Compare the advantages and disadvantages of exploring the Moon with astronauts as opposed to using mobile, unmanned instrument packages.

41. Describe how you would empirically test the idea that human behavior is related to the phases of the Moon. What problems are inherent in such testing?

42. How would our theories of the Moon's history have been affected if astronauts had discovered sedimentary rock on the Moon?

43. Imagine that you are planning a lunar landing mission. What type of landing site would you select? Where might you land to search for evidence of recent volcanic activity?

Web/CD-ROM Questions

44. Several unmanned missions to the Moon were in progress or under development as of early 2004. These include *LUNAR-A* and *SELENE* (Institute of Space and Astronautical Science, Japan) and *SMART-1* (European Space Agency). Search the World Wide Web for current information about these missions. When is each mission scheduled to be launched? What investigations will each mission make that have not been made before? What scientific issues may these missions resolve?

45. Determining the Size of the Planetesimal that Formed the Moon. Access the animation "The Formation of the Moon" in Chapter 10 of the *Universe* Web site or CD-ROM. By making measurements on the screen with a ruler, determine how many times larger the proto-Earth is than the impacting planetesimal. If Mars is about 50% the size of Earth, how does the planetesimal compare in size with present-day Mars?

Observing Projects

Observing tips and tools

If you do not have access to a telescope, you can learn a lot by observing the Moon through binoculars. Note that the Moon will appear right-side-up through binoculars but inverted through a telescope; if you are using a map of the Moon to aid your observations, you will need to take this into account. Inexpensive maps of the Moon can be purchased from most good bookstores or educational supply stores. You can determine the phase of the Moon either by looking at a calendar (most of which tell you the dates of new moon, first quarter, full moon, and third quarter), by checking the weather page of your newspaper, by consulting the current issue of *Sky & Telescope* or *Astronomy* magazine, by using the *Starry Night Backyard*™ program, or by using the World Wide Web.

46. Use a telescope or binoculars to observe the Moon. Compare the texture of the lunar surface you see on the maria with that of the lunar highlands. How does the visibility of details vary with distance from the terminator (the boundary between day and night on the Moon)?

47. Observe the Moon through a telescope every few nights over a period of two weeks between new moon and full moon. Make sketches of various surface features, such as craters, mountain ranges, and maria. How does the appearance of these features change with the Moon's phase? Which features are most easily seen at a low angle of illumination? Which features show up best with the Sun nearly overhead?

 48. Use the *Starry Night Backyard*™ program to observe the changing appearance of the Moon. Display the entire celestial sphere (select **Guides > Atlas** in the Go menu) and center on the Moon by using the **Find...** command in the Edit menu. Then use the zoom controls at the right-hand end of the Control Panel (at the top of the main window) to adjust your view so that you can see the Moon clearly. (Click on the binoculars to zoom in and on the person to zoom out.) Click on the control for the rate of time flow (immediately to the right of the date and time display) and set the discrete time step to 1 hour. Then click on the Forward button (a triangle that points to the right). (**a**) Describe how the phase of the Moon changes. (**b**) Look carefully at features near the left-hand and right-hand limbs (edges) of the Moon. Are these features always at the same position relative to the limb? Explain in terms of libration. (**c**) Note that the apparent size of the Moon changes somewhat over the course of a day (of simulated time). Explain in terms of the Earth's rotation. (*Hint:* In **Atlas** mode, *Starry Night Backyard*™ makes the Earth transparent so you can see objects that would normally be below the horizon. As the Earth rotates, your distance from the Moon changes.) (**d**) Change the discrete time step to 1 day and again click the Forward button. Does the apparent size of the Moon always stay the same, or does it vary? Explain what this tells you about the shape of the Moon's orbit.

 49. Use the *Deep Space Explorer*™ program to examine the Moon. In the left-hand part of the window under the heading **Solar System,** click on the triangle next to the word **Explore** and then click on **Moon.** You can zoom in and zoom out using the buttons at the upper left of the window (an upward-pointing triangle and a downward-pointing triangle). You can also rotate the Moon by putting the mouse cursor over the image, holding down the mouse button, and moving the mouse. (On a two-button mouse, hold down the left mouse button.) (**a**) Based *only* on what you can see in the image, what evidence can you find that the Moon is geologically inactive? Explain. (**b**) Spreading outward from some of the largest craters on the Moon are straight lines of light-colored material, called *rays,* that were caused by material ejected outward by the impact that caused the crater. Can you find any rays on the Moon? (Rotate the Moon around to see all of the illuminated surface.) Estimate the length of the rays by measuring their length on the screen with a ruler and comparing to the diameter of the Moon (see Table 10-1).

Collaborative Exercise

59. The image of Crater Clavius in Figure 10-4 reveals numerous craters. Using the idea that the Moon's landscape can only be changed by impacts, make a rough sketch showing ten of the largest craters and label them from oldest (those that showed up first) to youngest (the most recent ones). Explain your reasoning and any uncertainties.

11 Sun-Scorched Mercury

The images shown here depict two heavily cratered worlds: Mercury (on the left) and the Moon (on the right). While lunar craters have been known since Galileo first turned his telescope on the Moon in 1610, the surface of Mercury was not seen clearly until the 1970s.

Mercury is difficult to observe from Earth: It is too small and too close to the Sun. Earth-based visual observations reveal very few surface details, although radio and radar observations provide information about the planet's temperature and rotation. It was not until 1974 that the first (and so far only) spacecraft to visit Mercury gave us close-up images of the planet. The craters of Mercury, like those of the Moon, are the scars of countless impacts that occurred soon after the planets were formed.

Mercury is by no means a mere clone of the Moon, however. The lava flows on Mercury's surface lack the characteristic dark color of the lunar maria. The surface of Mercury has unusual wrinkles quite unlike anything found on the Moon or, for that matter, the Earth. And unlike the Moon, Mercury has a large iron core and a magnetic field. Mercury thus has a Moonlike surface but an Earthlike interior.

Mercury is also quite unlike our Moon in the way it spins on its axis. While the Moon makes one rotation during one orbit around the Earth, Mercury makes exactly 1½ rotations during one orbit around the Sun. No other planet or satellite rotates in this way.

Despite its small size, Mercury is an important planet to astronomers. If we can understand how Mercury formed and acquired its unique personality, we may be able to understand how the other terrestrial planets—including our own Earth—came to be.

Mercury | The Moon

 WEB LINK 11.1   Mercury and the Moon to scale. (NASA)   R I **V** U X G

Questions to Guide Your Reading

As you read the sections of this chapter, look for the answers to the following questions:

11-1   What makes Mercury such a difficult planet to see?

11-2   What is unique about Mercury's rotation?

11-3   How do the surface features on Mercury differ from those on the Moon?

11-4   Is Mercury's internal structure more like that of the Earth or the Moon?

11-1 | Earth-based optical observations of Mercury are difficult

Mercury is often one of the brightest objects in the sky. At its greatest brilliance, it appears brighter than any star, which is why Mercury has been known since ancient times. Its motions played a role in the religious beliefs of the ancient Mayans, Egyptians, Greeks, and Romans.

Although it may be bright, Mercury is so close to the Sun that it is quite difficult to observe. (It is said that Copernicus never saw it.) Mercury moves around the Sun at an average distance of only 0.387 AU (57.9 million kilometers, or 36 million miles) along an orbit that is more eccentric than that of any other planet except Pluto. **Figure 11-1** shows the orbits of Mercury and the Earth to scale. **Table 11-1** summarizes data about the planet.

Like all the planets, Mercury shines by reflected sunlight. Because it orbits so close to the Sun, it is exposed to sunlight that on average is about 7 times as intense as on Earth. Furthermore, Mercury is always less than 1.5 AU from Earth. These factors explain why Mercury appears bright, despite having a rather low albedo of about 0.12 (that is, it reflects only about 12% of the sunlight that falls on its surface, roughly comparable to weathered asphalt).

Mercury can best be seen with the naked eye when it is as far from the Sun in the sky as it can be, at its greatest eastern or western elongation (Section 4-2). For a few days near the time of **greatest eastern elongation**, Mercury appears low over the western horizon as an "evening star" for a short time after sunset. Near the time of **greatest western elongation**, Mercury can be glimpsed as a "morning star" that rises before the Sun in the brightening eastern sky.

CAUTION ! Remember that "greatest *western* elongation" and "greatest *eastern* elongation" refer to Mercury's position in the sky relative to the Sun. When Mercury is at greatest western elongation, it is west of the Sun in the sky; at dawn, when the Sun is rising in the east, Mercury has already risen. When Mercury is at greatest eastern elongation, it is east of the Sun in the sky and is still above the horizon when the Sun sets in the west.

Because its orbit is so close to the Sun, Mercury's maximum elongation is only 28°. The celestial sphere appears to rotate at 15° per hour (360° divided by 24 hours), so Mercury never rises more than 2 hours before sunrise or sets more than 2 hours after sunset. Unfortunately, Mercury's elliptical orbit and its inclination to the ecliptic often place Mercury much less than 28° from the horizon at sunset or sunrise. Hence, some elongations are more favorable for viewing Mercury than others, as **Figure 11-2** shows. A total of six or seven greatest elongations (both eastern and western) occur each year (**Table 11-2**), but usually only two of these will be favorable for viewing the planet.

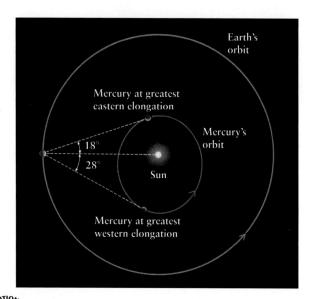

ANIMATION 11.1 **Figure 11-1**

Mercury's Orbit Mercury moves around the Sun every 88 days in a rather eccentric orbit. The distance between Mercury and the Sun varies from 70 million kilometers at aphelion to only 46 million kilometers at perihelion. As seen from the Earth, the angle between Mercury and the Sun at greatest eastern or western elongation can be as large as 28° (when Mercury is near aphelion) or as small as 18° (near perihelion).

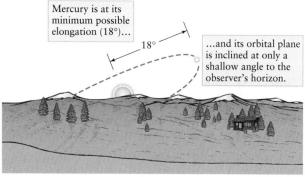

Figure 11-2

Favorable Versus Unfavorable Elongations The tilt of the Earth's axis, the inclination and eccentricity of Mercury's orbit, and the latitude of the observer combine to make an elongation either favorable (upper illustration) or unfavorable (lower illustration) for viewing Mercury.

table 11-1	Mercury Data
Average distance from Sun:	0.387 AU = 5.79 × 10⁷ km
Maximum distance from Sun:	0.467 AU = 6.98 × 10⁷ km
Minimum distance from Sun:	0.307 AU = 4.60 × 10⁷ km
Eccentricity of orbit:	0.206
Average orbital speed:	47.9 km/s
Orbital period:	87.969 days
Rotation period:	58.646 days
Inclination of equator to orbit:	0.5°
Inclination of orbit to ecliptic:	7° 00′ 16″
Diameter (equatorial):	4880 km = 0.383 Earth diameter
Mass:	3.302 × 10²³ kg = 0.0553 Earth mass
Average density:	5430 kg/m³
Escape speed:	4.3 km/s
Surface gravity (Earth = 1):	0.38
Albedo:	0.12
Average surface temperatures:	Day: 350°C = 662°F = 623 K
	Night: −170°C = −274°F = 103 K
Atmosphere:	Essentially none

 WEB LINK 11.2

(Astrogeology Team, U.S. Geological Survey)   R I **V** U X G

Naked-eye observations of Mercury are best made at dusk or dawn, but the best telescopic views are obtained in the daytime when the planet is high in the sky, far above the degrading atmospheric effects near the horizon. A yellow filter can eliminate much of the blue light from our sky. The photographs in Figure 11-3, which are among the finest Earth-based views of Mercury, were taken during daytime.

Mercury travels around the Sun faster than any other object in the solar system. Its sidereal period, or time to complete one full orbit, is only 88 days. Its synodic period, which is the time to go through a complete cycle of configurations as seen from the Earth (see Section 4-2), is 116 days or about one-third of a year. Thus, Mercury passes through inferior conjunction (when it is between the Earth and the Sun) at least three times a year. Therefore, you might expect to see Mercury occasionally silhouetted against the Sun. Such a passage in front of the Sun is called a **solar transit**. In fact, transits of Mercury across the Sun are not very common, because Mercury's orbit is tilted to the plane of the ecliptic by 7°. As a result, Mercury

table 11-2	Greatest Elongations of Mercury,* 2004–2008	
Year	Greatest western elongations	Greatest eastern elongations
2004	January 17, May 14, September 9, December 29	March 29, July 27, November 21
2005	April 26, August 23, December 12	March 12, July 9, November 3
2006	April 8, August 7, November 25	February 24, June 20, October 17
2007	March 22, July 20, November 8	February 7, June 2, September 29
2008	March 3, July 1, October 22	January 22, May 14, September 11

*Mercury can be seen in the eastern sky at dawn for a few days around greatest western elongation.
It can be seen in the western sky at dusk for a few days around greatest eastern elongation.

(a)   (b)

figure 11-3 R I V U X G

Earth-Based Views of Mercury These are two of the best photographs of Mercury ever produced with an Earth-based telescope. Hazy markings are faintly visible on the tiny planet. Note that Mercury goes through phases like the Moon. It appears larger when it is at first or third quarter, as shown in (a), than when it is nearly full, as shown in (b). Galileo's discovery that the planet Venus obeys this same relationship (see Figure 4-13) was compelling evidence that the planets orbit the Sun and not the Earth. (New Mexico State University Observatory)

usually lies well above or below the solar disk at the moment of inferior conjunction.

In order for a solar transit to take place, the Sun, Mercury, and the Earth must be in nearly perfect alignment. This arrangement can occur only in May and November, when the Earth is located near the line along which the planes of the Earth's orbit and Mercury's orbit intersect. Even then a solar transit will occur only if Mercury happens to be in inferior conjunction at that time. The first two transits of the twenty-first century are on November 8, 2006, and May 9, 2016. The longest transits occur in May, when Mercury is near aphelion and therefore is traveling comparatively slowly along its orbit. The maximum duration of a solar transit is 9 hours. The time-lapse photograph of a solar transit in Figure 11-4 is dramatic proof of how tiny Mercury is in comparison with the Sun!

figure 11-4 R I V U X G

A Solar Transit of Mercury This time-lapse photograph shows Mercury moving across the face of the Sun during the transit of May 7, 2003. The image shows a total elapsed time of 5 hours and 19 minutes. The dark sunspot visible near the middle of the Sun's disk is actually much larger than Mercury. (Dominique Dierick)

11-2 | Mercury rotates slowly and has an unusual spin-orbiting coupling

SCIENCE IN PROCESS *Radar observations coupled with an improved theory solved the mystery of Mercury's rotation*

During the 1880s, the Italian astronomer Giovanni Schiaparelli attempted to make the first map of Mercury. Unfortunately, even today's best telescopes reveal only a few faint, hazy markings. Schiaparelli's views of Mercury were so vague and indistinct that he made an understandable but major error, which went uncorrected for more than half a century. He erroneously concluded that Mercury always keeps the same side facing the Sun.

Many objects in our solar system are in synchronous rotation, so that their rotation period equals their period of revolution—a situation also called **1-to-1 spin-orbit coupling.** We saw in Section 4-8 how the Earth's tidal forces keep the Moon in synchronous rotation, so that it always keeps the same side toward the Earth. Tidal forces also keep the two moons of Mars with the same side facing their parent planet, and likewise many of the satellites of Jupiter and Saturn.

It had been suggested as early as 1865 that the Sun's tidal forces would keep Mercury in synchronous rotation. But this is not how Mercury rotates. The first clue about Mercury's true rotation period came in 1962, when astronomers detected radio radiation coming from the innermost planet.

As we learned in Sections 5-3 and 5-4, every dense object (such as a planet) emits electromagnetic radiation with a continuous spectrum. The dominant wavelength of this radiation depends on the object's temperature. Only at absolute zero does a dense object emit no radiation at all.

If Mercury exhibited synchronous rotation, one side of the planet would remain in perpetual, frigid darkness, quite near absolute zero. To test this idea, in 1962 the radio astronomer W. E. Howard and his colleagues at the University of Michigan began monitoring radiation from Mercury. (If the planet had one side that was very cold, its emission should be principally at quite long wavelengths. Hence, a radio telescope was used for these observations.) They were surprised to discover that the temperature on the planet's nighttime side is about 100 K (= −173°C = −280°F), not nearly as

Antenna at the focal point of the mirror   Concave disk (objective mirror)

figure 11-5 R I **V** U X G
The Arecibo Radio Telescope With a diameter of 305 m (1000 ft), this is the largest single-dish radio telescope on Earth. The dish is not movable, so it can be used only to examine objects near the zenith. The primary use of the telescope is to detect radio waves emitted by objects in space (compare Figure 6-25). To measure the rotation period of Mercury, radio waves were *emitted* by the telescope, and the waves that were reflected back from Mercury were detected. (Arecibo Observatory)

cold as expected. These observations were in direct contradiction to the idea of synchronous rotation.

One idea offered to explain Howard's observations was that Mercury rotates synchronously, but that winds in the planet's atmosphere carried warmth from the daytime side of the planet around to the nighttime side. But, in fact, there are no winds on Mercury. Its daytime temperature is too high and its gravity too weak to retain any substantial atmosphere (see Box 7-2).

The definitive observation was made in 1965, when Rolf B. Dyce and Gordon H. Pettengill used the giant 1000-ft radio telescope at the Arecibo Observatory in Puerto Rico (Figure 11-5) to bounce powerful radar pulses off Mercury. The outgoing radiation consisted of microwaves of a very specific wavelength. In the reflected signal echoed back from the planet, the wavelength had shifted as a result of the Doppler effect (see Section 5-9, especially Figure 5-24). As Mercury rotates, one side of the planet approaches the Earth, while the other side recedes from the Earth. Microwaves reflected from the planet's approaching side were shortened in wavelength, whereas those from its receding side were lengthened (Figure 11-6). Thus, the radar pulse went out at one specific wavelength, but it came back spread over a small wavelength range. From the width of this wavelength range, Dyce and Pettengill deduced that Mercury's rotation period is approximately 58.6 days.

Giuseppe Colombo, an Italian physicist with a long-standing interest in Mercury, found this number intriguing. Colombo noted that Dyce and Pettengill's result for the rotation period is very close to two-thirds of Mercury's sidereal period of 87.969 days:

$$^2/_3 \ (87.969 \text{ days}) = 58.646 \text{ days}$$

Colombo therefore boldly speculated that Mercury's true rotation period is exactly 58.646 days, or 58 days and 15½ hours. He

realized that this figure would mean that Mercury is locked into a **3-to-2 spin-orbit coupling:** The planet makes *three* complete rotations on its axis for every *two* complete orbits around the Sun. No other planet or satellite in the solar system has this curious relationship between its rotation and its orbital motion.

Figure 11-7 shows how gravitational forces from the Sun cause Mercury's 3-to-2 spin-orbit coupling. The force of gravity decreases with increasing distance, which explains how the Moon is able to raise a tidal bulge in the Earth's oceans (see Section 4-8).

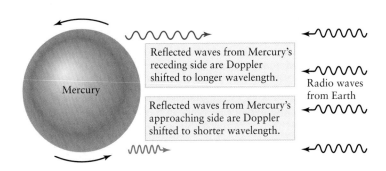

Reflected waves from Mercury's receding side are Doppler shifted to longer wavelength.

Radio waves from Earth

Reflected waves from Mercury's approaching side are Doppler shifted to shorter wavelength.

figure 11-6

Measuring Mercury's Rotation Period As Mercury rotates, one side of the planet moves away from the Earth and the other side toward the Earth. If radio waves of a single wavelength are beamed toward Mercury, waves reflected from these two sides will be Doppler shifted to longer and shorter wavelengths, respectively. By measuring the wavelength shift of the reflected radiation, astronomers deduced how rapidly Mercury rotates and thus determined its rotation period.

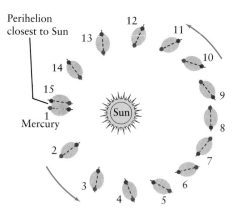

(a) **Mercury is slightly elongated:** The different gravitational forces on Mercury's two ends tend to twist the long axis to point toward the Sun.

(b) **If Mercury were in a circular orbit,** its long axis would always point toward the Sun: Mercury would be in synchronous rotation (1-to-1 spin-orbit coupling).

(c) **In fact Mercury is in an elliptical orbit,** and its long axis only points toward the Sun at perihelion: Mercury spins on its axis 1½ times during each complete orbit (3-to-2 spin-orbit coupling).

ANIMATION 11.2

Figure 11-7

Spin-Orbit Coupling (a) The Sun's gravitational force on the near side of Mercury's long axis is greater than the force on the far side. This tends to twist the long axis to point toward the Sun. (b) If Mercury were in a circular orbit, the twisting effect shown in (a) would always keep the same side of Mercury (shown by a red dot) facing the Sun. (c) Because Mercury's orbit is rather elongated, its rotation is more complex: The end of Mercury's long axis marked by a red dot faces the Sun at one perihelion (point 1), but at the next perihelion (point 15) the opposite end—marked by a blue dot—faces the Sun.

Mercury has no oceans, but it has a natural bulge of its own; thanks to the Sun's tidal forces, the planet is slightly elongated along one axis. The stronger gravitational force that the Sun exerts on the near side of the planet tends to twist the long axis to point toward the Sun, as Figure 11-7a shows. Indeed, Mercury's long axis would always point toward the Sun if its orbit were circular or nearly so. In this case the same side of Mercury would always face the Sun, and there would be synchronous rotation (Figure 11-7b). But Mercury's orbit has a rather high eccentricity, and the twisting effect on Mercury decreases rapidly as the planet moves away from the Sun. As a result, Mercury's long axis points toward the Sun only at perihelion (Figure 11-7c). From one perihelion to the next, alternate sides of Mercury face the Sun.

Colombo's enlightened guess was confirmed by further radar observations and, later, by the *Mariner 10* spacecraft. After making the first-ever close flyby of Mercury in the spring of 1974, *Mariner 10* was placed in a 176-day orbit around the Sun. This orbital period brings the spacecraft back to the planet every two Mercurian years, just enough time for the planet to complete three rotations. During the second and third flybys, the spacecraft's cameras showed that the same side of Mercury was facing the Sun as on the first flyby. Mercury does indeed rotate three times about its axis for every two orbits it makes of the Sun.

Because of the 3-to-2 spin-orbit coupling, the average time from sunrise to sunset on Mercury is just equal to its orbital period, or just under 88 days. This helps explain the tremendous difference between daytime and nighttime temperatures on Mercury. Not only is sunlight about 7 times more intense on Mercury than on the Earth

(because of the smaller size of Mercury's orbit), but that sunlight has almost three Earth months to heat up the surface. As a result, daytime temperatures at the equator are high enough—about 430°C (roughly 700 K, or 800°F)—to melt lead. (In comparison, a typical kitchen oven reaches only about 300°C or 550°F.)

The time from sunset to the next sunrise is also 88 days, and the surface thus has almost three Earth months of darkness during which to cool down. Therefore, nighttime temperatures reach to below –170°C (about 100 K, or –270°F). This is cold enough to freeze carbon dioxide and methane. The surface of Mercury is truly inhospitable!

While the Sun moves slowly across Mercury's sky, it does not always move from east to west as it does as seen from Earth. The reason is that Mercury's speed along its elliptical orbit varies substantially, in accordance with Kepler's second law (see Section 4-4, especially Figure 4-11). It moves fastest (59 km/s) at perihelion and slowest (39 km/s) at aphelion.

As seen from Mercury's surface, the Sun rises in the east and sets in the west, just as it does on the Earth. When Mercury is near perihelion, however, the planet's rapid motion along its orbit outpaces its leisurely rotation about its axis (one rotation in 58.646 days). Hence, the usual east-west movement of the Sun across Mercury's sky is interrupted, and the Sun actually stops and moves backward (from west to east) for a few Earth days. If you were standing on Mercury watching a sunset occurring at perihelion, the Sun would not simply set. It would dip below the western horizon and then come back up, only to set a second time a day or two later.

11-3 | Images from *Mariner 10* reveal Mercury's heavily cratered surface

Scientists acquired their first detailed knowledge about Mercury's surface during 1974 and 1975, when the unmanned *Mariner 10* spacecraft made three close flybys of the planet (**Figure 11-8a**). At each closest approach *Mariner 10* passed over Mercury's darkened, nighttime side, and so was not able to obtain images. The spacecraft's photography was therefore divided into two parts—incoming views, taken prior to closest approach, and outgoing views, taken as *Mariner 10* receded from the planet—which were synthesized to make the mosaic shown in Figure 11-8b.

The *Mariner 10* images show only one hemisphere of Mercury's surface, since the time interval between flybys was equal to twice the planet's orbital period and 3 times its rotation period. Hence, the same side of Mercury was illuminated by the Sun during each flyby.

As *Mariner 10* closed in on Mercury, scientists were struck by the Moonlike pictures appearing on their television monitors. It was obvious that Mercury, like the Moon, is a barren, heavily cratered world, with no evidence for plate tectonics. This is what we would expect based on Mercury's small size (see Table 7-1 and the figure that opens this chapter). As we saw in Section 7-6, a small world is expected to have little internal heat, and, hence, little or no geologic activity and an ancient, cratered surface. But craters are not the only features of Mercury's surface; there are also gently rolling plains, long, meandering cliffs, and an unusual sort of jumbled terrain. As

we will see, Mercury's distinctive surface shows that this planet is *not* merely a somewhat larger version of the Moon.

Figure 11-9 shows a typical close-up view sent back from *Mariner 10*. The consensus among astronomers is that most of the craters on both Mercury and the Moon were produced during the first 700 million years after the planets formed. Debris remaining after planet formation rained down on these young worlds, gouging out most of the craters we see today. We saw in Section 10-4 that the strongest evidence for this chronology comes from analysis and dating of Moon rocks. *Mariner 10* did not land on Mercury, so we are not able to make the same kind of direct analysis of rocks from the planet's surface.

The differences between Mercury and the Moon are as striking as the similarities. As an example, **Figure 11-10** shows the Moon's southern hemisphere. Notice how the lunar craters are densely packed, with one overlapping the next. In sharp contrast, Mercury's surface has extensive low-lying plains (examine the upper half of Figure 11-9). These large, smooth areas are about 2 km lower than the cratered terrain.

We saw in Section 10-4 that the lunar maria were produced by extensive lava flows between 3.1 and 3.8 billion years ago. Primordial lava flows probably also explain the plains of Mercury. As large meteoroids punctured the planet's thin, newly formed crust, lava welled up from the molten interior to flood low-lying areas.

From the number of craters that pit them, Mercury's plains appear to have been formed near the end of the era of heavy bombardment, a little more than 3.8 billion years ago. Mercury's plains are therefore older than most of the lunar maria. Analysis of

(a)

(b)

Figure 11-8 R I V U X G

Mariner 10 Views Mercury (a) The unmanned *Mariner 10* spacecraft was placed in a solar orbit that took it past Venus once and past Mercury three times during 1974 and 1975. (b) To make this mosaic of the side of Mercury viewed by *Mariner 10*, scientists combined

dozens of images recorded by the spacecraft as it flew toward and away from the planet. The blank swatch running down from Mercury's north pole is a region that was not imaged. (Astrogeology Team, U.S. Geological Survey)

figure 11-9 R I ▮ U X G
Mercurian Craters and Plains This view of Mercury's northern hemisphere was taken by *Mariner 10* at a range of 55,000 km (34,000 mi) from the planet's surface. Numerous craters and extensive intercrater plains appear in this photograph, which covers an area 480 km (300 mi) wide. (NASA)

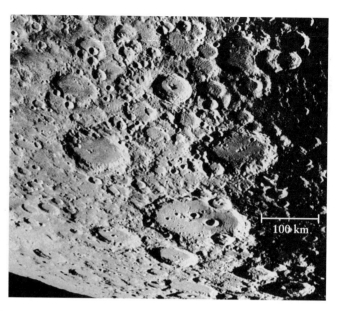

figure 11-10 R I ▮ U X G
Craters on the Moon This photograph taken from Earth shows a 600-km (370-mi) wide portion of the Moon's southern hemisphere during third quarter moon. Note how these lunar craters are more densely packed than those on the surface of Mercury, shown in Figure 11-9. (Carnegie Observatories)

Mariner 10 images made at different wavelengths also shows that the material in Mercury's plains has a lower iron content than the lunar maria. This is presumably why the plains of Mercury do not have the dark color of the Moon's maria. (Contrast the photographs of Mercury and the Moon in the figure that opens this chapter).

Mariner 10 also revealed numerous long cliffs, called **scarps**, meandering across Mercury's surface (**Figure 11-11**). Some scarps rise as much as 3 km (2 mi) above the surrounding plains and are 20 to 500 km long. These cliffs probably formed as the planet cooled and contracted.

ANALOGY The shrinkage of Mercury's crust is probably a result of a general property of most materials: They shrink when cooled and expand when heated. The lid on a jar of jam makes a useful analogy. When the jar is kept cold in the refrigerator, the metal lid contracts more than the glass jar. This makes a tight seal, but also makes the lid hard to unscrew. You can loosen the lid by running hot water over it; this makes the metal lid expand more than the glass jar, helping to break the seal.

If there had still been molten material beneath Mercury's surface when this contraction took place, lava would have leaked onto the surface around the scarps. This does not appear to have happened. Hence, the scarps must have formed relatively late in Mercury's history, after the lava flows had ended and after the planet had solidified to a substantial depth beneath the surface. With the exception of the scarps, there are no features on Mercury that resemble the boundaries of tectonic plates. Thus, we can regard Mercury's crust, like that of the Moon, as a single plate.

The most impressive feature discovered by *Mariner 10* was a huge impact scar called the Caloris Basin (*calor* is the Latin word for "heat"). The Sun is directly over the Caloris Basin during alternating perihelion passages, and thus it is the hottest place on the planet once every 176 days. During each of the *Mariner 10* flybys,

1. The floors of these craters were flooded by lava from Mercury's interior.

2. Some time after the lava cooled, Mercury's crust contracted to form this scarp.

3. This crater was distorted when the scarp formed.

50 km

figure 11-11 R I ▮ U X G
A Scarp A long cliff, or scarp, called Santa Maria Rupes runs across this *Mariner 10* image. This scarp is more than a kilometer high and extends for several hundred kilometers across a region near Mercury's equator. The area shown is approximately 200 km (120 mi) across. (NASA)

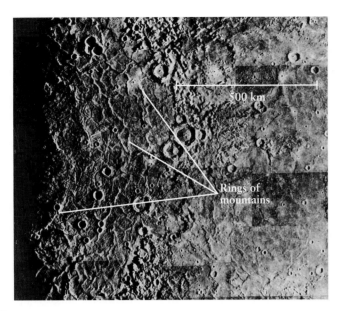

figure 11-12 R I **V** U X G

The Caloris Basin An ancient impact created this 1300-km (810-mi) diameter basin. Its center is hidden in the shadows just beyond the left side of the image. The impact fractured the surface extensively, forming several concentric chains of mountains. The mountains in the outermost ring are up to 2 km (6500 ft) high. (NASA)

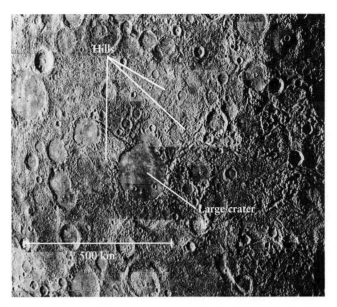

figure 11-13 R I **V** U X G

Unusual, Hilly Terrain The tiny, fine-grained wrinkles in this image are actually closely spaced hills on the side of Mercury opposite the Caloris Basin. The hills are about 5 to 10 km wide and have elevations between 100 and 1800 m (300 and 5900 ft). The large, smooth-floored crater near the center of this photograph is 170 km (106 mi) in diameter. (NASA)

the Caloris Basin happened to lie on the terminator (the line dividing day from night; see Section 10-1). Hence, in Figure 11-12 slightly more than half of this basin is hidden on the night side of the planet.

The Caloris Basin, which measures 1300 km (810 mi) in diameter, is both filled with and surrounded by smooth lava plains. The basin was probably gouged out by the impact of a large meteoroid that penetrated the planet's crust, allowing lava to flood out onto the surface and fill the basin. Because relatively few craters pockmark these lava flows, the Caloris impact must be relatively young. It must have occurred toward the end of the crater-making period that dominated the first 700 million years of our solar system's history.

The impact that created the huge Caloris Basin must have been so violent that it shook the entire planet. On the side of Mercury opposite the Caloris Basin, *Mariner 10* discovered a consequence of this impact—a jumbled, hilly region covering about 500,000 square kilometers, about twice the size of the state of Wyoming (Figure 11-13). Geologists theorize that seismic waves from the Caloris impact became focused as they passed through Mercury. As this concentrated seismic energy reached the far surface of the planet, it deformed the surface and created the hills shown in Figure 11-13.

Features similar to those of the Caloris Basin exist on our Moon. The best example is Mare Orientale, shown in Figure 11-14 (it can also be seen in Figure 10-7). Just as for the Caloris Basin, the impact that created Mare Orientale fractured the crust and produced encircling rings of mountains. The theory about the seismic upthrusting of jumbled hills on Mercury is supported by

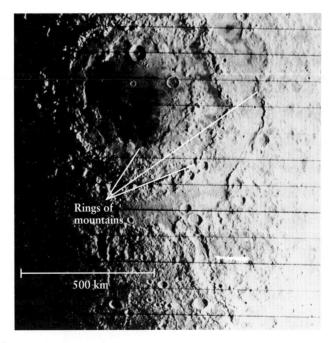

figure 11-14 R I **V** U X G

Mare Orientale on the Moon Mare Orientale is one of several large impact basins on the Moon that resemble the Caloris Basin on Mercury. The outermost ring of peaks forms a circle 900 km (560 mi) in diameter. This image was sent back in 1967 from the *Lunar Orbiter 4* spacecraft. Black lines interrupt the picture because the image was transmitted to the Earth in thin strips. (NASA)

figure 11-15 R I V U X G
Ice at Mercury's North Pole? This radar image is centered on Mercury's north pole. White indicates strong reflection of the radar signal, and darker shades indicate progressively weaker reflection. The bright dot at the center of the image, a region of very strong reflection, is thought to be a patch of frozen water (ice). Ice could survive on the floors of craters at Mercury's north pole, where it would remain permanently shaded from the Sun and reach temperatures as low as 150°C (about 125 K, or –235°F). (Courtesy of J. K. Harmon and Martin Slade, JPL/NASA)

the presence of similar (although less extensive) chaotic hills on the side of the Moon opposite Mare Orientale.

In the decades since the *Mariner 10* mission, astronomers have continued to study Mercury with both visible-light and radio telescopes. Radar observations have detected patches at Mercury's north and south poles that are unusually effective at reflecting radio waves (**Figure 11-15**). It has been suggested that these patches may be regions of water ice deep within craters where the Sun's rays never reach. (We saw in Section 10-2 that ice may also exist in shadowed regions at the Moon's poles.) If this is correct, then Mercury is truly a world of extremes, with perpetual ice caps at one or both poles but with midday temperatures at the equator that are high enough to melt lead.

11-4 | Mercury has an iron core and a surprising magnetic field

Mercury's average density of 5430 kg/m³ is quite similar to that of the Earth, which is 5515 kg/m³. We saw in Section 9-2 that typical rocks from the Earth's surface have a density of only about 3000 kg/m³ because they are composed primarily of lightweight, mineral-forming elements. The higher average density of our planet

is caused by the Earth's iron core. By studying how the Earth vibrates during earthquakes, geologists have deduced that our planet's iron core occupies about 17% of the Earth's volume.

Pressures inside a planet increase the density of rock by squeezing its atoms closer together. Because the Earth is 18 times more massive than Mercury, the weight of this larger mass compresses the Earth's core much more than Mercury's core. If the Earth were uncompressed, its density would be 4400 kg/m³, whereas Mercury's uncompressed density would still be a hefty 5300 kg/m³. This higher uncompressed density means that Mercury has a *larger* proportion of iron than the Earth: An iron core must occupy about 42% of Mercury's volume. Mercury is therefore the most iron-rich planet in the solar system. **Figure 11-16** is a scale drawing of the interior structures of Mercury and the Earth.

Several theories have been proposed to account for Mercury's high iron content. According to one theory, the inner regions of the primordial solar nebula were so warm that only those substances with high condensation temperatures—like iron-rich minerals—could have condensed into solids. Another theory suggests that a brief episode of very powerful solar winds could have stripped Mercury of its low-density mantle shortly after the Sun formed. A third possibility is that during the final stages of planet formation, Mercury was struck by a large planetesimal. Supercomputer simulations show that this cataclysmic collision would have ejected much of the lighter mantle, leaving a disproportionate amount of iron to reaccumulate to form the planet we see today (**Figure 11-17**).

An important clue to the structure of Mercury's iron core came from the *Mariner 10* magnetometers, which discovered that Mercury has a magnetic field similar to that of the Earth but only about

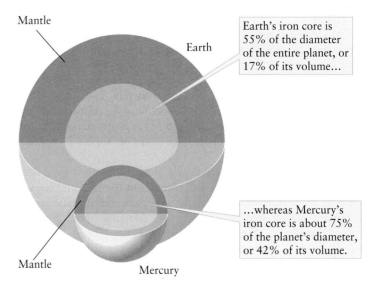

Mantle

Earth

Earth's iron core is 55% of the diameter of the entire planet, or 17% of its volume...

Mantle

Mercury

...whereas Mercury's iron core is about 75% of the planet's diameter, or 42% of its volume.

figure 11-16
The Internal Structures of Mercury and the Earth
Compared to the Earth's core, Mercury's iron-rich core takes up a much larger fraction of the planet's volume. Indeed, Mercury is the most iron-rich planet in the solar system. Surrounding Mercury's core is a 600-km-thick rocky mantle.

Proto-Mercury

Planetesimal ⅙ the mass of proto-Mercury

5000 km

Just before contact

Planetesimal plows into proto-Mercury's interior

5000 km

3 minutes after contact

Impact vaporizes the rocky mantles of both worlds

5000 km

6 minutes after contact

Mercury re-forms from iron-rich material within this banana-shaped region

Other material disperses into space

5000 km

30 minutes after contact

figure 11-17

Stripping Mercury's Mantle by a Collision To account for Mercury's high iron content, one theory proposes that a collision with a planet-sized object stripped Mercury of most of its rocky mantle. These four images show a supercomputer simulation of a head-on collision between proto-Mercury and a planetesimal one-sixth its mass. (Courtesy of W. Benz, A. G. W. Cameron, and W. Slattery)

1% as strong. We saw in Section 7-7 that for a planet to have a global magnetic field, it must have moving liquid material in its interior that conducts electricity. This moving material acts as a dynamo and generates the magnetic field. Hence, we conclude that (1) at least part of Mercury's core must be in a liquid state, and (2) there must be a source of energy to make material flow within this liquid region in order to produce electric currents. The second requirement is evidence that Mercury, like the Earth, also has a *solid* part of its core: As material deep in the liquid core cools and solidifies to join the solid portion of the core, it releases the energy needed to stir up the motions of the remaining liquid material. Mercury's magnetic field therefore suggests that its core has the same general structure as the Earth's core (recall Figure 9-9).

Mercury's magnetic field comes as a something of a surprise. The ancient, cratered surface of Mercury is evidence of a lack of geologic activity, which in turn shows that Mercury must have lost the internal heat that powers such activity. This is as we would expect for such a small planet (see Section 7-6). But the existence of a global magnetic field shows that Mercury is still partially molten inside, which is evidence that the planet *has* some internal heat. The evidence from cratering and from the global magnetic field are thus in contradiction with each other.

Two other observations further complicate our understanding of Mercury's interior. One is that the amount of electrically conducting molten material that remains within Mercury is thought to be too little to account for even the relatively weak global magnetic field. Another is that a planet must rotate fairly rapidly in order to stir its molten interior into the kind of motion that generates a magnetic field. But as we saw in Section 11-2, Mercury rotates very *slowly*: It takes 58.646 days to spin on its axis. We are forced to conclude that we still have much to learn about the interior of Mercury.

Whatever the origin of Mercury's magnetic field, we know that it interacts with the solar wind much as does the Earth's field, although on a much smaller scale (**Figure 11-18**). The solar wind is a constant flow of charged particles (mostly protons and electrons)

away from the outer layers of the Sun's upper atmosphere. A planet's magnetic field can repel and deflect the impinging particles, thereby forming an elongated "cavity" in the solar wind called a

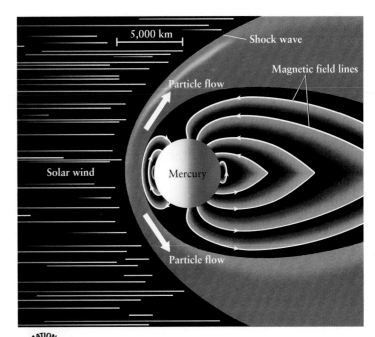

5,000 km

Shock wave

Magnetic field lines

Particle flow

Solar wind

Mercury

Particle flow

ANIMATION 11.4

figure 11-18

Mercury's Magnetosphere Mercury's weak magnetic field has only 1% the strength of the Earth's field. This is just strong enough to carve out a cavity in the solar wind, preventing the impinging particles from striking the planet's surface directly. Most of the particles of the solar wind are deflected around the planet in a turbulent region colored purple in this scale drawing. (Compare this with the Earth's much larger magnetosphere, shown in Figure 9-19.)

magnetosphere. (We discussed the Earth's magnetosphere in Section 9-4.)

Charged-particle detectors aboard *Mariner 10* mapped the structure of Mercury's magnetosphere. When the particles in the solar wind first encounter Mercury's magnetic field, they are abruptly slowed, producing a shock wave that marks the boundary where this sudden decrease in velocity occurs. Most of the particles from the solar wind are deflected around the planet, just as water is deflected to either side of the bow of a ship. Mercury's magnetosphere thus prevents the solar wind from reaching the planet's surface. Because Mercury's magnetic field is much weaker than that of the Earth, it is not able to capture particles from the solar wind. Hence, it has no structures like the Van Allen belts that surround the Earth (recall Figure 9-19).

Much about Mercury still remains unknown. What is the character of the side of the planet that *Mariner 10* did not see? What is the chemical composition of the surface rocks? Was the surface of Mercury once completely molten? What caused Mercury to have such a high iron content? What portion of the planet's core is liquid, and what portion is solid? Is there indeed ice perpetually frozen at the poles? Learning the answers will require extensive observations of Mercury by a spacecraft at close range.

 One such mission, *MESSENGER* (*ME*rcury *S*urface, *S*pace *EN*vironment, *GE*ochemistry, and *R*anging), is a U.S. spacecraft whose sophisticated instruments will make detailed measurements of surface composition and may help reveal the nature of Mercury's core (**Figure 11-19**). Because it will go into orbit around Mercury, *MESSENGER* will be able to make images of the side of the planet that *Mariner 10* did not observe. If all goes as planned, *MESSENGER* will arrive at Mercury in 2009. Another Mercury mission, *BepiColombo,* is being planned by the European Space Agency in collaboration with the Japanese Institute of Space and Astronautical Science. Named for the physicist who first proposed Mercury's 3-to-2 spin-orbit coupling, *BepiColombo* is actually three spacecraft: an orbiter that will study the planet itself, a smaller orbiter that will explore Mer-

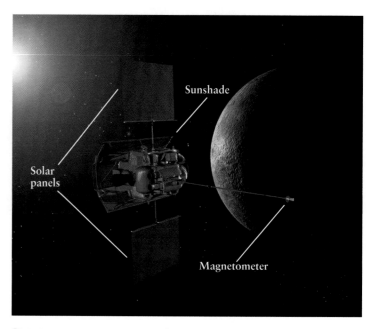

Figure 11-19

The *MESSENGER* Spacecraft   When the *MESSENGER* spacecraft arrives in Mercury orbit in 2009, a sunshade will keep the body of the spacecraft from overheating in the intense light close to the Sun. (The exposed solar panels will provide electrical power to the spacecraft.) The magnetometer at the end of the 3.6-meter (12-foot) boom will map Mercury's magnetic field and search for regions of magnetized rocks in the crust. (Johns Hopkins University/Applied Physics Laboratory)

cury's magnetosphere, and a lander that will touch down near one of the poles and make direct measurements of the surface and its properties. This mission may reach Mercury in 2011. Until this new generation of spacecraft begin returning new data, Mercury will remain one of the most enigmatic worlds of the solar system.

Key Words

1-to-1 spin-orbit coupling, p. 232

3-to-2 spin-orbit coupling, p. 233

greatest eastern elongation, p. 230

greatest western elongation, p. 230

scarp, p. 236

solar transit, p. 231

Key Ideas

Motions of Mercury in the Earth's Sky: At its greatest eastern and western elongations, Mercury is never more than 28° from the Sun, so it can be seen with the naked eye only briefly after sunset or before sunrise.

• Solar transits of Mercury occur about a dozen times per century.

Mercury's Rotation: Poor telescopic views of Mercury's surface led to the mistaken impression that the planet always keeps the same side toward the Sun, a configuration called synchronous rotation or 1-to-1 spin-orbit coupling.

• Radio and radar observations in the 1960s revealed that Mercury in fact has 3-to-2 spin-orbit coupling. For an observer on Mercury, the average time from sunset to sunrise (one-half of a solar day) is equal to the time for a complete orbit around the Sun (one Mercurian year).

The Surface of Mercury: The *Mariner 10* spacecraft made several passes near Mercury in the mid-1970s, providing pictures of its surface.

• The Mercurian surface is pocked with craters like those of the Moon, but there are extensive smooth plains between these craters.

• Long cliffs called scarps meander across the surface of Mercury. These scarps probably formed as the planet cooled, solidified, and shrank.

• The impact of a large object long ago formed the huge Caloris Basin and shoved up jumbled hills on the opposite side of the planet.

Mercury's Interior and Magnetic Field: Like the Earth, Mercury has an iron-rich core.

• The iron core of Mercury has a diameter equal to three-fourths of the planet's diameter, whereas the diameter of the Earth's core is only slightly more than one-half of the Earth's diameter.

• Mercury has a weak magnetic field, indicating that at least part of the iron core is liquid. The magnetic field produces a magnetosphere around Mercury that blocks the solar wind from reaching the surface of the planet.

Review Questions

1. Why is it impossible to see Mercury in the sky at midnight?

2. Why can't you see any surface features on Mercury when it is closest to the Earth?

3. Why are naked-eye observations of Mercury best made at dusk or dawn, while telescopic observations are best made during the day?

4. Table 11-2 shows that a greatest western elongation of Mercury is always followed by a greatest eastern elongation, and vice versa. Explain why.

5. What is a solar transit? Why are solar transits of Mercury relatively infrequent?

6. In his 1964 science fiction story "The Coldest Place," author Larry Niven described the "dark side" of Mercury as the coldest place in the solar system. What assumption did he make about the rotation of Mercury? Did this assumption turn out to be correct?

7. What is 3-to-2 spin-orbit coupling? How is the rotation period of an object exhibiting 3-to-2 spin-orbit coupling related to its orbital period? What aspects of Mercury's orbit cause it to exhibit 3-to-2 spin-orbit coupling? What telescopic observations proved this?

8. Explain why Mercury does not have a substantial atmosphere.

9. Explain why *Mariner 10* was able to photograph only one side of Mercury even though the spacecraft returned to the planet three times.

10. Compare the surfaces of Mercury and our Moon. How are they similar? How are they different?

11. What kind of surface features are found on Mercury? Why are they probably much older than most surface features on the Earth?

12. How could you tell which craters in Figure 11-10 are younger than the others?

13. How do we know that the scarps on Mercury are younger than the lava flows? How can you tell that the scarp in Figure 11-11 is younger than the vertically distorted crater at the center of the figure?

14. Explain why the Sun is directly over the Caloris Basin on Mercury only every other time that the planet is at perihelion.

15. Explain why mountains on Earth can be said to be formed from below, while the mountains that ring the Caloris Basin on Mercury and Mare Orientale on the Moon are best described as having formed from above.

16. Explain why there is unusual hilly terrain on the side of Mercury directly opposite the Caloris Basin.

17. If Mercury is the closest planet to the Sun and has such a high average surface temperature, how is it possible that ice might exist on its surface?

18. Why do astronomers think that Mercury has a very large iron core?

19. Briefly describe at least one possible history that would account for Mercury's large iron core.

20. Why is it surprising that Mercury has a global magnetic field?

21. Explain why the 58.646-day rotation period of Mercury implies that Mercury can have only a weak magnetic field.

22. How is Mercury's magnetosphere similar to that of the Earth? How is it different? Why do you suppose Mercury does not have Van Allen belts?

Advanced Questions

Questions preceded by an asterisk (*) involve topics discussed in the Boxes.

> Problem-solving tips and tools
>
> You may need to refresh your memory about the small-angle formula, found in Box 1-1; about Kepler's third law, described in Section 4-4; and about Wien's law and the Doppler effect, discussed in Section 5-4 and Section 5-9, respectively. The circumference of a circle of radius r is $2\pi r$. Box 7-2 discusses the criteria for a planet to be able to retain an atmosphere.

*23. Find the largest angular size that Mercury can have as seen from the Earth. In order for Mercury to have this apparent size, at what point in its orbit must it be?

24. Figure 11-1 shows Mercury with a greatest eastern elongation of 18° and a greatest western elongation of 28°. (**a**) Make a drawing like Figure 11-1 that shows how the Earth and Mercury would have to be positioned on their orbits to have Mercury at a greatest eastern elongation of 28° and a greatest western elongation of 18°. (**b**) On June 9, 2000, Mercury was at a greatest eastern elongation of 24°. Was Mercury at perihelion, aphelion, or some other point on its orbit? Explain.

*25. Suppose you have a superb telescope that can resolve features as small as 1 arcsec across. What is the size (in kilometers) of the smallest surface features you should be able to see on Mercury? How does your answer compare with the size of the Caloris Basin? (*Hint:* Assume that you choose to observe Mercury when it is at greatest elongation, 28° from the Sun. As Figure 11-1 shows, at this point in its orbit the Earth-Mercury distance is about the same as the Earth-Sun distance.)

26. (a) Explain why November solar transits of Mercury, which occur near the time of perihelion passage, are more common than May transits. (b) When a transit of Mercury takes place, can it be seen by observers anywhere on Earth? Why or why not?

27. Find the value of λ_{max} for blackbody radiation coming from the sunlit side of Mercury. In what part of the electromagnetic spectrum does this lie?

28. If the albedo of Mercury were increased, would the planet's surface temperature go up or down? Explain your answer.

29. (a) Mercury has a 58.646-day rotation period. What is the speed at which a point on the planet's equator moves due to this rotation? (*Hint:* Remember that speed is distance divided by time. What distance does a point on Mercury's equator travel as the planet makes one rotation?) (b) Use your answer to (a) to answer the following: As a result of rotation, what difference in wavelength is observed for a radio wave of wavelength 12.5 cm (such as is actually used in radar studies of Mercury) emitted from either the approaching or receding edge of the planet?

30. How much would an 80-kg person weigh on Mercury? How does that compare with that person's weight on the Moon? How much does that person weigh on the Earth?

31. When an impact crater is formed, material (called *ejecta*) is sprayed outward from the impact. (Look at the photograph of the Moon on the right-hand side of the figure that opens this chapter. At the lower right of this photograph, you can see light-colored ejecta surrounding the crater Stevinus.) While ejecta are found surrounding the craters on Mercury, they do not extend as far from the craters as do ejecta on the Moon. Explain why, using the difference in surface gravity between the Moon and Mercury.

*32. (a) Calculate the minimum molecular weight of a gas that could in theory be retained as an atmosphere by Mercury if the average daytime temperature were 620 K. (b) Are there any abundant gases that meet this minimum criterion? Why doesn't Mercury have an atmosphere of these gases? (*Hint:* See Box 7-2. The mass of a molecule is equal to its molecular weight, μ, multiplied by the mass of a hydrogen atom, $m_H = 1.67 \times 10^{-27}$ kg. The molecular weight of a molecule equals the sum of the atomic weights of its atoms, which can be looked up in a periodic table of the elements. Thus, for example, the molecular weight of CO_2 is 12.0 + 16.0 + 16.0 = 44.0, and so the molecule's mass is 44.0 m_H.)

33. The orbital period of *Mariner 10* is twice that of Mercury. Use this fact to calculate the length of the semimajor axis of the spacecraft's orbit.

34. Much of the area shown in Figure 11-11 is pockmarked with small impact craters. By contrast, there are very few small craters within the large crater at the upper right of this figure. Explain

why this suggests that the interior of this crater was flooded with lava while the area outside the crater was not.

35. Give at least two pieces of evidence that Mercury has undergone chemical differentiation.

36. Consider the idea that Mercury has a solid iron-bearing mantle that is permanently magnetized like a giant bar magnet. Using the fact that iron demagnetizes at temperatures above 770°C, present an argument against this explanation of Mercury's magnetic field.

Discussion Questions

37. Before about 350 B.C., the ancient Greeks did not realize that Mercury seen in the morning sky (which they called Apollo) and seen in the evening sky (which they called Hermes) were actually the same planet. Discuss why you think it took some time to realize this.

38. If you were planning a new mission to Mercury, what features and observations would be of particular interest to you?

39. What evidence do we have that the surface features on Mercury were not formed during recent geological history?

 ## Web/CD-ROM Questions

40. Search the World Wide Web for the latest information about new missions to Mercury. Which of these missions have been given funding so that they can proceed? When will they be launched, and when will they arrive in orbit around Mercury? What scientific experiments will they carry? What scientific issues are these instruments intended to resolve?

 41. Elongations of Mercury. Access the animation "Elongations of Mercury" in Chapter 11 of the *Universe* Web site or CD-ROM. (a) View the animation and notice the dates of the greatest eastern and greatest western elongations. Which time interval is greater: from a greatest eastern elongation to a greatest western elongation, or vice versa? (b) Based on what you observe in the animation, draw a diagram to explain your answer to the question in (a).

Observing Projects

Observing tips and tools

 Remember that Mercury is visible in the morning sky when it is at or near greatest western elongation and in the evening sky when at or near greatest eastern elongation. You can consult such magazines as *Sky & Telescope* and *Astronomy,* or the Web sites for these magazines, for more detailed information about when and where to look for Mercury during a given month. You can also use the *Starry Night Backyard*™ program on the CD-ROM that accompanies this textbook.

42. Refer to Table 11-2 to determine the dates of the next two or three greatest elongations of Mercury. Consult such magazines as *Sky & Telescope* and *Astronomy*, or the Web sites for these magazines, to determine if any of these greatest elongations is going to be a favorable one. If so, make plans to be one of those rare individuals who has actually seen the innermost planet of the solar system. Set aside several evenings (or mornings) around the date of the favorable elongation to reduce the chances of being "clouded out." Select an observing site that has a clear, unobstructed view of the horizon where the Sun sets (or rises). If possible, make arrangements to have a telescope at your disposal. Search for the planet on the dates you have selected, and make a drawing of its appearance through your telescope.

43. This observing project should be performed only under the direct supervision of an astronomer who knows how to point a telescope safely at Mercury. Make arrangements to view Mercury during broad daylight. This is best done by visiting an observatory where the coordinates (right ascension and declination) of Mercury's position can be used to point the telescope. **DO NOT LOOK AT THE SUN! Looking directly at the Sun can cause blindness.**

44. Use the *Starry Night Backyard™* program to observe solar transits of Mercury. Display the entire celestial sphere (select **Guides > Atlas** in the **Go** menu) and center on Mercury by using the **Find...** command in the **Edit** menu. Using the controls at the right-hand end of the Control Panel, zoom in until the field of view is 1°. (**a**) In the Control Panel, set the date and time to November 8, 2006, at 12:00:00 A.M. Then click on the control for the rate of time flow (immediately to the right of the date and time display) and set the discrete time step to 1 hour. Step backward or forward through time using the single-step buttons (the leftmost and the rightmost buttons) and record the times at which the solar transit begins and ends. What is the total duration of the solar transit? (**b**) Set the date and time to May 9, 2016, at 12:00:00 A.M. Again step backward or forward through time, record the times when the solar transit begins and ends, and find the total duration

of the solar transit. (**c**) The maximum duration of a solar transit of Mercury is 9 hours. Explain why the transits you observed in (a) and (b) last a substantially shorter time.

45. Use the *Deep Space Explorer™* program to examine Mercury. In the left-hand part of the window under the heading **Solar System,** click on the triangle next to the word **Explore** and then click on **Mercury.** You can zoom in and zoom out using the buttons at the upper left of the window (an upward-pointing triangle and a downward-pointing triangle). You can also rotate Mercury by putting the mouse cursor over the image, holding down the mouse button, and moving the mouse. (On a two-button mouse, hold down the left mouse button.) Estimate the diameter of the largest craters by measuring their size on the screen with a ruler and comparing to the diameter of Mercury (see Table 11-1).

Collaborative Exercises

46. If you were planning an evening picnic sometime in the next calendar year when Mercury would be visible, which date would you select and why? Use Table 11-2 to help.

47. Figures 11-7*b* and 11-7*c* show a planet in synchronous rotation and Mercury with a 3-to-2 spin-orbit coupling, respectively. Stand up and demonstrate how planets move in each of these rotations by "orbiting" around a stationary classmate who represents our Sun. How would Mercury's motion be different if it had a 4-to-2 spin-orbit coupling instead?

48. In the nineteenth century, French mathematician and astronomer U. J. J. Le Verrier led a failed search for a hypothetical planet named Vulcan (after the mythical blacksmith of the gods) orbiting closer to our Sun than Mercury. If Vulcan had an orbit with the same eccentricity as Mercury's orbit but only one-half the size, what would have been its maximum eastern and western elongations?

12 Cloud-Covered Venus

For decades Venus was a planet of mystery. It has nearly the same mass and diameter as the Earth, but is shrouded by clouds that perpetually hide the surface from view. What, scientists wondered, were conditions like below the clouds? Since Venus is closer to the Sun than Earth and hence hotter, might Venus be a hot, humid jungle world, perhaps teeming with animal life?

The answers to these tantalizing questions were provided by two new technologies: radar, which allowed astronomers to "see" through the clouds, and unmanned spacecraft, which orbited Venus at close range and even landed on its surface. Thanks to radar, we now know that the similarities between Venus and the Earth are deceptive. As the accompanying false-color topographic map shows, Venus has no true continents but merely highlands (shown in red) that rise gently above the planet's lower-lying areas (shown in blue). Much of the planet is covered with lava flows from volcanoes.

Thanks to observations from spacecraft, we also know that there can be no jungles on Venus. The thick atmosphere is composed almost entirely of carbon dioxide, producing such a strong greenhouse effect that the surface temperature is a searing 460°C (860°F). Furthermore, the clouds that envelop Venus contain droplets of corrosive sulfuric acid, perhaps the result of sulfurous emissions from the same volcanoes that spilled lava across the planet's surface.

The greenhouse effect has made Venus utterly hostile to living organisms. This same effect also acts in a milder form in our own atmosphere, and might someday cause our polar ice caps to melt. Hence, an understanding of our sister planet may give important insights into the future of our own world.

12-1 | The surface of Venus is hidden beneath a thick, highly reflective cloud cover

Venus, familiar for centuries as the "morning star" and "evening star," is easy to identify. Unlike Mercury, it is possible to see Venus without interference from the Sun's glare.

VIDEO 12.1

Radar reveals the topography of Venus. (NASA/JPL, MIT, and USGS) **R** I V U X G

Questions to Guide Your Reading

As you read the sections of this chapter, look for the answers to the following questions:

12-1 What makes Venus such a brilliant "morning star" or "evening star"?

12-2 What is strange about the rotation of Venus?

12-3 In what ways does Venus's atmosphere differ radically from our own?

12-4 Why do astronomers suspect that there are active volcanoes on Venus?

12-5 Why is there almost no water on Venus today? Why do astronomers think that water was once very common on Venus?

12-6 Does Venus have the same kind of active surface geology as the Earth?

Indeed, at times Venus is the brightest object in the night sky. These simple observations tell us quite a bit about Venus's orbit and atmosphere.

Venus can be seen fairly far from the Sun—at its greatest elongation, about 47° away (Figure 12-1)—because its orbit is almost twice as large as that of Mercury. At its greatest eastern elongation, when it is called an "evening star," Venus is seen high above the western horizon after sunset. At greatest western elongation, Venus is called the "morning star," because it rises nearly 3 hours before the Sun and is positioned high in the eastern sky at dawn. Table 12-1 summarizes some basic facts about Venus and its orbit, and Table 12-2 lists the dates of greatest eastern and western elongation for Venus from 2004 to 2008.

At its greatest brilliance, Venus is 16 times brighter than the brightest star and is outshone only by the Sun and the Moon. There are four reasons why: Venus is relatively large (almost the same size as the Earth), it is close to the Sun, it is close to the Earth, and it reflects 59% of the sunlight that falls on the planet (its albedo is 0.59).

Through Earth-based telescopes, Venus appears almost completely featureless, with neither continents nor mountains (see Figure 4-13). However, astronomers soon realized that they were not seeing the true surface of the planet. Rather, Venus must be covered by a thick, unbroken layer of clouds. A cloud layer would also

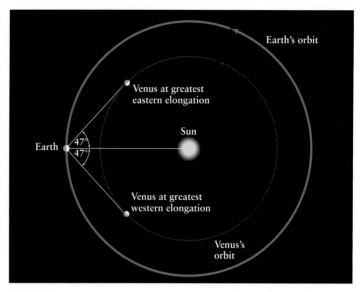

Figure 12-1

The Orbit of Venus Venus follows a nearly circular orbit around the Sun with a period of 224.7 days. At its greatest eastern elongation, Venus appears as a prominent "evening star"; at its greatest western elongation, it is a prominent "morning star."

table 12-1	Venus Data
Average distance from Sun:	0.723 AU = 1.082 × 10^8 km
Maximum distance from Sun:	0.728 AU = 1.089 × 10^8 km
Minimum distance from Sun:	0.718 AU = 1.075 × 10^8 km
Eccentricity of orbit:	0.0068
Average orbital speed:	35.0 km/s
Orbital period:	224.70 days
Rotation period:	243.01 days (retrograde)
Inclination of equator to orbit:	177.4°
Inclination of orbit to ecliptic:	3.39°
Diameter (equatorial):	12,104 km = 0.949 Earth diameter
Mass:	4.868 × 10^{24} kg = 0.815 Earth mass
Average density:	5243 kg/m^3
Escape speed:	10.4 km/s
Surface gravity (Earth = 1):	0.91
Albedo:	0.59
Average surface temperature:	460°C = 860°F = 733 K
Atmospheric composition (by number of molecules):	96.5% carbon dioxide (CO_2) 3.5% nitrogen (N_2), 0.003% water vapor (H_2O)

(NASA)   R I V **U** X G

table 12-2	Greatest Elongations of Venus,* 2004–2008	
Year	Greatest western elongations	Greatest eastern elongations
2004	August 17	March 29
2005	(none)	November 3
2006	March 25	(none)
2007	October 28	June 9
2008	(none)	(none)

*Venus can be seen in the eastern sky in the hours before dawn for several months around greatest western elongation. It can be seen in the western sky in the hours after dusk for several months around greatest eastern elongation.

explain why Venus reflects such a large fraction of the sunlight that falls on it. Direct evidence that Venus has a thick atmosphere came when nineteenth-century astronomers observed Venus near the time of inferior conjunction. This is when Venus lies most nearly between the Earth and the Sun so that we see the planet lit from behind. As Figure 12-2 shows, sunlight is scattered by Venus's atmosphere, producing a luminescent ring around the planet that would not otherwise be present.

One other aspect of Venus that can be observed in visible light is a solar transit, in which Venus passes directly in front of the Sun at inferior conjunction. This does not happen at every inferior conjunction, because the plane of the orbit of Venus is inclined by 3.39° to the ecliptic (the plane of the Earth's orbit). Indeed, solar transits of Venus are exceedingly rare; the last two were in 1874 and 1882, and not one transit of Venus occurred during the entire twentieth century. Venusian transits usually come in pairs separated by 8 years. The only pair of transits in the twenty-first century are on June 8, 2004, and June 6, 2012; the next pair will occur in 2117 and 2125. Because Mercury orbits closer to the Sun, it is more likely to pass in front of the Sun's disk as seen from the Earth, and transits of Mercury are more frequent (see Section 11-1).

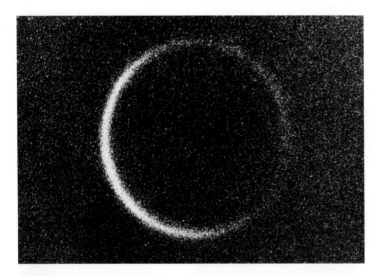

figure 12-2   R I 𝗩 U X G

Venus at Inferior Conjunction This photograph of Venus at inferior conjunction shows an illuminated atmosphere surrounding the planet. If Venus did not have an atmosphere to scatter sunlight, we would see a thin crescent rather than a ring. This photograph was specially processed to reveal the ring, which is too faint to show in the images in Figure 4-13. (Lowell Observatory)

| 12-2 | Venus's rotation is slow and retrograde

Venus rotates very slowly; the sidereal rotation period of the planet is 243.01 days, even longer than the planet's orbital period. To an imaginary inhabitant of Venus who could somehow see through the cloud cover, stars would move across the sky at a rate of only 1½° per Earth day. As seen from the Earth, by contrast, stars move 1½° across the sky in just 6 minutes.

Not only does Venus rotate slowly, it also rotates in an unusual direction. Most of the planets in the solar system spin on their axes in **prograde rotation.** Prograde means "forward," and planets with prograde rotation spin on their axes in the same direction in which they orbit the Sun (Figure 12-3a). As seen from the surface of a planet with prograde rotation, such as the Earth, the Sun and stars rise in the east and set in the west.

Venus is an exception to the rule: It spins in **retrograde rotation.** That is, the direction in which it spins on its axis is backward compared to the direction in which it orbits the Sun, as Figure 12-3b shows. (Recall that "retrograde" means "backward.") If an observer on the surface of Venus could see through the thick layer of clouds, the Sun and stars would rise in the west and set in the east!

 Be careful not to confuse retrograde *rotation* with retrograde *motion*. When Venus is near inferior conjunction, so that it is passing the Earth on its smaller, faster orbit

(a) Prograde rotation

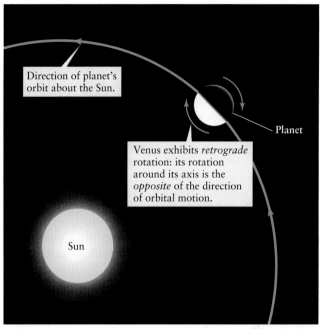

(b) Retrograde rotation

ANIMATION 12.1

Figure 12-3

Prograde and Retrograde Rotation (a) If you could view the solar system from a point several astronomical units above the Earth's north pole, you would see all of the planets orbiting the Sun in a counterclockwise direction. Most of the planets also rotate counterclockwise on their axes in prograde (forward) rotation. (b) Venus is an exception; from your perspective high above the plane of the solar system, you would see Venus rotating clockwise in retrograde (backward) rotation.

around the Sun, it appears to us to move from east to west on the celestial sphere from one night to the next. This apparent motion is called retrograde motion, because it is backward to the apparent west-to-east motion that Venus displays most of the time. (You may want to review the discussion of retrograde motion in Sections 4-1 and 4-2.) By contrast, the retrograde rotation of Venus means that it spins on its axis from east to west, rather than west to east like the Earth.

How do astronomers know how Venus rotates? One approach is to track features in the planet's cloud cover. But winds in the atmosphere cause such features to move relative to Venus's surface, so this approach does not give a direct measurement of how the planet itself rotates. It is necessary to watch the surface of Venus itself. As we have seen, visible light cannot penetrate the planet's perpetual cloud cover. Instead, astronomers must use a different form of electromagnetic radiation.

Clouds may contain gas, dust, haze, water droplets, or other small particles. In general, electromagnetic radiation can pass easily through such a cloud only if the wavelength is large compared to the size of the particles. We then say that the cloud is *transparent* to the radiation. For example, clouds in the Earth's atmosphere are made of water droplets with an average diameter of about 20 μm (2×10^{-5} m, or 20,000 nm). Visible light has wavelengths between 400 to 700 nm, which is less than the droplet size. Hence, visible light cannot easily pass through such clouds, which is why

cloudy days are darker than sunny days. But radio waves, with wavelengths of 0.1 m or more, and microwaves, with wavelengths from 10^{-3} m to 0.1 m, can pass through such a cloud with ease. (That is why your mobile phone, radio, or television works just as well on a cloudy or foggy day as on a clear one.)

Like clouds in the Earth's atmosphere, the clouds of Venus are transparent to radio waves and microwaves. Beginning in the early 1960s, advances in radio telescope technology made it possible to send microwave radiation to Venus and detect the waves reflected from its surface. Although they did not record surface features directly, these observations allowed measurements of Venus's rotation. The key was to measure how the wavelengths of the reflected radiation differed from the wavelength of the radiation originally sent to the planet.

If a beam of microwaves is sent from the Earth to Venus at one precise wavelength, the reflected waves come back spread over a small range of wavelengths. We saw in Section 11-2 that this is a consequence of the Doppler effect: Waves reflected from the approaching side of the rotating planet are shortened in wavelength, while waves reflected from the receding side undergo an increase in wavelength (see Figure 11-6). From the amount of the wavelength spread, astronomers can determine how rapidly Venus is rotating; by noticing which side of the planet is approaching and which is receding, they can tell whether the rotation is prograde or retrograde. Such measurements confirm that Venus's rotation is indeed retrograde and reveal the planet's slow rotation.

Why is Venus's retrograde rotation worthy of our attention? Here's why: It adds to the puzzle of how our solar system came to be. If you could view our solar system from a great distance above the Earth's north pole, you would see all the planets orbiting the Sun counterclockwise. Closer examination would reveal that the Sun and most of the planets also *rotate* counterclockwise on their axes; the exceptions are Venus, Uranus, and Pluto. Most of the satellites of the planets also move counterclockwise along their orbits and rotate in the same direction. Thus, the rotation of Venus is not just opposite to its orbital motion; it is opposite to most of the orbital and rotational motions in the solar system!

As we saw in Section 8-4, the Sun and planets formed from a solar nebula that was initially rotating. (Had it not been rotating, all of the material in the nebula would have fallen inward toward the protosun, leaving nothing to form the planets.) Because the planets formed from the material of this rotating cloud, they tend to rotate on their axes in that same direction. It is difficult to imagine how three planets could have bucked this trend. One theory suggests that the tidal forces exerted by the Sun tended to elongate Venus's atmosphere toward the Sun, much as the Moon's tidal forces elongate the Earth's oceans toward the Moon (see Section 4-8, especially Figure 4-24). Friction between the atmosphere and Venus's surface then acted to slow the planet's rotation. By themselves, these tidal effects would put Venus into synchronous prograde rotation. But in addition, the Sun's heat on the illuminated side of Venus generates strong winds in the planet's atmosphere, and the frictional forces of these winds drive Venus into slow retrograde rotation.

Venus's rotation has other puzzling aspects. The length of an apparent solar day on Venus—that is, from when the Sun is highest in the sky to when it is again highest in the sky (see Section 2-7)—is 116.8 Earth days. Remarkably, this is almost exactly one-fifth as long as the planet's 584-day synodic period. (We saw in Section 4-2 that this is the time between successive inferior conjunctions, when Venus and the Earth are closest together.) Is this curious ratio the result of gravitational interactions between Venus and the Earth? Or is it sheer coincidence? No one knows the answer.

12-3 | Venus has a hot, dense atmosphere and corrosive cloud layers

 SCIENCE IN PROCESS *Measuring Venus's microwave emissions revealed the nature of its atmosphere*

Venus is bathed in more intense sunlight than Earth because it is closer to the Sun. If Venus had no atmosphere, and if its surface had an albedo like that of Mercury or the Moon, heating by the Sun would bring its surface temperature to around 45°C (113°F)—comparable to that found in the hottest regions on Earth. What was unclear until the mid-twentieth century was how much effect Venus's atmosphere has on the planet's surface temperature.

We saw in Section 9-1 that gases such as water vapor (H_2O) and carbon dioxide (CO_2) in the Earth's atmosphere trap some of the infrared radiation that is emitted from our planet's surface. This elevates the temperatures of both the atmosphere and its surface, a phenomenon called the greenhouse effect. In 1932, Walter S. Adams

and Theodore Dunham Jr. at the Mount Wilson Observatory found absorption lines of CO_2 in the spectrum of sunlight reflected from Venus, indicating that carbon dioxide is also present in the Venusian atmosphere. (Section 7-3 describes how astronomers use spectroscopy to determine the chemical compositions of atmospheres.) Thus, Venus's surface, like the Earth's, should be warmed by the greenhouse effect.

However, Venus's perpetual cloud cover reflects back into space a substantial amount of the solar energy reaching the planet. This effect by itself acts to cool Venus's surface, just as clouds in our atmosphere can make overcast days cooler than sunny days. Some scientists suspected that this cooling kept the greenhouse effect in check, leaving Venus with surface temperatures below the boiling point of water. Venus might then have warm oceans and possibly even life in the form of tropical vegetation. But if the greenhouse effect were strong enough, surface temperatures could be so high that any liquid water would boil away. Then Venus would have a dry, desertlike surface with no oceans, lakes, or rivers.

This scientific controversy was resolved in 1962 when the unmanned U.S. spacecraft *Mariner 2* (Figure 12-4) made the first close flyby of Venus. As we learned in Sections 5-3 and 5-4, a dense object (like a planet's surface) emits radiation whose intensity and spectrum depend on the temperature of the object. *Mariner 2* carried instruments that measured radiation coming from the planet at two microwave wavelengths, 1.35 cm and 1.9 cm. Venus's atmosphere is transparent to both these wavelengths, so *Mariner 2* saw

WEB LINK 12.2 **Figure 12-4**
The *Mariner 2* Spacecraft *Mariner 2* was the first of all spacecraft from Earth to make a successful flyby of another planet. After coming within 34,773 km of Venus on December 14, 1962, it went into a permanent orbit around the Sun. (GSFC/NASA)

radiation that had been emitted by the planet's surface and then passed through the atmosphere like visible light through glass.

From the amount of radiation that *Mariner 2* detected at these two wavelengths, astronomers concluded that the surface temperature of Venus was more than 400°C—well above the boiling point of water and higher than even daytime temperatures on Mercury. Thus, there cannot be any liquid water on the planet's surface. Water vapor absorbs microwaves at 1.35 cm, so if there were substantial amounts of water vapor in Venus's atmosphere, it would have blocked this wavelength from reaching the detectors on board *Mariner 2*. In fact, the spacecraft did detect strong 1.35-cm emissions from Venus. Thus, the atmosphere, too, must be all but devoid of water.

 This picture of Venus as a dry, hellishly hot world was reinforced by subsequent spacecraft launched in the 1960s and 1970s. The United States sent a series of lightweight spacecraft past Venus and studied the planet and its environment with remote sensing devices. The Soviet Union, which had more powerful rockets, sent massive vehicles that plunged directly into the Venusian clouds.

Building spacecraft that could survive a descent into the Venusian atmosphere proved to be more challenging than anyone had expected. Finally, in 1970, the Soviet spacecraft *Venera 7* managed to transmit data for a few seconds directly from the Venusian surface. *Venera 7* and other successful Soviet landers during the early 1970s finally determined the surface temperature to be a nearly constant 460°C (860°F).

The success of the Soviet landers was a major technological achievement, for the pressure at the surface of Venus proved to be 90 atmospheres—that is, 90 times greater than the average air pressure at sea level on Earth. (One atmosphere, or 1 atm, is equal to 1.01×10^5 newtons per square meter, or 14.7 pounds per square inch.) This is about the same as the pressure at a depth of 1 kilometer below the surface of the Earth's oceans. The density of the atmosphere at the surface of Venus is likewise high, more than 50 times greater than the sea-level density of our atmosphere. The atmosphere is so massive that once heated by the Sun, it retains its heat throughout the long Venusian night. This explains why the temperatures on the day and night sides of Venus are almost identical.

Atmospheric probes revealed why Venus's surface temperatures are so high: 96.5% of the molecules in Venus's atmosphere are carbon dioxide, with nitrogen (N_2) making up most of the remaining 3.5% (see Table 9-4 for a comparison of the atmospheres of Venus, Earth, and Mars). Because Venus's atmosphere is so thick, and because most of the atmosphere is the greenhouse gas CO_2, the greenhouse effect has run wild. On Earth the greenhouse gases H_2O and CO_2 together make up only about 1% of our relatively sparse atmosphere, and the greenhouse effect has elevated the surface temperature by about 41°C (74°F). But on Venus, the dense shroud of CO_2 traps infrared radiation from the surface so effectively that it has raised the surface temperature by more than 400°C (720°F).

Soviet and American spacecraft also discovered that Venus's clouds are primarily confined to three high-altitude layers. An upper cloud layer lies at altitudes between 68 and 58 km, a denser and more opaque cloud layer from 58 to 52 km, and an even more dense and opaque layer between 52 and 48 km. Above and below the clouds are 20-km-thick layers of haze (**Figure 12-5**). Below the low-

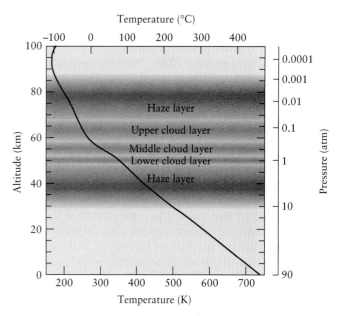

figure 12-5

Temperature and Pressure in the Venusian Atmosphere The scale on the right-hand side of the graph shows how pressure increases as you descend in the Venusian atmosphere. The atmosphere is so dense that the pressure 50 km above the surface is 1 atm, the same as at sea level on Earth. The curve shows how temperature varies with altitude (compare the curve for the Earth in Figure 9-22). Clouds in the Earth's atmosphere are seldom found above 12 km (8 mi); Venus's cloud layers lie at much higher altitudes.

est haze layer, the atmosphere is remarkably clear all the way down to the surface of Venus.

The Soviet and American probes all carried instruments that measured temperature and pressure as they descended to the planet's surface. Figure 12-5 shows how both pressure and temperature decrease smoothly with increasing altitude. As we saw in Section 9-6, the Earth's atmosphere has a much more complicated relationship between temperature and altitude (see Figure 9-22).

Spacecraft measurements show that sulfur, which is not found in any appreciable amount in our atmosphere, plays an important role in the Venusian atmosphere. Sulfur combines with other elements to form gases such as sulfur dioxide (SO_2) and hydrogen sulfide (H_2S), along with sulfuric acid (H_2SO_4), the same acid used in automobile batteries. While the Earth's clouds are composed of water droplets, Venusian clouds contain almost no water. Instead, they are composed of droplets of concentrated sulfuric acid. Thanks to the high temperatures on Venus, these droplets never rain down on the planet's surface; they simply evaporate at high altitude. (You can see a similar effect on Earth. On a hot day in the desert of the American Southwest, streamers of rain called *virga* appear out of the bottoms of clouds but evaporate before reaching the ground.)

The sulfuric acid in Venus's atmosphere makes the clouds a cauldron of chemical reactions hostile to metals and other solid materials. For example, reactions with fluorides and chlorides in surface rocks give rise to hydrofluoric acid (HF) and hydrochloric acid (HCl). Further reactions produce fluorosulfuric acid (HSO_3F), which is one of the most corrosive substances known to chemists, capable of dissolving lead, tin, and most rocks. It would be hard to imagine a less inviting place for humans to visit!

The Venusian atmosphere differs from our own not only in its pressures, temperatures, and chemical composition, but also in the patterns of its global circulation. Astronomers have been able to study these patterns using spacecraft in orbit around Venus. For example, in the early 1980s, the American *Pioneer Venus Orbiter* sent back numerous images of Venus taken at ultraviolet wavelengths, at which its atmospheric markings stand out best (**Figure 12-6**). By following individual cloud markings, scientists determined that the upper levels of Venus's atmosphere rotate around the planet in just 4 days. This confirmed earlier observations made with ground-based telescopes. The rapid motion of Venus's upper atmosphere is in sharp contrast to the slow rotation of the solid planet itself. Like the rotation of the planet, the motion of the atmosphere is in a retrograde direction (from east to west).

The *Pioneer Venus Multiprobe* spacecraft consisted of four probes that were dropped into the Venusian atmosphere in 1978. As they descended, they determined the atmosphere's dominant circulation patterns. Gases in the equatorial regions are warmed by the Sun, then rise upward and travel in the upper cloud layer toward the cooler polar regions. At the polar latitudes, the cooled gases sink to the lower cloud layer, in which they are transported back toward the equator. Recall from Section 9-1 that this process of heat transfer, whereby hot gases rise while cooler gases sink, is called *convection.*

The circulation of Venus's atmosphere is dominated by two huge convection cells, one in the northern hemisphere and another in the southern hemisphere, which circulate gases between the equatorial and polar regions of the planet (**Figure 12-7a**). (The convection

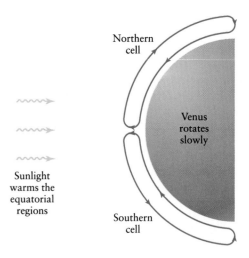

(a) Atmospheric circulation on Venus: two convection cells

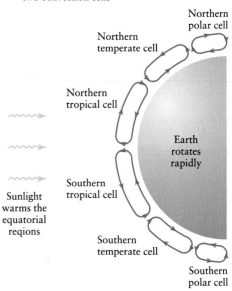

(b) Atmospheric circulation on Earth: six convection cells

Figure 12-7

Atmospheric Circulation on Venus and Earth Solar heating causes convection in the atmospheres of both Venus and Earth, with warm air rising at the equator and cold air descending at the poles. **(a)** Because Venus rotates only very slowly, it has little effect on the circulation. **(b)** Earth's rapid rotation distorts the atmospheric circulation into a more complex pattern (compare Figure 9-23).

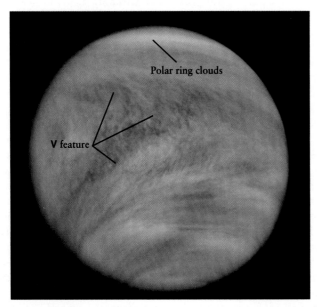

Figure 12-6 R I V U X G

Venus's Cloud Patterns The dark V feature in this false-color ultraviolet image of Venus is produced by the rapid motion of clouds around the planet's equator. The clouds move in the same retrograde direction as the rotation of the planet itself but at a much greater speed. The polar ring clouds are caused by atmospheric vortices near the pole (see Figure 12-7). (GSFC/NASA)

pattern in the Earth's atmosphere, shown in Figure 12-7*b* and Figure 9-23, is more complicated because the Earth's rapid rotation distorts the convection cells.) These convection cells, which are almost entirely contained within the main cloud layers, are called "driving cells" because they propel similar circulation cells above and below the main cloud deck (not shown in Figure 12-7*a*) somewhat like a meshed set of gears. (One effect of the secondary circulation is to produce a ring of clouds around the poles, shown in Figure 12-6.) The circulation is so effective at transporting heat around Venus's atmosphere that there is almost no temperature difference between the planet's equator and its poles.

As Venus's upper atmosphere makes its 4-day rotation around the planet, prevailing winds with speeds of 350 km/h (220 mi/h) blow from east to west at high altitude. These winds stretch out the driving cells and produce the characteristic V-shaped, chevronlike patterns visible in Figure 12-6. (In a similar way, the Earth's rotation stretches out the convection cells in our atmosphere to produce the complicated circulation patterns shown in Figure 9-23.)

On the Earth, friction between the atmosphere and the ground causes wind speeds at the surface to be much less than at high altitude. The same is true on Venus: The greatest wind speed measured by spacecraft on Venus's surface is only about 5 km/h (3 mi/h). Thus, only slight breezes disturb the crushing pressures and infernal temperatures found on Venus's dry, lifeless surface.

We have seen that the tremendous amount of carbon dioxide in the Venusian atmosphere explains why the surface is so hot. But where did all the CO_2 come from? And why do sulfur compounds play such an important role in Venus's clouds? We will find the answers to these questions by studying the present state and past history of Venus's surface.

12-4 | Volcanic eruptions are probably responsible for Venus's clouds

Sulfur compounds make up less than one part per billion of the Earth's atmosphere. By contrast, SO_2 and other sulfur compounds together make up about 0.015% of Venus's atmosphere as a whole and are the dominant component of the Venusian clouds. The apparent reason why sulfur compounds are relatively abundant on Venus is that sulfurous gases are being injected into the atmosphere by volcanic processes.

When *Pioneer Venus Orbiter* arrived at Venus in 1978, its ultraviolet spectrometer recorded unexpectedly high levels of sulfur dioxide and sulfuric acid, which steadily declined over the next several years. A similar, anomalously high abundance of haze particles may have occurred in the late 1950s, as indicated by telescopic observations made at that time from the Earth. In explanation, Larry W. Esposito of the University of Colorado proposed that in both the late 1950s and the late 1970s, energetic volcanic eruptions injected sulfur dioxide into Venus's upper atmosphere. It therefore seems that Venus's sulfur-rich clouds may be regularly replenished by active volcanoes.

This notion is based on the observation that volcanoes on Earth inject sulfurous gases into our atmosphere. For example, during the Mount St. Helens eruption of 1980 (**Figure 12-8**), geologists monitored the emission of substantial amounts of sulfuric acid and other sulfur compounds. Many of these substances are highly reactive and short-lived, forming sulfate compounds that become part of the planet's surface rocks. For these substances to be relatively abundant on Venus, they must be constantly replenished by new eruptions.

Volcanoes do indeed exist on Venus, and were first viewed by Earth-based radar observations. One of the most powerful tools for this task is the 305-m (1000-ft) dish at the Arecibo Observatory in Puerto Rico (see Figure 11-5). The Arecibo dish is used to transmit toward Venus a powerful, brief burst of microwaves, which easily penetrate the planet's clouds and reflect off its surface. Different

 Figure 12-8   R I **V** U X G
A Volcanic Eruption on Earth   The eruption of Mount St. Helens in Washington State on May 18, 1980, released an ash plume 40 km (25 mi) high. It injected into the atmosphere substantial amounts of sulfurous gases such as sulfur dioxide (SO_2), hydrogen sulfide (H_2S), and sulfuric acid (H_2SO_4). Figure 9-14 shows another volcanic eruption. (USGS)

types of terrain reflect microwaves more or less efficiently, so astronomers are able to construct a map of the Venusian surface by analyzing the reflected radiation.

Figure 12-9 is a radar view of a large Venusian volcano named Theia Mons. (Most features on Venus are named for women in history and legend. Theia was one of the Titans of Greek mythology.) It rises to an altitude of 6000 m (20,000 ft) and has gently sloping sides that extend over an area 1000 km (620 mi) in diameter. A volcano having this characteristic shape is called a **shield volcano,** because in profile it resembles an ancient Greek warrior's shield lying on the ground. Shield volcanoes also exist on both Earth and Mars; the best-known examples are the Hawaiian Islands. Shield volcanoes on Venus and Mars, as well as the Hawaiian Islands on the Earth, are thought to be the result of **hot-spot volcanism,** whereby a hot region beneath the planet's surface extrudes molten rock over a long period of time.

 By far the most detailed images of the Venusian volcanoes have come from the highly successful *Magellan* spacecraft, which went into orbit around Venus in

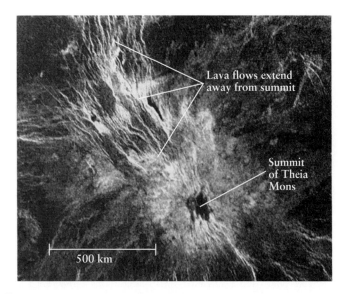

figure 12-9 ℝ I V U X G

An Earth-Based Radar Image of a Venusian Volcano This image shows solidified lava along the flanks of the Venusian volcano Theia Mons. Dark areas are where the surface is smooth and reflects microwaves only weakly; bright areas (such as the lava flows) have a rough-textured surface that reflects microwaves more strongly. The image covers an area roughly 3 times the size of Texas. (Courtesy of D. B. Campbell, Arecibo Observatory)

figure 12-10 ℝ I V U X G

A *Magellan* Radar Image of a Venusian Volcano The orbiting *Magellan* spacecraft provided detailed radar images of the Venusian surface. The smallest details visible are about 100 m across, about the size of a football stadium. The flat, 35 km- (22 mi-) wide summit of this volcano is surrounded by a number of ridges and valleys, which give it an insect-like appearance. (NASA)

1990. *Magellan* used sophisticated radar techniques to map more than 98% of the planet with a resolution of 100 m. It observed more than 1600 major volcanoes and volcanic features on Venus. Figure 12-10 shows a typical *Magellan* image of a volcano.

 In addition to mapping the planet with radar, *Magellan* used a radar altimeter that bounced microwaves off the ground directly below the spacecraft. By measuring the

time delay of the reflected waves, scientists could determine the height and depth of Venus's terrain. By combining all this information, scientists have constructed a detailed three-dimensional map of the Venusian surface. They have used this map and a supercomputer to produce vivid perspective views of the surface, like the one shown in Figure 12-11. Such views have provided scientists with a tremendous amount of insight into volcanism on Venus.

figure 12-11 ℝ I V U X G

Young Lava Flows on Venus This computer-generated perspective view shows the Venusian surface as you would see it from an altitude of 4 km (2.5 mi). The false color suggests the actual color of sunlight that penetrates Venus's thick clouds. As in Figure 12-9, the brighter color of the extensive lava flows indicates that they reflect radio waves more strongly. To emphasize the gently sloping volcanoes Sapas Mons and Maat Mons (named for ancient Phoenician and Egyptian goddesses respectively), the vertical scale has been exaggerated 10 times. (NASA, JPL Multimission Image Processing Laboratory)

Most of the volcanoes on Venus are probably inactive at present, just as is the case with most volcanoes on Earth. But *Magellan* found evidence of recent volcanic activity, some of which may be continuing today. The key to estimating the amount of recent volcanic activity is that the radar reflectivity of volcanic materials depends on whether the material is relatively fresh or relatively old. By mapping these reflectivity variations on Venus, *Magellan* found many areas with young lava flows. Some of the youngest material found by *Magellan* caps the volcano Maat Mons (see Figure 12-11). Geologists estimate that the topmost material is no more than 10 million years old and could be much younger. The presence of such young lava flows reinforces the idea that Venus, like the Earth but unlike the Moon or Mercury, has some present-day volcanic activity.

To verify the volcanic nature of the Venusian surface, it is necessary to visit the surface and examine rock samples. **Figure 12-12** is a panoramic view taken in 1981 by the Soviet spacecraft *Venera 13*, one of 10 unmanned spacecraft that the Soviet Union landed successfully on the surface of Venus. Russian scientists hypothesize that this region was covered with a thin layer of lava that fractured upon cooling to create the rounded, interlocking shapes seen in the photograph. This hypothesis agrees with information obtained from chemical analyses of surface material made by the space-

craft's instruments. These analyses indicate that the surface composition is similar to lava rocks called basalt, which are common on the Earth (see Figure 9-18*a*) and in the maria of the Moon (see Figure 10-15). The results from *Venera 13* and other landers are consistent with the picture of Venus as a world whose surface and atmosphere have been shaped by volcanic activity.

12-5 | The climate on Venus followed a different evolutionary path from that on Earth

Volcanic activity on Venus has not only shaped the planet's surface features but also influenced the Venusian atmosphere by adding new gases to it. Volcanoes on the Earth do the same thing, but the atmospheres of the two planets have evolved in very different ways. The Earth has abundant water in its oceans but very little carbon dioxide in its relatively thin atmosphere. By contrast, Venus is very dry and its thick atmosphere is mostly carbon dioxide. The challenge is to understand how these two planets, almost identical in size, ended up with atmospheres that are so different in chemical composition and density.

(a) Image from *Venera 13*

(b) Color-corrected image

Figure 12-12 R I **V** U X G

A Venusian Landscape (a) This wide-angle color photograph from *Venera 13* shows the rocky surface of Venus. The thick atmosphere absorbs the blue component of sunlight, giving the image an orange tint: The stripes on the spacecraft's color calibration bar that appear yellow are actually white in color. (b) This color-corrected view shows that the rocks are actually gray in color. The rocky plates covering the ground may be fractured segments of a thin layer of lava. (Courtesy of C. M. Pieters and the Russian Academy of Sciences)

The original atmospheres of both Venus and the Earth were derived at least in part from gases spewed forth, or outgassed, by volcanoes (see Section 9-5). The gases that emanate from present-day volcanoes on Earth, such as Mount St. Helens (see Figure 12-8), are predominantly water vapor (H_2O), carbon dioxide (CO_2), and sulfur dioxide (SO_2). These gases should therefore have been important parts of the original atmospheres of both Venus and the Earth. Much of the water on both planets is also thought to have come from impacts from comets, icy bodies formed in the outer solar system (see Section 7-5 and Section 8-5).

In fact, water probably once dominated the Venusian atmosphere. Venus and the Earth are similar in size and mass, so Venusian volcanoes may well have outgassed as much water vapor as on the Earth, and both planets would have had about the same number of comets strike their surfaces. Studies of how stars evolve suggest that the early Sun was only about 70% as luminous as it is now, so the temperature in Venus's early atmosphere must have been quite a bit lower. Water vapor may actually have been able to liquefy and form oceans on Venus.

But if water vapor and carbon dioxide were once so common in the atmospheres of both the Earth and Venus, what became of the Earth's CO_2? And what happened to the water on Venus?

The answer to the first question is that CO_2 is indeed found in abundance on the Earth. But, as we saw in Section 9-5, it is either dissolved in the oceans or chemically bound into carbonate rocks, such as the limestone and marble that formed in the oceans. If the Earth became as hot as Venus, much of its CO_2 would be boiled out of the oceans and baked out of the crust. Our planet would soon develop a thick, oppressive carbon dioxide atmosphere much like that of Venus.

To answer the question about Venus's lack of water, we must return to the early history of the planet. Just as on the present-day Earth, the oceans of Venus kept the amount of atmospheric CO_2 in check by dissolving it in the oceans and binding it up in carbonate rocks. Enough of the liquid water would have vaporized to create a thick cover of water vapor clouds. Since water vapor is a greenhouse gas, this humid atmosphere—perhaps denser than the Earth's present-day atmosphere, but far less dense than the atmosphere that surrounds Venus today—would have efficiently trapped heat from the Sun. At first, this would have had little effect on the oceans of Venus. Although the temperature would have climbed above 100°C, the boiling point of water at sea level on the Earth, the added atmospheric pressure from water vapor would have kept the water in Venus's oceans in the liquid state.

This hot and humid state of affairs may have persisted for several hundred million years. But as the Sun's energy output slowly increased over time, the temperature at the surface would eventually have risen above 374°C (647 K, or 705°F). Above this temperature, no matter what the atmospheric pressure, Venus's oceans would have begun to evaporate, and the added water vapor in the atmosphere would have increased the greenhouse effect. This would have made the temperature even higher and caused the oceans to evaporate faster, producing more water vapor. That in turn would have further intensified the greenhouse effect and made the temperature climb higher still. This is an example of a **runaway greenhouse effect,** in which an increase in temperature causes a further increase in temperature, and so on.

 A runaway greenhouse effect is like a house in which the thermostat has accidentally been connected backward. If the temperature in such a house gets above the value set on the thermostat, the heater comes on and makes the house even hotter.

Once Venus's oceans disappeared, so did the mechanism for removing carbon dioxide from the atmosphere. With no oceans to dissolve it, outgassed CO_2 began to accumulate in the atmosphere, making the greenhouse effect "run away" even faster. Temperatures eventually became high enough to "bake out" any CO_2 that was trapped in carbonate rocks. This liberated carbon dioxide formed the thick atmosphere of present-day Venus.

Sulfur dioxide (SO_2) is also a greenhouse gas. Although present in far smaller amounts than carbon dioxide, it would have contributed to Venus's rising temperatures. Like CO_2, it dissolves in water, a process that helps moderate the amount of sulfur dioxide in our atmosphere. But when the oceans evaporated on Venus, the amount of atmospheric SO_2 would have increased. (As we saw in Section 12-4, volcanic eruptions help to sustain the amount of SO_2 in Venus's atmosphere.)

The temperature in Venus's atmosphere would not have increased indefinitely. In time, solar ultraviolet radiation striking molecules of water vapor—the dominant cause of the runaway greenhouse effect—would have broken them into hydrogen and oxygen atoms. The lightweight hydrogen atoms would have then escaped into space (see Box 7-2 for a discussion of why lightweight atoms can escape a planet more easily than heavy atoms). The remaining atoms of oxygen, which is one of the most chemically active elements, would have readily combined with other substances in Venus's atmosphere. Eventually, almost all of the water vapor would have been irretrievably lost from Venus's atmosphere. With all the water vapor gone, and essentially all the carbon dioxide removed from surface rocks, the greenhouse effect would no longer have "run away," and the rising temperature would have leveled off.

Today, the infrared-absorbing properties of CO_2 have stabilized Venus's surface temperature at its present value of 460°C. Only minuscule amounts of water vapor—about 30 parts per million, or 0.003%—remain in the atmosphere. As on Earth, volcanic outgassing still adds small amounts of water vapor to the atmosphere, along with carbon dioxide and sulfur dioxide. But on Venus, these water molecules either combine with sulfur dioxide to form sulfuric acid clouds or break apart due to solar ultraviolet radiation. The great irony is that this state of affairs is the direct result of an earlier Venusian atmosphere that was predominantly water vapor!

Venus provides an object lesson about the importance of water on climate. On Earth, volcanic outgassing releases the greenhouse gases CO_2 and SO_2 into the atmosphere. But because these gases dissolve in rainwater and in the oceans, they are removed from the atmosphere, and their contribution to the greenhouse effect is moderated. On Venus, however, there are no oceans and no rainfall, and nothing moderates the greenhouse effect (Figure 12-13). As we will see in the next section, the absence of water on Venus also has major consequences for the planet's geology.

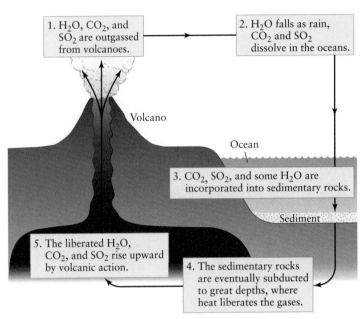

(a) Earth: H_2O, CO_2, and SO_2 are recycled

1. H_2O, CO_2, and SO_2 are outgassed from volcanoes.

2. H_2O falls as rain, CO_2 and SO_2 dissolve in the oceans.

3. CO_2, SO_2, and some H_2O are incorporated into sedimentary rocks.

4. The sedimentary rocks are eventually subducted to great depths, where heat liberates the gases.

5. The liberated H_2O, CO_2, and SO_2 rise upward by volcanic action.

Volcano

Ocean

Sediment

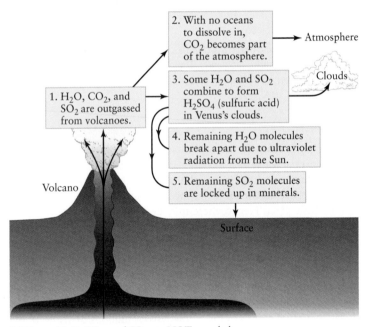

(b) Venus: H_2O, CO_2, and SO_2 are NOT recycled

1. H_2O, CO_2, and SO_2 are outgassed from volcanoes.

2. With no oceans to dissolve in, CO_2 becomes part of the atmosphere.

3. Some H_2O and SO_2 combine to form H_2SO_4 (sulfuric acid) in Venus's clouds.

4. Remaining H_2O molecules break apart due to ultraviolet radiation from the Sun.

5. Remaining SO_2 molecules are locked up in minerals.

Atmosphere

Clouds

Volcano

Surface

ANIMATION 12.3

figure 12-13

Chemical Cycles on Earth and Venus On both Earth and Venus, volcanic outgassing injects water vapor (H_2O), carbon dioxide (CO_2), and sulfur dioxide (SO_2) into the atmosphere. **(a)** The presence of oceans on Earth makes it possible to recycle H_2O, CO_2, and SO_2 into the Earth's interior. This helps keep their abundances in the atmosphere under control. **(b)** On Venus there are no oceans and no rainfall, so the outgassed H_2O, CO_2, and SO_2 follow different pathways than on Earth.

12-6 | The surface of Venus shows no evidence of plate tectonics

Venus is nearly the same size as Earth. Hence, we might expect it to have many of the same characteristics as the Earth: a molten interior, a magnetic field generated by fluid motions in its interior, and a crust that is reshaped by plate tectonics. But just as spacecraft revealed Venus's atmosphere to be different from what scientists had expected, they also provided many surprises about Venus's surface and interior.

The presence of volcanism on Venus (Section 12-4) strongly suggests that Venus, like Earth, has a molten interior. Unfortunately, we have no seismic data to back up this conclusion. None of the *Venera* spacecraft that landed on Venus carried seismometers. (In any event, the instruments on these spacecraft only lasted a few hours before succumbing to Venus's harsh environment.)

 Despite its molten interior, Venus has no planetwide magnetic field. One possible explanation is the planet's long rotation period. Venus rotates so slowly that the fluid material within the planet is hardly agitated at all, and so may not move in the fashion that generates a magnetic field. (We discussed the origin of planetary magnetic fields in Section 7-7.) Although it has no magnetic field and hence no magnetosphere, Venus is nonetheless shielded from the solar wind by ions in the planet's upper atmosphere.

Venus's crust also has many surprising characteristics. As we saw in Section 12-4, radar observations by the *Magellan* spacecraft have given us detailed three-dimensional information about the topography of Venus. The map in **Figure 12-14** summarizes this information, with a color code to denote elevation. (The image that opens this chapter is another such map.) Unlike Earth, where there is a sharp contrast in elevation between the ocean floors and the continents, about 60% of Venus's terrain lies within 500 m of the average elevation. More than 80% of Venus's surface is covered with volcanic plains that are the result of numerous lava flows.

Just a few large highlands rise above the planet's generally level surface. The highlands in the northern hemisphere (at the top of Figure 12-14), named Ishtar Terra, after the Babylonian goddess of love, are approximately the same size as Australia. Ishtar Terra is dominated by a high plateau ringed by high mountains; the highest, Maxwell Montes, rises to an altitude of 12 km (7 mi) above the average surface elevation. For comparison, the summit of Mount Everest on the Earth is 9 km (6 mi) above sea level. The most extensive Venusian highlands are Aphrodite Terra, which extends for 16,000 km around the equator. (Aphrodite was the Greek goddess whom the Romans called Venus.) Aphrodite Terra is 2000 km wide, giving it an area comparable to that of Africa. Radar images show that most of Aphrodite Terra is covered by networks of faults and fractures.

Before *Magellan*, geologists wondered if Venus's molten interior had given rise to plate tectonics, like those that have remolded the face of the Earth. If this were the case, then the same tectonic effects might also have shaped the surface of Venus. As we saw in Section 9-3, the Earth's hard outer shell, or lithosphere, is broken into about a dozen large plates that slowly shuffle across the globe. Long

Figure 12-14 R I V U X G
A Topographic Map of Venus Radar altimeter measurements by *Magellan* were used to produce this topographic map of Venus. Color indicates elevation—red corresponds to the highest, blue to the lowest. Gray areas were not mapped by *Magellan*. Flat plains of volcanic origin cover about 80% of the planet's surface, with only a few continentlike highlands. The volcanoes shown in Figure 12-11 lie at the far eastern end of one of these highland areas, Aphrodite Terra. (Peter Ford, MIT; NASA/JPL)

mountain ranges, like the Mid-Atlantic Ridge (see Figure 9-12), are created where fresh magma wells up from the Earth's interior to push the plates apart.

The images from *Magellan* show *no* evidence of Earthlike tectonics on Venus. On Earth, long chains of volcanic mountains (like the Cascades in North America or the Andes in South America) form along plate boundaries where subduction is taking place. Volcanic features on Venus, by contrast, do not appear in chains. Also absent on Venus is a type of faulting that always occurs with sea-floor spreading on the Earth. These faults, which are perpendicular to rifts on the ocean floor, give the Earth's ocean ridges an appearance that distinctly resembles a staircase (examine Figure 9-12). Because Venus lacks any evidence of subduction or of seafloor spreading, we know that there can have been only limited horizontal displacement of its lithosphere. Thus, like the Moon (see Section 10-1) and Mercury (see Section 11-3), Venus is a one-plate planet. But there have been local, small-scale deformations and reshaping of the surface. As an example, roughly a fifth of Venus's surface is covered by folded and faulted ridges.

Why is plate tectonics absent on Venus? One key reason may be the absence of water. Crustal motions on Earth are powered by large convection currents in the asthenosphere, the partially molten layer that lies between the lithosphere and the mantle (see Figure 9-15). This layer can flow because it contains water, which lowers the melting-point temperature of the asthenosphere's rock and makes it more plastic. (Water gets into the asthenosphere by being subducted along with descending crustal plates.) But Venus lost all the water in its atmosphere and from its surface during the era of the runaway greenhouse effect. Hence, there may be no asthenosphere at all beneath Venus's crust.

Another explanation may be that Venus's lithosphere is too plastic to sustain plate tectonics. The high surface temperature may soften rock so much that it is unable to sustain the kind of stresses associated with plate tectonics. Any large differences in elevation, such as are found on Earth between the continents and the sea-floors, would disappear after a few tens of millions of years. In this model, tall mountains such as Maxwell Montes must be very young, which reinforces the picture of Venus as a volcanically active world (see Section 12-4).

Figure 12-15 shows the difference between key geologic processes on Earth and on Venus. On Earth, new crust is formed at midocean ridges where material flows upward from the interior. This crust is eventually reprocessed into the interior at subduction zones. On Venus, by contrast, upwelling magma from the mantle pushes upward on the crust, elevating it and producing volcanic structures like those in Figures 12-9, 12-10, and 12-11. This hot-spot volcanism does not add much new material to the crust. Instead, the crust is simply pushed sideways. The result is that the

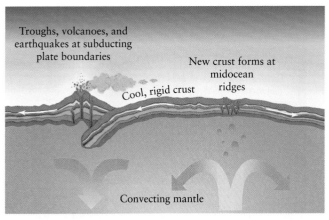

(a) Earth: The crust is cool and rigid, and so can move in large plates. New crust is created at midocean ridges and returned to the interior at subduction zones.

(b) Venus: The crust may be too hot and soft to move in rigid plates. The absence of water may limit motions beneath the crust.

Figure 12-15

Geologic Processes on Earth and Venus (a) Plate tectonics is active on Earth, but (b) Venus shows no evidence of plate tectonics. (Adapted from J. W. Head)

crust is compressed together over locations where magma is moving downward in the mantle. Such compression folds the surface into ridges and valleys. Features of this kind are found on Venus's broad lava plains (Figure 12-16).

Magellan has produced superb pictures of Venus's impact craters (Figure 12-17). Venus probably has about a thousand craters larger than a few kilometers in diameter, many more than have been found on Earth but only a small fraction of the number on the Moon or Mercury. We saw in Section 7-6 that the number of impact craters is a clue to the age of a planet's surface. Such craters formed at a rapid rate during the early history of the solar system, when considerable interplanetary debris still orbited the Sun. Since that time, impact rates have declined, as shown in Figure 10-18, and old craters have been obliterated as volcanism or tectonics renews a planet's surface (Figure 12-18). Consequently, the more craters a planet has, the older its surface. The age of the Venusian surface seems to be roughly 500 million years. This is about twice the age of the Earth's surface but much younger than the surfaces of the Moon or Mercury, each of which is billions of years old.

Surprisingly, Venus's craters are uniformly scattered across the planet's surface. We would expect that older regions on the surface—which have been exposed to bombardment for a longer time—would be more heavily cratered, while younger regions would be relatively free of craters. For example, the ancient highlands on the Moon are much more heavily cratered than the younger maria (see Section 10-4). Because such variations are not found on Venus, scientists conclude that the entire surface of the planet has essentially the *same* age. This is very different from the Earth, where geological formations of widely different ages can be found.

Scientists have proposed two competing hypotheses to explain the apparent youth of Venus's surface. According to the **equilibrium resurfacing hypothesis,** proposed by Roger J. Phillips of Washington

University, volcanic eruptions are continually covering up craters at about the same rate that craters are formed by impact. Central to this model is the observation that volcanism is widespread on Venus, so that craters are destroyed at roughly equal rates over all parts of the planet.

Figure 12-16 R I V U X G

Compressional Folding on Venus This *Magellan* radar image of the low-lying Lavinia Planitia region shows how compression of the surface, as suggested by Figure 12-15b, has formed a bright belt of ridges that run from lower left to upper right. Lava flows have flooded the surrounding terrain. (Black stripes are where data are missing.) (NASA/JPL)

figure 12-17 R I V U X G

A Venusian Crater This perspective view (constructed from *Magellan* data) shows an impact crater 48 km (30 mi) in diameter, about the same size as St. Louis, Missouri, and its suburbs. This crater is named for the Polish astronomer Maria Cunitz; the volcano Gula Mons is named after a Babylonian earth goddess. (NASA/JPL)

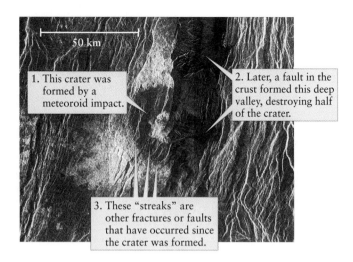

figure 12-18 R I V U X G

A Partially Obliterated Crater This *Magellan* image shows how the right half of an old impact crater 37 km (20 mi) in diameter was erased by a fault in the crust. This crater, named for the American economist Emily Balch, lies in Beta Regio (see the left side of Figure 12-14). (NASA/JPL)

An alternative model, put forth by Gerald G. Schraber of the U.S. Geological Survey and Robert G. Strom of the University of Arizona, proposes that an intense period of eruptions took place on Venus several hundred million years ago. According to this **global catastrophe hypothesis,** the entire globe was resurfaced over a short period with fresh lava, covering up any older craters. In one version of this hypothesis, upwelling material from the mantle solidifies on the underside of Venus's lithosphere. The increasingly thick lithosphere acts as an insulating blanket enveloping the planet's interior, trapping heat until the mantle becomes hot enough to break the lithosphere apart. The result is widespread volcanism that covers nearly the entire planet with lava. The lithosphere then cools and thickens, and the cycle begins anew. If this picture is correct, then every few hundred million years the entire face of Venus is erased at once, and there can be no very old rocks on the Venusian surface. By contrast, plate tectonics erases the Earth's surface in a piecemeal fashion, which is why some rocks on the Earth are billions of years old.

The intense eruptions proposed in the global catastrophe hypothesis would also have had tremendous effects on Venus's climate. Such eruptions would have outgassed tremendous amounts of carbon dioxide, water vapor, and sulfur dioxide. Simulations by Mark A. Bullock and David J. Grinspoon of the University of Colorado suggest that this fresh supply of greenhouse gases would eventually have raised Venus's surface temperature to 100°C higher than its present-day value. Such temperatures would have melted some of the solid materials that make up Venus's surface, which could explain certain riverlike features seen on the planet today (**Figure 12-19**). Atmospheric temperatures would also have increased, causing Venus's sulfuric acid clouds to evaporate. The atmosphere

 figure 12-19 R I V U X G

The Longest Channel in the Solar System This *Magellan* image shows a small portion of a long, meandering, 2 km- (1.2 mi-) wide channel named Baltis Vallis ("Baltis" is the Syrian word for the planet Venus). Carved out by molten lava, not water, Baltis Vallis has a total length of 6800 km (4200 mi). In comparison, the Nile—the longest river on the Earth—is some 6600 km (4100 mi) long. (NASA/JPL)

and surface would have returned to their present "mild" temperatures only after hundreds of millions of years.

Although Venus is nearly the Earth's twin in size, its surface and atmosphere have evolved in radically different ways than the Earth. In the long run, however, both planets may meet the same fate. Our Sun's luminosity will continue to increase as it ages, and in about 3½ billion years sunlight will be bright enough to begin evaporating the Earth's oceans. The atmosphere will experience the same sort of runaway greenhouse effect that occurred on Venus long ago. As temperatures drive toward 1000 K, vast amounts of carbon dioxide will be baked out of the Earth's rocks, creating a Venuslike atmosphere. By studying Venus, we may be looking into the distant future of our own world.

Key Words

equilibrium resurfacing
  hypothesis, p. 257
global catastrophe hypothesis,
  p. 258
hot-spot volcanism, p. 251

prograde rotation, p. 246
retrograde rotation, p. 246
runaway greenhouse effect,
  p. 254
shield volcano, p. 251

Key Ideas

Motions of Venus in the Earth's Sky: At its greatest eastern and western elongations, Venus is about 47° from the Sun, and it can be seen for several hours after sunset or before sunrise.

Comparison of the Earth and Venus: Venus is similar to the Earth in its size, mass, average density, and surface gravity, but it is covered by unbroken, highly reflective clouds that conceal its other features from Earth-based observers.

Venus's Rotation: Venus rotates slowly in a retrograde direction with a solar day of 117 Earth days and a rotation period of 243 Earth days. There are approximately two Venusian solar days in a Venusian year.

Venus's Atmosphere and Clouds: Spacecraft measurements reveal that 96.5% of the Venusian atmosphere is carbon dioxide. Most of the balance of the atmosphere is nitrogen.

• Venus's clouds consist of droplets of concentrated sulfuric acid. Active volcanoes on Venus may be a continual source of this sulfurous material.

• Venus's clouds are confined to altitudes between 48 and 68 km above the planet's surface. A haze layer extends down to an elevation of 30 km, beneath which the atmosphere is clear.

• The surface pressure on Venus is 90 atm, and the surface temperature is 460°C. Both temperature and pressure decrease as altitude increases.

• The circulation of the Venusian atmosphere is dominated by two huge convection currents in the cloud layers, one in the northern hemisphere and one in the southern hemisphere.

• The upper cloud layers of the Venusian atmosphere move rapidly around the planet in a retrograde direction, with a period of only about 4 Earth days.

The Greenhouse Effect: Venus's high temperature is caused by the greenhouse effect, as the dense carbon dioxide atmosphere traps and retains energy from sunlight.

• The early atmosphere of Venus contained substantial amounts of water vapor. This caused a runaway greenhouse effect that evaporated Venus's oceans and drove carbon dioxide out of the rocks and into the atmosphere. Almost all of the water vapor was eventually lost by the action of ultraviolet radiation on the upper atmosphere.

• The Earth has roughly as much carbon dioxide as Venus, but it has been dissolved in the Earth's oceans and chemically bound into its rocks.

The Surface of Venus: The surface of Venus is surprisingly flat, mostly covered with gently rolling hills. There are a few major highlands and several large volcanoes.

• The surface of Venus shows no evidence of the motion of large crustal plates, which plays a major role in shaping the Earth's surface.

• The density of craters suggests that the entire surface of Venus is no more than a few hundred million years old. According to the equilibrium resurfacing hypothesis, this happens because old craters are erased by ongoing volcanic eruptions. In the alternative global catastrophe hypothesis, all of Venus was resurfaced at essentially the same time.

Review Questions

1. As seen from the Earth, the brightness of Venus changes as it moves along its orbit. Describe the main factors that determine these variations in brightness.

2. Explain why the photograph in Figure 12-2 must have been made during the daytime.

3. If Venus did not have an atmosphere, how would its appearance as seen from Earth be different?

4. Why are solar transits of Venus so much rarer than solar transits of Mercury?

5. Why was it so difficult to determine the rate and direction of Venus's rotation? How were these finally determined?

6. Describe one explanation for the slow, retrograde rotation of Venus.

7. What roles does the greenhouse effect play in the atmospheres of Venus and the Earth?

8. Why was it difficult to determine Venus's surface temperature from Earth? How was this finally determined?

9. The *Mariner 2* spacecraft did not enter Venus's atmosphere, but it was nonetheless able to determine that the atmosphere is very dry. How was this done?

10. Why is it hotter on Venus than on Mercury?

11. How does the circulation of Venus's atmosphere differ from that of Earth's atmosphere?

12. What is the evidence for active volcanoes on Venus?

13. How can lava flows be distinguished from other types of terrain in radar images of the Venusian surface?

14. How might Venus's cloud cover change if all of Venus's volcanic activity suddenly stopped? How might these changes affect the overall Venusian environment?

15. Why are there no oceans on Venus? Where has Venus's water gone?

16. Why is there so much carbon dioxide in Venus's atmosphere while very little of this gas is present in the Earth's atmosphere?

17. What is the difference between the greenhouse effect as it exists on Venus today and the runaway greenhouse effect that existed in Venus's early atmosphere?

18. Describe the Venusian surface. What kinds of features would you see if you could travel around on the planet?

19. In what ways does the surface topography of Venus differ from that of the Earth?

20. Why do scientists think that Venus's surface was *not* molded by the kind of tectonic activity that shaped the Earth's surface?

21. Describe how Venus's lack of water and high surface temperatures may help explain the absence of plate tectonic activity.

22. Compare and contrast the kinds of geologic activity that occur on Venus with those that occur on Earth.

23. Describe two competing hypotheses that attempt to explain why Venus's surface is only a few hundred million years old.

| Advanced Questions |

Problem-solving tips and tools

Box 4-1 describes the relationship between Venus's sidereal orbital period and its synodic period (the time from one inferior conjunction to the next). You should recall that Wien's law (Section 5-4) relates the temperature of a blackbody to λ_{max}, its wavelength of maximum emission. Box 5-4 describes some of the physics of light scattering. Section 5-9 and Box 5-6 explain the Doppler effect and how to do calculations using it. We discussed the relationship between angular distance and linear distance in Box 1-1 and the idea of angular resolution in Section 6-3. The linear speed of a point on a planet's equator is the planet's circumference divided by its rotation period; Recall that the circumference of a circle of radius r is $2\pi r$.

24. Venus takes 440 days to move from greatest western elongation to greatest eastern elongation, but it needs only 144 days to go from greatest eastern elongation to greatest western elongation. With the aid of a diagram like Figure 12-1, explain why.

25. Before about 350 B.C. it was not generally understood that Venus seen in the morning sky (that is, at greatest western elongation) and Venus seen in the evening sky (at greatest eastern elongation) were actually the same planet. Construct a geocentric model of the planets (like that shown in Figure 4-15) in which the "morning Venus" and the "evening Venus" are two distinct planets.

26. (a) Use the data in Table 12-2 to determine how many days elapse from one greatest eastern elongation of Venus to the next. (b) Use Table 12-2 and your answer from part (a) to determine the date in 2009 when Venus will be at greatest eastern elongation. (c) If Venus has an orbital period of 224.7 days, why isn't the time interval you calculated in part (a) equal to 224.7 days? (*Hint:* See Box 4-1.)

27. Venus's sidereal rotation period is 243.01 days and its orbital period is 224.70 days. Use these data to prove that a solar day on Venus lasts 116.8 days. (*Hint:* Develop a formula relating Venus's solar day to its sidereal rotation period and orbital period similar to the first formula in Box 4-1.)

28. In Section 12-2 we described the relationship between the length of Venus's synodic period and the length of an apparent solar day on Venus. Using this and a diagram, explain why at each inferior conjunction the same side of Venus is turned toward the Earth.

29. If you aim microwaves of wavelength 1.9 cm at the entire face of Venus, what will be the spread in wavelength of the reflected waves caused by the planet's rotation? (See Figure 11-6.)

30. At what wavelength does Venus's surface emit the most radiation? Do astronomers have telescopes that can detect this radiation? Why can't we use such telescopes to view the planet's surface?

31. The *Mariner 2* spacecraft detected more microwave radiation when its instruments looked at the center of Venus's disk than when it looked at the edge, or limb, of the planet. (This effect is called *limb darkening*.) Explain how these observations show that the microwaves are emitted by the planet's surface rather than its atmosphere.

32. Explain how you could estimate the size of the droplets that make up Venus's clouds by beaming radio waves of different wavelengths through the clouds to a spacecraft on the planet's surface.

33. When the *Galileo* spacecraft flew past Venus in 1990 while on its way to Jupiter, it used its infrared camera to view lower-level clouds in the Venusian atmosphere. Why was it necessary to use infrared light to see these clouds?

34. In the classic Ray Bradbury science-fiction story "All Summer in a Day," human colonists on Venus are subjected to continuous rainfall except for one day every few years when the clouds part and the Sun comes out for an hour or so. Discuss how our understanding of Venus's atmosphere has evolved since this story was first published in 1954.

35. Suppose that Venus had no atmosphere at all. How would the albedo of Venus then compare with that of Mercury or the Moon? Explain your answer.

36. A hypothetical planet has an atmosphere that is opaque to visible light but transparent to infrared radiation. How would this

affect the planet's surface temperature? Contrast and compare this hypothetical planet's atmosphere with the greenhouse effect in Venus's atmosphere.

37. Water has a density of 1000 kg/m³, so a column of water n meters tall and 1 meter square at its base has a mass of $n \times 1000$ kg. On either the Earth or Venus, which have nearly the same surface gravity, a mass of 1 kg weighs about 9.8 newtons (2.2 lb). Calculate how deep you would have to descend into the Earth's oceans for the pressure to equal the atmospheric pressure on Venus's surface, 90 atm or 9×10^6 newtons per square meter.

38. Radar echoes become weaker as the distance increases between the radio source and the objected being observed. Use this to explain why the method of mapping the surface of Venus using radio waves emitted is most successful when Venus is near inferior conjunction.

39. The image in Figure 12-10 was made by the *Magellan* spacecraft from an altitude of 720 km (450 mi). Use the information given in the caption to Figure 12-10 to calculate the angular resolution of *Magellan*'s imaging system.

40. Marine organisms produce sulfur-bearing compounds, some of which escape from the oceans into the Earth's atmosphere. (These compounds are largely responsible for the characteristic smell of the sea.) Even more sulfurous gases are injected into our atmosphere by the burning of sulfur-rich fossil fuels, such as coal, in electric power plants. Both of these processes add more sulfur compounds to the atmosphere than do volcanic eruptions. On lifeless Venus, by contrast, volcanoes are the only source for sulfurous atmospheric gases. Why, then, are sulfur compounds so much rarer in our atmosphere than in the Venusian atmosphere?

41. The Hawaiian Islands lie along a nearly straight line (see the accompanying figure). A hot spot under the Pacific plate has

remained essentially stationary for 70 million years while the plate has moved to the northwest some 6000 km. The upwelling magma has thus produced a long chain of hot-spot volcanoes. (The oldest, at the northwest end of the chain, have eroded so much that they no longer rise above the ocean surface.) Hot-spot volcanoes are also found on Venus, but they do not form long chains. Explain what this tells us about tectonic activity on Venus.

Discussion Questions

42. Describe the apparent motion of the Sun during a "day" on Venus relative to (**a**) the horizon and (**b**) the background stars. (Assume that you can see through the cloud cover.)

43. If you were designing a space vehicle to land on Venus, what special features would you think necessary? In what ways would this mission and landing craft differ from a spacecraft designed for a similar mission to Mercury?

44. If you could examine rock samples from the surface of Venus, would you expect them to be the same as rock samples from Earth? Would you expect to find igneous, sedimentary, and metamorphic rocks like those found on Earth (see Section 9-3)? Explain your answers.

 ## Web/CD-ROM Questions

45. A number of European astronomers traveled to Asia and the Pacific islands to observe the transits of Venus in 1761 and 1769. Search the World Wide Web for information about these expeditions. Why were these events of such interest to astronomers? How definitive were the results of these observations?

46. Search the World Wide Web for the latest ideas about whether lightning occurs in the Venusian atmosphere. What was the original evidence for lightning on Venus? What recent observations either confirm or refute this evidence? Is there any evidence for other phenomena in the Venusian atmosphere that cause light to be emitted?

47. Search the World Wide Web for the latest information about proposed future missions to Venus, including a mission called *Venus Express*. Which of these missions have been given funding so that they can proceed? What scientific experiments will they carry? Which scientific issues are these instruments intended to resolve?

 48. Surface Temperature of Venus. Access the Active Integrated Media Module "Wien's Law" in Chapter 5 of the *Universe* Web site or CD-ROM. (**a**) Using the Wien's Law calculator, determine Venus's approximate temperature if it emits blackbody radiation with a peak wavelength of 3866 nm. (**b**) By trial and error, find the wavelength of maximum emission for a surface temperature of 733 K (for present-day Venus) and a surface temperature of 833 K (as it might be in the event of a global catastrophe that released more greenhouse gases into

Venus's atmosphere). In what part of the electromagnetic spectrum do these wavelengths lie?

Observing Projects

> Observing tips and tools
>
> Venus is visible in the morning sky when it is at or near greatest western elongation and in the evening sky when at or near greatest eastern elongation. These are also the times when Venus can be seen at the highest altitude above the horizon during the hours of darkness. Consult such magazines as *Sky & Telescope* and *Astronomy* or their Web sites for more detailed information about when and where to look for Venus during a given month. You can also use the *Starry Night Backyard*™ program on the CD-ROM that accompanies this textbook.

49. Refer to Table 12-2 to see if Venus is near a greatest elongation. If so, view the planet through a telescope. Make a sketch of the planet's appearance. From your sketch, can you determine if Venus is closer to us or farther from us than the Sun?

50. Using a small telescope, observe Venus once a week for a month and make a sketch of the planet's appearance on each occasion. From your sketches, can you determine whether Venus is approaching us or moving away from us?

51. This observing project should be performed only under the direct supervision of an astronomer who knows how to point a telescope safely at Venus. Make arrangements to view Venus during broad daylight. This is best done by visiting an observatory where the coordinates (right ascension and declination) of Venus's position can be used to point the telescope. **DO NOT LOOK AT THE SUN! Looking directly at the Sun can cause blindness.**

 52. Use the *Starry Night Backyard*™ program to observe solar transits of Venus. Display the entire celestial sphere (select **Guides > Atlas** in the **Go** menu) and center on Venus by using the **Find...** command in the **Edit** menu. Using the controls at the right-hand end of the Control Panel, zoom in until the field of view is 1°. (**a**) In the Control Panel, set the date and time to June 8, 2004, at 12:00:00 A.M., and set the time step to 1 hour. Step backward or forward through time using the single-step buttons (the leftmost and the rightmost buttons) and record the times at which the solar transit begins and ends. What is the total duration of the solar transit? (**b**) The ecliptic appears in *Starry Night Backyard*™ as a green line. During the transit, is Venus precisely on the ecliptic? If not, about how far off is it? (*Hint:* The Sun has an angular diameter of about 30 arcmin.) (**c**) Repeat parts (a) and (b) for the solar transit of Venus on June 6, 2012.

 53. Use the *Starry Night Backyard*™ program to observe the apparent motion of Venus on the celestial sphere. Display the entire celestial sphere (select **Guides > Atlas** in the **Go** menu) and center on the Sun by using the **Find...** command in the **Edit** menu. Using the controls at the right-hand end of the Control Panel, zoom out until the field of view is 100°. In the Control Panel, set the date and time to January 1, 2007, at 12:00:00 A.M., and set the time step to 1 day. (**a**) Using the buttons for "play forward" (a triangle pointing right) and "stop" (a solid square), find the first date after January 1, 2007 when Venus is as far to the right of the Sun as possible, and the first date after January 1, 2007 when Venus is as far to the left of the Sun as possible. Compare with Table 12-2. What is your interpretation of the two dates that you determined? (**b**) Set the date to December 1, 2007, and start the animation by clicking on the "play forward" button. Based on your observations, explain why Venus has neither a greatest western elongation nor a greatest eastern elongation during 2008.

 54. Use the *Deep Space Explorer*™ program to compare the orbits of Venus and the Earth. In the left-hand part of the window under the heading **Solar System,** click on the triangle next to the word **Explore** and then click on **Inner Solar System.** You can zoom in and zoom out using the buttons at the upper left of the window (an upward-pointing triangle and a downward-pointing triangle). You can also rotate the solar system by putting the mouse cursor over the image, holding down the mouse button, and moving the mouse. (On a two-button mouse, hold down the left mouse button.) Are the orbits of Venus and of the Earth in the same plane? At the time shown in the image, is Venus nearest to inferior conjunction, superior conjunction, greatest eastern elongation, or greatest western elongation as seen from Earth? Explain your answers.

Collaborative Exercises

55. In very general terms, Venus can be easily seen roughly two months before and two months after its greatest elongation. Using Table 12-2, will Venus be an evening star, a morning star, or likely not easily visible on your next birthday?

56. Figure 12-3 shows prograde and retrograde rotation. Stand up and demonstrate how Venus rotates as it orbits our Sun, using a classmate as our stationary Sun. How is this different from how Earth rotates in its orbit?

57. Surface features on Venus are given female names by the International Astronomical Union, mostly originating in ancient mythology. If the 10 largest mountains on Venus were to be named after famous females who are alive today, propose 10 names around a unified theme and explain why each is worthy of such a hypothetical honor.

13 Red Planet Mars

Mars Attacks! Invaders from Mars! Martians, Go Home! The fourth planet from the Sun has inspired many science-fiction films and novels about alien invasions. But why Mars? People have long speculated that life might exist there, because the red planet has many Earthlike characteristics. At the beginning of the twentieth century, some astronomers claimed to have seen networks of linear features on the Martian surface, perhaps "canals" built by an advanced civilization.

The Hubble Space Telescope image of Mars reproduced here shows no sign of canals, however. Indeed, over the past four decades spacecraft have confirmed that Mars has no canals, no signs of life, and no liquid water on its surface. But they have also found evidence that Mars was once a very different place. Spacecraft in Martian orbit have photographed ancient flood channels through which immense amounts of water once coursed, as well as gullies that may have been caused by smaller but still substantial flows. Data from spacecraft that have landed on Mars show that liquid water may have existed there for extended periods—perhaps long enough for Mars to have had large lakes and even an ocean.

Spacecraft have also revealed how the face of Mars has been reshaped. Internal forces have produced gigantic volcanoes and an immense valley that is as long as North America is wide. External forces, in the form of space debris that has collided with Mars, have strewn much of the planet's surface with craters of all sizes. The challenge to planetary scientists is to understand how Mars has evolved into the dry, desolate, and nearly airless planet that we see today.

A view of Mars during the opposition of 2003. (NASA; J. Bell, Cornell University; and M. Wolff, SSI) R I **V** U X G

Questions to Guide Your Reading

As you read the sections of this chapter, look for the answers to the following questions:

13-1 When is it possible to see Mars in the night sky?

13-2 Why was it once thought that there are canals on Mars?

13-3 How are the northern and southern hemispheres of Mars different from each other?

13-4 What is the evidence that there was once liquid water on Mars?

13-5 Why is the Martian atmosphere so thin?

13-6 What have we learned about Mars by sending spacecraft to land on its surface?

13-7 What causes the seasonal color changes on Mars?

13-8 As seen from Mars, how do the Martian moons move across the sky?

13-1 | Earth-based observations of Mars are best made during favorable oppositions

Mars is a relatively small planet, with just over half the diameter of the Earth. (Table 13-1 summarizes some basic information about Mars.) Nonetheless, Mars is the easiest of the other terrestrial planets to study with a small telescope. Unlike Mercury, which never appears very far from the Sun (see Section 11-1), Mars can be seen high in the night sky when it is at opposition. And unlike Venus, with its dense atmosphere and perpetual cloud cover (see Section 12-1), Mars has a thin, almost cloudless atmosphere that permits a clear view of the surface. Under the right conditions, a relatively small telescope can reveal substantial detail on the Martian surface (Figure 13-1).

We get the best views of Mars when it is at opposition, so that it is close to Earth and appears high in our night sky. Such oppositions occur at intervals of about 780 days. However, some oppositions provide better views of Mars than others. As Figure 13-2 shows, Mars has a noticeably elongated orbit. As a result, we get the best view of Mars when it is simultaneously at opposition and near the perihelion of its elliptical orbit, a configuration called a **favorable opposition.** The Earth-Mars distance can then be as small as 0.37 AU, or 56 million kilometers (35 million miles), and the angular diameter of Mars can be as large as 25.1 arcsec—about the same as a person's head seen at a distance of 1.6 km (1 mile). Under these conditions, Mars appears in the nighttime sky as a brilliant red object 3½ times brighter than Sirius, the brightest star in the sky. The image in Figure 13-1 shows Mars during the very favorable 2003 opposition, when Mars was closer to Earth than it had been in several thousand years.

Unfortunately, the time from one favorable opposition to the next is about 15 years. Hence, astronomers take advantage of all

figure 13-1   R I **V** U X G
Mars as Viewed from Earth This high-quality image of Mars was made using a 25-cm (10-in.) telescope with a CCD detector (see Section 6-4). Compare this with the Hubble Space Telescope image that opens this chapter, which was taken during the same month in 2003 and which shows the same face of Mars. (Courtesy of Sheldon Faworski and Donald Parker)

oppositions, even the not-so-favorable ones, to observe Mars. When Mars is at opposition but near aphelion, as will happen in 2012, the Earth-Mars distance can be as large as 0.68 AU, or 101 million kilometers (63 million miles). When viewed under these conditions through a small amateur telescope, Mars appears as a nearly featureless red dot.

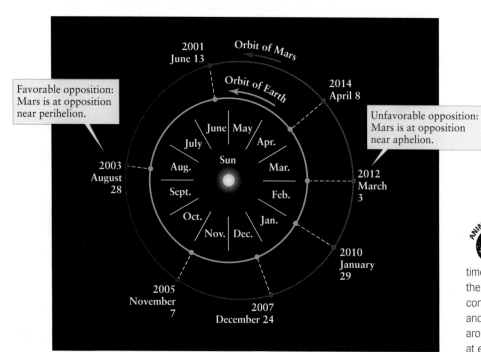

figure 13-2
The Orbits of Earth and Mars The best times to observe Mars are at opposition, when Mars is on the side of Earth opposite the Sun. The red and blue dots connected by dashed lines show the positions of Mars and Earth at various oppositions. The months shown around the Sun refer to the time of year when the Earth is at each position around its orbit.

table 13-1 | Mars Data

Average distance from Sun:	1.524 AU = 2.279×10^8 km
Maximum distance from Sun:	1.666 AU = 2.492×10^8 km
Minimum distance from Sun:	1.381 AU = 2.067×10^8 km
Eccentricity of orbit:	0.093
Average orbital speed:	24.1 km/s
Orbital period:	686.98 days = 1.88 years
Rotation period:	$24^h\ 37^m\ 22^s$
Inclination of equator to orbit:	25.19°
Inclination of orbit to ecliptic:	1.85°
Diameter (equatorial):	6794 km = 0.533 Earth diameter
Mass:	6.418×10^{23} kg = 0.107 Earth mass
Average density:	3934 kg/m³
Escape speed:	5.0 km/s
Surface gravity (Earth = 1):	0.38
Albedo:	0.15
Surface temperatures:	Maximum: 20°C = 70°F = 293 K
	Mean: –53°C = –63°F = 220 K
	Minimum: –140°C = –220°F = 133 K
Atmospheric composition (by number of molecules):	95.3% carbon dioxide (CO_2)
	2.7% nitrogen (N_2)
	0.03% water vapor (H_2O)
	2% other gases

WEB LINK 13.3   (NASA, USGS)   R I **V** U X G

⚠ **CAUTION** Although Mars appears bright red to Earth observers, its surface is actually reddish-brown. A sunlit brown surface, if surrounded by the blackness of space, is perceived by the human eye as having a red color.

Two Martian oppositions will take place between the years 2004 and 2007 (Table 13-2). Figure 4-2 shows the apparent path of Mars across the sky during the 2010 opposition.

13-2 | Earth-based observations were once thought to show evidence of intelligent life on Mars

WEB LINK 13.4   The first reliable record of surface features on Mars was made by the Dutch scientist Christiaan Huygens, who observed the planet during the opposition of 1659. He identified a prominent dark feature that we now call Syrtis Major.

table 13-2 | Oppositions of Mars, 2004–2008

	Earth-Mars distance		
Date of opposition	(AU)	(10^6 km)	Angular diameter (arcsec)
2005 November 7	0.464	69.4	20.2
2007 December 24	0.589	88.1	15.9

After observing this feature for several weeks, Huygens concluded that the rotation period of Mars is approximately 24 hours, the same as the Earth.

In 1666, the Italian astronomer Gian Domenico Cassini made more-refined observations and concluded that a Martian day is about 37½ minutes longer than a solar day on Earth. Cassini was also the first to see the Martian polar caps, which bear a striking superficial resemblance to the Arctic and Antarctic regions on Earth (see the figure that opens this chapter). More than a century later, the German-born English astronomer William Herschel suggested that the Martian polar caps might be made of ice or snow.

Herschel also found that Mars's axis of rotation is not perpendicular to the plane of the planet's orbit, but is tilted by about 25° away from the perpendicular. This is very close to the Earth's 23½° tilt (see Figure 2-12). This striking coincidence means that Mars experiences Earthlike seasons, with opposite seasons in the northern and southern Martian hemispheres (see Section 2-5). Because Mars takes nearly two (Earth) years to orbit the Sun, the Martian seasons last nearly twice as long as on Earth.

During spring and summer in a Martian hemisphere, the polar cap shrinks and the dark markings (which sometimes look greenish) become very distinct. Half a Martian year later, with the approach of fall and winter, the dark markings fade and the polar cap grows. These seasonal variations were widely held to mean that there is vegetation on Mars that changes seasonally, just as on the Earth. And if there was plant life, might there not also be intelligent beings?

Observations by the Italian astronomer Giovanni Schiaparelli during the favorable opposition of 1877 fueled speculations about life on Mars. He reported that during moments of "good seeing," when the air is exceptionally calm and clear, he was able to see 40 straight-line features crisscrossing the Martian surface (Figure 13-3). He called these dark linear features *canali*, an Italian word for "channels," which was soon mistranslated into English as "canals." The alleged discovery of canals implied that there were intelligent creatures on Mars capable of substantial engineering feats. This speculation helped motivate the American millionaire Percival Lowell to finance a major new observatory near Flagstaff, Arizona, primarily to study Mars. By the end of the nineteenth century, Lowell had reported observations of 160 Martian canals.

Although many astronomers observed Mars, not all saw the canals. In 1894, the American astronomer Edward Barnard, working at Lick Observatory in California, complained that "to save my soul I can't believe in the canals as Schiaparelli draws them." But objections by Barnard and others did little to sway the proponents of the canals.

As the nineteenth century drew to a close, speculation about Mars grew more and more fanciful. Perhaps the reddish-brown color of the planet meant that Mars was a desert world, and perhaps the Martian canals were an enormous planetwide irrigation network. From these ideas, it was a small leap to envision Mars as a dying world with canals carrying scarce water from melting polar caps to farmlands near the equator. The terrible plight of the Martian race formed the basis of inventive science fiction by Edgar Rice Burroughs, Ray Bradbury, and many others. It also led to the notion that the Martians might be a warlike race who schemed to invade the Earth for its abundant resources. (In Roman mythology, Mars was the god of war.) Stories of alien invasions, from H. G. Wells's

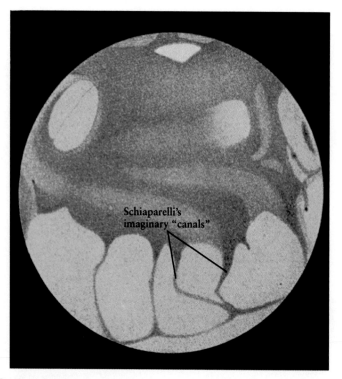

Schiaparelli's imaginary "canals"

Figure 13-3

The Mirage of the Martian Canals Giovanni Schiaparelli studied the Martian surface using a 20-cm (8-in.) telescope, the same size used by many amateur astronomers today. He recorded his observations in drawings like this one, which shows a network of linear features that Percival Lowell and others interpreted as irrigation canals. Later observations with larger telescopes revealed that the "canals" were mere illusions. (Michael Hoskin, ed., *The Cambridge Illustrated History of Astronomy,* Cambridge University Press, 1997, p. 286. Illustration by G. V. Schiaparelli. Courtesy Institute of Astronomy, University of Cambridge, UK)

1898 novel *The War of the Worlds* down to the present day, all owe their existence to the *canali* of Schiaparelli.

Modern observations of the Martian surface, such as the Hubble Space Telescope image that opens this chapter, show *no* sign whatsoever of linear canals. Why, then, were Schiaparelli, Lowell, and others so certain that the canals were there? One reason is that their observations were made from the Earth's surface, at the bottom of an often turbulent atmosphere that blurs images seen through a telescope. A second reason is that the human eye and brain can readily be deceived. The brain can connect together two dark streaks on the Martian surface to give the perception of a single "canal." In this way did astronomers such as Schiaparelli and Lowell delude themselves.

The episode of the Martian canals was not the first time that a number of scientists came to very wrong conclusions. Like any human activity, science will always involve human error and human frailties. But scientific investigation is fundamentally self-correcting, because new observations and experiments can always call into question the scientific theories of the past. Thus it was during the first decades of the twentieth century, when the Greek astronomer Eugène Antoniadi observed Mars with a state-of-the-art telescope and found that the "canals" were actually unconnected dark spots. And so it was again in the 1960s, when unmanned spacecraft from Earth first observed the Martian surface at close range.

13-3 | Unmanned spacecraft found craters, volcanoes, and canyons on Mars

Between 1964 and 1969, three American spacecraft—*Mariner 4*, *Mariner 6*, and *Mariner 7*—flew past Mars and sent back the first close-up pictures of the planet's surface. These images showed no trace of the canals of Schiaparelli and Lowell and absolutely no evidence of vegetation. The dark surface markings are just different-colored terrain.

Why did many Earth-based observers report that dark areas on Mars had the greenish hue of vegetation? The probable explanation is a quirk of human color vision. When a neutral gray area is viewed next to a bright red or brown area, the eye perceives the gray to have a blue-green color.

The most remarkable discovery of the first three *Mariner* missions to Mars was that the planet's surface is pockmarked with craters (see **Figure 13-4** and Figure 7-9*c*). At about the same time as these missions, scientists were learning that craters on the Moon were formed by interplanetary debris striking the lunar surface (see Section 10-4). The number of impacts tapered off about 3.8 billion years ago as the supply of such debris was used up, though mare-gouging impacts continued to about 3 billion years ago. The Martian craters, too, are the result of impacts from space. Since so many craters survive to the present day, at least part of the Martian surface must be extremely ancient (see Section 7-6).

Although many of the Martian craters are quite large, they had escaped detection from Earth because of the Martian atmosphere. This atmosphere is extremely thin by Earth standards; measurements from spacecraft reveal an average pressure at the Martian surface of only 0.0063 atmosphere, roughly the same as at an altitude of 35 km (115,000 ft) above the Earth's surface. (We saw in Section 9-6 and Section 12-3 that one atmosphere is the average sea-level pressure on the Earth.) But even this thin atmosphere obscures the Martian surface somewhat. That is why even the Hubble Space Telescope cannot see Martian craters distinctly (compare parts *a* and *b* of Figure 13-4), and why it took close-up images made by spacecraft near Mars to reveal these craters.

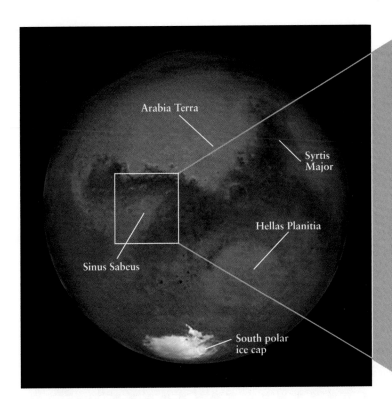

(a) Mars from the Hubble Space Telescope

(b) Closeup of Sinus Sabeus region

figure 13-4   R I **V** U X G

Martian Craters (a) This image, made during the favorable opposition of 2003, shows cratered regions on the side of Mars opposite to that in the image that opens this chapter. Arabia Terra is dotted with numerous flat-bottom craters. Syrtis Major was first identified by Christiaan Huygens in 1659. A single titanic impact carved out Hellas Planitia, which is 5 times the size of Texas. (b) This mosaic of images from the *Viking Orbiter 1* and *2* spacecraft shows an extensively cratered region located south of the Martian equator. (a: NASA; J. Bell, Cornell University; and M. Wolff, SSI; b: USGS)

Mariners 4, 6, and *7* transformed scientists' picture of Mars from that of an Earthlike world to that of a somewhat larger version of the Moon. But these three spacecraft observed just 10% of the Martian surface during their brief encounters. To get more complete observations required putting a spacecraft into orbit around Mars.

 Three such orbiters reached Mars in 1971: NASA's *Mariner 9,* which mapped the planet for a year, and the shorter-lived Soviet *Mars 2* and *Mars 3.* These were followed by NASA's two *Viking Orbiters,* which made images of Mars over a four-year period beginning in 1976; by *Phobos 2,* a Soviet spacecraft that carried out two months of observations in 1989; by NASA's *Mars Global Surveyor* and *Mars Odyssey,* which have been studying Mars at close range since 1997 and 2001, respectively; and the European Space Agency's *Mars Express,* which entered Mars orbit in 2003. (Several other orbital missions have been attempted, but failed either to achieve Mars orbit or to maintain contact with Earth. Later in this chapter we discuss missions that have landed on Mars.)

Each new mission used increasingly sophisticated technology to study the Martian surface. These missions have revealed Mars to be a remarkably diverse planet, with many geologic features unique in the solar system.

One of the most striking properties of the Martian surface is the difference between the northern and southern hemispheres. As the topographic map in **Figure 13-5** shows, the average elevation of the northern hemisphere is about 5 km (3 mi) lower than that of the southern hemisphere. Hence, planetary scientists refer to the **northern lowlands** and **southern highlands.** Furthermore, the northern lowlands are remarkably smooth and free of craters, while craters of many sizes are found across the southern highlands. Hence, the surface of the southern highlands must be relatively old, while that of the northern lowlands is relatively young (see Section 7-6). This implies that the northern lowlands have been resurfaced by some process that eradicated any ancient lowland craters. This set of striking differences between the two Martian hemispheres is called the **crustal dichotomy.**

Figure 13-5 shows that the dividing line between the northern lowlands and southern highlands is *not* precisely along the Martian equator. Thus, "northern hemisphere" is not exactly synonymous with "northern lowlands."

Several explanations for the Martian crustal dichotomy have been proposed. One idea is that a massive object struck the northern hemisphere relatively late in the planet's history. Like the impacts that formed the lunar maria (Sec-

 Figure 13-5

The Topography of Mars The color coding on this map of Mars shows elevations above (positive numbers) or below (negative numbers) the planet's average radius. The labeled features can be seen in Figures 13-4*a*, 13-6, 13-7, and 13-8. The landing sites for *Viking Landers 1* and *2* (VL1 and VL2), *Mars Pathfinder* (MP), the *Mars Exploration Rovers* (*Spirit* and *Opportunity*), and *Beagle 2* (B2; see Chapter 30) are each marked with an X. (MOLA Science Team, NASA/GSFC)

tion 10-4), such an impact would have excavated the crust to a depth of several kilometers and erased older craters. A smaller but still world-shaking impact of this kind produced Hellas Planitia, an immense basin 2300 km across and 6 km deep in the southern hemisphere that can actually be seen from Earth (see Figure 13-4*a* and Figure 13-5). One argument against this model is that the boundary between the northern lowlands and southern highlands is too irregular to be the wall of a giant impact crater.

Another idea is that Mars once had plate tectonic activity like the Earth, but that this activity came to an end long ago as Mars cooled. (We saw in Section 7-6 that small bodies tend to lose heat more rapidly than large bodies. Because Mars is substantially smaller than Earth, it should have lost more of its internal heat.) The young terrain in the northern lowlands could have formed by crustal spreading, such as happens along oceanic rifts on Earth like the Mid-Atlantic Ridge (see Figure 9-12).

Oceanic crust on Earth is about 30 km thinner than continental crust (see Table 9-3). Hence, if ancient plate tectonics produced the northern lowlands and southern highlands, we would expect a thinner crust in the northern hemisphere. One way to test this is to carefully monitor the motion of a spacecraft orbiting Mars. If there is a concentration of mass (such as a thicker crust) at one location on the planet, gravitational attraction will make the spacecraft speed up as it approaches the concentration and slow down as it moves away.

By analyzing the orbit of *Mars Global Surveyor* in just this way, a team of scientists headed by David Smith of NASA's Goddard Space Flight Center and Maria Zuber of MIT find that the Martian crust is about 40 km thick under the northern lowlands but about 70 km thick under the southern highlands. However, they find that the boundary between thin and thick crust does *not* line up with the boundary between high and low terrain, which contradicts the plate tectonic model.

It may be that the northern lowlands are the result of a combination of effects, including resurfacing by volcanic eruptions. Such resurfacing has taken place on Venus, where extensive lava flows have erased ancient craters (see Section 12-6). The first evidence that Mars, too, has had a volcanic past came when the *Mariner 9* mission discovered several enormous volcanoes. The largest of these, Olympus Mons, is the largest volcano in the solar system. It covers an area as big as the state of Missouri, and rises 24 km (15 mi) above the surrounding plains—nearly 3 times the height of Mount Everest (**Figure 13-6**). By comparison, the highest volcano on Earth, Mauna Loa in the Hawaiian Islands, has a summit only 8 km (5 mi) above the ocean floor.

Olympus Mons and other Martian volcanoes probably formed by *hot-spot volcanism*. In this process, magma wells upward from a hot spot in a planet's mantle, elevating the overlying surface and producing a volcano (see Section 12-4). On Earth, hot-spot volcanism is the origin of the Hawaiian Islands. These islands are part of a long chain of volcanoes that formed as the Pacific tectonic plate moved over a long-lived hot spot. On Mars, by contrast, the huge size of Olympus Mons strongly suggests an absence of plate tectonics. A single hot spot under this volcano probably pumped magma upward through the same vent for millions of years, producing one giant volcano rather than a long chain of smaller ones. As we saw in Section 12-4, the same process presumably gave rise to Maxwell Montes and other large volcanoes on Venus.

Figure 13-6 R I **V** U X G

Olympus Mons : This perspective view was created by combining a number of images from the two *Viking Orbiter* spacecraft. The base of Olympus Mons measures 600 km (370 mi) in diameter, and the scarps (cliffs) that surround the base are 6 km (4 mi) high in spots. The caldera, or volcanic crater, at the summit is approximately 70 km across, large enough to contain the state of Rhode Island. (© Calvin J. Hamilton)

If Olympus Mons is so gigantic, why didn't astronomers see it through telescopes long before *Mariner 9*? The explanation is that hot spots produce volcanoes with very gently sloping sides. Even when illuminated from the side, as at sunset or sunrise, such volcanoes cast only small shadows. Hence, Olympus Mons does not stand out from its surroundings as seen from Earth. (You may be able to see Olympus Mons as a light-colored circle in the upper part of the Hubble Space Telescope image that opens this chapter.)

Hot spots have produced a number of other features on Mars. A 2500-km- (1500-mi-) wide cluster of large volcanoes lies just east of Olympus Mons in a region called the Tharsis rise. It is actually a dome-shaped bulge that has been lifted 5 to 6 kilometers above the planet's average elevation (see Figure 13-5). Apparently, a massive plume of magma once welled upward from a hot spot under this region, producing the Tharsis rise and its volcanoes. Another hot spot on the opposite side of Mars produced a smaller bulge centered on the volcano Elysium Mons, shown in Figure 13-5. None of the Martian volcanoes are active today.

East of the Tharsis rise, *Mariner 9* discovered a vast chasm that runs roughly parallel to the Martian equator (**Figure 13-7**). If this canyon were located on Earth, it would stretch from Los Angeles to New York. In honor of the spacecraft that revealed so much of the Martian surface, this canyon has been designated Valles Marineris.

Valles Marineris begins with heavily fractured terrain in the west and ends with ancient cratered terrain in the east. Many geologists suspect that Valles Marineris was caused by the same upwelling of material that formed the Tharsis rise. As the Martian surface bulged upward at Tharsis, there would have been tremendous stresses on the crust, which would have caused extensive fracturing. Thus, Valles Marineris may be a **rift valley**, a feature created when a

figure 13-7 R I V U X G

Valles Marineris This mosaic of *Viking Orbiter* images shows the Tharsis rise, with its three large volcanoes, and the huge rift valley of Valles Marineris. The valley extends from west to east for more than 4000 km (2500 mi) and is 600 km (400 mi) wide at its center. Its deepest part is 8 km (5 mi) beneath the surrounding plateau. Compare this image with Figure 13-5. (USGS/NASA)

planet's crust breaks apart along a fault line. Rift valleys are found on Earth; two examples are the Red Sea (see Figure 9-16) and the Rhine River valley in Europe. Other, smaller rifts in the Martian crust are found all around the Tharsis rise.

While the motion of material in the mantle has had a large effect on the Martian surface, plate tectonics has not. Mars lacks the global network of ridges and subduction zones that plate tectonics has produced on the Earth. The reason is that the outer layers of the red planet have cooled more extensively than those of the Earth. The Martian crust is tens of kilometers thicker than the Earth's crust, making it impossible for one part of the crust to be subducted beneath another. Hence, the entire crust of Mars makes up a single tectonic plate, as is the case on the Moon (Section 10-1), Mercury (Section 11-3), and Venus (Section 12-6).

As we discussed in Section 7-7, Mars has no planetwide magnetic field but does have some weakly magnetized regions in the old southern highlands (see Figure 7-12). This suggests that Mars had a magnetic field when the planet was young and the highlands were still solidifying. Remarkably, different regions of the southern highlands are magnetized in different directions. This suggests that the Martian magnetic field reversed direction from time to time, just as the Earth's magnetic field does (see Section 9-4). There is no evidence for magnetized areas in impact basins such as Hellas Planitia, so the Martian magnetic field must have shut off by the time these basins were formed about 3.9 billion years ago.

Today, Mars probably has a core with a radius about half that of the planet. Unlike the Earth's core, however, the Martian core is thought to contain substantial amounts of sulfur in addition to iron. (We saw in Section 8-4 that planets can have different chemical compositions, depending on how far from the Sun they formed.) Temperatures in the core should be high enough to melt sulfur and sulfur compounds. However, the electrical properties of these substances are different from those of iron. Hence, while the flow of molten iron in the Earth's outer core causes electric currents that generate our planet's magnetic field, flows within the sulfur-rich Martian core do not.

13-4 | Surface features indicate that water once flowed on Mars

 SCIENCE IN PROCESS
Detecting subatomic particles emanating from Mars helps scientists locate water beneath the planet's surface

Spacecraft orbiting Mars have seen no evidence of rainfall or of liquid water anywhere on the planet's surface. But they have revealed many features that look like dried-up lakes and riverbeds (Figure 13-8). Intricate branched patterns and delicate channels meandering among flat-bottomed craters strongly suggest that water once flowed over the Martian surface. Such dried-up water channels are found throughout the old southern highlands but in relatively few places on the younger northern lowlands. Hence, large amounts of water flowed on the Martian surface in the distant past but not in the recent past.

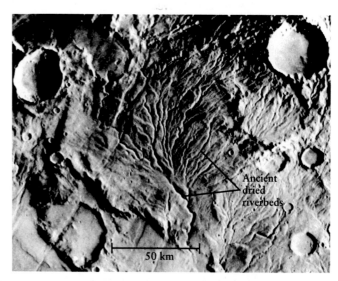

figure 13-8 R I V U X G

Ancient River Channels on Mars This *Viking Orbiter* image shows a network of dried riverbeds extending across the cratered southern highlands. Liquid water would evaporate in the sparse, present-day Martian atmosphere, so these channels may date from a time when the Martian atmosphere was thicker and its climate more Earthlike. (Michael Carr, USGS)

figure 13-9   R I **V** U X G

Signs of Ancient Floods on Mars These teardrop-shaped islands rise above the floor of an ancient Martian channel called Ares Valles. They were carved out by a torrent of water that once flowed from the bottom of this image toward the top. The size and shape of the islands suggest that the peak flow rate in these floods was more than 1000 times that of the Mississippi River. Similar flood-carved islands are found on Earth in eastern Washington State. (NASA/USGS)

Spacecraft images have also revealed evidence of powerful flash floods that long ago raged across the surface of Mars (**Figure 13-9**). In addition to flash floods, there is also evidence that water flowed for sustained periods. **Figure 13-10** shows a canyon that appears to have been shaped gradually by water erosion.

The features shown in Figures 13-8, 13-9, and 13-10 must all be quite old, for liquid water cannot exist anywhere on the Martian surface today. Water is liquid over only a limited temperature range: If the temperature is too low, water becomes ice, and if the temperature is too high, it becomes water vapor. What determines this temperature range is the atmospheric pressure above a body of water. If the pressure is very low, molecules easily escape from the liquid's surface, causing the water to vaporize. Thus, at low pressures, water more easily becomes water vapor. The average surface temperature on Mars is only 220 K (–53°C, or –63°F), and the average pressure is only 0.0063 atmosphere. With this combination of temperature and pressure, water can exist as a solid (ice) and as a gas (water vapor) but not as a liquid. You can see this same situation inside a freezer, where water vapor swirls around over ice cubes.

In order to keep water on Mars in the liquid state, it would be necessary to increase both the temperature (to keep water from freezing) and the pressure (to keep the liquid water from evaporating). Therefore, if liquid water existed for extended periods on Mars, the atmosphere must have been both thicker and warmer than it is today.

If water once flowed on Mars, where might that water be today? It is not in the Martian atmosphere, which is 95.3% carbon dioxide (CO_2) and contains only a trace amount of water vapor (see Table 13-1). Indeed, it never rains on Mars. If all the water vapor could somehow be squeezed out of the Martian atmosphere, it would not fill one of the five Great Lakes of North America.

Another place to look for the missing water is at the Martian polar ice caps, which you can see in the image that opens this chapter and in Figure 13-4a. However, these ice caps cannot simply be reservoirs of frozen water, because the winter temperature at the Martian poles is low enough (–140°C = –220°F) to freeze CO_2 as well. How much, then, of the ice caps is frozen CO_2 (dry ice) and how much is water ice?

Mariner 9 provided a partial answer to this question in 1972, as it observed spring and summer coming to the northern hemisphere of Mars. During the Martian spring, the north polar cap receded rapidly, consistent with a thin layer of CO_2 frost evaporating in the sunlight. With the arrival of summer, however, the rate of recession slowed abruptly. This indicated that a thicker and less easily evaporated layer of water ice had been exposed. Scientists concluded that the **residual polar caps**—the portions that survive the Martian summers—contain a large quantity of frozen water (Figure 13-11). The European *Mars Express* spacecraft confirmed this idea in 2004. It

figure 13-10   R I **V** U X G

Signs of Sustained Water Flow on Mars This *Mars Global Surveyor* image shows part of a 2.5-km- (1.6-mi-) wide canyon in the Martian southern highlands. The terraces within the canyon resemble structures found within riverbeds on Earth, suggesting that this canyon formed over time by water flow. (NASA/JPL/Malin Space Science Systems)

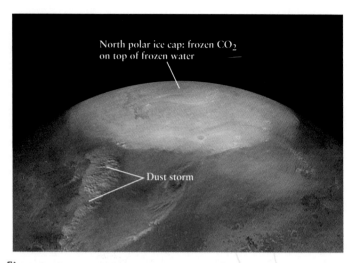

figure 13-11 R I V U X G

The North Polar Ice Cap This mosaic of *Mars Global Surveyor* images shows the Martian northern ice cap. During the northern winter, this region is covered by CO_2 frost. This begins to evaporate in the spring, exposing a residual polar cap made primarily of water ice. This evaporation also helps trigger dust storms (see Section 13-6). (NASA/JPL/Malin Space Science Systems)

used its infrared cameras to peer through the carbon dioxide layer at the south polar ice cap and detect the characteristic spectrum of water ice. Because we do not know the thickness of the layer of ice, the amount of frozen water stored in the Martian polar caps is still unknown.

The residual ice caps cannot be the only site of Martian water, however. Close-up views such as Figure 13-9 show that flash-flood features usually emerge from craters or from collapsed, jumbled terrain. It therefore seems likely that frozen water is also stored in **permafrost** under the Martian surface, similar to the layer beneath the tundra in the Earth's far northern regions (Figure 13-12). In the distant past, heat from a meteoroid impact or from volcanic activity occasionally caused a sudden melting of this subsurface ice. The ground then collapsed, and millions of tons of rock pushed the water to the surface.

Recent observations from the *Mars Odyssey* spacecraft have reinforced the idea that water is frozen just beneath the surface of Mars. **Cosmic rays** are fast-moving subatomic particles that enter our solar system from interstellar space. When cosmic rays strike and penetrate into the Martian soil, they collide with the nuclei of atoms in the soil and knock out neutrons. (Recall from Section 5-7 that atomic nuclei contain both positively charged protons and electrically neutral neutrons.) If these ejected neutrons originate within a meter or so of the surface, they can penetrate through the soil and escape into space. However, if there are water molecules (H_2O) in the surface, the nuclei of the hydrogen atoms in these molecules will absorb the neutrons and prevent them from escaping. An instrument on board *Mars Odyssey* detects neutrons coming from Mars; where the instrument finds a *deficiency* of neutrons is where the soil contains hydrogen and, presumably, water.

The results from the *Mars Odyssey* neutron measurements show that there is abundant water beneath the surface at both Martian poles (Figure 13-13). But they also show a surprising amount of subsurface water in two regions near the equator, the desert of Arabia Terra (see Figure 13-4*a* and Figure 13-5) and another region on the opposite side of the planet. One possible explanation is that about a million years ago, gravitational forces from other planets caused a temporary change in the tilt of Mars's rotation axis. The ice caps were then exposed to more direct sunlight, causing some of the water to evaporate and eventually refreeze elsewhere on the planet.

There are other surprising features on Mars related to the presence of water. A number of *Mars Global Surveyor* images show gullies apparently carved by water flowing down the walls of pits or craters (Figure 13-14). These could have formed when water trapped underground—where the pressure is greater and water can remain a liquid—seeped onto the surface. The gullies appear to be geologically young, so it is possible that even today some liquid water survives below the Martian surface. Alternatively, these gullies may have formed during periods when the Martian climate was colder than usual and the slopes of craters were covered with a layer of dust and snow. Melting snow on the underside of this layer could have carved out gullies, which were later exposed when the climate warmed and the snow evaporated.

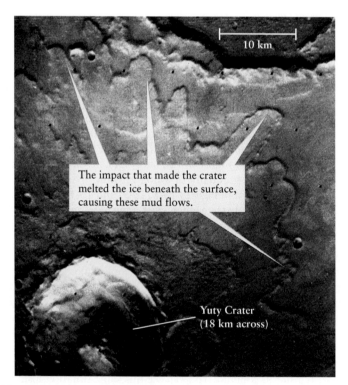

figure 13-12 R I V U X G

A Martian "Mud Splash" Most large, young Martian craters are surrounded by features that resemble mud flows. This is evidence for a subsurface layer of water ice, which would have been melted by a crater-forming impact. No such features appear around craters on the bone-dry Moon (see Figure 10-8). (GSFC/NASA)

Percent abundance of water (by mass)

High concentration of subsurface water in the polar regions

High concentration of subsurface water in some locations near the equator

figure 13-13

Water Beneath the Martian Surface This map of the abundance of water (probably in the form of ice) just below the surface of Mars is based on data collected from orbit by the *Mars Odyssey* spacecraft. These measurements can only reveal the presence of water or water ice to a depth of about a meter; there may be much more water at greater depths. (LANL)

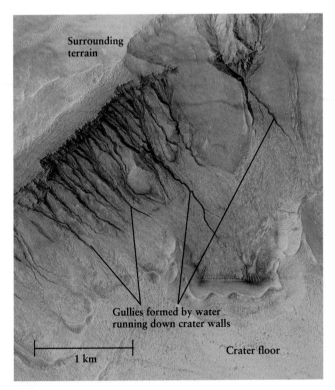

Surrounding terrain

Gullies formed by water running down crater walls

1 km

Crater floor

figure 13-14 R I V U X G

Martian Gullies When *Mars Global Surveyor* looked straight down into this crater in the Martian southern hemisphere, it saw a series of gullies along the crater wall. These gullies may have formed by subsurface water seeping out to the surface, or by the melting of snow that fell on crater walls. (NASA/JPL/Malin Space Science Systems)

The total amount of water on Mars is not known. But by examining flood channels at various locations around Mars, Michael Carr of the U.S. Geological Survey has estimated that there is enough water to cover the planet to a depth of 500 meters (1500 ft). (By comparison, Earth has enough water to cover our planet to a depth of 2700 meters, or 8900 ft.) Thus, while the total amount of water on Mars is less than on Earth, there may be enough to have once formed lakes or even oceans.

Was there ever a time when Mars was warmer and wetter than today, with liquid water on its surface? To find an answer, we must explore how the Martian atmosphere evolved.

13-5 | Earth and Mars began with similar atmospheres that evolved very differently

As nineteenth-century astronomers observed Mars through their telescopes, they were occasionally able to see cloud formations (Figure 13-15). This was the first evidence that Mars has an atmosphere (without an atmosphere, a planet can have no clouds). It was difficult to measure the Martian atmosphere's spectrum, however, and for many decades no one knew for sure what the atmosphere is made of. When the *Mariner 4* spacecraft flew past Mars in 1965, its measurements indicated that the planet's atmosphere is almost entirely carbon dioxide (CO_2). We now know that the extremely thin atmosphere is about 95.3% carbon dioxide and 2.7% nitrogen, with small amounts of argon, oxygen, carbon monoxide, and water vapor (see Table 13-1). The clouds themselves are made of tiny water ice crystals (as in wispy cirrus clouds on Earth) as well as crystals of CO_2 ice (dry ice).

Olympus Mons

Tharsis
volcanoes

1000 km

figure 13-15   R I **V** U X G

Clouds Top the Volcanoes of Mars On most Martian afternoons,
bluish-white clouds form at the summits of Olympus Mons and the
volcanoes of the Tharsis rise. (NASA/JPL/Malin Space Science Systems)

As we saw in Section 9-1, carbon dioxide (CO_2) and
water vapor (H_2O) both contribute to the greenhouse
effect. Because the atmospheres of the Earth and Mars
are transparent to sunlight, they are not heated directly by the Sun.
Instead, they are heated from below by energy reradiated from the
ground at longer, infrared wavelengths. Greenhouse gases such as
water vapor (H_2O) and carbon dioxide (CO_2) absorb these wavelengths, trapping the heat and warming the atmosphere. Without
the greenhouse effect, the average temperature on Earth would be
about 41°C lower than it is now, and our entire planet would be
frozen over. On Mars, the thin atmosphere permits much of the
infrared radiation to escape into space, so the greenhouse effect
increases the temperature by only about 5°C.

If the Martian atmosphere was thicker in the distant past, however, the greenhouse effect would have been stronger, and the atmospheric pressure and temperature could have been great enough for
liquid water to exist on the surface. Extinct Martian volcanoes suggest how such a thick, ancient atmosphere could have been created.
The terrestrial planets obtained their atmospheres primarily through
volcanic outgassing (see Section 9-5). Water vapor, carbon dioxide,
and nitrogen are among the most common gases in volcanic vapors.
These three gases probably constituted the bulk of the atmospheres
of both the Earth and Mars 4 billion years ago.

On Earth most of the water is in the oceans. Nitrogen (N_2),
which is not very reactive, is still in the atmosphere. Carbon dioxide, however, dissolves in water; thus, rain washed most of the CO_2

out of our planet's atmosphere long ago. The dissolved CO_2 reacted
with rocks in rivers and streams, and the residue was ultimately
deposited on the ocean floors, where it turned into carbonate
rocks, such as limestone. Consequently, most of the Earth's carbon
dioxide is tied up in the Earth's crust, and carbon dioxide makes up
only a small fraction of the Earth's atmosphere. (About 1 in every
3000 molecules in our atmosphere is a carbon dioxide molecule.)

The amount of carbon dioxide in our atmosphere is sustained in
part by plate tectonics. Tectonic activity causes carbonate rocks to
cycle through volcanoes, where they are heated and forced to liberate their trapped CO_2. This liberated gas then rejoins the atmosphere. Without volcanoes, rainfall would remove all the CO_2 from
our atmosphere in only a few thousand years. The truly unique feature of our atmosphere—the presence of oxygen—is the result of
biological activity, in particular, photosynthesis by plant life (Figure
13-16a).

If Mars had a thick atmosphere 4 billion years ago, rainfall
would have washed much of the planet's carbon dioxide from its
atmosphere, perhaps creating carbonate rocks in which the CO_2 is
today chemically bound. If so, these carbonate rocks are significant
reservoirs of the primordial CO_2 outgassed from ancient volcanoes.
Tectonic activity on Mars came to an end as the planet cooled, so its
carbonate rocks were not recycled through volcanoes as they are on
Earth. The depletion of carbon dioxide from the Martian atmosphere into the rocks would therefore have been permanent.

As the amount of atmospheric CO_2 declined, the greenhouse
effect on Mars would have weakened and temperatures begun to
fall. This temperature decrease would have caused more water
vapor to condense into raindrops and fall to the surface, taking even
more CO_2 with them and further weakening the greenhouse effect.
Thus, a decrease in temperature would have caused a further
decrease in temperature—a phenomenon sometimes called a **runaway icehouse effect**. (As we saw in Section 12-5, just the opposite
has taken place on Venus, where temperatures have spiraled upward
to a hellish level.) Ultimately rainfall would have removed both
water vapor and most of the carbon dioxide from the Martian
atmosphere. With only a thin CO_2 atmosphere remaining, surface
temperatures on Mars eventually stabilized at their present frigid
values (Figure 13-16b).

As both water vapor and carbon dioxide became depleted, ultraviolet light from the Sun could penetrate the thinning Martian
atmosphere to strip it of nitrogen. Nitrogen molecules (N_2) normally do not have enough thermal energy to escape from Mars, but
they can acquire that energy from ultraviolet photons, which break
the molecules in two. Ultraviolet photons can also split carbon dioxide and water molecules, giving their atoms enough energy to escape
(Figure 13-16b). Indeed, in 1971 the Soviet *Mars 2* spacecraft found
a stream of oxygen and hydrogen atoms (from the breakup of water
molecules) escaping into space. Oxygen atoms that did not escape
into space could have combined with iron-bearing minerals in the
surface, forming rustlike compounds. Such compounds have a characteristic reddish-brown color and may be responsible for the overall color of the planet.

Since Mars is being depleted of water in this way, the total
amount of water must have been greater in the past, perhaps
enough to flood the flat northern lowlands of Mars and form an
ocean. James Head of Brown University and his colleagues have

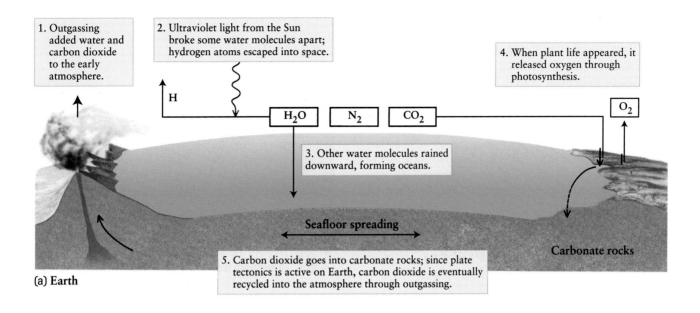

1. Outgassing added water and carbon dioxide to the early atmosphere.

2. Ultraviolet light from the Sun broke some water molecules apart; hydrogen atoms escaped into space.

4. When plant life appeared, it released oxygen through photosynthesis.

3. Other water molecules rained downward, forming oceans.

Seafloor spreading

5. Carbon dioxide goes into carbonate rocks; since plate tectonics is active on Earth, carbon dioxide is eventually recycled into the atmosphere through outgassing.

Carbonate rocks

(a) Earth

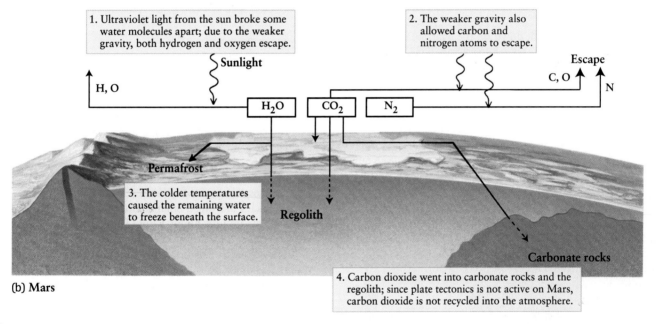

1. Ultraviolet light from the sun broke some water molecules apart; due to the weaker gravity, both hydrogen and oxygen escape.

2. The weaker gravity also allowed carbon and nitrogen atoms to escape.

Sunlight

Escape

3. The colder temperatures caused the remaining water to freeze beneath the surface.

Permafrost

Regolith

Carbonate rocks

4. Carbon dioxide went into carbonate rocks and the regolith; since plate tectonics is not active on Mars, carbon dioxide is not recycled into the atmosphere.

(b) Mars

figure 13-16

The Atmospheres of the Earth and Mars The evolution of the atmosphere followed different paths on **(a)** the geologically active Earth and **(b)** geologically inactive Mars. (Adapted from R. M. Haberle)

analyzed high-resolution topographical data from *Mars Global Surveyor* (see Figure 13-5) and have found evidence of an ancient shoreline that encircles the lowlands. Large channels leading from the highlands to the lowlands could have supplied a northern ocean with runoff from the higher terrain to the south.

How can we test this model of an ancient Mars that was warmer and wetter than today? One approach is to search for the carbonate rocks that should have been produced when an ancient, thick atmosphere cooled as shown in Figure 13-16*b*. Philip Christensen of Arizona State University and his collaborators used infrared detectors on board *Mars Odyssey* to look for the distinctive spectrum of such rocks on the Martian surface. While they did detect carbonates in Martian dust, they found *no* evidence of carbonate rocks anywhere on the planet.

Perhaps there are carbonate rocks on Mars, but they are buried beneath a thick layer of dust that obscures them from *Mars Odyssey*. Alternatively, the picture of a thick atmosphere on ancient Mars may be incorrect. The volcanoes on early Mars may have released only small amounts of carbon dioxide and water vapor, so that little CO_2 was available to form carbonates. Such a thin atmosphere would have generated only a feeble greenhouse effect, and the pressure and temperature might never have been high enough for large bodies of water to form. In this model, liquid water only existed on Mars on rare occasions, such as after a meteoroid impact melted some permafrost and caused a flash flood (see Figure 13-9).

 One tool that may help resolve the dilemma of how the Martian atmosphere evolved is *Mars Express,* placed in orbit in 2003 by the European Space Agency (ESA). *Mars Express* has instruments that can probe the subsurface structure of Mars to a depth of several kilometers and make high-precision measurements of the surface composition. The results from this spacecraft should help us better understand the full history of water on Mars.

13-6 | Landers have explored the surface of Mars

Spacecraft orbiting Mars have revolutionized our understanding of the planet. But nothing substitutes for data collected from a planet's surface, which is why several spacecraft have been sent to land on Mars.

The first spacecraft to touch down on the Martian surface were the *Viking Landers.* The two *Viking* spacecraft were launched in 1975 and arrived at Mars almost a year later. Each spacecraft consisted of two modules, an orbiter (introduced in Section 13-4) and a lander. After entering orbit about Mars, each lander separated from its orbiter and descended to the Martian surface with the aid of a heat shield, retrorockets, and a parachute. Each lander is roughly the size of a small automobile (**Figure 13-17**).

Viking Lander 1 touched down in July 1976 in a flat, moderately cratered plain called Chryse Planitia in the northern lowlands (see

 figure 13-18   R I **V** U X G
The View from *Viking Lander 1* The rocks of Chryse Planitia resemble volcanic rocks on Earth (see Figure 9-18*a*), and thus are thought to be pieces of an ancient lava flow that was broken apart by crater-forming impacts. Fine-grained debris covers the rocks and the spaces between them and has piled up to form dunes. (Dr. Edwin Bell II/NSSDC/GSFC/NASA)

Figure 13-5). From this site it radioed data back to Earth until November 1982. The spacecraft's cameras revealed a landscape littered with jagged rocks, as **Figure 13-18** shows. These are probably the wreckage of an ancient crater-forming impact, which ejected rocky debris in all directions.

Figure 13-18 also shows that the landscape around *Viking Lander 1* is covered with a fine-grained dust. These particles have piled up in places to form drifts that resemble sand dunes on Earth. The dust particles are much smaller than ordinary household dust on Earth: They are only about 1 μm (10^{-6} m) across, about the size of the particles in cigarette smoke.

Viking Lander 2 set down in Utopia Planitia in August 1976, on nearly the opposite side of the planet from *Viking Lander 1* but about 1500 kilometers (930 miles) closer to the north pole. (Figure 13-5 shows both landing sites.) It continued to transmit data to Earth until April 1980. The terrain around *Viking Lander 2* is remarkably crater-free, but is nonetheless strewn with rocks like those in Figure 13-18. It is thought that many of these rocks were ejected from a large crater about 200 kilometers east of the landing site. As at the *Viking Lander 1* site, the rocks in Utopia Planitia are covered with dust, though the dust there does not pile up in drifts.

Even before *Viking,* astronomers knew that dust was an important part of the Martian environment. Telescopic observers found that surface features on Mars are sometimes obscured by a yellowish haze. This haze was interpreted to be caused by windblown storms of dust. Because the Martian air is so tenuous, a 100-km/h (60-mi/h) wind on Mars is only about as strong as a 10-km/h (6-mi/h) breeze on Earth. Hence, only very small dust particles are carried aloft and spread over the surface by the Martian winds. Figure 13-11 shows a dust storm near the north polar ice cap; the dust storm shown in **Figure 13-19** covered the entire planet.

Airborne dust also affects the color of the Martian sky. The Earth's largely dust-free sky is blue because molecules in our

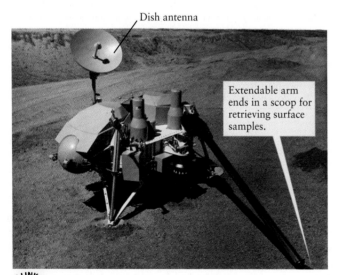

figure 13-17   R I **V** U X G
A *Viking Lander* Each *Viking Lander* is about 2 m (6 ft) tall from the base of its footpads to the top of its antenna. (This photograph shows a full-size replica under test on Earth.) The dish antenna transmitted data to one of the *Viking* orbiters overhead, which in turn relayed the information back to Earth. (NASA)

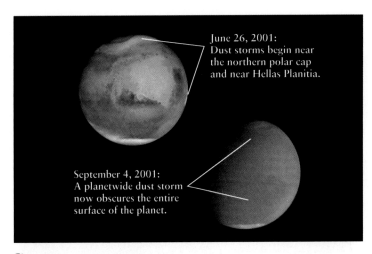

June 26, 2001:
Dust storms begin near
the northern polar cap
and near Hellas Planitia.

September 4, 2001:
A planetwide dust storm
now obscures the entire
surface of the planet.

Figure 13-19   R I **V** U X G

A Martian Dust Storm Dust storms on Mars can expand to cover the entire planet. These Hubble Space Telescope images show a particularly intense storm that took place in 2001. (NASA; J. Bell, Cornell University; M. Wolff, SSI; and the Hubble Heritage Team, STScI/AURA)

atmosphere scatter blue sunlight more effectively than other colors (see Box 5-4). On Mars, however, dust particles in the air do a good job of scattering all visible wavelengths, which by itself would make the sky appear white. But in addition, the dust particles contain a mineral called magnetite, which absorbs blue light. The result is that the Martian sky has a yellowish color reminiscent of butterscotch.

In addition to cameras, each *Viking Lander* had a scoop at the end of a mechanical arm to dig into the Martian regolith and retrieve samples for analysis (see Figure 13-17). Bits of the regolith clung to a magnet mounted on the scoop, indicating that the regolith contains iron. A device called an X-ray spectrometer measured the chemical composition of samples at both lander sites and showed them to be rich in iron, silicon, and sulfur. Thus, the Martian regolith can best be described as an iron-rich clay.

The *Viking Landers* each carried a miniature chemistry laboratory for analyzing the regolith. Curiously, when a small amount of water was added to a sample of regolith, the sample began to fizz and release oxygen. This experiment indicated that the regolith contains unstable chemicals called peroxides and superoxides, which break down in the presence of water to release oxygen gas.

The unstable chemistry of the Martian regolith probably comes from solar ultraviolet radiation that beats down on the planet's surface. (Unlike Earth, Mars has no ozone layer to screen out ultraviolet light.) Ultraviolet photons easily break apart molecules of carbon dioxide (CO_2) and water vapor (H_2O) by knocking off oxygen atoms, which then become loosely attached to chemicals in the regolith. Ultraviolet photons also produce highly reactive ozone (O_3) and hydrogen peroxide (H_2O_2), which also become incorporated in the regolith.

Among the most important scientific goals of the *Viking Landers* was to search for evidence of life on Mars. Although it was clear that Mars has neither civilizations nor vast fields of plants, it still

seemed possible that simple microorganisms could have evolved on Mars during its early brief epoch of Earthlike climate. If so, perhaps some species had survived to the present day. To test this idea, the *Viking Landers* looked for biologically significant chemical reactions in samples of the Martian regolith, but they failed to detect any reactions of this type. Perhaps life never existed on Mars, or perhaps it existed once but was destroyed by the presence of peroxides and superoxides in the regolith.

On Earth, hydrogen peroxide is commonly used as an antiseptic. When poured on a wound, it fizzes and froths as it reacts with organic material and destroys germs. Thus, the *Viking Landers* may have failed to detect any biochemistry on Mars because the regolith has literally been sterilized! Hence, we cannot yet tell whether life ever existed on Mars, or whether it might yet exist elsewhere on or beneath the planet's surface. (In Chapter 30 we discuss the *Viking Lander* experiments in detail, as well as the controversial proposal that fossil Martian microorganisms may have traveled to Earth aboard meteorites.)

One of the limitations of the *Viking Landers* was that they could collect and examine samples only within the reach of the spacecraft's mechanical arm. A more versatile unmanned explorer, *Mars Pathfinder,* landed on Mars on July 4, 1997. Unlike the *Viking Landers, Mars Pathfinder* did not have a heavy and expensive rocket engine to lower it gently to the Martian surface. Instead, the spacecraft was surrounded by airbags like those used in automobiles. Had there been anyone to watch *Mars Pathfinder* land, they would have seen an oversized beach ball hit the surface at 50 km/h (30 mi/h), then bounce for a kilometer before finally rolling to a stop. Unharmed by its wild ride, the spacecraft then deflated its airbags and began to study the geology of Mars.

On board *Mars Pathfinder* was a small, wheeled robot called *Sojourner* (Figure 13-20a). Following directions from Earth, *Sojourner* spent three months exploring the rocks around its landing site. Over those three months, the mission radioed to Earth more than 10 times as many images as had the two *Viking Landers* combined.

The X-ray spectrometer on each *Viking Lander* was located within the spacecraft body and could analyze only samples retrieved by the mechanical arm. *Sojourner,* by contrast, carried its X-ray spectrometer as well as its own camera to rocks several meters from *Mars Pathfinder,* as Figure 13-20a shows.

Mars Pathfinder landed about 800 km (500 mi) from *Viking Lander 1,* in an ancient flood channel in the northern lowlands called Ares Valles. (Figure 13-5 shows the landing site, and Figure 13-9 shows a portion of this plain.) Flood waters can carry rocks of all kinds great distances from their original locations, so scientists expected that this landing site would show great geologic diversity. They were not disappointed. Some rocks resemble the impact breccias found in the ancient highlands of the Moon (see Figure 10-17). Still others have a layered structure like sedimentary rocks on Earth (see Figure 9-18b). Such rocks form gradually at the bottom of bodies of water, so perhaps liquid water was stable in at least some regions of Mars for a substantial period of time.

Many of the rocks at the *Mars Pathfinder* landing site appear to be igneous rock produced by volcanic action. We have seen that Mars has volcanoes that were once active, so this was not a surprise. What was surprising was the chemical composition of these volcanic

Mars Pathfinder rover   Hill, 1 km (0.6 mi) away

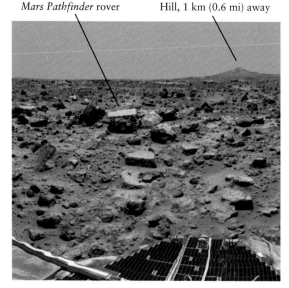

Cameras   Mars Exploration Rover   Mars Pathfinder rover

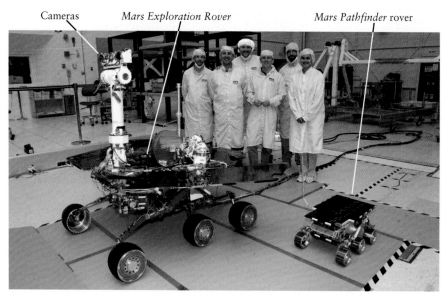

(a) A rover on Mars

(b) Two generations of rovers

WEB LINK 13.10

Figure 13-20 R I **V** U X G

Roving the Martian Surface (a) This view from the stationary part of *Mars Pathfinder* shows the rover, called *Sojourner*, deploying its X-ray spectrometer against a rock named Moe. Scientists gave other rocks whimsical names such as Jiminy Cricket and Pooh Bear. (b) The technicians in this photograph stand with one of the *Mars Exploration Rovers* launched in 2003 and with a replica of its ancestor *Sojourner*. (NASA/JPL)

rocks. Unlike basalts, the most common type of volcanic rock found on Earth (see Figure 9-18a), the rocks that *Sojourner* analyzed with its X-ray spectrometer have a relatively high concentration of silicon. Silicon-rich volcanic rocks, called **andesites,** are less dense than basalts. They are formed when crustal material gets so hot that it becomes molten, allowing less dense materials to float to the top.

Observations from *Mars Global Surveyor* indicate that andesites are found in broad regions of the young northern lowlands, including the *Mars Pathfinder* landing site. By contrast, basalts dominate in the older southern highlands. Thus, at some time in the past, something changed within Mars to alter the kind of lava that flowed from volcanoes. Why this happened is as yet unknown. Whatever the explanation, Mars has apparently had a more dynamic geologic past than heretofore suspected.

What *Mars Pathfinder* could not do was investigate the interiors of Martian rocks. This is part of the mission of the two *Mars Exploration Rovers*, which arrived on Mars in January 2004 (see Figure 13-20b; see also Figure 13-5 for the locations of the landing sites). Much larger and with a much greater range than *Sojourner*, the *Mars Exploration Rovers* (named *Spirit* and *Opportunity*) carry a tool for boring into rocks to determine their internal composition. This information reveals more about the conditions that existed when the rock was formed. In March 2004 scientists reported that the rocks at the *Opportunity* site are unusually rich in chemicals called sulfates. On Earth, sulfates are found only in rocks that either formed under water or have been exposed to water over long periods. This is compelling evidence that there was once liquid water on at least some parts of Mars.

13-7 | The Martian atmosphere changes dramatically with the seasons

The *Viking Lander* and *Mars Pathfinder* spacecraft all carried instruments to measure meteorological conditions on Mars. The thin, nearly cloudless Martian atmosphere provides little thermal insulation, and there are large swings in air temperature over the course of a Martian day. At the *Mars Pathfinder* site, temperatures ranged from 197 K (−76°C = −105°F) just before dawn to 263 K (−10°C = 14°F) in the "heat" of the afternoon.

Each afternoon, parcels of warm air rise from the heated surface and form whirlwinds called **dust devils.** (A similar phenomenon with the same name occurs in dry or desert terrain on Earth.) Martian dust devils can reach an altitude of 6 km (20,000 ft) and help spread dust particles across the planet's surface. All three landers measured changes in air pressure as dust devils swept past. Often several of these would pass in a single afternoon. Dust devils are large enough to be seen by spacecraft orbiting Mars (**Figure 13-21**). As the sun sets and afternoon turns to evening, the dust devils subside until the following day.

In addition to recording conditions over the course of a Martian day, the two long-lived *Viking Landers* measured how the weather changes over the course of a Martian year. (*Mars Pathfinder*, with its lifetime of just three months, could not do this.) *Viking Landers 1* and *2* both touched down in the northern hemisphere of Mars when it was spring in that hemisphere and autumn in the southern hemisphere. Initially, both spacecraft measured atmospheric pres-

WEB LINK 13.11

Figure 13-21 R I **V** U X G
A Martian Dust Devil This *Mars Global Surveyor* image shows a dust devil as seen from almost directly above. This tower of swirling air and dust casts a long shadow in the afternoon sun. The dust devil had been moving from right to left before the picture was taken, leaving a dark, curlicue-shaped trail in its wake. (NASA/JPL/Malin Space Science Systems)

Figure 13-22 R I **V** U X G
Winter in Mars's Northern Hemisphere This picture, taken during midwinter, shows a thin frost layer that lasted for about a hundred days at the *Viking Lander 2* site. Freezing carbon dioxide adheres to water-ice crystals and dust grains in the atmosphere, causing them to fall to the ground. (NASA/JPL)

sures around 0.008 atmosphere. After only a few weeks on the Martian surface, however, atmospheric pressure at both landing sites was dropping steadily. Mars seemed to be rapidly losing its atmosphere, and some scientists joked that soon all the air would be gone! The actual explanation proved to be quite straightforward: Winter was coming to the southern hemisphere.

During the southern hemisphere's winter, the Martian south pole becomes so cold that large amounts of carbon dioxide condense out of the atmosphere there, covering the ground with flakes of dry ice. The formation of this "snow" removes gas from the Martian atmosphere, thereby lowering the atmospheric pressure across the planet. Several months later, when spring comes to the southern hemisphere, the dry-ice snow evaporates rapidly and the atmospheric pressure returns to its prewinter levels. The pressure drops again in the southern hemisphere's summer, because it is then winter in the northern hemisphere, and dry-ice snow condenses at northern latitudes (**Figure 13-22**). Therefore, both temperature and atmospheric pressure change significantly with the Martian seasons.

As on the Earth, the seasons on Mars are caused by the tilt of the planet's rotation axis. When the north pole is tilted toward the Sun, it is summer in the northern hemisphere and winter in the southern hemisphere; when the south pole is tilted toward the Sun, the seasons are reversed (see Figures 2-12 and 2-13). But because Mars has a rather elliptical orbit, the seasonal variations in the planet's two hemispheres are somewhat different. Summer comes to the southern hemisphere of Mars at the same time that the planet makes its closest approach to the Sun. In the northern hemisphere, by contrast, summer comes when Mars is farthest from the Sun. Hence, the tem-

perature at the south pole during its summer should be higher than at the north pole when it is summer there. Paradoxically, the *Viking Orbiters* measured just the opposite! The north pole's temperature climbs to 200 K during the northern summer, while the midsummer temperature at the south pole never gets above 150 K.

Why is summer warmer in the northern hemisphere than in the southern hemisphere? This seemingly backward state of affairs is caused by dust in the atmosphere. When it is summer in the southern hemisphere, the Sun warms the soil surrounding the south polar ice cap, while the ice cap remains quite cold. This temperature contrast gives rise to strong winds that often produce swirling dust storms. A dusty atmosphere admits less sunlight than a clear one, so these storms help to keep the south polar ice cap cold even in midsummer. At the north pole, however, the solar heating during summer is weaker because the planet is far from the Sun. Hence, the temperature difference between the cold north polar ice cap and the surrounding sun-warmed soil is not as great, and the winds are not strong enough to produce dust storms that would cover the pole. (Note that the dust storm shown in Figure 13-11 is south of the north polar cap.) As a result, the north polar ice cap remains exposed to sunlight, the temperature there rises to levels never found at the south pole, and carbon dioxide frost is able to evaporate completely.

Seasonal winds can also explain why Mars changes in appearance over the course of a Martian year. As Figure 13-18 shows, small sand drifts were seen at the *Viking Lander 1* site. Careful scrutiny of pictures taken over several months reveals that these drifts are pushed around slightly by the seasonal winds. New rocks are occasionally exposed and others covered up. Therefore, as seen from Earth, large areas of the planet vary from dark to light with the passage of the seasons. These variations mimic the color changes that would be expected from vegetation, which is why early observers thought they saw signs of plant life on Mars (see Section 13-2).

13-8 | The two Martian moons resemble asteroids

Two moons move around Mars in orbits close to the planet's surface. These satellites are so tiny that they were not discovered until the favorable opposition of 1877. While Schiaparelli was seeing canals, the American astronomer Asaph Hall spotted the two moons, which orbit almost directly above its equator. He named them Phobos ("fear") and Deimos ("panic"), after the mythical horses that drew the chariot of the Greek god of war.

Phobos is the inner and larger of the Martian moons. It circles Mars in just 7 hours and 39 minutes at an average distance of only 9378 km (5828 mi) from the center of Mars. (Recall that our Moon orbits at an average distance of 384,400 km from the Earth's center.) This orbital period is much shorter than the Martian day; hence, Phobos rises in the *west* and gallops across the sky in only 5½ hours, as viewed by an observer near the Martian equator. During this time, Phobos appears several times brighter in the Martian sky than Venus does from the Earth.

Deimos, which is farther from Mars and somewhat smaller than Phobos, appears about as bright from Mars as Venus does from the Earth. Deimos orbits at an average distance of 23,460 km (14,580 mi) from the center of Mars, slightly beyond the right distance for a synchronous orbit—an orbit in which a satellite seems to hover above a single location on the planet's equator. As seen from the Martian surface, Deimos rises in the east and takes about three full days to creep slowly from one horizon to the other.

Images from spacecraft in Mars's orbit reveal Phobos and Deimos to be jagged and heavily cratered (**Figure 13-23**). Both moons are somewhat elongated from a spherical shape: Phobos is approxi-

mately 28 by 23 by 20 km in size, while Deimos is slightly smaller, roughly 16 by 12 by 10 km. Observations from the *Viking Orbiters* showed that Phobos and Deimos are both in synchronous rotation. The tidal forces that Mars exerts on these elongated moons cause them to always keep the same sides toward Mars, just as the Earth's tidal forces ensure that the same side of the Moon always faces us (see Section 4-8).

We saw in Section 10-5 that the Moon raises a tidal bulge on the Earth. Because the Earth rotates on its axis more rapidly than the Moon orbits the Earth, this bulge is dragged ahead of a line connecting the Earth and Moon (see Figure 10-19). The gravitational force of this bulge pulls the Moon into an ever larger orbit. In the same way, and for the same reason, Deimos is slowly moving away from Mars. But for fast-moving Phobos, the situation is reversed: Mars rotates on its axis more slowly than Phobos moves around Mars. This slow rotation holds back the tidal bulge that Phobos raises in the solid body of Mars, and this bulge thus is *behind* a line connecting the centers of Mars and Phobos. (Imagine that the bulge in Figure 10-19 is tilted clockwise rather than counterclockwise.) This bulge exerts a gravitational force on Phobos that pulls it back, slowing the moon's motion and making it slowly spiral inward toward Mars. This inward spiral will cease in about 40 million years when Phobos smashes into the Martian surface.

The origin of the Martian moons is unknown. One idea is that they are asteroids that Mars captured from the nearby asteroid belt (see Section 7-5). Like Phobos and Deimos, many asteroids reflect very little sunlight (typically less than 10%) because of a high carbon content. Phobos and Deimos also have very low densities (1900 kg/m^3 and 1760 kg/m^3 respectively), which are also characteristic of carbon-rich asteroids. Perhaps two of these asteroids wandered close enough to Mars to become permanently trapped by the planet's gravitational field. Alternatively, Phobos and Deimos may have formed in orbit around Mars out of debris left over from the formation of the planets. Some of this debris might have come from the asteroid belt, giving these two moons an asteroidlike character. Future measurements of the composition of the Martian moons—perhaps by landing spacecraft on their surfaces—may shed light on their origin.

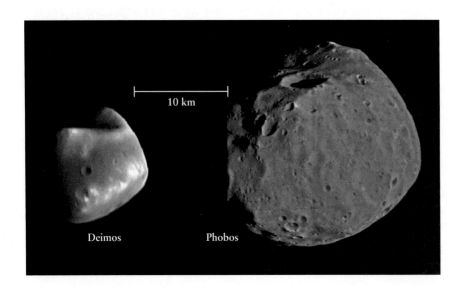

figure 13-23 R I **V** U X G

The Moons of Mars These *Viking Orbiter* images show Deimos and Phobos to the same scale. What appears to be a fold at the top of the image of Deimos is actually a side view of an impact crater 10 km across, comparable to the Moon's 22-km average diameter. An only slightly more powerful impact would have broken Deimos apart. (JPL/NASA)

The exploration of Mars and its two moons has really just begun. Recent discoveries from orbiters and landers have only whetted our appetite to know more about Mars. Fortunately, every 25 months the Earth and Mars are ideally placed for sending a spacecraft from one planet to another with the minimum thrust. For the next several years, one or more spacecraft will be launched toward Mars at each of these opportunities. By 2014, Mars will have been visited by a new generation of orbiters and landers from NASA, the European Space Agency (ESA), and the Japanese Institute of Space and Astronautical Science (ISAS). One of these spacecraft may land on Mars, collect samples with an advanced rover, and return those samples to Earth. Detailed analysis of Martian samples in laboratories on Earth could answer many open questions about the red planet, its formation, and its evolution.

It is ironic that at the end of the nineteenth century, the English novelist H. G. Wells envisioned a fleet of Martian spaceships coming to Earth on a mission of conquest. Today, at the beginning of the twenty-first century, it is we on Earth who are sending robotic spacecraft on a peaceful scientific "invasion" of Mars.

Key Words

andesite, p. 278	permafrost, p. 272
cosmic rays, p. 272	residual polar cap, p. 271
crustal dichotomy, p. 268	rift valley, p. 269
dust devil, p. 278	runaway icehouse effect,
favorable opposition, p. 264	p. 274
northern lowlands, p. 268	southern highlands, p. 268

Key Ideas

Earth-Based Observations of Mars: The best Earth-based views of Mars are obtained at favorable oppositions, when Mars is simultaneously at opposition and near perihelion.

• Earth-based observers found that a solar day on Mars is nearly the same length as on Earth, that Mars has polar caps that expand and shrink with the seasons, and that the Martian surface undergoes seasonal color changes.

• A few observers reported a network of linear features called canals. These observations, which proved to be illusions, led to many speculations about Martian life.

 Spacecraft Observation of Mars: Spacecraft have returned close-up views of the Martian surface. These show that the Martian surface has numerous craters, several huge volcanoes, a vast rift valley, and dried-up riverbeds—but no canals.

• The heavily cratered southern highlands are older and about 5 km higher in elevation than the smooth northern lowlands. The origin of this crustal dichotomy is not completely understood.

• Martian volcanoes and the Valles Marineris rift valley were formed by upwelling plumes of magma in the mantle. For reasons that are not understood, the chemical composition of ancient Martian lava is different from that of more recent lava.

• Mars has no planetwide magnetic field at present but may have had one in the ancient past.

Water on Mars: Flash-flood features and dried riverbeds on the Martian surface indicate that water has flowed on Mars at least occasionally.

• No liquid water can exist on the Martian surface today. But Mars's polar caps contain frozen water, a layer of permafrost may exist below the Martian regolith, and there may be liquid water beneath the surface.

• The Martian polar caps expand in winter as a thin layer of frozen carbon dioxide (dry ice) is deposited from the atmosphere.

• Mars's primordial atmosphere may have been thicker and warmer than the present-day atmosphere. It is unclear whether it contained enough carbon dioxide and water vapor to support a greenhouse effect that would permit liquid water to exist on the planet's surface.

The Martian Atmosphere: The Martian atmosphere is composed mostly of carbon dioxide. The atmospheric pressure on the surface is less than 1% that of the Earth and shows seasonal variations as carbon dioxide freezes onto and evaporates from the poles.

• Great dust storms sometimes blanket Mars. Fine-grained dust in its atmosphere gives the Martian sky a pinkish-orange tint.

• Seasonal winds blow dust across the face of Mars, covering and uncovering the underlying surface material and causing seasonal color changes. Afternoon dust devils help to transport dust from place to place.

Life on Mars: Chemical reactions in the Martian regolith, together with ultraviolet radiation from the Sun, appear to sterilize the Martian surface. Hence, the *Viking Lander* spacecraft found no evidence for present-day life on Mars.

The Moons of Mars: Mars has two small, football-shaped satellites that move in orbits close to the surface of the planet. They may be captured asteroids or may have formed in orbit around Mars out of solar system debris.

Review Questions

1. Why is it best to view Mars near opposition? Why are some oppositions better than others?

2. The image in Figure 13-1 shows a nearly full Mars. Is it possible to see Mars in a crescent phase when it is near opposition (Figure 13-2)? Why or why not?

3. Explain why Mars can best be viewed from Earth while it is undergoing retrograde motion.

4. Why are there seasons on Mars?

5. Why did Earth observers report that they had seen straight-line features ("canals") on Mars?

6. Why do scientists think that the surface of Mars is very old?

7. What is the Martian crustal dichotomy? What theories have been proposed to explain its origin? What are the arguments in favor of or against each theory?

8. Compare the volcanoes of Venus, the Earth, and Mars. Cite evidence that hot-spot volcanism is or was active on all three worlds.

9. Olympus Mons and Valles Marineris are respectively the largest volcano and the largest valley known in the solar system. How, then, could the three *Mariner* spacecraft that flew past Mars in the 1960s have missed seeing these features?

10. Giant impact basins like Hellas Planitia are found on Mars but not on Earth. Why not?

11. What geologic features (or lack thereof) on Mars have convinced scientists that plate tectonics did not significantly shape the Martian surface?

12. What geologic processes are thought to have created Valles Marineris?

13. If you were an astronaut on Mars, would it be useful to carry a compass for navigation? Why or why not?

14. Why is it impossible for liquid water to exist on Mars today? If liquid water existed on Mars in the past, what must have been different then?

15. Substantial quantities of liquid water have not been present on Mars for a very long time. Explain how we know this.

16. How was the *Mars Odyssey* spacecraft able to detect water beneath the Martian surface without landing on the planet?

17. Why is it reasonable to assume that the primordial atmospheres of the Earth and Mars were roughly the same?

18. Carbon dioxide accounts for about 95% of the present-day atmospheres of both Mars and Venus. Why, then, is there a strong greenhouse effect on Venus but only a weak greenhouse effect on Mars?

19. What is a runaway icehouse effect?

20. What did the *Mars Odyssey* spacecraft reveal about carbonate rocks on Mars? What do these results suggest about the history of water on Mars?

21. Why is Mars red?

22. Explain why the Martian sky has the color of butterscotch.

23. Were the *Viking Landers* able to determine whether life currently exists on Mars or whether it once existed there? Why or why not?

24. What discoveries did *Mars Pathfinder* make about Martian rocks? Why weren't the *Viking Landers* able to make these same discoveries 20 years earlier?

25. What is a dust devil? Why would you feel much less breeze from a Martian dust devil than from a dust devil on Earth?

26. Discuss the statement "On Mars, the winters are so cold that the atmosphere freezes." How accurate is this statement?

27. In which Martian hemisphere are the summers warmer? Why is there a difference between the hemispheres? As part of your explanation, make a drawing like Figure 2-12 that shows Mars in its orbit around the Sun.

28. Explain how the seasonal Martian winds fooled early observers into thinking that Mars had vegetation.

29. A full moon on Earth is bright enough to cast shadows. As seen from the Martian surface, would you expect a full Phobos or full Deimos to cast shadows? Why or why not?

Advanced Questions

> **Problem-solving tips and tools**
>
> You will need to remember the small angle formula (Box 1-1) and the formula for the angular resolution of a telescope (Section 6-3). For the relationship between a planet's sidereal and synodic periods, see Box 4-1. You may also need to review the form of Kepler's third law that explicitly includes mass (see Section 4-7 and Box 4-4). The volume of a sphere of radius r is $4\pi r^3/3$. The speed of light is given in Appendix 7.

30. Using Figure 13-2, explain why oppositions of Mars are most favorable when they occur in August.

31. Explain why Mars has the longest synodic period of all the planets, although its sidereal period is only 687 days.

32. **(a)** Use the data in Figure 13-2 to calculate the number of days that elapse between the oppositions in 2001 and 2003, between the oppositions in 2005 and 2007, and between the oppositions in 2010 and 2012. Are your three numbers identical? **(b)** According to Table 4-1, the synodic period of Mars—that is, the time between successive oppositions—is 780 days. How does this compare to the numbers of days you calculated in part (a)? Explain this apparent contradiction.

33. From the 2007 opposition of Mars to the 2010 opposition, how many orbits around the Sun does the Earth complete? How many orbits does Mars complete? (Your answers will *not* be whole numbers.)

34. When Mars is at opposition, what is its phase as seen from the Earth? What is the Earth's phase as seen from Mars? Explain with a diagram.

35. Explain why the image that opens this chapter shows much finer detail than Figure 13-1.

36. For a planet to appear to the naked eye as a disk rather than as a point of light, its angular size would have to be 1 arcmin, or 60 arcsec. (This is the same as the angular separation between lines in the bottom row of an optometrist's eye chart.) **(a)** How close would you have to be to Mars in order to see it as a disk with the naked eye? Does Mars ever get this close to Earth? **(b)** Would the Earth ever be visible as a disk to an astronaut on Mars? Would she be able to see the Earth and the Moon separately, or would they always appear as a single object? Explain.

37. (a) Suppose you have a telescope with an angular resolution of 1 arcsec. What is the size (in kilometers) of the smallest feature you could see on the Martian surface during the opposition of 2007 (see Table 13-2)? (b) Suppose you had access to the Hubble Space Telescope, which has an angular resolution of 0.1 arcsec. What is the size (in kilometers) of the smallest feature you could see on Mars with the HST during the 2007 opposition?

38. An amateur astronomer wants to purchase a telescope that will enable her to resolve Hellas Planitia (diameter 2300 km) during the 2005 opposition of Mars. What must be the minimum diameter of the telescope objective? Assume that she observes Mars using a yellow filter, so that the wavelength used is 550 nm. Ignore the blurring of the image by the Earth's atmosphere.

39. Imagine that you are part of the scientific team monitoring a spacecraft that has landed on Mars. At 5:00 P.M. in your control room on Earth, the spacecraft reports that the Sun is highest in the sky as seen from its location on Mars. When the Sun is next at its highest point as seen from the spacecraft, what time will it be in the control room on Earth?

40. The Earth's northern hemisphere is 39% land and 61% water, while its southern hemisphere is only 19% land and 81% water. Thus, the southern hemisphere could also be called the "water hemisphere." The Moon also has two distinct hemispheres, the near side (which has a number of maria) and the far side (which has almost none). How are these hemispheric differences on Earth and on the Moon similar to the Martian crustal dichotomy? How are they different?

41. For a group of properly attired astronauts equipped with oxygen tanks, a climb to the summit of Olympus Mons would actually be a relatively easy (albeit long) hike rather than a true mountain climb. Give two reasons why.

42. Mars Global Surveyor (MGS) is in a nearly circular orbit with an orbital period of 117 minutes. (a) Using the data in Table 13-1, find the radius of the orbit. (b) What is the average altitude of MGS above the Martian surface? (c) The orbit of MGS passes over the north and south poles of Mars. Explain how this makes it possible for the spacecraft to observe the entire surface of the planet.

43. On Mars, the difference in elevation between the highest point (the summit of Olympus Mons) and the lowest point (the bottom of the Hellas Planitia basin) is 30 km. On Earth, the corresponding elevation difference (from the peak of Mount Everest to the bottom of the deepest ocean) is only 20 km. Discuss why the maximum elevation difference is so much greater on Mars.

44. The elevations in Figure 13-5 were measured using an instrument called MOLA (Mars Orbiter Laser Altimeter) on board Mars Global Surveyor. MOLA fires a laser beam downward, then measures how long it takes for the beam to return to the spacecraft after reflecting off the surface. Suppose MOLA measures this round-trip time for the laser beam to reflect off the summit of Olympus Mons, as well as the round-trip time to reflect off the

bottom of Hellas Planitia (see Question 43). Which round-trip time is longer? How much longer is it? Do you need to know the distance from Mars Global Surveyor to the surface to answer this question?

45. (a) The Grand Canyon in Arizona was formed over 15 to 20 million years by the flowing waters of the Colorado River, as well as by rain and wind. Contrast this formation scenario to that of Valles Marineris. (b) Valles Marineris is sometimes called "the Grand Canyon of Mars." Is this an appropriate description? Why or why not?

46. The classic 1950 science-fiction movie Rocketship X-M shows astronauts on the Martian surface with oxygen masks for breathing but wearing ordinary clothing. Would this be a sensible choice of apparel for a walk on Mars? Why or why not?

47. Compare the evolution of the atmospheres of Venus, the Earth, and Mars. Explain how the atmospheres of these three planets turned out so differently from one another, even though they may all have started with roughly the same chemical composition.

48. Suppose Mars were moved into the Earth's orbit and the Earth were moved into Mars's orbit. What effect would this have on the atmospheres of the two planets?

49. Suppose that the only information you had about Mars was the image of the surface in Figure 13-18. Describe at least two ways that you could tell from this image that Mars has an atmosphere.

50. Although the Viking Lander 1 and Viking Lander 2 landing sites are 6500 km apart and have different geologic histories, the chemical compositions of the dust at both sites are nearly identical. (a) What does this suggest about the ability of the Martian winds to transport dust particles? (b) Would you expect that larger particles such as pebbles would also have identical chemical compositions at the two Viking Lander sites? Why or why not?

51. The landing site for Viking Lander 1 is in an ancient flood plain, while the site for Viking Lander 2 is near the southernmost extent of the north polar ice cap. The Viking mission planners selected these sites because they suspected water might lie beneath the surface there. Why were these good places to search for potential Martian microorganisms?

52. Is it reasonable to suppose that the polar regions of Mars might harbor life-forms, even though the Martian regolith is sterile at the Viking Lander sites? Explain.

53. Why do you suppose that Phobos and Deimos are not round like our Moon?

54. The orbit of Phobos has a semimajor axis of 9378 km. Use this information and the orbital period given in the text to calculate the mass of Mars. How does your answer compare with the mass of Mars given in Table 13-1?

55. Calculate the angular sizes of Phobos and Deimos as they pass overhead, as seen by an observer standing on the Martian equator. How do these sizes compare with that of the Moon seen from the Earth's surface? Would Phobos and Deimos appear as impressive in the Martian sky as the Moon does in our sky?

56. You are to put a spacecraft into a synchronous circular orbit around the Martian equator, so that its orbital period is equal to the planet's rotation period. Such a spacecraft would always be over the same part of the Martian surface. (**a**) Find the radius of the orbit and the altitude of the spacecraft above the Martian surface. (**b**) Suppose Mars had a third moon that was in a synchronous orbit. Would tidal forces make this moon tend to move toward Mars, away from Mars, or neither? Explain.

Discussion Questions

57. One explanation that has been proposed for the origin of Valles Marineris is that it is a place where plate tectonics started on Mars and where two parts of the crust began to move apart like tectonic plates. In this picture, tectonic activity died out on Mars because the planet's interior cooled too rapidly, so that convection in the mantle ceased. What sorts of observations of Mars would you make in an attempt to prove or disprove this explanation?

58. The total cost of the *Mars Global Surveyor* mission was about $154 million. (To put this number into perspective, in 2000 the U.S. Mint spent about $40 million to advertise its new $1 coin. Several recent Hollywood movies have had larger budgets than *Mars Global Surveyor*.) Does this expenditure seem reasonable to you? Why or why not?

59. Compare the scientific opportunities for long-term exploration offered by the Moon and Mars. What difficulties would be inherent in establishing a permanent base or colony on each of these two worlds?

60. Imagine you are an astronaut living at a base on Mars. Describe your day's activities, what you see, the weather, the spacesuit you are wearing, and so on.

61. It has been proposed to make conditions on Mars more Earth-like by using mirrors to direct sunlight onto the south polar ice cap, causing carbon dioxide to evaporate. Discuss how this could achieve the desired effect.

 ## Web/CD-ROM Questions

62. Search the World Wide Web for information about at least two of the following Mars missions: the Japanese orbiter *Nozomi*, the NASA *Mars Phoenix* mission, and the NASA *Mars Reconnaissance Orbiter* mission. What is the current status of these missions? What are their scientific goals?

63. In 1999 two NASA spacecraft—*Mars Climate Orbiter* and *Mars Polar Lander*—failed to reach their destinations. Search the World Wide Web for information about these missions. What were their scientific goals? How and why did the missions fail? Which future missions, if any, are intended to pick up where these missions left off?

64. Search the World Wide Web for information about possible manned missions to Mars. How long might such a mission take? How expensive would such a project be? What would be the advantages of a manned mission compared to an unmanned one?

 65. **Conjunctions of Mars.** Access and view the animation "The Orbits of Earth and Mars" in Chapter 13 of the *Universe* Web site or CD-ROM. (**a**) The animation highlights three dates when Mars is in opposition, so that the Earth lies directly between Mars and the Sun. By using the "Stop" and "Play" buttons in the animation, find two times during the animation when Mars is in *conjunction*, so that the Sun lies directly between Mars and the Earth (see Figure 4-6). For each conjunction, make a drawing showing the positions of the Sun, the Earth, and Mars, and record the month and year when the conjunction occurs. (*Hint:* See Figure 13-2.) (**b**) When Mars is in conjunction, at approximately what time of day does it rise as seen from Earth? At what time of day does it set? Is Mars suitably placed for telescopic observation when in conjunction?

Observing Projects

 Observing tips and tools
Mars is most easily seen around an opposition (see Table 13-2). At other times Mars may be visible only in the early morning hours before sunrise or in the early evening just after sunset. Consult such magazines as *Sky & Telescope* and *Astronomy* or their Web sites for more detailed information about when and where to look for Mars during a given month. You can also use the *Starry Night Backyard*™ program on the CD-ROM that accompanies this textbook.

66. If Mars is suitably placed for observation, arrange to view the planet through a telescope. Draw a picture of what you see. What magnifying power seems to give you the best image? Can you distinguish any surface features? Can you see a polar cap or dark markings? If not, can you offer an explanation for Mars's bland appearance?

 67. Use the *Starry Night Backyard*™ program to observe the appearance of Mars. Display the entire celestial sphere (select **Guides > Atlas** in the **Go** menu) and center on Mars by using the **Find...** command in the **Edit** menu. Using the controls at the right-hand end of the Control Panel, zoom in as far as you can. (**a**) In the Control Panel, set the date and time to August 28, 2003, at 12:00:00 A.M. to see Mars during a very favorable opposition, and set the time step to 1 hour. Then click on the "Forward" button (a triangle that points to the right). Describe what you see. (**b**) Click on the "Stop" button (a black square) in the Control Panel. Change the time step to 1 lunar month (1 "lunar m.") and click again on the Forward button. Describe what you see. Using a diagram like Figure 4-6, explain the changes in the apparent size of the planet.

68. Use the *Starry Night Backyard*™ program to observe the appearance of Mars. Display the entire celestial sphere (select **Guides > Atlas** in the **Go** menu) and center on Mars by using the **Find...** command in the **Edit** menu. Using the controls at the right-hand end of the Control Panel, zoom in as far as you can. In the Control Panel, set the date and time to August 28, 2003, at 12:00:00 A.M. to see Mars during a very favorable opposition. You will see then Mars and its two moons, Phobos and Deimos. (If you can't see the moons, click on the tab **View Options** at the left-hand edge of the window. Under "Solar System," make sure that "Planets-Moons" and "Labels" are both checked. Then click on the **View Options** tab again.) (**a**) In the Control Panel set the time step to 1 minute and click on the "Forward" button (a triangle that points to the right). Click on the "Stop" button (a black square) in the Control Panel. Record the date and time in the display, and note the position of Phobos (the inner moon). Step through time using the single-step button (the rightmost time control button) until Phobos returns to approximately the same position, then record the date and time in the display. From your observations, what is the orbital period of Phobos? How does your result compare with the orbital period given in Appendix 3? (**b**) Repeat part (a) for Deimos (the outer moon).

69. Use the *Deep Space Explorer*™ program to compare the orbits of Mars and the Earth. In the left-hand part of the window under the heading **Solar System**, click on the triangle next to the word **Explore** and then click on **Inner Solar System.** You can zoom in and zoom out using the buttons at the upper left of the window (an upward-pointing triangle and a downward-pointing triangle). You can also rotate the solar system by putting the mouse cursor over the image, holding down the mouse button, and moving the mouse. (On a two-button mouse, hold down the left mouse button.) (**a**) Is the time shown in the image a few weeks before an opposition or a few weeks after an opposition? (*Hint:* Arrowheads on the orbits show you which way each planet moves around its orbit.) (**b**) Are the orbits of Mars and of the Earth in the same plane? Use your answer to explain the shape of the apparent path of Mars on the celestial sphere in 2009–2010, shown in Figure 4-2.

Collaborative Exercises

70. Mars is generally visible in the night sky the two months before and two months after it is at opposition. Using Table 13-2, determine whether or not Mars will be easily visible on your next birthday. Create a simple sketch showing our Sun, the Earth, and Mars to illustrate your answer.

71. The image of Mars that opens this chapter is from the Hubble Space Telescope. Draw a circle on your paper roughly 5 cm in diameter and, taking turns, have each person in your group sketch a different region of Mars. How is your collaborative sketch different than Schiaparelli's drawing shown in Figure 13-3?

72. Water erosion features like those shown on Mars in Figure 13-10 are common on Earth. Exactly where would be the closest place to you that one could photograph similar features from an airplane?

BRUCE JAKOSKY

The Exploration of Mars

Bruce Jakosky is a professor in the Department of Geological Sciences and the Laboratory for Atmospheric and Space Physics at the University of Colorado at Boulder. His research interests are in the geology of planetary surfaces, the evolution of the Martian atmosphere and climate, the potential for life on Mars and elsewhere, and the societal and philosophical issues connected to life elsewhere. He has been involved with numerous spacecraft missions, and he heads up the University of Colorado Center for Astrobiology.

Mars sits at the intellectual center of NASA's space exploration program. This is due in large part to the tremendous scientific and public interest in the potential for life to exist there. However, you cannot just go and look for life. We are trying to explore Mars in a systematic way. We want to understand its geological history, in order to understand the availability of water and of the elements out of which life could be constructed. We want to understand the history of the Martian climate, in order to understand conditions at the surface and whether they ever could have supported life. We want to understand the geochemical processes within the crust, in order to determine the availability of energy from chemical reactions that could support a Martian metabolism. In upcoming spacecraft missions, we will try to determine whether there ever was liquid water. We will try to understand the "carbon chemistry" and the nature of organic molecules at the surface. Only then will a search for life make sense, and only then can the results be interpreted properly.

It's only been in the last decade, however, that the "life" questions have become so important. Yes, the *Viking* spacecraft in the mid-1970s searched for life. But they showed us how difficult such a search would be. With nearly three decades of hindsight, it is clear that we were incredibly naive about identifying evidence for life. For example, we can culture less than 1% of terrestrial microorganisms in labs here on Earth, yet we thought that we could go to another planet and successfully culture organisms about which we knew absolutely nothing! Although no evidence for life was found, it does not rule out the potential that Mars could have life. *Viking* was able to "search" only a very small fraction of the planet—and the least interesting areas, at that. And it was able to analyze only a very small fraction of the range of biological processes that could be operating.

Today, we know that life can occupy a much wider range of environments than we had previously thought, and that it can utilize a much wider range of metabolic processes. We have identified life at temperatures above the boiling point of water, or living within rocks and living off of chemical reactions. These types of environments almost certainly occur on Mars, begging the question of whether there could be life there today.

At the same time, the interest in life on Mars was rejuvenated in a very public discussion by a discovery of meteorites that had come from Mars. One of them, collected in Antarctica and known as ALH84001, was thought by some to have evidence for fossil life embedded within the rock. Although years of debate now lead us to believe that most of the original observations were either artifacts or ambiguous, that discussion stirred excitement within both the scientific community and the public.

My own interest in the potential for life on Mars has changed over time as well. At the time of the *Viking* missions, I was an undergraduate working on one of the experiments on the orbiter. Although *Viking* was a "biology" mission, the "real" planetary scientists focused on understanding the nature and history of the planet. For the next 15 years, my research emphasized the geology of its surface and the history of the atmosphere and climate. Through teaching undergraduates, though, I realized how deep the interest in the "life" questions is and that these were the issues that stirred individuals.

Slowly, I began to appreciate the broader questions of the relationship between biological and geological activity. Through these topics, we can bring together under a single umbrella issues such as the origin of our solar system and the objects in it; the evolution of planets, satellites, and smaller objects in our solar system; and the nature of stars and planets outside our solar system.

I think back to when I was 13 years old, watching the first human landings on the Moon on TV. Although the science was tremendous, I was engaged, like everybody else, by the human drama of exploring the place for the first time. That is what drives our exploration of space, and that is what unites the public and scientific interest in asking and then answering questions that cut to the core of what it means to be human. And that is why these questions, and the missions to address them, have become the intellectual drivers of the space exploration program.

14 Jupiter and Saturn: Lords of the Planets

Among the most remarkable sights in our solar system are the colorful, turbulent atmosphere of Jupiter and the ethereal beauty of Saturn's rings. Both of these giant worlds dwarf our own planet: More than 1200 Earths would fit inside Jupiter's immense bulk, and more than 700 Earths inside Saturn. Unlike the Earth, both Jupiter and Saturn are composed primarily of the lightweight elements hydrogen and helium, in abundances very similar to those in the Sun.

As these images show, the rapid rotations of both Jupiter and Saturn stretch their weather systems into colorful bands that extend completely around each planet. Both planets also have intense magnetic fields that are generated in their interiors, where hydrogen is so highly compressed that it becomes a metal.

Saturn is not merely a miniature version of Jupiter, however. Its muted colors and more flattened shape are clues that Saturn's atmosphere and interior have important differences from those of Jupiter. The most striking difference is Saturn's elaborate system of rings, composed of countless numbers of icy fragments orbiting in the plane of the planet's equator. The rings display a complex and elegant structure, which is shaped by subtle gravitational influences from Saturn's retinue of moons. Jupiter, too, has rings, but they are made of dark, dustlike particles that reflect little light. These systems of rings, along with the distinctive properties of the planets themselves, make Jupiter and Saturn essential destinations on our tour of the solar system.

Jupiter (left) and Saturn (right) to scale. (NASA, JPL, and University of Arizona; Hubble Heritage Team [STScI/AURA]/NASA) R I **V** U X G

Questions to Guide Your Reading

As you read the sections of this chapter, look for the answers to the following questions:

14-1 Why is the best month to see Jupiter different from one year to the next?

14-2 Why are there important differences between the atmospheres of Jupiter and Saturn?

14-3 What is going on in Jupiter's Great Red Spot?

14-4 What is the nature of the multicolored clouds of Jupiter and Saturn?

14-5 What does the chemical composition of Jupiter's atmosphere imply about the planet's origin?

14-6 How do astronomers know about the deep interiors of Jupiter and Saturn?

14-7 How do Jupiter and Saturn generate their intense magnetic fields?

14-8 Why would it be dangerous for humans to visit certain parts of the space around Jupiter?

14-9 How was it discovered that Saturn has rings?

14-10 Are Saturn's rings actually solid bands that encircle the planet?

14-11 How uniform and smooth are Saturn's rings?

14-12 How do Saturn's satellites affect the character of its rings?

table 14-1	Jupiter Data
Average distance from Sun:	5.203 AU = 7.783 × 10⁸ km
Maximum distance from Sun:	5.455 AU = 8.160 × 10⁸ km
Minimum distance from Sun:	4.950 AU = 7.406 × 10⁸ km
Eccentricity of orbit:	0.048
Average orbital speed:	13.1 km/s
Orbital period:	11.86 years
Rotation period:	9ʰ 50ᵐ 28ˢ (equatorial)
	9ʰ 55ᵐ 29ˢ (internal)
Inclination of equator to orbit:	3.12°
Inclination of orbit to ecliptic:	1.30°
Diameter:	142,984 km = 11.209 Earth diameters (equatorial)
	133,708 km = 10.482 Earth diameters (polar)
Mass:	1.899 × 10²⁷ kg = 317.8 Earth masses
Average density:	1326 kg/m³
Escape speed:	60.2 km/s
Surface gravity (Earth = 1):	2.36
Albedo:	0.44
Average temperature at cloudtops:	−108°C = −162°F = 165 K
Atmospheric composition (by number of molecules):	86.2% hydrogen (H_2), 13.6% helium (He), 0.2% methane (CH_4), ammonia (NH_3), water vapor (H_2O), and other gases

(NASA/JPL)   R I **V** U X G

14-1 | Jupiter and Saturn are the most massive planets in the solar system

Jupiter and Saturn are respectively the largest and second largest of the planets, and the largest objects in the solar system other than the Sun itself. Astronomers have known for centuries about the huge diameters and immense masses of these two planets. Given the distance to each planet and their angular sizes, they used the small-angle formula (Box 1-1) to calculate that Jupiter is about 11 times larger in diameter than Earth, while Saturn's diameter is about 9 times larger than Earth's (see Figure 7-2). By observing the orbits of Jupiter's four large moons and Saturn's large moon Titan (see Table 7-2) and applying Newton's form of Kepler's third law (see Section 4-7 and Box 4-4), astronomers also determined that Jupiter and Saturn are, respectively, 318 times and 95 times more massive than the Earth. In fact, Jupiter has 2½ times the combined mass of all the other planets, satellites, asteroids, meteoroids, and comets in the solar system. A visitor from interstellar space might well

describe our solar system as the Sun, Jupiter, and some debris! Table 14-1 and Table 14-2 list some basic data about Jupiter and Saturn.

As for any planet whose orbit lies outside the Earth's orbit, the best time to observe Jupiter or Saturn is when they are at opposition. At opposition, Jupiter can appear nearly 3 times brighter than Sirius, the brightest star in the sky. Only the Moon and Venus can outshine Jupiter at opposition. Saturn's brightness at opposition is only about 1/7 that of Jupiter, but it still outshines all of the stars except Sirius and Canopus.

Through a telescope, Jupiter at opposition presents a disk nearly 50 arcsec in diameter, approximately twice the angular diameter of Mars under the most favorable conditions (Figure 14-1a). Because Jupiter takes almost a dozen Earth years to orbit the Sun, it appears to meander slowly across the 12 constellations of the zodiac at the rate of approximately one constellation per year. Successive oppositions occur at intervals of about 13 months (Table 14-3).

Saturn's orbital period is even longer, more than 29 Earth years, so it moves even more slowly across the zodiac. If Saturn did not move at all, successive oppositions would occur exactly one Earth

table 14-2 | Saturn Data

Average distance from Sun:	9.572 AU = 1.432×10^9 km
Maximum distance from Sun:	10.081 AU = $1.508 \times 3\ 10^9$ km
Minimum distance from Sun:	9.063 AU = 1.356×10^9 km
Eccentricity of orbit:	0.053
Average orbital speed:	9.64 km/s
Orbital period:	29.37 years
Rotation period:	$10^h\ 13^m\ 59^s$ (equatorial)
	$10^h\ 39^m\ 25^s$ (internal)
Inclination of equator to orbit:	26.73°
Inclination of orbit to ecliptic:	2.48°
Diameter:	120,536 km = 9.449 Earth diameters (equatorial)
	108,728 km = 8.523 Earth diameters (polar)
Mass:	5.685×10^{26} kg = 95.16 Earth masses
Average density:	687 kg/m^3
Escape speed:	35.5 km/s
Surface gravity (Earth = 1):	0.92
Albedo:	0.46
Average temperature at cloudtops:	−180°C = −292°F = 93 K
Atmospheric composition (by number of molecules):	96.3% hydrogen (H_2), 3.3% helium (He), 0.4% methane (CH_4), ammonia (NH_3), water vapor (H_2O), and other gases

(STScI/Hubble Heritage Team)   R I **V** U X G

(a) Jupiter

(b) Saturn

figure 14-1   R I **V** U X G

Jupiter and Saturn as Viewed from Earth (a) The disk of Jupiter at opposition appears about 2½ times larger than (b) the disk of Saturn at opposition. Both planets display dark and light bands, though these are fainter on Saturn. (a: Courtesy of Stephen Larson; b: NASA)

table 14-3	Oppositions of Jupiter, 2004–2008		
	Earth-Jupiter distance		Angular diameter
Date of opposition	(AU)	(10⁶ km)	(arcsec)
2004 March 4	4.43	662	44.5
2005 April 3	4.46	667	44.2
2006 May 4	4.41	660	44.6
2007 June 5	4.30	644	45.7
2008 July 9	4.16	622	47.3

table 14-4	Oppositions of Saturn, 2004–2008		
	Earth-Saturn distance		Angular diameter
Date of opposition	(AU)	(10⁶ km)	(arcsec)
2005 January 13	8.08	1208	20.6
2006 January 27	8.13	1216	20.5
2007 February 10	8.20	1227	20.3
2008 February 24	8.29	1240	20.1

year apart; thanks to its slow motion, Saturn's oppositions occur at intervals of about one year and two weeks (Table 14-4). At opposition Saturn appears as a disk about 20 arcsec in diameter, with a dramatic difference from Jupiter: Saturn is surrounded by a magnificent system of rings (Figure 14-1b). Later in this chapter we will examine these rings in detail.

As seen through an Earth-based telescope, both Jupiter and Saturn display colorful bands that extend around each planet parallel to its equator (Figure 14-1). More detailed close-up views of Jupiter from a spacecraft (**Figure 14-2a**) show alternating dark and light bands parallel to Jupiter's equator in subtle tones of red, orange, brown, and yellow. The dark, reddish bands are called **belts**, and the light-colored bands are called **zones**. The Hubble Space Tele-

scope image in Figure 14-2b shows similar features on Saturn, but with colors that are much less pronounced. (We will see in Section 14-4 how differences between the atmospheres of Jupiter and Saturn can explain Saturn's washed-out appearance.)

In addition to these conspicuous stripes, a huge, red-orange oval called the **Great Red Spot** is often visible in Jupiter's southern hemisphere. This remarkable feature was first seen by the English scientist Robert Hooke in 1664 but may be much older. It appears to be an extraordinarily long-lived storm in the planet's dynamic atmosphere. Saturn has no long-lived storm systems of this kind. However, many careful observers have reported smaller spots and blemishes in the atmospheres of both Jupiter and Saturn that last for only a few weeks or months.

(a) Jupiter

(b) Saturn

  Figure 14-2 R I **V** U X G
Jupiter and Saturn as Viewed from Space
(a) This view of Jupiter is a composite of four images made by the *Cassini* spacecraft as it flew past Jupiter in 2000. (b) Images from the *Voyager 2* spacecraft as it flew past Saturn in 1981 were combined to show the planet in approximately natural color. (a: NASA/JPL/University of Arizona; b: USGS/NASA/JPL)

14-2 | Unlike the terrestrial planets, Jupiter and Saturn exhibit differential rotation

 SCIENCE IN PROCESS *Observations from spacecraft were needed to determine the chemical compositions of Jupiter and Saturn*

Observations of features like the Great Red Spot and smaller storms allow astronomers to determine how rapidly Jupiter and Saturn rotate. At its equator, Jupiter completes a full rotation in only 9 hours, 50 minutes, and 28 seconds, making it not only the largest and most massive planet in the solar system but also the one with the fastest rotation. However, Jupiter rotates in a strikingly different way from the Earth, the Moon, Mercury, Venus, or Mars. If Jupiter were a solid body like a terrestrial planet (or, for that matter, a billiard ball), all parts of Jupiter's surface would rotate through one complete circle in this same amount of time (Figure 14-3*a*). But by watching features in Jupiter's cloud cover, Gian Domenico Cassini discovered in 1690 that the polar regions of the planet rotate a little more slowly than do the equatorial regions. (You may recall this Italian astronomer from Section 13-2 as the gifted observer who first determined Mars's rate of rotation.) Near the poles, the rotation period of Jupiter's atmosphere is about 9 hours, 55 minutes, and 41 seconds. Saturn, too, has a longer rotation period near its poles (10 hours, 39 minutes, and 24 seconds) than at its equator (10 hours, 13 minutes, and 59 seconds).

 You can see this kind of rotation, called **differential rotation,** in the kitchen. As you stir the water in a pot, different parts of the liquid take different amounts of time to make one "rotation" around the center of the pot (Figure 14-3*b*). Differential rotation shows that neither Jupiter nor Saturn can be solid throughout their volumes: They must be at least partially fluid, like water in a pot.

If Jupiter and Saturn have partially fluid interiors, they cannot be made of the rocky materials that constitute the terrestrial planets. An important clue to the compositions of Jupiter and Saturn are their average densities, which are only 1326 kg/m³ for Jupiter and 687 kg/m³ for Saturn. (By comparison, the Earth's average density is 5515 kg/m³.) To explain these low average densities, Rupert Wildt of the University of Göttingen in Germany suggested in the 1930s that Jupiter and Saturn are composed mostly of hydrogen and helium atoms—the two lightest elements in the universe—held together by their mutual gravitational attraction to form a planet. Wildt was motivated in part by his observations of prominent absorption lines of methane and ammonia in Jupiter's spectrum. (We saw in Section 7-3 how spectroscopy plays an important role in understanding the planets.) A molecule of methane (CH_4) contains four hydrogen atoms, and a molecule of ammonia (NH_3) contains three. The presence of these hydrogen-rich molecules was strong, but indirect, evidence of abundant hydrogen in Jupiter's atmosphere.

Direct evidence for hydrogen and helium in Jupiter's atmosphere, however, was slow in coming. The problem was that neither gas produces prominent spectral lines in the visible sunlight reflected from the planet. To show the presence of these elements

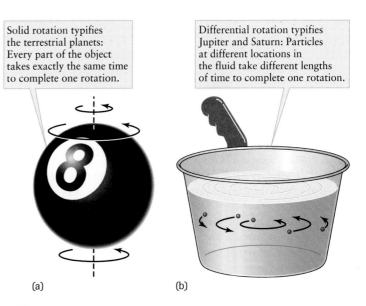

Solid rotation typifies the terrestrial planets: Every part of the object takes exactly the same time to complete one rotation.

Differential rotation typifies Jupiter and Saturn: Particles at different locations in the fluid take different lengths of time to complete one rotation.

(a)   (b)

Figure 14-3

Solid Rotation versus Differential Rotation (a) All parts of a solid object rotate together, but (b) a rotating fluid displays differential rotation. To see this, put some grains of sand, bread crumbs, or other small particles in a pot of water. Stir the water with a spoon to start it rotating, then take out the spoon. The particles near the center of the pot take less time to make a complete rotation than those away from the center.

conclusively, astronomers had to look for spectral lines in the ultraviolet part of the spectrum. These lines are very difficult to measure from the Earth, because almost no ultraviolet light penetrates our atmosphere (see Figure 6-28). Astronomers first detected the weak spectral lines of hydrogen molecules in Jupiter's spectrum in 1960. The presence of helium on Jupiter and Saturn was finally confirmed in the 1970s and 1980s, when spacecraft first flew past these planets and measured their hydrogen spectra in detail. (Collisions between helium and hydrogen atoms cause small but measurable changes in the hydrogen spectrum, which is what the spacecraft instruments detected.)

Today we know that the chemical composition of Jupiter's atmosphere is 86.2% hydrogen molecules (H_2), 13.6% helium atoms, and 0.2% methane, ammonia, water vapor, and other gases. The percentages in terms of *mass* are somewhat different because a helium atom is twice as massive as a hydrogen molecule. Hence, by mass, Jupiter's atmosphere is approximately 75% hydrogen, 24% helium, and 1% other substances, quite similar to that of the Sun. We will see evidence in Section 14-6 that Jupiter has a large rocky core made of heavier elements. It is estimated that the breakdown by mass of the planet as a whole (atmosphere plus interior) is approximately 71% hydrogen, 24% helium, and 5% all heavier elements.

Like Jupiter, Saturn is thought to have a large rocky core. But unlike Jupiter, data from Earth-based telescopes and spacecraft show that the atmosphere of Saturn has a serious helium deficiency: Its chemical composition is 96.3% hydrogen molecules, 3.3% helium, and 0.4% other substances (by mass, 92% hydrogen, 6% helium, and 2% other substances). This is a puzzle because Jupiter and Saturn are thought to have formed in similar ways from the gases of the solar nebula (see Section 8-4), and so both planets (and the Sun) should have essentially the same abundances of hydrogen and helium. So where did Saturn's helium go?

The explanation may be simply that Saturn is smaller than Jupiter, and so as a result Saturn probably cooled more rapidly. (We saw in Section 7-6 why a small world cools down faster than a large one.) This cooling would have triggered a process analogous to the way rain develops here on Earth. When the air is cool enough, humidity in the Earth's atmosphere condenses into raindrops that fall to the ground. On Saturn, however, it is droplets of liquid helium that condense within the planet's cold, hydrogen-rich outer layers. In this scenario, helium is deficient in Saturn's upper atmosphere simply because it has fallen farther down into the planet. By contrast, Jupiter's helium has not rained out because its upper atmosphere is warmer and the helium does not form droplets.

ANALOGY An analogy to helium "rainfall" within Saturn is what happens when you try to sweeten tea by adding sugar. If the tea is cold, the sugar does not dissolve well and tends to sink to the bottom of the glass even if you stir the tea with a spoon. But if the tea is hot, the sugar dissolves with only a little stirring. In the same way, it is thought that the descending helium droplets once again dissolve in hydrogen once they reach the warmer depths of Saturn's interior.

In this scenario, Jupiter and Saturn both have about the same overall chemical composition. But Saturn's smaller mass, less than a third that of Jupiter, means that there is less gravitational force tending to compress its hydrogen and helium. This explains why Saturn's density is only about half that of Jupiter, and is in fact the lowest of any planet in the solar system.

CAUTION Because Jupiter and Saturn are almost entirely hydrogen and helium, it would be impossible to land a spacecraft on either planet. An astronaut foolish enough to try would notice the hydrogen and helium around the spacecraft becoming denser, the temperature rising, and the pressure increasing as the spacecraft descended. But the hydrogen and helium would never solidify into a surface on which the spacecraft could touch down. Long before reaching the planet's rocky core, the pressure of the hydrogen and helium would reach such unimaginably high levels that any spacecraft, even one made of the strongest known materials, would be crushed.

14-3 | Spacecraft images show remarkable activity in the clouds of Jupiter and Saturn

Most of our detailed understanding of Jupiter and Saturn comes from a series of robotic spacecraft that have examined these remarkable planets at close range. They found striking evidence of stable, large-scale weather patterns in both planets' atmospheres, as well as evidence of dynamic changes on smaller scales.

 The first several spacecraft to visit Jupiter and Saturn each made a single flyby of the planet. *Pioneer 10* flew past Jupiter in December 1973; it was followed a year later by the nearly identical *Pioneer 11*, which went on to make the first-ever flyby of Saturn in 1979. Also in 1979 another pair of spacecraft, *Voyager 1* and *Voyager 2*, sailed past Jupiter. These spacecraft sent back spectacular close-up color pictures of Jupiter's dynamic atmosphere. Both *Voyagers* subsequently flew past Saturn. The first spacecraft to go into Jupiter orbit was *Galileo*, which carried out an extensive program of observations from 1995 to 2003. The *Cassini* spacecraft was scheduled to go into

(a) *Pioneer 11*, December 1974

(b) *Voyager 2*, July 1979

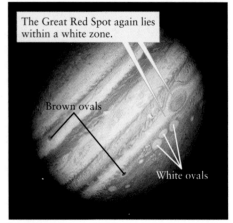

(c) HST, February 1995

Figure 14-4 R I **V** U X G

Three Views of Jupiter These images show major changes in the planet's upper atmosphere over a 20-year period. (a and b: NASA/JPL; c: Reta Beebe and Amy Simon, New Mexico State University, and NASA) orbit around Saturn in 2004 (see the image that opens Chapter 7); on the way to its destination it viewed Jupiter at close range, recording detailed images such as Figure 14-2*a*.

While the general pattern of Jupiter's atmosphere stayed the same during the four years between the *Pioneer* and *Voyager* flybys, there were some remarkable changes in the area surrounding the centuries-old Great Red Spot. During the *Pioneer* flybys, the Great Red Spot was embedded in a broad white zone that dominated the planet's southern hemisphere (**Figure 14-4***a*). By the time of the *Voyager* missions, a dark belt had broadened and encroached on the Great Red Spot from the north (Figure 14-4*b*). In the same way that colliding weather systems in our atmosphere can produce strong winds and turbulent air, the interaction in Jupiter's atmosphere between the belt and the Great Red Spot embroiled the entire region in turbulence (**Figure 14-5**). By 1995, when Jupiter was viewed by the Hubble Space Telescope, the Great Red Spot was once again centered within a white zone (Figure 14-4*c*).

Over the past three centuries, Earth-based observers have reported many long-term variations in the Great Red Spot's size and color. At its largest, it measured 40,000 by 14,000 km—so large that three Earths could fit side by side across it. At other times (as in 1976 and 1977), the spot almost faded from view. During the *Voyager* flybys of 1979, the Great Red Spot was comparable in size to the Earth (see Figure 14-5).

Clouds at different levels in Jupiter's atmosphere reflect different wavelengths of infrared light. Using this effect, astronomers used an infrared telescope on board the *Galileo* spacecraft to help clarify the vertical structure of the Great Red Spot. Most of the Great Red Spot is made of clouds at relatively high altitudes, surrounded by a collar of very low-level clouds about 50 km (30 miles, or 160,000 feet) below the high clouds at the center of the spot. This same kind of structure is seen in high-pressure areas in the Earth's atmosphere, although on a very much smaller scale.

Cloud motions in and around the Great Red Spot reveal that the spot rotates counterclockwise with a period of about six days. Furthermore, winds to the north of the spot blow to the west, and winds south of the spot move toward the east. The circulation around the Great Red Spot is thus like a wheel spinning between two oppositely moving surfaces (see Figure 14-5). Weather patterns on the Earth tend to change character and eventually dissipate when they move between plains and mountains or between land and sea. But because no solid surface or ocean underlies Jupiter's clouds, no such changes can occur for the Great Red Spot—which may explain how this wind pattern has survived for at least three centuries.

Other persistent features in Jupiter's atmosphere are the **white ovals**. Several white ovals are visible in Figure 14-4*b*. As in the Great Red Spot, wind flow in white ovals is counterclockwise. White ovals are also apparently long-lived; Earth-based observers have reported seeing them in the same location since 1938.

Most of the white ovals are observed in Jupiter's southern hemisphere, whereas **brown ovals** are more common in Jupiter's northern hemisphere. Brown ovals appear dark in a visible-light image like Figure 14-4*b*, but they appear bright in an infrared image. For this reason, brown ovals are understood to be holes in Jupiter's cloud cover. They permit us to see into the depths of the

Winds on the north side of the Great Red Spot flow westward.

Turbulence downwind of the Great Red Spot

Winds within the Great Red Spot circulate counterclockwise.

Winds on the south side of the Great Red Spot flow eastward.

20,000 km   Earth's diameter

 VIDEO 14.3 **figure 14-5**   R I **V** U X G
Circulation around the Great Red Spot This 1979 image from *Voyager 2* shows atmospheric turbulence and the direction of winds around the Great Red Spot (compare Figure 14-4*b*, taken at approximately the same time). Winds within the Great Red Spot itself make it spin counterclockwise, completing a full revolution in about six days. (NASA/JPL; adapted from A. P. Ingersoll)

Jovian atmosphere, where the temperature is higher and the atmosphere emits infrared light more strongly. White ovals, by contrast, have relatively low temperatures. They are areas with cold, high-altitude clouds that block our view of the lower levels of the atmosphere.

The *Voyager, Galileo*, and *Cassini* images might suggest a state of incomprehensible turmoil, but, surprisingly, there is also great regularity in the Jovian atmosphere. The computer-generated view in **Figure 14-6** shows how Jupiter would look to someone located directly over either the planet's north pole or its south pole. Note the regular spacing of such cloud features as ripples, plumes, and light-colored wisps. The regular spacing of white ovals in the southern hemisphere is quite apparent. These regularities are probably the result of stable, large-scale weather patterns that encircle Jupiter.

While Saturn has belts and zones like Jupiter, it has no storm systems as long-lived as Jupiter's Great Red Spot. But about every 30 years—roughly the orbital period of Saturn—Earth-based observers have reported seeing storms in Saturn's clouds that last for several days or even months. **Figure 14-7** shows one such "white spot" that appeared near Saturn's equator in 1994. Such storms are thought to form when warm gases in Saturn's atmosphere rise upward and cool, causing gaseous ammonia to solidify into crystals and form white clouds. (On Earth, similar rapid lifting of air

Brown ovals

Evenly spaced white ovals

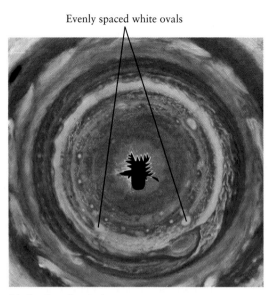

Evenly spaced light-colored plumes

(a) Northern hemisphere

(b) Southern hemisphere

figure 14-6   R I 🇻 U X G

Jupiter's Northern and Southern Hemispheres *Voyager* images were combined and computer processed to construct these views that look straight down onto Jupiter's north and south poles. Various cloud features are evenly spaced in longitude. The black spots at the poles themselves are areas not photographed by the spacecraft. (NASA/JPL)

occurs within thunderstorms and can cause water droplets to solidify into hailstones.) About 20 temporary white spots have been seen on Saturn over the past two centuries.

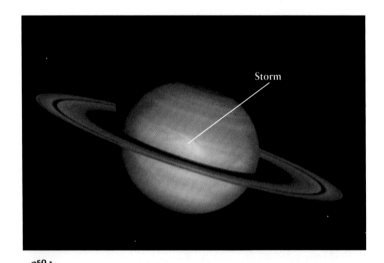

Storm

VIDEO 14.4

figure 14-7   R I 🇻 U X G

A New Storm on Saturn The white arrowhead-shaped feature near Saturn's equator is a giant storm the size of the Earth that appeared in August 1994. The storm lasted several months; it had changed little when this Hubble Space Telescope image was made in December 1994. (Reta Beebe, New Mexico State University; D. Gilmore and L. Bergeron, STScI; and NASA)

14-4 | The internal heat of Jupiter and Saturn has a major effect on the planets' atmospheres

Weather patterns on Earth are the result of the motions of air masses. Winter storms in the midwestern United States occur when cold, moist air moves southward from Canada, and tropical storms in the southwest United States are caused by the northeastward motion of hot, moist air from Hawaii. All such motions, as well as the overall global circulation of our atmosphere, are powered by the Sun. Sunlight is absorbed by the Earth's surface, and the heated surface in turn warms the atmosphere and stirs it into motion (see Sections 9-1 and 9-5). The energy of sunlight also powers the motions of the atmospheres of Venus (Section 12-3) and Mars (Section 13-5). On both Jupiter and Saturn, however, atmospheric motions are powered both by solar energy and by the internal energy of the planet.

In the late 1960s, astronomers using Earth-based telescopes made the remarkable discovery that Jupiter emits more energy in the form of infrared radiation than it absorbs from sunlight—in fact, about twice as much. (By comparison, the internal heat that the Earth radiates into space is only about 0.005% of what it absorbs from the Sun.) The excess heat presently escaping from Jupiter is thought to be energy left over from the formation of the planet 4.56 billion years ago. As gases from the solar nebula fell into the protoplanet, vast amounts of gravitational energy were converted into thermal energy that heated the planet's interior. Jupiter has been slowly radiating this energy into space ever since. Because Jupiter is so large, it has retained substantial thermal energy even after billions of years, and so still has plenty of energy to emit as infrared radiation.

Saturn, too, radiates into space more energy than it receives as sunlight. Because Saturn is smaller than Jupiter, it should have begun with less internal heat trapped inside, and it should have radiated that heat away more rapidly. Hence, we would expect Saturn to be radiating very little energy today. In fact, when we take Saturn's smaller mass into account, we find that Saturn releases about 25% *more* energy from its interior on a per-kilogram basis than does Jupiter. The explanation for this seeming paradox may be that helium is raining out of Saturn's upper atmosphere, as we described in Section 14-2. Friction between the falling helium droplets and the surrounding gases generates heat that eventually escapes from the planet's surface. This "raining out" of helium from Saturn's upper layers is calculated to have begun 2 billion years ago. The amount of thermal energy released by this process adequately accounts for the extra heat radiated by Saturn since that time.

Heat naturally flows from a hot place to a colder place, never the other way around. Hence, in order for heat to flow upward through the atmospheres of Jupiter and Saturn and radiate out into space, the temperature must be warmer deep inside the atmosphere than it is at the cloudtops. Infrared measurements have confirmed that the temperature within the atmospheres of both planets does indeed increase with increasing depth (**Figure 14-8**).

ANALOGY You can understand the statement that "heat naturally flows from a hot place to a colder place" by visualizing an ice cube placed on a hot frying pan. Experience tells you that the hot frying pan cools down a little while the ice warms up and melts. If heat were to flow the other way, the ice would get colder and the frying pan would get hotter—which never happens in the real world.

When a fluid is warm at the bottom and cool at the top, like water being warmed in a pot, one effective way for energy to be distributed through the fluid is by the up-and-down motion called *convection*. (We first discussed convection in Section 9-1; see, in particular, Figure 9-4.) Until recently it was generally thought that Jupiter's light-colored zones are regions where warm gas from low levels is rising and cooling, forming high-altitude clouds, while the dark-colored belts are regions where cool high-altitude gas is descending and being heated. However, this picture has been called into question by images made by *Cassini* during its 2000–2001 flyby of Jupiter, which showed details as small as 58 km (36 mi) across. The *Cassini* images revealed numerous white cloud cells welling upward in Jupiter's atmosphere, almost all of which appeared in the dark belts. These observations suggest that the traditional picture is incorrect, and that air is actually rising in the *belts* and sinking in the *zones*. *Cassini* will carry out similar observations

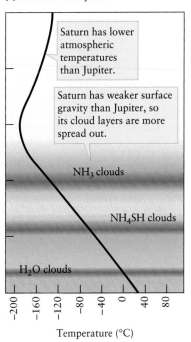

figure 14-8

The Upper Atmospheres of Jupiter and Saturn The black curves in these graphs show temperature versus altitude in each atmosphere, as well as the probable arrangements of the cloud layers. Zero altitude in each atmosphere is chosen to be the point where the pressure is 100 millibars, or one-tenth of Earth's atmospheric pressure. Beneath both planets' cloud layers, the atmosphere is composed almost entirely of hydrogen and helium. (Adapted from A. P. Ingersoll)

Zone

Belt

Great Red Spot

(a) Visible-light image   R I **V** U X G

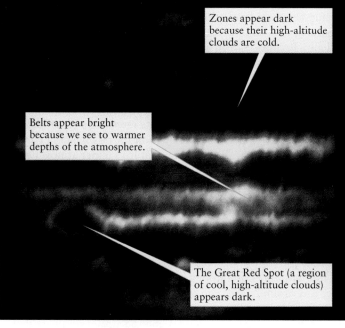

Zones appear dark because their high-altitude clouds are cold.

Belts appear bright because we see to warmer depths of the atmosphere.

The Great Red Spot (a region of cool, high-altitude clouds) appears dark.

(b) Infrared image   R **I** V U X G

figure 14-9

Jupiter's Warm Belts and Cool Zones (a) This visible-light image of Jupiter was made by the *Voyager 1* spacecraft. (b) Bright and dark areas in this Earth-based infrared image, taken at the same time as the image in (a), correspond to high and low temperatures, respectively. (NASA/JPL)

of the belts and zones of Saturn, and the results may clarify the true nature of convection in both planets' atmospheres.

One thing that is well known about Jupiter's atmosphere is that the belts are regions in which we can see into the atmosphere's lower levels. You can see this in **Figure 14-9**, which compares visible-light and infrared images of Jupiter. The cold Jovian clouds emit radiation primarily at infrared wavelengths, with an intensity that depends on the cloud temperature (see Section 5-4) and hence on the depth of the cloud within the atmosphere (see Figure 14-8*a*). The infrared image in Figure 14-9*b* shows stronger emission from the belts, where we are looking at lower-lying, warmer levels of Jupiter's atmosphere.

 In addition to the vertical motion of gases within the belts and zones, the very rapid rotation of both Jupiter and Saturn creates a global pattern of eastward and westward **zonal winds.** Wind speeds on Jupiter can exceed 500 km/h (300 mi/h); on Saturn, wind speeds near the equator approach 1800 km/h (1100 mi/h), approximately two-thirds the speed of sound in Saturn's atmosphere and faster than winds on any other planet. The Earth's atmospheric circulation has a similar pattern of eastward and westward flow (see Figure 9-23), but with slower wind speeds. The faster winds on Jupiter presumably result from the planet's more rapid rotation as well as the substantial flow of heat from the planet's interior. Why Saturn's winds should be even faster than Jupiter's is not yet completely understood.

The zonal winds on Jupiter are generally strongest at the boundaries between belts and zones. Within a given belt or zone, the wind reverses direction between the northern and southern boundaries. As Figure 14-5 shows, such reversals in wind direction are associated with circulating storms such as the Great Red Spot.

From spectroscopic observations and calculations of atmospheric temperature and pressure, scientists conclude that both Jupiter and Saturn have three main cloud layers of differing chemical composition. The uppermost cloud layer is composed of crystals of frozen ammonia. Deeper in the atmosphere, ammonia (NH_3) and hydrogen sulfide (H_2S)—a compound of hydrogen and sulfur—combine to produce ammonium hydrosulfide (NH_4SH) crystals. At even greater depths, the clouds are composed of crystals of frozen water.

Jupiter's strong surface gravity compresses these cloud layers into a region just 75 km deep in the planet's upper atmosphere (Figure 14-8*a*). But Saturn has a smaller mass and hence weaker surface gravity, so the atmosphere is less compressed and the same three cloud layers are spread out over a range of nearly 300 km (Figure 14-8*b*). The colors of Saturn's clouds are less dramatic than Jupiter's (see Figure 14-2) because deeper cloud layers are partially obscured by the hazy atmosphere above them.

The colors of clouds on Jupiter and Saturn depend on the temperatures of the clouds and, therefore, on the depth of the clouds within the atmosphere. Brown clouds are the warmest and are thus the deepest layers that we can see. Whitish clouds form the next layer up, followed by red clouds in the highest layer. The whitish zones on each planet are therefore somewhat higher than the brownish belts, while the red clouds in Jupiter's Great Red Spot are among the highest found anywhere on that planet.

Despite all the data from various spacecraft, we still do not know what gives the clouds of Jupiter and Saturn their colors. The crystals of NH_3, NH_4SH, and water ice in the three main cloud layers are all white. Thus, other chemicals must cause the browns, reds, and oranges. Certain molecules closely related to ammonium hydrosulfide, which can form into long chains that have a yellow-brown color, may play a role. Compounds of sulfur or phosphorus, which can assume many different colors depending on their temperature, might also be involved. The Sun's ultraviolet radiation may induce the chemical reactions that produce these colorful compounds.

14-5 | A space probe has explored Jupiter's deep atmosphere

 SCIENCE IN PROCESS *Measuring trace amounts of rare gases on Jupiter provides clues about the origin of the planets*

By observing Jupiter's and Saturn's cloud layers from a distance, we have been able to learn a great deal about the atmospheres beneath the clouds. But there is no substitute for making measurements on site, and for many years scientists planned to send a spacecraft to explore deep into Jupiter's atmosphere. While still 81.52 million kilometers (50.66 million miles) from Jupiter, the *Galileo* spacecraft released the *Galileo Probe*, a cone-shaped body about the size of an office desk. While *Galileo* itself later fired its rockets to place itself into an orbit around Jupiter, the *Galileo Probe* continued on a course that led it on December 7, 1995, to a point in Jupiter's clouds just north of the planet's equator.

A heat shield protected the *Galileo Probe* as air friction slowed its descent speed from 171,000 km/h (106,000 mi/h) to 40 km/h (25 mi/h) in just 3 minutes. The spacecraft then deployed a parachute and floated down through the atmosphere. For the next hour, the probe observed its surroundings and radioed its findings back to the main *Galileo* spacecraft, which in turn radioed them to Earth. The mission ceased at a point some 200 km (120 mi) below Jupiter's upper cloud layer, where the tremendous pressure (24 atmospheres) and high temperature (152°C = 305°F) finally overwhelmed the probe's electronics.

Although the *Galileo Probe* did not have a camera, it did carry a variety of instruments that made several new discoveries. A radio-emissions detector found evidence for lightning discharges that, while less frequent than on Earth, are individually much stronger than lightning bolts in our atmosphere. Other measurements showed that Jupiter's winds, which are brisk in the atmosphere above the clouds, are even stronger beneath the clouds: The *Galileo Probe* measured a nearly constant wind speed of 650 km/h (400 mi/h) throughout its descent. This shows convincingly that the energy source for the winds is Jupiter's interior heat. If the winds were driven primarily by solar heating, as is the case on Earth, the wind speed would have decreased with increasing depth.

A central purpose of the *Galileo Probe* mission was to test the three-layer cloud model depicted in Figure 14-8a by making direct measurements of Jupiter's clouds. But the probe saw only traces of the NH_3 and NH_4SH cloud layers and found no sign at all of the low-lying water clouds. One explanation of this surprising result is that the probe by sheer chance entered a **hot spot**, an unusually warm and cloud-free part of Jupiter's atmosphere. Indeed, observations from the Earth showed strong infrared emission from the probe entry site, as would be expected where a break in the cloud cover allows a view of Jupiter's warm interior (see Figure 14-9b).

The *Galileo Probe* also made measurements of the chemical composition of Jupiter's atmosphere. These revealed that the relative proportions of hydrogen and helium are almost exactly the same as in the Sun, in line with the accepted picture of how Jupiter formed from the solar nebula (see Section 8-4). But a number of heavy elements, including carbon, nitrogen, and sulfur, were found to be significantly more abundant on Jupiter than in the Sun. This was an expected result. Over the past several billion years, Jupiter's strong gravity should have pulled in many pieces of interplanetary debris, thus building up a small but appreciable abundance of heavy elements.

Water is a major constituent of small bodies in the outer solar system and thus should likewise be present inside Jupiter in noticeable quantities. Surprisingly, the amount of water actually detected by the *Galileo Probe* was less than half the expected amount. One proposed explanation for this disparity is that the hot spot where the probe entered the atmosphere was also unusually dry. If this explanation is correct, Jupiter's hot spots are analogous to hot, dry deserts on Earth. An alternative idea is that more water than expected has rained out of Jupiter's upper atmosphere and coalesced deep within the planet. For now, the enigma of Jupiter's missing water remains unresolved.

Another surprising and potentially important result from the *Galileo Probe* concerns three elements, argon (Ar), krypton (Kr), and xenon (Xe). These elements are three of the so-called **noble gases**, which do not combine with other atoms to form molecules. (The noble gases, which also include helium, appear in the rightmost column of the periodic table in Box 5-5.) All three of these elements appear in tiny amounts in the Earth's atmosphere and in the Sun, and so must also have been present in the solar nebula. But the *Galileo Probe* found that argon, krypton, and xenon are about 3 times as abundant in Jupiter's atmosphere as in the Sun's.

If these elements were incorporated into Jupiter directly from the gases of the solar nebula, they should be equally as abundant in Jupiter as in the Sun (which formed from the gas at the center of the solar nebula; see Section 8-3). So, the excess amounts of argon, krypton, and xenon must have entered Jupiter in the form of solid planetesimals, just as did carbon, nitrogen, and sulfur. But at Jupiter's distance from the Sun, the presumed temperature of the solar nebula was too high to permit argon, krypton, or xenon to solidify. How, then, did the excess amounts of these noble gases become part of Jupiter?

Several hypotheses have been proposed to explain these results. One idea is that icy bodies, including solid argon, krypton, and xenon, formed in an interstellar cloud even before the cloud collapsed to form the solar nebula. A second notion is that the solar nebula may actually have been colder than previously thought. If either of these ideas proves to be correct, it would mean that Jovian planets can form closer to their stars than had heretofore been thought. This would help to explain why astronomers find massive, presumably Jovian planets orbiting close to stars other than the Sun (see Section 8-5).

A third hypothesis proposes that our models of the solar nebula are correct, but that Jupiter originally formed much farther from the

Sun. Out in the cold, remote reaches of the solar nebula, argon, krypton, and xenon could have solidified into icy particles and fallen into the young Jupiter, thus enriching it in those elements. Interactions between Jupiter and the material of the solar nebula could then have forced the planet to spiral inward, eventually reaching its present distance from the Sun. As we saw in Section 8-5, a model of this kind has been proposed to explain why extrasolar planets orbit so close to their stars. More research will be needed to decide which of these hypotheses is closest to the truth.

The story of the *Galileo Probe* exemplifies two of the most remarkable aspects of scientific research. First, when new scientific observations are made, the results often pose many new questions at the same time that they provide answers to some old ones. And second, it sometimes happens that investigating what might seem to be trivial details—such as the abundances of trace elements in Jupiter's atmosphere—can reveal clues about truly grand issues, such as how the planets formed.

14-6 | The oblateness of Jupiter and Saturn reveals their rocky cores

The *Galileo Probe* penetrated only a few hundred kilometers into Jupiter's atmosphere before being crushed. To learn about Jupiter's structure at greater depths as well as that of Saturn, astronomers must use indirect clues. The shapes of Jupiter and Saturn are important clues, since they indicate the size of the rocky cores at the centers of the planets.

Even a casual glance through a small telescope shows that Jupiter and Saturn are not spherical but slightly flattened or **oblate**. (You can see this in Figures 14-1 and 14-2.) The diameter across Jupiter's equator (142,980 km) is 6.5% larger than its diameter from pole to pole (133,700 km). Thus, Jupiter is said to have an **oblateness** of 6.5%, or 0.065. Saturn has an even larger oblateness of 9.8%, or 0.098, making it the most oblate of all the planets. By comparison, the oblateness of the Earth is just 0.34%, or 0.0034.

If Jupiter and Saturn did not rotate, both would be perfect spheres. A massive, nonrotating object naturally settles into a spherical shape so that every atom on its surface experiences the same gravitational pull aimed directly at the object's center. Because Jupiter and Saturn do rotate, however, the body of each planet tends to fly outward and away from the axis of rotation.

ANALOGY You can demonstrate this effect for yourself. Stand with your arms hanging limp at your sides, then spin yourself around. Your arms will naturally tend to fly outward, away from the vertical axis of rotation of your body. Likewise, if you drive your car around a circular road, you feel thrown toward the outside of the circle; the car as a whole is rotating around the center of the circle, and you tend to move away from the rotation axis. (We discussed the physical principles behind this in Box 4-3.) For Jupiter and Saturn, this same effect deforms the planets into their nonspherical, oblate shapes.

The oblateness of a planet depends both on its rotation rate and on how the planet's mass is distributed over its volume. The challenge to planetary scientists is to create a model of a planet's mass

distribution that gives the right value for the oblateness. One such model for Jupiter suggests that 2.6% of its mass is concentrated in a dense, rocky core (Figure 14-10a). Although this core has about 8 times the mass of the Earth, the crushing weight of the remaining bulk of Jupiter compresses it to a diameter of just 11,000 km, slightly smaller than the Earth's diameter of 12,800 km. The pressure at the center of the core is estimated to be about 70 million atmospheres, and the central temperature is probably about 22,000 K. By contrast, the temperature at Jupiter's cloudtops is only 165 K.

If Jupiter formed by accretion of gases onto a rocky "seed," the present-day core is presumably that very seed. However, additional amounts of rocky material were presumably added later by meteoritic material that fell into the planet and sank to the center. (If Jupiter formed directly from the gases of the solar nebula, this additional material makes up the present-day core.) Material from icy planetesimals, too, would have sunk deep within Jupiter and added to the core. In the model we have been describing, these "ices"—

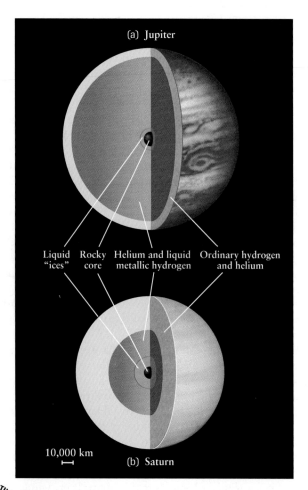

Figure 14-10
The Internal Structures of Jupiter and Saturn These diagrams of the interiors of Jupiter and Saturn are drawn to the same scale. Each planet's rocky core is surrounded by an outer core of liquid "ices," a layer of helium and liquid metallic hydrogen, and a layer of helium and ordinary molecular hydrogen (H₂). Saturn's rocky core contains a larger fraction of the planet's mass than does Jupiter's, while Saturn has a smaller volume of liquid metallic hydrogen than does Jupiter.

principally water (H_2O), methane (CH_4), ammonia (NH_3), and other molecules made by chemical reactions among these substances—form a layer some 3000 km thick that surrounds the rocky core. (Because these substances are less dense than rock, they "float" on top of the rocky core.) Despite the high pressures in this layer, the temperature is so high that the "ices" are probably in a liquid state.

Saturn rotates at about the same rate as Jupiter but has less mass, and therefore less gravity to pull its material inward. Hence, Saturn's rotation should cause material at its equator to bulge outward more than on Jupiter, giving Saturn a greater oblateness. In fact, if Saturn and Jupiter had the same internal structure, we would actually expect Saturn to be even more oblate than it really is. We can account for this discrepancy if Saturn has a different mass distribution than Jupiter. Detailed calculations suggest that about 10% of Saturn's mass is contained in its rocky core, compared to 2.6% for Jupiter's rocky core (see Figure 14-10b). As for Jupiter, the rocky core of Saturn is probably surrounded by an outer core of liquid "ices."

14-7 | Metallic hydrogen inside Jupiter and Saturn endows the planets with strong magnetic fields

The oblateness of Jupiter and Saturn gives information about the size of the planets' rocky cores. It does not, however, provide much insight into the planets' bulk between their cores and their colorful cloudtops. Most of our knowledge of these parts of the interiors of Jupiter and Saturn comes from observations with radio telescopes. By using these telescopes to detect radio waves from Jupiter and Saturn, astronomers have found evidence of electric currents flowing within both planets' hydrogen-rich interiors.

Astronomers began discovering radio emissions coming from Jupiter in the 1950s. A small portion of these emissions is **thermal radiation,** with the sort of spectrum that would be expected from an opaque object due to its temperature. (We discussed the electromagnetic radiation from heated objects in Sections 5-3 and 5-4.) However, most of the radio energy emitted from Jupiter is **nonthermal radiation.** This has a very different sort of spectrum from radiation from a heated object and is found in two distinct wavelength ranges.

At wavelengths of a few meters, Jupiter emits sporadic bursts of radio waves. This **decametric** (10-meter) **radiation** is probably caused by electrical discharges associated with powerful electric currents in Jupiter's ionosphere. These discharges result from electromagnetic interactions between Jupiter and its large satellite Io. At shorter wavelengths of a few tenths of a meter, Jupiter emits a nearly constant stream of electromagnetic radiation. The properties of this **decimetric** (tenth-meter) **radiation** show that it is emitted by electrons moving through a strong magnetic field at speeds comparable to the speed of light. The magnetic field deflects the electrons into spiral-shaped trajectories, and the spiraling electrons radiate energy like miniature radio antennas. Energy produced in this fashion is called **synchrotron radiation.** Astronomers now realize that synchrotron radiation is very important in a wide range of situations throughout the universe, from pulsars and quasars to the radio emissions of entire galaxies.

If some of Jupiter's radiation is indeed synchrotron radiation, then Jupiter must have a magnetic field. Measurements by the *Pioneer* and *Voyager* spacecraft verified this and indicated that the field

at Jupiter's equator is 14 times stronger than at the Earth's equator. Like the Earth's magnetic field, Jupiter's field is thought to be generated by motions of an electrically conducting fluid in the planet's interior. But wholly unlike Earth, the moving fluid within Jupiter is an exotic form of hydrogen called **liquid metallic hydrogen.**

As we discussed in Section 7-7, hydrogen only becomes a liquid metal when the pressure exceeds about 1.4 million atmospheres, which occurs at depths more than about 7000 km below Jupiter's cloudtops. Thus, the internal structure of Jupiter consists of four distinct regions, as shown in Figure 14-10a: a rocky core, a roughly 3000-km thick layer of liquid "ices," a layer of helium and liquid metallic hydrogen about 56,000 km thick, and a layer of helium and ordinary hydrogen about 7000 km thick. The colorful cloud patterns that we can see through telescopes are located in the outermost 100 km of the outer layer.

Figure 14-10a shows that much of the enormous bulk of Jupiter is composed of liquid metallic hydrogen. Jupiter's rapid rotation sets the liquid metallic hydrogen into motion, giving rise to a powerful dynamo 20,000 times stronger than Earth's and generating an intense magnetic field.

Saturn also has a substantial magnetic field, but it pales by Jupiter standards: Saturn's internal dynamo is only 3% as strong as Jupiter's. (To be fair, it is still 600 times stronger than Earth's dynamo.) This difference between the planets suggests that even though Saturn rotates nearly as rapidly as Jupiter, there must be far less liquid magnetic hydrogen within Saturn to be stirred up by the rotation and thereby generate a magnetic field. This makes sense: Compared to Jupiter, Saturn has less mass, less gravity, and less internal pressure to compress ordinary hydrogen into liquid metallic hydrogen. As a consequence, as Figure 10-14b shows, liquid metallic hydrogen probably makes up only a relatively small fraction of Saturn's bulk.

14-8 | Jupiter and Saturn have extensive magnetospheres

Jupiter's strong magnetic field surrounds the planet with a magnetosphere so large that it envelops the orbits of many of its moons. The *Pioneer* and *Voyager* spacecraft, which carried instruments to detect charged particles and magnetic fields, revealed the awesome dimensions of the Jovian magnetosphere. The shock wave that surrounds Jupiter's magnetosphere is nearly 30 million kilometers across. If you could see Jupiter's magnetosphere from the Earth, it would cover an area in the sky 16 times larger than the Moon. Figure 14-11 shows the structure of Jupiter's magnetosphere (compare Figure 9-19, which shows the Earth's much smaller magnetosphere).

The *Pioneer* and *Voyager* spacecraft reported the outer boundary of Jupiter's magnetosphere (on the left in Figure 14-11) at distances ranging from 3 to 7 million kilometers (2 to 4 million miles) above the planet. The "downstream" side of the magnetosphere (on the right in Figure 14-11) extends for more than a billion kilometers (650 million miles), even beyond the orbit of Saturn.

The inner regions of the Earth's magnetosphere are dominated by two huge Van Allen belts (recall Figure 9-19) that are filled with charged particles. In a similar manner, Jupiter entraps immense quantities of charged particles in belts of its own. An astronaut venturing into these belts would quickly receive a lethal dose of radiation. In addition, Jupiter's rapid rotation spews the charged particles

figure 14-11

Jupiter's Magnetosphere The shock wave marks where the supersonic solar wind is abruptly slowed to subsonic speeds. Most solar wind particles are deflected around Jupiter in the turbulent region (shown in purple) between the shock wave and magnetopause (where the pressure within the magnetosphere balances the pressure of the solar wind). Particles trapped inside Jupiter's magnetosphere are spread out into a vast current sheet by the planet's rapid rotation.

out into a huge **current sheet,** which lies close to the plane of Jupiter's magnetic equator (see Figure 14-11). Jupiter's magnetic axis is inclined 11° from the planet's axis of rotation, and the orientation of Jupiter's magnetic field is the reverse of the Earth's magnetic field—a compass would point toward the south on Jupiter.

Figure 14-12 shows radio emissions from charged particles in the densest regions of Jupiter's magnetosphere. These emissions vary slightly with a period of 9 hours, 55 minutes, and 30 seconds, as Jupiter's rotation changes the angle at which we view its magnetic field. Because the magnetic field is anchored deep within the planet, this variation reveals Jupiter's **internal rotation period.** This period, indicative of the rotation of the bulk of the planet's mass, is slightly slower than the atmospheric rotation at the equator (period 9 hours, 50 minutes, 28 seconds) but about the same as the atmospheric rotation at the poles (period 9 hours, 55 minutes, 41 seconds). The faster motion of the atmosphere at the equator relative to the interior is driven by Jupiter's internal heat.

The *Voyager* spacecraft discovered that the inner regions of Jupiter's magnetosphere contain a hot, gaslike mixture of charged particles called a **plasma.** A plasma is formed when a gas is heated to such extremely high temperatures that electrons are torn off the atoms of the gas. As a result, a plasma is a mixture of positively charged ions and negatively charged electrons. The plasma that envelops Jupiter consists primarily of electrons and protons, with some ions of helium, sulfur, and oxygen. A major source of these particles is Jupiter's volcanically active moon Io (see Table 7-2), which lies deep within the magnetosphere. Io ejects a ton of material into Jupiter's magnetosphere each second from its surface and volcanic plumes. These charged particles are caught up in Jupiter's rapidly rotating magnetic field and are accelerated to high speeds.

The fast-moving particles in Jupiter's plasma exert a substantial pressure that holds off the solar wind. The *Voyager* data suggest that the pressure balance between the solar wind and the hot plasma inside the Jovian magnetosphere is precarious. A gust in the solar wind can blow away some of the plasma, at which point the magnetosphere deflates rapidly to as little as one-half its original size. However, additional electrons and ions accelerated by Jupiter's rotating magnetic field soon replenish the plasma, and the magnetosphere expands again.

Saturn's magnetosphere is similar to Jupiter's, but is only about 10 to 20% as large and contains far fewer charged particles than

Jupiter's. There are two reasons for this deficiency. First, Saturn lacks a volcanic moon like Io to inject particles into the magnetosphere. Second, and more important, many charged particles are absorbed by the icy particles that make up Saturn's rings (see Section 14-10). The charged particles that do exist in Saturn's magnetosphere are concentrated in radiation belts similar to the Van Allen belts in the Earth's magnetosphere.

On Earth, particles trapped in our planet's magnetic field stream onto the north and south magnetic poles, creating the aurora (see Section 9-4). The same effect takes place on both Jupiter and Saturn. Images from the Hubble Space Telescope show glowing auroral rings centered on the north and south magnetic poles of Jupiter (Figure 14-13*a*) and of Saturn (Figure 14-13*b*). As on Earth, the emission comes from a region high in each planet's atmosphere, hundreds of kilometers above the cloudtops.

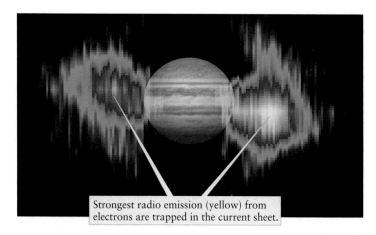

Strongest radio emission (yellow) from electrons are trapped in the current sheet.

figure 14-12 **R** **V** U X G

A Radio View of Jupiter Instruments on board the *Cassini* spacecraft recorded this false-color map of decimetric emission from Jupiter at a wavelength of 2.2 cm. The emission comes from a region elongated parallel to the planet's magnetic equator, which is tilted relative to its geographic equator. A visible-light image of Jupiter is superimposed on the radio image. (NASA/JPL)

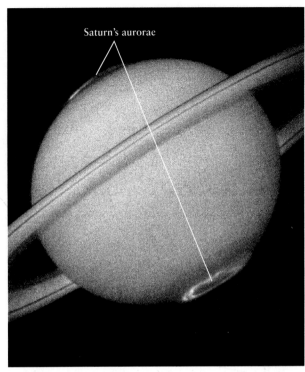

(a) Jupiter　R I **V** **U** X G

(b) Saturn　R I V **U** X G

figure 14-13

WEB LINK 14.5

Aurorae on Jupiter and Saturn On both Jupiter and Saturn, charged particles from the magnetosphere are funneled onto the planet's magnetic poles. When the particles collide with and excite molecules in the upper atmosphere, the molecules emit ultraviolet light. (The ultraviolet images in (a) are superimposed on a visible-light image of Jupiter.)

(a: J. Clarke, University of Michigan, and NASA; b: J. Trauger, JPL, and NASA)

14-9 | Earth-based observations reveal three broad rings encircling Saturn

Compared to Saturn, Jupiter is unquestionably larger, more massive, more colorful, and more dynamic in both its atmosphere and magnetosphere. But as seen through even a small telescope, Saturn stands out as perhaps the most beautiful of all the worlds of the solar system, thanks to its majestic rings (**Figure 14-14**). As we will see in Section 14-10, Jupiter also has rings, but they are so faint that they were not discovered until spacecraft could observe Jupiter at close range.

In 1610, Galileo became the first person to see Saturn through a telescope. He saw few details, but he did notice two puzzling lumps protruding from opposite edges of the planet's disk. Curiously, these lumps disappeared in 1612, only to reappear in 1613. Other observers saw similar appearances and disappearances over the next several decades.

In 1655, the Dutch astronomer Christiaan Huygens began to observe Saturn with a better telescope than was available to any of his predecessors. (We discussed in Section 13-2 how Huygens used this same telescope to observe Mars.) On the basis of his observations, Huygens suggested that Saturn was surrounded by a thin, flattened ring. At times this ring was edge-on as viewed from the Earth, making it almost impossible to see. At other times Earth observers

viewed Saturn from an angle either above or below the plane of the ring, and the ring was visible, as in Figure 14-1. (The lumps that Galileo saw were the parts of the ring to either side of Saturn, blurred by the poor resolution of his rather small telescope.) Astronomers confirmed this brilliant deduction over the next several years as they watched the ring's appearance change just as Huygens had predicted.

As the quality of telescopes improved, astronomers realized that Saturn's "ring" is actually a *system* of rings, as Figure 14-14 shows. In 1675, the Italian astronomer Gian Domenico Cassini (whom we also met in Section 13-2) discovered a dark division in the ring. The **Cassini division** is an apparent gap about 4500 kilometers wide that separates the outer **A ring** from the brighter **B ring** closer to the planet. In the mid-1800s, astronomers managed to identify the faint **C ring**, or *crepe* ring, that lies just inside the B ring.

Saturn and its rings are best seen when the planet is at or near opposition (see Table 14-4). A modest telescope gives a good view of the A and B rings, but a large telescope and excellent observing conditions are needed to see the C ring.

Earth-based views of the Saturnian ring system change as Saturn slowly orbits the Sun. Huygens was the first to understand that this change occurs because the rings lie in the plane of Saturn's equator, and this plane is tilted 27° from the plane of Saturn's orbit. As Saturn orbits the Sun, its rotation axis and the plane of its equator maintain the same orientation in space. (The same is

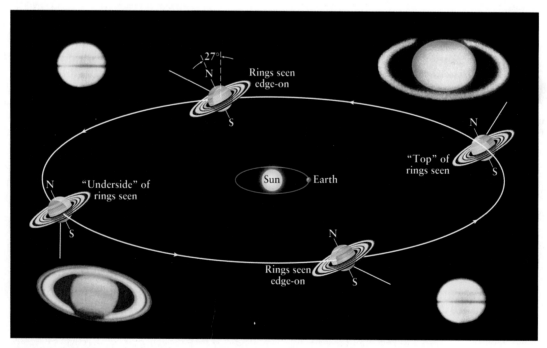

figure 14-14   R I **V** U X G

Saturn's System of Rings This *Voyager 1* image shows many details of Saturn's rings. Saturn's equatorial diameter is labeled, as are the diameters of the inner and outer edges of the rings. The C ring is so faint that it is almost invisible in this view. Saturn is visible through the rings (look near the bottom of the image), which shows that the rings are not solid. (NASA/JPL)

true for the Earth; see Figure 2-12.) Hence, over the course of a Saturnian year, an Earth-based observer views the rings from various angles (**Figure 14-15**). At certain times Saturn's north pole is tilted toward the Earth and the observer looks "down" on the "top side" of the rings. Half a Saturnian year later, Saturn's south pole is tilted toward us and the "underside" of the rings is exposed to our Earth-based view.

When our line of sight to Saturn is in the plane of the rings, the rings are viewed edge-on and seem to disappear entirely (see the images at the upper left and lower right corners of Figure 14-15). This disappearance indicates that the rings are very thin. In fact, they are thought to be only a few tens of meters thick. In proportion to their diameter, Saturn's rings are thousands of times thinner than the sheets of paper used to print this book.

ANIMATION 14.1

figure 14-15   R I **V** U X G
The Changing Appearance of Saturn's Rings
Saturn's rings are aligned with its equator, which is tilted 27° from Saturn's orbital plane. As Saturn moves around its orbit, the rings maintain the same orientation in space, so observers on Earth see the rings at various angles.

The accompanying Earth-based photographs show Saturn at various points in its orbit. Note that the rings seem to disappear entirely when viewed edge-on, which occurs about every 15 years. (Lowell Observatory)

The last edge-on presentation of Saturn's rings was in 1995–1996, and the next will occur in 2008–2009. Until 2008, Earth observers will see the "underside" of the rings (as in the image at the lower left corner of Figure 14-15).

14-10 | Saturn's rings are composed of numerous icy fragments, while Jupiter's rings are made of small rocky particles

WEB LINK 14.6 Astronomers have long known that Saturn's rings could not possibly be solid sheets of matter. In 1857, the Scottish physicist James Clerk Maxwell (whom we last encountered in Section 5-2) proved theoretically that if the rings were solid, differences in Saturn's gravitational pull on different parts of the rings would cause the rings to tear apart. He concluded that Saturn's rings are made of "an indefinite number of unconnected particles."

In 1895, James Keeler at the Allegheny Observatory in Pittsburgh became the first to confirm by observation that the rings are not rigid. He did this by photographing the spectrum of sunlight reflected from Saturn's rings. As the rings orbit Saturn, the spectral lines from the side approaching us are blueshifted by the Doppler effect (recall Section 5-9, especially Figure 5-24). At the same time, spectral lines from the receding side of the rings are redshifted. Keeler noted that the size of the wavelength shift increased inward across the rings—the closer to the planet, the greater the shift. This variation in Doppler shift proved that the inner portions of Saturn's rings are moving around the planet more rapidly than the outer portions. Indeed, the orbital speeds across the rings are in complete agreement with Kepler's third law: The square of the orbital period about Saturn at any place in the rings is proportional to the cube of the distance from Saturn's center (see Section 4-7 and Box 4-4). This is exactly what would be expected if the rings consisted of numerous tiny "moonlets," or **ring particles,** each individually circling Saturn.

Saturn's rings are quite bright; they reflect 80% of the sunlight that falls on them. (By comparison, Saturn itself reflects 46% of incoming sunlight.) Astronomers therefore long suspected that the ring particles are made of ice and ice-coated rock. This hunch was confirmed in the 1970s, when the American astronomers Gerard P. Kuiper and Carl Pilcher identified absorption features of frozen water in the rings' near-infrared spectrum. The *Voyager 1* and *Voyager 2* spacecraft made even more detailed infrared measurements as they flew past Saturn in 1980 and 1981, respectively. These indicate that the temperature of the rings ranges from –180°C (–290°F) in the sunshine to less than –200°C (–330°F) in Saturn's shadow. Water ice is in no danger of melting or evaporating at these temperatures.

To determine the sizes of the particles that make up Saturn's rings, astronomers analyzed the radio signals received from the *Voyager* spacecraft as the spacecraft passed behind the rings. How easily radio waves can travel through the rings depends on the relationship between the wavelength and the particle size. The results showed that most of the particles range in size from pebble-sized fragments about 1 cm in diameter to chunks about 5 m across, the size of large boulders. Most abundant are snowball-sized particles about 10 cm in diameter.

All this material may be ancient debris that failed to accrete (fall together) into satellites. The total amount of material in the rings is quite small. If Saturn's entire ring system were compressed together to make a satellite, it would be no more than 100 km (60 mi) in diameter. But, in fact, the ring particles are so close to Saturn that they will never be able to form moons.

To see why, imagine a collection of small particles orbiting a planet. Gravitational attraction between neighboring particles tends to pull the particles together. However, because the various particles are at differing distances from the parent planet, they also experience different amounts of gravitational pull from the planet. This difference in gravitational pull is a **tidal force** that tends to keep the particles separated. (We discussed tidal forces in detail in Section 4-8. You may want to review that section, and in particular Figure 4-23.)

The closer a pair of particles is to the planet, the greater the tidal force that tries to pull the pair apart. At a certain distance from the planet's center, called the **Roche limit,** the disruptive tidal force is just as strong as the gravitational force between the particles. (The concept of this limit was developed in the mid-1800s by the French mathematician Edouard Roche.) Inside the Roche limit, the tidal force overwhelms the gravitational pull between neighboring particles, and these particles cannot accrete to form a larger body. Instead, they tend to spread out into a ring around the planet. Indeed, most of Saturn's system of rings visible in Figure 14-14 lies within the planet's Roche limit.

All large planetary satellites are found outside their planet's Roche limit. If any large satellite were to come inside its planet's Roche limit, the planet's tidal forces would cause the satellite to break up into fragments. (We will see in Chapter 16 that such a catastrophic tidal disruption may be the eventual fate of Neptune's large satellite, Triton.)

CAUTION It may seem that it would be impossible for any object to hold together inside a planet's Roche limit. But the ring particles inside Saturn's Roche limit survive and do not break apart. The reason is that the Roche limit applies only to objects held together by the *gravitational* attraction of each part of the object for the other parts. By contrast, the forces that hold a rock or a ball of ice together are *chemical* bonds between the object's atoms and molecules. These chemical forces are much stronger than the disruptive tidal force of a nearby planet, so the rock or ball of ice does not break apart. In the same way, people walking around on the Earth's surface (which is inside the Earth's Roche limit) are in no danger of coming apart, because we are held together by comparatively strong chemical forces rather than gravity.

When the *Voyager 1* spacecraft flew past Jupiter in 1979, it trained its cameras not just on the planet but also on the space around the planet's equator. It discovered that Jupiter, too, has a system of rings that lies within its Roche limit (**Figure 14-16**). These rings differ from Saturn's in two important ways. First, Jupiter's rings are composed of tiny particles of rock with an average size of only about 1 μm (= 0.001 mm = 10^{-6} m) and that reflect less

Figure 14-16   R I **V** U X G

Jupiter's Main Ring   This *Galileo* image shows Jupiter's main ring almost edge-on. The ring's outer radius, about 129,000 km from the center of Jupiter, is close to the orbit of Adrastea, the second closest (after Metis) of Jupiter's moons. Not shown is an even larger and even more tenuous pair of "gossamer rings," one of which extends out to 181,000 km and the other out to 222,000 km. (NASA/JPL; Cornell University)

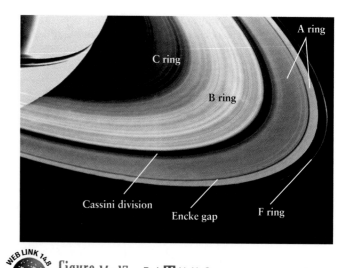

Figure 14-17   R I **V** U X G

Details of Saturn's Rings   The colors in this *Voyager 1* image have been enhanced to emphasize small differences between different portions of the rings. The broad Cassini division is clearly visible, as is the narrow Encke gap in the outer A ring. The very thin F ring lies just beyond the outer edge of the A ring. (NASA/JPL)

than 5% of the sunlight that falls on them. Second, there is very little material in the rings of Jupiter, less than 1/100,000 (10^{-5}) the amount of material in Saturn's rings. As a result, Jupiter's rings are extremely faint, which explains why their presence was first revealed by a spacecraft rather than an Earth-based telescope. The ring particles are thought to originate from meteorite impacts on Jupiter's four small, inner satellites, two of which are visible in Figure 14-16.

We will see in Chapter 16 that the rings of Uranus and Neptune are also made of many individual particles orbiting inside each planet's Roche limit. Like the rings of Jupiter, these rings are quite dim and difficult to see from Earth.

14-11 Saturn's rings consist of thousands of narrow, closely spaced ringlets

The photograph in Figure 14-1*b*, which was made in 1974, indicates what astronomers understood about the structure of Saturn's rings in the mid-1970s. Each of the A, B, and C rings appeared to be rather uniform, with little or no evidence of any internal structure. Within a few years, however, close-up observations from spacecraft revealed the true complexity of the rings, as well as their chemical composition.

Three spacecraft—*Pioneer 11*, *Voyager 1*, and *Voyager 2*—flew past Saturn in 1979, 1980, and 1981, respectively, after each had first made a close flyby of Jupiter. (*Pioneer 10* also flew past Jupiter but did not visit Saturn. Instead, it was aimed so that Jupiter's gravity deflected its path out of the plane of the solar system and into interstellar space.) *Pioneer 11* had only a relatively limited capability to make images, but cameras on board the two *Voyager* spacecraft sent back a number of pictures showing the detailed structure of Saturn's rings.

Figure 14-17 shows one such *Voyager* image. Some of the features seen in this image were expected, including the **Encke gap**, a 270-km-wide division in the outer A ring observed by the German

astronomer Johann Franz Encke in 1838. But to the amazement of scientists, images like the one in Figure 14-17 revealed that the broad A, B, and C rings are not uniform at all but instead consist of hundreds upon hundreds of closely spaced bands or **ringlets**. These ringlets are arrangements of ring particles that have evolved from the combined gravitational forces of neighboring particles, of Saturn's moons, and of the planet itself.

The *Voyager* cameras also sent back high-quality pictures of the narrow **F ring**, which was first detected by *Pioneer 11*. The F ring, which you can see in Figure 14-17, is only about 100 km wide and lies 4000 km beyond the outer edge of the A ring. Close-up views showed that the F ring often consists of several intertwined strands (Figure 14-18). One *Voyager 1* image displayed a total of five strands, each about 10 km across. Planetary scientists suspect that gravitational forces from tiny, unseen moons within the F ring produce this curious and complex structure.

Spacecraft have done more than show Saturn's rings in greater detail; they have also viewed the rings from perspectives not possible from Earth. Through Earth-based telescopes, we can see only the sunlit side of the rings. From this perspective, the B ring appears very bright, the A ring moderately bright, the C ring dim, and the Cassini division dark (see Figure 14-14). The fraction of sunlight reflected back toward the Sun (the *albedo*) is directly related to the concentration and size of the particles in the ring. The B ring is bright because it has a high concentration of relatively large, icy, reflective particles, whereas the darker Cassini division has a lower concentration of such particles.

The *Voyagers* expanded our knowledge of the ring particles by imaging the shaded side of the rings, which cannot be seen from Earth. Figure 14-19 shows such an image. Because the spacecraft was looking back toward the Sun, this picture shows sunlight that has passed through the rings. The B ring looks darkest in this image

figure 14-18 R I **V** U X G

Details of Saturn's F Ring This *Voyager 1* image shows several strands that together constitute Saturn's F ring (compare Figure 14-17). The total width of the F ring is about 100 km. The narrowness of the F ring is the result of gravitational forces exerted by two small satellites (see Section 14-11). (NASA/JPL)

tively small number of particles. (In the same way, dust particles in the air in your room become visible when a shaft of sunlight passes through them.)

The process by which light bounces off particles in its path is called **light scattering**. The proportion of light scattered in various directions depends upon both the size of the particles and the wavelength of the light. (See Box 5-4 for examples of light scattering on Earth.) Because visible light has a much smaller wavelength than radio waves, light scattering allows scientists to measure the sizes of particles too small to detect using the radio technique described in Section 14-10. As an example, by measuring the amount of light scattered from the rings at various wavelengths and from different angles as the two *Voyager* spacecraft sped past Saturn, scientists showed that the F ring contains a substantial number of tiny particles about 1 μm (= 10^{-6} m) in diameter. This is about the same size as the particles found in smoke.

There are also subtle differences in color from one ring to the next, as the computer-enhanced image in Figure 14-20 shows. Although the main chemical constituent of the ring particles is frozen water, trace amounts of other chemicals—perhaps coating the surfaces of the ice particles—are probably responsible for the different colors. These trace chemicals have not yet been identified.

The color variations in Figure 14-20 suggest that the icy particles do not migrate substantially from one ringlet to another. Had such migration taken place, the color differences would have been smeared out over time. The color differences may also indicate that different sorts of material were added to the rings at different times. In this scenario, the rings did not all form at the same time as Saturn but were added to over an extended period. New ring material could have come from small satellites that shattered after being hit by a stray asteroid or comet.

because little sunlight gets through its dense collection of particles. If the Cassini division were completely empty, it would look black, because we would then see through it to the blackness of empty space. But the Cassini division looks *bright* in Figure 14-19. This shows that the Cassini division is not empty but contains a rela-

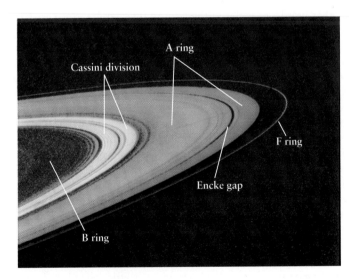

figure 14-19 R I **V** U X G

The View from the Far Side of Saturn's Rings This false-color view of the side of the rings away from the Sun was taken by *Voyager 1*. The Cassini division appears white, not black like the empty gap between the A and F rings. Hence, the Cassini division cannot be empty, but must contain a number of relatively small particles that scatter sunlight like dust motes. (NASA/JPL)

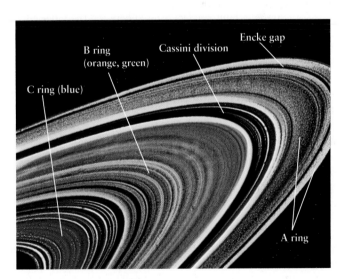

figure 14-20 R I **V** U X G

Color Variations in Saturn's Rings Computer processing has severely exaggerated the subtle color variations in this *Voyager 2* view of the sunlit side of Saturn's rings (actually a composite of visible-light and ultraviolet images). These color variations are indicative of slight differences in chemical composition among particles in different parts of the rings. (NASA/JPL)

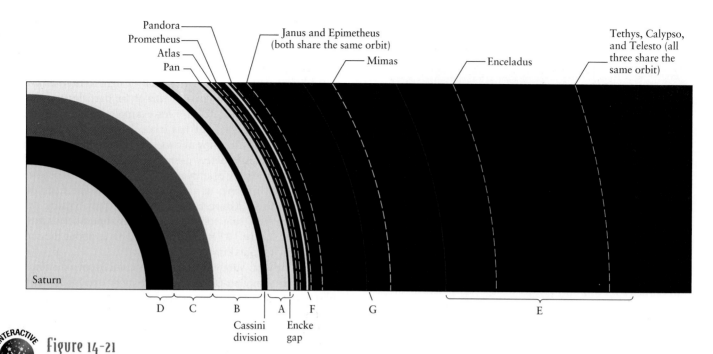

Pandora
Prometheus
Atlas
Pan

Janus and Epimetheus
(both share the same orbit)

Mimas

Enceladus

Tethys, Calypso,
and Telesto (all
three share the
same orbit)

Saturn

D C B A F G E

Cassini
division

Encke
gap

Figure 14-21
The Arrangement of Saturn's Rings This scale drawing shows the location of Saturn's rings along with the orbits of the 11 inner satellites. Only the A, B, and C rings can readily be seen from the Earth. The remaining rings are very faint, and their existence was only confirmed by spacecraft flybys.

In addition to revealing new details about the A, B, C, and F rings, the *Voyager* cameras also discovered three new ring systems: the D, E, and G rings. The drawing in Figure 14-21 shows the layout of all of Saturn's known rings along with the orbits of some of Saturn's satellites. The **D ring** is Saturn's innermost ring system. It consists of a series of extremely faint ringlets located between the inner edge of the C ring and the Saturnian cloudtops. The **E ring** and the **G ring** both lie far from the planet, well beyond the narrow F ring. Both of these outer ring systems are extremely faint, fuzzy, and tenuous. Each lacks the ringlet structure so prominent in the main ring systems. The E ring encloses the orbit of Enceladus, one of Saturn's icy satellites. Some scientists suspect that water geysers on Enceladus are the source of ice particles in the E ring, much as Io's volcanoes supply material to Jupiter's magnetosphere (see Section 14-8).

14-12 | Saturn's inner satellites affect the appearance and structure of its rings

Why do Saturn's rings have such a complex structure? The answer is that, like any object in the universe, the particles that make up the rings are affected by the force of gravity. Saturn's gravitational pull keeps the ring particles in orbit around the planet. But the satellites of Saturn also exert gravitational forces on the rings. These forces can shape the orbits of the ring particles and help give rise to the rings' structure.

Astronomers have long known that one of Saturn's moons, Mimas, has a gravitational effect on its ring system. Mimas is a moderate-sized satellite that orbits Saturn every 22.6 hours. According to Kepler's third law, ring particles in the Cassini division should orbit Saturn approximately every 11.3 hours. Consequently, a given set of ring particles in the Cassini division will find itself between Saturn and Mimas on every second orbit (that is, every 22.6 hours). In other words, particles in the Cassini division are in a 2-to-1 resonance with Mimas. Because of these repeated alignments, the combined gravitational forces of Saturn and Mimas cause small particles to deviate from their original orbits, sweeping them out of the Cassini division. This helps explain why the Cassini division is relatively empty (although not completely so, as Figure 14-19 shows).

While Mimas tends to spread ring particles apart, two other satellites exert gravitational forces that herd particles together. The *Voyager* spacecraft discovered these satellites—both less than 150 km across—orbiting on either side of the narrow F ring (Figure 14-22). Pandora, the outer of the two satellites, moves around Saturn at a slightly slower speed than the particles in the ring. As ring particles pass Pandora, they experience a tiny gravitational tug that slows them down slightly. Hence, these particles lose a little energy and fall into orbits a bit closer to Saturn. At the same time, Prometheus, the inner satellite, orbits the planet somewhat faster than the F ring particles. The gravitational pull of Prometheus makes the ring particles speed up a little, giving them some extra energy and nudging them into slightly higher orbits. The competition between the pulls exerted by these two satellites focuses the F ring particles into a well-defined, narrow band. Because of their confining influence, Prometheus and Pandora are called **shepherd satellites.** We will see in Chapter 16 that Uranus is encircled by narrow rings that are likewise kept confined by shepherd satellites.

figure 14-22 R I **V** U X G

Saturn's F Ring and Its Two Shepherds The tiny satellites Prometheus and Pandora orbit Saturn on either side of its F ring, shepherding the ring particles into a band about 100 km wide. At the instant shown in this *Voyager 2* image, the two satellites were less than 1800 km (1100 mi) apart; Prometheus passed Pandora about 2 hours later. The satellites lap each other every 25 days. (NASA/JPL)

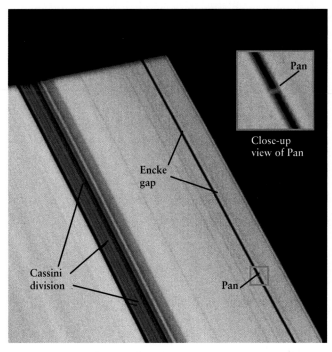

figure 14-23 R I **V** U X G

Pan and the Encke Gap This *Voyager 2* image shows Saturn's tiny moon Pan orbiting within the Encke gap (see also Figure 14-17). Pan's existence was predicted theoretically from the properties of the Encke gap; it was discovered in 1990 by Mark Showalter of NASA after a computer search of 30,000 *Voyager* images made a decade earlier. (Calvin J. Hamilton and NASA/JPL)

The Encke gap in Saturn's A ring, shown in Figure 14-17, actually has a satellite orbiting within it (Figure 14-23). This miniature moon, Pan, is just 20 km (12 mi) in diameter. In a fashion similar to the F ring's shepherds, Pan's gravitational forces cause ring particles outside its orbit to move farther out and cause those inside its orbit to move farther inward toward Saturn. In this way, Pan helps to maintain a gap in the rings that is 270 km (170 mi) wide, many times wider than Pan itself. Gravitational forces from Pan also cause wavelike disturbances in the ring material on either side of the gap. Other small satellites presumably orbit within Saturn's rings and cause small gaps of their own, but their existence has yet to be confirmed.

We should learn much more about the rings of Saturn from the *Cassini* spacecraft, which was scheduled to go into orbit around the ringed planet in July 2004 (see the image that opens Chapter 7). A joint project involving NASA, the European Space Agency, the Italian Space Agency, and scientists from 17 nations, *Cassini* will monitor Saturn's rings and atmosphere for at least four years. The results of this long-term observation program will help scientists better understand this beautiful and complex planet.

Cassini will also make detailed observations of many of Saturn's satellites. Jupiter and Saturn together have an exotic collection of moons, including Saturn's moon Titan, which may have hydrocarbon seas beneath its thick atmosphere; Jupiter's satellite Io, the most volcanic world in the solar system; and Saturn's satellite Tethys, which may have been resurfaced by an exotic "lava" made of water and ammonia. We will examine these alien worlds in Chapter 15.

| Key Words |

Key Ideas

Composition and Structure: Jupiter and Saturn are both much larger than Earth. Each is composed of 71% hydrogen, 24% helium, and 5% all other elements by mass. Both planets have a higher percentage of heavy elements than does the Sun.

• Jupiter probably has a rocky core several times more massive than the Earth. The core is surrounded by a layer of liquid "ices" (water, ammonia, methane, and associated compounds). On top of this is a layer of helium and liquid metallic hydrogen and an outermost layer composed primarily of ordinary hydrogen and helium. All of Jupiter's visible features are near the top of this outermost layer.

• Saturn's internal structure is similar to that of Jupiter, but its core makes up a larger fraction of its volume and its liquid metallic hydrogen mantle is shallower than that of Jupiter.

• Jupiter and Saturn both rotate so rapidly that the planets are noticeably flattened. The rotation of both planets' interiors is revealed by variations in the radio emissions from the planets.

• Both Jupiter and Saturn emit more energy than they receive from the Sun. Presumably both planets are still cooling.

Atmospheres: The visible "surfaces" of Jupiter and Saturn are actually the tops of their clouds. The rapid rotation of the planets twists the clouds into dark belts and light zones that run parallel to the equator. Strong zonal winds run along the belts and zones.

• The outer layers of both planets' atmospheres show differential rotation: The equatorial regions rotate slightly faster than the polar regions. For both Jupiter and Saturn, the polar rotation rate is nearly the same as the internal rotation rate.

• The colored ovals visible in the Jovian atmosphere represent gigantic storms. Some, such as the Great Red Spot, are quite stable and persist for many years. Storms in Saturn's atmosphere seem to be shorter-lived.

• There are presumed to be three cloud layers in the atmospheres of Jupiter and Saturn. The reasons for the distinctive colors of these different layers are not yet known. The cloud layers in Saturn's atmosphere are spread out over a greater range of altitude than those of Jupiter, giving Saturn a more washed-out appearance.

• Saturn's atmosphere contains less helium than Jupiter's atmosphere. This lower abundance may be the result of helium raining downward into the planet. Helium "rainfall" may also account for Saturn's surprisingly strong heat output.

Magnetic Fields and Magnetospheres: Jupiter has a strong magnetic field created by currents in the metallic hydrogen layer. Its huge magnetosphere contains a vast current sheet of electrically charged particles. Charged particles in the densest portions of Jupiter's magnetosphere emit synchrotron radiation at radio wavelengths.

• The Jovian magnetosphere encloses a low-density plasma of charged particles. The magnetosphere exists in a delicate balance between pressures from the plasma and from the solar wind. When this balance is disturbed, the size of the magnetosphere fluctuates drastically.

• Saturn's magnetic field and magnetosphere are much less extensive than Jupiter's.

Rings: Saturn is circled by a system of thin, broad rings lying in the plane of the planet's equator. This system is tilted away from the plane of Saturn's orbit, which causes the rings to be seen at various angles by an Earth-based observer over the course of a Saturnian year.

Structure of the Rings: Saturn has three major, broad rings (A, B, and C) that can be seen from Earth. Other, fainter rings were observed by the *Voyager* spacecraft.

• The principal rings of Saturn are composed of numerous particles of ice and ice-coated rock ranging in size from a few micrometers to about 10 m. Most of the rings exist inside the Roche limit of Saturn, where disruptive tidal forces are stronger than the gravitational forces attracting the ring particles to each other.

• Each of Saturn's major rings is composed of a great many narrow ringlets. The faint F ring, which is just outside the A ring, is kept narrow by the gravitational pull of shepherd satellites.

• Jupiter's faint rings are composed of a relatively small amount of small, dark, rocky particles that reflect very little light.

Review Questions

1. Which planet, Mars or Jupiter, passes closer to the Earth? On which planet is it easier to see details with an Earth-based telescope? Explain your answers.

2. Saturn is the most distant of the planets visible without a telescope. Is there any way we could infer this from naked-eye observations? Explain. (*Hint:* Think about how Saturn's position on the celestial sphere must change over the course of weeks or months.)

3. In what ways are the motions of Jupiter's atmosphere like the motion of water stirred in a pot (see Figure 14-3*b*)? In what ways are they different?

4. Compare Jupiter's chemical composition with that of the Sun. What does this tell us about Jupiter's formation?

5. Astronomers can detect the presence of hydrogen in stars by looking for the characteristic absorption lines of hydrogen in the star's visible spectrum (Figure 5-19). They can also detect hydrogen in glowing gas clouds by looking for hydrogen's characteristic emission lines (Figure 5-16). Explain why neither of these techniques helped Earth-based astronomers to detect hydrogen in Jupiter's atmosphere.

6. Is the chemical composition of Jupiter as a whole the same as that of its atmosphere? Explain any differences.

7. On a warm, humid day, water vapor remains in the atmosphere. But if the temperature drops suddenly, the water vapor forms droplets, clouds appear, and it begins to rain. Relate this observation to why there is relatively little helium in Saturn's atmosphere compared to the atmosphere of Jupiter.

8. What are the belts and zones in the atmospheres of Jupiter and Saturn? Is the Great Red Spot more like a belt or a zone? Explain your answer.

9. Give one possible explanation why weather systems on Jupiter are longer-lived than weather systems on Earth.

10. What are white ovals and brown ovals? What can we infer about them from infrared observations?

11. Compare and contrast the source of energy for motions in the Earth's atmosphere with the energy source for motions in the atmospheres of Jupiter and Saturn.

12. Both Jupiter and Saturn emit more energy than they receive from the Sun in the form of sunlight. Compare the internal energy sources of the two planets that produce this emission.

13. What observations from the *Cassini* spacecraft contradict the accepted picture of convection in Jupiter's atmosphere?

14. Compare the atmospheres of Jupiter and Saturn. Why does Saturn's atmosphere look "washed out" in comparison to that of Jupiter?

15. Which data from the *Galileo Probe* were in agreement with astronomers' predictions? Which data were surprising?

16. Fewer than one in every 10^5 atoms in Jupiter's atmosphere is an argon atom, and fewer than one in 10^8 is an atom of krypton or xenon. If these atoms are so rare, why are scientists concerned about them? How do the abundances of these elements in Jupiter's atmosphere compare to the abundances in the Sun? What hypotheses have been offered to explain these observations?

17. Why is Jupiter oblate? What do astronomers learn from the value of Jupiter's oblateness?

18. What is liquid metallic hydrogen? What is its significance for Jupiter?

19. Describe Jupiter's internal structure, and compare it with the internal structure of the Earth.

20. What data and what techniques have been used to determine the internal structure of Jupiter?

21. Compare the interiors of Jupiter and Saturn. What are the similarities? What are the differences?

22. Explain why Saturn is more oblate than Jupiter, even though Saturn rotates more slowly.

23. If Jupiter does not have any observable solid surface and its atmosphere rotates differentially, how are astronomers able to determine the planet's internal rotation rate?

24. Compare and contrast Jupiter's magnetosphere with the magnetosphere of a terrestrial planet like Earth. Why is the size of the Jovian magnetosphere highly variable, while that of the Earth's magnetosphere is not?

25. What is a plasma? Where are plasmas found in the vicinity of Jupiter?

26. Why is Saturn's magnetosphere less extensive than Jupiter's?

27. When Saturn is at different points in its orbit, we see different aspects of its rings because the planet has a 27° tilt. If the tilt angle were different, would it be possible to see the upper and lower sides of the rings at all points in Saturn's orbit? If so, what would the tilt angle have to be? Explain your answers.

28. What observations of Saturn's rings proved that they are not solid?

29. If Saturn's rings are not solid, why do they look solid when viewed through a telescope?

30. Although the *Voyager* spacecraft did not collect any samples of Saturn's ring particles, measurements from these spacecraft allowed scientists to determine the sizes of the particles. Explain how this was done.

31. The Space Shuttle and other spacecraft orbit the Earth well within the Earth's Roche limit. Explain why these spacecraft are not torn apart by tidal forces.

32. How do Jupiter's rings differ from those of Saturn?

33. Describe the structure of Saturn's rings. What evidence is there that ring particles do not migrate significantly between ringlets?

34. During the planning stages for the *Pioneer 11* mission, when relatively little was known about Saturn's rings, it was proposed to have the spacecraft fly through the Cassini division. Why would this have been a bad idea?

35. What is the relationship between Saturn's satellite Mimas and the Cassini division?

36. Why is the term "shepherd satellite" appropriate for the objects so named? Explain how a shepherd satellite operates.

Advanced Questions

The question preceded by an asterisk (*) involves topics discussed in Box 7-2.

Problem-solving tips and tools

Box 4-4 describes how to use Newton's form of Kepler's third law. Newton's universal law of gravitation, discussed in Section 4-7, is the basic equation from which you can calculate a planet's surface gravity. Box 7-2 discusses escape speed, and Sections 5-3 and 5-4 discuss the properties of thermal radiation (including the Stefan-Boltzmann law, which relates the temperature of a body to the amount of thermal radiation that it emits).

---

37. The angular diameter of Jupiter at opposition varies little from one opposition to the next (see Table 14-3). By contrast, the angular diameter of Mars at opposition is quite variable (see Table 13-2). Explain why there is a difference between these two planets.

38. Jupiter was at opposition on March 4, 2004. On that date Jupiter appeared to be in the constellation Leo. Approximately when will Jupiter next be at opposition in this same region of the celestial sphere? Explain your answer.

39. Jupiter's equatorial diameter and the rotation period at Jupiter's equator are both given in Table 14-1. Use these data to calculate the speed at which an object at the cloudtops along Jupiter's equator moves around the center of the planet.

40. Using orbital data for a Jovian satellite of your choice (see Appendix 3), calculate the mass of Jupiter. How does your answer compare with the mass quoted in Table 14-1?

41. It has been claimed that Saturn would float if one had a large enough bathtub. Using the mass and size of Saturn given in Table 14-2, confirm that the planet's average density is about 690 kg/m^3, and comment on this somewhat fanciful claim.

42. Roughly speaking, Jupiter's composition (by mass) is three-quarters hydrogen and one-quarter helium. The mass of a single hydrogen atom is given in Appendix 7; the mass of a single helium atom is about 4 times greater. Use these numbers to calculate how many hydrogen atoms and how many helium atoms there are in Jupiter.

43. Use the information given in Section 14-3 to estimate the wind velocities in the Great Red Spot, which rotates with a period of about six days.

44. If Jupiter emitted just as much energy per second (as infrared radiation) as it receives from the Sun, the average temperature of the planet's cloudtops would be about 107 K. Given that Jupiter actually emits twice this much energy per second, calculate what the average temperature must actually be.

45. An astronaut floating above Saturn's cloudtops would see a blue sky, even though Saturn's atmosphere has a very different chemical composition than Earth's. Explain why. (*Hint:* See Box 5-4.)

46. (a) From Figure 14-8a, by how much does the temperature increase as you descend from the 100-millibar level in Jupiter's atmosphere to an altitude 100 km below that level? (b) Use Figure 14-8b to answer the same question for Saturn's atmosphere. (c) In the Earth's troposphere (see Section 9-5), the air temperature increases by 6.4°C for each kilometer that you descend. In which planet's atmosphere—Earth, Jupiter, or Saturn—does the temperature increase most rapidly with decreasing altitude?

***47.** Consider a hypothetical future spacecraft that would float, suspended from a balloon, for extended periods in Jupiter's upper atmosphere. If we want this spacecraft to return to Earth after completing its mission, calculate the speed at which the spacecraft's rocket motor would have to accelerate it in order to escape Jupiter's gravitational pull. Compare with the escape speed from the Earth, equal to 11.2 km/s.

48. The *Galileo Probe* had a mass of 339 kg. On Earth, its weight (the gravitational force exerted on it by Earth) was 3320 newtons, or 747 lb. What was the gravitational force that Jupiter exerted on the *Galileo Probe* when it entered Jupiter's clouds?

49. From the information given in Section 14-6, calculate the average density of Jupiter's rocky core. How does this compare with the average density of the Earth? With the average density of the Earth's solid inner core? (See Table 9-1 and Table 9-2 for data about the Earth.)

50. In the outermost part of Jupiter's outer layer (shown in yellow in the upper part of Figure 14-10), hydrogen is principally in the form of molecules (H$_2$). Deep within the liquid metallic hydrogen layer (shown in orange in Figure 14-10), hydrogen is in the form of single atoms. Recent laboratory experiments suggest that there is a gradual transition between these two states, and that the transition layer overlaps the boundary between the ordinary hydrogen and liquid metallic hydrogen layers. Use this information to redraw the upper part of Figure 14-10 and to label the following regions in Jupiter's interior: (i) ordinary (nonmetallic) hydrogen molecules; (ii) nonmetallic hydrogen with a mixture of atoms and molecules; (iii) liquid metallic hydrogen with a mixture of atoms and molecules; (iv) liquid metallic hydrogen atoms.

51. As seen from Earth, the intervals between successive edge-on presentations of Saturn's rings alternate between about 13 years, 9 months, and about 15 years, 9 months. Why do you think these two intervals are not equal?

52. (a) Use Newton's form of Kepler's third law to calculate the orbital periods of particles at the outer edge of Saturn's A ring and at the inner edge of the B ring. (b) Saturn's rings orbit in the same direction as Saturn's rotation. If you were floating along with the cloudtops at Saturn's equator, would the outer edge of the A ring and the inner edge of the B ring appear to move in the same or opposite directions? Explain.

53. This *Voyager 2* close-up image of Saturn's rings shows a number of dark, straight features called *spokes*. As these features orbit around Saturn, they tend to retain their shape like the rigid spokes on a rotating bicycle wheel. (The black dots were added by the *Voyager* camera system to help scientists calibrate the electronic image.) The spokes rotate at the

(NASA/JPL)   R I **V** U X G

same rate as Saturn's magnetic field and are thought to be clouds of tiny, electrically charged particles that are kept in orbit by magnetic forces. Explain why the spokes could *not* maintain their shape if they were kept in orbit by gravitational forces alone.

54. The *Cassini* division involves a 2-to-1 resonance with Mimas. Does the location of the Encke gap—133,500 km from Saturn's center—correspond to a resonance with one of the other satellites? If so, which one? (See Appendix 3.)

Discussion Questions

55. Describe some of the semipermanent features in Jupiter's atmosphere. Compare and contrast these long-lived features with some of the transient phenomena seen in Jupiter's clouds.

56. Suppose you were asked to design a mission to Jupiter involving an unmanned airplanelike vehicle that would spend many days (months?) flying through the Jovian clouds. What observations, measurements, and analyses should this aircraft be prepared to make? What dangers might the aircraft encounter, and what design problems would you have to overcome?

57. What sort of experiment or space mission would you design in order to establish definitively whether Jupiter has a rocky core?

58. The classic science-fiction films *2001: A Space Odyssey* and *2010: The Year We Make Contact* both involve manned spacecraft in orbit around Jupiter. What kinds of observations could humans make on such a mission that cannot be made by robotic spacecraft? What would be the risks associated with such a mission? Do you think that a manned Jupiter mission would be as worthwhile as a manned mission to Mars? Explain your answers.

59. Suppose that Saturn were somehow moved to an orbit around the Sun with semimajor axis 1 AU, the same as the Earth's. Discuss what long-term effects this would have on the planet and its rings.

 ## Web/CD-ROM Questions

60. On Jupiter, the noble gases argon, krypton, and xenon provide important clues about Jupiter's past. On Earth, xenon is used in electronic strobe lamps because it emits a very white light when excited by an electric current. Argon, by contrast, is one of the gases used to fill ordinary incandescent lightbulbs. Search the World Wide Web for information about why argon is used in this way. Why do premium, long-life lightbulbs use krypton rather than argon?

 61. Moving Weather Systems on Jupiter. Access and view the video "The Great Red Spot" in Chapter 14 of the *Universe* Web site or CD-ROM. (**a**) Near the bottom of the video window you will see a white oval moving from left to right. By stepping through the video one frame at a time, estimate how long it takes this oval to move a distance equal to its horizontal dimension. (*Hint:* You can keep track of time by noticing

how many frames it takes a feature in the Great Red Spot, at the center of the video window, to move in a complete circle around the center of the spot. The actual time to move in such a circle is about six days.) (**b**) The horizontal dimension of the white oval is about 4000 km. At what approximate speed (in kilometers per hour) does the white oval move?

62. Search the World Wide Web, especially the Web sites for NASA's Jet Propulsion Laboratory and the European Space Agency, for information about the current status of the *Cassini* mission. What recent discoveries has *Cassini* made? What ideas are being considered for the *Cassini* extended mission, to begin in 2008?

63. The two *Voyager* spacecraft were launched from Earth along a trajectory that took them directly to Jupiter. The force of Jupiter's gravity then gave the two spacecraft a "kick" that helped push them onward to Saturn. The much larger *Cassini* spacecraft, by contrast, was first launched on a trajectory that took it past Venus. Search the World Wide Web, especially the Web sites for NASA's Jet Propulsion Laboratory and the European Space Agency, for information about the trajectory that *Cassini* is taking through the solar system. Explain why this trajectory is so different than that of the *Voyagers*.

 64. The Rotation Rate of Saturn. Access and view the video "Saturn from the Hubble Space Telescope" in Chapter 14 of the *Universe* Web site or CD-ROM. The total amount of time that actually elapses in this video is 42.6 hours. Using this information, identify and follow an atmospheric feature and determine the rotation period of Saturn. How does your answer compare with the value given in Table 14-2?

Observing Projects

 Observing tips and tools

Like Mars, Jupiter and Saturn are most easily seen around opposition. The dates of Jupiter's and Saturn's oppositions are listed in Tables 14-3 and 14-4. Each planet is visible in the night sky for several months before or after its opposition. At other times, Jupiter and Saturn may be visible in either the predawn morning sky or the early evening sky. Consult such magazines as *Sky & Telescope* and *Astronomy* or their Web sites for more detailed information about when and where to look for Jupiter during a given month. You can also use the *Starry Night Backyard*™ program on the CD-ROM that accompanies this textbook. A relatively small telescope with an objective diameter of 15 to 20 cm (6 to 8 inches), used with a medium-power eyepiece to give a magnification of 25X or so, should enable you to see some of the dark belts on Jupiter and the rings of Saturn.

65. If Jupiter is visible in the night sky, make arrangements to view the planet through a telescope. What magnifying power seems to give you the best view? Draw a picture of what you see. Can you see any belts and zones? How many? Can you see the Great Red Spot?

66. Make arrangements to view Jupiter's Great Red Spot through a telescope. Consult the *Sky & Telescope* Web site, which lists the times when the center of the Great Red Spot passes across Jupiter's central meridian. The Great Red Spot is well placed for viewing for 50 minutes before and after this happens. You will need a refractor with an objective of at least 15 cm (6 in.) diameter or a reflector with an objective of at least 20 cm (8 in.) diameter. Using a pale blue or green filter can increase the color contrast and make the spot more visible. For other useful hints, see the article "Tracking Jupiter's Great Red Spot" by Alan MacRobert (*Sky & Telescope*, September 1997).

67. View Saturn through a small telescope. Make a sketch of what you see. Estimate the angle at which the rings are tilted to your line of sight. Can you see the Cassini division? Can you see any belts or zones in Saturn's clouds? Is there a faint, starlike object near Saturn that might be Saturn's satellite Titan? What observations could you perform to test whether the starlike object is a Saturnian satellite?

 68. Use the *Starry Night Backyard*™ program to observe the appearance of Jupiter. Display the entire celestial sphere (select **Guides > Atlas** in the **Go** menu) and center on Jupiter by using the **Find...** command in the **Edit** menu. Using the controls at the right-hand end of the Control Panel, zoom in as far as you can. In the Control Panel, set the date and time to March 4, 2004, at 12:00:00 A.M. to see Jupiter at opposition, and set the time step to 1 minute. Then click on the "Forward" button (a triangle that points to the right). Step through enough time to determine the rotation period of the planet. (*Hint:* You may want to track the motion of an easily recognizable feature, such as the Great Red Spot.) How does your answer compare with the rotation period given in Table 14-1?

 69. Use the *Starry Night Backyard*™ program to observe the appearance of Saturn. Display the entire celestial sphere (select **Guides > Atlas** in the **Go** menu) and center on Saturn by using the **Find...** command in the **Edit** menu. Using the controls at the right-hand end of the Control Panel,

zoom in as far as you can. In the Control Panel, set the time step to 1 month ("lunar m.") and click on the "Forward" button (a triangle that points to the right). Describe how Saturn's appearance changes over time. Explain what causes these changes.

 70. Use the *Deep Space Explorer*™ program to examine Jupiter and Saturn. (**a**) In the left-hand part of the window under the heading **Solar System,** click on the triangle next to the word **Explore** and then click on **Jupiter.** You can zoom in and zoom out using the buttons at the upper left of the window (an upward-pointing triangle and a downward-pointing triangle). You can also rotate Jupiter by putting the mouse cursor over the image, holding down the mouse button, and moving the mouse. (On a two-button mouse, hold down the left mouse button.) How many white ovals and brown ovals can you count in Jupiter's southern hemisphere (the hemisphere in which the Great Red Spot is located)? How many white ovals and brown ovals can you count in the northern hemisphere? What general rule can you state about the abundance of these storms in the two hemispheres? (**b**) In the left-hand part of the window under the heading **Solar System,** click on **Saturn.** Rotate Saturn so that you are looking straight down on the plane of the rings. Draw a copy of what you see and label the different rings and divisions. (Compare Figure 14-21.)

Collaborative Exercises

71. Using a ruler with millimeter markings on five various images of Jupiter in the text, Figures 14-1*a*, 14-2*a*, and 14-4, determine the ratio of the longest width of the Great Red Spot to the full diameter of Jupiter. Each group member should measure a different image and all values should be averaged.

72. The text provides different years that spacecraft have flown by Jupiter and Saturn. List these dates and create a time line by listing one important event that was occurring on Earth during each of those years.

73. If the largest circle you can draw on a piece of paper represents the largest diameter of Saturn's rings, about how large would Saturn be if scaled appropriately? Which item in a group member's backpack is closest to this size?

15 Jupiter and Saturn's Satellites of Fire and Ice

The world shown here is neither the creation of an abstract artist nor the product of a science fiction writer's imagination. It is Europa, a satellite of Jupiter whose icy surface is crisscrossed by fractures that suggest a dynamic geologic history. Beneath the ice may lie a worldwide ocean, within which there might possibly be single-celled organisms.

Europa is one of four large satellites discovered by Galileo in 1610. Astronomers discovered these four satellites anew in the late twentieth century thanks to the *Voyager* flybys in 1979 and the orbiting *Galileo* spacecraft in the late 1990s. They found that Io, Europa, Ganymede, and Callisto are entirely different from any world previously seen or even imagined.

Io, the most geologically active world in the solar system, has a surface covered with colorful sulfur compounds deposited by dozens of active volcanoes. Like Europa, Ganymede and Callisto are each covered with a thick layer of ice; unlike Europa, Callisto has a cratered, geologically dead surface, while Ganymede has a substantial magnetic field of its own.

No less exotic is Saturn's large satellite Titan, which is enveloped by a thick, nitrogen-rich atmosphere from which liquid hydrocarbons may fall as rain. Saturn also has medium-sized moons that show evidence of volcanic activity, but with frigid "lavas" made of water and ammonia. In this chapter we will explore these and other exotic aspects of Jupiter and Saturn's extensive collection of satellites.

 A color-enhanced view of Jupiter's satellite Europa. (NASA/JPL/University of Arizona) R **I** **V** U X G

Questions to Guide Your Reading

As you read the sections of this chapter, look for the answers to the following questions:

15-1   What is special about the orbits of Jupiter's Galilean satellites?

15-2   Are all the Galilean satellites made of rocky material, like the Earth's moon?

15-3   What could account for differences between the inner and outer Galilean satellites?

15-4   Why does Io have active volcanoes? How does Io's volcanic activity differ from that on Earth?

15-5   How does Io act like an electric generator?

15-6   What is the evidence that Europa has an ocean beneath its surface?

15-7   What is unusual about the magnetic fields of Ganymede and Callisto?

15-8   How is it possible for Saturn's moon Titan to have an atmosphere?

15-9   Why do some of Jupiter's moons orbit in the "wrong" direction?

15-10 What kinds of geologic activity are seen on Saturn's medium-sized satellites?

15-1 | Jupiter's Galilean satellites are easily seen with Earth-based telescopes

When Galileo trained his telescope on Jupiter in January 1610, he became the first person to observe satellites orbiting another planet (see Section 4-5). He called them the Medicean Stars, in order to curry the favor of a wealthy Florentine patron of the arts and sciences. We now call them the **Galilean satellites**. Individually, they are named Io, Europa, Ganymede, and Callisto, after four mythical lovers and companions of the god whom the Greeks called Zeus and the Romans called Jupiter.

When viewed through an Earth-based telescope, the Galilean satellites look like pinpoints of light (**Figure 15-1**). All four satellites orbit Jupiter in nearly the same plane as the planet's equator. From Earth, we always see that plane nearly edge-on, so the satellites appear to move back and forth relative to Jupiter (see Figure 4-17). The orbital periods are fairly short, ranging from 1.8 (Earth) days for Io to 16.7 days for Callisto, so we can easily see the satellites move from one night to the next and even during a single night. These motions led Galileo to realize that he was seeing objects orbiting around Jupiter, in much the same way that Copernicus said that planets move around the Sun (see Section 4-2).

Each of the four Galilean satellites is bright enough to be visible to the naked eye. Why, then, did Galileo need a telescope to discover them? The reason is that as seen from the Earth, the angular separation between Jupiter and the satellites is quite small—never more than 10 arcminutes for Callisto and even less for the other three Galilean satellites. To the naked eye, these satellites are lost in the overwhelming glare of Jupiter. But a small telescope or even binoculars increases the apparent angular separation by enough to make the Galilean satellites visible.

The brightness of each satellite varies slightly as it orbits Jupiter. This is because the satellites also spin on their axes as they orbit Jupiter, and dark and light areas on their surfaces are alternately exposed to and hidden from our view. Remarkably, each Galilean satellite goes through one complete cycle of brightness during one orbital period. This tells us that each satellite is in *synchronous rota-*

tion, so that it rotates exactly once on its axis during each trip around its orbit (see Figure 3-4*b*). As we saw in Section 4-8, our Moon's synchronous rotation is the result of gravitational forces exerted on the Moon by Earth. Likewise, Jupiter's gravitational forces keep the Galilean satellites in synchronous rotation.

 Synchronous rotation means that the rotation period and orbital period are in a 1-to-1 ratio for each Galilean satellite. Remarkably, there is also a simple ratio of the different orbital periods of the three inner Galilean satellites, Io, Europa, and Ganymede. During the 7.155 days that Ganymede takes to complete one orbit around Jupiter, Europa makes two orbits and Io makes four orbits. Thus, the orbital periods of Io, Europa, and Ganymede are in the ratio 1:2:4, which you can verify from the data in Table 15-1.

This special relationship among the satellites' orbits is maintained by the gravitational forces that they exert on one another. Those forces can be quite strong, because the three inner Galilean satellites pass relatively close to one another; at their closest approach, Io and Europa are separated by only two-thirds the distance from the Earth to the Moon. Such a close approach occurs once for every two of Io's orbits, so Europa's gravitational pull acts on Io in a rhythmic way. Indeed, Io, Europa, and Ganymede all exert rhythmic gravitational tugs on one another. Just as a drummer's rhythm keeps musicians on the same beat, this gravitational rhythm maintains the simple ratio of orbital periods among the satellites.

By contrast, Callisto orbits at a relatively large distance from the other three large satellites. Hence, the gravitational forces on Callisto from the other satellites are relatively weak, and there is no simple relationship between Callisto's orbital period and the period of Io, Europa, or Ganymede.

Because the orbital planes of the Galilean satellites are nearly edge-on to our line of sight, we see these satellites undergoing transits, eclipses, and occultations. In a transit, a satellite passes between us and Jupiter, and we see the satellite's shadow as a black dot against the planet's colorful cloudtops. (Figure 14-2*a* shows Europa's shadow on Jupiter.) In an eclipse, one of the satellites disappears and then reappears as it passes into and out of Jupiter's enormous shadow. In an **occultation,** a satellite passes completely behind Jupiter, and the satellite is temporarily blocked from our Earth-based view. (The word *occultation* comes from a Latin verb meaning "to cover.")

Before spacecraft first ventured to Jupiter, astronomers used eclipses to estimate the diameters of the Galilean satellites. When a satellite emerges from Jupiter's shadow, it does not blink on instantly. Instead, there is a brief interval during which the satellite gets progressively brighter as more of its surface is exposed to sunlight. The satellite's diameter is calculated from the duration of this interval and the satellite's orbital speed (which is known from Kepler's laws).

Occultations can also be used to determine the diameter of a satellite by measuring the time it takes to disappear behind Jupiter or to reappear from behind the planet. In addition, the Galilean satellites occasionally occult one another, because all four orbit Jupiter in the plane of the planet's equator. Every six years, when the Earth passes through this equatorial plane, there is a period of a few months during which Earth observers can see mutual occultations of the satellites. Once again, timing the occultations enables astronomers to calculate the diameters of the satellites.

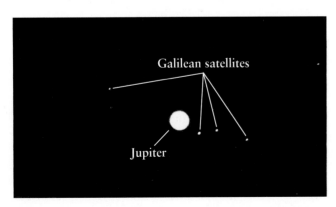

Figure 15-1 R I **V** U X G

The Galilean Satellites through a Small Telescope This photograph, taken by an amateur astronomer with a small telescope, shows the four Galilean satellites alongside an overexposed image of Jupiter. (Courtesy of J. Jenkins)

table 15-1	The Galilean Satellites Compared with the Moon, Mercury, and Mars						
	Average distance from Jupiter (km)	Orbital period (days)	Diameter (km)	Mass		Average density (kg/m³)	Albedo
				(kg)	(Moon = 1)		
Io	421,600	1.769	3642	8.932×10^{22}	1.22	3529	0.63
Europa	670,900	3.551	3120	4.791×10^{22}	0.65	3018	0.64
Ganymede	1,070,000	7.155	5268	1.482×10^{23}	2.02	1936	0.43
Callisto	1,883,000	16.689	4800	1.077×10^{23}	1.47	1851	0.17
Moon	—	—	3476	7.349×10^{22}	1.00	3344	0.11
Mercury	—	—	4880	3.302×10^{23}	4.49	5430	0.12
Mars	—	—	6794	6.419×10^{23}	8.73	3934	0.15

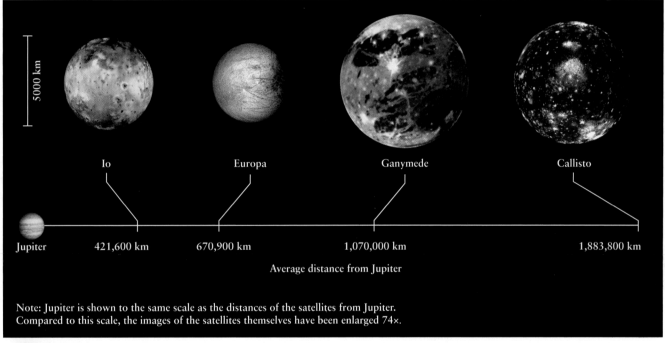

Note: Jupiter is shown to the same scale as the distances of the satellites from Jupiter. Compared to this scale, the images of the satellites themselves have been enlarged 74×.

WEB LINK 15.2 (NASA/JPL)   R I **V** U X G

As Table 15-1 shows, the smallest Galilean satellite (Europa) is slightly smaller than our Moon; the largest satellite (Ganymede) is larger than Mercury and more than three-quarters the size of Mars. Thus, the Galilean satellites truly are worlds in their own right.

15-2 | Data from spacecraft reveal the unique properties of the Galilean satellites

WEB LINK 15.3 Even the finest images made with Earth-based telescopes have revealed relatively few details about the Galilean satellites. Almost everything we know about these satellites has come from observations made at close range by spacecraft.

The first close-range observations of the Galilean satellites were made by the *Pioneer 10* and *Pioneer 11* spacecraft as they flew past Jupiter in 1973 and 1974. The images from these missions were of relatively low resolution, however, and much more information has come from three subsequent missions. *Voyager 1* and *Voyager 2* flew past Jupiter in 1979 and recorded tens of thousands of images, one of which is shown in Figure 15-2. An even more extensive investigation of Jupiter's satellites was carried out by the *Galileo* spacecraft, which orbited Jupiter from 1995 to 2003.

One key goal of these missions was to determine the densities of the Galilean satellites to very high accuracy. Given the density of a planet or satellite, astronomers can draw conclusions about its chemical composition and internal structure. To find the densities, scientists first determined the satellite masses. They did this by

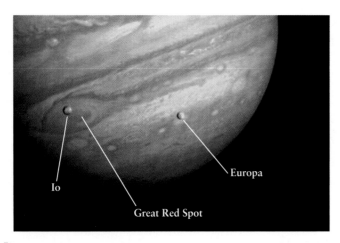

Io

Europa

Great Red Spot

Figure 15-2 R I **V** U X G

Io and Europa from *Voyager 1* *Voyager 1* recorded this image of Jupiter and the Galilean satellites Io and Europa in February 1979, when it was 20 million kilometers (12.4 million miles) from the giant planet. The satellites are actually smaller than some of Jupiter's cloud features, such as the Great Red Spot (see Section 14-3). *Voyager 1* and *Voyager 2* also recorded many images of the Galilean satellites at closer range. (NASA/JPL)

measuring how the gravity of Io, Europa, Ganymede, and Callisto deflected the trajectories of the *Voyager* and *Galileo* spacecraft. Given the masses and diameters of the satellites, they then calculated each satellite's average density (mass divided by volume). Table 15-1 lists our current knowledge about the sizes and masses of the Galilean satellites.

Of the four Galilean satellites, Europa proves to be the least massive: Its mass is only two-thirds that of our Moon. Ganymede is by far the most massive of the four, with more than double the mass of our Moon. In fact, Ganymede is the most massive satellite anywhere in the solar system. In second place is Callisto, with about 1½ times the Moon's mass.

It would be quite incorrect, however, to think of Ganymede and Callisto as merely larger versions of our Moon. The reason is that both Ganymede and Callisto have very low average densities of less than 2000 kg/m³. By comparison, typical rocks in the Earth's crust have densities around 3000 kg/m³, and the average density of our Moon is 3344 kg/m³. The low densities of Ganymede and Callisto mean that these satellites cannot be composed primarily of rock. Instead, they are thought to be made of roughly equal parts of rock and water ice. Water ice is substantially less dense than rock, but it becomes as rigid as rock under high pressure (such as is found in the interiors of the Galilean satellites). Furthermore, water molecules are relatively common in the solar system. Enough water would have been available in the early solar system to make up a substantial portion of a large satellite such as Ganymede or Callisto.

Water ice cannot be a major constituent of the two inner Galilean satellites, however. The innermost satellite, Io, has the highest average density, 3529 kg/m³, slightly greater than the density of our Moon. The next satellite out, Europa, has an average density of

3018 kg/m³. Both of these values are close to the densities of typical rocks in the Earth's crust. Hence, it is reasonable to suppose that both Io and Europa are made primarily of rocky material.

The most definitive evidence for water ice on Jupiter's satellites has come from spectroscopic observations. The spectra of sunlight reflected from Europa, Ganymede, and Callisto show absorption at the infrared wavelengths characteristic of water ice molecules. You can see this in Figure 7-4, which compares the spectrum of Europa to the spectrum of ice. Because Europa's density shows that it is composed mostly of rock, its ice must be limited to the satellite's outer regions. Among the Galilean satellites, only Io shows no trace of water ice on its surface or in its interior.

15-3 | The Galilean satellites formed like a solar system in miniature

As Table 15-1 shows, the more distant a Galilean satellite is from Jupiter, the greater the proportion of ice within the satellite and the lower its average density. But why should this be? An important clue is that the average densities of the planets show a similar trend: Moving outward from the Sun, the average density of the planets steadily decreases, from more than 5000 kg/m³ for Mercury to less than 1000 kg/m³ for Saturn (recall Table 7-1). The best explanation for this similarity is that the Galilean satellites formed around Jupiter in much the same way that the planets formed around the Sun, although on a much smaller scale.

We saw in Section 8-4 that dust grains in the solar nebula played an important role in planet formation. In the inner parts of the nebula, close to the glowing protosun, only dense, rocky grains were able to survive. These accumulated over time into the dense, rocky inner planets. But in the cold outer reaches of the nebula, dust grains were able to retain icy coatings of low-density material like water and ammonia. Hence, when the Jovian planets formed in the outer solar nebula and incorporated these ice-coated grains, the planets ended up with both rock and ices in their cores as well as substantial amounts of ammonia and water throughout their interiors and atmospheres (see Section 14-1 and Section 14-6).

As Jupiter coalesced, the gas accumulating around it formed a rotating "Jovian nebula." The central part of this nebula became the huge envelope of hydrogen and helium that makes up most of Jupiter's bulk. But in the outer parts of this nebula, dust grains could have accreted to form small solid bodies. These grew to become the Galilean satellites.

Although the "Jovian nebula" was far from the protosun, not all of it was cold. Jupiter, like the protosun, must have emitted substantial amounts of radiation due to Kelvin-Helmholtz contraction, which we discussed in Section 8-3. (We saw in Section 14-4 that even today, Jupiter emits more energy due to its internal heat than it absorbs from sunlight.) Hence, temperatures very close to Jupiter must have been substantially higher than at locations farther from the planet. Calculations by NASA scientists James Pollack, Fraser Fanale, and collaborators show that only rocky material would condense at the orbital distances of Io and Europa, but frozen water could be retained and incorporated into satellites at the distances of Ganymede and Callisto. In this way Jupiter ended up with two distinct classes of Galilean satellites (**Figure 15-3**).

figure 15-3

Formation of the Galilean Satellites Heat from the protosun made it impossible for icy grains to survive within the innermost 2.5 to 4 AU of the solar nebula. In the same way, Jupiter's heat evaporated any icy grains that were too close to the center of the "Jovian nebula." Hence, the two inner Galilean satellites were formed primarily from rock, while the outer two incorporated both rock and ice.

The analogy between the solar nebula and the "Jovian nebula" is not exact. Unlike the Sun, Jupiter is a "failed star." Its internal temperatures and pressures never became high enough to ignite nuclear reactions that convert hydrogen into helium. (Jupiter's mass would have had to be about 80 times larger for these reactions to have begun.) Furthermore, the icy worlds that formed in the outer region of the "Jovian nebula"—namely, Ganymede and Callisto—were of relatively small mass. Hence, they were unable to attract gas

from the nebula and become Jovian planets in their own right. Nonetheless, it is remarkable how the main processes of "planet" formation seem to have occurred twice in our solar system: once in the solar nebula around the Sun, and once in microcosm in the gas cloud around Jupiter.

The diverse densities of the Galilean satellites suggest that these four worlds may be equally diverse in their geology. This turns out to be entirely correct. Because each of the Galilean satellites is unique, we devote the next several sections to each satellite in turn.

15-4 | Io is covered with colorful sulfur compounds ejected from active volcanoes

SCIENCE IN PROCESS *The theory of tidal forces helped predict Io's volcanic nature*

Before the two *Voyager* spacecraft flew past Jupiter, the nature of Io was largely unknown. On the basis of Io's size and density, however, it was expected that Io would be a world much like our own Moon—geologically dead, with little internal heat available to power tectonic or volcanic activity (see Section 7-6). Hence, it was thought that Io's surface would be extensively cratered, because there would have been little geologic activity to erase those craters over the satellite's history.

Io proved to be utterly different from these naive predictions. On March 5, 1979, *Voyager 1* came within 21,000 kilometers (13,000 miles) of Io and began sending back a series of bizarre and unexpected pictures of the satellite (Figure 15-4). These images showed that Io has *no* impact craters at all! Instead, the surface is pockmarked by irregularly shaped pits and is blotched with color, giving Io an appearance totally unlike that of any other world in the solar system. Baffled by what they were seeing, scientists at the Jet Propulsion Laboratory (JPL) in Pasadena, California (where the pictures were being received and enhanced by computer processing), jokingly

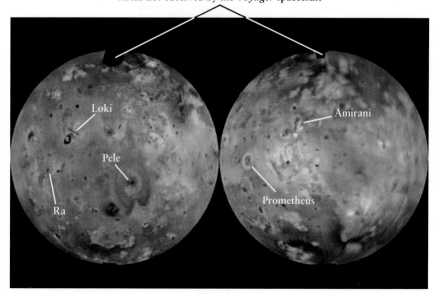

figure 15-4 R I **V** U X G

Io These mosaics of Io's two hemispheres were built up from individual *Voyager* images. The extraordinary colors are probably caused by deposits of sulfur and sulfur compounds ejected from Io's numerous volcanoes. The labels show several of these volcanoes, which are named for sun gods and fire gods of different cultures. (NASA/JPL) compared Io to a pizza or a rotten orange. In fact, the scientists were seeing confirmation of a brilliant prediction published only a few days earlier: Io's surface is the result of intense volcanic activity.

Three days before *Voyager 1* flew past Io, the journal *Science* published an article by Stanton Peale of the University of California, Santa Barbara, and Patrick Cassen and Ray Reynolds of NASA's Ames Research Center. In this article, they reported their conclusion that Io's interior must be kept hot by Jupiter's tidal forces.

As we saw in Section 4-8, tidal forces are differences in the gravitational pull on different parts of a planet or satellite. These forces tend to deform the shape of the planet or satellite (see Figure 4-24). Io, for example, is somewhat distorted from a spherical shape by tidal forces from Jupiter. If Io were in a perfectly circular orbit so that it traveled around Jupiter at a steady speed, Io's rotation would always be in lockstep with its orbital motion. In this case, Io would be in perfect synchronous rotation, with its long axis always pointed directly toward Jupiter. But as Io moves around its orbit, Europa and Ganymede exert gravitational tugs on it in a regular, rhythmic fashion, thanks to the 1:2:4 ratio among the orbital periods of these three satellites. These tugs distort Io's orbit into an ellipse, and so its speed varies as it moves around its orbit. Consequently, Io's long axis "nods" back and forth by about ½° as seen from Jupiter. Due to this "nodding," the tidal stresses that Jupiter exerts on Io vary rhythmically as Io moves around its orbit. These varying tidal stresses alternately squeeze and flex Io. Just as a ball of clay or bread dough gets warm as you knead it between your fingers, this squeezing and flexing causes **tidal heating** of Io's interior. (The Earth's Moon also "nods" as it orbits our planet, but experiences almost no tidal heating because the Earth's tidal forces are very weak compared to those of massive Jupiter.)

Tidal heating adds energy to Io at a rate of about 10^{14} watts, equivalent to 24 tons of TNT exploding every second. As this energy makes its way to the satellite's surface, it provides about 2.5 watts of power to each square meter of Io's surface. By comparison, the average global heat flow through the Earth's crust is 0.06 watts per square meter. Only in volcanically active areas on Earth do we find heat flows that even come close to Io's average. Thus, Peale, Cassen, and Reynolds predicted "widespread and recurrent surface volcanism" on Io.

No one expected that *Voyager 1* would obtain images of erupting volcanoes on Io. After all, a spacecraft making a single trip past the Earth would be highly unlikely to catch a large volcano actually erupting. But several days after the Jupiter flyby, Linda Morabito, a navigation engineer at JPL, noticed a large umbrella-shaped cloud protruding from Io in one image (**Figure 15-5a**). She had discovered a plume of gas from a volcanic eruption. Careful reexamination of the close-up images revealed seven more giant volcanic plumes. More recently, the *Galileo* spacecraft, which came as close as 200 km to Io's surface during its eight years in orbit around Jupiter, returned detailed images of several such plumes (Figure 15-5b). These observations resoundingly confirm the Peale-Cassen-Reynolds prediction: Io is by far the most volcanic world in the solar system.

Io's volcanic plumes rise to astonishing heights of 70 to 280 km above Io's surface. To reach such altitudes, the material must emerge from volcanic vents with speeds between 300 and 1000 m/s (1100 to 3600 km/h, or 700 to 2200 mi/h). Even the most violent terrestrial volcanoes, like Vesuvius, Krakatoa, and Mount St. Helens (see Figure 12-8), have eruption speeds of only around 100 m/s (360 km/h, or 220 mi/h). Scientists thus began to suspect that Io's volcanic activity must be fundamentally different from volcanism here on Earth.

An important clue about volcanism on Io came from the infrared spectrometers aboard *Voyager 1*, which detected abundant sulfur and sulfur dioxide in Io's volcanic plumes. This led to the

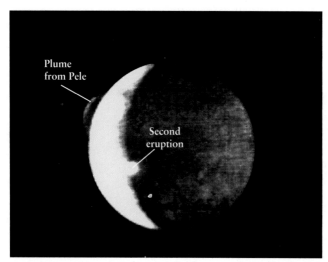

(a) *Voyager 1*, March 1979

(b) *Galileo*, November 1997

Figure 15-5 R I **V** U X G

Volcanic Plumes on Io (a) Io's volcanic eruptions were first discovered on this *Voyager 1* image. The plume at upper left, which extends 260 km (150 mi) above Io's surface, came from the volcano Pele (see Figure 15-4). (b) This *Galileo* image shows plumes from the volcanoes Prometheus (see Figure 15-4) and Zamama. The plumes rise to about 100 km (60 mi) above Io's surface and are about 250 km (150 mi) wide. (a: NASA; b: Planetary Image Research Laboratory/University of Arizona/JPL/NASA)

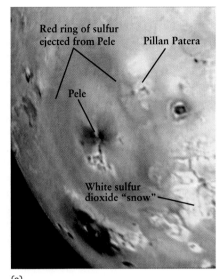

Red ring of sulfur
ejected from Pele

Pillan Patera

Pele

White sulfur
dioxide "snow"

(a)

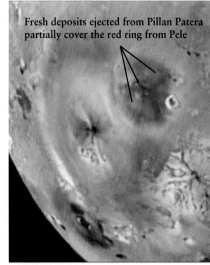

Fresh deposits ejected from Pillan Patera
partially cover the red ring from Pele

(b)

figure 15-6 R I **V** U X G

Rapid Changes on Io (a) This *Galileo* image shows a portion of Io's southern hemisphere. Sulfur ejected from the active volcanic vent Pele produced a red ring with diffuse edges that resembles paint sprayed from a can. (b) A few months later, a volcanic plume erupted from a vent called Pillan Patera, spraying dark material over an area some 400 km (250 mi) in diameter (about the size of Arizona). Other than Earth, Io is the only world in the solar system that shows such noticeable changes over short time spans. (NASA/JPL)

idea that the plumes are actually more like geysers than volcanic eruptions. In a geyser on Earth, water seeps down to volcanically heated rocks, where it changes to steam and erupts explosively through a vent. The planetary geologists Susan Kieffer, Eugene Shoemaker, and Bradford Smith suggested that sulfur dioxide rather than water could be the principal propulsive agent driving volcanic plumes on Io. Sulfur dioxide is a solid at the frigid temperatures found on most of Io's surface, but it should be molten at depths of only a few kilometers. Just as the explosive conversion of water into steam produces a geyser on the Earth, the conversion of liquid sulfur dioxide into a high-pressure gas could result in eruption velocities of up to 1000 m/s.

Io's dramatic coloration (see Figure 15-4) is probably due to sulfur and sulfur dioxide, which is ejected in volcanic plumes and later falls back to the surface. Sulfur is normally bright yellow, which explains the dominant color of Io's surface. But if sulfur is heated and suddenly cooled, as would happen if it were ejected from a volcanic vent and allowed to fall to the surface, it can assume a range of colors from orange and red to black. Indeed, these colors are commonly found around active volcanic vents (**Figure 15-6**). Whitish surface deposits (examine Figure 15-4 and Figure 15-6*a*), by contrast, are probably due to sulfur dioxide (SO_2). Volcanic vents on Earth commonly discharge SO_2 in the form of an acrid gas. But on Io, when hot SO_2 gas is released by an eruption into the cold vacuum of space, it crystallizes into white snowflakes. This sulfur dioxide "snow" then rains back onto Io's surface.

The plumes are not the whole story of volcanism on Io, however. The sources of the plumes appear in *Voyager* images as black spots 10 to 50 km in diameter, which are actually volcanic vents (see Fig-

ure 15-4). Many of these black spots, which cover 5% of Io's surface, are surrounded by dark lava flows. **Figure 15-7** is a *Galileo* image of one of Io's most active volcanic regions, where in 1999 lava was found spouting to altitudes of thousands of meters along a fissure 40 km (25 mi) in length. The glow from this lava fountain was so intense that it was visible with Earth-based telescopes.

Lava flows

Volcanic
calderas

Curtain of lava

50 km

figure 15-7 R I **V** U X G

Io's Lava Flows and a Curtain of Fire This mosaic of *Galileo* spacecraft images shows a nested chain of volcanic calderas (pits) with black lava flows. The red streak at upper left is a fissure in the surface about 40 km (25 mi) long, through which lava is erupting upward in a vertical sheet or curtain. The erupting lava reaches the amazing height of 1500 m (5000 ft). (University of Arizona/JPL/NASA)

Thus, Io has two distinct styles of volcanic activity: unique geyserlike plumes and lava flows that are more dramatic versions of volcanic flows on Earth.

While sulfur and sulfur compounds are the key constituents of Io's volcanic plumes, the lava flows must be made of something else. Infrared measurements from the *Galileo* spacecraft (Figure 15-8) show that lava flows on Io are at temperatures of 1700 to 2000 K (about 1450 to 1750°C, or 2600 to 3150°F). At these temperatures sulfur could not remain molten but would evaporate almost instantly. Io's lavas are also unlikely to have the same chemical composition as typical lavas on Earth, which have temperatures of only 1300 to 1450 K. Instead, lavas on Io are probably **ultramafic lavas.** These are enriched in magnesium and iron, which give the lava a higher melting temperature.

Solidified ultramafic lavas are found on Earth, but primarily in lava beds that formed billions of years ago when the Earth's interior was much hotter than today. The presence of molten ultramafic lava on Io suggests that its interior, too, is substantially hotter than that of the present-day Earth. One model proposes that Io has a 100-km (60-mi) thick crust floating atop a worldwide ocean of liquid magma 800 km (500 mi) deep. (As we saw in Section 9-2, the mantle that lies underneath the Earth's crust is a plastic solid rather than a true liquid.) The global extent of this "magma ocean" would explain why volcanic activity is found at all points on Io's surface, rather than being concentrated in pockets as on Earth (see Section 9-3). It could also explain the origin of Io's mountains, which reach heights up to 10 km (30,000 ft). In this model, these mountains are

blocks of Io's crust that have tilted before sinking into the depths of the magma ocean.

Io's worldwide volcanic activity is remarkably persistent. When *Voyager 2* flew through the Jovian system in July 1979, four months after *Voyager 1,* almost all of the volcanic plumes seen by *Voyager 1* were still active. And when *Galileo* first viewed Io in 1996, about half of the volcanoes seen by the *Voyager* spacecraft were still ejecting material, while other, new volcanoes had become active.

Io may have as many as 300 active volcanoes, each of which ejects an estimated 10,000 tons of material per second. Altogether, volcanism on Io may eject as much as 10^{13} tons of matter each year. This is sufficient to cover the satellite's entire surface to a depth of 1 meter in a century, or to cover an area of 1000 square kilometers in a few weeks (see Figure 15-6). Thanks to this continual "repaving" of the surface, there are probably no long-lived features on Io, and any impact craters are quickly obliterated.

15-5 | Jupiter's magnetic field makes electric currents flow through Io

Most of the material ejected from Io's volcanoes falls back onto the satellite's surface. But some does not, by virtue of Io's location deep within Jupiter's magnetosphere. The magnetosphere contains charged particles, some of which collide with Io and its volcanic plumes. The impact of these collisions knocks ions out of the plumes and off the surface, and these ions become part of Jupiter's magnetosphere. Indeed, material from Io is the main source for Jupiter's magnetosphere as a whole. Some of the ejected material forms the **Io torus,** a huge doughnut-shaped ring of ionized gas, or plasma (see Section 14-8), that circles Jupiter at the distance of Io's orbit. The plasma's glow can be detected from Earth (Figure 15-9a).

Jupiter's magnetic field has other remarkable effects on Io. As Jupiter rotates, its magnetic field rotates with it, and this field sweeps over Io at high speed. This generates a voltage of 400,000 volts across the satellite, which causes 5 million amperes of electric current to flow through Io.

 Whenever you go shopping with a credit card, you use the same physical principle that generates an electric current within Io. The credit card's number is imprinted as a magnetic code in the dark stripe on the back of the card. The salesperson who swipes your card through the card reader is actually moving the card's magnetic field past a coil of wire within the reader. This generates within the coil an electric current that carries the same coded information about your credit card number. That information is transmitted to the credit card company, which (it is hoped) approves your purchase. An electric generator at a power plant works in the same way. A coil of wire is moved through a strong magnetic field, which makes current flow in the coil. This current is delivered to transmission lines and eventually to your home.

In fact, a current flows not only through Io but also through the sea of charged particles in Jupiter's magnetosphere and through Jupiter's atmosphere. This forms a gigantic, oval-shaped electric circuit that connects Io and Jupiter in the same way that

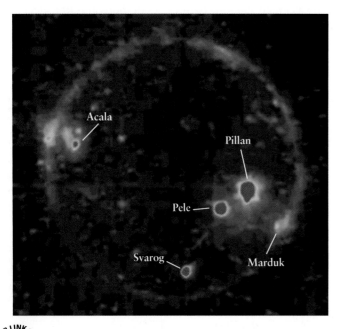

WEB LINK 15.6

figure 15-8   R **I** **V** U X G

Io's Glowing Volcanoes This false-color *Galileo* image (made while Io was in the darkness of Jupiter's shadow) shows the infrared and visible glow from several volcanic regions, including Pele and Pillan (see Figure 15-6). These regions are substantially hotter than typical lava flows on Earth. (Planetary Image Research Laboratory/University of Arizona/JPL/NASA)

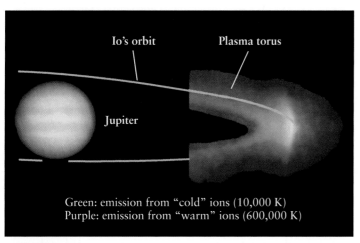

Io's orbit   Plasma torus

Jupiter

Green: emission from "cold" ions (10,000 K)
Purple: emission from "warm" ions (600,000 K)

(a) R **I** **V** U X G

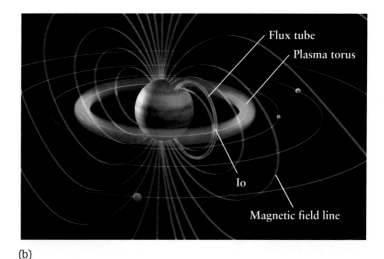

Flux tube
Plasma torus

Io

Magnetic field line

(b)

Figure 15-9

The Io Torus (a) An Earth-based telescope recorded this false-color infrared image of Io's plasma torus. Because of Jupiter's glare, only the outer edge of the torus could be photographed; an artist added the visible-light picture of Jupiter and the yellow line indicating the rest of the torus. (b) A flux tube (a region of concentrated magnetic field) carries plasma between Jupiter and Io, forming an immense electric circuit. (a: Courtesy of J. Trauger; b: Alfred T. Kamajian and Torrence V. Johnson, "The Galileo Mission to Jupiter and Its Moons," *Scientific American*, February 2000, p. 44)

wires connect the battery and the lightbulb in a flashlight (Figure 15-9*b*). Jupiter's aurora, which we discussed in Section 14-8, is particularly strong at the locations where this current strikes the upper atmosphere of Jupiter.

Part of the electric current that flows between Io and Jupiter is made up of electrons that spiral around Jupiter's magnetic field lines. As they spiral, the electrons act like miniature antennas and emit radio waves. This is the source of Jupiter's bursts of decametric radio radiation, which we discussed in Section 14-7.

The electric current within and near Io also creates a weak magnetic field, which was first detected by *Voyager 1*. Remarkably, measurements made by *Galileo* during its close approaches to Io suggest that the magnetic field near Io may be too strong—comparable to that of Mercury, although weaker than that of Earth—to be generated by Io's electric current alone. This has led scientists to speculate that Io generates its own magnetic field through the motions of molten material in its interior, just as the Earth does (see Section 7-7 and Section 9-4). If these speculations are correct, Io is the smallest world in the solar system to generate its own magnetic field.

If Io has undergone enough tidal heating to melt its interior, chemical differentiation must have taken place and the satellite should have a dense core. (We described chemical differentiation in Box 7-1 and Section 8-4.) To test for this, scientists measured how Io's gravity deflected the trajectory of *Galileo* as the spacecraft flew past. From these measurements, they could determine not only Io's mass but also the satellite's oblateness (how much it deviates from a spherical shape because of its rotation). The amount of oblateness indicates the size of the core; the greater the fraction of the satellite's mass contained in its core, the less oblate the satellite will be. The *Galileo* observations suggest that Io has a dense core composed of iron and iron sulfide (FeS), with a radius

of about 900 km (about half the satellite's overall radius). Surrounding the core is a mantle of partially molten rock, on top of which is Io's visible crust.

15-6 | Europa is covered with a smooth layer of ice that may cover a worldwide ocean

Europa, the second of the Galilean satellites, is the smoothest body in the solar system. There are no mountains and no surface features greater than a few hundred meters high (Figure 15-10). There are almost no craters, indicating a young surface that has been reprocessed by geologic activity. The dominant surface feature is a worldwide network of stripes and cracks (see Figure 15-11 and the image that opens this chapter). Like Io, Europa is an exception to the general rule that a small world should be cratered and geologically dead (see Section 7-6).

Neither *Voyager 1* nor *Voyager 2* flew very near Europa, so most of our knowledge of this satellite comes from the close passes made by *Galileo*. (Both Figure 15-10 and Figure 15-11 are *Galileo* images.) But even before these spacecraft visited Jupiter, spectroscopic observations from Earth indicated that Europa's surface is almost pure frozen water (see Figure 7-4). This was confirmed by instruments on board *Galileo*, which showed that Europa's infrared spectrum is a close match to that of a thin layer of fine-grained water ice frost on top of a surface of pure water ice. (The brown areas in Figure 15-10 show where the icy surface contains deposits of rocky material from meteoritic impacts, from Europa's interior, or from a combination of these sources.)

The purity of Europa's ice suggests that water is somehow brought upward from the moon's interior to the surface, where it

figure 15-10 R ■ ▼ U X G

Europa Dark lines crisscross Europa's smooth, icy surface in this composite of visible and infrared images from *Galileo*. These are fractures in Europa's crust that can be as much as 20 to 40 km (12 to 25 mi) wide. Only one recent impact crater can be seen in this image. This indicates that Europa has a very young surface on which all older craters have been erased. (NASA/JPL)

Some features on Europa's surface, such as the fracture patterns shown in Figure 15-11, may be the direct result of the crust being stretched and compressed by tidal flexing. Figure 15-12 shows other features, such as networks of ridges and a young, very smooth circular area, that were probably caused by the internal heat that tidal flexing generates. The rich variety of terrain depicted in Figure 15-12, with stress ridges going in every direction, shows that Europa has a complex geologic history.

Among the unique structures found on Europa's surface are **ice rafts**. The area shown in Figure 15-13*a* was apparently subjected to folding, producing the same kind of linear features as those in Figure 15-12. But a later tectonic disturbance broke the surface into small chunks of crust a few kilometers across, which then "rafted" into new positions. A similar sort of rafting happens in the Earth's Arctic Ocean every spring, when the winter's accumulation of surface ice breaks up into drifting ice floes (Figure 15-13*b*). The existence of such structures on Europa strongly suggests that there is a subsurface layer of liquid water or soft ice over which the ice rafts can slide with little resistance.

The ridges, faults, ice rafts, and other features strongly suggest that Europa has substantial amounts of internal heat. This heat could prevent water from freezing beneath Europa's surface, creating a worldwide ocean beneath the crust. If this picture is correct, geologic processes on Europa may be an exotic version of those on Earth, with the roles of solid rock and molten magma being played by solid ice and liquid water.

A future spacecraft may use radar to penetrate through Europa's icy crust to search for definitive proof of liquid water beneath the

solidifies to make a fresh, smooth layer of ice. Indeed, some *Galileo* images show what appear to be lava flows on Europa's surface, although the "lava" in this case is mostly ice. This idea helps to explain why Europa has very few craters (any old ones have simply been covered up) and why its surface is so smooth. Europa's surface may thus represent a water-and-ice version of plate tectonics.

CAUTION Although Europa's surface is almost pure water ice, keep in mind that Europa is *not* merely a giant ice ball. The satellite's density shows that rocky material makes up about 85 to 90% of Europa's mass. Hence, only a small fraction of the mass, about 10 to 15%, is water ice. Because the surface is icy, we can conclude that the rocky material is found within Europa's interior.

Europa is too small to have retained much of the internal heat that it had when it first formed. But something is powering the geologic processes that erase craters and bring fresh water to Europa's surface. If not internal heat, what is the source of this energy? The most likely answer, just as for Io, is Jupiter's tidal forces. The rhythmic gravitational tugs exerted by Io and Ganymede on Europa deform its orbit into an ellipse; the varying speed of Europa as it goes around its orbit causes Europa to "nod" back and forth, causing variations in tidal stresses that make Europa flex. But because it is farther from Jupiter, tidal effects on Europa are only about one-fourth as strong as those on Io, which may explain why no ongoing volcanic activity has yet been seen on Europa.

Smooth ice plains with a network of fractures

Linear ridges (dark colors are caused by minerals in the ice)

500 km

figure 15-11 R ■ ▼ U X G

Europa's Fractured Crust Fractures and dark, linear features crisscross Europa's icy surface in this false-color composite of visible and infrared images from *Galileo*. The smooth ice plains (shown in blue) are the basic terrain found on Io. (NASA/JPL)

Ridges produced by folding and faulting of the surface.

5 km

Smooth area formed by fluid erupting onto the surface.

Rugged patch created by a subsurface disturbance.

WEB LINK 15.8

Figure 15-12 R I **V** U X G
Europa in Close-up This high-resolution *Galileo* image shows a network of overlapping ridges, part of which has been erased to leave a smooth area and part of which has been jumbled into a rugged patch of terrain. Europa's interior must have enough energy to power this complex geologic activity. (NASA/JPL)

surface. But magnetic field measurements made by the *Galileo* spacecraft have already provided some key evidence favoring this picture. Unlike the Earth or Jupiter, Europa does not seem to create a steady magnetic field of its own. But as Europa moves through

Jupiter's intense magnetic field, electric currents are induced within the satellite's interior, just as they are within Io (see Section 15-5), and these currents generate a weak but measurable field. (The strength and direction of this induced field vary as Europa moves through different parts of the Jovian magnetosphere, which would not be the case if Europa generated its field by itself.)

To explain these observations, there must be an electrically conducting fluid beneath Europa's crust—a perfect description of an underground ocean of water with dissolved minerals. (Pure water is a very poor conductor of electricity, so other substances must be present to make the water conducting.) If some of this water should penetrate upward to Europa's surface through cracks in the crust, it would vaporize and spread the dissolved minerals across the terrain. This could explain the reddish-brown colors in Figure 15-13a.

By combining measurements of Europa's induced magnetic field, gravitational pull, and oblateness, scientists conclude that Europa's outermost 100 to 200 km are ice and water. (It is not clear how much of this is liquid and how much solid.) Within this outer shell is a rocky mantle surrounding a metallic core some 600 km (400 mi) in radius.

Remarkably, Europa also has an extremely thin atmosphere of oxygen. (By Earth standards, this atmosphere would qualify as a near-vacuum.) Oxygen in the Earth's atmosphere is produced by plants through photosynthesis (see Section 9-5). But Europa's oxygen atmosphere is probably the result of ions from Jupiter's magnetosphere striking the satellite's icy surface. These collisions break apart water molecules, liberating atoms of hydrogen (which escape into space) and oxygen.

Ice rafts (fractured segments of Europa's icy crust)

Linear features show how rafts once fit together.

10 km

(a) Ice rafts on Europa

Ice floes

Akpatok Island

10 km

(b) Ice floes on Earth

WEB LINK 15.9

Figure 15-13 R I **V** U X G
Moving Ice on Europa and Earth (a) Some time after a series of ridges formed in this region of Europa's surface, the icy crust broke into "rafts" that were moved around by an underlying liquid or plastic layer. The colors in this *Galileo* image may be due to minerals that were released from beneath the surface after the crust broke apart. (b) Europa's ice rafts are analogous to ice floes created when pack ice breaks up, as in this spacecraft view of part of the Canadian arctic. (a: NASA/JPL; b: USGS and NASA)

The existence of a warm, subsurface ocean on Europa, if proved, would make Europa the only world in the solar system other than the Earth on which there is liquid water. This would have dramatic implications. On Earth, water and warmth are essentials for the existence of life. Perhaps single-celled organisms have evolved in the water beneath Europa's crust, where they would use dissolved minerals and organic compounds as food sources. In light of this possibility, NASA has taken steps to prevent biological contamination of Europa. At the end of the *Galileo* mission in 2003, the spacecraft (which may have carried traces of organisms from Earth) was sent to burn up in Jupiter's atmosphere, rather than remaining in orbit where it might someday crash into Europa. An appropriately sterilized spacecraft may one day visit Europa and search for evidence of life within this exotic satellite.

15-7 | Liquid water may also lie beneath the cratered surfaces of Ganymede and Callisto

Unlike Io and Europa, the two outer Galilean satellites have cratered surfaces. In this respect, Ganymede and Callisto bear a superficial resemblance to our own Moon (see the illustration with Table 15-1). But unlike the Moon's craters, the craters on both Ganymede and Callisto are made primarily of ice rather than rock, and Ganymede has a number of surface features that indicate a geologically active past.

 Ganymede is the largest satellite in our solar system and is even larger than the planet Mercury. Like our Moon, Ganymede has two distinct kinds of terrain, called **dark terrain** and **bright terrain** (Figure 15-14). On the Moon, the dark maria are younger than the light-colored lunar highlands (see Section 10-1). On Ganymede, by contrast, the dark terrain is older, as indicated by its higher density of craters. The bright terrain is much less cratered and therefore younger. Because ice is more reflective than rock, even Ganymede's dark terrain is substantially brighter than the lunar surface.

Voyager and *Galileo* images show noticeable differences between young and old impact craters on Ganymede. The youngest craters are surrounded by bright rays of freshly exposed ice that has been excavated by the impact (see the lower left of Figure 15-14). Older craters are darker, perhaps because of chemical processes triggered by exposure to sunlight.

The close-up *Galileo* image in **Figure 15-15** shows the contrast between dark and bright terrain. The area on the left-hand side of this image is strewn with large and small craters, reinforcing the idea that the dark terrain is billions of years old. The dark terrain is also marked by long, deep furrows. These are deformations of Ganymede's crust that have partially erased some of the oldest craters. Since other craters lie on top of the furrows, the stresses that created the furrows must have occurred long ago. Since geologic activity is a result of a planet or satellite's internal heat, this shows that Ganymede must have had a warm interior in the distant past.

Linear features are also seen in the bright terrain on the right in Figure 15-15. The *Voyager* missions discovered that the bright terrain is marked by long grooves, some of which extend for hundreds of kilometers and are as much as a kilometer deep. Images from

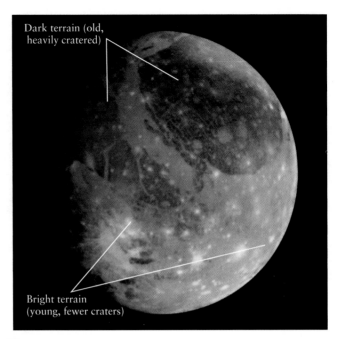

Dark terrain (old, heavily cratered)

Bright terrain (young, fewer craters)

Figure 15-14 R I **V** U X G

Ganymede Two distinct types of terrain—one dark, heavily cratered, and hence old, the other bright, less cratered, and hence younger—are visible in this *Galileo* image of the hemisphere of Ganymede that always faces away from Jupiter. Craters in general appear bright, suggesting that the impacts that formed the craters excavated the surface to reveal ice underneath. (NASA/JPL)

Galileo show that some adjacent grooves run in different directions, indicating that the bright terrain has been subjected to a variety of tectonic stresses. *Galileo* also found a number of small craters overlying the grooves, some of which are visible in Figure 15-15. The density of craters shows that the bright terrain, while younger than the dark terrain, is at least a billion years old.

After the *Voyager* missions, it was thought that the bright terrain represented fresh ice that had flooded through cracks in Ganymede's crust, analogous to the maria on our Moon (see Section 10-1). However, high-resolution *Galileo* images do not show the kinds of surface features that would be expected if this terrain was volcanic in origin. Yet there are several low-lying areas within the bright terrain that appear to have been flooded by a watery fluid that subsequently froze. These low-lying areas may be *rift valleys* where the crust has been pulled apart, like Valles Marineris on Mars (see Section 13-3). In order to flood the floors of these valleys, there must have been liquid water beneath Ganymede's surface when the valleys formed. This is additional evidence that Ganymede must have had a warm interior in the past, and that this internal warmth persisted at least until the era when the bright terrain formed a billion years ago. This is rather surprising, since Ganymede is too small to have retained much of the heat left over from its formation or from radioactive elements, and it orbits too far from Jupiter to have any substantial tidal heating.

An equally great surprise from *Galileo* observations was the discovery that Ganymede has its own magnetic field, and that this

Figure 15-15 R I **V** U X G

Evidence for Geologic Activity on Ganymede Stresses in Ganymede's crust have produced linear features that extend for hundreds of kilometers across both the dark and bright terrain. These features must be quite ancient, since numerous impact craters have formed on top of them. (NASA)

field is twice as strong as that of Mercury. This field is sufficiently strong to trap charged particles, giving Ganymede its own "mini-magnetosphere" within Jupiter's much larger magnetosphere. The presence of a magnetic field shows that electrically conducting material must be in motion within Ganymede to act as a dynamo, which means that the satellite must have substantial internal heat even today. A warm interior causes differentiation, and *Galileo* measurements indeed showed that Ganymede is highly differentiated. It has a metallic core some 500 km in radius, surrounded by a rocky mantle and by an outer shell of ice some 800 km thick.

An even more surprising result is that Ganymede's magnetic field varies somewhat as it orbits Jupiter, and that these variations are correlated with Ganymede's location within Jupiter's magnetic field. One explanation for this variation is that Ganymede has a layer of salty, electrically conducting liquid water some 170 km beneath its surface. As the satellite moves through Jupiter's magnetic field, electric currents are induced in this layer just as happens within Europa (see Section 15-6), and these currents generate a weak field that adds to the stronger field produced by Ganymede's dynamo.

A possible explanation for Ganymede's internal heat is that gravitational forces from the other Galilean satellites may have affected its orbit. While Ganymede's present-day orbit around Jupiter is quite circular, calculations show that the orbit could have been more elliptical in the past. Just as for Io and Europa, such an elliptical orbit would have given rise to substantial tidal heating of Ganymede's interior. It may have retained a substantial amount of that heat down to the present day.

WEB LINK 15.11 In contrast to Io, Europa, and Ganymede is Callisto, Jupiter's outermost Galilean satellite. Images from *Voyager* showed that Callisto has numerous impact craters scattered over an icy crust (**Figure 15-16**). Callisto's ice is not as reflective as that on Europa or Ganymede, however; the surface appears to be covered with some sort of dark mineral deposit.

CAUTION Figure 15-16, as well as the images that accompany Table 15-1, shows that Callisto is the darkest of the Galilean satellites. Nonetheless, its surface is actually more than twice as reflective as that of our own Moon. Callisto's surface may be made of "dirty" ice, but even dirty ice reflects more light than the gray rock found on the Moon. If you could somehow replace our own Moon with Callisto, a full moon as seen from Earth would be 4 times brighter than it is now (thanks also in part to Callisto's larger size).

The *Voyager* images led scientists to conclude that Callisto is a dead world, with an ancient surface that has never been reshaped by

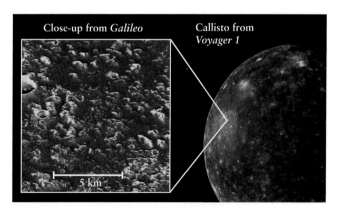

Figure 15-16 R I **V** U X G

Callisto Numerous craters pockmark Callisto's icy surface. The inset shows a portion of a huge impact basin called Valhalla. Most of the very smallest craters in this close-up view have been completely obliterated, and the terrain between the craters is blanketed by dark, dusty material. (Left: Arizona State University/JPL/NASA; right: JPL/NASA)

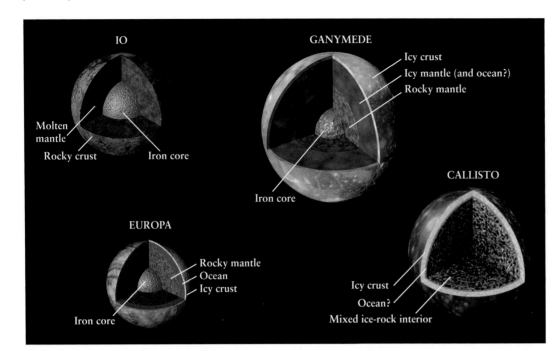

INTERACTIVE EXERCISE 15.1

figure 15-17
Interiors of the Galilean Satellites

These cross-sectional diagrams show the probable internal structures of the four Galilean satellites of Jupiter, based on information from the *Galileo* mission. (NASA/JPL)

geologic processes. There is no simple relationship between Callisto's orbital period and those of the other Galilean satellites, so it would seem that Callisto never experienced any tidal heating and hence was unable to power any geologic processes at its surface. But once the *Galileo* spacecraft went into orbit around Jupiter and began returning high-resolution images, scientists realized that Callisto's nature is not so simple. In fact, Callisto proves to be the most perplexing of the Galilean satellites.

One curious discovery from *Galileo* is that while Callisto has numerous large craters, there are very few with diameters less than 1 km. Such small craters are abundant on Ganymede, which has probably had the same impact history as Callisto. This implies that most of Callisto's small craters have been eroded away. How this could have happened is not known. Another surprising result is that Callisto's surface is covered by a blanket of dark, dusty material (see the inset in Figure 15-16). Where this material came from, and how it came to be distributed across the satellite's surface, is not understood.

Callisto's most unexpected feature is that, like Europa and Ganymede, it appears to have an induced magnetic field that varies as the satellite moves through different parts of Jupiter's magnetosphere. Callisto must have a layer of electrically conducting material in order to generate such a field. One model suggests that this material is in the form of a subsurface ocean like Europa's but only about 10 km (6 mi) deep. However, without tidal heating, Callisto's interior is probably so cold that a layer of liquid water would quickly freeze. One proposed explanation is that Callisto's ocean is a mixture of water and a form of "antifreeze." In cold climates on Earth, antifreeze is added to the coolant in automobile radiators to lower the liquid's freezing temperature and prevent it from solidifying. Ammonia in Callisto's proposed subsurface ocean could play the same role.

The presence of a liquid layer beneath Callisto's surface is difficult to understand. Even with ammonia acting as an antifreeze, this layer cannot be too cold or it would solidify. This argues for a relatively *warm* interior. But *Galileo* measurements suggest that Cal-

listo's interior is not highly differentiated. This implies that the interior is *cold*, and may never have been heated enough to undergo complete chemical differentiation. The full story of Callisto's interior, it would seem, has yet to be told.

Although Ganymede and Callisto have many differences, one feature they share is an extremely thin and tenuous atmosphere. The molecules that make up Ganymede's atmosphere are oxygen (O_2) and ozone (O_3). These are thought to be released from gas bubbles trapped in Ganymede's surface ice. On Callisto, by contrast, the atmosphere is composed of carbon dioxide (CO_2). This gas may have evaporated from Callisto's poles, where temperatures are low enough to form CO_2 ice.

The diagrams in **Figure 15-17** summarize the probable interior structures of the Galilean satellites. They demonstrate the remarkable variety of these terrestrial worlds.

15-8 | Titan has a thick, opaque atmosphere rich in methane, nitrogen, and hydrocarbons

WEB LINK 15.2

Unlike Jupiter, Saturn has only one large satellite that is comparable in size to our own Moon. This satellite, Titan, has a diameter of 5150 km. Of all the satellites of the solar system, only Jupiter's satellite Ganymede (Section 15-7) is larger. Like Ganymede, Titan has a low density that suggests it is made of a mixture and ice and rock. (See Table 7-2 for a comparison between Titan and the six other large satellites of our solar system.)

Titan (**Figure 15-18**) was discovered by Christiaan Huygens in 1665, but its unique features were not revealed until the twentieth century. By the early 1900s, scientists had begun to suspect that Titan might have an atmosphere, because it is cool enough and massive enough to retain heavy gases. (Box 7-2 explains the criteria for a planet or satellite to retain an atmosphere.) These suspi-

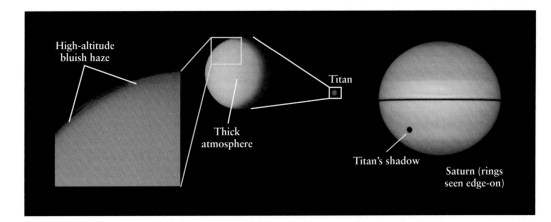

Figure 15-18 R **V** U X G

Titan The Hubble Space Telescope image at far right shows Titan and the shadow it casts on Saturn's clouds. The *Voyager* image in the middle shows how Titan's atmosphere gives it a nearly featureless appearance. At far left is a color-enhanced *Voyager* image that shows a haze extending well above the satellite's visible edge, or limb. (Left: JPL/NASA; center: NASA; right: Erich Karkoschka, LPL/STScI/NASA)

cions were confirmed in 1944, when Gerard Kuiper discovered the spectral lines of methane in sunlight reflected from Titan. (We discussed Titan's spectrum and its interpretation in Section 7-3.) Titan is now known to be the only satellite in the solar system with an appreciable atmosphere.

Because of its atmosphere, Titan was a primary target for the *Voyager* missions. To the chagrin of mission scientists and engineers, however, *Voyager 1* spent hour after precious hour of its limited observation time sending back featureless images like the one at the center of Figure 15-18. Titan's atmosphere is so thick—about 200 km (120 mi) deep, or about 10 times the depth of the Earth's troposphere—that it blocks any view of the surface. The haze surrounding Titan is so dense that little sunlight penetrates to the ground; noon on Titan is probably no brighter than a moonlit midnight on Earth.

By examining how the radio signal from *Voyager 1* was affected by passing through Titan's atmosphere, scientists inferred that the atmospheric pressure at Titan's surface is 50% greater than the atmospheric pressure at sea level on the Earth. Titan's surface gravity is weaker than that of Earth, so to produce this pressure Titan must have considerably more gas in its atmosphere than Earth.

Indeed, calculations show that about 10 times more gas (by mass) lies above a square meter of Titan's surface than above a square meter of Earth.

Voyager data suggest that roughly 90% of Titan's atmosphere is nitrogen. Most of this nitrogen probably came from ammonia (NH_3), a compound of nitrogen and hydrogen that is quite common in the outer solar system. Ammonia is easily broken into hydrogen and nitrogen atoms by the Sun's ultraviolet radiation. Titan's gravity is too weak to retain hydrogen atoms (see Box 7-2), so these atoms escape into space and leave the more massive nitrogen atoms behind.

The second most abundant gas on Titan is methane (CH_4), which is the principal component of the "natural gas" used on Earth as fuel. The interaction of methane with ultraviolet light from the Sun produces a variety of other carbon-hydrogen compounds, or **hydrocarbons**. For example, the *Voyager* infrared spectrometers detected small amounts of ethane (C_2H_6), acetylene (C_2H_2), ethylene (C_2H_4), and propane (C_3H_8) in Titan's atmosphere. During summer on one of Titan's hemispheres, these chemicals can be lifted by convection and form clouds large enough to be visible from Earth (**Figure 15-19**).

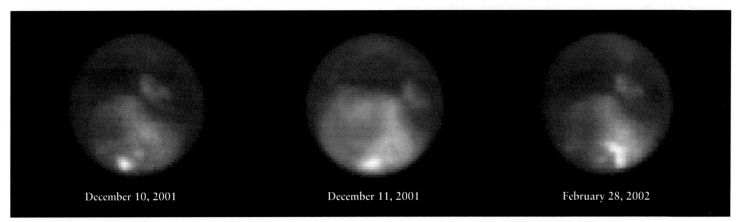

December 10, 2001   December 11, 2001   February 28, 2002

Figure 15-19 R **I** V U X G

Summer Clouds on Titan These false-color infrared images, made using adaptive optics (see Section 6-4), show summer in Titan's southern hemisphere (at the bottom of each image). The white blotches are hydrocarbon clouds, which may form in the same way as summer thunderstorm clouds on Earth. The largest cloud, in the February 28, 2002, image, is over 1400 km (900 mi) long. (Courtesy of Michael Brown, Caltech).

Ethane is the most abundant hydrocarbon produced from methane in Titan's atmosphere. It condenses into droplets as it is produced and falls to Titan's surface. Over the history of the solar system, enough ethane may have been produced to create rivers and lakes. To test this idea, astronomers have made observations at infrared wavelengths (which can penetrate all the way through Titan's atmosphere) in an effort to map the satellite's surface. Images produced in this way suggest that there may indeed be bodies of liquid hydrocarbons on Titan (**Figure 15-20**). These images also confirm that Titan, like most satellites in the solar system, is in synchronous rotation.

Nitrogen in Titan's atmosphere can combine with airborne hydrocarbons to produce other compounds, such as hydrogen cyanide (HCN). Hydrogen cyanide, along with other molecules, can join together in long, repeating molecular chains to form substances called **polymers**. Droplets of some polymers remain sus-

pended in the atmosphere, forming an **aerosol**. (Common aerosols on the Earth include fog, mist, and paint sprayed from a can.) The polymers in Titan's airborne aerosol are thought to be responsible for the reddish-brown color seen in Figure 15-18. The heavier polymer particles do not remain airborne but settle down onto Titan's solid surface, probably covering it with a thick layer of sticky, tarlike goo.

Some of the compounds of hydrogen, nitrogen, and carbon found in Titan's atmosphere and on its surface are the building blocks of the organic molecules on which life is based. There is little reason to suspect that life exists on Titan, however; its surface temperature of 95 K (−178°C = −288°F) is prohibitively cold. Nevertheless, a more detailed study of Titan's chemistry may shed light on the origins of life on the Earth.

Scientists hope to learn much more about Saturn's largest satellite from *Huygens*, a probe designed to enter Titan's atmosphere. *Huygens* was launched in 1997 as part of the *Cassini* mission (see Section 14-3), and was to be released from *Cassini* before the parent spacecraft entered Saturn orbit in 2004. (When these words were written in early 2004, *Cassini* and *Huygens* were still en route.) The *Huygens* mission plan included a 2½-hour descent on a parachute through Titan's atmosphere. During the descent, the spacecraft was to radio its findings about the atmosphere back to *Cassini* to be relayed on to Earth. After landing at 25 km/h (15 mi/h), *Huygens* was to give an on-site report on the nature of Titan's surface—be it water ice, solid ammonia, liquid ethane, or hydrocarbon "goo."

15-9 | Jupiter has dozens of small satellites that have different origins

 Besides the Galilean satellites, Jupiter has at least 59 other small satellites. Each is named for a mythological character associated with Jupiter. Appendix 3 gives details of some of the satellites' orbits. It is helpful to sort Jupiter's 63 known satellites (as of early 2004) into three groups according to their sizes and orbits.

Four planet-sized satellites, Io, Europa, Ganymede, and Callisto, can be thought of as four of the terrestrial worlds of the solar system. All are comparable in size to the planet Mercury (see Table 15-1). The semimajor axes of their orbits range from 5.9 to 26 times the radius of Jupiter. (By comparison, the distance from the Earth to the Moon is about 60 times the Earth's radius.)

Four small inner satellites, Metis, Adrastea, Amalthea, and Thebe, have orbital semimajor axes between 1.8 and 3.1 times the radius of Jupiter. As **Figure 15-21** shows, these satellites have irregular shapes much like a potato (or an asteroid). The largest of the inner satellites, Amalthea, measures about 270 by 150 km (170 by 95 mi)—about 10 times larger than the moons of Mars, but only about one-tenth the size of the Galilean satellites. Discovered in 1892, Amalthea has a distinct reddish color due to sulfur that Jupiter's magnetosphere removed from Io's volcanic plumes, then deposited onto Amalthea's surface. The three other inner satellites are even smaller than Amalthea. They were first discovered in images from *Voyager 1* and *Voyager 2*.

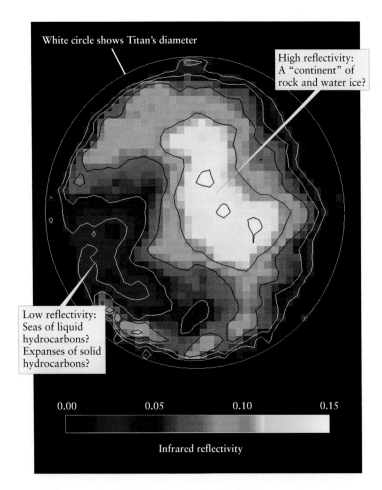

White circle shows Titan's diameter

High reflectivity: A "continent" of rock and water ice?

Low reflectivity: Seas of liquid hydrocarbons? Expanses of solid hydrocarbons?

| 0.00 | 0.05 | 0.10 | 0.15 |

Infrared reflectivity

figure 15-20 R■VUXG

Hydrocarbon Seas on Titan? The false colors in this infrared view of Titan indicate what fraction of infrared light from the Sun is reflected by different parts of the surface. The area in dark blue, with a reflectivity of less than 0.03 (3%), may be a region where hydrocarbons have rained down and collected on the surface. (S. G. Gibbard, B. Macintosh, and C. Max, IGPP/LLNL)

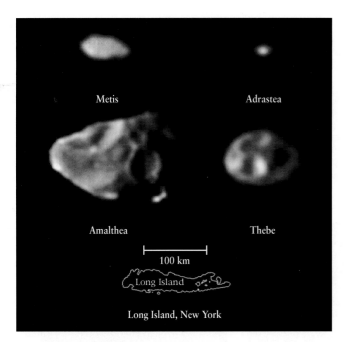

figure 15-21   R I **V** U X G

Jupiter's Four Inner Satellites First observed in 1892, Amalthea was the last satellite of any planet to be discovered without the aid of photography. Metis, Adrastea, and Thebe, by contrast, are so small that they were not discovered until 1979. These images from the *Galileo* spacecraft are the first to show these three satellites as anything more than points of light. (NASA/JPL; Cornell University)

Fifty-five small outer satellites have diameters from 184 km for the largest to about 1 km for the smallest. All orbit far beyond Callisto, with semimajor axes that range from about 100 to 360 times the radius of Jupiter. (Twenty-three of these satellites were discovered in 2003 alone, and even more may be discovered as telescope technology continues to evolve.)

The Galilean satellites and the four inner satellites all orbit Jupiter in the plane of the planet's equator. (The same is true of Jupiter's rings, discussed in Section 14-10.) Furthermore, these are all **prograde orbits,** which means that these objects orbit Jupiter in the same direction as Jupiter's rotation. This is what we would expect if these satellites formed from the same primordial, rotating cloud as Jupiter itself. (In a similar way, the planets all orbit the Sun in nearly the same plane and in the same direction as the Sun's rotation. As we saw in Section 8-3, this reinforces the idea that the Sun and planets formed from the same solar nebula.)

In contrast, the 55 outer satellites all circle Jupiter along orbits that are inclined at steep angles to the planet's equatorial plane. These satellites are thought not to have formed along with Jupiter; rather, they are probably wayward asteroids captured by Jupiter's powerful gravitational field. One piece of evidence for this is that 48 of the outer satellites are in **retrograde orbits:** They orbit Jupiter in a direction opposite the planet's rotation. Calculations show that it is easier for Jupiter to capture a satellite into a retrograde orbit than a prograde one.

15-10 | The icy surfaces of Saturn's six moderate-sized moons provide clues to their histories

Saturn, too, has an extensive collection of satellites: 31 were known as of early 2004. As we did for Jupiter's satellites in Section 15-9, we can subdivide the satellites of Saturn into three categories based on their sizes and their orbits. (Appendix 3 has information about the orbits of all of Saturn's satellites.) We will see that while Saturn does not have a set of large moons like the Galilean satellites of Jupiter, it does have a remarkably diverse collection of moderate-sized satellites.

One planet-sized satellite, Titan, is actually one of the terrestrial worlds of our solar system. With its diameter of 5150 km, Titan is intermediate in size between Mercury and Mars. Its prograde orbit has a semimajor axis of just over 20 times the radius of Saturn and lies in nearly the same plane as Saturn's equator and rings.

Six moderate-sized satellites were all discovered before 1800. These six satellites have properties and diameters that lead us to group them in pairs: Mimas and Enceladus (400 to 500 km in diameter), Tethys and Dione (roughly 1000 km in diameter), and Rhea and Iapetus (about 1500 km in diameter). The semimajor axes of their orbits range from 3.1 to 59 times the radius of Saturn. Like Titan, all of their orbits are prograde and close to the plane of Saturn's equator.

Twenty-four small satellites have sizes from 7 to 266 km. Some are in prograde orbits as small as 2.2 times the radius of Saturn, and lie within or close to the rings (see Figure 14-21). These may be jagged fragments of ice and rock produced by impacts and collisions. The shepherd satellites that shape Saturn's F ring (Section 14-12) may be such fragments. Others are in retrograde orbits with semimajor axes as large as 380 times Saturn's radius, and may be captured asteroids.

The most curious and mysterious of these worlds may be Saturn's six moderate-sized satellites (**Figure 15-22**). These six satellites have very low average densities of less than 1400 kg/m³. This implies that unlike the satellites of the inner solar system or the Galilean satellites of Jupiter, they are composed primarily of ices (mostly frozen water and ammonia) with very little rock. The motions of these mid-sized satellites are not surprising: All six orbit the planet in the plane of Saturn's equator and in the direction of the planet's rotation, and all six exhibit synchronous rotation. What is surprising is the curious variety of surface features that the *Voyager* spacecraft discovered on these satellites.

Mimas, the innermost of the six middle-sized moons, has the heavily cratered surface we would expect on such a small world. Its most remarkable feature is a crater so large (Figure 15-22*a*) that the impact that formed it must have come close to shattering Mimas into fragments.

 In contrast to Mimas's heavily cratered surface, its neighbor and near twin, Enceladus, has extensive crater-free regions (Figure 15-22*b*). Apparently some form of geologic activity has resurfaced these areas within the past 100 million years, and may still be under way. Geologic activity may also explain why Enceladus has an amazingly high albedo of 0.95, making it the most reflective large object in our solar system. Some process has freed Enceladus's

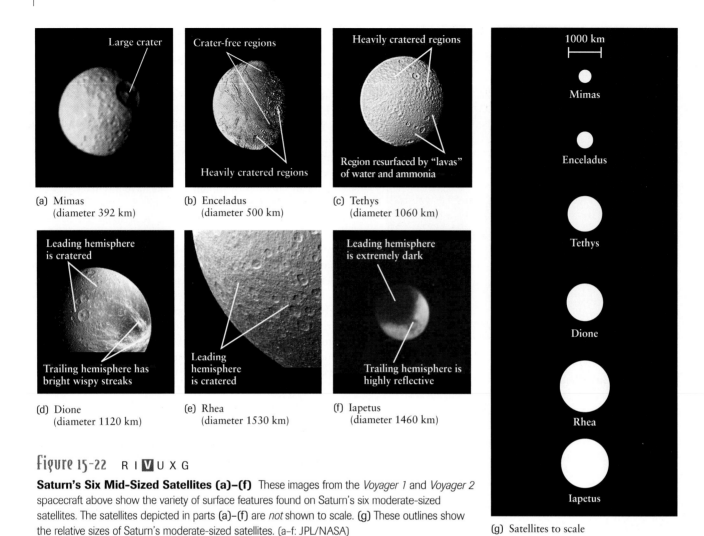

(a) Mimas
(diameter 392 km)

(b) Enceladus
(diameter 500 km)

(c) Tethys
(diameter 1060 km)

(d) Dione
(diameter 1120 km)

(e) Rhea
(diameter 1530 km)

(f) Iapetus
(diameter 1460 km)

(g) Satellites to scale

Figure 15-22 R I V U X G

Saturn's Six Mid-Sized Satellites (a)–(f) These images from the *Voyager 1* and *Voyager 2* spacecraft above show the variety of surface features found on Saturn's six moderate-sized satellites. The satellites depicted in parts (a)–(f) are *not* shown to scale. (g) These outlines show the relative sizes of Saturn's moderate-sized satellites. (a–f: JPL/NASA)

icy surface of rock and dust, whose presence on the other icy satellites is the likeliest explanation for their lower albedos. Evidence for even more dramatic geologic activity comes from studies of Saturn's E ring, which suggest that some kind of eruptions on Enceladus supply particles to this ring (see Section 14-11). However, the *Voyager* spacecraft did not see any such eruptions.

What could be powering the geologic activity on Enceladus? One guess is tidal heating, which provides the internal heat for the inner Galilean satellites of Jupiter (see Section 15-4). The satellite Dione orbits Saturn in 65.7 hours, almost exactly twice Enceladus's orbital period of 32.9 hours, and the 1:2 ratio of orbital periods should set Enceladus up to be caught in a tidal tug-of-war between Saturn and Dione. Calculations suggest that the amount of tidal heating produced in this way is modest, but should be enough to melt the ices of Enceladus at least part of the time. Curiously, there is also a 1:2 ratio between the orbital periods of Mimas and Saturn's moon Tethys, but the cratered surface of Mimas shows no sign that this moon was ever tidally heated.

Tethys and Dione are the next-largest pair of Saturn's six moderate-sized satellites. Most of Tethys, like Mimas, is heavily cratered. But Tethys also has a curious, roughly circular plain where craters are more sparse than on the rest of the satellite

(Figure 15-22c). Like the lunar maria (see Section 10-1), this region has apparently been resurfaced by lava flows. The difference is that the "lava" on Tethys is probably a mixture of water and ammonia, which flows like a viscous fluid at the ultracold temperatures found on Saturn's icy satellites. Dione, slightly larger than Tethys, also has a strange surface dichotomy: The surfaces of its leading hemisphere (that is, the hemisphere facing toward its direction of orbital motion) and trailing hemisphere are quite different. Dione's leading hemisphere is heavily cratered, but its trailing hemisphere is covered by a network of strange, wispy markings (Figure 15-22d). These wisps may be troughs and valleys in the icy surface, or they may be fresh deposits of frozen water formed by volcanic outgassing from Dione's interior.

Because Tethys and Dione have about double the diameter of Enceladus, the energy to power their geologic activity probably came from within. These larger satellites retain heat more effectively, including heat from the decay of radioactive materials in each satellite's interior. (Such materials are scarce within the icy satellites of Saturn, but only relatively little heat is needed to power ice volcanism.)

Rhea and Iapetus are the largest and most distant of Saturn's moderate-sized moons. Rhea resembles Dione in that its leading hemisphere is heavily cratered (Figure 15-22*e*); its trailing hemisphere is covered by a network of odd wispy markings. Whatever process produced the wisps on one side of Dione may also have been active on the corresponding side of Rhea.

Iapetus has been known to be unusual ever since its discovery in 1671: Unlike the other satellites, its brightness varies greatly as it orbits Saturn. Images from the *Voyager* spacecraft show extreme differences between the leading and trailing hemispheres of Iapetus (Figure 15-22*f*). The leading hemisphere is as black as asphalt (albedo = 0.05), but the trailing hemisphere is highly reflective (albedo = 0.50), like the surfaces of the other moderate-sized moons.

What could account for the two-sided nature of Iapetus? One possibility is that the dark material covering Iapetus's leading hemi-sphere may have come from Phoebe, a smaller satellite that orbits about 3½ times farther from Saturn than Iapetus. Phoebe moves in a retrograde direction along an orbit tilted well away from the plane of Saturn's equator. This motion leads scientists to suspect that Phoebe is a captured asteroid, and its dark surface does indeed resemble the surfaces of a particular class of carbon-rich asteroids. Bits and pieces of this dark charcoal-like material drifting toward Saturn may have been swept up onto the leading hemisphere of Iapetus. However, a sticking point with this scenario is that the dark material on Iapetus is reddish in color, while Phoebe's surface is not. For now, the curious coloration of Iapetus remains unexplained. Observations from the *Cassini* spacecraft may cast light on this and other mysteries of Saturn's satellites.

Key Words

aerosol, p. 328
bright terrain (Ganymede), p. 324
dark terrain (Ganymede), p. 324
Galilean satellites, p. 314
hydrocarbon, p. 327

ice rafts (Europa), p. 322
Io torus, p. 320
occultation, p. 314
polymer, p. 328
prograde orbit, p. 329

retrograde orbit, p. 329
tidal heating, p. 318
ultramafic lava, p. 320

Key Ideas

Nature of the Galilean Satellites: The four Galilean satellites orbit Jupiter in the plane of its equator. All are in synchronous rotation.

• The orbital periods of the three innermost Galilean satellites, Io, Europa, and Ganymede, are in the ratio 1:2:4.

• The two innermost Galilean satellites, Io and Europa, have roughly the same size and density as our Moon. They are composed principally of rocky material. The two outermost Galilean satellites, Ganymede and Callisto, are roughly the size of Mercury. Lower in density than either the Moon or Mercury, they are made of roughly equal parts ice and rock.

• The Galilean satellites probably formed in a similar fashion to our solar system but on a smaller scale.

Io: Io is covered with a colorful layer of sulfur compounds deposited by frequent explosive eruptions from volcanic vents. These eruptions resemble terrestrial geysers.

• The energy to heat Io's interior and produce the satellite's volcanic activity comes from tidal forces that flex the satellite. This tidal flexing is aided by the 1:2:4 ratio of orbital periods among the inner three Galilean satellites.

• The Io torus is a ring of electrically charged particles circling Jupiter at the distance of Io's orbit. Interactions between this ring and Jupiter's magnetic field produce strong radio emissions. Io may also have a magnetic field of its own.

Europa: While composed primarily of rock, Europa is covered with a smooth layer of water ice.

• The surface has hardly any craters, indicating a geologically active history. Other indications are a worldwide network of long cracks and ice rafts that indicate a subsurface layer of liquid water or soft ice. As for Io, tidal heating is responsible for Europa's internal heat.

• An ocean may lie beneath Europa's frozen surface. Minerals dissolved in this ocean may explain Europa's induced magnetic field.

Ganymede: Two types of terrain are found on the icy surface of Ganymede: areas of dark, ancient, heavily cratered surface and regions of heavily grooved, lighter-colored, younger terrain.

• Ganymede is highly differentiated, and probably has a metallic core. It has a surprisingly strong magnetic field and a magnetosphere of its own.

• While there is at present little tidal heating of Ganymede, it may have been heated in this fashion in the past. An induced magnetic field suggests that it, too, has a layer of liquid water beneath the surface.

Callisto: Callisto has a heavily cratered crust of water ice. The surface shows little sign of geologic activity, because there was never any significant tidal heating of Callisto. However, some unknown processes have erased the smallest craters and blanketed the surface with a dark, dusty substance.

• Magnetic field data seem to suggest that Callisto has a shallow subsurface ocean.

Titan: The largest Saturnian satellite, Titan, is a terrestrial world with a dense nitrogen atmosphere. A variety of hydrocarbons are produced there by the interaction of sunlight with methane. These compounds form an aerosol layer in Titan's atmosphere and possibly cover some of its surface with lakes of ethane.

Other Satellites: As of early 2004, Jupiter has a total of 63 known satellites and Saturn has a total of 31.

• In addition to the Galilean satellites, Jupiter has four small inner satellites that lie inside Io's orbit. Like the Galilean satellites, these orbit in the plane of Jupiter's equator. The remaining satellites are small and move in much larger orbits that are noticeably inclined to the plane of Jupiter's equator. Many of these orbit in the direction opposite to Jupiter's rotation.

• In addition to Titan, six moderate-sized moons circle Saturn in regular orbits: Mimas, Enceladus, Tethys, Dione, Rhea, and Iapetus. They are probably composed largely of ice, but their surface features and histories vary significantly. The other, smaller moons include shepherd satellites that control the shapes of Saturn's rings and captured asteroids in large retrograde orbits.

Review Questions

1. Why can't the Galilean satellites be seen with the naked eye?

2. During the time it takes Ganymede to complete one orbit, how many orbits do Io and Europa complete?

3. In what ways does the system of Galilean satellites resemble our solar system? In what ways is it different?

4. No spacecraft from Earth has ever landed on any of the Galilean satellites. How, then, can we know anything about the chemical compositions of these satellites?

5. In what ways did the formation of the Galilean satellites mimic the formation of the planets? In what ways were the two formation processes different?

6. All of the Galilean satellites orbit Jupiter in the same direction. Furthermore, the planes of their orbits all lie within 0.5° of Jupiter's equatorial plane. Explain why these observations are consistent with the idea that the Galilean satellites formed from a "Jovian nebula."

7. What is the source of energy that powers Io's volcanoes? How is it related to the orbits of Io and the other Galilean satellites?

8. Io has no impact craters on its surface, while our Moon is covered with craters. What is the explanation for this difference?

9. Despite all the gases released from its interior by volcanic activity, Io does not possess a thick atmosphere. Explain why not.

10. Long before the *Voyager* flybys, Earth-based astronomers reported that Io appeared brighter than usual for the few hours after it emerged from Jupiter's shadow. From what we know about the material ejected from Io's volcanoes, suggest an explanation for this brief brightening of Io.

11. How do lavas on Io differ from typical lavas found on Earth? What does this difference tell us about Io's interior?

12. What accounts for the different colors found on Io's surface?

13. What is the Io torus? What is its source?

14. What is the origin of the electric current that flows through Io?

15. How was the *Galileo* spacecraft used to determine the internal structure of Io and the other Galilean satellites?

16. What surface features on Europa provide evidence for geologic activity?

17. What is the evidence for an ocean of liquid water beneath Europa's icy surface? What is the evidence that substances other than water are dissolved in this ocean?

18. What aspects of Europa lead scientists to speculate that life may exist there? What precautions have been taken to prevent contaminating Europa with Earth organisms?

19. Why is ice an important constituent of Ganymede and Callisto, but not of the Earth's Moon?

20. How do scientists know that the dark terrain on Ganymede is younger than the bright terrain?

21. In what ways is Ganymede like our own Moon? In what ways is it different? What are the reasons for the differences?

22. Why were scientists surprised to learn that Ganymede has a magnetic field? What does this field tell us about Ganymede's history?

23. What leads scientists to suspect that there is liquid water beneath Ganymede's surface?

24. Why are numerous impact craters found on Ganymede and Callisto but not on Io or Europa?

25. Describe the surprising aspects of Callisto's surface and interior that were revealed by the *Galileo* spacecraft. Why did these come as a surprise?

26. Compare and contrast the surface features of the four Galilean satellites. Discuss their relative geological activity and the evolution of these four satellites.

27. The larger the orbit of a Galilean satellite, the less geologic activity that satellite has. Explain why.

28. Explain how the 1:2:4 ratio of the orbital periods of Io, Europa, and Ganymede is related to the geologic activity on these satellites.

29. Describe Titan's atmosphere. What effect has the Sun's ultraviolet radiation had on Titan's atmosphere?

30. The images of Titan's clouds in Figure 15-19 were made using a detector tuned to an infrared wavelength of 2.11 μm. By contrast, images of Titan's surface like Figure 15-20 are made at a wavelength of 2.06 μm. What does this tell you about the transparency of Titan's lower atmosphere to these two wavelengths of infrared light?

31. Why do scientists suspect that there may be liquid hydrocarbons on Titan?

32. How would you account for the existence of the satellites of Jupiter other than the Galilean ones?

33. Which of Saturn's moderate-sized satellites show evidence of geologic activity? What might be the energy source for this activity?

34. Explain why debris from Phoebe would be expected to pile up only on the leading hemisphere of Iapetus. (*Hint:* How do the orbits of these two satellites compare? How does the orbital motion of debris falling slowly inward toward Saturn compare with the orbital motion of Iapetus?)

35. Saturn's equator is tilted by 27° from the ecliptic, while Jupiter's equator is tilted by only 3°. Use these data to explain why we see fewer transits, eclipses, and occultations of Saturn's satellites than of the Galilean satellites.

Advanced Questions

Questions preceded by an asterisk (*) involve topics discussed in Box 1-1 or Box 7-2.

> **Problem-solving tips and tools**
>
> Newton's form of Kepler's third law (Box 4-4) relates the masses of two objects in orbit around each other to the period and size of the orbit. The small-angle formula is discussed in Box 1-1. The best seeing conditions on the Earth give a limiting angular resolution of ¼ arcsec. Because the orbits of the Galilean satellites are almost perfect circles, you can easily calculate the orbital speeds of these satellites from the data listed in Table 15-1. Data about Jupiter itself are given in Table 14-1 and data about all of the satellites of Jupiter and Saturn can be found in Appendix 3. For a discussion of escape speed and how planets retain their atmospheres, see Box 7-2.

36. Using the orbital data in Table 15-1, demonstrate that the Galilean satellites obey Kepler's third law.

*37. What is the size of the smallest feature you should be able to see on a Galilean satellite through a large telescope under conditions of excellent seeing when Jupiter is near opposition? How does this compare with the best *Galileo* images, which have resolutions around 25 meters?

*38. Astronomers who observed Io with the Hubble Space Telescope in 1992 claimed that they could see features as small as 150 km across. When the observations were made, Io was 4.45 AU from the Earth. What was the angular resolution of the Hubble Space Telescope at this time? (Subsequent repair missions to the Hubble Space Telescope improved its resolution.)

39. Using the diameter of Io (3642 km) as a scale, estimate the height to which the plume of Pele rises above the surface of Io in Figure 15-5a. (You will need to make measurements on this figure using a ruler.) Compare your answer to the value given in the figure caption.

40. Explain why the volcanic plumes in Figure 15-5b have a bluish color. (*Hint:* See Box 5-4.)

41. Jupiter, its magnetic field, and the charged particles that are trapped in the magnetosphere all rotate together once every 10 hours. Io takes 1.77 days to complete one orbit. Using a diagram, explain why particles from Jupiter's magnetosphere hit Io primarily from behind (that is, on the side of Io that trails as it orbits the planet).

42. Assuming material is ejected from Io into Jupiter's magnetosphere at the rate of 1 ton per second (1000 kg/s), how long will it be before Io loses 10% of its mass? How does your answer compare with the age of the solar system?

43. Volcanic eruptions from Io add about 1 ton of material to Jupiter's magnetosphere every second. Based on the information given in Section 15-4, what fraction of the material ejected from Io's volcanoes goes into Jupiter's magnetosphere?

44. How long does it take for Ganymede to entirely enter or entirely leave Jupiter's shadow? Assume that the shadow has a sharp edge.

45. Figure 15-15 shows that Ganymede's dark terrain has both more craters and larger craters than the bright terrain. Explain what this tells you about the sizes of meteoroids present in the solar system in the distant past and in the more recent past.

46. Put a few cubes of ice in a glass, then fill the glass with warm water to make the ice cubes crack. The fracture lines in each cube are clearly visible because they reflect light. Use this observation to suggest why Ganymede's icy bright terrain, which may have been badly fractured by the stresses that produced the grooves (Figure 15-15), is 25% more reflective than the dark terrain.

*47. Find the escape speed on Titan. What is the limiting molecular weight of gases that could be retained by Titan's gravity? (*Hint:* Use the ideas presented in Box 7-2 and assume an average atmospheric temperature of 95 K.)

48. Many of the gases in the atmosphere of Titan, such as methane, ethane, and acetylene, are highly flammable. Why, then, doesn't Titan's atmosphere catch fire? (*Hint:* What gas in our atmosphere is needed to make wood, coal, or gasoline burn?)

49. Suppose the Earth's Moon were removed and replaced in its orbit by Titan. What changes would you expect to occur in Titan's atmosphere? Would solar eclipses be more or less common as seen from Earth? Explain your answers.

50. At infrared wavelengths, the Hubble Space Telescope can see details on Titan's surface as small as 580 km (360 mi) across. Determine the angular resolution of the Hubble Space Telescope using infrared light. If visible light is used, is the angular resolution better, worse, or the same? Explain your answer.

51. (a) To an observer floating in Jupiter's Great Red Spot, Io would rise in the east and set in the west, but Metis would rise in the *west* and set in the *east*. Explain how this is possible. (b) At what distance from the center of Jupiter would a satellite have to orbit so that it would neither rise nor set as seen by the observer in (a)?

52. (a) To an observer on Enceladus, what is the time interval between successive oppositions of Dione? Explain your answer. (b) As seen from Enceladus, what is the angular diameter of Dione at opposition? How does this compare to the angular diameter of the Moon as seen from Earth (about ½°)?

Discussion Questions

53. If you could replace our Moon with Io, and if Io could maintain its present amount of volcanic activity, what changes would this cause in our nighttime sky? Do you think that Io could in fact remain volcanically active in this case? Why or why not?

54. In the classic science-fiction film *2010: The Year We Make Contact,* an alien intelligence causes Jupiter to contract so much that nuclear reactions begin at its center. As a result, Jupiter becomes a star like the Sun. Is this possible in principle? Explain your answer.

55. Speculate on the possibility that Europa, Ganymede, or Callisto might harbor some sort of life. Explain your reasoning.

56. Suppose you were planning four missions that would land a spacecraft on each of the Galilean satellites. What kinds of questions would you want these missions to answer? What kinds of data would you want your spacecraft to send back? In view of the different environments on the four satellites, how would the designs of the four spacecraft differ? Be specific about the possible hazards and problems each spacecraft might encounter in landing on the four satellites. If only one of these missions could be funded, which one would you choose? Why?

57. Comment on the suggestion that Titan may harbor life-forms.

58. Jupiter's satellite Io and Saturn's satellite Enceladus are both geologically active, and both are in 2-to-1 resonances with other satellites. However, the amount of geologic activity of Enceladus is far less than on Io. Discuss some possible reasons for this difference.

59. Imagine that you are in charge of planning the *Cassini* orbital tour of the Saturnian system. In your opinion, which satellites in the system should be examined, what data should be collected, and what kinds of questions should the mission attempt to answer?

 Web/CD-ROM Questions

60. Various spacecraft missions have been proposed to explore Europa in greater detail. Search the World Wide Web for information about these. How would these missions test for the presence of an ocean beneath Europa's surface?

61. Twenty-one new satellites of Jupiter were discovered in 2003. Search the World Wide Web for information about how these satellites were discovered. How was it determined that these satellites are actually in orbit around Jupiter?

62. In 2003 astronomers reported the discovery of a new moon of Saturn, given the interim designation S/2003 S1. Search the World Wide Web for information about this satellite. How did astronomers discover it? How do they infer the satellite's diameter?

 63. The Surface of Ganymede. Access and view the video "Jupiter's Moon Ganymede" in Chapter 15 of the *Universe* Web site or CD-ROM. Describe the different surface features that you see, and explain how each type of feature was formed.

Observing Projects

Observing tips and tools

You can easily find the apparent positions of the Galilean satellites for any date and time using the *Starry Night Backyard*™ software on the CD-ROM that accompanies this textbook. For even more detailed information about satellite positions, consult the "Satellites of Jupiter" section in the *Astronomical Almanac* for the current year. If your goal is to view Saturn's satellites, consult the section entitled "Satellites of Saturn" in the *Astronomical Almanac* for the current year. This includes a diagram showing the orbits of Mimas, Enceladus, Tethys, Dione, Rhea, Titan, and Hyperion. Plan your observing session by looking up the dates and times of the most recent greatest eastern elongations of the various satellites. You will have to convert from universal time (UT), which is the same as Greenwich Mean Time, to your local time zone. Then, using the tick marks along the orbits in the diagram, estimate the positions of the satellites relative to Saturn at the time you will be at the telescope. Another useful resource is the "Celestial Calendar" section of *Sky & Telescope*. During months when Saturn is visible in the night sky, this section of the magazine includes a chart of Saturn's satellites.

64. Observe Jupiter through a pair of binoculars. Can you see all four Galilean satellites? Make a drawing of what you observe. If you look again after an hour or two, can you see any changes?

65. Observe Jupiter through a small telescope on three or four consecutive nights. Make a drawing each night showing the positions of the Galilean satellites relative to Jupiter. Record the time and date of each observation. Consult the sources listed above in the "Observing tips and tools" to see if you can identify the satellites by name.

66. Make arrangements to observe an eclipse, a transit, or an occultation of one of the Galilean satellites. Consult a listing of such phenomena in the "Satellites of Jupiter" section in the *Astronomical Almanac* for the current year. Choose the phenomenon you would like to see and calculate its scheduled time by converting the universal time given in the *Astronomical Almanac* to your local time.

67. If you are fortunate enough to have access to a large telescope with a primary mirror 1 meter or more in diameter, make arrangements to view Jupiter through that telescope. Describe what you see. Under conditions of excellent seeing when Jupiter is near opposition, the Galilean satellites should look like tiny discs rather than starlike pinpoints of light.

68. If you have access to a moderately large telescope, make arrangements to observe several of Saturn's satellites. At the telescope, you should have no trouble identifying Titan. Tethys, Dione, and Rhea are about one-sixth as bright as Titan and should be the next easiest satellites to find. Can you confidently identify any of the other satellites?

 69. Use the *Starry Night Backyard*™ program to observe the Galilean satellites of Jupiter. Display the entire celestial sphere (select **Guides > Atlas** in the **Go** menu) and center on Jupiter by using the **Find...** command in the **Edit** menu. Using the controls at the right-hand end of the Control Panel, zoom in as far as you can. In the Control Panel, set the date and time to March 4, 2004, at 12:00:00 A.M. to see Jupiter at opposition, and set the time step to 1 hour. Then click on the "Forward" button (a triangle that points to the right). You will see the four Galilean satellites orbiting Jupiter. (**a**) Are all four satellites ever on the same side of Jupiter? (**b**) Observe the satellites passing in front of and behind Jupiter. (Zoom in as needed.) Explain how your observations tell you that all four satellites orbit Jupiter in the same direction.

 70. Use the *Starry Night Backyard*™ program to view Saturn from its satellite Enceladus. First select **Viewing Location...** in the **View** menu. At the top of the window that appears, click on the word **Earth** and scroll down to **Enceladus** (listed under Saturn). In the list of locations on Enceladus, scroll to "Ahmad Crater" and then click on **Set Location.** To see Saturn from this location on Enceladus, center on Saturn by using the **Find...** command in the **Edit** menu. (**a**) In the Control Panel, set the discrete time step to 1 minute, then click on the "Forward" button (a triangle that points to the right). How do the stars appear to move as seen from this location on Enceladus? How does Saturn appear to move? What do these observations tell you about the relationship between the orbital period and rotation period of Enceladus? (**b**) Set the discrete time step to 1 hour. By stepping forward through time using the rightmost single-step button, determine how much time elapses from when Saturn appears full from this location to when it next appears full. Explain why this is the same as the orbital period of Enceladus, and compare to the value of the orbital period given in Appendix 3.

 71. Use the *Deep Space Explorer*™ program to examine the satellites of Saturn. (**a**) In the left-hand part of the window under the heading **Solar System,** click on the triangle next to the word **Explore** and then click on **Saturn.** You can rotate Saturn by putting the mouse cursor over the image, holding down the mouse button, and moving the mouse. (On a two-button mouse, hold down the left mouse button.) Use this technique to rotate Saturn so that you are viewing the rings edge-on. Then zoom in and zoom out using the buttons at the upper left of the window (an upward-pointing triangle and a downward-pointing triangle). As you zoom, the distant stars will remain stationary but the satellites of Saturn will appear to move. You should be able to see at least eight satellites. Which satellites are these? (Move the mouse to center the cursor on a satellite and its name will appear. If you can't see all eight, try zooming out farther.) (**b**) The plane of Saturn's rings is the same as the plane of Saturn's equator. Which satellites appear to be the farthest from this plane?

Collaborative Exercises

72. Imagine that scientists are proposing to send a robotic lander to Jupiter's satellite Callisto. Create a 100-word written proposal describing a robotic lander mission to another of the Galilean satellites, explaining why your group found it to be the most interesting and why the government should allocate the money for your alternative project. In your proposal, be sure to demonstrate your knowledge of Callisto and the other satellite.

73. From the data and the accompanying images for Table 15-1, "The Galilean Satellites Compared to the Moon, Mercury, and Mars," use someone's shoe to represent the 150,000 km diameter of Jupiter and determine about how many "shoes" away would each of the Galilean satellites be from Jupiter.

16 The Outer Worlds

Beyond Saturn, in the cold, dark recesses of the solar system, orbit three planets that have long been shrouded in mystery. These planets—Uranus, Neptune, and Pluto (all shown here to the same scale as the Earth)—are so distant, so dimly lit by the Sun, and so slow in their motion against the stars that they were unknown to ancient astronomers and were discovered only after the invention of the telescope. Even then, little was known about Uranus and Neptune until *Voyager 2* flew past these planets during the 1980s.

Surrounded by a system of small moons and thin, dark rings, Uranus is tipped on its side so that its axis of rotation lies nearly in its orbital plane. This remarkable orientation suggests that Uranus may have been the victim of a staggering impact by a massive planetesimal. Neptune is, at first glance, a denser, more massive version of Uranus, but it is a far more active world. It has an internal energy source that Uranus seemingly lacks, as well as atmospheric bands and storm activity resembling those on Jupiter.

Neptune also has dark rings, small, icy moons, and an intriguing large satellite, Triton, which is nearly devoid of impact craters and has geysers that squirt nitrogen-rich vapors. Triton's retrograde orbit suggests that this strange world may have been gravitationally captured by Neptune.

A close cousin of Triton is Pluto, the smallest and most remote of the planets. Pluto still harbors many mysteries, because it has not yet been visited by a spacecraft. It is just one of thousands of small, icy worlds that orbit the Sun at the outskirts of the solar system.

Earth   Uranus   Neptune   Pluto

The Earth and the three outer planets to scale. (Alan Stern, Southwest Research Institute; Mark Buie, Lowell Observatory; NASA; and ESA)   R I **V** U X G

Questions to Guide Your Reading

As you read the sections of this chapter, look for the answers to the following questions:

16-1   How did Uranus and Neptune come to be discovered?

16-2   What gives Uranus its distinctive greenish-blue color?

16-3   Why are the clouds on Neptune so much more visible than those on Uranus?

16-4   Are Uranus and Neptune merely smaller versions of Jupiter and Saturn?

16-5   What is so unusual about the magnetic fields of Uranus and Neptune?

16-6   Why are the rings of Uranus and Neptune so difficult to see?

16-7   Do the moons of Uranus show any signs of geologic activity?

16-8   What makes Neptune's moon Triton unique in the solar system?

16-9   Are there other planets beyond Pluto?

16-1 | Uranus was discovered by chance, but Neptune's existence was predicted by applying Newtonian mechanics

Uranus was discovered quite by accident by William Herschel, a German-born musician who moved to England in 1757 and became fascinated by astronomy. Using a telescope that he built himself, Herschel was systematically surveying the sky on March 13, 1781, when he noticed a faint, fuzzy object that he first thought to be a distant comet. By the end of 1781, however, his observations had revealed that the object's orbit was relatively circular and was larger than Saturn's orbit. Comets, by contrast, can normally be seen only when they follow elliptical orbits that bring them much closer to the Sun.

Herschel had discovered the seventh planet from the Sun. In doing so, he had doubled the radius of the known solar system from 9.5 AU (the semimajor axis of Saturn's orbit) to 19.2 AU (the distance from the Sun to Uranus). Herschel originally named his discovery Georgium Sidus (Latin for "Georgian star") in honor of the reigning monarch, George III. The name Uranus—in Greek mythology, the personification of Heaven—came into currency only some decades later.

Although Herschel received the credit for discovering Uranus, he was by no means the first person to have seen it. At opposition, Uranus is just barely bright enough to be seen with the naked eye under good observing conditions, so it was probably seen by the ancients. Many other astronomers with telescopes had sighted this planet before Herschel; it is plotted on at least 20 star charts drawn between 1690 and 1781. But all these other observers mistook Uranus for a dim star. Herschel was the first to track its motion relative to the stars and recognize it as a planet. This was no small task, because Uranus moves very slowly on the celestial sphere, just over 4° in the space of a year (compared to about 12° for Saturn and about 35° for Jupiter).

It was by carefully tracking Uranus's slow motions that astronomers were led to discover Neptune. By the beginning of the nineteenth century, it had become painfully clear to astronomers that they could not accurately predict the orbit of Uranus using Newtonian mechanics. By 1830 the discrepancy between the planet's predicted and observed positions had become large enough (2 arcmin) that some scientists suspected that Newton's law of gravitation might not be accurate at great distances from the Sun.

By the mid-1840s, two scientists working independently—the French astronomer Urbain Jean Joseph Le Verrier and the English mathematician John Couch Adams—were exploring an earlier and sounder suggestion. Perhaps the gravitational pull of an as yet

table 16-1	Uranus Data
Average distance from Sun:	19.194 AU = 2.871×10^9 km
Maximum distance from Sun:	20.017 AU = 2.995×10^9 km
Minimum distance from Sun:	18.371 AU = 2.748×10^9 km
Eccentricity of orbit:	0.0429
Average orbital speed:	6.83 km/s
Orbital period:	84.099 years
Rotation period (internal):	17.24 hours
Inclination of equator to orbit:	97.86°
Inclination of orbit to ecliptic:	0.77°
Diameter:	51,118 km = 4.007 Earth diameters (equatorial)
Mass:	8.682×10^{25} kg = 14.53 Earth masses
Average density:	1318 kg/m³
Escape speed:	21.3 km/s
Surface gravity (Earth = 1):	0.90
Albedo:	0.56
Average temperature at cloudtops:	−218°C = −360°F = 55 K
Atmospheric composition (by number of molecules):	82.5% hydrogen (H_2), 15.2% helium (He), 2.3% methane (CH_4)

WEB LINK 16.2 (NASA/JPL) R I **V** U X G

table 16-2	Neptune Data
Average distance from Sun:	30.066 AU = 4.498×10^9 km
Maximum distance from Sun:	30.367 AU = 4.543×10^9 km
Minimum distance from Sun:	29.765 AU = 4.453×10^9 km
Eccentricity of orbit:	0.010
Minimum distance from Sun:	29.765 AU = 4.453×10^9 km
Eccentricity of orbit:	0.010
Average orbital speed:	5.5 km/s
Orbital period:	164.86 years
Rotation period (internal):	16.11 hours
Inclination of equator to orbit:	29.56°
Inclination of orbit to ecliptic:	1.77°
Diameter:	49,528 km = 3.883 Earth diameters (equatorial)
Mass:	1.024×10^{26} kg = 17.15 Earth masses
Average density:	1638 kg/m³
Escape speed:	23.5 km/s
Surface gravity (Earth = 1):	1.1
Albedo:	0.51
Average temperature at cloudtops:	–218°C = –360°F = 55 K
Atmospheric composition (by number of molecules):	79% hydrogen (H_2), 18% helium (He), 3% methane (CH_4)

(NASA/JPL)   R I **V** U X G

undiscovered planet was causing Uranus to deviate slightly from its predicted orbit. Calculations by both scientists concluded that Uranus had indeed caught up with and had passed a more distant planet. Uranus had accelerated slightly as it approached the unknown planet, then decelerated slightly as it receded from the planet.

Astronomers at Cambridge University Observatory undertook a six-week search for Adams's proposed new planet in the summer of 1846. Recently unearthed historical evidence suggests that Adams's predictions of the new planet's position were rather imprecise, ranging over as much as 20° of the sky (40 times the angular diameter of the Moon). As a result, Cambridge astronomers were unsuccessful in their search.

Meanwhile, Le Verrier wrote to Johann Gottfried Galle at the Berlin Observatory with a more accurate calculation. Galle received the letter on September 23, 1846, and sighted the planet that very night after just a half-hour of searching. Working with Heinrich d'Arrest, Galle located an uncharted star with the expected brightness in the predicted location. Subsequent observations confirmed that this new "star" showed a planetlike motion with respect to other stars.

Le Verrier proposed that the planet be called Neptune. After years of debate between English and French astronomers, the credit for its discovery came to be divided equally between Adams and Le Verrier. Neptune is thus the only planet whose existence was revealed by calculation rather than chance discovery.

At opposition, Neptune can be bright enough to be visible through small telescopes. The first person thus to have seen it was probably Galileo. His drawings from January 1613, when he was using his telescope to observe Jupiter and its four large satellites, show a "star" less than 1 arcmin from Neptune's location. Galileo even noted in his observation log that on one night this star seemed to have moved in relation to the other stars. But Galileo would have been hard pressed to identify Neptune as a planet, because its motion against the background stars is so slow (just over 2° per year).

Through a large, modern telescope, both Uranus and Neptune are dim, uninspiring sights. Each planet appears as a hazy, featureless disk with a faint greenish-blue tinge. Although Uranus and Neptune are both about 4 times larger in diameter than Earth, they are so distant that their angular diameters as seen from the Earth are tiny—no more than 4 arcsec for Uranus and just over 2 arcsec for Neptune. To an Earth-based observer, Uranus is roughly the size of a golf ball seen at a distance of 1 kilometer.

From 2004 through 2008, Uranus and Neptune are less than 30° apart in the sky in the adjacent constellations of Aquarius and

Capricornus (the Sea Goat). During these years, both planets are at opposition in either August or September, which are thus the best months to view them with a telescope. Table 16-1 and Table 16-2 give basic data about Uranus and Neptune, respectively.

16-2 | Uranus is nearly featureless and has an unusually tilted axis of rotation

Scientists had hoped that *Voyager 2* would reveal cloud patterns in Uranus's atmosphere when it flew past the planet in January 1986. But even images recorded at close range showed Uranus to be remarkably featureless (Figure 16-1). Faint cloud markings became visible in these images only after extreme computer enhancement.

Voyager 2 data confirmed that the Uranian atmosphere is dominated by hydrogen (82.5%) and helium (15.2%), similar to the atmospheres of Jupiter and Saturn. Uranus differs, however, in that 2.3% of its atmosphere is methane (CH_4), which is 5 to 10 times the percentage found on Jupiter and Saturn. In fact, Uranus has a higher percentage of all heavy elements—including carbon atoms, which are found in molecules of methane—than Jupiter and Saturn. (In Section 16-4 we will investigate how this could have come about.)

Methane preferentially absorbs the longer wavelengths of visible light, so sunlight reflected from Uranus's upper atmosphere is depleted of its reds and yellows. This gives the planet its distinct greenish-blue appearance. As on Saturn's moon Titan (see Section 15-8), ultraviolet light from the Sun turns some of the methane gas into a hydrocarbon haze, making it difficult to see the lower levels of the atmosphere.

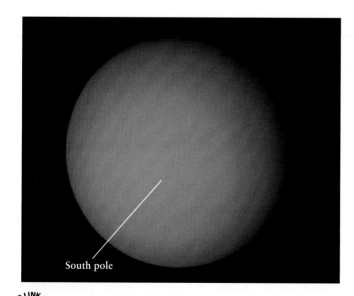

South pole

Figure 16-1 R I **V** U X G

Uranus from *Voyager 2* This image looks nearly straight down onto Uranus's south pole, which was pointed almost directly at the Sun when *Voyager 2* flew past in 1986. None of the *Voyager 2* images of Uranus shows any pronounced cloud patterns. The color is due to methane in the planet's atmosphere, which absorbs red light but reflects green and blue. (JPL/NASA)

Ammonia (NH_3), which constitutes 0.01 to 0.03% of the atmospheres of Jupiter and Saturn, is almost completely absent from the Uranian atmosphere. The reason is that Uranus is colder than Jupiter or Saturn: The temperature in its upper atmosphere is only 55 K (−218°C = −360°F). Ammonia freezes at these very low temperatures, so any ammonia has long since precipitated out of the atmosphere and into the planet's interior. For the same reason, Uranus's atmosphere is also lacking in water. Hence, the substances that make up the clouds on Jupiter and Saturn—ammonia, ammonium hydrosulfide (NH_4SH), and water—are not available on Uranus. This helps to explain the bland, uniform appearance of the planet shown in Figure 16-1.

The few clouds found on Uranus are made primarily of methane, which condenses into droplets only if the pressure is sufficiently high. Therefore, methane clouds form only at lower levels within the atmosphere, where they are difficult to see.

Voyager 2 also confirmed that Uranus's rotation axis is tilted in a unique and bizarre way. Herschel found the first evidence of this in 1787, when he discovered two moons orbiting Uranus in a plane that is almost perpendicular to the plane of the planet's orbit around the Sun. Because the large moons of Jupiter and Saturn were known to orbit in the same plane as their planet's equator and in the same direction as their planet's rotation, it was thought that the same must be true for the moons of Uranus. Thus, Uranus's equator must be almost perpendicular to the plane of its orbit, and its rotation axis must lie very nearly in that plane (Figure 16-2).

Careful measurement shows that Uranus's axis of rotation is tilted by 98°, as compared to 23½° for Earth (compare Figure 16-2 with Figure 2-12). A tilt angle greater than 90° means that Uranus exhibits retrograde (backward) rotation like that shown in Figure 12-3*b*. Along with Venus and Pluto, it is one of only three planets in the solar system to do so. Modern astronomers suspect that Uranus acquired its large tilt angle billions of years ago, when another massive body collided with Uranus while the planet was still forming.

The radical tilt of its axis means that as Uranus moves along its 84-year orbit, its north and south poles alternately point toward or away from the Sun. This produces highly exaggerated seasonal changes. For example, when *Voyager 2* flew by in 1986, Uranus's south pole was pointed toward the Sun. Most of the planet's southern hemisphere was bathed in continuous sunlight, while most of the northern hemisphere was subjected to a continuous, frigid winter night. But over the following quarter of a Uranian year later (21 of our years), sunlight has gradually been returning to the northern hemisphere, triggering immense storms there. Figure 16-3 shows some of these storms, which are much more visible at infrared wavelengths than with visible light.

By following the motions of clouds and storm systems on Uranus, scientists find that the planet's winds flow to the east—that is, in the same direction as the planet's rotation—at northern and southern latitudes, but to the west near the equator. This is quite unlike the situation on Jupiter and Saturn, where the zonal winds alternate direction many times between the north and south poles (see Section 14-4). The fastest Uranian winds (about 700 km/h, or 440 mi/h) are found at the equator.

Although Uranus's equatorial region was receiving little sunlight at the time of the *Voyager 2* flyby, the atmospheric temperature there (about 55 K = −218°C = −359°F) was not too different from

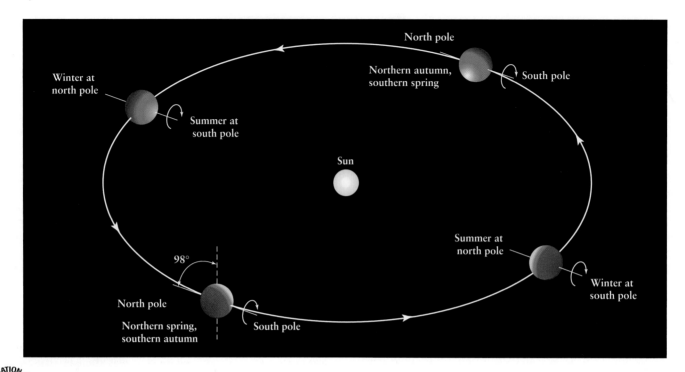

Figure 16-2

Exaggerated Seasons on Uranus For most planets, the rotation axis is roughly perpendicular to the plane of the planet's orbit around the Sun. But for Uranus, the rotation axis is tilted by 98° from the perpendicular, which causes severely exaggerated seasons. For example, during midsummer at Uranus's south pole, the Sun appears nearly overhead for many Earth years, while the planet's northern regions are in continuous darkness. Half an orbit later, the seasons are reversed.

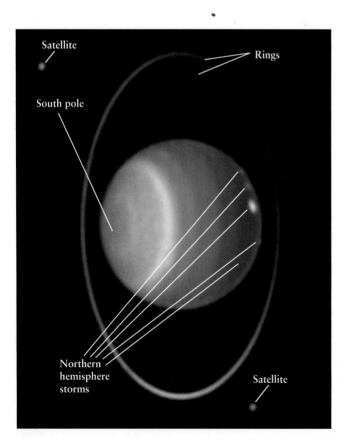

that at the sunlit pole. Heat must therefore be efficiently transported from the poles to the equator. This north-south heat transport, which is perpendicular to the wind flow, may have mixed and homogenized the atmosphere to make Uranus nearly featureless.

The rotation period of Uranus's atmosphere is about 16 hours. Like Jupiter and Saturn, Uranus rotates differentially, so this period depends on the latitude. This can be measured by tracking the motions of clouds. To determine the rotation period for the underlying body of the planet, scientists looked to Uranus's magnetic field, which is presumably anchored in the planet's interior, or at least in the deeper and denser layers of its atmosphere. Data from *Voyager 2* indicate that Uranus's internal period of rotation is 17.24 hours.

Figure 16-3 R **I** V U X G

Sunlight Produces Uranian Storms When the Hubble Space Telescope took this false-color infrared image of Uranus in 1998, sunlight was returning to the planet's northern hemisphere. (Figure 16-1 shows Uranus 12 years earlier.) Solar heating of the northern hemisphere triggered a series of storms, each more than 1000 km (550 mi) across. This image also shows Uranus's rings (see Section 16-6) and two of its satellites (see Section 16-7). (Erich Karkoschka/STScI/NASA)

16-3 | Neptune is a cold, bluish world with Jupiterlike atmospheric features

At first glance, Neptune appears to be the twin of Uranus. (See the image that opens this chapter, and compare Tables 16-1 and 16-2.) But these two planets are by no means identical. While Neptune and Uranus have almost the same diameter, Neptune is 18% more massive. Neptune's axis of rotation also has a more modest 29½° tilt. When *Voyager 2* flew past Neptune in August 1989, it revealed that the planet has a more active and dynamic atmosphere than Uranus. This activity suggests a powerful source of energy within Neptune.

The *Voyager 2* data showed that Neptune has essentially the same atmospheric composition as Uranus: 79% hydrogen, 18% helium, 3% methane, and almost no ammonia or water vapor. As for Uranus, the presence of methane gives Neptune a characteristic bluish-green color. The temperature in the upper atmosphere is also the same as on Uranus, about 55 K. That this should be so, even though Neptune is much farther from the Sun, is further evidence that Neptune has a strong internal heat source.

Unlike Uranus, however, Neptune has clearly visible cloud patterns in its atmosphere. At the time that *Voyager 2* flew past Nep-tune, the most prominent feature in the planet's atmosphere was a giant storm called the Great Dark Spot (**Figure 16-4**). The Great Dark Spot had a number of similarities to Jupiter's Great Red Spot (see Sections 14-1 and 14-3). The storms on both planets were comparable in size to the Earth's diameter, both appeared at about the same latitude in the southern hemisphere, and the winds in both storms circulated in a counterclockwise direction (see Figure 14-5). But Neptune's Great Dark Spot appears not to have been as long-lived as the Great Red Spot on Jupiter. When the Hubble Space Telescope first viewed Neptune in 1994, the Great Dark Spot had disappeared. Another dark storm appeared in 1995 in Neptune's northern hemisphere.

Voyager 2 also saw a few conspicuous whitish clouds on Neptune. These clouds are thought to be produced when winds carry methane gas into the cool, upper atmosphere, where it condenses into crystals of methane ice. *Voyager 2* images show these high-altitude clouds casting shadows onto lower levels of Neptune's atmosphere (**Figure 16-5**). Images from the Hubble Space Telescope also show the presence of high-altitude clouds (**Figure 16-6**).

Thanks to its greater distance from the Sun, Neptune receives less than half as much energy from the Sun as Uranus. But with less solar energy available to power atmospheric motions, why are there high-altitude clouds and huge, dark storms on Neptune but

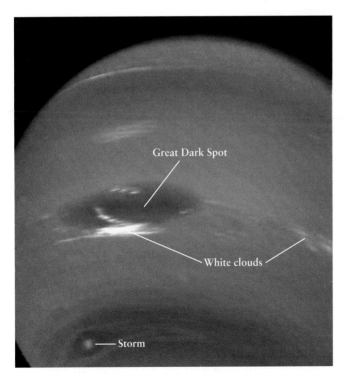

figure 16-4 R I **V** U X G

Neptune from *Voyager 2* When this picture of Neptune's southern hemisphere was taken in 1989, the Great Dark Spot measured about 12,000 by 8000 km, comparable in size to the Earth. (A smaller storm appears at the lower left.) The white clouds are thought to be composed of crystals of methane ice. The color contrast in this image has been exaggerated to emphasize the differences between dark and light areas. (NASA/JPL)

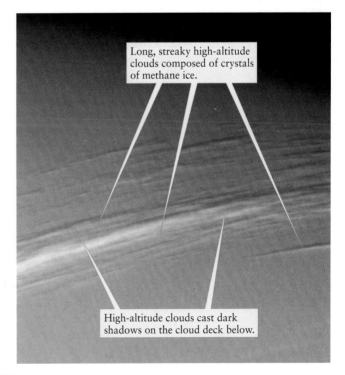

figure 16-5 R I **V** U X G

Cirrus Clouds over Neptune *Voyager 2* recorded this image of clouds near Neptune's terminator (the border between day and night on the planet). Like wispy, high-altitude cirrus clouds in the Earth's atmosphere, these clouds are thought to be made of ice crystals. The difference is that Neptune's cirrus clouds are probably methane ice, not water ice as on Earth. (NASA/JPL)

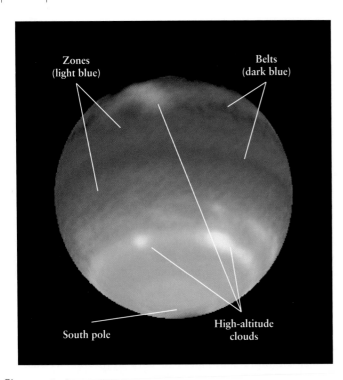

Figure 16-6 R I **V** U X G

Neptune's Banded Structure This enhanced-color Hubble Space Telescope image shows Neptune's belts and zones. White areas denote high-altitude clouds; the very highest clouds (near the top of the image) are shown in yellow-red. The atmosphere absorbs blue light in the green belt near the south pole, perhaps indicating a different chemical composition there. (Lawrence Sromovsky, University of Wisconsin-Madison; STScI/NASA)

not on Uranus? At least part of the answer is probably that Neptune is still slowly contracting, thus converting gravitational energy into thermal energy that heats the planet's core. (The same is true for Jupiter and Saturn, as we saw in Section 14-4.) The evidence for this is that Neptune, like Jupiter and Saturn, emits more energy than it receives from the Sun. The combination of a warm interior and a cold outer atmosphere can cause convection in Neptune's atmosphere, producing the up-and-down motion of gases that generates clouds and storms.

Uranus, by contrast, appears to have little or no internal source of thermal energy; measurements to date show that it radiates just as much energy into space as it receives from the Sun. Therefore, Uranus lacks the dynamic atmospheric activity found on Neptune and has a much blander appearance (compare Figure 16-1 and Figure 16-4). It is not known why Uranus should be different from the other Jovian planets.

Neptune also resembles Jupiter in having faint belts and zones parallel to the planet's equator (see Figure 16-6). As on Jupiter, the light-colored zones are regions where clouds have been lifted to relatively high altitudes, while the dark belts are where we see to lower regions of the atmosphere.

Like those on Uranus, most of Neptune's clouds are probably made of droplets of liquid methane. Because these droplets form fairly deep within the atmosphere, the clouds are more difficult to see than the ones on Jupiter. Hence, Neptune's belts and zones are less pronounced than Jupiter's, although more so than those on Uranus (thanks to the extra cloud-building energy from Neptune's interior). As described above, Neptune's high-altitude clouds (see Figure 16-5) are probably made of *frozen* methane.

Although Neptune displays more evidence of up-and-down motion in its atmosphere than Uranus, the global pattern of east and west winds is almost identical on the two planets. This is rather strange. The two planets are heated very differently by the Sun, thanks to their different distances from the Sun and the different tilts of their axes of rotation, so we might have expected the wind patterns on Uranus and Neptune also to be different. Perhaps the explanation of these wind patterns will involve understanding how heat is transported not only within the atmospheres of Uranus and Neptune but within their interiors as well.

16-4 | Uranus and Neptune contain a higher proportion of heavy elements than Jupiter and Saturn

 SCIENCE IN PROCESS *Explaining the origins of Uranus and Neptune poses a challenge to planetary scientists*

At first glance, Uranus and Neptune might seem to be simply smaller and less massive versions of Jupiter or Saturn (see Table 7-1). But like many first impressions, this one is misleading because it fails to take into account what lies beneath the surface. We saw in Section 14-2 that the interiors of both Jupiter and Saturn are composed primarily of hydrogen and helium, in nearly the same abundance as the Sun. But Uranus and Neptune must have a different composition, because their average densities are too high.

If Uranus and Neptune had solar abundances of the elements, the smaller masses of these planets would produce less gravitational compression, and Uranus and Neptune would both have lower average densities than Jupiter or Saturn. In fact, however, Uranus and Neptune have average densities (1320 kg/m^3 and 1640 kg/m^3, respectively) that are comparable to or greater than those of Jupiter (1330 kg/m^3) or Saturn (690 kg/m^3). Therefore, we can conclude that Uranus and Neptune contain greater proportions of the heavier elements than do Jupiter or Saturn.

This picture is not what we might expect. According to our discussion in Section 8-4 of how the solar system formed, hydrogen and helium should be relatively more abundant as distance from the vaporizing heat of the Sun increases. But Uranus and Neptune contain a greater percentage of heavy elements, and, therefore, a *smaller* percentage of hydrogen and helium, than Jupiter and Saturn.

A related problem is explaining the masses of Uranus and Neptune. At the locations of Uranus (19 AU from the Sun) and Neptune (30 AU from the Sun) the solar nebula was probably so sparse that these planets would have taken tens of hundreds of millions of years to grow to their present sizes around a core of icy planetesimals. But protoplanetary disks observed around other stars (see Section 8-3) do not appear to survive for that long before they dissipate. In other

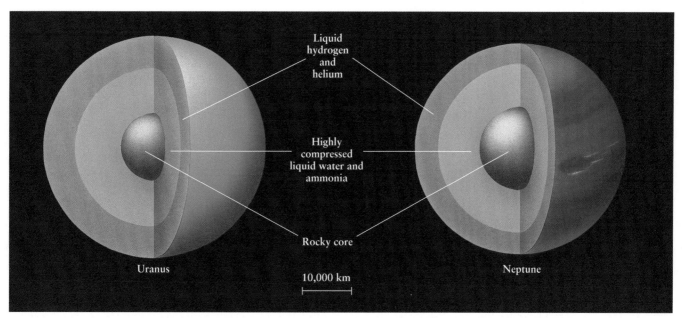

figure 16-7

The Internal Structures of Uranus and Neptune In the model shown here, both Uranus and Neptune have a rocky core, resembling a terrestrial planet; a mantle of liquid water with ammonia dissolved in it; and an outer layer of liquid molecular hydrogen and liquid helium. The atmospheres are very thin shells on top of the hydrogen-helium layer. Uranus and Neptune have about the same diameter, but since Neptune is more massive it may have a somewhat larger core.

words, the problem is not how to explain why Uranus and Neptune are smaller than Jupiter and Saturn; it is how to explain why Uranus and Neptune should exist at all!

One possible solution is that Uranus and Neptune formed in denser regions of the solar nebula between 4 and 10 AU from the Sun, where they could have grown rapidly to their present sizes. Had they remained at these locations, they could eventually have accumulated enough hydrogen and helium to become as large as Jupiter or Saturn. But before that could happen, gravitational interactions with Jupiter and Saturn would have pushed Uranus and Neptune outward into their present large orbits. Because the solar nebula was very sparse at those greater distances, Uranus and Neptune stopped growing and remained at the sizes we see today.

An alternative idea is that Uranus and Neptune formed in their present locations, not around a core of planetesimals but directly from the gases of the solar nebula. (We introduced this idea in Section 8-4.) Once a ball of gas formed, icy particles within the ball would have settled to its center, forming a solid core. This process could have formed Uranus and Neptune in only a few hundred years. In this model, Uranus and Neptune would actually have ended up larger than they are today, with substantially more hydrogen and helium. But if there happened to be a hot star relatively near the solar nebula, it would have emitted substantial amounts of ultraviolet radiation (see Section 5-3 and Figure 5-10). Ultraviolet photons have high energy (see Section 5-5), and a shower of them could have stripped Uranus and Neptune of many of their lightweight hydrogen and helium atoms, leaving shrunken planets with a larger percentage of heavy elements. (In this picture, Jupiter and Saturn

were protected from the star's ultraviolet radiation because they were shrouded by the thicker clouds of material in their part of the solar nebula.) More research will be needed to better determine the origins of Uranus and Neptune.

Figure 16-7 shows one model for the present-day internal structures of Uranus and Neptune. (Compare this to the internal structures of Jupiter and Saturn, shown in Figure 14-10.) In this model each planet has a rocky core roughly the size of Earth, although more massive. Each planet's core is surrounded by a mantle of liquid water and ammonia. (This means that the mantle is chemically similar to household window cleaning fluid.) Around the mantle is a layer of liquid molecular hydrogen and liquid helium, with a small percentage of liquid methane. This layer is relatively shallow compared to those on Jupiter and Saturn, and the pressure is not high enough to convert the liquid hydrogen into liquid metallic hydrogen.

16-5 | The magnetic fields of both Uranus and Neptune are oriented at unusual angles

Astronomers were quite surprised by the data sent back from *Voyager 2*'s magnetometer as the spacecraft sped past Uranus and Neptune. These data, as well as radio emissions from charged particles in their magnetospheres, showed that the magnetic fields of both Uranus and Neptune are tilted at steep angles to their axes of rotation. Uranus's **magnetic axis,** the line connecting its north and south magnetic poles, is inclined by 59° from its axis of rotation;

INTERACTIVE EXERCISE 16.1

figure 16-8

The Magnetic Fields of Five Planets This drawing shows how the magnetic fields of the Earth, Jupiter, Saturn, Uranus, and Neptune are tilted relative to their rotation axes (shown in red). For both Uranus and Neptune, the magnetic fields are offset from the planet's center and steeply inclined to the rotation axis.

Neptune's magnetic axis is tilted by 47°. By contrast, the magnetic axes of Earth, Jupiter, and Saturn are all nearly aligned with their rotation axes; the angle between their magnetic and rotation axes is 12° or less (Figure 16-8). Scientists were also surprised to find that the magnetic fields of Uranus and Neptune are offset from the centers of the planets.

CAUTION The drawing of the Earth's magnetic field at the far left of Figure 16-8 may seem to be mislabeled, because it shows the *south* pole of a magnet at the Earth's *North* Pole. But, in fact, this is correct, as you can understand by thinking about how magnets work. On a magnet that is free to swivel, like the magnetized needle in a compass, the "north pole" is called that because it tends to point north on Earth. Likewise, a compass needle's south pole points toward the south on Earth. Furthermore, opposite magnetic poles attract. If you take two magnets and put them next to each other, they try to align themselves so that one magnet's north pole is next to the other magnet's south pole. Now, if you think of a compass needle as one magnet and the entire Earth as the other magnet, it makes sense that the compass needle's north pole is being drawn toward a magnetic *south* pole—which happens to be located near the Earth's geographic North Pole. The Earth's magnetic pole nearest its geographic North Pole is called the "magnetic north pole." Note that the magnets drawn inside Jupiter, Saturn, and Neptune are oriented opposite to the Earth's. On any of these planets, the north pole of a compass needle would point south, not north!

Why are the magnetic axes and axes of rotation of Uranus and Neptune so badly misaligned? And why are the magnetic fields offset from the centers of the planets? One possibility is that their magnetic fields might be undergoing a reversal; geological data show that the Earth's magnetic field has switched north to south and back again many times in the past. Another possibility is that the misalignments resulted from catastrophic collisions with planet-sized bodies. As we discussed in Section 16-2, the tilt of Uranus's rotation axis and its system of moons suggest that such collisions occurred long ago. As we will see in Section 16-7, Neptune may have gravitationally captured its largest moon, Triton, but no one knows if that incident was responsible for offsetting Neptune's magnetic axis.

Because neither Uranus nor Neptune is massive enough to compress hydrogen to a metallic state, their magnetic fields cannot be generated in the same way as those of Jupiter and Saturn. Instead, under the high pressures found in the watery mantles of Uranus and Neptune, dissolved molecules such as ammonia lose one or more electrons and become electrically charged (that is, they become ionized; see Section 5-8). Water is a good conductor of electricity when it has such electrically charged molecules dissolved in it, and electric currents in this fluid are probably the source of the magnetic fields of Uranus and Neptune.

16-6 | Uranus and Neptune each have a system of thin, dark rings

 SCIENCE IN PROCESS *The search for information about Uranus's atmosphere led to the discovery of its rings*

Like the planet itself, Uranus's rings were discovered by accident. On March 10, 1977, Uranus was scheduled to move in front of a faint star, as seen from the Indian Ocean. A team of astronomers headed by James L. Elliot of Cornell University observed this event, called an **occultation,** from a NASA airplane equipped with a telescope. They hoped that by measuring how long the star was hidden, they could accurately measure Uranus's size. In addition, by carefully measuring how the starlight faded when Uranus passed in front of the star, they planned to deduce important properties of Uranus's upper atmosphere.

To everyone's surprise, the background star briefly blinked on and off several times just before the star passed behind Uranus and again immediately after (Figure 16-9). The astronomers concluded that Uranus must be surrounded by a series of nine narrow rings. During the *Voyager 2* flyby, two more rings were discovered.

Unlike Saturn's rings, the rings of Uranus are dark and narrow: Most are less than 10 km wide. Typical particles in Saturn's rings are chunks of reflective ice the size of snowballs (a few centimeters across), but typical particles in Uranus's rings are 0.1 to 10 meters in size and are no more reflective than lumps of coal. As a result, the Uranian rings are very dark; they reflect only a few percent of the sunlight that falls on them. It is not surprising that these narrow, dark rings escaped detection for so long. Figure 16-10 is a *Voyager 2* image of the rings from the side of the planet away from the Sun, where light

scattering from exceptionally small ring particles makes them more visible. (We discussed light scattering in Section 5-6 and Box 5-4.) Figure 16-3 shows the Sun-facing side of the rings.

All of Uranus's rings are located less than 2 Uranian radii from the planet's center, well within the planet's Roche limit (see Section 14-10). Some sort of mechanism, possibly one involving shepherd satellites, efficiently confines particles to their narrow orbits. (In Section 14-12 we discussed how shepherd satellites help keep planetary rings narrow.) *Voyager 2* searched for shepherd satellites but found only two. The others may be so small and dark that they have simply escaped detection. Or perhaps they aren't there, and our ideas need revision.

Like Uranus, Neptune is surrounded by a system of thin, dark rings that were first detected in stellar occultations. Figure 16-11 is a *Voyager 2* image of Neptune's rings. The ring particles vary in size from a few micrometers (1 μm = 10^{-6} m) to about 10 meters.

It is so cold at Uranus and Neptune that the ring particles can retain methane ice. Scientists speculate that eons of impacts by electrons trapped in the magnetospheres of the planets have converted

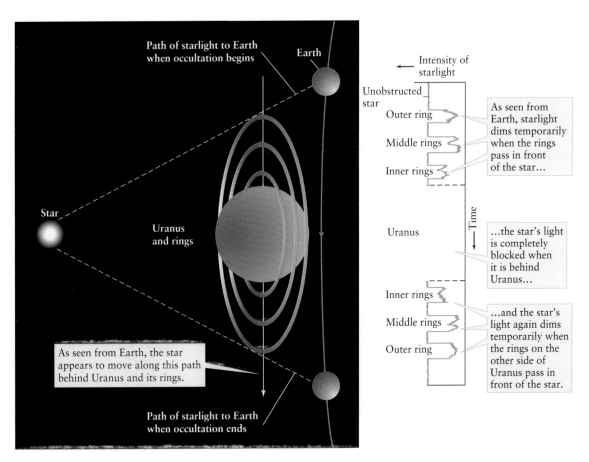

figure 16-9
How the Rings of Uranus Were Discovered As seen from Earth, Uranus occasionally appears to move in front of a distant star. Such an event is called an occultation. The star's light is completely blocked when it is behind the planet. But before and after the occultation by Uranus, the starlight dims temporarily as the rings pass in front of the star.

figure 16-10 R I **V** U X G

The Shaded Side of Uranus's Rings This view, taken when *Voyager 2* was in Uranus's shadow, looks back toward the Sun. Numerous fine-grained dust particles gleam in the spaces between the main rings. This dust is probably debris from collisions between larger particles in the main rings. The short horizontal streaks are star images blurred by the spacecraft's motion during the exposure. (NASA/JPL)

figure 16-11 R I **V** U X G

Neptune's Rings The two main rings of Neptune and a faint, inner ring can easily be seen in this composite of two *Voyager 2* images. There is also a faint sheet of particles whose outer edge is located between the two main rings and extends inward toward the planet. The overexposed image of Neptune itself has been blocked out. (NASA)

this methane ice into dark carbon compounds. This **radiation darkening** can account for the low reflectivity of the rings around both Uranus and Neptune.

16-7 | Some of Uranus's satellites show evidence of past tidal heating

Before the *Voyager 2* flyby of Uranus, five moderate-sized satellites—Titania, Oberon, Ariel, Umbriel, and Miranda—were known to orbit the planet (**Figure 16-12**). Most are named after sprites and spirits in Shakespeare's plays. They range in diameter from about 1600 km (1000 mi) for Titania and Oberon to less than 500 km (300 mi) for Miranda. All these moons have average densities around 1500 kg/m³, which is consistent with a mixture of ice and rock.

Voyager 2 discovered 10 other small Uranian satellites, most of which have diameters of less than 100 km (60 mi); 12 more were found using ground-based telescopes and the Hubble Space Telescope between 1997 and 2003. Only Jupiter and Saturn have more known satellites than Uranus. **Figure 16-13** shows several of the

1000 km

Miranda   Ariel   Umbriel   Titania   Oberon

figure 16-12 R I **V** U X G

Uranus's Principal Satellites This "family portrait" (a montage of five *Voyager 2* images) shows Uranus's five largest moons to the same scale and correctly displays their respective reflectivities.

(The darkest satellite, Umbriel, is actually more reflective than the Earth's Moon.) All five satellites have grayish surfaces, with only slight variations in color. (NASA/JPL)

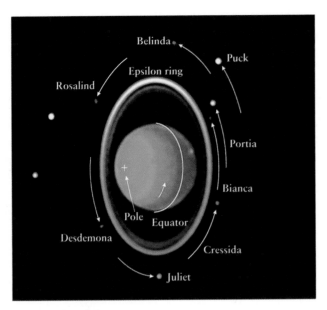

figure 16-13   R ▣ V U X G

Uranus's Rings and Small Satellites All eight satellites shown in this false-color infrared image from the Hubble Space Telescope were discovered by *Voyager 2* when it flew past Uranus in 1986. All lie within 86,000 km of the planet's center (only about one-fifth of the distance from the Earth to our Moon) and all are less than 160 km (100 mi) in diameter. The arcs show how far each satellite moves around its orbit in 90 minutes. (Erich Karkoschka, University of Arizona; and NASA)

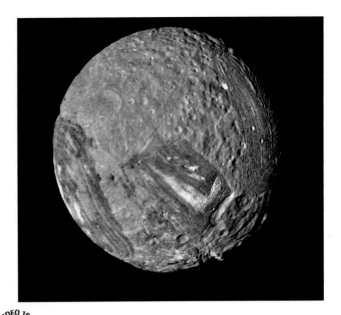

figure 16-14   R I ▣ U X G

Miranda This composite of *Voyager 2* images shows that part of Miranda's surface is ancient and heavily cratered, while other parts are dominated by parallel networks of valleys and ridges. At the very bottom of the image—where a "bite" seems to have been taken out of Miranda—is a range of enormous cliffs that jut upward to an elevation of 20 km, twice as high as Mount Everest. (NASA/JPL)

smaller satellites. Appendix 3 summarizes information about all 27 known moons of Uranus.

Umbriel and Oberon both appear to be geologically dead worlds, with surfaces dominated by impact craters. By contrast, Ariel's surface appears to have been cracked at some time in the past, allowing some sort of ice lava to flood low-lying areas. A similar process appears to have taken place on Titania. This geologic activity may be due to a combination of the satellites' internal heat and tidal heating like that which powers volcanism on Jupiter's satellite Io (see Section 15-4). Io's tidal heating is only possible because of the 1:2:4 ratio of the orbital periods of Io, Europa, and Ganymede. While there are no such simple ratios between the present-day orbital periods of Uranus's satellites, there may have been in the past. If so, tidal heating could have helped reshape the surfaces of some of the satellites.

Unique among Uranus's satellites is Miranda, which has a landscape unlike that of any other world in the solar system. Much of the surface is heavily cratered, as we would expect for a satellite only 470 km in diameter, but several regions have unusual and dramatic topography (Figure 16-14). Detailed analysis of Miranda's geology suggests that this satellite's orbital period was once in a whole-number ratio with that of more massive Umbriel or Ariel. The resulting tidal heating melted Miranda's interior, causing dense rocks in some locations on the surface to settle toward the satellite's center as blocks of less dense ice were forced upward toward the surface, thus creating Miranda's resurfaced terrain. Tidal heating must have ceased before this process could run its

full course, which explains why some ancient, heavily cratered regions remain on Miranda's surface.

16-8 | Triton is a frigid, icy world with a young surface and a tenuous atmosphere

Neptune has 13 known satellites, listed in Appendix 3. They are named for mythological beings related to bodies of water. (Neptune itself is named for the Roman god of the sea.) Most of these worlds are small, icy bodies, probably similar to the smaller satellites of Uranus. The one striking exception is Triton, Neptune's largest satellite (see Table 7-2). In many ways Triton is quite unlike any other world in the solar system.

Triton is in a retrograde orbit: It goes around Neptune oppositely to the direction in which the planet rotates. Furthermore, the plane of Triton's orbit is inclined by 23° from the plane of Neptune's equator. It is difficult to imagine how a satellite might form out of the same cloud of material as a planet and end up orbiting in a direction opposite the planet's rotation and in such a tilted plane. Hence, Triton probably formed elsewhere in the solar system, collided long ago with a now-vanished satellite of Neptune, and was captured by Neptune's gravity.

Triton is not the only satellite in the solar system that is thought to have been captured into a retrograde orbit; a few of the small outer satellites of Jupiter and Saturn have retrograde orbits and are probably captured asteroids (see Section 14-10 and Section 14-11). But with a diameter of 2706 km, a bit smaller than our Moon, Triton is certainly the largest such captured satellite.

Figure 16-15 shows the icy, reflective surface of Triton as imaged by *Voyager 2*. There is a conspicuous absence of large craters, which immediately tells us that Triton has a young surface on which the scars of ancient impacts have largely been erased by tectonic activity. There are areas that resemble frozen lakes and may be the calderas of extinct ice volcanoes. Some of Triton's surface features resemble the long cracks seen on Europa (Section 15-6) and Ganymede (Section 15-7). Still other features are unique to Triton. For example, in the upper portion of Figure 16-15 you can see dimpled, wrinkled terrain that resembles the skin of a cantaloupe.

Triton's tectonically active history is probably related to its having been captured into orbit around Neptune. After its capture, Triton most likely started off in a highly elliptical orbit, but today the satellite's orbit is quite circular. The satellite's original elliptical orbit would have been made circular by tidal forces exerted on Triton by Neptune's gravity. These tidal forces would also have stretched and flexed Triton, causing enough tidal heating to melt much of the interior. The resulting volcanic activity (with lavas made of ice rather than molten rock) would have obliterated Triton's original surface features, including craters.

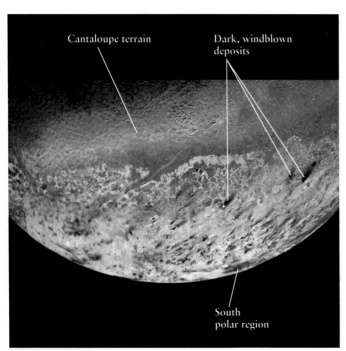

Cantaloupe terrain

Dark, windblown deposits

South polar region

Figure 16-15 R I **V** U X G

Triton Several high-resolution *Voyager 2* images were assembled to create this mosaic. The pinkish material surrounding Triton's south polar region is probably nitrogen frost. Some of this presumably evaporates when summer comes to the south pole; the northward flow of the evaporated gas may cause the dark surface markings. Farther north is a brown area of "cantaloupe terrain" that resembles the skin of a melon. (NASA/JPL)

There may still be warmth in Triton's interior today. *Voyager 2* observed plumes of dark material being ejected from the surface to a height of 8 km (5 mi). These plumes may have been generated from a hot spot far below Triton's surface, similar to geysers on the Earth. Alternatively, the energy source for the plumes may be sunlight that warms the surface, producing subsurface pockets of gas and creating fissures in the icy surface through which the gas can escape.

Triton's surface temperature is only 38 K (−235°C = −391°F), the lowest of any world yet visited by spacecraft. This temperature is low enough to solidify nitrogen, and indeed the spectrum of sunlight reflected from Triton's surface shows absorption lines due to nitrogen ice as well as methane ice. But Triton is also warm enough to allow some nitrogen to evaporate from the surface, like the steam that rises from ice cubes when you first take them out of the freezer. *Voyager 2* confirmed that Triton has a very thin nitrogen atmosphere with a surface pressure of only 1.6×10^{-5} atmosphere, about the same as at an altitude of 100 km above the Earth's surface. Despite its thinness, Triton's atmosphere has noticeable effects. *Voyager 2* saw areas on Triton's surface where dark material has been blown downwind by a steady breeze (see Figure 16-15). It also observed dark material ejected from the geyserlike plumes being carried as far as 150 km by high-altitude winds.

Just as tidal forces presumably played a large role in Triton's history, they also determine its future. Triton raises a tidal bulge on Neptune, just as our Moon distorts the Earth (recall Figure 10-19). In the case of the Earth-Moon system, the gravitational pull of the Earth's tidal bulge causes the Moon to spiral away from the Earth. But because Triton's orbit is retrograde, the tidal bulge on Neptune exerts a force on Triton that makes the satellite slow down rather than speed up. (In Figure 10-19, imagine that the moon is orbiting toward the bottom of the figure rather than toward the top.) This is causing Triton to spiral gradually in toward Neptune. In approximately 100 million years, Triton will be inside Neptune's Roche limit, and the satellite will eventually be torn to pieces by tidal forces. When this happens, the planet will develop a spectacular ring system—overshadowing by far its present-day set of narrow rings—as rock fragments gradually spread out along Triton's former orbit.

Prior to *Voyager 2*, only one other satellite was known to orbit Neptune. Nereid, which was first sighted in 1949, is in a prograde orbit. Hence, it orbits Neptune in the direction opposite to Triton. Nereid also has the most eccentric orbit of any satellite in the solar system; its distance from Neptune varies from 1.4 million to 9.7 million kilometers. One possible explanation is that when Triton was captured by Neptune's gravity, the interplay of gravitational forces exerted on Nereid by both Neptune and Triton moved Nereid from a relatively circular orbit (like those of Neptune's other, smaller moons) into its present elliptical one.

16-9 | Pluto and its moon, Charon, may be typical of a thousand icy objects that orbit far from the Sun

Speculations about a ninth planet date back to the late 1800s, when a few astronomers suggested that Neptune's orbit was being perturbed by an unknown object. Encouraged by the fame of Adams

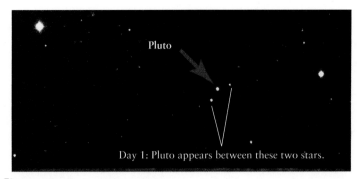

figure 16-16   R I **V** U X G

Pluto's Motion across the Sky   Pluto was discovered in 1930 by searching for a dim, starlike object that moves slowly in relation to the background stars. These two photographs were taken one day apart. Even when its apparent motion on the celestial sphere is fastest, Pluto moves only about 1 arcmin per day relative to the stars. (Lick Observatory)

and Le Verrier, several people set out to discover "Planet X." Two prosperous Boston gentlemen, William Pickering and Percival Lowell, were prominent in this effort. (Lowell also enthusiastically promoted the idea of canals of Mars; see Section 13-2.) Modern calculations show that there are, in fact, no unaccounted perturbations of Neptune's orbit. It is thus not surprising that no planet was found at the positions predicted by Pickering, Lowell, and others. Yet the search continued.

Before he died in 1916, Lowell urged that a special camera with a wide field of view be constructed to help search for Planet X. After many delays, the camera was finished in 1929 and installed at the Lowell Observatory in Flagstaff, Arizona, where a young astronomer, Clyde W. Tombaugh, had joined the staff to carry on the project. On February 18, 1930, Tombaugh finally discovered the long-sought planet. It was disappointingly faint—a thousand times dimmer than the dimmest object visible with the naked eye and 250 times dimmer than Neptune at opposition—and presented no discernible disk. The planet was named for Pluto, the mythological god of the underworld, whose name has Percival Lowell's initials as its first two letters. The discovery was publicly announced on March 13, 1930, the 149th anniversary of the discovery of Uranus. The two photographs in Figure 16-16 show one day's motion of Pluto.

Pluto's orbit about the Sun is more elliptical and more steeply inclined to the plane of the ecliptic than the orbit of any other planet (see Figure 7-1). In fact, Pluto's orbit is so eccentric that this planet is sometimes closer to the Sun than Neptune. This was the case from 1979 until 1999; indeed, when Pluto was at perihelion in 1989, it was more than 10^8 km closer to the Sun than was Neptune. Table 16-3 lists additional data about Pluto.

Pluto is so far away that it subtends only a very small angle of 0.15 arcsec. Hence, it is extraordinarily difficult to resolve any surface features on the planet. But by observing Pluto with the Hubble Space Telescope over the course of a solar day on Pluto (6.3872 Earth days) and using computer image processing, Alan Stern and Marc Buie generated the maps of Pluto's surface shown in Figure 16-17. These maps show bright polar ice caps as well as regions of different reflectivity near the planet's equator. Observations of Pluto's rotation confirm that the planet's rotation axis is tipped by more than 90°, so that Pluto has retrograde rotation like Uranus.

One of the most remarkable discoveries about Pluto was made simply by examining its image in a photograph. In 1978, while examining some photographs of Pluto, James W. Christy of the U.S. Naval Observatory noticed that the image of the planet on a photographic plate appeared slightly elongated. Pluto seemed to have a

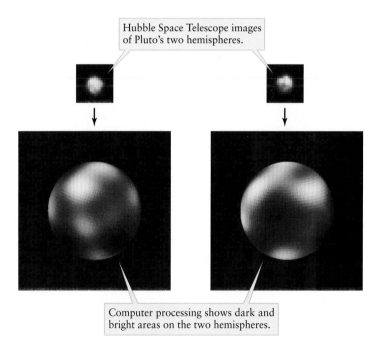

Hubble Space Telescope images of Pluto's two hemispheres.

Computer processing shows dark and bright areas on the two hemispheres.

figure 16-17   R I **V** **U** X G

Pluto's Surface   The bright regions at the top and bottom of each of Pluto's hemispheres may be polar ice caps. The bright regions nearer the equator may be impact basins where more reflective subsurface ice has been exposed. (Alan Stern, Southwest Research Institute; Marc Buie, Lowell Observatory; NASA; and ESA)

table 16-3	Pluto Data
Average distance from Sun:	39.537 AU = 5.915×10^9 km
Maximum distance from Sun:	49.425 AU = 7.394×10^9 km
Minimum distance from Sun:	29.649 AU = 4.435×10^9 km
Eccentricity of orbit:	0.2501
Average orbital speed:	4.7 km/s
Orbital period:	248.60 years
Rotation period:	6.387 days
Inclination of equator to orbit:	122.52°
Inclination of orbit to ecliptic:	17.146°
Diameter:	about 2300 km = 0.18 Earth diameter
Mass:	1.3×10^{22} kg = 0.0021 Earth mass
Average density:	about 1900 kg/m³
Escape speed:	1.2 km/s
Surface gravity (Earth = 1):	0.07
Albedo:	0.5
Average surface temperature:	−233°C = −387°F = 40 K

(Alan Stern, Southwest Research Institute; Marc Buie, Lowell Observatory; NASA; and ESA) R I **V** **U** X G

lump on one side, as shown in Figure 16-18. (This image was made with a telescope just a few kilometers from the Lowell Observatory, where Pluto itself was discovered.) He promptly examined a number of other photographs of Pluto, and found a series that showed the lump moving clockwise around Pluto with a period of about 6 days.

Christy concluded that the lump was actually a satellite of Pluto. He proposed that the newly discovered moon be christened Charon (pronounced KAR-en), after the mythical boatman who ferried souls across the River Styx to Hades, the domain ruled by Pluto. (Christy also chose the name because of its similarity to Charlene, his wife's name.) The average distance between Charon and Pluto is a scant 19,640 km, less than 5% of the distance between the Earth and our Moon. The best pictures of Pluto and Charon have been obtained using the Hubble Space Telescope (Figure 16-19).

Observations show that Charon's orbital period of 6.3872 days is the same as the rotational period of Pluto *and* the rotational period of Charon. In other words, both Pluto and Charon rotate synchronously with their orbital motion, and so both always keep the same face toward each other. As seen from the Charon-facing side of Pluto, the satellite remains perpetually suspended above the horizon. Likewise, Pluto neither rises nor sets as seen from Charon.

Soon after the discovery of Charon, astronomers witnessed an alignment that occurs only once every 124 years. From 1985 through 1990, Charon's orbital plane appeared nearly edge-on as seen from the Earth, allowing astronomers to view mutual eclipses of Pluto and its moon. As the bodies passed in front of each other, their combined brightness diminished in ways that revealed their sizes and surface characteristics. Data from the eclipses gave Pluto's diameter as about 2300 km and Charon's as about 1200 km. For comparison, our Moon's diameter (3476 km) nearly equals the diameters of Pluto and Charon added together. The brightness variations during eclipses suggest that Charon has a bright southern polar cap.

The average densities of Pluto and Charon, at about 2000 kg/m³, are essentially the same as that of Triton (2070 kg/m³). All three worlds are therefore probably composed of a mixture of rock and ice, as we might expect for small bodies that formed in the cold outer reaches of the solar nebula.

Pluto's spectrum shows absorption lines of various solid ices that cover the planet's surface, including nitrogen (N_2), methane (CH_4), and carbon monoxide (CO). Stellar occultation measurements have shown that Pluto, like Triton, has a very thin atmosphere. Exposed to a daytime temperature of around 40 K, N_2 and CO ices turn to gas more easily than frozen CH_4. For this reason, most of Pluto's tenuous atmosphere probably consists of N_2 and CO. Charon's weaker gravity has apparently allowed most of the N_2, CH_4, and CO to escape into space; only water ice is found on its surface.

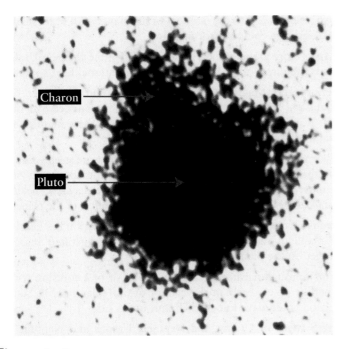

figure 16-18 R I **V** U X G

The Discovery of Pluto's Moon Charon was discovered in 1978 when James Christy noticed a "lump" protruding from the top of this greatly magnified image of Pluto. (This picture is a photographic negative, so black and white are reversed.) The small irregular blobs surrounding Pluto and Charon are part of the photographic emulsion. (U.S. Naval Observatory)

  figure 16-19 R I **V** U X G

Pluto and Charon—A "Double Planet" Pluto and Charon resemble each other in mass and size more than any other planet-satellite pair in the solar system. The distance between planet and satellite is also the smallest in the solar system. When this Hubble Space Telescope image was made, Pluto and Charon's angular separation as seen from Earth was only 0.9 arcsec. (R. Albrecht, ESA/ESO Space Telescope Coordinating Facility; and NASA)

Pluto and Charon are remarkably like each other in mass, size, and density. Throughout the rest of the solar system, planets are many times larger and more massive than any of their satellites. The exceptional similarities between Pluto and Charon suggest to some astronomers that this binary system may have formed when Pluto collided with a similar body. Perhaps chunks of matter were stripped from the second body, leaving behind a mass, now called Charon, that was captured into orbit by Pluto's gravity. Alternatively, Charon may have been captured by Pluto during a close encounter between the two worlds.

For either of these scenarios to be feasible, there must have been many Plutolike objects in the outer regions of the solar system. Some astronomers estimate that there must have been at least a thousand Plutos in order for a collision or close encounter between two of them to have occurred at least once since the solar system formed some 4.56 billion years ago.

In 1992 astronomers David Jewitt and Jane Luu found the first evidence for this picture when they discovered a small object 42 AU from the Sun. Named 1992 QB$_1$, its diameter is estimated to be only 240 km. Like Pluto, 1992 QB$_1$ has a reddish color, possibly because frozen methane has been degraded by eons of radiation exposure. As of this writing (2004), about 1000 such icy objects have been discovered orbiting beyond Neptune in what is called the **Kuiper belt**, which lies roughly between 30 and 500 AU from the Sun (see Section 7-5). At least nine of these objects have satellites of their own, suggesting that the process that gave rise to Pluto and Charon was quite common in the Kuiper belt.

The similarity between the Pluto-Charon "double planet" and Kuiper belt objects suggests that Pluto and Charon, too, should be thought of as Kuiper belt objects. (In this sense, the first Kuiper belt object to be discovered was not 1992 QB$_1$ but Pluto itself.) This reclassification means that Pluto may not truly merit the title of "planet." A controversy continues to simmer among planetary scientists about how best to pigeonhole this icy world.

Classifying Pluto and Charon as members of the Kuiper belt also suggests that there might be other icy, Pluto-sized objects in the cold regions beyond Neptune. In 2002, Chadwick Trujillo and Michael Brown of Caltech discovered one such object orbiting the Sun at an average distance of 43 AU. Named Quaoar ("kwah-o-wahr") after a creation deity of the Native Americans who once dominated the Los Angeles basin, this object is 1300 km in diameter, slightly larger than Charon. Astronomers are discovering new objects in the Kuiper Belt at the rate of about 100 a year, and it is likely that a handful of these may be as large or larger than Quaoar.

Excitement about the Plutolike worlds of the Kuiper belt has motivated the development of a spacecraft called *New Horizons* (Figure 16-20). If launched on schedule in 2006, *New Horizons* will swing by Jupiter in 2007 and make the first-ever flyby of Pluto and Charon in 2015. It will then be aimed toward one or two other targets in the Kuiper belt. The choice of targets will not be made until a year before the Pluto-Charon flyby, giving astronomers extra time to discover new members of the Kuiper belt that might be suitable destinations. The high-resolution images and other data to be returned by *New Horizons* promise to revolutionize our understanding of these most remote members of the solar system.

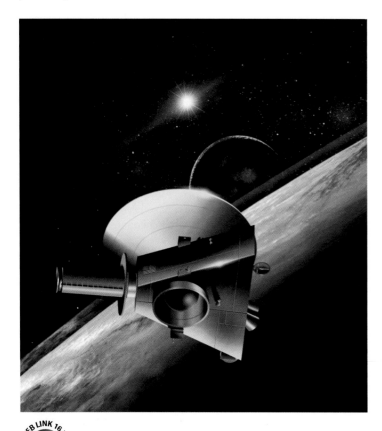

WEB LINK 16.10

Figure 16-20

New Horizons The first spacecraft to visit worlds beyond Neptune will carry a suite of instruments to examine the surfaces, atmospheres, and environments of Pluto, Charon, and targets of opportunity in the Kuiper belt. This illustration shows *New Horizons* during the flyby of Pluto and Charon. (Johns Hopkins University, Applied Physics Laboratory; Southwest Research Institute)

| Key Words |

Kuiper belt, p. 351

magnetic axis, p. 343

occultation, p. 344

radiation darkening, p. 346

| Key Ideas |

Discovery of the Outer Planets: Uranus was discovered by chance, while Neptune was discovered at a location predicted by applying Newtonian mechanics. Pluto was discovered after a long search.

Atmospheres of Uranus and Neptune: Both Uranus and Neptune have atmospheres composed primarily of hydrogen, helium, and a few percent methane.

• Methane absorbs red light, giving Uranus and Neptune their greenish-blue color.

• No white ammonia clouds are seen on Uranus or Neptune. Presumably the low temperatures have caused almost all the ammo-

nia to precipitate into the interiors of the planets. All of these planets' clouds are composed of methane.

• Much more cloud activity is seen on Neptune than on Uranus. This is because Uranus lacks a substantial internal heat source.

Interiors and Magnetic Fields of Uranus and Neptune: Both Uranus and Neptune may have a rocky core surrounded by a mantle of water and ammonia. Electric currents in these mantles may generate the magnetic fields of the planets.

• The magnetic axes of both Uranus and Neptune are steeply inclined from their axes of rotation. The magnetic and rotational axes of all the other planets are more nearly parallel. The magnetic fields of Uranus and Neptune are also offset from the centers of the planets.

Uranus's Unusual Rotation: Uranus's axis of rotation lies nearly in the plane of its orbit, producing greatly exaggerated seasonal changes on the planet.

• This unusual orientation may be the result of a collision with a planetlike object early in the history of our solar system. Such a collision could have knocked Uranus on its side.

Ring Systems of Uranus and Neptune: Uranus and Neptune are both surrounded by systems of thin, dark rings. The low reflectivity of the ring particles may be due to radiation-darkened methane ice.

Satellites of Uranus and Neptune: Uranus has five satellites similar to the moderate-sized moons of Saturn, plus at least 22 more small satellites. Neptune has 13 satellites, one of which (Triton) is comparable in size to our Moon or the Galilean satellites of Jupiter.

• Triton has a young, icy surface indicative of tectonic activity. The energy for this activity may have been provided by tidal heating that occurred when Triton was captured by Neptune's gravity into a retrograde orbit.

• Triton has a tenuous nitrogen atmosphere.

Pluto and Charon: Pluto and its moon, Charon, move together in a highly elliptical orbit steeply inclined to the plane of the ecliptic. They are the only worlds in the solar system not yet visited by spacecraft.

• Several hundred small, icy worlds have been discovered beyond Neptune. Pluto and Charon are part of this population.

| Review Questions |

1. Could astronomers in antiquity have seen Uranus? If so, why was it not recognized as a planet?

2. Why do you suppose that the discovery of Neptune is rated as one of the great triumphs of science, whereas the discoveries of Uranus and Pluto are not?

3. Why is it so difficult to see features in the atmosphere of Uranus?

4. (a) Draw a figure like Figure 16-2, and indicate on it where Uranus was in 1986 and 1998. Explain your reasoning. (*Hint:* See Figure 16-1.) (b) In approximately what year will the Sun next be highest in the sky as seen from Uranus's south pole? Explain your reasoning.

5. Why do you suppose the tilt of Uranus's rotation axis was deduced from the orbits of its satellites and not by observing the rotation of the planet itself?

6. Describe the seasons on Uranus. In what ways are the Uranian seasons different from those on Earth?

7. Explain the statement "Methane is to Uranus's atmosphere as water is to Earth's atmosphere."

8. A number of storms in the Uranian atmosphere can be seen in Figure 16-3, but none are visible in Figure 16-1. How can you account for the difference?

9. Why are Uranus and Neptune distinctly blue-green in color, while Jupiter or Saturn are not?

10. Why are fewer white clouds seen on Uranus and Neptune than on Jupiter and Saturn?

11. Why do Uranus and Neptune have higher densities than Jupiter and Saturn?

12. Discuss some competing explanations of why Uranus and Neptune are substantially smaller than Jupiter and Saturn.

13. How do the orientations of Uranus's and Neptune's magnetic axes differ from those of other planets?

14. Briefly describe the evidence supporting the idea that Uranus was struck by a large planetlike object several billion years ago.

15. Compare the rings that surround Jupiter, Saturn, Uranus, and Neptune. Briefly discuss their similarities and differences.

16. The 1977 occultation that led to the discovery of Uranus's rings was visible from the Indian Ocean. Explain why it could not be seen from other parts of the Earth's night side.

17. As *Voyager 2* flew past Uranus, it produced images only of the southern hemispheres of the planet's satellites. Why do you suppose this was?

18. Why do astronomers think that the energy needed to resurface parts of Miranda came from tidal heating rather than the satellite's own internal heat?

19. Using the data in Appendix 3, explain why Uranus's satellites Caliban and Sycorax (both discovered in 1997) were probably captured from space rather than having formed at the same time as the planet itself.

20. Briefly describe the evidence supporting the idea that Triton was captured by Neptune.

21. If you were floating in a balloon in Neptune's upper atmosphere, in what part of the sky would you see Triton rise? Explain your reasoning.

22. Why is it reasonable to suppose that Neptune will someday be surrounded by a broad system of rings, perhaps similar to those that surround Saturn?

23. How can astronomers distinguish a faint solar system object like Pluto from background stars within the same field of view?

24. Describe the evidence supporting the idea that Pluto is one of many similar icy worlds that once occupied the outer regions of the solar system.

Advanced Questions

Questions preceded by an asterisk (*) involve topics discussed in Box 1-1.

> **Problem-solving tips and tools**
>
> See Box 1-1 for the small-angle formula. Recall that the volume of a sphere of radius r is $4\pi r^3/3$. Section 4-7 discusses the gravitational force between two objects and Newton's form of Kepler's third law. You will find discussions of the original form of Kepler's third law in Section 4-4, tidal forces in Section 4-8, Wien's law for blackbody radiation in Section 5-4, and the transparency of the Earth's atmosphere to various wavelengths of light in Section 6-7.

25. For which configuration of the Sun, Uranus, and Neptune is the gravitational force of Neptune on Uranus at a maximum? For this configuration, calculate the gravitational force exerted by the Sun on Uranus and by Neptune on Uranus. Then calculate the fraction by which the sunward gravitational pull on Uranus is reduced by Neptune at that configuration. Based on your calculations, do you expect that Neptune has a relatively large or relatively small effect on Uranus's orbit?

26. At certain points in its orbit, a stellar occultation by Uranus would *not* reveal the existence of the rings. What points are those? How often does this circumstance arise? Explain using a diagram.

27. According to one model for the internal structure of Uranus, the rocky core and the surrounding shell of water and methane ices together make up 80% of the planet's mass. This interior region extends from the center of Uranus to about 70% of the planet's radius. **(a)** Find the average density of this interior region. **(b)** How does your answer to (a) compare with the average density of Uranus as a whole? Is this what you would expect? Why?

28. Uranus's epsilon (ε) ring has a radius of 51,150 km. **(a)** How long does it take a particle in the ε ring to make one complete orbit of Uranus? **(b)** If you were riding on one of the particles in the ε ring and watching a cloud near Uranus's equator, would the cloud appear to move eastward or westward as Uranus rotates? Explain your answer.

29. **(a)** Show that the ratio of the orbital periods of Neptune and Pluto is very close to 2:3. (This ratio is thought to result from gravitational interactions between Neptune and Pluto. These interactions prevent Neptune and Pluto from ever getting very close to each other.) **(b)** Several Kuiper belt objects have been discovered with the same orbital period and hence the same semimajor axis as Pluto. Explain how these objects, called *Plutinos,* can avoid colliding with Pluto.

30. Suppose you were standing on Pluto. Describe the motions of Charon relative to the Sun, the stars, and your own horizon. Would you ever be able to see a total eclipse of the Sun? (*Hint:* You will need to calculate the angles subtended by Charon and by the Sun as seen by an observer on Pluto.) In what circumstances would you *never* see Charon?

31. It is thought that Pluto's tenuous atmosphere may become even thinner as the planet moves toward aphelion (which it will reach in 2113), then regain its present density as it again moves toward perihelion. Why should this be?

32. The brightness of sunlight is inversely proportional to the square of the distance from the Sun. For example, at a distance of 4 AU from the Sun, sunlight is only $(^1/_4)^2 = ^1/_{16} = 0.0625$ as bright as at 1 AU. Compared with the brightness of sunlight on the Earth, what is its brightness **(a)** on Pluto at perihelion and **(b)** on Pluto at aphelion? **(c)** How much brighter is it on Pluto at perihelion compared with aphelion? (Even this brightness is quite low. Noon on Pluto is about as dim as it is on the Earth a half hour after sunset on a moonless night.)

*33. If Earth-based telescopes can resolve angles down to 0.25 arcsec, how large could a Kuiper belt object be at Pluto's average distance from the Sun and still not present a resolvable disk?

34. The observations of Pluto shown in Figure 16-18 were made using blue and ultraviolet light. What advantages does this have over observations made with red or infrared light?

*35. Calculate the maximum angular separation between Pluto and Charon as seen from Earth. (Assume that Pluto is at its minimum distance from the Sun and that Pluto is at opposition as seen from Earth.) Compare your answer with the angular separation given in the caption to Figure 16-19.

36. Presumably Pluto and Charon raise tidal bulges on each other. Explain why the average distance between Pluto and Charon is probably constant, rather than increasing like the Earth-Moon distance or decreasing like the Neptune-Triton distance. Include a diagram like Figure 10-19 as part of your answer.

37. Suppose you wanted to search for planets beyond Pluto. Why might it be advantageous to do your observations at infrared rather than visible wavelengths? (*Hint:* At visible wavelengths, the light we see from planets is reflected sunlight. At what wavelengths would you expect distant planets to *emit* their own light most strongly? Use Wien's law to calculate the wavelength range best suited for your search.) Could such observations be done at an observatory on the Earth's surface? Explain.

38. The *New Horizons* spacecraft will swing by Jupiter to get a boost from that planet's gravity, enabling it to reach Pluto relatively quickly. To see what would happen if this technique were not used, consider a spacecraft trajectory that is an elliptical orbit around the Sun. The perihelion of this orbit is at 1 AU from the Sun (at the Earth) and the aphelion is at 30 AU (at Pluto's position). Calculate how long it would take a spacecraft in this orbit to make the one-way trip from Earth to Pluto. Based on the information in Section 16-9, how much time is saved by making a swing by Jupiter instead?

Discussion Questions

39. Discuss the evidence presented by the outer planets that suggests that catastrophic impacts of planetlike objects occurred during the early history of our solar system.

40. Some scientists are discussing the possibility of placing spacecraft in orbit about Uranus and Neptune. What kinds of data should be collected, and what questions would you like to see answered by these missions?

41. If Triton had been formed along with Neptune rather than having been captured, would you expect it to be in a prograde or retrograde orbit? Would you expect the satellite to show signs of tectonic activity? Explain your answers.

42. Would you expect the surfaces of Pluto and Charon to be heavily cratered? Explain why or why not.

43. Imagine that you are in charge of planning the *New Horizons* flyby of Pluto and Charon. In your opinion, what data should be collected and what kinds of questions should the mission attempt to answer?

 ## Web/CD-ROM Questions

 44. Miranda. Access and view the video "Uranus's Moon Miranda" in Chapter 16 of the *Universe* Web site or CD-ROM. Discuss some of the challenges that would be involved in launching a spacecraft from Earth to land on the surface of Miranda.

45. The discovery image of Charon (Figure 16-18) was made by an astronomer at the U.S. Naval Observatory. Why do you suppose the U.S. Navy carries out work in astronomy? Search the World Wide Web for the answer.

46. Search the World Wide Web for a list of Kuiper belt objects. What are the largest and smallest objects of this sort that have so far been found, and how large are they? Have any objects larger than Quaoar been found?

 47. Separation of Pluto and Charon. Pluto is located about 4.5 billion km from Earth and has a maximum observable separation from Charon of about 0.9 arcseconds. Access the AIMM (Active Integrated Media Module) called "Small-Angle Toolbox" in Chapter 1 of the *Universe* Web site or CD-ROM. Use this AIMM and the above data to determine the distance between Pluto and Charon. How does your answer compare with the value given in the text?

Observing Projects

Observing tips and tools

 During the period 2004–2008, Uranus will be at opposition in late August and early September, Neptune in August, and Pluto in June. You can find Uranus and Neptune with binoculars if you know where to look (a good star chart is essential), but Pluto is so dim that it can be a challenge to spot even with a 25-cm (10-inch) telescope. Each year, star charts that enable you to find these planets are printed in the issue of *Sky & Telescope* for the month in which each planet is first visible in the nighttime sky. You can also locate the outer planets using the *Starry Night Backyard™* program on the CD-ROM that accompanies this textbook.

48. Make arrangements to view Uranus through a telescope. The planet is best seen at or near opposition. Use a star chart at the telescope to find the planet. Are you certain that you have found Uranus? Can you see a disk? What is its color?

49. If you have access to a large telescope, make arrangements to view Neptune. Like Uranus, Neptune is best seen at or near opposition and can most easily be found using a star chart. Can you see a disk? What is its color?

50. If you have access to a large telescope (at least 25 cm in diameter), make arrangements to view Pluto. Using the star chart from *Sky & Telescope* referred to above, view the part of the sky where Pluto is expected to be seen and make a careful sketch of all of the stars that you see. Repeat this process on a later night. Can you identify the "star" that has moved?

51. Use the *Starry Night Backyard*™ program to observe the five large satellites of Uranus. Display the entire celestial sphere (select **Guides > Atlas** in the **Go** menu) and center on Uranus by using the **Find...** command in the **Edit** menu. Using the controls at the right-hand end of the Control Panel, zoom in as far as you can. In the Control Panel, set the year to 1986 and the time step to 1 hour. Then click on the "Forward" button (a triangle that points to the right). (**a**) Describe how the satellites move, and relate your observations to Kepler's third law (see Sections 4-4 and 4-7). (**b**) Set the year to 2007 and again click on the "Forward" button. How do the orbits look different than in (a)? Explain any differences.

52. Use the *Starry Night Backyard*™ program to observe Neptune and its satellites Triton and Nereid. Display the entire celestial sphere (select **Guides > Atlas** in the **Go** menu) and center on Neptune by using the **Find...** command in the **Edit** menu. Using the controls at the right-hand end of the Control Panel, zoom in as far as you can. In the Control Panel, set the time step to 1 hour, then click on the "Forward" button (a triangle that points to the right). (**a**) The satellite you see orbiting close to Neptune is Triton. In what direction (clockwise or counterclockwise) does it orbit Neptune? (**b**) Nereid is in a rather elongated orbit that takes it far from Neptune. To see all of its orbit, first use the **Find...** command in the **Edit** menu to center on Nereid. Then set the time step to 1 day and click on the "Forward" button. In what direction (clockwise or counterclockwise) does Nereid orbit Neptune? When it is closest to Neptune, is it inside or outside Triton's orbit?

53. Use the *Starry Night Backyard*™ program to observe Pluto and Charon. First select **Solar System > Pluto** in the **Go** menu. In the Control Panel will be displayed the distance from Pluto from which we are observing it. Click on this distance to bring up a new menu and select **Hover as Pluto Rotates.** Using the controls at the right-hand end of the Control Panel, zoom out as far as you can. In the Control Panel, set the time step to 1 hour, then click on the "Forward" button (a triangle that points to the right). (**a**) Estimate Charon's orbital period. (**b**) By following a spot on Pluto's surface, estimate Pluto's rotation period. How does it compare to your answer in part (a)? (**c**) Using the **Find...** command in the **Edit** menu, center on the Sun. Then zoom out as far as you can, set the time step to 1 year, and click on the "Forward" button to see the apparent motion of the Sun as seen from Pluto. Observe the motion for several centuries of simulated time. Does the Sun always appear to move at the same speed? Use the properties of Pluto's orbit to explain why or why not.

54. Use the *Deep Space Explorer*™ program to examine the satellites of Uranus. (**a**) In the left-hand part of the window under the heading **Solar System,** click on the triangle next to the word **Explore** and then click on **Uranus.** Then zoom in and zoom out using the buttons at the upper left of the window (an upward-pointing triangle and a downward-pointing triangle). As you zoom, the distant stars will remain stationary but the satellites of Uranus will appear to move. You should be able to see five satellites. Which satellites are these? (Move the mouse to center the cursor on a satellite and its name will appear. If you can't see all five, try zooming out farther.) (**b**) You can rotate Uranus by putting the mouse cursor over the image, holding down the mouse button, and moving the mouse. (On a two-button mouse, hold down the left mouse button.) Use this technique to rotate Uranus until you are seeing the plane of the satellites' orbits edge-on. Do all of the satellites appear to lie in the same plane? How do you imagine that this plane relates to the plane of Uranus's equator? Why do you suspect this is so?

Collaborative Exercise

55. Sir William Herschel, a British astronomer, discovered Uranus in 1781 and named it Georgium Sidus, after the reigning monarch, George III. What name might Uranus have been given in 1781 if an American had discovered it and why? What if it had been discovered in the United States in 1881? In 1991?

17 Vagabonds of the Solar System

On the Tuesday night after Easter in the year 1066, a strange new star appeared in the European sky. Seemingly trailing fire, this "star" hung in the night sky for weeks. Some, including the English king Harold, may have taken it as an ill omen; others, such as Harold's rival William, Duke of Normandy, may have regarded it as a sign of good fortune. Perhaps they were both right, because by the end of the year Harold was dead and William was installed on the English throne.

Neither man ever knew that he was actually seeing the trail of a city-sized ball of ice and dust, part of which evaporated as it rounded the Sun to produce a long tail. They were seeing a comet—remarkably, the same Comet Halley that last appeared in 1986 and will next be seen in 2061. Although we now understand that comets are natural phenomena rather than supernatural omens, they still have the power to awe and inspire us, as did Comet Hale-Bopp (shown here) in 1997.

While comets are made of ice and dust, asteroids are rocky bits of the solar nebula that never formed into a full-sized planet. Some asteroids break into fragments, as do some comets, and some of these fragments fall to the Earth as meteorites. On extraordinarily rare occasions, an entire asteroid strikes Earth. Such an asteroid impact may have led to the extinction of the dinosaurs some 65 million years ago. Thus, these minor members of our solar system can have major consequences for our planet.

 Comet Hale-Bopp, the great comet of 1997. (Courtesy of Johnny Horne) R I **V** U X G

Questions to Guide Your Reading

As you read the sections of this chapter, look for the answers to the following questions:

17-1 How and why were the asteroids first discovered?

17-2 Why didn't the asteroids coalesce to form a single planet?

17-3 What do asteroids look like?

17-4 How might an asteroid have caused the extinction of the dinosaurs?

17-5 What are the differences among meteoroids, meteors, and meteorites?

17-6 What do meteorites tell us about the way in which the solar system formed?

17-7 Why do comets have tails?

17-8 Where do comets come from?

17-9 What is the connection between comets and meteor showers?

17-1 | A search for a planet between Mars and Jupiter led to the discovery of asteroids

After William Herschel's discovery of Uranus in 1781 (which we described in Section 16-1), many astronomers began to wonder if there were other, as yet undiscovered, planets. If these planets were too dim to be seen by the naked eye, they might still be visible through telescopes. These planets would presumably be found close to the plane of the ecliptic, because all the other planets orbit the Sun in or near that plane (see Section 7-1). But how far from the Sun might these additional planets be found?

Astronomers in the late eighteenth century had a simple rule of thumb, called the *Titius-Bode Law,* relating the sizes of planetary orbits: From one planet to the next, the semimajor axis of the orbit increases by a factor between approximately 1.4 and 2. For example, the semimajor axis of Mercury's orbit is 0.39 AU. Venus's orbit has a semimajor axis of 0.72 AU, which is larger by a factor of (0.72)/(0.39) = 1.85. (Unlike Newton's laws, the Titius-Bode "Law" is not a fundamental law of nature. It is probably just a reflection of how our solar system happened to form.)

The glaring exception is Jupiter (semimajor axis 5.20 AU), which is more than *3 times* farther from the Sun than Mars (semimajor axis 1.52 AU). Table 7-1 depicts this large space between the orbits of Mars and Jupiter. If the rule of thumb is correct, astronomers reasoned, there should be a "missing planet" with an orbit about 1.4 to 2 times larger than that of Mars. This planet should therefore have a semimajor axis between 2 and 3 AU.

Six German astronomers, who jokingly called themselves the "Celestial Police," organized an international group to begin a careful search for this missing planet. Before their search got under way, however, surprising news reached them from Giuseppe Piazzi, a Sicilian astronomer. Piazzi had been carefully mapping faint stars in the constellation of Taurus when on January 1, 1801, the first night of the nineteenth century, he noticed a dim, previously uncharted star. This star's position shifted slightly over the next several nights. Suspecting that he might have found the "missing planet," Piazzi excitedly wrote to Johann Bode, the director of the Berlin Observatory and a member of the Celestial Police.

Unfortunately, Piazzi's letter did not reach Bode until late March. By that time, the Earth had moved around its orbit so that Piazzi's object appeared too near the Sun to be visible in the night sky. With no way of knowing where to look after it emerged from the Sun's glare, astronomers feared that Piazzi's object might have been lost.

Upon hearing of this dilemma, the brilliant young German mathematician Karl Friedrich Gauss took up the challenge. He developed a general method of computing an object's orbit from only three separate measurements of its position on the celestial sphere. (With slight modifications, this same method is used by astronomers today.) In November 1801, Gauss predicted that Piazzi's object would be found in the predawn sky in the constellation of Virgo. And indeed Piazzi's object was sighted again on December 31, 1801, only a short distance from the position Gauss had calculated. Piazzi named the object Ceres (pronounced SEE-reez), after the patron goddess of Sicily in Roman mythology.

Ceres orbits the Sun once every 4.6 years at an average distance of 2.77 AU, just where astronomers had expected to find the missing planet. But Ceres is very small; its diameter is estimated to be a scant 918 km (**Figure 17-1**). Hence, Ceres reflects only a little sunlight, which is why it cannot be seen with the naked eye even at opposition. (It can be seen with binoculars, but it looks like just another faint star.) Because of its small size, it is known today as a **minor planet,** or **asteroid.**

On March 28, 1802, Heinrich Olbers discovered another faint, starlike object that moved against the background stars. He called it Pallas, after the Greek goddess of wisdom. Like Ceres, Pallas orbits the Sun every 4.6 years at an average distance of 2.77 AU, but its orbit is more steeply inclined from the plane of the ecliptic and is somewhat more eccentric. Pallas is even smaller and dimmer than Ceres, with an estimated diameter of only 522 km. Obviously, Pallas was also not the missing planet.

Did the missing planet even exist? Some astronomers speculated that perhaps there had once been such a full-size planet, but it had somehow broken apart or exploded to produce a population of asteroids orbiting between Mars and Jupiter. The search was on to discover this population. Two more such asteroids were discovered in the next few years, Juno in 1804 and Vesta in 1807. Several hundred more were discovered beginning in the mid-1800s, by which time telescopic equipment and techniques had improved.

The next major breakthrough came in 1891, when the German astronomer Max Wolf began using photographic techniques to search for asteroids. Before this, asteroids had to be painstakingly discovered by scrutinizing the skies for faint, uncharted, starlike objects whose positions move slightly from one night to the next. With photography, an astronomer simply aims a camera-equipped telescope at the stars and takes a long exposure. If an asteroid happens to be in the field of view, its orbital motion during the long

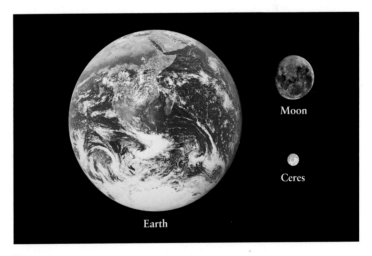

Figure 17-1   R I **V** U X G

Ceres Compared with the Earth and Moon A drawing of Ceres, the largest of the asteroids and the first one discovered, is shown here to the same scale as the Earth and the Moon. Ceres is too small to be considered a planet, and it cannot be regarded as a moon because it does not orbit any other body. To denote their status, asteroids are also called minor planets. (NASA) exposure leaves a distinctive trail on the photographic plate (Figure 17-2). Using this technique, Wolf alone discovered 228 asteroids.

 Today, more than 50,000 asteroids are known and tens or hundreds more are discovered every month, some by amateur astronomers. After the discoverer reports a new find to the Minor Planet Center of the Smithsonian Astrophysical Observatory, the new asteroid is first given a provisional designation. For example, asteroid 1980 JE was the fifth ("E") to be discovered during the second half of May (the tenth half-month of the year, "J") 1980. If the asteroid is located again on at least four succeeding oppositions (a process that can take decades), the asteroid is assigned an official sequential number (Ceres is 1, Pallas is 2, and so forth), and the discoverer is given the privilege of suggesting a name for the asteroid. The suggested name must then be approved by the International Astronomical Union. For example, the asteroid 1980 JE was officially named 3834 Zappafrank (after the American musician Frank Zappa) in 1994.

Ceres—or, in modern nomenclature, 1 Ceres—is unquestionably the largest asteroid. With a diameter of 934 km, 1 Ceres accounts for about 30% of the mass of all the asteroids combined. Only three asteroids—1 Ceres, 2 Pallas, and 4 Vesta—have diameters greater than 300 km. Thirty other asteroids have diameters between 200 and 300 km, and 200 more are bigger than 100 km across.

There may be many hundreds of thousands of asteroids bright enough to be detected by Earth-based telescopes. The vast majority are less than 1 km across. Like Ceres, Pallas, Vesta, and Juno, most asteroids orbit the Sun at distances between 2 and 3.5 AU. This region of our solar system between the orbits of Mars and Jupiter is called the **asteroid belt** (Figure 17-3).

 You may wonder how spacecraft such as *Voyager 1, Voyager 2,* and *Galileo* were able to cross the asteroid belt to reach Jupiter. Aren't asteroids hazards to space navigation,

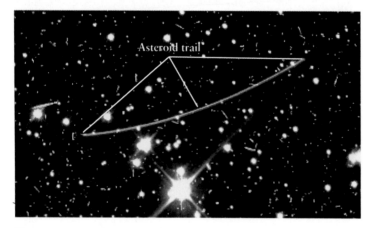

figure 17-2 R I **V** U X G

The Trail of an Asteroid Telescopes used to photograph the stars are motorized to follow the apparent motion of the celestial sphere. Because asteroids orbit the Sun, their positions change with respect to the stars, and they leave blurred trails on time exposure images of the stars. This asteroid trail was recorded quite by accident by the Hubble Space Telescope. (R. Evans and K. Stapelfeldt, Jet Propulsion Laboratory; and NASA)

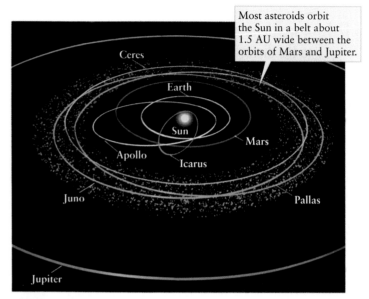

Most asteroids orbit the Sun in a belt about 1.5 AU wide between the orbits of Mars and Jupiter.

 ## figure 17-3

The Asteroid Belt Most asteroids, including the large asteroids 1 Ceres, 2 Pallas, and 3 Juno, have roughly circular orbits that lie within the asteroid belt. By contrast, some asteroids, such as 1862 Apollo and 1566 Icarus, have eccentric orbits that carry them inside the orbit of Earth.

as they are sometimes depicted in science-fiction movies? Wouldn't these spacecraft have been likely to run into an asteroid? Happily, the answer to both questions is no! While it is true that there are more than 10^5 asteroids, they are spread over a belt with a total area (as seen from above the plane of the ecliptic) of about 10^{17} square kilometers—about 10^8 times greater than the Earth's entire surface area. Hence, the average distance between asteroids in the ecliptic plane is about 10^6 kilometers, about twice the distance between the Earth and Moon. Furthermore, many asteroids have orbits that are tilted out of the ecliptic. The *Galileo* spacecraft did pass within a few thousand kilometers of two asteroids, but these passes were intentional and required that the spacecraft be carefully aimed.

What of the nineteenth-century notion of a missing planet? This idea has long since been discarded, because the combined matter of all the asteroids (including an estimate for those not yet officially known) would produce an object barely 1500 km in diameter. This is considerably smaller than anything that could be considered a missing planet. We now understand that asteroids are actually debris left over from the formation of the solar system.

17-2 | Jupiter's gravity helped shape the asteroid belt

Where did the asteroid belt come from? Why did asteroids, rather than a planet, form in the solar nebula (see Section 8-5) in the region between Mars and Jupiter? Important insights into these questions come from supercomputer simulations of the formation of the terrestrial planets, like the one depicted in Figure 8-9 but more elaborate. A typical simulation starts off with a billion (10^9) or so planetesimals, each with a mass of 10^{15} kg or more, so that their combined mass equals that of the terrestrial planets. The computer then follows these planetesimals as they collide and accrete to form the planets. By selecting different speeds and positions for the planetesimals at the start of the simulation, a scientist can study alternative scenarios of the development of the inner solar system.

If the effects of Jupiter's gravity are not included in a simulation, an Earth-sized planet usually forms in the asteroid belt, giving us five terrestrial planets instead of four. But when Jupiter's gravity is added, this fifth terrestrial planet is less likely to form. Jupiter's strong pull "clears out" the asteroid belt by disrupting the orbits of planetesimals in the belt, ejecting most of them from the solar system altogether. As a result, the asteroid belt becomes depleted of planetesimals before a planet has a chance to form. The few planetesimals that remain in the simulation—only a small fraction of those that originally orbited between Mars and Jupiter—become the asteroids that we see today.

Jupiter's gravitational influence, however, probably cannot explain why asteroid orbits have the wide variety of semimajor axes, eccentricities, and inclinations to the ecliptic shown in Figure 17-3. Even with the influence of Jupiter, collisions between planetesimals in the solar nebula would have left the asteroids in roughly circular orbits close to the ecliptic plane, like the orbits of the planets. Therefore, something else must have "stirred up" the asteroids to put them into their current state. Recent analyses suggest that one or more Mars-sized objects—that is, terrestrial planets intermediate in size between the Earth and the Moon—did in fact form within the asteroid belt. Planetesimals that passed within close range of these Mars-sized objects would have experienced strong gravitational forces, and these forces could have deflected the planetesimals into the eccentric or inclined orbits that many asteroids follow today.

Mars-sized objects would also have helped clear out the asteroid belt by deflecting planetesimals into orbits that crossed Jupiter's orbit. Once a planetesimal wandered close to Jupiter, it could be ejected from the solar system by the giant planet's gravitational force.

In this scheme, the Mars-sized objects disappeared when gravitational forces from Jupiter either accelerated them out of the solar system entirely or caused them to slow down and fall into the Sun. A Mars-sized object colliding with the Earth is thought to have produced the Moon (see Figure 10-20). Perhaps this object was an "escapee" from the asteroid belt.

Jupiter's gravity continues to influence the asteroid belt down to the present day. The American astronomer Daniel Kirkwood found the first evidence for this in 1867, when he discovered gaps in the asteroid belt. These features, today called **Kirkwood gaps,** can best be seen in a histogram like **Figure 17-4,** which shows asteroid orbital periods. The gaps in this histogram show regions where there are relatively few asteroids. Curiously, these gaps occur for asteroid orbits whose periods are simple fractions (such as $^1/_3$, $^2/_5$, $^3/_7$, and $^1/_2$) of Jupiter's orbital period.

To understand why the Kirkwood gaps exist, imagine an asteroid within the belt that circles the Sun once every 5.93 years, exactly half of Jupiter's orbital period. On every second trip around the Sun, the asteroid finds itself lined up between Jupiter and the Sun, always at the same location and with the same orientation. Because of these repeated alignments, called a *2-to-1 resonance,* Jupiter's gravity

Figure 17-4

The Kirkwood Gaps This graph displays the numbers of asteroids at various distances from the Sun. Very few asteroids are found in orbits whose orbital periods are equal to the simple fractions $^1/_3$, $^2/_5$, $^3/_7$, and $^1/_2$ of Jupiter's orbital period. The Trojan asteroids (at far right) follow the same orbit as Jupiter and thus have the same orbital period. We describe these in Section 17-4.

deflects the asteroid from its original 5.93-year orbit, ultimately ejecting it from the asteroid belt. Similar resonances clear out the Cassini division in Saturn's rings (see Section 14-12) and drive the tidal heating of Jupiter's moon Io (see Section 15-4). According to Kepler's third law, a period of 5.93 years corresponds to a semimajor axis of 3.28 AU. Because of Jupiter's gravity, almost no asteroids orbit the Sun at this average distance.

Another Kirkwood gap corresponds to an orbital period of one-third Jupiter's period, or 3.95 years (a *3-to-1 resonance*). Additional gaps exist for other simple ratios between the periods of asteroids and Jupiter. Any Mars-sized bodies that existed in the asteroid belt were probably ejected after wandering into one of the Kirkwood gaps.

17-3 | Asteroids occasionally collide with one another

Although the asteroid belt is mostly empty space, collisions between asteroids should occur from time to time. Asteroids move in orbits with a variety of different eccentricities and inclinations (see Figure 17-3), and some of these orbits intersect each other. A collision between asteroids must be an awe-inspiring event. Typical collision speeds are estimated to be from 1 to 5 km/s (3600 to 18,000 km/h, or 2000 to 11,000 mi/h), which is more than sufficient to shatter rock. Recent observations show that these titanic impacts play an important role in determining the nature of asteroids.

State-of-the-art optical telescopes have good enough resolution to reveal surface details on large asteroids such as 1 Ceres, 2 Pallas, and 4 Vesta (Figure 17-5). These asteroids resemble miniature terrestrial planets. They have enough mass, and hence enough gravity, to pull themselves into roughly spherical shapes. Vesta has a surface of basaltic rock resulting from ancient lava flows, suggesting that it underwent chemical differentiation (see Box 7-1 and Section 8-5). These large asteroids are likely to have extensively cratered surfaces due to collisions with other asteroids in the asteroid belt.

Optical telescopes reveal much less detail about the smaller asteroids. But by studying how the brightness of an asteroid changes as it rotates, astronomers can infer the asteroid's shape. An elongated asteroid appears dimmest when its long axis points toward the Sun so that it presents less of its surface to reflect sunlight. By contrast, a rotating spherical asteroid will have a more constant brightness. Such observations show that most smaller asteroids are not spherical at all. Instead, they retain the odd shapes produced by previous collisions with other asteroids.

An important recent innovation in studying asteroids has been the use of radar. An intense radio beam is sent toward an asteroid, and the reflected signal is detected and analyzed to determine the asteroid's shape (Figure 17-6). The amount of the radio signal that is reflected also gives clues about the texture of the surface.

The most powerful technique for studying asteroids is to send a spacecraft to make observations at close range. The Jupiter-bound *Galileo* spacecraft gave us our first close-up view of an asteroid in 1991, when it flew past 951 Gaspra. Two years later *Galileo* made a close pass by 243 Ida.

More recently, the *NEAR Shoemaker* spacecraft made a close approach to 253 Mathilde in 1997 (Figure 17-7), and *Deep Space 1* came within 26 km (16 mi) of 9969 Braille in 1999. *NEAR Shoemaker* went into orbit around 433 Eros

figure 17-6 R I V U X G
A Radar View of a Medium-Sized Asteroid This radar image reveals the curious dog-bone shape of the asteroid 216 Kleopatra. The Arecibo radio telescope (see Figure 11-5) was used both to send radio waves toward Kleopatra and to detect the waves that were reflected back. Scientists then created this image by computer analysis of the reflected waves. (Arecibo Observatory, JPL/NASA)

in 2000 (see Figure 7-7) and touched down on Eros the following year. It was the first spacecraft to orbit and land on an asteroid.

What have spacecraft and Earth-based radar told us about the asteroids? To appreciate the answer, we should first consider what scientists expected to find with these new tools. A few years ago, the consensus among planetary scientists was that the smaller asteroids were solid, rocky objects. The surfaces were presumed to be cratered by impacts, but only relatively small craters were expected. The reason is that an impact forceful enough to make a very large

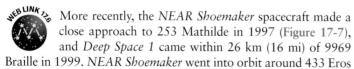

figure 17-5 R I V U X G
A Telescope View of a Large Asteroid With a diameter of 530 km (330 mi), 4 Vesta is the third largest of all the asteroids. This Hubble Space Telescope image shows dark areas that are presumably lava flows, similar to the Moon's maria. The asteroid's nonspherical shape may be the result of a massive impact that tore away a substantial portion of Vesta. (B. Zellner, Georgia Southern University; P. Thomas, Cornell University; and STScI/NASA)

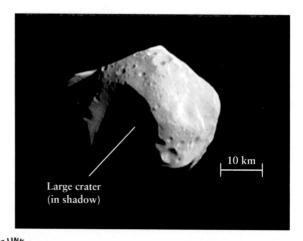

figure 17-7 R I V U X G
A Spacecraft View of a Small Asteroid The asteroid 253 Mathilde has an albedo of only 0.04, making it half as reflective as a charcoal briquette. It has an irregular shape, with dimensions 66 × 48 × 46 km. The partially shadowed crater indicated by the label is about 20 km across and 10 km deep. The *NEAR Shoemaker* spacecraft imaged Mathilde at close range during its close flyby in June 1997. (Johns Hopkins University, Applied Physics Laboratory)

crater would probably fracture a rocky asteroid into two or more smaller asteroids.

The truth proves to be rather different. It is now suspected that most asteroids larger than about a kilometer are not entirely solid, but are actually composed of several pieces. An extreme example is Mathilde (see Figure 17-7). This asteroid has such a low density (1300 kg/m³) that it cannot be made of solid rock, which typically has densities of 2500 kg/m³ or higher. Instead, it is probably a "rubble pile" of small fragments that fit together loosely. Billions of years of impacts have totally shattered this asteroid. But the impacts were sufficiently gentle that the resulting fragments drifted away relatively slowly, only to be pulled back onto the asteroid by gravitational attraction.

Another surprising discovery is that some asteroids have extremely large craters, comparable in size to the asteroid itself. An example is the immense crater on Mathilde, shown in Figure 17-7. The object that formed this crater probably collided with Mathilde at a speed of 1 to 5 km/s (3600 to 18,000 km/h, or 2000 to 11,000 mi/h), which is more than sufficient to shatter rock. Why didn't this collision break the asteroid completely apart, sending the pieces flying in all directions? The explanation may be that the asteroid had already been broken into a loose collection of fragments before this major impact took place. Such a fragmented asteroid can more easily absorb the energy of a collision than can a solid, rocky one.

 If you fire a rifle bullet at a wine glass, the glass will shatter into tiny pieces. But if you fire the same bullet into a bag full of sand (which has much the same chemical composition as glass), the main damage will be a hole in the bag. In a similar way, an asteroid that is a collection of small pieces can survive a major impact more easily than can a solid asteroid.

Relatively low-speed collisions between asteroids probably produce rubble piles like Mathilde. But higher-speed collisions may shatter an asteroid permanently. Evidence for this was pointed out in 1918 by the Japanese astronomer Kiyotsugu Hirayama, who drew attention to families of asteroids that share nearly identical orbits. Each of these groupings, now called **Hirayama families,** presumably resulted from a parent asteroid that was broken into fragments by a high-speed, high-energy collision with another asteroid.

17-4 | Asteroids are found outside the asteroid belt— and have struck the Earth

Scientists have found a possible connection between asteroid impacts and biological extinctions

While Jupiter's gravitational pull depletes certain orbits in the asteroid belt, it actually captures asteroids at two locations much farther from the Sun. The gravitational forces of the Sun and Jupiter work together to hold asteroids at two locations called the **stable Lagrange points.** One of these points is one-sixth of the way around Jupiter's orbit ahead of the planet, while the other point is the same distance behind the planet (Figure 17-8). The French

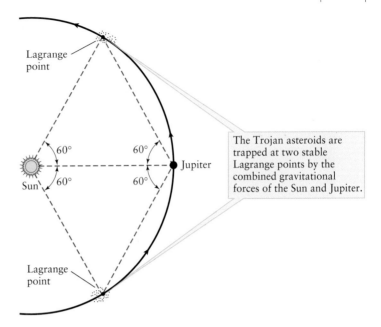

The Trojan asteroids are trapped at two stable Lagrange points by the combined gravitational forces of the Sun and Jupiter.

figure 17-8

The Trojan Asteroids The combined gravitational forces of Jupiter and the Sun tend to trap asteroids at the two stable Lagrange points along Jupiter's orbit. Note that the Sun, Jupiter, and either one of these points lie at the vertices of an equilateral triangle. Asteroids at these points are named after Homeric heroes of the Trojan War.

mathematician Joseph Louis Lagrange predicted the existence of these points in 1772; asteroids were first discovered there in 1906.

The asteroids trapped at Jupiter's Lagrange points are called **Trojan asteroids,** named individually after heroes of the Trojan War. Over 1600 Trojan asteroids have been catalogued so far. An asteroid has also been discovered at one of the Lagrange points of Neptune.

Some asteroids have highly elliptical orbits that bring them into the inner regions of the solar system. Asteroids that cross Mars's orbit, or whose orbits lie completely within that of Mars, are called **near-Earth objects** or **NEOs.**

Occasionally, a near-Earth object may pass relatively close to the Earth. On October 30, 1937, Hermes passed the Earth at a distance of 900,000 km (560,000 mi), only a little more than twice the distance to the Moon. One of the closest calls to date occurred on December 9, 1994, when an asteroid called 1994 XM$_1$ passed within a mere 105,000 km (60,000 mi) of Earth (Figure 17-9). Because 1994 XM$_1$ is only about 10 meters across (about the size of a house), even if it had been aimed directly at the Earth it would probably have broken up in the atmosphere before reaching the surface.

More than 2500 near-Earth objects are known, and there may be hundreds or thousands more that we have not yet detected. Fortunately, space is a big place and the Earth is a rather small target, so asteroids strike the Earth only very rarely.

However, collisions between asteroids produce numerous smaller chunks of rock, and many of these do eventually rain down on the terrestrial planets. Fortunately for us, the majority of these fragments, usually called **meteoroids,** are quite small. But on rare

Asteroid trail

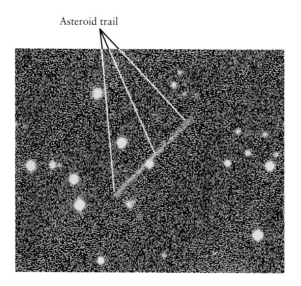

figure 17-9   R I **V** U X G

The Near-Earth Object 1994 XM₁ This image was made on December 9, 1994, 12 hours before asteroid 1994 XM₁ made its closest approach to the Earth. The asteroid was moving so rapidly during the exposure that it appears as a streak rather than as a point of light. (J. V. Scotti, The Spacewatch Project, University of Arizona)

occasions a large fragment collides with our planet. When such a collision takes place, the result is an impact crater whose diameter depends on the mass and speed of the impinging object.

A relatively young, pristine terrestrial impact crater is the Barringer Crater near Winslow, Arizona (Figure 17-10). This crater was formed 50,000 years ago when an iron-rich meteoroid, measuring approximately 50 m (160 ft) across, struck the ground at a speed of more than 11 km/s (40,000 km/h, or 25,000 mi/h). The resulting blast was equivalent to the detonation of a 20-megaton hydrogen bomb and left a crater 1.2 km (3/4 mi) wide and 200 m (650 ft) deep at its center. (Figure 7-9b shows a much larger but much older crater in Canada.)

Iron is an important constituent of asteroids and meteoroids. Another element, iridium, is common in iron-rich minerals but rare in other types of rocks. Measurements of iridium in the Earth's crust can thus tell us the rate at which meteoritic material has been deposited on the Earth over the ages. The geologist Walter Alvarez and his physicist father, Luis Alvarez, from the University of California, Berkeley, made such measurements in the late 1970s.

Working at a site of exposed marine limestone in the Apennine Mountains in Italy, the Alvarez team discovered an exceptionally high abundance of iridium in a dark-colored layer of clay between limestone strata (Figure 17-11). Since this discovery in 1979, a comparable layer of iridium-rich material has been uncovered at numerous sites around the world. In every case, geological dating reveals that this apparently worldwide layer of iridium-rich clay was deposited about 65 million years ago.

Paleontologists were quick to realize the significance of this date, because it was 65 million years ago that all the dinosaurs rather suddenly became extinct. In fact, more than 75% of all the species on the Earth disappeared within a relatively brief span of time.

The Alvarez discovery suggests a startling explanation for this dramatic mass extinction: Perhaps an asteroid hit the Earth at that time. An asteroid 10 km in diameter slamming into our planet would have produced a fiery plume that ignited tremendous wildfires, killing untold numbers of animals and consuming much of the Earth's vegetation. The dust thrown up from the impact and soot from wildfires hung in the atmosphere, blackening the sky for months after the impact and before finally drifting to the ground and depositing a layer of iridium-rich clay around the world. (Evidence for this picture is that a layer of soot 65 million years old is also

figure 17-10   R I **V** U X G

The Barringer Crater This 1.2-km- (3/4-mi-) wide crater was formed by a meteoroid impact about 50,000 years ago. Fewer than 200 other impact craters between 3 and 150 km across have been identified on the Earth. None of these is more than 500 million years old, because the reshaping of the Earth's surface by erosion and by tectonic processes tends to erase craters over time. (Meteor Crater Enterprises)

10 cm

figure 17-11   R I **V** U X G

Iridium-Rich Clay This photograph of strata in Italy's Apennine Mountains shows a dark-colored layer of iridium-rich clay. This is sandwiched between older layers of white limestone (lower right) and younger layers of grayish limestone (upper left). The iridium-rich layer is thought to be the result of an asteroid impact 65 million years ago. The coin is the size of a U.S. quarter. (Courtesy of W. Alvarez) found at various sites around the world. From the thickness of this layer, scientists calculate that the wildfires produced nearly 70 *billion* tons of soot.) Temperatures would have plummeted due to the blocking out of sunlight by dust and soot, only to increase later due to global warming brought on by greenhouse gases (see Section 9-1 and Section 9-7) released by the wildfires. Tiny rodentlike creatures capable of ferreting out seeds and nuts were among the animals that managed to survive this catastrophe, setting the stage for the rise of mammals in the eons that followed. As the dominant mammals on the planet, we may owe our very existence to an ancient asteroid.

In 1992 a team of geologists suggested that this asteroid crashed into a site in Mexico. They based this conclusion on glassy debris and violently shocked grains of rock ejected from the 180-km-diameter Chicxulub Crater on the Yucatán Peninsula. From the known rate at which radioactive potassium decays (see Section 8-3 and Box 8-1), the scientists find that the asteroid struck 64.98 million years ago, in remarkable agreement with the age of the iridium-rich layer of clay.

Some geologists and paleontologists are not convinced that an asteroid impact led to the extinction of the dinosaurs. But many scientists agree that this hypothesis fits the available evidence better than any other explanation that has been offered so far.

17-5 | Meteorites are classified as stones, stony irons, or irons, depending on their composition

A meteoroid, like an asteroid, is a chunk of rock in space. There is no official dividing line between meteoroids and asteroids, but the term *asteroid* is generally applied only to objects larger than a few hundred meters across.

A **meteor** is the brief flash of light (sometimes called a *shooting star*) that is visible at night when a meteoroid enters the Earth's atmosphere (Figure 17-12). Frictional heat is generated as the meteoroid plunges through the atmosphere, leaving a fiery trail across the night sky.

Figure 17-13 R I **V** U X G

A Meteorite "Fender-Bender" On the evening of October 9, 1992, Michelle Knapp of Peekskill, New York, heard a noise from outside that sounded like a car crash. She discovered that the trunk of her car had been smashed by a 12-kilogram (27-pound) meteorite, which was lying beside the car and was still warm to the touch. Meteorite damage is not covered by automobile insurance, but Ms. Knapp was offered several tens of thousands of dollars for the meteorite (and the car). (R. A. Langheinrich Meteorites, Ilion, N.Y.)

 If a piece of rock survives its fiery descent through the atmosphere and reaches the ground, it is called a **meteorite**. As incredible as it may seem, an estimated total of 300 tons of extraterrestrial matter falls on the Earth each day. Most of this is dust rather than meteorites, however, and is hardly noticeable. It is fortunate that large meteorite falls are rather rare events, because these fast-moving rock fragments can cause substantial damage when they strike the ground (Figure 17-13). Because they are so rare, meteorites are prized by scientists and collectors alike. Almost all meteorites are asteroid fragments and thus provide information about the chemical composition of asteroids. A very few meteorites have been identified as pieces of the Moon or Mars that were blasted off the surfaces of those worlds by asteroid impacts and eventually landed on the Earth.

People have been finding specimens of meteorites for thousands of years, and descriptions of them appear in ancient Chinese, Greek, and Roman literature. Many civilizations have regarded meteorites as objects of veneration. The sacred Black Stone, enshrined at the Ka'aba in the Great Mosque at Mecca, may be a relic of a meteorite impact.

The extraterrestrial origin of meteorites was hotly debated until as late as the eighteenth century. Upon hearing a lecture by two Yale professors, President Thomas Jefferson is said to have remarked, "I could more easily believe that two Yankee professors could lie than that stones could fall from Heaven." Although several falling meteorites had been widely witnessed and specimens from them had been collected, many scientists were reluctant to accept the idea that rocks could fall to the Earth from outer space. Conclusive evidence for the extraterrestrial origin of meteorites came on April 26, 1803, when fragments pelted the French town of L'Aigle. This event was

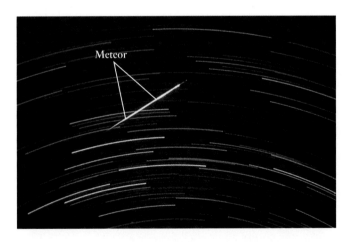

Figure 17-12 R I **V** U X G

A Meteor A meteor is produced when a meteoroid—a piece of interplanetary rock or dust—strikes the Earth's atmosphere at high speed (typically 5 to 30 km/s). The heat of air friction makes the meteoroid glow as it disintegrates. Most meteoroids burn up completely at altitudes of 100 km (60 mi) or so. This long exposure (notice the star trails) shows an exceptionally bright meteor called a fireball. (Courtesy of R. A. Oriti)

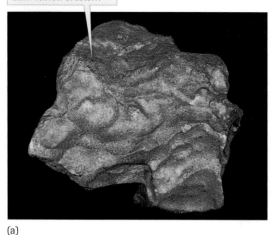

Many stony meteorites are coated with dark fusion crusts...

(a)

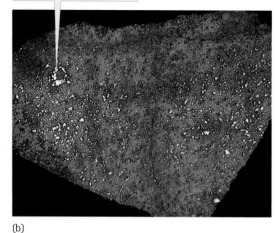

...but when cut and polished they reveal tiny specks of iron in the rock.

(b)

figure 17-14 R I **V** U X G

Stony Meteorites Of all meteorites that fall on the Earth, 95% are stones. (a) Many freshly discovered specimens, like the one shown here, are coated with dark fusion crusts. This particular stone fell in Texas. (b) Some stony meteorites are found to contain tiny specks of iron mixed in the rock. This specimen was discovered in California. (From the collection of R. A. Oriti)

analyzed by the noted physicist Jean-Baptiste Biot, and his conclusions helped finally to convince scientists that meteorites were indeed extraterrestrial.

Meteorites are classified into three broad categories: stones, stony irons, and irons. As their name suggests, **stony meteorites,** or **stones,** look like ordinary rocks at first glance, but they are sometimes covered with a **fusion crust** (Figure 17-14a). This crust is produced by the melting of the meteorite's outer layers during its fiery descent through the atmosphere. When a stony meteorite is cut in two and polished, tiny flecks of iron can sometimes be found in the rock (Figure 17-14b).

Although stony meteorites account for about 95% of all the meteoritic material that falls to the Earth, they are the most difficult meteorite specimens to find. If they lie undiscovered and become exposed to the weather for a few years, they look almost indistinguishable from common terrestrial rocks. Meteorites with a high iron content can be found much more easily, because they can be located with a metal detector. Consequently, iron and stony iron meteorites dominate most museum collections.

Stony iron meteorites consist of roughly equal amounts of rock and iron. Figure 17-15, for example, shows the mineral olivine suspended in a matrix of iron. Only about 1% of the meteorites that fall to the Earth are stony irons.

Iron meteorites (Figure 17-16), or **irons,** account for about 4% of the material that falls on the Earth. Iron meteorites usually contain no stone, but many contain from 10% to 20% nickel. Before humans began extracting iron from ores around 2000 B.C., the only source of the metal was from iron meteorites. Because these are so rare, iron was regarded as a precious metal like gold or silver.

In 1808, Count Alois von Widmanstätten, director of the Imperial Porcelain Works in Vienna, discovered a conclusive test for the most common type of iron meteorite. About 75% of all iron meteorites have a unique crystalline structure. These **Widmanstätten patterns** (pronounced VIT-mahn-shtetten) become visible when the meteorite is cut, polished, and briefly dipped into a dilute solution of acid (Figure 17-16b). Nickel-iron crystals of this type can form only if the molten metal cools slowly over many millions of years. Therefore, Widmanstätten patterns are never found in counterfeit meteorites, or "meteorwrongs."

Widmanstätten patterns suggest that some asteroids were originally at least partly molten inside and remained partly molten long

after they were formed. As soon as an asteroid accreted from planetesimals some 4.56 billion years ago, rapid decay of short-lived radioactive isotopes could have heated the asteroid's interior to temperatures above the melting point of rock. If the asteroid was large enough—at least 200 to 400 km in diameter—to insulate the interior and prevent cooling, the interior would have remained molten over the next few million years and chemical differentiation would have occurred (see Box 7-1 and Section 8-5). Iron and nickel would have sunk toward the asteroid's center, forcing less dense rock upward toward the asteroid's surface. Like a terrestrial planet, such a **differentiated asteroid** would have a distinct core and crust. After a differentiated asteroid cooled and its core solidified, collisions with other asteroids would have fragmented the parent body into meteoroids. Iron meteorites are therefore specimens from a differentiated asteroid's core, while stony irons come from regions between a differentiated asteroid's core and its crust.

Unlike irons and stony irons, stony meteorites may or may not come from differentiated asteroids. If an asteroid was too small, it would not have retained its internal heat long enough for chemical

figure 17-15 R I **V** U X G

A Stony Iron Meteorite Stony irons account for about 1% of all the meteorites that fall to the Earth. This particular specimen, a variety of stony iron called a pallasite, fell in Chile. (Chip Clark)

Iron meteorites are composed of nickel-iron minerals and are characterized by a surface covered with depressions...

(a)

...and when cut and polished, by interlocking crystals in a Widmanstätten pattern.

(b)

Figure 17-16 R I **V** U X G

Iron Meteorites (a) The surface of a typical iron is covered with small depressions caused by ablation (removal by melting) during its high-speed descent through the atmosphere. Note the one-cent coin for scale. (b) When cut, polished, and etched with a weak acid solution, most iron meteorites exhibit interlocking crystals in designs called Widmanstätten patterns. Both of these iron meteorites were found in Australia. (From the collection of R. A. Oriti)

differentiation to occur. Some stony meteorites are fragments of such **undifferentiated asteroids** (also called *primitive* asteroids), while others are relics of the crust of differentiated asteroids.

17-6 | Some meteorites retain traces of the early solar system

A class of rare stony meteorites called **carbonaceous chondrites** shows no evidence of ever having been melted as part of asteroids. As suggested by their name, carbonaceous chondrites contain substantial amounts of carbon and carbon compounds, including complex organic molecules and as much as 20% water bound into the minerals. These compounds would have been broken down and the water driven out if these meteorites had been subjected to heating and melting. Carbonaceous chondrites may therefore be samples of the original material from which our solar system was created. The asteroid 253 Mathilde, shown in Figure 17-7, has a very dark gray color and the same sort of spectrum as a carbonaceous chondrite. It, too, is likely composed of material that predates the formation of the solar system.

Amino acids, the building blocks of proteins upon which terrestrial life is based, are among the organic compounds occasionally found inside carbonaceous chondrites. Interstellar organic material has thus probably been falling on our planet since its formation. Some scientists suspect that carbonaceous chondrites may have played a role in the origin of life on Earth.

Shortly after midnight on February 8, 1969, the night sky around Chihuahua, Mexico, was illuminated by a brilliant blue-white light moving across the heavens. As the light crossed the sky, it disintegrated in a spectacular, noisy explosion that dropped thousands of rocks and pebbles over terrified onlookers. Within hours, teams of scientists were on their way to collect specimens of a carbonaceous chondrite, collectively named the Allende meteorite after the locality (Figure 17-17).

Specimens were scattered in an elongated ellipse approximately 50 km long by 10 km wide. Most fragments were coated with a fusion crust, but minerals immediately beneath the crust showed no signs of damage. Surface material is peeled away as it becomes heated during flight through the atmosphere; this process forms the fusion crust. Because the heat has little time to penetrate the meteorite's interior, compounds there are left intact.

One of the most striking discoveries to come from the Allende meteorite was evidence suggesting that a nearby star exploded into a **supernova** about 4.6 billion years ago. In a supernova, a massive star reaches the end of its life cycle and blows itself apart in a cataclysm that hurls matter outward at tremendous speeds. During this detonation, violent collisions between nuclei produce a host of radioactive elements, including ^{26}Al, a radioactive isotope of aluminum.

Researchers found clear evidence for the former presence of ^{26}Al in the Allende meteorite. Chemical analyses revealed a high abundance of a stable isotope of magnesium (^{26}Mg), which is produced by the radioactive decay of ^{26}Al. Some astronomers interpret this as evidence for a supernova in our vicinity at about the time the Sun was born. Indeed, by compressing interstellar gas and dust, the supernova's shock wave may have triggered the birth of our solar system.

Figure 17-17 R I **V** U X G

A Piece of the Allende Meteorite This carbonaceous chondrite fell near Chihuahua, Mexico, in February 1969. The meteorite's dark color is due to its high abundance of carbon. Radioactive age-dating indicates that this meteorite is 4.56 billion years old, suggesting that this meteorite is a specimen of primitive planetary material that predates the formation of the planets. The ruler is 15 cm (6 in.) long. (Courtesy of J. A. Wood)

17-7 | A comet is a dusty chunk of ice that partially vaporizes as it passes near the Sun

WEB LINK 17.1 Just as heat from the protosun produced two classes of planets, the terrestrial and the Jovian, two main types of small bodies formed in the solar system. Near the Sun, interplanetary debris consists of the rocky objects called asteroids or meteoroids. Far from the Sun, where temperatures in the early solar system were low enough to permit ices of water, methane, ammonia, and carbon dioxide to form, interplanetary debris took the form of loose collections of ices and small rocky particles. These objects are like icy versions of "rubble pile" asteroids such as 253 Mathilde (see Section 17-3). They could be referred to as "dirty snowballs," a name coined by the Harvard astronomer Fred Whipple. More commonly, however, they are called **comets.**

Asteroids travel around the Sun along roughly circular orbits that are largely confined to the asteroid belt and that lie close to the plane of the ecliptic. In sharp contrast, many comets travel around the Sun along highly elliptical orbits inclined at random angles to the ecliptic. Comets are just a few kilometers across, so when they are far from the Sun they are difficult to spot with even a large telescope. As a comet approaches the Sun, however, solar heat begins to vaporize the comet's ices, liberating gases as well as dust particles. The liberated gases begin to glow, producing a fuzzy, luminous ball called a **coma** that is typically 1 million km in diameter. Some of these luminous gases stream outward into a long, flowing **tail.** This tail, which can be more than 100 million kilometers in length—comparable to the distance from the Earth to the Sun—is one of the most awesome sights that can be seen in the night sky (see Figure 17-18 and the image that opens this chapter).

LOOKING DEEPER 17-2 Many comets are discovered when they are still far from the Sun, long before their tails grow to full size. Most of these discoveries are made by amateur astronomers, many of whom use only binoculars or small telescopes to aid in their search.

Figure 17-19 depicts the structure of a comet. The solid part of the comet, from which the coma and tail emanate, is called the **nucleus.** It is a mixture of ice and dust that typically measures a few kilometers across. Before 1986, no one had seen the nucleus of a comet. It is small, dim, and buried in the glare of the coma. In that year, the first close-up pictures of a comet's nucleus were obtained by a spacecraft that flew past Comet Halley (Figure 17-20). Comet Halley's potato-shaped nucleus is darker than coal, reflecting only about 4% of the light that strikes it. This dark color is probably caused by a layer of carbon-rich compounds and dust that is left behind as the comet's ice evaporates.

Several 15-km-long jets of dust particles were seen emanating from bright areas on Halley's nucleus. These seem to be active only when exposed to the Sun. The bright areas, which probably cover about 10% of the surface of the nucleus, are presumably places where the dark layer covering the nucleus is particularly thin. When exposed to the Sun, these areas evaporate rapidly, producing the jets. When the nucleus's rotation brings the bright areas into darkness, away from the Sun, the jets shut off.

A part of a comet that is not visible to the human eye is the **hydrogen envelope,** a huge sphere of tenuous gas surrounding the

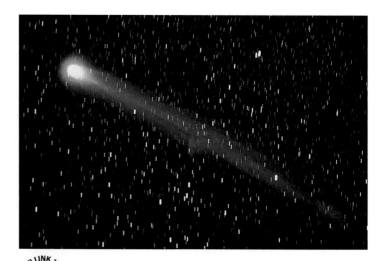

WEB LINK 17.2 figure 17-18   R I **V** U X G
Comet Hyakutake Using binoculars, Japanese amateur astronomer Yuji Hyakutake first noticed this comet on the morning of January 30, 1996. Two months later, Comet Hyakutake came within 0.1 AU (15 million km, or 9 million mi) of Earth and became one of the brightest comets of the twentieth century. This image was made in April 1996, when the comet's tail extended more than 30° across the sky (more than 60 times the diameter of the full moon). The stars appear as short streaks because the camera followed the comet's motion during the exposure. (Courtesy of Gary Goodman)

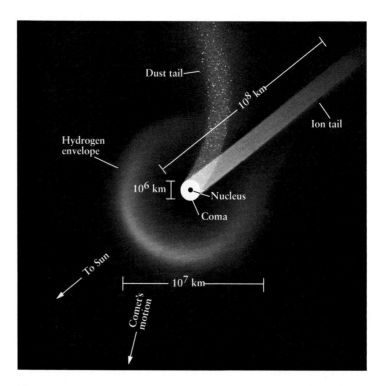

figure 17-19

The Structure of a Comet The solid part of a comet is the nucleus. It is typically about 10 km in diameter. The coma typically measures 1 million kilometers in diameter, and the hydrogen envelope usually spans 10 million kilometers. A comet's tail can be as long as 1 AU.

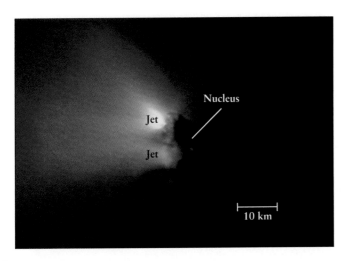

figure 17-20 R I **V** U X G

The Nucleus of Comet Halley This close-up picture, taken in March 1986 by a camera on board the European Space Agency spacecraft *Giotto,* shows the dark, potato-shaped nucleus of Comet Halley. Sunlight from the left illuminates half of the nucleus, which is about 15 km long and 8 km wide. Two bright jets of dust extend from the nucleus toward the Sun. This suggests that most of the vaporization takes place on the sunlit side. (Max-Planck-Institut für Aeronomie)

A visible-light image shows the comet's tail.

An ultraviolet image reveals the comet's immense hydrogen envelope.

R I **V** U X G   R I V **U** X G

figure 17-21

A Comet and Its Hydrogen Envelope These visible-light and ultraviolet images of Comet Kohoutek, which made an appearance in 1974, are reproduced to the same scale. The hydrogen envelope is only visible in the ultraviolet image. (Johns Hopkins University, Naval Research Laboratory)

nucleus. This hydrogen comes from water molecules (H_2O) that escape from the comet's evaporating ice and then break apart when they absorb ultraviolet photons from the Sun. The hydrogen atoms also absorb solar ultraviolet photons, which excite the atoms from the ground state into an excited state (see Section 5-8). When the atoms return to the ground state, they emit ultraviolet photons.

Unfortunately for astronomers, these emissions from a comet's hydrogen envelope cannot be detected by Earth-based telescopes because our atmosphere is largely opaque to ultraviolet light (see Section 6-7). Instead, cameras above the Earth's atmosphere must be used. Figure 17-21 shows two views of the same comet, one as it appeared in visible light to Earth-based observers and one as photographed by an ultraviolet camera aboard a rocket. The ultraviolet image shows the enormous extent of the hydrogen envelope, which can span 10 million kilometers.

As the diagram in Figure 17-19 suggests, comet tails always point away from the Sun. This is true regardless of the direction of the comet's motion (Figure 17-22). The implication that something from the Sun was "blowing" the comet's gases radially outward led Ludwig Biermann to predict the existence of a *solar wind*, a stream of particles rushing away from the Sun (see Section 8-5). A decade later, in 1962, Biermann's prediction was confirmed when the solar wind was detected for the first time by instruments on the *Mariner 2* spacecraft.

In fact, the Sun usually produces *two* comet tails—an **ion tail** and a **dust tail** (Figure 17-23). Ionized atoms and molecules—that is, atoms and molecules missing one or more electrons—are swept directly away from the Sun by the solar wind to form the relatively straight ion tail. The distinct blue color of the ion tail is caused by

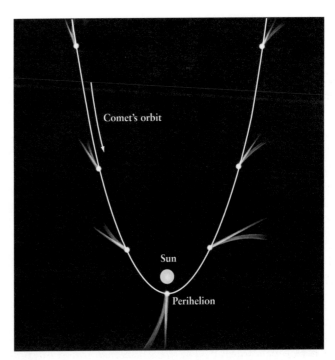

figure 17-22

The Orbit and Tail of a Comet The solar wind and the pressure of sunlight blow a comet's dust particles and ionized atoms away from the Sun. Consequently, a comet's tail points generally away from the Sun. In particular, the tail does *not* always stream behind the nucleus. At the upper right of this figure, the comet is moving upward along its orbit and is literally chasing its own tail.

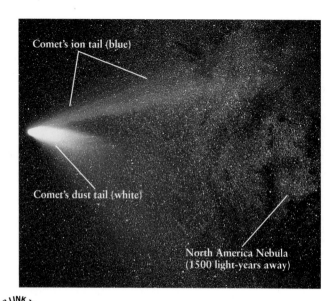

figure 17-23 R I **V** U X G
The Two Tails of Comet Hale-Bopp A comet's dust tail is white because it reflects sunlight, while the molecules in the ion tail emit their own light with a characteristic blue color. When this picture was taken on March 8, 1997, the ion tail extended more than 10° across the sky. The red object to the right is the North America Nebula, a star-forming region some 1500 light-years beyond the solar system. (Courtesy of Tony and Daphne Hallas Astrophotos)

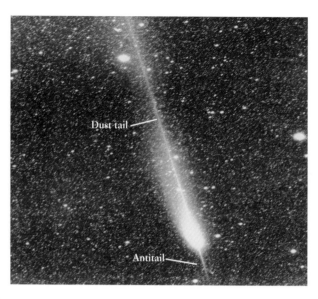

figure 17-24 R I **V** U X G
The Antitail of Comet Hale-Bopp In January 1998, nine months after passing perihelion, Comet Hale-Bopp was oriented in such a way that the end of its arched dust tail looked like a spike sticking out of the comet's head. This spike is called an "antitail" because it appears to protrude in front of the comet. (European Southern Observatory)

emissions from carbon-bearing molecules such as CN and C_2. The dust tail is formed when photons of light strike dust particles freed from the evaporating nucleus. (Figure 8-6 shows a dust particle of this type.) Light exerts a pressure on any object that absorbs or reflects it. This pressure, called **radiation pressure,** is quite weak but

is strong enough to make fine-grained dust particles in a comet's coma drift away from the comet, thus producing a dust tail. The solar wind has less of an effect on dust particles than on (much smaller) ions, so the dust tail ends up being curved rather than straight. On rare occasions a comet is oriented in such a way that its dust tail appears to stick out in front of the comet (Figure 17-24).

After the comet passes perihelion, it recedes from the Sun back into the cold regions of the outer solar system. The ices stop vaporizing, the coma and tail disappear, and the comet goes back into an inert state—until the next time its orbit takes it toward the inner solar system. An object in an elliptical orbit spends most of its time far from the Sun, so it is only during a relatively brief period before and after perihelion that a comet can have a prominent tail. An example is Comet Halley, which orbits the Sun along a highly elliptical path that stretches from just inside the Earth's orbit to slightly beyond the orbit of Neptune (Figure 17-25). The comet's tail is visible to the naked eye or with binoculars only during a few months around perihelion, which last occurred in 1986 and will happen again in 2061.

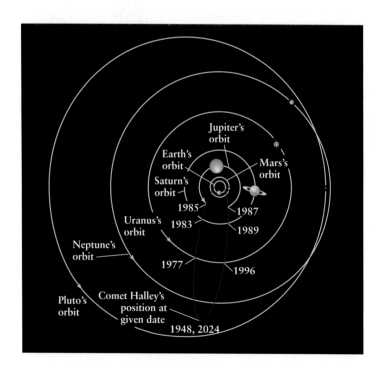

figure 17-25
Comet Halley's Eccentric Orbit Like most comets, Comet Halley has an elongated orbit and spends most of its time far from the Sun. Figure 4-22 shows the comet's tail, which appears only when the comet is close to perihelion. Comet Halley has been observed at intervals of about 76 years—the period of its orbit—since 88 B.C. (see Section 4-7).

Other comets have orbits that are larger and more elongated, with even longer periods. The orbit of Comet Hyakutake (see Figure 17-18) takes it to a distance of about 2000 AU from the Sun, 50 times the size of Pluto's orbit, with an orbital period of around 30,000 years!

17-8 | Comets originate either from a belt beyond Pluto or from a vast cloud in near interstellar space

Why do comets have such elliptical orbits, while asteroids have relatively circular ones? The reason cannot have anything to do with comets being icy and asteroids being rocky, because the gravitational forces that shape an object's orbit do not depend on its chemical composition. Instead, the answer has to do with how the gravitational pulls of the planets helped to shape the orbits of asteroids and comets in different ways.

As we saw in Section 17-2, the asteroids formed in the space between Mars and Jupiter. The relatively small terrestrial planets, with their correspondingly small masses, exert only weak gravitational forces on asteroids. Hence, most asteroids, like the planets, remained in fairly circular orbits. (The exceptions are those few asteroids that wandered too close to Mars, whose gravitational force deflected these asteroids into becoming the near-Earth objects described in Section 17-4.) Jupiter's gravity acted primarily to bunch the asteroids away from the Kirkwood gaps.

Comets, by contrast, first formed in the outer solar system just beyond Neptune, where temperatures were low enough to permit ices to condense into chunks several kilometers across. Because Neptune is much more massive than the terrestrial planets, its gravitational forces had a major effect, deflecting these icy bodies in a wide variety of directions.

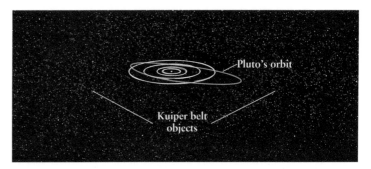

Figure 17-26

The Kuiper Belt The Kuiper belt of comets lies close to the ecliptic, like the asteroid belt, and extends from around the orbit of Pluto to about 500 AU from the Sun. Observations show that the combined mass of the objects in the Kuiper belt is far greater than the combined mass of all of the asteroids in the asteroid belt.

There are now two large reservoirs of comets, the Kuiper belt and the Oort cloud. The **Kuiper belt** (see Section 7-5 and Section 16-9) lies close to the plane of the ecliptic and extends from about 30 to 500 AU from the Sun (Figure 17-26). The existence of this belt was first proposed by the American astronomer Gerard Kuiper in 1951; the first Kuiper belt object to be discovered was seen in 1992. (This was the object 1992 QB_1, which we described in Section 16-9.) About 1000 of these objects have now been observed (Figure 17-27). It is suspected that there could be more than 100,000 objects in the Kuiper belt with diameters of 100 km or greater and many more smaller objects. Like other planetesimals, most Kuiper belt objects probably formed at the same distance from the Sun as we find them today.

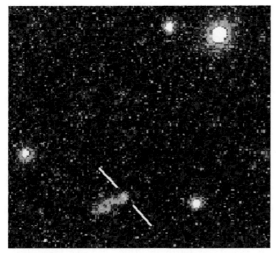

(a) Kuiper belt object 1993 SC (between the white lines)

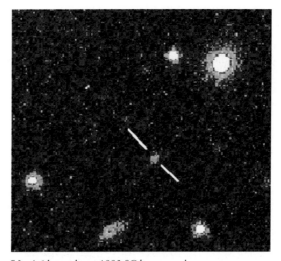

(b) 4.6 hours later, 1993 SC has moved against the background of stars

Figure 17-27 R I **V** U X G

A Kuiper Belt Object Like asteroids, Kuiper belt objects are discovered by looking for objects that slowly change positions relative to the stars. In these false-color images—the first to show 1993 SC, the sixth Kuiper belt object to be discovered—blue represents dark areas, while green, yellow, red, and white denote successively brighter areas. (Alan Fitzsimmons, Queen's University of Belfast)

Jupiter-family comets, which orbit the Sun in fewer than 20 years and return again and again to the inner solar system at predictable intervals, are thought to come from the Kuiper belt. Collisions between Kuiper belt objects can break off relatively small chunks, and gravitational perturbations from Neptune can occasionally launch one of these chunks into a highly elliptical orbit that takes it close to the Sun. When one of these chunks approaches the Sun, it develops a visible tail and appears as a comet. A few of these perturbed objects from the Kuiper belt are thought to have been captured by the gravitational pull of Saturn, becoming that planet's small outer satellites (see Section 15-10).

The majority of comets are **intermediate-period comets,** with orbital periods between 20 and 200 years (including Comet Halley), and **long-period comets,** which take roughly 1 to 30 million years to complete one orbit of the Sun. Long-period comets travel along extremely elongated orbits and consequently spend most of their time at distances of roughly 10^4 to 10^5 AU from the Sun, or about one-fifth of the way to the nearest star. These orbits extend far beyond the Kuiper belt. Furthermore, intermediate-period and long-period comets have orbital planes that are often steeply inclined to the plane of the ecliptic. (Comet Halley's orbital plane is inclined by 162° to the ecliptic, which means that it orbits the Sun in the opposite direction to the planets.)

The best explanation for these observations is that intermediate-period and long-period comets come from a reservoir that extends from the Kuiper belt to some 50,000 AU from the Sun. This reservoir, first hypothesized by Dutch astronomer Jan Oort in 1950 and now called the **Oort cloud,** does not lie in or near the ecliptic; rather, it is a spherical distribution centered on the Sun. This explains why many intermediate-period and long-period comets have steeply inclined orbits.

Because astronomers discover long-period comets at the rate of about one per month, it is reasonable to suppose that there is an enormous population of comets in the Oort cloud. Estimates of the number of "dirty snowballs" in the Oort cloud range as high as 5 trillion (5×10^{12}). Only such a large reservoir of comet nuclei would explain why we see so many long-period comets, even though each one takes several million years to travel once around its orbit. Because the Oort cloud is so distant, it has not yet been possible to detect objects in the Oort cloud directly.

The Oort cloud was probably created 4.56 billion years ago from numerous icy planetesimals that orbited the Sun in the vicinity of the newly formed Jovian planets. During near-collisions with the giant planets, many of these chunks of ice and dust were catapulted by gravity into highly elliptical orbits, in much the same way that *Pioneer* and *Voyager* spacecraft were flung far from the Sun during their flybys of the Jovian planets. Gravitational perturbations from nearby stars could have tilted the planes of the orbits in all directions, giving the Oort cloud its spherical shape.

The distinction between long-period, intermediate-period, and Jupiter-family comets can be blurred by the effects of gravitational perturbations. During a return trip toward the Sun, an encounter with a Jovian planet may force a comet into a much larger orbit. Alternatively, such an encounter can move a long-period comet into a smaller orbit (**Figure 17-28**). Several comets have been perturbed in this way into orbits that always remain within the inner solar system. None of these has a prominent tail, however.

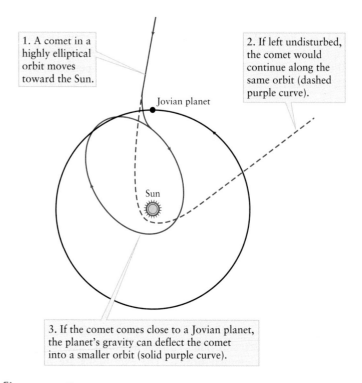

1. A comet in a highly elliptical orbit moves toward the Sun.

2. If left undisturbed, the comet would continue along the same orbit (dashed purple curve).

Jovian planet

Sun

3. If the comet comes close to a Jovian planet, the planet's gravity can deflect the comet into a smaller orbit (solid purple curve).

Figure 17-28

Transforming a Comet's Orbit The gravitational force of a planet can deflect a comet from a highly elliptical orbit into a less elliptical one. In some cases a comet ends up with an orbit that keeps it within the inner solar system.

Our description of cometary origins suggests that different comets are of different ages. Computer simulations suggest that many Jupiter-family comets were produced no more than a few hundred million years ago by collisions between larger icy objects in the Kuiper belt. Comets emanating from the Oort cloud may date from the very early solar system, but may have originated at somewhat different times at different distances from the center of the solar nebula before being catapulted into the Oort cloud. This variety of ages is one reason why comets are of particular interest to planetary scientists: Comets carry information about the history of the cold, outer regions of the solar system, but are brought into the inner solar system where we can examine them at closer range using telescopes.

 Both NASA and the European Space Agency are exploring comets at even closer range using unmanned spacecraft. NASA's *Deep Space 1*, launched in 1998, came within 2000 km of Comet Borrelly in 2001. A second NASA spacecraft, *Stardust*, encountered Comet Wild 2 in 2004 and collected a sample of material from the comet's coma. It is scheduled to return this sample to Earth in 2006. NASA's *Deep Impact* mission was scheduled for a 2004 launch and a 2005 rendezvous with Comet Tempel 1. The spacecraft was designed to launch a 500-kg metal cylinder at the comet's nucleus, excavating a large crater and revealing its internal composition. Most ambitious of all is the European Space Agency's *Rosetta* mission, which was scheduled to fly past at least one asteroid on its way to intercept Comet 67P/Churyumov-Gerasimenko in 2014. If all goes well, *Rosetta* will go into orbit around the comet nucleus and release a small probe that will actually land on the nucleus.

17-9 | Comets eventually break apart, and their fragments give rise to meteor showers

Because they evaporate away part of their mass each time they pass near the Sun, comets cannot last forever. A typical comet may lose about 0.5% to 1% of its ice each time it passes near the Sun. Hence, the ice completely vaporizes after about 100 or 200 perihelion passages, leaving only a swarm of dust and pebbles. Astronomers have observed some comet nuclei in the process of fragmenting (Figure 17-29).

One remarkable comet, called Shoemaker-Levy 9, broke apart for a different reason. As the comet swung by Jupiter in July 1992, the giant planet's tidal forces tore the nucleus into more than 20 fragments. Jupiter's gravity then deflected the trajectories of these fragments so that they plummeted into the planet's atmosphere.

As a comet's nucleus evaporates, residual dust and rock fragments form a **meteoritic swarm,** a loose collection of debris that continues to circle the Sun along the comet's orbit (Figure 17-30). If the Earth's orbit happens to pass through this swarm, a **meteor shower** is seen as the dust particles strike the Earth's upper atmosphere.

Nearly a dozen meteor showers can be seen each year (Table 17-1). Note that the **radiants** for these showers—that is, the places among the stars from which the meteors appear to come—are not confined to the constellations of the zodiac. This means that the swarms do not all orbit in the plane of the ecliptic. The reason is

Figure 17-29   R I **V** U X G

The Fragmentation of Comet LINEAR The Lincoln Near Earth Asteroid Research (LINEAR) project discovered this comet in 1999. As it passed the perihelion of its orbit in July 2000, Comet LINEAR broke into more than a dozen small fragments. These continue to orbit the Sun along the same trajectory that the comet followed prior to its breakup. This image was made using the Very Large Telescope (see Figure 6-18). (European Southern Observatory)

that the orbits of their parent comets were inclined at random angles to the plane of the ecliptic (see Figure 17-30).

On June 30, 1908, a spectacular explosion occurred over the Tunguska region of Siberia that released about 10^{15} joules of energy, equivalent to a nuclear detonation of several hundred kilotons. The blast knocked a man off his porch some 60 km away and could be heard more than 1000 km away. Millions of tons of dust were injected into the atmosphere, darkening the sky as far away as California.

Preoccupied with political upheaval and World War I, Russia did not send a scientific expedition to the site until 1927. Researchers

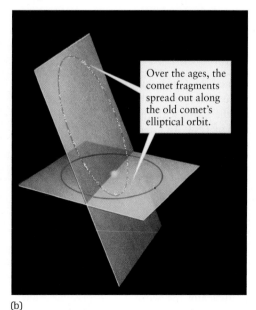

(a)   (b)

Figure 17-30

Meteoritic Swarms Rock fragments and dust from "burned out" comets continue to circle the Sun. **(a)** The most spectacular meteor showers occur when the Earth passes through compact swarms of debris from recently extinct comets. **(b)** Meteor showers associated with older comets are more predictable, because the Earth passes through the evenly distributed swarm on each trip around the Sun.

table 17-1 | Prominent Yearly Meteor Showers

Shower name	Date of maximum intensity*	Typical hourly rate	Average speed (km/s)	Radiant constellation
Quadrantids	January 3	40	40	Boötes
Lyrids	April 22	15	50	Lyra
Eta Aquarids	May 4	20	64	Aquarius
Delta Aquarids	July 30	20	40	Aquarius
Perseids	August 12	50	60	Perseus
Orionids	October 21	20	66	Orion
Taurids	November 4	15	30	Taurus
Leonids	November 16	15	70	Leo
Geminids	December 13	50	35	Gemini
Ursids	December 22	15	35	Ursa Minor

*The date of maximum intensity is the best time to observe a particular shower, although good displays can often be seen a day or two before or after the maximum. The typical hourly rate is given for an observer under optimum viewing conditions. The average speed refers to how fast the meteoroids are moving when they strike the atmosphere.

found that trees had been seared and felled radially outward in an area about 50 km in diameter (Figure 17-31). There was no clear evidence of a crater. In fact, the trees at "ground zero" were left standing upright, although they were completely stripped of branches and leaves. Because no significant meteorite samples were found, for many years it was assumed that a small comet had struck Earth.

Recently, however, several teams of astronomers have argued that the Tunguska explosion was actually caused by a stony asteroid traveling at supersonic speed. They arrived at this conclusion after assessing the effects of various impactor sizes, speeds, and compositions. Even data from above-ground nuclear detonations of the 1940s and 1950s were worked into the calculations. The Tunguska event is well-matched by a stony asteroid about 80 m (260 ft) in diameter entering the Earth's atmosphere at 22 km/s (80,000 km/h = 50,000 mph). A small comet would have broken up too high in the atmosphere to cause significant damage on the ground.

Large objects, such as that which formed the Chicxulub Crater and may have led to the demise of the dinosaurs (see Section 17-3), occasionally strike the Earth with the destructive force of as much as a million megatons of TNT. An asteroid 1 or 2 kilometers in diameter, striking the Earth at 30 km/s, would destroy an area the size of California. Such an impact would throw more than 10^{13} kg of microscopic dust particles into the atmosphere, which would decrease the amount of sunlight reaching the Earth's surface by enough to threaten the health of the world's agriculture. The loss of a year's crops could lead to the demise of a quarter of the world's population and would place our civilization—although perhaps not the survival of our species—in grave jeopardy.

The good news is that such large objects strike the Earth *very* infrequently. An asteroid 1 km in diameter might be expected to hit the Earth only once every 300,000 years. Statistically, the probability that an average person will die because of such an event is about 1 in 20,000, or only about 1% as great as the probability of

dying in an automobile accident. While the threat of asteroid or comet impact exists, a catastrophic impact is very unlikely to occur in our lifetimes.

figure 17-31 R I V U X G

Aftermath of the Tunguska Event In 1908 a stony asteroid with a mass of about 10^8 kilograms entered the Earth's atmosphere over the Tunguska region of Siberia. Its passage through the atmosphere made a blazing trail in the sky some 800 km long. The asteroid apparently exploded before reaching the surface, blowing down trees for hundreds of kilometers around "ground zero." (Courtesy of Sovfoto)

Key Words

amino acids, p. 365

asteroid, p. 357

asteroid belt, p. 358

carbonaceous chondrite, p. 365

coma (of a comet), p. 366

comet, p. 366

differentiated asteroid, p. 364

dust tail, p. 367

fusion crust, p. 364

Hirayama family, p. 361

hydrogen envelope, p. 366

intermediate-period comet, p. 370

iron meteorite (iron), p. 364

ion tail, p. 367

Jupiter-family comet, p. 370

Kirkwood gaps, p. 359

Kuiper belt, p. 369

long-period comet, p. 370

meteor, p. 363

meteor shower, p. 371

meteorite, p. 363

meteoritic swarm, p. 371

meteoroid, p. 361

minor planet, p. 357

near-Earth object (NEO), p. 361

nucleus (of a comet), p. 366

Oort cloud, p. 370

radiant (of a meteor shower), p. 371

radiation pressure, p. 368

stable Lagrange points, p. 361

stony iron meteorite, p. 364

stony meteorite (stone), p. 364

supernova, p. 365

tail (of a comet), p. 366

Trojan asteroid, p. 361

undifferentiated asteroid, p. 365

Widmanstätten patterns, p. 364

Key Ideas

Discovery of the Asteroids: Astronomers first discovered the asteroids while searching for a "missing planet."

• Thousands of asteroids with diameters ranging from a few kilometers up to 1000 kilometers orbit within the asteroid belt between the orbits of Mars and Jupiter.

Origin of the Asteroids: The asteroids are the relics of planetesimals that failed to accrete into a full-sized planet, thanks to the effects of Jupiter and other Mars-sized objects.

• Even today, gravitational perturbations by Jupiter deplete certain orbits within the asteroid belt. The resulting gaps, called Kirkwood gaps, occur at simple fractions of Jupiter's orbital period.

• Jupiter's gravity also captures asteroids in two locations, called Lagrangian points, along Jupiter's orbit.

Asteroid Collisions: Asteroids undergo collisions with each other, causing them to break up into smaller fragments.

• Some asteroids, called near-Earth objects, move in elliptical orbits that cross the orbits of Mars and Earth. If such an asteroid strikes the Earth, it forms an impact crater whose diameter depends on both the mass and the speed of the asteroid.

• An asteroid may have struck the Earth 65 million years ago, possibly causing the extinction of the dinosaurs and many other species.

Meteoroids, Meteors, and Meteorites: Small rocks in space are called meteoroids. If a meteoroid enters the Earth's atmosphere, it produces a fiery trail called a meteor. If part of the object survives the fall, the fragment that reaches the Earth's surface is called a meteorite.

• Meteorites are grouped into three major classes, according to composition: iron, stony iron, and stony meteorites. Irons and stony irons are fragments of the core of an asteroid that was large enough and hot enough to have undergone chemical differentiation, just like a terrestrial planet. Some stony meteorites come from the crust of such differentiated meteorites, while others are fragments of small asteroids that never underwent differentiation.

• Rare stony meteorites called carbonaceous chondrites may be relatively unmodified material from the solar nebula. These meteorites often contain organic material and may have played a role in the origin of life on Earth.

• Analysis of isotopes in certain meteorites suggests that a nearby supernova may have triggered the formation of the solar system 4.56 billion years ago.

Comets: A comet is a chunk of ice with imbedded rock fragments that generally moves in a highly elliptical orbit about the Sun.

• As a comet approaches the Sun, its icy nucleus develops a luminous coma, surrounded by a vast hydrogen envelope. An ion tail and a dust tail extend from the comet, pushed away from the Sun by the solar wind and radiation pressure, respectively.

• Fragments of "burned out" comets produce meteoritic swarms. A meteor shower is seen when the Earth passes through a meteoritic swarm.

 Origin of Comets: Comets are thought to originate from two regions, the Kuiper belt and the Oort cloud.

• The Kuiper belt lies in the plane of the ecliptic at distances between 30 and 500 AU from the Sun. It is thought to contain many tens of thousands of comet nuclei. Many Jupiter-family comets probably come from the Kuiper belt, and hundreds of larger objects have been observed in the Kuiper belt.

• The Oort cloud contains billions of comet nuclei in a spherical distribution that extends out to 50,000 AU from the Sun. Intermediate-period and long-period comets are thought to originate in the Oort cloud. As yet no objects in the Oort cloud have been detected directly.

Review Questions

1. What led astronomers to suspect that there were members of the solar system that orbit between Mars and Jupiter?

2. How did the first asteroid come to be discovered? What role did theoretical calculations play in confirming the discovery? How did this discovery differ from what astronomers had expected to find?

3. How do modern astronomers discover new asteroids?

4. Describe the asteroid belt. Does it lie completely within the plane of the ecliptic? What are its inner and outer radii?

5. What are Kirkwood gaps? What causes them?

6. Compare the explanation of the Kirkwood gaps in the asteroid belt to the way in which Saturn's moons help produce divisions in that planet's rings (see Section 14-11).

7. What is the evidence that some asteroids are made of a loose conglomeration of smaller pieces?

8. The asteroid 243 Ida, which was viewed by the *Galileo* spacecraft, is a member of a Hirayama family. Discuss what this tells us about the history of this asteroid. Where might you look to find other members of the same family?

9. What are the Trojan asteroids, and where are they located? What holds them in this location?

10. What are near-Earth objects? What is the evidence that the Earth has been struck by these objects?

11. What is the evidence that an asteroid impact could have contributed to the demise of the dinosaurs?

12. What is the difference between a meteoroid, a meteor, and a meteorite?

13. Is there anywhere on the Earth where you might find large numbers of stony meteorites that are not significantly weathered? If so, where? If not, why not?

14. Scientists can tell that certain meteorites came from the interior of an asteroid rather than from its outer layers. Explain how this is done.

15. Why are some asteroids differentiated while others are not?

16. Suppose you found a rock you suspect might be a meteorite. Describe some of the things you could do to determine whether it was a meteorite or a "meteorwrong."

17. What is the evidence that carbonaceous chondrites are essentially unaltered relics of the early solar system? What do they suggest about how the solar system may have formed?

18. With the aid of a drawing, describe the structure of a comet.

19. Why is the phrase "dirty snowball" an appropriate characterization of a comet's nucleus?

20. Why do the ion tail and dust tail of a comet point in different directions?

21. Why do comets have prominent tails for only a short time during each orbit?

22. Why is it that Jupiter and Saturn can be seen in the night sky every year, while seeing specific comets such as Halley and Hyakutake is a once-in-a-lifetime event?

23. What is the Kuiper belt? How does it compare with the asteroid belt? What are the similarities? What are the differences? What is the relationship between the Kuiper belt and comets?

24. What is the Oort cloud? How might it be related to planetesimals left over from the formation of the solar system?

25. Why are comets more likely to break apart at perihelion than at aphelion?

26. Why do astronomers think that meteorites come from asteroids, while meteor showers are related to comets?

27. Why are asteroids, meteorites, and comets all of special interest to astronomers who want to understand the early history and subsequent evolution of the solar system?

Advanced Questions

Questions preceded by an asterisk (*) involve topics discussed in Box 1-1 or Box 7-2.

> **Problem-solving tips and tools**
>
> The small-angle formula is described in Box 1-1. We discussed retrograde motion in Section 4-1 and described its causes in Section 4-2. You will need to use Kepler's third law, described in Section 4-4 and Box 4-2, in some of the problems below. Box 7-2 discusses the concept of escape speed. A spherical object of radius r intercepts an amount of sunlight proportional to its cross-sectional area, equal to πr^2. The volume of a sphere of radius r is $4\pi r^3/3$.

28. When Olbers discovered Pallas in March 1802, the asteroid was moving from east to west relative to the stars. At what time of night was Pallas highest in the sky over Olbers's observatory? Explain your reasoning.

29. Consider the Kirkwood gap whose orbital period is two-fifths of Jupiter's 11.86-year period. Calculate the distance from the Sun to this gap. Does your answer agree with Figure 17-4?

*30. When the image in Figure 17-5 was made, the asteroid 4 Vesta was 1.77×10^8 km from the Hubble Space Telescope. (a) What was the angular size of the asteroid as a whole? (b) You can see individual pixels in the image shown in Figure 17-5. Using a ruler and the scale bar in the figure, determine how many kilometers on the surface of Vesta are contained in the width of one pixel. (c) What is the angular width of each pixel in arcseconds?

31. Suppose that a binary asteroid (two asteroids orbiting each other) is observed in which one member is 16 times brighter than the other. Suppose that both members have the same albedo and that the larger of the two is 120 km in diameter. What is the diameter of the other member?

32. The accompanying image from the *Galileo* spacecraft shows the asteroid 243 Ida, which has dimensions $56 \times 24 \times 21$ km. *Galileo* discovered a tiny moon called Dactyl, just $1.6 \times 1.4 \times 1.2$ km in size, which orbits Ida at a distance of about 100 km. (In Greek mythology, the Dactyli were beings who lived on the slopes of

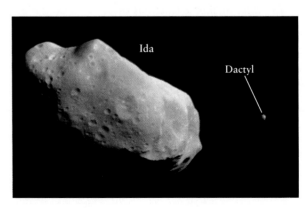

(JPL/NASA)   R I **V** U X G

Mount Ida.) Describe a scenario that could explain how Ida came to have a moon.

*33. Assume that Ida's tiny moon Dactyl (see Question 32) has a density of 2500 kg/m^3. (a) Calculate the mass of Dactyl in kilograms. For simplicity, assume that Dactyl is a sphere 1.4 km in diameter. (b) Calculate the escape speed from the surface of Dactyl. If you were an astronaut standing on Dactyl's surface, could you throw a baseball straight up so that it would never come down? Professional baseball pitchers can throw at speeds around 40 m/s (140 km/h, or 90 mi/h); your throwing speed is probably a bit less.

34. Imagine that you are an astronaut standing on the surface of a Trojan asteroid. How will you see the phase of Jupiter change with the passage of time? How will you see Jupiter move relative to the distant stars? Explain your answers.

35. Use the percentages of stones, irons, and stony iron meteorites that fall to Earth to estimate what fraction of their parent asteroids' interior volume consisted of an iron core. Assume that the percentages of stones and irons that fall to the Earth indicate the fractions of a parent asteroid's interior volume occupied by rock and iron, respectively. How valid do you think this assumption is?

36. On March 8, 1997, Comet Hale-Bopp was 1.39 AU from the Earth and 1.00 AU from the Sun. Use this information and that given in the caption to Figure 17-23 to estimate the length of the comet's ion tail on that date. Give your answer in kilometers and astronomical units.

37. *Sun-grazing comets* come so close to the Sun that their perihelion distances are essentially zero. Find the orbital periods of Sun-grazing comets whose aphelion distances are (a) 100 AU, (b) 1000 AU, (c) 10,000 AU, and (d) 100,000 AU. Assuming that these comets can survive only a hundred perihelion passages, calculate their lifetimes. (*Hint:* Remember that the semimajor axis of an orbit is one-half the length of the orbit's long axis.)

38. Comets are generally brighter a few weeks after passing perihelion than a few weeks before passing perihelion. Explain why might this be. (*Hint:* Water, including water ice, does an excellent job of retaining heat.)

39. The hydrogen clouds of comets are especially bright at an ultraviolet wavelength of 122 nm. Use Figure 5-22 to explain why.

40. A *very* crude model of a typical comet nucleus is a cube of ice (density 1000 kg/m^3) 10 km on a side. (a) What is the mass of this nucleus? (b) Suppose 1% of the mass of the nucleus evaporates away to form the comet's tail. Suppose further that the tail is 100 million (10^8) km long and 1 million (10^6) km wide. Estimate the average density of the tail (in kg/m^3). For comparison, the density of the air you breathe is about 1.2 kg/m^3. (c) In 1910 the Earth actually passed through the tail of Comet Halley. At the time there was some concern among the general public that this could have deleterious effects on human health. Was this concern justified? Why or why not?

Discussion Questions

41. From the abundance of craters on the Moon and Mercury, we know that numerous asteroids and meteoroids struck the inner

planets early in the history of our solar system. Is it reasonable to suppose that numerous comets also pelted the planets 3.5 to 4.5 billion years ago? Speculate about the effects of such a cometary bombardment, especially with regard to the evolution of the primordial atmospheres on the terrestrial planets.

42. In the 1998 movie *Armageddon*, an asteroid "the size of Texas" is on a collision course with Earth. The asteroid is first discovered by astronomers just 18 days prior to impact. To avert disaster, a team of astronauts blasts the asteroid into two pieces just 4 hours before impact. Discuss the plausibility of this scenario. (*Hint:* On average, the state of Texas extends for about 750 km from north to south and from east to west. How does this compare with the size of the largest known asteroids?)

43. Suppose astronomers discovered that the near-Earth object 1994 XM$_1$ had been disturbed in such a way as to put it on a collision course with Earth. Describe what humanity could do within the framework of present technology to counter such a catastrophe.

 Web/CD-ROM Questions

44. A NASA spacecraft called *Dawn* is intended to go into orbit around two asteroids, 1 Ceres and 4 Vesta. Search the World Wide Web for information about this mission. Why were these two particular asteroids selected for study? When is the spacecraft to be launched, and when will it arrive at each asteroid?

45. Search the World Wide Web to find out why some scientists disagree with the idea that a tremendous impact led to the demise of the dinosaurs. (They do not dispute that the impact took place, only what its consequences were.) What are their arguments? From what you learn, what is your opinion?

46. Several scientific research programs are dedicated to the search for near-Earth objects (NEOs), especially those that might someday strike our planet. Search the World Wide Web for information about at least one of these programs. How does this program search for NEOs? How many NEOs has this program discovered? Will any of these pose a threat in the near future?

 47. Estimating the Speed of a Comet. Access and view the video "Two Comets and an Active Sun" in Chapter 17 of the *Universe* Web site or CD-ROM. (a) Why don't the comets reappear after passing the Sun? (b) The white circle shows the size of the Sun, which has the diameter 1.39×10^6 km. Using this to set the scale, step through the video and measure how the position of one of the comets changes. Use your measurements and the times displayed in the video to estimate the comet's speed in km/h and km/s. (Assume that the comet moved in the same plane as that shown in the video.) As part of your answer, explain the technique and calculations you used. (c) How does your answer in (b) compare with the orbital speed of Mercury, the innermost and fastest-moving planet (see Table 11-1)? Why is there a difference? (d) If the comet's motion was not in the same plane as that shown in the video, was its actual speed more or less than your estimate in (b)? Explain.

Observing Projects

48. Make arrangements to view a meteor shower. Table 17-1 lists the dates of major meteor showers. Choose a shower that will occur near the time of a new moon. Set your alarm clock for the early morning hours (1 to 3 A.M.). Get comfortable in a reclining chair or lie on your back so that you can view a large portion of the sky. Record how long you observe, how many meteors you see, and what location in the sky they seem to come from. How well does your observed hourly rate agree with published estimates, such as those in Table 17-1? Is the radiant of the meteor shower apparent from your observations?

49. If a comet is visible with a telescope at your disposal, make arrangements to view it. Can you distinguish the comet from background stars? Can you see its coma? Can you see a tail?

50. Make arrangements to view an asteroid. Observe the asteroid on at least two occasions, separated by a few days. On each night, draw a star chart of the objects in your telescope's field of view. Has the position of one starlike object shifted between observing sessions? Does the position of the moving object agree with the path plotted on published star charts? Do you feel confident that you have in fact seen the asteroid?

 51. Use the *Starry Night Backyard*™ program to observe two asteroids. Display the entire celestial sphere (select **Guides > Atlas** in the **Go** menu) and center on Ceres by using the **Find...** command in the **Edit** menu. In the Control Panel, set the time step to 1 day. Then click on the "Forward" button (a triangle that points to the right). (a) Watch Ceres for at least two years of simulated time. Describe how Ceres moves. How can you tell that

Ceres orbits the Sun in the same direction as the planets? Does Ceres remain as close to the ecliptic (shown as a solid green line) as the planets do? (If you don't see a green line representing the ecliptic, select **Show View Options Panel** in the **View** menu, then make sure the box "The Ecliptic" is checked.) (b) Click on the "Stop" button (a black square) and use the **Find...** command in the **Edit** menu to center on Pallas. Again click on the "Forward" button, and watch how Pallas moves for at least two years of simulated time. How does the motion of Pallas differ from that of Ceres? How are the two motions similar? Which asteroid's orbit is more steeply inclined to the plane of the Earth's orbit? How can you tell?

 52. Use the *Starry Night Backyard*™ program to study the motion of a comet. First set up the field of view so that you are observing the inner solar system from a distance (select **Solar System > Inner Solarsystem** in the **Go** menu). In the Control Panel, click on the "Stop" button (a black square) to halt the animation, and then set the date to January 1, 1995, and the time step to 1 day. Center on Comet Hyakutake by using the **Find...** command in the **Edit** menu to search for *Hyakutake*. Using the controls at the right-hand end of the Control Panel, zoom in until the field of view is about 25 degrees. Then click on the "Forward" button (a triangle that points to the right). (a) Watch the motion of Comet Hyakutake for at least two years of simulated time. Describe what you see. Is the comet's orbit in about the same plane as the orbits of the inner planets, or is it steeply inclined to that plane? How does the comet's speed vary as it moves along its orbit? During which part of the orbit is the tail visible? In what direction does the tail point? (b) Click on the "Stop" button to halt the animation, and set up the field of view so that you are observing from Earth (select **Go Home** in the **View** menu, then select **Guides > Atlas** in the **Go** menu). Again set the date to January 1, 1995, and the time step to 1 day, and again center on Comet Hyakutake. Use the controls at the right-hand end of the Control Panel to zoom out as far as possible. Then click on the "Forward" button" and watch the comet's motion for at least two years of simulated time. Describe the motion, and explain why it is more complicated than the motion you observed in part (a). (c) Stop the animation, set the date to today's date, set the time step to 1 month ("lunar m."), and restart the animation. Comet Hyakutake is currently moving almost directly away from the Sun, so as seen from the Sun its position on the celestial sphere should not change. Is this what you see in *Starry Night Backyard*™? Explain any differences. (*Hint:* You are observing from the Earth, not the Sun.)

 53. Use the *Deep Space Explorer*™ program to examine Comet Halley. In the left-hand part of the window under the heading **Solar System,** click on the triangle next to the word **Explore** and then click on **Halley's Comet.** You can rotate the comet by putting the mouse cursor over the image, holding down the mouse button, and moving the mouse. (On a two-button mouse, hold down the left mouse button.) (a) In what direction does the tail of the comet point? Explain why. (b) Based on the direction of the comet's tail, can you tell in which direction the comet is moving at the instant shown in *Deep Space Explorer*™? If so, what is that direction, and how can you tell? If not, explain why not.

18 Our Star, the Sun

The Sun is by far the brightest object in the sky. By earthly standards, the temperature of its glowing surface is remarkably high, about 5800 K. Yet astronomers have found regions of the Sun that reach temperatures of tens of thousands or even millions of kelvins. Gases at such temperatures emit ultraviolet light and can be seen in an image like this one from an ultraviolet telescope in space. Some of the most energetic regions on the Sun spawn solar flares like the one shown here. These powerful eruptions can propel solar material across space to reach the Earth and other planets.

In recent decades, astronomers have used tools such as space telescopes to make a wealth of discoveries about the Sun. We have learned that the Sun shines because at its core hundreds of millions of tons of hydrogen are converted to helium every second. We have found the by-products of this transmutation—strange, ethereal particles called neutrinos—streaming outward from the Sun into space. We have discovered that the Sun has a surprisingly violent atmosphere, with a host of features such as sunspots whose numbers rise and fall on a predictable 11-year cycle. By studying the Sun's vibrations, we have begun to probe beneath its surface into hitherto unexplored realms. And we have just begun to investigate how changes in the Sun's activity can affect the Earth's environment as well as our technological society.

An ultraviolet image of a solar flare erupting from the Sun's surface. (*SOHO*/EIT/ESA/NASA) R I V U X G

Questions to Guide Your Reading

As you read the sections of this chapter, look for the answers to the following questions:

18-1 What is the source of the Sun's energy?

18-2 What is the internal structure of the Sun?

18-3 How can astronomers measure the properties of the Sun's interior?

18-4 How can we be sure that thermonuclear reactions are happening in the Sun's core?

18-5 Does the Sun have a solid surface?

18-6 Since the Sun is so bright, how is it possible to see its dim outer atmosphere?

18-7 Where does the solar wind come from?

18-8 What are sunspots? Why do they appear dark?

18-9 What is the connection between sunspots and the Sun's magnetic field?

18-10 What causes eruptions in the Sun's atmosphere?

18-1 | The Sun's energy is generated by thermonuclear reactions in its core

SCIENCE IN PROCESS
Ideas from relativity and nuclear physics led to an understanding of how the Sun shines

The Sun is the largest member of the solar system, with almost a thousand times more mass than all the planets, moons, asteroids, comets, and meteoroids put together. But the Sun is also a star. In fact, it is a remarkably typical star, with a mass, size, surface temperature, and chemical composition that are approximately midway between the extremes exhibited by the myriad other stars in the heavens. Table 18-1 lists essential data about the Sun.

For most people, what matters most about the Sun is the energy that it radiates into space. Without the Sun's warming rays, our atmosphere and oceans would freeze into an icy layer coating a desperately cold planet, and life on Earth would be impossible. To understand why we are here, we must understand the nature of the Sun.

What makes the Sun such an important source of energy? An important part of the answer is that the Sun has a far higher surface temperature than any of the planets or moons. The Sun's spectrum is close to that of an idealized blackbody with a temperature of 5800 K (see Sections 5-3 and 5-4, especially Figure 5-11). Thanks to this high temperature, each square meter of the Sun's surface emits a tremendous amount of radiation, principally at visible wavelengths. Indeed, the Sun is the only object in the solar system that emits substantial amounts of visible light. The light that we see from the Moon and planets is actually sunlight that struck those worlds and was reflected toward Earth.

The Sun's size also helps us explain its tremendous energy output. Because the Sun is so large, the total number of square meters of radiating surface—that is, its surface area—is immense. Hence, the total amount of energy emitted by the Sun each second, called its **luminosity,** is very large indeed: about 3.9×10^{26} watts, or 3.9×10^{26} joules of energy emitted per second. (We discussed the relation among the Sun's surface temperature, radius, and luminosity in Box 5-2.) Astronomers denote the Sun's luminosity by the symbol $L_\odot$. A circle with a dot in the center is the astronomical symbol for the Sun and was also used by ancient astrologers.

These ideas lead us to a more fundamental question: What keeps the Sun's visible surface so hot? Or, put another way, what is the fundamental source of the tremendous energies that the Sun radiates into space? For centuries, this was one of the greatest mysteries in science. The mystery deepened in the nineteenth century, when geologists and biologists found convincing evidence that life had existed

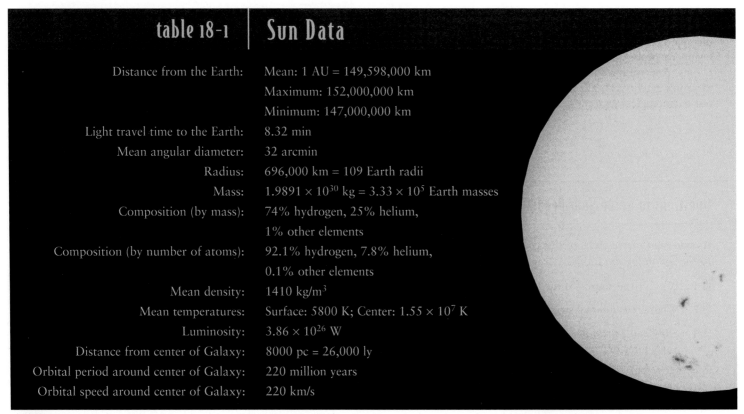

table 18-1	Sun Data
Distance from the Earth:	Mean: 1 AU = 149,598,000 km
	Maximum: 152,000,000 km
	Minimum: 147,000,000 km
Light travel time to the Earth:	8.32 min
Mean angular diameter:	32 arcmin
Radius:	696,000 km = 109 Earth radii
Mass:	1.9891×10^{30} kg = 3.33×10^5 Earth masses
Composition (by mass):	74% hydrogen, 25% helium, 1% other elements
Composition (by number of atoms):	92.1% hydrogen, 7.8% helium, 0.1% other elements
Mean density:	1410 kg/m³
Mean temperatures:	Surface: 5800 K; Center: 1.55×10^7 K
Luminosity:	3.86×10^{26} W
Distance from center of Galaxy:	8000 pc = 26,000 ly
Orbital period around center of Galaxy:	220 million years
Orbital speed around center of Galaxy:	220 km/s

Figure 18-1 R I ☑ U X G

The Sun The Sun's visible surface has a temperature of about 5800 K. At this temperature, all solids and liquids vaporize to form gases. It was only in the twentieth century that scientists discovered what has kept the Sun so hot for billions of years: the thermonuclear fusion of hydrogen nuclei in the Sun's core. (Jeremy Woodhouse/PhotoDisc)

on Earth for at least several hundred million years. (We now know that the Earth is 4.56 billion years old and that life has existed on it for most of its history.) Since life as we know it depends crucially on sunlight, the Sun must be as old. This posed a severe problem for physicists. What source of energy could have kept the Sun shining for so long (Figure 18-1)?

One attempt to explain solar energy was made in the mid-1800s by the English physicist Lord Kelvin (for whom the temperature scale is named) and the German scientist Hermann von Helmholtz. They argued that the tremendous weight of the Sun's outer layers should cause the Sun to contract gradually, compressing its interior gases. Whenever a gas is compressed, its temperature rises. (You can demonstrate this with a bicycle pump: As you pump air into a tire, the temperature of the air increases and the pump becomes warm to the touch.) Kelvin and Helmholtz thus suggested that gravitational contraction could cause the Sun's gases to become hot enough to radiate energy out into space.

This process, called *Kelvin-Helmholtz contraction,* actually does occur during the earliest stages of the birth of a star like the Sun (see Section 8-4). But Kelvin-Helmholtz contraction cannot be the major source of the Sun's energy today. If it were, the Sun would have had to be much larger in the relatively recent past. Helmholtz's own calculations showed that the Sun could have started its initial collapse from the solar nebula no more than about 25 million years ago. But the geological and fossil record shows that the Earth is far older than that, and so the Sun must be as well. Hence, this model of a Sun that shines because it shrinks cannot be correct.

On Earth, a common way to produce heat and light is by burning fuel, such as a log in a fireplace or coal in a power plant. Is it possible that a similar process explains the energy released by the Sun? The answer is no, because this process could not continue for a long enough time to explain the age of the Earth. The chemical reactions involved in burning release roughly 10^{-19} joule of energy per atom. Therefore, the number of atoms that would have to undergo chemical reactions each second to generate the Sun's luminosity of 3.9×10^{26} joules per second is approximately

$$\frac{3.9 \times 10^{26} \text{ joules per second}}{10^{-19} \text{ joule per atom}} = 3.9 \times 10^{45} \text{ atoms per second}$$

From its mass and chemical composition, we know that the Sun contains about 10^{57} atoms. Thus, the length of time required to consume the entire Sun by burning would be

$$\frac{10^{57} \text{ atoms}}{3.9 \times 10^{45} \text{ atoms per second}} = 3 \times 10^{11} \text{ seconds}$$

There are about 3×10^7 seconds in a year. Hence, in this model, the Sun would burn itself out in a mere 10,000 (10^4) years! This is far shorter than the known age of the Earth, so chemical reactions also cannot explain how the Sun shines.

The source of the Sun's luminosity could be explained if there were a process that was like burning but released much more energy per atom. Then the rate at which atoms would have to be consumed would be far less, and the lifetime of the Sun could be long enough to be consistent with the known age of the Earth. Albert Einstein discovered the key to such a process in 1905.

According to his *special theory of relativity,* a quantity *m* of mass can in principle be converted into an amount of energy *E* according to a now-famous equation:

Einstein's mass-energy equation

$$E = mc^2$$

m = quantity of mass, in kg

c = speed of light = 3×10^8 m/s

E = amount of energy into which the mass can be converted, in joules

The speed of light *c* is a large number, so c^2 is huge. Therefore, a small amount of matter can release an awesome amount of energy.

Inspired by Einstein's ideas, astronomers began to wonder if the Sun's energy output might come from the conversion of matter into energy. The Sun's low density of 1410 kg/m³ indicates that it must be made of the very lightest atoms, primarily hydrogen and helium. In the 1920s, the British astronomer Arthur Eddington showed that temperatures near the center of the Sun must be so high that atoms become completely ionized. Hence, at the Sun's center we expect to find hydrogen nuclei and electrons flying around independent of each other.

Another British astronomer, Robert Atkinson, suggested that under these conditions hydrogen nuclei could fuse together to produce helium nuclei in a *nuclear reaction* that transforms a tiny amount of mass into a large amount of energy. Experiments in the laboratory using individual nuclei show that such reactions can indeed take place. The process of converting hydrogen into helium is called **hydrogen fusion.** (It is also sometimes called *hydrogen burning,* even though nothing is actually burned in the conventional sense. Ordinary burning involves chemical reactions that rearrange the outer electrons of atoms but have no effect on the atoms' nuclei.) Hydrogen fusion provides the devastating energy released in a hydrogen bomb (see Figure 1-4).

The fusing together of nuclei is also called **thermonuclear fusion,** because it can take place only at extremely high temperatures. The reason is that all nuclei have a positive electric charge and so tend to repel one another. But in the extreme heat and pressure at the Sun's center, hydrogen nuclei (protons) are moving so fast that they can overcome their electric repulsion and actually touch one another. When that happens, thermonuclear fusion can take place.

ANALOGY You can think of protons as tiny electrically charged spheres that are coated with a very powerful glue. If the spheres are not touching, the repulsion between their charges pushes them apart. But if the spheres are forced into contact, the strength of the glue "fuses" them together.

CAUTION Be careful not to confuse thermonuclear fusion with the similar-sounding process of *nuclear fission.* In nuclear fusion, energy is released by joining together nuclei of lightweight atoms such as hydrogen. In nuclear fission, by contrast, the nuclei of very massive atoms such as uranium or plutonium

release energy by fragmenting into smaller nuclei. Nuclear power plants produce energy using fission, not fusion. (Generating power using fusion has been a goal of researchers for decades, but no one has yet discovered a commercially viable way to do this.)

We learned in Section 5-8 that the nucleus of a hydrogen atom (H) consists of a single proton. The nucleus of a helium atom (He) consists of two protons and two neutrons. In the nuclear process that Atkinson described, four hydrogen nuclei combine to form one helium nucleus, with a concurrent release of energy:

$$4\,H \rightarrow He + energy$$

In several separate reactions, two of the four protons are changed into neutrons, and eventually combine with the remaining protons to produce a helium nucleus (**Figure 18-2**). Each time this process takes place, a small fraction (0.7%) of the combined mass of the hydrogen nuclei does not show up in the mass of the helium nucleus. This "lost" mass is converted into energy. **Box 18-1** describes how to use Einstein's mass-energy equation to calculate the amount of energy released.

CAUTION You may have heard the idea that mass is always conserved (that is, it is neither created nor destroyed), or that energy is always conserved in a reaction. Einstein's ideas show that neither of these statements is quite correct, because mass can be converted into energy and vice versa. A more accurate statement is that the total amount of mass *plus* energy is conserved. Hence, the destruction of mass in the Sun does not violate any laws of nature.

For every four hydrogen nuclei converted into a helium nucleus, 4.3×10^{-12} joule of energy is released. This may seem like only a tiny amount of energy, but it is about 10^7 times larger than the amount of energy released in a typical chemical reaction, such as occurs in ordinary burning. Thus, thermonuclear fusion can explain how the Sun could have been shining for billions of years.

To produce the Sun's luminosity of 3.9×10^{26} joules per second, 6×10^{11} kg (600 million metric tons) of hydrogen must be converted into helium each second. This rate is prodigious, but there is a literally astronomical amount of hydrogen in the Sun. In particular, the Sun's core contains enough hydrogen to have been giving off energy at the present rate for as long as the solar system has existed, about 4.56 billion years, and to continue doing so for more than 6 billion years into the future.

 The Sun's energy actually comes from a series of nuclear reactions, called the **proton-proton chain,** in which hydrogen nuclei combine to form helium (see Figure 18-2 and Box 18-1). The proton-proton chain is also the energy source for many of the stars in the sky. In stars with central temperatures that are much hotter than that of the Sun, however, hydrogen fusion proceeds according to a different set of nuclear reactions, called the **CNO cycle,** in which carbon, nitrogen, and oxygen nuclei absorb protons to produce helium nuclei. Still other thermonuclear reactions, such as helium fusion, carbon fusion, and oxygen fusion, occur late in the lives of many stars.

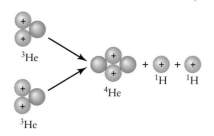

(a) Step 1:

- Two protons (hydrogen nuclei, ^{1}H) collide.
- One of the protons changes into a neutron (shown in blue), a neutral, nearly massless neutrino (ν), and a positively charged electron, or positron (e⁺).
- The proton and neutron form a hydrogen isotope (^{2}H).
- The positron encounters an ordinary electron (e⁻), annihilating both particles and converting them into gamma-ray photons (γ).

(b) Step 2:

- The ^{2}H nucleus from the first step collides with a third proton.
- A helium isotope (^{3}He) is formed and another gamma-ray photon is released.

(c) Step 3:

- Two ^{3}He nuclei collide.
- A different helium isotope with two protons and two neutrons (^{4}He) is formed and two protons are released.

ANIMATION 18.1

figure 18-2

Fusing Hydrogen into Helium Hydrogen fusion in the Sun usually takes place in three steps, each of which releases energy (see Box 18-1). In the first two steps, three hydrogen nuclei (^{1}H) fuse to make one ^{3}He nucleus. In the third step, two such ^{3}He nuclei fuse to make a ^{4}He nucleus, and two ^{1}H nuclei are released. Since two of the original six ^{1}H nuclei are returned to their original state, we can summarize the process as 4 ^{1}H → ^{4}He + energy.

box 18-1   TOOLS OF THE ASTRONOMER'S TRADE

Converting Mass into Energy

Figure 18-2 shows the steps involved in the thermonuclear fusion of hydrogen at the Sun's center. In these steps, four protons are converted into a single nucleus of ^{4}He, an isotope of helium with two protons and two neutrons. (As we saw in Box 5-5, different isotopes of the same element have the same number of protons but different numbers of neutrons.) The reaction depicted in Figure 18-2a also produces a neutral, nearly massless particle called the *neutrino*. Neutrinos respond hardly at all to ordinary matter, so they travel almost unimpeded through the Sun's massive bulk. Hence, the energy that neutrinos carry is quickly lost into space. This loss is not great, however, because the neutrinos carry relatively little energy. (See Section 18-4 for more about these curious particles.)

Most of the energy released by thermonuclear fusion appears in the form of gamma-ray photons. The energy of these photons remains trapped within the Sun for a long time, thus maintaining the Sun's intense internal heat. Some gamma-ray photons are produced by the reaction in Figure 18-2b. Others appear when an electron in the Sun's interior annihilates a positively charged electron, or **positron**, which is a by-product

of the reaction in Figure 18-2a. An electron and a positron are respectively matter and antimatter, and they convert entirely into energy when they meet. (You may have thought that "antimatter" was pure science fiction, but it is being created and annihilated in the Sun as you read these words.)

We can summarize the thermonuclear fusion of hydrogen as follows:

$$4 \; {}^{1}\text{H} \rightarrow {}^{4}\text{He} + \text{neutrinos} + \text{gamma-ray photons}$$

To calculate how much energy is released in this process, we use Einstein's mass-energy formula: The energy released is equal to the amount of mass consumed multiplied by c^2, where c is the speed of light. To see how much mass is consumed, we compare the combined mass of four hydrogen atoms (the ingredients) to the mass of one helium atom (the product):

4 hydrogen atoms	=	6.693×10^{-27} kg
−1 helium atom	=	6.645×10^{-27} kg
Mass lost	=	0.048×10^{-27} kg

(continued)

TOOLS OF THE ASTRONOMER'S TRADE continued

Thus, a small fraction (0.7%) of the mass of the hydrogen going into the nuclear reaction does not show up in the mass of the helium. This lost mass is converted into an amount of energy $E = mc^2$:

$$E = mc^2 = (0.048 \times 10^{-27} \text{ kg}) (3 \times 10^8 \text{ m/s})^2$$
$$= 4.3 \times 10^{-12} \text{ joule}$$

This is the amount of energy released by the formation of a single helium atom. It would light a 10-watt lightbulb for almost one-half of a trillionth of a second.

EXAMPLE: How much energy is released when 1 kg of hydrogen is converted to helium?

Situation: We are given the initial mass of hydrogen. We know that a fraction of the mass is lost when the hydrogen undergoes fusion to make helium; our goal is to find the quantity of energy into which this lost mass is transformed.

Tools: We use the equation $E = mc^2$ and the result that 0.7% of the mass is lost when hydrogen is converted into hydrogen.

Answer: When one kilogram of hydrogen is converted to helium, the amount of mass lost is 0.7% of 1 kg, or 0.007 kg. (This means that 0.993 kg of helium comes out.) Using Einstein's equation, we find that this missing 0.007 kg of matter is transformed into an amount of energy equal to

$$E = mc^2 = (0.007 \text{ kg}) (3 \times 10^8 \text{ m/s})^2 = 6.3 \times 10^{14} \text{ joules}$$

Review: This equals the energy released by burning 20,000 metric tons (2×10^7 kg) of coal! Hydrogen fusion is a *much* more efficient energy source than ordinary burning.

The Sun's luminosity is 3.9×10^{26} joules per second. To generate this much power, hydrogen must be consumed at a rate of

$$\frac{3.9 \times 10^{26} \text{ joules per second}}{6.3 \times 10^{14} \text{ joules per kilogram}}$$

$$= 6 \times 10^{11} \text{ kilograms per second}$$

That is, the Sun converts 600 million metric tons of hydrogen into helium every second.

18-2 | A theoretical model of the Sun shows how energy gets from its center to its surface

While thermonuclear fusion is the source of the Sun's energy, this process cannot take place everywhere within the Sun. As we have seen, extremely high temperatures—in excess of 10^7 K—are required for atomic nuclei to fuse together to form larger nuclei. The temperature of the Sun's visible surface, about 5800 K, is too low for these reactions to occur there. Hence, thermonuclear fusion can be taking place only within the Sun's interior. But precisely where does it take place? And how does the energy produced by fusion make its way to the surface, where it is emitted into space in the form of photons?

To answer these questions, we must understand conditions in the Sun's interior. Ideally, we would send an exploratory spacecraft to probe deep into the Sun; in practice, the Sun's intense heat would vaporize even the sturdiest spacecraft. Instead, astronomers use the laws of physics to construct a theoretical model of the Sun. (We discussed the use of models in science in Section 1-1.) Let's see what ingredients go into building a model of this kind.

Note first that the Sun is not undergoing any dramatic changes. The Sun is not exploding or collapsing, nor is it significantly heat-ing or cooling. The Sun is thus said to be in both *hydrostatic equilibrium* and *thermal equilibrium*.

To understand what is meant by **hydrostatic equilibrium**, imagine a slab of material in the solar interior (Figure 18-3a). In equilibrium, the slab on average will move neither up nor down. (In fact, there are upward and downward motions of material inside the Sun, but these motions average out.) Equilibrium is maintained by a balance among three forces that act on this slab:

1. The downward pressure of the layers of solar material above the slab.

2. The upward pressure of the hot gases beneath the slab.

3. The slab's weight—that is, the downward gravitational pull it feels from the rest of the Sun.

The pressure from below must balance both the slab's weight and the pressure from above. Hence, the pressure below the slab must be greater than that above the slab. In other words, pressure has to increase with increasing depth. For the same reason, pressure increases as you dive deeper into the ocean (Figure 18-3b) or as you move toward lower altitudes in our atmosphere.

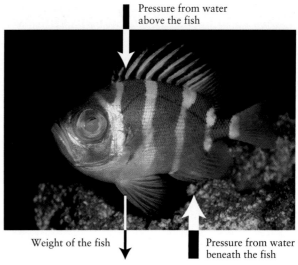

(a) Material inside the sun is in hydrostatic equilibrium, so forces balance

(b) A fish floating in water is in hydrostatic equilibrium, so forces balance

figure 18-3

Hydrostatic Equilibrium (a) Material in the Sun's interior tends to move neither up nor down. The upward forces on a slab of solar material (due to pressure of gases below the slab) must balance the downward forces (due to the slab's weight and the pressure of gases above the slab). Hence, the pressure must increase with increasing depth. (b) The same principle applies to a fish floating in water. In equilibrium, the forces balance and the fish neither rises nor sinks. (Ken Usami/PhotoDisc)

Hydrostatic equilibrium also tells us about the density of the slab. If the slab is too dense, its weight will be too great and it will sink; if the density is too low, the slab will rise. To prevent this, the density of solar material must have a certain value at each depth within the solar interior. (The same principle applies to objects that float beneath the surface of the ocean. Scuba divers wear weight belts to increase their average density so that they will neither rise nor sink but will stay submerged at the same level.)

Another consideration is that the Sun's interior is so hot that it is completely gaseous. Gases compress and become more dense when you apply greater pressure to them, so density must increase along with pressure as you go to greater depths within the Sun. Furthermore, when you compress a gas, its temperature tends to rise, so the temperature must also increase as you move toward the Sun's center.

While the temperature in the solar interior is different at different depths, the temperature at each depth remains constant in time. This principle is called **thermal equilibrium.** For the Sun to be in thermal equilibrium, all the energy generated by thermonuclear reactions in the Sun's core must be transported to the Sun's glowing surface, where it can be radiated into space. If too much energy flowed from the core to the surface to be radiated away, the Sun's interior would cool down; alternatively, the Sun's interior would heat up if too little energy flowed to the surface.

But exactly how is energy transported from the Sun's center to its surface? There are three methods of energy transport: *conduction, convection,* and *radiative diffusion.* Only the last two are important inside the Sun.

If you heat one end of a metal bar with a blowtorch, energy flows to the other end of the bar so that it too becomes warm. The effi-ciency of this method of energy transport, called **conduction,** varies significantly from one substance to another. For example, copper is a good conductor of heat, but wood is not (which is why copper pots often have wooden handles). Conduction is not an efficient means of energy transport in substances with low average densities, including the gases inside stars like the Sun.

Inside stars like our Sun, energy moves from center to surface by two other means: convection and radiative diffusion. **Convection** is the circulation of fluids—gases or liquids—between hot and cool regions. Hot gases rise toward a star's surface, while cool gases sink back down toward the star's center. This physical movement of gases transports heat energy outward in a star, just as the physical movement of water boiling in a pot transports energy from the bottom of the pot (where the heat is applied) to the cooler water at the surface.

In **radiative diffusion,** photons created in the thermonuclear inferno at a star's center diffuse outward toward the star's surface. Individual photons are absorbed and reemitted by atoms and electrons inside the star. The overall result is an outward migration from the hot core, where photons are constantly created, toward the cooler surface, where they escape into space.

To construct a model of a star like the Sun, astrophysicists express the ideas of hydrostatic equilibrium, thermal equilibrium, and energy transport as a set of equations. To ensure that the model applies to the particular star under study, they also make use of astronomical observations of the star's surface. (For example, to construct a model of the Sun, they use the data that the Sun's surface temperature is 5800 K, its luminosity is 3.9×10^{26} W, and the gas pressure and density at the surface are almost zero.) The astrophysicists then use a computer to solve their set of equations and calculate conditions layer by layer in toward the star's center. The

table 18-2 | A Theoretical Model of the Sun

Distance from the Sun's center (solar radii)	Fraction of luminosity	Fraction of mass	Temperature ($\times 10^6$ K)	Density (kg/m^3)	Pressure (relative to pressure at center)
0.0	0.00	0.00	15.5	160,000	1.00
0.1	0.42	0.07	13.0	90,000	0.46
0.2	0.94	0.35	9.5	40,000	0.15
0.3	1.00	0.64	6.7	13,000	0.04
0.4	1.00	0.85	4.8	4,000	0.007
0.5	1.00	0.94	3.4	1,000	0.001
0.6	1.00	0.98	2.2	400	0.0003
0.7	1.00	0.99	1.2	80	4×10^{-5}
0.8	1.00	1.00	0.7	20	5×10^{-6}
0.9	1.00	1.00	0.3	2	3×10^{-7}
1.0	1.00	1.00	0.006	0.00030	4×10^{-13}

Note: The distance from the Sun's center is expressed as a fraction of the Sun's radius ($R_\odot$). Thus, 0.0 is at the center of the Sun and 1.0 is at the surface. The fraction of luminosity is that portion of the Sun's total luminosity produced within each distance from the center; this is equal to 1.00 for distances of 0.25 $R_\odot$ or more, which means that all of the Sun's nuclear reactions occur within 0.25 solar radius from the Sun's center. The fraction of mass is that portion of the Sun's total mass lying within each distance from the Sun's center. The pressure is expressed as a fraction of the pressure at the center of the Sun.

result is a model of how temperature, pressure, and density increase with increasing depth below the star's surface.

Table 18-2 and Figure 18-4 show a theoretical model of the Sun that was calculated in just this way. Different models of the Sun use slightly different assumptions, but all models give essentially the same results as those shown here. From such computer models we

have learned that at the Sun's center the density is 160,000 kg/m^3 (14 times the density of lead!), the temperature is 1.55×10^7 K, and the pressure is 3.4×10^{11} atm. (One atmosphere, or 1 atm, is the average atmospheric pressure at sea level on Earth.)

Table 18-2 and Figure 18-4 show that the solar luminosity rises to 100% at about one-quarter of the way from the Sun's center to

figure 18-4
A Theoretical Model of the Sun's Interior These graphs depict what percentage of the Sun's total luminosity is produced within each distance from the center (upper left), what percentage of the

total mass lies within each distance from the center (lower left), the temperature at each distance (upper right) and the density at each distance (lower right). (See Table 18-2 for a numerical version of this model.)

its surface. In other words, the Sun's energy production occurs within a volume that extends out only to 0.25 $R_\odot$. (The symbol $R_\odot$ denotes the solar radius, or radius of the Sun as a whole, equal to 696,000 km.) Outside 0.25 $R_\odot$, the density and temperature are too low for thermonuclear reactions to take place. Also note that 94% of the total mass of the Sun is found within the inner 0.5 $R_\odot$. Hence, the outer 0.5 $R_\odot$ contains only a relatively small amount of material.

How energy flows from the Sun's center toward its surface depends on how easily photons move through the gas. If the solar gases are comparatively transparent, photons can travel moderate distances before being scattered or absorbed, and energy is thus transported by radiative diffusion. If the gases are comparatively opaque, photons cannot get through the gas easily and heat builds up. Convection then becomes the most efficient means of energy transport. The gases start to churn, with hot gas moving upward and cooler gas sinking downward.

From the center of the Sun out to about 0.71 $R_\odot$, energy is transported by radiative diffusion. Hence, this region is called the **radiative zone.** Beyond about 0.71 $R_\odot$, the temperature is low enough (a mere 2×10^6 K or so) for electrons and hydrogen nuclei to join into hydrogen atoms. These atoms are very effective at absorbing photons, much more so than free electrons or nuclei, and this absorption chokes off the outward flow of photons. Therefore, beyond about 0.71 $R_\odot$, radiative diffusion is not an effective way to transport energy. Instead, convection dominates the energy flow in this outer region, which is why it is called the **convective zone.** Figure 18-5 shows these aspects of the Sun's internal structure.

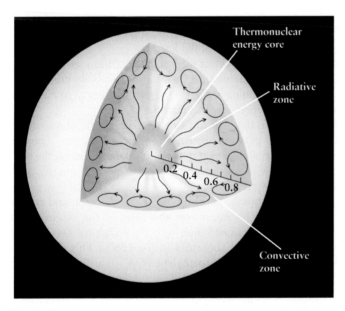

figure 18-5

The Sun's Internal Structure Thermonuclear reactions occur in the Sun's core, which extends out to a distance of 0.25 $R_\odot$ from the center. Energy is transported outward, via radiative diffusion, to a distance of about 0.71 $R_\odot$. In the outer layers between 0.71 $R_\odot$ and 1.00 $R_\odot$, energy flows outward by convection.

Although energy travels through the radiative zone in the form of photons, the photons have a difficult time of it. Table 18-2 shows that the material in this zone is extremely dense, so photons from the Sun's core take a long time to diffuse through the radiative zone. As a result, it takes approximately 170,000 years for energy created at the Sun's center to travel 696,000 km to the solar surface and finally escape as sunlight. The energy flows outward at an average rate of 50 centimeters per hour, or about 20 times slower than a snail's pace.

Once the energy escapes from the Sun, it travels much faster—at the speed of light. Thus, solar energy that reaches you today took only 8 minutes to travel the 150 million kilometers from the Sun's surface to the Earth. But this energy was actually produced by thermonuclear reactions that took place in the Sun's core hundreds of thousands of years ago.

18-3 | Astronomers probe the solar interior using the Sun's own vibrations

We have described how astrophysicists construct models of the Sun. But since we cannot see into the Sun's opaque interior, how can we check these models to see if they are accurate? What is needed is a technique for probing the Sun's interior. A very powerful technique of just this kind involves measuring vibrations of the Sun as a whole. This field of solar research is called **helioseismology.**

Vibrations are a useful tool for examining the hidden interiors of all kinds of objects. Food shoppers test whether melons are ripe by tapping on them and listening to the vibrations. Geologists can determine the structure of the Earth's interior by using seismographs to record vibrations during earthquakes.

Although there are no true "sunquakes," the Sun does vibrate at a variety of frequencies, somewhat like a ringing bell. These vibrations were first noticed in 1960 by Robert Leighton of the California Institute of Technology, who made high-precision Doppler shift observations of the solar surface. These measurements revealed that parts of the Sun's surface move up and down about 10 meters every 5 minutes. Since the mid-1970s, several astronomers have reported slower vibrations, having periods ranging from 20 to 160 minutes. The detection of extremely slow vibrations has inspired astronomers to organize networks of telescopes around and in orbit above the Earth to monitor the Sun's vibrations on a continuous basis.

The vibrations of the Sun's surface can be compared with sound waves. If you could somehow survive within the Sun's outermost layers, you would first notice a deafening roar, somewhat like a jet engine, produced by turbulence in the Sun's gases. But superimposed on this noise would be a variety of nearly pure tones. You would need greatly enhanced hearing to detect these tones, however; the strongest has a frequency of just 0.003 hertz, 13 octaves below the lowest frequency audible to humans. (Recall from Section 5-2 that one hertz is one oscillation per second.)

In 1970, Roger Ulrich, at UCLA, pointed out that sound waves moving upward from the solar interior would be reflected back inward after reaching the solar surface. However, as a reflected

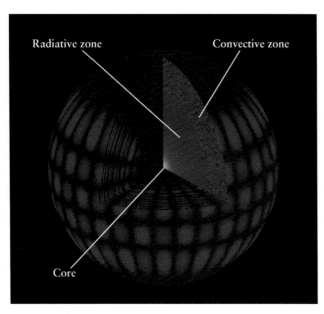

Figure 18-6

A Sound Wave Resonating in the Sun This computer-generated image shows one of the millions of ways in which the Sun's interior vibrates. The regions that are moving outward are colored blue, those moving inward, red. As the cutaway shows, these oscillations are thought to extend into the Sun's radiative zone (compare Figure 18-5). (National Solar Observatory)

sound wave descends back into the Sun, the increasing density and pressure bend the wave severely, turning it around and aiming it back toward the solar surface. In other words, sound waves bounce back and forth between the solar surface and layers deep within the Sun. These sound waves can reinforce each other if their wavelength is the right size, just as sound waves of a particular wavelength resonate inside an organ pipe.

The Sun oscillates in millions of ways as a result of waves resonating in its interior. **Figure 18-6** is a computer-generated illustration of one such mode of vibration. Helioseismologists can deduce information about the solar interior from measurements of these oscillations. For example, they have been able to set limits on the amount of helium in the Sun's core and convective zone and to determine the thickness of the transition region between the radiative zone and convective zone. They have also found that the convective zone is thicker than previously thought.

18-4 | Neutrinos reveal information about the Sun's core—and have surprises of their own

SCIENCE IN PROCESS *Scientists use the most ethereal of subatomic particles to learn about the Sun, and vice versa*

We have seen circumstantial evidence that thermonuclear fusion is the source of the Sun's power. To be certain, however, we need more definitive evidence. How can we show that thermonuclear fusion really is taking place in the Sun's core? Unfortunately, helioseismology is of little assistance. The vibrations that astronomers see on the Sun's surface do not penetrate that far into the interior.

Although the light energy that we receive from the Sun originates in the core, it provides few clues about conditions there. The problem is that this energy has changed form repeatedly during its passage from the core: It appeared first as photons diffusing through the radiative zone, then as heat transported through the outer layers by convection, and then again as photons emitted from the Sun's glowing surface. As a result of these transformations, much of the information that the Sun's radiated energy once carried about conditions in the core has been lost.

 ANALOGY If you make a photocopy of a photocopy of a photocopy of an original document, the final result may be so blurred as to be unreadable. In an analogous way, because solar energy is transformed many times while en route to Earth, the story it could tell us about the Sun's core is hopelessly blurred.

But there is a way for scientists to learn about conditions in the Sun's core and to get direct evidence that thermonuclear fusion really does happen there. The trick is to detect the subatomic byproducts of thermonuclear fusion reactions.

As part of the process of hydrogen fusion, protons change into neutrons and release **neutrinos** (see Figure 18-2a and Box 18-1). Like photons, neutrinos are particles that have no electric charge. Unlike photons, however, neutrinos interact only very weakly with matter. Even the vast bulk of the Sun offers little impediment to their passage, so neutrinos must be streaming out of the core and into space. Indeed, the conversion of hydrogen into helium at the Sun's center produces 10^{38} neutrinos each second. Every second, about 10^{14} neutrinos from the Sun—that is, **solar neutrinos**—must pass through each square meter of the Earth.

If it were possible to detect these solar neutrinos, we would have direct evidence that thermonuclear reactions really do take place in the Sun's core. Beginning in the 1960s, scientists began to build neutrino detectors for precisely this purpose.

The challenge is that neutrinos are exceedingly difficult to detect. Just as neutrinos pass unimpeded through the Sun, they also pass through the Earth almost as if it were not there. We stress the word "almost," because neutrinos can and do interact with matter, albeit infrequently.

On rare occasions a neutrino will strike a neutron and convert it into a proton. This effect was the basis of the original solar neutrino detector, designed and built by Raymond Davis of the Brookhaven National Laboratory in the 1960s. This device used 100,000 gallons of perchloroethylene (C_2Cl_4), a fluid used in dry cleaning, in a huge tank buried deep underground. Most of the solar neutrinos that entered Davis's tank passed right through it with no effect whatsoever. But occasionally a neutrino struck the nucleus of one of the chlorine atoms (^{37}Cl) in the cleaning fluid and converted one of its neutrons into a proton, creating a radioactive atom of argon (^{37}Ar).

The rate at which argon is produced is related to the neutrino flux—that is, the number of neutrinos from the Sun arriving at the Earth per square meter per second. By counting the number of newly created argon atoms, Davis was able to determine the neutrino flux from the Sun. (Other subatomic particles besides neutrinos can also induce reactions that create radioactive atoms. By placing the experiment deep underground, however, the body of the Earth absorbs essentially all such particles—with the exception of neutrinos.)

Davis and his collaborators found that solar neutrinos created one radioactive argon atom in the tank every three days. But this rate corresponds to only one-third of the neutrino flux predicted from standard models of the Sun. This troubling discrepancy between theory and observation, called the **solar neutrino problem,** motivated scientists around the world to conduct further experiments to measure solar neutrinos.

One key question was whether the neutrinos that Davis had detected had really come from the Sun. (The Davis experiment had no way to determine the direction from which neutrinos had entered the tank of cleaning fluid.) This was resolved by an experiment in Japan called Kamiokande, which was designed by the physicist Masatoshi Koshiba. A large underground tank containing 3000 tons of water was surrounded by 1100 light detectors. From time to time, a high-energy solar neutrino struck an electron in one of the water molecules, dislodging it and sending it flying like a pin hit by a bowling ball. The recoiling electron produced a flash of light, which was sensed by the detectors. By analyzing the flashes, scientists could tell the direction from which the neutrinos were coming and confirmed that they emanated from the Sun. These results in the late 1980s gave direct evidence that thermonuclear fusion is indeed occurring in the Sun's core. (Davis and Koshiba both received the 2002 Nobel Prize in Physics for their pioneering research on solar neutrinos.)

Like Davis's experiment, however, Koshiba and his colleagues at Kamiokande detected only a fraction of the expected flux of neutrinos. Where, then, were the missing solar neutrinos?

One proposed solution had to do with the energy of the detected neutrinos. The vast majority of neutrinos from the Sun are created during the first step in the proton-proton chain, in which two protons combine to form a heavy isotope of hydrogen (see Figure 18-2a). But these neutrinos have too little energy to convert chlorine into argon. Both Davis's and Koshiba's experiments responded only to high-energy neutrinos produced by reactions that occur only part of the time near the end of the proton-proton chain. (Figure 18-2 does not show these reactions.) Could it be that the discrepancy between theory and observation would go away if the flux of low-energy neutrinos could be measured?

To test this idea, two teams of physicists constructed neutrino detectors that used several tons of gallium (a liquid metal) rather than cleaning fluid. Low-energy neutrinos convert gallium (^{71}Ga) into a radioactive isotope of germanium (^{71}Ge). By chemically separating the germanium from the gallium and counting the radioactive atoms, the physicists were able to measure the flux of low-energy solar neutrinos. These experiments—GALLEX in Italy and SAGE (Soviet-American Gallium Experiment) in Russia—detected only 50% to 60% of the expected neutrino

flux. Hence, the solar neutrino problem is a discrepancy between theory and observation for neutrinos of all energies.

Another proposed solution to the neutrino problem was that the Sun's core is cooler than predicted by solar models. If the Sun's central temperature were only 10% less than the current estimate, fewer neutrinos would be produced and the neutrino flux would agree with experiments. However, a lower central temperature would cause other obvious features, such as the Sun's size and surface temperature, to be different from what we observe.

 Only very recently has the solution to the neutrino problem been found. The answer lies not in how neutrinos are produced, but rather in what happens to them between the Sun's core and detectors on the Earth. Physicists have found that there are actually three types of neutrinos. Only one of these types is produced in the Sun, and it is only this type that can be detected by the experiments we have described. But if some of the solar neutrinos change in flight into a different type of neutrino, the detectors in these experiments would record only a fraction of the total neutrino flux. This effect is called **neutrino oscillation.** In June 1998, scientists at the Super-Kamiokande neutrino observatory (a larger and more sensitive device than Kamiokande) revealed evidence that neutrino oscillation does indeed take place.

The best confirmation of this idea has come from the Sudbury Neutrino Observatory (SNO) in Canada. Like Kamiokande and Super-Kamiokande, SNO uses an large tank of water placed deep underground (**Figure 18-7**). But unlike those earlier experiments, SNO can detect all of the three types of neutrinos. It does this by using heavy water, in which the hydrogen atoms in an H_2O molecule are the isotope ^2H shown in Figure 18-2a, with both a proton and a neutron in its nucleus. (Each hydrogen atom in ordinary, or "light," water has a solitary proton as its nucleus.) If a high-energy solar neutrino of any type passes through SNO's tank of heavy water, it can knock the neutron out of one of the ^2H nuclei. The ejected neutron can then be captured by another nucleus, and this capture releases energy that manifests itself as a tiny burst of light. As in Kamiokande, detectors around SNO's water tank record these light flashes.

What SNO has observed as of early 2004 is that the combined flux of all three types of neutrinos coming from the Sun is *equal* to the theoretical prediction. Together with the results from earlier neutrino experiments, this strongly suggests that the Sun is indeed producing neutrinos at the predicted rate as a by-product of thermonuclear reactions. But before these neutrinos can reach the Earth, about two-thirds of them undergo an oscillation and change their type.

The story of the solar neutrino problem illustrates how two different branches of science—in this case, studies of the solar interior and investigations of subatomic particles—can sometimes interact, to the mutual benefit of both. While there is still much we do not understand about the Sun and about neutrinos, a new generation of neutrino detectors in Japan, Canada, and elsewhere promises to further our knowledge of these exotic realms of astronomy and physics. The essay by John Bahcall that follows this chapter has more to say about the connections between solar astronomy and neutrino physics.

Hollow transparent sphere
to hold heavy water

Frame for light detectors

Workers

Figure 18-7 R I **V** U X G
The Sudbury Neutrino Observatory Under Construction The transparent acrylic sphere holds 1000 tons of heavy water. Any of the three types of solar neutrino produces a flash of light when it interacts with the heavy water. The flash is sensed by 9600 light detectors surrounding the tank. (The detectors were not all installed when this photograph was taken.) (Photo courtesy of SNO)

18-5 | The photosphere is the lowest of three main layers in the Sun's atmosphere

Although the Sun's core is hidden from our direct view, we can easily see sunlight coming from the high-temperature gases that make up the Sun's atmosphere. These outermost layers of the Sun prove to be the sites of truly dramatic activity, much of which has a direct impact on our planet. By studying these layers, we gain further insight into the character of the Sun as a whole.

A visible-light photograph like **Figure 18-8** makes it appear that the Sun has a definite surface. This is actually an illusion; the Sun is gaseous throughout its volume because of its high internal temperature, and the gases simply become less and less dense as you move farther away from the Sun's center.

Why, then, does the Sun appear to have a sharp, well-defined surface? The reason is that essentially all of the Sun's visible light emanates from a single, thin layer of gas called the **photosphere** ("sphere of light"). Just as you can see only a certain distance through the Earth's atmosphere before objects vanish in the haze, we can see only about 400 km into the photosphere. This distance is so small compared with the Sun's radius of 696,000 km that the photosphere appears to be a definite surface.

The photosphere is actually the lowest of the three layers that together constitute the solar atmosphere. Above it are the *chromosphere* and the *corona*, both of which are transparent to visible light. We can see them only using special techniques, which we discuss later in this chapter. Everything below the photosphere is called the *solar interior*.

The photosphere is heated from below by energy streaming outward from the solar interior. Hence, temperature should decrease

as you go upward in the photosphere, just as in the solar interior (see Table 18-2 and Figure 18-4). We know this is the case because the photosphere appears darker around the edge, or *limb*, of the Sun than it does toward the center of the solar disk, an effect called **limb darkening** (examine Figure 18-8). This happens because when

Figure 18-8 R I **V** U X G
The Photosphere The photosphere is the layer in the solar atmosphere from which the Sun's visible light is emitted. Note that the Sun appears darker around its limb, or edge; here we are seeing the upper photosphere, which is relatively cool and thus glows less brightly. (The dark sunspots, which we discuss in Section 18-8, are also relatively cool regions.) (Celestron International)

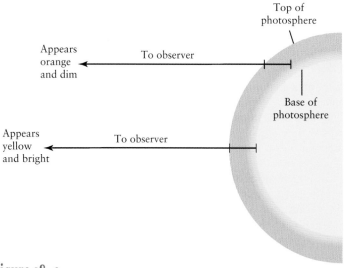

figure 18-9

The Origin of Limb Darkening Light from the Sun's limb and light from the center of its disk both travel about the same straight-line distance through the photosphere to reach us. Because of the Sun's curved shape, light from the limb comes from a greater height within the photosphere, where the temperature is lower and the gases glow less brightly. Hence, the limb appears darker and more orange.

we look near the Sun's limb, we do not see as deeply into the photosphere as we do when we look near the center of the disk (**Figure 18-9**). The high-altitude gas we observe at the limb is not as hot and thus does not glow as brightly as the deeper, hotter gas seen near the disk center.

The spectrum of the Sun's photosphere confirms how its temperature varies with altitude. As we saw in Section 18-1, the photosphere shines like a nearly perfect blackbody with an average temperature of about 5800 K. However, superimposed on this spectrum are many dark absorption lines (see Figure 5-12). As discussed in Section 5-6, we see an *absorption line spectrum* of this sort whenever we view a hot, glowing object through a relatively cool gas. In this case, the hot object is the lower part of the photosphere; the cooler gas is in the upper part of the photosphere, where the temperature declines to about 4400 K. All the absorption lines in the Sun's spectrum are produced in this relatively cool layer, as atoms selectively absorb photons of various wavelengths streaming outward from the hotter layers below.

CAUTION! You may find it hard to think of 4400 K as "cool." But keep in mind that the ratio of 4400 K to 5800 K, the temperature in the lower photosphere, is the same as the ratio of the temperature on a Siberian winter night to that of a typical day in Hawaii.

We can learn still more about the photosphere by examining it with a telescope—but only when using special dark filters to prevent eye damage. ***Looking directly at the Sun without the correct filter, whether with the naked eye or with a telescope, can cause permanent blindness!*** Under good observing conditions, astronomers using such filter-equipped telescopes can often see a blotchy pattern in the photosphere, called **granulation** (**Figure 18-10**). Each light-colored **granule** measures about 1000 km (600 mi) across and is surrounded by a darkish boundary. The difference in brightness between the center and the edge of a granule corresponds to a temperature drop of about 300 K.

VIDEO 18.1   VIDEO 18.2

figure 18-10   R I **V** U X G

Solar Granulation High-resolution photographs of the Sun's surface reveal a blotchy pattern called granulation. Granules are convection cells about 1000 km (600 mi) wide in the Sun's photosphere. **Inset:** Rising hot gas produces bright granules. Cooler gas sinks downward along the boundaries between granules; this gas glows less brightly, giving the boundaries their dark appearance. This convective motion transports heat from the Sun's interior outward to the solar atmosphere. (MSFC/NASA; inset: Goran Scharmer, Lund Observatory)

Granulation is caused by convection of the gas in the photosphere. The inset in Figure 18-10 shows how gas from lower levels rises upward in granules, cools off, spills over the edges of the granules, and then plunges back down into the Sun. This can occur only if the gas is heated from below, like a pot of water being heated on a stove (see Section 18-2). Along with limb darkening and the Sun's absorption line spectrum, granulation shows that the upper part of the photosphere must be cooler than the lower part.

Time-lapse photography reveals more of the photosphere's dynamic activity. Granules form, disappear, and reform in cycles lasting only a few minutes. At any one time, about 4 million granules cover the solar surface. A typical granule occupies about 106 square kilometers, equal to the areas of Texas and Oklahoma combined.

Superimposed on the pattern of granulation are even larger convection cells called **supergranules** (Figure 18-11). As in granules, gases rise upward in the middle of a supergranule, move horizontally outward toward its edge, and descend back into the Sun. The difference is that a typical supergranule is about 35,000 km in diameter, large enough to enclose several hundred granules. This large-scale convection moves at only about 0.4 km/s (1400 km/h, or 900 mi/h), about one-tenth the speed of gases churning in a granule. A given supergranule lasts about a day.

Blue: areas of rising gas

Red: areas of sinking gas

figure 18-11 R I **V** U X G

Supergranules and Large-Scale Convection Supergranules display relatively little contrast between their center and edges, so they are hard to observe in ordinary images. But they can be seen in a false-color Doppler image like this one. Light from gas that is approaching us (that is, rising) is shifted toward shorter wavelengths, while light from receding gas (that is, descending) is shifted toward longer wavelengths (see Section 5-9). (David Hathaway, MSFC/NASA)

ANALOGY Similar patterns of large-scale and small-scale convection can be found in the Earth's atmosphere. On the large scale, air rises gradually at a low-pressure area, then sinks gradually at a high-pressure area, which might be hundreds of kilometers away. This is analogous to the flow in a supergranule. Thunderstorms in our atmosphere are small but intense convection cells within which air moves rapidly up and down. Like granules, they last only a relatively short time before they dissipate.

Although the photosphere is a very active place, it actually contains relatively little material. Careful examination of the spectrum shows that it has a density of only about 10^{-4} kg/m³, roughly 0.01% the density of the Earth's atmosphere at sea level. The photosphere is made primarily of hydrogen and helium, the most abundant elements in the solar system (see Figure 8-3).

Despite being such a thin gas, the photosphere is surprisingly opaque to visible light. If it were not so opaque, we could see into the Sun's interior to a depth of hundreds of thousands of kilometers, instead of the mere 400 km that we can see down into the photosphere. What makes the photosphere so opaque is that its hydrogen atoms sometimes acquire an extra electron, becoming **negative hydrogen ions**. The extra electron is only loosely attached and can be dislodged if it absorbs a photon of any visible wavelength. Hence, negative hydrogen ions are very efficient light absorbers, and there are enough of these light-absorbing ions in the photosphere to make it quite opaque. Because it is so opaque, the photosphere's spectrum is close to that of an ideal blackbody.

18-6 | The chromosphere is characterized by spikes of rising gas

An ordinary visible-light image such as Figure 18-8 gives the impression that the Sun ends at the top of the photosphere. But during a total solar eclipse, the Moon blocks the photosphere from our view, revealing a glowing, pinkish layer of gas above the photosphere (Figure 18-12). This is the tenuous **chromosphere** ("sphere of color"), the second of the three major levels in the Sun's atmosphere. The chromosphere is only about one ten-thousandth (10^{-4}) as dense as the photosphere, or about 10^{-8} as dense as our own atmosphere. No wonder it is normally invisible!

Unlike the photosphere, which has an absorption line spectrum, the chromosphere has a spectrum dominated by emission lines. An emission line spectrum is produced by the atoms of a hot, thin gas (see Section 5-6 and Section 5-8). As their electrons fall from higher to lower energy levels, the atoms emit photons.

One of the strongest emission lines in the chromosphere's spectrum is the H_α line at 656.3 nm, which is emitted by a hydrogen atom when its single electron falls from the $n = 3$ level to the $n = 2$ level (recall Figure 5-21b). This wavelength is in the red part of the spectrum, which gives the chromosphere its characteristic pinkish color. The spectrum also contains emission lines of singly ionized calcium, as well as lines due to ionized helium and ionized metals. In fact, helium was originally discovered in the chromospheric spectrum in 1868, almost 30 years before helium gas was first isolated on Earth.

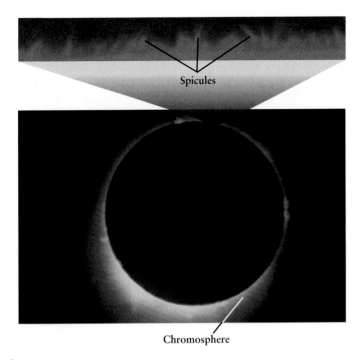

The Chromosphere During a total solar eclipse, the Sun's glowing chromosphere can be seen around the edge of the Moon. It appears pinkish because its hot gases emit light at only certain discrete wavelengths, principally the H_α emission of hydrogen at a red wavelength of 656.3 nm. The expanded area above shows spicules, jets of chromospheric gas that surge upward into the Sun's outer atmosphere. (NOAO)

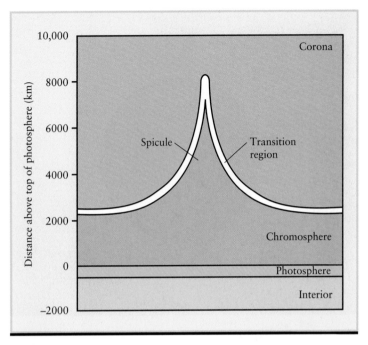

figure 18-13

The Solar Atmosphere This schematic diagram shows the three layers of the solar atmosphere. The lowest, the photosphere, is about 400 km thick. The chromosphere extends about 2000 km higher, with spicules jutting up to nearly 10,000 km above the photosphere. Above a transition region is the Sun's outermost layer, the corona, which we discuss in Section 18-7. It extends many millions of kilometers out into space. (Adapted from J. A. Eddy)

Analysis of the chromospheric spectrum shows that temperature *increases* with increasing height in the chromosphere. This is just the opposite of the situation in the photosphere, where temperature decreases with increasing height. The temperature is about 4400 K at the top of the photosphere; 2000 km higher, at the top of the chromosphere, the temperature is nearly 25,000 K.

The photospheric spectrum is dominated by absorption lines at certain wavelengths, while the spectrum of the chromosphere has emission lines at these same wavelengths. In other words, the photosphere appears dark at the wavelengths at which the chromosphere emits most strongly, such as the H_α wavelength of 656.3 nm. By viewing the Sun through a special filter that is transparent to light only at the wavelength of H_α, astronomers can screen out light from the photosphere and make the chromosphere visible. (The same technique can be used with other wavelengths at which the chromosphere emits strongly, including nonvisible wavelengths.) This makes it possible to see the chromosphere at any time, not just during a solar eclipse.

The top photograph in Figure 18-12 is a high-resolution image of the Sun's chromosphere taken through an H_α filter. This image shows numerous vertical spikes, which are actually jets of rising gas called **spicules.** A typical spicule lasts just 15 minutes or so: It rises at the rate of about 20 km/s (72,000 km/h, or 45,000 mi/h), can reach a height of several thousand kilometers, and then collapses and fades away (Figure 18-13). Approximately 300,000 spicules exist at any one time, covering about 1% of the Sun's surface.

Spicules are generally located directly above the edges of supergranules (see Figure 18-11). This is a surprising result, because chromospheric gases are rising in a spicule while photospheric gases are *descending* at the edge of a supergranule. What, then, is pulling gases upward to form spicules? As we will see in Section 18-10, the answer proves to be the Sun's intense magnetic field. But before we delve into how this happens, let us complete our tour of the solar atmosphere by exploring its outermost, least dense, most dynamic, and most bizarre layer—the region called the corona.

18-7 | The corona ejects mass into space to form the solar wind

The **corona,** or outermost region of the Sun's atmosphere, begins at the top of the chromosphere. It extends out to a distance of several

figure 18-14   R I **V** U X G

The Solar Corona This striking photograph of the corona was taken during the total solar eclipse of July 11, 1991. Numerous streamers extend for millions of kilometers above the solar surface. The unearthly light of the corona is one of the most extraordinary aspects of experiencing a solar eclipse. (Courtesy of R. Christen and M. Christen, Astro-Physics, Inc.)

In this narrow transition region between the chromosphere and corona, the temperature rises abruptly by about a factor of 100.

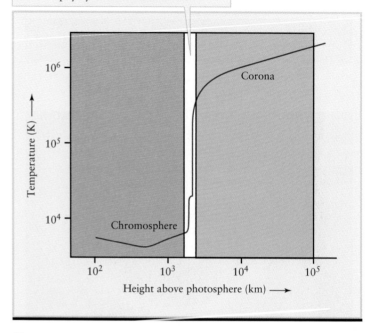

figure 18-15

Temperatures in the Sun's Upper Atmosphere This graph shows how temperature varies with altitude in the Sun's chromosphere and corona and in the narrow transition region between them. In order to show a large range of values, both the vertical and horizontal scales are nonlinear. (Adapted from A. Gabriel)

million kilometers. Despite its tremendous extent, the corona is only about one-millionth (10^{-6}) as bright as the photosphere—no brighter than the full moon. Hence, the corona can be viewed only when the light from the photosphere is blocked out, either by use of a specially designed telescope or during a total eclipse.

Figure 18-14 is an exceptionally detailed photograph of the Sun's corona taken during a solar eclipse. It shows that the corona is not merely a spherical shell of gas surrounding the Sun. Rather, numerous streamers extend in different directions far above the solar surface. The shapes of these streamers vary on time scales of days or weeks. (For another view of the corona during a solar eclipse, see Figure 3-11.)

The corona has an emission line spectrum, characteristic of a hot, thin gas. When this spectrum was first measured in the nineteenth century, astronomers found a number of emission lines at wavelengths that had never been seen in the laboratory. Their explanation was that the corona contained elements that had not yet been detected on the Earth. However, laboratory experiments in the 1930s revealed that these unusual emission lines were in fact caused by the same atoms found elsewhere in the universe—but in highly ionized states. For example, a prominent green line at 530.3 nm is caused by highly ionized iron atoms, each of which has been stripped of 13 of its 26 electrons. In order to strip that many electrons from atoms, temperatures in the corona must reach 2 million kelvins (2×10^6 K) or even higher. **Figure 18-15** shows how temperature in both the chromosphere and corona varies with altitude.

CAUTION! The corona is actually not very "hot"—that is, it contains very little thermal energy. The reason is that the corona is nearly a vacuum. In the corona there are only about 10^{11} atoms per cubic meter, compared with about 10^{23} atoms per cubic meter in the Sun's photosphere and about 10^{25} atoms per cubic meter in the air that we breathe. Because of the corona's high temperature, the atoms there are moving at very high speeds. But because there are so few atoms in the corona, the total amount of energy in these moving atoms (a measure of how "hot" the gas is) is rather low. If you flew a spaceship into the corona, you would have to worry about becoming overheated by the intense light coming from the photosphere, but you would notice hardly any heating from the corona's ultrathin gas.

ANALOGY The situation in the corona is similar to that inside a conventional oven that is being used for baking. Both the walls of the oven and the air inside the oven are at the same high temperature, but the air contains very few atoms and thus carries little energy. If you put your hand in the oven momentarily, the lion's share of the heat you feel is radiation from the oven walls.

The low density of the corona explains why it is so dim compared with the photosphere. In general, the higher the temperature of a gas, the brighter it glows. But because there are so few atoms in the corona, the net amount of light that it emits is very feeble compared with the light from the much cooler, but also much denser, photosphere.

The Earth's gravity keeps our atmosphere from escaping into space. In the same way, the Sun's powerful gravitational attraction keeps most of the gases of the photosphere, chromosphere, and corona from escaping. But the corona's high temperature means that its atoms and ions are moving at very high speeds, around a million kilometers per hour. As a result, some of the coronal gas does escape. This outflow of gas, which we first encountered in Section 8-5, is called the **solar wind.**

Each second the Sun ejects about a million tons (10^9 kg) of material into the solar wind. But the Sun is so massive that, even over its entire lifetime, it will eject only a few tenths of a percent of its total mass. The solar wind is composed almost entirely of electrons and nuclei of hydrogen and helium. About 0.1% of the solar wind is made up of ions of more massive atoms, such as silicon, sulfur, calcium, chromium, nickel, iron, and argon. The aurorae seen at far northern or southern latitudes on Earth are produced when electrons and ions from the solar wind enter our upper atmosphere.

Special telescopes enable astronomers to see the origin of the solar wind. To appreciate what sort of telescopes are needed, note that because the temperature of the coronal gas is so high, ions in the corona are moving very fast (see Box 7-2). When ions collide, the energy of the impact is so great that the ion's electrons are boosted to very high energy levels. As the electrons fall back to lower levels, they emit high-energy photons in the ultraviolet and X-ray portions of the spectrum—wavelengths at which the photosphere and chromosphere are relatively dim. Hence, telescopes sensitive to these short wavelengths are ideal for studying the corona and the flow of the solar wind.

The Earth's atmosphere is opaque to ultraviolet light and X rays, so telescopes for these wavelengths must be placed on board spacecraft (see Section 6-7, especially Figure 6-28). **Figure 18-16** shows an ultraviolet view of the corona from the spacecraft *SOHO* (*Solar and Heliospheric Observatory*), a joint project of the European Space Agency (ESA) and NASA.

Figure 18-16 reveals that the corona is not uniform in temperature or density. The densest, highest-temperature regions appear bright, while the thinner, lower-temperature regions are dark. Note the large dark area, called a **coronal hole** because it is almost devoid of luminous gas. Particles streaming away from the Sun can most easily flow outward through these particularly thin regions. Therefore, it is thought that coronal holes are the main corridors through which particles of the solar wind escape from the Sun.

Evidence in favor of this picture has come from the *Ulysses* spacecraft, another joint ESA/NASA mission. In 1994 and 1995, *Ulysses* became the first spacecraft to fly over the Sun's north and south poles, where there are apparently permanent coronal holes. The spacecraft indeed measured a stronger solar wind emanating from these holes.

The temperatures in the corona and the chromosphere are not at all what we would expect. Just as you feel warm if you stand close to a campfire but cold if you move away, we would expect that the temperature in the corona and chromosphere would *decrease* with increasing altitude and, hence, increasing distance

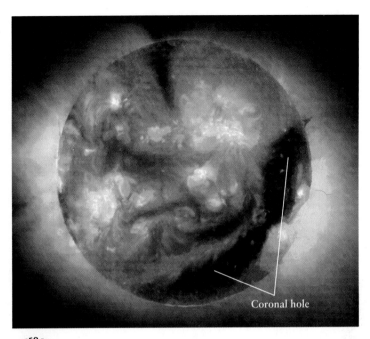

Coronal hole

figure 18-16   R I **V** U X G
The Ultraviolet Corona   The *SOHO* spacecraft recorded this false-color ultraviolet view of the solar corona. The dark feature running across the Sun's disk from the top is a coronal hole, a region where the coronal gases are thinner than elsewhere. Such holes are often the source of strong gusts in the solar wind. (*SOHO*/EIT/ESA/NASA)

from the warmth of the Sun's photosphere. Why, then, does the temperature in these regions *increase* with increasing altitude? This has been one of the major unsolved mysteries in astronomy for the past half-century. As astronomers have tried to resolve this dilemma, they have found important clues in one of the Sun's most familiar features—sunspots.

18-8 | Sunspots are low-temperature regions in the photosphere

Granules, supergranules, spicules, and the solar wind occur continuously. These features are said to be aspects of the *quiet* Sun. But other, more dramatic features appear periodically, including massive eruptions and regions of concentrated magnetic fields. When these are present, astronomers refer to the *active* Sun. The features of the active Sun that can most easily be seen with even a small telescope (although only with an appropriate filter attached) are sunspots.

Sunspots are irregularly shaped dark regions in the photosphere. Sometimes sunspots appear in isolation (**Figure 18-17a**), but frequently they are found in sunspot groups (Figure 18-17b; see also Figure 18-8). Although sunspots vary greatly in size, typical ones measure a few tens of thousands of kilometers across—comparable to the diameter of the Earth. Sunspots are not permanent features of the photosphere but last between a few hours and a few months.

(a)

(b)

 figure 18-17 R I ⓥ U X G

Sunspots **(a)** This high-resolution photograph of the photosphere shows a mature sunspot. The dark center of the spot is called the umbra. It is bordered by the penumbra, which is less dark and has a

featherlike appearance. **(b)** In this view of a typical sunspot group, several sunspots are close enough to overlap. In both images you can see granulation in the surrounding, undisturbed photosphere. (NOAO)

Each sunspot has a dark central core, called the *umbra*, and a brighter border called the *penumbra*. We used these same terms in Section 3-4 to refer to different parts of the Earth's or the Moon's shadow. But a sunspot is not a shadow; it is a region in the photosphere where the temperature is relatively low, which makes it appear darker than its surroundings. If the surrounding photosphere is blocked from view, a sunspot's umbra appears red and the penumbra appears orange. As we saw in Section 5-4, Wien's law relates the color of a blackbody (which depends on the wavelength at which it emits the most light) to the blackbody's temperature. The colors of a sunspot indicate that the temperature of the umbra is typically 4300 K and that of the penumbra is typically 5000 K. While high by earthly standards, these temperatures are quite a bit lower than the average photospheric temperature of 5800 K.

The Stefan-Boltzmann law (see Section 5-4) tells us that the energy flux from a blackbody is proportional to the fourth power of its temperature. This law lets us compare the amounts of light energy emitted by a square meter of a sunspot's umbra and by a square meter of undisturbed photosphere. The ratio is:

$$\frac{\text{flux from umbra}}{\text{flux from photosphere}} = \left(\frac{4300 \text{ K}}{5800 \text{ K}}\right)^4 = 0.30$$

That is, the umbra emits only 30% as much light as an equally large patch of undisturbed photosphere. This is why sunspots appear so dark.

Occasionally, a sunspot group is large enough to be seen without a telescope. Chinese astronomers recorded such sightings 2000 years ago, and huge sunspot groups visible to the naked eye (with an appropriate filter) were seen in 1989 and 2003. But it was not until

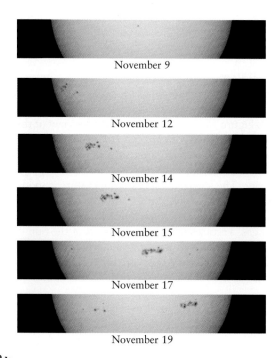

November 9

November 12

November 14

November 15

November 17

November 19

figure 18-18 R I ⓥ U X G

Tracking the Sun's Rotation with Sunspots This series of photographs taken in 1999 shows the rotation of the Sun. By observing the same group of sunspots from one day to the next, Galileo found that the Sun rotates once in about four weeks. (The equatorial regions of the Sun actually rotate somewhat faster than the polar regions.) Notice how the sunspot group shown here changed its shape. (The Carnegie Observatories)

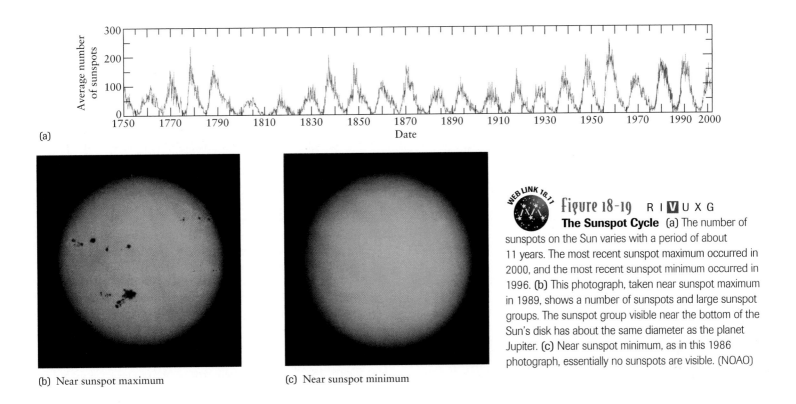

(a)

(b) Near sunspot maximum

(c) Near sunspot minimum

figure 18-19 R I **V** U X G
The Sunspot Cycle (a) The number of sunspots on the Sun varies with a period of about 11 years. The most recent sunspot maximum occurred in 2000, and the most recent sunspot minimum occurred in 1996. (b) This photograph, taken near sunspot maximum in 1989, shows a number of sunspots and large sunspot groups. The sunspot group visible near the bottom of the Sun's disk has about the same diameter as the planet Jupiter. (c) Near sunspot minimum, as in this 1986 photograph, essentially no sunspots are visible. (NOAO)

Galileo introduced the telescope into astronomy (see Section 4-5) that anyone was able to examine sunspots in detail.

Galileo discovered that he could determine the Sun's rotation rate by tracking sunspots as they moved across the solar disk (**Figure 18-18**). He found that the Sun rotates once in about four weeks. A typical sunspot group lasts about two months, so a specific one can be followed for two solar rotations.

Further observations by the British astronomer Richard Carrington in 1859 demonstrated that the Sun does not rotate as a rigid body. Instead, the equatorial regions rotate more rapidly than the polar regions. This phenomenon is known as **differential rotation.** Thus, while a sunspot near the solar equator takes only 25 days to go once around the Sun, a sunspot at 30° north or south of the equator takes 27½ days. The rotation period at 75° north or south is about 33 days, while near the poles it may be as long as 35 days.

The average number of sunspots on the Sun is not constant, but varies in a predictable **sunspot cycle** (**Figure 18-19a**). This phenomenon was first reported by the German astronomer Heinrich Schwabe in 1843 after many years of observing. As Figure 18-19a shows, the average number of sunspots varies with a period of about 11 years. A period of exceptionally many sunspots is a **sunspot maximum** (Figure 18-19b), as occurred in 1979, 1989, and 2000. Conversely, the Sun is almost devoid of sunspots at a **sunspot minimum** (Figure 18-19c), as occurred in 1976, 1986, and 1996 and is projected to occur in 2007.

The locations of sunspots also vary with the same 11-year sunspot cycle. At the beginning of a cycle, just after a sunspot minimum, sunspots first appear at latitudes around 30° north and south of the solar equator (**Figure 18-20**). Over the succeeding years, the sunspots occur closer and closer to the equator.

figure 18-20

Variations in the Average Latitude of Sunspots The dots in this graph record how far north or south of the Sun's equator sunspots were observed. At the beginning of each sunspot cycle, most sunspots are found near latitudes 30° north or south. As the cycle goes on, sunspots typically form closer to the equator. (NASA Marshall Space Flight Center)

18-9 | Sunspots are produced by a 22-year cycle in the Sun's magnetic field

Why should the number of sunspots vary with an 11-year cycle? Why should their average latitude vary over the course of a cycle? And why should sunspots exist at all? The first step toward answering these questions came in 1908, when the American astronomer George Ellery Hale discovered that sunspots are associated with intense magnetic fields on the Sun.

When Hale focused a spectroscope on sunlight coming from a sunspot, he found that many spectral lines appear to be split into several closely spaced lines (Figure 18-21). This "splitting" of spectral lines is called the **Zeeman effect,** after the Dutch physicist Pieter Zeeman who first observed it in his laboratory in 1896. Zeeman showed that a spectral line splits when the atoms are subjected to an intense magnetic field. The more intense the magnetic field, the wider the separation of the split lines.

Hale's discovery showed that sunspots are places where the hot gases of the photosphere are bathed in a concentrated magnetic field. Because of the high temperature of the Sun's atmosphere, many of the atoms there are ionized. The solar atmosphere is thus a gaseous mixture called a **plasma,** in which electrically charged ions and electrons can move freely. Like any moving, electrically charged objects, they can be deflected by magnetic fields. **Figure 18-22** shows how a magnetic field in the laboratory bends a beam of fast-moving electrons into a curved trajectory. Similarly, the paths of moving ions and electrons in the photosphere are deflected by the Sun's magnetic field. In particular, magnetic forces act on the hot plasma that rises from the Sun's interior due to convection. Where the magnetic field is particularly strong, these forces push the hot plasma away. The result is a localized region where the gas is relatively cool and thus glows less brightly—in other words, a sunspot.

To get a fuller picture of the Sun's magnetic fields, astronomers take images of the Sun at two wavelengths, one just less than and one just greater than the wavelength of a magnetically split spectral line. From the difference between these two images, they can construct a picture called a **magnetogram,** which displays the magnetic fields in the solar atmosphere. **Figure 18-23a** is an ordinary white-light photograph of the Sun taken at the same time as the magnetogram in Figure 18-23b. In the magnetogram, dark blue indicates areas of the photosphere with one magnetic polarity (north), and yellow indicates areas with the opposite (south) magnetic polarity. This image shows that many sunspot groups have roughly comparable areas covered by north and south magnetic polarities (see also Figure 18-23c). Thus, a sunspot group resembles a giant bar

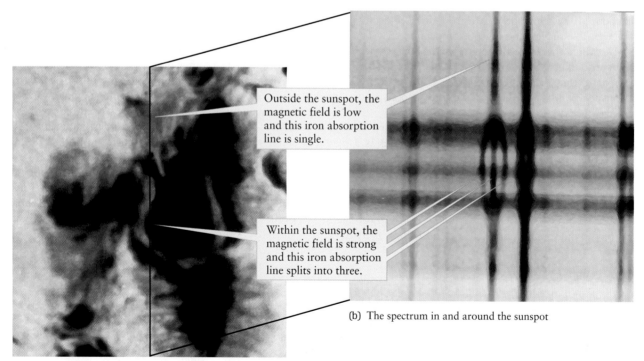

Outside the sunspot, the magnetic field is low and this iron absorption line is single.

Within the sunspot, the magnetic field is strong and this iron absorption line splits into three.

(a) A sunspot

(b) The spectrum in and around the sunspot

Figure 18-21 R I V U X G

Sunspots Have Strong Magnetic Fields (a) A black line in this image of a sunspot shows where the slit of a spectrograph was aimed. (b) This is a portion of the resulting spectrum, including a dark absorption line caused by iron atoms in the photosphere. The splitting of this line by the sunspot's magnetic field can be used to calculate the field strength. Typical sunspot magnetic fields are over 5000 times stronger than the Earth's field at its north and south poles. (NOAO)

Figure 18-22 R I **V** U X G

Magnetic Fields Deflect Moving, Electrically Charged Objects In this laboratory experiment, a beam of negatively charged electrons (shown by a blue arc) is aimed straight upward from the center of the apparatus. The entire apparatus is inside a large magnet, and the magnetic field deflects the beam into a curved path. (Courtesy of Central Scientific Company)

magnet, with a north magnetic pole at one end and a south magnetic pole at the other.

If different sunspot groups were unrelated to one another, their magnetic poles would be randomly oriented, like a bunch of compass needles all pointing in random directions. As Hale discovered, however, there is a striking regularity in the magnetization of sunspot groups. As a given sunspot group moves with the Sun's rotation, the sunspots in front are called the "preceding members" of the group. The spots that follow behind are referred to as the "following members." Hale compared the sunspot groups in the two solar hemispheres, north or south of the Sun's equator. He found that the preceding members in one solar hemisphere all have the same magnetic polarity, while the preceding members in the other hemisphere have the opposite polarity. Furthermore, in the hemisphere where the Sun has its north magnetic pole, the preceding members of all sunspot groups have north magnetic polarity. In the opposite hemisphere, where the Sun has its south magnetic pole, the preceding members all have south magnetic polarity.

Along with his colleague Seth B. Nicholson, Hale also discovered that the Sun's polarity pattern completely reverses itself every 11 years—the same interval as the time from one solar maximum to the next. The hemisphere that has preceding north magnetic poles during one 11-year sunspot cycle will have preceding south magnetic poles during the next 11-year cycle, and vice versa. The north and south magnetic poles of the Sun itself also reverse every 11 years. Thus, the Sun's magnetic pattern repeats itself only after two sunspot cycles, which is why astronomers speak of a **22-year solar cycle.**

In 1960, the American astronomer Horace Babcock proposed a description that seems to account for many features of this 22-year solar cycle. Babcock's scenario, called a **magnetic-dynamo model,**

(a) Visible-light image (b) Magnetogram (c) Magnetogram of a sunspot group

Figure 18-23 R I **V** U X G

Mapping the Sun's Magnetic Field (a) This visible-light image and (b) this false-color magnetogram were recorded at the same time. Dark blue and yellow areas in the magnetogram have north and south magnetic polarity, respectively; blue-green regions have weak magnetic fields. The highly magnetized regions in (b) correlate with the sunspots in (a). (c) The two ends of this large sunspot group have opposite magnetic polarities (colored blue and yellow), like the ends of a giant bar magnet. (NOAO)

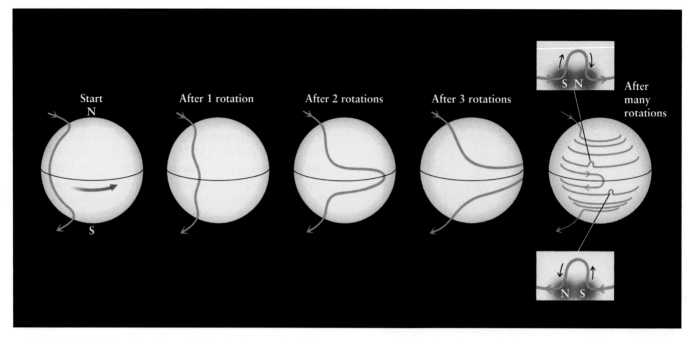

Figure 18-24

Babcock's Magnetic Dynamo Model Magnetic field lines tend to move along with the plasma in the Sun's outer layers. Because the Sun rotates faster at the equator than near the poles, a field line that starts off running from the Sun's north magnetic pole (N) to its south magnetic pole (S) ends up wrapped around the Sun like twine wrapped around a ball. The insets on the far right show how sunspot groups appear where the concentrated magnetic field rises through the photosphere.

makes use of two basic properties of the Sun's photosphere—differential rotation and convection. Differential rotation causes the magnetic field in the photosphere to become wrapped around the Sun (Figure 18-24). As a result, the magnetic field becomes concentrated at certain latitudes on either side of the solar equator. Convection in the photosphere creates tangles in the concentrated magnetic field, and "kinks" erupt through the solar surface. Sunspots appear where the magnetic field protrudes through the photosphere. The theory suggests that sunspots should appear first at northern and southern latitudes and later form nearer to the equator. This is just what is observed (see Figure 18-20). Note also that as shown on the far right in Figure 18-24, the preceding member of a sunspot group has the same polarity (N or S) as the Sun's magnetic pole in that hemisphere. This is just as Hale observed.

Differential rotation eventually undoes the twisted magnetic field. The preceding members of sunspot groups move toward the Sun's equator, while the following members migrate toward the poles. Because the preceding members from the two hemispheres have opposite magnetic polarities, their magnetic fields cancel each other out when they meet at the equator. The following members in each hemisphere have the opposite polarity to the Sun's pole in that hemisphere; hence, when they converge on the pole, the following members first cancel out and then reverse the Sun's overall magnetic field. The fields are now completely relaxed. Once again, differential rotation begins to twist the Sun's magnetic field, but now with all magnetic polarities reversed. In this way, Babcock's model helps to explain the change in field direction every 11 years.

Recent discoveries in helioseismology (Section 18-3) offer new insights into the Sun's magnetic field. By comparing the speeds of sound waves that travel with and against the Sun's rotation, helioseismologists have been able to determine the Sun's rotation rate at different depths and latitudes. As shown in Figure 18-25, the Sun's surface pattern of differential rotation persists through the convective zone. Farther in, within the radiative zone, the Sun seems to rotate like a rigid object with a period of 27 days at all latitudes. Astronomers suspect that the Sun's magnetic field originates in a relatively thin layer where the radiative and convective zones meet and slide past each other due to their different rotation rates.

Much about sunspots and solar activity remains mysterious. The best calculations predict that a sunspot should break up and disperse fairly quickly, but, in fact, sunspots can persist for many weeks. There are also perplexing irregularities in the solar cycle. For example, the overall reversal of the Sun's magnetic field is often piecemeal and haphazard. One pole may reverse polarity long before the other. For several weeks the Sun's surface may have two north magnetic poles and no south magnetic pole at all.

What is more, there seem to be times when all traces of sunspots and the sunspot cycle vanish for many years. For example, virtually no sunspots were seen from 1645 through 1715. Curiously, during these same years Europe experienced record low temperatures, often referred to as the Little Ice Age, whereas the western United States was subjected to severe drought. By contrast, there was apparently a period of increased sunspot activity during the eleventh and twelfth centuries, during which the Earth was warmer than it is today. Thus, variations in solar activity appear to affect climates on

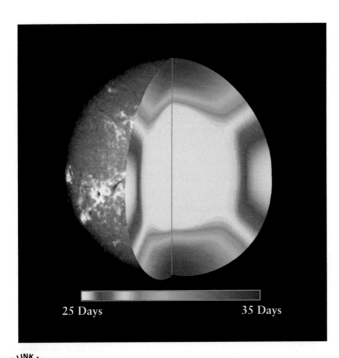

Figure 18-25

Rotation of the Solar Interior This cutaway picture of the Sun shows how the solar rotation period (shown by different colors) varies with depth and latitude. The surface and the convective zone have differential rotation (a short period at the equator and longer periods near the poles). Deeper within the Sun, the radiative zone seems to rotate like a rigid sphere. (Courtesy of K. Libbrecht, Big Bear Solar Observatory)

the Earth. The origin of this Sun-Earth connection is a topic of ongoing research.

18-10 | The Sun's magnetic field also produces other forms of solar activity

Astronomers now understand that the Sun's magnetic field does more than just explain the presence of sunspots. It is also responsible for the existence of spicules, as well as a host of other dramatic phenomena in the chromosphere and corona.

In a plasma, magnetic field lines and the material of the plasma tend to move together. This means that as convection pushes material toward the edge of a supergranule, it pushes magnetic field lines as well. The result is that vertical magnetic field lines pile up around a supergranule. Plasma that "sticks" to these magnetic field lines thus ends up lifted upward, forming a spicule (see Figures 18-12 and 18-13).

The tendency of plasma to follow the Sun's magnetic field may also explain why the temperature of the corona is so high. Spacecraft observations show magnetic field arches extending tens of thousands of kilometers into the corona, with streamers of electrically charged particles moving along each arch (Figure 18-26). If two arches with oppositely directed magnetic fields come into proximity, their magnetic fields can cancel each other out in a process called **magnetic reconnection.** The tremendous amount of energy stored in the magnetic field is then released into the corona. (A single arch contains as much energy as a hydroelectric power plant would generate in a million years.) The amount of energy released in this way appears to be more than enough to maintain the corona's temperature.

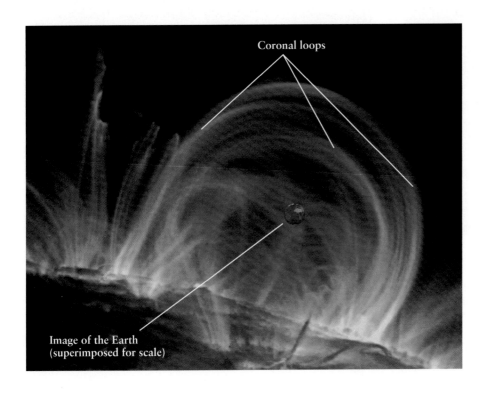

Coronal loops

Image of the Earth
(superimposed for scale)

Figure 18-26 R I V **U** X G

Magnetic Arches in the Corona This false-color ultraviolet image from the *TRACE* spacecraft (*Transition Region and Coronal Explorer*) shows magnetic field loops suspended high above the solar surface. (Compare these to the picture of the Earth, which is shown to scale.) The loops are made visible by the glowing gases trapped within them. (Stanford-Lockheed Institute for Space Research; *TRACE*; and NASA)

Magnetic heating can also explain why the parts of the corona that are most prominent in ultraviolet images are often those that lie on top of sunspots. (Some examples are the bright regions in Figure 18-16.) The intense magnetic field of the sunspots helps give the overlying coronal gas such a high temperature that it emits copious amounts of high-energy ultraviolet photons, as well as even more energetic X-ray photons.

Spicules and coronal heating occur even when the Sun is quiet. But magnetic fields can also explain many aspects of the active Sun in addition to sunspots. Figure 18-27 is an image of the chromosphere made with an H$_\alpha$ filter during a sunspot maximum. The bright areas are called **plages** (from the French word for "beach"). These are bright, hot regions in the chromosphere that tend to form just before the appearance of new sunspots. They are probably created by magnetic fields that push upward from the Sun's interior, compressing and heating a portion of the chromosphere. The dark streaks, called **filaments,** are relatively cool and dense parts of the chromosphere that have been pulled along with magnetic field lines as they arch to high altitudes.

When seen from the side, so that they are viewed against the dark background of space, filaments appear as bright, arching columns of gas called **prominences** (Figure 18-28). They can extend for tens of thousands of kilometers above the photosphere. Some prominences last for only a few hours, while others persist for many months. The most energetic prominences break free of the magnetic fields that confined them and burst into space.

 Violent, eruptive events on the Sun, called **solar flares,** occur in complex sunspot groups. Within only a few minutes, temperatures in a compact region may soar to 5×10^6 K and vast quantities of particles and radiation are blasted out into space. (The image that opens this chapter shows such a flare.) These eruptions can also cause disturbances that spread outward in the solar atmosphere, like the ripples that appear when you drop a rock into a pond.

The most energetic flares carry as much as 10^{30} joules of energy, equivalent to 10^{14} one-megaton nuclear weapons being exploded at once! However, the energy of a solar flare does not come from thermonuclear fusion in the solar atmosphere; instead, it appears to be released from the intense magnetic field around a sunspot group.

As energetic as solar flares are, they are dwarfed by **coronal mass ejections** (Figure 18-29a). In such an event, more than 10^{12} kilo-

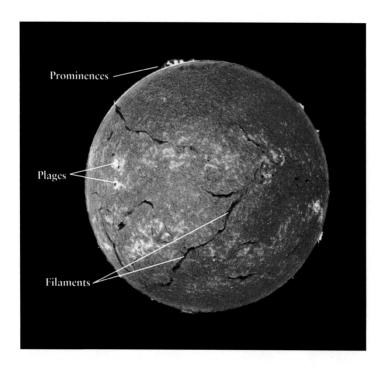

figure 18-27 R I **V** U X G

The Active Sun Seen through an H$_\alpha$ Filter This image was made using a red filter that only passes light at a wavelength of 656 nm. The spectrum of the photosphere has an absorption line at this wavelength and so appears dark. Hence, this filter reveals the photosphere and corona. Prominences, plages, and filaments are associated with strong magnetic fields. (NASA)

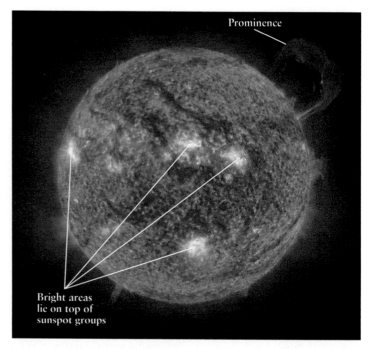

figure 18-28 R I V **U** X G

A Solar Prominence A huge prominence arches above the solar surface in this ultraviolet image from the *SOHO* spacecraft. The image was recorded using light at a wavelength of 30.4 nm, emitted by singly ionized helium atoms at a temperature of about 60,000 K. By comparison, the material within the arches in Figure 18-26 reaches temperatures in excess of 2×10^6 K. (*SOHO* EIT/ESA/NASA)

Material ejected
from the corona

Ejected material encounters
the Earth's magnetosphere

Earth

(a) A coronal mass ejection

(b) Two to four days later

figure 18-29   R I V U X G

A Coronal Mass Ejection (a) *SOHO* recorded this coronal mass ejection in an X-ray image. (The image of the Sun itself was made at ultraviolet wavelengths.) (b) Within two to four days the fastest-moving ejected material reaches a distance of 1 AU from the Sun. Most particles are deflected by the Earth's magnetosphere, but some are able to reach the Earth. (The ejection shown in (a) was not aimed toward the Earth and did not affect us.) (*SOHO*/EIT/LASCO/ESA/NASA)

grams (a billion tons) of high-temperature coronal gas is blasted into space at speeds of hundreds of kilometers per second. A typical coronal mass ejection lasts a few hours. These explosive events seem to be related to large-scale alterations in the Sun's magnetic field. Such events occur every few months; smaller eruptions may occur almost daily.

If a solar flare or coronal mass ejection happens to be aimed toward Earth, a stream of high-energy electrons and nuclei reaches us a few days later (Figure 18-29b). When this plasma arrives, it can interfere with satellites, pose a health hazard to astronauts in orbit, and disrupt electrical and communications equipment on the Earth's surface. Telescopes on Earth and on board spacecraft now monitor the Sun continuously to provide warnings of dangerous levels of solar particles.

The numbers of plages, filaments, solar flares, and coronal mass ejections all vary with the same 11-year cycle as sunspots. But unlike sunspots, coronal mass ejections never completely cease, even when the Sun is at its quietest. Astronomers are devoting substantial effort to understanding these and other aspects of our dynamic Sun.

Key Words

The term preceded by an asterisk (*) is discussed in Box 18-1.

22-year solar cycle, p. 397
chromosphere, p. 390
CNO cycle, p. 380
conduction, p. 383
convection, p. 383
convective zone, p. 385
corona, p. 391
coronal hole, p. 393
coronal mass ejection, p. 400
differential rotation, p. 395
filament, p. 400
granulation, p. 389
granule, p. 389
helioseismology, p. 385
hydrogen fusion, p. 380

hydrostatic equilibrium, p. 382
limb darkening, p. 388
luminosity (of the Sun), p. 378
magnetic-dynamo model, p. 397
magnetogram, p. 396
magnetic reconnection, p. 399
negative hydrogen ion, p. 390
neutrino, p. 386
neutrino oscillation, p. 387
photosphere, p. 388
plage, p. 400
plasma, p. 396
*positron, p. 381
prominence, p. 400

Key Ideas

Hydrogen Fusion in the Sun's Core: The Sun's energy is produced by hydrogen fusion, a sequence of thermonuclear reactions in which four hydrogen nuclei combine to produce a single helium nucleus.

• The energy released in a nuclear reaction corresponds to a slight reduction of mass according to Einstein's equation $E = mc^2$.

• Thermonuclear fusion occurs only at very high temperatures; for example, hydrogen fusion occurs only at temperatures in excess of about 10^7 K. In the Sun, fusion occurs only in the dense, hot core.

Models of the Sun's Interior: A theoretical description of a star's interior can be calculated using the laws of physics.

• The standard model of the Sun suggests that hydrogen fusion takes place in a core extending from the Sun's center to about 0.25 solar radius.

• The core is surrounded by a radiative zone extending to about 0.71 solar radius. In this zone, energy travels outward through radiative diffusion.

• The radiative zone is surrounded by a rather opaque convective zone of gas at relatively low temperature and pressure. In this zone, energy travels outward primarily through convection.

Solar Neutrinos and Helioseismology: Conditions in the solar interior can be inferred from measurements of solar neutrinos and of solar vibrations.

• Neutrinos emitted in thermonuclear reactions in the Sun's core have been detected, but in smaller numbers than expected. Recent neutrino experiments explain why this is so.

• Helioseismology is the study of how the Sun vibrates. These vibrations have been used to infer pressures, densities, chemical compositions, and rotation rates within the Sun.

 The Sun's Atmosphere: The Sun's atmosphere has three main layers: the photosphere, the chromosphere, and the corona. Everything below the solar atmosphere is called the solar interior.

• The visible surface of the Sun, the photosphere, is the lowest layer in the solar atmosphere. Its spectrum is similar to that of a blackbody at a temperature of 5800 K. Convection in the photosphere produces granules.

• Above the photosphere is a layer of less dense but higher-temperature gases called the chromosphere. Spicules extend upward from the photosphere into the chromosphere along the boundaries of supergranules.

• The outermost layer of the solar atmosphere, the corona, is made of very high-temperature gases at extremely low density. Activity in the corona includes coronal mass ejections and coronal holes. The solar corona blends into the solar wind at great distances from the Sun.

The Active Sun: The Sun's surface features vary in an 11-year cycle. This is related to a 22-year cycle in which the surface magnetic field increases, decreases, and then increases again with the opposite polarity.

• Sunspots are relatively cool regions produced by local concentrations of the Sun's magnetic field. The average number of sunspots increases and decreases in a regular cycle of approximately 11 years, with reversed magnetic polarities from one 11-year cycle to the next. Two such cycles make up the 22-year solar cycle.

• The magnetic-dynamo model suggests that many features of the solar cycle are due to changes in the Sun's magnetic field. These changes are caused by convection and the Sun's differential rotation.

• A solar flare is a brief eruption of hot, ionized gases from a sunspot group. A coronal mass ejection is a much larger eruption that involves immense amounts of gas from the corona.

Review Questions

1. What is Kelvin-Helmholtz contraction? Why is it ruled out as a source of the present-day Sun's energy?

2. Why is it impossible for the burning of substances like coal to be the source of the Sun's energy?

3. What is hydrogen fusion? Why is hydrogen fusion fundamentally unlike the burning of a log in a fireplace?

4. Why do thermonuclear reactions occur only in the Sun's core, not in its outer regions?

5. Describe how the net result of the reactions in Figure 18-2 is the conversion of four protons into a single helium nucleus. What other particles are produced in this process? How many of each particle are produced?

6. Give an everyday example of hydrostatic equilibrium. Give an example of thermal equilibrium. Explain how these equilibrium conditions apply to each example.

7. If thermonuclear fusion in the Sun were suddenly to stop, what would eventually happen to the overall radius of the Sun? Justify your answer using the ideas of hydrostatic equilibrium and thermal equilibrium.

8. Give some everyday examples of conduction, convection, and radiative diffusion.

9. Describe the Sun's interior. Include references to the main physical processes that occur at various depths within the Sun.

10. Suppose thermonuclear fusion in the Sun stopped abruptly. Would the intensity of sunlight decrease just as abruptly? Why or why not?

11. Explain how studying the oscillations of the Sun's surface can give important, detailed information about physical conditions deep within the Sun.

12. What is a neutrino? Why is it useful to study neutrinos coming from the Sun? What do they tell us that cannot be learned from other avenues of research?

13. Unlike all other types of telescopes, neutrino detectors are placed deep underground. Why?

14. What is the solar neutrino problem? What solution to this problem is suggested by the results from the Sudbury Neutrino Observatory?

15. Describe the dangers in attempting to observe the Sun. How have astronomers learned to circumvent these observational problems?

16. Briefly describe the three layers that make up the Sun's atmosphere. In what ways do they differ from each other?

17. What is solar granulation? Describe how convection gives rise to granules.

18. High-resolution spectroscopy of the photosphere reveals that absorption lines are blueshifted in the spectrum of the central, bright regions of granules but are redshifted in the spectrum of the dark boundaries between granules. Explain how these observations show that granulation is due to convection.

19. What is the difference between granules and supergranules?

20. What are spicules? Where are they found? How can you observe them? What causes them?

21. How do astronomers know that the temperature of the corona is so high?

22. How do astronomers know when the next sunspot maximum and minimum will occur?

23. Why do astronomers say that the solar cycle is really 22 years long, even though the number of sunspots varies over an 11-year period?

24. Explain how the magnetic-dynamo model accounts for the solar cycle.

25. Describe one explanation for why the corona has a higher temperature than the chromosphere.

26. Why should solar flares and coronal mass ejections be a concern for businesses that use telecommunication satellites?

Advanced Questions

Questions preceded by an asterisk (*) involve the topic discussed in Box 18-1.

> **Problem-solving tips and tools**
>
> You may have to review Wien's law and the Stefan-Boltzmann law, which are the subject of Section 5-4. Section 5-5 discusses the properties of photons. As we described in Box 5-2, you can simplify calculations by taking ratios, such as the ratio of the flux from a sunspot to the flux from the undisturbed photosphere. When you do this, all the cumbersome constants cancel out. Figure 5-7 shows the various parts of the electromagnetic spectrum. We introduced the Doppler effect in Section 5-9 and Box 5-6. For information about the planets, see Table 7-1.

27. Calculate how much energy would be released if each of the following masses were converted *entirely* into their equivalent energy: (**a**) a carbon atom with a mass of 2×10^{-26} kg, (**b**) 1 kilogram, and (**c**) a planet as massive as the Earth (6×10^{24} kg).

28. Use the luminosity of the Sun (given in Table 18-1) and the answers to the previous question to calculate how long the Sun must shine in order to release an amount of energy equal to that produced by the complete mass-to-energy conversion of (**a**) a carbon atom, (**b**) 1 kilogram, and (**c**) the Earth.

29. Assuming that the current rate of hydrogen fusion in the Sun remains constant, what fraction of the Sun's mass will be converted into helium over the next 5 billion years? How will this affect the overall chemical composition of the Sun?

30. (**a**) Estimate how many kilograms of hydrogen the Sun has consumed over the past 4.56 billion years, and estimate the amount of mass that the Sun has lost as a result. Assume that the Sun's luminosity has remained constant during that time. (**b**) In fact, the Sun's luminosity when it first formed was only about 70% of its present value. With this in mind, explain whether your answers to part (a) are an overestimate or an underestimate.

*31. To convert one kilogram of hydrogen (^{1}H) into helium (^{4}He) as described in Box 18-1, you must start with 1.5 kg of hydrogen. Explain why, and explain what happens to the other 0.5 kg. (*Hint: How many ^{1}H nuclei are used to make the two ^{3}He nuclei in Figure 18-2c? How many of these ^{1}H nuclei end up being incorporated into the ^{4}He nucleus in Figure 18-2c?*)

*32. (**a**) A positron has the same mass as an electron (see Appendix 7). Calculate the amount of energy released by the annihilation of an electron and positron. (**b**) The products of this annihilation are two photons, each of equal energy. Calculate the wavelength of each photon, and confirm from Figure 5-7 that this wavelength is the gamma-ray range.

33. Sirius is the brightest star in the night sky. It has a luminosity of 23.5 L$_\odot$, that is, it is 23.5 times as luminous as the Sun and burns hydrogen at a rate 23.5 times greater than the Sun. How many kilograms of hydrogen does Sirius convert into helium each second?

34. (Refer to the preceding question.) Sirius has 2.3 times the mass of the Sun. Do you expect that the lifetime of Sirius will be longer, shorter, or the same length as that of the Sun? Explain your reasoning.

35. (a) If the Sun were not in a state of hydrostatic equilibrium, would its diameter remain the same? Explain your reasoning. (b) If the Sun were not in a state of thermal equilibrium, would its luminosity remain the same? What about its surface temperature? Explain your reasoning.

36. Using the mass and size of the Sun given in Table 18-1, verify that the average density of the Sun is 1410 kg/m^3. Compare your answer with the average densities of the Jovian planets.

37. Use the data in Table 18-2 to calculate the average density of material within 0.1 R$_\odot$ of the center of the Sun. (You will need to use the mass and radius of the Sun as a whole, given in Table 18-1.) Explain why your answer is not the same as the density at 0.1 R$_\odot$ given in Table 18-2.

38. In a typical solar oscillation, the Sun's surface moves up or down at a maximum speed of 0.1 m/s. An astronomer sets out to measure this speed by detecting the Doppler shift of an absorption line of iron with wavelength 557.6099 nm. What is the maximum wavelength shift that she will observe?

39. Explain why the results from the Sudbury Neutrino Observatory (SNO) only provide an answer to the solar neutrino problem for relatively high-energy neutrinos. (*Hint:* Can SNO detect solar neutrinos of all energies?)

40. The amount of energy required to dislodge the extra electron from a negative hydrogen ion is 1.2 × 10^{-19} J. (a) The extra electron can be dislodged if the ion absorbs a photon of sufficiently short wavelength. (Recall from Section 5-5 that the higher the energy of a photon, the shorter its wavelength.) Find the longest wavelength (in nm) that can accomplish this. (b) In what part of the electromagnetic spectrum does this wavelength lie? (c) Would a photon of visible light be able to dislodge the extra electron? Explain. (d) Explain why the photosphere, which contains negative hydrogen ions, is quite opaque to visible light but is less opaque to light with wavelengths longer than the value you calculated in (a).

41. Astronomers often use an H$_\alpha$ filter to view the chromosphere. Explain why this can also be accomplished with filters that are transparent only to the wavelengths of the H and K lines of ionized calcium. (*Hint:* The H and K lines are dark lines in the spectrum of the photosphere.)

42. Calculate the wavelengths at which the photosphere, chromosphere, and corona emit the most radiation. Explain how the results of your calculations suggest the best way to observe these regions of the solar atmosphere. (*Hint:* Treat each part of the atmosphere as a perfect blackbody. Assume average temperatures

of 50,000 K and 1.5 × 10^6 K for the chromosphere and corona, respectively.)

43. On November 15, 1999, the planet Mercury passed in front of the Sun as seen from Earth. The *TRACE* spacecraft made these time-lapse images of this event using ultraviolet light (top) and visible light (bottom). (Mercury moved from left to right in these images. The time between successive views of Mercury is 6 to 9 minutes.) Explain why the Sun appears somewhat larger in the ultraviolet image than in the visible-light image.

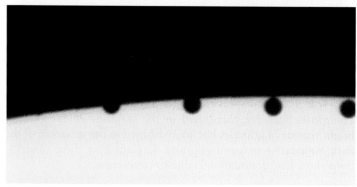

(K. Schrijver, Stanford-Lockheed Institute for Space Research; *TRACE*; and NASA)   R I **V** U X G

44. The moving images on a television set are made by a fast-moving electron beam that sweeps over the back of the screen. Explain why placing a strong magnet next to the screen distorts the picture. (*Caution:* Don't try this with your television! You can do permanent damage to your set.)

45. Find the wavelength of maximum emission of the umbra of a sunspot and the wavelength of maximum emission of a sunspot's penumbra. In what part of the electromagnetic spectrum do these wavelengths lie?

46. (a) Find the ratio of the energy flux from a patch of a sunspot's penumbra to the energy flux from an equally large patch of undisturbed photosphere. Which patch is brighter? (b) Find the ratio of the energy flux from a patch of a sunspot's penumbra to the energy flux from an equally large patch of umbra. Again, which patch is brighter?

47. Suppose that you want to determine the Sun's rotation rate by observing its sunspots. Is it necessary to take the Earth's orbital motion into account? Why or why not?

48. (a) Using a ruler to make measurements of Figure 18-26, determine how far the arches in that figure extend above the Sun's surface. The diameter of the Earth is 12,756 km. (b) In Figure 18-26 (an ultraviolet image) the photosphere appears dark compared to the arches. Explain why.

49. The amount of visible light emitted by the Sun varies only a little over the 11-year sunspot cycle. But the amount of X rays emitted by the Sun can be 10 times greater at solar maximum than at solar minimum. Explain why these two types of radiation should be so different in their variability.

Discussion Questions

50. Discuss the extent to which cultures around the world have worshiped the Sun as a deity throughout history. Why do you suppose there has been such widespread veneration?

51. In the movie *Star Trek IV: The Voyage Home,* the starship *Enterprise* flies on a trajectory that passes close to the Sun's surface. What features should a real spaceship have to survive such a flight? Why?

52. Discuss some of the difficulties in correlating solar activity with changes in the Earth's climate.

53. Describe some of the advantages and disadvantages of observing the Sun (a) from space and (b) from the Earth's south pole. What kinds of phenomena and issues might solar astronomers want to explore from these locations?

Web/CD-ROM Questions

54. Search the World Wide Web for the latest information from the neutrino detectors at the Super-Kamiokande Observatory and the Sudbury Neutrino Observatory. What are the most recent results from these detectors? What is the current thinking about the solar neutrino problem? What is the status of a detector called Borexino? How does it differ from other neutrino experiments?

55. Search the World Wide Web for information about features in the solar atmosphere called *sigmoids.* What are they? What causes them? How might they provide a way to predict coronal mass ejections?

56. Determining the Lifetime of a Solar Granule. Access and view the video "Granules on the Sun's Surface" in Chapter 18 of the *Universe* Web site or CD-ROM. Your task is to determine the approximate lifetime of a solar granule on the photosphere. Select an area, then slowly and rhythmically repeat "start, stop, start, stop" until you can consistently predict the appearance and disappearance of granules. While keeping your rhythm, move to a different area of the video and continue monitoring the appearance and disappearance of granules. When you are confident you have the timing right, move your eyes (or use a partner) to look at the clock shown in the video. Determine the length of time between the appearance and disappearance of the granules and record your answer.

Observing Projects

Observing tips and tools

At the risk of repeating ourselves, we remind you to *never look directly at the Sun, because it can easily cause permanent blindness.* You can view the Sun safely without a telescope just by using two pieces of white cardboard. First, use a pin to poke a small hole in one piece of cardboard; this will be your "lens," and the other piece of cardboard will be your "viewing screen." Hold the "lens" piece of cardboard so that it is face-on to the Sun and sunlight can pass through the hole. With your other hand, hold the "viewing screen" so that the sunlight from the "lens" falls on it. Adjust the distance between the two pieces of cardboard so that you see a sharp image of the Sun on the "viewing screen." This image is perfectly safe to view. It is actually possible to see sunspots with this low-tech apparatus.

For a better view, use a telescope with a solar filter that fits on the front of the telescope. A standard solar filter is a piece of glass coated with a thin layer of metal to give it a mirror-like appearance. This coating reflects almost all the sunlight that falls on it, so that only a tiny, safe amount of sunlight enters the telescope. An H_α filter, which looks like a red piece of glass, keeps the light at a safe level by admitting only a very narrow range of wavelengths. (Filters that fit on the back of the telescope are *not* recommended. The telescope focuses concentrated sunlight on such a filter, heating it and making it susceptible to cracking—and if the filter cracks when you are looking through it, your eye will be ruined instantly and permanently.)

To use a telescope with a solar filter, first aim the telescope away from the Sun, then put on the filter. Keep the lens cap on the telescope's secondary wide-angle "finder scope" (if it has one), because the heat of sunlight can fry the finder scope's optics. Next, aim the telescope toward the Sun, using the telescope's shadow to judge when you are pointed in the right direction. You can then safely look through the telescope's eyepiece. When you are done, make sure you point the telescope away from the Sun before removing the filter and storing the telescope.

Note that the amount of solar activity that you can see (sunspots, filaments, flares, prominences, and so on) will depend on where the Sun is in its 11-year sunspot cycle.

57. Use a telescope with a solar filter to observe the surface of the Sun. Do you see any sunspots? Sketch their appearance. Can you distinguish between the umbrae and penumbrae of the sunspots? Can you see limb darkening? Can you see any granulation?

58. If you have access to an H$_\alpha$ filter attached to a telescope especially designed for viewing the Sun safely, use this instrument to examine the solar surface. How does the appearance of the Sun differ from that in white light? What do sunspots look like in H$_\alpha$? Can you see any prominences? Can you see any filaments? Are the filaments in the H$_\alpha$ image near any sunspots seen in white light? (Note that the amount of activity that you see will be much greater at some times during the solar cycle than at others.)

59. Use the *Starry Night Backyard*™ program to measure the Sun's rotation. Display the entire celestial sphere (select **Guides > Atlas** in the **Go** menu) and center on the Sun by using the **Find...** command in the **Edit** menu. Using the controls at the right-hand end of the Control Panel, zoom in until you can clearly see details on the Sun's surface. Set the time step in the Control Panel to 1 day. Using the "Step time forward" and "Step time backward" buttons in the Control Panel, step through enough time to determine the rotation period of the Sun. Which part of the actual Sun's surface rotates at the rate shown in *Starry Night Backyard*™? (The program does not show the Sun's differential rotation.)

60. Use the *Deep Space Explorer*™ program to examine the Sun. In the left-hand part of the window under the heading **Solar System,** click on the triangle next to the word **Explore** and then click on **Inner Solar System.** Zoom in on the Sun using the buttons at the upper left of the window (an upward-pointing triangle and a downward-pointing triangle). You can rotate the Sun by putting the mouse cursor over the image, holding down the mouse button, and moving the mouse. (On a two-button mouse, hold down the left mouse button.) (**a**) The Sun's equator lies close to the plane of the ecliptic. Where do most of the sunspots visible on the image of the Sun lie relative to the solar equator? (**b**) Based on your observations in (a), does the image in *Deep Space Explorer*™ show the Sun near the beginning, middle, or end of the 11-year sunspot cycle? Explain your reasoning.

Collaborative Exercises

61. Variations in the average latitude of sunspots are shown in Figure 18-20. Estimate the average latitude of sunspots in the year you were born and estimate the average latitude on your twenty-first birthday. Make a rough sketch to illustrate your answers.

62. Create a diagram showing a sketch of how limb darkening on the Sun would look different if the Sun had either a thicker or thinner photosphere. Be sure to include a caption explaining your diagram.

63. Solar granules, shown in Figure 18-10, are about 1000 km across. What city is about that distance away from where you are right now? What city is that distance from the birthplace of each group member?

64. Magnetic arches in the corona are shown in Figure 18-26. How many Earths high are these arches, and how many Earths could fit inside one arch?

JOHN N. BAHCALL

Searching for Neutrinos Beyond the Textbooks

After graduate school at Harvard University, in 1961 **John Bahcall** joined the faculty of the California Institute of Technology, where he remained until 1971. He then became a professor of natural sciences at the Institute for Advanced Study, in Princeton, New Jersey. Bahcall was president of the American Astronomical Society in 1990–1992 and has received the Warner, Heinmann, and Russell prizes from the society for his theoretical work on solar neutrinos, quasars, and dark matter. In 1992, he received the NASA Distinguished Public Service Medal for his observations and leadership with the Hubble Space Telescope. In 1998, Bahcall received the Presidential Medal of Science (from President Bill Clinton) for his theoretical work on solar neutrinos and for his role in the development of the Hubble Space Telescope. He was also awarded the first Hans Bethe prize from the American Physical Society in 1998, and in 2003 received the Franklin Medal for Physics (which he shared with Raymond David and M. Koshiba), the Dan Davis Prize, and the Gold Medal of the Royal Astronomical Society.

Bahcall chaired the National Academy Decade Survey Committee for Astronomy and Astrophysics in the 1990s, which successfully set priorities for astrophysical research projects. He was elected vice president of the American Physical Society in 2004 and will become in 2006 the first person to have served as president of both the American Astronomical Society and the American Physical Society.

In collaboration with Raymond Davis Jr., Bahcall proposed in 1964 that neutrinos from the Sun could be detected via a practical chlorine detector. In the subsequent four decades, Bahcall refined theoretical predictions and interpretations of solar neutrino detectors. This joint work was recognized in 2003 by the Fermi Award (shared with Davis). Bahcall's other areas of expertise include models of the Galaxy, dark matter, atomic and nuclear physics applied to astronomical systems, stellar evolution, and quasar emission and absorption lines.

When Bahcall went in search of solar neutrinos, he applied atomic and nuclear physics to the stars. In his many areas of expertise, in fact, Dr. Bahcall studies just that interplay—between the theories of physics and our understanding of the heavens. At the Institute for Advanced Study, he has developed models of our Galaxy, dark matter, stellar evolution, and the spectra of quasars.

I n attempting to understand how the Sun shines, physicists, chemists, and astronomers have been confronted with a mystery—the case of the missing neutrinos. These exotic particles travel at essentially the speed of light. They are so elusive that they can traverse a thousand light-years of lead before being stopped.

Physicists and astronomers believe that the Sun shines because of the conversion of hydrogen nuclei (protons) into helium nuclei (alpha particles), with the subsequent release of a substantial amount of nuclear energy. The same basic process, *nuclear fusion*, produces the explosion of a hydrogen bomb. We think that about 600 million tons (or 6×10^{11} kg) of hydrogen

are converted to helium every second in the Sun's central regions, providing the energy that we know as sunlight and making life on Earth possible.

In the early 1960s, Ray Davis and I proposed to test the theory of how the Sun shines. Ray, a chemist at Brookhaven National Laboratory, had developed a neutrino detector that uses a cleaning fluid containing chlorine. Using standard theories of physics and astronomy, I calculated the rate at which neutrinos are produced in the Sun. I could then predict the rate at which neutrinos should be captured in the largest detector Ray could build. If my calculations matched the experiment, they would confirm that the Sun shines by nuclear fusion in its interior. In effect, we proposed using neutrinos to look into the center of the Sun in much the same way your doctor uses X rays or ultrasound to look into the interior of your body.

The actual experiment used 100,000 gallons of perchloroethylene, about enough to fill an Olympic swimming pool. Ray Davis and his collaborators put their detector in a deep gold mine, to shield it from other particles that hit the surface of the Earth. To everyone's surprise, Ray's chlorine detector captured many fewer neutrinos than I had predicted. The results were challenged and checked repeatedly over the following three decades, but always with the same result: Many neutrinos appear to be missing! The case of the missing neutrinos grew stronger with time. By the early 1990s, three other experiments, each with a different type of detector, searched for neutrinos from the Sun. They all found fewer than I predicted.

What was wrong? Where have the neutrinos gone? There are three possibilities: Either the experiments are wrong, the standard model of how the Sun (and other stars) shine is wrong, or something happens to the neutrinos after they are produced. Since different experiments all showed that fewer neutrinos were detected than I predicted, it seemed that the culprit was either the standard model of physics or the standard model of the Sun.

In the 1990s, seismic measurements made with optical light on the surface of the Sun confirmed predictions of the standard solar model to high precision. The speed of sound within the solar interior that was predicted by the standard solar model agreed with the helioseismological measurements to extraordinary accuracy, to better than 0.1%. Many physicists began to acknowledge that our understanding of basic physics needed to be corrected, rather than our understanding of how the Sun shines.

Finally, in the last decade of the twentieth century and in the first few years of the twenty-first century, the mystery of the missing neutrinos was decided definitely by new experiments in Japan and Canada. The Canadian and Japanese solar neutrino experiments show that neutrinos created in the solar interior change into neutrinos that are more difficult to detect as they pass out of the Sun and travel to the Earth. The neutrinos change their personalities, so to speak! Most easily detected are the so-called electron-type neutrinos; more difficult to detect are *muon*-type and *tau*-type neutrinos. The new neutrino detectors in Japan and Canada provide clear experimental demonstrations of physical processes, neutrino personality changes, that cannot be described correctly by the standard model of particle physics.

The peculiar behavior of neutrinos, their oscillation between different neutrino types, offers a clue that may help lead us to new laws of particle physics, the laws governing the smallest scales of matter. The observed changes from one neutrino type to another imply that neutrinos have tiny masses. Since ordinary telescopes that detect light cannot see neutrinos, we know that neutrinos account for some of the mysterious "dark matter" in the universe. Unfortunately, the simplest interpretations of the solar neutrino experiments suggest that neutrinos constitute only about 0.1% of the dark matter.

However, the real message of the experiments on solar neutrinos is even more remarkable. The history of solar neutrino research shows that if you work on the frontier of science, you may, while looking for something quite different, be lucky enough to stumble across something that is beautiful and unexpected.

19 The Nature of the Stars

To the unaided eye, the night sky is spangled with several thousand stars, each appearing as a bright pinpoint of light. With a pair of binoculars, you can see some 10,000 other, fainter stars; with a 15-cm (6-in.) telescope, the total rises to more than 2 million. Astronomers now know that there are in excess of 100 billion (10^{11}) stars in our Milky Way Galaxy alone.

But what are these distant pinpoints? To the great thinkers of ancient Greece, the stars were bits of light embedded in a vast sphere with the Earth at its center. They thought the stars were composed of a mysterious "fifth element," quite unlike anything found on Earth.

Today, we know that the stars are made of the same chemical elements found on Earth. We know their sizes, their temperatures, their masses, and something of their internal structures. We understand, too, why the stars in the accompanying image come in a range of beautiful colors: Blue stars have high surface temperatures, while the surface temperatures of red and yellow stars are relatively low.

How have we learned these things? How can we know the nature of the stars, objects so distant that their light takes years or centuries to reach us? In this chapter, we will learn about the measurements and calculations that astronomers make to determine the properties of stars. We will also take a first look at the Hertzsprung-Russell diagram, an important tool that helps astronomers systematize the wealth of available information about the stars. In later chapters, we will use this diagram to help us understand how stars are born, evolve, and eventually die.

Infrared and visible-light images were combined to show these stars in the cluster NGC 6397. (NASA and the Hubble Heritage Team [AURA/STScI]) R **I** **V** U X G

Questions to Guide Your Reading

As you read the sections of this chapter, look for the answers to the following questions:

19-1 How far away are the stars?

19-2 What evidence do astronomers have that the Sun is a typical star?

19-3 What is meant by a "first-magnitude" or "second-magnitude" star?

19-4 Why are some stars red and others blue?

19-5 What are the stars made of?

19-6 As stars go, is our Sun especially large or small?

19-7 What are giant, supergiant, and white dwarf stars?

19-8 How do we know the distances to remote stars?

19-9 Why are binary star systems important in astronomy?

19-10 How can a star's spectrum show whether it is actually a binary star system?

19-11 What do astronomers learn from stars that eclipse each other?

19-1 | Careful measurements of the parallaxes of stars reveal their distances

The vast majority of stars are objects very much like the Sun. This understanding followed from the discovery that the stars are tremendously far from us, at distances so great that their light takes years to reach us. Because the stars at night are clearly visible to the naked eye despite these huge distances, it must be that the **luminosity** of the stars—that is, how much energy they emit into space per second—is comparable to or greater than that of the Sun. Just as for the Sun, the only explanation for such tremendous luminosities is that thermonuclear reactions are occurring within the stars (see Section 18-1).

Clearly, then, it is important to know how distant the stars are. But how do we measure these distances? You might think this is done by comparing how bright different stars appear. Perhaps the star Betelgeuse in the constellation Orion appears bright because it is relatively close, while the dimmer and less conspicuous star Polaris (the North Star, in the constellation Ursa Minor) is farther away.

But this line of reasoning is incorrect: Polaris is actually closer to us than Betelgeuse! How bright a star appears is *not* a good indicator of its distance. If you see a light on a darkened road, it could be a motorcycle headlight a kilometer away or a person holding a flashlight just a few meters away. In the same way, a bright star might be extremely far away but have an unusually high luminosity, and a dim star might be relatively close but have a rather low luminosity. Astronomers must use other techniques to determine the distances to the stars.

The most straightforward way of measuring stellar distances uses an effect called **parallax.** This is the apparent displacement of an object because of a change in the observer's point of view (**Figure 19-1**). To see how parallax works, hold your arm out straight in front of you. Now look at the hand on your outstretched arm, first with your left eye closed, then with your right eye closed. When you close one eye and open the other, your hand appears to shift back and forth against the background of more distant objects.

The closer the object you are viewing, the greater the parallax shift. To see this, repeat the experiment with your hand held closer to your face. Your brain analyzes such parallax shifts constantly as it compares the images from your left and right eyes, and in this way determines the distances to objects around you. This is the origin of depth perception.

To measure the distance to a star, astronomers measure the parallax shift of the star using two points of view that are as far apart as possible—at opposite sides of the Earth's orbit. The direction from Earth to a nearby star changes as our planet orbits the Sun, and the nearby star appears to move back and forth against the background of more distant stars (**Figure 19-2**). This motion is called **stellar parallax.** The parallax (*p*) of a star is equal to half the angle through which the star's apparent position shifts as the Earth moves from one side of its orbit to the other. The larger the parallax *p*, the smaller the distance *d* to the star (compare Figure 19-2*a* with Figure 19-2*b*).

It is convenient to measure the distance *d* in **parsecs.** A star with a parallax angle of 1 second of arc (*p* = 1 arcsec) is at a distance of

figure 19-1

Parallax Imagine looking at some nearby object (a tree) against a distant background (mountains). When you move from one location to another, the nearby object appears to shift with respect to the distant background scenery. This familiar phenomenon is called parallax.

1 parsec (*d* = 1 pc). (The word "parsec" is a contraction of the phrase "the distance at which a star has a *par*allax of one arc*sec*ond." Recall from Section 1-7 that 1 parsec equals 3.26 light-years, 3.09×10^{13} km, or 206,265 AU; see Figure 1-12.) If the angle *p* is measured in arcseconds, then the distance *d* to the star in parsecs is given by the following equation:

Relation between a star's distance and its parallax

$$d = \frac{1}{p}$$

d = distance to a star, in parsecs

p = parallax angle of that star, in arcseconds

This simple relationship between parallax and distance in parsecs is one of the main reasons that astronomers usually measure cosmic distances in parsecs rather than light-years. For example, a star whose parallax is *p* = 0.1 arcsec is at a distance *d* = 1/(0.1) = 10 parsecs from the Earth. Barnard's star, named for the American astronomer Edward E. Barnard, has a parallax of 0.547 arcsec. Hence, the distance to this star is:

$$d = \frac{1}{p} = \frac{1}{0.547} = 1.83 \text{ pc}$$

Because 1 parsec is 3.26 light-years, this can also be expressed as

$$d = 1.83 \text{ pc} \times \frac{3.26 \text{ ly}}{1 \text{ pc}} = 5.96 \text{ ly}$$

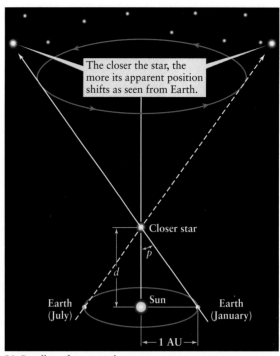

(a) Parallax of a nearby star

(b) Parallax of an even closer star

Figure 19-2
ANIMATION 19.1

Stellar Parallax (a) As the Earth orbits the Sun, a nearby star appears to shift its position against the background of distant stars. The parallax (p) of the star is equal to the angular radius of the Earth's orbit as seen from the star. (b) The closer the star is to us, the greater the parallax angle p. The distance d to the star (in parsecs) is equal to the reciprocal of the parallax angle p (in arcseconds): $d = 1/p$.

All known stars have parallax angles less than one arcsecond. In other words, the closest star is more than 1 parsec away. It was not until 1838 that the first successful parallax measurements were made by the German astronomer and mathematician Friedrich Wilhelm Bessel. He found the parallax angle of the star 61 Cygni to be just $^1/_3$ arcsec—equal to the angular diameter of a dime at a distance of 11 kilometers, or 7 miles. He thus determined that this star is about 3 pc from the Earth. (Modern measurements give a slightly smaller parallax angle, which means that 61 Cygni is actually more than 3 pc away.) The star Proxima Centauri has the largest known parallax angle, 0.772 arcsec, and hence is the closest known star (other than the Sun); its distance is 1/(0.772) = 1.30 pc.

WEB LINK 19.3
Appendix 4 at the back of this book lists all the stars within 4 pc of the Sun, as determined by parallax measurements. Most of these stars are far too dim to be seen with the naked eye, which is why their names are probably unfamiliar to you. By contrast, the majority of the familiar, bright stars in the nighttime sky (listed in Appendix 5) are so far away that their parallaxes cannot be measured from the Earth's surface. They appear bright not because they are close, but because they are far more luminous than the Sun. The brightest stars in the sky are *not* necessarily the nearest stars!

Parallax angles smaller than about 0.01 arcsec are extremely difficult to measure from the Earth, in part because of the blurring effects of the atmosphere. Therefore, the parallax method used with ground-based telescopes can give fairly reliable distances only for stars nearer than about 1/0.01 = 100 pc. But an observatory in space is unhampered by the atmosphere. Observations made from spacecraft therefore permit astronomers to measure even smaller parallax angles and thus determine the distances to more remote stars.

WEB LINK 19.4
In 1989 the European Space Agency (ESA) launched the satellite *Hipparcos,* an acronym for *Hi*gh *P*recision *Pa*rallax *C*ollecting *S*atellite (and a commemoration of the ancient Greek astronomer Hipparchus, who created one of the first star charts). Over more than three years of observations, the telescope aboard *Hipparcos* was used to measure the parallaxes of 118,000 stars with an accuracy of 0.001 arcsecond. This has enabled astronomers to determine stellar distances out to several hundred parsecs, and with much greater precision than has been possible with ground-based observations. In the years to come, astronomers will increasingly turn to space-based observations to determine stellar distances.

Unfortunately, most of the stars in the Galaxy are so far away that their parallax angles are too small to measure even with an orbiting telescope. Later in this chapter, we will discuss a technique that can be used to find the distances to these more remote stars. In Chapters 26 and 28 we will learn about other techniques that astronomers use to determine the much larger distances to galaxies

figure 19-3 R I **V** U X G
The Motion of Barnard's Star Three photographs taken over a four-year period were combined to show the motion of Barnard's star, which lies 1.82 parsecs away in the constellation Ophiuchus. Over this time interval, Barnard's star moved more than 41 arcseconds on the celestial sphere (about 0.69 arcminutes, or 0.012°), more than any other star. (John Sanford/Science Photo Library)

beyond the Milky Way. These techniques also help us understand the overall size, age, and structure of the universe.

Because it can be used only on relatively close stars, stellar parallax might seem to be of limited usefulness. But parallax measurements are the cornerstone for all other methods of finding the distances to remote objects. These other methods require a precise and accurate knowledge of the distances to nearby stars, as determined by stellar parallax. Hence, any inaccuracies in the parallax angles for nearby stars can translate into substantial errors in measurement for the whole universe. For this reason, astronomers are continually trying to perfect their parallax-measuring techniques.

Stellar parallax is an *apparent* motion of stars caused by the Earth's orbital motion around the Sun. But stars are not fixed objects and actually do move through space. As a result, stars change their positions on the celestial sphere (Figure 19-3), and they move either toward or away from the Sun. These motions are sufficiently slow, however, that changes in the positions of the stars are hardly noticeable over a human lifetime. Box 19-1 describes how astronomers study these motions and what insights they gain from these studies.

19-2 | If a star's distance is known, its luminosity can be determined from its brightness

All the stars you can see in the nighttime sky shine by thermonuclear fusion, just as the Sun does (see Section 18-1). But they are by no means merely identical copies of the Sun. Stars differ in their luminosity (L), the amount of light energy they emit each second. Lumi-

nosity is usually measured either in watts (1 watt, or 1 W, is 1 joule per second) or as a multiple of the Sun's luminosity ($L_\odot$, equal to 3.86×10^{26} W). Most stars are less luminous than the Sun, but some blaze forth with a million times the Sun's luminosity. Knowing a star's luminosity is essential for determining the star's history, its present-day internal structure, and its future evolution.

To determine the luminosity of a star, we first note that as light energy moves away from its source, it spreads out over increasingly larger regions of space. Imagine a sphere of radius d centered on the light source, as in Figure 19-4. The amount of energy that passes each second through a square meter of the sphere's surface area is the total luminosity of the source (L) divided by the total surface area of the sphere (equal to $4\pi d^2$). This quantity is called the **apparent brightness** of the light, or just **brightness** (b), because how bright a light source appears depends on how much light energy per second enters through the area of a light detector (such as your eye). Apparent brightness is measured in watts per square meter (W/m²).

Inverse-square law relating apparent brightness and luminosity

$$b = \frac{L}{4\pi d^2}$$

b = apparent brightness of a star's light, in W/m²

L = star's luminosity, in W

d = distance to star, in meters

This relationship is called the **inverse-square law,** because the apparent brightness of light that an observer can see or measure is inversely proportional to the square of the observer's distance (d)

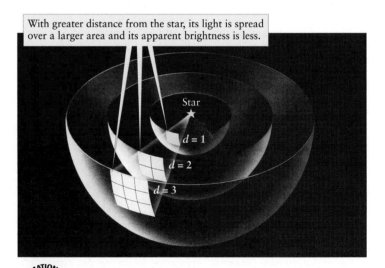

With greater distance from the star, its light is spread over a larger area and its apparent brightness is less.

figure 19-4
The Inverse-Square Law Radiation from a light source illuminates an area that increases as the square of the distance from the source. Hence, the apparent brightness decreases as the square of the distance. The brightness at $d = 2$ is $1/(2^2) = 1/4$ of the brightness at $d = 1$, and the brightness at $d = 3$ is $1/(3^2) = 1/9$ of that at $d = 1$.

box 19-1 TOOLS OF THE ASTRONOMER'S TRADE

Stellar Motions

Stars can move through space in any direction. The **space velocity** of a star describes how fast and in what direction it is moving. As the accompanying figure shows, a star's space velocity v can be broken into components parallel and perpendicular to our line of sight.

The component perpendicular to our line of sight—that is, across the plane of the sky—is called the star's **tangential velocity** (v_t). To determine it, astronomers must know the distance to a star (d) and its **proper motion** (μ, the Greek letter mu), which is the number of arcseconds that the star appears to move per year on the celestial sphere. Proper motion does not repeat itself yearly, so it can be distinguished from the apparent back-and-forth motion due to parallax. In terms of a star's distance and proper motion, its tangential velocity (in km/s) is

$$v_t = 4.74\mu d$$

where μ is in arcseconds per year and d is in parsecs. For example, Barnard's star (Figure 19-3) has a proper motion of 10.358 arcseconds per year and a distance of 1.83 pc. Hence, its tangential velocity is

$$v_t = 4.74(10.358)(1.83) = 89.8 \text{ km/s}$$

The component of a star's motion parallel to our line of sight—that is, either directly toward us or directly away from us—is its **radial velocity** (v_r). It can be determined from measurements of the Doppler shifts of the star's spectral lines (see Section 5-9 and Box 5-6). If a star is approaching us, the wavelengths of all of its spectral lines are decreased (blueshifted); if the star is receding from us, the wavelengths are increased (redshifted). The radial velocity is related to the wavelength shift by the equation

$$\frac{\lambda - \lambda_0}{\lambda_0} = \frac{v_r}{c}$$

In this equation, λ is the wavelength of light coming from the star, λ_0 is what the wavelength would be if the star were not moving, and c is the speed of light. As an illustration, a particular spectral line of iron in the spectrum of Barnard's star has a wavelength (λ) of 516.445 nm. As measured in a laboratory on the Earth, the same spectral line has a wavelength (λ_0) of 516.629 nm. Thus, for Barnard's star, our equation becomes

$$\frac{516.445 \text{ nm} - 516.629 \text{ nm}}{516.629 \text{ nm}} = -0.000356 = \frac{v_r}{c}$$

v = the star's space velocity
v_r = the star's radial velocity
v_t = the star's tangential velocity

d = distance from Earth to the star

This side is also equal to v_r

This side is also equal to v_t

Star

Earth

Solving this equation for the radial velocity v_r, we find

$$v_r = (-0.000356)\,c = (-0.000356)(3.00 \times 10^5 \text{ km/s})$$
$$= -107 \text{ km/s}$$

The negative sign means that Barnard's star is moving toward us. You can check this interpretation by noting that the wavelength λ = 516.445 nm received from Barnard's star is less than the laboratory wavelength λ_0 = 516.629 nm; hence, the light from the star is blueshifted, which indeed means that the star is approaching. If the star were receding, its radial velocity would be positive.

The illustration shows that the tangential velocity and radial velocity form two sides of a right triangle. The long side (hypotenuse) of this triangle is the space velocity (v). From the Pythagorean theorem, the space velocity is

$$v = \sqrt{v_t^2 + v_r^2}$$

For Barnard's star, the space velocity is

$$v = \sqrt{(89.8 \text{ km/s})^2 + (-107 \text{ km/s})^2} = 140 \text{ km/s}$$

Therefore, Barnard's star is moving through space at a speed of 140 km/s (503,000 km/h, or 312,000 mi/h) relative to the Sun.

Determining the space velocities of stars is essential for understanding the structure of the Galaxy. Studies show that the stars in our local neighborhood are moving in wide orbits around the center of the Galaxy, which lies some 8000 pc (26,000 light-years) away in the direction of the constellation Sagittarius (the Archer). While many of the orbits are roughly circular and lie in nearly the same plane, others are highly elliptical or steeply inclined to the galactic plane. We will see in Chapter 25 how the orbits of stars and gas clouds reveal the Galaxy's spiral structure.

from the source. If you double your distance from a light source, its radiation is spread out over an area 4 times larger, so the apparent brightness you see is decreased by a factor of 4. Similarly, at triple the distance, the apparent brightness is $1/9$ as great (see Figure 19-4).

We can apply the inverse-square law to the Sun, which is 1.50×10^{11} m from Earth. Its apparent brightness ($b_\odot$) is

$$b_\odot = \frac{3.86 \times 10^{26}\ \text{W}}{4\pi(1.50 \times 10^{11}\ \text{m})^2} = 1370\ \text{W/m}^2$$

That is, a solar panel with an area of 1 square meter receives 1370 watts of power from the Sun.

Astronomers measure the apparent brightness of a star using a telescope with an attached light-sensitive instrument, similar to the light meter in a camera that determines the proper exposure. Measuring a star's apparent brightness is called **photometry.**

The inverse-square law says that we can find a star's luminosity if we know its distance and its apparent brightness. To do this, it is convenient to express this law in a somewhat different form. We first rearrange the above equation:

$$L = 4\pi d^2 b$$

We then apply this equation to the Sun. That is, we write a similar equation relating the Sun's luminosity ($L_\odot$), the distance from the Earth to the Sun ($d_\odot$, equal to 1 AU), and the Sun's apparent brightness ($b_\odot$):

$$L_\odot = 4\pi d_\odot{}^2 b_\odot$$

If we take the ratio of these two equations, the unpleasant factor of 4π drops out and we are left with the following:

Determining a star's luminosity from its apparent brightness

$$\frac{L}{L_\odot} = \left(\frac{d}{d_\odot}\right)^2 \frac{b}{b_\odot}$$

$L/L_\odot$ = ratio of the star's luminosity to the Sun's luminosity

$d/d_\odot$ = ratio of the star's distance to the Earth-Sun distance

$b/b_\odot$ = ratio of the star's apparent brightness to the Sun's apparent brightness

We need to know just two things to find a star's luminosity: the distance to a star as compared to the Earth-Sun distance (the ratio $d/d_\odot$), and how that star's apparent brightness compares to that of the Sun (the ratio $b/b_\odot$). Then we can use the above equation to find how luminous that star is compared to the Sun (the ratio $L/L_\odot$).

In other words, this equation gives us a general rule relating the luminosity, distance, and apparent brightness of a star:

We can determine the luminosity of a star from its distance and apparent brightness. For a given distance, the brighter the star, the more luminous that star must be. For a given apparent brightness, the more distant the star, the more luminous it must be to be seen at that distance.

Box 19-2 shows how to use the above equation to determine the luminosity of the nearby star ε (epsilon) Eridani, the fifth brightest star in the constellation Eridanus (named for a river in Greek mythology). Parallax measurements indicate that ε Eridani is 3.23 pc away, and photometry shows that the star appears only 6.73×10^{-13} as bright as the Sun; using the above equation, we find that ε Eridani has only 0.30 times the luminosity of the Sun.

Calculations of this kind show that stars come in a wide variety of different luminosities, with values that range from about $10^6\ L_\odot$ (a million times the Sun's luminosity) to only about $10^{-4}\ L_\odot$ (a mere ten-thousandth of the Sun's light output). The most luminous star emits roughly 10^{10} times more energy each second than the least luminous! (To put this number in perspective, about 10^{10} human beings have lived on the Earth since our species first evolved.)

As stars go, our Sun is neither extremely luminous nor extremely dim; it is a rather ordinary, garden-variety star. It is somewhat more luminous than most stars, however. Of more than 30 stars within 4 pc of the Sun (see Appendix 4), only three (α Centauri, Sirius, and Procyon) have a greater luminosity than the Sun.

To better characterize a typical population of stars, astronomers count the stars out to a certain distance from the Sun and plot the number of stars that have different luminosities. The resulting graph is called the **luminosity function.** Figure 19-5 shows the luminosity function for stars in our part of the Milky Way

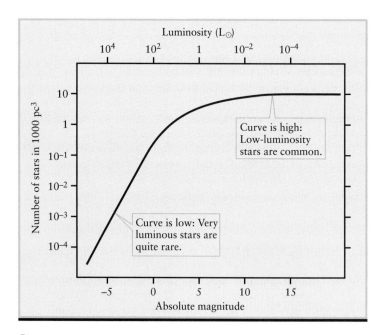

Figure 19-5

The Luminosity Function This graph shows how many stars of a given luminosity lie within a representative 1000 cubic-parsec volume. The scale at the bottom of the graph shows absolute magnitude, an alternative measure of a star's luminosity (described in Section 19-3). (Adapted from J. Bahcall and R. Soneira)

BOX 19-2 | TOOLS OF THE ASTRONOMER'S TRADE

Luminosity, Distance, and Apparent Brightness

The inverse-square law (Section 19-2) relates a star's luminosity, distance, and apparent brightness to the corresponding quantities for the Sun:

$$\frac{L}{L_\odot} = \left(\frac{d}{d_\odot}\right)^2 \frac{b}{b_\odot}$$

We can use a similar equation to relate the luminosities, distances, and apparent brightnesses of *any* two stars, which we call star 1 and star 2:

$$\frac{L_1}{L_2} = \left(\frac{d_1}{d_2}\right)^2 \frac{b_1}{b_2}$$

EXAMPLE: The star ε (epsilon) Eridani is 3.23 pc from Earth. As seen from Earth, this star appears only 6.73×10^{-13} as bright as the Sun. What is the luminosity of ε Eridani compared with that of the Sun?

Situation: We are given the distance to ε Eridani ($d = 3.23$ pc) and this star's brightness compared to that of the Sun ($b/b_\odot = 6.73 \times 10^{-13}$). Our goal is to find the ratio of the luminosity of ε Eridani to that of the Sun, that is, the quantity $L/L_\odot$.

Tools: Since we are asked to compare this star to the Sun, we use the first of the two equations given above, $L/L_\odot = (d/d_\odot)^2(b/b_\odot)$, to solve for $L/L_\odot$.

Answer: Our equation requires the ratio of the star's distance to the Sun's distance, $d/d_\odot$. The distance from the Earth to the Sun is $d_\odot = 1$ AU. To calculate the ratio $d/d_\odot$, we must express both distances in the same units. There are 206,265 AU in 1 parsec, so we can write the distance to e Eridani as $d = (3.23$ pc$)(206{,}265$ AU/pc$) = 6.66 \times 10^5$ AU. Hence, the ratio of distances is $d/d_\odot = (6.66 \times 10^5$ AU$)/(1$ AU$) = 6.66 \times 10^5$. Then we find that the ratio of the luminosity of ε Eridani (L) to the Sun's luminosity ($L_\odot$) is

$$\frac{L}{L_\odot} = \left(\frac{d}{d_\odot}\right)^2 \frac{b}{b_\odot} = (6.66 \times 10^5)^2 \times (6.73 \times 10^{-13}) = 0.30$$

Review: This result means that ε Eridani is only 0.30 as luminous as the Sun; that is, its power output is only 30% as great.

EXAMPLE: Suppose star 1 is at half the distance of star 2 (that is, $d_1/d_2 = \frac{1}{2}$) and that star 1 appears twice as bright as star 2 (that is, $b_1/b_2 = 2$). How do the luminosities of these two stars compare?

Situation: For these two stars, we are given the ratio of distances (d_1/d_2) and the ratio of apparent brightnesses (b_1/b_2). Our goal is to find the ratio of their luminosities (L_1/L_2).

Tools: Since we are comparing two stars, neither of which is the Sun, we use the second of the two equations above: $L_1/L_2 = (d_1/d_2)^2(b_1/b_2)$.

Answer: Plugging in to our equation, we find

$$\frac{L_1}{L_2} = \left(\frac{1}{2}\right)^2 \times 2 = \frac{1}{2}$$

Review: This result says that star 1 has only one-half the luminosity of star 2. Despite this, star 1 appears brighter than star 2 because it is closer to us.

The two equations above are also useful in the method of *spectroscopic parallax*, which we discuss in Section 19-8. It turns out that a star's luminosity can be determined simply by analyzing the star's spectrum. If the star's apparent brightness is also known, the star's distance can be calculated. The inverse-square law can be rewritten as an expression for the ratio of the star's distance from the Earth (d) to the Earth-Sun distance ($d_\odot$):

$$\frac{d}{d_\odot} = \sqrt{\frac{(L/L_\odot)}{(b/b_\odot)}}$$

We can also use this formula as a relation between the properties of any two stars, 1 and 2:

$$\frac{d_1}{d_2} = \sqrt{\frac{(L_1/L_2)}{(b_1/b_2)}}$$

EXAMPLE: The star Pleione in the constellation Taurus is 190 times as luminous as the Sun but appears only 3.19×10^{-13} as bright as the Sun. How far is Pleione from Earth?

Situation: We are told the ratio of Pleione's luminosity to that of the Sun ($L/L_\odot = 190$) and the ratio of their apparent brightnesses ($b/b_\odot = 3.19 \times 10^{-13}$). Our goal is to find the distance d from the Earth to Pleione.

Tools: Since we are comparing Pleione to the Sun, we use the first of the two equations above.

Answer: Our equation tells us the ratio of the Earth-Pleione distance to the Earth-Sun distance:

(continued)

TOOLS OF THE ASTRONOMER'S TRADE continued

$$\frac{d}{d_\odot} = \sqrt{\frac{(L/L_\odot)}{(b/b_\odot)}} = \sqrt{\frac{190}{3.19 \times 10^{-13}}} = \sqrt{5.95 \times 10^{14}} = 2.44 \times 10^7$$

Since the distance from Earth to Sun is $d_\odot = 1$ AU, this means that Pleione is 2.44×10^7 AU from Earth. Using the conversion 206,265 AU = 1 pc, the star's distance is $d = (2.44 \times 10^7 \text{ AU})(1 \text{ pc}/206,265 \text{ AU}) = 118$ pc.

Review: We can check our result by comparing with the above example about the star ε Eridani. Pleione has a much greater luminosity than ε Eridani (190 times the Sun's luminosity versus 0.30 times), but Pleione appears dimmer than ε Eridani (3.19×10^{-13} times as bright as the Sun compared to 6.73×10^{-13} times). For this to be true, Pleione must be much farther away from Earth than is ε Eridani, which is just what our results show ($d = 118$ pc for Pleione compared to $d = 3.23$ pc for ε Eridani).

EXAMPLE: The star δ (delta) Cephei, which lies 300 pc from the Earth, is thousands of times more luminous than the Sun. Thanks to this great luminosity, stars like δ Cephei can be seen in galaxies millions of parsecs away. As an example, the Hubble Space Telescope has detected stars like δ Cephei within the galaxy NGC 3351, which lies in the direction of the constellation Leo. These stars appear only 9×10^{-10} as bright as δ Cephei. What is the distance to NGC 3351?

Situation: To determine the distance we want, we need to find the distance to a star within NGC 3351. We are told that certain stars within this galaxy are identical to δ Cephei but appear only 9×10^{-10} as bright.

Tools: We use the equation $d_1 / d_2 = \sqrt{(L_1/L_2)/(b_1/b_2)}$ to relate two stars, one within NGC 3351 (call this star 1) and the identical star δ Cephei (star 2). Our goal is to find d_1.

Answer: Since the two stars are identical, they have the same luminosity ($L_1 = L_2$, or $L_1/L_2 = 1$). The brightness ratio is $b_1/b_2 = 9 \times 10^{-10}$, so our equation tells us that

$$\frac{d_1}{d_2} = \sqrt{\frac{(L_1/L_2)}{(b_1/b_2)}} = \sqrt{\frac{1}{9 \times 10^{-10}}} = \sqrt{1.1 \times 10^9} = 33,000$$

Hence, NGC 3351 is 33,000 times farther away than δ Cephei, which is 300 pc from Earth. The distance from the Earth to NGC 3351 is therefore $(33,000)(300 \text{ pc}) = 10^7$ pc, or 10 megaparsecs (10 Mpc).

Review: This example illustrates one technique that astronomers use to measure extremely large distances. We will learn more about stars like δ Cephei in Chapter 21, and in Chapter 26 we will explore further how they are used to determine the distances to remote galaxies.

Galaxy. The curve declines very steeply for the most luminous stars toward the left side of the graph, indicating that they are quite rare. For example, this graph shows that stars like the Sun are about 10,000 times more common than stars like Spica (which has a luminosity of 2100 $L_\odot$).

The exact shape of the curve in Figure 19-5 applies only to the vicinity of the Sun and similar regions in our Milky Way Galaxy. Other locations have somewhat different luminosity functions. In stellar populations in general, however, low-luminosity stars are much more common than high-luminosity ones.

19-3 | Astronomers often use the magnitude scale to denote brightness

Because astronomy is among the most ancient of sciences, some of the tools used by modern astronomers are actually many centuries old. One such tool is the **magnitude scale,** which astronomers frequently use to denote the brightness of stars. This scale was introduced in the second century B.C. by the Greek astronomer Hipparchus, who called the brightest stars first-magnitude stars. Stars

about half as bright as first-magnitude stars were called second-magnitude stars, and so forth, down to sixth-magnitude stars, the dimmest ones he could see. After telescopes came into use, astronomers extended Hipparchus's magnitude scale to include even dimmer stars.

The magnitudes in Hipparchus's scale are properly called **apparent magnitudes,** because they describe how bright an object *appears* to an Earth-based observer. Apparent magnitude is directly related to apparent brightness.

CAUTION ! The magnitude scale can be confusing because it works "backward." Keep in mind that the *greater* the apparent magnitude, the *dimmer* the star. A star of apparent magnitude +3 (a third-magnitude star) is dimmer than a star of apparent magnitude +2 (a second-magnitude star).

In the nineteenth century, astronomers developed better techniques for measuring the light energy arriving from a star. These measurements showed that a first-magnitude star is about 100 times brighter than a sixth-magnitude star. In other words, it would take 100 stars of magnitude +6 to provide as much light energy as we receive from a single star of magnitude +1. To make computations easier, the magnitude scale was redefined so that a magnitude difference of 5 corresponds exactly to a factor of 100 in brightness. A magnitude difference of 1 corresponds to a factor of 2.512 in brightness, because

$$2.512 \times 2.512 \times 2.512 \times 2.512 \times 2.512 = (2.512)^5 = 100$$

Thus, it takes 2.512 third-magnitude stars to provide as much light as we receive from a single second-magnitude star.

Figure 19-6 illustrates the modern apparent magnitude scale. The dimmest stars visible through a pair of binoculars have an apparent magnitude of +10, and the dimmest stars that can be photographed in a one-hour exposure with the Keck telescopes (see Section 6-2) or the Hubble Space Telescope have apparent magnitude +30. Modern astronomers also use negative numbers to extend Hipparchus's scale to include very bright objects. For example, Sirius, the brightest star in the sky, has an apparent magnitude of −1.43. The Sun, the brightest object in the sky, has an apparent magnitude of −26.7.

Apparent magnitude is a measure of a star's apparent brightness as seen from Earth. A related quantity that measures a star's true energy output—that is, its luminosity—is called **absolute magnitude**. This is the apparent magnitude a star would have if it were located exactly 10 parsecs from Earth.

ANALOGY If you wanted to compare the light output of two different lightbulbs, you would naturally place them side by side so that both bulbs were the same distance from you. In the absolute magnitude scale, we imagine doing the same thing with stars to compare their luminosities.

If the Sun were moved to a distance of 10 parsecs from the Earth, it would have an apparent magnitude of +4.8. The absolute magnitude of the Sun is thus +4.8. The absolute magnitudes of the stars range from approximately +15 for the least luminous to −10 for the most luminous. (*Note:* Like apparent magnitudes, absolute magnitudes work "backward": The *greater* the absolute magnitude, the *less luminous* the star.) The Sun's absolute magnitude is about in the middle of this range.

There is a mathematical relationship between absolute magnitude and luminosity, which astronomers use to convert one to the other as they see fit. It is also possible to rewrite the inverse-square law, which we introduced in Section 19-2, as an equation that relates a star's apparent magnitude (a measure of its apparent brightness), its absolute magnitude (a measure of its luminosity), and its distance. Box 19-3 describes these relationships.

Because the "backward" magnitude scales can be confusing, we will use them only occasionally in this book. We will usually speak of a star's luminosity rather than its absolute magnitude and will describe a star's appearance in terms of apparent brightness rather than apparent magnitude. But if you go on to learn more about astronomy, you will undoubtedly make frequent use of apparent magnitude and absolute magnitude.

(a) Some apparent magnitudes

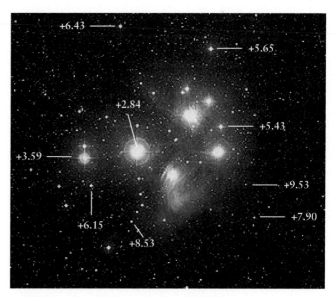

(b) Apparent magnitudes of stars in the Pleiades

Figure 19-6

The Apparent Magnitude Scale (a) Astronomers denote the apparent brightness of objects in the sky by their apparent magnitudes. The greater the apparent magnitude, the dimmer the object. (b) This photograph of the Pleiades cluster, located about 120 pc away in the constellation Taurus, shows the apparent magnitudes of some of its stars. Most are too faint to be seen by the naked eye. (David Malin/Anglo-Australian Observatory)

TOOLS OF THE ASTRONOMER'S TRADE

Apparent Magnitude and Absolute Magnitude

Astronomers commonly express a star's apparent brightness in terms of apparent magnitude (denoted by a lowercase m), and the star's luminosity in terms of absolute magnitude (denoted by a capital M). While we do not use these quantities extensively in this book, it is useful to know a few simple relationships involving them.

Consider two stars, labeled 1 and 2, with apparent magnitudes m_1 and m_2 and brightnesses b_1 and b_2, respectively. The *ratio* of their apparent brightnesses (b_1/b_2) corresponds to a *difference* in their apparent magnitudes ($m_2 - m_1$). As we learned in Section 19-3, each step in magnitude corresponds to a factor of 2.512 in brightness; we receive 2.512 times more energy per square meter per second from a third-magnitude star than from a fourth-magnitude star. This idea was used to construct the table that follows.

Apparent magnitude difference ($m_2 - m_1$)	Ratio of apparent brightness (b_1/b_2)
1	2.512
2	$(2.512)^2 = 6.31$
3	$(2.512)^3 = 15.85$
4	$(2.512)^4 = 39.82$
5	$(2.512)^5 = 100$
10	$(2.512)^{10} = 10^4$
15	$(2.512)^{15} = 10^6$
20	$(2.512)^{20} = 10^8$

A simple equation relates the difference between two stars' apparent magnitudes to the ratio of their brightnesses:

Magnitude difference related to brightness ratio

$$m_2 - m_1 = 2.5 \log \left(\frac{b_1}{b_2} \right)$$

$m_1, m_2 =$ apparent magnitudes of stars 1 and 2
$b_1, b_2 =$ apparent brightnesses of stars 1 and 2

In this equation, $\log (b_1/b_2)$ is the logarithm of the brightness ratio. The logarithm of $1000 = 10^3$ is 3, the logarithm of $10 = 10^1$ is 1, and the logarithm of $1 = 10^0$ is 0.

EXAMPLE: At their most brilliant, Venus has a magnitude of about −4 and Mercury has a magnitude of about −2. How many times brighter are these than the dimmest stars visible to the naked eye, with a magnitude of +6?

Situation: In each case we want to find a ratio of two apparent brightnesses (the brightness of Venus or Mercury compared to that of the dimmest naked-eye stars).

Tools: In each case we will convert a *difference* in apparent magnitude between the planet and the naked-eye star into a *ratio* of their brightnesses.

Answer: The magnitude difference between Venus and the dimmest stars visible to the naked eye is +6 − (−4) = 10. From the table, this difference corresponds to a brightness ratio of $(2.512)^{10} = 10^4 = 10,000$, so Venus at it most brilliant is 10,000 times brighter than the dimmest naked-eye stars.

The magnitude difference between Mercury and the dimmest naked-eye stars is +4 − (−4) = 8. While this value is not in the table, you can see that the corresponding ratio of brightnesses is $(2.512)^8 = (2.512)^{5+3} = (2.512)^5 \times (2.512)^3$. From the table, $(2.512)^5 = 100$ and $(2.512)^3 = 15.85$. Hence, the ratio of brightnesses is $100 \times 15.85 = 1585$. Hence, Mercury at its most brilliant is 1585 times brighter than the dimmest stars visible to the naked eye.

Review: Can you show that when at their most brilliant, Venus is 6.31 times brighter than Mercury? (*Hint:* No multiplication or division is required—just notice the difference in apparent magnitude between Venus and Mercury, and consult the table.)

EXAMPLE: The variable star RR Lyrae in the constellation Lyra (the Harp) periodically doubles its light output. By how much does its apparent magnitude change?

Situation: We are given a ratio of two brightnesses (the star at its maximum is twice as bright as at its minimum). Our goal is to find the corresponding difference in apparent magnitude.

Tools: We let 1 denote the star at its maximum brightness and 2 denote the same star at its dimmest, so the ratio of brightnesses is $b_1/b_2 = 2$. We then use the equation $m_2 - m_1 = 2.5 \log (b_1/b_2)$ to solve for the apparent magnitude difference $m_2 - m_1$.

Answer: Using a calculator, we find $m_2 - m_1 = 2.5 \log (2) = 2.5 \times 0.30 = 0.75$. RR Lyrae therefore varies periodically in brightness by 0.75 magnitude.

Review: Our answer means that at its dimmest, RR Lyrae has an apparent magnitude m_2 that is 0.75 *greater* than its apparent magnitude m_1 when it is brightest. (Remember that a greater value of apparent magnitude means the star is dimmer, not brighter!)

TOOLS OF THE ASTRONOMER'S TRADE continued

The inverse-square law relating a star's apparent brightness and luminosity can be rewritten in terms of the star's apparent magnitude (m), absolute magnitude (M), and distance from the Earth (d). This can be expressed as an equation:

Relation between a star's apparent magnitude and absolute magnitude

$$m - M = 5 \log d - 5$$

m = star's apparent magnitude
M = star's absolute magnitude
d = distance from the Earth to the star in parsecs

In this expression $m - M$ is called the **distance modulus,** and $\log d$ means the logarithm of the distance d in parsecs. For convenience, the following table gives the values of the distance d corresponding to different values of $m - M$.

Distance modulus $m - M$	Distance d (pc)
−4	1.6
−3	2.5
−2	4.0
−1	6.3
0	10
1	16
2	25
3	40
4	63
5	100
10	10^3
15	10^4
20	10^5

This table shows that if a star is less than 10 pc away, its distance modulus $m - M$ is negative. That is, its apparent magnitude (m) is less than its absolute magnitude (M). If the star is more than 10 pc away, $m - M$ is positive and m is greater than M. As an example, the star ε (epsilon) Indi, which is in the direction of the southern constellation Indus, has apparent magnitude $m = +4.7$. It is 3.6 pc away, which is less than 10 pc, so its apparent magnitude is less than its absolute magnitude.

EXAMPLE: Find the absolute magnitude of ε Indi.

Situation: We are given the distance to ε Indi ($d = 3.6$ pc) and its apparent magnitude ($m = +4.7$). Our goal is to find the star's absolute magnitude M.

Tools: We use the formula $m - M = 5 \log d - 5$ to solve for M.

Answer: Since $d = 3.6$ pc, we use a calculator to find $\log d = \log 3.6 = 0.56$. Therefore the star's distance modulus is $m - M = 5(0.56) - 5 = -2.2$, and the star's absolute magnitude is $M = m - (-2.2) = +4.7 + 2.2 = +6.9$.

Review: As a check on our calculations, note that this star's distance modulus $m - M = -2.2$ is less than zero, as it should be for a star less than 10 pc away. Note that our Sun has absolute magnitude +4.8; ε Indi has a greater absolute magnitude, so it is less luminous than the Sun.

EXAMPLE: Suppose you were viewing the Sun from a planet orbiting another star 100 pc away. Could you see it without using a telescope?

Situation: We learned in the preceding examples that the Sun has absolute magnitude $M = +4.8$ and that the dimmest stars visible to the naked eye have apparent magnitude $m = +6$. Our goal is to determine whether the Sun would be visible to the naked eye at a distance of 100 pc.

Tools: We use the relationship $m - M = 5 \log d - 5$ to find the Sun's apparent magnitude at $d = 100$ pc. If this is greater than +6, the Sun would not be visible at that distance. (Remember that the greater the apparent magnitude, the dimmer the star.)

Answer: From the table, at $d = 100$ pc the distance modulus is $m - M = 5$. So, as seen from this distant planet, the Sun's apparent magnitude would be $m = M + 5 = +4.8 + 5 = +9.8$. This is greater than the naked-eye limit $m = +6$, so the Sun could not be seen.

Review: The Sun is by far the brightest object in the Earth's sky. But our result tells us that to an inhabitant of a planetary system 100 pc away—a rather small distance in a galaxy that is thousands of parsecs across—our own Sun would be just another insignificant star, visible only through a telescope.

The magnitude system is also used by astronomers to express the colors of stars as seen through different filters, as we describe in Section 19-4. For example, rather than quantifying a star's color by the color ratio b_V/b_B (a star's apparent brightness as seen through a V filter divided by the brightness through a B filter), astronomers commonly use the *color index* B–V, which is the difference in the star's apparent magnitude as measured with these two filters. We will not use this system in this book, however (but see Advanced Questions 51 and 52).

19-4 | A star's color depends on its surface temperature

The image that opens this chapter shows that stars come in different colors. You can see these colors even with the naked eye. For example, you can easily see the red color of Betelgeuse, the star in the "armpit" of the constellation Orion, and the blue tint of Bellatrix at Orion's other "shoulder" (see Figure 2-2). Colors are most evident for the brightest stars, because your color vision does not work well at low light levels.

It's true that the light from a star will appear redshifted if the star is moving away from you and blueshifted if it's moving toward you. But for even the fastest stars, these color shifts are so tiny that it takes sensitive instruments to measure them. The red color of Betelgeuse and the blue color of Bellatrix are not due to their motions; they are the actual colors of the stars.

We saw in Section 5-3 that a star's color is directly related to its surface temperature. The intensity of light from a relatively cool star peaks at long wavelengths, making the star look red (**Figure 19-7a**). A hot star's intensity curve peaks at shorter wavelengths, so the star looks blue (Figure 19-7c). For a star with an intermediate temperature, such as the Sun, the intensity peak is near the middle of the visible spectrum. This gives the star a yellowish color (Figure 19-7b). This leads to an important general rule about star colors and surface temperatures:

Red stars are relatively cold, with low surface temperatures; blue stars are relatively hot, with high surface temperatures.

Figure 19-7 shows that astronomers can accurately determine the surface temperature of a star by carefully measuring its color. To do this, the star's light is collected by a telescope and passed through one of a set of differently colored filters. The filtered light is then collected by a light-sensitive device such as a CCD (see Section 6-4).

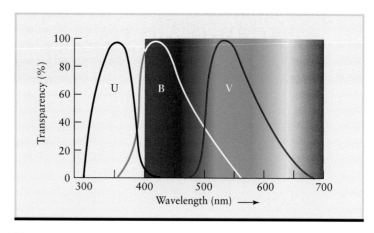

figure 19-8

U, B, and V Filters This graph shows the wavelengths to which the standard filters are transparent. The U filter is transparent to near-ultraviolet light. The B filter is transparent to violet, blue, and green light, while the V filter is transparent to green and yellow light. By measuring the apparent brightness of a star with each of these filters and comparing the results, an astronomer can determine the star's surface temperature.

The process is then repeated with each of the filters in the set. The star's image will have a different brightness through each colored filter, and by comparing these brightnesses astronomers can find the wavelength at which the star's intensity curve has its peak—and hence the star's temperature.

Let's look at this procedure in more detail. The most commonly used filters are called U, B, and V, and the technique that uses them is called **UBV photometry**. Each filter is transparent in a different band of wavelengths: the ultraviolet (U), the blue (B), and the yellow-green (V, for visual) region of the visible spectrum (**Figure 19-8**).

(a)

(b)

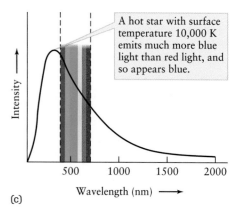

(c)

figure 19-7

Temperature and Color These graphs show the intensity of light emitted by three hypothetical stars plotted against wavelength (compare with Figure 5-10). The rainbow band indicates the range of visible wavelengths. The star's apparent color depends on whether the intensity curve has larger values at the short-wavelength or long-wavelength end of the visible spectrum.

table 19-1	Colors of Selected Stars			
Star	**Surface temperature (K)**	b_V/b_B	b_B/b_U	**Apparent color**
Bellatrix (γ Orionis)	21,500	0.81	0.45	Blue
Regulus (α Leonis)	12,000	0.90	0.72	Blue-white
Sirius (α Canis Majoris)	9400	1.00	0.96	Blue-white
Megrez (δ Ursae Majoris)	8630	1.07	1.07	White
Altair (α Aquilae)	7800	1.23	1.08	Yellow-white
Sun	5800	1.87	1.17	Yellow-white
Aldebaran (α Tauri)	4000	4.12	5.76	Orange
Betelgeuse (α Orionis)	3500	5.55	6.66	Red

Source: J.-C. Mermilliod, B. Hauck, and M. Mermilliod, University of Lausanne.

The transparency of the V filter mimics the sensitivity of the human eye.

To determine a star's temperature using UBV photometry, the astronomer first measures the star's brightness through each of the filters individually. This gives three apparent brightnesses for the star, designated b_U, b_B, and b_V. The astronomer then compares the intensity of starlight in neighboring wavelength bands by taking the ratios of these brightnesses: b_V/b_B and b_B/b_U. Table 19-1 gives values for these **color ratios** for several stars with different surface temperatures.

If a star is very hot, its radiation is skewed toward short, ultraviolet wavelengths as in Figure 19-7c. This makes the star dim through the V filter, brighter through the B filter, and brightest through the U filter. Hence, for a hot star b_V is less than b_B, which in turn is less than b_U, and the ratios b_V/b_B and b_B/b_U are both less than 1. One such star is Bellatrix (see Table 19-1), which has a surface temperature of 21,500 K.

In contrast, if a star is cool, its radiation peaks at long wavelengths as in Figure 19-7a. Such a star appears brightest through the V filter, dimmer through the B filter, and dimmest through the U filter (see Figure 19-8). In other words, for a cool star b_V is greater than b_B, which in turn is greater than b_U. Hence, the ratios b_V/b_B and b_B/b_U will both be greater than 1. The star Betelgeuse (surface temperature 3500 K) is an example.

You can see these differences between hot and cool stars in parts *a* and *c* of Figure 6-30, which show the constellation Orion at ultraviolet wavelengths (a bit shorter than those transmitted by the U filter) and at visible wavelengths that approximate the transmission of a V filter. The hot star Bellatrix is brighter in the ultraviolet image (Figure 6-30a) than at visible wavelengths (Figure 6-30c). (Figure 6-30d shows the names of the stars.) The situation is reversed for the cool star Betelgeuse: It is bright at visible wavelengths, but at ultraviolet wavelengths it is too dim to show up in the image.

Figure 19-9 graphs the relationship between a star's b_V/b_B color ratio and its temperature. If you know the value of the b_V/b_B color ratio for a given star, you can use this graph to find the star's surface temperature. As an example, for the Sun b_V/b_B equals 1.87, which corresponds to a surface temperature of 5800 K.

CAUTION! As we will see in Chapter 20, tiny dust particles that pervade interstellar space cause distant stars to appear redder than they really are. (In the same way, particles in the Earth's atmosphere make the setting Sun look redder; see Box 5-4.) Astronomers must take this reddening into account whenever they attempt to determine a star's surface temperature from its color ratios. A star's spectrum provides a more precise measure of a star's surface temperature, as we will see next. But it is quicker and easier to observe a star's colors with a set of U, B, and V filters than it is to take the star's spectrum with a spectrograph.

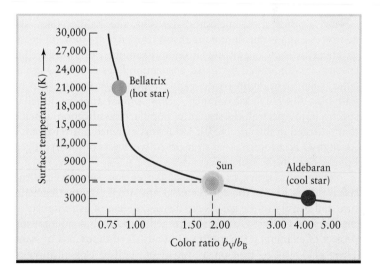

figure 19-9

Temperature, Color, and Color Ratio The b_V/b_B color ratio is the ratio of a star's apparent brightnesses through a V filter and through a B filter. This ratio is small for hot, blue stars but large for cool, red stars. After measuring a star's brightness with the B and V filters, an astronomer can estimate the star's surface temperature from a graph like this one.

19-5 | The spectra of stars reveal their chemical compositions as well as surface temperatures

SCIENCE IN PROCESS

Deciphering the information in starlight took the painstaking work of generations of astronomers

We have seen how the color of a star's light helps astronomers determine its surface temperature. To determine the other properties of a star, astronomers must analyze the spectrum of its light in more detail. This technique of *stellar spectroscopy* began in 1817 when Joseph Fraunhofer, a German instrument maker, attached a spectroscope to a telescope and pointed it toward the stars. Fraunhofer had earlier observed that the Sun has an absorption line spectrum—that is, a continuous spectrum with dark absorption lines (see Section 5-6). He found that stars have the same kind of spectra, which reinforces the idea that our Sun is a rather typical star. But Fraunhofer also found that the pattern of absorption lines is different for different stars.

We see an absorption line spectrum when a cool gas lies between us and a hot, glowing object (recall Figure 5-14). The light from the hot, glowing object itself has a continuous spectrum. In the case of a star, light with a continuous spectrum is produced at low-lying levels of the star's atmosphere where the gases are hot and dense. The absorption lines are created when this light flows outward through the upper layers of the star's atmosphere. Atoms in these cooler, less dense layers absorb radiation at specific wavelengths, which depend on the specific kinds of atoms present—hydrogen, helium, or other elements—and on whether or not the atoms are ionized. Absorption lines in the Sun's spectrum are produced in this same way (see Section 18-5).

WEB LINK 19.8
Some stars have spectra in which the Balmer absorption lines of hydrogen are prominent. But in the spectra of other stars, including the Sun, the Balmer lines are nearly absent and the dominant absorption lines are those of heavier elements such as calcium, iron, and sodium. Still other stellar spectra are dominated by broad absorption lines caused by molecules, such as titanium oxide, rather than single atoms. To cope with this diversity, astronomers group similar-appearing stellar spectra into **spectral classes**. In a popular classification scheme that emerged in the late 1890s, a star was assigned a letter from A through O according to the strength or weakness of the hydrogen Balmer lines in the star's spectrum.

Nineteenth-century science could not explain why or how the spectral lines of a particular chemical are affected by the temperature and density of the gas. Nevertheless, a team of astronomers at the Harvard College Observatory forged ahead with a monumental project of examining the spectra of hundreds of thousands of stars. Their goal was to develop a system of spectral classification in which all spectral features, not just Balmer lines, change smoothly from one spectral class to the next.

The Harvard project was financed by the estate of Henry Draper, a wealthy New York physician and amateur astronomer who in 1872 became the first person to photograph stellar absorption lines. Researchers on the project included Edward C. Pickering, Williamina Fleming, Antonia Maury, and Annie Jump Cannon

(Figure 19-10). As a result of their efforts, many of the original A-through-O classes were dropped and others were consolidated. The remaining spectral classes were reordered in the sequence **OBAFGKM**. You can remember this sequence with the mnemonic: "Oh, Be A Fine Girl (or Guy), Kiss Me!"

Cannon refined the original OBAFGKM sequence into smaller steps called **spectral types**. These steps are indicated by attaching an integer from 0 through 9 to the original letter. For example, the spectral class F includes spectral types F0, F1, F2, ... , F8, F9, which are followed by the spectral types G0, G1, G2, ... , G8, G9, and so on.

Figure 19-11 shows representative spectra of several spectral types. The strengths of spectral lines change gradually from one spectral type to the next. For example, the Balmer absorption lines of hydrogen become increasingly prominent as you go from spectral type B0 to A0. From A0 onward through the F and G classes, the hydrogen lines weaken and almost fade from view. The Sun, whose spectrum is dominated by calcium and iron, is a G2 star.

The Harvard project culminated in the *Henry Draper Catalogue,* published between 1918 and 1924. It listed 225,300 stars, each of which Cannon had personally classified. Meanwhile, physicists had been making important discoveries about the structure of atoms. Ernest Rutherford had shown that atoms have nuclei (recall

WEB LINK 19.9

figure 19-10   R I **V** U X G
Classifying the Spectra of the Stars The modern scheme of classifying stars by their spectra was developed at the Harvard College Observatory in the late nineteenth century. A team of women astronomers led by Edward C. Pickering and Williamina Fleming (standing) analyzed hundreds of thousands of spectra. Social conventions of the time prevented most women astronomers from using research telescopes or receiving salaries comparable to men's. (Harvard College Observatory)

figure 19-11   R I **V** U X G

Principal Types of Stellar Spectra Stars of different spectral classes and different surface temperatures have spectra dominated by different absorption lines. Notice how the Balmer lines of hydrogen (H_α, H_β, H_γ and H_δ) are strongest for hot stars of spectral class A, while absorption lines due to calcium (Ca) are strongest in medium-temperature G and K stars. The spectra of M stars have broad, dark bands caused by molecules of titanium oxide (TiO), which can only exist at relatively low temperatures. A roman numeral after a chemical symbol shows whether the absorption line is caused by un-ionized atoms (roman numeral I) or by atoms that have lost one electron (roman numeral II). (R. Bell, University of Maryland, and M. Briley, University of Wisconsin at Oshkosh)

Figure 5-17), and Niels Bohr made the remarkable hypothesis that electrons circle atomic nuclei along discrete orbits (see Figure 5-20). These advances gave scientists the conceptual and mathematical tools needed to understand stellar spectra.

In the 1920s, the Harvard astronomer Cecilia Payne and the Indian physicist Meghnad Saha demonstrated that the OBAFGKM spectral sequence is actually a sequence in temperature. The hottest stars are O stars. Their absorption lines can occur only if these stars have surface temperatures above 25,000 K. M stars are the coolest stars. The spectral features of M stars are consistent with stellar surface temperatures of about 3000 K.

To see why the appearance of a star's spectrum is profoundly affected by the star's surface temperature, consider the Balmer lines of hydrogen. Hydrogen is by far the most abundant element in the universe, accounting for about three-quarters of the mass of a typical star. Yet the Balmer lines do not necessarily show up in a star's spectrum. As we saw in Section 5-8, Balmer absorption lines are produced when an electron in the $n = 2$ orbit of hydrogen is lifted into a higher orbit by absorbing a photon with the right amount of energy (see Figure 5-22). If the star is much hotter than 10,000 K, the photons pouring out of the star's interior have such high energy that they easily knock electrons out of hydrogen atoms in the star's atmosphere. This process ionizes the gas. With its only electron torn away, a hydrogen atom cannot produce absorption lines. Hence, the Balmer lines will be relatively weak in the spectra of such hot stars, such as the hot O and B2 stars in Figure 19-11.

Conversely, if the star's atmosphere is much cooler than 10,000 K, almost all the hydrogen atoms are in the lowest ($n = 1$) energy state. Most of the photons passing through the star's atmosphere possess too little energy to boost electrons up from the $n = 1$ to the $n = 2$ orbit of the hydrogen atoms. Hence, very few of these atoms will have electrons in the $n = 2$ orbit, and only these few can absorb the photons characteristic of the Balmer lines. As a result, these lines are nearly absent from the spectrum of a cool star. (You can see this in the spectra of the cool M0 and M2 stars in Figure 19-11.)

For the Balmer lines to be prominent in a star's spectrum, the star must be hot enough to excite the electrons out of the ground state but not so hot that all the hydrogen atoms become ionized. A stellar surface temperature of about 9000 K produces the strongest hydrogen lines; this is the case for the stars of spectral types A0 and A5 in Figure 19-11.

 Every other type of atom or molecule also has a characteristic temperature range in which it produces prominent absorption lines in the observable part of the spectrum. Figure 19-12 shows the relative strengths of absorption lines produced by different chemicals. By measuring the details of these lines in a given star's spectrum, astronomers can accurately determine that star's surface temperature.

For example, the spectral lines of neutral (that is, un-ionized) helium are strong around 25,000 K. At this temperature, photons have enough energy to excite helium atoms without tearing away the electrons altogether. In stars hotter than about 30,000 K, helium atoms become singly ionized, that is, they lose one of their two electrons. The remaining electron produces a set of spectral lines that is recognizably different from those of neutral helium. Hence, when the spectral lines of singly ionized helium appear in a star's spectrum, we know that the star's surface temperature is greater than 30,000 K.

Astronomers use the term **metals** to refer to all elements other than hydrogen and helium. (This idiosyncratic use of the term "metal" is quite different from the definition used by chemists and other scientists. To a chemist, sodium and iron are metals but carbon and oxygen are not; to an astronomer, all of these substances are metals.) In this terminology, metals dominate the spectra of stars cooler than 10,000 K. Ionized metals are prominent for surface temperatures between 6000 and 8000 K, while neutral metals are strongest between approximately 5500 and 4000 K.

Below 4000 K, certain atoms in a star's atmosphere combine to form molecules. (At higher temperatures atoms move so fast that when they collide, they bounce off each other rather than "sticking together" to form molecules.) As these molecules vibrate and rotate, they produce bands of spectral lines that dominate the star's spectrum. Most noticeable are the lines of titanium oxide (TiO), which are strongest for surface temperatures of about 3000 K.

 Since 1995 astronomers have found a number of stars with surface temperatures even lower than those of spectral class M. Strictly speaking, these are not stars but **brown dwarfs**, which we introduced in Section 8-6. Brown dwarfs are too small to sustain thermonuclear fusion in their cores. Instead, these "substars" glow primarily from the heat released by Kelvin-Helmholtz contraction (see Section 18-1). (They do undergo fusion reactions for a brief period during their evolution.) Brown dwarfs are so cold that they are best observed with infrared telescopes (see Figure 8-14). Such observations reveal that brown dwarf spectra have a rich variety of absorption lines due to molecules. Some of these molecules actually form into solid grains in a brown dwarf's atmosphere.

To describe brown dwarf spectra, astronomers have defined two new spectral classes, L and T. Thus, the modern spectral sequence of stars and brown dwarfs is OBAFGKMLT. (Can you think of a new mnemonic that includes L and T?) Table 19-2 summarizes the relationship between the temperature and spectra of stars and brown dwarfs.

When the effects of temperature are accounted for, astronomers find that *all* stars have essentially the same chemical composition. We can state the results as a general rule:

By mass, almost all stars (including the Sun) and brown dwarfs are about three-quarters hydrogen, one-quarter helium, and 1% or less metals.

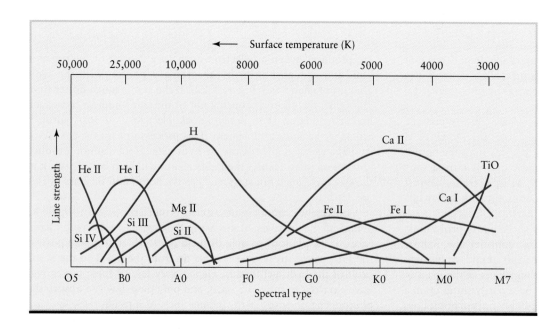

figure 19-12

The Strengths of Absorption Lines
Each curve in this graph peaks at the stellar surface temperature for which that chemical's absorption line is strongest. For example, hydrogen (H) absorption lines are strongest in A stars with surface temperatures near 10,000 K. Roman numeral I denotes neutral, un-ionized atoms; II, III, and IV denote atoms that are singly, doubly, or triply ionized (that is, have lost one, two, or three electrons).

table 19-2	**The Spectral Sequence**			
Spectral class	Color	Temperature (K)	Spectral lines	Examples
O	Blue-violet	30,000–50,000	Ionized atoms, especially helium	Naos (ζ Puppis), Mintaka (δ Orionis)
B	Blue-white	11,000–30,000	Neutral helium, some hydrogen	Spica (α Virginis), Rigel (β Orionis)
A	White	7500–11,000	Strong hydrogen, some ionized metals	Sirius (α Canis Majoris), Vega (α Lyrae)
F	Yellow-white	5900–7500	Hydrogen and ionized metals such as calcium and iron	Canopus (α Carinae), Procyon (α Canis Minoris)
G	Yellow	5200–5900	Both neutral and ionized metals, especially ionized calcium	Sun, Capella (α Aurigae)
K	Orange	3900–5200	Neutral metals	Arcturus (α Boötis), Aldebaran (α Tauri)
M	Red-orange	2500–3900	Strong titanium oxide and some neutral calcium	Antares (α Scorpii), Betelgeuse (α Orionis)
L	Red	1300–2500	Neutral potassium, rubidium, and cesium, and metal hydrides	Brown dwarf Teide 1
T	Red	below 1300	Strong neutral potassium and some water (H_2O)	Brown dwarf Gliese 229B

19-6 | Stars come in a wide variety of sizes

With even the best telescopes, stars appear as nothing more than bright points of light. On a photograph or CCD image, brighter stars appear larger than dim ones (see Figures 19-3 and 19-6b), but these apparent sizes give no indication of the star's actual size. To determine the size of a star, astronomers combine information about its luminosity (determined from its distance and apparent brightness) and its surface temperature (determined from its spectral type). In this way, they find that some stars are quite a bit smaller than the Sun, while others are a thousand times larger.

The key to finding a star's radius from its luminosity and surface temperature is the Stefan-Boltzmann law (see Section 5-4). This law says that the amount of energy radiated per second from a square meter of a blackbody's surface—that is, the energy flux (F)—is proportional to the fourth power of the temperature of that surface (T), as given by the equation $F = \sigma T^4$. This equation applies very well to stars, whose spectra are quite similar to that of a perfect blackbody. (Absorption lines, while important for determining the star's chemical composition and surface temperature, make only relatively small modifications to a star's blackbody spectrum.)

A star's luminosity is the amount of energy emitted per second from its entire surface. This equals the energy flux F multiplied by the total number of square meters on the star's surface (that is, the star's surface area). We expect that most stars are nearly spherical, like the Sun, so we can use the formula for the surface area of a sphere. This is $4\pi R^2$, where R is the star's radius (the distance from its center to its surface). Multiplying together the formulas for energy flux and surface area, we can write the star's luminosity as follows:

Relationship between a star's luminosity, radius, and surface temperature

$$L = 4\pi R^2 \sigma T^4$$

L = star's luminosity, in watts

R = star's radius, in meters

σ = Stefan-Boltzmann constant = 5.67×10^{-8} W m^{-2} K^{-4}

T = star's surface temperature, in kelvins

This equation says that a relatively cool star (low surface temperature T), for which the energy flux is quite low, can nonetheless be very luminous if it has a large enough radius R. Alternatively, a relatively hot star (large T) can have a very low luminosity if the star has only a little surface area (small R).

Box 19-4 describes how to use the above equation to calculate a star's radius if its luminosity and surface temperature are known. We can express the idea behind these calculations in terms of the following general rule:

We can determine the radius of a star from its luminosity and surface temperature. For a given luminosity, the greater the surface temperature, the smaller the radius must be. For a given surface temperature, the greater the luminosity, the larger the radius must be.

box 19-4   TOOLS OF THE ASTRONOMER'S TRADE

Stellar Radii, Luminosities, and Surface Temperatures

Because stars emit light in almost exactly the same fashion as blackbodies, we can use the Stefan-Boltzmann law to relate a star's luminosity (L), surface temperature (T), and radius (R). The relevant equation is

$$L = 4\pi R^2 \sigma T^4$$

As written, this equation involves the Stefan-Boltzmann constant σ, which is equal to 5.67×10^{-8} W m^{-2} K^{-4}. In many calculations, it is more convenient to relate everything to the Sun, which is a typical star. Specifically, for the Sun we have $L_\odot = 4\pi R_\odot^2 \sigma T_\odot^4$, where $L_\odot$ is the Sun's luminosity, $R_\odot$ is the Sun's radius, and $T_\odot$ is the Sun's surface temperature (equal to 5800 K). Dividing the general equation for L by this specific equation for the Sun, we obtain

$$\frac{L}{L_\odot} = \left(\frac{R}{R_\odot}\right)^2 \left(\frac{T}{T_\odot}\right)^4$$

This is an easier formula to use because the constant σ has cancelled out. We can also rearrange terms to arrive at a useful equation for the radius (R) of a star:

Radius of a star related to its luminosity and surface temperature

$$\frac{R}{R_\odot} = \left(\frac{T_\odot}{T}\right)^2 \sqrt{\frac{L}{L_\odot}}$$

$R/R_\odot$ = ratio of the star's radius to the Sun's radius

$T_\odot/T$ = ratio of the Sun's surface temperature to the star's surface temperature

$L/L_\odot$ = ratio of the star's luminosity to the Sun's luminosity

EXAMPLE: The bright reddish star Betelgeuse in the constellation Orion (see Figure 2-2 or Figure 6-30c) is 60,000 times more luminous than the Sun and has a surface temperature of 3500 K. What is its radius?

Situation: We are given the star's luminosity $L = 60{,}000$ $L_\odot$ and its surface temperature $T = 3500$ K. Our goal is to find the star's radius R.

Tools: We use the above equation to find the ratio of the star's radius to the radius of the Sun, $R/R_\odot$. Note that we also know the Sun's surface temperature, $T_\odot = 5800$ K.

Answer: Substituting these data into the above equation, we get

$$\frac{R}{R_\odot} = \left(\frac{5800 \text{ K}}{3500 \text{ K}}\right)^2 \sqrt{6 \times 10^4} = 670$$

Review: Our result tells us that Betelgeuse's radius is 670 times larger than that of the Sun. The Sun's radius is 6.96×10^5 km, so we can also express the radius of Betelgeuse as $(670)(6.96 \times 10^5 \text{ km}) = 4.7 \times 10^8$ km. This is more than 3 AU. If Betelgeuse were located at the center of our solar system, it would extend beyond the orbit of Mars!

EXAMPLE: Sirius, the brightest star in the sky, is actually two stars orbiting each other (a binary star). The less luminous star, Sirius B, is a white dwarf that is too dim to see with the naked eye. Its luminosity is 0.0025 $L_\odot$ and its surface temperature is 10,000 K. How large is Sirius B compared to the Earth?

Situation: Again we are asked to find a star's radius from its luminosity and surface temperature.

Tools: As in the preceding example, we use the equation given above.

Answer: The ratio of the radius of Sirius B to the Sun's radius is

$$\frac{R}{R_\odot} = \left(\frac{5800 \text{ K}}{10{,}000 \text{ K}}\right)^2 \sqrt{0.0025} = 0.017$$

Since the Sun's radius is $R_\odot = 6.96 \times 10^5$ km, the radius of Sirius B is $(0.017)(6.96 \times 10^5 \text{ km}) = 12{,}000$ km. From Table 7-1, the Earth's radius (half its diameter) is 6378 km. Hence, this star is only about twice the radius of the Earth.

Review: Sirius B's radius would be large for a terrestrial planet, but it is minuscule for a star. The name *dwarf* is well deserved!

The radii of some stars have been measured with other techniques (see Section 19-11). These other methods yield values consistent with those calculated by the methods we have just described.

figure 19-13
Finding Key Properties of a Nearby Star
This flowchart shows how astronomers determine the properties of a relatively nearby star (one close enough that its parallax can be measured). The rounded purple boxes show the measurements that must be made of the star, the blue ovals show the key equations that are used (from Sections 19-2, 19-5, and 19-6), and the green rectangles show the inferred properties of the stars. A different procedure is followed for more distant stars (see Section 19-8, especially Figure 19-16).

ANALOGY In a similar way, a roaring campfire can emit more light than a welder's torch. The campfire is at a lower temperature than the torch, but has a much larger surface area from which it emits light.

Using this general rule as shown in Box 19-4, astronomers find that stars come in a wide range of sizes. The smallest stars visible through ordinary telescopes, called *white dwarfs,* are about the same size as the Earth. Although their surface temperatures can be very high (25,000 K or more), white dwarfs have so little surface area that their luminosities are very low (less than $0.01 \ L_\odot$). The largest stars, called *supergiants,* are a thousand times larger in radius than the Sun (and 10^5 times larger than white dwarfs). If our own Sun were replaced by one of these supergiants, the Earth's orbit would lie completely inside the star!

Figure 19-13 summarizes how astronomers determine the distance from Earth, luminosity, surface temperature, chemical composition, and radius of a star close enough to us so that its parallax can be measured. Remarkably, all of these properties can be deduced from just a few measured quantities: the star's parallax angle, apparent brightness, and spectrum.

19-7 | Hertzsprung-Russell (H-R) diagrams reveal the different kinds of stars

Astronomers have collected a wealth of data about the stars, but merely having tables of numerical data is not enough. Like all scientists, astronomers want to analyze their data to look for trends

and underlying principles. One of the best ways to look for trends in any set of data, whether it comes from astronomy, finance, medicine, or meteorology, is to create a graph showing how one quantity depends on another. For example, investors consult graphs of stock market values versus dates, and weather forecasters make graphs of temperature versus altitude to determine whether thunderstorms will form. Astronomers have found that a particular graph of stellar properties shows that stars fall naturally into just a few categories. This graph, one of the most important in all astronomy, will in later chapters help us understand how stars form, evolve, and eventually die.

Which properties of stars should we include in a graph? Most stars have about the same chemical composition, but two properties of stars—their luminosities and surface temperatures—differ substantially from one star to another. Stars also come in a wide range of radii, but a star's radius is a secondary property that can be found from the luminosity and surface temperature (as we saw in Section 19-6 and Box 19-4). We also relegate the positions and space velocities of stars to secondary importance. (In a similar way, a physician is more interested in your weight and blood pressure than in where you live or how fast you drive.) We can then ask the following question: What do we learn when we graph the luminosities of stars versus their surface temperatures?

The first answer to this question was given in 1911 by the Danish astronomer Ejnar Hertzsprung. He pointed out that a regular pattern appears when the absolute magnitudes of stars (which measure their luminosities) are plotted against their colors (which measure their surface temperatures). Two years later, the American astronomer Henry Norris Russell independently discovered a similar regularity in a graph using spectral types (another measure of surface temperature) instead of colors. In recognition of their originators, graphs of this kind are today known as **Hertzsprung-Russell diagrams,** or **H-R diagrams** (Figure 19-14).

Figure 19-14*a* is a typical Hertzsprung-Russell diagram. Each dot represents a star whose spectral type and luminosity have been determined. The most luminous stars are near the top of the diagram, the least luminous stars near the bottom. Hot stars of spectral classes O and B are toward the left side of the graph and cool stars of spectral class M are toward the right.

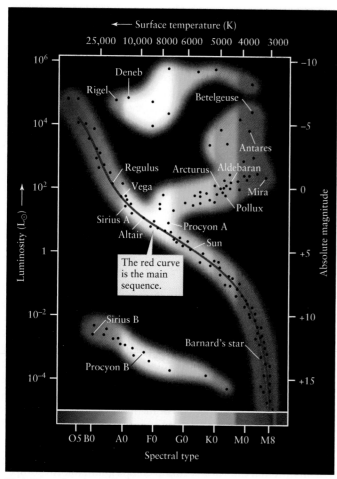

(a) A Hertzsprung-Russell (H-R) diagram

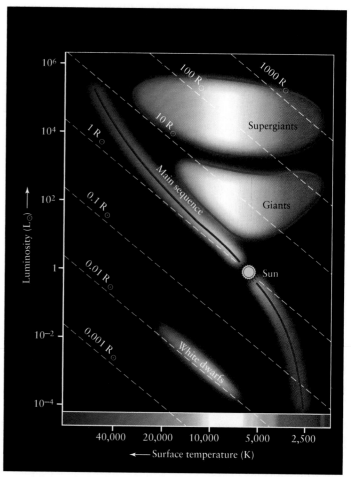

(b) The sizes of stars on an H-R diagram

figure 19-14

Hertzsprung-Russell (H-R) Diagrams On an H-R diagram, the luminosities (or absolute magnitudes) of stars are plotted against their spectral types (or surface temperatures). **(a)** The data points are grouped in just a few regions on the graph, showing that luminosity and spectral type are correlated. Most stars lie along the red curve called the main sequence. Giants like Arcturus as well as supergiants like Rigel and Betelgeuse are above the main sequence, and white dwarfs

like Sirius B are below it. **(b)** The blue curves on this H-R diagram enclose the regions of the diagram in which different types of stars are found. The dashed diagonal lines indicate different stellar radii. For a given stellar radius, as the surface temperature increases (that is, moving from right to left in the diagram), the star glows more intensely and the luminosity increases (that is, moving upward in the diagram). Note that the Sun is intermediate in luminosity, surface temperature, and radius.

CAUTION You are probably accustomed to graphs in which the numbers on the horizontal axis increase as you move to the right. (For example, the business section of a newspaper includes a graph of stock market values versus dates, with later dates to the right of earlier ones.) But on an H-R diagram the temperature scale on the horizontal axis increases toward the *left*. This practice stems from the original diagrams of Hertzsprung and Russell, who placed hot O stars on the left and cool M stars on the right. This arrangement is a tradition that no one has seriously tried to change.

The most striking feature of the H-R diagram is that the data points are not scattered randomly over the graph but are grouped in

a few distinct regions. The luminosities and surface temperatures of stars do *not* have random values; instead, these two quantities are related!

The band stretching diagonally across the H-R diagram includes about 90% of the stars in the night sky. This band, called the **main sequence,** extends from the hot, luminous, blue stars in the upper left corner of the diagram to the cool, dim, red stars in the lower right corner. A star whose properties place it in this region of an H-R diagram is called a **main-sequence star.** The Sun (spectral type G2, luminosity 1 $L_\odot$, absolute magnitude +4.8) is such a star. We will find that all main-sequence stars are like the Sun in that *hydrogen fusion*—thermonuclear reactions that convert hydrogen into helium (see Section 18-1)—is taking place in their cores.

The upper right side of the H-R diagram shows a second major grouping of data points. Stars represented by these points are both luminous and cool. From the Stefan-Boltzmann law, we know that a cool star radiates much less light per unit of surface area than a hot star. In order for these stars to be as luminous as they are, they must be huge (see Section 19-6), and so they are called **giants**. These stars are around 10 to 100 times larger than the Sun. You can see this in Figure 19-14b, which is an H-R diagram to which dashed lines have been added to represent stellar radii. Most giant stars are around 100 to 1000 times more luminous than the Sun and have surface temperatures of about 3000 to 6000 K. Cooler members of this class of stars (those with surface temperatures from about 3000 to 4000 K) are often called **red giants** because they appear reddish. The bright yellowish stars in the image that opens this chapter are red giants. (Notice how they outshine the blue stars in this image, which are at the same distance from Earth as the red giants but are smaller and less luminous.) A number of red giants can easily be seen with the naked eye, including Aldebaran in the constellation Taurus and Arcturus in Boötes.

A few rare stars are considerably bigger and brighter than typical red giants, with radii up to 1000 $R_\odot$. Appropriately enough, these superluminous stars are called **supergiants**. Betelgeuse in Orion (see Box 19-4) and Antares in Scorpius are two supergiants you can find in the nighttime sky. Together, giants and supergiants make up about 1% of the stars in the sky.

Both giants and supergiants have thermonuclear reactions occurring in their interiors, but the character of those reactions and where in the star they occur can be quite different than for a main-sequence star like the Sun. We will study these stars in more detail in Chapters 21 and 22.

The remaining 9% of stars form a distinct grouping of data points toward the lower left corner of the Hertzsprung-Russell diagram. Although these stars are hot, their luminosities are quite low; hence, they must be small. They are appropriately called **white dwarfs**. These stars, which are so dim that they can be seen only with a telescope, are approximately the same size as the Earth. As we will learn in Chapter 22, no thermonuclear reactions take place within white dwarf stars. Rather, like embers left from a fire, they are the still-glowing remnants of what were once giant stars.

By contrast, *brown* dwarfs (which lie at the extreme lower right of the main sequence, off the bottom and right-hand edge of Figure 19-14a or Figure 19-14b) are objects that will never become stars. They are comparable in radius to the planet Jupiter (that is, intermediate in size between the Earth and the Sun; see Figure 7-2). The study of brown dwarfs is still in its infancy, but it appears that there may be twice as many brown dwarfs as there are "real" stars.

 ANALOGY You can think of white dwarfs as "has-been" stars whose days of glory have passed. In this analogy, a brown dwarf is a "never-will-be."

The existence of fundamentally different types of stars is the first important lesson to come from the H-R diagram. In later chapters we will find that these different types represent various stages in the lives of stars. We will use the H-R diagram as an essential tool for understanding how stars evolve.

19-8 | Details of a star's spectrum reveal whether it is a giant, a white dwarf, or a main-sequence star

A star's surface temperature largely determines which lines are prominent in its spectrum. Therefore, classifying stars by spectral type is essentially the same as categorizing them by surface temperature. But, as the H-R diagram in Figure 19-14b shows, stars of the same surface temperature can have very different luminosities. As an example, a star with surface temperature 5800 K could be either a white dwarf, a main-sequence star, a giant, or a supergiant, depending on its luminosity. By examining the details of a star's spectrum, however, astronomers can determine to which of these categories a star belongs. This gives astronomers a tool to determine the distances to stars millions of parsecs away, far beyond the maximum distance that can be measured using stellar parallax.

Figure 19-15 compares the spectra of two stars of the same spectral type but different luminosity: a B8 supergiant and a B8 main-sequence star. Note that the Balmer lines of hydrogen are narrow in the spectrum of the very luminous supergiant but quite broad in the spectrum of the less luminous main-sequence star. In general, for stars of spectral types B through F, the more luminous the star, the narrower its hydrogen lines.

Fundamentally, these differences between stars of different luminosity are due to differences between the stars' atmospheres, where absorption lines are produced. Hydrogen lines in particular are affected by the density and pressure of the gas in a star's atmosphere. The higher the density and pressure, the more frequently hydrogen atoms collide and interact with other atoms and ions in the atmosphere. These collisions shift the energy levels in the hydrogen atoms and thus broaden the hydrogen spectral lines.

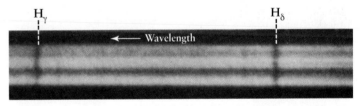

(a) A supergiant star has a low-density, low-pressure atmosphere: its spectrum has narrow absorption lines

(b) A main-sequence star has a denser, higher-pressure atmosphere: its spectrum has broad absorption lines

Figure 19-15   R I **V** U X G

How Luminosity Affects a Star's Spectrum These are the spectra of two stars of the same spectral type (B8) and surface temperature (13,400 K) but different luminosities: (a) the B8 supergiant Rigel (luminosity 58,000 $L_\odot$) in Orion, and (b) the B8 main-sequence star Algol (luminosity 100 $L_\odot$) in Perseus. (From W. W. Morgan, P. C. Keenan, and E. Kellman, *An Atlas of Stellar Spectra*)

In the atmosphere of a luminous giant star, the density and pressure are quite low because the star's mass is spread over a huge volume. Atoms and ions in the atmosphere are relatively far apart; hence, collisions between them are sufficiently infrequent that hydrogen atoms can produce narrow Balmer lines. A main-sequence star, however, is much more compact than a giant or supergiant. In the denser atmosphere of a main-sequence star, frequent interatomic collisions perturb the energy levels in the hydrogen atoms, thereby producing broader Balmer lines.

In the 1930s, W. W. Morgan and P. C. Keenan of the Yerkes Observatory of the University of Chicago developed a system of **luminosity classes** based upon the subtle differences in spectral lines. When these luminosity classes are plotted on an H-R diagram (Figure 19-16), they provide a useful subdivision of the star types in the upper right of the diagram. Luminosity classes Ia and Ib are composed of supergiants; luminosity class V includes all the main-sequence stars. The intermediate classes distinguish giant stars of various luminosities. Note that for stars of a given surface temperature (that is, a given spectral type), the *higher* the number of the luminosity class, the *lower* the star's luminosity.

As we will see in Chapters 21 and 22, different luminosity classes represent different stages in the evolution of a star. White dwarfs are not given a luminosity class of their own; as we mentioned in Section 19-7, they represent a final stage in stellar evolution in which no thermonuclear reactions take place.

Astronomers commonly use a shorthand description that combines a star's spectral type and its luminosity class. For example, the Sun is said to be a G2 V star. The spectral type indicates the star's surface temperature, and the luminosity class indicates its luminosity. Thus, an astronomer knows immediately that any G2 V star is a main-sequence star with a luminosity of about 1 $L_\odot$ and a surface temperature of about 5800 K. Similarly, a description of Aldebaran as a K5 III star tells an astronomer that it is a red giant with a luminosity of around 370 $L_\odot$ and a surface temperature of about 4000 K.

A star's spectral type and luminosity class, combined with the information on the H-R diagram, enable astronomers to estimate the star's distance from the Earth. As an example, consider the star Pleione in the constellation Taurus. Its spectrum reveals Pleione to be a B8 V star (a hot, blue, main-sequence star, like the one in Figure 19-15b). Using Figure 19-16, we can read off that such a star's luminosity is 190 $L_\odot$. Given the star's luminosity and its apparent brightness—in the case of Pleione, 3.9×10^{-13} of the apparent brightness of the Sun—we can use the inverse-square law to determine its distance from the Earth. The mathematical details are worked out in Box 19-2.

This method for determining distance, in which the luminosity of a star is found using spectroscopy, is called **spectroscopic parallax**. (The name is a bit misleading, because no parallax angle is involved. A better name for this method, although not the one used by astronomers, would be "spectroscopic distance determination.") **Figure 19-17** summarizes the method of spectroscopic parallax.

Spectroscopic parallax is an incredibly powerful technique. No matter how remote a star is, this technique allows astronomers to determine its distance, provided only that its spectrum and apparent brightness can be measured. Box 19-2 gives an example of how spectroscopic parallax has been used to find the distance to stars in other galaxies tens of millions of parsecs away. By contrast, we saw in Section 19-1 that "real" stellar parallaxes can be measured only for stars within a few hundred parsecs.

Unfortunately, spectroscopic parallax has its limitations; distances to individual stars determined using this method are only accurate to at best 10%. The reason is that the luminosity classes shown in Figure 19-16 are not thin lines on the H-R diagram but are moderately broad bands. Hence, even if a star's spectral type and luminosity class are known, there is still some uncertainty in the luminosity that we read off an H-R diagram. Nonetheless, spectroscopic parallax is often the only means that an astronomer has to estimate the distance to remote stars.

What has been left out of this discussion is *why* different stars have different spectral types and luminosities. One key factor, as we shall see, turns out to be the mass of the star.

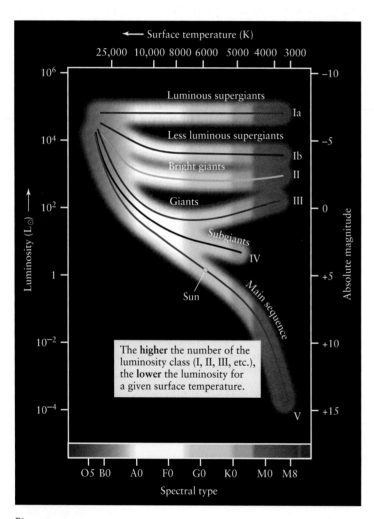

Figure 19-16

Luminosity Classes The H-R diagram is divided into regions corresponding to stars of different luminosity classes. (White dwarfs do not have their own luminosity class.) A star's spectrum reveals both its spectral type and its luminosity class; from these, the star's luminosity can be determined.

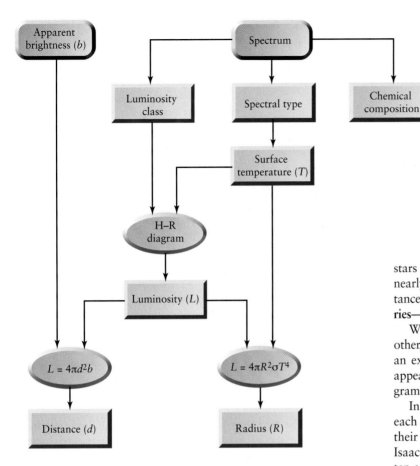

Figure 19-17

The Method of Spectroscopic Parallax If a star is too far away, its parallax angle is too small to allow a direct determination of its distance. This flowchart shows how astronomers deduce the properties of such a distant star. Note that the H-R diagram plays a central role in determining the star's luminosity from its spectral type and luminosity class. Just as for nearby stars (see Figure 19-13), the star's chemical composition is determined from its spectrum, and the star's radius is calculated from the luminosity and surface temperature.

stars are **optical double stars,** which are two stars that lie along nearly the same line of sight but are actually at very different distances from us. But many double stars are true **binary stars,** or **binaries**—pairs of stars that actually orbit each other (Figure 19-18).

When astronomers can actually see the two stars orbiting each other, a binary is called a **visual binary.** By observing the binary over an extended period, astronomers can plot the orbit that one star appears to describe around the other, as shown in the center diagram in Figure 19-18.

In fact, *both* stars in a binary system are in motion: They orbit each other because of their mutual gravitational attraction, and their orbital motions obey Kepler's third law as formulated by Isaac Newton (see Section 4-7 and Box 4-4). This law can be written as follows:

Kepler's third law for binary star systems

$$M_1 + M_2 = \frac{a^3}{P^2}$$

M_1, M_2 = masses of two stars in binary system, in solar masses

a = semimajor axis of one star's orbit around the other, in AU

P = orbital period, in years

Here a is the semimajor axis of the elliptical orbit that one star appears to describe around the other, plotted as in the center diagram in Figure 19-18. As this equation indicates, if we can measure this semimajor axis (a) and the orbital period (P), we can learn something about the masses of the two stars.

In principle, the orbital period of a visual binary is easy to determine. All you have to do is see how long it takes for the two stars to revolve once about each other. The two stars shown in Figure 19-18 are quite close, so the orbital period is only a few weeks. Many binary systems have much larger separations, however, and the period may be so long that more than one astronomer's lifetime is needed to complete the observations.

Determining the semimajor axis of an orbit can also be a challenge. The *angular* separation between the stars can be determined by observation. To convert this angle into a physical distance between the stars, we need to know the distance between the binary

19-9 | Binary star systems provide crucial information about stellar masses

We now know something about the sizes, temperatures, and luminosities of stars. To complete our picture of the physical properties of stars, we need to know their masses. In this section, we will see that stars come in a wide range of masses. We will also discover an important relationship between the mass and luminosity of main-sequence stars. This relationship is crucial to understanding why some main-sequence stars are hot and luminous, while others are cool and dim. It will also help us understand what happens to a star as it ages and evolves.

Determining the masses of stars is not trivial, however. The problem is that there is no practical, direct way to measure the mass of an isolated star. Fortunately for astronomers, about half of the visible stars in the night sky are not isolated individuals. Instead, they are multiple-star systems, in which two or more stars orbit each other. By carefully observing the motions of these stars, astronomers can glean important information about their masses.

A pair of stars located at nearly the same position in the night sky is called a **double star.** The Anglo-German astronomer William Herschel made the first organized search for such pairs. Between 1782 and 1821, he published three catalogs listing more than 800 double stars. Late in the nineteenth century, his son, John Herschel, discovered 10,000 more doubles. Some of these double

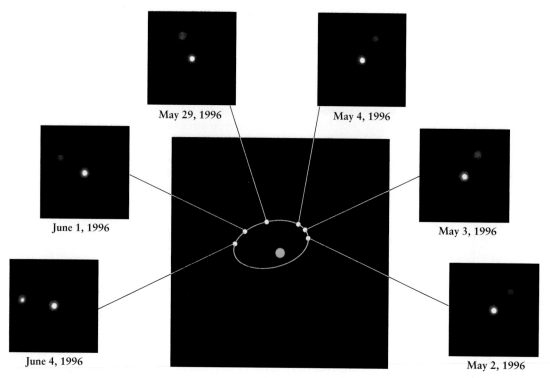

May 29, 1996

May 4, 1996

June 1, 1996

May 3, 1996

June 4, 1996

May 2, 1996

Figure 19-18 R I **V** U X G

A Binary Star System The two stars that make up the binary system called ζ (zeta) Ursae Majoris are separated by only about 0.01 arcsecond. The images surrounding the center diagram show the relative positions of the two stars over half of their orbital period. For simplicity, the diagram shows one star as remaining stationary; in reality, both stars move around their common center of mass. (Navy Prototype Optical Interferometer, Flagstaff, Arizona; Courtesy of Dr. Christian A. Hummel)

and the Earth. This can be found from parallax measurements or by using spectroscopic parallax. The astronomer must also take into account how the orbit is tilted to our line of sight.

Once both P and a have been determined, Kepler's third law can be used to calculate $M_1 + M_2$, the sum of the masses of the two stars in the binary system. But this analysis tells us nothing about the *individual* masses of the two stars. To obtain these, more information about the motions of the two stars is needed.

Each of the two stars in a binary system actually moves in an elliptical orbit about the **center of mass** of the system. Imagine two children sitting on opposite ends of a seesaw (Figure 19-19a). For the seesaw to balance properly, they must position themselves so that their center of mass—an imaginary point that lies along a line connecting their two bodies—is at the fulcrum, or pivot point of the seesaw. If the two children have the same mass, the center of mass lies midway between them, and they should sit equal dis-

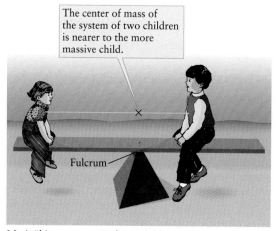

The center of mass of the system of two children is nearer to the more massive child.

Fulcrum

(a) A "binary system" of two children

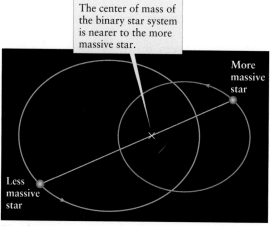

The center of mass of the binary star system is nearer to the more massive star.

More massive star

Less massive star

(b) A binary star system

ANIMATION 19.3

Figure 19-19
Center of Mass in a Binary Star System (a) A seesaw balances if the fulcrum is at the center of mass of the two children. (b) The members of a binary star system orbit around the center of mass of the two

stars. Although their elliptical orbits cross each other, the two stars are always on opposite sides of the center of mass and thus never collide.

The mass-luminosity relation shows that the main sequence on an H-R diagram is a progression in mass as well as in luminosity and surface temperature (Figure 19-21). The hot, bright, bluish stars in the upper left corner of an H-R diagram are the most massive main-sequence stars. Likewise, the dim, cool, reddish stars in the lower right corner of an H-R diagram are the least massive. Main-sequence stars of intermediate temperature and luminosity also have intermediate masses.

The mass of a main-sequence star also helps determine its radius. Referring back to Figure 19-14b, we see that if we go along the main sequence from low luminosity to high luminosity, the radius of the star increases. Thus we have the following general rule for main-sequence stars:

The greater the mass of a main-sequence star, the greater its luminosity, its surface temperature, and its radius.

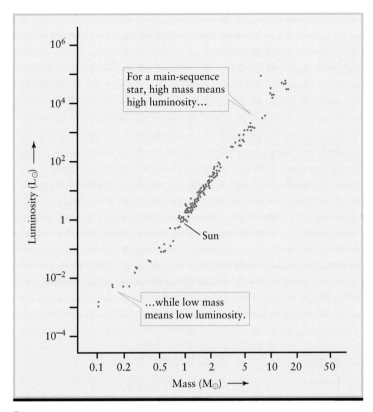

Figure 19-20

The Mass-Luminosity Relation For main-sequence stars, there is a direct correlation between mass and luminosity—the more massive a star, the more luminous it is. A main-sequence star of mass 10 $M_\odot$ (that is, 10 times the Sun's mass) has roughly 3000 times the Sun's luminosity (3000 $L_\odot$); one with 0.1 $M_\odot$ has a luminosity of only about 0.001 $L_\odot$.

tances from the fulcrum. If their masses are different, the center of mass is closer to the heavier child.

Just as the seesaw naturally balances at its center of mass, the two stars that make up a binary system naturally orbit around their center of mass (Figure 19-19b). The center of mass always lies along the line connecting the two stars and is closer to the more massive star.

The center of mass of a visual binary is located by plotting the separate orbits of the two stars, as in Figure 19-19b, using the background stars as reference points. The center of mass lies at the common focus of the two elliptical orbits. Comparing the relative sizes of the two orbits around the center of mass yields the ratio of the two stars' masses, M_1/M_2. The sum $M_1 + M_2$ is already known from Kepler's third law, so the individual masses of the two stars can then be determined.

Years of careful, patient observations of binaries have slowly yielded the masses of many stars. As the data accumulated, an important trend began to emerge: For main-sequence stars, there is a direct correlation between mass and luminosity. The more massive a main-sequence star, the more luminous it is. Figure 19-20 depicts this **mass-luminosity relation** as a graph. The range of stellar masses extends from less than 0.1 of a solar mass to more than 50 solar masses. The Sun's mass lies between these extremes.

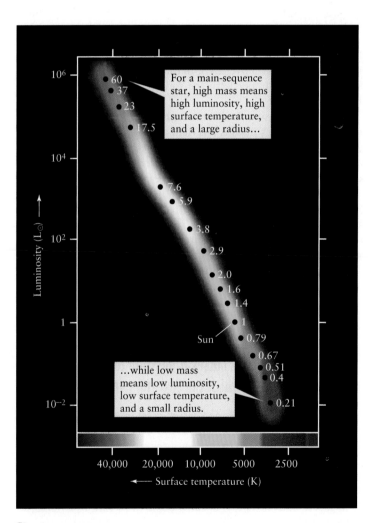

Figure 19-21

The Main Sequence and Masses On this H-R diagram, each dot represents a main-sequence star. The number next to each dot is the mass of that star in solar masses ($M_\odot$). As you move up the main sequence from the lower right to the upper left of the H-R diagram, the mass, luminosity, and surface temperature of main-sequence stars all increase.

Why is mass the controlling factor in determining the properties of a main-sequence star? The answer is that all main-sequence stars are objects like the Sun, with essentially the same chemical composition as the Sun but with different masses. Like the Sun, all main-sequence stars shine because thermonuclear reactions at their cores convert hydrogen to helium and release energy. The greater the total mass of the star, the greater the pressure and temperature at the core, the more rapidly thermonuclear reactions take place in the core, and the greater the energy output—that is, the luminosity—of the star. In other words, the greater the mass of a main-sequence star, the greater its luminosity. This statement is just the mass-luminosity relation, which we can now recognize as a natural consequence of the nature of main-sequence stars.

Like the Sun, main-sequence stars are in a state of both hydrostatic equilibrium and thermal equilibrium (see Section 18-2). Calculations using models of a main-sequence star's interior (like the solar models we discussed in Section 18-2) show that to maintain equilibrium, a more massive star must have a larger radius and a higher surface temperature. This is just what we see when we plot the curve of the main sequence on an H-R diagram (see Figure 19-14b). As you move up the main sequence from less massive stars (at the lower right in the H-R diagram) to more massive stars (at the upper left), the radius and surface temperature both increase.

Calculations using hydrostatic and thermal equilibrium also show that if a star's mass is less than about $0.08 M_{\odot}$, the core pressure and temperature are too low for thermonuclear reactions to take place. The "star" is then a brown dwarf. Brown dwarfs also obey a mass-luminosity relation: The greater the mass, the faster the brown dwarf contracts because of its own gravity, the more rapidly it radiates energy into space, and, hence, the more luminous the brown dwarf is.

There are no simple mass-luminosity relations for giant, supergiant, or white dwarf stars. Why these stars lie where they do on an H-R diagram will become apparent when we study the evolution of stars in Chapters 21 and 22. We will find that main-sequence stars evolve into giant and supergiant stars, and that some of these eventually end their lives as white dwarfs.

19-10 | Spectroscopy makes it possible to study binary systems in which the two stars are close together

We have described how the masses of stars can be determined from observations of visual binaries, in which the two stars can be distinguished from each other. But if the two stars in a binary system are too close together, the images of the two stars can blend to produce the semblance of a single star. Happily, in many cases we can use spectroscopy to decide whether a seemingly single star is in fact a binary system. Spectroscopic observations of binaries provide additional useful information about star masses.

Some binaries are discovered when the spectrum of a star shows incongruous spectral lines. For example, the spectrum of what appears to be a single star may include both strong hydrogen lines (characteristic of a type A star) and strong absorption bands of titanium oxide (typical of a type M star). Because a single star cannot have the differing physical properties of these two spectral types,

such a star must actually be a binary system that is too far away for us to resolve its individual stars. A binary system detected in this way is called a **spectrum binary**.

Other binary systems can be detected using the Doppler effect. If a star is moving toward the Earth, its spectral lines are displaced toward the short-wavelength (blue) end of the spectrum. Conversely, the spectral lines of a star moving away from us are shifted toward the long-wavelength (red) end of the spectrum. The upper portion of **Figure 19-22** applies these ideas to a hypothetical binary star system with an orbital plane that is edge-on to our line of sight.

As the two stars move around their orbits, they periodically approach and recede from us. Hence, the spectral lines of the two stars are alternately blueshifted and redshifted. The two stars in this hypothetical system are so close together that they appear through a telescope as a single star with a single spectrum. Because one star shows a blueshift while the other is showing a redshift, the spectral lines of the binary system appear to split apart and rejoin periodically. Stars whose binary character is revealed by such shifting spectral lines are called **spectroscopic binaries.**

To analyze a spectroscopic binary, astronomers measure the wavelength shift of each star's spectral lines and use the Doppler shift formula (introduced in Section 5-9 and Box 5-6) to determine the **radial velocity** of each star—that is, how fast and in what direction it is moving along our line of sight. The lower portion of Figure 19-22 shows a graph of the radial velocity versus time, called a **radial velocity curve,** for the binary system HD 171978. ("HD" refers to the *Henry Draper Catalogue*, mentioned in Section 19-5.) Each of the two stars alternately approaches and recedes as it orbits around the center of mass. The pattern of the curves repeats every 15 days, which is the orbital period of the binary.

Figure 19-23 shows two spectra of the spectroscopic binary κ (kappa) Arietis taken a few days apart. In Figure 19-23a, two sets of spectral lines are visible, offset slightly in opposite directions from the normal positions of these lines. This corresponds to stage 1 or stage 3 in Figure 19-22; one of the orbiting stars is moving toward the Earth and has its spectral lines blueshifted, and the other star is moving away from the Earth and has its lines redshifted. A few days later, the stars have progressed along their orbits so that neither star is moving toward or away from the Earth, corresponding to stage 2 or stage 4 in Figure 19-22. At this time there are no Doppler shifts, and the spectral lines of both stars are at the same positions. That is why only one set of spectral lines appears in Figure 19-23b.

CAUTION It is important to emphasize that the Doppler effect applies only to motion along the line of sight. Motion perpendicular to the line of sight does not affect the observed wavelengths of spectral lines. Hence, the ideal orientation for a spectroscopic binary is to have the stars orbit in a plane that is edge-on to our line of sight. (By contrast, a *visual* binary is best observed if the orbital plane is face-on to our line of sight.) For the Doppler shifts to be noticeable, the orbital speeds of the two stars should be at least a few kilometers per second.

The binaries depicted in Figures 19-22 and 19-23 are called *double-line* spectroscopic binaries, because the spectral lines of both stars in the binary system can be seen. Most spectroscopic binaries,

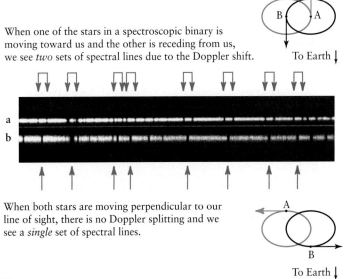

Figure 19-22 caption text to the right of the graph.

figure 19-22

Radial Velocity Curves The lower graph displays the radial velocity curves of the binary system HD 171978. The drawings at the top indicate the positions of the stars (labeled A and B) and the spectra of the binary at four selected moments (stages 1, 2, 3, and 4) during an orbital period. Note that at stages 1 and 3, the Doppler effect splits apart the absorption lines from stars A and B; compare with Figure 19-23. The entire binary star system is moving away from us at 12 km/s, which is why the entire pattern of radial velocity curves is displaced upward from the zero-velocity line.

however, are *single-line* spectroscopic binaries: One of the stars is so dim that its spectral lines cannot be detected. The star is obviously a binary, however, because its spectral lines shift back and forth, thereby revealing the orbital motions of two stars about their center of mass.

As for visual binaries, spectroscopic binaries allow astronomers to learn about stellar masses. From a radial velocity curve, one can find the *ratio* of the masses of the two stars in a binary. The *sum* of the masses is related to the orbital speeds of the two stars by Kepler's laws and Newtonian mechanics. If both the ratio of the masses and their sum are known, the individual masses can be determined using algebra. However, determining the sum of the masses requires that we know how the binary orbits are tilted from our line of sight. This is because the Doppler shifts reveal only the radial velocities of the stars rather than their true orbital speeds. This tilt is often impossible to determine, because we cannot see the individual stars in the binary. Thus, the masses of stars in spectroscopic binaries tend to be uncertain.

There is one important case in which we can determine the orbital tilt of a spectroscopic binary. If the two stars are observed to eclipse each other periodically, then we must be viewing the orbit nearly edge-on. As we will see next, individual stellar masses—as well as other useful data—can be determined if a spectroscopic binary also happens to be such an *eclipsing* binary.

When one of the stars in a spectroscopic binary is moving toward us and the other is receding from us, we see *two* sets of spectral lines due to the Doppler shift.

To Earth ↓

When both stars are moving perpendicular to our line of sight, there is no Doppler splitting and we see a *single* set of spectral lines.

To Earth ↓

figure 19-23   R I **V** U X G

A Spectroscopic Binary The visible-light spectrum of the double-line spectroscopic binary κ (kappa) Arietis has spectral lines that shift back and forth as the two stars revolve about each other. (Lick Observatory)

19-11 | Light curves of eclipsing binaries provide detailed information about the two stars

Some binary systems are oriented so that the two stars periodically eclipse each other as seen from the Earth. These **eclipsing binaries** can be detected even when the two stars cannot be resolved visually as two distinct images in the telescope. The apparent brightness of the image of the binary dims briefly each time one star blocks the light from the other.

Using a sensitive detector at the focus of a telescope, an astronomer can measure the incoming light intensity quite accurately and create a **light curve** (Figure 19-24). The shape of the light curve for an eclipsing binary reveals at a glance whether the eclipse is partial or total (compare Figures 19-24a and 19-24b). Figure 19-24d shows an observation of a binary system undergoing a total eclipse.

In fact, the light curve of an eclipsing binary can yield a surprising amount of information. For example, the ratio of the surface temperatures can be determined from how much their combined light is diminished when the stars eclipse each other. Also, the duration of a mutual eclipse tells astronomers about the relative sizes of the stars and their orbits.

If the eclipsing binary is also a double-line spectroscopic binary, an astronomer can calculate the mass and radius of each star from the light curves and the velocity curves. Unfortunately, very few binary stars are of this ideal type. Stellar radii determined in this way agree well with the values found using the Stefan-Boltzmann law, as described in Section 19-6.

The shape of a light curve can reveal many additional details about a binary system. In some binaries, for example, the gravitational pull of one star distorts the other, much as the Moon distorts the Earth's oceans in producing tides (see Figure 4-24). Figure 19-24c shows how such tidal distortion gives the light curve a different shape than in Figure 19-24b.

Information about stellar atmospheres can also be derived from light curves. Suppose that one star of a binary is a luminous main-sequence star and the other is a bloated red giant. By observing exactly how the light from the bright main-sequence star is gradu-

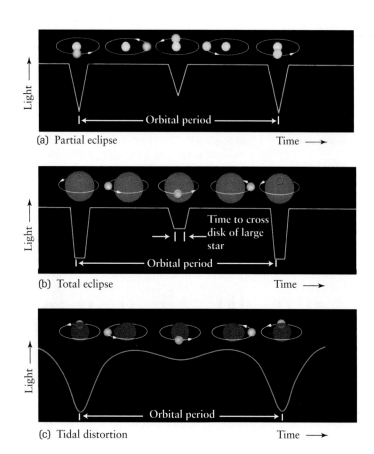

(a) Partial eclipse

(b) Total eclipse

(c) Tidal distortion

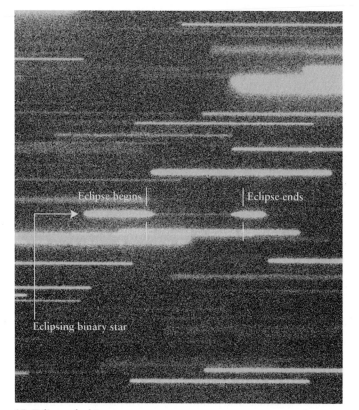

(d) Eclipse of a binary star

figure 19-24 R I **V** U X G

Representative Light Curves of Eclipsing Binaries (a), (b), (c) The shape of the light curve of an eclipsing binary can reveal many details about the two stars that make up the binary. (d) This image shows the binary star NN Serpens (indicated by the arrow) undergoing a total eclipse. The telescope was moved during the exposure so that the sky drifted slowly from left to right across the field of view. During the 10.5-minute duration of the eclipse, the dimmer star of the binary system (an M6 main-sequence star) passed in front of the other, more luminous star (a white dwarf). The binary became so dim that it almost disappeared. (European Southern Observatory) ally cut off as it moves behind the edge of the red giant during the beginning of an eclipse, astronomers can infer the pressure and density in the upper atmosphere of the red giant.

Binary systems are tremendously important because they enable astronomers to measure stellar masses as well as other key properties of stars. In the next several chapters, we will use this information to help us piece together the story of *stellar evolution*—how stars are born, evolve, and eventually die.

Key Words

Terms preceded by an asterisk (*) are discussed in the Boxes.

absolute magnitude, p. 417	OBAFGKM, p. 422
apparent brightness (brightness), p. 412	optical double star, p. 431
	parallax, p. 410
apparent magnitude, p. 416	parsec, p. 410
binary star (binary), p. 431	photometry, p. 414
brown dwarf, p. 424	*proper motion, p. 413
center of mass, p. 432	*radial velocity, pp. 413, 434
color ratio, p. 421	radial velocity curve, p. 434
*distance modulus, p. 419	red giant, p. 429
double star, p. 431	*space velocity, p. 413
eclipsing binary, p. 436	spectral classes, p. 422
giant, p. 429	spectral types, p. 422
Hertzsprung-Russell diagram (H-R diagram), p. 427	spectroscopic binary, p. 434
	spectroscopic parallax, p. 430
inverse-square law, p. 412	spectrum binary, p. 434
light curve, p. 436	stellar parallax, p. 410
luminosity, p. 410	supergiant, p. 429
luminosity class, p. 430	*tangential velocity, p. 413
luminosity function, p. 414	UBV photometry, p. 420
magnitude scale, p. 416	visual binary, p. 431
main sequence, p. 428	white dwarf, p. 429
main-sequence star, p. 428	
mass-luminosity relation, p. 433	
metals, p. 424	

Key Ideas

Measuring Distances to Nearby Stars: Distances to the nearer stars can be determined by parallax, the apparent shift of a star against the background stars observed as the Earth moves along its orbit.

• Parallax measurements made from orbit, above the blurring effects of the atmosphere, are much more accurate than those made with Earth-based telescopes.

• Stellar parallaxes can only be measured for stars within a few hundred parsecs.

The Inverse-Square Law: A star's luminosity (total light output), apparent brightness, and distance from the Earth are related by the inverse-square law. If any two of these quantities are known, the third can be calculated.

The Population of Stars: Stars of relatively low luminosity are more common than more luminous stars. Our own Sun is a rather average star of intermediate luminosity.

The Magnitude Scale: The apparent magnitude scale is an alternative way to measure a star's apparent brightness.

• The absolute magnitude of a star is the apparent magnitude it would have if viewed from a distance of 10 parsecs. A version of the inverse-square law relates a star's absolute magnitude, apparent magnitude, and distance.

Photometry and Color Ratios: Photometry measures the apparent brightness of a star. The color ratios of a star are the ratios of brightness values obtained through different standard filters, such as the U, B, and V filters. These ratios are a measure of the star's surface temperature.

Spectral Types: Stars are classified into spectral types (subdivisions of the spectral classes O, B, A, F, G, K, and M), based on the major patterns of spectral lines in their spectra. The spectral class and type of a star is directly related to its surface temperature: O stars are the hottest and M stars are the coolest.

• Most brown dwarfs are in even cooler spectral classes called L and T. Unlike true stars, brown dwarfs are too small to sustain thermonuclear fusion.

Hertzsprung-Russell Diagram: The Hertzsprung-Russell (H-R) diagram is a graph plotting the absolute magnitudes of stars against their spectral types—or, equivalently, their luminosities against surface temperatures.

• The positions on the H-R diagram of most stars are along the main sequence, a band that extends from high luminosity and high surface temperature to low luminosity and low surface temperature.

• On the H-R diagram, giant and supergiant stars lie above the main sequence, while white dwarfs are below the main sequence.

• By carefully examining a star's spectral lines, astronomers can determine whether that star is a main-sequence star, giant, supergiant, or white dwarf. Using the H-R diagram and the inverse-square law, the star's luminosity and distance can be found without measuring its stellar parallax.

Binary Stars: Binary stars, in which two stars are held in orbit around each other by their mutual gravitational attraction, are surprisingly common. Those that can be resolved into two distinct star images by an Earth-based telescope are called visual binaries.

• Each of the two stars in a binary system moves in an elliptical orbit about the center of mass of the system.

• Binary stars are important because they allow astronomers to determine the masses of the two stars in a binary system. The masses can be computed from measurements of the orbital period and orbital dimensions of the system.

Mass-Luminosity Relation for Main-Sequence Stars: Main-sequence stars are stars like the Sun but with different masses.

• The mass-luminosity relation expresses a direct correlation between mass and luminosity for main-sequence stars. The greater the mass of a main-sequence star, the greater its luminosity (and also the greater its radius and surface temperature).

Spectroscopic Observations of Binary Stars: Some binaries can be detected and analyzed, even though the system may be so distant or the two stars so close together that the two star images cannot be resolved.

• A spectrum binary appears to be a single star but has a spectrum with the absorption lines for two distinctly different spectral types.

• A spectroscopic binary has spectral lines that shift back and forth in wavelength. This is caused by the Doppler effect, as the orbits of the stars carry them first toward then away from the Earth.

• An eclipsing binary is a system whose orbits are viewed nearly edge-on from the Earth, so that one star periodically eclipses the other. Detailed information about the stars in an eclipsing binary can be obtained from a study of the binary's radial velocity curve and its light curve.

Review Questions

1. Explain the difference between a star's apparent brightness and its luminosity.

2. Describe how the parallax method of finding a star's distance is similar to binocular (two-eye) vision in humans.

3. Why are measurements of stellar parallax difficult to make? What are the advantages of making these measurements from orbit?

4. What is the inverse-square law? Use it to explain why an ordinary lightbulb can appear brighter than a star, even though the lightbulb emits far less light energy per second.

5. Briefly describe how you would determine the luminosity of a nearby star. Of what value is knowing the luminosity of various stars?

6. Which are more common, stars more luminous than the Sun or stars less luminous than the Sun?

7. Why is the magnitude scale called a "backward" scale? What is the difference between apparent magnitude and absolute magnitude?

8. The star Zubenelgenubi (from the Arabic for "scorpion's southern claw") has apparent magnitude +2.75, while the star Sulafat (Arabic for "tortoise") has apparent magnitude +3.25. Which star appears brighter? From this information alone, what can you conclude about the luminosities of these stars? Explain.

9. Explain why the color ratios of a star are related to the star's surface temperature.

10. Would it be possible for a star to appear bright when viewed through a U filter or a V filter, but dim when viewed through a B filter? Explain.

11. Which gives a more accurate measure of a star's surface temperature, its color ratios or its spectral lines? Explain.

12. Menkalinan (Arabic for "shoulder of the rein-holder") is an A2 star in the constellation Auriga (the Charioteer). What is its spectral class? What is its spectral type? Which gives a more precise description of the spectrum of Menkalinan?

13. A fellow student expresses the opinion that since the Sun's spectrum has only weak absorption lines of hydrogen, this element cannot be a major constituent of the Sun. How would you enlighten this person?

14. What are the most prominent absorption lines you would expect to find in the spectrum of a star with a surface temperature of (**a**) 35,000 K, (**b**) 2800 K, and (**c**) 5800 K (like the Sun)? Briefly describe why these stars have such different spectra even though they have essentially the same chemical composition.

15. If a red star and a blue star both have the same radius and both are the same distance from the Earth, which one looks brighter in the night sky? Explain why.

16. If a red star and a blue star both appear equally bright and both are the same distance from the Earth, which one has the larger radius? Explain why.

17. Sketch a Hertzsprung-Russell diagram. Indicate the regions on your diagram occupied by (**a**) main-sequence stars, (**b**) red giants, (**c**) supergiants, (**d**) white dwarfs, and (**e**) the Sun.

18. Most of the bright stars in the night sky (see Appendix 5) are giants and supergiants. How can this be, if giants and supergiants make up only 1% of the population of stars?

19. Explain why the dashed lines in Figure 19-14*b* slope down and to the right.

20. Some giant and supergiant stars are of the same spectral type (G2) as the Sun. What aspects of the spectrum of a G2 star would you concentrate on to determine the star's luminosity class? Explain what you would look for.

21. Briefly describe how you would determine the distance to a star whose parallax is too small to measure.

22. What information about stars do astronomers learn from binary systems that cannot be learned in any other way? What measurements do they make of binary systems to garner this information?

23. Suppose that you want to determine the temperature, diameter, and luminosity of an isolated star (not a member of a binary system). Which of these physical quantities require you to know the distance to the star? Explain.

24. What is the mass-luminosity relation? Does it apply to stars of all kinds?

25. Use Figure 19-20 to (**a**) estimate the mass of a main-sequence star that is 1000 times as luminous as the Sun, and (**b**) estimate the luminosity of a main-sequence star whose mass is one-fifth that of the Sun. Explain your answers.

26. Which is more massive, a red main-sequence star or a blue main-sequence star? Which has the greater radius? Explain your answers.

27. How do white dwarfs differ from brown dwarfs? Which are more massive? Which are larger in radius? Which are denser?

28. Sketch the radial velocity curves of a binary consisting of two identical stars moving in circular orbits that are (**a**) perpendicular to and (**b**) parallel to our line of sight.

29. Give two reasons why a visual binary star is unlikely to also be a spectroscopic binary star.

30. Sketch the light curve of an eclipsing binary consisting of two identical stars in highly elongated orbits oriented so that (**a**) their major axes are pointed toward the Earth and (**b**) their major axes are perpendicular to our line of sight.

Advanced Questions

Questions preceded by an asterisk (*) involve topics discussed in the Boxes.

> ### Problem-solving tips and tools
>
> Look carefully at the worked examples in Boxes 19-1, 19-2, 19-3, and 19-4 before attempting these exercises. For data on the solar system, see Table 7-1 or Appendices 1 and 2 at the back of this book. Remember that a telescope's light-gathering power is proportional to the area of its objective or primary mirror. The volume of a sphere of radius r is $4\pi r^3/3$. Make use of the H-R diagrams in this chapter to answer questions involving spectroscopic parallax. As Box 19-3 shows, some of the problems concerning magnitudes may require facility with logarithms.

31. Find the average distance from the Sun to Pluto in parsecs. Compared to Pluto, how many times farther away from the Sun is Proxima Centauri?

32. Suppose that a dim star were located 2 million AU from the Sun. Find (**a**) the distance to the star in parsecs and (**b**) the parallax angle of the star. Would this angle be measurable with present-day techniques?

33. The star GJ 1156 has a parallax angle of 0.153 arcsec. How far away is the star?

*34. Kapteyn's star (named after the Dutch astronomer who found it) has a parallax of 0.255 arcsec, a proper motion of 8.67 arcsec per year, and a radial velocity of +246 km/s. (**a**) What is the star's tangential velocity? (**b**) What is the star's actual speed relative to the Sun? (**c**) Is Kapteyn's star moving toward the Sun or away from the Sun? Explain.

*35. How far away is a star that has a proper motion of 0.08 arcseconds per year and a tangential velocity of 40 km/s? For a star at this distance, what would its tangential velocity have to be in order for it to exhibit the same proper motion as Barnard's star (see Box 19-1)?

*36. The space velocity of a certain star is 120 km/s and its radial velocity is 48 km/s. Find the star's tangential velocity.

*37. In the spectrum of a particular star, the Balmer line H_α has a wavelength of 656.41 nm. The laboratory value for the wavelength of H_α is 656.28 nm. (**a**) Find the star's radial velocity. (**b**) Is this star approaching us or moving away? Explain. (**c**) Find the wavelength at which you would expect to find H_α in the spectrum of this star, given that the laboratory wavelength of H_α is 486.13 nm. (**d**) Do your answers depend on the distance from the Sun to this star? Why or why not?

*38. Derive the equation given in Box 19-1 relating proper motion and tangential velocity. (*Hint:* See Box 1-1.)

39. How much dimmer does the Sun appear from Uranus than from Earth? (*Hint:* The average distance between a planet and the Sun equals the semimajor axis of the planet's orbit.)

40. Stars *A* and *B* are both equally bright as seen from Earth, but *A* is 12 pc away while *B* is 48 pc away. Which star has the greater luminosity? How many times greater is it?

41. Stars *C* and *D* both have the same luminosity, but *C* is 60 pc from Earth while *D* is 12 pc from Earth. Which star appears brighter as seen from Earth? How many times brighter is it?

42. Suppose two stars have the same apparent brightness, but one star is 9 times farther away than the other. What is the ratio of their luminosities? Which one is more luminous, the closer star or the farther star?

43. The *solar constant*, equal to 1370 W/m^2, is the amount of light energy from the Sun that falls on 1 square meter of the Earth's surface in 1 second (see Section 19-2). What would the distance between the Earth and the Sun have to be in order for the solar constant to be 1 watt per square meter (1 W/m^2)?

44. The star Procyon in Canis Minor (the Small Dog) is a prominent star in the winter sky, with an apparent brightness 1.3×10^{-11} that of the Sun. It is also one of the nearest stars, being only 3.50 parsecs from Earth. What is the luminosity of Procyon? Express your answer as a multiple of the Sun's luminosity.

45. The star HIP 92403 (also called Ross 154) is only 2.97 parsecs from Earth but can be seen only with a telescope, because it is 60 times dimmer than the dimmest star visible to the unaided eye. How close to us would this star have to be in order for it to be visible without a telescope? Give your answer in parsecs and in AU. Compare with the semimajor axis of Pluto's orbit around the Sun.

*46. The star HIP 72509 has an apparent magnitude of +12.1 and a parallax angle of 0.222 arcsecond. (**a**) Determine its absolute magnitude. (**b**) Find the approximate ratio of the luminosity of HIP 72509 to the Sun's luminosity.

*47. Suppose you can just barely see a twelfth-magnitude star through an amateur's 6-inch telescope. What is the magnitude of the dimmest star you could see through a 60-inch telescope?

*48. A certain type of variable star is known to have an average absolute magnitude of 0.0. Such stars are observed in a particular star cluster to have an average apparent magnitude of +14.0. What is the distance to that star cluster?

*49. (a) Find the absolute magnitudes of the brightest and dimmest of the labeled stars in Figure 19-6b. Assume that all of these stars are 110 pc from Earth. (b) If a star in the Pleiades cluster is just bright enough to be seen from Earth with the naked eye, what is its absolute magnitude? Is such a star more or less luminous than the Sun? Explain.

50. (a) On a copy of Figure 19-8, sketch the intensity curve for a blackbody at a temperature of 3000 K. Note that this figure shows a smaller wavelength range than Figure 19-7a. (b) Repeat part (a) for a blackbody at 12,000 K (see Figure 19-7c). (c) Use your sketches from parts (a) and (b) to explain why the color ratios b_V/b_B and b_B/b_U are less than 1 for very hot stars but greater than 1 for very cool stars.

*51. Astronomers usually express a star's color using apparent magnitudes. The star's apparent magnitude as viewed through a B filter is called m_B, and its apparent magnitude as viewed through a V filter is m_V. The difference $m_B - m_V$ is called the *B–V color index* ("B minus V"). Is the B–V color index positive or negative for very hot stars? What about very cool stars? Explain your answers.

*52. (See Question 51.) The B–V color index is related to the color ratio b_V/b_B by the equation

$$m_B - m_V = 2.5 \log \left(\frac{b_V}{b_B} \right)$$

(a) Explain why this equation is correct. (b) Use the data in Table 19-1 to calculate the B–V color indices for Bellatrix, the Sun, and Betelgeuse. From your results, describe a simple rule that relates the value of the B–V color index to a star's color.

53. The bright star Rigel in the constellation Orion has a surface temperature about 1.6 times that of the Sun. Its luminosity is about 64,000 $L_\odot$. What is Rigel's radius compared to the radius of the Sun?

54. The Sun's surface temperature is 5800 K. Using Figure 19-12, arrange the following absorption lines in the Sun's spectrum from the strongest to the weakest, and explain your reasoning: (i) neutral calcium; (ii) singly ionized calcium; (iii) neutral iron; (iv) singly ionized iron.

55. (See Figure 19-12.) What temperature and spectral classification would you give to a star with equal line strengths of hydrogen (H) and neutral helium (He I)? Explain.

56. Star P has double the radius of star Q. Stars P and Q have surface temperatures 4000 K and 8000 K, respectively. Which star has the greater luminosity? How many times greater is it?

57. Star X has 12 times the luminosity of star Y. Stars X and Y have surface temperatures 3500 K and 7800 K, respectively. Which star has the larger radius? How many times larger is it?

58. Suppose a star experiences an outburst in which its surface temperature doubles but its average density (its mass divided by its volume) decreases by a factor of 8. The mass of the star stays the same. By what factors do the star's radius and luminosity change?

59. The Sun experiences solar flares (see Section 18-10). The amount of energy radiated by even the strongest solar flare is not enough to have an appreciable effect on the Sun's luminosity. But when a flare of the same size occurs on a main-sequence star of spectral class M, the star's brightness can increase by as much as a factor of 2. Why should there be an appreciable increase in brightness for a main-sequence M star but not for the Sun?

60. The bright star Zubeneschmali (β Librae) is of spectral type B8 and has a luminosity of 130 $L_\odot$. What is the star's approximate surface temperature? How does its radius compare to that of the Sun?

61. Castor (α Geminorum) is an A1 V star with an apparent brightness of 4.4×10^{-12} that of the Sun. Determine the approximate distance from the Earth to Castor (in parsecs).

62. A brown dwarf called CoD–33°7795 B has a luminosity of 0.0025$L_\odot$. It has a relatively high surface temperature of 2550 K, which suggests that it is very young and has not yet had time to cool down by emitting radiation. (a) What is this brown dwarf's spectral class? (b) Find the radius of CoD–33°7795 B. Express your answer in terms of the Sun's radius and in kilometers. How does this compare to the radius of Jupiter? Is the name "dwarf" justified?

63. The visual binary 70 Ophiuchi (see the accompanying figure) has a period of 87.7 years. The parallax of 70 Ophiuchi is 0.2 arcsec, and the apparent length of the semimajor axis as seen through a telescope is 4.5 arcsec. (a) What is the distance to 70 Ophiuchi in parsecs? (b) What is the actual length of the semimajor axis in AU? (c) What is the sum of the masses of the two stars? Give your answer in solar masses.

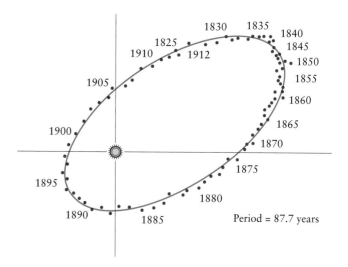

64. An astronomer observing a binary star finds that one of the stars orbits the other once every 5 years at a distance of 10 AU. (a) Find the sum of the masses of the two stars. (b) If the mass ratio of the system (M_1/M_2) is 4.0, find the individual masses of the stars. Give your answers in terms of the mass of the Sun.

Discussion Questions

*65. As seen from the starship *Enterprise* in the *Star Trek* television series and movies, stars appear to move across the sky due to the starship's motion. How fast would the *Enterprise* have to move in order for a star 1 pc away to appear to move 1° per second? (*Hint:* The speed of the star as seen from the *Enterprise* is the same as the speed of the *Enterprise* relative to the star.) How does this compare with the speed of light? Do you think the stars appear to move as seen from an orbiting space shuttle, which moves at about 8 km/s?

66. From its orbit around the Earth, the *Hipparcos* satellite could measure stellar parallax angles with acceptable accuracy only if the angles were larger than about 0.002 arcsec. Discuss the advantages or disadvantages of making parallax measurements from a satellite in a large solar orbit, say at the distance of Jupiter from the Sun. If this satellite can also measure parallax angles of 0.002 arcsec, what is the distance of the most remote stars that can be accurately determined? How much bigger a volume of space would be covered compared to the Earth-based observations? How many more stars would you expect to be contained in that volume?

67. It is desirable to be able to measure the radial velocity of stars (using the Doppler effect) to an accuracy of 1 km/s or better. One complication is that radial velocities refer to the motion of the star relative to the Sun, while the observations are made using a telescope on the Earth. Is it important to take into account the motion of the Earth around the Sun? Is it important to take into account the Earth's rotational motion? To answer this question, you will have to calculate the Earth's orbital speed and the speed of a point on the Earth's equator (the part of the Earth's surface that moves at the greatest speed because of the planet's rotation). If one or both of these effects are of importance, how do you suppose astronomers compensate for them?

Web/CD-ROM Questions

68. Search the World Wide Web for information about *Gaia,* a European Space Agency (ESA) spacecraft planned to extend the work carried out by *Hipparcos.* When is the spacecraft planned to be launched? How will *Gaia* compare to *Hipparcos?* For how many more stars will it be able to measure parallaxes? What other types of research will it carry out?

69. Search the World Wide Web for recent discoveries about brown dwarfs. Are all brown dwarfs found orbiting normal stars, as in Figure 8-14, or are they also found in isolation? The Sun experiences flares (see Section 18-10), as do other normal stars; is there any evidence that brown dwarfs also experience flares? If so, is there anything unusual about these flares?

70. Distances to Stars Using Parallax. Access the Active Integrated Media Module "Using Parallax to Determine Distance" in Chapter 19 of the *Universe* Web site or CD-ROM. Use this to determine the distance in parsecs and in light-years to each of the following stars: (**a**) Betelgeuse (parallax $p = 0.00763$ arcsecond); (**b**) Vega ($p = 0.129$ arcsecond); (**c**) Antares ($p = 0.00540$ arcsecond); (**d**) Sirius ($p = 0.379$ arcsecond).

Observing Projects

> **Observing tips and tools**
>
> Even through a telescope, the colors of stars are sometimes subtle and difficult to see. To give your eye the best chance of seeing color, use the "averted vision" trick: When looking through the telescope eyepiece, direct your vision a little to one side of the star you are most interested in. This places the light from that star on a more sensitive part of your eye's retina.

71. The accompanying table on this page lists five well-known red stars. It includes their right ascension and declination (celestial coordinates described in Box 2-1), apparent magnitudes, and color ratios. As their apparent magnitudes indicate, all these stars are somewhat variable. Observe at least two of these stars both by eye and through a small telescope. Is the reddish color of the stars readily apparent, especially in contrast to neighboring stars? (The Jesuit priest and astronomer Angelo Secchi named Y Canum Venaticorum "La Superba," and μ Cephei is often called William Herschel's "Garnet Star.")

Star	Right ascension	Declination	Apparent magnitude	b_V/b_B
Betelgeuse	$5^h 55.2^m$	+7° 24′	0.4–1.3	5.5
Y Canum Venaticorum	12 45.1	+45 26	5.5–6.0	10.4
Antares	16 29.4	−26 26	0.9–1.8	5.4
μ Cephei	21 43.5	+58 47	3.6–5.1	8.7
TX Piscium	23 46.4	+3 29	5.3–5.8	11.0

Note: The right ascensions and declinations are given for epoch 2000.

Star	Right ascension	Declination	Apparent magnitudes	Angular separation (arcseconds)	Spectral types
55 Piscium	0^h 39.9^m	+21° 26′	5.4 and 8.7	6.5″	K0 and F3
γ Andromedae	2 03.9	+42 20	2.3 and 4.8	9.8″	K3 and A0
32 Eridani	3 54.3	− 2 57	4.8 and 6.1	6.8″	G5 and A2
ι Cancri	8 46.7	+28 46	4.2 and 6.6	30.5″	G5 and A5
γ Leonis	10 20.0	+19 51	2.2 and 3.5	4.4″	K0 and G7
24 Coma Berenicis	12 35.1	+18 23	5.2 and 6.7	20.3″	K0 and A3
ν Boötis	14 45.0	+27 04	2.5 and 4.9	2.8″	K0 and A0
α Herculis	17 14.6	+14 23	3.5 and 5.4	4.7″	M5 and G5
59 Serpentis	18 27.2	+ 0 12	5.3 and 7.6	3.8″	G0 and A6
β Cygni	19 30.7	+27 58	3.1 and 5.1	34.3″	K3 and B8
δ Cephei	22 29.2	+58 25	4* and 7.5	20.4″	F5 and A0

Note: The right ascensions and declinations are given for epoch 2000.
The brighter star in the δ Cephei binary system is a variable star of approximately the fourth magnitude.

72. The above table of double stars includes vivid examples of contrasting star colors. The table lists the angular separation between the stars of each double. Observe at least four of these double stars through a telescope. Use the spectral types listed to estimate the difference in surface temperature of the stars in each pair you observe. Does the double with the greatest difference in temperature seem to present the greatest color contrast? From what you see through the telescope and on what you know about the H-R diagram, explain why all the cool stars (spectral types K and M) listed are probably giants or supergiants.

73. Observe the eclipsing binary Algol (β Persei), using nearby stars to judge its brightness during the course of an eclipse. Algol has an orbital period of 2.87 days, and, with the onset of primary eclipse, its apparent magnitude drops from 2.1 to 3.4. It remains this faint for about 2 hours. The entire eclipse, from start to finish, takes about 10 hours. Consult the "Celestial Calendar" section of the current issue of *Sky & Telescope* for the predicted dates and times of the minima of Algol. Note that the schedule is given in Universal Time (the same as Greenwich Mean Time), so you will have to convert the time to that of your own time zone. Algol is normally the second brightest star in the constellation of Perseus. Because of its position on the celestial sphere (R.A. = 3^h 08.2^m, Decl. = 40° 57′), Algol is readily visible from northern latitudes during the fall and winter months.

74. Use the *Starry Night Backyard*™ program to investigate the brightest stars. Select **Go Home** in the **View** menu to show the sky as seen from your location. (If the program doesn't place you at your true location, use the **Viewing Location...** command in the **View** menu.) In the Control Panel, set the date to today's date and the time to midnight (12:00:00 A.M.). Then select **Show View Options Panel** in the **View** menu, and under "Constellations" click the check boxes for "Boundaries" and "Labels." Then select **Hide View Options Panel** in the **View** menu. You will now see outlines of the constellations. (a) Scroll around the sky and identify at least five of the brighter stars by clicking the cursor on them. (Brighter stars are shown as larger dots.) Which stars did you select? In which constellation does each of these stars lie? Which of these stars are listed in Appendix 5? Of these, which is the most luminous? Which is the most distant? (b) In the Control Panel, set the date to six months from today, and again set the time to 12:00:00 A.M. Which of the stars that you selected in part (a) are visible? (You can use the **Find...** command in the **Edit** menu.) Which are not? Explain why the passage of six months should make a difference.

75. Use the *Deep Space Explorer*™ program to examine the nearby stars. In the left-hand part of the window under the heading **Milky Way,** click on the triangle next to the word **Explore** and then click on **Local Stars.** You can zoom in or out using the buttons at the upper left of the window (an upward-pointing triangle and a downward-pointing triangle). You can rotate the swarm of stars by putting the mouse cursor over the image, holding down the mouse button, and moving the mouse. (On a two-button mouse, hold down the left mouse button.) If you move the cursor over a star, you will see the star's apparent magnitude as seen from Earth and its distance from Earth. (a) Find at least 5 stars within 50 light-years of the Sun. Which of these stars are visible from Earth with the naked eye from a dark location? Which are visible with the naked eye from a brightly lit city? (Hint: The naked eye can see stars as faint as apparent magnitude $m = +6$ from a dark location, but only as faint as $m = +4$ from an inner city.) (b) Find at least 5 stars more than 500 light-years from the Sun. Which of these stars are visible from Earth with the naked eye from a dark location? Are the naked-eye stars more likely to be giants or supergiants, or are they more likely to be main-sequence stars? Explain your answer.

20 The Birth of Stars

The stars that illuminate our nights seem eternal and unchanging. But this permanence is an illusion. Each of the stars visible to the naked eye shines due to thermonuclear reactions and has only a finite amount of fuel available for these reactions. Hence, stars cannot last forever: They form from material in interstellar space, evolve over millions or billions of years, and eventually die. In this chapter our concern is with how stars are born and become part of the main sequence.

Stars form within cold, dark clouds of gas and dust that are scattered abundantly throughout our Galaxy. One such cloud appears as a dark area on the far right-hand side of the accompanying photograph. Perhaps a dark cloud like this encounters one of the Galaxy's spiral arms, or perhaps a supernova detonates nearby. From the shock of events like these, the cloud begins to contract under the pull of gravity, forming protostars—the fragments that will one day become stars. As a protostar develops, its internal pressure builds and its temperature rises. In time, hydrogen fusion begins, and a star is born. The hottest, bluest, and brightest young stars, like those in the accompanying image, emit ultraviolet radiation that excites the surrounding interstellar gas. The result is a beautiful glowing nebula, which typically has the red color characteristic of excited hydrogen (as shown in the photograph).

In Chapters 21 and 22 we will see how stars mature and grow old. Some even blow themselves apart in death throes that enrich interstellar space with the material for future generations of stars. Thus, like the mythical phoenix, new stars arise from the ashes of the old.

 A region of star formation about 1400 pc (4000 ly) from Earth in the southern constellation Ara (the Altar). (European Southern Observatory) R I **V** U X G

Questions to Guide Your Reading

As you read the sections of this chapter, look for the answers to the following questions:

20-1 Why do astronomers think that stars evolve?

20-2 What kind of matter exists in the spaces between the stars?

20-3 In what kind of nebulae do new stars form?

20-4 What steps are involved in forming a star like the Sun?

20-5 When a star forms, why does it end up with only a fraction of the available matter?

20-6 What do star clusters tell us about the formation of stars?

20-7 Where in the Galaxy does star formation take place?

20-8 How can the death of one star trigger the birth of many other stars?

20-1 | Understanding how stars evolve requires both observation and ideas from physics

Over the past several decades, astronomers have labored to develop an understanding of **stellar evolution:** that is, how stars are born, live their lives, and finally die. Our own Sun provides evidence that stars are not permanent. The energy radiated by the Sun comes from thermonuclear reactions in its core, which consume 6×10^{11} kg of hydrogen each second and convert it into helium (see Section 18-1). While the amount of hydrogen in the Sun's core is vast, it is not infinite; therefore, the Sun cannot always have been shining, nor can it continue to shine forever. The same is true for all other main-sequence stars, which are fundamentally the same kinds of objects as the Sun but with different masses (see Section 19-9). Thus, stars must have a beginning as well as an end.

Stars last very much longer than the lifetime of any astronomer—indeed, far longer than the entire history of human civilization. Thus, it is impossible to watch a single star go through its formation, evolution, and eventual demise. Rather, astronomers have to piece together the evolutionary history of stars by studying different stars at different stages in their life cycles.

 To see the magnitude of this task, imagine that you are a biologist from another planet who sets out to understand the life cycles of human beings. You send a spacecraft to fly above the Earth and photograph humans in action. Unfortunately, the spacecraft fails after collecting only 20 seconds of data, but during that time its sophisticated equipment sends back observations of thousands of different humans. From this brief snapshot of life on the Earth—only 10^{-8} (a hundred-millionth) of a typical human lifetime—how would you decide which were the young humans and which were the older ones? Without a look inside our bodies to see the biological processes that shape our lives, could you tell how humans are born and how they die? And how could you deduce the various biological changes that humans undergo as they age?

Astronomers, too, have data spanning only a tiny fraction of any star's lifetime. A star like the Sun can last for about 10^{10} years, whereas astronomers have been observing stars in detail for only about a century—as in our analogy, roughly 10^{-8} of the life span of a typical star. Astronomers are also frustrated by being unable to see the interiors of stars. For example, we cannot see the thermonuclear reactions that convert hydrogen into helium. But astronomers have an advantage over the biologist in our story. Unlike humans, stars are made of relatively simple substances, primarily hydrogen and helium, that are found almost exclusively in the form of gases. Of the three phases of matter—gas, liquid, and solid—gases are by far the simplest to understand.

Astronomers use our understanding of gases to build theoretical models of the interiors of stars, like the model of the Sun we saw in Section 18-2. Models help to complete the story of stellar evolution. In fact, like all great dramas, the story of stellar evolution can be regarded as a struggle between two opposing and unyielding forces: Gravity continually tries to make a star shrink, while the star's internal pressure tends to make the star expand. When these two opposing forces are in balance, the star is in a state of hydrostatic equilibrium (see Figure 18-3).

But what happens when changes within the star cause either pressure or gravity to predominate? The star must then either expand or contract until it reaches a new equilibrium. In the process, it will change not only in size but also in luminosity and color.

In the following chapters, we will find that giant and supergiant stars are the result of pressure gaining the upper hand over gravity. Both giants and supergiants turn out to be aging stars that have become tremendously luminous and ballooned to hundreds or thousands of times their previous size. White dwarfs, by contrast, are the result of the balance tipping in gravity's favor. These are even older stars that have collapsed to a fraction of the size they had while on the main sequence. In this chapter, however, we will see how the combat between gravity and pressure explains the birth of stars. We start our journey within the diffuse clouds of gas and dust that permeate our galaxy.

20-2 | Interstellar gas and dust pervade the Galaxy

Where do stars come from? As we saw in Section 8-4, our Sun condensed from a solar nebula, a collection of gas and dust in interstellar space. Observations suggest that other stars originate in a similar way (see Figure 8-5). To understand the formation of stars, we must first understand the substance between the stars from which they form.

At first glance, the space between the stars seems to be empty. On closer inspection, we find that it is filled with a thin gas laced with microscopic dust particles. This combination of gas and dust is called the **interstellar medium.** Evidence for this medium includes interstellar clouds of various types, curious lines in the spectra of binary star systems, and an apparent dimming and reddening of distant stars.

You can see evidence for the interstellar medium with the naked eye. Look carefully at the constellation Orion (**Figure 20-1a**), visible on winter nights in the northern hemisphere and summer nights in the southern hemisphere. While most of the stars in the constellation appear as sharply defined points of light, the middle "star" in Orion's sword has a fuzzy appearance. This becomes more obvious with binoculars or a telescope. As Figure 20-1b shows, this "star" is actually not a star at all, but the Orion Nebula—a cloud in interstellar space. Any interstellar cloud is called a **nebula** (plural **nebulae**) or **nebulosity.**

The Orion Nebula emits its own light, with the characteristic emission line spectrum of a hot, thin gas. For this reason it is called an **emission nebula.** Many emission nebulae can be seen with a small telescope. Figure 20-2 shows some of these nebulae in a different part of the constellation Orion. Emission nebulae are direct evidence of gas atoms in the interstellar medium.

Typical emission nebulae have masses that range from about 100 to about 10,000 solar masses. Because this mass is spread over a huge volume that is light-years across, the density is quite low by Earth standards, only a few thousand hydrogen atoms per cubic centimeter. (By comparison, the air you are breathing contains more than 10^{19} atoms per cm^3.)

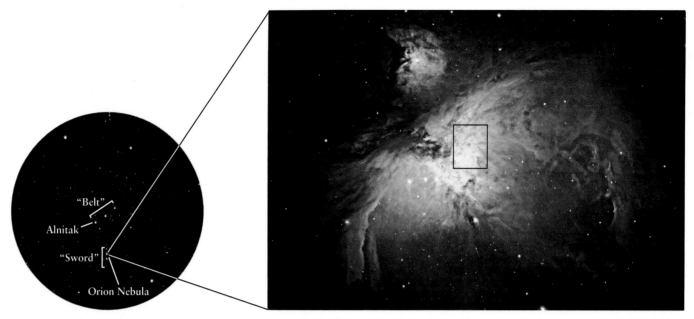

(a) A wide-angle view of Orion.   (b) A closeup of the Orion Nebula.

figure 20-1   R I **V** U X G

The Orion Nebula (a) The middle "star" of the three that make up Orion's sword is actually an interstellar cloud called the Orion Nebula. (b) The nebula is about 450 pc (1500 ly) from Earth and contains about 300 solar masses of material. Within the area shown by the box are four hot, massive stars called the Trapezium. They produce the ultraviolet light that makes the nebula glow. Figure 20-10a shows an even closer view of the area within the box. (a: R. C. Mitchell, Central Washington University; b: Anglo-Australian Observatory)

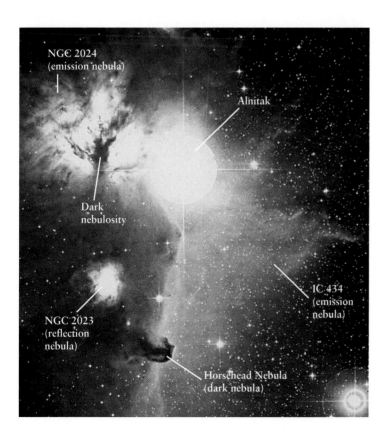

Emission nebulae are found near hot, luminous stars of spectral types O and B. Such stars emit copious amounts of ultraviolet radiation. When atoms in the nearby interstellar gas absorb these energetic ultraviolet photons, the atoms become ionized. Indeed, emission nebulae are composed primarily of ionized hydrogen atoms, that is, free protons (hydrogen nuclei) and electrons. Astronomers use the notation H I for neutral, un-ionized hydrogen atoms and H II for ionized hydrogen atoms, which is why emission nebulae are also called **H II regions.**

H II regions emit visible light when some of the free protons and electrons get back together to form hydrogen atoms, a process called **recombination.** When an atom forms by recombination, the electron is typically captured into a high-energy orbit. As the electron cascades downward through the atom's energy levels toward the ground state, the atom emits visible-light photons. Particularly

figure 20-2   R I **V** U X G

Emission, Reflection, and Dark Nebulae in Orion
A variety of different nebulae appear in the sky around Alnitak, the easternmost star in Orion's belt (see Figure 20-1a). All the nebulae lie approximately 500 pc (1600 ly) from Earth. They are actually nowhere near Alnitak, which is only 250 pc (820 ly) distant. This photograph shows an area of the sky about 1.5° across. (Royal Observatory, Edinburgh)

box 20-1   THE HEAVENS ON THE EARTH

Fluorescent Lights

The light that comes from glowing interstellar clouds is, quite literally, otherworldly. But the same principles that explain how such clouds emit light are also at the heart of light phenomena that we see here on the Earth.

A fluorescent lamp produces light in a manner not too different from an emission nebula (H II region). In both cases, the physical effect is called **fluorescence**: High-energy ultraviolet photons are absorbed, and the absorbed energy is reradiated as lower-energy photons of visible light.

Within the glass tube of a fluorescent lamp is a small amount of the element mercury. When you turn on the lamp, an electric current passes through the tube, vaporizing the mercury and exciting its atoms. This excited mercury vapor radiates light with an emission-line spectrum, including lines in the ultraviolet. The white fluorescent coating on the inside of the glass tube absorbs these ultraviolet photons, exciting the coating's molecules to high energy levels.

The molecules then cascade down through a number of lower levels before reaching the ground state. During this cascade, visible-light photons of many different wavelengths are emitted, giving an essentially continuous spectrum and a very white light. (By comparison, the hydrogen atoms in an H II region emit at only certain discrete wavelengths, because the spectrum of hydrogen is much simpler than that of the fluorescent coating's molecules. Another difference is that the molecules in the fluorescent tube never become ionized.)

Many common materials display fluorescence. Among them are teeth, fingernails, and certain minerals; when illuminated with ultraviolet light, these materials glow with a blue or green color. (Most natural history museums and science museums have an exhibit showing fluorescent minerals.) Laundry detergent also contains fluorescent material. After washing, your laundry absorbs ultraviolet light from the Sun and glows faintly, making it appear "whiter than white."

---

important is the transition from $n = 3$ to $n = 2$. It produces H_α photons with a wavelength of 656 nm, in the red portion of the visible spectrum (see Section 5-8, especially Figure 5-21). These photons give H II regions their distinctive reddish color.

For each high-energy, ultraviolet photon absorbed by a hydrogen atom to ionize it, several photons of lower energy are emitted when a proton and electron recombine. As **Box 20-1** describes, a similar effect takes place in a fluorescent light bulb. In this sense, H II regions are immense fluorescent light fixtures!

Further evidence of interstellar gas comes from the spectra of binary star systems. As the two stars that make up a binary system orbit their common center of mass, their spectral lines shift back and forth (see Figure 19-23). But certain calcium and sodium lines are found to remain at fixed wavelengths. These **stationary absorption lines** are therefore not associated with the binary star. Instead, they must be caused by interstellar gas between us and the binary system.

In addition to the presence of gas atoms in H II regions, Figure 20-2 also shows two kinds of evidence for larger bits of matter, called **dust grains,** in the interstellar medium. These dust grains make their appearance in *dark nebulae* and *reflection nebulae*.

A **dark nebula** is so opaque that it blocks any visible light coming from stars that lie behind it. One such dark nebula, called the Horsehead Nebula for its shape, protrudes in front of one of the H II regions in Figure 20-2. **Figure 20-3** shows another dark nebula, called Barnard 86. These nebulae have a relatively dense concentration of microscopic dust grains, which scatter and absorb light much more efficiently than single atoms. Typical dark nebulae have temperatures between 10 K to 100 K, which is low enough for hydrogen atoms to form molecules. Such nebulae contain from 10^4 to 10^9 particles (atoms, molecules, and dust grains) per cubic

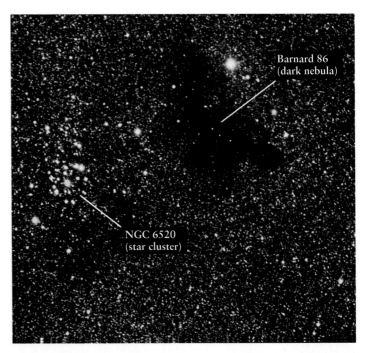

figure 20-3   R I **V** U X G

A Dark Nebula When first discovered in the late 1700s, dark nebulae were thought to be "holes in the heavens" where very few stars are present. In fact, they are opaque regions that block out light from the stars beyond them. The few stars that appear to be within Barnard 86 lie between us and the nebula. Barnard 86 is in the constellation Sagittarius and has an angular diameter of 4 arcminutes, about $1/7$ the angular diameter of the full moon. (Anglo-Australian Observatory) centimeter. Although tenuous by Earth standards, dark nebulae are large enough—typically many light-years deep—that they block the passage of light. In the same way, a sufficient depth of haze or smoke in our atmosphere can make it impossible to see distant mountains.

The other evidence for dust in Figure 20-2 is the bluish haze surrounding the star immediately above and to the left of the Horsehead Nebula. **Figure 20-4** shows a similar haze around a different set of stars. A haze of this kind, called a **reflection nebula,** is caused by fine grains of dust in a lower concentration than that found in dark nebulae. The light we see coming from the nebula is starlight that has been scattered and reflected by these dust grains. The grains are only about 500 nm across, no larger than a typical wavelength of visible light, and they scatter short-wavelength blue light more efficiently than long-wavelength red light. Hence, reflection nebulae have a characteristic blue color. Box 5-4 explains how a similar process—the scattering of sunlight in our atmosphere—gives rise to the blue color of the sky.

In the 1930s, the American astronomer Robert Trumpler discovered two other convincing pieces of evidence for the existence of interstellar matter—*interstellar extinction* and *interstellar reddening*. While studying the brightness and distances of certain star clusters, Trumpler noticed that remote clusters seem to be dimmer than would be expected from their distance alone. His observations demonstrated that the intensity of light from remote stars is reduced as the light passes through material in interstellar space. This process is called **interstellar extinction.** (In the same way, the headlights of oncoming cars appear dimmer when there is smoke or fog in the air.) The light from remote stars is also reddened as it passes through the interstellar medium, because the blue component of their starlight is scattered and absorbed by interstellar dust. This effect, shown in **Figure 20-5,** is called **interstellar reddening.** The same effect makes the setting Sun look red (see Box 5-4).

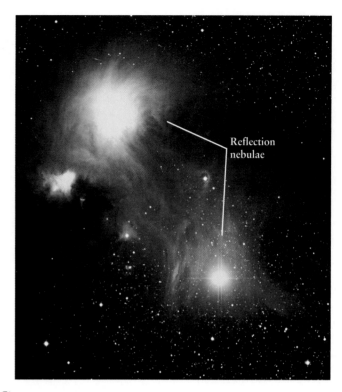

Figure 20-4   R I **V** U X G

Reflection Nebulae Wispy reflection nebulae called NGC 6726-27-29 surround several stars in the constellation Corona Australis (the Southern Crown). Unlike emission nebulae, reflection nebulae do not emit their own light, but scatter and reflect light from the stars that they surround. This scattered starlight is quite blue in color. The region shown here is about 23 arcminutes across. (David Malin/Anglo-Australian Observatory)

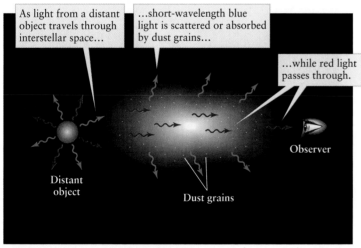

(a) How dust causes interstellar reddening

Figure 20-5   R I **V** U X G

Interstellar Reddening (a) Dust grains in interstellar space scatter or absorb blue light more than red light. Thus light from a distant object appears redder than it really is. (b) The emission nebulae NGC 3603 and NGC 3576 are different distances from Earth. Light from the more distant nebula must

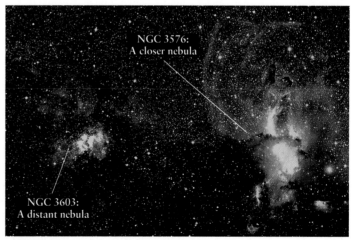

(b) Reddening depends on distance

pass through more interstellar dust to reach us, so more interstellar reddening occurs and NGC 3603 is a ruddier shade of red. The two nebulae are about 1° apart in the sky. (Anglo-Australian Observatory)

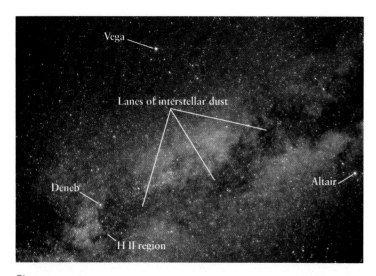

figure 20-6 R I **V** U X G

Gas and Dust in the Milky Way Glowing gas clouds (emission nebulae or H II regions) and dark, dusty regions are concentrated close to the midplane of the Milky Way Galaxy, of which our Sun is part. This wide-angle photograph also shows the three bright stars that make up the "summer triangle" (see Figure 2-8). (© Jerry Lodriguss/Photo Researchers).

Interstellar reddening is a subtle effect that can be detected only in telescopic images. If a star looks red as seen with the naked eye or through even a moderately large telescope, the cause is not the interstellar medium between that star and Earth. Instead, the red color means that star has a low surface temperature (see Figure 19-7). It is also important to understand the distinction between interstellar reddening and the Doppler effect. If an object is moving away from us, the Doppler effect *shifts* all of that object's light toward longer wavelengths (a redshift). Interstellar reddening, by contrast, makes objects appear red not by shifting wavelengths but by filtering out short wavelengths. The effect is the same as if we looked at an object through red-colored glasses, which let red light pass but block out blue light. The one similarity between Doppler shifts and interstellar reddening is that neither has any discernible effect on what you see with the naked eye.

Observations indicate that interstellar gas and dust are largely confined to the plane of the Milky Way—that faint, hazy band of stars you can see stretching across the sky on a dark, moonless night. **Figure 20-6** shows glowing emission nebulae and dark lanes of dust along the Milky Way.

As we will see in Chapter 25, the band of the Milky Way is actually our inside view of our Galaxy, which is a flat, disk-shaped collection of several hundred billion stars about 50,000 pc (160,000 ly) in diameter. The Sun is located about 8000 pc (26,000 ly) from our Galaxy's center. Astronomers know these dimensions because they can map our Galaxy from the locations of bright stars and nebulae and also by using radio telescopes. Observations indicate that bright stars, gas, and glowing nebulae

are mostly located within a few hundred parsecs of the midplane of our Galaxy, and that they lie along arching spiral arms that wind outward from the Galaxy's center. If we could view our Galaxy from a great distance, it would look somewhat like one of the galaxies shown in **Figure 20-7**.

Interstellar gas and dust are the raw material from which stars are made. The disk of our Galaxy, where most of this matter is concentrated, is therefore the site of ongoing star formation. Our next goal is to examine the steps by which stars are formed.

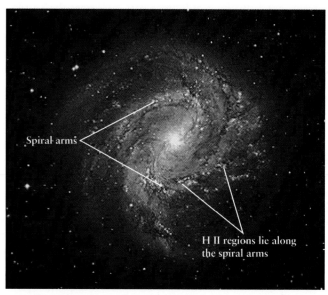

(a) We see spiral galaxy M83 nearly face-on

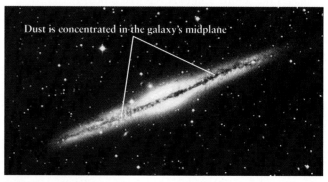

(b) We see spiral galaxy NGC 891 nearly edge-on

figure 20-7 R I **V** U X G

Spiral Galaxies Spiral galaxies, like our own Milky Way Galaxy, consist of stars, gas, and dust that are largely confined to a flattened, rotating disk. (a) This face-on view of a galaxy shows luminous stars and H II regions along the spiral arms. (b) This edge-on view of a different galaxy shows a dark band caused by dust in this galaxy's interstellar medium. Although in different parts of the sky, both M83 and NGC 891 are about 7 million pc (23 million ly) from Earth and have angular diameters of about 13 arcminutes. (a: David Malin/Anglo-Australian Observatory; b: Instituto de Astrofísica de Canarias/Royal Greenwich Observatory/David Malin)

20-3 | Protostars form in cold, dark nebulae

How do stars form in the interstellar medium? In this section, we discuss how a contracting cloud gives birth to a clump called a *protostar*—a future main-sequence star.

In order for interstellar material to condense and form a star, the force of gravity—which tends to draw interstellar material together—must overwhelm the internal pressure pushing the material apart. This means that stars will most easily form in regions where the interstellar material is relatively dense, so that atoms and dust grains are close together and gravitational attraction is enhanced.

To assist star formation, the pressure of the interstellar medium should be relatively low. This means that the interstellar medium should be as cold as possible, because the pressure of a gas goes down as the gas temperature decreases. (Conversely, increasing the temperature makes the pressure increase. That's why the air pressure inside automobile tires is higher after the auto has been driven a while and the tires have warmed up.)

The only parts of the interstellar medium with high enough density and low enough temperature for stars to form are the dark nebulae. Many of these, such as the Horsehead Nebula (see Figure 20-2) and Barnard 86 (see Figure 20-3), were discovered and catalogued around 1900 by Edward Barnard and are known as **Barnard objects.** (The Horsehead Nebula is also known as Barnard 33.) Other relatively small, nearly spherical dark nebulae are known as **Bok globules,** after the Dutch-American astronomer Bart Bok, who first called attention to them during the 1940s (**Figure 20-8**). A Bok globule resembles the inner core of a Barnard object with the outer, less dense portions stripped away. The density of the gas and dust within a Barnard object or Bok globule is indeed quite high by cosmic standards, in the range from 100 to 10,000 particles per cm^3; by comparison, most of the interstellar medium contains only 0.1 to 20 particles per cm^3. Radio emissions from molecules within these clouds indicate that their internal temperatures are very low, only about 10 K.

A typical Barnard object contains a few thousand solar masses of gas and dust spread over a volume roughly 10 pc (30 ly) across; a typical Bok globule is about one-tenth as large. The chemical composition of this material is the standard "cosmic abundance" of about 74% (by mass) hydrogen, 25% helium, and 1% heavier elements (review Figure 8-3). Within these clouds, the densest portions can contract under their own mutual gravitational attraction and form clumps called **protostars.** Each protostar will eventually evolve into a main-sequence star. Because dark nebulae contain many solar masses of material, it is possible for a large number of protostars to form out of a single such nebula. Thus, we can think of dark nebulae as "stellar nurseries."

In the 1950s, astrophysicists such as Louis Henyey in the United States and C. Hayashi in Japan performed calculations that enabled them to describe the earliest stages of a protostar. At first, a protostar is merely a cool blob of gas several times larger than our solar system. The pressure inside the protostar is too low to support all this cool gas, and so the protostar contracts. As it does, gravitational energy is converted into thermal energy, making the gases heat up and start glowing. (We discussed this process, called Kelvin-Helmholtz contraction, in Section 8-4 and Section 18-1.) Energy

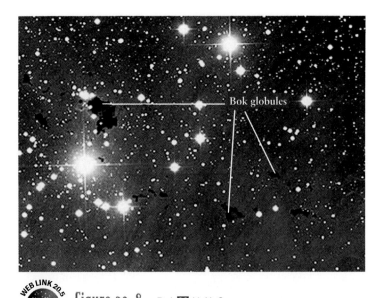

Figure 20-8   R I **V** U X G

Bok Globules   The black spots on this photograph of a glowing H II region are not imperfections in the image, but clouds of dust and gas called Bok globules. A typical Bok globule is a parsec or less in size and contains from one to a thousand solar masses of material. The stars in front of the H II region are part of the cluster IC 2944, which lies in the constellation Centaurus (the Centaur). The image shows an area about 8 × 10 arcminutes. (Anglo-Australian Observatory)

from the interior of the protostar is transported outward by convection, warming its surface.

After only a few thousand years of gravitational contraction, the protostar's surface temperature reaches 2000 to 3000 K. At this point the protostar is still quite large, so its glowing gases produce substantial luminosity. (The greater a star's radius and surface temperature, the greater its luminosity. You can review the details in Section 19-6 and Box 19-4.) For example, after only 1000 years of contraction, a protostar of 1 solar mass is 20 times larger in radius than the Sun and has 100 times the Sun's luminosity. Unlike the Sun, however, the luminosity of a young protostar is not the result of thermonuclear fusion, because the protostar's core is not yet hot enough for fusion reactions to begin. Instead, the radiated energy comes exclusively from the heating of the protostar as it contracts.

In order to determine the conditions inside a contracting protostar, astrophysicists use computers to solve equations similar to those used for calculating the structure of the Sun (see Section 18-2). The results tell how the protostar's luminosity and surface temperature change at various stages in its contraction. This information, when plotted on a Hertzsprung-Russell diagram, provides a protostar's **evolutionary track.** The track shows us how the protostar's appearance changes because of changes in its interior. These theoretical tracks agree quite well with actual observations of protostars.

CAUTION   When astronomers refer to a star "following an evolutionary track" or "moving on the H-R diagram," they mean that the star's luminosity, surface temperature, or both

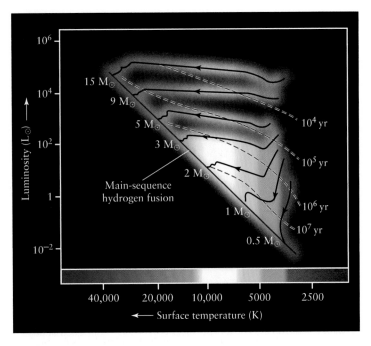

figure 20-9

Pre–Main-Sequence Evolutionary Tracks As a protostar evolves, its luminosity and surface temperature both change. The tracks shown here depict these changes for protostars of seven different masses. Each dashed red line shows the age of a protostar when its evolutionary track crosses that line. We will see in Section 20-5 that protostars lose quite a bit of mass as they evolve: The mass shown for each track is the value when the protostar finally settles down as a main-sequence star.

change. Hence, the point that represents the star on an H-R diagram changes its position. This is completely unrelated to how the star physically moves through space!

Figure 20-9 shows evolutionary tracks for protostars of seven different masses, ranging from 0.5 $M_\odot$ to 15 $M_\odot$. Because protostars are relatively cool when they begin to shine at visible wavelengths, these tracks all begin near the right (low-temperature) side of the H-R diagram. The subsequent evolution is somewhat different, however, depending on the protostar's mass.

Observing the evolution of a protostar can be quite a challenge. Indeed, while Figure 20-9 shows that young protostars are much more luminous than the Sun, it is quite unlikely that you have ever seen a protostar shining in the night sky. The reason is that protostars form within clouds that contain substantial amounts of interstellar dust. The dust in a protostar's immediate surroundings, called its **cocoon nebula**, absorbs the vast amounts of visible light emitted by the protostar and makes it very hard to detect using visible wavelengths.

 Protostars can be seen, however, using infrared wavelengths. Because it absorbs so much energy from the protostar that it surrounds, a dusty cocoon nebula becomes heated to a few hundred kelvins. The warmed dust then

reradiates its thermal energy at infrared wavelengths, to which the cocoon nebula is relatively transparent. So, by using infrared telescopes, astronomers can see protostars within the "stellar nursery" of a dark nebula.

Figure 20-10 shows visible-light and infrared views of a stellar nursery within the Orion Nebula. The visible-light view (Figure 20-10a) shows glowing gas clouds that lie in front of a dark, dusty nebula that almost completely blocks visible light. The infrared image (Figure 20-10b) allows us to see through the dust, revealing a number of newly formed protostars within the dark nebula. The properties of these protostars agree well with the theoretical ideas outlined in this section.

figure 20-10

Newborn Stars in the Orion Nebula (a) This visible-light view shows the inner regions of the Orion Nebula (the boxed area in Figure 20-1b). (b) This false-color infrared image of the outlined area in (a) reveals stars hidden by dust. The diagonal dimension of the image is about 0.1 parsec (0.4 light-year, or 20,000 AU). Yellow-orange denotes emission from young stars and dust; blue denotes emission from excited hydrogen molecules. Several curved arcs are thought to be shock waves caused by material flowing out of protostars faster than the speed of sound waves in the nebula. These shock waves can trigger the formation of even more stars (see Section 20-7). (a: C. R. O'Dell, S. K. Wong, and NASA; b: R. Thompson, M. Rieke, G. Schneider, S. Stolovy, E. Erickson, D. Axon, and NASA)

20-4 | Protostars evolve into main-sequence stars

Evolutionary tracks show how a protostar matures into a star as its gases contract. The details of this evolution depend on the star's mass. To follow these details, remember the basic principle that luminosity is proportional to the square of radius and to the fourth power of surface temperature (see Section 19-6).

For a protostar with the same mass as our Sun (1 $M_\odot$), the outer layers are cool and quite opaque (for the same reasons that the Sun's photosphere is opaque; see Section 18-5). This means that energy released from the shrinking inner layers in the form of radiation cannot reach the surface. Instead, energy flows outward by the slower and less effective method of convection. Hence, the surface temperature of the contracting protostar stays roughly constant, the luminosity decreases as the radius decreases, and the evolutionary track moves downward on the H-R diagram in Figure 20-9.

Although its surface temperature changes relatively little, the internal temperature of the shrinking protostar increases. After a time, the interior becomes ionized, which makes it less opaque. Energy is then conveyed outward by radiation in the interior and by convection in the opaque outer layers, just as in the present-day Sun (see Section 18-2, especially Figure 18-5). This makes it easier for energy to escape from the protostar, so the luminosity increases. As a result, the evolutionary track bends upward (higher luminosity) and to the left (higher surface temperature, caused by the increased energy flow).

In time, the 1-$M_\odot$ protostar's interior temperature reaches a few million kelvins, hot enough for thermonuclear reactions to begin converting hydrogen into helium. As we saw in Section 18-1 and Box 18-1, these reactions release enormous amounts of energy.

Eventually, these reactions provide enough heat and internal pressure to stop the star's gravitational contraction, and hydrostatic equilibrium is reached. The protostar's evolutionary track has now led it to the main sequence, and the protostar has become a full-fledged main-sequence star.

More massive protostars evolve a bit differently. If its mass is more than about 4 $M_\odot$, a protostar contracts and heats more rapidly, and hydrogen fusion begins quite early. As a result, the luminosity quickly stabilizes at nearly its final value, while the surface temperature continues to increase as the star shrinks. Thus, the evolutionary tracks of massive protostars traverse the H-R diagram roughly horizontally (signifying approximately constant luminosity) in the direction from right to left (from low to high surface temperature). You can see this most easily for the 9-$M_\odot$ and 15-$M_\odot$ evolutionary tracks in Figure 20-9.

Greater mass means greater pressure and temperature in the interior, which means that a massive star has an even larger temperature difference between its core and its outer layers than the Sun. This causes convection deep in the interior of a massive star (Figure 20-11). By contrast, a massive star's outer layers are of such low density that energy flows through them more easily by radiation than by convection. Therefore, main-sequence stars with masses more than about 4 $M_\odot$ have convective interiors but radiative outer layers (Figure 20-11a). By contrast, less massive main-sequence stars such as the Sun have radiative interiors and convective outer layers (Figure 20-11b).

The internal structure is also different for main-sequence stars of very low mass. As such a star forms from a protostar, the interior temperature is never high enough to fully ionize the interior. The interior remains too opaque for radiation to flow efficiently, so energy is transported by convection throughout the volume of the star (Figure 20-11c).

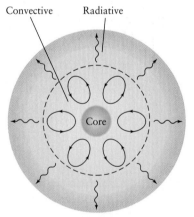

(a) Mass more than about 4 $M_\odot$: Energy flows by convection in the inner regions and by radiation in the outer regions.

(b) Mass between about 4 $M_\odot$ and 0.8 $M_\odot$: Energy flows by radiation in the inner regions and by convection in the outer regions.

(c) Mass less than 0.8 $M_\odot$: Energy flows by convection throughout the star's interior.

Figure 20-11
Main-Sequence Stars of Different Masses Stellar models show that when a protostar evolves into a main-sequence star, its internal structure depends on its mass. *Note:* The three stars shown here are *not* drawn to scale. Compared with a 1-$M_\odot$ main-sequence star like that shown in (b), a 6-$M_\odot$ main-sequence star like that in (a) has more than 4 times the radius, and a 0.2-$M_\odot$ main-sequence star like that in (c) has only one-third the radius.

All the protostar evolutionary tracks in Figure 20-9 end on the main sequence. The main sequence therefore represents stars in which thermonuclear reactions are converting hydrogen into helium. For most stars, this is a stable situation. For example, our Sun will remain on or very near the main sequence, quietly fusing hydrogen into helium at its core, for a total of some 10^{10} years. Each evolutionary track ends at a point along the main sequence that agrees with the mass-luminosity relation (recall Figure 19-21 and Figure 19-22). The most massive stars are the most luminous, while the least massive stars are the least luminous.

The theory of how protostars evolve helps explain why the main sequence has both an upper mass limit and a lower mass limit. As we saw in Section 19-5, protostars less massive than about 0.08 $M_\odot$ can never develop the necessary pressure and temperature to start hydrogen fusion in their cores. Instead, such "failed stars" end up as *brown dwarfs*, which shine faintly by Kelvin-Helmholtz contraction (see Figure 8-14).

Protostars with masses greater than about 200 solar masses also do not become main-sequence stars. Such a protostar rapidly becomes very luminous, resulting in tremendous internal pressures. This pressure is so great that it overwhelms the effects of gravity, expelling the outer layers into space and disrupting the star. Main-sequence stars therefore have masses between about 0.08 and 200 $M_\odot$, although the high-mass stars are extremely rare.

CAUTION! Two words of caution are in order here. First, while the evolutionary tracks of protostars begin in the red giant region of the H-R diagram (the upper right), protostars are *not* red giants. As we will see in Chapter 21, red giant stars represent a stage in the evolution of stars that comes *after* being a main-sequence star. Second, it is worth remembering that stars live out most of their lives on the main sequence, after only a relatively brief period as protostars. A 15-$M_\odot$ protostar takes only 20,000 years to become a main-sequence star, and a 1-$M_\odot$ protostar takes about 2×10^7 years. By contrast, the Sun has been a main-sequence star for about 4.56×10^9 years. By astronomical standards, pre–main-sequence stars are quite transitory.

20-5 | During the birth process, stars both gain and lose mass

After reading the previous section, you may think that a main-sequence star forms simply by collapsing inward. In fact, much of the material of a cold, dark nebula is ejected into space and never incorporated into stars. As it is ejected, this material may help sweep away the dust surrounding a young star, making the star observable at visible wavelengths.

Mass ejection into space is a hallmark of **T Tauri stars**. These are protostars with emission lines as well as absorption lines in their spectra and whose luminosity can change irregularly on time scales of a few days. T Tauri stars have masses less than about 3 $M_\odot$ and ages around 10^6 years, so on an H-R diagram such as Figure 20-9 they appear above the right-hand end of the main sequence. The emission lines show that these protostars are surrounded by a thin, hot gas. The Doppler shifts of these emission lines suggest that the protostars eject gas at speeds around 80 km/s (300,000 km/h, or 180,000 mi/h).

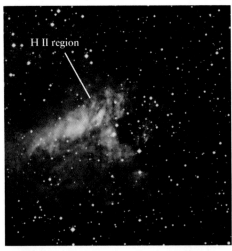

(a) Visible-light image   R I **V** U X G

H II region

(b) False-color infrared image   R **I** V U X G

Young stars lie within the dark nebula around the H II region.

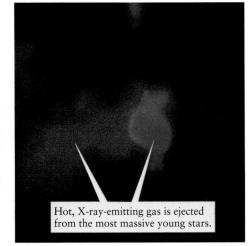

(c) False-color X-ray image   R I V U **X** G

Hot, X-ray-emitting gas is ejected from the most massive young stars.

WEB LINK 20.8

Figure 20-12

Mass Loss from Young, Massive Stars (a) The Omega Nebula, also known as M17, is a region of star formation in the constellation Sagittarius about 1700 pc (5500 ly) from Earth. (b) This infrared image allows us to see through dust, revealing recently formed stars that cannot be seen in (a). (c) The most massive young stars eject copious amounts of hot gas. Red indicates X-ray emission from gas at a temperature of 1.5×10^6 K; blue indicates even hotter gas at a temperature of 7×10^6 K. Astronomers do not see such X-ray emission from the Orion nebula (Figure 20-1), which has many young stars but very few massive ones. (a: Palomar Observatory DSS; b: 2MASS/UMass/IPAC-Caltech/NASA/NSF; c: NASA/CXC/PSU/L. Townsley et al.)

On average, T Tauri stars eject about 10^{-8} to 10^{-7} solar masses of material per year. This may seem like a small amount, but by comparison the present-day Sun loses only about 10^{-14} $M_\odot$ per year. The T Tauri phase of a protostar may last 10^7 years or so, during which time the protostar may eject roughly a solar mass of material. Thus, the mass of the final main-sequence star is quite a bit less than that of the cloud of gas and dust from which the star originated. (The stellar masses shown in Figure 20-9 are those of the final, main-sequence stars.)

Young stars that are more massive than about 3 $M_\odot$ do not vary in luminosity like T Tauri stars. They do lose mass, however, because the pressure of radiation at their surfaces is so strong that it blows gas into space. One place this can be seen is in the Omega Nebula (**Figure 20-12a**), where new stars are being formed (Figure 20-12b). The most massive of these stars eject gases of such high temperature that they emit X rays (Figure 20-12c).

 In the early 1980s, it was discovered that many young stars, including T Tauri stars, also lose mass by ejecting gas along two narrow, oppositely directed jets—a phenomenon called **bipolar outflow.** As this material is ejected into space at speeds of several hundred kilometers per second, it collides with the surrounding interstellar medium and produces knots of hot, ionized gas that glow with an emission-line spectrum. These glowing knots are called **Herbig-Haro objects** after the two astronomers, George Herbig in the United States and Guillermo Haro in Mexico, who discovered them independently.

Figure 20-13 is a Hubble Space Telescope image of the Herbig-Haro objects HH 1 and HH 2, which are produced by the two jets from a single young star in the constellation Orion. Herbig-Haro objects like these change noticeably in position, size, shape, and brightness from year to year, indicating the dynamic character of bipolar outflows.

Observations suggest that most protostars eject material in the form of jets at some point during their evolution. These bipolar outflows are very short-lived by astronomical standards, a mere 10^4 to 10^5 years, but they are so energetic that they typically eject into space more mass than ends up in the final protostar.

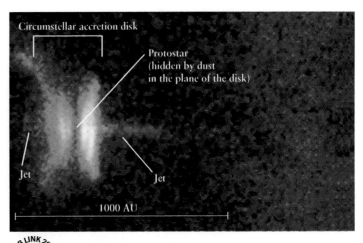

Figure 20-14 R I **V** U X G
A Circumstellar Accretion Disk and Jets This false-color image shows a star surrounded by an accretion disk, which we see nearly edge-on. Red denotes emission from ionized gas, while green denotes starlight scattered from dust particles in the disk. The midplane of the accretion disk is so dusty and opaque that it appears dark. Two oppositely directed jets flow away from the star, perpendicular to the disk and along the disk's rotation axis. This star lies 140 pc (460 ly) from Earth. (C. Burrows, the WFPC-2 Investigation Definition Team, and NASA)

Protostars slowly add mass to themselves at the same time that they rapidly eject it into space. In fact, the two processes are related. As a protostar's nebula contracts, it spins faster and flattens into a disk with the protostar itself at the center. The same flattening took place in the solar nebula, from which the Sun and planets formed (see Section 8-4). Particles orbiting the protostar within this disk collide with each other, causing them to lose energy, spiral inward onto the protostar, and add to the protostar's mass. This is a process of **accretion,** and the disk of material being added to the protostar in this way is called a **circumstellar accretion disk. Figure 20-14** is

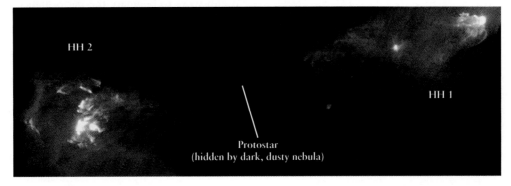

Figure 20-13 R I **V** U X G
Bipolar Outflow and Herbig-Haro Objects The two bright knots of glowing, ionized gas called HH 1 and HH 2 are Herbig-Haro objects. They are created when fast-moving gas ejected from a protostar slams into the surrounding interstellar medium, heating the gas to high temperature. HH 1 and HH 2 are 0.34 parsec (1.1 light-year) apart and lie 470 pc (1500 ly) from Earth in the constellation Orion. (J. Hester, the WFPC-2 Investigation Definition Team, and NASA)

(a)

(b)

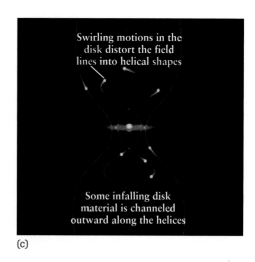

(c)

Figure 20-15

A Magnetic Model for Bipolar Outflow (a) Observations suggest that circumstellar accretion disks are threaded by magnetic field lines, as shown here. (b), (c) The contraction and rotation of the disk make the magnetic field lines distort and twist into helices. These helices steer some of the disk material into jets that stream perpendicular to the plane of the disk, as in Figure 20-14. (Adapted from Alfred T. Kamajian/Thomas P. Ray, "Fountain of Youth: Early Days in the Life of a Star," *Scientific American*, August 2000)

an edge-on view of a circumstellar accretion disk, showing two oppositely directed jets emanating from a point at or near the center of the disk (where the protostar is located).

What causes some of the material in the disk to be blasted outward in a pair of jets? One model involves the magnetic field of the dark nebula in which the star forms (Figure 20-15). As material in the circumstellar accretion disk falls inward, it drags the magnetic field lines along with it. (We saw in Section 18-9 how a similar mechanism may explain the Sun's 22-year cycle.) Parts of the disk at different distances from the central protostar orbit at different speeds, and this can twist the magnetic field lines into two helix shapes, one on each side of the disk. The helices then act as channels that guide infalling material away from the protostar, forming two opposing jets.

Many astronomers suspect that interactions among the protostar, the accretion disk, and the jets help to slow the protostar's rotation. If so, this would explain why main-sequence stars generally spin much more slowly than protostars of the same final mass.

 In the 1990s, astronomers using the Hubble Space Telescope discovered many examples of disks around newly formed stars in the Orion Nebula (see Figure 20-1), one of the most prominent star-forming regions in the northern sky. Figure 8-5 shows a number of these **protoplanetary disks,** or **proplyds,** that surround young stars within the nebula. As the name suggests, protoplanetary disks are thought to contain the material from which planets form around stars. They are what remains of a circumstellar accretion disk after much of the material has either fallen onto the star or been ejected by bipolar outflows.

Not all stars are thought to form protoplanetary disks; the exceptions probably include stars with masses in excess of about 3 $M_\odot$, as well as many stars in binary systems. But surveys of the Orion Nebula show that disks are found around most young, low-mass stars. Thus, disk formation may be a natural stage in the birth of many stars.

20-6 | Young star clusters give insight into star formation and evolution

Dark nebulae contain tens or hundreds of solar masses of gas and dust, enough to form many stars. As a consequence, these nebulae tend to form groups or **clusters** of young stars. Such clusters typically include stars with a range of different masses, all of which began to form out of the parent nebula at roughly the same time. Hence, star clusters are valuable laboratories for comparing the evolution of different stars. (In an analogous way, foot races are useful ways to compare the performance of sprinters because all the competitors start the race at the same time.) Star clusters are also objects of great natural beauty. One example is M16, shown in Figure 20-16; another is NGC 6520, depicted in Figure 20-3.

All the stars in a cluster may begin to form nearly simultaneously, but they do not all become main-sequence stars at the same time. As you can see from their evolutionary tracks (see Figure 20-9), high-mass stars evolve more rapidly than low-mass stars. The more massive the protostar, the sooner it develops the central pressures and temperatures needed for steady hydrogen fusion to begin.

Upon reaching the main sequence, *high-mass* protostars become hot, ultraluminous stars of spectral types O and B. As we saw in Section 20-2, these are the sorts of stars whose ultraviolet radiation ionizes the surrounding interstellar medium to produce an H II region. Figure 20-16 shows such an H II region, called the Eagle Nebula, surrounding the young star cluster M16. A few hundred thousand years ago, this region of space would have had

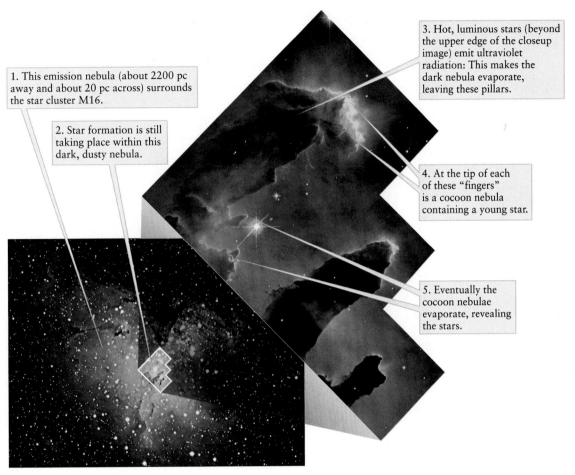

1. This emission nebula (about 2200 pc away and about 20 pc across) surrounds the star cluster M16.

2. Star formation is still taking place within this dark, dusty nebula.

3. Hot, luminous stars (beyond the upper edge of the closeup image) emit ultraviolet radiation: This makes the dark nebula evaporate, leaving these pillars.

4. At the tip of each of these "fingers" is a cocoon nebula containing a young star.

5. Eventually the cocoon nebulae evaporate, revealing the stars.

figure 20-16 R I **V** U X G

A Star Cluster with an H II Region The star cluster M16 is thought to be no more than 800,000 years old, and star formation is still taking place within adjacent dark, dusty globules. The inset shows three dense, cold pillars of gas and dust silhouetted against the glowing background of the red emission nebula (called the Eagle Nebula for its shape). The pillar at the upper left extends about 0.3 parsec (1 light-year) from base to tip, and each of its "fingers" is somewhat broader than our entire solar system. (Anglo-Australian Observatory; J. Hester and P. Scowen, Arizona State University; NASA)

a far less dramatic appearance. It was then a dark nebula, with protostars just beginning to form. Over the intervening millennia, mass ejection from these evolving protostars swept away the obscuring dust. The exposed young, hot stars heated the relatively thin remnants of the original dark nebula, creating the H II region that we see today.

When the most massive protostars to form out of a dark nebula have reached the main sequence, other *low-mass* protostars are still evolving nearby within their dusty cocoons. The evolution of these low-mass stars can be disturbed by their more massive neighbors. As an example, the inset in Figure 20-16 is a close-up of part of the Eagle Nebula. Within these opaque pillars of cold gas and dust, protostars are still forming. At the same time, however, the pillars are being eroded by intense ultraviolet light from hot, massive stars that have already shed their cocoons. As each pillar evaporates, the embryonic stars within have their surrounding material stripped away prematurely, limiting the total mass that these stars can accrete.

Star clusters tell us still more about how high-mass and low-mass stars evolve. **Figure 20-17a** shows the young star cluster NGC 2264 and its associated emission nebula. Astronomers have measured each star's apparent brightness and color ratio. Knowing the distance to the cluster, they have deduced the luminosities and surface temperatures of the stars (see Section 19-2 and Section 19-4). Figure 20-17b shows all these stars on an H-R diagram. Note that the hottest and most massive stars, with surface temperatures around 20,000 K, are on the main sequence. Stars cooler than about 10,000 K, however, have not yet quite arrived at the main sequence. These are less massive stars in the final stages of pre–main-sequence contraction and are just now beginning to ignite thermonuclear reactions at their centers. To find the ages of these stars, we can compare Figure 20-17b with the theoretical

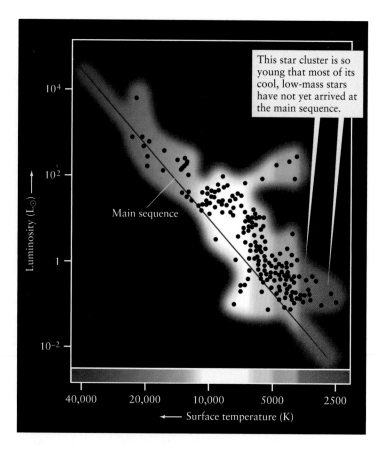

This star cluster is so young that most of its cool, low-mass stars have not yet arrived at the main sequence.

Main sequence

figure 20-17 R I V U X G

A Young Star Cluster and Its H-R Diagram (a) This photograph shows an H II region and the young star cluster NGC 2264 in the constellation Monoceros (the Unicorn). It lies about 800 pc (2600 ly) from Earth. (b) Each dot plotted on this H-R diagram represents a star in

NGC 2264 whose luminosity and surface temperature have been determined. This star cluster probably started forming only 2 million years ago. (Anglo-Australian Observatory)

calculations of protostar evolution in Figure 20-9. It turns out that this particular cluster is probably about 2 million years old.

Figure 20-18a shows another young star cluster called the Pleiades. The photograph shows that the Pleiades must be older than NGC 2264, the cluster in Figure 20-17a. The gas that must once have formed an H II region around this cluster has dissipated into interstellar space, leaving only traces of dusty material that forms reflection nebulae around the cluster's stars. The H-R diagram for the Pleiades in Figure 20-18b bears out this idea. In contrast to the H-R diagram for NGC 2264, nearly all the stars in the Pleiades are on the main sequence. The cluster's age is about 50 million years, which is how long it takes for the least massive stars to finally begin hydrogen fusion in their cores.

CAUTION Note that the data points for the most massive stars in the Pleiades (at the upper left of the H-R diagram in Figure 20-18b) lie above the main sequence. This is *not* because these stars have yet to arrive at the main sequence. Rather, these stars were the first members of the cluster to arrive at the main sequence some time ago and are now the first members to leave it. They have used up the hydrogen in their cores, so the steady process

of core hydrogen fusion that characterizes main-sequence stars cannot continue. In Chapter 21 we will see why massive stars spend a rather short time as main-sequence stars and will study what happens to stars after the main-sequence phase of their lives.

WEB LINK 20.12 A loose collection of stars such as NGC 2264 or the Pleiades is referred to as an **open cluster** (or *galactic cluster*, since such clusters are usually found in the plane of the Milky Way Galaxy). Open clusters possess barely enough mass to hold themselves together by gravitation. Occasionally, a star moving faster than average will escape, or "evaporate," from an open cluster. Indeed, by the time the stars are a few billion years old, they may be so widely separated that a cluster no longer exists.

If a group of stars is gravitationally unbound from the very beginning—that is, if the stars are moving away from one another so rapidly that gravitational forces cannot keep them together—then the group is called a **stellar association**. Because young stellar associations are typically dominated by luminous O and B main-sequence stars, they are also called **OB associations**. The image that opens this chapter shows part of an OB association in the southern constellation Ara (the Altar).

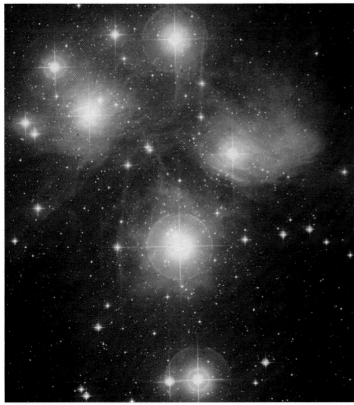

(a) The Pleiades star cluster

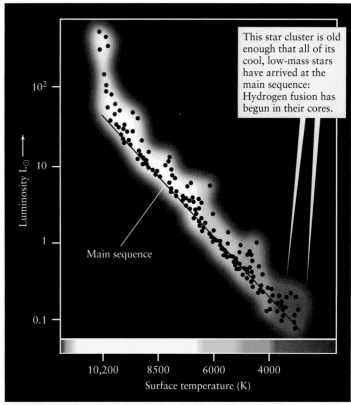

This star cluster is old enough that all of its cool, low-mass stars have arrived at the main sequence: Hydrogen fusion has begun in their cores.

Main sequence

(b) An H-R diagram of the stars in the Pleiades

Figure 20-18   R I **V** U X G

The Pleiades and Its H-R Diagram (a) The Pleiades star cluster is 117 pc (380 ly) from Earth in the constellation Taurus, and can be seen with the naked eye. (b) Each dot plotted on this H-R diagram represents a star in the Pleiades whose luminosity and surface temperature have been measured. (*Note:* The scales on this H-R diagram are different from those in Figure 20-17*b*.) The Pleiades is about 50 million (5×10^7) years old. (Anglo-Australian Observatory)

20-7 | Star birth can begin in giant molecular clouds

SCIENCE IN PROCESS

Observing the Galaxy at millimeter wavelengths reveals the cold gas that spawns new stars

We have seen that star formation takes place within dark nebulae. But where within our Galaxy are these dark nebulae found? Does star formation take place everywhere within the Milky Way, or only in certain special locations? The answers to such questions can enhance our understanding of star formation and of the nature of our home Galaxy.

Dark nebulae are a challenge to locate simply because they *are* dark—they do not emit visible light. Nearby dark nebulae can be seen silhouetted against background stars or H II regions (see Figure 20-2), but sufficiently distant dark nebulae are impossible to see with visible light because of interstellar extinction. They can, however, be detected using longer-wavelength radiation that can pass unaffected through interstellar dust. In fact, dark nebulae actually emit radiation at millimeter wavelengths.

Such emission takes place because in the cold depths of interstellar space, atoms combine to form molecules. The laws of quan-

tum mechanics predict that just as electrons within atoms can occupy only certain specific energy levels (see Section 5-8), molecules can vibrate and rotate only at certain specific rates. When a molecule goes from one vibrational state or rotational state to another, it either emits or absorbs a photon. (In the same way, an atom emits or absorbs a photon as an electron jumps from one energy level to another.) Most molecules are strong emitters of radiation with wavelengths of around 1 to 10 millimeters (mm). Consequently, observations with radio telescopes tuned to millimeter wavelengths make it possible to detect interstellar molecules of different types. More than 100 different kinds of molecules have so far been discovered in interstellar space, and the list is constantly growing.

Hydrogen is by far the most abundant element in the universe. Unfortunately, in cold nebulae much of it is in a molecular form (H_2) that is difficult to detect. The reason is that the hydrogen molecule is symmetric, with two atoms of equal mass joined together, and such molecules do not emit many photons at radio frequencies. In contrast, asymmetric molecules that consist of two atoms of unequal mass joined together, such as carbon monoxide (CO), are easily detectable at radio frequencies. When a carbon monoxide molecule makes a transition from one rate of rotation to another, it emits a photon at a wavelength of 2.6 mm or shorter.

The ratio of carbon monoxide to hydrogen in interstellar space is reasonably constant: For every CO molecule, there are about 10,000 H_2 molecules. As a result, carbon monoxide is an excellent "tracer" for molecular hydrogen gas. Wherever astronomers detect strong emission from CO, they know molecular hydrogen gas must be abundant.

The first systematic surveys of our Galaxy looking for 2.6-mm CO radiation were undertaken in 1974 by the American astronomers Philip Solomon and Nicholas Scoville. In mapping the locations of CO emission, they discovered huge clouds, now called **giant molecular clouds,** that must contain enormous amounts of hydrogen. These clouds have masses in the range of 10^5 to 2×10^6 solar masses and diameters that range from about 15 to 100 pc (50 to 300 ly). Inside one of these clouds, there are about 200 hydrogen molecules per cubic centimeter. This is several thousand times greater than the average density of matter in the disk of our Galaxy, yet only 10^{-17} as dense as the air we breathe. Astronomers now estimate that our Galaxy contains about 5000 of these enormous clouds.

Figure 20-19 is a map of radio emissions from carbon monoxide in the constellations Orion and Monoceros. Note the extensive

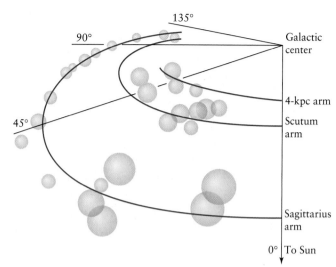

figure 20-20

Giant Molecular Clouds in the Milky Way This perspective drawing shows the locations of giant molecular clouds in an inner part of our Galaxy as seen from a vantage point above the Sun. These clouds lie primarily along the Galaxy's spiral arms, shown by red arcs. The distance from the Sun to the galactic center is about 8000 pc (26,000 ly). (Adapted from T. M. Dame and colleagues)

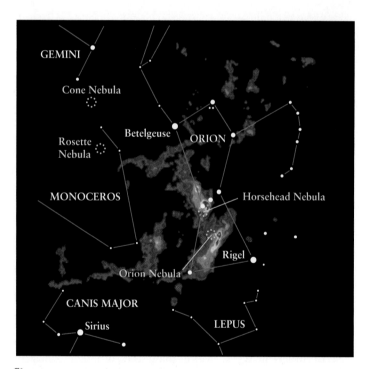

figure 20-19 R I V U X G

Mapping Molecular Clouds A radio telescope was tuned to a wavelength of 2.6 mm to detect emissions from carbon monoxide (CO) molecules in the constellations Orion and Monoceros. The result was this false-color map, which shows a 35° × 40° section of the sky. The star chart overlay shows the locations of four prominent star-forming nebulae. The Orion and Horsehead nebulae are located at sites of intense CO emission (shown in red and yellow), indicating the presence of a particularly dense molecular cloud at these sites of star formation. (Courtesy of R. Maddalena, M. Morris, J. Moscowitz, and P. Thaddeus)

areas of the sky covered by giant molecular clouds. This part of the sky is of particular interest because it includes several star-forming regions. By comparing the radio map with the star chart overlay, you can see that the areas where CO emission is strongest, and, thus, where giant molecular clouds are densest, are sites of star formation. Therefore, giant molecular clouds are associated with the formation of stars. Particularly dense regions within these clouds form dark nebulae, and within these stars are born.

By using CO emissions to map out giant molecular clouds, astronomers can find the locations in our Galaxy where star formation occurs. These investigations reveal that molecular clouds clearly outline our Galaxy's spiral arms, as Figure 20-20 shows. (Figure 20-7a shows the arrangement of spiral arms in another, similar galaxy.) These clouds lie 1000 pc (3000 ly) apart, strung along the spiral arms like beads on a string. This arrangement resembles the spacing of H II regions along the arms of other spiral galaxies, such as the one shown in Figure 20-7a. The presence of both molecular clouds and H II regions shows that spiral arms are sites of ongoing star formation.

In Chapter 25 we will learn that spiral arms are locations where matter "piles up" temporarily as it orbits the center of the Galaxy. You can think of matter in a spiral arm as analogous to a freeway traffic jam. Just as cars are squeezed close together when they enter a traffic jam, a giant molecular cloud is compressed when it passes through a spiral arm. When this happens, vigorous star formation begins in the cloud's densest regions.

As soon as massive O and B stars form, they emit ultraviolet light that ionizes the surrounding hydrogen, and an H II region is born. An H II region is thus a small, bright "hot spot" in a giant molecu-

Radiation and stellar winds from this massive, luminous star...

...may have triggered the formation of these stars.

Figure 20-21   R **I** V U X G

How O and B Stars Trigger Star Formation Stellar winds and ultraviolet radiation from young O and B stars produce a shock wave that compresses gas farther into the giant molecular cloud. This stimulates star formation, producing more O and B stars, which stimulate still more star formation, and so on. Meanwhile, older stars are left behind. The inset shows a massive star that has spawned other, smaller stars in this way. These stars are about 770 pc (2500 ly) from Earth in the Cone Nebula, a star-forming region in the constellation Monoceros. The younger stars are just 0.04 to 0.08 ly from the central star. (Adapted from C. Lada, L. Blitz, and B. Elmegreen; inset: R. Thompson, M. Rieke, G, Schneider, and NASA)

lar cloud. An example is the Orion Nebula, shown in Figure 20-1b. Four hot, luminous O and B stars at the heart of the nebula produce the ionizing radiation that makes the surrounding gases glow (see Figure 20-10a). The Orion Nebula is embedded in the edge of a giant molecular cloud whose mass is estimated at 500,000 $M_\odot$. The H II regions in Figure 20-2 are located at a different point on the edge of the same molecular cloud, some 25 pc (80 ly) from the Orion Nebula.

Once star formation has begun and an H II region has formed, the massive O and B stars at the core of the H II region induce star formation in the rest of the giant molecular cloud. Ultraviolet radiation and vigorous stellar winds from the O and B stars carve out a cavity in the cloud, and the H II region, heated by the stars, expands into it. These winds travel faster than the speed of sound in the gas—that is, they are **supersonic.** Just as an airplane creates a shock wave (a sonic boom) if it flies faster than sound waves in our atmosphere, a shock wave forms where the expanding H II region pushes at supersonic speed into the rest of the giant molecular cloud. This shock wave compresses the gas through which it passes, stimulating more star birth. Newborn O and B stars further expand the H II region into the giant molecular cloud. Meanwhile, the older O and B stars, which were left behind, begin to disperse (**Figure 20-21**). In this way, an OB association "eats into" a giant molecular cloud, leaving stars in its wake.

Infrared observations reveal many features that resemble protostars in the swept-up layer immediately behind the shock wave from an OB association. For example, Figure 20-10 shows both optical and infrared views of the core of the Orion Nebula. Four O and B stars, called the Trapezium, and glowing gas and dust dominate

the view at visible wavelengths (see Figure 20-10a). Infrared observations penetrate this obscuring material to reveal dozens of infrared objects (see Figure 20-10b). These are thought to be cocoons of warm dust enveloping newly formed stars. Shock waves from the Trapezium may have helped to trigger the formation of these stars. In turn, these newly formed stars generate shock waves of their own, which you can see in Figure 20-10b.

20-8 | Supernovae compress the interstellar medium and can trigger star birth

Spiral arms are not the only mechanism for triggering the birth of stars. Presumably, anything that compresses interstellar clouds will do the job. The most dramatic is a *supernova*, caused by the violent death of a massive star after it has left the main sequence. As we will see in Chapter 22, the core of the doomed star collapses suddenly, releasing vast quantities of particles and energy that blow the star apart. The star's outer layers are blasted into space at speeds of several thousand kilometers per second.

Astronomers have found many nebulae across the sky that are the shredded funeral shrouds of these dead stars. Such nebulae, like the one shown in Figure 20-22, are known as **supernova remnants.** Many supernova remnants have a distinctly circular or arched appearance, as would be expected for an expanding shell of gas. This wall of gas is typically moving away from the dead star faster than sound waves can travel through the interstellar medium. As we

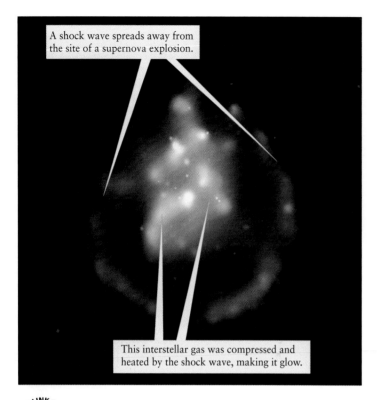

A shock wave spreads away from the site of a supernova explosion.

This interstellar gas was compressed and heated by the shock wave, making it glow.

Figure 20-22   R I V U X G

A Supernova Remnant The Chandra X-ray Observatory recorded this image of a 10,000-year-old supernova remnant named SNR 0103-72.6 for its coordinates on the celestial sphere (see Box 2-1). In 10,000 years, the shock wave has expanded outward about 23 pc (75 ly) in all directions from the site of the supernova explosion. The false colors denote the temperature of the gas: Red is coolest and blue is hottest. (NASA/CXC/PSU/S. Park et al.)

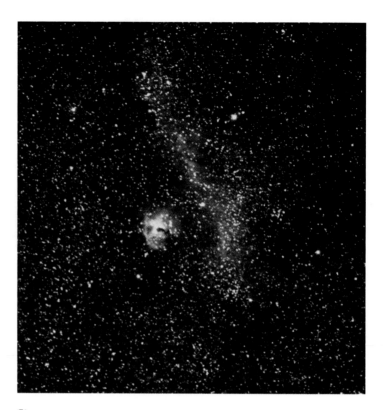

Figure 20-23   R I V U X G

The Canis Major R1 Association This luminous arc of gas, about 30 pc (100 ly) long, is studded with numerous young stars. Both the luminous arc and the young stars can be traced to the same source, a supernova explosion. The shock wave from the supernova explosion is exciting the gas and making it glow; the same shock wave also compresses the interstellar medium through which it passes, triggering star formation. (Courtesy of H. Vehrenberg)

saw in Section 20-7, such supersonic motion produces a shock wave that abruptly compresses the medium through which it passes. When a gas is compressed rapidly, its temperature rises, and this temperature rise causes the gas to glow as shown in Figure 20-22.

When the expanding shell of a supernova remnant slams into an interstellar cloud, it squeezes the cloud, stimulating star birth. This kind of star birth is taking place in the stellar association seen in Figure 20-23. This stellar nursery is located along a luminous arc of gas about 30 pc (100 ly) in length that is presumably the remnant of an ancient supernova explosion. In fact, this arc is part of an almost complete ring of glowing gas with a diameter of about 60 pc (200 ly). Spectroscopic observations of stars along this arc reveal substantial T Tauri activity. This activity results from newborn stars undergoing mass loss in their final stages of contraction before they become main-sequence stars.

Supernovae produce a variety of atomic nuclei, including some that are not produced in any other way. These nuclei are dispersed into space by the explosion. Some of these telltale nuclei have been discovered in meteorites that have fallen to Earth. Since meteorites formed very early in the history of our solar system (see Section 8-3), this suggests that a supernova occurred nearby when our solar system was very young. Some astronomers have used this idea to propose that the Sun was once a member of a loose stellar association created by a supernova. Individual stellar motions soon carry the stars of such an association in various directions away from their birthplaces. About 4.56 billion years have passed since the birth of our star, so if the Sun was once part of a stellar association, its brothers and sisters are now widely scattered across the Galaxy.

Many other processes can also trigger star formation. For example, a collision between two interstellar clouds can create new stars. Compression occurs at the interface between the two colliding clouds and vigorous star formation follows. Similarly, stellar winds from a group of O and B stars may exert strong enough pressure on interstellar clouds to cause compression, followed by star formation. (This process is similar to the one depicted in Figure 20-21.)

Our understanding of star birth has improved dramatically in recent years, primarily through infrared- and millimeter-wavelength observations. Nevertheless, many puzzles and mysteries remain. One problem is that different modes of star birth tend to produce different percentages of different kinds of stars. For example, the passage of a spiral arm through a giant molecular cloud tends to produce an abundance of massive O and B stars. In contrast, the shock wave from a supernova seems to produce fewer O and B stars but many more of the less massive A, F, G, and K stars. We do not yet know why this is so.

Despite these unanswered questions, it is now clear that star birth involves mechanisms on a colossal scale, from the deaths of massive stars to the rotation of an entire galaxy. In many respects, we have just begun to appreciate these cosmic processes. The study of cold, dark stellar nurseries will be an active and exciting area of astronomical research for many years to come.

In the chapters that follow, we will learn that the interstellar medium is both the birthplace of new stars and a dumping ground for dying stars. At the end of its life, a star can shed most of its mass in an outburst that enriches interstellar space with new chemical elements. The interstellar medium is therefore both nursery and graveyard. Because of this intimate relationship with stars, the interstellar medium evolves as successive generations of stars live out their lives. Understanding the details of this cosmic symbiosis is one of the challenges of modern astronomy.

Key Words

The term preceded by an asterisk (*) is discussed in Box 20-1.

accretion, p. 453

Barnard object, p. 449

bipolar outflow, p. 453

Bok globule, p. 449

circumstellar accretion disk, p. 453

cluster (of stars), p. 454

cocoon nebula, p. 450

dark nebula, p. 446

dust grains, p. 446

emission nebula, p. 444

evolutionary track, p. 449

*fluorescence, p. 446

giant molecular cloud, p. 458

H II region, p. 445

Herbig-Haro object, p. 453

interstellar extinction, p. 447

interstellar medium, p. 444

interstellar reddening, p. 447

nebula (*plural* nebulae), p. 444

nebulosity, p. 444

OB association, p. 456

open cluster, p. 456

protoplanetary disk (proplyd), p. 454

protostar, p. 449

recombination, p. 445

reflection nebula, p. 447

stationary absorption lines, p. 446

stellar association, p. 456

stellar evolution, p. 444

supernova remnant, p. 459

supersonic, p. 459

T Tauri star, p. 452

Key Ideas

Stellar Evolution: Because stars shine by thermonuclear reactions, they have a finite life span. The theory of stellar evolution describes how stars form and change during that life span.

 The Interstellar Medium: Interstellar gas and dust, which make up the interstellar medium, are concentrated in the disk of the Galaxy. Clouds within the interstellar medium are called nebulae.

• Dark nebulae are so dense that they are opaque. They appear as dark blots against a background of distant stars.

• Emission nebulae, or H II regions, are glowing, ionized clouds of gas. Emission nebulae are powered by ultraviolet light that they absorb from nearby hot stars.

• Reflection nebulae are produced when starlight is reflected from dust grains in the interstellar medium, producing a characteristic bluish glow.

Protostars: Star formation begins in dense, cold nebulae, where gravitational attraction causes a clump of material to condense into a protostar.

• As a protostar grows by the gravitational accretion of gases, Kelvin-Helmholtz contraction causes it to heat and begin glowing. Its relatively low temperature and high luminosity place it in the upper right region on an H-R diagram.

• Further evolution of a protostar causes it to move toward the main sequence on the H-R diagram. When its core temperatures become high enough to ignite steady hydrogen burning, it becomes a main-sequence star.

• The more massive the protostar, the more rapidly it evolves.

Mass Loss by Protostars: In the final stages of pre–main-sequence contraction, when thermonuclear reactions are about to begin in its core, a protostar may eject large amounts of gas into space.

• Low-mass stars that vigorously eject gas are called T Tauri stars.

• A circumstellar accretion disk provides material that a young star ejects as jets. Clumps of glowing gas called Herbig-Haro objects are sometimes found along these jets and at their ends.

Star Clusters: Newborn stars may form an open or galactic cluster. Stars are held together in such a cluster by gravity. Occasionally a star moving more rapidly than average will escape, or "evaporate," from such a cluster.

• A stellar association is a group of newborn stars that are moving apart so rapidly that their gravitational attraction for one another cannot pull them into orbit about one another.

O and B Stars and Their Relation to H II Regions: The most massive protostars to form out of a dark nebula rapidly become main sequence O and B stars. They emit strong ultraviolet radiation that ionizes hydrogen in the surrounding cloud, thus creating the reddish emission nebulae called H II regions.

• Ultraviolet radiation and stellar winds from the O and B stars at the core of an H II region create shock waves that move outward through the gas cloud, compressing the gas and triggering the formation of more protostars.

Giant Molecular Clouds: The spiral arms of our Galaxy are laced with giant molecular clouds, immense nebulae so cold that their constituent atoms can form into molecules.

• Star-forming regions appear when a giant molecular cloud is compressed. This can be caused by the cloud's passage through one of the spiral arms of our Galaxy, by a supernova explosion, or by other mechanisms.

| Review Questions |

1. Why is it more difficult to observe the life cycles of stars than the life cycles of planets or animals?

2. If an interstellar medium fills the space between the stars, how is that we are able to see the stars at all?

3. Summarize the evidence that interstellar space contains (**a**) gas and (**b**) dust.

4. What are H II regions? Near what kinds of stars are they found? Why do only these stars give rise to H II regions?

5. What are stationary absorption lines? In what sort of spectra are they seen? How do they give evidence for the existence of the interstellar medium?

6. In Figure 20-2, what makes the Horsehead Nebula dark? What makes IC 434 glow?

7. Why is the daytime sky blue? Why are distant mountains purple? Why is the Sun red when seen near the horizon at sunrise or sunset? In what ways are your answers analogous to the explanations for the bluish color of reflection nebulae and the process of interstellar reddening?

8. To see the constellation Coma Berenices (Berenice's Hair) you must look perpendicular to the plane of The Milky Way. By contrast, the Milky Way passes through the constellation Cassiopeia (named for a mythical queen). Would you expect H II regions to be more abundant in Coma Berenices or in Cassiopeia? Explain your reasoning.

9. Why are low temperatures necessary in order for protostars to form inside dark nebulae?

10. Compare and contrast Barnard objects and Bok globules. How many Sun-sized stars could you make out of a Barnard object? Out of a Bok globule?

11. The interior of a dark nebula is billions of times less dense than the air that you breathe. How, then, are dark nebulae able to block out starlight?

12. Describe the energy source that causes a protostar to shine. How does this source differ from the energy source inside a main-sequence star?

13. What is an evolutionary track? How can evolutionary tracks help us interpret the H-R diagram?

14. What happens inside a protostar to slow and eventually halt its gravitational contraction?

15. Why are the evolutionary tracks of high-mass stars different from those of low-mass stars? For which kind of star is the evolution more rapid? Why?

16. Why are protostars more easily seen with an infrared telescope than with a visible-light telescope?

17. In what ways is the internal structure of a 1-$M_\odot$ main-sequence star different from that of a 5-$M_\odot$ main-sequence star? From that of a 0.5-$M_\odot$ main-sequence star? What features are common to all these stars?

18. What sets the limits on the maximum and minimum masses of a main-sequence star?

19. What are Herbig-Haro objects? Why are they often found in pairs?

20. Why do disks form around contracting protostars? What is the connection between disks and bipolar outflows?

21. Young open clusters like those shown in Figures 20-17 and 20-18 are found only in the plane of the Galaxy. Explain why this should be.

22. Why are observations at millimeter wavelengths so much more useful in exploring interstellar clouds than observations at visible wavelengths?

23. What are giant molecular clouds? What role do these clouds play in the birth of stars?

24. Giant molecular clouds are among the largest objects in our Galaxy. Why, then, were they discovered only relatively recently?

25. Consider the following stages in the evolution of a young star cluster: (i) H II region; (ii) dark nebula; (iii) formation of O and B stars; (iv) giant molecular cloud. Put these stages in the correct chronological order and discuss how they are related.

26. Briefly describe four mechanisms that compress the interstellar medium and trigger star formation.

| Advanced Questions |

> Problem-solving tips and tools
>
> You may find it helpful to review Box 19-4, which describes the relationship among a star's luminosity, radius, and surface temperature. The small-angle formula is described in Box 1-1. Orbital periods are described by Kepler's third law, which we discussed in Boxes 4-2 and 4-4. Remember that the Stefan-Boltzmann law (Box 5-2) relates the temperature of a blackbody to its energy flux. Remember, too, that the volume of a sphere of radius r is $4\pi r^3/3$.

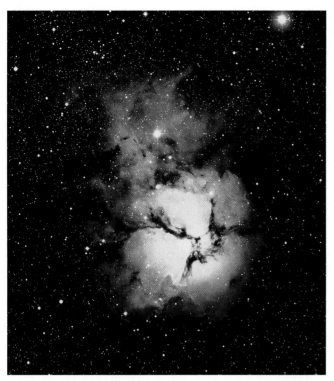

(Anglo-Australian Observatory)   R I **V** U X G

27. The visible-light photograph above shows the Trifid Nebula in the constellation Sagittarius. Label the following features on this photograph: (**a**) reflection nebulae (and the star or stars whose light is being reflected); (**b**) dark nebulae; (**c**) H II regions; (**d**) regions where star formation may be occurring. Explain how you identified each feature.

28. If you looked at the spectrum of a reflection nebula, would you see absorption lines, emission lines, or no lines? Explain your answer. As part of your explanation, describe how the spectrum demonstrates that the light was reflected from nearby stars.

29. In the direction of a particular star cluster, interstellar extinction allows only 15% of a star's light to pass through each kiloparsec (1000 pc) of the interstellar medium. If the star cluster is 3.0 kiloparsecs away, what percentage of its photons survive the trip to the Earth?

30. Find the density (in atoms per cubic centimeter) of a Bok globule having a radius of 1 light-year and a mass of 100 $M_\odot$. How does your result compare with the density of a typical H II region, between 80 and 600 atoms per cm³? (Assume that the globule is made purely of hydrogen atoms.)

31. The infrared-bright object at the center of Figure 20-10*b* is called the *Becklin-Neugebauer object* after its discoverers. Like the other bright spots in that image, it is a newly formed star. Assuming that all these stars began the process of formation of the same time, what can you conclude about the mass of the Becklin-Neugebauer object compared with those of the other stars? Does your conclusion depend on whether or not the stars have reached the main sequence? Explain your reasoning.

32. The two false-color images below show a portion of the Trifid Nebula (see Question 27). The reddish-orange view is a false-color infrared image, while the bluish picture (shown to the same scale) was made with visible light. Explain why the dark streaks in the visible-light image appear bright in the infrared image.

33. At one stage during its birth, the protosun had a luminosity of 1000 $L_\odot$ and a surface temperature of about 1000 K. At this time, what was its radius? Express your answer in three ways: as a multiple of the Sun's present-day radius, in kilometers, and in astronomical units.

34. A newly formed protostar and a red giant are both located in the same region on the H-R diagram. Explain how you could distinguish between these two.

35. (**a**) Determine the radius of the circumstellar accretion disk in Figure 20-14. (You will need to measure this image with a ruler. Note the scale bar in this figure.) Give your answer in astronomical units and in kilometers. (**b**) Assume that the young star at the

R **I** V U X G   R I **V** U X G

(ESA/ISO, ISOCAM, and J. Cernicharo et al.; IAC, Observatorio del Teide, Tenerife)

center of this disk has a mass of 1 $M_\odot$. What is the orbital period (in years) of a particle at the outer edge of the disk? (c) Using your ruler again, determine the length of the jet that extends to the right of the circumstellar disk in Figure 20-14. At a speed of 200 km/s, how long does it take gas to traverse the entire visible length of the jet?

36. The star cluster NGC 2264 (Figure 20-17) contains numerous T Tauri stars, while the Pleiades (Figure 20-18) contains none. Explain why there is a difference.

37. The concentration or abundance of ethyl alcohol in a typical molecular cloud is about 1 molecule per 10^8 cubic meters. What volume of such a cloud would contain enough alcohol to make a martini (about 10 grams of alcohol)? A molecule of ethyl alcohol has 46 times the mass of a hydrogen atom (that is, ethyl alcohol has a molecular weight of 46).

38. (a) The supernova remnant shown in Figure 20-22, SNR 0103-72.6, actually lies outside the Milky Way Galaxy, about 5800 pc (190,000 ly) away in a smaller galaxy called the Small Magellanic Cloud. For this supernova remnant to be visible at that distance, there must be very little dust between Earth and the Small Magellanic Cloud. Explain why. (b) The radius (half the diameter) of SNR 0103-72.6 is 23 pc. Calculate its angular diameter in arcminutes as seen from Earth.

39. From the information given in the caption for Figure 20-22, calculate the average speed at which the shock wave has spread away from the site of the supernova explosion. Give your answer in kilometers per second and as a fraction of the speed of light. (*Hint:* There are 3.16×10^7 seconds in a year and the speed of light is 3.00×10^5 km/s.)

Discussion Questions

40. Some science-fiction movies show stars suddenly becoming dramatically brighter when they are "born" (that is, when thermonuclear fusion reactions begin in their cores). Discuss whether this is a reasonable depiction.

41. Suppose that the electrons in hydrogen atoms were not as strongly attracted to the nuclei of those atoms, so that these atoms were easier to ionize. What consequences might this have for the internal structure of main-sequence stars? Explain your reasoning.

42. What do you think would happen if our solar system were to pass through a giant molecular cloud? Do you think the Earth has ever passed through such clouds?

43. Many of the molecules found in giant molecular clouds are *organic* molecules (that is, they contain carbon). Speculate about the possibility of life-forms and biological processes occurring in giant molecular clouds. In what ways might the conditions existing in giant molecular clouds favor or hinder biological evolution?

44. Speculate on why a shock wave from a supernova seems to produce relatively few high-mass O and B stars, compared to the lower-mass A, F, G, and K stars.

 Web/CD-ROM Questions

45. In recent years astronomers have been able to learn about the character of the interstellar medium in the vicinity of the Sun. Search the World Wide Web for information about aspects of the nearby interstellar medium, including features called the Local Interstellar Cloud and the Local Bubble. How do astronomers study the nearby interstellar medium? What makes these studies difficult? Is the interstellar medium relatively uniform in our neighborhood, or is it clumpy? If the latter, is our solar system in a relatively thin or thick part of the interstellar medium? How is our solar system moving through the interstellar medium?

46. Search the World Wide Web for recent discoveries about how brown dwarfs form. Do they tend to form in the same locations as "real" stars? Do they form in relatively small or relatively large numbers compared to "real" stars? What techniques are used to make these discoveries?

 47. Measuring a Stellar Jet. Access the animation "A Stellar Jet in the Trifid Nebula" in Chapter 20 of the *Universe* Web site or CD-ROM. (a) The Trifid Nebula as a whole has an angular diameter of 28 arcmin. By stepping through the animation, estimate the angular size of the stellar jet shown at the end of the animation. (b) The Trifid Nebula is about 2800 pc (9000 ly) from Earth. Estimate the length of the jet in light-years. (c) If gas in the jet travels at 200 km/s, how long does it take to traverse the length of the jet? Give your answer in years. (*Hint:* There are 3.16×10^7 seconds in a year.)

Observing Projects

> **Observing tips and tools**
>
> After looking at the beautiful color photographs of nebulae in this chapter, you may find the view through a telescope a bit disappointing at first, but fear not. You can see a great deal with even a small telescope. To get the best view of a dim nebula using a telescope, use the same "averted vision" trick we described in "Observing tips and tools" in Chapter 19: If you direct your vision a little to one side of the object that you are looking at, the light from that object will go onto a more sensitive part of the retina.

48. Use a telescope to observe at least two of the following H II regions. In each case, can you guess which stars are probably responsible for the ionizing radiation that causes the nebula to glow? Can you see any obscuration or silhouetted features that suggest the presence of interstellar dust? Draw a picture of what you see through the telescope and compare it with a photograph of the object. Take note of which portions of the nebula were not visible through your telescope.

Nebula	Right ascension	Declination
M42 (Orion)	5^h 35.4^m	−5° 27′
M43	5 35.6^m	−5 16
M20 (Trifid)	18 02.6^m	−23 02
M8 (Lagoon)	18 03.8^m	−24 23
M17 (Omega)	18 20.8^m	−16 11

Note: The right ascensions and declinations are given for epoch 2000.

49. On an exceptionally clear, moonless night, use a telescope to observe at least one of the following dark nebulae. These nebulae are very difficult to find, because they are recognizable only by the absence of stars in an otherwise starry part of the sky. Are you confident that you actually saw the dark nebula? Does the pattern of background stars suggest a particular shape to the nebula?

Nebula	Right ascension	Declination
Barnard 72 (the Snake)	17^h 23.5^m	−23° 38′
Barnard 86	18 02.7	−27 50
Barnard 133	19 06.1	−6 50
Barnard 142 and 143	19 40.7	−10 57

Note: The right ascensions and declinations are given for epoch 2000.

50. A few fine objects cover such large regions of the sky that they are best seen with binoculars. If you have access to a high-quality pair of binoculars, observe the North American Nebula in Cygnus and the Pipe Nebula in Ophiuchus. Both nebulae are quite faint, so you should attempt to observe them only on an exceptionally dark, clear, moonless night. The North America Nebula is a cloud of glowing hydrogen gas located about 3° east of Deneb, the brightest star in Cygnus. While searching for the North America Nebula, you may glimpse another diffuse H II region, the Pelican Nebula, located about 2° southeast of Deneb. The Pipe Nebula is a 7°-long, meandering, dark nebula to the south and to the east of the star θ (theta) Ophiuchi, which is in a section of Ophiuchus that extends southward between the constellations of Scorpius and Sagittarius. Located about 12° east of the bright red star Antares, you can locate θ Ophiuchi using the *Starry Night Backyard*™ software on the CD-ROM that accompanies this book.

51. Use the *Starry Night Backyard*™ program to investigate star-forming regions. Display the entire celestial sphere (select **Guides > Atlas** in the **Go** menu) and center on M20 (the Trifid Nebula, shown in the figure that accompanies Question 27) by using the **Find...** command in the **Edit** menu.

Using the controls at the right-hand end of the Control Panel, zoom out as far as possible. Set the time step to 1 month ("lunar m.") and step through time in one-month intervals. (**a**) In which month is M20 highest in the sky at noon? Explain how you determined this. (**b**) In which month is M20 highest in the sky at midnight, so that it is best placed for observing with a telescope? Explain how you determined this.

52. Use the *Deep Space Explorer*™ program to examine the Milky Way Galaxy. In the left-hand part of the window under the heading **Milky Way**, click on the triangle next to the word **Explore** and then click on **Sun in Milky Way.** You can zoom in or out using the buttons at the upper left of the window (an upward-pointing triangle and a downward-pointing triangle). You can rotate the Galaxy by putting the mouse cursor over the image, holding down the mouse button, and moving the mouse. (On a two-button mouse, hold down the left mouse button.) (**a**) You can identify H II regions by their characteristic magenta color. Describe where in the Galaxy you find these. Are most found in the inner part of the Galaxy or in its outer regions? (**b**) Where do you find dark lanes of dust, in the inner part of the Galaxy or in its outer regions? Do you see any connection between the locations of dust and of H II regions? If there is a connection, what do you think causes it? If there is not a connection, why is this the case?

Collaborative Exercises

53. Imagine that your group walks into a store that specializes in selling antiquated clothing. Prepare a list of observable characteristics that you would look for to distinguish which items were from the early, middle, and late twentieth century. Also, write a paragraph that specifically describes how this task is similar to how astronomers understand the evolution of stars.

54. Consider advertisement signs visible at night in your community and provide specific examples of ones that are examples of the three different types of nebulae that astronomers observe and study. If an example doesn't exist in your community, creatively design an advertisement sign that could serve as an example.

55. The pre–main-sequence evolutionary tracks shown in Figure 20-9 describe the tracks of seven protostars of different masses. Imagine a new sort of H-R diagram that plots a human male's increasing age versus decreasing hair density on the head instead of increasing luminosity versus decreasing temperature. Create and carefully label a sketch of this imaginary HaiR diagram showing both the majority of the U.S. male population and a few oddities. Finally, draw a line that clearly labels your sketch to show how a typical male undergoing male-pattern baldness might slowly change position on the HaiR diagram over the course of a human life span.

21 Stellar Evolution: After the Main Sequence

Imagine a world like Earth, but orbiting a star more than 100 times larger and 2000 times more luminous than our Sun. Bathed in the star's intense light, the surface of this world is utterly dry, airless, and hot enough to melt iron. If you could somehow survive on the daytime side of this world, you would see the star filling almost the entire sky.

This bizarre planet is not a creation of science fiction—it is our own Earth some 7.6 billion years from now. The bloated star is our own Sun, which in that remote era will have become a red-giant star.

A main-sequence star evolves into a red giant when all the hydrogen in its core is consumed. The star's core contracts and heats up, but its outer layers expand and cool. In the hot, compressed core, helium fusion becomes a new energy source. The more massive a star, the more rapidly it consumes its core's hydrogen and the sooner it evolves into a giant.

The interiors of stars are hidden from our direct view, so much of the story in this chapter is based on theory. We back up those calculations with observations of star clusters, which contain stars of different masses but roughly the same age. (An example is the cluster shown here, many of whose stars have evolved into luminous red giants.) Other observations show that some red-giant stars actually pulsate, and that stars can evolve along very different paths if they are part of a binary star system.

WEB LINK 21.1 The globular cluster NGC 6093 has numerous bright red giant stars. (Hubble Heritage Team, AURA/STScI/NASA)
R I **V** U X G

Questions to Guide Your Reading

As you read the sections of this chapter, look for the answers to the following questions:

21-1   How will our Sun change over the next few billion years?

21-2   Why are red giants larger than main-sequence stars?

21-3   Do all stars evolve into red giants at the same rate?

21-4   How do we know that many stars lived and died before our Sun was born?

21-5   Why do some giant stars pulsate in and out?

21-6   Why do stars in some binary systems evolve in unusual ways?

21-1 | When core hydrogen fusion ceases, a main-sequence star becomes a red giant

At the core, main-sequence stars are all fundamentally alike. As we saw in Section 20-4, it is in their cores that all such stars convert hydrogen into helium by thermonuclear reactions. This process is called **core hydrogen fusion.** The total time that a star will spend fusing hydrogen into helium in its core, and thus the total time that it will spend as a main-sequence star, is called its **main-sequence lifetime.** For our Sun, the main-sequence lifetime is about 12 billion (1.2×10^{10}) years. Hydrogen fusion has been going on in the Sun's core for the past 4.56 billion (4.56×10^9) years, so our Sun is less than halfway through its main-sequence lifetime.

What happens to a star after the core hydrogen has been used up, so that it is no longer a main-sequence star? As we will see, it expands dramatically to become a red giant. To understand why this happens, it is useful to first look at how a star evolves *during* its main-sequence lifetime.

A protostar becomes a main-sequence star when steady hydrogen fusion begins in its core and it achieves *hydrostatic equilibrium*—a balance between the inward force of gravity and the outward pressure produced by hydrogen fusion (see Section 18-2 and Section 20-4). Such a freshly formed main-sequence star is called a **zero-age main-sequence star.**

We make the distinction between "main sequence" and "zero-age main sequence" because a star undergoes noticeable changes in luminosity, surface temperature, and radius during its main-sequence lifetime. These changes are a result of core hydrogen fusion, which alters the chemical composition of the core. As an example, when our Sun first formed, its composition was the same at all points throughout its volume: by mass, about 74% hydrogen,

25% helium, and 1% heavy elements. But as **Figure 21-1** shows, the Sun's core now contains a greater mass of helium than of hydrogen. (There is still enough hydrogen for another 7 billion years or so of core hydrogen fusion.)

Thanks to core hydrogen fusion, the total number of atomic nuclei in a star's core decreases with time: In each reaction, four hydrogen nuclei are converted to a single helium nucleus (see Figure 18-2 and Box 18-1). With fewer particles bouncing around to provide the core's internal pressure, the core contracts slightly under the weight of the star's outer layers. Compression makes the core denser and increases its temperature. (Box 21-1 gives some everyday examples of how the temperature of a gas changes when it compresses or expands.) As a result of these changes in density and temperature, the pressure in the compressed core is actually higher than before.

As the star's core shrinks, its outer layers expand and shine more brightly. Here's why: As the core's density and temperature increase, hydrogen nuclei in the core collide with one another more frequently, and the rate of core hydrogen fusion increases. Hence, the star's luminosity increases. The radius of the star as a whole also increases slightly, because increased core pressure pushes outward on the star's outer layers. The star's surface temperature changes as well, because it is related to the luminosity and radius (see Section 19-6 and Box 19-4). As an example, theoretical calculations indicate that over the past 4.56×10^9 years, our Sun has become 40% more luminous, grown in radius by 6%, and increased in surface temperature by 300 K (**Figure 21-2**).

As a main-sequence star ages and evolves, the increase in energy outflow from its core also heats the material immediately surrounding the core. As a result, hydrogen fusion can begin in this surrounding material. By tapping this fresh supply of hydrogen, a star manages to eke out a few million more years of main-sequence existence.

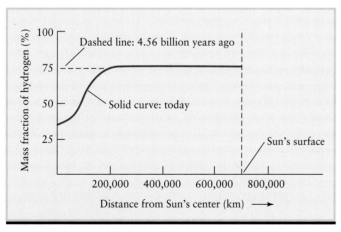

(a) Hydrogen in the Sun's interior

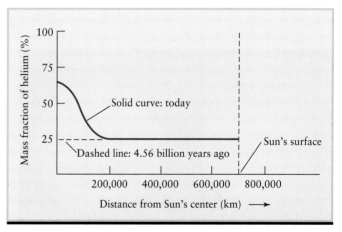

(b) Helium in the Sun's interior

Figure 21-1

Changes in the Sun's Chemical Composition These graphs show the percentage by mass of (a) hydrogen and (b) helium at different points within the Sun's interior. The dashed horizontal lines show that these percentages were the same throughout the Sun's volume when it first formed. As the solid curves show, over the past 4.56×10^9 years, thermonuclear reactions at the core have depleted hydrogen in the core and increased the amount of helium in the core.

box 21-1 THE HEAVENS ON THE EARTH

Compressing and Expanding Gases

As a star evolves, various parts of the star either contract or expand. When this happens, the gases behave in much the same way as gases here on the Earth when they are forced to compress or allowed to expand.

When a gas is compressed, its temperature rises. You know this by personal experience if you have ever had to inflate a bicycle tire with a hand pump. As you pump, the compressed air gets warm and makes the pump warm to the touch. The same effect happens on a larger scale in southern California during Santa Ana winds or downwind from the Rocky Mountains when there are Chinook winds. Both of these strong winds blow from the mountains down to the lowlands. Even though the mountain air is cold, the winds that reach low elevations can be very hot. (Chinook winds have been known to raise the temperature by as much as 27°C, or 49°F, in only two minutes!) The explanation is compression. Air blown downhill by the winds is compressed by the greater air pressure at lower altitudes, and this compression raises the temperature of the air.

Expanding gases tend to drop in temperature. When you open a bottle of carbonated beverage, the gases trapped in the bottle expand and cool down. The cooling can be so great that a little cloud forms within the neck of the bottle. Clouds form in the atmosphere in the same way. Rising air cools as it goes to higher altitudes, where the pressure is lower, and the cooling makes water in the air condense into droplets.

Here's an experiment you can do to feel the cooling of expanding gases. Your breath is actually quite warm, as you can feel if you open your mouth wide, hold the back of your hand next to your mouth, and exhale. But if you bring your lips together to form an "o" and again blow on your hand, your breath feels cool. In the second case, your exhaled breath has to expand as it passes between your lips to the outside air. That makes its temperature drop.

A star's main-sequence lifetime depends critically on its mass. As Table 21-1 shows, massive stars have short main-sequence lifetimes because they are also very luminous (see Section 19-9, and particularly Figure 19-21). To emit energy so rapidly, these stars must be depleting the hydrogen in their cores at a prodigious rate. Hence, even though a massive O or B main-sequence star contains much more hydrogen fuel in its core than a less-massive M star, it exhausts its hydrogen much sooner. High-mass O and B stars gobble up their hydrogen fuel in only a few million years, but low-mass M stars take hundreds of billions of years to use up their hydrogen. Thus, a main-sequence star's mass determines not only its luminosity and spectral type, but also how long it can remain a main-sequence star (see Box 21-2 for details).

We saw in Section 20-4 how more-massive stars evolve more quickly through the protostar phase to become main-sequence stars (see Figure 20-9). In general, the more massive the star, the more rapidly it goes through *all* the phases of its life. Still, most of the stars we are able to detect are in their main-sequence phase, because this phase lasts so much longer than other luminous phases. (In Chapters 22, 23, and 24 we will explore the final phases of a star's existence, when it ceases to have an appreciable luminosity.)

At the end of a star's main-sequence lifetime, all the hydrogen in its core has been used up and hydrogen fusion ceases there. In this new stage, hydrogen fusion continues only in the hydrogen-rich material just outside the core, a situation called **shell hydrogen fusion**. At first, this process occurs only in the hottest region just outside the core, where the hydrogen fuel has not yet been exhausted.

Strangely enough, the end of core hydrogen fusion *increases* the core's temperature. When thermonuclear reactions first cease in the core, nothing remains to generate heat there. Hence, the core starts to cool and the pressure there starts to decrease. This pressure decrease allows the star's core to again compress under the weight of the outer layers. As the core contracts, its temperature again increases, and heat begins to flow outward from the core even though no nuclear reactions are taking place there. (Technically, gravitational energy is converted into thermal energy, as in Kelvin-Helmholtz contraction; see Section 8-4 and Section 18-1).

This new flow of heat warms the gases around the core, increasing the rate of shell hydrogen fusion and making the shell eat further

figure 21-2

The Zero-Age Sun and Today's Sun Over the past 4.56×10^9 years, much of the hydrogen in the Sun's core has been converted into helium, the core has contracted a bit, and the Sun's luminosity has gone up by about 40%. These changes in the core have made the Sun's outer layers expand in radius by 6% and increased the surface temperature from 5500 K to 5800 K.

box 21-2   TOOLS OF THE ASTRONOMER'S TRADE

Main-Sequence Lifetimes

Hydrogen fusion converts a portion of a star's mass into energy. We can use Einstein's famous equation relating mass and energy to calculate how long a star will remain on the main sequence.

Suppose that M is the mass of a star and f is the fraction of the star's mass that is converted into energy by hydrogen fusion. The total energy E supplied by the hydrogen fusion can be expressed as

$$E = fMc^2$$

In this equation c is the speed of light.

This energy from hydrogen fusion is released gradually over millions or billions of years. If L is the star's luminosity (energy released per unit time) and t is the star's main-sequence lifetime (the total time over which the hydrogen fusion occurs), then we can write

$$L = \frac{E}{t}$$

(Actually, this equation is only an approximation. A star's luminosity is not quite constant over its entire main-sequence lifetime. But the variations are not important for our purposes.) We can rewrite this equation as

$$E = Lt$$

From this equation and $E = fMc^2$, we see that

$$Lt = fMc^2$$

which can be rewritten as

$$t = \frac{fMc^2}{L}$$

Thus, a star's lifetime on the main sequence is proportional to its mass (M) divided by its luminosity (L). Using the symbol $\propto$ to denote "is proportional to," we write

$$t \propto \frac{M}{L}$$

We can carry this analysis further by recalling that main-sequence stars obey the mass-luminosity relation (see Section 19-9, especially Figure 19-21). The distribution of data on the graph in Figure 19-21 tells us that a star's luminosity is roughly proportional to the 3.5 power of its mass:

$$L \propto M^{3.5}$$

Substituting this relationship into the previous proportionality, we find that

$$t \propto \frac{M}{M^{3.5}} = \frac{1}{M^{2.5}} = \frac{1}{M^2 \sqrt{M}}$$

This approximate relationship can be used to obtain rough estimates of how long a star will remain on the main sequence. It is often convenient to relate these estimates to the Sun (a typical 1-$M_\odot$ star), which will spend 1.2×10^{10} years on the main sequence.

EXAMPLE: How will a star whose mass is 4 $M_\odot$ remain on the main sequence?

Situation: Given the mass of a star, we are asked to determine its main-sequence lifetime.

Tools: We use the relationship $t \propto 1/M^{2.5}$.

Answer: The star has 4 times the mass of the Sun, so it will be on the main sequence for approximately

$$\frac{1}{4^{2.5}} = \frac{1}{4^2\sqrt{4}} = \frac{1}{32} \text{ times the Sun's main-sequence lifetime}$$

Thus, a 4-$M_\odot$ main-sequence star will fuse hydrogen in its core for about $\frac{1}{32} \times 1.2 \times 10^{10}$ years, or about 4×10^8 (400 million) years.

Review: Our result makes sense: A star more massive than the Sun must have a shorter main-sequence lifetime.

outward into the surrounding matter. Helium produced by reactions in the shell falls down onto the core, which continues to contract and heat up as it gains mass. Over the course of hundreds of millions of years, the core of a 1-$M_\odot$ star compresses to about one-third of its original radius, while its central temperature increases from about 15 million (1.5×10^7) K to about 100 million (10^8) K.

During this post–main-sequence phase, the star's outer layers expand just as dramatically as the core contracts. As the hydrogen-fusing shell works its way outward, egged on by heat from the contracting core, the star's luminosity increases substantially. This increases the star's internal pressure and makes the star's outer layers expand to many times their original radius. This tremendous expansion causes those layers to cool down, and the star's surface temperature drops (see Box 21-1). Once the temperature of the star's bloated surface falls to about 3500 K, the gases glow with a reddish hue, in accordance with Wien's law (see Figure 19-7a). The star is then appropriately called a **red giant** (Figure 21-3). Thus, we see that red-giant stars are former main-sequence stars that have evolved into a new stage of existence. We can summarize these observations as a general rule:

table 21-1	Approximate Main-Sequence Lifetimes			
Mass (M$_\odot$)	Surface temperature (K)	Spectral class	Luminosity (L$_\odot$)	Main-sequence lifetime (10^6 years)
25	35,000	O	80,000	4
15	30,000	B	10,000	15
3	11,000	A	60	800
1.5	7000	F	5	4500
1.0	6000	G	1	12,000
0.75	5000	K	0.5	25,000
0.50	4000	M	0.03	700,000

The main-sequence lifetimes were estimated using the relationship $t \propto 1/M^{2.5}$ (see Box 21-2).

Stars join the main sequence when they begin hydrogen fusion in their cores. They leave the main sequence and become giant stars when the core hydrogen is depleted.

Red-giant stars undergo substantial **mass loss** because of their large diameters and correspondingly weak surface gravity. This makes it relatively easy for gases to escape from the red giant into space. Mass loss can be detected in a star's spectrum, because gas escaping from a red giant toward a telescope on the Earth produces narrow absorption lines that are slightly blueshifted by the Doppler effect (review Figure 5-24). Typical observed blueshifts correspond to

a speed of about 10 km/s. A typical red giant loses roughly 10^{-7} M$_\odot$ of matter per year. For comparison, the Sun's present-day mass loss rate is only 10^{-14} M$_\odot$ per year. Hence, an evolving star loses a substantial amount of mass as it becomes a red giant. **Figure 21-4** shows a star losing mass in this way.

We can use these ideas to peer into the future of our planet and our solar system. The Sun's luminosity will continue to increase as it goes through its main-sequence lifetime, and the temperature of the Earth will increase with it. One and a half billion years from now the Earth's average surface temperature will be 50°C (122°F). By 3½ billion years from now the surface temperature of the Earth

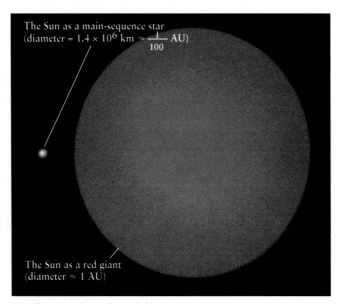

(a) The Sun today and as a red giant

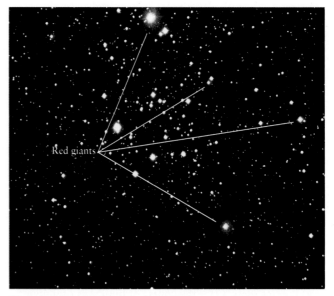

(b) Red giant stars in the star cluster M50

Figure 21-3 R I **V** U X G

Red Giants (a) The present-day Sun produces energy in a hydrogen-fusing core about 100,000 km in diameter. Some 7.6 billion years from now, when the Sun becomes a red giant, its energy source will be a shell only about 30,000 km in diameter within which hydrogen fusion will take place at a furious rate. The Sun's luminosity will be about 2000 times greater than today, and the increased luminosity will make the Sun's outer layers expand to approximately 100 times their present size. (b) This composite of visible and infrared images shows bright red giant stars in the open cluster M50 in the constellation Monoceros (the Unicorn). (T. Credner and S. Kohle, Astronomical Institutes of the University of Bonn)

figure 21-4 R I **V** U X G

A Mass-Loss Star As stars age and become giant stars, they expand tremendously and shed matter into space. This star, HD 65750, is losing matter at a high rate and is surrounded by a reflection nebula (IC 2220) caused by starlight reflecting from dust grains. These dust grains may have condensed from material shed by the star. (Anglo–Australian Observatory)

will exceed the boiling temperature of water. All the oceans will boil away, and the Earth will become utterly incapable of supporting life. These increasingly hostile conditions will pose the ultimate challenge to whatever intelligent beings might inhabit the Earth in the distant future.

About 7 billion years from now, our Sun will finish converting hydrogen into helium at its core. As the Sun's core contracts, its atmosphere will expand to envelop Mercury and perhaps reach to the orbit of Venus. Roughly 700 million years after leaving the main sequence, our red-giant Sun will have swollen to a diameter of about 1 AU—roughly 100 times larger than its present size—and its surface temperature will have dropped to about 3500 K. Shell hydrogen fusion will proceed at such a furious rate that our star will shine with the brightness of 2000 present-day Suns. Some of the inner planets will be vaporized, and the thick atmospheres of the outer planets will evaporate away to reveal tiny, rocky cores. Thus, in its later years, the aging Sun may destroy the planets that have accompanied it since its birth.

21-2 | Fusion of helium into carbon and oxygen begins at the center of a red giant

When a star first becomes a red giant, its hydrogen-fusing shell surrounds a small, compact core of almost pure helium. In a moderately low-mass red giant, which the Sun will become 7 billion years from now, the dense helium core is about twice the diameter of the Earth.

Helium, the "ash" of hydrogen fusion, is a potential nuclear fuel: **Helium fusion,** the thermonuclear fusion of helium nuclei to make heavier nuclei, releases energy. But this reaction does not take place within the core of our present-day Sun because the temperature there is too low. Each helium nucleus contains two protons, so it has twice the positive electric charge of a hydrogen nucleus, and there is a much stronger electric repulsion between two helium nuclei than between two hydrogen nuclei. For helium nuclei to overcome this repulsion and get close enough to fuse together, they must be moving

at very high speeds, which means that the temperature of the helium gas must be very high. (For more on the relationship between the temperature of a gas and the speed of atoms in the gas, see Box 7-2.)

When a star first becomes a red giant, the temperature of the contracted helium core is still too low for helium nuclei to fuse. But as the hydrogen-fusing shell adds mass to the helium core, the core contracts even more, further increasing the star's central temperature. When the central temperature finally reaches 100 million (10^8) K, **core helium fusion**—that is, thermonuclear fusion of helium in the core—begins. As a result, the aging star again has a central energy source for the first time since leaving the main sequence (**Figure 21-5**).

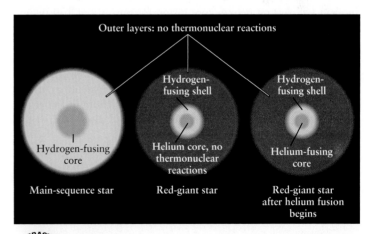

figure 21-5

INTERACTIVE EXERCISE 21.1

Stages in Stellar Evolution In a main-sequence star (left), hydrogen is converted into helium in the core. When the core hydrogen is exhausted, hydrogen fusion continues in a shell, and the star expands to become a red giant (center). When the temperature in the red giant's core becomes high enough because of contraction, core helium fusion begins (right). (These three pictures are *not* drawn to scale. The star is much larger in its red-giant phase than in its main-sequence phase, then shrinks somewhat when core helium fusion begins.)

Helium fusion occurs in two steps. First, two helium nuclei combine to form a beryllium nucleus:

$$^4He + {}^4He \rightarrow {}^8Be$$

This particular beryllium isotope, which has four protons and four neutrons, is very unstable and breaks into two helium nuclei soon after it forms. However, in the star's dense core a third helium nucleus may strike the ^8Be nucleus before it has a chance to fall apart. Such a collision creates a stable isotope of carbon and releases energy in the form of a gamma-ray photon (γ):

$$^8Be + {}^4He \rightarrow {}^{12}C + \gamma$$

This process of fusing three helium nuclei to form a carbon nucleus is called the **triple alpha process,** because helium nuclei (^4He) are also called **alpha particles** by nuclear physicists. Some of the carbon nuclei created in this process can fuse with an additional helium nucleus to produce a stable isotope of oxygen and release more energy:

$$^{12}C + {}^4He \rightarrow {}^{16}O + \gamma$$

Thus, both carbon and oxygen make up the "ash" of helium fusion. It is interesting to note that ^{12}C and ^{16}O are the most common isotopes of carbon and oxygen, respectively; the vast majority of the carbon atoms in your body are ^{12}C, and almost all the oxygen you breathe is ^{16}O. We will discuss the significance of this in Section 21-4.

The second step in the triple alpha process and the process of oxygen formation both release energy. The onset of these reactions reestablishes thermal equilibrium and prevents any further gravitational contraction of the star's core. A mature red giant fuses helium in its core for a much shorter time than it spent fusing hydrogen in its core as a main-sequence star. For example, in the distant future the Sun will sustain helium core fusion for only about 100 million years. (While this is going on, hydrogen fusion is still continuing in a shell around the core.)

How helium fusion begins at a red giant's center depends on the mass of the star. In high-mass stars (greater than 2 to 3 $M_\odot$), helium fusion begins gradually as temperatures in the star's core approach 10^8 K. In low-mass stars (less than 2 to 3 $M_\odot$), helium fusion begins explosively and suddenly, in what is called the **helium flash.** Table 21-2 summarizes these differences.

The helium flash occurs because of unusual conditions that develop in the core of a low-mass star as it becomes a red giant. To appreciate these conditions we must first understand how an ordinary gas behaves. Then we can explore how the densely packed electrons at the star's center alter this behavior.

When a gas is compressed, it usually becomes denser and warmer. To describe this process, scientists use the convenient concept of an **ideal gas,** which has a simple relationship between pressure, temperature, and density. Specifically, the pressure exerted by an ideal gas is directly proportional to both the density and the temperature of the gas. Many real gases actually behave like an ideal gas over a wide range of temperatures and densities.

Under most circumstances, the gases inside a star act like an ideal gas. If the gas is compressed, it heats up, and if it expands, it cools down (see Box 21-1). This behavior serves as a safety valve, ensuring that the star does not explode. For example, if energy production overheats the star's core, the core expands, cooling the gases and slowing the rate of thermonuclear reactions. Conversely, if too little energy is being created to support the star's overlying layers, the core compresses, increasing the temperature and thus speeding up the thermonuclear reactions to increase the energy output.

In a low-mass red giant, however, the core behaves very differently from an ideal gas. The core must be compressed tremendously in order to become hot enough for helium fusion to begin. At these extreme pressures and temperatures, the atoms are completely ionized, and most of the core consists of nuclei and detached electrons. Eventually, the free electrons become so closely crowded that a limit to further compression is reached, as predicted by a remarkable law of quantum mechanics called the **Pauli exclusion principle.** Formulated in 1925 by the Austrian physicist Wolfgang Pauli, this principle states that two electrons cannot simultaneously occupy the same quantum state. A quantum state is a particular set of circumstances concerning locations and speeds that are available to a particle. In the submicroscopic world of electrons, the Pauli exclusion principle is analogous to saying that you can't have two things in the same place at the same time.

Just before the onset of helium fusion, the electrons in the core of a low-mass star are so closely crowded together that any further compression would violate the Pauli exclusion principle. Because the electrons cannot be squeezed any closer together, they produce a powerful pressure that resists further core contraction.

This phenomenon, in which closely packed particles resist compression as a consequence of the Pauli exclusion principle, is called **degeneracy.** Astronomers say that the electrons in the helium-rich core of a low-mass red giant are "degenerate," and that the core is supported by **degenerate-electron pressure.** This degenerate pressure, unlike the pressure of an ideal gas, does not depend on temperature. Remarkably, you can find degenerate electrons on Earth in an ordinary piece of metal (Figure 21-6).

When the temperature in the core of a low-mass red giant reaches the high level required for the triple alpha process, energy begins to be released. The helium heats up, which makes the triple alpha process happen even faster. However, the pressure provided by the degenerate electrons is independent of the temperature, so the pressure does not change. Without the "safety valve" of increasing pressure, the star's core cannot expand and cool. The rising temperature causes the helium to fuse at an ever-increasing rate, producing the helium flash.

Eventually, the temperature becomes so high that the electrons in the core are no longer degenerate. The electrons then behave like an

table 21-2	How Helium Core Fusion Begins in Different Red Giants	
Mass of star	**Onset of helium burning in core**	
Less than 2–3 solar masses	Explosive (helium flash)	
More than 2–3 solar masses	Gradual	

Figure 21-6   R I ⊻ U X G

Degenerate Electrons The electrons in a piece of metal, like the chrome grille on this classic car, are so close together that they are affected by the Pauli exclusion principle. The resulting degenerate-electron pressure helps make metals strong and difficult to compress. A more powerful version of this same effect happens inside the cores of low-mass red-giant stars. (Santokh Kochar/PhotoDisc)

ideal gas and the star's core expands, terminating the helium flash. These events occur so rapidly that the helium flash is over in seconds, after which the star's core settles down to a steady rate of helium fusion.

CAUTION ❗ The term "helium flash" might give you the impression that a star emits a sudden flash of light when the helium flash occurs. If this were true, it would be an incredible sight. During the brief time interval when the helium flash occurs, the helium-fusing core is 10^{11} times more luminous than the present-day Sun, comparable to the total luminosity of all the stars in the Milky Way Galaxy! But in fact the helium flash has no immediately visible consequences—for two reasons. First, much of the energy released during the helium flash goes into heating the core and terminating the degenerate state of the electrons. Second, the energy that does escape the core is largely absorbed by the star's outer layers, which are quite opaque (just like the Sun's present-day interior; see Section 18-2). Therefore, the explosive drama of the helium flash takes place where it cannot be seen directly.

Whether a helium flash occurs or not, the onset of core helium fusion actually causes a *decrease* in the luminosity of the star. This is the opposite of what you might expect—after all, turning on a new energy source should make the luminosity greater, not less. What happens is that after the onset of core helium fusion, a star's superheated core expands like an ideal gas. (If the star is of sufficiently low mass to have had a degenerate core, the increased temperature after the helium flash makes the core too hot to remain degenerate. Hence, these stars also end up with cores that behave like ideal gases.) Temperatures drop around the expanding core, so the hydrogen-fusing shell reduces its energy output and the star's

luminosity decreases. This allows the star's outer layers to contract and heat up. Consequently, a post–helium-flash star is less luminous, hotter at the surface, and smaller than a red giant.

Core helium fusion lasts for only a relatively short time. Calculations suggest that a 1-$M_\odot$ star like the Sun sustains core hydrogen fusion for about 12 billion (1.2×10^{10}) years, followed by about 250 million (2.5×10^8) years of shell hydrogen fusion leading up to the helium flash. After the helium flash, such a star can fuse helium in its core (while simultaneously fusing hydrogen in a shell around the core) for only 100 million (10^8) years, a mere 1% of its main-sequence lifetime. Figure 21-7 summarizes these evolutionary stages. In Chapter 22 we will take up the story of what happens after a star has consumed all the helium in its core.

Here is the story of post–main-sequence evolution in its briefest form: Before the beginning of core helium fusion, the star's core compresses and the outer layers expand, and just after core helium fusion begins, the core expands and the outer layers compress. We will see in Chapter 22 that this behavior, in which the inner and outer regions of the star change in opposite ways, occurs again and again in the final stages of a star's evolution.

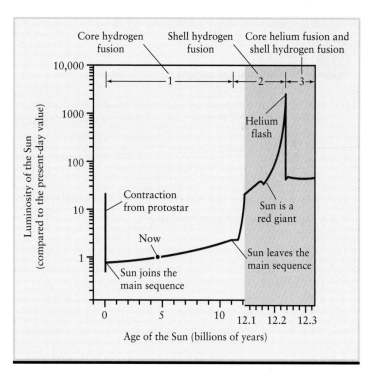

Figure 21-7

Stages in the Evolution of the Sun This diagram shows how the luminosity of the Sun (a 1-$M_\odot$ star) changes over time. The Sun began as a protostar whose luminosity decreased rapidly as the protostar contracted. Once established as a main-sequence star with core hydrogen fusion, the Sun's luminosity increases slowly over billions of years. The post–main-sequence evolution is much more rapid, so a different time scale is used in the right-hand portion of the graph. (Adapted from Mark A. Garlick, based on calculations by I.-Juliana Sackmann and Kathleen E. Kramer)

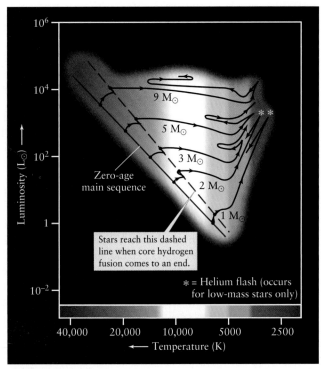

(a) Post–main-sequence evolutionary tracks of five stars with different mass

(b) H-R diagram of 20,853 stars—note the width of the main sequence

Figure 21-8
Stellar Evolution on and off the Main Sequence

(a) The two lowest-mass stars shown here (1 M$_\odot$ and 2 M$_\odot$) undergo a helium flash at their centers, as shown by the asterisks. In the high-mass stars, core helium fusion ignites more gradually where the evolutionary tracks make a sharp downward turn in the red-giant region of the diagram.

(b) Data from the *Hipparcos* satellite (see Section 19-1) was used to create this H-R diagram. The thickness of the main sequence is due in large part to stars evolving during their main-sequence lifetimes. (a: Adapted from I. Iben; b: Adapted from M. A. C. Perryman)

21-3 | H-R diagrams and observations of star clusters reveal how red giants evolve

To see how stars evolve during and after their main-sequence lifetimes, it is helpful to follow them on a Hertzsprung-Russell (H-R) diagram. On such a diagram, zero-age main-sequence stars lie along a line called the **zero-age main sequence**, or **ZAMS** (**Figure 21-8a**). These stars have just emerged from their protostar stage, are steadily fusing hydrogen into helium in their cores, and have attained hydrostatic equilibrium. With the passage of time, hydrogen in a main-sequence star's core is converted to helium, the luminosity slowly increases, the star slowly expands, and the star's position on the H-R diagram inches away from the ZAMS. As a result, the main sequence on an H-R diagram is a fairly broad band rather than a narrow line (Figure 21-8b).

The dashed line in Figure 21-8a denotes stars whose cores have been exhausted of hydrogen and in which core hydrogen fusion has ceased. From there, the points representing high-mass stars (3 M$_\odot$, 5 M$_\odot$, and 9 M$_\odot$) move rapidly from left to right across the H-R diagram. This means that, although the star's surface temperature is decreasing, its surface area is increasing at a rate that keeps its overall luminosity roughly constant. During this transition, the star's core contracts and its outer layers expand as energy flows outward from the hydrogen-fusing shell.

Just before core helium fusion begins, the evolutionary tracks of high-mass stars turn upward in the red-giant region of the H-R diagram (to the upper right of the main sequence). After core helium fusion begins, however, the cores of these stars expand, the outer layers contract, and the evolutionary tracks back away from these temporary peak luminosities. The tracks then wander back and forth in the red-giant region while the stars readjust to their new energy sources.

Figure 21-8a also shows the evolutionary tracks of two low-mass stars (1 M$_\odot$ and 2 M$_\odot$). The onset of core helium fusion in these stars occurs with a helium flash, indicated by the red asterisks in the figure. As we saw in the previous section, after the helium flash, these stars shrink and become less luminous. The decrease in size is proportionally greater than the decrease in luminosity, and so the surface temperatures increase. Hence, after the helium flash, the evolutionary tracks move down and to the left.

We can summarize our understanding of stellar evolution from birth through the onset of helium fusion by following the evolution of a hypothetical cluster of stars. The eight H-R diagrams in **Figure 21-9** are from a computer simulation of the evolution of 100 stars that differ only in initial mass. In this simulation all stars form at the same

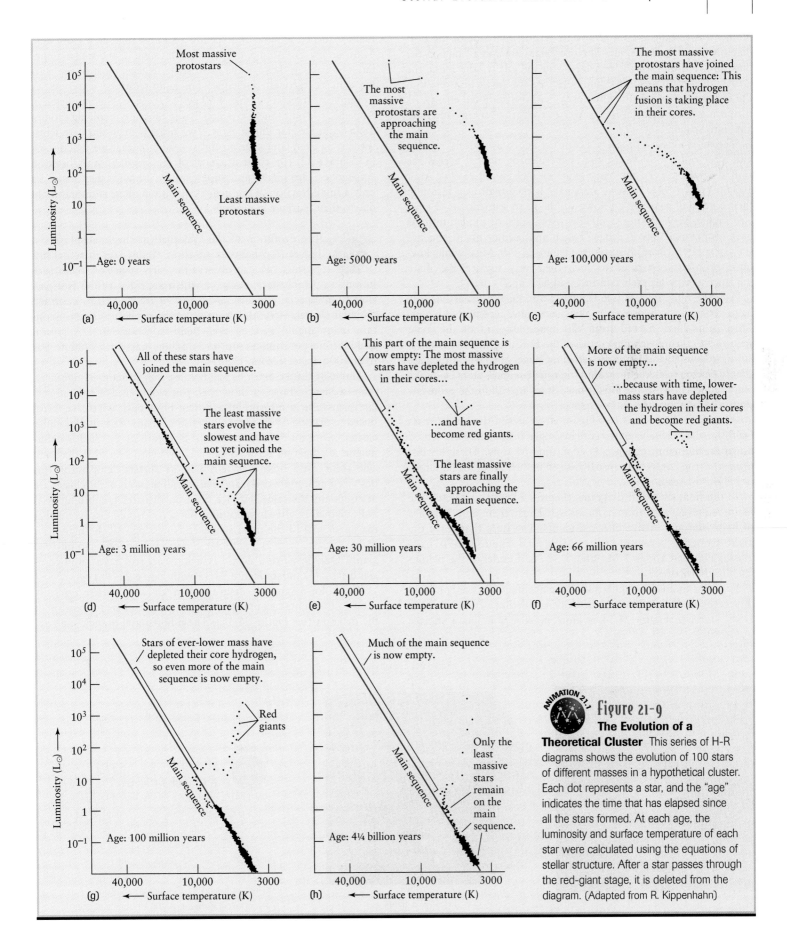

(a) Age: 0 years

Most massive protostars

Least massive protostars

Main sequence

(b) Age: 5000 years

The most massive protostars are approaching the main sequence.

Main sequence

(c) Age: 100,000 years

The most massive protostars have joined the main sequence: This means that hydrogen fusion is taking place in their cores.

Main sequence

(d) Age: 3 million years

All of these stars have joined the main sequence.

The least massive stars evolve the slowest and have not yet joined the main sequence.

Main sequence

(e) Age: 30 million years

This part of the main sequence is now empty: The most massive stars have depleted the hydrogen in their cores...

...and have become red giants.

The least massive stars are finally approaching the main sequence.

Main sequence

(f) Age: 66 million years

More of the main sequence is now empty...

...because with time, lower-mass stars have depleted the hydrogen in their cores and become red giants.

Main sequence

(g) Age: 100 million years

Stars of ever-lower mass have depleted their core hydrogen, so even more of the main sequence is now empty.

Red giants

Main sequence

(h) Age: 4¼ billion years

Much of the main sequence is now empty.

Only the least massive stars remain on the main sequence.

Main sequence

Luminosity (L_☉)

← Surface temperature (K)

figure 21-9
ANIMATION 21.1
The Evolution of a Theoretical Cluster This series of H-R diagrams shows the evolution of 100 stars of different masses in a hypothetical cluster. Each dot represents a star, and the "age" indicates the time that has elapsed since all the stars formed. At each age, the luminosity and surface temperature of each star were calculated using the equations of stellar structure. After a star passes through the red-giant stage, it is deleted from the diagram. (Adapted from R. Kippenhahn)

moment. All begin as cool protostars on the right side of the H-R diagram (see Figure 21-9*a*). The protostars are spread out on the diagram according to their masses, and the greater the mass, the greater the protostar's initial luminosity. As we saw in Section 20-3, the source of a protostar's luminosity is its gravitational energy. As the protostar contracts, this is converted to thermal energy and radiated into space.

The most massive protostars contract and heat up very rapidly, and after only 5000 years, they have already moved across the H-R diagram toward the main sequence (see Figure 21-9*b*). After 100,000 years, these massive stars have ignited hydrogen fusion in their cores and have settled down on the main sequence as O stars (see Figure 21-9*c*). After 3 million years, stars of moderate mass have also ignited core hydrogen fusion and become main-sequence stars of spectral classes B and A (see Figure 21-9*d*). Meanwhile, low-mass protostars continue to inch their way toward the main sequence as they leisurely contract and heat up.

After 30 million years (see Figure 21-9*e*), the most massive stars have depleted the hydrogen in their cores and become red giants. These stars have moved from the upper left end of the main sequence to the upper right corner of the H-R diagram. (This simulation follows stars only to the red-giant stage, after which they are simply deleted from the diagram.) Intermediate-mass stars lie on the main sequence, while the lowest-mass stars are still in the protostar stage and lie above the main sequence.

After 66 million years (see Figure 21-9*f*), even the lowest-mass protostars have finally ignited core hydrogen fusion and have settled down on the main sequence as cool, dim, M stars. These lowest-mass stars can continue to fuse hydrogen in their cores for hundreds of billions of years.

In the final two H-R diagrams (Figures 21-9*g* and 21-9*h*), the main sequence gets shorter as stars exhaust their core supplies of hydrogen and evolve into red giants. The stars that leave the main sequence between Figure 21-9*g* and Figure 21-9*h* have masses between about 1 $M_\odot$ and 3 $M_\odot$ and undergo the helium flash in their cores.

For all stars in this simulation, the giant stage lasts only a brief time compared to the star's main-sequence lifetime. Compared to a 1-$M_\odot$ star (see Figure 21-7), a more massive star has a shorter main-sequence lifetime *and* spends a shorter time as a giant star. Thus, at any given time, only a small fraction of the stellar population is passing through the giant stage. Hence, most of the stars we can see through telescopes are main-sequence stars. As an example, of the stars within 4.00 pc (13.05 ly) of the Sun listed in Appendix 4, only one—Procyon A—is presently evolving from a main-sequence star into a giant. Two other nearby stars are white dwarfs, an even later stage in stellar evolution that we will discuss in Chapter 22. (By contrast, most of the *brightest* stars listed in Appendix 5 are giants and supergiants. Although they make up only a small fraction of the stellar population, these stars stand out due to their extreme luminosity.)

 The evolution of the hypothetical cluster displayed in Figure 21-9 helps us interpret what we see in actual star clusters. We can observe the early stages of stellar evolution in open clusters, many of which are quite young (see Section 20-6, especially Figure 20-17 and Figure 20-18). By contrast, **globular clusters**—so called because of their spherical shape—contain many highly evolved post–main-sequence stars. A typical globular cluster contains up to 1 million stars in a volume less than 100 parsecs across (**Figure 21-10**).

Globular clusters must be old, because they contain no high-mass main-sequence stars. To show this, you would measure the apparent magnitude (a measure of apparent brightness, which we introduced in Section 19-3) and color ratio of many stars in a globular cluster, then plot the data as shown in **Figure 21-11**. Such a **color-magnitude diagram** for a cluster is equivalent to an H-R diagram. The color ratio of a star tells you its surface temperature (as described in Section 19-4), and because all the stars in the cluster are at essentially the same distance from us, their relative brightnesses indicate their relative luminosities. What you would discover is that the upper half of the main sequence is missing. All the high-mass main-sequence stars in a globular cluster evolved long ago into red giants. Only low-mass, slowly evolving stars still have core hydrogen fusion. (Compare Figure 21-11 with Figure 21-9*h*.)

The color-magnitude diagram of a globular cluster typically shows a horizontal grouping of stars in the left-of-center portion of the diagram (see Figure 21-11). These **horizontal-branch stars** are

Horizontal-branch stars

Red giants

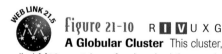

Figure 21-10   R **I** **V** U X G
A Globular Cluster This cluster, called M10, contains a few hundred thousand stars within a diameter of only 20 pc (70 ly). It lies approximately 5000 pc (16,000 ly) from the Earth in the constellation Ophiuchus. Most of the stars in this image are either red giants or blue, horizontal-branch stars with both core helium fusion and shell hydrogen fusion. (T. Credner and S. Kohle, Astronomical Institutes of the University of Bonn)

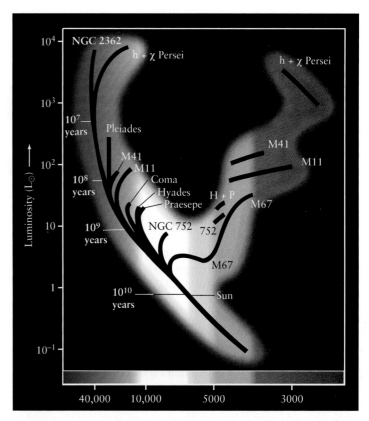

figure 21-11

A Color-Magnitude Diagram of a Globular Cluster Each dot in this diagram represents the apparent visual magnitude (a measure of the brightness as seen through a V filter) and surface temperature (as measured by the color ratio b_V/b_B) of a star in the globular cluster M55 in Sagittarius. Because all the stars in M55 are at essentially the same distance from the Earth (about 6000 pc or 20,000 ly), their apparent visual magnitudes are a direct measure of their luminosities. Note that the upper half of the main sequence is missing. (Adapted from D. Schade, D. VandenBerg, and F. Hartwick)

figure 21-12

An H-R Diagram for Open Star Clusters The black bands indicate where stars from various open clusters fall on the H-R diagram. The age of a cluster can be estimated from the location of the cluster's turnoff point, where the cluster's most massive stars are just now leaving the main sequence. The times for these turnoff points are listed alongside the main sequence. For example, the Pleiades cluster turnoff point is near the 10^8-year point, so this cluster is about 10^8 years old. (Adapted from A. Sandage)

post–helium-flash, low-mass stars with luminosities of about 50 L$_\odot$ in which there is both core helium fusion and shell hydrogen fusion. In years to come, these stars will move back toward the red-giant region as their fuel is devoured. The inset in Figure 21-10 shows some of these stars. Our own Sun will go through a horizontal-branch phase in the distant future; this is the phase labeled by the number 3 at the far right of Figure 21-7.

An H-R diagram can be used to determine the age of a cluster. In the H-R diagram for a very young cluster, all the stars are on or near the main sequence. (An example is the open cluster NGC 2264, shown in Figure 20-17.) As a cluster gets older, however, stars begin to leave the main sequence. The high-mass, high-luminosity stars are the first to become red giants. Over the years, the main sequence gets shorter and shorter, like a candle burning down.

The age of a cluster can be found from the **turnoff point,** which is the top of the surviving portion of the main sequence on the cluster's H-R diagram (see Figure 21-11). The stars at the turnoff point are just now exhausting the hydrogen in their cores, so their main-

sequence lifetime is equal to the age of the cluster. For example, in the case of the globular cluster M55 plotted in Figure 21-11, 0.8-M$_\odot$ stars have just left the main sequence, indicating that the cluster's age is more than 12 billion (1.2×10^{10}) years (see Table 21-1).

Figure 21-12 shows data for several star clusters plotted on a single H-R diagram. This graph also shows turnoff-point times from which the ages of the clusters can be estimated.

21-4 | Stellar evolution has produced two distinct populations of stars

Studies of star clusters reveal a curious difference between the youngest and oldest stars in our Galaxy. Stars in the youngest clusters (those with most of their main sequences still intact) are said to be **metal rich,** because their spectra contain many prominent spectral lines of heavy elements. (Recall from Section 19-5 that

The spectrum of this Population II star shows absorption lines of hydrogen (such as H_γ and H_δ) but only very weak absorption lines of metals ... such a star is **metal-poor.**

H_δ

Wavelength ⟶

H_γ

(a)

(b)

The spectrum of this Population I star has stronger absorption lines of metals ... such a star is **metal-rich.**

Figure 21-13   R I **V** U X G

Spectra of a Metal-Poor Star and a Metal-Rich Star The abundance of metals (elements heavier than hydrogen and helium) in a star can be inferred from its spectrum. These spectra compare (a) a metal-poor, Population II star and (b) a metal-rich, Population I star (the Sun) of the same surface temperature. We described the hydrogen absorption lines H_γ (wavelength 434 nm) and H_δ (wavelength 410 nm) in Section 5-8. (Lick Observatory)

astronomers use the term "metal" to denote any element other than hydrogen and helium, which are the two lightest elements.) Such stars are also called **Population I stars.** The Sun is a relatively young, metal-rich, Population I star.

By contrast, the spectra of stars in the oldest clusters show only weak lines of heavy elements. These ancient stars are thus said to be **metal poor,** because heavy elements are only about 3% as abundant in these stars as in the Sun. They are also called **Population II stars.** The stars in globular clusters are metal poor, Population II stars. **Figure 21-13** shows the difference in spectra between a metal-poor, Population II star and the Sun (a metal-rich, Population I star).

 Note that "metal rich" and "metal poor" are relative terms. In even the most metal-rich star known, metals make up just a few percent of the total mass of the star.

To explain why there are two distinct populations of stars, we must go back to the Big Bang, the explosive origin of the universe that took place some 13.7 billion years ago. As we will discuss in Chapter 29, the early universe consisted almost exclusively of hydrogen and helium, with almost no heavy elements (metals). The first stars to form were likewise metal poor. The least massive of these have survived to the present day and are now the ancient stars of Population II.

The more massive of the original stars evolved more rapidly and no longer shine. But as these stars evolved, helium fusion in their cores produced metals—carbon and oxygen. In the most massive stars, as we will learn in Chapter 23, further thermonuclear reactions produced even heavier elements. As these massive original stars aged and died, they expelled their metal-enriched gases into space. (The star shown in Figure 21-4 is going through such a mass-loss phase.) This expelled material joined the interstellar medium and was eventually incorporated into a second generation of stars that have a higher concentration of heavy elements. These metal-rich members of the second stellar generation are the Population I stars, of which our Sun is an example.

 Be careful not to let the designations of the two stellar populations confuse you. Population *I* stars are members of a *second* stellar generation, while Population *II* stars belong to an older *first* generation.

The relatively high concentration of heavy elements in the Sun means that the solar nebula, from which both the Sun and planets formed (see Section 8-4), must likewise have been metal rich. The Earth is composed almost entirely of heavy elements, as are our bodies. Thus, our very existence is intimately linked to the Sun's being a Population I star. A planet like the Earth probably could not have formed from the metal-poor gases that went into making Population II stars.

The concept of two stellar populations provides insight into our own origins. Recall from Section 21-2 that helium fusion in red-giant stars produces the same isotopes of carbon (^{12}C) and oxygen (^{16}O) that are found most commonly on the Earth. The reason is that the Earth's carbon and oxygen atoms actually *were* produced by helium fusion. These reactions occurred billions of years ago within an earlier generation of stars that died and gave up their atoms to the interstellar medium—the same atoms that later became part of our solar system, our planet, and our bodies. We are literally children of the stars.

21-5 | Many mature stars pulsate

 SCIENCE IN PROCESS   *By studying variable stars, astronomers gain insight into late stages of stellar evolution*

We saw in Section 18-3 that the surface of our Sun vibrates in and out, although by only a small amount. But other stars undergo substantial changes in size, alternately swelling and shrinking. As these stars pulsate, they also vary dramatically in brightness. We now understand that these **pulsating variable stars** are actually evolved, post–main-sequence stars.

Pulsating variable stars were first discovered in 1595 by the Dutch minister and amateur astronomer David Fabricius. He noticed that the star o (omicron) Ceti is sometimes bright enough to be easily seen with the naked eye but at other times fades to invisibility (Figure 21-14). By 1660, astronomers realized that these brightness variations repeated with a period of 332 days. Seventeenth-century astronomers were so enthralled by this variable that they renamed the star Mira ("wonderful").

Mira is an example of a class of pulsating stars called **long-period variables.** These stars are cool red giants that vary in brightness by a factor of 100 or more over a period of months or years. With surface temperatures of about 3500 K and average luminosities that range from about 10 to 10,000 $L_\odot$, they occupy the upper right side of the H-R diagram (Figure 21-15). Some, like Mira, are periodic, but others are irregular. Many eject large amounts of gas and dust into space.

Astronomers do not fully understand why some cool red giants become long-period variables. It is difficult to calculate accurate stellar models to describe such huge stars with extended, tenuous atmospheres.

Astronomers have a much better understanding of other pulsating stars, called **Cepheid variables,** or simply Cepheids. A Cepheid variable is recognized by the characteristic way in which its light output varies—rapid brightening followed by gradual dimming. They are named for δ (delta) Cephei, an example of this type of star discovered in 1784 by John Goodricke, a deaf, mute, 19-year-old English amateur astronomer. He found that at its most brilliant, δ Cephei is 2.3 times as bright as at its dimmest. The cycle of brightness variations repeats every 5.4 days. (Sadly, Goodricke paid for his discoveries with his life; he caught pneumonia while making his nightly observations and died before his twenty-second birthday.) The surface temperatures and luminosities of the Cepheid variables place them in the upper middle of the H-R diagram (see Figure 21-15).

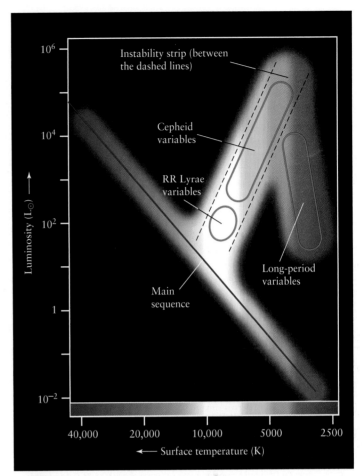

Figure 21-15

Variable Stars on the H-R Diagram Pulsating variable stars are found in the upper right of the H-R diagram. Long-period variables like Mira are cool red giant stars that pulsate slowly, changing their brightness in a semi-regular fashion over months or years. Cepheid variables and RR Lyrae variables are located in the instability strip, which lies between the main sequence and the red-giant region. A star passing through this strip along its evolutionary track becomes unstable and pulsates.

(a)   (b)

Figure 21-14   R I **V** U X G

Mira—A Long-Period Variable Star Mira, or o (omicron) Ceti, is a variable star whose luminosity varies with a 332-day period. At its dimmest, as in (a) (photographed in December 1961), Mira is less than 1% as bright as when it is at maximum, as in (b) (January 1965). These brightness variations occur because Mira pulsates. (Lowell Observatory)

After core helium fusion begins, mature stars move across the middle of the H-R diagram. Figure 21-8 shows the evolutionary tracks of high-mass stars crisscrossing the H-R diagram. Post–helium-flash, low-mass stars also cross the middle of the H-R diagram between the red-giant region and the horizontal branch.

During these transitions across the H-R diagram, a star can become unstable and pulsate. In fact, there is a region on the H-R diagram between the upper main sequence and the red-giant branch called the **instability strip** (see Figure 21-15). When an evolving star passes through this region, the star pulsates and its brightness varies periodically. **Figure 21-16a** shows the brightness variations of δ Cephei, which lies within the instability strip.

A Cepheid variable brightens and fades because the star's outer envelope cyclically expands and contracts. The first to observe this

(a) The light curve of δ Cephei (a graph of brightness versus time)

(b) Radial velocity versus time for δ Cephei
(positive: star is contracting; negative: star is expanding)

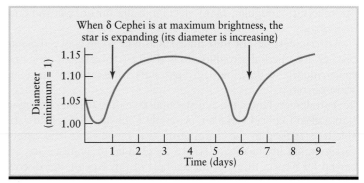

(c) Surface temperature versus time for δ Cephei

(d) Diameter versus time for δ Cephei

Figure 21-16

δ Cephei—A Pulsating Star (a) As δ Cephei pulsates, it brightens quickly (the light curve moves upward sharply) but fades more slowly (the curve declines more gently). The increases and decreases in brightness are nearly in step with variations in (b) the star's radial velocity

(positive when the star contracts and the surface moves away from us, negative when the star expands and the surface approaches us) as well as in (c) the star's surface temperature. (d) The star is still expanding when it is at its brightest and hottest (compare with parts **a** and **b**).

was the Russian astronomer Aristarkh Belopol'skii, who noticed in 1894 that spectral lines in the spectrum of δ Cephei shift back and forth with the same 5.4-day period as that of the magnitude variations. From the Doppler effect, we can translate these wavelength shifts into radial velocities and draw a velocity curve (Figure 21-16b). Negative velocities mean that the star's surface is expanding toward us; positive velocities mean that the star's surface is receding. Note that the light curve and velocity curve are mirror images of each other. The star is brighter than average while it is expanding and dimmer than average while contracting.

When a Cepheid variable pulsates, the star's surface oscillates up and down like a spring. During these cyclical expansions and contractions, the star's gases alternately heat up and cool down. Figure 21-16c shows the resulting changes in the star's surface temperature. Figure 21-16d graphs the periodic changes in the star's diameter.

Just as a bouncing ball eventually comes to rest, a pulsating star would soon stop pulsating without something to keep its oscillations going. In 1914, the British astronomer Arthur Eddington suggested that a Cepheid pulsates because the star is more opaque when compressed than when expanded. When the star is compressed,

trapped heat increases the internal pressure, which pushes the star's surface outward. When the star expands, the heat escapes, the internal pressure drops, and the star's surface falls inward.

In the 1960s, the American astronomer John Cox followed up on Eddington's idea and proved that helium is what keeps Cepheids pulsating. Normally, when a star's helium is compressed, the gas increases in temperature and becomes more transparent. But in certain layers near the star's surface, compression may ionize helium (remove one of its electrons) instead of raising its temperature. Ionized helium gas is quite opaque, so these layers effectively trap heat and make the star expand, as Eddington suggested. This expansion cools the outer layers and makes the helium ions recombine with electrons, which makes the gas more transparent and releases the trapped energy. The star's surface then falls inward, recompressing the helium, and the cycle begins all over again.

Cepheid variables are important because they have two properties that allow astronomers to determine the distances to very remote objects. First, Cepheids can be seen even at distances of millions of parsecs. This is because they are very luminous, ranging from a few hundred times solar luminosity to more than 10^4 $L_\odot$. Second, there is a direct relationship between a Cepheid's period and its average luminosity: The dimmest Cepheid variables pulsate rapidly, with periods of 1 to 2 days, while the most luminous Cepheids pulsate with much slower periods of about 100 days.

Figure 21-17 shows this **period-luminosity relation**. By measuring the period of a distant Cepheid's brightness variations and using a graph like Figure 21-17, an astronomer can determine the star's luminosity. By also measuring the star's apparent brightness, the distance to the Cepheid can be found by using the inverse-square law (see Section 19-2). By applying the period-luminosity relation in this way to Cepheids in other galaxies, astronomers have been able to calculate the distances to those galaxies with great accuracy. (Box 19-2 gives an example of such a calculation.) As we will see in Chapters 26 and 28, such measurements play an important role in determining the overall size and structure of the universe.

How a Cepheid pulsates depends on the amount of heavy elements in the star's outer layers, because even trace amounts of these elements can have a large effect on how opaque the stellar gases are. Hence, Cepheids are classified according to their metal content. If the star is a metal-rich, Population I star, it is called a **Type I Cepheid**; if it is a metal-poor, Population II star, it is called a **Type II Cepheid**. As Figure 21-17 shows, these two types of Cepheids exhibit different period-luminosity relations. In order to know which period-luminosity relation to apply to a given Cepheid, an astronomer must determine the star's metal content from its spectrum (see Figure 21-13).

The evolutionary tracks of mature, high-mass stars pass back and forth through the upper end of the instability strip on the H-R diagram. These stars become Cepheids when helium ionization occurs at just the right depth to drive the pulsations. For stars on the high-temperature (left) side of the instability strip, helium ionization occurs too close to the surface and involves only an insignificant fraction of the star's mass. For stars on the cool (right) side of the instability strip, convection in the star's outer layers prevents the storage of the energy needed to drive the pulsations. Thus, Cepheids exist only in a narrow temperature range on the H-R diagram.

Stars of lower mass do not become Cepheids. Instead, after undergoing the helium flash, their evolutionary tracks pass through the lower end of the instability strip as they move along the horizontal branch. Some of these stars become **RR Lyrae variables,** named for their prototype in the constellation Lyra. RR Lyrae variables all have periods shorter than one day and roughly the same average luminosity as horizontal-branch stars, about 100 $L_\odot$. In fact, the RR Lyrae region of the instability strip (see Figure 21-15) is actually a segment of the horizontal branch. RR Lyrae stars are all metal-poor, Population II stars. Many have been found in globular clusters, and they have been used to determine the distances to those clusters in the same way that Cepheids are used to find the distances to other galaxies. In Chapter 25 we will see how RR Lyrae stars helped astronomers determine the size of the Milky Way Galaxy.

In some cases the expansion speed of a pulsating star exceeds the star's escape speed. When this happens, the star's outer layers are ejected completely. We will see in Chapter 22 that dying stars eject significant amounts of mass in this way, renewing and enriching the interstellar medium for future generations of stars.

21-6 | Mass transfer can affect the evolution of close binary star systems

We have outlined what happens when a main-sequence star evolves into a red giant. What we have ignored is that more than half of all stars are members of multiple-star systems, including binaries. If the stars in such a system are widely separated, the individual stars follow the same course of evolution as if they were isolated. In a **close binary,** however, when one star expands to become a red giant, its outer layers can be gravitationally captured by the nearby companion star. In other words, a bloated red giant in a close binary system can dump gas onto its companion, a process called **mass transfer.**

Our modern understanding of mass transfer in close binaries is based on the work of French mathematician Edouard Roche. In the mid-1800s, Roche studied how rotation and mutual tidal interaction affect the stars in a binary system. Tidal forces cause the two stars in a close binary to keep the same sides facing each other, just as our Moon keeps its same side facing the Earth (see Section 4-8). But because stars are gaseous, not solid, rotation and tidal forces can have significant effects on their shapes.

In widely separated binaries, the stars are so far apart that tidal effects are small, and, therefore, the stars are nearly perfect spheres. In close binaries, where the separation between the stars is not much greater than their sizes, tidal effects are strong, causing the stars to be somewhat egg-shaped.

Figure 21-17

Period-Luminosity Relations for Cepheids The greater the average luminosity of a Cepheid variable, the longer its period and the slower its pulsations. Note that there are actually two distinct period-luminosity relations—one for Type I Cepheids and one for the less luminous Type II Cepheids. (Adapted from H. C. Arp)

Roche discovered a surface that marks the gravitational domain of each star in a close binary. **Figure 21-18a** shows the outline of this surface as a dashed line. The two halves of this surface, each of which encloses one of the stars, are known as **Roche lobes.** The more massive star is always located inside the larger Roche lobe. If gas from a star leaks over its Roche lobe, it is no longer bound by gravity to that star. This escaped gas is free either to fall onto the companion star or to escape from the binary system.

The point where the two Roche lobes touch, called the **inner Lagrangian point,** is a kind of balance point between the two stars in a binary. It is here that the effects of gravity and rotation cancel each other. When mass transfer occurs in a close binary, gases flow through the inner Lagrangian point from one star to the other.

In many binaries, the stars are so far apart that even during their red-giant stages the surfaces of the stars remain well inside their Roche lobes. As a result, little mass transfer can occur and each star lives out its life as if it were single and isolated. A binary system of this kind is called a **detached binary** (Figure 21-18a).

However, if the two stars are close enough, when one star expands to become a red giant, it may fill or overflow its Roche lobe. Such a system is called a **semidetached binary** (Figure 21-18b). If both stars fill their Roche lobes, the two stars actually touch and the system is called a **contact binary** (Figure 21-18c). It is quite

unlikely, however, that both stars exactly fill their Roche lobes at the same time. It is more likely that they overflow their Roche lobes, giving rise to a common envelope of gas. Such a system is called an **overcontact binary** (Figure 21-18d).

The binary star system Algol (from an Arabic term for "demon") provided the first clear evidence of mass transfer in close binaries. Also called β (beta) Persei, Algol can easily be seen with the naked eye. Ancient astronomers knew that Algol varies periodically in brightness by a factor of more than 2. In 1782, John Goodricke (the discoverer of δ Cephei's variability) first suggested that these brightness variations take place because Algol is an *eclipsing* binary, discussed in Section 19-11. Because the orbital plane of the two stars that make up the binary system is nearly edge-on to our line of sight, one star periodically eclipses the other. Algol's light curve (**Figure 21-19a**) and spectrum show that Goodricke's brilliant hypothesis is correct, and that Algol is a semidetached binary. The detached star (on the right in Figure 21-19a) is a luminous blue main-sequence star, while its less massive companion is a dimmer red giant that fills its Roche lobe.

According to stellar evolution theory, the more massive a star, the more rapidly it should evolve. But in Algol and similar binaries, the more massive star (on the right in Figure 21-19a) is still on the main sequence, whereas the less massive star (on the left in Figure 21-19a)

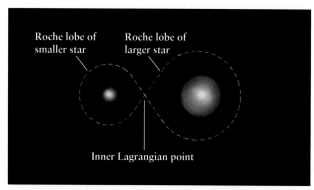

(a) Detached binary: Neither star fills its Roche lobe.

(b) Semi-detached binary: One star fills its Roche lobe.

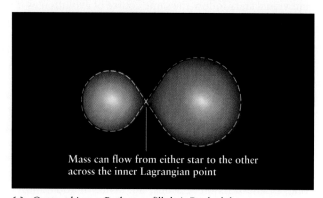

(c) Contact binary: Both stars fill their Roche lobes.

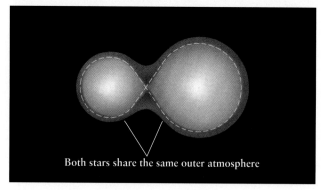

(d) Overcontact binary: Both stars overfill their Roche lobes.

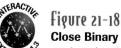

Figure 21-18

Close Binary Star Systems The gravitational domain of a star in a close binary system is called its Roche lobe. The two Roche lobes meet at the inner Lagrangian point. The sizes of the stars relative to their Roche lobes determine whether the system is **(a)** a detached binary, **(b)** a semidetached binary, **(c)** a contact binary, or **(d)** an overcontact binary.

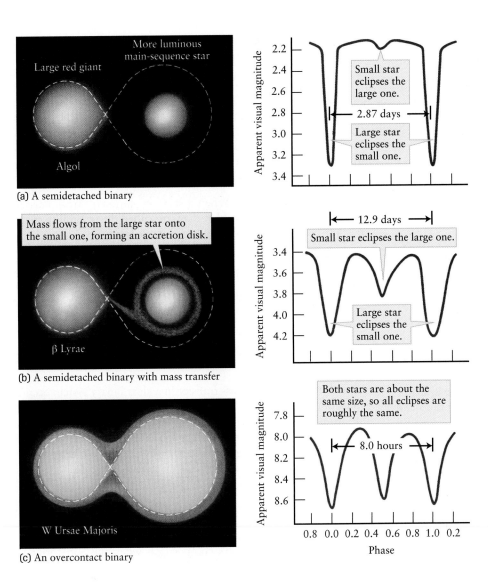

(a) A semidetached binary

Small star eclipses the large one.

2.87 days

Large star eclipses the small one.

Mass flows from the large star onto the small one, forming an accretion disk.

β Lyrae

(b) A semidetached binary with mass transfer

12.9 days

Small star eclipses the large one.

Large star eclipses the small one.

W Ursae Majoris

(c) An overcontact binary

Both stars are about the same size, so all eclipses are roughly the same.

8.0 hours

Phase

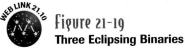

Figure 21-19

Three Eclipsing Binaries

Compare the light curves of these three eclipsing binaries to those in Figure 19-25. **(a)** Algol is a semidetached binary. The deep eclipse occurs when the large red-giant star blocks the light from the smaller but more luminous main-sequence star. (Compare Figure 19-24*b*.) **(b)** β Lyrae's light curve is also at its lowest when the larger star completely eclipses the smaller one. Half an orbital period later, the smaller star partially eclipses the larger one, making a shallower dip in the light curve. **(c)** W Ursae Majoris is an overcontact binary in which both stars overfill their Roche lobes. The extremely short period of this binary indicates that the two stars are very close to each other.

---

has evolved to become a red giant. This seems backward, since the two stars in a binary system form simultaneously and thus are the same age. The explanation for this apparent contradiction is that the red giant in Algol-type binaries was originally the more massive star. As it left the main sequence to become a red giant, this star overflowed its Roche lobe, dumping gas onto the originally less massive companion. Because of the resulting mass transfer, that companion became the more massive star.

Mass transfer is also important in another class of semidetached binaries, called β (beta) Lyrae variables, after their prototype in the constellation Lyra (the Harp). As with Algol, the less massive star in β Lyrae (on the left in Figure 21-19*b*) fills its Roche lobe. Unlike Algol, however, the more massive detached star (on the right in Figure 21-19*b*) is the dimmer of the two stars. Apparently, this detached star is enveloped in a rotating *accretion disk* of gas captured from its bloated companion. This disk partially blocks the light coming from the detached star, making it appear dimmer.

What is the fate of an Algol or β Lyrae system? If the detached star is massive enough, it will evolve rapidly, expanding to also fill its Roche lobe. The result will be an overcontact binary in which the

two stars share the gases of their outer layers. Such binaries are sometimes called W Ursae Majoris stars, after the prototype of this class (Figure 21-19*c*). Mass transfer can also continue even after nuclear reactions cease in one of the stars in a close binary. In Chapters 22 and 23 we will see how mass transfer onto dead stars produces some of the most unusual objects in the sky.

Key Words

alpha particle, p. 472

Cepheid variable, p. 479

close binary, p. 481

color-magnitude diagram, p. 476

contact binary, p. 482

core helium fusion, p. 471

core hydrogen fusion, p. 467

degeneracy, p. 472

degenerate-electron pressure, p. 472

detached binary, p. 482

globular cluster, p. 476

helium flash, p. 472

helium fusion, p. 471

horizontal-branch star, p. 476

Key Ideas

The Main-Sequence Lifetime: The duration of a star's main-sequence lifetime depends on the amount of hydrogen in the star's core and the rate at which the hydrogen is consumed.

• The more massive a star, the shorter is its main-sequence lifetime. The Sun has been a main-sequence star for about 4.56 billion years and should remain one for about another 7 billion years.

• During a star's main-sequence lifetime, the star expands somewhat and undergoes a modest increase in luminosity.

Becoming a Red Giant: Core hydrogen fusion ceases when the hydrogen has been exhausted in the core of a main-sequence star. This leaves a core of nearly pure helium surrounded by a shell through which hydrogen fusion works its way outward in the star. The core shrinks and becomes hotter, while the star's outer layers expand and cool. The result is a red giant star.

• As a star becomes a red giant, its evolutionary track moves rapidly from the main sequence to the red-giant region of the H-R diagram. The more massive the star, the more rapidly this evolution takes place.

Helium Fusion: When the central temperature of a red giant reaches about 100 million K, helium fusion begins in the core. This process, also called the triple alpha process, converts helium to carbon and oxygen.

• In a more massive red giant, helium fusion begins gradually; in a less massive red giant, it begins suddenly, in a process called the helium flash.

• After the helium flash, a low-mass star moves quickly from the red-giant region of the H-R diagram to the horizontal branch.

Star Clusters and Stellar Populations: The age of a star cluster can be estimated by plotting its stars on an H-R diagram.

• The cluster's age is equal to the age of the main-sequence stars at the turnoff point (the upper end of the remaining main sequence).

• As a cluster ages, the main sequence is "eaten away" from the upper left as stars of progressively smaller mass evolve into red giants.

• Relatively young Population I stars are metal rich; ancient Population II stars are metal poor. The metals (heavy elements) in Population I stars were manufactured by thermonuclear reactions in an earlier generation of Population II stars, then ejected into space and incorporated into a later stellar generation.

Pulsating Variable Stars: When a star's evolutionary track carries it through a region in the H-R diagram called the instability strip, the star becomes unstable and begins to pulsate.

• Cepheid variables are high-mass pulsating variables. There is a direct relationship between their periods of pulsation and their luminosities.

• RR Lyrae variables are low-mass, metal-poor pulsating variables with short periods.

• Long-period variable stars also pulsate but in a fashion that is less well understood.

Close Binary Systems: Mass transfer in a close binary system occurs when one star in a close binary overflows its Roche lobe. Gas flowing from one star to the other passes across the inner Lagrangian point. This mass transfer can affect the evolutionary history of the stars that make up the binary system.

Review Questions

1. How does the chemical composition of the present-day Sun's core compare to the core's composition when the Sun formed? What caused the change?

2. Why do high-mass main-sequence stars have shorter lifetimes than those of lower mass?

3. On what grounds are astronomers able to say that the Sun has about 7×10^9 years remaining in its main-sequence stage?

4. Which regions of the Sun are denser today than they were a billion years ago? Which regions are less dense? What has caused these changes?

5. What will happen inside the Sun 7 billion years from now, when it begins to mature into a red giant?

6. Explain why the Earth is expected to become inhospitable to life long before the Sun becomes a red giant.

7. Explain how it is possible for the core of a red giant to contract at the same time that its outer layers expand.

8. Why does helium fusion require much higher temperatures than hydrogen fusion?

9. How is a degenerate gas different from ordinary gases?

10. What is the helium flash? Why does it happen in some stars but not in others?

11. Why does a star's luminosity decrease after helium fusion begins in its core?

12. What does it mean when an astronomer says that a star "moves" from one place to another on an H-R diagram?

13. Explain why the majority of the stars visible through telescopes are main-sequence stars.

14. On an H-R diagram, main-sequence stars do not lie along a single narrow line but are spread out over a band (see Figure 21-8*b*). On the basis of how stars evolve during their main-sequence lifetime, explain why this should be so.

15. Explain how and why the turnoff point on the H-R diagram of a cluster is related to the cluster's age.

16. There is a good deal of evidence that our universe is no more than about 14 billion years old (see Chapter 28). Explain why no main-sequence stars of spectral class M have yet evolved into red-giant stars.

17. How do astronomers know that globular clusters are made of old stars?

18. Red giant stars appear more pronounced in composites of infrared images and visible-light images, like those in Figure 21-3*b* and Figure 21-10. Explain why.

19. The horizontal-branch stars in Figure 21-10 appear blue. (a) Explain why this is consistent with the color-magnitude diagram shown in Figure 21-11. (b) All horizontal-branch stars were once red giants. Explain what happened to these stars to change their color.

20. What is the difference between Population I and Population II stars? In what sense can the stars of one population be regarded as the "children" of the other population?

21. Both diamonds and graphite (the material used in pencils to make marks on paper) are crystalline forms of carbon. Most of the carbon atoms in these substances have nuclei with 6 protons and 6 neutrons (^{12}C). Where did these nuclei come from?

22. Why do astronomers attribute the observed Doppler shifts of a Cepheid variable to pulsation, rather than to some other cause, such as orbital motion?

23. Why do Cepheid stars pulsate? Why are these stars important to astronomers who study galaxies beyond the Milky Way?

24. What is a Roche lobe? What is the inner Lagrangian point? Why are Roche lobes important in close binary star systems?

25. What is the difference between a detached binary, a semidetached binary, a contact binary, and an overcontact binary?

26. Massive main-sequence stars turn into red giants before less massive stars. Why, then, is the more massive star in Algol a main-sequence star and the less massive star a red giant?

Advanced Questions

Questions preceded by an asterisk (*) involve topics discussed in the Boxes in this chapter, in the Boxes in Chapter 19, or in Box 7-2.

> **Problem-solving tips and tools**
>
> Recall from Section 18-1 that 6×10^{11} kg of hydrogen is converted into helium each second at the Sun's center. Recall also that you must use absolute (Kelvin) temperatures when using the Stefan-Boltzmann law. You may find it helpful to review the discussion of apparent magnitude, absolute magnitude, and luminosity in Box 19-3. Section 19-4 discusses the connection between the surface temperatures and colors of stars. Newton's form of Kepler's third law (see Section 19-9) describes the orbits of stars in binary systems. Box 7-2 gives the formula for escape speed and for the average speeds of gas molecules.

27. The radius of the Sun has increased over the past several billion years. Over the same time period, the size of the Moon's orbit around the Earth has also increased. A few billion years ago, were annular eclipses of the Sun (see Figure 3-12) more or less common than they are today? Explain.

28. The Sun has increased in radius by 6% over the past 4.56 billion years. Its present-day radius is 696,000 km. What was its radius 4.56 billion years ago? (*Hint:* The answer is *not* 654,000 km.)

*29. Calculate the escape speed from (a) the surface of the present-day Sun and (b) the surface of the Sun when it becomes a red giant, with essentially the same mass as today but with a radius that is 100 times larger. (c) Explain how your results show that a red-giant star can lose mass more easily than a main-sequence star.

*30. Calculate the average speed of a hydrogen atom (mass 1.67×10^{-27} kg) (a) in the atmosphere of the present-day Sun, with temperature 5800 K, and (b) in the atmosphere of a 1-$M_\odot$ red giant, with temperature 3500 K. (c) Compare your results with the escape speeds that you calculated in Question 27. Use this comparison to discuss how well the present-day Sun and a 1-$M_\odot$ red giant can retain hydrogen in their atmospheres.

31. Use the value of the Sun's luminosity to calculate what mass of hydrogen the Sun will convert into helium during its entire main-sequence lifetime of 1.2×10^{10} years. (Assume that the Sun's luminosity remains nearly constant during the entire 1.2×10^{10} years.) What fraction does this represent of the total mass of hydrogen that was originally in the Sun?

32. (a) The main-sequence stars Sirius (spectral type A1), Vega (A0), Spica (B1), Fomalhaut (A3), and Regulus (B7) are among the 20 brightest stars in the sky. Explain how you can tell that all these stars are younger than the Sun. (b) The third-brightest star in the sky, although it can be seen only south of 29° north latitude, is α (alpha) Centauri A. It is a main-sequence star of spectral type G2, the same as the Sun. Can you tell from this whether α Centauri A is younger than the Sun, the same age, or older? Explain your reasoning.

33. Using the same horizontal and vertical scales as in Figure 21-8a, make points on an H-R diagram for each of the stars listed in Table 21-1. Label each point with the star's mass and its main-sequence lifetime. Which of these stars will remain on the main sequence after 10^9 years? After 10^{11} years?

*34. Calculate the main-sequence lifetimes of (a) a 9-$M_\odot$ star and (b) a 0.25-$M_\odot$ star. Compare these lifetimes with that of the Sun.

*35. The earliest fossil records indicate that life appeared on the Earth about a billion years after the formation of the solar system. What is the most mass that a star could have in order that its lifetime on the main sequence is long enough to permit life to form on one or more of its planets? Assume that the evolutionary processes would be similar to those that occurred on the Earth.

36. As a red giant, the Sun's luminosity will be about 2000 times greater than it is now, so the amount of solar energy falling on the Earth will increase to 2000 times its present-day value. Hence, to maintain thermal equilibrium, each square meter of the Earth's surface will have to radiate 2000 times as much energy into space as it does now. Use the Stefan-Boltzmann law to determine what the Earth's surface temperature will be under these conditions. (*Hint:* The present-day Earth has an average surface temperature of 14°C.)

37. When the Sun becomes a red giant, its luminosity will be about 2000 times greater than it is today. Assuming that this luminosity is caused *only* by fusion of the Sun's remaining hydrogen, calculate how long our star will be a red giant. (In fact, only a fraction of the remaining hydrogen will be consumed, and the luminosity will vary over time as shown in Figure 21-7.)

38. What observations would you make of a star to determine whether its primary source of energy is hydrogen fusion or helium fusion?

39. The star whose spectrum is shown in Figure 21-13a has a lower percentage of heavy elements than the Sun, whose spectrum is shown in Figure 21-13b. Hence, the star in Figure 21-13a has a higher percentage of hydrogen. Why, then, isn't the H_δ absorption line of hydrogen noticeably darker for the star in Figure 21-13a?

40. Would you expect the color of a Cepheid variable star (see Figure 21-16) to change during the star's oscillation period? If not, why not? If so, describe why the color should change, and describe the color changes you would expect to see during an oscillation period.

41. The brightness of a certain Cepheid variable star increases and decreases with a period of 10 days. (a) What must this star's luminosity be if its spectrum has strong absorption lines of hydrogen and helium, but no strong absorption lines of heavy elements? (b) Repeat part (a) for the case in which the star's spectrum also has strong absorption lines of heavy elements.

42. The star X Arietis is an RR Lyrae variable. Its apparent brightness varies between 2.0×10^{-15} and 4.9×10^{-15} that of the Sun with a period of 0.65 day. Interstellar extinction dims the star by 37%. Approximately how far away is the star?

43. The apparent brightness of δ Cephei (a Type I Cepheid variable) varies with a period of 5.4 days. Its average apparent brightness is 5.1×10^{-13} that of the Sun. Approximately how far away is δ Cephei? (Ignore interstellar extinction.)

44. Suppose you find a binary star system in which the more massive star is a red giant and the less massive star is a main-sequence star. Would you expect that mass transfer between the stars has played an important role in the evolution of these stars? Explain your reasoning.

45. The larger star in the Algol binary system (see Figure 21-19a) is of spectral class K, while the smaller star is of spectral class B. Discuss how the color of Algol changes as seen through a small telescope (through which Algol appears as a single star). What is the color during a deep eclipse, when the large star eclipses the small one? What is the color when the small star eclipses the large one?

46. Suppose the detached star in β Lyrae (Figure 21-19b) did not have an accretion disk. Would the deeper dips in the light curve be deeper, shallower, or about the same? What about the shallower dip? Explain your answers.

47. The two stars that make up the overcontact binary W Ursae Majoris (Figure 21-19c) have estimated masses of 0.99 $M_\odot$ and 0.62 $M_\odot$. (a) Find the average separation between the two stars. Give your answer in kilometers. (b) The radii of the two stars are estimated to be 1.14 $R_\odot$ and 0.83 $R_\odot$. Show that these values and your result in part (a) are consistent with the statement that this is an overcontact binary.

48. The stars that make up the binary system W Ursae Majoris (see Figure 21-19c) have particularly strong magnetic fields. Explain how astronomers could have discovered this. (*Hint:* See Section 18-9.)

 49. Consult recent issues of *Sky & Telescope* and *Astronomy* to find out when Mira will next reach maximum brightness. Look up the star's location in the sky using the *Starry Night Backyard*™ program on the CD-ROM that comes with this book. (Use the **Find...** command in the **Edit** menu to search for Omicron Ceti.) Why is it unlikely that you will be able to observe Mira at maximum brightness?

Discussion Questions

50. Eventually the Sun's luminosity will increase to the point where the Earth can no longer sustain life. Discuss what measures a future civilization might take to preserve itself from such a calamity.

51. The half-life of the ^{8}Be nucleus is 2.6×10^{-16} second, which is the average time that elapses before this unstable nucleus decays into two alpha particles. How would the universe be different if instead the ^{8}Be half-life were zero? How would the universe be different if the ^{8}Be nucleus were stable?

52. Discuss how H-R diagrams of star clusters could be used to set limits on the age of the universe. Could they be used to set lower limits on the age? Could they be used to set upper limits? Explain your reasoning.

Web/CD-ROM Questions

53. Suppose that an oxygen nucleus (^{16}O) were fused with a helium nucleus (^4He). What element would be formed? Look up the relative abundance of this element in, for example, the *Handbook of Chemistry and Physics* or on the World Wide Web. Based on the abundance, comment on whether such a process is likely.

54. Although Polaris, the North Star, is a Cepheid variable, it pulsates in a somewhat different way than other Cepheids. Search the World Wide Web for information about this star's pulsations and how they have been measured by astronomers at the U.S. Naval Observatory. How does Polaris pulsate? How does this differ from other Cepheids?

55. Observing Stellar Evolution. Step through the animation "The Hertzsprung-Russell Diagram and Stellar Evolution" in Chapter 21 of the *Universe* Web site or CD-ROM. Use this animation to answer the following questions. (a) How does a 1-$M_\odot$ star move on the H-R diagram during its first 4.56 billion (4560 million) years of existence? Compare this with the discussion in Section 21-1 of how the Sun has evolved over the past 4.56 billion years. (b) What is the zero-age spectral class of a 2-$M_\odot$ star? At what age does such a star evolve into a red giant of spectral class K? (c) What is the approximate zero-age luminosity of a 1.3-$M_\odot$ star? What is its approximate luminosity when it becomes a red giant? (d) Suppose a star cluster has no main-sequence stars of spectral classes O or B. What is the approximate age of the cluster? (e) Approximately how long do the most massive stars of spectral class B live before leaving the main sequence? What about the most massive stars of spectral class F?

Observing Projects

> **Observing tips and tools**
>
> An excellent resource for learning how to observe variable stars is the Web site of the American Association of Variable Star Observers. A wealth of data about specific variable stars can be found on the *Sky & Telescope* Web site and in the three volumes of *Burnham's Celestial Handbook: An Observer's Guide to the Universe Beyond the Solar System* (Dover, 1978). This book also provides useful information about star clusters, as does the Web site for SEDS (Students for the Exploration and Development of Space).

56. Observe several of the red giants and supergiants listed below with the naked eye and through a telescope. (Note that γ Andromedae, α Tauri, and ε Pegasi are all multiple-star systems. The spectral type and luminosity class refer to the brightest star in these systems.) You can locate these stars in the sky using the *Starry Night Backyard*™ program on the CD-ROM that comes with this book. Is the reddish color of these stars apparent when they are compared with neighboring stars?

Star	Spectral type	R.A.	Decl.
Almach (γ Andromedae)	K3 II	$2^h\ 03.9^m$	+42° 20″
Aldebaran (α Tauri)	K5 III	4 35.9	+16 31
Betelgeuse (α Orionis)	M2 I	5 55.2	+07 24
Arcturus (α Boötis)	K2 III	14 15.7	+19 11
Antares (α Scorpii)	M1 I	16 29.5	−26 26
Eltanin (γ Draconis)	K5 III	17 56.7	+51 29
Enif (ε Pegasi)	K2 I	21 44.2	+09 52

Note: The right ascensions and declinations are given for epoch 2000.

57. Several of the open clusters referred to in Figure 21-12 can be seen with a good pair of binoculars. Observe as many of these clusters as you can, using both a telescope and binoculars. (Some are actually so large that they will not fit in the field of view of many telescopes.) You can locate these clusters in the sky using the *Starry Night Backyard*™ program on the CD-ROM that comes with this book. Note that in the following table, the M prefix refers to the Messier Catalog, NGC refers to the New General Catalogue, and Mel refers to the Melotte Catalog. In making your own observations, note the overall distribution of stars in each cluster. Which clusters are seen better through binoculars than through a telescope? Which clusters can you see with the naked eye?

Star cluster	Constellation	R.A.	Decl.
h Persei (NGC 869)	Perseus	$2^h\ 19.0^m$	+57° 09′
χ Persei (NGC 884)	Perseus	2 22.4	+57 07
Pleiades (M45)	Taurus	3 47.0	+24 07
Hyades (Mel 25)	Taurus	4 27	+16 00
Praesepe (M44)	Cancer	8 40.1	+19 59
Coma (Mel 111)	Coma Berenices	12 25	+26 00
Wild Duck (M11)	Scutum	18 51.1	−06 16

Note: The right ascensions and declinations are given for epoch 2000.

58. There are many beautiful globular clusters scattered around the sky that can be easily seen with a small telescope. Several of the brightest and nearest globulars are listed below. You can locate these clusters in the sky using the *Starry Night Backyard*™ program on the CD-ROM that comes with this book. Observe as many of these globular clusters as you can. Can you distinguish individual stars toward the center of each cluster? Do you notice any differences in the overall distribution of stars between clusters?

Star cluster	Constellation	R.A.	Decl.
M3 (NGC 5272)	Canes Venatici	13^h 42.2^m	+28° 23′
M5 (NGC 5904)	Serpens	15 18.6	+ 2 05
M4 (NGC 6121)	Scorpius	16 23.6	−26 32
M13 (NGC 6205)	Hercules	16 41.7	+36 28
M12 (NGC 6218)	Ophiuchus	16 47.2	− 1 57
M28 (NGC 6626)	Sagittarius	18 24.5	−24 52
M22 (NGC 6656)	Sagittarius	18 36.4	−23 54
M55 (NGC 6809)	Sagittarius	19 40.0	−30 58
M15 (NGC 7078)	Pegasus	21 30.0	+12 10

Note: The right ascensions and declinations are given for epoch 2000.

59. Use the *Starry Night Backyard*™ program to view some of the objects described in this chapter. First select **Go Home** in the **View** menu to show the sky as seen from your location. (If the program doesn't place you at your true location, use the **Viewing Location...** command in the **View** menu.) In the Control Panel, set the date to today's date and the time to midnight (12:00:00 A.M.). Click on the "Info" tab at the left-hand side of the main *Starry Night Backyard*™ window (this will present a display with information about what you are viewing). Then locate the giant star Aldebaran, the open cluster M44, and the globular cluster M12. (For each object, use the **Find...** command in the **Edit** menu.) (**a**) Which of these are visible from your location tonight at midnight? (**b**) For each object that is visible, use the zoom controls at the upper right of the Control Panel to get the best possible view. Describe the appearance of the object. (**c**) For each object that is visible, in which direction of the compass would you have to look at midnight to see it (that is, what is its *azimuth*)? How far above the horizon would you have to look (that is, what is its *altitude*)? (Look in the **Info** panel at the left-hand side of the screen for this information.)

60. Use the *Deep Space Explore*™ program to look for signs of stellar evolution in our Milky Way Galaxy. In the left-hand part of the window under the heading **Milky Way,** click on the triangle next to the word **Explore** and then click on **Sun in Milky Way.** You can zoom in or out using the buttons at the upper left of the window (an upward-pointing triangle and a downward-pointing triangle). You can rotate the Galaxy by putting the mouse cursor over the image, holding down the mouse button, and moving the mouse. (On a two-button mouse, hold down the left mouse button.) (**a**) What is the color of the central part of the Galaxy? Based on the color of this region, do you expect that there are many massive main-sequence stars there? Would you expect to find many young stars there? Explain your reasoning. (**b**) Repeat part (**a**) for the outer regions of the Galaxy.

Collaborative Exercises

61. The inverse mass-lifetime relationship for stars is sometimes likened to automobiles in that the more massive vehicles, such as commercial semi–tractor-trailer trucks, need to consume significantly more fuel to travel at highway speeds than more lightweight and economical vehicles. As a group, create a table called "Maximum Vehicle Driving Distances," much like Table 21-1, "Main-Sequence Lifetimes," by making estimates for any five vehicles of your groups' choosing. The table's column headings should be: (1) vehicle make and model; (2) estimated gas tank size; (3) cost to fill tank; (4) estimated mileage (in miles per gallon); and (5) number of miles driven on a single fill-up.

62. Consider Figure 21-17, showing the period-luminosity relation for Cepheids. If the length of time it has been since someone in your group last purchased milk or juice at the grocery store was identical to the pulsation period of a Type I Cepheid, how much longer a pulsation period would a Type II Cepheid need in order to reach the same luminosity as the Type I Cepheid in this time frame?

63. Figure 21-16a shows a light curve of apparent magnitude versus time in days for δ Cephei—a pulsating star that reaches maximum brightness every 5.4 days. Create a new sketch of apparent magnitude versus time in days showing three different stars: (1) δ Cephei; (2) a slightly smaller pulsating star; and (3) a slightly larger pulsating star, all of which have about the same total change in apparent magnitude.

22 Stellar Evolution: The Deaths of Stars

When a star reaches the end of its main-sequence lifetime and becomes a red giant, it comes to have a compressed core and a bloated atmosphere. Finally, it devours its remaining nuclear fuels and begins to die. The character of its death depends crucially on the star's mass.

A star of relatively low mass—such as our own Sun—ends its evolution by gently expelling its outer layers into space. These ejected gases form a glowing cloud called a *planetary nebula* such as the one shown here in the upper image. The burned-out core that remains is called a *white dwarf*.

In contrast, a high-mass star ends its life in almost inconceivable violence. At the end of its short life, the core of such a star collapses suddenly. This triggers a powerful *supernova* explosion that can be as luminous as an entire galaxy of stars. A white dwarf, too, can become a supernova if it accretes gas from a companion star in a close binary system.

Thermonuclear reactions in supernovae produce a wide variety of heavy elements, which are ejected into the interstellar medium. (The supernova remnant shown here in the lower image is rich in these elements.) Such heavy elements are essential building blocks for terrestrial worlds like our Earth. Thus the deaths of massive stars can provide the seeds for planets orbiting succeeding generations of stars.

A planetary nebula

R I **V** U X G

A supernova remnant

R I V U **X** G

WEB LINK 22.1 The planetary nebula M57 (the Ring Nebula) and the supernova remnant Cassiopeia A. (Hubble Heritage Team, AURA/STScI/NASA; NASA/CXC/SAO/ Rutgers/J. Hughes)

Questions to Guide Your Reading

As you read the sections of this chapter, look for the answers to the following questions:

22-1 What kinds of nuclear reactions occur within a star like the Sun as it ages?

22-2 Where did the carbon atoms in our bodies come from?

22-3 What is a planetary nebula, and what does it have to do with planets?

22-4 What is a white dwarf star?

22-5 Why do high-mass stars go through more evolutionary stages than low-mass stars?

22-6 What happens within a high-mass star to turn it into a supernova?

22-7 Why was SN 1987A an unusual supernova?

22-8 What was learned by detecting neutrinos from SN 1987A?

22-9 How do white dwarf stars give rise to certain types of supernovae?

22-10 What vestiges are left after a supernova explosion?

22-1 | Low-mass stars go through two distinct red-giant stages

As we saw in Chapter 21, a main-sequence star eventually uses up the hydrogen in its core. The aged star then leaves the main sequence. Let's examine what happens next to a star of relatively low mass, less than about 4 $M_\odot$.

We can describe a star's post–main-sequence evolution using an evolutionary track on a H-R diagram. **Figure 22-1** shows the track for a 1-$M_\odot$ star like the Sun. Once core hydrogen fusion ceases, the core shrinks, heating the surrounding hydrogen and triggering shell hydrogen fusion. The new outpouring of energy causes the star's outer layers to expand and cool, and the star becomes a red giant. As the luminosity increases and the surface temperature drops, the post–main-sequence star moves up and to the right along the **red-giant branch** on an H-R diagram (Figure 22-1a).

Next, the helium-rich core of the star shrinks and heats until eventually **core helium fusion** begins. This second post–main-sequence stage begins gradually in stars more massive than about 2–3 $M_\odot$, but for less massive stars it comes suddenly—in a *helium flash*. During core helium fusion, the surrounding hydrogen-fusing shell still provides most of the red giant's luminosity.

As we learned in Section 21-2, the core expands when core helium fusion begins, which makes the core cool down a bit. (We saw in

Box 21-1 that letting a gas expand tends to lower its temperature, while compressing a gas tends to increase its temperature.) The cooling of the core also cools the surrounding hydrogen-fusing shell, so that the shell releases energy more slowly. Hence, the luminosity goes down a bit after core helium fusion begins.

The slower rate of energy release also lets the star's outer layers contract. As they contract, they heat up, so the star's surface temperature increases and its evolutionary track moves to the left on the H-R diagram in Figure 22-1b. The luminosity changes relatively little during this stage, so the evolutionary track moves almost horizontally, along a path called the **horizontal branch**. Horizontal-branch stars have helium-fusing cores surrounded by hydrogen-fusing shells. Figure 21-10 shows horizontal-branch stars in a globular cluster, and Figure 21-7 shows the evolution of the luminosity of a 1-$M_\odot$ star up to this point in its history.

Helium fusion produces nuclei of carbon and oxygen. After about a hundred million (10^8) years of core helium fusion, essentially all the helium in the core of a 1-$M_\odot$ star has been converted into carbon and oxygen, the fusion of helium in the core ceases. (This corresponds to the right-hand end of the graph in Figure 21-7.) Without thermonuclear reactions to maintain the core's internal pressure, the core again contracts, until it is stopped by degenerate-electron pressure (described in Section 21-2). This contraction releases heat into the surrounding helium-rich gases, and a new stage of helium fusion begins in a thin shell around the core. This process is called **shell helium fusion.**

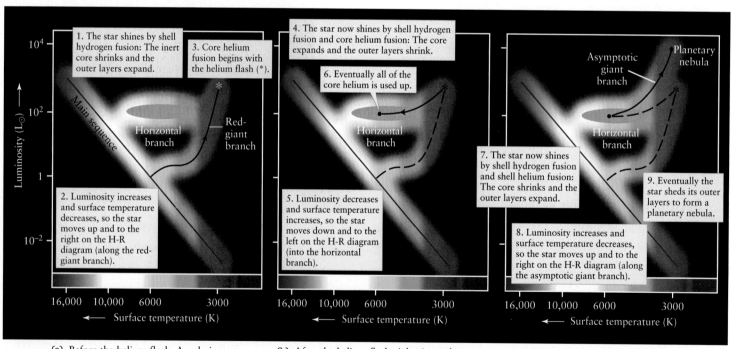

(a) Before the helium flash: A red-giant star

(b) After the helium flash: A horizontal-branch star

(c) After core helium fusion ends: An AGB star

Figure 22-1
The Post–Main-Sequence Evolution of a 1-M$_\odot$ Star
These H-R diagrams show the evolutionary track of a star like the Sun as it goes through the stages of being (a) a red-giant star, (b) a horizontal-branch star, and (c) an asymptotic giant branch (AGB) star. The star eventually evolves into a planetary nebula (described in Section 22-3).

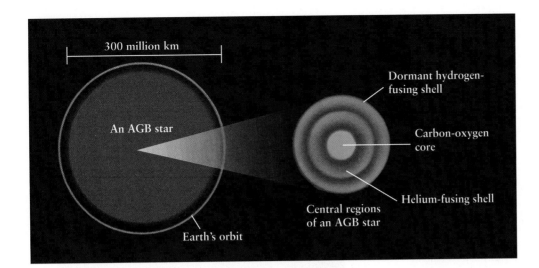

figure 22-2

The Structure of an Old, Low-Mass AGB Star Near the end of its life, a star like the Sun becomes an immense, red, asymptotic giant branch (AGB) star. The star's inert core, active helium-fusing shell, and dormant hydrogen-fusing shell are all contained within a volume roughly the size of the Earth. Thermonuclear reactions in the helium-fusing shell are so rapid that the star's luminosity is thousands of times that of the present-day Sun.

History now repeats itself—the star enters a *second* red-giant phase. A star first becomes a red giant at the end of its main-sequence lifetime, when the outpouring of energy from shell hydrogen fusion makes the star's outer layers expand and cool. In the same way, the outpouring of energy from shell *helium* fusion causes the outer layers to expand again. The low-mass star ascends into the red-giant region of the H-R diagram for a second time (Figure 22-1c), but now with even greater luminosity than during its first red-giant phase.

Stars in this second red-giant phase are commonly called **asymptotic giant branch stars**, or **AGB stars**, and their evolutionary tracks follow what is called the **asymptotic giant branch**. (*Asymptotic* means "approaching," and, indeed, a star on the asymptotic giant branch approaches the red-giant branch from the left on an H-R diagram.)

When a low-mass star first becomes an AGB star, it consists of an inert, degenerate carbon-oxygen core and a helium-fusing shell, both inside a hydrogen-fusing shell, all within a volume approximately the size of the Earth. This small, dense central region is surrounded by an enormous hydrogen-rich envelope about as big as Earth's orbit. After a while, the expansion of the star's outer layers causes the hydrogen-fusing shell to also expand and cool, and thermonuclear reactions in this shell temporarily cease. This leaves the aging star's structure as shown in Figure 22-2.

We saw in Section 21-1 that the more massive a star, the shorter the amount of time it remains on the main sequence. Similarly, the greater the star's mass, the more rapidly it goes through the stages of post–main-sequence evolution. Hence, we can see all of these stages by studying star clusters, which contain stars that are all the same age but that have a range of masses (see Section 21-3). Figure 22-3 shows a color-magnitude diagram for the globular cluster M55, which is at least 13 billion years old. The least massive stars in this cluster are still on the main sequence. Progressively more massive stars have evolved to the red giant branch, the horizontal branch, and the asymptotic giant branch.

A 1-$M_\odot$ AGB star can reach a maximum luminosity of nearly 10^4 $L_\odot$, as compared with approximately 10^3 $L_\odot$ when it reached the helium flash and a relatively paltry 1 $L_\odot$ during its main-sequence

figure 22-3

Stellar Evolution in a Globular Cluster In the old globular cluster M55, stars with masses less than about 0.8 $M_\odot$ are still on the main sequence, converting hydrogen into helium in their cores. Slightly more massive stars have consumed their core hydrogen and are ascending the red-giant branch; even more massive stars have begun helium core fusion and are found on the horizontal branch. The most massive stars (which still have less than 4 $M_\odot$) have consumed all the helium in their cores and are ascending the asymptotic giant branch. (Compare with Figure 21-11.) (Adapted from D. Schade, D. VandenBerg, and F. Hartwick) lifetime. When the Sun becomes an AGB star some 12.3 billion years from now, this tremendous increase in luminosity will cause Mars and the Jovian planets to largely evaporate away. The Sun's bloated outer layers will reach to the Earth's orbit. Mercury and perhaps Venus will simply be swallowed whole.

22-2 | Dredge-ups bring the products of nuclear fusion to a giant star's surface

As we saw in Section 18-2, energy is transported outward from a star's core by one of two processes—radiative diffusion or convection. The first is the passage of energy in the form of electromagnetic radiation, and it dominates only when a star's gases are relatively transparent. The second involves up-and-down movement of the star's gases. Convection plays a very important role in giant stars, because it helps supply the cosmos with the elements essential to life.

 In the Sun, convection dominates only the outer layers, from around 0.71 solar radius (measured from the center of the Sun) up to the photosphere (recall Figure 18-5). During the final stages of a star's life, however, the convective zone can become so broad that it extends down to the star's core. At these times, convection can "dredge up" the heavy elements produced in and around the core by thermonuclear fusion, transporting them all the way to the star's surface.

 The *first* dredge-up takes place after core hydrogen fusion stops, when the star becomes a red giant for the first time. Convection dips so deeply into the star that material processed by the CNO cycle of hydrogen fusion (see Section 18-1) is carried up to the star's surface, changing the relative abundances of carbon, nitrogen, and oxygen. A *second* dredge-up occurs after core helium fusion ceases, further altering the abundances of carbon, nitrogen, and oxygen. Still later, during the AGB stage, a *third* dredge-up can occur if the star has a mass greater than about 2 $M_\odot$. This third dredge-up transports large amounts of freshly synthesized carbon to the star's surface, and the star's spectrum thus exhibits prominent absorption bands of carbon-rich molecules like C_2, CH, and CN. For this reason, an AGB star that has undergone a third dredge-up is called a **carbon star.**

All AGB stars have very strong stellar winds that cause them to lose mass at very high rates, up to 10^{-4} $M_\odot$ per year (a thousand times greater than that of a red giant, and 10^{10} times greater than the rate at which our present-day Sun loses mass). The surface temperature of AGB stars is relatively low, around 3000 K, so any ejected carbon-rich molecules can condense to form tiny grains of soot. Indeed, carbon stars are commonly found to be obscured in sooty cocoons of ejected matter (**Figure 22-4**).

Carbon stars are important because they enrich the interstellar medium with carbon and some nitrogen and oxygen. The triple alpha process that occurs in helium fusion is the *only* way that carbon can be made, and carbon stars are the primary avenue by which this element is dispersed into interstellar space. Indeed, most of the carbon in your body was produced long ago inside a star by the triple alpha process (see Section 21-2), dredged up to the star's surface, ejected into space, and later incorporated into

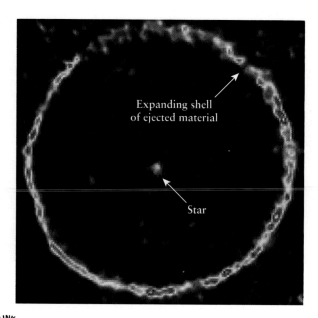

Figure 22-4 R I V U X G
A Carbon Star TT Cygni is an AGB star in the constellation Cygnus that ejects some of its carbon-rich outer layers into space. Some of the ejected carbon combines with oxygen to form molecules of carbon monoxide (CO), whose emissions can be detected with a radio telescope. This radio image shows the CO emissions from a shell of material that TT Cygni ejected some 7000 years ago. Over that time, the shell has expanded to a diameter of about ½ light-year. (H. Olofsson [Stockholm Observatory] et al./NASA)

the solar nebula from which our Earth—and all of the life on it—eventually formed.

22-3 | Low-mass stars die by gently ejecting their outer layers, creating planetary nebulae

The AGB stage in the evolution of a star of low mass (less than about 4 $M_\odot$) is a dramatic turning point. Before this stage, a star loses mass only gradually through steady stellar winds. But as it evolves during its AGB stage, a star divests itself completely of its outer layers. The aging star undergoes a series of bursts in luminosity, and in each burst it ejects a shell of material into space. (The shell around the AGB star TT Cygni, shown in Figure 22-4, was probably created in this way.) Eventually, all that remains of a low-mass star is a fiercely hot, exposed core, surrounded by glowing shells of ejected gas. This late stage in the life of a star is called a **planetary nebula** (see the upper image on the opening page of this chapter).

To understand how this process works, consider the internal structure of an AGB star as shown in Figure 22-2. As the helium in the helium-fusing shell is used up, the pressure that holds up the dormant hydrogen-fusing shell decreases. Hence, the dormant hydrogen shell contracts and heats up, and hydrogen fusion begins anew. This revitalized hydrogen fusion creates helium, which rains downward onto the temporarily dormant helium-fusing shell. As the helium shell gains mass, it shrinks and heats up. When the temperature of the helium shell reaches a certain critical value, it reignites in a **helium shell flash** that is similar to (but less intense than) the helium flash that occurred earlier in the evolution of a low-mass star (see Section 21-2). The released energy pushes the hydrogen-fusing shell outward, making it cool off, so that hydrogen fusion ceases and this shell again becomes dormant. The process then starts over again.

When a helium shell flash occurs, the luminosity of an AGB star increases substantially in a relatively short-lived burst called a **thermal pulse. Figure 22-5**, which is based on a theoretical calculation of the evolution of a 1-$M_\odot$ star, shows that thermal pulses begin when the star is about 12.365 billion years old. The calculations predict that thermal pulses occur at ever-shorter intervals of about 100,000 years.

During these thermal pulses, the dying star's outer layers can separate completely from its carbon-oxygen core. As the ejected material expands into space, dust grains condense out of the cooling gases. Radiation pressure from the star's hot, burned-out core acts on the specks of dust, propelling them further outward, and the star sheds its outer layers altogether. In this way an aging 1-$M_\odot$ star loses as much as 40% of its mass. More massive stars eject even greater fractions of their original mass.

As a dying star ejects its outer layers, the star's hot core becomes exposed. With a surface temperature of about 100,000 K, this exposed core emits ultraviolet radiation intense enough to ionize and excite the expanding shell of ejected gases. These gases therefore glow and emit visible light through the process of fluorescence (see Box 20-1), producing a planetary nebula like those shown in **Figure 22-6**.

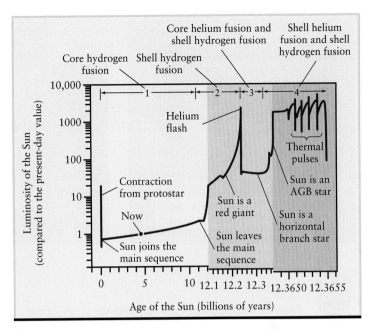

figure 22-5

Further Stages in the Evolution of the Sun This diagram, which shows how the luminosity of the Sun (a 1-$M_\odot$ star) changes over time, is an extension of Figure 21-7. We use different scales for the final stages because the evolution is so rapid. During the AGB stage there are brief periods of runaway helium fusion, causing spikes in luminosity called thermal pulses. (Adapted from Mark A. Garlick, based on calculations by I.-Juliana Sackmann and Kathleen E. Kramer)

(a) R I **V** U X G

(b) R I **V** U X G

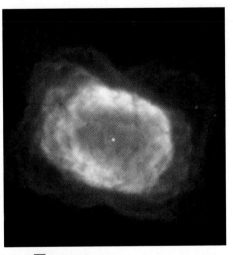

(c) R **I** V U X G

figure 22-6
WEB LINK 22.4

Planetary Nebulae (a) The pinkish blob is a planetary nebula surrounding a star in the globular cluster M15, about 10,000 pc (33,000 ly) from Earth in the constellation Pegasus. (b) The Helix Nebula is the closest of all known planetary nebulae. It lies 140 pc (450 ly) from Earth in the constellation Aquarius and has an angular diameter of about ½° (the same as the full Moon). The "ring" is actually a spherical shell; we see a substantial thickness of the shell only when we look near its rim. (c) This infrared image of the planetary nebula NGC 7027 shows wispy shells of cooler gas surrounding a more luminous shell of hot gas, suggesting a complex evolutionary history. NGC 7027 is about 900 pc (3000 ly) from Earth in the constellation Cygnus and is roughly 14,000 AU across. (a: NASA/Hubble Heritage Team, STScI/AURA; b: Anglo-Australian Observatory; c: William B. Latter, SIRTF Science Center/Caltech, and NASA)

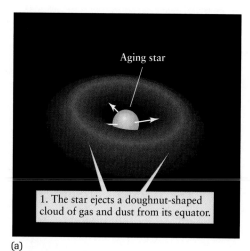

1. The star ejects a doughnut-shaped cloud of gas and dust from its equator.

Aging star

(a)

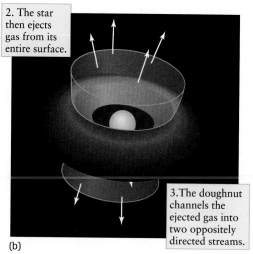

2. The star then ejects gas from its entire surface.

3. The doughnut channels the ejected gas into two oppositely directed streams.

(b)

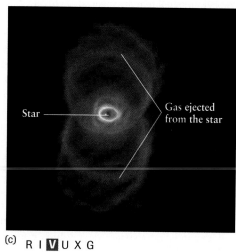

Star

Gas ejected from the star

(c)   R I **V** U X G

WEB LINK 22.5

Figure 22-7

Making an Elongated Planetary Nebula (a), (b) These illustrations show one proposed explanation for why many planetary nebulae have an elongated shape. (c) The planetary nebula MyCn18, shown here in false color, may have acquired its elongated shape in this way. It lies some 2500 pc (8000 ly) from Earth in the constellation Musca (the Fly). (R. Sahai and J. Trauger, Jet Propulsion Laboratory; the WFPC-2 Science Team; and NASA)

CAUTION

Despite their name, planetary nebulae have nothing to do with planets. This misleading term was introduced in the nineteenth century because these glowing objects looked like distant Jovian planets when viewed through the small telescopes then available. The difference between planets and planetary nebulae became obvious with the advent of spectroscopy: Planets have *absorption* line spectra (see Section 7-3), but the excited gases of planetary nebulae have *emission* line spectra.

Planetary nebulae are quite common. Astronomers estimate that there are 20,000 to 50,000 planetary nebulae in our Galaxy alone. Many planetary nebulae, such as those in Figure 22-6, are more or less spherical in shape. This is a result of the symmetrical way in which the gases were ejected. But if the rate of expansion is not the same in all directions, the resulting nebula takes on an hourglass or dumbbell appearance (**Figure 22-7**).

Spectroscopic observations of planetary nebulae show emission lines of ionized hydrogen, oxygen, and nitrogen. From the Doppler shifts of these lines, astronomers have concluded that the expanding shell of gas moves outward from a dying star at speeds from 10 to 30 km/s. For a shell expanding at such speeds to have attained the typical diameter of a planetary nebula, about 1 light-year, it must have begun expanding about 10,000 years ago. Thus, by astronomical standards, the planetary nebulae we see today were created only very recently.

We do not observe planetary nebulae that are more than about 50,000 years old. After this length of time, the shell has spread out so far from the cooling central star that its gases cease to glow and simply fade from view. The nebula's gases then mix with the surrounding interstellar medium. Astronomers estimate that all the planetary nebulae in the Galaxy return a total of about 5 $M_\odot$ to the interstellar medium each year. This amounts to 15% of all the

matter expelled by all the various sorts of stars in the Galaxy each year. Because of this significant contribution, planetary nebulae play an important role in the chemical evolution of the Galaxy as a whole.

22-4 | The burned-out core of a low-mass star cools and contracts until it becomes a white dwarf

Stars less massive than about 4 $M_\odot$ never develop the necessary central pressures or temperatures to ignite thermonuclear reactions that use carbon or oxygen as fuel. Instead, as we have seen, the process of mass ejection just strips away the star's outer layers and leaves behind the hot carbon-oxygen core. With no thermonuclear reactions taking place, the core simply cools down like a dying ember. These burnt-out relics of a star's former glory are called **white dwarfs**. White dwarfs are quite small by stellar standards—approximately the same size as the Earth (see Section 19-7).

You might think that without thermonuclear reactions to provide internal heat and pressure, a white dwarf should keep on shrinking under the influence of its own gravity as it cools. Actually, however, a cooling white dwarf maintains its size, because the burnt-out stellar core is so dense that most of its electrons are degenerate (see Section 21-2). Thus, degenerate-electron pressure supports the star against further collapse. This pressure does not depend on temperature, so it continues to hold up the star even as the temperature drops.

Many white dwarfs are found in the solar neighborhood, but all are too faint to be seen with the naked eye. One of the first white dwarfs to be discovered is a companion to Sirius, the brightest star in the night sky. In 1844 the German astronomer Friedrich

Bessel noticed that Sirius was moving back and forth slightly, as if it was being orbited by an unseen object. This companion, designated Sirius B (Figure 22-8), was first glimpsed in 1862 by the American astronomer Alvan Clark. Recent satellite observations at ultraviolet wavelengths, where hot white dwarfs emit most of their light, show that the surface temperature of Sirius B is about 30,000 K.

Observations of white dwarfs in binary systems like Sirius allow astronomers to determine the mass, radius, and density of these stars (see Sections 19-9, 19-10, and 19-11). Such observations show that the density of the degenerate matter in a white dwarf is typically 10^9 kg/m^3 (a million times denser than water). A teaspoonful of white dwarf matter brought to Earth would weigh nearly 5.5 tons—as much as an elephant!

As we learned in Section 21-2, degenerate matter has a very different relationship between its pressure, density, and temperature than that of ordinary gases. Consequently, white dwarf stars have an unusual **mass-radius relation:** The more massive a white dwarf star, the *smaller* it is.

Figure 22-9 displays the mass-radius relation for white dwarfs. Note that the more degenerate matter you pile onto a white dwarf, the smaller it becomes. However, there is a limit to how much pressure degenerate electrons can produce. As a result, there is an upper limit to the mass that a white dwarf can have. This maximum mass is called the **Chandrasekhar limit,** after the Indian-American scientist Subrahmanyan Chandrasekhar, who pioneered theoretical studies of white dwarfs in the 1930s. (The orbiting Chandra X-ray Observatory, described in

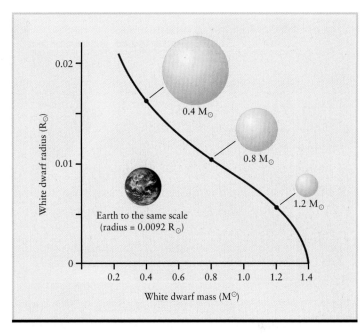

Figure 22-9

The Mass-Radius Relationship for White Dwarfs The more massive a white dwarf is, the smaller its radius. (The drawings of white dwarfs of different mass are drawn to the same scale as the image of the Earth.) This unusual relationship is a result of the degenerate-electron pressure that supports the star. The maximum mass of a white dwarf, called the Chandrasekhar limit, is 1.4 M$_\odot$.

Section 6-7, is named in his honor.) The Chandrasekhar limit is equal to 1.4 M$_\odot$, meaning that all white dwarfs must have masses less than 1.4 M$_\odot$.

The material inside a white dwarf consists mostly of ionized carbon and oxygen atoms floating in a sea of degenerate electrons. As the dead star cools, the particles in this material slow down, and electric forces between the ions begin to prevail over the random thermal motions. The ions no longer move freely. Instead, they arrange themselves in orderly rows, like an immense crystal lattice. From this time on, you could say that the star is "solid." The degenerate electrons move around freely in this crystal material, just as electrons move freely through an electrically conducting metal like copper or silver. A diamond is also crystallized carbon, so a cool carbon-oxygen white dwarf resembles an immense spherical diamond!

Figure 22-10 shows the evolutionary tracks followed by three burned-out stellar cores as they become white dwarfs. The initial red giants had masses of 0.8, 1.5, and 3.0 M$_\odot$. Mass ejection strips these dying stars of up to 60% of their matter. During their final spasms, the luminosity and surface temperature of these stars change quite rapidly. The points representing these stars on an H-R diagram race along their evolutionary tracks, sometimes executing loops corresponding to thermal pulses. Finally, as the ejected nebulae fade and the stellar cores cool, the evolutionary tracks of these dying stars take a sharp turn toward the white dwarf region of the H-R diagram.

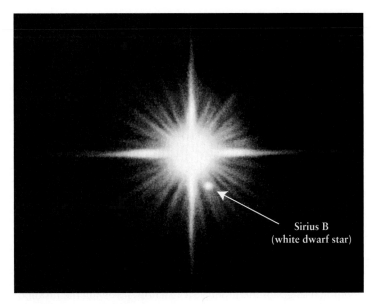

Figure 22-8   R I **V** U X G

Sirius and Its White Dwarf Companion Sirius, the brightest-appearing star in the sky, is actually a binary star: The secondary star, called Sirius B, is a white dwarf. In this photograph Sirius B is almost obscured by the glare of the primary star. The spikes and rays around Sirius are the result of optical effects within the telescope. (Courtesy of R. B. Minton)

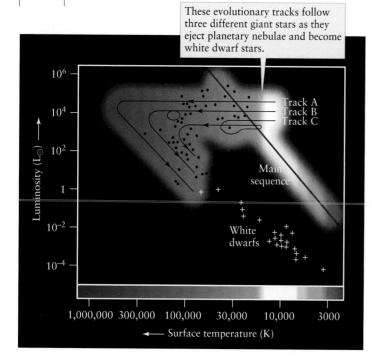

These evolutionary tracks follow three different giant stars as they eject planetary nebulae and become white dwarf stars.

Evolutionary track	Mass ($M_\odot$)		
	Giant star	Ejected nebula	White dwarf
A	3.0	1.8	1.2
B	1.5	0.7	0.8
C	0.8	0.2	0.6

Figure 22-10

Evolution from Giants to White Dwarfs This H-R diagram shows the evolutionary tracks of three low-mass giant stars as they eject planetary nebulae. The table gives the extent of mass loss in each case. The dots represent the central stars of planetary nebulae whose surface temperatures and luminosities have been determined. The crosses are white dwarfs for which similar data exist. (Adapted from B. Paczynski)

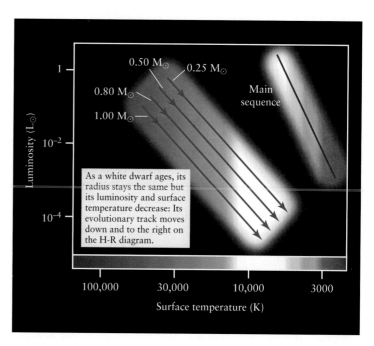

Figure 22-11

White Dwarf "Cooling Curves" As white dwarf stars radiate their internal energy into space, they become dimmer and cooler. The blue lines show the evolutionary tracks of four white dwarfs of different mass: The more massive a white dwarf, the smaller and hence fainter it is. Compare these "cooling curves" with the lines of constant radius in Figure 19-15.

Although a white dwarf maintains the same size as it cools, its luminosity and surface temperature both decrease with time. Consequently, the evolutionary tracks of aging white dwarfs point toward the lower right corner of the H-R diagram. You can see this in Figure 22-10; Figure 22-11 shows it in more detail. The energy that the white dwarf radiates into space comes only from the star's internal heat, which is a relic from the white dwarf's past existence as a stellar core. Over billions of years, white dwarfs grow dimmer and dimmer as their surface temperatures drop toward absolute zero.

After ejecting much of its mass into space, our own Sun will eventually evolve into a white dwarf star about the size of the Earth and with perhaps one-tenth of its present luminosity. It will become even dimmer as it cools. After 5 billion years as a white dwarf, the Sun will radiate with no more than one ten-thousandth of its present brilliance. With the passage of eons, our Sun will simply fade into obscurity.

22-5 | High-mass stars create heavy elements in their cores

During the entire lifetime of a low-mass star (mass less than about 4 $M_\odot$), just two kinds of thermonuclear reactions take place—hydrogen fusion and helium fusion. The heaviest elements manufactured by these reactions are carbon and oxygen.

The life story of a *high-mass* star (with a zero-age mass greater than about 4 $M_\odot$) begins with these same reactions. But theoretical calculations show that high-mass stars can also go through several additional stages of thermonuclear reactions involving the fusion of carbon, oxygen, and other heavy nuclei. As a result, high-mass stars end their lives quite differently from low-mass stars.

Why is fusion of heavy nuclei possible only in a high-mass star? The reason is that such a star has more gravity trying to pull it together and more pressure at its center. Hence, the star's central temperature is also higher—high enough to fuse heavy elements. Heavy-nucleus fusion reactions require high temperatures, because these nuclei have large electric charges, which means that there are strong electric forces that tend to keep the nuclei apart. Only at the great speeds associated with very high temperatures can the nuclei travel fast enough to overcome their mutual electric repulsion and fuse together.

As we discussed in Section 21-1, when most main-sequence stars use up their core hydrogen, they begin shell hydrogen fusion and enter a red-giant phase. These stars then begin core helium fusion when the core temperature becomes high enough. (The exceptions are stars with mass less than 0.4 $M_\odot$. Convection in their interiors prevents the ignition of shell hydrogen fusion, and they have too little internal pressure to trigger core helium fusion.) The differences between low-mass and high-mass stars become pronounced after helium core fusion ends, when the core is composed primarily of carbon and oxygen.

Let us consider how the late stages in the evolution of a high-mass star differ from those of a low-mass star. In low-mass stars, as we saw in Section 22-4, the carbon-oxygen core eventually becomes exposed and becomes a white dwarf. But in stars whose overall mass is more than about 4 $M_\odot$, the carbon-oxygen core is more massive than the Chandrasekhar limit of 1.4 $M_\odot$, so degenerate-electron pressure cannot prevent the core from contracting and heating. Hence, a high-mass star is able to enter a new round of core thermonuclear reactions. When the central temperature of such a high-mass star reaches 600 million kelvins (6×10^8 K), the first of the new thermonuclear reactions, **carbon fusion,** begins. Carbon fusion consumes carbon and produces oxygen, neon, sodium, and magnesium.

If a star has an even larger main-sequence mass of about 8 $M_\odot$ or so (before mass ejection), even more thermonuclear reactions can take place. After the cessation of carbon fusion, the core will again contract, and the star's central temperature can rise to 1 billion kelvins (10^9 K). At this temperature **neon fusion** begins. This uses up the neon accumulated from carbon fusion and further increases the concentrations of oxygen and magnesium in the star's core.

After neon fusion ends, the core will again contract, and **oxygen fusion** will begin when the central temperature of the star reaches about 1.5 billion kelvins (1.5×10^9 K). The principal product of oxygen fusion is sulfur. Once oxygen fusion is over, the core will contract yet again. If the central temperature reaches 2.7 billion kelvins (2.7×10^9 K), **silicon fusion** begins, producing a variety of nuclei from sulfur to iron.

Between each new stage of core thermonuclear reactions comes a period of shell fusion and a new red-giant phase. This means that the evolutionary tracks of high-mass stars go through a number of back-and-forth gyrations on an H-R diagram. When each new stage of core fusion begins, the luminosity and radius decrease and the surface temperature increases; when that stage ceases and only shell fusion is taking place, the luminosity and radius increase, the surface temperature decreases, and the star again becomes a giant. While all of this is going on, the star is also losing mass at a rapid rate (**Figure 22-12**).

As a high-mass star consumes increasingly heavier nuclei, the thermonuclear reactions produce a wider variety of products. For example, oxygen fusion also produces silicon, phosphorus, and magnesium. Some thermonuclear reactions that create heavy elements also release neutrons. A neutron is like a proton except that

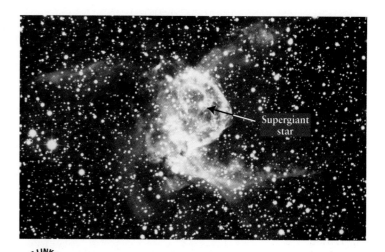

WEB LINK 22.7

Figure 22-12 R I **V** U X G

Mass Loss from a Supergiant Star At the heart of this nebulosity, called NGC 2359, lies a supergiant star with a mass in excess of 40 $M_\odot$. This star is losing mass at a rapid rate in a strong stellar wind. As this wind collides with the surrounding interstellar gas and dust, it creates the "bubble" shown here. NGC 2359, which has an angular size of 10 by 5 arcmin, lies in the constellation Canis Major. (Anglo-Australian Observatory)

it carries no electric charge. Therefore, neutrons are not repelled by positively charged nuclei, and so can easily collide and combine with them. This absorption of neutrons by nuclei, called **neutron capture,** creates many elements and isotopes that are not produced directly in fusion reactions.

Each stage of thermonuclear reactions in a high-mass star helps to trigger the succeeding stage. In each stage, when the star exhausts a given variety of nuclear fuel in its core, gravitational contraction takes the core to ever-higher densities and temperatures, thereby igniting the "ash" of the previous fusion stage—and possibly the outlying shell of unburned fuel as well.

The increasing density and temperature of the core make each successive thermonuclear reaction more rapid than the one that preceded it. As an example, Stanford Woosley at the University of California, Santa Cruz, and Thomas Weaver at Lawrence Livermore National Laboratory have made detailed calculations for a star with a zero-age mass of 25 $M_\odot$. They find that carbon fusion in such a star lasts for 600 years, neon fusion for 1 year, and oxygen fusion for only 6 months. The last, and briefest, stage of nuclear reactions is silicon fusion. The entire core supply of silicon in a 25-$M_\odot$ star is used up in only one day! Table 22-1 summarizes the evolutionary stages in the life of such a 25-$M_\odot$ star.

Each stage of core fusion in a high-mass star generates a new shell of material around the core. After several such stages, the internal structure of a truly massive star—say, 25 $M_\odot$ or greater—resembles that of an onion, as Figure 22-13 shows. Because thermonuclear reactions can take place simultaneously in several shells, energy is released at such a rapid rate that the star's outer layers expand tremendously. The result is a **supergiant** star, whose luminosity and radius are much larger than those of a giant (see Section 19-7).

table 22-1	Evolutionary Stages of a 25-M$_\odot$ Star		
Stage	Core temperature (K)	Core density (kg/m^3)	Duration of stage
Hydrogen fusion	4×10^7	5×10^3	7×10^6 years
Helium fusion	2×10^8	7×10^5	7×10^5 years
Carbon fusion	6×10^8	2×10^8	600 years
Neon fusion	1.2×10^9	4×10^9	1 year
Oxygen fusion	1.5×10^9	10^{10}	6 months
Silicon fusion	2.7×10^9	3×10^{10}	1 day
Core collapse	5.4×10^9	3×10^{12}	¼ second
Core bounce	2.3×10^{10}	4×10^{15}	milliseconds
Explosive (supernova)	about 10^9	varies	10 seconds

Several of the brightest stars in the sky are supergiants, including Betelgeuse and Rigel in the constellation Orion and Antares in the constellation Scorpius. (Figure 19-14 shows the locations of these stars on an H-R diagram.) They appear bright not because they are particularly close, but because they are extraordinarily luminous.

A supergiant star cannot keep adding shells to its "onion" structure forever, because the sequence of thermonuclear reactions cannot go on indefinitely. In order for an element to serve as a

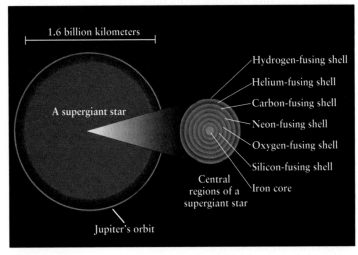

figure 22-13

The Structure of an Old High-Mass Star Near the end of its life, a high-mass star becomes a supergiant whose overall size can be as large as Jupiter's orbit around the Sun. The star's energy comes from six concentric fusing shells, all contained within a volume roughly the same size as the Earth. Thermonuclear reactions cannot occur within the iron core, because fusion reactions that involve iron absorb energy rather than release it.

thermonuclear fuel, energy must be given off when its nuclei collide and fuse. This released energy is a result of the strong nuclear force of attraction that draws nucleons (neutrons and protons) together. However, protons also repel one another by the weaker electric force. As a result of this electric repulsion, adding extra protons to nuclei larger than iron, which has 26 protons, requires an *input* of energy rather than causing energy to be released. Nuclei of this size or larger cannot act as fuel for thermonuclear reactions. Hence, the sequence of fusion stages ends with silicon fusion. One of the products of silicon fusion is iron, and the result is a star with an iron-rich core in which no thermonuclear reactions take place (see Figure 22-13).

Shell fusion in the layers surrounding the iron-rich core consumes the star's remaining reserves of fuel. At this stage the entire energy-producing region of the star is contained in a volume no bigger than the Earth, some 10^6 times smaller in radius than the overall size of the star. This state of affairs will soon come to an end, because the buildup of an inert, iron-rich core signals the impending violent death of a massive supergiant star.

22-6 | High-mass stars violently blow apart in supernova explosions

Our present understanding is that all stars of about 8 M$_\odot$ or less divest most of their mass in the form of planetary nebulae. The burned-out core that remains settles down to become a white dwarf star. But the truly massive stars—stellar heavyweights that begin their lives with more than 8 solar masses of material—do not pass through a planetary nebula phase. Instead, they die in spectacular *supernova explosions*.

To understand what happens in a supernova explosion, we must look deep inside a massive star at the end of its life. Of course, we cannot do this in actuality, because the interiors of stars are opaque. But astronomers have developed theoretical models based on what we know about the behavior of gases and atomic nuclei. The story that follows, while largely theoretical, describes our observations of supernovae fairly well. And as we will see in Section 22-8, a special kind of "telescope" has allowed us to glimpse the interior of at least one famous supernova.

The core of an aging, massive star gets progressively hotter as it contracts to ignite successive stages of thermonuclear fusion. Wien's law (Section 5-4) and Planck's law (Section 5-5) together tell us that as the temperature of an object like a star increases, so does the energy of the photons it emits. When the temperature in the core of a massive star reaches a few hundred million kelvins, the photons are energetic enough to initiate a host of nuclear reactions that create neutrinos. These neutrinos, which carry off energy, escape from the star's core, just as solar neutrinos flow freely out of the Sun (see Section 18-4 and the essay "Searching for Neutrinos Beyond the Textbooks" at the end of Chapter 18).

To compensate for the energy drained by the neutrinos, the star must provide energy either by consuming more thermonuclear fuel, by contracting, or both. But when the star's core is converted into iron (as shown in Figure 22-13), no more energy-producing thermonuclear reactions are possible, and the only source of energy is contraction and rapid heating.

Once a star with an original mass of about 8 $M_\odot$ or more develops an iron-rich core, the core contracts very rapidly, so that the core temperature skyrockets to 5×10^9 K within a tenth of a second. The gamma-ray photons emitted by the intensely hot core have so much energy that when they collide with iron nuclei, they begin to break the iron nuclei down into much smaller helium nuclei (^{4}He). This process is called **photodisintegration**. As Table 22-1 shows, it takes a high-mass star millions of years and several stages of thermonuclear reactions to build up an iron core; within a fraction of a second, photodisintegration undoes much of those millions of years of effort.

Within another tenth of a second, the core becomes so dense that the negatively charged electrons within it are forced to combine with the positively charged protons to produce electrically neutral neutrons. This process also releases a flood of neutrinos, denoted by the Greek letter ν (nu):

$$e^- + p \rightarrow n + \nu$$

Although neutrinos interact only very weakly with matter (see Section 18-4), the core is now so dense that even neutrinos cannot escape from it immediately. But because these neutrinos carry away a substantial amount of energy as they escape from the core, the core cools down and condenses even further.

At about 0.25 second after its rapid contraction begins, the core is less than 20 km in diameter and its density is in excess of 4×10^{17} kg/m^3. This is **nuclear density,** the density with which neutrons and protons are packed together inside nuclei. (If the Earth were compressed to this density, it would be only 300 meters, or 1000 feet, in diameter.)

Matter at nuclear density or higher is extraordinarily difficult to compress. Thus, when the density of the neutron-rich core begins to exceed nuclear density, the core suddenly becomes very stiff and rigid. The core's contraction comes to a sudden halt, and the innermost part of the core actually bounces back and expands somewhat.

This *core bounce* sends a powerful wave of pressure, like an unimaginably intense sound wave, outward into the outer core.

During this critical stage, the cooling of the core has caused the pressure to decrease profoundly in the regions surrounding the core. Without pressure to hold it up against gravity, the material from these regions plunges inward at speeds up to 15% of the speed of light. When this inward-moving material crashes down onto the rigid core, it encounters the outward-moving pressure wave. In just a fraction of a second, the material that fell onto the core begins to move back out toward the star's surface, propelled in part by the flood of neutrinos trying to escape from the star's core.

The outward-moving wave accelerates as it encounters less and less resistance and soon reaches a speed greater than the speed of sound waves in the star's outer layers. When this happens, the wave becomes a shock wave, like the sonic boom produced by a supersonic airplane.

After a few hours, the shock wave reaches the star's surface, by which time the star's outer layers have begun to lift away from the core. The energy released in this titanic event is an incomprehensibly large 10^{46} joules—a hundred times more energy than the Sun has emitted over its entire 4.56-billion year history. When the star's outer layers thin out sufficiently, a portion of this energy escapes in a torrent of light. The star has become a **supernova** (plural **supernovae**).

Supercomputer simulations provide many insights into the violent, complex, and rapidly changing conditions deep inside a star as it is torn apart by a supernova explosion (**Figure 22-14**). For example, Figure 22-14a and Figure 22-14b show snapshots of the first 20 milliseconds after the stiffening of the inner core of a 25-$M_\odot$ star. The simulation predicts turbulent swirls and eddies that grow behind the shock wave as it moves outward from the star's core. Evidence in favor of such turbulence comes from images of the remnants of long-ago supernovae (Figure 22-14c). Such images show that material is ejected from the supernova not in uniform shells but in irregular clumps. This is just what would be expected from a turbulent explosion. (We discuss supernova remnants further in Section 22-10.)

Detailed computer calculations suggest that a 25-$M_\odot$ star ejects about 96% of its material to the interstellar medium for use in producing future generations of stars. Less massive stars eject proportionally less of their mass into space when they become supernovae.

Before this material is ejected into space, it is compressed so much by the passage of the shock wave through the star's outer layers that a new wave of thermonuclear reactions sets in. These reactions produce many more chemical elements, including *all* the elements heavier than iron. These reactions require a tremendous input of energy, and thus cannot take place during the star's presupernova lifetime.

The energy-rich environment of a supernova shock wave is almost the only place in the universe where such heavy elements as zinc, silver, tin, gold, mercury, lead, and uranium can be produced. (In Chapter 23 we will see another, even more exotic mechanism for producing these heaviest elements.) Remarkably, all of these elements are found on the Earth. This means that our solar system, our Earth, and our bodies include material that long ago was part of a star that lived, evolved, and died as a supernova.

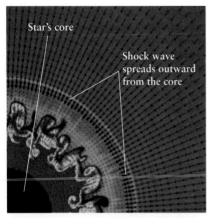

Star's core

Shock wave
spreads outward
from the core

(a) 10 milliseconds after the core "bounce"

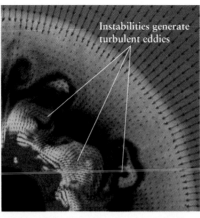

Instabilities generate
turbulent eddies

(b) 20 milliseconds after the core "bounce"

figure 22-14
Turbulence in a Supernova

(a), (b) These images from a supercomputer simulation show the inner regions of a massive star during the first 20 milliseconds of a supernova explosion. The colors indicate temperatures that range from 2×10^{10} K (red) to 6×10^{8} K (blue). (c) Turbulence causes material to be ejected from the supernova in irregular "blobs," as shown by these images of the supernova remnant Cassiopeia A. Each image was made using an X-ray wavelength emitted by a particular element. (A false-color image of Cassiopeia A made using several X-ray wavelengths opens this chapter; see also Figure 22-23.) (a, b: Courtesy of Adam Burrows, University of Arizona; and Bruce Fryxell, NASA/GSFC; c: U. Hwang et al., NASA/GSFC)

Silicon

Calcium

Iron

(c) Material was ejected in "blobs" from the supernova that produced the Cassiopeia A supernova remnant   R I V U **X** G

22-7 | In 1987 a nearby supernova gave us a close-up look at the death of a massive star

Supernovae have peak luminosities as great as 10^{9} L$_{\odot}$, rivaling the light output of an entire galaxy. This makes it possible to see supernovae in galaxies far beyond our own Milky Way Galaxy, and indeed hundreds of these distant supernovae are observed each year. But one recent and unusually close supernova has provided astronomers with a unique opportunity to check the theoretical ideas presented in Section 22-6.

On February 23, 1987, a supernova was discovered in the Large Magellanic Cloud (LMC), a companion galaxy to our Milky Way some 50,000 pc (160,000 ly) from Earth. The supernova, designated SN 1987A because it was the first discovered that year, occurred near an enormous H II region in the LMC called the Tarantula Nebula (Figure 22-15). The supernova was so bright that observers in the southern hemisphere could see it without a telescope.

Such a bright supernova is a rare event. In the last thousand years, only five other supernovae—in 1006, 1054, 1181, 1572, and 1604—have been bright enough to be seen with the naked eye, and all of these occurred in our own Milky Way Galaxy. (As we described in Section 4-3, the supernova of 1572 had a major influence on Tycho Brahe's ideas about the heavens.) Although outside our Galaxy, SN 1987A occurred relatively close to us in a part of the heavens that is obscured only slightly by the Milky Way's interstellar dust. Furthermore, not long after SN 1987A appeared, several new orbiting telescopes were placed into service. These enabled astronomers to study the supernova's evolution with unprecedented resolution and in wavelength ranges not accessible from the Earth's surface (see Section 6-7). As a result, SN 1987A has given astronomers an unprecedented view of the violent death of a massive star.

The light from a supernova such as SN 1987A does not all come in a single brief flash; the outer layers continue to glow as they expand into space. For the first 20 days after the detonation of SN 1987A, this glow was powered primarily by the tremendous heat that the shock wave deposited in the star's outer layers. As the expanding gases cooled, the light energy began to be provided by a different source—the decay of radioactive isotopes of cobalt, nickel, and titanium produced in the supernova explosion.

Astronomers have been able to pinpoint the specific isotopes involved because different radioactive nuclei emit gamma rays of different wavelengths when they decay. These emissions have

The supernova remained visible to the naked eye for several months after the detonation.

Ideally, SN 1987A would have confirmed the theories of astronomers about typical supernovae. But SN 1987A was *not* typical. Its luminosity peaked at roughly 10^8 $L_\odot$, only a tenth of the maximum luminosity observed for other, more distant supernovae. Fortunately, the doomed star had been observed prior to becoming a supernova, and these observations helped explain why SN 1987A was an exceptional case.

The star that exploded and became SN 1987A, called the **progenitor star,** was identified as a blue B3 I supergiant (**Figure 22-16**). When this star was on the main sequence its mass was about 20 $M_\odot$, although by the time it exploded—some 10^7 years after it first formed—it probably had shed a few solar masses. The evolutionary track for an aging 20-$M_\odot$ star wanders back and forth across the top of the H-R diagram as the star alternates between being a hot (blue) supergiant and a cool (red) supergiant. The star's size changes significantly as the surface temperature changes. A blue supergiant is 10 times larger in diameter than the Sun, but a red supergiant of the same luminosity has a diameter 1000 times that of the Sun.

Because the progenitor star was relatively small when it exploded, the star's outer layers were close to the core and thus held more strongly by the core's gravitational attraction. When the detonation occurred, a relatively large fraction of the shock wave's energy had to be used against this gravitational attraction to push the outer layers into space. Hence, the amount of shock wave energy available to be converted into light was smaller than for most supernovae. The result was that SN 1987A was only a tenth as bright as an exploding red supergiant.

Three and a half years after SN 1987A exploded, astronomers used the newly launched Hubble Space Telescope to obtain a picture of the supernova. To their surprise, the image showed a ring of

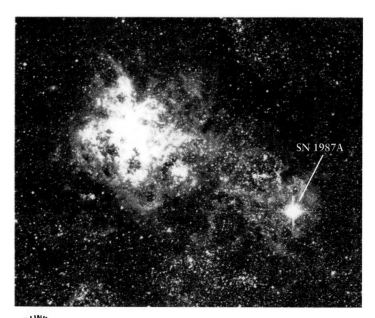

WEB LINK 22.9

figure 22-15 R I **V** U X G
Supernova 1987A This photograph, taken soon after the discovery of SN 1987A, shows a portion of the Large Magellenic Cloud that includes the supernova and a huge H II region called the Tarantula Nebula. Although it was 50,000 pc from Earth, SN 1987A was bright enough to be seen without a telescope. (European Southern Observatory)

been detected by orbiting gamma-ray telescopes (see Figure 6-33). Thanks to these radioactive decays, the brightness of SN 1987A actually *increased* for the first 85 days after the detonation, then settled into a slow decline as the radioactive isotopes were used up.

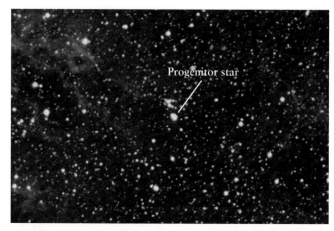

(a) Before the star exploded

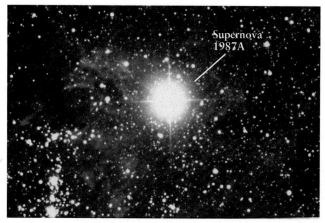

(b) After the star exploded

WEB LINK 22.10

figure 22-16 R I **V** U X G
SN 1987A—Before and After (a) This photograph shows a small section of the Large Magellanic Cloud as it appeared before the explosion of SN 1987A. The supernova's progenitor star was a B3 blue supergiant. (b) This image shows a somewhat larger region of the sky a few days after the supernova exploded into brightness. (Anglo-Australian Observatory)

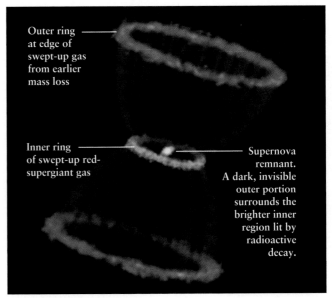

(a) Supernova 1987A seen in 1996 R I **V** U X G

(b) An explanation of the rings

figure 22-17

SN 1987A and Its "Three-Ring Circus" (a) This true-color view from the Hubble Space Telescope shows three bright rings around SN 1987A. (b) This drawing shows the probable origin of the rings. A wind from the progenitor star (see Figure 22-16a) formed an hourglass-shaped shell surrounding the star. (Compare with Figure 22-7.) Ultraviolet light from the supernova explosion ionized ring-shaped regions in the shell, causing them to glow. (Robert Kirshner and Peter Challis, Harvard-Smithsonian Center for Astrophysics; STScI)

glowing gas around the exploded star. After the optics of the Hubble Space Telescope were repaired in 1994 (see Section 6-7), SN 1987A was observed again and a set of *three* glowing rings was revealed (**Figure 22-17a**).

These rings are relics of a hydrogen-rich outer atmosphere that was ejected by gentle stellar winds from the progenitor star when it was a red supergiant, about 20,000 years ago. This diffuse gas expanded in a hourglass shape (Figure 22-17b), because it was blocked from expanding around the star's equator either by a pre-existing ring of gas or by the orbit of an as yet unseen companion star. (Figure 22-7 shows a similar model used to explain the shapes of certain planetary nebulae.) The outer rings in Figure 22-17a are parts of the hourglass that were ionized by the initial flash of ultraviolet radiation from the supernova; as electrons recombine with the ions, the rings emit visible-light photons. (We described this process of *recombination* in Section 20-2.)

 By the early years of the twenty-first century, the shock wave from the supernova was beginning to collide with the "waist" of the hourglass shown in Figure 22-17b. This collision is making the hourglass glow more brightly in visible wavelengths—though not enough, unfortunately, to make the supernova again visible to the naked eye—and emit copious radiation at X-ray and ultraviolet wavelengths.

By studying this cosmic collision, astronomers hope to learn more about the shock wave and the supernova explosion that spawned it. They hope as well to learn about the matter in the inner ring, which will give us insight into the stellar winds that blew from the progenitor star thousands of years ago. Because it provides a unique laboratory for studying the evolution of a supernova, SN 1987A will be carefully monitored by astronomers for decades to come.

22-8 | Neutrinos emanate from supernovae like SN 1987A

 Scientists have developed "neutrino telescopes" that can single out neutrinos coming from supernovae

In addition to electromagnetic radiation, supernovae also emit a brief but intense burst of neutrinos from their collapsing cores. In fact, theory suggests that *most* of the energy released by the exploding star is in the form of neutrinos. If it were possible to detect the flood of neutrinos from an exploding star, astronomers would have direct evidence of the nuclear processes that occur within the star during its final seconds before becoming a supernova.

Unfortunately, detecting neutrinos is difficult because under most conditions matter is transparent to neutrinos (see Section 18-4). Consequently, when supernova neutrinos encounter the Earth, almost all of them pass completely through the planet as if it were not there. The challenge to scientists is to detect the tiny fraction of neutrinos that *do* interact with the matter through which they pass.

During the 1980s, two "neutrino telescopes" uniquely suited for detecting supernova neutrinos went into operation—the Kamiokande detector in Japan (a joint project of the University of Tokyo and the University of Pennsylvania), and the IMB detector (a collaboration of the University of California at Irvine, the University of Michigan, and Brookhaven National Laboratory). Both detectors consisted of large tanks containing thousands of tons of water. On the rare occasion when a neutrino collided with one of the water molecules, it produced a brief flash of light. Any light flash was recorded by photomultiplier tubes lining the walls of the tank. Because other types of subatomic particles besides neutrinos could also produce similar light flashes, the detectors were placed deep underground so that hundreds of meters of earth would screen out almost all particles except neutrinos. (The more recent Sudbury Neutrino Observatory, shown in Figure 18-7, has a similar design.)

What causes the light flashes? And how do we know whether a neutrino comes from a supernova? The key is that a single supernova neutrino carries a relatively large amount of energy, typically 20 MeV or more. (We introduced the unit of energy called the electron volt, or eV, in Section 5-5. One MeV is equal to 10^6 electron volts. Only the most energetic nuclear reactions produce particles with energies of more than 1 MeV.) If such a high-energy neutrino hits a proton in the water-filled tank of a neutrino telescope, the collision produces a positron (see Box 18-1). The positron then recoils at a speed greater than the speed of light in water, which is 2.3×10^5 km/s. (Such motion does not violate the ultimate speed limit in the universe, 3×10^5 km/s, which is the speed of light in a *vacuum*.)

Just as an airplane that flies faster than sound produces a shock wave (a sonic boom), a positron that moves through a substance such as water faster than the speed of light in that substance produces a shock wave of light. This shock wave is called **Cerenkov radiation,** after the Russian physicist Pavel A. Cerenkov, who first observed it in 1934. It is this radiation that is detected by the photomultiplier tubes that line the detector walls.

By measuring the properties of the Cerenkov radiation from a recoiling positron, scientists can determine the positron's energy and, therefore, the energy of the neutrino that created the positron. This allows them to tell the difference between the high-energy neutrinos from supernovae and neutrinos from the Sun, which typically have energies of 1 MeV or less.

Both the Kamiokande and IMB detectors were operational on February 23, 1987, when SN 1987A was first observed in the Large Magellanic Cloud. Soon afterward, the physicists working with these detectors excitedly reported that they had detected Cerenkov flashes from a 12-second burst of neutrinos that reached the Earth 3 hours before astronomers saw the light from the exploding star. Only a few neutrinos were seen: The Kamiokande detector saw flashes from 12 neutrinos at about the same time that 8 were recorded by the IMB detector. But when the physicists factored in the sensitivity of their detectors, they calculated that Kamiokande and IMB had actually been exposed to a torrent of more than 10^{16} neutrinos.

Given the flux of neutrinos measured by the detectors and the 160,000–light-year distance from SN 1987A to the Earth, physicists used the inverse-square law to determine the total number of neutrinos that had been emitted from the supernova. (This law applies to neutrinos just as it does to electromagnetic radiation; see Section 19-2.) They found that over a 10-second period, SN 1987A emitted 10^{58} neutrinos with a total energy of 10^{46} joules. This is more than 100 times as much energy as the Sun has emitted in its entire history and more than 100 times the amount of energy that the supernova emitted in the form of electromagnetic radiation. Indeed, for a few seconds the supernova's neutrino luminosity—that is, the *rate* at which it emitted energy in the form of neutrinos—was 10 times greater than the total luminosity in electromagnetic radiation of all of the stars in the observable universe! Such comparisons give a hint of the incomprehensible violence with which a supernova explodes.

Why did the neutrinos from SN 1987A arrive 3 hours *before* the first light was seen? As we saw in Section 22-6, neutrinos are produced when thermonuclear reactions cease in the core of a massive star and the core collapses. These neutrinos encounter little delay as they pass through the volume of the star. The tremendous increase in the star's light output, by contrast, occurs only when the shock wave reaches the star's outermost layers (which are thin enough to allow light to pass through them). It took 3 hours for this shock wave to travel outward from the star's core to its surface, by which time the neutrino burst was already billions of kilometers beyond the dying star. For the next 160,000 years, the neutrinos that would eventually produce light flashes in Kamiokande and IMB remained in front of the photons emitted from the star's surface, and so the neutrinos were detected before the supernova's light. Thus, the neutrino data from SN 1987A gave astronomers direct confirmation of theoretical ideas about how supernova explosions take place.

Kamiokande and IMB have both been replaced by a new generation of neutrino telescopes (see Section 18-4). While these new detectors are intended primarily to observe neutrinos from the Sun, they are also fully capable of measuring neutrino bursts from nearby supernovae. Astronomers have identified a number of supergiant stars in our Galaxy that are likely to explode into supernovae, among them the bright red supergiant Betelgeuse in the constellation Orion (see Figure 2-2 and Figure 6-30).

Unfortunately, astronomers do not yet know how to predict precisely when such stars will explode into supernovae. It may be many thousands of years before Betelgeuse explodes. Then again, it could happen tomorrow. If it does, neutrino telescopes will be ready to record the collapse of its massive core.

22-9 | White dwarfs in close binary systems can also become supernovae

Astronomers discover dozens of supernovae in distant galaxies every year (Figure 22-18), but not all of these are the result of massive stars dying violently. A totally different type of supernova occurs when a white dwarf star in a binary system blows itself completely apart. The first clue that two entirely distinct chains of events could produce supernovae was rather subtle: Some supernovae have prominent hydrogen emission lines in their spectra but others do not.

Supernovae with hydrogen emission lines, called **Type II supernovae,** are the sort we described in Section 22-6. They are caused by the death of highly evolved massive stars that still have ample hydrogen in their atmospheres when they explode. When the star explodes, the hydrogen atoms are excited and glow prominently,

(a) Spiral galaxy M81

Progenitor star

(b) Before the explosion

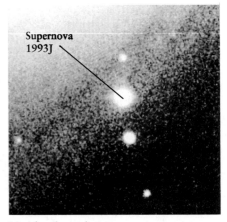

Supernova 1993J

(c) After the explosion

WEB LINK 22.12

Figure 22-18 R I **V** U X G

A Supernova in a Distant Galaxy On the night of March 28, 1993, Francisco Garcia Diaz, a Spanish amateur astronomer, discovered supernova SN 1993J in the galaxy M81 in Ursa Major. (a) M81 lies some 3.6 million pc (12 million ly) from Earth. Its angular size is about half that of the full moon. (b) The progenitor star that later exploded into SN 1993J was a K0 I red supergiant. (c) This image shows the same part of the sky as (b). Like SN 1987A, SN 1993J resulted from the explosion of a massive star. (a: Palomar Observatory; b, c: D. Jones and E. Telles, Isaac Newton Telescope)

producing hydrogen emission lines. SN 1987A, the topic of Sections 22-7 and 22-8, was a Type II supernova.

Hydrogen lines are missing in the spectrum of a **Type I supernova,** which tells us that little or no hydrogen is left in the debris from the explosion. Type I supernovae are further divided into three important subclasses. **Type Ia** supernovae have spectra that include a strong absorption line of ionized silicon. **Type Ib** and **Type Ic** supernovae both lack the ionized silicon line. The difference between them is that the spectra of Type Ib supernovae have a strong helium absorption line, while those of Type Ic supernovae do not. **Figure 22-19** shows these different supernova spectra.

Astronomers suspect that Type Ib and Ic supernovae are caused by dying massive stars, just like Type II supernovae. The difference is that the progenitor stars of Type Ib and Ic supernovae have been stripped of their outer layers before they explode. A star can lose its outer layers to a strong stellar wind (see Figure 22-12) or, if it is part of a close binary system, by transferring mass to its companion star (see Figure 21-19b). If enough mass remains for the star's core to collapse, the star dies as a Type Ib supernova. Because the outer layers of hydrogen are absent, the supernova's spectrum exhibits no hydrogen lines but many helium lines (Figure 22-19b). Type Ic supernovae have apparently undergone even more mass loss prior to their explosion; their spectra show that they have lost much of their helium as well as their hydrogen (Figure 22-19c).

The evidence that Type II, Type Ib, and Type Ic supernovae all begin as massive stars is that all three types are found only near sites of recent star formation. The life span of a massive star from its formation to its explosive death as a supernova is only about 10^7 years, less time than it took our own Sun to condense from a protostar to a main-sequence star (see Figure 20-9) and a mere blink of an eye on the time scale of stellar evolution. Because massive stars live such a short time, it makes sense that they should meet their demise very close to where they were formed.

Type Ia supernovae, by contrast, are found even in galaxies where there is no ongoing star formation. Hence, they are probably *not* the death throes of massive supergiant stars. Instead, Type Ia supernovae may result from the thermonuclear explosion of a white dwarf star. This may sound contradictory, because we saw in Section 22-4 that white dwarf stars have no thermonuclear reactions going on in their interiors. But these reactions *can* occur if a carbon-oxygen–rich white dwarf is in a close, semidetached binary system with a red giant star (see Figure 21-18b).

As the red giant evolves and its outer layers expand, it overflows its Roche lobe and dumps gas from its outer layers onto the white dwarf (Figure 22-19a). When the total mass of the white dwarf approaches the Chandrasekhar limit, the increased pressure applied to the white dwarf's interior causes carbon fusion to begin there. Hence, the interior temperature of the white dwarf increases.

ANIMATION 22.4

If the white dwarf were made of ordinary matter, the temperature increase would cause a further increase in pressure, the white dwarf would expand and cool, and the carbon-fusing reactions would abate. But because the white dwarf is composed of degenerate matter, this "safety valve" between temperature and pressure does not operate. Instead, the increased temperature just makes the reactions proceed at an ever-increasing rate, in a catastrophic runaway process reminiscent of the helium flash in low-mass stars. Soon the temperature gets so high that the electrons are no longer degenerate and the white dwarf blows apart, dispersing all of its mass into space.

Before exploding, the white dwarf contained primarily carbon and oxygen and almost no hydrogen or helium, which explains the absence of hydrogen and helium lines in the spectrum of the resulting supernova. Silicon is a by-product of the carbon-fusing reaction and gives rise to the silicon absorption line characteristic of Type Ia supernovae.

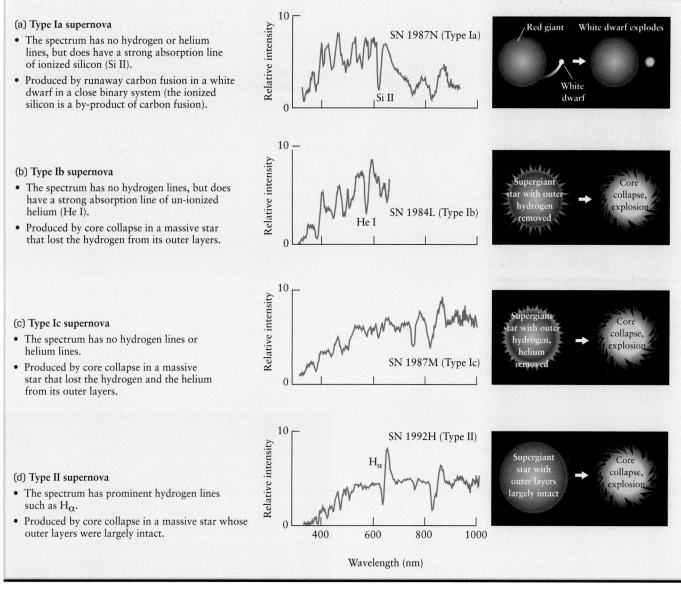

(a) Type Ia supernova
- The spectrum has no hydrogen or helium lines, but does have a strong absorption line of ionized silicon (Si II).
- Produced by runaway carbon fusion in a white dwarf in a close binary system (the ionized silicon is a by-product of carbon fusion).

(b) Type Ib supernova
- The spectrum has no hydrogen lines, but does have a strong absorption line of un-ionized helium (He I).
- Produced by core collapse in a massive star that lost the hydrogen from its outer layers.

(c) Type Ic supernova
- The spectrum has no hydrogen lines or helium lines.
- Produced by core collapse in a massive star that lost the hydrogen and the helium from its outer layers.

(d) Type II supernova
- The spectrum has prominent hydrogen lines such as H_α.
- Produced by core collapse in a massive star whose outer layers were largely intact.

Figure 22-19

Supernova Types These illustrations show the characteristic spectra and the probable origins of supernovae of **(a)** Type Ia, **(b)** Type Ib, **(c)** Type Ic, and **(d)** Type II. (Spectra courtesy of Alexei V. Filippenko, University of California, Berkeley)

 Different types of supernovae have fundamentally different energy sources. Type II, Ib, and Ic supernovae are powered by *gravitational* energy. The star's iron-rich core collapses inward under the influence of its own gravity, releasing a flood of energy in the form of neutrinos; the outer layers of the star also fall inward under gravity, providing the energy to power the nuclear reactions that generate the supernova's electromagnetic radiation. Type Ia supernovae, by contrast, are powered by *nuclear* energy released in the explosive thermonuclear fusion of a white dwarf star. While Type Ia supernovae typically emit more energy in the form of electromagnetic radiation than supernovae of other types, they do not emit copious numbers of neutrinos because there is no core col-

lapse. If we include the energy emitted in the form of neutrinos, the most luminous supernovae by far are those of Type II.

In addition to the differences in their spectra, different types of supernovae can be distinguished by their light curves (**Figure 22-20**). All supernovae begin with a sudden rise in brightness that occurs in less than a day. After reaching peak luminosity, Type Ia, Ib, and Ic supernovae settle into a steady, gradual decline in luminosity. (An example is the supernova of 1006, which is thought to have been of Type Ia. This supernova, which at its peak was more than 200 times brighter than any other star in the sky, took three years to fade into invisibility.) By contrast, the Type II light curve has a

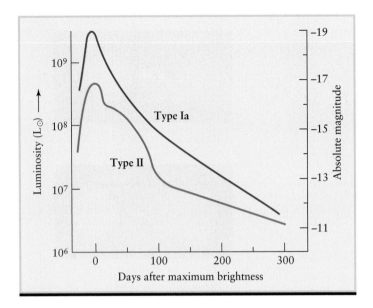

figure 22-20

Supernova Light Curves A Type Ia supernova reaches maximum brightness in about a day, followed by a gradual decline in brightness. A Type II supernova reaches a maximum brightness only about one-fourth that of a Type Ia supernova and usually has alternating intervals of steep and gradual declines.

steplike appearance caused by alternating periods of steep and gradual declines in brightness.

For all supernova types, the energy source during the period of declining brightness is the decay of radioactive isotopes produced during the supernova explosion. Because a different set of thermonuclear reactions occurs for each type of supernova, each type produces a unique set of isotopes that decay at different rates. This helps explains the distinctive light curves for different supernova types.

For the same reason, each type of supernova ejects a somewhat different mix of elements into the interstellar medium. As an example, Type Ia supernovae are primarily responsible for the elements near iron in the periodic table, because they generate these elements in more copious quantities than Type II supernovae.

A number of astronomers are now measuring the distances to remote galaxies by looking for Type Ia supernovae in those galaxies. This is possible because there is a simple relationship between the rate at which a Type Ia supernova fades away and its peak luminosity: The slower it fades, the greater its luminosity. Hence, by observing how rapidly a distant Type Ia supernova fades, astronomers can determine its peak luminosity. A measurement of the supernova's peak apparent brightness then tells us (through the inverse-square law) the distance to the supernova, and, therefore, the distance to the supernova's host galaxy. The tremendous luminosity of Type Ia supernovae allows this method to be used for galaxies more than 10^9 light-years distant. In Chapter 28 we will learn what such studies tell us about the size and evolution of the universe as a whole.

22-10 | A supernova remnant can be detected at many wavelengths for centuries after the explosion

Astronomers find the debris of supernova explosions, called **supernova remnants,** scattered across the sky. A beautiful example is the Veil Nebula, shown in Figure 22-21. The doomed star's outer layers were blasted into space with such violence that they are still traveling through the interstellar medium at supersonic speeds 15,000 years later. As this expanding shell of gas plows through space, it collides with atoms in the interstellar medium, exciting the gas and making it glow.

A few nearby supernova remnants cover sizable areas of the sky. The largest is the Gum Nebula, named after the astronomer Colin Gum, who first noticed its faint glowing wisps on photographs of the southern sky (Figure 22-22). Its 40° angular diameter is centered on the constellation Vela.

The Gum Nebula looks big simply because it is quite close to us. Its center is only about 460 pc (1300 ly) from Earth, and its near side is just 100 pc (330 ly) away. Studies of the nebula's expansion rate suggest that this supernova exploded around 9000 B.C. At maximum brilliance, the exploding star probably was as bright as the Moon at first quarter. Like the first quarter moon, it would have been visible in the daytime!

figure 22-21   R I **V** U X G

The Veil Nebula—A Supernova Remnant This nebulosity is a portion of the Cygnus Loop, which is the roughly spherical remnant of a supernova that exploded about 15,000 years ago. The distance to the nebula is about 800 pc (2600 ly), and the overall diameter of the loop is about 35 pc (120 ly). (Palomar Observatory)

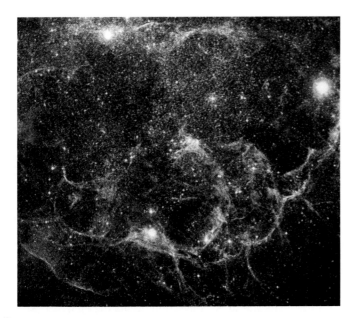

Figure 22-22 R I V U X G

The Gum Nebula—A Supernova Remnant This nebula has the largest angular size (40°) of any known supernova remnant. Only the central regions of the Gum Nebula are shown here. The supernova explosion occurred about 11,000 years ago, and the remnant now has a diameter of about 700 pc (2300 ly). (Royal Observatory, Edinburgh)

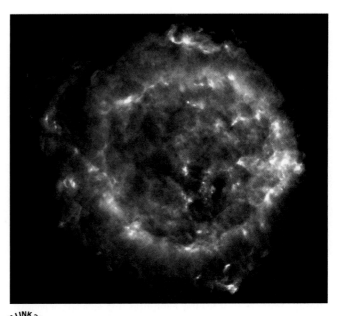

Figure 22-23 R I V U X G

Cassiopeia A—A Supernova Remnant This false-color radio image of Cassiopeia A was produced by the Very Large Array (see Figure 6-26). The impact of supernova material on the interstellar medium causes ionization, and the liberated electrons generate radio waves as they move. Cassiopeia A is roughly 3300 pc (11,000 ly) from Earth. (Image courtesy of NRAO/AUI)

Many supernova remnants are virtually invisible at optical wavelengths. However, when the expanding gases collide with the interstellar medium, they radiate energy at a wide range of wavelengths, from X rays through radio waves. For example, Figure 22-23 shows a radio image of the supernova remnant Cassiopeia A; Figure 22-14*c* and the lower image on the page that opens this chapter show Cassiopeia A at X-ray wavelengths. Optical photographs of this part of the sky reveal only a few small, faint wisps. In fact, radio searches for supernova remnants are more fruitful than optical searches. Only two dozen supernova remnants have been found in visible-light images, but more than 100 remnants have been discovered by radio astronomers. Other remnants have been observed by the new generation of orbiting X-ray telescopes (see Figure 20-22).

From the expansion rate of Cassiopeia A, astronomers conclude that this supernova explosion occurred about 300 years ago. Although telescopes were in wide use by the late 1600s, no one saw the outburst (and no one today knows why). The last supernova seen in our Galaxy, which occurred in 1604, was observed by Johannes Kepler. In 1572, Tycho Brahe also recorded the sudden appearance of an exceptionally bright star in the sky. To find any other accounts of nearby bright supernovae, we must delve into astronomical records that are almost 1000 years old.

At first glance, this apparent lack of nearby supernovae may seem puzzling. From the frequency with which supernovae occur in distant galaxies, it is reasonable to suppose that a galaxy such as our own should have as many as five supernovae per century. Where have they been?

As we will learn from the study of galaxies in Chapters 25 and 26, the plane of our Galaxy is where massive stars are born and supernovae explode. This region is so rich in interstellar dust, however, that we simply cannot see very far into space in the directions occupied by the Milky Way (see Section 20-2). In other words, supernovae probably do in fact erupt every few decades in remote parts of our Galaxy, but their detonations are hidden from our view by intervening interstellar matter.

A supernova remnant may be all that is left after some supernovae explode. But for supernovae of Types II, Ib, and Ic, which are caused by core collapse in a massive star, the core itself may also remain. If there is a relic of the core, it may be either a *neutron star* or a *black hole*, depending on the mass of the core and the conditions within it during the collapse. Neutron stars, as the name suggests, are made primarily of neutrons. Wholly unlike anything we have studied so far, these exotic objects are the subject of Chapter 23. We will study black holes, which are far stranger even than neutron stars, in Chapter 24.

CAUTION! Although neutron stars and black holes can be part of the debris from a supernova explosion, they are *not* called "supernova remnants." That term is applied exclusively to the gas that spreads away from the site of the supernova explosion.

We have seen in this chapter that only the most massive stars end their lives as supernovae of Types II, Ib, or Ic. We learned earlier that mass plays a central role in determining the speed with which a star

Figure 22-24

Pathways of Stellar Evolution The evolution of an isolated star (one that is not part of a close multiple-star system) depends on the star's mass. The more massive the star, the more rapid its evolution. The scale on the left gives the mass of the star when it is on the main sequence, and the scale at the right shows the mass of the resulting stellar corpse. Stars whose original masses are less than about 8 $M_\odot$ eject enough mass over their lifetimes to end up as white dwarfs (which requires that the corpse be less massive than the Chandrasekhar limit of 1.4 $M_\odot$). High-mass stars can evolve into supernovae of Type Ib, Ic, or II and become neutron stars or black holes. It is also possible that a high-mass star can leave no corpse at all.

forms and joins the main sequence (see Figure 20-9), the star's luminosity and surface temperature while on the main sequence (see Figures 19-21 and 19-22), and how long a star can remain on the main sequence (see Table 21-1). Now we see that mass also determines the eventual fate of a star—whether it ends in the whimper of a planetary nebula, as do low-mass stars like the Sun, or follows the evolutionary path of a high-mass star and ends with an almost inconceivably intense bang (**Figure 22-24**).

Key Words

asymptotic giant branch, p. 491	helium shell flash, p. 493
asymptotic giant branch star (AGB star), p. 491	horizontal branch, p. 490
	mass-radius relation, p. 495
carbon fusion, p. 497	neon fusion, p. 497
carbon star, p. 492	neutron capture, p. 497
Cerenkov radiation, p. 503	nuclear density, p. 499
Chandrasekhar limit, p. 495	oxygen fusion, p. 497
core helium fusion, p. 490	photodisintegration, p. 499
dredge-up, p. 492	planetary nebula, p. 492

progenitor star, p. 501	thermal pulse, p. 493
red-giant branch, p. 490	Type I supernova, p. 504
shell helium fusion, p. 490	Type Ia supernova, p. 504
silicon fusion, p. 497	Type Ib supernova, p. 504
supergiant, p. 497	Type Ic supernova, p. 504
supernova (*plural* supernovae), p. 499	Type II supernova, p. 503
	white dwarf, p. 494
supernova remnant, p. 506	

Key Ideas

Late Evolution of Low-Mass Stars: A low-mass star becomes a red giant when shell hydrogen fusion begins, a horizontal-branch star when core helium fusion begins, and an asymptotic giant branch (AGB) star when the helium in the core is exhausted and shell helium fusion begins.

• As a low-mass star ages, convection occurs over a larger portion of its volume. This takes heavy elements formed in the star's interior and distributes them throughout the star.

Planetary Nebulae and White Dwarfs: Helium shell flashes in an old, low-mass star produce thermal pulses during which more than half the star's mass may be ejected into space. This exposes the hot carbon-oxygen core of the star.

• Ultraviolet radiation from the exposed core ionizes and excites the ejected gases, producing a planetary nebula.

• No further nuclear reactions take place within the exposed core. Instead, it becomes a degenerate, dense sphere about the size of the Earth and is called a white dwarf. It glows from thermal radiation; as the sphere cools, it becomes dimmer.

Late Evolution of High-Mass Stars: Unlike a low-mass star, a high-mass star undergoes an extended sequence of thermonuclear reactions in its core and shells. These include carbon fusion, neon fusion, oxygen fusion, and silicon fusion.

• In the last stages of its life, a high-mass star has an iron-rich core surrounded by concentric shells hosting the various thermonuclear reactions. The sequence of thermonuclear reactions stops here, because the formation of elements heavier than iron requires an input of energy rather than causing energy to be released.

The Deaths of Massive Stars: A high-mass star dies in a violent cataclysm in which its core collapses and most of its matter is ejected into space at high speeds. The luminosity of the star increases suddenly by a factor of around 10^8 during this explosion, producing a supernova.

• More than 99% of the energy from such a supernova is emitted in the form of neutrinos from the collapsing core.

• The matter ejected from the supernova, moving at supersonic speeds through interstellar gases and dust, glows as a nebula called a supernova remnant.

Other Types of Supernovae: An accreting white dwarf in a close binary system can also become a supernova when carbon fusion ignites explosively throughout such a degenerate star.

• Type Ia supernovae are those produced by accreting white dwarfs in close binaries. Type II supernovae are created by the deaths of massive stars, as are supernovae of Type Ib and Type Ic; these latter types occur when the star has lost a substantial part of its outer layers before exploding.

• Most supernovae occurring in our Galaxy are hidden from our view by interstellar dust and gases.

Review Questions

1. What is the horizontal branch? Where is it located on an H-R diagram? How do stars on the horizontal branch differ from red giants or main-sequence stars?

2. Horizontal-branch stars are sometimes referred to as "helium main-sequence stars." In what sense is this true?

3. Some blue main-sequence stars in our region of the Galaxy have the same luminosity and surface temperature as the horizontal-branch stars in the globular cluster M55 (see Figure 22-3). How do we know that the horizontal-branch stars in M55 are not main-sequence stars?

4. What is the asymptotic giant branch? Where is it located on an H-R diagram? How do asymptotic giant branch stars differ from red giants or main-sequence stars?

5. Is a carbon star a star that is made of carbon? Explain.

6. What is the connection between dredge-ups in old stars and life on Earth?

7. What are thermal pulses in AGB stars? What causes them? What effect do they have on the luminosity of the star?

8. How is a planetary nebula formed?

9. How can an astronomer tell the difference between a planetary nebula and a planet?

10. Why do we not observe planetary nebulae that are more than about 50,000 years old?

11. What is a white dwarf? Does it produce light in the same way as a star like the Sun?

12. How does the radius of a white dwarf depend on its mass? How is this different from other types of stars?

13. What is the significance of the Chandrasekhar limit?

14. On an H-R diagram, sketch the evolutionary track that the Sun will follow from when it leaves the main sequence to when it becomes a white dwarf. Approximately how much mass will the Sun have when it becomes a white dwarf? Where will the rest of the mass go?

15. What prevents thermonuclear reactions from occurring at the center of a white dwarf? If no thermonuclear reactions are occurring in its core, why doesn't the star collapse?

16. Why do you suppose that all the white dwarfs known to astronomers are relatively close to the Sun?

17. Why does the mass of a star play such an important role in determining the star's evolution?

18. Why is the temperature in a star's core so important in determining which nuclear reactions can occur there?

19. What is the difference between a red giant and a red supergiant?

20. In what way does the structure of an aging supergiant resemble that of an onion?

21. What is nuclear density? Why is it significant when a star's core reaches this density?

22. Why is SN 1987A so interesting to astronomers? In what ways was it not a typical supernova?

23. Why are neutrinos emitted by supernovae? How can these neutrinos be detected? How can they be distinguished from solar neutrinos?

24. What are the differences among Type Ia, Type Ib, Type Ic, and Type II supernovae? Which type is most unlike the other three, and why?

25. How can a supernova continue to shine for many years after it explodes?

26. How do supernova remnants produce radiation at nonvisible wavelengths?

27. There may have been recent supernovae in our Galaxy that have not been observable even though they are incredibly luminous. How is it this possible?

28. Is our own Sun likely to become a supernova? Why or why not?

Advanced Questions

Questions preceded by an asterisk (*) involve topics discussed in the Boxes in Chapter 1, Chapter 7, or Chapter 19.

> **Problem-solving tips and tools**
>
> The small-angle formula is given in Box 1-1. You may find it useful to review Box 19-4, which discusses stellar radii and their relationship to temperature and luminosity. Section 4-7 explains the formula for gravitational force, and Box 7-2 explains the concept of escape speed. Sections 5-2 and 5-4 describe some key properties of light, especially blackbody radiation. We discussed the relationship among luminosity, apparent brightness, and distance in Box 19-2. The relationship among absolute magnitude, apparent magnitude, and distance was the topic of Box 19-3.

29. The globular cluster M15 depicted in Figure 22-6a contains 30,000 old stars, but only one of these stars is presently in the planetary nebula stage of its evolution. Explain why planetary nebulae are not more prevalent in M15.

30. Explain why astronomers had to use infrared light to detect the faint wisps in planetary nebula NGC 7027 (Figure 22-6c).

31. The central star in a newly formed planetary nebula has a luminosity of 1000 $L_\odot$ and a surface temperature of 100,000 K. What is the star's radius? Give your answer as a multiple of the Sun's radius.

32. You want to determine the age of a planetary nebula. What observations should you make, and how would you use the resulting data?

33. Figure 22-6*b* shows the Helix Nebula. Use the information given in the figure caption to calculate the diameter of the nebula in pc and in AU.

*34. The Ring Nebula in the constellation Lyra has an angular size of 1.4 arcmin × 1.0 arcmin. It is expanding at the rate of about 20 km/s. Approximately how long ago did the central star shed its outer layers? Assume that the nebula is 2,700 ly from Earth.

35. The accompanying image shows the planetary nebula IC 418 in the constellation Lepus (the Hare). (**a**) The image shows a small shell of glowing gas (shown in blue) within a larger glowing gas shell (shown in orange). Discuss how IC 418 could have acquired this pair of gas shells. (**b**) Explain why the outer shell looks thicker around the edges than near the middle.

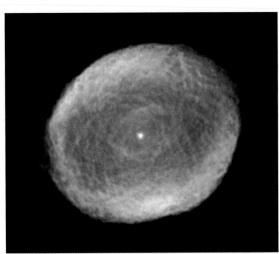

(NASA and Hubble Heritage Team, STScI/AURA)   R I **V** U X G

36. (**a**) Calculate the wavelength of maximum emission of the white dwarf Sirius B. In what part of the electromagnetic spectrum does this wavelength lie? (**b**) In a visible-light photograph such as Figure 22-8, Sirius B appears much fainter than its primary star. But in an image made with an X-ray telescope, Sirius B is the brighter star. Explain the difference.

37. (**a**) Find the average density of a 1-$M_\odot$ white dwarf having the same diameter as the Earth. (**b**) What speed is required to eject gas from the white dwarf's surface? (This is also the speed with which interstellar gas falling from a great distance would strike the star's surface.)

38. In the classic 1960s science-fiction comic book *The Atom*, a physicist discovers a basketball-sized meteorite (about 10 cm in radius) that is actually a fragment of a white dwarf star. With

some difficulty, he manages to hand-carry the meteorite back to his laboratory. Estimate the mass of such a fragment, and discuss the plausibility of this scenario.

39. (**a**) What kinds of stars would you monitor if you wished to observe a supernova explosion from its very beginning? (**b**) Examine Appendices 4 and 5, which list the nearest and brightest stars, respectively. Which, if any, of these stars are possible supernova candidates? Explain.

*40. Consider a high-mass star just prior to a supernova explosion, with a core of diameter 20 km and density 4×10^{17} kg/m³. (**a**) Calculate the mass of the core. Give your answer in kilograms and in solar masses. (**b**) Calculate the force of gravity on a 1-kg object at the surface of the core. How many times larger is this than the gravitational force on such an object at the surface of the Earth, about 10 newtons? (**c**) Calculate the escape speed from the surface of the star's core. Give your answer in m/s and as a fraction of the speed of light. What does this tell you about how powerful a supernova explosion must be in order to blow material away from the star's core?

41. The shock wave that traveled through the progenitor star of SN 1987A took 3 hours to reach the star's surface. (**a**) Given the size of a blue supergiant star (see Section 22-7), estimate the speed with which the shock wave traveled through the star's outer layers. (The core of the progenitor star was very small, so you may consider the shock wave to have started at the very center of the star.) Give your answer in meters per second. (**b**) Compare your answer with the speed of sound waves in our atmosphere, about 340 m/s, and with the speed of light. (**c**) A shock wave traveling through a gas is a special case of a sound wave. In general, sound waves travel faster through denser, less easily compressed materials. Thus, sound travels faster through water (about 1500 m/s) than through our atmosphere, and faster still through steel (about 5900 m/s). Use this idea to compare the gases within the progenitor star of SN 1987A with the gases in our atmosphere in terms of their average density and how easily they are compressed.

42. The neutrinos from SN 1987A arrived 3 hours before the visible light. While they were en route to the Earth, what was the distance between the neutrinos and the first photons from SN 1987A? Assume that neutrinos are massless and thus travel at the speed of light. Give your answers in kilometers and in AU.

*43. Compared to SN 1987A (see Figure 22-15), the supernova SN 1993J (see Figure 22-18) had a maximum apparent brightness only 9.1×10^{-4} as great. Using the distances from Earth to each of these supernovae, determine the ratio of the maximum luminosity of SN 1993J to that of SN 1987A. Which of the two supernovae had the greater maximum luminosity?

*44. Suppose that the brightness of a star becoming a supernova increases by 20 magnitudes. Show that this corresponds to an increase of 10^8 in luminosity.

*45. Suppose that the red-supergiant star Betelgeuse, which lies some 425 light-years from the Earth, becomes a Type II supernova. (**a**) At the height of the outburst, how bright would it appear in the sky? Give your answer as a fraction of the brightness of the Sun ($b_\odot$). (**b**) How would it compare with the brightness of Venus (about 10^{-9} $b_\odot$)?

*46. In July 1997, a supernova named SN 1997cw exploded in the galaxy NGC 105 in the constellation Cetus (the Whale). It reached an apparent magnitude of +16.5 at maximum brilliance, and its spectrum showed an absorption line of ionized silicon. Use this information to find the distance to NGC 105. (*Hint:* Inspect the light curves in Figure 22-20 to find the *absolute* magnitudes of typical supernovae at peak brightness.)

47. Figure 22-21 shows a portion of the Veil Nebula in Cygnus. Use the information given in the caption to find the average speed at which material has been moving away from the site of the supernova explosion over the past 15,000 years. Express your answer in km/s and as a fraction of the speed of light.

Discussion Questions

48. Suppose that you discover a small, glowing disk of light while searching the sky with a telescope. How would you decide if this object is a planetary nebula? Could your object be something else? Explain.

49. Suppose the convective zone in AGB stars did *not* reach all the way down into their carbon-rich cores. How might this have affected the origin and evolution of life on Earth?

50. Imagine that our Sun was somehow replaced by a 1-$M_\odot$ white dwarf star, and that our Earth continued in an orbit of semi-major axis 1 AU around this star. Discuss what effects this would have on our planet. What would the white dwarf look like as seen from Earth? Could you look at it safely with the unaided eye? Would the Earth's surface temperature remain the same as it is now?

51. The major final product of silicon fusion is ^{56}Fe, an isotope of iron with 26 protons and 30 neutrons. This is also the most common isotope of iron found on Earth. Discuss what this tells you about the origin of the solar system.

52. SN 1987A did not agree with the theoretical picture outlined in Section 22-6. Does this mean that the theory was wrong? Discuss.

 ## Web/CD-ROM Questions

53. It has been claimed that the Dogon tribe in western Africa has known for thousands of years that Sirius is a binary star. Search the World Wide Web for information about these claims. What is the basis of these claims? Why are scientists skeptical, and how do they refute these claims?

54. Search the World Wide Web for recent information about SN 1993J. Has the shape of the supernova's light curve been adequately explained? Has the supernova produced any surprises?

55. Search the World Wide Web for information about SN 1994I, a supernova that occurred in the galaxy M51 (NGC 5194). Why

was this supernova unusual? Was it bright enough to have been seen by amateur astronomers?

 56. Convection Inside a Giant Star. Access and view the animation "Convection Inside a Giant Star" in Chapter 22 of the *Universe* Web site or CD-ROM. Describe the motion of material in the interior of the star. In what ways is this motion similar to convection within the present-day Sun (see Section 18-2)? In what ways is it different? Is a dredge-up taking place in this animation? How can you tell?

  57. Types of Supernovae. Access and view the animations "In the Heart of a Type II Supernova" and "A Type Ia Supernova" in Chapter 22 of the *Universe* Web site or CD-ROM. Describe how these two types of supernova are fundamentally different in their origin.

Observing Projects

> **Observing tips and tools**
>
> While planetary nebulae are rather bright objects, their brightness is spread over a relatively large angular size, which can make seeing them a challenge for the beginning observer. For example, the Helix Nebula (see Figure 22-6*b*) has the largest angular size of any planetary nebula but is also one of the most difficult to see. To improve your view, make your observations on a dark, moonless night from a location well shielded from city lights. Another useful trick, mentioned in Chapters 19 and 20, is to use "averted vision." Once you have the nebula centered in the telescope, you will get a brighter and clearer image if you look at the nebula out of the corner of your eye. The so-called Blinking Planetary in Cygnus affords an excellent demonstration of this effect; the nebula seems to disappear when you look straight at it, but it reappears as soon as you look toward the side of your field of view.
>
> Another useful tip is to view the nebula through a green filter (a #58, or O III, filter available from telescope supply houses). Green light is emitted by excited, doubly ionized oxygen atoms, which are common in planetary nebulae but not in most other celestial objects. Using such a filter can make a planetary nebula stand out more distinctly against the sky. As a side benefit, it also helps to block out stray light from street lamps. The same tips also apply to observing supernova remnants.

58. Although they represent a fleeting stage at the end of a star's life, planetary nebulae are found all across the sky. Some of the brightest are listed in the accompanying table on page 512. Note that the distances to most of these nebulae are quite uncertain. Observe as many of these planetary nebulae as you can on a clear, moonless night using the largest telescope at your disposal. Note and compare the various shapes of the different nebulae. In how many cases can you see the central star? The central star in the

Eskimo Nebula is supposed to be the "nose" of an Eskimo wearing a parka. Can you see this pattern?

Planetary nebula	Distance (light-years)	Angular size	Constellation	Right Ascension	Declination
Dumbbell (M27, NGC 6853)	490–3500	8.0 × 5.7	Vulpecula	19h 59.6m	+22° 43′
Ring (M57, NGC 6720)	1300–4100	1.4 × 1.0	Lyra	18 53.6	+33 02
Little Dumbbell (M76, NGC 650)	1700–15,000	2.7 × 1.8	Perseus	01 42.4	+51 34
Owl (M97, NGC 3587)	1300–12,000	3.4 × 3.3	Ursa Major	11 14.8	+55 01
Saturn (NGC 7009)	1600–3900	0.4 × 1.6	Aquarius	21 04.2	–11 22
Helix (NGC 7293)	450	41 × 41	Aquarius	22 29.6	–20.48
Eskimo (NGC 2392)	1400–10,000	0.5 × 0.5	Gemini	7 29.2	+20 55
Blinking Planetary (NGC 6826)	3300(?)	2.2 × 0.5	Cygnus	19 44.8	+50 31

59. Northern hemisphere observers with modest telescopes can see two supernova remnants, one in the winter sky and the other in the summer sky. Both are quite faint, however, so you should schedule your observations for a moonless night. The winter sky contains the Crab Nebula, which is discussed in detail in Chapter 23. The coordinates are R.A. = 5^h 34.5^m and Decl. = +22° 00′, which places the object near the star marking the eastern horn of Taurus (the Bull). Whereas the entire Crab Nebula easily fits in the field of view of an eyepiece, the Veil or Cirrus Nebula in the summer sky is so vast that you can see only a small fraction of it at a time. The easiest way to find the Veil Nebula is to aim the telescope at the star 52 Cygni (R.A. = 20^h 45.7^m and Decl. = +30° 43′), which lies on one of the brightest portions of the nebula. If you then move the telescope slightly north or south until 52 Cygni is just out of the field of view, you should see faint wisps of glowing gas.

60. Use a telescope to observe the remarkable triple star 40 Eridani, whose coordinates are R.A. = 4^h 15.3^m and Decl. = –7° 39′. The primary, a 4.4-magnitude yellowish star like the Sun, has a 9.6-magnitude white dwarf companion, the most easily seen white dwarf in the sky. On a clear, dark night with a moderately large telescope, you should also see that the white dwarf has an 11th-magnitude companion, which completes this most interesting trio.

 61. The red supergiant Betelgeuse in the constellation Orion will someday explode as a supernova. Use the *Starry Night Backyard*™ program to investigate how the supernova might appear if this was to happen tonight. Select **Go Home** in the **View** menu to show the sky as seen from your location at the present time. (If the program doesn't place you at your true location, use the **Viewing Location...** command in the **View** menu.) Click on the "Info" tab at the left-hand side of the main *Starry Night Backyard*™ window (this will present a display with information about what you are viewing). Then locate Betelgeuse using the **Find...** command in the **Edit** menu. If Betelgeuse is below the horizon, allow the program to reset the time to when it is visible. (**a**) At what time does Betelgeuse rise on today's date? At what time does it set? (**b**) If Betelgeuse became a supernova today, would it be visible in the daytime? How would it appear at night? Do you think it would cast shadows? (**c**) Are Betelgeuse and the Moon both in the night sky tonight? (Use the **Find...** command in the **Edit** menu to locate the Moon.) If they are, what kinds of shadows might they both cast if Betelgeuse became a supernova?

 62. Use the *Deep Space Explorer*™ program to help visualize the location of Supernova 1987A. In the left-hand part of the window under the heading **Milky Way**, click on the triangle next to the word **Explore** and then click on **Local Neighbourhood**. You can zoom in or out using the buttons at the upper left of the window (an upward-pointing triangle and a downward-pointing triangle). You can rotate the Milky Way Galaxy and its neighbor galaxies by putting the mouse cursor over the image, holding down the mouse button, and moving the mouse. (On a two-button mouse, hold down the left mouse button.) (**a**) Center the field of view on the Sun (use the **Find...** command in the **Edit** menu). Describe the position of the Large Magellanic Cloud (LMC), within which SN 1987A lies, relative to the Milky Way Galaxy and to our solar system. (**b**) Use the **Find...** command in the **Edit** menu to center on the LMC. You should be able to locate the Tarantula Nebula, shown in Figure 22-15. Is SN 1987A near the center or the edge of the LMC? (Note that although *Deep Space Explorer*™ depicts the LMC as being rather flat, it is thought to be an irregular blob of stars with some thickness.)

Collaborative Exercise

63. Imagine that a supernova originating from a close binary star system began (as seen from Earth) on the most recent birthday of the youngest person in your group. Using the light curves in Figure 22-21, what would its new luminosity be today and how bright would it appear in the sky (apparent magnitude) if it were located 10 parsecs (32.6 light-years) away? How would your answers change if you were to discover that the supernova actually originated from an isolated star with a mass 15 times greater than our Sun?

23 Neutron Stars

Almost all the light we see from objects in the night sky is due to thermonuclear fusion reactions. Such reactions are what make stars shine, and in turn make any surrounding nebulae glow. Even the light we see from the Moon and planets—which is simply reflected sunlight—can be traced back to thermonuclear reactions in the Sun's core.

By contrast, the exotic object shown here—the Crab Nebula in Taurus—is undergoing no thermonuclear fusion at all. Despite this, the object is so energetic that it is emitting copious amounts of X rays. This image shows tilted, glowing rings more than a light-year across, with oppositely directed jets flowing outward perpendicular to the plane of the rings. At the object's very center, too small to pick out in this image, is the source of all this dynamic activity: a neutron star, the leftover core of what was once a supergiant star.

Before 1967 neutron stars were considered pure speculation. But today we know of more than 1300 of these exotic stellar corpses, and hundreds of thousands more may be strewn across the Milky Way Galaxy. As we will discover in this chapter, neutron stars have powerful magnetic fields that sweep beams of radiation across the sky. We detect these beams as pulsating radio signals. We will find that neutron stars in close binary systems display even more outrageous behavior, including bouts of explosive helium fusion that yield a vast outpouring of X rays. And we will see that a similar process involving a white dwarf in a close binary system produces a short-lived intense burst of visible light called a nova.

 A composite image of the Crab Nebula using X-ray (blue) and visible (red) wavelengths. (X-ray: NASA/CXC/ASU/ J. Hester et al.; optical: NASA/CXC/ASU/J. Hester et al.)
R I **V** U **X** G

Questions to Guide Your Reading

As you read the sections of this chapter, look for the answers to the following questions:

23-1 What led scientists to the idea of a neutron star?

23-2 What are pulsars, and how were they discovered?

23-3 How did astronomers determine the connection between pulsars and neutron stars?

23-4 How can a neutron star supply energy to a surrounding nebula?

23-5 What are conditions like inside a neutron star?

23-6 How are some neutron stars able to spin several hundred times per second?

23-7 Why do some pulsars emit fantastic amounts of X rays?

23-8 Are X-ray bursters and novae similar to supernovae?

23-9 How massive can a neutron star be?

23-1 | Scientists first proposed the existence of neutron stars in the 1930s

On the morning of July 4, 1054, Yang Wei-T'e, the imperial astronomer to the Chinese court, made a startling discovery. Just a few minutes before sunrise, a new and dazzling object ascended above the eastern horizon. This "guest star," as Yang called it, was far brighter than Venus and more resplendent than any star he had ever seen.

Yang's records show that the "guest star" was so brilliant that it could easily be seen during broad daylight for the rest of July. Records from Constantinople (now Istanbul, Turkey) also describe this object, and works of art made by the Anasazi culture in the American Southwest suggest that they may have seen it as well (Figure 23-1). Over the next 21 months, however, the "guest star" faded to invisibility.

We now know that the "guest star" of 1054 was actually a remarkable stellar transformation: A massive star some 6500 light-years away perished in a supernova explosion, leaving behind both a supernova remnant (see Figure 1-6 and Section 22-10) and a bizarre object called a *neutron star*. But it was not until 1932 that the English physicist James Chadwick discovered the subatomic particle called the neutron. The neutron has about the same mass as the positively charged proton, but has no electric charge. Within a

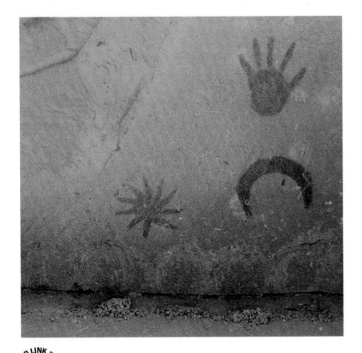

WEB LINK 23.2 **Figure 23-1** R I **V** U X G
A Supernova Pictograph? This drawing in an eleventh-century structure in New Mexico shows a ten-pointed star next to a crescent. It may depict the scene on the morning of July 5, 1054, when a "guest star" appeared next to the waning crescent moon. (Courtesy of National Parks Service)

year of Chadwick's discovery, two astronomers were inspired to predict the existence of neutron stars.

As we saw in Section 22-6, a proton and an electron can combine under high pressure to form a neutron. Could such pressures be found within the core of a dying star? If so, they could transform the core into a sphere of neutrons—a **neutron star.** This is just what was proposed by Fritz Zwicky at the California Institute of Technology (Caltech) and his colleague Walter Baade at Mount Wilson Observatory. "With all reserve," Zwicky and Baade theorized, "we advance the view that supernovae represent the transition from ordinary stars into neutron stars, which in their final stages consist of extremely closely packed neutrons." In other words, there could be at least two types of stellar corpses—white dwarfs and neutron stars.

Zwicky and Baade realized that once a neutron star had formed in response to tremendous pressures, it would be able to resist any further compression. The neutrons themselves would provide a counterbalancing outward pressure. Motivated by the idea that white dwarfs are supported by degenerate electron pressure, Zwicky and Baade proposed that a highly compact ball of neutrons would similarly produce a **degenerate neutron pressure.** (Like electrons, neutrons obey the Pauli exclusion principle that we described in Section 21-2.) This powerful pressure could support a stellar corpse—perhaps one even more massive than a white dwarf, because degenerate neutron pressure can be much stronger than degenerate electron pressure. So, while a white dwarf collapses if its mass is above the Chandrasekhar limit (see Section 22-4), a neutron star might not.

Although Zwicky and Baade's proposal would prove to be very close to the mark, most scientists politely ignored it for years. After all, a neutron star must be a rather weird object. If brought to the Earth's surface, a single thimbleful of neutron star matter would weigh 100 million tons!

A star compacted to such densities must be very small. A 1.4-$M_\odot$ neutron star would have a diameter of only 20 km (12 mi), about the size of a moderate-sized city on Earth. Its surface gravity would be so strong that an object would have to travel at one-half the speed of light to escape into space. These conditions seemed outrageous until the late 1960s, when astronomers discovered pulsating radio sources.

23-2 | The discovery of pulsars in the 1960s stimulated interest in neutron stars

 As a young graduate student at Cambridge University, Jocelyn Bell spent many months helping construct an array of radio antennas covering 4½ acres of the English countryside. The instrument was completed by the summer of 1967, and Bell and her colleagues in Anthony Hewish's research group began using it to scrutinize radio emissions from the sky. They were looking for radio sources that "twinkle" like stars; that is, they looked for random small fluctuations in brightness caused by the motion of gas between the source and the observer. While searching for such random flickering, Bell noticed that the antennas had detected regular pulses of radio noise from one particular location in the sky. These radio pulses were arriving at regular intervals of 1.3373011 seconds.

The pulsations were much more rapid than those of any other astronomical object known at that time. Indeed, they were so rapid and regular that the Cambridge team at first suspected that they might not be of natural origin. Instead, they might be signals from an advanced alien civilization. That possibility had to be discarded within a few months after several more of these pulsating radio sources, which came to be called **pulsars,** were discovered across the sky. In all cases, the periods were extremely regular, ranging from about 0.25 second for the fastest to about 1.5 seconds for the slowest (**Figure 23-2**).

What could these objects possibly be? As is often done in science, astronomers first decided what pulsars could *not* be. It was clear that pulsars were not ordinary stars or nebulae, because while these objects emit some radio waves, their emissions do not pulsate (that is, their light output does not increase and decrease rhythmically). One type of star that *appears* to pulsate is an eclipsing binary star system, which appears dimmer when one star in the binary system passes in front of the other (see Section 19-11). However, in order to vary in brightness with a typical pulsar period of 1 second or less, an eclipsing binary would have to have an equally short orbital period. Newton's form of Kepler's third law (see Section 4-7) implies that the two stars would have to orbit only 1000 kilometers or so from each other. This is less than the diameter of any ordinary star, even a white dwarf star. Hence, the eclipsing binary hypothesis can be ruled out, because it leads to the highly implausible picture of two stars that overlap!

Astronomers also ruled out the idea that pulsars might be variable stars, which vary in luminosity as they increase and decrease in size (see Section 21-5). Pulsating variables have periods of days or weeks, while pulsars have periods of seconds or less. Just as the broad wings of an eagle could never flap as fast as a hummingbird's wings without breaking, an ordinary star or even a white dwarf is too large to pulsate in and out in less than a second.

Another possibility, however, was given serious consideration. Perhaps pulsars were rapidly rotating white dwarf stars with some sort of radio-emitting "hot spots" on their surfaces. But even this model had its limits, because a white dwarf that rotated on its axis in just 1 second would be on the verge of flying apart. What finally ruled out the spinning white dwarf model was the discovery of a pulsar in the middle of the Crab Nebula, the supernova remnant left behind by the supernova that Yang Wei-T'e saw in 1054 (see Figure 1-6). That discovery clarified the true nature of pulsars.

At the time of its discovery, the Crab pulsar was the fastest pulsar known to astronomers. Its period is 0.0333 second, which means that it flashes 1/0.0333 = 30 times each second. It was immediately apparent to astrophysicists that white dwarfs are too big and bulky to generate 30 signals per second; calculations demonstrated that a white dwarf could not rotate that fast without tearing itself apart. Hence, the existence of the Crab pulsar means that the stellar corpse at the center of the Crab Nebula has to be much smaller and more compact than a white dwarf.

This was a truly unsettling realization, because most astronomers in the mid-1960s thought that *all* stellar corpses are white dwarfs. The number of white dwarfs in the sky seemed to account for all the stars that must have died since our Galaxy was formed. It was generally assumed that all dying stars—even the most massive ones, which produce supernovae—somehow manage to eject enough matter so that their corpses do not exceed the Chandrasekhar limit. But the discovery of the Crab pulsar showed these conservative opinions to be in error.

It was now clear that pulsars had to be something far more exotic than rotating white dwarfs. As Thomas Gold of Cornell University emphasized, astronomers would now have to seriously consider the existence of neutron stars.

23-3 | Pulsars are rapidly rotating neutron stars with intense magnetic fields

SCIENCE IN PROCESS

The neutron-star model of pulsars had to pass several stringent tests to be accepted by astronomers

Science is, by its very nature, based on skepticism. If a scientist proposes a new and exotic explanation for an unexplained phenomenon, that explanation will be accepted only if it satisfies two demanding criteria. First, there must be excellent reasons why more conventional explanations cannot work. Second, the new explanation must help scientists understand a range of other phenomena.

A good example is the Copernican idea that the Earth orbits the Sun, which gained general acceptance only after Galileo's observations (see Section 4-5). First, Galileo showed the older geocentric explanation of celestial motions to be untenable, and, second, it became clear that the Copernican model could also explain the motions of Jupiter's moons.

We have seen how the discovery of the Crab pulsar ruled out the idea that pulsars were white dwarfs. But astronomers soon accepted the more exotic idea that pulsars were actually neutron stars. To understand why, we must assume the skeptical outlook of

WEB LINK 23.4

figure 23-2

A Recording of a Pulsar This chart recording shows the intensity of radio emission from one of the first pulsars to be discovered. (The designation means "pulsar at a right ascension of 03ʰ 29ᵐ and a declination of +54°".) Note that the interval between pulses is very regular, even though some pulses are weak and others are strong. (Adapted from R. N. Manchester and J. H. Taylor) the scientist and ask some probing questions about this idea. Why should neutron stars emit radiation at all? In particular, why should they emit radio waves? Why should the emissions be pulsed? And how could the pulses occur as rapidly as 30 times per second, as in the Crab pulsar? The answers to all of these questions are intimately related to the small size of neutron stars.

 Because neutron stars are very small, they should also rotate rapidly. A typical star, such as our Sun, takes nearly a full month to rotate once about its axis. But just as an ice skater doing a pirouette speeds up when she pulls in her arms, a collapsing star also speeds up as its size shrinks. If our Sun were compressed to the size of a neutron star, it would spin about 1000 times per second! Because neutron stars are so small and dense, they can spin this rapidly without flying apart.

The small size of neutron stars also implies that they have intense magnetic fields. It seems safe to say that every star possesses some magnetic field, but typically the strength of this field is quite low. The magnetic field of a main-sequence star is spread out over billions of square kilometers of the star's surface. However, if such a star collapses down to a neutron star, its surface area (which is proportional to the square of its radius) shrinks by a factor of about 10^{10}. The magnetic field, which is bonded to the star's ionized gases (see Section 18-9), becomes concentrated onto an area 10^{10} times smaller than before the collapse, and thus the field strength increases by a factor of 10^{10}.

A star's magnetic field splits its spectral lines into two or more lines whose spacing reveals the field's strength (see Figure 18-21). The strength of a magnetic field is usually measured in the unit called the *gauss* (G), named after the famous German mathematician and astronomer Karl Friedrich Gauss. For example, the magnetic splitting of lines in the spectra of main-sequence stars reveals field strengths in the range from about 1 G (the value for the Sun) to several thousand gauss. The splitting of lines in the spectra of certain white dwarfs corresponds to field strengths in excess of 10^6 G, comparable to the strongest fields ever produced in a laboratory. But the magnetic fields surrounding typical neutron stars are a million times stronger still, on the order of 1 trillion gauss (10^{12} G). Some neutron stars may have fields as strong as 10^{15} G.

The magnetic field of a neutron star makes it possible for the star to radiate pulses of energy toward our telescopes. A number of different models have been proposed for how rotating neutron stars generate radiation; we'll discuss just one leading model. The idea behind this model is that the magnetic axis of a neutron star, the line connecting the north and south magnetic poles, is likely to be inclined at an angle to the rotation axis (**Figure 23-3**). After all, there is no fundamental reason for these two axes to coincide. (Indeed, these two axes do not coincide for any of the planets of our solar system.)

In 1969, Peter Goldreich at Caltech pointed out that the combination of such a powerful magnetic field and rapid rotation would act like a giant electric generator, creating very strong electric fields near the neutron star's surface. Indeed, these fields are so intense that part of their energy is used to create electrons and positrons out of nothingness in a process called **pair production.** The powerful electric fields push these charged particles out from the neutron star's surface and into its curved magnetic field, as sketched in Figure 23-3.

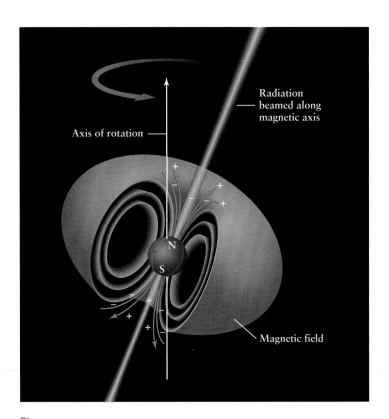

Figure 23-3

A Rotating, Magnetized Neutron Star Charged particles are accelerated near a magnetized neutron star's magnetic poles (labeled N and S), producing two oppositely directed beams of radiation. If the star's magnetic axis (a line that connects the north and south magnetic poles) is tilted at an angle from the axis of rotation, as shown here, the beams sweep around the sky as the star rotates. If the Earth happens to lie in the path of one of the beams, we detect radiation that appears to pulse on and off.

As the particles spiral along the curved field, they are accelerated and emit energy in the form of electromagnetic radiation. (A radio transmission antenna works in a similar way. Electrons are accelerated back and forth along the length of the antenna, producing radio waves.) The result is that two narrow beams of radiation pour out of the neutron star's north and south magnetic polar regions.

 A rotating, magnetized neutron star is somewhat like a lighthouse beacon. As the star rotates, the beams of radiation sweep around the sky. If at some point during the rotation one of those beams happens to point toward the Earth, we will detect a brief flash as the beam sweeps over us. Hence, a radio telescope will detect regular pulses of radiation, with one pulse being received for each rotation of the neutron star.

 The name *pulsar* may lead you to think that the source of radio waves is actually pulsing. But in the model just described, this is not the case at all. Instead, beams of radiation are emitted continuously from the magnetic poles of the neutron star. The pulsing that astronomers detect here on the

Earth is simply a result of the rapid rotation of the neutron star, which brings one of the beams periodically into our line of sight. In this sense, the analogy between a pulsar and a lighthouse beacon is a very close one.

 Soon after the rotating neutron star model of pulsars was proposed in the late 1960s, a team of astronomers at the University of Arizona began wondering if pulsars might also emit pulses at wavelengths other than the radio part of the spectrum. To investigate, they began a search for *visible-light* pulses from the Crab pulsar (**Figure 23-4**). They aimed a telescope at the center of the Crab Nebula (Figure 23-4a), then used a spinning disk with a slit in it to "chop," or interrupt, incoming light at rapid intervals. To their surprise and delight, they found that one of the stars at the center of the nebula appears to be flashing on and off 30 times each second (Figure 23-4b)—exactly the same pulse rate as observed with radio waves.

This observation was strong evidence in favor of pulsars being rotating, magnetized neutron stars, because charged particles accelerating in a magnetic field can radiate strongly over a very wide range of wavelengths. By contrast, the blackbody radiation from an ordinary star, which has a surface temperature in thousands of degrees, is very weak at radio wavelengths. (The Sun's weak radio emissions come not from its glowing surface but rather from charged particles spiraling in the solar magnetic field.)

Since the pioneering days of the 1960s, periodic flashes of radiation have been detected from the Crab pulsar at still other wavelengths, including X rays (Figure 23-4c). The pulsar period is the same at all wavelengths, just as we would expect if the emissions are coming from a portion of the neutron star, which comes into view periodically as the star rotates.

The many successes of the neutron star model of pulsars led to the rapid acceptance of the model. Since 1968 radio astronomers have discovered more than 1300 pulsars scattered across the sky, and it is estimated that perhaps 10^5 more are strewn around the disk of the Milky Way Galaxy. Each one is presumed to be the neutron star corpse of an extinct, massive star. Radio telescopes have detected pulsars with a wide variety of pulse periods, from a relatively sluggish 8.51 s to a blindingly fast 0.00156 s. In each case the pulse period is thought to be the same as the rotation period of the neutron star.

Supernovae have been seen throughout history, but they did not necessarily produce pulsars. For example, those observed by Tycho Brahe in 1572 and by Johannes Kepler in 1604 were probably Type Ia supernovae (see Section 22-9), and it is difficult to imagine how

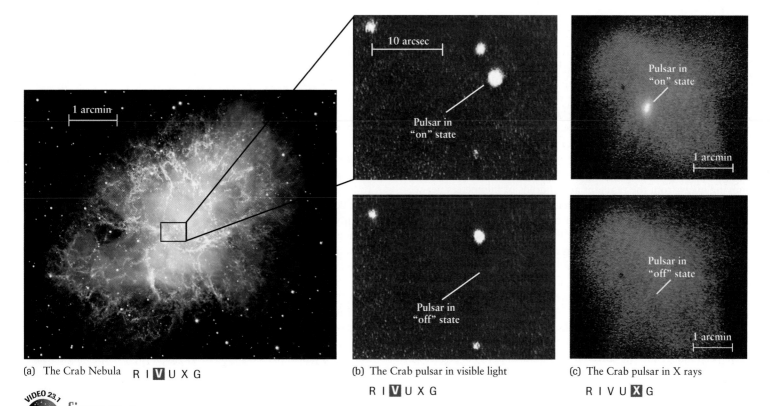

(a) The Crab Nebula   R I **V** U X G

(b) The Crab pulsar in visible light
R I **V** U X G

(c) The Crab pulsar in X rays
R I V U **X** G

figure 23-4

The Crab Pulsar (a) A pulsar is located at the center of the Crab Nebula, which is about 2000 pc (6500 ly) from Earth and about 3 pc (10 ly) across. The boxed area is shown in closeup in part (b). (b) We see a flash of visible light when the rotating pulsar's beam is directed toward us (the "on" state). The pulsar fades (the "off" state) when the beam is aimed elsewhere. (c) The Crab pulsar also pulses "on" and "off" at X-ray wavelengths. The radio pulses, visual flashes, and X-ray flashes all have a period of 0.033 s. (a: The FORS Team, VLT, European Southern Observatory; b: Lick Observatory; c: Einstein Observatory, Harvard-Smithsonian Center for Astrophysics)

figure 23-5 R I V U X G
A Fast-Moving Pulsar When the progenitor star of this supernova remnant exploded, it ejected the star's core at such a high speed that the core overtook the blast wave and moved beyond the boundary of the supernova remnant. The ejected core became a pulsar, denoted B1757-24. Tens of thousands of years after the explosion, the pulsar is still moving at about 600 km/s (350 mi/s). In these false-color radio images, red denotes areas of strongest radio emission. (National Radio Astronomy Observatory)

an exploding white dwarf might become a neutron star. Indeed, no pulsar has been found at the locations of either of those supernovae. The primary sources of pulsars are thought to be Type II supernovae, in which the cores of massive stars collapse (as described in Section 22-6). The supernova that left behind the Crab Nebula and the Crab pulsar is thought to have been of Type II.

The Crab pulsar is not far from the center of the Crab Nebula. This is what we might expect: When the progenitor star of the supernova of 1054 exploded, its core remained at the center of the explosion and became a rotating neutron star, while the star's outer layers were blasted into space to form the nebula. However, Figure 23-5 shows a pulsar that lies *outside* its supernova remnant and is moving away from the remnant at hundreds of kilometers per second. It appears that when the progenitor star exploded, the explosion was much more powerful on one side of the core than the other. As a result, the core was sent flying at immense speed. Many such fast-moving pulsars have been observed, which suggests that supernova explosions are often asymmetrical. In this way studies of pulsars have given us new insights into the nature of supernovae.

The most intensely studied of all recent Type II supernovae is the great supernova of 1987 (SN 1987A, described in Section 22-7). Astronomers have been carefully observing its remains for signs of a pulsar. To date, however, no confirmed observations of pulses have been made, and the search for a pulsar in SN 1987A goes on.

23-4 | Pulsars gradually slow down as they radiate energy into space

 As remarkable as the Crab pulsar is, the Crab Nebula that surrounds it is even more so. Although the nebula is 2000 pc (6500 ly) away, it can be spotted in dark

skies with even a small telescope. It shines with a luminosity 75,000 times that of the Sun (and some 10^8 times the luminosity of the pulses from the Crab pulsar).

But why is the Crab Nebula so luminous? Certain other nebulae—the H II regions described in Section 20-2—can also be seen at great distances. But H II regions contain young, hot stars that give off ultraviolet radiation. This radiation ionizes the surrounding gas and provides the energy that the H II region emits into space. The Crab Nebula, however, is different from H II regions: It contains *no* young stars. What, then, is the source of this nebula's prodigious energy output? As we will see, the Crab pulsar provides the energy not only for radio pulses but for the entire Crab Nebula.

In 1966, two years before the discovery of pulsars, John A. Wheeler at Princeton University and Franco Pacini in Italy had already speculated that the ultimate source of the Crab Nebula's immense energy output might be a spinning neutron pulsar. This speculation was quickly accepted as the correct explanation of the Crab Nebula's luminosity. What confirmed their idea was the discovery that the Crab pulsar is slowing down.

Although pulsars were first noted for their very regular pulses, careful measurements by radio telescopes soon revealed that the rotation period of a typical pulsar—that is, the time for it to spin once on its axis—increases by a few billionths of a second each day. The Crab pulsar, which is one of the most rapidly rotating pulsars, is also slowing more quickly than most—its period increases by 3×10^{-8} second each day. Thirty billionths of a second may sound trivial, but there is so much energy in the rotation of a rapidly spinning neutron star that even the slightest slowdown corresponds to a tremendous loss of energy. In fact, the observed rate of slowing for the Crab pulsar corresponds to a rate of energy loss equal to the entire luminosity of the Crab Nebula.

The interpretation is that the energy lost by the Crab pulsar as it slows down is transferred to the surrounding nebula. But *how* does this transfer of energy take place, and how is the energy radiated into space? The answer lies in the diffuse part of the Crab Nebula, which shines with an eerie bluish light (see Figure 23-4a). This same type of light was first produced on the Earth in 1947 in a particle accelerator in Schenectady, New York. This machine, called a *synchrotron,* was built by the General Electric Company to accelerate electrons to nearly the speed of light for experiments in nuclear physics. (The electrons are said to be *relativistic,* because their velocities are near the speed of light and Einstein's theory of relativity must be applied to understand their motions.)

As a beam of electrons whirled along a circular path within the cyclotron, held in orbit by powerful magnets, scientists noticed that it emitted a strange light. It is now known that this light, called **synchrotron radiation,** is emitted whenever high-speed electrons move along curved paths through a magnetic field.

The total energy output of the Crab Nebula in synchrotron radiation is 3×10^{31} watts, compared to the relatively paltry 4×10^{26} watts emitted by the Sun. Thus, the Crab Nebula must contain quite a large number of relativistic electrons spiraling in an extensive magnetic field. **Figure 23-6** shows where these electrons come from:

They are ejected from the neutron star's vicinity, forming immense jets that emanate from the star's poles and rings that spread outward from its equator. (Figure 20-15 shows a related process that produces jets from young stars.) As the Crab pulsar slows, the energy of its rotation is transferred to these electrons through the magnetic field and is then emitted by them in the form of synchrotron radiation. So much energy is transferred in this way that the Crab Nebula shines very brightly.

Because a spinning neutron star slows down as it radiates away its rotational energy, it follows that an old neutron star should be spinning more slowly than a young one. Slower pulsars are thought to have been formed in supernova explosions that occurred hundreds of thousands of years ago. Over the ages, the supernova remnants have dispersed into the interstellar medium, and the pulsars left behind have slowed to periods of a second or more. (More than 400 such old pulsars are known.) Only relatively young, rapidly rotating pulsars like the Crab pulsar have enough energy to emit flashes of visible light along with radio pulses (see Figure 23-4b). We can summarize these ideas as a general rule:

An isolated pulsar slows down as it ages, so its pulse period increases.

From a pulsar's period (which increases as it ages) and the rate at which its period is increasing (which decreases as it ages), astronomers can estimate how old a pulsar is.

Like all general rules, the one relating a pulsar's age and its period has exceptions. In Section 23-6 we will see how some old pulsars that are *not* isolated—that is, are members of binary star systems—can be reaccelerated to truly dizzying rotation speeds.

23-5 | Superfluidity and superconductivity are among the strange properties of neutron stars

The radiation from pulsars gives us information about the surroundings of neutron stars, as well as how they rotate. To deduce what conditions are like *inside* a neutron star, however, astrophysicists must construct theoretical models. We saw examples of such models for the Sun in Section 18-2 and for post–main-sequence stars in Sections 22-1 and 22-5.

The models for neutron stars are very different. For one thing, neutron stars are thought to have a solid crust on their surface. Furthermore, the interior of a neutron star is a sea of densely packed, degenerate neutrons, with properties quite unlike those of ordinary gases or even degenerate electrons. Detailed calculations strongly suggest that degenerate neutron matter can flow without any friction whatsoever, a phenomenon called **superfluidity.**

Superfluidity can be observed in the laboratory by cooling liquid helium to temperatures near absolute zero. Because it is frictionless, superfluid liquid helium exhibits strange properties such as being able to creep up the walls of a container in apparent defiance of gravity. Within a neutron star, elongated, friction-free whirlpools of superfluid neutrons may form. Their interaction with the star's crust may be the cause of sudden changes in a pulsar's rotation.

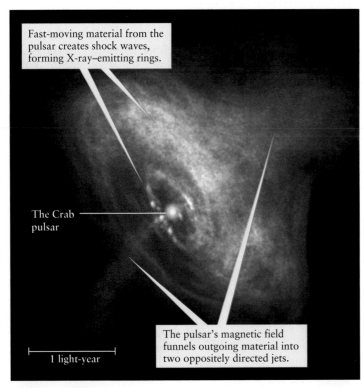

Fast-moving material from the pulsar creates shock waves, forming X-ray–emitting rings.

The Crab pulsar

1 light-year

The pulsar's magnetic field funnels outgoing material into two oppositely directed jets.

  Figure 23-6 R I V U X G
The Dynamic Heart of the Crab Nebula This image from the Chandra X-ray Observatory shows material being expelled from the Crab pulsar into the surrounding nebula. (The image that opens this chapter combines this X-ray view with a visible-light view.) Some of the ejected material travels at more than half the speed of light. (NASA/CXC/ASU/J. Hester et al.)

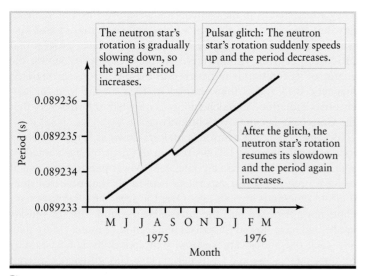

figure 23-7

A Pulsar Glitch This graph shows how the period of the Vela pulsar varied during 1975 and 1976. Most of the time the neutron star's rotation slowed as its rotational energy was converted into radiation. In September 1975, however, a sudden speedup (glitch) occurred. A number of pulsars have been observed to undergo multiple glitches at intervals of a few years.

In addition to the general slowing down of pulsars, astronomers have found that they sometimes exhibit a sudden speedup, sometimes called a **glitch.** For example, Figure 23-7 shows measurements of the period of the Vela pulsar (named for the constellation in which it lies) during 1975 and 1976. On this graph, the pulsar's gradual slowdown is shown as a steady increase in its period. In September 1975, however, there was an abrupt speedup, after which the pulsar continued to slow down at its usual rate. Similar glitches are seen for other pulsars.

Current opinion among pulsar theorists is that glitches are caused by the superfluid neutrons within the neutron star. As a rotating neutron star radiates energy into space, the rotation of its crust slows down, but the neutron whirlpools in the star's interior continue to rotate with the same speed. Some of these whirlpools cling to the crust, as though they were bungee cords with one end attached to the crust and the other to the star's interior. As the crust slows down relative to the interior, these superfluid "bungee cords" stretch out. When the tension in them gets too great, they deliver a sharp jolt that makes the crust speed up suddenly. (An older model, in which pulsar glitches were caused by a "starquake" in the neutron star's settling crust, does not appear to agree with the accumulated observational data on glitches.)

Superfluidity is not the only exotic property of neutron star interiors. Models of the internal structure of a neutron star strongly suggest that the protons in the core can move around without experiencing any electrical resistance whatsoever. This phenomenon, called **superconductivity,** also occurs on the Earth with certain substances at low temperatures. Once you start an electric current moving in a superconducting material, it keeps moving forever.

You may have been surprised to read in the preceding paragraph about *protons* in a *neutron* star. While a neutron star is made up predominantly of neutrons, some protons and electrons must be scattered throughout the star's interior. Indeed, a pulsar's magnetic field must be anchored to the neutron star by charged particles. Neutrons are electrically neutral, so without the protons and electrons in its interior a neutron star would rapidly lose its magnetic field.

As Figure 23-8 shows, the structure of a neutron star probably consists of a core with superfluid neutrons and superconducting protons, a mantle of superfluid neutrons in which some of the neutrons combine with protons to form nuclei, and a brittle crust less than a kilometer thick. At the center of a neutron star, the density is on the order of 10^{18} kg/m^3. This is some 10 trillion (10^{13}) times greater than the central density of our present-day Sun and approximately twice the density of an atomic nucleus. It has been speculated that under these conditions the neutrons and protons dissolve into more fundamental particles called *quarks*. To date, however, astronomers have not found conclusive observational evidence of such extreme states of matter.

While the deep interiors of neutron stars remain mysterious, astronomers have recently found evidence that the crusts of neutron stars are surrounded by a highly unusual atmosphere (Figure 23-9). The evidence comes from the neutron star 1E 1207.4-5209, which like the Crab pulsar emits X rays and lies at the center of a supernova remnant (Figure 23-9a). The dashed curve in Figure 23-9b

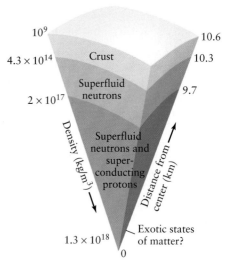

figure 23-8

A Model of a Neutron Star This theoretical model for a 1.4-M$_\odot$ neutron star has a superconducting, superfluid core 9.7 km in radius. The core is surrounded by a 0.6-km thick mantle containing nuclei, electrons, and superfluid neutrons. The star's crust is only 0.3 km thick (the length of three football fields) and is composed of heavy nuclei (such as iron) and free electrons. (The thicknesses of the layers are not shown to scale.)

(a) X-ray image

(b) X-ray spectrum

Figure 23-9 R I V U **X** G

Evidence for a Neutron-Star Atmosphere (a) This image from XMM-Newton (see Figure 6-32*b*) shows a neutron star roughly 2000 pc (7000 ly) from Earth in the constellation Centaurus. Like the Crab pulsar (Figure 23-4*a*), this neutron star is an X-ray–emitting pulsar. (b) The dashed curve shows the X-ray spectrum that would be expected if the neutron star had no atmosphere. The observed dips are evidence that an atmosphere is present. (a: ESA/CESR/G. Bignami et al.)

shows the expected spectrum of the emitted X rays. The actual spectrum, shown by the solid curve in Figure 23-9*b*, has two pronounced dips. The interpretation of these dips is that they are caused by absorption of certain X-ray wavelengths by an atmosphere around the neutron star. (Compare Figure 23-9*b* with Figure 7-3.) The atoms that make up such an atmosphere must be dramatically different from those around you: In the intense magnetic field of a young neutron star, ordinarily spherical atoms are deformed into elongated cigar shapes and have radically altered spectral lines.

The bizarre properties of neutron star interiors and atmospheres are of great interest to astronomers and physicists precisely because they are so bizarre. Just as physicians learn about health by studying disease, scientists gain insight into the fundamental nature of matter by studying its behavior under extreme conditions.

23-6 | The fastest pulsars were probably created by mass transfer in close binary systems

In 1982 radio astronomers discovered a pulsar whose period is only 1.558 ms. (One millisecond, or 1 ms, equals a thousandth of a second). This remarkable pulsar, called PSR 1937+21 after

its coordinates in the sky, is a neutron star spinning 642 times per second—about 3 times faster than the blades of a kitchen blender! Such incredibly rapid rotation is a clue to this pulsar's unusual history.

As we have seen, pulsars should slow down as they age. The rapid spin rate of PSR 1937+21 suggests that it has hardly aged at all and so must be very young. A young pulsar should also be slowing down very rapidly. The faster it rotates, the more rapidly it can transfer rotational energy to its surroundings and the faster the slowdown. But PSR 1937+21 is actually slowing down at a very *gradual* rate, millions of times more gradually than the Crab pulsar. Such a gradual slowdown is characteristic of a pulsar hundreds of millions of years old. But if PSR 1937+21 is so old, how can it be spinning so rapidly?

PSR 1937+21 is not the only pulsar with unusually rapid rotation. Since 1982, astronomers have discovered more than 80 very fast pulsars, which are now called **millisecond pulsars**. All have periods between 1 and 10 ms, which means that these neutron stars are spinning at rates of 100 to 1000 rotations per second.

An important clue to understanding millisecond pulsars is that the majority are in close binary systems, with only a small separation between the spinning neutron star and its companion. (We know the separation must be small because the orbital periods of these binary systems are short, between 10 and 100 days. Kepler's third law for binary star systems, described in Section 19-9, tells us that the shorter the period, the smaller the distance between the two stars.) This fact suggests a scenario for how millisecond pulsars acquired their very rapid rotation.

Imagine a binary system consisting of a high-mass star and a low-mass star. The high-mass star evolves more rapidly than the low-mass star, and within a few million years becomes a Type II supernova that creates a neutron star. Like most newborn neutron stars, it spins several times per second, and, thus, initially we see a pulsar rather like the Crab or Vela pulsar.

 Over the next few billion years, the pulsar slows down as it radiates energy into space. Meanwhile, the slowly evolving low-mass star begins to expand as it evolves away from the main sequence to become a red giant. When the red giant gets big enough to fill its Roche lobe, it starts to spill gas over the inner Lagrangian point onto the neutron star. (See Section 21-6 for a review of close binary systems.) The infalling gas strikes the neutron star's surface at high speed and at an angle that causes the star to spin faster. In this way, a slow, aging pulsar is "spun up" by mass transfer from its bloated companion.

What about the few millisecond pulsars, like PSR 1937+21, that are *not* members of close binaries? It may be that solitary millisecond pulsars were once part of close binary systems, but the companion stars have been eroded away by the high-energy particles emitted by the pulsar after it was spun up. The Black Widow pulsar—named for the species of spider whose females kill and eat their mates—may be caught in the act of destroying its companion in just such a process (Figure 23-10).

The Black Widow pulsar (a neutron star spinning 622 times per second) and its companion orbit each other with a period of 9.16 hours. The orbital plane of the system is nearly edge-on to our line of sight, so this system is an eclipsing binary (described in Section 19-11): The companion blocks out the pulsar's radio signals for 45 minutes during each eclipse. Just before and after an eclipse, however, the pulsar's signals are delayed substantially, as if the radio waves were being slowed as they pass through a cloud of ionized gas surrounding the companion star. This circumstellar material is probably the star's outer layers, dislodged by radiation and by fast-moving particles emitted from the pulsar. In a few hundred million years or so, the companion of the pulsar will have completely disintegrated, leaving behind only a spun-up, solitary millisecond pulsar.

It is also possible that both stars in a close binary system might evolve into neutron stars, forming a pair of pulsars. Several such *binary pulsars* are known. Because the two pulsars are so massive and move around each other in tight orbits, they lose energy through a process called *gravitational radiation*. (We will discuss this process, which is a prediction of Einstein's general theory of relativity, in more detail in Chapter 24.) As a result the two neutron stars spiral toward each other and eventually collide. Computer simulations of such a collision show that some of the neutrons are ejected into space, where many of them decay into protons and combine to form various heavy elements. In addition to the heavy elements produced directly by supernova collisions, the elements produced by such neutron star collisions may play an important role in seeding the interstellar medium with the building blocks of planets (see Section 22-6).

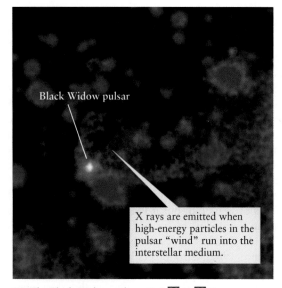

(a) The Black Widow pulsar R I **V** U **X** G

Black Widow pulsar

X rays are emitted when high-energy particles in the pulsar "wind" run into the interstellar medium.

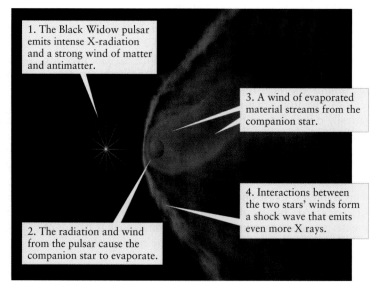

(b) An illustration of the pulsar and its companion

1. The Black Widow pulsar emits intense X-radiation and a strong wind of matter and antimatter.

3. A wind of evaporated material streams from the companion star.

4. Interactions between the two stars' winds form a shock wave that emits even more X rays.

2. The radiation and wind from the pulsar cause the companion star to evaporate.

Figure 23-10

The Black Widow Pulsar (a) This false-color composite of optical and X-ray images shows pulsar B1957+20, called the Black Widow. As the pulsar moves through space at over 200 km/s, it emits a wind that streams behind it. (b) This artist's conception of the pulsar and its doomed companion shows an area far smaller than the white dot in (a) labeled "Black Widow pulsar." (a: NASA/CXC/ASTRON/B. Stappers et al. (X-ray), AAO/J. Bland-Hawthorn & H. Jones (visible); b: CXC/M. Weiss)

23-7 | Pulsating X-ray sources are also neutron stars in close binary systems

Millisecond pulsars are not the only result of having a neutron star in a close binary system with an ordinary star. Even more exotic are **pulsating X-ray sources,** in which material from the companion star is drawn onto the magnetic poles of the neutron star, producing intense hot spots that emit breathtaking amounts of X rays. As the neutron star spins and the X-ray beams are directed alternately toward us and away from us, we see the X rays flash on and off. Such pulsating X-ray sources, while similar in many ways to ordinary pulsars, are far more luminous.

Pulsating X-ray sources were discovered in 1971 by the *Uhuru* spacecraft, the first to give astronomers a comprehensive look at the X-ray sky. (As we saw in Section 6-7, the Earth's atmosphere is opaque to X rays.) The first X-ray source found to emit pulses was Centaurus X-3, the third X-ray source found in the southern constellation Centaurus (**Figure 23-11**). The X-ray pulses have a regular period of 4.84 s. A few months later, similar pulses were discovered coming from a source designated Hercules X-1, which had a period of 1.24 s. Because the periods of these two X-ray sources are so short, astronomers began to suspect that they were actually observing rapidly rotating neutron stars.

It soon became clear that Centaurus X-3 and Hercules X-1 are members of binary systems. One piece of evidence is that every 2.087 days, Centaurus X-3 appears to turn off for almost 12 hours. Apparently, Centaurus X-3 is a neutron star in an eclipsing binary system, and it takes nearly 12 hours for the X-ray source to pass behind its companion star. Hercules X-1, too, appears to turn off periodically, corresponding to a 6-hour eclipse every 1.7 days by its visible companion, the star HZ Herculis. Moreover, careful timing of its X-ray pulses shows a periodic Doppler shifting every 1.7 days, which is direct evidence of orbital motion about a companion star. When the X-ray source is approaching us, the pulses from Hercules X-1 are separated by slightly less than 1.24 seconds. When the source is receding from us, slightly more than 1.24 seconds elapse between the pulses.

Putting all the pieces together, astronomers now realize that pulsating X-ray sources like Centaurus X-3 and Hercules X-1 are binary systems in which one of the stars is a neutron star. All these binary systems have very short orbital periods, which means that the distance between the two stars is quite small. The neutron star can therefore capture gases escaping from its ordinary companion (**Figure 23-12**). More than 20 such exotic binary systems, also called *X-ray binary pulsars,* are known. A typical rate of mass transfer from the ordinary star to the neutron star is roughly 10^{-9} solar masses per year.

Because of its strong gravity, a neutron star in a pulsating X-ray source easily captures much of the gas escaping from its companion. But like an ordinary pulsar, the neutron star is rotating rapidly and has a powerful magnetic field inclined to the axis of rotation (recall Figure 23-3). As the gas falls toward the neutron star, its magnetic field funnels the incoming matter down onto the star's north and south magnetic polar regions. The neutron star's gravity is so strong that the gas is traveling at nearly half the speed of light by the time it crashes onto the star's surface. This violent impact creates spots at both poles with searing temperatures of about 10^8 K. Material raised to these temperatures emits abundant X rays, and because so much material is dumped onto these hot spots the X-ray luminosity reaches roughly 10^{31} watts. This is much greater than the X-ray luminosity of isolated pulsars like the Crab (see Figure 23-4*b*) or 1E 1207.4-5209 (see Figure 23-9*a*), whose surface temperatures are a relatively chilly 10^6 K or so. It is also nearly 100,000 times greater than the combined luminosity of the Sun at all wavelengths.

As the neutron star rotates, the beams of X rays from its magnetic poles sweep around the sky. If the Earth happens to be in the path of one of the two beams, we can observe a pulsating X-ray source. We detect a pulse once per rotation period, which means that the neutron star in Hercules X-1 spins at the rate of once every 1.24 seconds. If the orbital plane of the binary system is nearly edge-on to our line of sight, as for Centaurus X-3 and Hercules X-1, the pulsating X rays appear to turn off when the neutron star is eclipsed by the companion star. Other pulsating X-ray sources, such as Scorpius X-1, do not undergo this sort of turning off. The orbital planes of these systems are oriented more nearly face-on to our line of sight, so no eclipses occur.

The neutron stars in most pulsating X-ray sources have rotation periods (and hence pulse periods) of a few seconds. But as time passes, the neutron star accretes more and more mass from

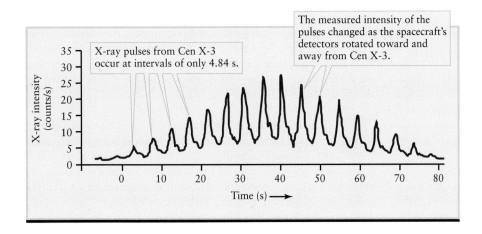

The measured intensity of the pulses changed as the spacecraft's detectors rotated toward and away from Cen X-3.

X-ray pulses from Cen X-3 occur at intervals of only 4.84 s.

figure 23-11

X-Ray Pulses from Centaurus X-3 This graph shows the intensity of X rays detected by the *Uhuru* spacecraft as Centaurus X-3 (often abbreviated as Cen X-3) moved across the satellite's field of view. Because the pulse period is so short, Centaurus X-3 is in all probability a rotating neutron star. (Adapted from R. Giacconi and colleagues)

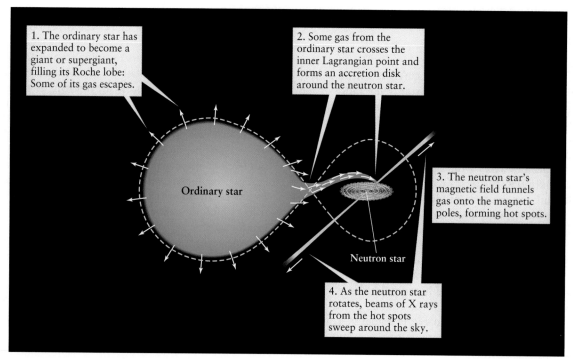

1. The ordinary star has expanded to become a giant or supergiant, filling its Roche lobe: Some of its gas escapes.

2. Some gas from the ordinary star crosses the inner Lagrangian point and forms an accretion disk around the neutron star.

Ordinary star

3. The neutron star's magnetic field funnels gas onto the magnetic poles, forming hot spots.

Neutron star

4. As the neutron star rotates, beams of X rays from the hot spots sweep around the sky.

Figure 23-12
A Model of a Pulsating X-Ray Source Pulsating X-ray sources are close, semidetached binary systems (compare with Figure 21-18*b*) in which one member is a rotating neutron star. Gas from the ordinary star flows onto the neutron star and creates hot spots that emit beams of X-rays. If one of the beams is oriented so that it sweeps over the Earth as the neutron star rotates, we see a pulse of X rays on each sweep.

its companion. In the process, it will spin up and may eventually become a millisecond pulsar (Section 23-6). In 1998 the *Rossi X-ray Timing Explorer* spacecraft first observed an evolutionary "missing link" between pulsating X-ray sources and millisecond pulsars. This object, designated SAX J1808.4-3658, is an X-ray–emitting millisecond pulsar in a close binary system with a pulse period of just 2.5 ms. Apparently the pulsar has accreted a substantial amount of mass from its companion star. Like the Black Widow pulsar (see Figure 23-10), it has also blown much of the companion's mass into space. Within a billion years, the companion star may disappear altogether.

23-8 | Explosive thermonuclear processes on white dwarfs and neutron stars produce novae and bursters

Still other exotic phenomena occur when a stellar corpse is part of a close binary system. One example is a **nova** (plural **novae**), in which a faint star suddenly brightens by a factor of 10^4 to 10^8 over a few days or hours, reaching a peak luminosity of about 10^5 $L_\odot$. By contrast, a *supernova* has a peak luminosity of about 10^9 $L_\odot$.

A nova's abrupt rise in brightness is followed by a gradual decline that may stretch out over several months or more (Figure 23-13). Every year two or three novae are seen in our Galaxy, and several dozen more are thought to take place in remote regions of the Galaxy that are obscured from our view by interstellar dust.

In the 1950s, painstaking observations of numerous novae by Robert Kraft, Merle Walker, and their colleagues at the University of California's Lick Observatory led to the conclusion that all novae are members of close binary systems containing a white dwarf. Gradual mass transfer from the ordinary companion star, which presumably fills its Roche lobe, deposits fresh hydrogen onto the white dwarf (see Figure 21-19*b* for a schematic diagram of this sort of mass transfer).

Because of the white dwarf's strong gravity, this hydrogen is compressed into a dense layer covering the hot surface of the white dwarf. As more gas is deposited and compressed, the temperature in the hydrogen layer increases. When the temperature reaches about 10^7 K, hydrogen fusion ignites throughout the gas layer, embroiling the white dwarf's surface in a thermonuclear holocaust that we see as a nova (Figure 23-14).

CAUTION It is important to understand the similarities and differences between novae and Type Ia supernovae. Both kinds of celestial explosions are thought to occur in close binary systems where one of the stars is a white dwarf. But, as befits their name, supernovae are much more energetic. A Type Ia supernova explosion radiates 10^{44} joules of energy into space, while the corresponding figure for a typical nova is 10^{37} joules.

(a) Nova Herculis 1934 shortly after peak brightness

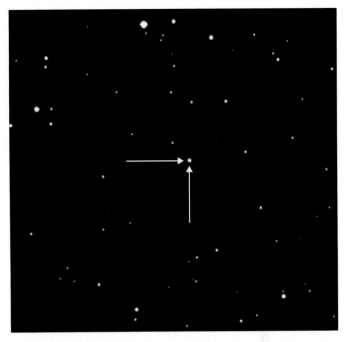

(b) Two months later

figure 23-13   R I **V** U X G

Nova Herculis 1934 These two pictures show a nova (a) shortly after peak brightness as a star of apparent magnitude +3, bright enough to be seen easily with the naked eye, and (b) two months later, when it had faded by a factor of 4000 in brightness to apparent magnitude +12. (See Figure 19-6 for a description of the apparent magnitude scale.) Novae are named after the constellation and year in which they appear. (Lick Observatory)

(To be fair to novae, this relatively paltry figure is as much energy as our Sun emits in 1000 years.) The difference is thought to be that in a Type Ia supernova, the white dwarf accretes much more mass from its companion. This added mass causes so much compression that nuclear reactions can take place *inside* the white dwarf. Eventually, these reactions blow the white dwarf completely apart. In a nova, by contrast, nuclear reactions occur only within the accreted material. The reaction is more sedate because it takes place only on the white dwarf's surface (perhaps because the accretion rate is less or because the white dwarf had less mass in the first place).

Because the white dwarf itself survives a nova explosion, it is possible for the same star to undergo more than one nova. As an example, the star T Coronae Borealis erupted as a nova in 1866, then put in a repeat performance in 1946. By contrast, a given star can only be a supernova once.

A surface explosion similar to a nova also occurs with neutron stars. In 1975 it was discovered that some objects in the sky emit sudden, powerful bursts of X rays. **Figure 23-15** shows the record of a typical burst. The source emits X rays at a constant low level until suddenly, without warning, there is an abrupt increase in X rays, followed by a gradual decline. An entire burst typically lasts for only 20 s. Unlike pulsating X-ray sources, there is a fairly long interval of hours or days between bursts. Sources that behave

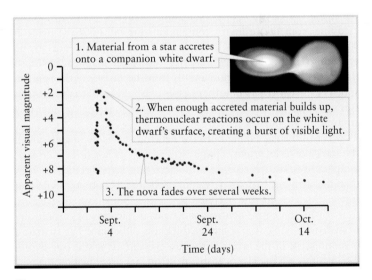

1. Material from a star accretes onto a companion white dwarf.

2. When enough accreted material builds up, thermonuclear reactions occur on the white dwarf's surface, creating a burst of visible light.

3. The nova fades over several weeks.

figure 23-14

The Light Curve of a Nova This illustration and graph show the history of Nova Cygni 1975, a typical nova. Its rapid rise and gradual decline in apparent brightness are characteristic of all novae. This nova, also designated V1500 Cyg, was easily visible to the naked eye (that is, was brighter than an apparent magnitude of +6) for nearly a week. (Illustration courtesy CXC/M. Weiss)

1. Material from a star accretes onto a companion neutron star.

2. When enough accreted material builds up, thermonuclear reactions occur on the neutron star's surface, creating a burst of X rays.

3. The X-ray burster fades within seconds.

Figure 23-15

The Light Curve of an X-Ray Burster This illustration and graph show the history of a typical X-ray burster. A burster emits a constant low intensity of X rays interspersed with occasional powerful X-ray bursts. This burst was recorded on September 28, 1975, by an Earth-orbiting X-ray telescope. Contrast this figure with Figure 23-14, which shows a typical nova. (Data adapted from W. H. G. Lewin; illustration courtesy CXC/M. Weiss)

in this fashion are known as **X-ray bursters.** Several dozen X-ray bursters have been discovered in our Galaxy.

X-ray bursters, like novae, are thought to involve close binaries whose stars are engaged in mass transfer. With a burster, however, the stellar corpse is a neutron star rather than a white dwarf. Gases escaping from the ordinary companion star fall onto the neutron star. The X-ray burster's magnetic field is probably not strong enough to funnel the falling material toward the magnetic poles, so the gases are distributed more evenly over the surface of the neutron star. The energy released as these gases crash down onto the neutron star's surface produces the low-level X rays that are continuously emitted by the burster.

Most of the gas falling onto the neutron star is hydrogen, which the star's powerful gravity compresses against its hot surface. In fact, temperatures and pressures in this accreting layer become so high that the arriving hydrogen is converted into helium by hydrogen fusion. As a result, the accreted gases develop a layered structure that covers the entire neutron star, with a few tens of centimeters of hydrogen lying atop a similar thickness of helium. The structure is reminiscent of the layers within an evolved giant star (see Figure 22-2), although the layers atop a neutron star are much more compressed, thanks to the star's tremendous surface gravity.

When the helium layer is about 1 m thick, helium fusion ignites explosively and heats the neutron star's surface to about 3×10^7 K. At this temperature the surface predominantly emits X rays, but the emission ceases within a few seconds as the surface cools. Hence, we observe a sudden burst of X rays only a few seconds in duration. New hydrogen then flows onto the neutron star, and the whole process starts over. Indeed, X-ray bursters typically emit a burst every few hours or days.

Whereas explosive *hydrogen* fusion on a white dwarf produces a nova, explosive *helium* fusion on a neutron star produces an X-ray burster. In both cases, the process is explosive, because the fuel is compressed so tightly against the star's surface that it becomes degenerate, like the star itself. As with the helium flash inside red giants (described in Section 21-2), the ignition of a degenerate thermonuclear fuel involves a sudden thermal runaway. This is because an increase in temperature does not produce a corresponding increase in pressure that would otherwise relieve compression of the gases and slow the nuclear reactions.

One of the great puzzles in modern astronomy has been the nature of an even more remarkable class of events called *gamma-ray bursters.* As the name suggests, these objects emit sudden, intense bursts of high-energy gamma rays. During the short duration of its burst, a gamma-ray burster can be 10^6 to 10^9 times brighter than a supernova at the same distance! We will discuss these exotic objects further in Chapter 27.

23-9 | Like a white dwarf, a neutron star has an upper limit on its mass

A white dwarf will collapse if its mass is greater than the Chandrasekhar limit of 1.4 $M_\odot$. At that point, degenerate electron pressure cannot support the overpowering weight of the star's matter, which presses inward from all sides (see Section 22-4). The mass of a neutron star also has an upper limit. However, the pressure within a neutron star is harder to analyze, because it comes from *two* sources. One is the degenerate nature of the neutrons, and the other is the strong nuclear force that acts between the neutrons themselves.

The strong nuclear force is what holds protons and neutrons together in atomic nuclei. Neutrons exert strong nuclear forces on one another only when they are almost touching. This force behaves somewhat like the force that billiard balls exert on one another when they touch: It strongly resists further compression. (Try squeezing two billiard balls together and see how much success you have.) Hence, the strong nuclear force is a major contributor to the star's internal pressure. Unfortunately, there is a good deal of uncertainty about the details of this force. This translates into uncertainties about how much weight the neutron star's internal pressure can support—that is, the neutron star's maximum mass. Theoretical estimates of this maximum mass range from 2 to 3 $M_\odot$.

Before pulsars were discovered, most astronomers believed all dead stars to be white dwarfs. Dying stars were thought to somehow eject enough material so that their corpses could be below the Chandrasekhar limit. The discovery of neutron stars proved this idea incorrect. Inspired by this lesson, astronomers soon began wondering what might happen if a dying massive star failed to eject enough matter to get below the upper limit for a neutron star. For example, what might a 5-$M_\odot$ stellar corpse be like?

The gravity associated with a neutron star is so strong that the escape speed from it is roughly one-half the speed of light. But if a stellar corpse has a mass greater than 3 $M_\odot$, so much matter is crushed into such a small volume that the escape speed actually *exceeds* the speed of light. Because nothing can travel faster than light, nothing—not even light—can leave this dead star. Its gravity is so powerful that it leaves a hole in the fabric of space and time. Thus, the discovery of neutron stars inspired astrophysicists to examine seriously one of the most bizarre and fantastic objects ever predicted by modern science, the black hole. We take up its story in the next chapter.

Key Words

degenerate neutron pressure, p. 514

glitch, p. 520

millisecond pulsar, p. 521

neutron star, p. 514

nova (*plural* novae), p. 524

pair production, p. 516

pulsar, p. 515

pulsating X-ray source, p. 523

superconductivity, p. 520

superfluidity, p. 519

synchrotron radiation, p. 519

X-ray burster, p. 526

Key Ideas

Neutron Stars: A neutron star is a dense stellar corpse consisting primarily of closely packed degenerate neutrons.

• A neutron star typically has a diameter of about 20 km, a mass less than 3 $M_\odot$, a magnetic field 10^{12} times stronger than that of the Sun, and a rotation period of roughly 1 second.

• A neutron star consists of a superfluid, superconducting core surrounded by a superfluid mantle and a thin, brittle crust.

• Intense beams of radiation emanate from regions near the north and south magnetic poles of a neutron star. These beams are produced by streams of charged particles moving in the star's intense magnetic field.

Pulsars: A pulsar is a source of periodic pulses of radio radiation. These pulses are produced as beams of radio waves from a neutron star's magnetic poles sweep past the Earth.

• The pulse rate of many pulsars is slowing steadily. This reflects the gradual slowing of the neutron star's rotation as it radiates energy into space. Sudden speedups of the pulse rate, called glitches, may be caused by interactions between the neutron star's crust and its superfluid interior.

Neutron Stars in Close Binary Systems: If a neutron star is in a close binary system with an ordinary star, tidal forces will draw gas from the ordinary star onto the neutron star.

• The transfer of material onto the neutron star can make it rotate extremely rapidly, giving rise to a millisecond pulsar.

• Magnetic forces can funnel the gas onto the neutron star's magnetic poles, producing hot spots. These hot spots then radiate intense beams of X rays. As the neutron star rotates, the X-ray beams appear to flash on and off. Such a system is called a pulsating X-ray variable.

Novae and Bursters: Material from an ordinary star in a close binary can fall onto the surface of the companion white dwarf or neutron star to produce a surface layer in which thermonuclear reactions can explosively ignite.

• Explosive hydrogen fusion may occur in the surface layer of a companion white dwarf, producing the sudden increase in luminosity that we call a nova. The peak luminosity of a nova is only 10^{-4} of that observed in a supernova.

• Explosive helium fusion may occur in the surface layer of a companion neutron star. This produces a sudden increase in X-ray radiation, which we call a burster.

Review Questions

1. What are neutron stars? What led scientists to propose their existence?

2. What is degenerate neutron pressure? How does it make it possible for a neutron star to be more massive than the Chandrasekhar limit for white dwarfs?

3. How were pulsars discovered? How do they differ from variable stars?

4. Why do astronomers think that pulsars are rapidly rotating neutron stars?

5. During the weeks immediately following the discovery of the first pulsar, one suggested explanation was that the pulses might be signals from an extraterrestrial civilization. Why did astronomers soon discard this idea?

6. Why do neutron stars rotate so much more rapidly than ordinary stars? Why do they have such strong magnetic fields?

7. How are rotating neutron stars able to produce pulses of radiation as seen by an observer on Earth?

8. Do all supernova remnants contain pulsars? Are all pulsars found within supernova remnants? For each question, explain why or why not.

9. Is our Sun likely to end up as a neutron star? Why or why not?

10. Why are some neutron stars seen moving through space at hundreds of kilometers per second?

11. What is synchrotron radiation? How is it involved in making the Crab Nebula glow?

12. Why does an isolated pulsar rotate more slowly as time goes by?

13. Astronomers have deduced that the Vela pulsar is about 11,000 years old. How do you suppose they did this?

14. What is the difference between superconductivity and superfluidity?

15. What is a pulsar glitch? How does a glitch affect a pulsar's period? What could be the cause of glitches?

16. Compare the internal structure of a white dwarf to that of a neutron star. What are the similarities? What are the differences?

17. What is the evidence that neutron stars have atmospheres?

18. Why do astronomers think that millisecond pulsars are very old?

19. If millisecond pulsars are formed in close binary systems, why are some found without companion stars?

20. Describe a pulsating X-ray source like Hercules X-1 or Centaurus X-3. What produces the pulsation?

21. What is the connection between pulsating X-ray sources and millisecond pulsars?

22. What is the difference between a nova and a Type Ia supernova?

23. What are the similarities between novae and X-ray bursters? What are the differences?

24. What are the similarities between pulsating X-ray sources and X-ray bursters? What are the differences?

25. Why is the maximum mass of neutron stars not known as accurately as the Chandrasekhar limit for white dwarfs?

Advanced Questions

> **Problem-solving tips and tools**
>
> The volume of a sphere of radius r is $4\pi r^3/3$. The small-angle formula is given in Box 1-1. Section 5-4 describes how to relate the temperature of a blackbody to its wavelength of maximum emission. Section 19-6 gives the formula relating a star's luminosity, surface temperature, and radius (see Box 19-4 for worked examples). Appendix 6 gives the conversion between seconds and years as well as the radius of the Sun.

26. Using a diagram like Figure 23-3, explain why the number of pulsars that we observe in nearby space is probably quite a bit less than the number of rotating, magnetized neutron stars in nearby space.

27. There are many more main-sequence stars of low mass (less than 8 $M_\odot$) than of high mass (8 $M_\odot$ or more). Use this fact to explain why white dwarf stars are far more common than neutron stars.

28. The distance to the Crab Nebula is about 2000 parsecs. In what year did the star actually explode? Explain your answer.

29. How do we know that the Crab pulsar is really embedded in the Crab Nebula and not simply located at a different distance along the same line of sight?

30. The Crab Nebula has an apparent size of about 5 arcmin, and this size is increasing at a rate of 0.23 arcsec per year. (a) Assume that the expansion rate has been constant over the entire history of the Crab Nebula. Based on this assumption, in what year would Earth observers have seen the supernova explosion that formed the nebula? (b) Does your answer to part (a) agree with the known year of the supernova, 1054 A.D.? If not, can you point to assumptions you made in your computations that led to the discrepancies? Or do you think your calculations suggest additional physical

effects are at work in the Crab Nebula, over and above a constant rate of expansion?

31. Emission lines in the spectrum of the Crab Nebula exhibit a Doppler shift, which indicates that gas in the nebula is moving toward us at 1450 km/s. (a) Assume that the expanding gas has been moving at the same speed since the original supernova explosion, observed in 1054 A.D., and calculate what radius and what diameter (in light-years) we should observe the nebula to have today. (b) Compare your result in part (a) to the actual size of the nebula, given in the caption to Figure 23-4.

32. The supernova remnant G5.4-1.2 shown in Figure 23-5 lies about 5000 pc (16,000 ly) from Earth in the constellation Sagittarius. (a) The green arc in the large left-hand image in Figure 23-5 represents part of the outer edge of this spherical supernova remnant. Estimate the diameter of this remnant in parsecs. (*Hint:* To calculate this, you will need to make measurements on the image.) (b) How far (in parsecs) did the neutron star travel from where it was formed (at the position of the supernova's progenitor star, presumably at the center of the present-day remnant) to the position shown in Figure 23-5? Explain your answer.

33. To determine accurately the period of a pulsar, astronomers must take into account the Earth's orbital motion about the Sun. (a) Explain why. (b) Knowing that the Earth's orbital velocity is 30 km/s, calculate the maximum correction to a pulsar's period because of the Earth's motion. Explain why the size of the correction is greatest for pulsars located near the ecliptic.

34. If a pulsar has period P (in seconds) and its period is increasing at a rate R (in seconds per second), an approximate formula for the age T of the pulsar (in seconds) is $T = P/2R$. For the Crab pulsar, $P = 0.0333$ s and $R = 4.21 \times 10^{-13}$ s/s. (a) Calculate the approximate age of the Crab pulsar in years. (b) Based on the information given in Section 23-1, is your result in (a) an underestimate or an overestimate? Explain your answer.

35. A neutron has a mass of about 1.7×10^{-27} kg and a radius of about 10^{-15} m. (a) Compare the density of matter in a neutron with the average density of a neutron star. (b) If the neutron star's density is more than that of a neutron, the neutrons within the star are overlapping; if it is less, the neutrons are not overlapping. Which of these seems to be the case? What do you think is happening at the center of the neutron star, where densities are higher than average?

36. X-ray pulsars are speeding up but ordinary (radio) pulsars are slowing down. Propose an explanation for this difference.

37. If the model for Hercules X-1 discussed in the text is correct, at what orientation of the binary system do we see its maximum optical brightness? Explain your answer.

38. Explain why heavy elements that are produced by neutron star collisions can still be regarded as having been processed through a supernova.

39. In an X-ray burster, the surface of a neutron star 10 km in radius is heated to a temperature of 3×10^7 K. (a) Determine the wavelength of maximum emission of the heated surface (which you may treat as a blackbody). In what part of the electromagnetic

spectrum does this lie? (See Figure 5-7.) (b) Find the luminosity of the heated neutron star. Give your answer in watts and in terms of the luminosity of the Sun, given in Table 18-1. How does this compare with the peak luminosity of a nova? Of a Type Ia supernova?

40. The nearest neutron star, called RX J185635-3754, is just 60 pc (200 ly) from Earth. It is thought to be the relic of a star that underwent a supernova explosion about 1 million years ago. The explosion ejected the neutron star at high speed, so it is now moving through nearly empty space. (a) RX J185635-3754 is *not* a pulsar, that is, it does not emit pulses of radiation. Suggest why this might be so. (b) The neutron star has a surface temperature of 600,000 K. Find the wavelength at which it emits most strongly, and explain why the neutron star appears as a steady, nonpulsating object in an X-ray telescope. (c) RX J185635-3754 has a total luminosity at all wavelengths of about $0.046L_\odot$. Calculate its radius, and explain why astronomers conclude that it is a neutron star.

Discussion Questions

41. Imagine that were somehow able to stand (and survive) at one of the magnetic poles of the Crab pulsar. Describe what you would see. How would the stars appear to move in the sky? What would you see if you looked straight up? What factors make this location a very unhealthy place to visit?

42. Accretion disks in close binary systems are too small to be seen directly with even the highest-resolution telescopes. How, then, can astronomers detect the presence of such accretion disks?

43. When neutrons are very close to one another, they repel one another through the strong nuclear force. If this repulsion were made even stronger, what effect might this have on the maximum mass of a neutron star? Explain your answer.

Web/CD-ROM Questions

44. Search the World Wide Web for information about the latest observations of the stellar remnant at the center of SN 1987A. Has a pulsar been detected? If so, how fast is it spinning? Has the supernova's debris thinned out enough to give a clear view of the neutron star?

45. Search the World Wide Web for information about a class of neutron stars called *magnetars*. How do they differ from ordinary neutron stars? What is their connection to objects called soft gamma-ray repeaters?

46. Monitoring the Crab Pulsar. Access the video "The Crab Pulsar" in Chapter 23 of the *Universe* Web site or CD-ROM. View the video and use it to answer the following questions. For each part, explain how you determined your answer. (a) How many rotations does the neutron star complete during the duration of the video? (b) Is the neutron star visible at the beginning of the video? If not, explain why not. (c) How does

the peak brightness of the Crab pulsar compare to the steady brightness of the nearby star? (d) What total amount of time is depicted in the video?

Observing Projects

47. If you did not take the opportunity to observe the Crab Nebula as part of the exercises in Chapter 22, do so now. The Crab Nebula is visible in the night sky from October through March. Its epoch 2000 coordinates are R.A. = 5^h 34.5^m and Decl. = 22° 00′, which is near the star marking the eastern horn of Taurus (the Bull). Be sure to schedule your observations for a moonless night. The larger the telescope you use, the better, because the Crab Nebula is quite dim.

48. Consult the World Wide Web to see if any novae have been sighted recently. If by good fortune one has been sighted, what is its apparent magnitude? Is it within reach of a telescope at your disposal? If so, arrange to observe it. Draw what you see through the eyepiece, noting the object's brightness in comparison with other stars in the field of view. If possible, observe the same object a few weeks or months later to see how its brightness has changed.

49. Use the *Starry Night Backyard*™ program to observe the sky in July 1054, when the supernova that spawned the Crab Nebula was visible from the American Southwest. Select **Viewing Location...** in the **View** menu, click on the "Latitude/Longitude" tab, enter 36° N for latitude and 109° W for longitude, and click on the "Set Location" button. You are now near the location of the pictograph shown in Figure 23-1. In the Control Panel, set the date to 7/5/1054 A.D. (July 5, 1054) and the time to 5:00 A.M. Use **Find...** in the **Edit** menu to find and center on the Crab Nebula. Zoom in or out until you can see both the position of the nebula and the Moon. You may find it helpful to turn daylight on or off (select **Show Daylight** or **Hide Daylight** in the **View** menu). (a) What is the phase of the Moon? (b) Investigate how the relative positions of the Moon and the Crab Nebula change when you set the date to July 4, 1054, or July 6, 1054. On which date do the relative positions of the Moon and the Crab Nebula give the best match to the pictograph in Figure 23-1?

50. Use the *Deep Space Explorer*™ program to locate the Small Magellanic Cloud (SMC), the site of a large concentration of X-ray pulsars. In the left-hand part of the window under the heading **Milky Way,** click on the triangle next to the word **Explore** and then click on **Local Neighbourhood.** You can zoom in or out using the buttons at the upper left of the window (an upward-pointing triangle and a downward-pointing triangle). You can rotate the Milky Way Galaxy and its neighbor galaxies by putting the mouse cursor over the image, holding down the mouse button, and moving the mouse. (On a two-button mouse, hold down the left mouse button.) (a) Center the field of view on the Sun (use the **Find...** command in the **Edit** menu). Describe the position of the SMC relative to the Milky Way Galaxy and to our solar system. (b) Use the **Find...** command in the **Edit** menu to center on the SMC. Pulsars are produced by supernovae, and only certain types of stars become supernovae. What evidence do you see that these types of stars are present in the SMC? (Note that although *Deep Space Explorer*™ depicts the SMC as being rather flat, it is thought to be an irregular blob of stars with some thickness.)

Collaborative Exercises

51. Consider the graph showing a recording of a pulsar in Figure 23-2. Sketch and label similar graphs that your group estimates for:

(1) a rapidly spinning, professional ice skater holding a flashlight; (2) an emergency signal on an ambulance; and (3) a rotating beacon at an airport.

52. As stars go, pulsars are tiny, only about 20 km across. Name three specific things or places that have a size or a separation of about 20 km.

53. If the Crab Nebula is slowing such that its period is increasing at a rate of 3×10^{-8} seconds per day, how much slower is it going today than on the day the youngest member of your group was born?

24 Black Holes

Imagine a swirling disk of gas and dust, orbiting around an object that has more mass than the Sun but is so dark that it cannot be seen. Imagine the material in this disk being compressed and heated as it spirals into the unseen object, reaching temperatures so high that the material emits X rays. And imagine that the unseen object has such powerful gravity that any material that falls into it simply disappears, never to be seen again.

Such hellish maelstroms, like the one shown in the illustration, really do exist. The disks are called *accretion disks,* and the unseen objects with immensely strong gravity are called *black holes.*

The matter that makes up a black hole has been so greatly compressed that it violently warps space and time. If you get too close to a black hole, the speed you would need to escape exceeds the speed of light. Because nothing can travel faster than light, nothing—not even light—can escape from a black hole.

Black holes, whose existence was predicted by Einstein's general theory of relativity (our best description of what gravity is and how it behaves), are both strange and simple. The structure of a black hole is completely specified by only three quantities—its mass, electric charge, and angular momentum. Perhaps strangest of all, there is compelling evidence that black holes really exist. In recent years, astronomers have found that certain binary star systems contain black holes. Even more remarkable is the discovery that black holes of more than a million solar masses may lie at the centers of many galaxies, including our own Milky Way.

 An artist's impression of a rotating black hole with an accretion disk. (XMM-Newton/ESA/NASA)

Questions to Guide Your Reading

As you read the sections of this chapter, look for the answers to the following questions:

24-1 What are the two central ideas behind Einstein's special theory of relativity?

24-2 How do astronomers search for black holes?

24-3 What are supermassive black holes, and where are they found?

24-4 In what sense is a black hole "black"?

24-5 In what way are black holes actually simpler than any other objects in astronomy?

24-6 What happens to an object that falls into a black hole?

24-7 Why do some pulsars emit fantastic amounts of X rays?

24-8 Do black holes last forever?

24-1 | The special theory of relativity changes our conceptions of space and time

The idea of a black hole is a truly strange one. To appreciate it, we must discard some "commonsense" notions about the nature of space and time.

According to the classical physics of Newton, space is perfectly uniform and fills the universe like a rigid framework. Similarly, time passes at a monotonous, unchanging rate. It is always possible to know exactly how fast you are moving through this rigid fabric of space and time.

Those ideas were upset forever in 1905 when Albert Einstein proposed his **special theory of relativity**. This theory describes how motion affects our measurements of distance and time. It is "special" in the sense of being specialized. In particular, it does not include the effects of gravity. The word "relativity" is used because one of the key ideas of the theory is that all measurements are made relative to an observer. Contrary to Newton, there is *no* absolute framework of space and time. In particular, the distance between two points is not an absolute, nor is the time interval between two events. Instead, the values that you measure for these qualities depend on how you are moving, and are relative to you. Someone moving in a different way would measure different values for these quantities.

Remarkably, Einstein's theory is based on just two basic principles. The first is quite simple:

Your description of physical reality is the same regardless of the constant velocity at which you move.

In other words, if you are moving in a straight line at a constant speed, you experience the same laws of physics as you would if you were moving at any other constant speed and in any other direction. As an example, suppose you were inside a closed railroad car moving due north in a straight line at 100 km/h. Any measurements you make inside the car—for example, how long your thumb is or how much time elapses between ticks of your watch—will give exactly the same results as if the car were moving in any other direction or at any other speed, or were not moving at all.

Einstein's second principle is much more bizarre:

Regardless of your speed or direction of motion, you always measure the speed of light to be the same.

To see what this implies, imagine that you are in a spaceship moving toward a flashlight. Even if you are moving at 99% of the speed of light, you will measure the photons from the flashlight to be moving at the same speed ($c = 3 \times 10^8$ m/s $= 3 \times 10^5$ km/s) as if your spaceship were motionless. This statement is in direct conflict with the Newtonian view that a stationary person and a moving person should measure different speeds (**Figure 24-1**).

Speed involves both distance and time. Since speed has a very different behavior in the special theory of relativity than in Newton's physics, it follows that both space and time behave differently as well. Indeed, in relativity time proves to be so intimately intertwined with the three dimensions of space that we regard them as a single *four*-dimensional entity called **spacetime**.

(a)

Incorrect Newtonian description:
As seen by the astronaut in spaceship, the light is approaching her at $(3 \times 10^8$ m/s$) + (1 \times 10^8$ m/s$) = 4 \times 10^8$ m/s.

Light
3×10^8 m/s 1×10^8 m/s

Correct Einsteinian description:
As seen by the astronaut in spaceship, the light is approaching her at 3×10^8 m/s.

(b)

Figure 24-1

The Speed of Light Is the Same to All Observers (a) The speed you measure for ordinary objects depends on how you are moving. Thus, the batter sees the ball moving at 30 m/s, but the outfielder sees it moving at 40 m/s relative to her. (b) Einstein showed that this commonsense principle does not apply to light. No matter how fast or in what direction the astronaut in the spaceship is moving, she and the astronaut holding the flashlight will always measure light to have the same speed.

Einstein expressed his ideas about spacetime in a mathematical form and used this to make a number of predictions about nature. All these predictions have been verified in innumerable experiments. One prediction is that the length you measure an object to have depends on how that object is moving; the faster it moves, the shorter its length along its direction of motion (**Figure 24-2**). This is called **length contraction**. In other words, if a railroad car moves past you at high speed, from your perspective on the ground you will actually measure it to be shorter than if it were at rest (Figure 24-2*a*). However, if you are on board the railroad car and moving with it, you will measure its length to be the same as measured on the ground when it was at rest. The word "relativity" emphasizes the importance of the relative speed between the observer and the object being measured.

If the idea of length contraction seems outrageous, it's because this effect is noticeable only at very high speeds, comparable to that of light. But even the fastest spacecraft ever built by humans travels at a mere 1/25,000 of the speed of light. At this speed, a spacecraft 10 m long would be contracted in length by only 8 nm—a distance equal to the width of a single protein molecule! For moving cars, trains, and airplanes, length contraction is far too small to measure.

This train is at rest relative to you.

The same train is now moving relative to you.

(a) Length contraction

The clock moves
relative to you

(b) Time dilation

figure 24-2

Length Contraction and Time Dilation (a) The faster an object moves, the shorter it becomes along its direction of motion. Speed has no effect on the object's dimensions perpendicular to the direction of motion. (b) The faster a clock moves, the slower it runs. This graph shows how many seconds (as measured on your clock) it takes a clock that moves relative to you to tick off one second. The effect is pronounced only for speeds near the speed of light c.

As Box 24-1 describes, however, we can easily see the effects of length contraction for subatomic particles that do indeed travel at nearly the speed of light.

A second result of relativity is no less strange: A clock moving past you runs more slowly than a clock that is at rest. Like length contraction, this **time dilation** is a very small effect unless the clock is moving at extremely high speeds (Figure 24-2b). Nonetheless, physicists have confirmed the existence of time dilation by using an extremely accurate atomic clock carried on a jet airliner. When the airliner landed, they found that the on-board clock had actually ticked off slightly less time than an identical, stationary clock on the ground. For the passengers on board the airliner during this experiment, however, time flowed at a normal rate: The on-board clock ticked off the seconds as usual, their hearts beat at a normal rate, and so on. You only measure a clock (or a beating heart) to be running slow if it is *moving* relative to you. Box 24-1 discusses time dilation in more detail.

The special theory of relativity predicts that the astronaut with the flashlight in Figure 24-1b sees the astronaut flying in the spaceship as shortened along the direction of motion and as having slowly ticking clocks. Remarkably, it also says that the *flying* astronaut sees the astronaut with the flashlight (who is moving relative to her) as being shortened and as having slowly ticking clocks! This may seem contradictory, but it is not: Each spaceship is moving relative to the other, and in the special theory of relativity only relative motion matters. Each astronaut's measurements are correct relative to his or her frame of reference.

 You may have the idea that the special theory of relativity implies that there are *no* absolutes in nature and that everything is relative. But, in fact, the special theory is based on the principles that the laws of physics and the speed of light in a vacuum *are* absolutes. Only certain quantities, such as distance and time, depend on your state of motion. You may also have the idea that length contraction and time dilation are just optical illusions caused by high speeds. But these effects are *real*. A moving spaceship doesn't just look shorter or seem to be shorter—it really *is* shorter. Likewise, a moving clock doesn't merely look like it's ticking slowly, or seem to be ticking slowly—it really *does* tick more slowly. Einstein's theory is supported by every experiment designed to test it. Relativity is very strange, but it is also very real!

Einstein's special theory of relativity predicts another famous relationship: Any object with mass (m) has energy (E) embodied in its mass. This is expressed by the well-known formula $E = mc^2$, where c is the speed of light. In thermonuclear reactions, part of the energy embodied in the mass of atomic nuclei is converted into other forms of energy. This released energy is what makes the Sun and the stars shine (see Section 18-1). The sunlight we see by day and the starlight that graces our nights are a resounding verification of the special theory of relativity.

The special theory of relativity also explains why it is impossible for a spaceship to move at the speed of light. If it could, then a light beam traveling in the same direction as the spaceship would appear to the ship's crew to be stationary. But this would contradict the second of Einstein's principles, which says that all observers, including those on a spaceship, must see light traveling at the speed of light. Therefore, it cannot be possible for a spaceship, or indeed any material object, to travel at the speed of light. This, too, has been verified by experiments. Subatomic particles can be made to travel at 99.99999999% of the speed of light c, but they can never make it all the way to c.

box 24-1 TOOLS OF THE ASTRONOMER'S TRADE

Time Dilation and Length Contraction

The special theory of relativity describes how motion affects measurements of time and distance. By using the two basic principles of his theory—that no matter how fast you move, the laws of physics and the speed of light are the same—Einstein concluded that these measurements must depend on how the person making the measurements is moving.

To describe how measurements depend on motion, Einstein derived a series of equations named the **Lorentz transformations,** after the famous Dutch physicist Hendrik Antoon Lorentz. (Lorentz was a contemporary of Einstein who developed these equations independently but did not grasp their true meaning.) These equations tell us exactly how moving clocks slow down and how moving objects shrink.

To appreciate the Lorentz transformations, imagine two observers named Sergio and Majeeda. Sergio is at rest on Earth, while Majeeda is flying past in her spaceship at a speed v. Sergio and Majeeda both observe the same phenomenon on Earth—say, the beating of Sergio's heart or the ticking of Sergio's watch—which appears to occur over an interval of time. According to Sergio's clock (which is not moving relative to the phenomenon), the phenomenon lasts for T_0 seconds. This is called the **proper time** of the phenomenon. But according to Majeeda's clock (which *is* moving relative to the phenomenon), the same phenomenon lasts for a different length of time, T seconds. These two time intervals are related as follows:

Lorentz transformation for time

$$T = \frac{T_0}{\sqrt{1 - (v/c)^2}}$$

T = time interval measured by an observer moving relative to the phenomenon

T_0 = time interval measured by a observer *not* moving relative to the phenomenon (proper time)

v = speed of the moving observer relative to the phenomenon

c = speed of light

EXAMPLE: Sergio heats a cup of water in a microwave oven for 1 minute. According to Majeeda, who is flying past Sergio at 98% of the speed of light, how long does it take to heat the water?

Situation: The phenomenon in question is heating the water in the microwave oven. Sergio is not moving relative to this phenomenon (so he measures the *proper* time interval T_0). Majeeda is moving at speed $v = 0.98c$ relative to this

phenomenon (so she measures a different time interval T). Our goal is to determine the value of T.

Tools: We use the Lorentz transformation for time to calculate T.

Answer: We have $v/c = 0.98$, so

$$T = \frac{T_0}{\sqrt{1 - (v/c)^2}} = \frac{1 \text{ minute}}{\sqrt{1 - (0.98)^2}} = 5 \text{ minutes}$$

Review: A phenomenon that lasts for $T_0 = 1$ minute on Sergio's clock is stretched out to $T = 5T_0 = 5$ minutes as measured on Majeeda's clock moving at 98% of the speed of light. Other phenomena are affected in the same way: As measured by Majeeda, it takes 5 seconds for Sergio's wristwatch to tick off one second, and the minute hand on Sergio's wristwatch takes 5 hours to make a complete sweep. The converse is also true. As measured by *Sergio*, the minute hand on *Majeeda's* wristwatch will take 5 hours to make a complete sweep. This phenomenon, in which events moving relative to an observer happen at a slower pace, is called *time dilation*.

The Lorentz transformation for time is plotted in Figure 24-2b, which shows how 1 second measured on a stationary clock (say, Sergio's) is stretched out when measured using a clock carried by a moving observer (such as Majeeda). For speeds less than about half the speed of light, the mathematical factor $\sqrt{1 - (v/c)^2}$ of is not too different from 1, and there is little difference between the recordings of the stationary and moving clocks. At the fastest speeds that humans have ever traveled (en route from the Earth to the Moon), the difference between 1 and the factor $\sqrt{1 - (v/c)^2}$ is less than 10^{-9}. So, for any speed associated with human activities, stationary and slowly moving clocks tick at essentially the same rate. As the next example shows, however, time dilation is important for subatomic particles that travel at speeds comparable to c.

EXAMPLE: When unstable particles called *muons* (pronounced "mew-ons") are produced in experiments on Earth, they decay into other particles in an average time of 2.2×10^{-6} s. Muons are also produced by fast-moving protons from interstellar space when they collide with atoms in the Earth's upper atmosphere. These muons typically move at 99.9% of the speed of light and are formed at an altitude of 10 km. How long do such muons last before they decay?

Situation: The phenomenon in question is the life of a muon, which lasts a time $T_0 = 2.2 \times 10^{-6}$ s as measured by an observer not moving with respect to the muon. Our goal is to find the time interval T measured by an observer on Earth, which is moving at $v = 0.999c$ relative to the muon (the same speed at which the muon is moving relative to the Earth).

Tools: As in the previous example, we use the Lorentz transformation for time.

Answer: Using $v/c = 0.999$,

$$T = \frac{2.2 \times 10^{-6}\,\text{s}}{\sqrt{1 - (0.999)^2}} = 4.9 \times 10^{-5}\,\text{s}$$

Review: At this speed, the muon's lifetime is slowed down by time dilation by a factor of more than 22. Note that as measured by an observer on Earth, the time that it takes a muon produced at an altitude of 10 km = 10,000 m to reach the Earth's surface is

$$\frac{\text{distance}}{\text{speed}} = \frac{10,000\,\text{m}}{0.999 \times 3.00 \times 10^8\,\text{m/s}} = 3.3 \times 10^{-5}\,\text{s}$$

Were there no time dilation, such a muon would decay in just 2.2×10^{-6} s and would never reach the Earth's surface. But in fact, these muons *are* detected by experiments on the Earth's surface, because a muon moving at $0.999c$ lasts more than 3.3×10^{-5} s. The detection at the Earth's surface of muons from the upper atmosphere is compelling evidence for the reality of time dilation.

In the same terminology as "proper time," we say that a ruler at rest measures **proper length** or **proper distance** (L_0). According to the Lorentz transformations, distances perpendicular to the direction of motion are unaffected. However, a ruler of proper length L_0 held parallel to the direction of motion shrinks to a length L, given by

Lorentz transformation for length

$$L = L_0 \sqrt{1 - (v/c)^2}$$

L = length of a moving object along the direction of motion

L_0 = length of the same object at rest (proper length)

v = speed of the moving object

c = speed of light

EXAMPLE: Again, imagine that Majeeda is traveling at 98% of the speed of light relative to Sergio. If Majeeda holds a 1-meter ruler parallel to the direction of motion, how long is this ruler as measured by Sergio?

Situation: The ruler is at rest relative to Majeeda, so she measures its *proper* length $L_0 = 1$ m. Our goal is to determine its length L as measured by Sergio, who is moving at $v = 0.98c$ relative to Majeeda and her ruler.

Tools: We use the Lorentz transformation for length.

Answer: With $v/c = 0.98$, we find

$$L = L_0 \sqrt{1 - (v/c)^2} = (1\,\text{m}) \sqrt{1 - (0.98)^2}$$
$$= (1\,\text{m}) \times (0.20) = 0.20\,\text{m} = 20\,\text{cm}$$

Review: We saw in the first example that according to Sergio, Majeeda's clocks are ticking only one-fifth as fast as his. This example shows that he also measures Majeeda's 1-meter ruler to be only one-fifth as long (20 cm) when held parallel to the direction of motion. Note that the converse is also true: If *Sergio* holds a 1-m ruler parallel to the direction of relative motion, *Majeeda* measures it to be only 20 cm long. This shrinkage of length is called *length contraction*.

EXAMPLE: Consider again the above example about muons created 10 km above the Earth's surface. If a muon is traveling straight down, what is the distance to the surface as measured by an observer riding along with the muon?

Situation: Imagine a ruler that extends straight up from the Earth's surface to where the muon is formed. This ruler is at rest relative to the Earth, so its length of 10 km is the proper length L_0. Our goal is to find the length L of this ruler as measured by the observer riding with the muon.

Tools: As in the previous example, we use the Lorentz transformation for length.

Answer: With $v/c = 0.999$, we calculate

$$L = (1\,\text{km}) \sqrt{1 - (0.999)^2} = 0.45\,\text{km} = 450\,\text{m}$$

Review: The distance is contracted tremendously as measured by an observer riding with the muon. This result gives us another way to explain how such muons are able to reach the Earth's surface. As measured by the muon, the time required to travel the contracted distance is

$$\frac{\text{distance}}{\text{speed}} = \frac{450\,\text{m}}{0.999 \times 3.00 \times 10^8\,\text{m/s}} = 1.5 \times 10^{-6}\,\text{s}$$

This is less time than the 2.2×10^{-6} s that an average muon takes to decay. Hence, muons can successfully reach the Earth's surface.

24-2 | The general theory of relativity predicts black holes

SCIENCE IN PROCESS *Careful experiments have verified the key ideas of Einstein's theory of gravity*

Einstein's special theory of relativity is a comprehensive description of the behavior of light and, by extension, of electricity and magnetism. (Recall from Section 5-2 that light is both electric and magnetic in nature.) Einstein's next goal was to develop an even more comprehensive theory that also explained gravity. This was the **general theory of relativity**, which he published in 1915.

According to Newton's theory of gravity, an apple falls to the floor because the force of gravity pulls the apple down. But Einstein pointed out that the apple would appear to behave in exactly the same way in space, far from the gravitational influence of any planet or star, if the floor were to accelerate upward (in other words, if the floor came up to meet the apple).

In Figure 24-3, two famous gentlemen are watching an apple fall toward the floor of their closed compartments. They have no way of telling who is at rest on the Earth and who is in the hypothetical elevator moving upward through empty space at a constantly increasing speed. This is an example of Einstein's **equivalence principle**, which states that in a small volume of space, the downward pull of gravity can be accurately and completely duplicated by an upward acceleration of the observer.

The equivalence principle is the key to the general theory of relativity. It allowed Einstein to focus entirely on motion, rather than

force, in discussing gravity. A hallmark of gravity is that it causes the same acceleration no matter the mass of the object. For example, a baseball and a cannon ball have very different masses, but if you drop them side by side in a vacuum, they accelerate downward at exactly the same rate. To explain this, Einstein envisioned gravity as being caused by a *curvature of space*. In fact, his general theory of relativity describes gravity entirely in terms of the geometry of both space *and* time, that is, of spacetime. Far from a source of gravity, like a planet or a star, spacetime is "flat" and clocks tick at their normal rate. Closer to a source of gravity, however, space is curved and clocks slow down.

A useful analogy is to picture the spacetime near a massive object such as the Sun as being curved like the surface in Figure 24-4. Imagine a ball rolling along this surface. Far from the "well" that represents the Sun, the surface is fairly flat and the ball moves in a straight line. If the ball passes near the well, however, it curves in toward it. If the ball is moving at an appropriate speed, it might move in an orbit around the sides of the well. The curvature of the well has the same effect on a ball of any size, which explains why gravity produces the same acceleration on objects of different mass.

Einstein's general theory of relativity and its picture of curved spacetime have been tested in a variety of different ways. In what follows we discuss some key experimental tests of the theory.

Experimental Test 1: The gravitational bending of light. Light rays naturally travel in straight lines. But if the space through which

This compartment is at rest in the Earth's gravitational field.

(a) The apple hits the floor of the compartment because the Earth's gravity accelerates the apple downward.

This compartment is moving in a gravity-free environment.

(b) The apple hits the floor of the compartment because the compartment accelerates upward.

Figure 24-3
The Equivalence Principle The equivalence principle asserts that you cannot distinguish between (a) being at rest in a gravitational field and (b) being accelerated upward in a gravity-free

environment. This idea was an important step in Einstein's quest to develop the general theory of relativity.

figure 24-4

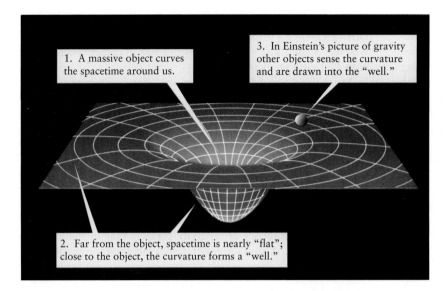

1. A massive object curves the spacetime around us.

3. In Einstein's picture of gravity other objects sense the curvature and are drawn into the "well."

2. Far from the object, spacetime is nearly "flat"; close to the object, the curvature forms a "well."

The Gravitational Curvature of Spacetime According to Einstein's general theory of relativity, spacetime becomes curved near a massive object. To help you visualize the curvature of four-dimensional spacetime, this figure shows the curvature of a *two*-dimensional space around a massive object.

the rays travel is curved, as happens when light passes near the surface of a massive object like the Sun, the paths of the rays will likewise be curved (Figure 24-5). In other words, gravity should bend light rays, an effect not predicted by Newtonian mechanics because light has no mass. This prediction was first tested in 1919 during a total solar eclipse. During totality, when the Moon blocked out the Sun's disk, astronomers photographed the stars around the Sun. Careful measurements afterward revealed that the stars were shifted from their usual positions by an amount consistent with Einstein's theory.

Experimental Test 2: The precession of Mercury's orbit. As the planet Mercury moves along its elliptical orbit, the orbit itself slowly changes orientation or *precesses* (Figure 24-6). Most of Mercury's precession is caused by the gravitational pull of the other planets, as explained by Newtonian mechanics. But once the effects of all the other planets had been accounted for, there remained an unexplained excess rotation of Mercury's major axis of 43 arcsec per century. Although this discrepancy may seem very small, it frustrated astronomers for half a century. Some astronomers searched for a missing planet even closer to the Sun that might be tugging on Mercury; none has ever been found. Einstein showed that at Mercury's position close to the Sun, the general theory of relativity predicts a small correction to Newton's description of gravity. This correction is just enough to account for the excess precession.

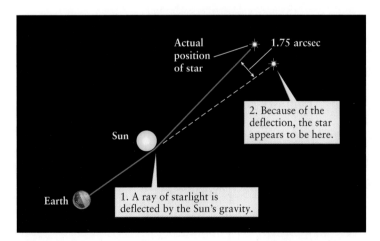

Actual position of star

1.75 arcsec

2. Because of the deflection, the star appears to be here.

Sun

Earth

1. A ray of starlight is deflected by the Sun's gravity.

figure 24-5

The Gravitational Deflection of Light Light rays are deflected by the curved spacetime around a massive object like the Sun. The maximum deflection is very small, only 1.75 arcsec for a light ray grazing the Sun's surface. By contrast, Newton's theory of gravity predicts *no* deflection at all. The deflection of starlight by the Sun was confirmed during a solar eclipse in 1919.

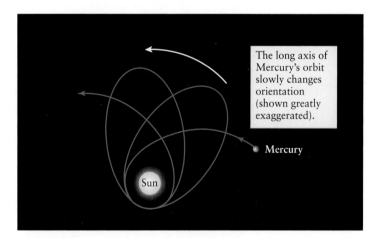

The long axis of Mercury's orbit slowly changes orientation (shown greatly exaggerated).

Mercury

Sun

figure 24-6

The Precession of Mercury's Orbit Mercury's orbit changes orientation at a rate of 574 arcsec (about one-sixth of a degree) per century. Newtonian mechanics predicts that the gravitational influences of the other planets should make the orientation change by only 531 arcsec per century. Einstein was able to explain the excess 43 arcsec per century using his general theory of relativity.

Experimental Test 3: The gravitational slowing of time and the gravitational red shift (**Figure 24-7**). In the general theory of relativity, a massive object such as the Earth warps time as well as space. Einstein predicted that clocks on the ground floor of a building should tick slightly more slowly than clocks on the top floor, which are farther from the Earth (Figure 24-7*a*).

A light wave can be thought of as a clock; just as a clock makes a steady number of ticks per minute, an observer sees a steady number of complete cycles of a light wave passing by each second. If a light beam is aimed straight up from the ground floor of a building, an observer at the top floor will measure a "slow-ticking" light wave with a lower frequency, and thus a longer wavelength, than will an observer on the ground floor (Figure 24-7*b*). The increase in wavelength means that a photon reaching the top floor has less energy than when it left the ground floor. (You may want to review the discussion of frequency and wavelength in Section 5-2 as well as the description of photons in Section 5-5.) These effects, which have no counterpart in Newton's theory of gravity, are called the **gravitational redshift.**

 Be careful not to confuse the gravitational redshift with a Doppler shift. In the Doppler effect, redshifts are caused by a light source moving away from an observer. Gravitational redshifts, by contrast, are caused by time flowing at different rates at different locations. No motion is involved.

The American physicists Robert Pound and Glen Rebka first measured the gravitational redshift in 1960 using gamma rays fired between the top and bottom of a shaft 20 meters tall. Because the Earth's gravity is relatively weak, the redshift that they measured was very small ($\Delta\lambda/\lambda = 2.5 \times 10^{-15}$) but was in complete agreement with Einstein's prediction.

Much larger shifts are seen in the spectra of white dwarfs, whose spectral lines are redshifted as light climbs out of the white dwarf's intense surface gravity. As an example, the gravitational redshift of the spectral lines of the white dwarf Sirius B (see Figure 22-8) is $\Delta\lambda/\lambda = 3.0 \times 10^{-4}$, which also agrees with the general theory of relativity.

 Experimental Test 4: Gravitational waves. Electric charges oscillating up and down in a radio transmitter's antenna produce electromagnetic radiation. In a similar way, the general theory of relativity predicts that oscillating massive objects should produce **gravitational radiation,** or **gravitational waves.** (Newton's theory of gravity makes no such prediction.)

Gravitational radiation is exceedingly difficult to detect, because it is by nature much weaker than electromagnetic radiation. Although physicists have built a number of sensitive "antennas" for gravitational radiation, no confirmed detections have been made as of this writing (2004). But compelling *indirect* evidence for the existence of gravitational radiation has come from a binary system of two neutron stars. Russell Hulse and Joseph Taylor at the University of Massachusetts discovered that these stars are slowly spiraling toward each

(a) The gravitational slowing of time

Compared to a clock on the top floor, a clock on the ground floor is deeper in the Earth's gravitational field and so ticks more slowly.

As a light wave climbs in a gravitational field, its frequency decreases and its wavelength increases.

Light source

(b) The gravitational redshift

Figure 24-7

The Gravitational Slowing of Time and the Gravitational Redshift
(a) Clocks at different heights in a gravitational field tick at different rates.
(b) The oscillations of a light wave constitute a type of clock. As a light wave climbs from the ground floor toward the top floor, its oscillation frequency becomes lower and its wavelength becomes longer.

other and losing energy in the process (see Section 23-6). The rate at which they lose energy is just what would be expected if the two neutron stars are emitting gravitational radiation as predicted by Einstein. Hulse and Taylor shared the 1993 Nobel Prize in physics for their discovery.

The general theory of relativity has never made an incorrect prediction. It now stands as our most accurate and complete description of gravity. Einstein demonstrated that Newtonian mechanics is accurate only when applied to low speeds and weak gravity. If extremely high speeds or powerful gravity are involved, only a calculation using relativity will give correct answers.

Perhaps the most dramatic prediction of the general theory of relativity concerns what happens when a large amount of matter is concentrated in a small volume. We have seen that if a dying star is not too massive, it ends up as a white dwarf star. If the dying star is more massive than the Chandrasekhar limit of about 1.4 $M_\odot$, it cannot exist as a stable white dwarf star and, instead, shrinks down to form a neutron star. But if the dying star has more mass than the maximum permissible for a neutron star, about 2 to 3 $M_\odot$, not even the internal pressure of neutrons can hold the star up against its own gravity, and the star contracts rapidly.

As the star's matter becomes compressed to enormous densities, the strength of gravity at the surface of this rapidly shrinking sphere also increases dramatically. According to the general theory of relativity, the space immediately surrounding the star becomes so highly curved that it closes on itself (**Figure 24-8**). Photons flying outward at an angle from the star's surface arc back inward, while photons

that fly straight outward undergo such a strong gravitational redshift that they lose all their energy and cease to exist.

Ordinary matter can never travel as fast as light. Hence, if light cannot escape from the collapsing star, neither can anything else. An object from which neither matter nor electromagnetic radiation can escape is called a **black hole**. In a sense, a hole is punched in the fabric of the universe, and the dying star disappears into this cavity (**Figure 24-9**).

None of the star's mass is lost when it collapses to form a black hole, however. This mass gives the spacetime around the black hole its strong curvature. Thanks to this curvature, the black hole's gravitational influence can still be felt by other objects.

CAUTION! Some low-quality science-fiction movies and books suggest that black holes are evil things that go around gobbling up everything in the universe. Not so! The bizarre effects created by highly warped spacetime are limited to a region quite near the hole. For example, the effects of the general theory of relativity predominate only within 1000 km of a 10-$M_\odot$ black hole. Beyond 1000 km, gravity is weak enough that Newtonian physics can adequately describe everything. If our own Sun somehow turned into a black hole (an event that, happily, seems to be quite impossible), the orbits of the planets would hardly be affected at all.

Perhaps the most remarkable aspect of black holes is that they really exist! As we will see, astronomers have located a number of black holes with masses a few times that of the Sun. What is truly

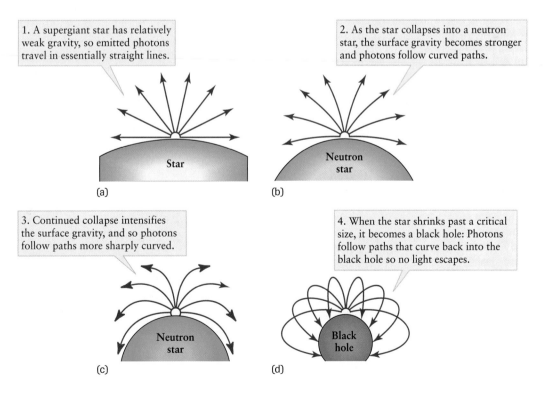

figure 24-8

The Formation of a Black Hole (a)–(c) These illustrations show four steps leading up to the formation of a black hole from a dying star. (d) When the star becomes a black hole, not even photons emitted directly upward from the surface can escape; they undergo an infinite gravitational redshift and disappear.

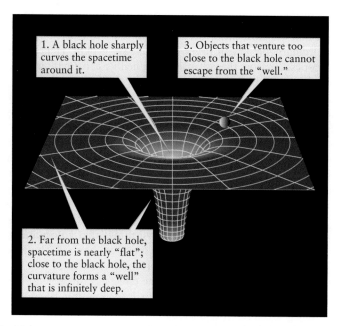

1. A black hole sharply curves the spacetime around it.

3. Objects that venture too close to the black hole cannot escape from the "well."

2. Far from the black hole, spacetime is nearly "flat"; close to the black hole, the curvature forms a "well" that is infinitely deep.

Figure 24-9

Curved Spacetime around a Black Hole This diagram suggests how spacetime is distorted by a black hole's mass. As in Figure 24-4, spacetime is represented as a two-dimensional surface. Unlike the situation in Figure 24-4, a black hole's gravitational "well" is infinitely deep.

few seconds (see Section 23-7). Instead, the X-ray emissions from Cygnus X-1 are highly variable and irregular; they flicker on time scales as short as one-hundredth of a second. One of the fundamental concepts in physics is that nothing can travel faster than the speed of light (recall Section 24-1). Because of this limitation, an object cannot flicker faster than the time required for light to travel across the object. Because light travels 3000 km in a hundredth of a second, Cygnus X-1 can be no more than 3000 km across, or about a quarter the size of the Earth.

Cygnus X-1 occasionally emits bursts of radio radiation, and in 1971 radio astronomers used these bursts to show that the source was at the same location in the sky as the star HDE 226868 (Figure 24-10). Spectroscopic observations revealed that HDE 226868 is a B0 supergiant with a surface temperature of about 31,000 K. Such stars do not emit significant amounts of X rays, so HDE 226868 alone cannot be the X-ray source Cygnus X-1. Because binary stars are very common, astronomers began to suspect that the visible star and the X-ray source are in orbit about each other.

Further spectroscopic observations soon showed that the lines in the spectrum of HDE 226868 shift back and forth with a period of 5.6 days. This behavior is characteristic of a single-line spectroscopic binary (see Section 19-10); the companion of HDE 226868 is too dim to produce its own set of spectral lines. The clear implication is that HDE 226868 and Cygnus X-1 are the two components of a binary star system.

amazing is that they have also discovered many truly immense black holes containing millions or billions of solar masses. These discoveries are a resounding confirmation of the ideas of the general theory of relativity. In the next two sections we will see how astronomers hunt down black holes in space.

24-3 | Certain binary star systems probably contain black holes

Finding black holes is a difficult business. Because light cannot escape from inside the black hole, you cannot observe one directly in the same way that you can observe a star or a planet. The best you can hope for is to detect the effects of a black hole's powerful gravity.

Close binary star systems offer one way to find **stellar-mass black holes** (that is, ones with masses is comparable to those of ordinary stars). Imagine that one of the stars in a binary system evolves into a black hole. If the black hole orbits close enough to the other, ordinary star in the system, tidal forces can draw matter from the ordinary star onto the black hole. If we can detect radiation coming from this "stolen" matter, we can infer the presence of the black hole.

The first sign of such emissions from a binary system with a black hole came shortly after the launch of the *Uhuru* X-ray–detecting satellite in 1971. Astronomers became intrigued with an X-ray source designated Cygnus X-1. This source is quite unlike pulsating X-ray sources, which emit regular bursts of X rays every

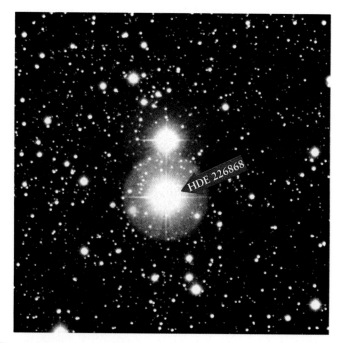

Figure 24-10   R I [V] U X G

HDE 226868 This photograph from the 5-m telescope on Palomar Mountain shows HDE 226868, the B0 blue supergiant star at the location of the X-ray source Cygnus X-1. This star is located about 2500 pc (87000 ly) from Earth. The bright star directly above HDE 226868 is not part of the binary system. (Courtesy of J. Kristian, Carnegie Observatories)

From what we know about the masses of other supergiant stars, HDE 226868 is estimated to have a mass of roughly 30 $M_\odot$. As a result, the unseen member of the binary system must have a mass of about 7 $M_\odot$ or more. Otherwise, it would not exert enough gravitational pull to make the B0 star wobble by the amount deduced from the periodic Doppler shifting of its spectral lines. Because the unseen companion does not emit visible light, it cannot be an ordinary star. Furthermore, because 7 solar masses is too large for either a white dwarf or a neutron star, Cygnus X-1 is likely to be a black hole.

The case for Cygnus X-1 being a black hole is not airtight. HDE 226868 might have a low mass for its spectral type, which would imply a somewhat lower mass for Cygnus X-1. In addition, uncertainties in the distance to the binary system could further reduce estimates of the mass of Cygnus X-1. If all these uncertainties combined in just the right way, the estimated mass of Cygnus X-1 could be pushed down to about 3 $M_\odot$. Thus, there is a slim chance that Cygnus X-1 might contain the most massive possible neutron star rather than a black hole.

If Cygnus X-1 does contain a black hole, the X rays do not come from the black hole itself. Gas captured from HDE 226868 goes into orbit about the hole, forming an accretion disk about 4 million kilometers in diameter (Figure 24-11). As material in the disk gradually spirals in toward the hole, friction heats the gas to temperatures approaching 2×10^6 K. In the final 200 kilometers above the hole, these extremely hot gases emit the X rays that our satellites detect. Presumably the X-ray flickering is caused by small hot spots on the rapidly rotating inner edge of the accretion disk. In this way, the black hole's existence is announced by doomed gases just before they plunge to oblivion.

Astronomers have found more than 20 other black hole candidates like Cygnus X-1. All are compact X-ray sources orbiting ordinary stars in a spectroscopic binary system. One of the best candidates is V404 Cygni in the constellation Cygnus. Doppler shift measurements reveal that as the visible star moves around its 6.47-day orbit, its radial velocity (see Section 19-10) varies by more than 400 km/s. These data give a firm lower limit of 6.26 $M_\odot$ for the mass of the unseen companion. It is probably impossible for a neutron star to be more massive than 3 $M_\odot$, so V404 Cygni must almost certainly be a black hole.

Another particularly convincing case is the flickering X-ray source A0620-00 in the constellation Monoceros. (The A refers to the British satellite *Ariel 5*, which discovered the source; the numbers denote its position in the sky.) The visible companion to A0620-00 is an orange K5 main-sequence star called V616 Monocerotis, which orbits the X-ray source every 7.75 hours. Because this star is relatively faint, it is possible to observe the shifting spectral lines from *both* the visible companion and the X-ray source as they orbit each other. With this more complete information about the orbits, astronomers have determined the mass of A0620-00 to be more than 3.2 $M_\odot$, and more probably about 9 $M_\odot$.

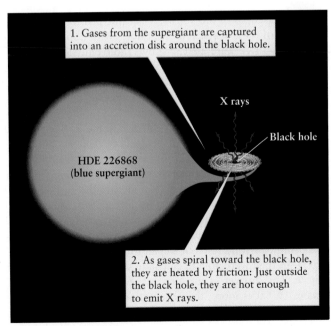

(a) A schematic diagram of Cygnus X-1

(b) An artist's impression of Cygnus X-1

Figure 24-11
The Cygnus X-1 System (a) The larger member of the Cygnus X-1 system is a B0 supergiant of about 30 $M_\odot$. The other, unseen member of the system has a mass of at least 7 $M_\odot$ and is probably a black hole. (b) This illustration shows how the Cygnus X-1 system might look at close range. At even closer range, the black hole and its immediate surroundings might appear as shown in Figure 24-12. (Courtesy of D. Norton, Science Graphics)

Astronomers have seen jets of hot, glowing material extending several light-years from some black hole candidates. The ejected material emerges from the vicinity of the black hole at speeds approaching the speed of light. It is thought that these jets are formed by strong electric and magnetic fields in the material around a *rotating* black hole, much as occurs for rotating protostars (see Figure 20-15) and for rotating neutron stars (see Figure 23-6). Figure 24-12 is an artist's impression of the immediate vicinity of such a rotating black hole. (We will discuss the curious features of rotating black holes in Section 24-6.)

If there really are black holes in close binary systems, how did they get there? One possibility is that one member of the binary explodes as a Type II supernova, leaving a burned-out core whose mass exceeds 3 $M_\odot$. This core undergoes gravitational collapse to form a black hole.

Another possibility is that a white dwarf or a neutron star in a binary system can become a black hole if it accretes enough matter from its companion star. This transformation can occur when the companion star becomes a red giant and dumps a significant part of its mass over its Roche lobe (see Figure 21-19b).

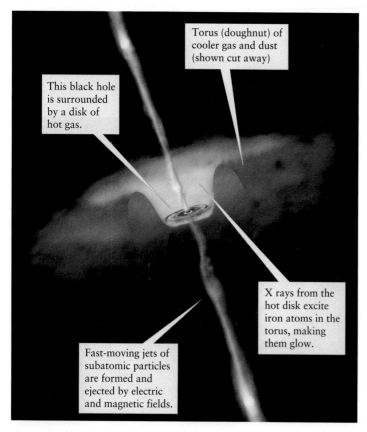

This black hole is surrounded by a disk of hot gas.

Torus (doughnut) of cooler gas and dust (shown cut away)

X rays from the hot disk excite iron atoms in the torus, making them glow.

Fast-moving jets of subatomic particles are formed and ejected by electric and magnetic fields.

Figure 24-12

The Environment of an Accreting Black Hole If a black hole is rotating, it can generate strong electric and magnetic fields in its immediate vicinity. These fields draw material from the accretion disk around the black hole and accelerate it into oppositely directed jets along the black hole's rotation axis. This illustration also shows other features of the material surrounding such a black hole. (CXC/M. Weiss)

Another possibility is two dead stars coalescing to form a black hole. For example, imagine two neutron stars orbiting each other, like the binary systems discussed in Section 23-6 and Section 24-2. Because the two stars emit gravitational radiation, they gradually spiral in toward each other and eventually merge. If their total mass exceeds 2 to 3 $M_\odot$, the entire system may become a black hole.

24-4 | Supermassive black holes exist at the centers of most galaxies

Stellar evolution makes it possible to form black holes with masses several times that of the Sun. But calculations suggest other ways to form black holes that are either extremely large or extraordinarily small.

As we will discuss in Chapter 26, galaxies formed in the early universe by the coalescence of gas clouds. During this formation process, some of a galaxy's gas could have plunged straight toward the center of the galaxy, where it collected and compressed together under its own gravity. If this central clump became sufficiently dense, it could have formed a black hole. Typical galaxies have masses of 10^{11} $M_\odot$, so even a tiny percentage of the galaxy's gas collected at its center would give rise to a **supermassive black hole** with a truly stupendous mass.

Since the 1970s, astronomers have found evidence suggesting the existence of supermassive black holes in other galaxies. With the advent of a new generation of optical telescopes in the 1990s, it became possible to see the environments of these suspected black holes with unprecedented detail (Figure 24-13). Figure 24-13a shows radio emissions from two jets that extend some 15,000 pc (50,000 ly) from the center of the galaxy NGC 4261. These resemble an immensely magnified version of the jets seen emerging from some stellar-mass black holes (see Section 24-3). The highly magnified Hubble Space Telescope image in Figure 24-13b shows a disk some 250 pc (800 ly) in diameter at the very center of this galaxy. The jets are perpendicular to the plane of the disk, just as in Figure 24-12.

By measuring the Doppler shifts of light coming from the two sides of the disk, astronomers found that the disk material was orbiting around the bright object at the center of the disk at speeds of hundreds of kilometers per second. (By comparison, the Earth orbits the Sun at 30 km/s.) Given the size of the material's orbit, and using Newton's form of Kepler's third law (see Section 4-7 and Box 4-4), they were able to calculate the mass of the central bright object. The answer is an amazing 1.2 *billion* (1.2×10^9) solar masses! What is more, the observations show that this object can be no larger than our solar system. The only possible explanation is that the object at the center of NGC 4261 is a black hole.

Dozens of other black holes in the centers of galaxies have been identified by their gravitational effect on surrounding gas and dust. Surveys of galaxies have shown that supermassive black holes are not at all unusual; most large galaxies appear to have them at their centers. As we will see in the next chapter, a black hole with several million solar masses lies at the center of our own Milky Way Galaxy, some 8000 pc (26,000 ly) from Earth.

Jet (radio image)

Galaxy NGC 4261 (visible-light image)

Jet (radio image)

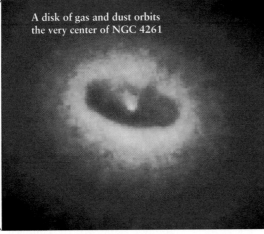

A disk of gas and dust orbits the very center of NGC 4261

(a) Galaxy NGC 4261

R I **V** U X G

(b) Evidence for a supermassive black hole in NGC 4261

R I **V** U X G

ANIMATION 24.2

figure 24-13
A Supermassive Black Hole The galaxy NGC 4261 lies some 30 million pc (100 million ly) from Earth in the constellation Virgo. **(a)** This composite view superimposes a visible-light image of the galaxy (white) with a radio image of the galaxy's immense jets (orange). **(b)** This Hubble Space Telescope image shows a disk of gas and dust about 250 pc (800 ly) across at the center of NGC 4261. Observations indicate that a supermassive black hole is at the center of the disk. (NASA, ESA)

Black holes seem to fall into two very different groups depending on their size. Supermassive black holes have masses from 10^6 to 10^9 $M_\odot$, while stellar-mass black holes like those discussed in Section 24-3 have masses around 10 $M_\odot$. In 2000 a team of astronomers from Japan, the United States, and Great Britain reported finding a "missing link" between these two groups.

 WEB LINK 24.8

The new discovery was a fluctuating X-ray source located some 200 pc (600 ly) from the center of the galaxy M82 (**Figure 24-14**). Images do not reveal material orbiting around this source, so we cannot determine the source's mass using Newton's form of Kepler's third law. But by relating the source's X-ray luminosity to the rate at which matter

would have to fall onto the source to produce that luminosity, it can be shown that the source must have a mass of at least 500 $M_\odot$ or so. The rapid fluctuations of the source show that it has a very small diameter, which suggests that it is a black hole (see Section 24-3). Hence the source in M82 may be a **mid-mass black hole**. Additional mid-mass black hole candidates have been found in other galaxies. Such black holes might have formed by the coalescence of many normal stars or by the direct merger of stellar-mass black holes.

During the early 1970s, Stephen Hawking of Cambridge University proposed the existence of yet another type of black hole. He suggested that during the Big Bang some 13.7 billion years ago, local pockets of the universe could have been dense enough and under

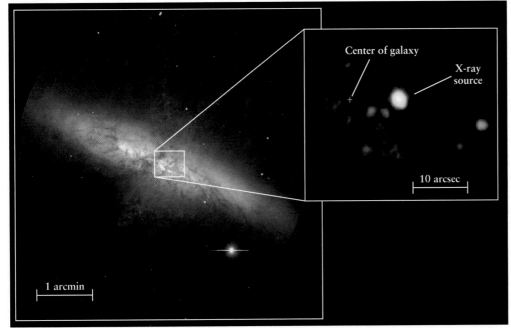

Center of galaxy

X-ray source

10 arcsec

1 arcmin

R I V U **X** G

VIDEO 24.1

figure 24-14
A Mid-Mass Black Hole?
M82 is an unusual galaxy in the constellation Ursa Major. The inset shows an image of the central region of this galaxy from the Chandra X-ray Observatory. The bright, compact X-ray source varies in its light output over a period of months. The properties of this source suggest that it may be a black hole of roughly 500 solar masses or more. (Subaru Telescope, National Astronomical Observatory of Japan; inset: NASA/SAO/CXC)

R I **V** U X G

sufficient pressure to form small black holes. According to his theory, these **primordial black holes** could have had masses as large as the Earth or as small as 5×10^{-8} kg, about 1/40 the mass of a single raindrop. To date, however, no evidence has been found for the existence of primordial black holes.

24-5 | A nonrotating black hole has only a "center" and a "surface"

Understanding in detail how black holes form is a challenging task. But once a black hole forms, it has a remarkably simple structure.

Surrounding a black hole, where the escape speed from the hole just equals the speed of light, is the **event horizon.** You can think of this sphere as the "surface" of the black hole, although the black hole's mass all lies well inside this "surface." Once a massive dying star collapses to within its event horizon, it disappears permanently from the universe. The term "event horizon" is quite appropriate, because this surface is like a horizon beyond which we cannot see any events.

If the dying star is not rotating before it collapses, the black hole will likewise not be rotating. The distance from the center of a nonrotating black hole to its event horizon is called the **Schwarzschild radius** (denoted R_{Sch}), after the German physicist Karl Schwarzschild who first determined its properties. **Box 24-2** describes how to calculate the Schwarzschild radius, which depends only on the black hole's mass. The more massive the black hole, the larger its event horizon.

Once a nonrotating star has contracted inside its event horizon, nothing can prevent its complete collapse. The star's entire mass is crushed to zero volume—and hence infinite density—at a single point, known as the **singularity,** at the center of the black hole. We now can see that the structure of a nonrotating black hole is quite simple. As drawn in **Figure 24-15,** it has only two parts: a singularity at the center surrounded by a spherical event horizon.

To understand why the complete collapse of such a doomed star is inevitable, think about your own life on the Earth, far from any black holes. You have the freedom to move as you wish through the three dimensions of space: up and down, left and right, or forward and back. But you do not have the freedom to move at will through the dimension of time. Whether we like it or not, we are all carried inexorably from the cradle to the grave.

box 24-2 | TOOLS OF THE ASTRONOMER'S TRADE

The Schwarzschild Radius

The Schwarzschild radius (R_{Sch}) is the distance from the center of a nonrotating black hole to its event horizon. You can think of it as the "size" of the black hole. The Schwarzschild radius is related to the mass M of the black hole by

Schwarzschild radius

$$R_{Sch} = \frac{2GM}{c^2}$$

R_{Sch} = Schwarzschild radius of a black hole

G = universal constant of gravitation

M = mass of black hole

c = speed of light

When using this formula, be careful to express the mass in kilograms, not solar masses. That is because of the units in which G and c are commonly expressed: $G = 6.67 \times 10^{-11}$ N · m²/kg² and $c = 3.00 \times 10^8$ m/s. If you use M in kilograms, the answer that you get for R_{Sch} will be in meters.

EXAMPLE: Find the Schwarzschild radius (in kilometers) of a black hole with 10 times the mass of the Sun.

Situation: We are given the mass of the black hole (10 $M_\odot$) and wish to determine its Schwarzschild radius R_{Sch}.

Tools: We use the above formula for R_{Sch}, being careful to first convert the mass from solar masses to kilograms.

Answer: One solar mass is 1.99×10^{30} kg, so in this case $M = 10 \times 1.99 \times 10^{30}$ kg $= 1.99 \times 10^{31}$ kg. The Schwarzschild radius of this black hole is then

$$R_{Sch} = \frac{2GM}{c^2} = \frac{2(6.67 \times 10^{-11})(1.99 \times 10^{31})}{(3.00 \times 10^8)^2}$$
$$= 3.0 \times 10^4 \text{ m}$$

One kilometer is 10^3 meters, so the Schwarzschild radius of this 10-$M_\odot$ black hole is

$$R_{Sch} = (3.0 \times 10^4 \text{ m}) \times \frac{1 \text{ km}}{10^3 \text{ m}} = 30 \text{ km, or } 19 \text{ mi}$$

Review: The Schwarzschild radius is directly proportional to the mass of the black hole. Thus, a black hole with 10 times the mass (100 $M_\odot$) would have a Schwarzschild radius 10 times larger ($R_{Sch} = 10 \times 30$ km = 300 km); a black hole with half the mass (5 $M_\odot$) would have half as large a Schwarzschild radius ($R_{Sch} = \frac{1}{2} \times 30$ km = 15 km), and so on.

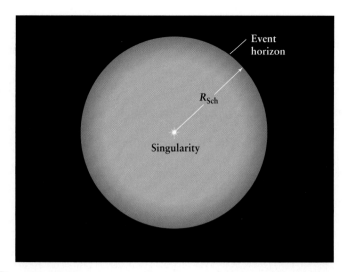

figure 24-15

The Structure of a Nonrotating (Schwarzschild) Black Hole
A nonrotating black hole has only two parts: a singularity, where all of the mass is located, and a surrounding event horizon. The distance from the singularity to the event horizon is called the Schwarzschild radius (R_{Sch}). The event horizon is a one-way surface: Things can fall in, but nothing can get out.

Inside a black hole, gravity distorts the structure of spacetime so severely that the directions of space and time become interchanged. In a sense, inside a black hole you acquire a limited ability to affect the passage of time. This seeming gain does you no good, however, because you lose a corresponding amount of freedom to move through space. Whether you like it or not, you will be dragged inexorably from the event horizon toward the singularity. Just as no force in the universe can prevent the forward march of time from past to future outside a black hole, no force in the universe can prevent the inward march of space from event horizon to singularity inside a black hole. Once an object dropped into a black hole crosses the event horizon, it is gone forever, like an object dropped into a bottomless pit.

The same is true for a light beam aimed at the black hole. Because all this light will be absorbed and none reflected back, a black hole is totally and utterly black. (By contrast, even the blackest object on Earth reflects *some* light when you shine a flashlight on it. A black hole would reflect none.)

At a black hole's singularity, the strength of gravity is infinite, so the curvature of spacetime there is infinite. Space and time at the singularity are thus all jumbled up. They do not exist as separate, distinctive entities.

This confusion of space and time has profound implications for what goes on inside a black hole. All the laws of physics require a clear, distinct background of space and time. Without this identifiable background, we cannot speak rationally about the arrangement of objects in space or the ordering of events in time. Because space and time are all jumbled up at the center of a black hole, the singularity there does not obey the laws of physics. The singularity behaves in a random, capricious fashion, totally devoid of rhyme or reason.

Fortunately, we are shielded from the singularity by the event horizon. In other words, although random things do happen at the singularity, none of their effects manages to escape beyond the event horizon. Consequently, the outside universe remains understandable and predictable.

The chaotic, random behavior of the singularity has been so disturbing to physicists that in 1969 the British mathematician Roger Penrose and his colleagues proposed the **law of cosmic censorship:** "Thou shalt not have naked singularities." In other words, every singularity must be completely surrounded by an event horizon, because an exposed singularity could affect the universe in an unpredictable and random way.

24-6 | Just three numbers completely describe the structure of a black hole

In addition to shielding us from singularities, the event horizon prevents us from ever knowing much about anything that falls into a black hole. A black hole is, in fact, an "information sink." Many properties of matter falling into a black hole, such as its chemical composition, texture, color, shape, and size, simply vanish as soon as the matter crosses the event horizon.

Because this information has completely vanished, it cannot affect the structure or properties of the hole. For example, consider two hypothetical black holes, one made from the gravitational collapse of 10 $M_\odot$ of iron and the other made from the gravitational collapse of 10 $M_\odot$ of peanut butter. Obviously, quite different substances went into the creation of the two holes. Once the event horizons of these two black holes have formed, however, both the iron and the peanut butter will have permanently disappeared from the universe. As seen from the outside, the two holes are absolutely identical, making it impossible for us to tell which ate the peanut butter and which ate the iron. A black hole is thus unaffected by the information it destroys.

Because a black hole is indeed an information sink, it is reasonable to wonder whether we can determine anything at all about a black hole. What properties *do* characterize a black hole?

First, we can measure the *mass* of a black hole. One way to do this would be by placing a satellite into orbit around the hole. After measuring the size and period of the satellite's orbit, we could use Newton's form of Kepler's third law (recall Section 4-7 and Box 4-4) to determine the mass of the black hole. This is equal to the total mass of all the material that has gone into the hole.

Second, we can also measure the total *electric charge* possessed by a black hole. Like gravity, the electric force acts over long distances—it is a long-range interaction that is felt in the space around the hole. Appropriate equipment on a space probe passing near the hole could measure the strength of the electric force, and the electric charge could thus be determined.

In actuality, we would not expect a black hole to possess much, if any, electric charge. For example, if a hole did happen to start off with a sizable positive charge, it would vigorously attract vast numbers of negatively charged electrons from the interstellar medium, which would soon neutralize the hole's charge. For this reason, astronomers usually neglect electric charge when discussing real black holes.

Although a black hole might theoretically have a tiny electric charge, it can have no magnetic field of its own whatsoever. (We discussed in Section 24-3 how magnetic fields are involved in producing jets from black holes. However, these fields are associated with the accretion disk around the black hole, not the black hole itself.)

When a black hole is created, however, the collapsing star from which it forms may possess an appreciable magnetic field. The star must therefore radiate this magnetic field away before it can settle down inside its event horizon. Theory predicts that the star does this by emitting electromagnetic and gravitational waves. As we saw in Section 24-2, gravitational waves are ripples in the overall geometry of space. Various experiments soon to be in operation may detect bursts of gravitational radiation emitted by massive stars as they collapse to form black holes.

 Third, we can detect the effects of a black hole's rotation, that is, measure its *angular momentum*. An object's angular momentum is related to how fast it rotates and how the object's mass is distributed over its volume. As a dead star collapses into a black hole, its rotation naturally speeds up as its mass moves toward the center, just as a figure skater rotates faster when she pulls her arms and legs in. This same effect explains the rotation of the solar nebula (see Section 8-4), as well as why neutron stars spin so fast (see Section 23-3). We expect a black hole that forms from a rotating star to be spinning very rapidly.

When the matter that collapses to form a black hole is rotating, that matter does not compress to a point. Instead, it collapses into a ring-shaped singularity located between the center of the hole and the event horizon (**Figure 24-16**). The structure of such rotating black holes was first worked out in 1963 by the New Zealand mathematician Roy Kerr. Most rotating black holes should be spinning thousands of times per second, even faster than the most rapid pulsars (see Section 23-6).

 If a rotating black hole is surrounded by an accretion disk with a magnetic field, it may be possible for the field to steal some of the rotational energy and angular momentum from the black hole and transfer it to the disk. The magnetic field acts as a "brake" that retards the black hole's rotation and makes the disk's rotation speed up. (On Earth, magnetic braking is used to slow locomotives, amusement park rides, and hybrid automobiles to a smooth stop without generating excess heat.) The illustration that opens this chapter shows an arching magnetic field that connects an accretion disk with a rotating black hole in just this way. This process may be taking place with the supermassive black hole at the center of the galaxy MCG–6-15-30. Observations with the XMM-Newton telescope (see Section 6-7) indicate that unusually intense X-rays are coming from the accretion disk around this black hole. The suspicion is that the energy to power this radiation may be extracted from the black hole's rotation.

 Even a rotating black hole without an accretion disk can transfer energy to other objects. This is possible according to Einstein's general theory of relativity, which makes the startling prediction that a rotating body drags spacetime around it. (Very precise spacecraft measurements indicate that the rotating Earth drags spacetime in just this way.) Sur-

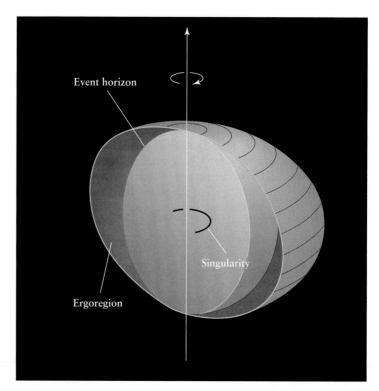

figure 24-16

The Structure of a Rotating (Kerr) Black Hole The singularity of a rotating black hole is an infinitely thin ring centered on the geometrical center of the hole. (It appears as an arc in this cutaway diagram.) Outside the spherical event horizon is the doughnut-shaped ergoregion, where the dragging of spacetime around the hole is so severe that nothing can remain at rest. The tan-colored surface marks the outer boundary of the ergoregion.

rounding the event horizon of every rotating black hole is a region where this dragging of space and time is so severe that it is impossible to stay in the same place. No matter what you do, you get pulled around the hole, along with the rotating geometry of space and time. This region, where it is impossible to be at rest, is called the **ergoregion** (see Figure 24-16).

To measure a black hole's angular momentum, we could hypothetically place two satellites in orbit about the hole. Suppose that one satellite circles the hole in the same direction the hole rotates and the other in the opposite direction. One satellite is thus carried along by the geometry of space and time, but the other is constantly fighting its way "upstream." The two satellites will thus have different orbital periods. By comparing these two periods, astronomers can deduce the total angular momentum of the hole.

Because the ergoregion is outside the event horizon, this bizarre region is accessible to us, and spacecraft could travel through it without disappearing into the black hole. According to detailed calculations, objects grazing the ergoregion could be catapulted back out into space at tremendous speeds. In other words, the ejected object could leave the ergoregion with more energy than it had initially, having extracted added energy from the hole's rotation. This is called the *Penrose process*, after Roger Penrose, the British mathematician who proposed it.

Mass, charge, and angular momentum are the only three properties that a black hole possesses. This simplicity is the essence of the **no-hair theorem**, first formulated in the early 1970s: "Black holes have no hair." Once matter has fallen into the hole, any and all additional properties ("hair") carried by the matter have disappeared from the universe. Hence, these properties can have no effect on the structure of the hole.

24-7 | Falling into a black hole is an infinite voyage

WEB LINK 24.11 Imagine that you are on board a spacecraft a safe distance from a 5-M$_\odot$ black hole. A distance of 1000 Schwarzschild radii, or 15,000 km, would suffice. You now release a space probe and let it fall into the black hole. To make the probe easier to track, it is coated with a phosphorescent paint that emits a blue glow. You also equip the probe with a video camera that transmits images of the view of the approaching black hole. What will you see as the probe falls?

As Figure 24-17a shows, at 1000 Schwarzschild radii from the black hole the video camera would send back a rather normal view of space. But as the probe approaches the black hole, the bending of light by the black hole (see Section 24-2) becomes more pronounced. Light rays passing close to the back hole are deflected

so severely that at close range, the camera will actually see multiple images of the same stars (Figure 24-17b).

How will the probe itself look from your vantage point at a safe distance? You might expect that as the probe falls, its speed should continue to increase. This is true up to a point. But as the probe approaches the event horizon, where the black hole's gravity is extremely strong, the gravitational slowing of time becomes so pronounced that the probe will appear to slow down! From your point of view, the spacecraft takes an infinite amount of time to reach the event horizon, where it will appear to remain suspended for all eternity.

To watch these effects, however, you will need special equipment. The reason is the gravitational redshift: As the probe falls, the light reaching you from the probe is shifted to longer and longer wavelengths. The probe's blue glow will turn green, then yellow, then red, and eventually fade into infrared wavelengths that your unaided eye cannot detect.

If you could view the falling probe with infrared goggles, you would see it come to an unpleasant end. Near the event horizon, the strength of the black hole's gravity increases dramatically as the probe moves just a short distance closer to the hole. In fact, the side of the probe nearest the black hole feels a much stronger gravitational pull than the side opposite the hole. These *tidal forces* are like those that the Moon exerts on the Earth (see Section 4-8), but tremendously stronger. Furthermore, the sides of the probe are pulled together, since the hole's gravity makes them fall in straight lines aimed at the center of the hole. The net effect is that the probe will be stretched out along the line pointing toward the hole, and

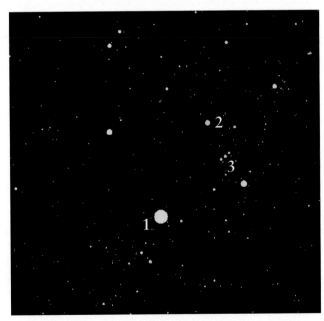

(a) Looking directly toward the black hole from a distance of 1000 Schwarzschild radii: Note positions of stars 1, 2, and 3.

(b) Looking directly toward the black hole from a distance of 10 Schwarzschild radii: Light bending causes multiple images.

Figure 24-17

The View Approaching a Black Hole These images are taken from a computer simulation of the appearance of the sky behind a black hole. **(a)** The bending of light by the black hole is negligible at large distances, but **(b)** becomes evident as you approach the black hole. (Courtesy Robert J. Nemiroff, Michigan Technological University)

Probe far from black hole   Probe approaching black hole

Black hole

(a)   (b)   (c)   (d)

Event horizon

Figure 24-18

Falling into a Black Hole (a) A cube-shaped probe is dropped from a distance of 1000 Schwarzschild radii from a 5-$M_\odot$ black hole. (b), (c), (d) As the probe approaches the event horizon, it is distorted into a long, thin shape by the black hole's extreme gravity. A distant observer sees the probe change color as photons from the probe undergo a strong gravitational redshift.

squeezed together along the perpendicular directions (**Figure 24-18**). The stretching is so great that it will rip the probe into atoms, and even rip the atoms apart.

Someone foolhardy enough to ride along with the probe would have a very different view of the fall than yours. From this astronaut's point of view, there is no slowing of time, and the probe continues its fall through the event horizon into the singularity. The astronaut is unlikely to enjoy the ride, however, since he will be disintegrated by tidal forces as he falls inward toward the singularity.

Could a black hole somehow be connected to another part of spacetime, or even some other universe? General relativity predicts that such connections, called **wormholes,** can exist for rotating black holes. **Box 24-3** discusses the possibility that such wormholes really exist.

---

box 24-3   THE HEAVENS ON THE EARTH

Wormholes and Time Machines

Wormholes have become a staple of science fiction. They are certainly a convenient plot device: The gallant crew flies their spaceship into one end of a wormhole, and a short time later they emerge many light-years away. Time machines, too, have been part of science fiction ever since H. G. Wells published his classic novel *The Time Machine* at the end of the nineteenth century. What would it be like to go back in time and watch famous historical events—or even change history?

Wormholes and time machines challenge our normal ideas about space and time. So, too, does the general theory of relativity. Could it be that this theory really makes it possible to travel through wormholes and to travel in time?

In the 1930s, Einstein and his colleague Nathan Rosen discovered that a black hole could possibly connect our universe with a second domain of space and time that is separate from ours. The first diagram shows this connection, called an *Einstein-Rosen bridge*. You could think of the upper surface as our universe and the lower surface as a "parallel universe." Alternatively, the upper and lower surfaces could be different regions of our own universe. In that case, an Einstein-Rosen bridge would connect our universe with itself, forming a wormhole, shown in the second diagram.

A wormhole may seem like a shortcut to distant places in our universe, but detailed calculations reveal a major obstacle. The powerful gravity of a black hole causes the wormhole to collapse almost as soon as it forms. As a result, to get from one side of a wormhole to the other, you would have to travel faster than the speed of light, which is not possible.

Caltech physicists Kip Thorne, Michael Morris, and Ulvi Yurtsever have proposed a scheme that might get around the difficulty of a collapsing wormhole. According to general relativity, they point out, pressure as well as mass can be a source of gravity. Normally we do not see the gravitational effects of pressure because they are so small. Thorne and his colleagues speculate that a technologically advanced civilization might someday be able to use pressure to produce *antigravity* strong enough to keep the wormhole open.

If a wormhole could be held open, it could also be a time machine. To see how, imagine you take one end of a wormhole and move it around for a while at speeds very near the speed of light. As we saw in Box 24-1, such motion causes clocks to slow down. Thus, when you stop moving that end of the wormhole, you find that it has not aged as much as the stationary end. In other words, one side of the wormhole has a different time from the other. As a result, you could go into one end of the wormhole at a late time and come out at an early time. For example, you might go in at 10 A.M. and come out at 9 A.M.

Time machines challenge ordinary logic. If you could get back from a trip an hour before you left, you could meet yourself and tell yourself what a nice journey you had. Then

24-8 | Black holes evaporate

With all the mass of a black hole hidden behind its event horizon and collapsed into a singularity, it may seem that there is no way of getting mass from the black hole back out into the universe. But in fact there is, as Stephen Hawking pointed out in the 1970s. To understand how this is possible, we must look at the behavior of matter at the submicroscopic scale of nuclei, and electrons.

The **Heisenberg uncertainty principle** is a basic tenet of submicroscopic physics. This principle states that you cannot determine precisely both the position and the speed of a subatomic particle. Over extremely short distances or times, a certain amount of "fuzziness" is built into the nature of reality.

The Heisenberg uncertainty principle leads to the concept of **virtual pairs**: At every point in space, pairs of particles and antiparticles are constantly being created and destroyed. An antiparticle is quite like an ordinary particle except that it has an opposite electric charge and can annihilate an ordinary particle so that both disappear, usually leaving a pair of photons in their place. In the case of virtual pairs, the process of creation and annihilation occurs over such incredibly brief time intervals that these virtual particles and antiparticles cannot be observed directly.

Think about a tiny black hole. Furthermore, think about the momentary creation of a virtual pair of an electron and a positron (the antiparticle of an electron) just outside the hole's event horizon. It may happen that one of the particles falls into the black hole. Its partner is then deprived of a counterpart with which to annihilate and must therefore become a real particle. To accomplish this conversion, some of the energy of the black hole's gravity must be converted into matter, according to $E = mc^2$. This decreases the mass of the black hole by a corresponding amount, and the particle is free to escape from the hole. In this way, particles can quantum-mechanically "leak" out of a black hole, carrying some of the hole's mass with them (**Figure 24-19**).

The less mass a black hole has, the more easily particles can leak out through its event horizon. Jacob Bekenstein and Stephen Hawking proved mathematically that you can speak of the *temperature* of a black hole as a way of describing the amounts of energy carried away by particles leaking out of it. For example, a 1-trillion-ton (10^{15} kg) black hole emits particles and energy as if it were a blackbody with a temperature of nearly 10^9 K.

For stellar-mass black holes, such as Cygnus X-1, this effect is negligible over time spans of billions of years. The reason is that the temperature of a black hole is inversely proportional to the black hole's mass. Compared to a trillion-ton black hole, a 5-$M_\odot$ (10^{31} kg)

THE HEAVENS ON THE EARTH continued

both of you could take the trip. If you and your twin return just before you both left, there would be four of you. And all four of you could take the trip again. And then all eight of you. Then all sixteen of you.

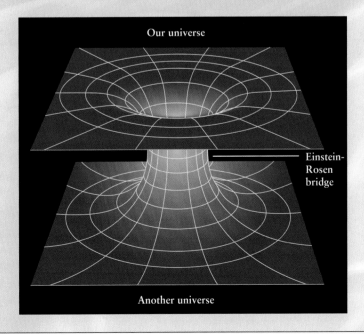

Our universe

Einstein-Rosen bridge

Another universe

Making copies of yourself is an example of how time machines violate *causality*, the notion that effects must follow their causes. To date, we have never seen a violation of causality. Many scientists would therefore like to show that time machines cannot exist. The British astrophysicist Stephen Hawking points to one strong bit of observational evidence against time machines: We are not being visited by hordes of tourists from the future. If we could discover why nature seems to preclude time machines, we would have a much deeper understanding of the nature of space and time.

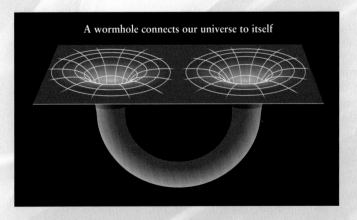

A wormhole connects our universe to itself

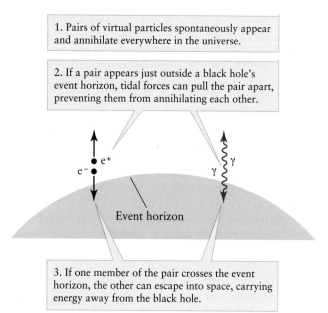

1. Pairs of virtual particles spontaneously appear and annihilate everywhere in the universe.

2. If a pair appears just outside a black hole's event horizon, tidal forces can pull the pair apart, preventing them from annihilating each other.

e^- • • e^+ γ γ

Event horizon

3. If one member of the pair crosses the event horizon, the other can escape into space, carrying energy away from the black hole.

Figure 24-19

Evaporation of a Black Hole This illustration shows two pairs of virtual particles—an electron (e^-) and a positron (e^+), and a pair of photons (γ)—appearing just outside the event horizon of a black hole. If one member of the pair escapes and carries energy away from the black hole, the black hole decreases in mass and the event horizon shrinks.

hole has a mass that is 10^{16} times greater and a temperature that is only 10^{-16} as much:

$$10^{-16} \times 10^9 \text{ K} = 10^{-7} \text{ K}$$

This temperature is barely above absolute zero. In other words, ordinary black holes have such low temperatures that particles hardly ever manage to escape from them.

The story is different for low-mass black holes, such as the primordial black holes we discussed in Section 24-4. As particles escape from a small black hole, the mass of the black hole decreases, making its temperature go up. As its temperature rises, still more particles escape, further decreasing the hole's mass and forcing the temperature still higher. This runaway process of **black hole evaporation** finally causes the hole to vanish altogether. During its final seconds of evaporation, the hole gives up the last of its mass in a violent burst of energy equal to the detonation of a billion megaton hydrogen bombs.

A 10^{10}-kg primordial black hole (comparable to the mass of Mount Everest) would take about 15 billion years to evaporate. This is close to the present age of the universe. If primordial black holes were formed in the Big Bang, we would expect to see some of them going through the explosive final stages of evaporation. To date, however, there is no compelling evidence that we have seen such evaporations taking place, and many astronomers suspect that there are no primordial black holes.

By contrast, a 5-$M_\odot$ black hole would take more than 10^{62} years to evaporate, and a supermassive black hole of 5 million solar masses would take more than 10^{80} years. These time spans are far longer than the age of the universe. We can safely predict that the black holes we have observed to date will remain as black holes for the foreseeable future.

Key Words

Terms preceded by an asterisk (*) are discussed in the Boxes.

black hole, p. 539
black hole evaporation, p. 550
equivalence principle, p. 536
ergoregion, p. 546
event horizon, p. 544
general theory of relativity, p. 536
gravitational radiation (gravitational waves), p. 538
gravitational redshift, p. 538
Heisenberg uncertainty principle, p. 549
law of cosmic censorship, p. 545
length contraction, p. 532
*Lorentz transformations, p. 534
mid-mass black hole, p. 543

no-hair theorem, p. 547
primordial black hole, p. 544
*proper length (proper distance), p. 535
*proper time, p. 534
Schwarzschild radius (R_{Sch}), p. 544
singularity, p. 544
spacetime, p. 532
special theory of relativity, p. 532
stellar-mass black hole, p. 540
supermassive black hole, p. 542
time dilation, p. 533
virtual pairs, p. 549
wormhole, p. 548

Key Ideas

The Special Theory of Relativity: This theory, published by Einstein in 1905, is based on the notion that there is no such thing as absolute space or time.

• The speed of light is the same to all observers, no matter how fast they are moving.

• An observer will note a slowing of clocks and a shortening of rulers that are moving with respect to the observer. This effect becomes significant only if the clock or ruler is moving at a substantial fraction of the speed of light.

• Space and time are not wholly independent of each other, but are aspects of a single entity called spacetime.

The General Theory of Relativity: Published by Einstein in 1915, this is a theory of gravity. Any massive object causes space to curve and time to slow down, and these effects manifest themselves as a gravitational force. These distortions of space and time are most noticeable in the vicinity of large masses or compact objects.

• The general theory of relativity is our most accurate description of gravitation. It predicts a number of phenomena, including the bending of light by gravity and the gravitational redshift, whose existence has been confirmed by observation and experiment.

• The general theory of relativity also predicts the existence of gravitational waves, which are ripples in the overall geometry of space and time produced by moving masses. Gravitational waves have been detected indirectly, and specialized antennas are under construction to make direct measurement of the gravitational waves from cosmic cataclysms.

Black Holes: If a stellar corpse has a mass greater than about 2 to 3 $M_\odot$, gravitational compression will overwhelm any and all forms of internal pressure. The stellar corpse will collapse to such a high density that its escape speed exceeds the speed of light.

Observing Black Holes: Black holes have been detected using indirect methods.

• Some binary star systems contain a black hole. In such a system, gases captured from the companion star by the black hole emit detectable X rays.

• Many galaxies have supermassive black holes at their centers. These are detected by observing the motions of material around the black hole.

Properties of Black Holes: The entire mass of a black hole is concentrated in an infinitely dense singularity.

• The singularity is surrounded by a surface called the event horizon, where the escape speed equals the speed of light. Nothing—not even light—can escape from inside the event horizon.

• A black hole has only three physical properties: mass, electric charge, and angular momentum.

• A rotating black hole (one with angular momentum) has an ergoregion around the outside of the event horizon. In the ergoregion, space and time themselves are dragged along with the rotation of the black hole.

• Black holes can evaporate, but in most cases at an extremely slow rate.

Review Questions

1. If you drop a ball inside a car traveling at a steady 50 km/h in a straight line on a smooth road, does it fall in the same way as if the car was stationary? How does this question relate to Einstein's special theory of relativity?

2. In Einstein's special theory of relativity, two different observers moving at different speeds will measure the same value of the speed of light. Will these same observers measure the same value of, say, the speed of an airplane? Explain.

3. Serena flies past Michael in her spaceship at nearly the speed of light. According to Michael, Serena's clock runs slow. According to Serena, does Michael's clock run slow, fast, or at the normal rate? Explain.

4. Ole flies past Lena in a spherical spaceship at nearly the speed of light. According to Lena, how does the distance from the front to the back of Ole's spaceship (that is, measured along the direction of motion) compare to the distance from the top to the bottom (that is, measured perpendicular to the direction of motion)? Explain.

5. Why does the speed of light represent an ultimate speed limit?

6. Why is Einstein's general theory of relativity a better description of gravity than Newton's universal law of gravitation? Under what circumstances is Newton's description of gravity adequate?

7. Describe two different predictions of the general theory of relativity and how these predictions were tested experimentally. Do the results of the experiments agree with the theory?

8. How does a gravitational redshift differ from a Doppler shift?

9. In what circumstances are degenerate electron pressure and degenerate neutron pressure incapable of preventing the complete gravitational collapse of a dead star?

10. Should we worry about the Earth being pulled into a black hole? Why or why not?

11. How does rapid flickering of an X-ray source provide evidence that the source is small?

12. All the stellar-mass black hole candidates mentioned in the text are members of very short-period binary systems. Explain how this makes it possible to detect the presence of the black hole.

13. Astronomers cannot actually see the black hole candidates in close binary systems. How, then, do they know that these candidates are not white dwarfs or neutron stars?

14. Describe two ways in which a member of a binary star system could become a black hole.

15. How do astronomers locate supermassive black holes in galaxies?

16. What is a mid-mass black hole? How are such objects thought to form?

17. When we say that the Moon has a radius of 1738 km, we mean that this is the smallest radius that encloses all of the Moon's material. In this sense, is it correct to think of the Schwarzschild radius as the radius of a black hole? Why or why not?

18. A twenty-third-century instructor at Starfleet Academy tells her students, "If someday your starship falls into a black hole, it'll be your own fault." Explain why it would require careful piloting to direct a spacecraft into a black hole.

19. In what way is a black hole blacker than black ink or a black piece of paper?

20. If the Sun suddenly became a black hole, how would the Earth's orbit be affected? Explain.

21. According to the general theory of relativity, why can't some sort of yet-undiscovered degenerate pressure prevent the matter inside a black hole from collapsing all the way down to a singularity?

22. What is the law of cosmic censorship?

23. Is it possible to tell which chemical elements went into a black hole? Why or why not?

24. Why is it unlikely that a black hole has an electric charge?

25. What kind of black hole is surrounded by an ergoregion? What happens inside the ergoregion?

26. What is the no-hair theorem?

27. As seen by a distant observer, how long does it take an object dropped from a great distance to fall through the event horizon of a black hole? Explain.

28. If even light cannot escape from a black hole, how is it possible for black holes to evaporate?

Advanced Questions

Questions preceded by an asterisk (*) involve topics discussed in the Boxes.

Problem-solving tips and tools

Remember that the time to travel a certain distance is equal to the distance divided by the speed, and that the density of an object is its mass divided by its volume. The volume of a sphere of radius r is $4\pi r^3/3$. Section 4-7 describes Newton's law of universal gravitation. Box 4-4 shows how to use Newton's formulation of Kepler's third law, which explicitly includes masses; when using this formula, note that the period P must be expressed in seconds, the semimajor axis a in meters, and the masses in kilograms. For another version of this formula, in which period is in years, semimajor axis in AU, and masses in solar masses, see Section 19-9.

---

*29. A spaceship flies from Earth to a distant star at a constant speed. Upon arrival, a clock on board the spaceship shows a total elapsed time of 8 years for the trip. An identical clock on the Earth shows that the total elapsed time for the trip was 10 years. What was the speed of the spaceship relative to the Earth?

*30. An unstable particle called a positive pion (pronounced "pie-on") decays in an average time of 2.6×10^{-8} s. On average, how long will a positive pion last if it is traveling at 95% of the speed of light?

*31. How fast should a meter stick be moving in order to appear to be only 60 cm long?

*32. An astronaut flies from the Earth to a distant star at 80% of the speed of light. As measured by the astronaut, the one-way trip takes 15 years. (**a**) How long does the trip take as measured by an observer on the Earth? (**b**) What is the distance from the Earth to the star (in light-years) as measured by an Earth observer? As measured by the astronaut?

33. In the binary system of two neutron stars discovered by Hulse and Taylor (Section 24-2), one of the neutron stars is a pulsar. The distance between the two stars varies between 1.1 and 4.8 times the radius of the Sun. The time interval between pulses from the pulsar is not constant: It is greatest when the two stars are closest to each other and least when the two stars are farthest apart. Explain why this is consistent with the gravitational slowing of time (Figure 24-7a).

34. Find the total mass of the neutron star binary system discovered by Hulse and Taylor (Section 24-2), for which the orbital period is 7.75 hours and the average distance between the neutron stars is 2.8 solar radii. Is your result reasonable for a pair of neutron stars? Explain.

35. Estimate how long it will be until the two neutron stars that make up the binary system discovered by Hulse and Taylor collide with each other. Assume that the distance between the two stars will continue to decrease at its present rate of 3 mm every 7.75 hours, and use the data given in Question 34. (You can assume that the two stars are very small, so they will collide when the distance between them is equal to zero.)

36. The orbital period of the binary system containing A0620-00 is 0.32 day, and Doppler shift measurements reveal that the radial velocity of the X-ray source peaks at 457 km/s (about 1 million miles per hour). (**a**) Assuming that the orbit of the X-ray source is a circle, find the radius of its orbit in kilometers. (This is actually an estimate of the semimajor axis of the orbit.) (**b**) By using Newton's form of Kepler's third law, prove that the mass of the X-ray source must be at least 3.1 times the mass of the Sun. (*Hint:* Assume that the mass of the K5V visible star—about 0.5 $M_\odot$ from the mass-luminosity relationship—is negligible compared to that of the invisible companion.)

37. Find the orbital period of a star moving in a circular orbit of radius 500 AU around the supermassive black hole in the galaxy NGC 4261 (Section 24-4).

*38. Find the Schwarzschild radius for an object having a mass equal to that of the planet Saturn.

*39. What is the Schwarzschild radius of a black hole whose mass is that of (**a**) the Earth, (**b**) the Sun, (**c**) the supermassive black hole in NGC 4261 (Section 24-4)? In each case, also calculate what the density would be if the matter were spread uniformly throughout the volume of the event horizon.

*40. What is the mass in kilograms of a black hole whose Schwarzschild radius is 11 km?

*41. To what density must the matter of a dead 8-$M_\odot$ star be compressed in order for the star to disappear inside its event horizon? How does this compare with the density at the center of a neutron star, about 3×10^{18} kg/m^3?

*42. Prove that the density of matter needed to produce a black hole is inversely proportional to the square of the mass of the hole. If you wanted to make a black hole from matter compressed to the density of water (1000 kg/m^3), how much mass would you need?

Discussion Questions

43. The speed of light is the same for all observers, regardless of their motion. Discuss why this requires us to abandon the Newtonian view of space and time.

44. Describe the kinds of observations you might make in order to locate and identify black holes.

45. Speculate on the effects you might encounter on a trip to the center of a black hole (assuming that you could survive the journey).

Web/CD-ROM Questions

46. Search the World Wide Web for information about a stellar-mass black hole candidate named V4641 Sgr. In what ways does it resemble other black hole candidates such as Cygnus X-1 and V404 Cygni? In what ways is it different and more dramatic? How do astronomers explain why V4641 Sgr is different?

47. Search the World Wide Web for information about the mid-mass black hole candidate in M82. Is this still thought to be a mid-mass black hole? What new evidence has been used to either support or oppose the idea that this object is a mid-mass black hole?

48. The Equivalence Principle. Access the animation "The Equivalence Principle" in Chapter 24 of the *Universe* Web site or CD-ROM. View the animation and answer the following questions. (**a**) Describe what is happening as viewed from the frame of reference of the elevator. What causes the apple to fall to the floor of each elevator? (**b**) Describe what is happening as viewed from the frame of reference of the stars. What causes the apple to fall to the floor of each elevator? (**c**) Think of another experiment you could perform with the apple. Describe what would happen during this experiment as seen by Newton (in the left-hand box) and by Einstein (in the right-hand box).

Observing Projects

49. You cannot see a black hole with a telescope. Nevertheless, you might want to observe the visible companion of Cygnus X-1. The epoch 2000 coordinates of this ninth-magnitude star are R.A. = 19^h 58.4^m and Decl. = +35° 12′, which is quite near the bright star η (eta) Cygni. Compare what you see with the photograph in Figure 24-10.

50. Use the *Starry Night Backyard*™ program to plan observations of Cygnus X-1. Center the field of view on Cygnus X-1 (select **Find...** in the **Edit** menu). If Cygnus X-1 is below the horizon, allow the program to reset the time to when it can best be seen. Using the controls at the right-hand end of the Control Panel, zoom out to the maximum field of view. (**a**) Using the time controls in the Control Panel, step through time and determine when Cygnus X-1 rises and sets on today's date from your location. (**b**) At approximately what time on today's date is Cygnus X-1 highest in the sky? Is tonight a good night for observing this star with a visible-light telescope? Would it be better placed in the sky for observation six months from now? Explain how you determined this.

51. Use the *Deep Space Explorer*™ program to examine galaxies with supermassive black holes. In the left-hand part of the window under the heading **Milky Way**, click on the triangle next to the word **Explore** and then click on **Sun in Milky Way**. Use the **Find...** command in the **Edit** menu to examine each of the following galaxies: (i) NGC 4261; (ii) M87; (iii) M31. Do you see any common features of these three galaxies? Suggest why advances in telescope technology were required before supermassive black holes could be discovered in these galaxies.

Collaborative Exercise

52. Using Einstein's theory of relativity, estimate (1) the length of your pencil or pen at constant speed at the speeds of a bicycle rider, a car traveling on the highway and a commerical jet liner at cruising altitude; and (2) the speed of a light beam emitted by a spaceshift traveling at 200,000 kilometers per second toward another spaceship traveling at the same speed.

25 Our Galaxy

On a clear, moonless night, away from the glare of city lights, you can often see a hazy, luminous band stretching across the sky. This band, called the Milky Way, extends all the way around the celestial sphere. The accompanying photograph is centered on the brightest part of the Milky Way, in the constellation Sagittarius.

Galileo, the first person to view the Milky Way with a telescope, discovered that it is composed of countless dim stars. Today, we realize that the Milky Way is actually a disk tens of thousands of parsecs across containing hundreds of billions of stars—one of which is our own Sun—as well as vast quantities of gas and dust.

Recently, astronomers have discovered that most of the Milky Way's mass is not in its stars, but in a halo of *dark matter* that emits no measurable radiation. The character of this dark matter remains mysterious. Even more mysterious is the very center of the Milky Way, where radio, infrared, and X-ray observations reveal the presence of a black hole with a mass of 3.7 million Suns.

The Milky Way is just one of myriad *galaxies,* or systems of stars and interstellar matter, that are spread across the observable universe. By studying our home Milky Way Galaxy, we begin to explore the universe on a grand scale. Instead of focusing on individual stars, we look at the overall arrangement and history of a huge stellar community of which the Sun is a member. In this way, we gain insights into galaxies in general and prepare ourselves to ask fundamental questions about the cosmos.

 This wide-angle image shows the Milky Way from the constellation Cygnus (at left) to the constellation Crux (at right). (Dirk Hoppe)   R I **V** U X G

Questions to Guide Your Reading

As you read the sections of this chapter, look for the answers to the following questions:

25-1 What is our Galaxy? How do astronomers know where we are located within it?

25-2 What is the shape and size of our Galaxy?

25-3 How do we know that our Galaxy has spiral arms?

25-4 What is most of the Galaxy made of? Is it stars, gas, dust, or something else?

25-5 What is the nature of the spiral arms?

25-6 What lies at the very center of our Galaxy?

25-1 | The Sun is located in the disk of our Galaxy, about 8000 parsecs from the galactic center

SCIENCE IN PROCESS

Observations of pulsating variable stars revealed the immense size of the Milky Way

Eighteenth-century astronomers were the first to suspect that because the Milky Way completely encircles us, all the stars in the sky are part of an enormous disk of stars—the **Milky Way Galaxy**. As we learned in Section 1-4, a **galaxy** is an immense collection of stars and interstellar matter, far larger than a star cluster. We have an edge-on view from inside the disk of our own Milky Way Galaxy, which is why the Milky Way appears as a band around the sky (**Figure 25-1**).

But where within this disk is our own Sun? Until the twentieth century, the prevailing opinion was that the Sun and planets lie at the Galaxy's center. One of the first to come to this conclusion was the eighteenth-century English astronomer William Herschel, who discovered the planet Uranus and was a pioneering cataloger of binary star systems (see Section 19-9). Herschel's approach to determining the Sun's position within the Galaxy was to count the number of stars in each of 683 regions of the sky. He reasoned that he should see the greatest number of stars toward the Galaxy's center and a lesser number toward the Galaxy's edge.

Herschel found approximately the same density of stars all along the Milky Way. Therefore, he concluded that we are at the center of our Galaxy (**Figure 25-2**). In the early 1900s, the Dutch astronomer Jacobus Kapteyn came to essentially the same conclusion by analyzing the brightness and proper motions of a large number of stars. According to Kapteyn, the Milky Way is about 17 kpc (17 kiloparsecs = 17,000 parsecs, or 55,000 light-years) in diameter, with the Sun near its center.

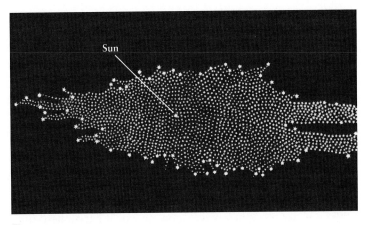

figure 25-2

Herschel's Map of Our Galaxy In a paper published in 1785, the English astronomer William Herschel presented this map of the Milky Way Galaxy. He determined the Galaxy's shape by counting the numbers of stars in various parts of the sky. Herschel's conclusions were flawed because interstellar dust blocked his view of distant stars, leading him to the erroneous idea that the Sun is at the center of the Galaxy. (Yerkes Observatory)

Both Herschel and Kapteyn were wrong about the Sun being at the center of our Galaxy. The reason for their mistake was finally discerned in 1930 by Robert J. Trumpler of Lick Observatory. While studying star clusters, Trumpler discovered that the more remote clusters appear unusually dim—more so than would be expected from their distances alone. As a result, Trumpler concluded that interstellar space must not be a perfect vacuum: It must contain dust that absorbs or scatters light from distant stars.

(a)

(b)

R I **V** U X G

figure 25-1

Our View of the Milky Way (a) The Milky Way Galaxy is a disk-shaped collection of stars. When we look out at the night sky in the plane of the disk, the stars appear as a band of light that stretches all the way around the sky. When we look perpendicular to the plane of the Galaxy, we see only those relatively few stars that lie between us and the "top" or "bottom" of the disk. (b) This wide-angle photograph shows a 180° view of the Milky Way centered on the constellation Sagittarius (compare with the photograph that opens this chapter). The dark streaks across the Milky Way are due to interstellar dust in the plane of our Galaxy. (Courtesy of Dennis di Cicco)

Like the stars themselves, interstellar dust is concentrated in the plane of the Galaxy (see Section 20-2). As a result, it obscures our view within the plane and makes distant objects appear dim, an effect called **interstellar extinction**. Great patches of interstellar dust are clearly visible in wide-angle photographs such as the one that opens this chapter. Thanks to interstellar extinction, Herschel and Kapteyn were actually seeing only the nearest stars in the Galaxy. Hence, they had no idea of either the enormous size of the Galaxy or of the vast number of stars concentrated around the galactic center.

ANALOGY Herschel and Kapteyn faced much the same dilemma as a lost motorist on a foggy night. Unable to see more than a city block in any direction, he would have a hard time deciding what part of town he was in. If the fog layer were relatively shallow, however, our motorist would be able to see the lights from tall buildings that extend above the fog, and in that way he could determine his location (**Figure 25-3a**).

The same principle applies to our Galaxy. While interstellar dust in the plane of our Galaxy hides the sky covered by the Milky Way, we have an almost unobscured view out of the plane (that is, to either side of the Milky Way). To find our location in the Galaxy, we need to locate bright objects that are part of the Galaxy but lie outside its plane in unobscured regions of the sky.

WEB LINK 25.2 Fortunately, bright objects like this do exist. They are the **globular clusters**, a class of star clusters associated with the Galaxy but which lie outside its plane (Figure 25-3b). As we saw in Section 21-3, a typical globular cluster is a spherical distribution of roughly 10^6 stars packed in a volume only a few hundred light-years across (see Figure 21-10).

However, to use globular clusters to determine our location in the Galaxy, we must first determine the distances from the Earth to these clusters. (Think again of our lost motorist—glimpsing the lights of a skyscraper through the fog is useful to the motorist only if he can tell how far away the skyscraper is.) Pulsating variable stars in globular clusters provide the distances, giving astronomers the key to the dimensions of our Galaxy.

In 1912, the American astronomer Henrietta Leavitt reported her important discovery of the period-luminosity relation for Cepheid variables. As we saw in Section 21-5, Cepheid variables are pulsating stars that vary periodically in brightness (see Figure 21-16). Leavitt studied numerous Cepheids in the Small Magellanic Cloud (a small galaxy near the Milky Way) and found their periods to be directly related to their average luminosities. As **Figure 25-4** shows, the longer a Cepheid's period, the greater its luminosity.

The period-luminosity law is an important tool in astronomy because it can be used to determine distances. For example, suppose you find a Cepheid variable in the sky. By measuring its period and using a graph like Figure 25-4, you can determine the star's average luminosity. Knowing the star's average luminosity, you can find out how far away the star must be in order to give the observed brightness. (Box 19-2 explains how this is done.)

Shortly after Leavitt's discovery of the period-luminosity law, Harlow Shapley, a young astronomer at the Mount Wilson Observatory in California, began studying a family of pulsating stars closely related to Cepheid variables called **RR Lyrae variables**. The light curve of an RR Lyrae variable is similar to that of a Cepheid, but RR Lyrae variables have shorter pulsation periods and lower peak luminosities (see Figure 25-4).

The tremendous importance of RR Lyrae variables is that they are commonly found in globular clusters (**Figure 25-5**). By using the period-luminosity relationship for these stars, Shapley was able to

(a) Determining your position in the fog

> A thin layer of fog makes it difficult to see very far horizontally, but you can still see objects above the fog layer.

Fog

(b) Determining your position in the Galaxy

> Interstellar dust makes it difficult to see very far within the plane of the Galaxy, but you can still see objects move above the galactic plane.

Globular cluster

Sun

Earth

Disk of the Milky Way Galaxy

Center of Galaxy

figure 25-3
Finding the Center of the Galaxy (a) A motorist lost on a foggy night can determine his location by looking for tall buildings that extend above the fog. (b) In the same way, astronomers determine our location in the

Galaxy by observing globular clusters that are part of the Galaxy but lie outside the obscuring material in the galactic disk. The globular clusters form a spherical halo centered on the center of the Galaxy.

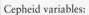

Cepheid variables:
• Found throughout the Galaxy
• Pulsation periods of 1 to 50 days
• Luminosity related to pulsation period

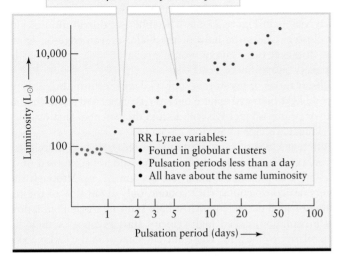

RR Lyrae variables:
• Found in globular clusters
• Pulsation periods less than a day
• All have about the same luminosity

figure 25-4
Period and Luminosity for Cepheid and RR Lyrae Variables
This graph shows the relationship between period and luminosity for
Cepheid variables and RR Lyrae variables. Cepheids come in a broad
range of luminosities: The more luminous the Cepheid, the longer its
pulsation period. By contrast, RR Lyrae variables are horizontal branch
stars that all have roughly the same average luminosity of about 100 $L_\odot$.

figure 25-5   R I **V** U X G
RR Lyrae Variables in a Globular Cluster The arrows
point to three RR Lyrae variables in the globular cluster M55, located in the
constellation Sagittarius. From the average apparent brightness (as seen in
this photograph) and average luminosity (known to be roughly 100 $L_\odot$) of
these variable stars, astronomers have deduced that the distance to M55 is
6500 pc (20,000 ly). (Harvard Observatory)

determine the distances to the 93 globular clusters then known. He
found that some of them were more than 100,000 light-years from
Earth. The large values of these distances immediately suggested that
the Galaxy was much larger than Herschel or Kapteyn had thought.

Another striking property of globular clusters is how they are distributed across the sky. Ordinary stars and open clusters of stars (see
Section 20-6, especially Figures 20-17 and 20-18) are rather uniformly spread along the Milky Way. However, the majority of the
93 globular clusters that Shapley studied are located in one-half of
the sky, widely scattered around the portion of the Milky Way that
is in the constellation Sagittarius.

From the directions to the globular clusters and their distances
from us, Shapley mapped out the three-dimensional distribution of
these clusters in space. In 1920 he concluded that the globular clusters form a huge spherical distribution centered not on the Earth but
rather about a point in the Milky Way several kiloparsecs away in
the direction of Sagittarius (see Figure 25-3*b*). This point, reasoned
Shapley, must coincide with the center of our Galaxy, because of
gravitational forces between the disk of the Galaxy and the "halo"
of globular clusters. Therefore, by locating the center of the distribution of globular clusters, Shapley was in effect measuring the location of the galactic center.

Since Shapley's pioneering observations, many astronomers have
measured the distance from the Sun to the **galactic nucleus,** the center of our Galaxy. Shapley's estimate of this distance was too large
by about a factor of 2, because he did not take into account the
effects of interstellar extinction (which were not well understood at
the time). Today, the generally accepted distance to the galactic

nucleus is about 8 kpc (26,000 ly); the actual distance could be
greater or less than that value by about 1 kpc (3300 ly).

Just as Copernicus and Galileo showed that the Earth was not
at the center of the solar system, Shapley and his successors
showed that the solar system lies nowhere near the center of the
Galaxy. We see that the Earth indeed occupies no special position
in the universe.

25-2 | Observations at nonvisible wavelengths reveal the shape of the Galaxy

At visible wavelengths, light suffers so much interstellar extinction
that the galactic nucleus is totally obscured from view. But the
amount of interstellar extinction is roughly inversely proportional
to wavelength. In other words, the longer the wavelength, the farther radiation can travel through interstellar dust without being
scattered or absorbed. As a result, we can see farther into the plane
of the Milky Way at infrared wavelengths than at visible wavelengths, and radio waves can traverse the Galaxy freely. For this
reason, telescopes sensitive to these nonvisible wavelengths are
important tools for studying the structure of our Galaxy.

Infrared light is particularly useful for tracing the location
of interstellar dust in the Galaxy. Starlight warms the dust
grains to temperatures in the range of 10 to 90 K; thus,
in accordance with Wien's law (see Section 5-4), the dust emits
radiation predominately at wavelengths from about 30 to 300 μm.

These are called **far-infrared** wavelengths, because they lie in the part of the infrared spectrum most different in wavelength from visible light (see Figure 5-7). At these wavelengths, interstellar dust radiates more strongly than stars, so a far-infrared view of the sky is principally a view of where the dust is. In 1983 the Infrared Astronomical Satellite (IRAS) scanned the sky with a 60-cm reflecting telescope at far-infrared wavelengths, giving the panoramic view of the Milky Way's dust shown in Figure 25-6a.

In 1990 an instrument on the Cosmic Background Explorer (COBE) satellite scanned the sky at **near-infrared** wavelengths, that is, relatively short wavelengths closer to the visible spectrum. Figure 25-6b shows the resulting near-infrared view of the plane of the Milky Way. At near-infrared wavelengths, interstellar dust does not emit very much light. Hence, the light sources in Figure 25-6b are stars, which do emit strongly in the near-infrared (especially the cool stars, such as red giants and supergiants). Because interstellar dust causes little interstellar extinction in the near-infrared, many of the stars whose light is recorded in Figure 25-6b lie deep within the Milky Way.

Observations such as those shown in Figure 25-6, along with the known distance to the center of the Galaxy, have helped astronomers to establish the dimensions of the Galaxy. The **disk** of our Galaxy is about 50 kpc (160,000 ly) in diameter and about 0.6 kpc (2000 ly) thick, as shown in Figure 25-7. The center of the Galaxy is surrounded by a distribution of stars, called the **central bulge**, which is about 2 kpc (6500 ly) in diameter. This central bulge is clearly visible in Figure 25-6b. The spherical distribution of globular clusters traces the **halo** of the Galaxy. Seen edge-on from a great distance, our Galaxy would probably look somewhat like the galaxy shown in Figure 25-8.

Observations suggest that our Galaxy's central bulge is not shaped like a flattened sphere but may be elongated somewhat into a bar or peanut shape. We will discuss the potential significance of this elongated shape in Section 25-5.

Different kinds of stars are found in the various components of our Galaxy. The globular clusters in the halo are composed of old, metal-poor, Population II stars (see Section 21-4). Although these clusters are conspicuous, they contain only about 1% of the total number of stars in the halo. Most halo stars are single Population II stars in isolation, called **high-velocity stars** because of their high speed relative to the Sun. These ancient stars orbit the Galaxy along paths tilted at random angles to the plane of the Milky Way.

The stars in the disk are mostly young, metal-rich, Population I stars like the Sun. The disk of a galaxy like the Milky Way appears bluish because its light is dominated by radiation from hot O and B main-sequence stars. Such stars have very short main-sequence lifetimes (see Section 21-1, especially Table 21-1), so they must be quite young by astronomical standards. Hence, their presence shows that

Dust lies mostly in the plane of the Galaxy (seen edge-on)

(a) Infrared emission from dust at wavelengths of 25, 60, and 100 μm

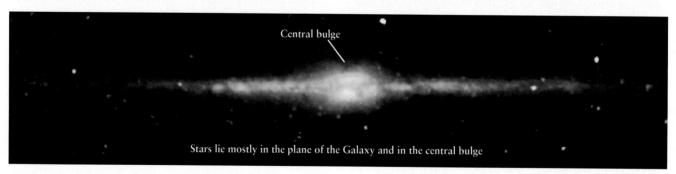

Central bulge

Stars lie mostly in the plane of the Galaxy and in the central bulge

(b) Infrared emission from dust at wavelengths of 1.2, 2.2, and 3.4 μm

Figure 25-6 R **I** V U X G

The Infrared Milky Way (a) This view was constructed from observations made at far-infrared wavelengths by the IRAS spacecraft. Interstellar dust, which is mostly confined to the plane of the Galaxy, is the principal source of radiation in this wavelength range. (b) Observing at near-infrared wavelengths, as in this composite of COBE data, allows us to see much farther through interstellar dust than we can at visible wavelengths. Light in this wavelength range comes mostly from stars in the plane of the Galaxy and in the bulge at the Galaxy's center. (NASA)

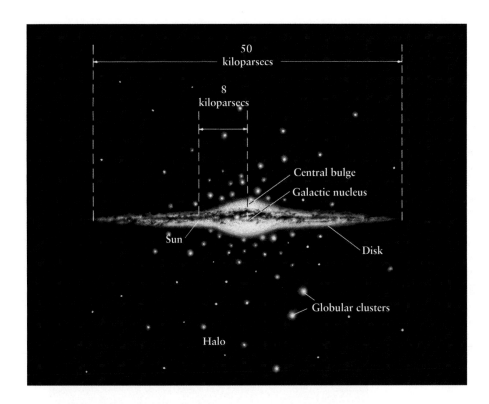

50 kiloparsecs

8 kiloparsecs

Central bulge
Galactic nucleus
Sun
Disk
Globular clusters
Halo

figure 25-7
Our Galaxy (Schematic Edge-on View) There are three major components of our Galaxy: a disk, a central bulge, and a halo. The disk contains gas and dust along with metal-rich (Population I) stars. The halo is composed almost exclusively of old, metal-poor (Population II) stars. The central bulge is a mixture of Population I and Population II stars.

there must be active star formation in the galactic disk. By contrast, no O or B stars are present in the halo, which implies that star formation ceased there long ago.

The central bulge contains both Population I and Population II stars, suggesting that some of the stars in the bulge are quite ancient, whereas others were created more recently. The central bulge looks

yellowish or reddish because it contains many red giants and red supergiants (see Figure 1-7).

Why are there such different populations of stars in the halo, disk, and central bulge? Why has star formation stopped in some regions of the Galaxy but continues in other regions? The answers to these questions are related to the way that stars, as well as the gas and dust from which stars form, move within the Galaxy.

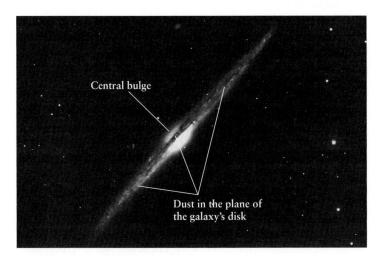

Central bulge

Dust in the plane of the galaxy's disk

figure 25-8 R I **V** U X G
Edge-on View of the Galaxy NGC 4565 If we could view our Galaxy edge-on from a great distance, it would probably look like this galaxy in the constellation Coma Berenices. A layer of obscuring dust is clearly visible in the plane of the galaxy, as is the central bulge. NGC 4565 is about 15 million pc (50 million ly) from Earth. (NOAO/AURA/NSF)

25-3 | Observations of star-forming regions reveal that our Galaxy has spiral arms

Because interstellar dust obscures our visible-light view in the plane of our Galaxy, a detailed understanding of the structure of the galactic disk had to wait until the development of radio astronomy. Thanks to their long wavelengths, radio waves can penetrate the interstellar medium even more easily than infrared light and can travel without being scattered or absorbed. As we shall see in this section, both radio and optical observations reveal that our Galaxy has **spiral arms**, spiral-shaped concentrations of gas and dust that extend outward from the center in a shape reminiscent of a pinwheel.

Hydrogen is by far the most abundant element in the universe (see Section 8-2, especially Figure 8-3). Hence, by looking for concentrations of hydrogen gas, we should be able to detect important clues about the distribution of matter in our Galaxy. Unfortunately, ordinary visible-light telescopes are of little use in this quest. This is so because hydrogen atoms can only emit visible light if they are first excited to high energy levels (see Section 5-8, and especially Figure 5-22). This is quite unlikely to occur in the cold depths of interstellar space. Furthermore, even if there are some hydrogen atoms that glow strongly at visible wavelengths, interstellar extinction due to dust (see Section 26-1) would make it impossible to see this glow from distant parts of the Galaxy.

What makes it possible to map out the distribution of hydrogen in our Galaxy is that even cold hydrogen clouds emit *radio* waves. As we saw in Section 25-2, radio waves can easily penetrate the interstellar medium, so we can detect the radio emission from such cold clouds no matter where in the Galaxy they lie. The hydrogen in these clouds is neutral—that is, not ionized—and is called **H I**. (This distinguishes it from ionized hydrogen, which is designated H II.) To understand how H I clouds can emit radio waves, we must probe a bit more deeply into the structure of protons and electrons, the particles of which hydrogen atoms are made.

In addition to having mass and charge, particles such as protons and electrons possess a tiny amount of angular momentum (that is, rotational motion) commonly called **spin**. Very roughly, you can visualize a proton or electron as a tiny, electrically charged sphere that spins on its axis. Because electric charges in motion generate magnetic fields, a proton or electron behaves like a tiny magnet with a north pole and a south pole (**Figure 25-9**).

If you have ever played with magnets, you know that two magnets attract when the north pole of one magnet is next to the south pole of the other and repel when two like poles (both north or both south) are next to each other (Figure 25-9a). In other words, the energy of the two magnets is least when opposite poles are together and highest when like poles are together. Hence, as shown in Figure 25-9b, the energy of a hydrogen atom is slightly different depending on whether the spins of the proton and electron are in the same direction or opposite directions. (According to the laws of quantum mechanics, these are the only two possibilities; the spins cannot be at random angles.)

If the spin of the electron changes its orientation from the higher-energy configuration to the lower-energy one—called a **spin-flip transition**—a photon is emitted. The energy difference between the two spin configurations is very small, only about 10^{-6} as great as those between different electron orbits (see Figure 5-22). Therefore, the photon emitted in a spin-flip transition between these configurations has only a small energy, and thus its wavelength is a relatively long 21 cm—a radio wavelength.

The spin-flip transition in neutral hydrogen was first predicted in 1944 by the Dutch astronomer Hendrik van de Hulst. His calculations suggested that it should be possible to detect the **21-cm radio emission** from interstellar hydrogen, although a very sensitive radio telescope would be required. In 1951, Harold Ewen and Edward Purcell at Harvard University first succeeded in detecting this faint emission from hydrogen between the stars.

Figure 25-10 shows the results of a more recent 21-cm survey of the entire sky. Neutral hydrogen gas (H I) in the plane of the Milky Way stands out prominently as a bright band across the middle of this image.

 The distribution of gas in the Milky Way is not uniform but is actually quite frothy. In fact, our Sun lies near the edge of an irregularly shaped region within which the

(a) The magnetic energy of two bar magnets depends on their relative orientation

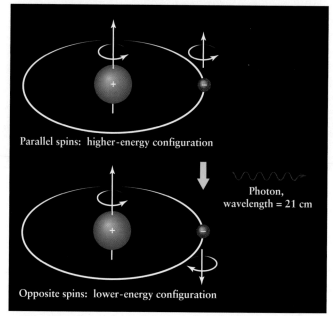

(b) The magnetic energy of a proton and electron depends on their relative spin orientation

Figure 25-9

Magnetic Interactions in the Hydrogen Atom (a) The energy of a pair of magnets is high when their north poles or their south poles are near each other, and low when they have opposite poles near each other. (b) Thanks to their spin, electrons and protons are both tiny magnets. When the electron flips from the higher-energy configuration (with its spin in the same direction the proton's spin) to the lower-energy configuration (with its spin opposite to the proton's spin), the atom loses a tiny amount of energy and emits a radio photon with a wavelength of 21 cm.

interstellar medium is very thin but at very high temperatures (about 10^6 K). This region, called the **Local Bubble,** is several hundred parsecs across. The Local Bubble may have been carved out by a supernova that exploded nearby some 300,000 years ago.

Remarkably, spin-flip transitions are used not only to map our Galaxy but also to map the internal structure of the human body. **Box 25-1** discusses this application, called magnetic resonance imaging.

21-cm emission shows that hydrogen gas is
concentrated along the plane of the Galaxy

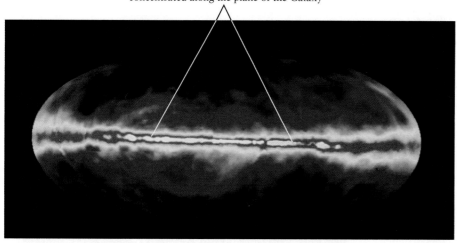

Figure 25-10   R I V U X G

The Sky at 21 Centimeters This image was
made by mapping the sky with radio telescopes
tuned to the 21-cm wavelength emitted by neutral
interstellar hydrogen (H I). The entire sky has been
mapped onto an oval, and the plane of the Galaxy
extends horizontally across the image. Black and blue
represent the weakest emission, and red and white
the strongest. (Courtesy of C. Jones and W. Forman,
Harvard-Smithsonian Center for Astrophysics)

box 25-1   THE HEAVENS ON THE EARTH

Spin-Flip Transitions in Medicine

WEB LINK 25.6

Thanks to their spin, protons and electrons act
like microscopic bar magnets. In a hydrogen
atom, the interaction between the magnetism of
the electron and that of the proton gives rise to the 21-cm
radio emission. But these particles can also interact with
outside magnetic fields, such as that produced by a large
electromagnet. This is the physical principle behind **magnetic
resonance imaging (MRI)**, an important diagnostic tool of
modern medicine.

Much of the human body is made of water. Every water
molecule has two hydrogen atoms, each of which has a nucleus
made of a single proton. If a person's body is placed in a strong
magnetic field, the spins of the protons in the hydrogen atoms of
their body can either be in the same direction as the field
("aligned") or in the direction opposite to the field ("opposed").
The aligned orientation has lower energy, and therefore the
majority of protons end up with their spins in this orientation.
But if a radio wave of just the right wavelength is now sent
through the person's body, an aligned proton can absorb a radio
photon and flip its spin into the higher-energy, opposed
orientation. How much of the radio wave is absorbed depends
on the number of protons in the body, which in turns depends
on how much water (and, thus, how much water-containing
tissue) is in the body.

In magnetic resonance imaging, a magnetic field is used whose
strength varies from place to place. The difference in energy
between the opposed and aligned orientations of a proton
depends on the strength of the magnetic field, so radio waves will
only be absorbed at places where this energy difference is equal to
the energy of a radio photon. (This equality is called *resonance*,
which is how magnetic resonance imaging gets its name.) By
varying the magnetic field strength over the body and the

wavelength of the radio waves, and by measuring how much of
the radio wave is absorbed by different parts of the body, it is
possible to map out the body's tissues. The accompanying false-
color image shows such a map of a patient's head.

Unlike X-ray images, which show only the densest parts of
the body, such as bones and teeth, magnetic resonance imaging
can be used to view less dense (but water-containing) soft tissue.
Just as the 21-cm radio emission has given astronomers a clear
view of what were hidden regions of our Galaxy, magnetic
resonance imaging allows modern medicine to see otherwise
invisible parts of the human body.

(Scott Camazine/Photo Researchers)   R I V U X G

The detection of 21-cm radio radiation was a major breakthrough that permitted astronomers to probe the galactic disk. **Figure 25-11** shows how this was done. Suppose that you aim a radio telescope along a particular line of sight across the Galaxy. Your radio receiver, located at *S* (the position of the solar system), picks up 21-cm emission from H I clouds at points 1, 2, 3, and 4. However, the radio waves from these various clouds are Doppler shifted by slightly different amounts, because the clouds are moving at different speeds as they travel with the rotating Galaxy.

It is important to remember that the Doppler shift reveals only motion along the line of sight (review Figure 5-24). In Figure 25-11, cloud 2 has the highest speed along our line of sight, because it is moving directly toward us. Consequently, the radio waves from cloud 2 exhibit a larger Doppler shift than those from the other three clouds along our line of sight. Because clouds 1 and 3 are at the same distance from the galactic center, they have the same orbital speed. The fraction of their velocity parallel to our line of sight is also the same, so their radio waves exhibit the same Doppler shift, which is less than the Doppler shift of cloud 2. Finally, cloud 4 is the same distance from the galactic center as the Sun. This cloud is thus orbiting the Galaxy at the same speed as the Sun, resulting in no net motion along the line of sight. Radio waves from cloud 4, as well as from hydrogen gas near the Sun, are not Doppler shifted at all.

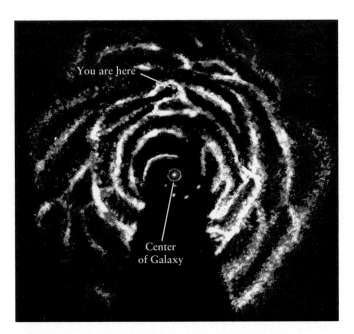

fiqure 25-12 **R** I V U X G

A Map of Neutral Hydrogen in Our Galaxy This map, constructed from radio-telescope surveys of 21-cm radiation, shows the distribution of hydrogen gas in a face-on view of our Galaxy. The map suggests a spiral structure. Details in the blank, wedge-shaped region at the bottom of the map are unknown. Gas in this part of the Galaxy is moving perpendicular to our line of sight and thus does not exhibit a detectable Doppler shift. (Courtesy of G. Westerhout)

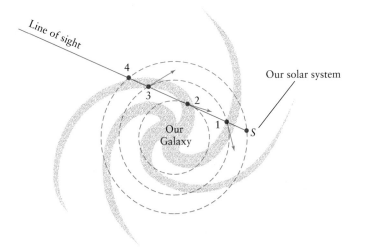

• Hydrogen clouds 1 and 3 are approaching us: They have a moderate blueshift.

• Hydrogen cloud 2 is approaching us at a faster speed: It has a larger blueshift.

• Hydrogen cloud 4 is neither approaching nor receding: It has no redshift or blueshift.

fiqure 25-11

A Technique for Mapping Our Galaxy If we look within the plane of our Galaxy from our position at *S*, hydrogen clouds at different locations (shown as 1, 2, 3, and 4) along our line of sight are moving at slightly different speeds relative to us. As a result, radio waves from these various gas clouds are subjected to slightly different Doppler shifts. This permits radio astronomers to sort out the gas clouds and thus map the Galaxy.

These various Doppler shifts cause radio waves from gases in different parts of the Galaxy to arrive at our radio telescopes with wavelengths slightly different from 21 cm. It is therefore possible to sort out the various gas clouds and thus produce a map of the Galaxy like that shown in **Figure 25-12**.

Figure 25-12 shows that neutral hydrogen gas is not spread uniformly around the disk of the Galaxy but is concentrated into numerous arched lanes. Similar features are seen in other galaxies beyond the Milky Way. As an example, the galaxy in **Figure 25-13a** has prominent spiral arms outlined by hot, luminous, blue main-sequence stars and the red emission nebulae (H II regions) found near many such stars. Stars of this sort are very short-lived, so these features indicate that spiral arms are sites of active, ongoing star formation. The 21-cm radio image of this same galaxy, shown in Figure 25-13b, shows that spiral arms are also regions where neutral hydrogen gas is concentrated, similar to the structures in our own Galaxy visible in Figure 25-12. This similarity is a strong indication that our Galaxy also has spiral arms.

 Photographs such as Figure 25-13a can lead to the impression that there are very few stars between the spiral arms of a galaxy. Nothing could be further from the truth! In fact, stars are distributed rather uniformly throughout the disk of a galaxy like the one in Figure 25-13a; the density of stars in the spiral arms is only about 5% higher than in the rest of

H II regions (red) are in the spiral arms.

Hot luminous, young stars (blue) are in the spiral arms.

(a) Visible-light view of M83 R I **V** U X G

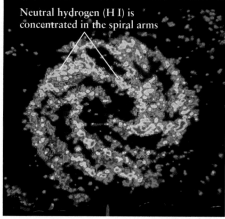

Neutral hydrogen (H I) is concentrated in the spiral arms

(b) 21-cm radio view of M83 **R** I V U X G

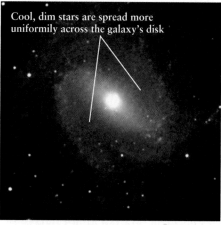

Cool, dim stars are spread more uniformily across the galaxy's disk

(c) Near-infrared view of M83 R **I** V U X G

WEB LINK 25.7

figure 25-13

A Spiral Galaxy The galaxy M83 lies in the southern constellation Hydra about 5 million pc (15 million ly) from Earth. (a) This visible-light image clearly shows the spiral arms. The presence of young stars and H II regions indicates that star formation takes place in spiral arms. (b) This radio view at a wavelength of 21 cm shows the emission from neutral interstellar hydrogen gas (H I). Note that essentially the same pattern of spiral arms is traced out in this image as in the visible-light photograph. (c) M83 has a much smoother appearance in this near-infrared view. This shows that cooler stars, which emit strongly in the infrared, are spread more uniformly across the galaxy's disk. Note the elongated bar shape of the central bulge. (a: Anglo-Australian Observatory; b: VLA, NRAO; c: S. Van Dyk/IPAC)

the disk. The spiral arms stand out nonetheless because they are where hot, blue O and B stars are found. One such star is about 10^4 times more luminous than an average star in the disk, so the light from O and B stars completely dominates the visible appearance of a spiral galaxy. An infrared image such as Figure 25-13c gives a better impression of how stars of all kinds are distributed through a spiral galaxy's disk.

Figure 25-13a suggests that we can confirm the presence of spiral structure in our own Galaxy by mapping the locations of star-forming regions. Such regions are marked by OB associations, H II regions, and molecular clouds (see Section 20-7). Unfortunately, the first two of these are best observed using visible light, and interstellar extinction limits the range of visual observations in the plane of the Galaxy to less than 3 kpc (10,000 ly) from the Earth. But there are enough OB associations and H II regions within this range to plot the spiral arms in the vicinity of the Sun.

Molecular clouds are easier to observe at great distances, because molecules of carbon monoxide (CO) in these clouds emit radio waves that are relatively unaffected by interstellar extinction. Hence, the positions of molecular clouds have been plotted even in remote regions of the Galaxy (see Figure 20-20). (We saw in Section 20-7 that CO molecules in molecular clouds emit more strongly than the hydrogen atoms do, even though hydrogen is the principal constituent of these clouds.)

Taken together, all these observations demonstrate that our Galaxy has four major spiral arms and several short arm segments (Figure 25-14). The Sun is located just outside a relatively short arm segment called the Orion arm, which includes the Orion Nebula and neighboring sites of vigorous star formation in that constellation.

Two major spiral arms border either side of the Sun's position. The Sagittarius arm is on the side toward the galactic center. This is the arm you see on June and July nights when you look at the portion of the Milky Way stretching across Scorpius and Sagittarius

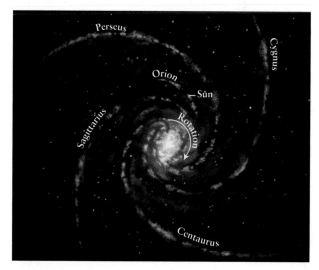

figure 25-14

Our Galaxy (Face-on View) Our Galaxy has four major spiral arms and several shorter arm segments. The Sun is located just outside the Orion arm, between two major spiral arms. The Galaxy's diameter is about 50,000 pc (160,000 ly), and the Sun is about 8000 pc (26,000 ly) from the galactic center. Note that the central bulge is somewhat elongated, as is also the case for M83 (see Figure 25-13).

(see the photograph that opens this chapter). In December and January, when our nighttime view is directed away from the galactic center, we see the Perseus arm. The remaining two major spiral arms are usually referred to as the Centaurus arm and the Cygnus arm.

Why are the young stars, star-forming regions, and clouds of neutral hydrogen in our Galaxy all found predominantly in the spiral arms? To answer this question, we must understand why spiral arms exist at all. Spiral arms are essentially cosmic "traffic jams," places where matter piles up as it orbits around the center of the Galaxy. This orbital motion, which is essential to grasping the significance of spiral arms, is our next topic as we continue the exploration of the Galaxy.

25-4 | The rotation of our Galaxy reveals the presence of dark matter

The spiral arms in the disk of our Galaxy suggest that the disk rotates. This means that the stars, gas, and dust in our Galaxy are all orbiting the galactic center. Indeed, if this were not the case, mutual gravitational attraction would cause the entire Galaxy to collapse into the galactic center. In the same way, the Moon is kept from crashing into the Earth and the planets from crashing into the Sun because of their motion around their orbits (see Section 4-7).

Measuring the rotation of our Galaxy accurately is a difficult business. But such challenging measurements have been made, as we shall see, and the results lead to a remarkable conclusion: Most of the mass of the Galaxy is in the form of *dark matter*, a mysterious sort of material that emits no light at all.

Radio observations of 21-cm radiation from hydrogen gas provide important clues about our Galaxy's rotation. Doppler shift measurements of this radiation indicate that stars and gas all orbit in the same direction around the galactic center, just as the planets all orbit in the same direction around the Sun. Measurements also show that the orbital speed of stars and gas about the galactic center is fairly uniform throughout much of the Galaxy's disk (**Figure 25-15**). As a result, stars inside the Sun's orbit complete a trip around the galactic center more quickly than the Sun, because the stars have a shorter distance to travel. Conversely, stars outside the Sun's orbit take longer to go once around the galactic center because they have farther to travel. As seen by Earth-based astronomers moving along with the Sun, stars inside the Sun's orbit overtake and pass us, while we overtake and pass stars outside the Sun's orbit (Figure 25-15*a*).

Figure 25-15

The Rotation of Our Galaxy (a) This schematic diagram shows three stars (the Sun and two others) orbiting the center of the Galaxy. Although they start off lined up, the stars become increasingly separated as they move along their orbits. Stars inside the Sun's orbit overtake and move ahead of the Sun, while stars far from the galactic center lag behind the Sun. (b) The stars would remain lined up if the Galaxy rotated like a solid disk. This is not what is observed. (c) If stars orbited the galactic center in the same way that planets orbit the Sun, stars inside the Sun's orbit would overtake us faster than they are observed to do.

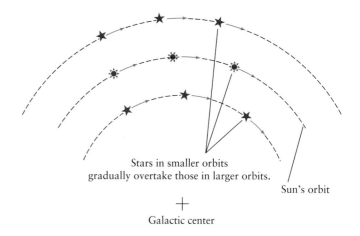

(a) The orbital speed of stars and gas around the galactic center is nearly uniform throughout most of our Galaxy.

Stars in smaller orbits gradually overtake those in larger orbits.

Sun's orbit

Galactic center

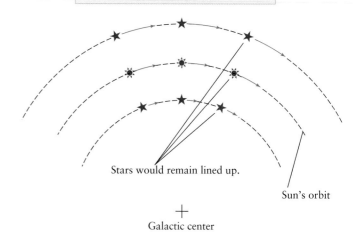

(b) If our Galaxy rotated like a solid disk, the orbital speed would be greater for stars and gas in larger orbits.

Stars would remain lined up.

Sun's orbit

Galactic center

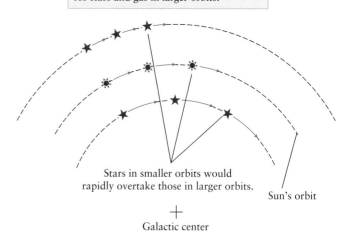

(c) If the Sun and stars obeyed Kepler's third law, the orbital speed would be less for stars and gas in larger orbits.

Stars in smaller orbits would rapidly overtake those in larger orbits.

Sun's orbit

Galactic center

! **CAUTION** Note that when we say that objects in different parts of the Galaxy orbit at the same speed, we do *not* mean that the Galaxy rotates like a solid disk. All parts of a rotating solid disk—a CD or DVD, for example—take the same time to complete one rotation. Because the outer part of the disk has to travel around a larger circle than the inner part, the speed (distance per time) is greater in the outer part (Figure 25-15*b*). By contrast, the orbital speed of material in our Galaxy is roughly the same at all distances from the galactic center.

The most familiar examples of orbital motion are the motions of the planets around the Sun. As we saw in Section 4-7, the farther a planet is from the Sun, the less gravitational force it experiences and the slower the speed it needs to have to remain in orbit. The same would be true for the orbits of stars and gas in the Galaxy *if* they were held in orbit by a single, massive object at the galactic center (Figure 25-15*c*). Hence, the 21-cm observations of our Galaxy, which show that the speed does *not* decrease with increasing distance from the galactic center, demonstrate that there is no such single, massive object holding objects in their galactic orbits.

Instead, what keeps a star in its orbit around the center of the Galaxy is the combined gravitational force exerted on it by *all* of the mass (including stars, gas, and dust) that lies within the star's orbit. (It turns out that the gravitational force from matter *outside* a star's orbit has little or no net effect on the star's motion around the galactic center.) This gives us a tool for determining the Galaxy's mass and how that mass is distributed.

An important example is the orbital motion of the Sun (and the solar system) around the center of the Galaxy. If we know the semimajor axis and period of the Sun's orbit, we can use Newton's form of Kepler's third law (described in Section 4-7) to determine the mass of that portion of the Galaxy that lies within the orbit. We saw in Section 25-2 that the Sun is about 8000 pc (26,000 ly) from the galactic center; this is the semimajor axis of the Sun's orbit. The orbit is in fact nearly circular, so we can regard 8000 pc as the radius r of the orbit. In one complete trip around the Galaxy, the Sun travels a distance equal to the circumference of its orbit, which is $2\pi r$. The time required for one orbit, or orbital period P, is equal to the distance traveled divided by the Sun's orbital speed v:

Period of the Sun's orbit around the galactic center

$$P = \frac{2\pi r}{v}$$

P = orbital period of the Sun

r = distance from the Sun to the galactic center

v = orbital speed of the Sun

Unfortunately, we cannot tell the Sun's orbital speed from 21-cm observations, since these reveal only how fast things are moving relative to the Sun. Instead, we need to measure how the Sun is moving relative to a background that is not rotating along with the rest of the Galaxy. Such a background is provided by distant galaxies beyond the Milky Way and by the globular clusters. (Since globular clusters lie outside the plane of the Galaxy, they do not take part in the rotation of the disk.) By measuring the Doppler shifts of these objects and averaging their velocities, astronomers deduce that the Sun is moving along its orbit around the galactic center at about 220 km/s—about 790,000 kilometers per hour or 490,000 miles per hour!

Using this information, we find that the Sun's orbital period is

$$P = \frac{2\pi \times 8000 \text{ pc}}{220 \text{ km/s}} \times \frac{3.09 \times 10^{13} \text{ km}}{1 \text{ pc}}$$
$$= 7.1 \times 10^{15} \text{ s} = 2.2 \times 10^{8} \text{ years}$$

Traveling at 790,000 kilometers per hour, it takes the Sun about 220 million years to complete one trip around the Galaxy. (In the 65 million years since the demise of the dinosaurs, our solar system has traveled less than a third of the way around its orbit.) The Galaxy is a very large place!

Box 25-2 shows how to combine the radius and period of the Sun's orbit to calculate the total mass of all the matter that lies inside the Sun's orbit. Such calculations give an answer of 9.0×10^{10} M$_\odot$ (90 million solar masses). As Figure 25-7 shows, the Galaxy extends well beyond the Sun's orbit, so the mass of the entire Galaxy must be larger than this.

In recent years, astronomers have been astonished to discover how much matter may lie outside the Sun's orbit. The clues come from 21-cm radiation emitted by hydrogen in spiral arms that extend to the outer reaches of the Galaxy. Because we know the true speed of the Sun, we can convert the Doppler shifts of this radiation into actual speeds for the spiral arms. This calculation gives us a **rotation curve**, a graph of the speed of galactic rotation measured outward from the galactic center (**Figure 25-16**). We would expect that for gas clouds beyond the confines of most of the Galaxy's mass, the orbital speed should decrease with increasing distance from the Galaxy's center, just as the orbital speeds of the planets decrease with increasing distance from the Sun (see Figure 25-15*c*). But as Figure 25-16 shows, the Galaxy's rotation curve is quite flat, indicating roughly uniform orbital speeds well beyond the visible edge of the galactic disk.

To explain these nearly uniform orbital speeds in the outer parts of the Galaxy, astronomers conclude that a large amount of matter must lie outside the Sun's orbit. When this matter is included, the total mass of our Galaxy could exceed 10^{12} M$_\odot$ or more, of which about one-tenth is in the form of stars. This implies that our Galaxy contains roughly 200 billion stars.

 These observations lead to a profound mystery. Stars, gas, and dust account for only about 10% of the Galaxy's total mass. What, then, makes up the remaining 90% of the matter in our Galaxy? Whatever it is, it is dark. It does not show up on photographs, nor indeed in images made in any part of the electromagnetic spectrum. This unseen material, which is by far the predominant constituent of our Galaxy, is called **dark matter.** We sense its presence only through its gravitational influence on the orbits of stars and gas clouds.

 CAUTION Be careful not to confuse dark *matter* with dark *nebulae*. A dark nebula like the one in Figure 20-3 emits no visible light, but does radiate at longer wavelengths. By contrast,

box 25-2 | TOOLS OF THE ASTRONOMER'S TRADE

Estimating the Mass Inside the Sun's Orbit

The force that keeps the Sun in orbit around the center of the Galaxy is the gravitational pull of all the matter interior to the Sun's orbit. We can estimate the total mass of all of this matter using Newton's form of Kepler's third law (see Section 4-7 and Box 4-4):

$$P^2 = \frac{4\pi^2 a^3}{G(M + M_\odot)}$$

In this equation P is the orbital period of the Sun, a is the semimajor axis of the Sun's orbit around the galactic center, G is the universal constant of gravitation, M is the amount of mass inside the Sun's orbit, and $M_\odot$ is the mass of the Sun.

Because the Sun is only one of more than 10^{11} stars in the Galaxy, the Sun's mass is minuscule compared to M. Hence, we can safely replace the sum $M + M_\odot$ in the above equation by simply M. If we now assume that the Sun's orbit is a circle, the semimajor axis a of the orbit is just the radius of this circle, which we call r. As we saw in Section 25-4, the period P of the orbit is equal to $2\pi r/v$, where v is the Sun's orbital speed. You can then show that

$$M = \frac{rv^2}{G}$$

(We leave the derivation of this equation as an exercise at the end of this chapter.)

Now we can insert known values to obtain the mass inside the Sun's orbit. Being careful to express distance in meters and speed in meters per second, we have $v = 220$ km/s = 2.2×10^5 m/s, $G = 6.67 \times 10^{-11}$ newton $\cdot$ m^2/kg^2, and

$$r = 26{,}000 \text{ light-years} \times \frac{9.46 \times 10^{12} \text{ km}}{1 \text{ light-year}} \times \frac{10^3 \text{ m}}{1 \text{ km}}$$

$$= 2.5 \times 10^{20} \text{ m}$$

Hence, we find that

$$M = \frac{2.5 \times 10^{20} \times (2.2 \times 10^5)}{6.67 \times 10^{-11}} = 1.8 \times 10^{41} \text{ kg}$$

or, in terms of the mass of the Sun,

$$M = 1.8 \times 10^{41} \text{ kg} \times \frac{1 \text{ M}_\odot}{1.99 \times 10^{30} \text{ kg}} = 9.0 \times 10^{10} \text{ M}_\odot$$

This estimate involves only mass that is interior to the Sun's orbit. Matter outside the Sun's orbit has no net gravitational effect on the Sun's motion and thus does not enter into Kepler's third law. (This is strictly true only if the matter outside our orbit is distributed over a sphere rather than a disk. In fact, the dark matter that dominates our galaxy seems to have a spherical distribution.)

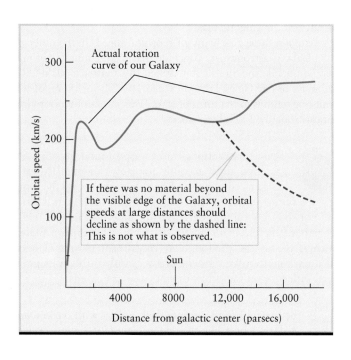

no electromagnetic radiation of any kind has yet been discovered coming from dark matter.

Observations of star groupings outside the Milky Way suggest that our Galaxy's dark matter forms a spherical halo centered on the galactic nucleus, like the halo of globular clusters and high-velocity stars shown in Figure 25-7. However, the dark matter halo is much larger; it may extend to a distance of 100–200 kpc from the center of our Galaxy, some 2 to 4 times the extent of the visible halo. Analysis of the rotation curve in Figure 25-16 shows that the den-

figure 25-16

The Galaxy's Rotation Curve The blue curve shows the orbital speeds of stars and gas in the disk of the Galaxy out to a distance of 18,000 parsecs from the galactic center. (Very few stars are found beyond this distance.) The dashed red curve indicates how this orbital speed should decline beyond the confines of most of the Galaxy's visible mass. Because there is no such decline, there must be an abundance of invisible dark matter that extends to great distances from the galactic center.

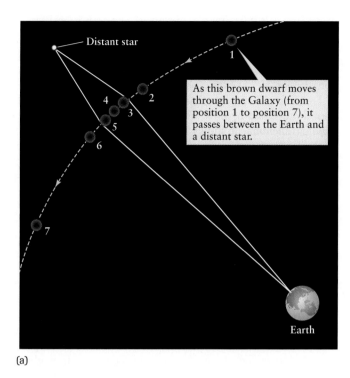

As this brown dwarf moves through the Galaxy (from position 1 to position 7), it passes between the Earth and a distant star.

(a)

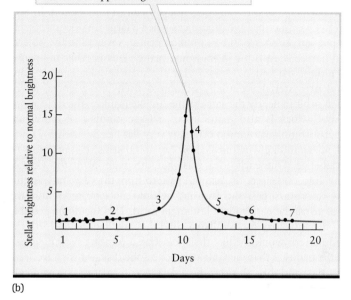

When the brown dwarf is directly between us and the distant star [near position 4 in (a)], it acts as a gravitational lens and makes the distant star appear brighter.

(b)

figure 25-17

Microlensing by Dark Matter in the Galactic Halo (a) If a dense object such as a brown dwarf or black hole passes between the Earth and a distant star, the gravitational curvature of space around the dense object deflects the starlight and focuses it in our direction. This effect is called microlensing. (b) This light curve shows the gravitational microlensing of light from a star in the Galaxy's central bulge. Astronomers do not know the nature of the object that passed between the Earth and this star to cause the microlensing. (Courtesy of the MACHO and GMAN Collaborations)

sity of the dark matter halo decreases with increasing distance from the center of the Galaxy.

 What is the nature of this mysterious dark matter? One proposal is that the dark matter halo is composed, at least in part, of dim objects with masses less than 1 $M_\odot$. These objects, which could include brown dwarfs, white dwarfs, or black holes, are called **massive compact halo objects,** or **MACHOs.** Astronomers have searched for MACHOs by monitoring the light from distant stars. If a MACHO passes between us and the star, its gravity will bend the light coming from the star. (In Section 24-2 we described how gravity can bend starlight.) As **Figure 25-17** shows, the MACHO's gravity acts like a lens that focuses the light from the star. This effect, called **microlensing,** makes the star appear to brighten substantially for a few days.

Astronomers have indeed detected MACHOs in this way, but not enough to completely solve the dark matter mystery. MACHOs with very low mass (10^{-6} to 0.1 $M_\odot$ each) do not appear to be a significant part of the dark matter halo. MACHOs of roughly 0.5 $M_\odot$ are more prevalent, but account for only about half of the dark matter halo.

The remainder of the dark matter is thought to be much more exotic. One candidate is a neutrino with a small amount of mass. If these neutrinos are sufficiently massive, and if enough of them are present in the halo of the Galaxy, they might constitute a reasonable fraction of the dark matter. As we saw in Section 18-4, one type of neutrino can transform into another. These transformations can take place only if neutrinos have a nonzero amount of mass. Thus, neutrinos must comprise at least part of the dark matter, though it is not known how much.

Another speculative possibility is a new class of subatomic particle called **weakly interacting massive particles,** or **WIMPs.** These particles, whose existence is suggested by certain theories but has not yet been confirmed experimentally, would not emit or absorb electromagnetic radiation. Physicists are attempting to detect these curious particles, which would have masses 10 to 10,000 times greater than a proton or neutron, by using a large crystal cooled almost to absolute zero. If a WIMP should enter this crystal and collide with one of its atoms, the collision will deposit a tiny but measurable amount of heat in the crystal.

As yet, the true nature of dark matter remains a mystery. Furthermore, this mystery is not confined to our own Galaxy. In Chapter 26 we will find that other galaxies have the same sort of rotation curve as in Figure 25-16, indicating that they also contain vast amounts of dark matter. Indeed, dark matter appears to make up most of the mass in our universe. Hence, the quest to understand dark matter is one of the most important in modern astronomy.

25-5 | Spiral arms are caused by density waves that sweep around the Galaxy

The disk shape of our Galaxy is not difficult to understand. In Section 8-4 we described what happens when a large number of objects are put into orbit around a common center: Over time the objects tend naturally to orbit in the same plane. This is what happened when our solar system formed from the solar nebula. There a giant cloud of material eventually organized itself into planets, all of which orbit in nearly the same plane. In like fashion, the disk of our Galaxy, which is also made up of a large number of individual objects orbiting a common center, is very flat (see Figure 25-7).

Understanding why our Galaxy has spiral arms presents more of a challenge. One early idea was that the material in the Galaxy somehow condensed into a spiral pattern from the very start. In this view, once stars, gas, and dust had become concentrated within the spiral arms, the pattern would remain fixed. This would be possible only if the Galaxy rotated like a solid disk (see Figure 25-15b); the fixed pattern would be like the spokes on a rotating bicycle wheel. But the reality is that the Galaxy is not a solid disk. As we have seen, stars, gas, and dust all orbit the galactic center with approximately the same speed, as shown in Figure 25-15a. Let us see why this makes it impossible for a rigid spiral pattern to persist.

Imagine four stars, A, B, C, and D, that originally lie on a line extending outward from the galactic center (**Figure 25-18a**). In a given amount of time, each of the stars travels the same distance around its orbit. But because the innermost star has a smaller orbit than the others, it takes less time to complete one orbit. As a result, a line connecting the four stars is soon bent into a spiral (Figure 25-18b). Moreover, the spiral becomes tighter and tighter with the passage of time (Figures 25-18c and 25-18d). This "winding up" of the spiral arms causes the spiral structure to disappear completely after a few hundred million years—a very brief time compared to the age of our Galaxy, thought to be about 12.5 billion (1.25×10^{10}) years.

Figure 25-18 suggests that the Milky Way's spiral arms ought to have disappeared by now. The fact that they have not is called the **winding dilemma.** It shows that the spiral arms cannot simply be assemblages of stars and interstellar matter that travel around the Galaxy together, like a troop of soldiers marching in formation around a flagpole. In other words, the spiral arms cannot be made of anything *material*. What, then, can the spiral arms be?

In the 1940s, the Swedish astronomer Bertil Lindblad proposed that the spiral arms of a galaxy are actually a pattern that moves through the Galaxy like ripples on water. This idea was greatly enhanced and embellished in the 1960s by the American astronomers Chia Chiao Lin and Frank Shu. In this picture, spiral arms are a kind of wave, like the waves that move across the surface of a pond when you toss a stone into the water. Water molecules pile up at a crest of the wave but spread out again when the crest passes. By analogy, Lindblad, Lin, and Shu pictured a pattern of **density waves** sweeping around the Galaxy. These waves make matter pile up in the spiral arms, which are the crests of the waves. Individual parts of the Galaxy's material are compressed only temporarily when they pass through a spiral arm. The pattern of spiral arms persists, however, just as the waves made by a stone dropped in the water can persist for quite awhile after the stone has sunk.

To understand better how a density wave operates in a galaxy, think again about a water wave in a pond. If one part of the pond is disturbed by dropping a stone into it, the molecules in that part will be displaced a bit. They will nudge the molecules next to them,

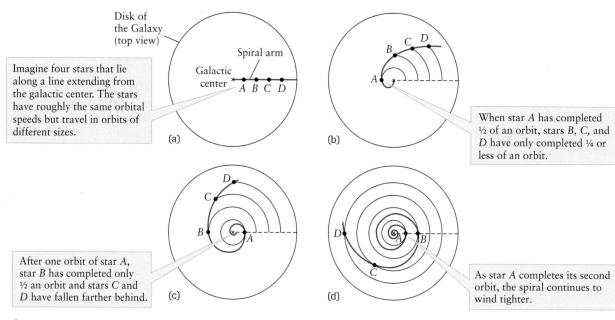

Figure 25-18

The Winding Dilemma This series of drawings shows that spiral arms in galaxies like the Milky Way cannot simply be assemblages of stars. If they were, the spiral arms would "wind up" and disappear in just a few hundred million years.

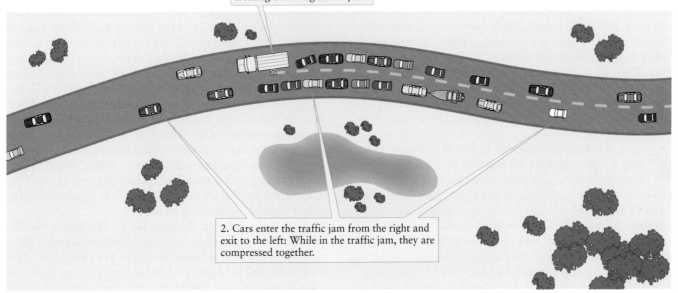

1. A crew of painters moves slowly along the highway, creating a moving traffic jam.

2. Cars enter the traffic jam from the right and exit to the left: While in the traffic jam, they are compressed together.

Figure 25-19

A Density Wave on the Highway A density wave in a spiral galaxy is analogous to a crew of painters moving slowly along the highway, creating a moving traffic jam. Like such a traffic jam, a density wave in a spiral galaxy is a slow-moving region where stars, gas, and dust are more densely packed than in the rest of the galaxy. As the material of the galaxy passes through the density wave, it is compressed. This triggers star formation, as Figure 25-20 shows.

causing those molecules to be displaced and to nudge the molecules beyond them. In this way the wave disturbance spreads throughout the pond.

In a galaxy, stars play the role of water molecules. Although stars and interstellar clouds of gas and dust are separated by vast distances, they can nonetheless exert forces on each other because they are affected by each other's gravity. If a region of above-average density should form, its gravitational attraction will draw nearby material into it. The displacement of this material will change the gravitational force that it exerts on other parts of the galaxy, causing additional displacements. In this way a spiral-shaped density wave can travel around the disk of a galaxy.

ANALOGY A key feature of density waves is that they move more slowly around a galaxy than do stars or interstellar matter. To visualize this, imagine workers painting a line down a busy freeway (**Figure 25-19**). The cars normally cruise along the freeway at high speed, but the crew of painters is moving much more slowly. When the cars come up on the painters, they must slow down temporarily to avoid hitting anyone. As seen from the air, cars are jammed together around the painters. An individual car spends only a few moments in the traffic jam before resuming its usual speed, but the traffic jam itself lasts all day, inching its way along the road as the painters advance.

A similar crowding takes place when interstellar matter enters a spiral arm. This crowding plays a key role in the formation of stars and the recycling of the interstellar medium. As interstellar gas and dust moves through a spiral arm, it is compressed into new nebulae (**Figure 25-20**). This compression begins the process by which new stars form, which we described in Section 20-3.

Figure 25-20

Star Formation in the Density-Wave Model A spiral arm is a region where the density of material is higher than in the surrounding parts of a galaxy. Interstellar matter moves around the galactic center rapidly (shown by the red arrows) and is compressed as it passes through the slow-moving spiral arms (whose motion is shown by the blue arrows). This compression triggers star formation in the interstellar matter, so that new stars appear on the "downstream" side of the densest part of the spiral arms.

These freshly formed stars continue to orbit around the center of their galaxy, just like the matter from which they formed. The most luminous among these are the hot, massive, blue O and B stars, which may have emission nebulae (H II regions) associated with them. These stars have main-sequence lifetimes of only 3 to 15 million years (see Table 21-1), which is very short compared to the 220 million years required for the Sun to make a complete orbit around the Galaxy.

 As a result, these luminous O and B stars can travel only a relatively short distance before dying off. Therefore, these stars, and their associated H II regions, are only seen in or close to the spiral arm in which they formed. (Figure 25-13a illustrates this for the spiral galaxy M83.) Less massive stars have much longer main-sequence lifetimes, and thus their orbits are able to carry them all around the galactic disk. These less-luminous stars are found throughout the disk, including between the spiral arms (see Figure 25-13c).

The density-wave model of spiral arms explains why the disk of our Galaxy is dominated by metal-rich Population I stars. Because the material left over from the death of ancient stars is enriched in heavy elements, new generations of stars formed in spiral arms are likely to be more metal-rich than their ancestors.

The density-wave model is still under development. One problem is finding a driving mechanism that keeps density waves going in spiral galaxies. After all, density waves expend an enormous amount of energy to compress the interstellar gas and dust. Hence, we would expect that density waves should eventually die away, just as do ripples on a pond. The American astronomers Debra and Bruce Elmegreen have suggested that gravity can supply that needed energy. As mentioned in Section 25-2, the central bulge of our Galaxy may be elongated somewhat into a bar shape, much like the central bulge of the galaxy M83 (see Figure 25-13a). The asymmetric gravitational field of such a bar pulls on the stars and interstellar matter of a galaxy to generate density waves. Another factor that may help to generate and sustain spiral arms is the gravitational interactions *between* galaxies. We will discuss this in Chapter 26.

Spiral density waves may not be the whole story behind spiral arms in our Galaxy and other galaxies. The reason is that spiral density waves should produce very well defined spiral arms. We do indeed see many so-called **grand-design spiral galaxies** (Figure 25-21a), with thin, graceful, and well-defined spiral arms. But in some galaxies, called **flocculent spiral galaxies** (Figure 25-21b), the spiral arms are broad, fuzzy, chaotic, and poorly defined. ("Flocculent" means "resembling wool.")

To explain such flocculent spirals, M. W. Mueller and W. David Arnett in 1976 proposed a theory of **self-propagating star formation.** Imagine that star formation begins in a dense interstellar cloud within the disk of a galaxy that does not yet have spiral arms. As soon as hot, massive stars form, their radiation and stellar winds compress nearby matter, triggering the formation of additional stars in that gas. When massive stars become supernovae, they produce shock waves that further compress the surrounding interstellar medium, thus encouraging still more star formation.

Although all parts of this broad, star-forming region have approximately the same orbital speed about the galaxy's center, the inner regions have a shorter distance to travel to complete one orbit than the outer regions. As a result, the inner edges of the star-forming region move ahead of the outer edges as the Galaxy rotates. The bright O and B stars and their nearby glowing nebulae soon become stretched out in the form of a spiral arm. These spiral arms come and go essentially at random across a galaxy. Bits and pieces

(a) Grand-design spiral galaxy

(b) Flocculent spiral galaxy

Figure 25-21   R I **V** U X G

Variety in Spiral Arms The differences from one spiral galaxy to another suggest that more than one process can create spiral arms. (a) NGC 628 is a grand-design spiral galaxy with thin, well-defined spiral arms. (b) NGC 7793 is a flocculent spiral galaxy with fuzzy, poorly defined spiral arms. (Courtesy of P. Seiden, D. Elmegreen, B. Elmegreen, and A. Mobarak; IBM) of spiral arms appear where star formation has recently begun but fade and disappear at other locations where all the massive stars have died off. Self-propagating star formation therefore tends to produce flocculent spiral galaxies that have a chaotic appearance with poorly defined spiral arms, like the galaxy in Figure 25-21b.

The two theories presented here are very different in character. In the density-wave model, star formation is caused by the spiral arms; in the self-propagating star formation model, by contrast, the spiral arms are caused by star formation. The correct description of spiral arms in our Galaxy remains a topic of active research.

25-6 | Infrared, radio, and X-ray observations are used to probe the galactic center

The innermost region of our Galaxy is an active, crowded place. If you lived on a planet near the galactic center, you could see a million stars as bright as Sirius, the brightest single star in our own night sky. The total intensity of starlight from all those nearby stars would be equivalent to 200 of our full moons. In effect, night would never really fall on a planet near the center of the Milky Way. At the center of this empire of light, however, lies the darkest of all objects in the universe—a black hole millions of times more massive than the Sun.

Because of the severe interstellar absorption at visual wavelengths, most of our information about the galactic center comes from infrared and radio observations. **Figure 25-22** shows three infrared views of the center of our Galaxy. Figure 25-22a is a wide-angle view covering a 50° segment of the Milky Way from Sagittarius through Scorpius. (The photograph that opens this chapter shows this same region at visible wavelengths, viewed at a different angle.) The prominent reddish band through the center of this false-color infrared image is a layer of dust in the plane of the Galaxy. Figure 25-22b is an IRAS view of the galactic center. It is surrounded by numerous streamers of dust (shown in blue). The strongest infrared emission (shown in white) comes from a grouping of several powerful sources of radio waves. One of these sources, **Sagittarius A*** (say "A star"), lies at the very center of the Galaxy. (Its position, pinpointed with simultaneous observations by radio telescopes scattered around the world, seems to be very near the gravitational center of the Galaxy.) The high-resolution infrared view in Figure 25-22c, made using adaptive optics (see Section 6-3), shows hundreds of stars crowded within 1 ly (0.3 pc) of Sagittarius A*. Compare this to our region of the Galaxy, where the average distance between stars is more than a light-year.

Sagittarius A* itself does not appear in infrared images. Nonetheless, astronomers have used infrared observations to make truly startling discoveries about this object. Since the 1990s, two research groups—one headed by Reinhard Genzel of the Max Planck Institute for Extraterrestrial Physics in Garching, Germany, and another led by Andrea Ghez at the University of California, Los Angeles—have been using infrared detectors to monitor the motions of stars in the immediate vicinity of Sagittarius A*. They have found a number of stars orbiting around Sagittarius A* at speeds in excess of

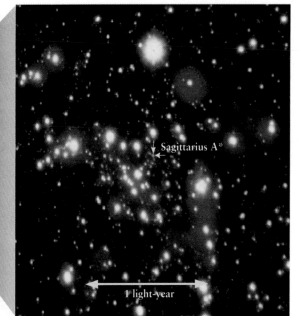

(a) A wide-angle (50°) infrared view

(b) A close-up view shows a more luminous region at the galactic center

(c) An extreme close-up view centered on Sagittarius A*, a radio source at the very center of the Milky Way Galaxy, shows hundreds of stars within 1 ly (0.3 pc)

Figure 25-22   R **I** V U X G

The Galactic Center (a) In this false-color infrared image, the reddish band is dust in the plane of the Galaxy and the fainter bluish blobs are interstellar clouds heated by young O and B stars. (b) This close-up infrared view covers the area outlined by the white rectangle in (a). (c) Adaptive optics reveals stars densely packed around the galactic center. (a, b: NASA; c: R. Schödel et al., MPE/ESO)

1500 km/s. (By comparison, the Earth orbits the Sun at a lack-adaisical 30 km/s.) In 2002, the Garching group observed one such star, called S2, as its elliptical orbit brought it within a mere 120 AU (three times the distance from the Sun to Pluto) from Sagittarius A* (Figure 25-23). At its closest approach, S2 was traveling at a breath-taking speed of 5000 km/s, or nearly 2% of the speed of light!

In order to keep stars like S2 in such small, rapid orbits, Sagittarius A* must exert a powerful gravitational force and hence must be very massive. By applying Newton's form of Kepler's third law to the motions of these stars around Sagittarius A*, the UCLA group calculates the mass of Sagittarius A* to be a remarkable 3.7 *million* solar masses (3.7×10^6 M$_\odot$). Furthermore, the small separation between S2 and Sagittarius A* at closest approach shows that Sagittarius A* can be no more than 120 AU in radius. An object this massive and this compact can only be one thing: a supermassive black hole (see Section 24-4).

Evidence in favor of this picture comes from the Chandra X-ray Observatory, which has observed X-ray flares coming from Sagittarius A*. The flares brighten dramatically over the space of just 10 minutes, which shows that the size of the flare's source can be no larger than the distance that light travels in 10 minutes. (We used a similar argument in Section 24-3 to show that the flickering X-ray source Cygnus X-1 must be very small.) In 10 minutes light travels a distance of 1.8×10^8 km or 1.2 AU, and only a black hole could pack a mass of 3.7×10^6 M$_\odot$ into a volume that size or smaller. The X-ray flares were presumably emitted by blobs of material that were compressed and heated as they fell into the black hole (see Section 24-3).

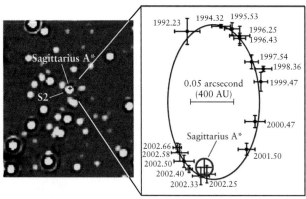

figure 25-23 R **I** V U X G
Stars Orbiting Sagittarius A* The diagram shows the results of ten years' observations of S2, one of several stars that orbit an unseen object at the position of the Sagittarius A* radio source. (The date of each observation is given in fractions of a year: for example, 2001.50 is July 1, 2001.) Note how rapidly S2 moved during 2002, when it was closest to Sagittarius A*. The actual plane of the orbit is not face-on to us, so S2 did not come as close to Sagittarius A* as the diagram might suggest. (ESO)

The X-ray flares from Sagittarius A* are relatively feeble, which suggests that the supermassive black hole is swallowing only relatively small amounts of material. But the region around Sagittarius A* is nonetheless an active and dynamic place. **Figure 25-24a** is a

(a) A radio view of the galactic center   R I V U X G

(b) An X-ray view of the galactic center   R I V U **X** G

figure 25-24
The Energetic Center of the Galaxy (a) The area shown in this radio image has the same angular size as the full moon. Sagittarius A*, at the very center of the Galaxy, is one of the brightest radio sources in the sky. Magnetic fields shape nearby interstellar gas into immense, graceful arches.

(b) This composite of images at X-ray wavelengths from 0.16 to 0.62 nm shows lobes of gas on either side of Sagittarius A*. The character of the X-ray emission shows that the gas temperature is as high as 2×10^7 K. (a: NRAO/VLA/F. Zadeh et al.; b: NASA/CXC/MIT/F. K. Baganoff et al.)

wide-angle radio image of the galactic center covering an area more than 60 pc (200 ly) across. Huge filaments of gas stretch for 20 pc (65 ly) northward of the galactic center (to the right and upward in Figure 25-24*a*), then abruptly arch southward (down and to the left in the figure). The orderly arrangement of these filaments is reminiscent of prominences on the Sun (see Section 18-10, especially Figure 18-28). This suggests that, as on the Sun, there is ionized gas at the galactic center that is being controlled by a powerful magnetic field. Indeed, much of the radio emission from the galactic center is synchrotron radiation: As we saw in Section 23-4, such radiation is produced by high-energy electrons spiraling in a magnetic field.

The false-color X-ray image in Figure 25-24*b* shows the immediate vicinity of Sagittarius A*. The black hole is flanked by lobes of hot, ionized, X-ray–emitting gas that extend for dozens of light years. These are thought to be the relics of immense explosions that may have taken place over the past several thousand years. Perhaps these past explosions cleared away much of the material around

Sagittarius A*, leaving only small amounts to fall into the black hole. This could explain why the X-ray flares from Sagittarius A* are so weak.

The supermassive black hole at the center of our Galaxy is not unique. Observations show that such titanic black holes are a feature of most large galaxies. In Chapter 27 we will see how black holes of this kind power *quasars*, the most luminous sustained light sources in the cosmos.

 Astronomers are still groping for a better understanding of the galactic center. With future developments in very-long-baseline interferometry (described in Section 6-6), it may be possible to actually obtain a picture of the supermassive black hole lurking there. During the coming years, observations from Earth-orbiting satellites as well as from radio and infrared telescopes on the ground will certainly add to our knowledge of the core of the Milky Way.

Key Words

The term preceded by an asterisk (*) is discussed in Box 25-1.

central bulge (of a galaxy), p. 558

dark matter, p. 565

density wave, p. 568

disk (of a galaxy), p. 558

far-infrared, p. 558

flocculent spiral galaxy, p. 570

galactic nucleus, p. 557

galaxy, p. 555

globular cluster, p. 556

grand-design spiral galaxy, p. 570

H I, p. 560

halo (of a galaxy), p. 558

high-velocity star, p. 558

interstellar extinction, p. 556

Local Bubble, p. 560

*magnetic resonance imaging (MRI), p. 561

massive compact halo object (MACHO), p. 567

microlensing, p. 567

Milky Way Galaxy, p. 555

near-infrared, p. 558

rotation curve, p. 565

RR Lyrae variable, p. 556

Sagittarius A*, p. 571

self-propagating star formation, p. 570

spin (of a particle), p. 560

spin-flip transition, p. 560

spiral arm, p. 559

21-cm radio emission, p. 560

weakly interacting massive particle (WIMP), p. 567

winding dilemma, p. 568

Key Ideas

The Shape and Size of the Galaxy: Our Galaxy has a disk about 50 kpc (160,000 ly) in diameter and about 600 pc (2000 ly) thick, with a high concentration of interstellar dust and gas in the disk.

• The galactic center is surrounded by a large distribution of stars called the central bulge. This bulge is not perfectly symmetrical, but may have a bar or peanut shape.

• The disk of the Galaxy is surrounded by a spherical distribution of globular clusters and old stars, called the galactic halo.

• There are about 200 billion (2×10^{11}) stars in the Galaxy's disk, central bulge, and halo.

The Sun's Location in the Galaxy: Our Sun lies within the galactic disk, some 8000 pc (26,000 ly) from the center of the Galaxy.

• Interstellar dust obscures our view at visible wavelengths along lines of sight that lie in the plane of the galactic disk. As a result,

the Sun's location in the Galaxy was unknown for many years. This dilemma was resolved by observing parts of the Galaxy outside the disk.

• The Sun orbits around the center of the Galaxy at a speed of about 790,000 km/h. It takes about 220 million years to complete one orbit.

The Rotation of the Galaxy and Dark Matter: From studies of the rotation of the Galaxy, astronomers estimate that the total mass of the Galaxy is about 10^{12} $M_\odot$. Only about 10% of this mass is in the form of visible stars, gas, and dust. The remaining 90% is in some nonvisible form, called dark matter, that extends beyond the edge of the luminous material in the Galaxy.

• Our Galaxy's dark matter may be a combination of MACHOs (dim, star-sized objects), massive neutrinos, and WIMPs (relatively massive subatomic particles).

The Galaxy's Spiral Structure: OB associations, H II regions, and molecular clouds in the galactic disk outline huge spiral arms.

• Spiral arms can be traced from the positions of clouds of atomic hydrogen. These can be detected throughout the galactic disk by the 21-cm radio waves emitted by the spin-flip transition in hydrogen. These emissions easily penetrate the intervening interstellar dust.

Theories of Spiral Structure: There are two leading theories of spiral structure in galaxies.

• According to the density-wave theory, spiral arms are created by density waves that sweep around the Galaxy. The gravitational field of this spiral pattern compresses the interstellar clouds through which it passes, thereby triggering the formation of the OB associations and H II regions that illuminate the spiral arms.

• According to the theory of self-propagating star formation, spiral arms are caused by the birth of stars over an extended region in a galaxy. Differential rotation of the galaxy stretches the star-forming region into an elongated arch of stars and nebulae.

The Galactic Nucleus: The innermost part of the Galaxy, or galactic nucleus, has been studied through its radio, infrared, and X-ray emissions (which are able to pass through interstellar dust).

• A strong radio source called Sagittarius A* is located at the galactic center. This marks the position of a supermassive black hole with a mass of about 3.7×10^6 $M_\odot$.

Review Questions

1. Why do the stars of the Galaxy appear to form a bright band that extends around the sky?

2. How did interstellar extinction mislead astronomers into believing that we are at the center of our Galaxy?

3. How did observations of globular clusters help astronomers determine our location in the Galaxy?

4. What are RR Lyrae stars? Why are they useful for determining the distance from our solar system to the center of the Galaxy?

5. Why are infrared telescopes useful for exploring the structure of the Galaxy? Why is it important to make observations at both near-infrared and far-infrared wavelengths?

6. The galactic halo is dominated by Population II stars, whereas the galactic disk contains predominantly Population I stars. In which of these parts of the Galaxy has star formation taken place recently? Explain.

7. What must happen within a hydrogen atom for it to emit a photon of wavelength 21 cm?

8. Most interstellar hydrogen atoms emit only radio waves at a wavelength of 21 cm, but some hydrogen clouds emit profuse amounts of visible light (see, for example, Figure 20-1 and Figure 20-2). What causes this difference?

9. How do astronomers determine the distances to H I (neutral hydrogen) clouds?

10. The radio map in Figure 25-12 has a large gap on the side of the Galaxy opposite to ours. Why is this?

11. In a spiral galaxy, are stars in general concentrated in the spiral arms? Why are spiral arms so prominent in visible-light images of spiral galaxies?

12. Many classic black-and-white photographs of spiral galaxies were taken using film that was most sensitive to blue light. Explain why the spiral arms were particularly prominent in such photographs.

13. What kinds of objects (other than H I clouds) do astronomers observe to map out the Galaxy's spiral structure? What is special about these objects? Which of these can be observed at great distances?

14. Why don't astronomers detect 21-cm radiation from the hydrogen in giant molecular clouds?

15. In what way are the orbits of stars in the galactic disk different from the orbits of planets in our solar system? What does this difference imply about the way that matter is distributed in the Galaxy?

16. How do astronomers determine how fast the Sun moves in its orbit around the Galaxy? How does this speed tell us about the amount of mass inside the Sun's orbit? Does this speed tell us about the amount of mass outside the Sun's orbit?

17. How do astronomers conclude that vast quantities of dark matter surround our Galaxy? How is this dark matter distributed in space?

18. Another student tells you that the Milky Way Galaxy is made up "mostly of stars." Is this statement accurate? Why or why not?

19. What is the difference between dark matter and dark nebulae?

20. What proposals have been made to explain the nature of dark matter? What experiments or observations have been made to investigate these proposals? What are the results of this research?

21. What is the winding dilemma? What does it tell us about the nature of spiral arms?

22. Do density waves form a stationary pattern in a galaxy? If not, do they move more rapidly, less rapidly, or at the same speed as stars in the disk?

23. In our Galaxy, why are stars of spectral classes O and B found in or near the spiral arms? Is the same true for stars of other spectral classes? Explain why or why not.

24. Compare the kinds of spiral arms produced by density waves with those produced by self-propagating star formation. By examining Figure 25-14, cite evidence that both processes may occur in our Galaxy.

25. What is the evidence that there is a supermassive black hole at the center of our Galaxy? How is it possible to determine the mass of this black hole?

Advanced Questions

Questions preceded by an asterisk (*) involve topics discussed in the Boxes.

Problem-solving tips and tools

Several of the following questions make extensive use of Newton's form of Kepler's third law, and you might find it helpful to review Box 4-4. Another useful version of Kepler's third law is given in Section 19-9. Box 19-2 discusses the relationship between luminosity, apparent brightness, and distance. We discussed the relationship between the energy and wavelength of a photon in Section 5-5. According to the Pythagorean theorem, an isosceles right triangle has a hypotenuse that is longer than its sides by a factor of $\sqrt{2} = 1.414$. The formula for the Schwarzschild radius of a black hole is given in Box 24-2. You will find the small-angle formula in Box 1-1 useful. It is also helpful to remember that an object 1 AU across viewed at a distance of 1 parsec has an angular size of 1 arcsecond. Remember, too, that the volume of a cylinder is equal to its height multiplied by the area of its base, and that the area of a circle of radius r is πr^2. You can find other geometrical formulae in Appendix 8.

---

26. Discuss how the Milky Way would appear to us if the Sun were relocated to (**a**) the edge of the Galaxy; (**b**) the galactic halo; (**c**) the galactic bulge.

27. Explain why globular clusters spend most of their time in the galactic halo, even though their eccentric orbits take them close to the galactic center.

28. The disk of the Galaxy is about 50 kpc in diameter and 600 pc thick. (**a**) Find the volume of the disk in cubic parsecs. (**b**) Find the volume (in cubic parsecs) of a sphere 300 pc in radius centered on the Sun. (**c**) If supernovae occur randomly throughout the volume of the Galaxy, what is the probability that a given supernova will occur within 300 pc of the Sun? If there are about three supernovae each century in our Galaxy, how often, on average, should we expect to see one within 300 pc of the Sun?

*29. An RR Lyrae star whose peak luminosity is 100 $L_\odot$ is in a globular cluster. At its peak luminosity, this star appears from Earth to be only 1.47×10^{-18} as bright as the Sun. Determine the distance to this globular cluster (**a**) in AU and (**b**) in parsecs.

30. A typical hydrogen atom in interstellar space undergoes a spin-flip transition only once every 10^7 years. How, then, is it at all possible to detect the 21-cm radio emission from interstellar hydrogen?

31. Calculate the energy of the photon emitted when a hydrogen atom undergoes a spin-flip transition. How many such photons would it take to equal the energy of a single H_α photon of wavelength 656.3 nm?

32. Suppose you were to use a radio telescope to measure the Doppler shift of 21-cm radiation in the plane of the Galaxy. (**a**) If you observe 21-cm radiation from clouds of atomic hydrogen at an angle of 45° from the galactic center, you will see the highest Dop-

pler shift from a cloud that is as far from the galactic center as it is from the Sun. Explain this statement using a diagram. (**b**) Find the distance from the Sun to the particular cloud mentioned in (a).

33. Calculate approximately how many times our solar system has orbited the center of our Galaxy since the Sun and planets were formed 4.56×10^9 years ago.

34. Sketch the rotation curve you would obtain if the Galaxy were rotating like a rigid body.

35. The mass of our Galaxy interior to the Sun's orbit is calculated from the radius of the Sun's orbit and its orbital speed. By how much would this estimate be in error if the calculated distance to the galactic center were off by 10%? By how much would this estimate be in error if the calculated orbital velocity were off by 10%? Explain your reasoning.

36. A gas cloud located in the spiral arm of a distant galaxy is observed to have an orbital velocity of 400 km/s. If the cloud is 20,000 pc from the center of the galaxy and is moving in a circular orbit, find (**a**) the orbital period of the cloud and (**b**) the mass of the galaxy contained within the cloud's orbit.

37. According to the Galaxy's rotation curve in Figure 25-16, a star 16 kpc from the galactic center has an orbital speed of about 270 km/s. Calculate the mass within that star's orbit.

38. Speculate on the reasons for the rapid rise in the Galaxy's rotation curve (see Figure 25-16) at distances close to the galactic center.

*39. Show that the form of Kepler's third law stated in Box 25-2, $P^2 = 4\pi^2 a^3 / G(M + M_\odot)$, is equivalent to $M = rv^2/G$, provided the orbit is a circle. (*Hint:* The mass of the Sun ($M_\odot$) is much less than the mass of the Galaxy inside the Sun's orbit (M).)

40. The accompanying image shows the spiral galaxy M74, located about 55 million light-years from the Earth in the constellation

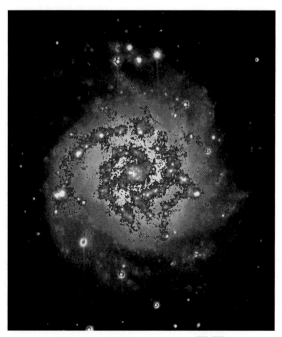

(Astronomical Society of the Pacific)   R I **V** **U** X G

Pisces (the Fish). It is actually a superposition of two false-color images: The red portion is an optical image taken at visible wavelengths, while the blue portion is an ultraviolet image made by NASA's Ultraviolet Imaging Telescope, which was carried into orbit by the space shuttle *Columbia* during the *Astro-1* mission in 1990. Compare the visible and ultraviolet images and, from what you know about stellar evolution and spiral structure, explain the differences you see.

41. (a) Calculate the Schwarzschild radius of a supermassive black hole of mass 3.7×10^6 M$_\odot$, the estimated mass of the black hole at the galactic center. Give your answer in both kilometers and astronomical units. (b) What is the angular diameter of such a black hole as seen at a distance of 8 kpc, the distance from the Earth to the galactic center? Give your answer in arcseconds. Observing an object with such a small angular size will be a challenge indeed! (c) What is the angular diameter of such a black hole as seen from a distance of 120 AU, the closest that the star S2 comes to Sagittarius A? Again, give your answer in arcseconds. Would it be discernible to the naked eye at that distance? (A normal human eye can see details as small as about 60 arcseconds.)

42. (a) The diagram in Figure 25-23 shows that at the distance of Sagittarius A, a length of 400 AU has an angular size of 0.05 arcsecond. Use this information to calculate the distance to Sagittarius A*. (b) Using the data given in Figure 25-23, estimate the orbital period of the star S2. (*Hint:* How long does S2 appear to take to complete half an orbit?) Given the period and the semimajor axis of the star's orbit, is it possible to calculate the mass of S2 itself? If it is, explain how this could be done; if not, explain why not.

43. The stars S0-1 and S0-2 orbit Sagittarius A with orbital periods of 63 and 17 years, respectively. (a) Assuming that the supermassive black hole in Sagittarius A* has a mass of 3.7×10^6 M$_\odot$, determine the semimajor axes of the orbits of these two stars. Give your answers in AU. (b) Calculate the angular size of each orbit's semimajor axis as seen from Earth. (See Section 25-1 for the distance from the Earth to the center of the Galaxy.) Explain why extremely high-resolution infrared images are required to observe the motions of these stars.

44. Consider a star that orbits around Sagittarius A* in a circular orbit of radius 530 AU. (a) If the star's orbital speed is 2500 km/s, what is its orbital period? Give your answer in years. (b) Determine the sum of the masses of Sagittarius A* and the star. Give your answer in solar masses. (Your answer is an estimate of the mass of Sagittarius A*, because the mass of a single star is negligibly small by comparison.)

Discussion Questions

45. From what you know about stellar evolution, the interstellar medium, and the density-wave theory, explain the appearance and structure of the spiral arms of grand-design spiral galaxies.

46. What observations would you make to determine the nature of the dark matter in our Galaxy's halo?

47. Describe how the appearance of the night sky might change if dark matter were visible to our eyes.

48. Discuss how a supermassive black hole could have formed at the center of our Galaxy.

Web/CD-ROM Questions

49. Some scientists have suggested that the rotation curve of the Galaxy can be explained by modifying the laws of physics rather than by positing the existence of dark matter. Search the World Wide Web for information about these proposed modifications, called MOND (for *MO*dified *N*ewtonian *D*ynamics). Under what circumstances do MOND and conventional physics predict different behavior? What evidence is there that MOND might be correct?

50. **Fast-Moving Stars at the Galactic Center.** Access and view the video "Fast-Moving Stars at the Galactic Center" in Chapter 25 of the *Universe* Web site or CD-ROM. Explain how you can tell which of the stars in the video are actually close to Sagittarius A* and which just happen to lie along our line of sight to the center of the Galaxy.

Observing Projects

51. Use the *Starry Night Backyard*™ program to observe the Milky Way. (a) Display the entire celestial sphere (select **Guides > Atlas** in the **Go** menu). Using the "hand" cursor, look at different parts of the Milky Way. Can you identify the direction toward the galactic nucleus? In this direction the Milky Way appears broadest. Check your identification by centering the field of view on the constellation Sagittarius (use the **Find...** command in the **Edit** menu). (b) Select **Go Home** in the **View...** menu to make the horizon visible, then set the local time to midnight (12:00:00 A.M.). Then set the date to January 1, then February 1, and so on. In which month is the galactic nucleus highest in the sky at midnight, so that it is most easily seen?

52. Use the *Starry Night Backyard*™ program to determine how the plane of the Galaxy is oriented on the celestial sphere. Display the entire celestial sphere (select **Guides > Atlas** in the **Go** menu). Next, use the hand cursor to adjust your field of view so that the celestial equator, shown as a red line, runs across the middle of the screen. Finally, move your field of view until you can see where the Milky Way crosses the celestial equator. (a) Measure the angle between the Milky Way and the celestial equator either on the screen or on a printout of the screen. (Some estimation is needed here.) How well aligned with each other are the plane of the Milky Way and the plane of

the Earth's equator? (**b**) A third plane of interest is the plane of the ecliptic, which is shown as a green line. (If you can't see this, select **Show View Options Panel** in the **View** menu and check the box for "The Ecliptic.") Adjust your field of view so that the ecliptic appears as a straight line rather than as a curve (which means you are viewing in a direction that lies in the ecliptic plane) and so that you can see where the ecliptic crosses the Milky Way. Measure the angle between the Milky Way and the ecliptic either on the screen or on a printout of the screen. (Again, some estimation will be needed.) How well aligned with each other are the plane of the Milky Way and the plane of the Earth's orbit around the Sun?

 53. Use the *Deep Space Explorer*™ program to measure the dimensions of the Milky Way Galaxy. In the left-hand part of the window under the heading **Milky Way,** click on the triangle next to the word **Explore** and then click on **Sun in Milky Way.** You can zoom in or out using the buttons at the upper left of the window (an upward-pointing triangle and a downward-pointing triangle). You can rotate the Milky Way Galaxy by putting the mouse cursor over the image, holding down the mouse button, and moving the mouse. (On a two-button mouse, hold down the left mouse button.) (**a**) Rotate the Galaxy so that you are seeing it face-on. Using a ruler, measure the distance on the screen from the center of the Galaxy to the position of the Sun. According to this chapter, what is this actual distance in kiloparsecs? (**b**) Now rotate the Galaxy so that you are seeing it edge-on. Using your ruler, measure the diameter of the central bulge of the Galaxy. Using your measurements in part (a) to set the scale, calculate the actual diameter of the central bulge in kiloparsecs.

Collaborative Exercises

54. Student book bags often contain a wide collection of odd-shaped objects. Each person in your group should rummage through their own book bags and find one object that is most similar to the Milky Way Galaxy in shape. List the items from each group member's belongings and describe what about the items is similar to the shape of our Galaxy and what about the items is not similar, then indicate which of the items is the closest match.

55. One strategy for identifying a central location is called *triangulation*. In triangulation, a central position can be pinpointed by knowing the distance from each of three different places. First, on a piece of paper, create a rough map showing where each person in your group lives. Second, create a circle around each person's home that has a radius equal to the distance that each home is from your classroom. Label the place where the circles intersect as your classroom. Why can you not identify the position of the classroom with only two people's circles?

56. A technique for mapping our Galaxy, shown in Figure 25-11, shows how emission spectra from hydrogen clouds would be shifted due to their motion around the Galaxy. Create a similar sketch showing an oval automobile racetrack with four cars moving on the track and a stationary observer outside the track at one end. Position and label the four moving cars all sounding their horns: (1) one that would have its horn sound shifted to longer wavelengths; (2) one that would have its horn sound shifted to shorter wavelengths; and (3) two cars moving in opposite directions so that their horn sounds would have no Doppler shift at all.

26 Galaxies

A century ago, most astronomers thought that the entire universe was only a few thousand light-years across and that nothing lay beyond our Milky Way Galaxy. One of the most important discoveries of the twentieth century was that this picture was utterly wrong. We now understand that the Milky Way is just one of billions of galaxies strewn across billions of light-years. The accompanying image shows three of them—the spiral galaxy M31 in the constellation Andromeda and two smaller galaxies (M32 and M110) that orbit it.

Some galaxies are spirals like M31 or the Milky Way, with arching spiral arms that are active sites of star formation. Others, like M32 and M110, are featureless, ellipse-shaped agglomerations of stars, virtually devoid of interstellar gas and dust. Some galaxies are only one-hundredth the size and one ten-thousandth the mass of the Milky Way. Others are giants, with 5 times the size and 50 times the mass of the Milky Way. Only about 10% of a typical galaxy's mass emits radiation of any kind; the remainder is made up of the mysterious dark matter.

Just as most stars are found within galaxies, most galaxies are located in groups and clusters. These clusters of galaxies stretch across the universe, forming huge, lacy patterns. Remarkably, remote clusters of galaxies are receding from us; the greater their distance, the more rapidly they are moving away. This relationship between distance and recessional velocity, called the Hubble law, reveals that our immense universe is expanding. In Chapters 28 and 29 we will learn what this implies about the distant past and remote future of the universe.

 The Great Galaxy in Andromeda (M31) and its two satellite galaxies. (Bill and Sally Fletcher/Tom Stack & Associates) R I **V** U X G

Questions to Guide Your Reading

As you read the sections of this chapter, look for the answers to the following questions:

26-1 How did astronomers first discover other galaxies?

26-2 How did astronomers first determine the distances to galaxies?

26-3 Do all galaxies have spiral arms, like the Milky Way?

26-4 How do modern astronomers tell how far away galaxies are?

26-5 How do the spectra of galaxies tell astronomers that the universe is expanding?

26-6 Are galaxies isolated in space, or are they found near other galaxies?

26-7 What happens when galaxies collide with each other?

26-8 Is dark matter found in galaxies beyond the Milky Way?

26-9 How do astronomers think galaxies formed?

26-1 | When galaxies were first discovered, it was not clear that they lie far beyond the Milky Way

As early as 1755, the German philosopher Immanuel Kant suggested that vast collections of stars lie outside the confines of the Milky Way. Less than a century later, an Irish astronomer observed the structure of some of the "island universes" that Kant proposed.

William Parsons, the third Earl of Rosse, was a wealthy amateur astronomer who used his fortune to build immense telescopes. His pièce de résistance, completed in February 1845, had an objective mirror 1.8 meters (6 feet) in diameter (Figure 26-1). The mirror was mounted at one end of a 60-foot tube controlled by cables, straps, pulleys, and cranes (Figure 26-1a). For many years, this triumph of nineteenth-century engineering was the largest telescope in the world.

With this new telescope, Rosse examined many of the nebulae that had been discovered and cataloged by William Herschel. He observed that some of these nebulae have a distinct spiral structure. One of the best examples is M51, also called NGC 5194.

Lacking photographic equipment, Lord Rosse had to make drawings of what he saw. Figure 26-1b shows a drawing he made of M51, which compares favorably with modern photographs (Figure 26-2). Views such as this inspired Lord Rosse to echo Kant's proposal that such nebulae are actually "island universes."

Many astronomers of the nineteenth century disagreed with this notion of island universes. A considerable number of nebulae are in fact scattered throughout the Milky Way. (Figures 20-2 and 20-3 show some examples.) It therefore seemed reasonable that "spiral nebulae," even though they are very different in shape from other sorts of nebulae, could also be components of our Galaxy.

 The astronomical community became increasingly divided over the nature of the spiral nebulae. In April 1920, two opposing ideas were presented before the National Academy of Sciences in Washington, D.C. On one side was Harlow

M31 (NGC 5194)

NGC 5195

Red H II regions

Figure 26-2   R I **V** U X G

A Modern View of the Spiral Galaxy M51   This galaxy, also called NGC 5194, is some 8.5 Mpc (8.5 million parsecs, or 28 million ly) from Earth and is about 20 kpc (20,000 parsecs, or 65,000 ly) in diameter. Its spiral arms are outlined by glowing H II regions, which reveal the sites of star formation (see Section 20-2). One spiral arm extends toward the companion galaxy NGC 5195. (CFHT)

Shapley from the Mount Wilson Observatory, renowned for his recent determination of the size of the Milky Way Galaxy (see Section 25-1). Shapley thought the spiral nebulae were relatively small, nearby objects scattered around our Galaxy like the globular clusters he had studied. Opposing Shapley was Heber D. Curtis of the University of California's Lick Observatory. Curtis championed the

Telescope tube

(a) Rosse's "Leviathan of Parsonstown"   R I **V** U X G

(b) M51 as viewed through the "Leviathan"

Figure 26-1

A Pioneering View of Another Galaxy   (a) Built in 1845, this structure housed the largest telescope of its day. The telescope itself (the black cylinder pointing at a 45° angle above the horizontal) was restored to its original state during 1996–1998. (b) Using his telescope, Lord Rosse made this sketch of spiral structure in M51. This galaxy, whose angular size is 8 × 11 arcminutes (about a third the angular size of the full moon), is today called the Whirlpool Galaxy because of its distinctive appearance. (a: Birr Castle Demesne; b: Courtesy of Lund Humphries) island universe theory, arguing that each of these spiral nebulae is a rotating system of stars much like our own Galaxy.

The Shapley-Curtis "debate" generated much heat but little light. Nothing was decided, because no one could present conclusive evidence to demonstrate exactly how far away the spiral nebulae are. Astronomy desperately needed a definitive determination of the distance to a spiral nebula. Such a measurement became the first great achievement of a young man who studied astronomy at the Yerkes Observatory, near Chicago. His name was Edwin Hubble.

26-2 | Hubble proved that the spiral nebulae are far beyond the Milky Way

After completing his studies, Edwin Hubble joined the staff of the Mount Wilson Observatory in Pasadena, California. On October 6, 1923, he took a historic photograph of the Andromeda "Nebula," one of the spiral nebulae around which controversy raged. (The image that opens this chapter is a modern photograph of this object.)

Hubble carefully examined his photographic plate and discovered what he at first thought to be a nova. Referring to previous plates of that region, he soon realized that the object was actually a Cepheid variable star. Further scrutiny of additional plates over the next several months revealed several more Cepheids. **Figure 26-3** shows modern observations of a Cepheid in another "spiral nebula."

 Figure 26-3 R I **V** U X G
A Cepheid Variable in a Distant Galaxy By observing Cepheid variable stars in M100, the galaxy shown here, astronomers have found that it is about 17 Mpc (56 million ly) from Earth. The insets show one of the Cepheids in M100 at different stages in its brightness cycle, which lasts several weeks. (Wendy L. Freedman, Carnegie Institution of Washington, and NASA)

As we saw in Section 21-5, Cepheid variables help astronomers determine distances. An astronomer begins by carefully measuring the variations in apparent brightness of a Cepheid variable, then recording the results in the form of a plot of brightness versus time, or light curve (see Figure 21-16a). This graph gives the variable star's period and average brightness. Given the star's period, the astronomer then uses the period-luminosity relation shown in Figure 21-17 to find the Cepheid's average luminosity. (The astronomer must also examine the spectrum of the star to determine whether it is a metal-rich Type I Cepheid or a metal-poor Type II Cepheid. As Figure 21-17 shows, these have somewhat different period-luminosity relations.) Knowing both the apparent brightness and luminosity of the Cepheid, the astronomer can then use the inverse-square law to calculate the distance to the star (see Box 19-2). **Box 26-1** presents an example of how this is done.

Cepheid variables are intrinsically quite luminous, with average luminosities that can exceed $10^4 L_\odot$. Hubble realized that for these luminous stars to appear as dim as they were on his photographs of the Andromeda "Nebula," they must be extremely far away. Straightforward calculations using modern data reveal that M31 is some 750 kiloparsecs (2.5 million light-years) from Earth. Based on its angular size, M31 has a diameter of 70 kiloparsecs—larger than the diameter of our own Milky Way Galaxy!

These results prove that the Andromeda "Nebula" is actually an enormous stellar system, far beyond the confines of the Milky Way. Today, this system is properly called the Andromeda Galaxy. (Under good observing conditions, you can actually see this galaxy's central bulge with the naked eye. If you could see the entire Andromeda Galaxy, it would cover an area of the sky roughly 5 times as large as the full moon.)

Hubble's results, which were presented at a meeting of the American Astronomical Society on December 30, 1924, settled the Shapley-Curtis "debate" once and for all. The universe was recognized to be far larger and populated with far bigger objects than anyone had seriously imagined. Hubble had discovered the realm of the galaxies.

CAUTION In everyday language, many people use the words "galaxy" and "universe" interchangeably. It is true that before Hubble's discoveries our Milky Way Galaxy was thought to constitute essentially the entire universe. But we now know that the universe contains literally billions of galaxies. A single galaxy, vast though it may be, is just a tiny part of the entire observable universe.

26-3 | Galaxies are classified according to their appearance

Millions of galaxies are visible across every unobscured part of the sky. Although all galaxies are made up of large numbers of stars, they come in a variety of shapes and sizes.

WEB LINK 26.4 Hubble classified galaxies into four broad categories based on their appearance. These categories form the basis for the **Hubble classification**, a scheme that is still

box 26-1 | TOOLS OF THE ASTRONOMER'S TRADE

Cepheids and Supernovae as Indicators of Distance

Because their periods are directly linked to their luminosities, Cepheid variables are one of the most reliable tools astronomers have for determining the distances to galaxies. To this day, astronomers use this link—much as Hubble did back in the 1920s—to measure intergalactic distances. More recently, they have begun to use Type Ia supernovae, which are far more luminous and thus can be seen much farther away, to determine the distances to very remote galaxies.

EXAMPLE: In 1992 a team of astronomers observed Cepheid variables in a galaxy called IC 4182 to deduce this galaxy's distance from the Earth. One such Cepheid has a period of 42.0 days and an average apparent magnitude (*m*) of +22.0. (See Box 19-3 for an explanation of the apparent magnitude scale.) By comparison, the dimmest star you can see with the naked eye has *m* = +6; this Cepheid in IC 4182 appears less than one one-millionth as bright. The star's spectrum shows that it is a metal-rich Type I Cepheid variable.

According to the period-luminosity relation shown in Figure 21-17, such a Type I Cepheid with a period of 42.0 days has an average luminosity of 33,000 $L_\odot$. This can expressed by saying that this Cepheid has an average absolute magnitude (*M*) of –6.5. (This compares to *M* = +4.8 for the Sun.) Use this information to determine the distance to IC 4182.

Situation: We are given the apparent magnitude *m* = +22.0 and the absolute magnitude *M* = –6.5 of the Cepheid variable star in IC 4182. Our goal is to calculate the distance to this star, and hence the distance to the galaxy of which it is part.

Tools: We use the relationship between apparent magnitude, absolute magnitude, and distance given in Box 19-3.

Answer: In Box 19-3, we saw that the apparent magnitude of a star is related to its absolute magnitude and distance in parsecs (*d*) by

$$m - M = 5 \log d - 5$$

This can be rewritten as

$$d = 10^{(m - M + 5)/5} \text{ parsecs}$$

We have *m* – *M* = (+22.0) – (–6.5) = 22.0 + 6.5 = 28.5. (Recall from Box 19-3 that *m* – *M* is called the *distance modulus*.) Hence, our equation becomes

$$d = 10^{(28.5 + 5)/5} \text{ parsecs} = 10^{6.7} \text{ parsecs} = 5 \times 10^6 \text{ parsecs}$$

(A calculator is needed to calculate the quantity $10^{6.7}$.)

Review: Our result tells us that the galaxy is 5 million parcsecs, or 5 Mpc, from Earth (1 Mpc = 10^6 pc). This distance can also be expressed as 16 million light-years.

EXAMPLE: Astronomers are interested in IC 4182 because a Type Ia supernova was observed there in 1937. All Type Ia supernovae are exploding white dwarfs that reach nearly the same maximum brightness at the peak of their outburst (see Section 22-9). Once astronomers know the peak absolute magnitude of Type Ia supernovae, they can use these supernovae as distance indicators. Because the distance to IC 4182 is known from its Cepheids, the 1937 observations of the supernova in that galaxy help us calibrate Type Ia supernovae as distance indicators.

At maximum brightness, the 1937 supernova reached an apparent magnitude of *m* = +8.6. What was its absolute magnitude at maximum brightness?

Situation: We are given the supernova's apparent magnitude *m*, and we know its distance from the previous example. Our goal is to calculate its absolute magnitude *M*.

Tools: We again use the relationship *m* – *M* = 5 log *d* – 5.

Answer: We could plug in the value of *d* found in the previous example. But it is simpler to note that the distance modulus *m* – *M* has the same value no matter whether it refers to a Cepheid, a supernova, or any other object, just so it is at the same distance *d*. From the Cepheid example we have *m* – *M* = 28.5 for IC 4182, so

$$M = m - (m - M) = 8.6 - (28.5) = -19.9$$

This absolute magnitude corresponds to a remarkable peak luminosity of 10^{10} $L_\odot$.

Review: Whenever astronomers find a Type Ia supernova in a remote galaxy, they can combine this absolute magnitude with the observed maximum apparent magnitude to get the galaxy's distance modulus, from which the galaxy's distance can be easily calculated (just as we did above for the Cepheids in IC 4182). This technique has been used to determine the distances to galaxies hundreds of millions of parsecs away (see Section 26-4).

table 26-1	Some Properties of Galaxies		
	Spiral (S) and barred spiral (SB) galaxies	Elliptical galaxies (E)	Irregular galaxies (Irr)
Mass ($M_\odot$)	10^9 to 4×10^{11}	10^5 to 10^{13}	10^8 to 3×10^{10}
Luminosity ($L_\odot$)	10^8 to 2×10^{10}	3×10^5 to 10^{11}	10^7 to 10^9
Diameter (kpc)	5 to 250	1 to 200	1 to 10
Stellar populations	Spiral arms: young Population I Nucleus and throughout disk: Population II and old Population I	Population II and old Population I	mostly Population I
Percentage of observed galaxies	77%	20%*	3%

This percentage does not include dwarf elliptical galaxies that are as yet too dim and distant to detect. Hence, the actual percentage of galaxies that are ellipticals may be higher than shown here.

used today. The four classes of galaxies are the spirals, classified S; barred spirals, or SB; ellipticals, E; and irregulars, Irr. Table 26-1 summarizes some key properties of each class. These various types of galaxies differ not only in their shapes but also in the kinds of processes taking place within them.

M31 (shown in the image that opens this chapter), M51 (see Figure 26-2), and M100 (see Figure 26-3) are examples of **spiral galaxies.** Figure 26-4 shows that spiral galaxies are characterized by arched lanes of stars, just as is our own Milky Way Galaxy (see Section 25-3). The spiral arms contain young, hot, blue stars and their associated H II regions, indicating ongoing star formation.

Thermonuclear reactions within stars create *metals*, that is, elements heavier than hydrogen or helium (see Section 19-5). These metals are dispersed into space as the stars evolve and die. So, if new stars are being formed from the interstellar matter in spiral galaxies, they will incorporate these metals and be Population I stars (see Section 21-4 and Section 25-2). Indeed, the visible-light spectrum of the disk of a spiral galaxy has strong metal absorption lines. Such a spectrum is a composite of the spectra of many stars and shows that the stars in the disk are principally of Population I. By contrast, there is relatively little star formation in the central bulges of spiral galaxies, and these regions are dominated by old Population II stars that have a low metal content. This also explains why the central bulges of spiral galaxies have a yellowish or reddish color; as a pop-

(a) Sa (NGC 1357)

(b) Sb (M81)

(c) Sc (NGC 4321)

figure 26-4 R I **V** U X G

Spiral Galaxies Edwin Hubble classified spiral galaxies according to the texture of their spiral arms and the relative size of their central bulges. Sa galaxies have smooth, broad spiral arms and the largest central bulges, while Sc galaxies have narrow, well-defined arms and the smallest central bulges. (a: STScI Digital Sky Survey; b: N. A. Sharp/AURA/NOAO/NSF; c: AngloAustralian Observatory) ulation of stars ages, the massive, luminous blue stars die off first, leaving only the longer-lived, low-mass red stars.

Hubble further classified spiral galaxies according to the texture of the spiral arms and the relative size of the central bulge. Spirals with smooth, broad spiral arms and a fat central bulge are called Sa galaxies, for spiral type *a* (Figure 26-4*a*); those with moderately well-defined spiral arms and a moderate-sized central bulge, like M31 and M51, are Sb galaxies (Figure 26-4*b*); and galaxies with narrow, well-defined spiral arms and a tiny central bulge are Sc galaxies (Figure 26-4*c*).

The differences between Sa, Sb, and Sc galaxies may be related to the relative amounts of gas and dust they contain. Observations with infrared telescopes (which detect the emission from interstellar dust) and radio telescopes (which detect radiation from interstellar gases such as hydrogen and carbon monoxide) show that about 4% of the mass of a Sa galaxy is in the form of gas and dust. This percentage is 8% for Sb galaxies and 25% for Sc galaxies.

Interstellar gas and dust is the material from which new stars are formed, so an Sc galaxy has a greater proportion of its mass involved in star formation than an Sb or Sa galaxy. Hence, a Sc galaxy has a large disk (where star formation occurs) and a small central bulge (where there is little or no star formation). By comparison, a Sa galaxy, which has relatively little gas and dust, and thus less material from which to form stars, has a large central bulge and only a small star-forming disk.

In **barred spiral galaxies,** such as those shown in Figure 26-5, the spiral arms originate at the ends of a bar-shaped region running through the galaxy's nucleus rather than from the nucleus itself. As with ordinary spirals, Hubble subdivided barred spirals according to the relative size of their central bulge and the character of their spiral arms. A SBa galaxy has a large central bulge and thin, tightly wound spiral arms (Figure 26-5*a*). Likewise, a SBb galaxy is a barred spiral with a moderate central bulge and moderately wound spiral arms (Figure 26-5*b*), while a SBc galaxy has lumpy, loosely

wound spiral arms and a tiny central bulge (Figure 26-5*c*). As for ordinary spiral galaxies, the difference between SBa, SBb, and SBc galaxies may be related to the amount of gas and dust in the galaxy.

 Bars appear to form naturally in many spiral galaxies. This conclusion comes from computer simulations of galaxies, which set hundreds of thousands of simulated "stars" into orbit around a common center. As the "stars" orbit and exert gravitational forces on one another, a bar structure forms in most cases. Indeed, barred spiral galaxies outnumber ordinary spirals by about two to one. (As we saw in Section 25-2, there is evidence that the Milky Way Galaxy is a barred spiral. The bar points roughly toward the Earth so that we see it almost end-on.)

Why don't all spiral galaxies have bars? According to calculations by Jeremiah Ostriker and P. J. E. Peebles of Princeton University, a bar will not develop if a galaxy is surrounded by a sufficiently massive halo of nonluminous *dark matter*. (In Section 25-4 we saw evidence that our Milky Way Galaxy is surrounded by such a dark matter halo.) The difference between barred spirals and ordinary spirals may thus lie in the amount of dark matter the galaxy possesses. In Section 26-8 we will see compelling evidence for the existence of dark matter in spiral galaxies.

Elliptical galaxies, so named because of their distinctly elliptical shapes, have no spiral arms. Hubble subdivided these galaxies according to how round or flattened they look. The roundest elliptical galaxies are called E0 galaxies and the flattest, E7 galaxies. Elliptical galaxies with intermediate amounts of flattening are given designations between these extremes (Figure 26-6).

CAUTION Unlike the designations for spirals and barred spirals, the classifications E0 through E7 may not reflect the true shape of elliptical galaxies. An E1 or E2 galaxy might actually be a very flattened disk of stars that we just happen to view face-on, and a cigar-shaped E7 galaxy might look spherical if

(a) SBa (NGC 4650)   (b) SBb (M83)   (c) SBc (NGC 1365)

Figure 26-5 R I **V** U X G

Barred Spiral Galaxies As with spiral galaxies, Hubble classified barred spirals according to the texture of their spiral arms (which correlates to the sizes of their central bulges). SBa galaxies have the smoothest spiral arms and the largest central bulges, while SBc galaxies have narrow, well-defined arms and the smallest central bulges. (a: AURA/NOAO/NSF; b, c: Anglo-Australian Observatory)

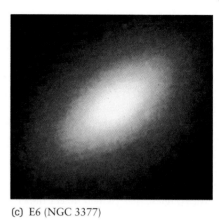

(a) E0 (M105)

(b) E3 (NGC 4365)

(c) E6 (NGC 3377)

figure 26-6 R I **V** U X G

Elliptical Galaxies Hubble classified elliptical galaxies according to how round or flattened they look. A galaxy that appears round is labeled E0, and the flattest-appearing elliptical galaxies are designated E7. (Courtesy of J. D. Wray; McDonald Observatory)

seen end-on. The Hubble scheme classifies galaxies entirely by how they *appear* to us on the Earth.

Elliptical galaxies look far less dramatic than their spiral and barred spiral cousins. The reason, as shown by radio and infrared observations, is that ellipticals are virtually devoid of interstellar gas and dust. Consequently, there is little material from which stars could have recently formed, and indeed there is no evidence of young stars in most elliptical galaxies. For the most part, star formation in elliptical galaxies ended long ago. Hence, these galaxies are composed of old, red, Population II stars with only small amounts of metals.

Elliptical galaxies come in a wide range of sizes and masses. Both the largest and the smallest galaxies in the known universe are elliptical. **Figure 26-7** shows two **giant elliptical galaxies** that are about 20 times larger than an average galaxy. These giant ellipticals are located near the middle of a large cluster of galaxies in the constellation Virgo. (We will discuss this cluster in Section 26-6.)

Giant ellipticals are rather rare, but **dwarf elliptical galaxies** are quite common. Dwarf ellipticals are only a fraction the size of their normal counterparts and contain so few stars—only a few million, compared to more than 100 billion (10^{11}) stars in our Milky Way Galaxy—that these galaxies are completely transparent. You can actually see straight through the center of a dwarf galaxy and out the other side, as **Figure 26-8** shows.

The visible light from a galaxy is emitted by its stars, so the visible-light spectrum of a galaxy has absorption lines. But because a galaxy's stars are in motion, with some approaching us and others moving away, the Doppler effect smears out and broadens the absorption lines. The average motions of stars in a galaxy can be deduced from the details of this broadening.

For elliptical galaxies, studies of this kind show that star motions are quite random. In a very round (E0) elliptical galaxy, this randomness is **isotropic,** meaning "equal in all directions." Because the stars are whizzing around equally in all directions, the galaxy is genuinely spherical. In a flattened (E7) elliptical galaxy, the randomness of the stellar motions is **anisotropic,** which means that the range of star speeds is different in different directions.

Edwin Hubble summarized his classification scheme for spiral, barred spiral, and elliptical galaxies in a diagram, now called the **tuning fork diagram** for its shape (**Figure 26-9**). According to this scheme, galaxies midway in appearance between ellipticals and the

M86

M84

figure 26-7 R I **V** U X G

Giant Elliptical Galaxies The Virgo cluster is a rich, sprawling collection of more than 2000 galaxies about 15 Mpc (50 million light-years) from the Earth. Only the center of this huge cluster appears in this photograph. The two largest members of this cluster are the giant elliptical galaxies M84 and M86. These galaxies have angular sizes of 5 to 7 arcmin. (Royal Observatory, Edinburgh)

figure 26-8 R I **V** U X G

A Dwarf Elliptical Galaxy This diffuse cloud of stars is a nearby E4 dwarf elliptical called Leo I. It actually orbits the Milky Way at a distance of about 180 kpc (600,000 ly). Leo I is about 1 kpc (3000 ly) in diameter but contains so few stars that you can see through the galaxy's center. (Anglo-Australian Observatory)

two kinds of spirals are denoted as S0 and SB0 galaxies, also called **lenticular galaxies.** Although they look somewhat elliptical, lenticular ("lens-shaped") galaxies have both a central bulge and a disk like spiral galaxies, but no discernible spiral arms. They are therefore sometimes referred to as "armless spirals." The giant elliptical galaxies M84 and M86, shown in Figure 26-7, may both actually be lenticular S0 galaxies.

CAUTION ! When Hubble first drew his tuning fork diagram, he had the idea that it represented an evolutionary sequence. He thought that galaxies evolved over time from the left to the right of the diagram, beginning as ellipticals and eventually becoming either spiral or barred spiral galaxies. We now understand that this is not the case at all! For one thing, elliptical galaxies have little or no overall rotation, while spiral and barred spiral galaxies have a substantial amount of overall rotation. There is no way that an elliptical galaxy could suddenly start rotating, which means that it could not evolve into a spiral galaxy.

A more modern interpretation of the Hubble tuning fork diagram is that it is an arrangement of galaxies according to their overall rotation. A rapidly rotating collection of matter in space tends to form a disk, while a slowly rotating collection does not. Thus, the elliptical galaxies at the far left of the tuning fork diagram have little internal rotation and hence no disk. Sa and SBa galaxies have enough overall rotation to form a disk, though their central bulges are still dominant. The galaxies with the greatest amount of overall rotation are Sc and SBc galaxies, in which the central bulges are small and most of the gas, dust, and stars are in the disk.

Galaxies that do not fit into the scheme of spirals, barred spirals, and ellipticals are usually referred to as **irregular galaxies.** They are generally rich in interstellar gas and dust, and have both young and old stars. For lack of any better scheme, the irregular galaxies are sometimes placed between the ends of the tines of the Hubble tuning fork diagram, as in Figure 26-9.

Hubble defined two types of irregulars. Irr I galaxies have only hints of organized structure, and have many OB associations and H II regions. The best-known examples of Irr I galaxies are the Large Magellanic Cloud (Figure 26-10) and the Small Magellanic Cloud. Both are nearby companions of our Milky Way and can be seen with the naked eye from southern latitudes. Both these galaxies contain substantial amounts of interstellar gas. Tidal forces exerted on these irregular galaxies by the Milky Way help

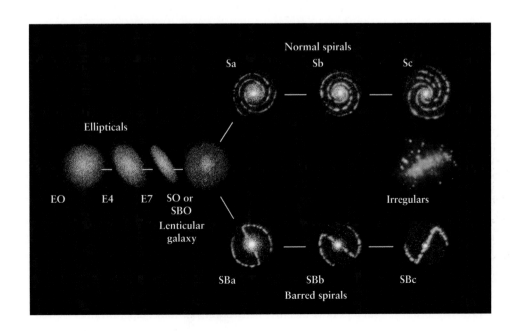

figure 26-9

Hubble's Tuning Fork Diagram Edwin Hubble's classification of regular galaxies is shown in his tuning fork diagram. An elliptical galaxy is classified by how flattened it appears. A spiral or barred spiral galaxy is classified by the texture of its spiral arms and the size of its central bulge. An S0 or SB0 galaxy, also called a lenticular galaxy, is intermediate between ellipticals and spirals. Irregular galaxies do not fit into this simple classification scheme.

Tarantula Nebula
(an H II region)

Large Magellanic Cloud

 figure 26-10 R I V U X G

The Large Magellanic Cloud (LMC) At a distance of only 55 kpc (179,000 ly), this Irr I galaxy is the third closest known companion of our Milky Way Galaxy. About 19 kpc (62,000 ly) across, the LMC spans 22° across the sky, or about 50 times the size of the full moon. One sign that star formation is ongoing in the LMC is the Tarantula Nebula, whose diameter of 250 pc (800 ly) and mass of 5×10^6 M$_\odot$ make it the largest known H II region. (Anglo-Australian Observatory)

to compress the gas, which is why both of the Magellanic Clouds are sites of active star formation.

The other type of irregulars, called Irr II galaxies, have asymmetrical, distorted shapes that seem to have been caused by collisions with other galaxies or by violent activity in their nuclei. M82, shown in Figure 24-14, is an example of an Irr II galaxy.

26-4 | Astronomers use various techniques to determine the distances to remote galaxies

 SCIENCE IN PROCESS *By studying pulsating stars and supernova explosions, astronomers gain essential tools for learning about galaxies*

A key question that astronomers ask about galaxies is "How far away are they?" Unfortunately, many of the techniques that are used to measure distances within our Milky Way Galaxy cannot be used for the far greater distances to other galaxies. The extremely accurate parallax method (see Section 19-1) can be used only for stars within about 500 pc. Beyond that distance, parallax angles become too small to measure. Spectroscopic parallax, in which the distance to a star or star cluster is found with the help of the H-R diagram (see Section 19-8), is accurate only out to roughly 10 kpc from Earth; more distant stars or clusters are too dim to give reliable results.

To determine the distance to a remote galaxy, astronomers look instead for a **standard candle**—an object, such as a star, that lies within that galaxy and for which we know the luminosity (or, equivalently, the absolute magnitude, described in Section 19-3). By measuring how bright the standard candle appears, astronomers can calculate its distance—and hence the distance to the galaxy of which it is part—using the inverse-square law.

The challenge is to find standard candles that are luminous enough to be seen across the tremendous distances to galaxies. To be useful, standard candles should have four properties:

1. They should be luminous, so we can see them out to great distances.

2. We should be fairly certain about their luminosities, so we can be equally certain of any distance calculated from a standard candle's apparent brightness and luminosity.

3. They should be easily identifiable—for example, by the shape of the light curve of a variable star.

4. They should be relatively common, so that astronomers can use them to determine the distances to many different galaxies.

For nearby galaxies, Cepheid variable stars make reliable standard candles. These variables can be seen out to about 30 Mpc (100 million ly) using the Hubble Space Telescope, and their luminosity can be determined from their period through the period-luminosity relation (Figure 21-17). Box 26-1 gives an example of using Cepheid variables to determine distances. RR Lyrae stars, which are Population II variable stars often found in globular clusters, can be used as standard candles in a similar way. (We saw in Section 25-1 how RR Lyrae variables helped determine the size of our Galaxy.) Because they are less luminous than Cepheids, RR Lyrae variables can be seen only out to 100 kpc (300,000 ly).

Beyond about 30 Mpc even the brightest Cepheid variables, which have luminosities of about 2×10^4 L$_\odot$, fade from view. Astronomers have tried to use even more luminous stars such as blue supergiants to serve as standard candles. However, this idea is based on the assumption that there is a fixed upper limit on the luminosities of stars, which may not be the case. Hence, these standard candles are not very "standard," and distances measured in this way are somewhat uncertain.

One class of standard candles that astronomers have used beyond 30 Mpc are Type Ia supernovae (see Section 22-9). These occur when a white dwarf in a close binary system accretes enough matter from its companion to blow itself apart in a thermonuclear conflagration. A Type Ia supernova can reach a maximum luminosity of about 3×10^9 L$_\odot$ (Figure 26-11). If a Type Ia supernova is seen in a distant galaxy and its maximum apparent brightness measured, the inverse-square law can be used to find the galaxy's distance (see Box 26-1).

One complication is that not all Type Ia supernovae are equally luminous. Fortunately, there is a simple relationship between the peak luminosity of a Type Ia supernova and the rate at which the luminosity decreases after the peak: The more slowly the brightness decreases, the more luminous the supernova. Using this relationship, astronomers have measured distances to supernovae more than 1000 Mpc (3 billion ly) from Earth.

Unfortunately, this technique can be used only for galaxies in which we happen to observe a Type Ia supernova. But telescopic surveys now identify many dozens of these supernovae every year,

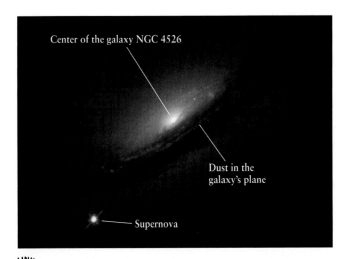

Center of the galaxy NGC 4526

Dust in the galaxy's plane

Supernova

figure 26-11 R I **V** U X G

A Supernova in a Spiral Galaxy This image from the Hubble Space Telescope shows a Type Ia supernova observed in 1994 in the outskirts of spiral galaxy NGC 4526. Note that the supernova is comparable in brightness to the entire galaxy. Such supernovae, which can be seen at extreme distances, are important standard candles used to determine the distances to faraway galaxies. (Harvard-Smithsonian Center for Astrophysics)

so the number of galaxies whose distances can be measured in this way is continually increasing.

Other methods for determining the distances to galaxies do not make use of standard candles. One was discovered in the 1970s by the astronomers Brent Tully and Richard Fisher. They found that the width of the hydrogen 21-cm emission line of a spiral galaxy (see Section 25-3) is related to the galaxy's luminosity. This correlation is the **Tully-Fisher relation**—the broader the line, the more luminous the galaxy.

Such a relationship exists because radiation from the approaching side of a rotating galaxy is blueshifted while that from the galaxy's receding side is redshifted. Thus, the 21-cm line is Doppler broadened by an amount directly related to how fast a galaxy is rotating. Rotation speed is related to the galaxy's mass by Newton's form of Kepler's third law. The more massive a galaxy, the more stars it contains and thus the more luminous it is. Consequently, the width of a galaxy's 21-cm line is directly related to its luminosity.

Because line widths can be measured quite accurately, astronomers can use the Tully-Fisher relation to determine the luminosity of a distant spiral galaxy. By combining this information with measurements of apparent brightness, they can calculate the distance to the galaxy. This technique can be used to measure distances of 100 Mpc or more.

Elliptical galaxies do not rotate, so the Tully-Fisher relation cannot be used to determine their distances. But in 1987 the American astronomers Marc Davis and George Djorgovski pointed out a correlation between the size of an elliptical galaxy, the average motions of its stars, and how the galaxy's brightness appears distributed over its surface.

In geometry, three points define a plane, so the relationship among size, motion, and brightness is called the **fundamental plane.** By measuring the last two quantities, an astronomer can use the fundamental plane relationship to determine a galaxy's actual size. And by comparing this to the galaxy's apparent size, the astronomer can calculate the distance to the galaxy using the small-angle formula (Box 1-1). Ellipticals can be substantially larger and more luminous than spirals, so the fundamental plane can be used at somewhat greater distances than the Tully-Fisher relation.

Figure 26-12 shows the ranges of applicability of several important means of determining astronomical distances. Because

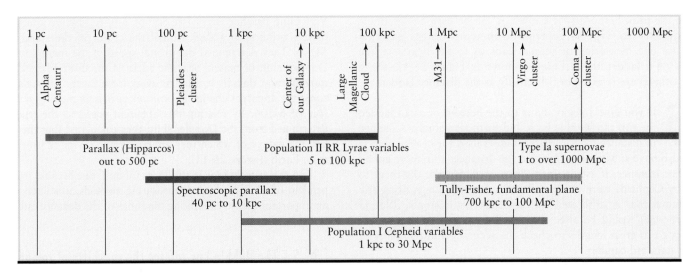

figure 26-12

The Distance Ladder Astronomers employ a variety of techniques for determining the distances to objects beyond the solar system. Because their ranges of applicability overlap, one technique can be used to calibrate another. The arrows indicate distances to several important objects. Note that each division on the scale indicates a tenfold increase in distance, such as from 1 to 10 Mpc.

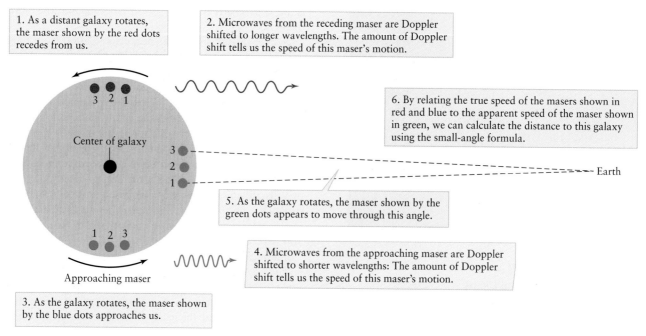

1. As a distant galaxy rotates, the maser shown by the red dots recedes from us.

2. Microwaves from the receding maser are Doppler shifted to longer wavelengths. The amount of Doppler shift tells us the speed of this maser's motion.

6. By relating the true speed of the masers shown in red and blue to the apparent speed of the maser shown in green, we can calculate the distance to this galaxy using the small-angle formula.

Center of galaxy

Earth

5. As the galaxy rotates, the maser shown by the green dots appears to move through this angle.

4. Microwaves from the approaching maser are Doppler shifted to shorter wavelengths: The amount of Doppler shift tells us the speed of this maser's motion.

Approaching maser

3. As the galaxy rotates, the maser shown by the blue dots approaches us.

Figure 26-13

Measuring the Distance to a Galaxy Using Masers This drawing shows interstellar clouds called masers (the colored dots) moving from position 1 to 2 to 3 as they orbit the center of a galaxy. The redshift and blueshift of microwaves from the masers shown in red and blue tell us their orbital speed. By relating this to the angle through which the masers shown in green appear to move in a certain amount of time, we can calculate the distance to the galaxy.

these ranges overlap, one technique can be used to calibrate another. As an example, astronomers have studied Cepheids in nearby galaxies that have been host to Type Ia supernovae. The Cepheids provide the distances to these nearby galaxies, making it possible to determine the peak luminosity of each supernova using its maximum apparent brightness and the inverse-square law. Once the peak luminosity is known, it can be used to determine the distance to Type Ia supernovae in more distant galaxies. Because one measuring technique leads us to the next one like rungs on a ladder, the techniques shown in Figure 26-12 (along with others) are referred to collectively as the **distance ladder.**

If you give a slight shake to the bottom of a tall ladder, the top can wobble back and forth alarmingly. A change in distance-measuring techniques used for nearby objects can also have substantial effects on the distances to remote galaxies. For instance, if astronomers discovered that the distances to nearby Cepheids were in error, distance measurements using any technique that is calibrated by Cepheids (for example, Type Ia supernovae) would be affected as well. For this reason, astronomers go to great lengths to check the accuracy and reliability of their standard candles.

WEB LINK 26.10 One distance-measuring technique that has broken free of the distance ladder uses observations of molecular clouds called **masers.** ("Maser" is an acronym for "microwave amplification by stimulated emission of radiation.")

Just as an electric current stimulates a laser to emit an intense beam of visible light, nearby luminous stars can stimulate water molecules in a maser to emit intensely at microwave wavelengths. This radiation is so intense that masers can be detected millions of parsecs away.

During the 1990s, Jim Herrnstein and his collaborators used the Very Long Baseline Array (see Section 6-6) to observe a number of masers orbiting in a disk around the center of the spiral galaxy M106. They determined the orbital speed of the masers from the Doppler shift of masers near the edges of the disk, where they are moving most directly toward or away from Earth. They also measured the apparent change in position of masers moving across the face of M106. By relating this apparent speed to the true speed determined from the Doppler shift, they calculated that the masers and the galaxy of which they are part are 7.2 Mpc (23 million ly) from Earth (Figure 26-13).

The maser technique is still in its infancy. But because it is independent of all other distance measuring methods, it is likely to play an important role in calibrating the rungs of the distance ladder.

26-5 | The Hubble law relates the redshifts of remote galaxies to their distances from the Earth

Whenever an astronomer finds an object in the sky that can be seen or photographed, the natural inclination is to attach a spectrograph to a telescope and record the spectrum. As long ago as 1914, Vesto M. Slipher, working at the Lowell Observatory in Arizona, began taking spectra of "spiral nebulae." He was surprised to discover that of the 15 spiral nebulae he studied, the spectral lines of 11 were shifted toward the red end of the spectrum, indicating that they were moving away from the Earth.

This marked dominance of redshifts was presented by Curtis in the 1920 Shapley-Curtis "debate" as evidence that these spiral nebulae could not be ordinary nebulae in our Milky Way Galaxy. It was only later that astronomers realized that the redshifts of spiral nebulae—that is, galaxies—reveal a basic law of our expanding universe.

 During the 1920s, Edwin Hubble and Milton Humason photographed the spectra of many galaxies with the 100-inch (2.5-meter) telescope on Mount Wilson in California. By observing the apparent brightnesses and pulsation periods of Cepheid variables in these galaxies, they were also able to measure the distance to each galaxy (see Section 26-2). Hubble and Humason found that most galaxies show a redshift in their spectrum. They also found a direct correlation between the distance to a galaxy and its redshift:

The more distant a galaxy, the greater its redshift and the more rapidly it is receding from us.

In other words, nearby galaxies are moving away from us slowly, and more distant galaxies are rushing away from us much more rapidly. **Figure 26-14** shows this relationship for five representative elliptical galaxies. This universal recessional movement is referred to as the **Hubble flow.**

Hubble estimated the distances to a number of galaxies and the redshifts of those galaxies. The **redshift,** denoted by the symbol z, is found by taking the wavelength (λ) observed for a given spectral line, subtracting from it the ordinary, unshifted wavelength of that line (λ_0) to get the wavelength difference ($\Delta\lambda$), and then dividing that difference by λ_0:

Redshift of a receding object

$$z = \frac{\lambda - \lambda_0}{\lambda_0} = \frac{\Delta\lambda}{\lambda_0}$$

z = redshift of an object

λ_0 = ordinary, unshifted wavelength of a spectral line

λ = wavelength of that spectral line that is actually observed from the object

From the redshifts, Hubble used the Doppler formula to calculate the speed at which these galaxies are receding from us. **Box 26-2** describes how this is done. Plotting the data on a graph of distance versus speed, Hubble found that the points lie near a straight line. **Figure 26-15** is a modern version of Hubble's graph based on recent data.

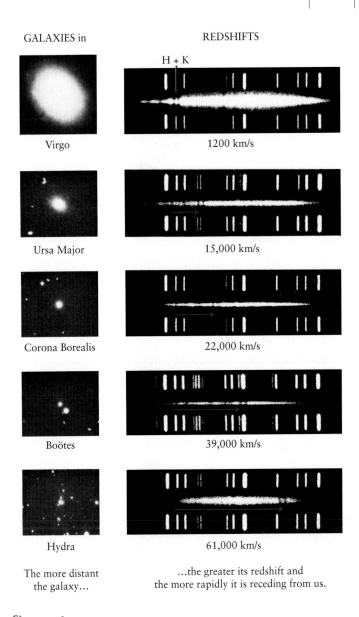

GALAXIES in   **REDSHIFTS**

Virgo — 1200 km/s

Ursa Major — 15,000 km/s

Corona Borealis — 22,000 km/s

Boötes — 39,000 km/s

Hydra — 61,000 km/s

The more distant the galaxy...   ...the greater its redshift and the more rapidly it is receding from us.

Figure 26-14   R I **V** U X G

Relating the Distances and Redshifts of Galaxies These five galaxies are arranged, from top to bottom, in order of increasing distance from us. All are shown at the same magnification. Each galaxy's spectrum is a bright band with dark absorption lines; the bright lines above and below it are a comparison spectrum of a light source at the observatory on Earth. The horizontal red arrows show how much the H and K lines of singly ionized calcium are redshifted in each galaxy's spectrum. Below each spectrum is the recessional velocity calculated from the redshift. The more distant a galaxy is, the greater its redshift. (Carnegie Observatories)

This relationship between the distances to galaxies and their redshifts was one of the most important astronomical discoveries of the twentieth century. As we will see in Chapter 28, this relationship tells us that we are living in an expanding universe. In 1929,

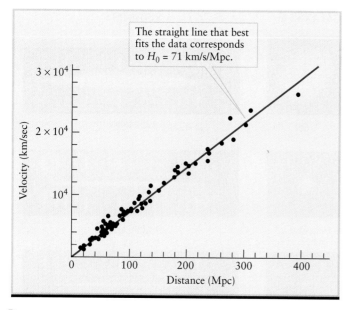

The straight line that best fits the data corresponds to $H_0 = 71$ km/s/Mpc.

figure 26-15

The Hubble Law This graph plots the distances and recessional velocities of a sample of galaxies. The straight line is the best fit to the data. This linear relationship between distance and recessional velocity is called the Hubble law.

Hubble published this discovery, which is now known as the **Hubble law**. The Hubble law is most easily stated as a formula:

The Hubble law

$$v = H_0 d$$

v = recessional velocity of a galaxy

H_0 = Hubble constant

d = distance to the galaxy

This formula is the equation for the straight line displayed in Figure 26-15, and the **Hubble constant** H_0 is the slope of this straight line. From the data plotted on this graph we find that $H_0 = 71$ km/s/Mpc (say "71 kilometers per second per megaparsec"). In other words, for each million parsecs to a galaxy, the galaxy's speed away from us increases by 71 km/s. For example, a galaxy located 100 million parsecs from the Earth should be rushing away from us with a speed of 7100 km/s. (In other books you may see the units of the Hubble constant written with exponents: 71 km s^{-1} Mpc^{-1}.)

 A common misconception about the Hubble law is that *all* galaxies are moving away from the Milky Way. The reality is that galaxies have their own motions relative to

---

box 26-2 TOOLS OF THE ASTRONOMER'S TRADE

The Hubble Law and the Relativistic Redshift

Suppose that you aim a telescope at an extremely distant galaxy. You take a spectrum of the galaxy and find that the spectral lines are shifted toward the red end of the spectrum. For example, a particular spectral line whose normal wavelength is λ_0 appears in the galaxy's spectrum at a longer wavelength λ. The spectral line has thus been shifted by an amount $\Delta\lambda = \lambda - \lambda_0$. The redshift of the galaxy, z, is given by

$$z = \frac{\lambda - \lambda_0}{\lambda_0} = \frac{\Delta\lambda}{\lambda_0}$$

The redshift means that the galaxy is receding from us. According to the Hubble law, the recessional velocity v of a galaxy is related to its distance d from the Earth by

$$v = H_0 d$$

where H_0 is the Hubble constant. We can rewrite this as

$$d = \frac{v}{H_0}$$

Given the value of H_0, we can find the distance d to the galaxy if we know how to determine the recessional velocity v from the redshift z.

If the redshift and recessional velocity are not too great, we can ignore the effects of the special theory of relativity and use the following equation:

$$z = \frac{v}{c} \quad \text{(valid for low speeds only)}$$

where c is the speed of light. For example, a 5% shift in wavelength ($z = 0.05$) corresponds to a recessional velocity of 5% of the speed of light ($v = 0.05c$). For redshifts greater than about 0.1, however, we must use the relativistic equation for the Doppler shift:

$$z = \sqrt{\frac{c + v}{c - v}} - 1$$

A useful way to rewrite this relationship is as follows:

$$\frac{v}{c} = \frac{(z + 1)^2 - 1}{(z + 1)^2 + 1} \quad \text{valid for all speeds}$$

TOOLS OF THE ASTRONOMER'S TRADE continued

The accompanying graph displays this relationship. Note that z approaches infinity as v approaches the speed of light.

EXAMPLE: When measured in a laboratory on Earth, the so-called K line of singly ionized calcium has a wavelength $\lambda_0 = 393.3$ nm. But when you observe the spectrum of the giant elliptical galaxy NGC 4889, you find this spectral line at $\lambda = 401.8$ nm. Using $H_0 = 71$ km/s/Mpc, find the distance to this galaxy.

Situation: We are given the values of λ and λ_0 for a line in this galaxy's spectrum. Our goal is to determine the galaxy's distance d.

Tools: We use the relationship $z = (\lambda - \lambda_0)/\lambda_0$ to determine the redshift. We then use the appropriate formula to determine the galaxy's recessional velocity v, and finally use the Hubble law to determine the distance to the galaxy.

Answer: The redshift of this galaxy is

$$z = \frac{401.8 \text{ nm} - 393.3 \text{ nm}}{393.3 \text{ nm}} = 0.0216$$

This is substantially less than 0.1, so we can safely use the low-speed relationship between recessional speed and redshift: $z = v/c$. So NGC 4889 is moving away from us with a speed

$$v = zc = (0.0216)(3 \times 10^5 \text{ km/s}) = 6480 \text{ km/s}$$

Using $H_0 = 71$ km/s/Mpc in the Hubble law, the distance to this galaxy is

$$d = \frac{v}{H_0} = \frac{6480 \text{ km/s}}{71 \text{ km/s/Mpc}} = 90 \text{ Mpc}$$

Review: This galaxy is receding from us at 0.0216 (2.16%) of the speed of light, and is 90 megaparsecs (290 million light-years) away. Thus the light we see from NGC 4889 today left the galaxy 290 million years ago, even before the first dinosaurs appeared on Earth.

EXAMPLE: In late 1997 astronomers observed a Type Ia supernova (called SN 1997ff) with a redshift $z = 1.7$. Use the Hubble law to find the distance to this supernova.

Situation: We are to use the redshift z to determine the distance d.

Tools: Since the redshift is large, we use the relativistic formula relating recessional velocity v and redshift z to calculate the value of v. We then use the Hubble law to determine the galaxy's distance d.

Answer: From the above relativistic formula,

$$\frac{v}{c} = \frac{(1.7 + 1)^2 - 1}{(1.7 + 1)^2 + 1} = \frac{(2.7)^2 - 1}{(2.7)^2 + 1} = \frac{6.29}{8.29} = 0.76$$

The supernova's recessional velocity is $v = 0.76c = (0.76)(3 \times 10^5 \text{ km/s}) = 2.3 \times 10^5$ km/s. From the Hubble law, the distance to the supernova is

$$d = \frac{v}{H_0} = \frac{2.3 \times 10^5 \text{ km/s}}{71 \text{ km/s/Mpc}} = 3200 \text{ Mpc} = 10^{10} \text{ ly}$$

This supernova is a remarkable 10 *billion* light-years away.

Review: For this supernova the value of z is greater than 1. Had we used the low-speed relationship $z = v/c$, we would have come to the erroneous conclusion that $v = 1.7c$—that is, that the supernova is receding from us faster than the speed of light. Using the correct relativistic formula tells us that the recessional velocity is only 76% of the speed of light.

In Chapter 28 we will see how observations of very distant supernovae like SN 1997ff reveal important deviations from the Hubble law. In particular, these observations show that the expansion of the universe is not proceeding at a constant rate, but is actually accelerating.

one another, thanks to their mutual gravitational attraction. For nearby galaxies, the speed of the Hubble flow is small compared to these intrinsic velocities. Hence, some of the nearest galaxies, including M31 (shown in the figure that opens this chapter), are actually approaching us and have blueshifts rather than redshifts. But for distant galaxies, the Hubble speed $v = H_0 d$ is much greater than any intrinsic motion that the galaxies might have. Even if the intrinsic velocity of such a distant galaxy is toward the Milky Way, the fast-moving Hubble flow sweeps that galaxy away from us.

The exact value of the Hubble constant has been a topic of heated debate among astronomers for several decades. The problem is that while redshifts are relatively easy to measure in a reliable way, distances to galaxies (especially remote galaxies) are not, as we saw in Section 26-4. Hence, astronomers who use different methods of determining galactic distances have obtained different values of H_0. To see why this is so, it is helpful to rewrite the Hubble law as

$$H_0 = \frac{v}{d}$$

This shows that if galaxies of a given recessional velocity (v) are far away (so d is large), the Hubble constant H_0 must be relatively small. But if these galaxies are relatively close (so d is small), then H_0 must have a larger value.

In the past, astronomers who used Type Ia supernovae for determining galactic distances found galaxies to be farther away than their colleagues who employed the Tully-Fisher relation. Therefore, the supernova adherents found values of H_0 in the range from about 40 to 65 km/s/Mpc, while the values from the Tully-Fisher relation ranged from about 80 to 100 km/s/Mpc.

In the past few years, the Hubble Space Telescope has been used to observe Cepheid variables with unprecedented precision and in galaxies as far away as 30 Mpc (100 million ly). These observations and others suggest a value of the Hubble constant of about 71 km/s/Mpc, with an uncertainty of no more than 10%. At the same time, reanalysis of the supernova and Tully-Fisher results have brought the values of H_0 from these techniques closer to the Hubble Space Telescope Cepheid value. We will adopt the value $H_0 = 71$ km/s/Mpc in this book.

Determining the value of H_0 has been an important task of astronomers for a very simple reason: The Hubble constant is one of the most important numbers in all astronomy. It expresses the rate at which the universe is expanding and, as we will see in Chapter 28, even helps give the age of the universe. Furthermore, the Hubble law can be used to determine the distances to extremely remote galaxies and quasars. If the redshift of a galaxy is known, the Hubble law can be used to determine its distance from the Earth, as Box 26-2 shows. Thus, the value of the Hubble constant helps determine the distances of the most remote objects in the universe that astronomers can observe.

Because the value of H_0 remains somewhat uncertain, astronomers often express the distance to a remote galaxy simply in terms of its redshift z (which can be measured very accurately). Given the redshift, the distance to this galaxy can be calculated from the Hubble law, but the distance obtained in this way will depend on the particular value of H_0 adopted. Rather than going through these calculations, an astronomer might simply say that a certain galaxy is "at $z = 0.128$." From the Hubble law relating redshift and dis-

tance, this makes it clear that the galaxy in question is farther away than one at $z = 0.120$ but not as far as one at $z = 0.130$. When astronomers use redshift to describe distance, they are making use of the following general rule:

The greater the redshift of a distant galaxy, the greater its distance.

26-6 | Galaxies are grouped into clusters and superclusters

Galaxies are not scattered randomly throughout the universe but are found in **clusters**. Figure 26-16 shows one such cluster. Like stars within a star cluster, the members of a cluster of galaxies are in continual motion around one another. They appear to be at rest only because they are so distant from us.

A cluster is said to be either **poor** or **rich**, depending on how many galaxies it contains. Poor clusters, which far outnumber rich ones, are often called **groups**. For example, the Milky Way Galaxy, the Andromeda Galaxy (M31), and the Large and Small Magellanic Clouds belong to a poor cluster familiarly known as the **Local Group**. The Local Group contains more than 40 galaxies, most of which are dwarf ellipticals (see Figure 26-8). Figure 26-17 is a map of a portion of the Local Group.

In recent years, astronomers have discovered several previously unknown dwarf elliptical galaxies in the Local Group. As of this writing (early 2004) the nearest and most recently discovered is the

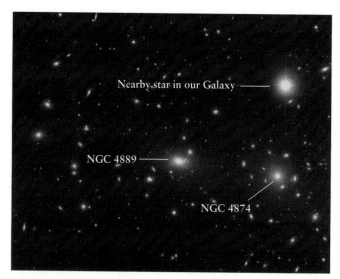

Figure 26-16 R I **V** U X G

The Coma Cluster of Galaxies This rich, regular cluster is about 90 Mpc (300 million light-years) from the Earth. Almost all of the spots of light in this image are individual galaxies of the cluster. Two giant elliptical galaxies, NGC 4889 and NGC 4874, dominate the center of the cluster. The bright star at the upper right is within our own Milky Way Galaxy, a million times closer than any of the galaxies shown here. (O. Lopez-Cruz and I. K. Shelton, University of Toronto; KPNO)

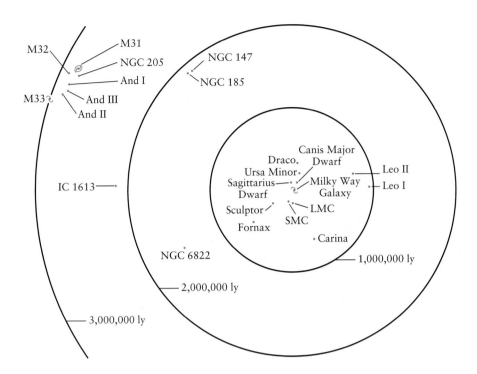

Figure 26-17

The Local Group This map shows the distribution of about 3/4 of the 40-plus galaxies that comprise the Local Group, a poor, irregular cluster of which our Galaxy is part. (The galaxies are actually distributed in three-dimensional space. In this figure they have been drawn in a plane to represent their distances from the Milky Way.) The largest and most massive galaxy in the Local Group is M31, the Andromeda Galaxy; in second place is the Milky Way. Both large galaxies are surrounded by a number of small satellite galaxies.

Canis Major Dwarf, so named for the constellation in whose direction it lies as seen from Earth. It is about 13 kpc (42,000 ly) from the center of the Milky Way Galaxy and a mere 8 kpc (25,000 ly) from Earth (about the same as the distance from Earth to the center of the Milky Way). Tidal forces exerted by the Milky Way on the Canis Major Dwarf are causing this dwarf galaxy to gradually disintegrate and leave a trail of debris behind it (**Figure 26-18**).

We may never know the total number of galaxies in the Local Group, because dust in the plane of the Milky Way obscures our view over a considerable region of the sky. Nevertheless, we can be certain that no additional large spiral galaxies are hidden by the Milky Way. Radio astronomers would have detected 21-cm radiation from them, even though their visible light is completely absorbed by interstellar dust.

The nearest fairly rich cluster is the Virgo cluster, a collection of more than 2000 galaxies covering a 10° × 12° area of the sky. Measurements of Cepheid variables in the spiral galaxy M100, a member of the Virgo cluster, give a distance of about 15 Mpc (50 million ly); the Tully-Fisher relation and observations of Type Ia supernovae give similar distances to this cluster. The overall diameter of the cluster is about 3 Mpc (9 million ly).

The center of the Virgo cluster is dominated by three giant elliptical galaxies, two of which can be seen in Figure 26-7. The diameter of each of these enormous galaxies is comparable to the 750-kpc distance between the Milky Way and M31. In other words, one giant elliptical is approximately the same size as the entire Local Group!

Astronomers also categorize clusters of galaxies as regular or irregular, depending on the overall shape of the cluster. The Virgo cluster, for example, is called an **irregular cluster,** because its galaxies are scattered throughout a sprawling region of the sky. Our own Local Group is also an irregular cluster. In contrast, a **regular cluster** has a distinctly spherical appearance, with a marked concentration of galaxies at its center.

The nearest example of a rich, regular cluster is the Coma cluster, located about 90 Mpc (300 million light-years) from us toward the constellation Coma Berenices (Berenice's Hair). Despite its great distance, telescopic images of this cluster show more than

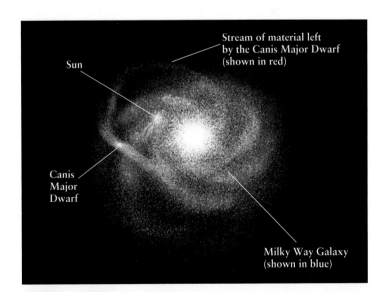

Figure 26-18

The Canis Major Dwarf Discovered in 2003, this dwarf elliptical galaxy is actually slightly closer to Earth than is the center of the Milky Way Galaxy. This illustration shows the stream of material left behind by the Canis Major Dwarf as it orbits the Milky Way. This material is torn away by the Milky Way's tidal forces (see Section 4-8). (R. Ibata et al., Observatoire de Strasbourg/Université Louis Pasteur; 2MASS; and NASA)

1000 galaxies (see Figure 26-16). The Coma cluster almost certainly contains many thousands of dwarf ellipticals, so the total membership of the cluster may be as many as 10,000 galaxies. The core of the Coma cluster is dominated by two giant ellipticals surrounded by many normal-sized galaxies.

The overall shape of a cluster is related to the dominant types of galaxies it contains. Rich, regular clusters contain mostly elliptical and lenticular galaxies. For example, about 80% of the brightest galaxies in the Coma cluster (see Figure 26-16) are ellipticals; only a few spiral galaxies are scattered around the cluster's outer regions. Irregular clusters, such as the Virgo cluster and the Hercules cluster (Figure 26-19), have a more even mixture of galaxy types.

Clusters of galaxies are themselves grouped together in huge associations called **superclusters.** A typical supercluster contains dozens of individual clusters spread over a region of space up to 45 Mpc (150 million ly) across. Figure 26-20 shows the distribution of clusters in our part of the universe. The nearer ones, out to the Virgo cluster, are members of the *Local Supercluster.* The other clusters shown in Figure 26-20 belong to other superclusters. The most massive cluster in the local universe is called the *Great Attractor;* its gravity is so great that the Milky Way as well as the rest of the Local Supercluster is moving toward it at speeds of several hundred kilometers per second.

Observations indicate that unlike clusters, superclusters are not bound together by gravity. That is, most clusters in each supercluster are drifting away from most of the other clusters in that same supercluster. Furthermore, the superclusters are all moving away from each other due to the Hubble flow.

Since the 1980s, astronomers have been working to understand how superclusters are distributed in space. Some of this structure is revealed by maps such as that shown in Figure 26-21, which displays the positions on the sky of 1.6 million galaxies. Such maps reveal that superclusters are not randomly distributed, but seem to

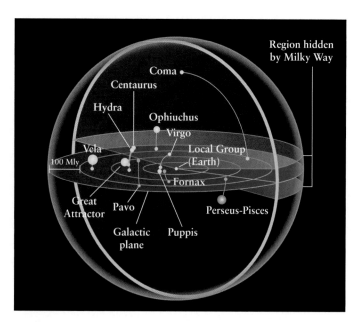

figure 26-20

Nearby Clusters of Galaxies This illustration shows a sphere of space 800 million ly (250 Mpc) in diameter centered on the Earth in the Local Group. Each spherical dot represents a cluster of galaxies. To better see the three-dimensionality of this figure, colored arcs are drawn from each cluster to the green plane, which is an extension of the plane of the Milky Way outward into the universe. Note that clusters of galaxies are unevenly distributed here, as they are elsewhere in the universe.

lie along filaments. But to comprehend more fully the distribution of superclusters, it is necessary to map their positions in three dimensions. This is done by measuring both the position of a galaxy on the sky as well as the galaxy's redshift. Using the Hubble law (see Section 26-5), astronomers can use each galaxy's redshift to estimate its distance from Earth and thus its position in three-dimensional space.

The first three-dimensional map of this kind, released in 1980 by astronomers at the Harvard-Smithsonian Center for Astrophysics, was based on measurements of 2400 galaxies as far away as 200 Mpc (700 million ly). This map required many years of telescopic observations. Technology for astronomy has advanced tremendously since then, and it is now possible to measure the redshifts of 400 galaxies in a single hour!

The most extensive galaxy maps available at this writing (early 2004) are those from the Sloan Digital Sky Survey, a joint project of astronomers from the United States, Japan, and Germany, and from the Two Degree Field Galactic Redshift Survey (2dFGRS), a collaboration between Australian, British, and U.S. astronomers. (The name refers to the 2° field of view of the telescope used for the observations, which is unusually wide for a research telescope.) Figure 26-22a shows a map made from 2dFGRS measurements of more than 60,000 galaxies. This particular map encompasses two wedge-shaped slices of the universe, one on either side of the plane of the Milky Way (Figure 26-22b). The

figure 26-19   R I **V** U X G

The Hercules Cluster This irregular cluster, which is about 200 Mpc (650 million ly) from Earth, is much less dense than the Coma cluster shown in Figure 26-16. The Hercules cluster contains many spiral galaxies, often associated in pairs and small groups. (NOAO)

figure 26-21 R I V U X G

Structure in the Nearby Universe This composite infrared image from the 2MASS (Two-Micron All-Sky Survey) project shows the light from 1.6 million galaxies. In this image, the entire sky is projected onto an oval; the blue band running vertically across the center of the image is light from the plane of the Milky Way. Note that galaxies form a lacy, filamentary structure. Note also the large, dark voids that contain few galaxies. (2MASS; IPAC/Caltech; and University of Massachusetts)

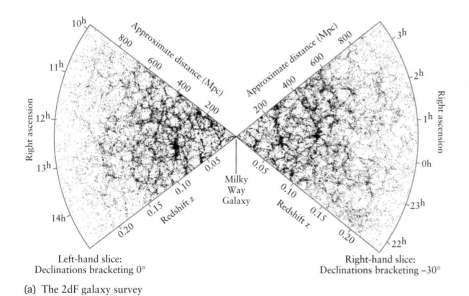

(a) The 2dF galaxy survey

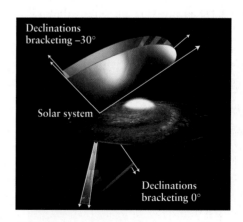

(b) Fields of view in the 2dF survey

figure 26-22

The Large-Scale Distribution of Galaxies (a) This map shows the distribution of 62,559 galaxies in two wedges extending out to redshift $z = 0.25$. (For an explanation of right ascension and declination, see Box 2-1.) Note the prominent voids surrounded by thin regions full of galaxies. (b) The two wedges shown in (a) lie roughly perpendicular to the plane of the Milky Way. These were chosen to avoid the obscuring dust that lies in our Galaxy's plane. (Courtesy the 2dF Galaxy Redshift Survey Team/Anglo-Australian Observatory)

Earth (in the Milky Way) is at the apex of the wedge-shaped map; each dot represents a galaxy. The measurements used to create this map included galaxies with redshifts as large as $z = 0.25$, corresponding to a recessional velocity of 66,000 km/s. With a Hubble constant of 71 km/s/Mpc, this means that the map in Figure 26-22a extends out to a distance of nearly 1000 Mpc, or 3 *billion* light-years from Earth.

Maps such as that shown in Figure 26-22a reveal enormous **voids** where exceptionally few galaxies are found. These voids are roughly spherical and measure 30 to 120 Mpc (100 million to 400 million ly) in diameter. They are not entirely empty, however. There is evidence for hydrogen clouds in some voids, while others may be subdivided by strings of dim galaxies.

Figure 26-22a shows that most galaxies are concentrated in sheets on the surfaces between voids. These sheets can be more than 100 Mpc long and several megaparsecs thick. This pattern is similar to that of soapsuds in a kitchen sink, with sheets of soap film (analogous to galaxies) surrounding air bubbles (analogous to voids). These titanic sheets of galaxies are the largest structures known in the universe: On scales much larger than 100 Mpc, the distribution of galaxies in the universe appears to be roughly uniform. As we will see in Chapter 29, this pattern of sheets and voids contains important clues about how clusters of galaxies formed in the early universe.

26-7 | Colliding galaxies produce starbursts, spiral arms, and other spectacular phenomena

Occasionally, two galaxies within a cluster or from adjacent clusters can collide with each other. Past collisions have hurled vast numbers of stars into intergalactic space. In some cases, we can even observe a collision in progress, a cosmic catastrophe that gives birth to new stars. And astronomers can predict collisions that will not take place for billions of years, such as the collision that is fated to occur between the galaxy M31 and our own Milky Way Galaxy.

When two galaxies collide at high speed, the huge clouds of interstellar gas and dust in the galaxies slam into each other and can be completely stopped in their tracks. In this way, two colliding galaxies can be stripped of their interstellar gas and dust.

The best evidence that such collisions take place is that many rich clusters of galaxies are strong sources of X rays (**Figure 26-23**). This emission reveals the presence of substantial amounts of hot **intracluster gas** (that is, gas within the cluster) at temperatures between 10^7 and 10^8 K. The only way that such large amounts of gas could be heated to such extremely high temperatures is in violent collisions between galaxies.

CAUTION Although galaxies can and do collide, it is highly unlikely that the *stars* from two colliding galaxies actually run into each other. The reason is that the stars within a galaxy are very widely separated from one another, with a tremendous amount of space between them.

In a less violent collision or a near-miss between two galaxies, the compressed interstellar gas may have more time to cool, allowing many protostars to form. Such collisions may account for **starburst galaxies** such as M82 (**Figure 26-24**), which blaze with the light of numerous newborn stars. These galaxies have bright centers surrounded by clouds of warm interstellar dust, indicating recent, vigorous star birth. Their warm dust is so abundant that starburst galaxies are among the most luminous objects in the universe at infrared wavelengths.

The starburst galaxy M82 shown in Figure 26-24 also shows the effects of strong winds from young, luminous stars. It also contains a number of luminous globular clusters. Unlike the globular clusters in our Galaxy, whose stars are about 12.5 billion years old, those in M82 are no more than 600 million years old. These young globular clusters are another sign of recent star formation.

M82 is one member of a nearby cluster of galaxies that includes the beautiful spiral galaxy M81 and a fainter elliptical companion called NGC 3077 (**Figure 26-25a**). Radio surveys of that region of the sky reveal enormous streams of hydrogen gas connecting the

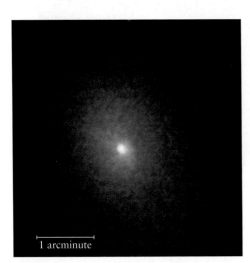

(a) An X-ray image of Abell 2029 shows emission from hot gas. R I V U **X** G

(b) A visible-light image of Abell 2029 shows the cluster's galaxies. R I **V** U X G

Figure 26-23
X-ray Emission from a Cluster of Galaxies (a) An X-ray image of this cluster of galaxies shows emission from hot gas between the galaxies. The gas was heated by collisions between galaxies within the cluster. (b) The galaxies themselves are too dim at X-ray wavelengths to be seen in (a), but are apparent at visible wavelengths. This cluster, one of many cataloged by the UCLA astronomer George O. Abell, is about 300 Mpc (1 billion ly) from Earth in the constellation Serpens. (a: NASA, CXC, and University of California, Irvine/A. Lewis et al. b: Palomar Observatory DSS)

WEB LINK 26.16

R **V** U X G

R **I** **V** U X G

figure 26-24

A Starburst Galaxy Prolific star formation is occurring at the center of the irregular galaxy M82, which lies about 3.6 Mpc (12 million ly) from Earth in the constellation Ursa Major. M82 also contains an unusual X-ray source, shown in Figure 24-14. (Left: N. A. Sharp/ AURA, NOAO, and NSF; right: NASA, ESA, and R. de Grijs/Institute of Astronomy, Cambridge)

(a) R **I** **V** U X G (b) **R** I V U X G

figure 26-25

The M81 Group (a) The starburst galaxy M82 (see Figure 26-24) is part of a cluster of about a dozen galaxies. This wide-angle visible-light photograph shows the three brightest galaxies of the cluster. The area shown is about 1° across. (b) This false-color radio image of the same region, created from data taken by the Very Large Array, shows streamers of hydrogen gas that connect the three bright galaxies as well as several dim ones. (a: Palomar Sky Survey; b: M. S. Yun, VLA and Harvard) three galaxies (Figure 26-25*b*). The loops and twists in these streamers suggest that the three galaxies have had several close encounters over the ages. A similar stream of hydrogen gas connects our Galaxy with its second nearest neighbor, the Large Magellanic Cloud (see Figure 26-10), suggesting a history of close encounters between our Galaxy and the LMC.

Tidal forces between colliding galaxies can deform the galaxies from their original shapes, just as the tidal forces of the Moon on the Earth deform the oceans and help give rise to the tides (see Section 4-8, especially Figure 4-26). The galactic deformation is so great that thousands of stars can be hurled into intergalactic space along huge, arching streams. (This same effect has stripped material away from the Canis Major Dwarf galaxy as it orbits the Milky Way, as shown in Figure 26-18.) Supercomputer simulations of such collisions show that while some of the stars are flung far and wide, other stars slow down and the galaxies may merge.

 Figure 26-26 shows one such simulation. As the two galaxies pass through each other, they are severely distorted by gravitational interactions and throw out a pair of extended tails. The actual colliding galaxies shown in Figure 26-27 have tails of just this kind. The interaction also prevents the galaxies in the simulation from continuing on their original paths. Instead, they fall back together for a second encounter (at 625 million years). The simulated galaxies merge soon thereafter, leaving a single object.

 Our own Milky Way Galaxy is expected to undergo a galactic collision like that shown in Figure 26-27. The Milky Way and the Andromeda Galaxy, shown in the image that opens this chapter, are actually approaching each other and should collide in another 6 billion years or so. (Recall that our solar system is only 4.56 billion years old.) When this happens, the sky will light up with a plethora of newly formed stars, followed in rapid succession by a string of supernovae, as the most massive of these stars complete their life spans. Any inhabitants of either galaxy will see a night sky far more dramatic and tempestuous than ours.

 When two galaxies merge, the result is a bigger galaxy. If this new galaxy is located in a rich cluster, it may capture and devour additional galaxies, growing to enormous dimensions by **galactic cannibalism.** Cannibalism differs from mergers in that the galaxy that does the devouring is bigger than its "meal," whereas merging galaxies are about the same size.

Many astronomers suspect that galactic cannibalism is the reason that giant ellipticals are so huge. As we have seen, giant galaxies typically occupy the centers of rich clusters. In many cases, smaller galaxies are located around these giants. As they pass through the extended halo of a giant elliptical, these smaller galaxies slow down and are eventually devoured by the larger galaxy.

Close encounters between galaxies provide a third way of forming spiral arms (in addition to density waves and self-propagating star formation, discussed in Sec-

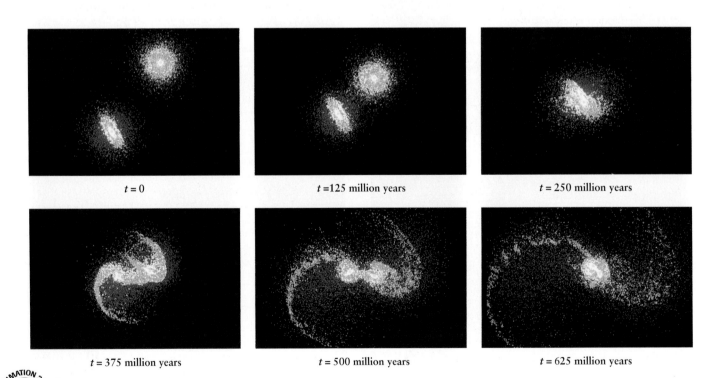

t = 0	*t* =125 million years	*t* = 250 million years
t = 375 million years	*t* = 500 million years	*t* = 625 million years

 Figure 26-26

A Simulated Collision Between Two Galaxies These frames from a supercomputer simulation show the collision and merger of two galaxies accompanied by an ejection of stars into intergalactic space. Stars in the disk of each galaxy are colored blue, while stars in their central bulges are yellow-white. Red indicates dark matter that surrounds each galaxy. The frames progress at 125-million-year intervals. Compare the bottom frames with Figure 26-27. (Courtesy of J. Barnes, University of Hawaii)

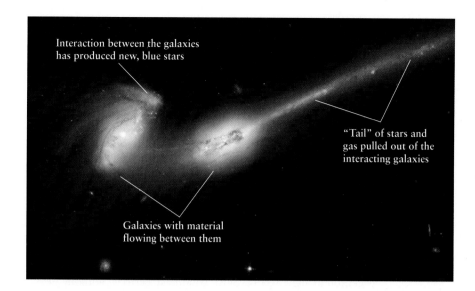

Interaction between the galaxies
has produced new, blue stars

"Tail" of stars and
gas pulled out of the
interacting galaxies

Galaxies with material
flowing between them

Figure 26-27   R I **V** U X G
Two Colliding Galaxies—
"The Mice" Known collectively as NGC 4676, these two galaxies are strongly deformed by their tidal influences on each other. Note how the interaction has compressed the material of the galaxies, making a new generation of hot, blue stars. Supercomputer simulations suggest that we are seeing these galaxies about 160 million years after their closest approach, and that they will eventually coalesce into a single galaxy. (NASA, H. Ford/JHU, G. Illingworth/UCSC/Lick, M.Clampin/STScI, G. Hartig/STScI, the ACS Science Team, and ESA)

tion 25-5). Supercomputer simulations clearly demonstrate that spiral arms can be created during a collision, either by drawing out long streamers of stars or by compressing clouds of interstellar gas. For example, the spiral arms of M51 (examine Figure 26-2) may have been produced by a close encounter with a second galaxy. The disruptive galaxy, NGC 5195, is now located at the end of one of the spiral arms created by the collision.

The very fact of our existence may be intimately related to interactions between galaxies. Some astronomers argue that the spiral arms of our Milky Way Galaxy were similarly produced by a close encounter with the Large Magellanic Cloud. As we saw in Section 25-5, spiral arms compress the interstellar medium in the Milky Way's disk to form Population I stars like our own Sun, which have enough heavy elements to produce Earthlike planets. Thus, the chain of events that led to the formation of our Sun, our solar system, and ourselves may have been initiated by a long-ago interaction between two galaxies.

26-8 | Most of the matter in the universe has yet to be discovered

A cluster of galaxies must be held together by gravity. In other words, there must be enough matter in the cluster to prevent the galaxies from wandering away. Nevertheless, careful examination of a rich cluster, like the Coma cluster shown in Figure 26-16, reveals that the mass of the visually luminous matter (principally the stars in the galaxies) is not at all sufficient to bind the cluster gravitationally. The observed line-of-sight speeds of the galaxies, measured by Doppler shifts, are so large that the cluster should have broken apart long ago. Considerably more mass than is visible is needed to keep the galaxies bound in orbit about the center of the cluster.

We encountered a similar situation in studying our own Milky Way Galaxy in Section 25-4: The total mass of our Galaxy is more than the amount of visible mass. As for our Galaxy, we conclude that clusters of galaxies must contain significant amounts of nonlu-

minous *dark matter*. If this dark matter were not there, the galaxies would have long ago dispersed in random directions and the cluster would no longer exist today. Analyses demonstrate that the total mass needed to bind a typical rich cluster is about 10 times greater than the mass of material that shows up on visible-light images.

As for our Galaxy, the problem is to determine what form the invisible mass takes. A partial solution to this **dark-matter problem,** which dates from the 1930s, was provided by the discovery in the late 1970s of hot, X-ray-emitting gas within clusters of galaxies (see Figure 26-23a). By measuring the amount of X-ray emission, astronomers find that the total mass of intracluster gas in a typical rich cluster can be greater than the combined mass of all the stars in all the cluster's galaxies. This is sufficient to account for only about 10% of the invisible mass, however. The remainder is dark matter of unknown composition.

Although we do not know what dark matter is made of, it is possible to investigate how dark matter is distributed in galaxies and clusters of galaxies. It appears that dark matter lies within and immediately surrounding galaxies, not in the vast spaces between galaxies. The evidence for this comes principally from observations of the rotation curves of galaxies and of the gravitational bending of light by clusters of galaxies.

As we saw in Section 25-4, a rotation curve is a graph that shows how fast stars in a galaxy are moving at different distances from that galaxy's center. For example, Figure 25-16 is the rotation curve for our Galaxy. As Figure 26-28 illustrates, many other spiral galaxies have similar rotation curves that remain remarkably flat out to surprisingly great distances from each galaxy's center. In other words, the orbital speed of the stars remains roughly constant out to the visible edges of these galaxies.

This observation tells us that we still have not detected the *true* edges of these galaxies (and many similar ones). Near the true edge of a galaxy we should see a decline in orbital speed, in accordance with Kepler's third law (see Figure 25-16). Because this decline has not been observed, astronomers conclude that there must be a considerable amount of dark material that extends well beyond the visible portion of the disk.

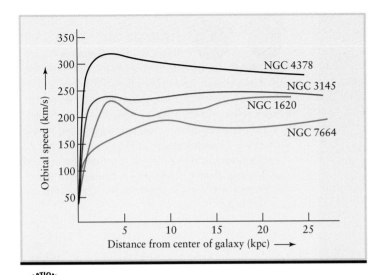

ANIMATION 26.7

figure 26-28

The Rotation Curves of Four Spiral Galaxies This graph shows how the orbital speed of material in the disks of four spiral galaxies varies with the distance from the center of each galaxy. If most of each galaxy's mass were concentrated near its center, these curves would fall off at large distances. But these and many other galaxies have flat rotation curves that do not fall off. This indicates the presence of extended halos of dark matter. (Adapted from V. Rubin and K. Ford)

Further evidence about how dark matter is distributed comes from the gravitational bending of light rays, which we described in Section 24-2. As Figure 24-5 shows, the gravity of a single star like the Sun can deflect light by only a few arcseconds. But a more massive object such as a galaxy can produce much greater deflections, and the amount of this deflection can be used to determine the galaxy's mass. For example, suppose that the Earth, a massive galaxy, and a background light source (such as a more distant galaxy) are in nearly perfect alignment, as sketched in Figure 26-29a. Because of the warped space around the massive galaxy, light from the background source curves around the galaxy as it heads toward us. As a result, light rays can travel along two paths from the background source to us here on the Earth. Thus, we should see two images of the background source.

A powerful source of gravity that distorts background images is called a **gravitational lens.** For gravitational lensing to work, the alignment between the Earth, the massive galaxy, and a remote background light source must be almost perfect. Without nearly perfect alignment, the second image of the background star is too faint to be noticeable.

Beginning in 1979, astronomers have discovered a great number of examples of gravitational lensing. The example shown in Figure 26-29b is almost exactly like the ideal situation depicted in Figure 26-29a. If the alignment is very slightly off, the image of the distant galaxy is distorted into an arc as shown in Figure 26-29c. Figure 26-29d shows a more complicated example of lensing that

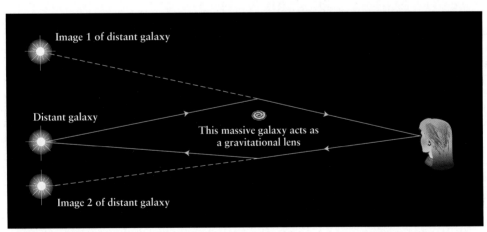

(a) How gravitational lensing happens

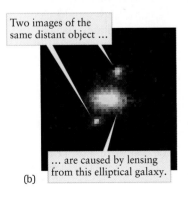

Two images of the same distant object …

… are caused by lensing from this elliptical galaxy.

(b)

This bluish arc is a distorted image of a distant galaxy …

… produced by lensing from this red galaxy.

(c)

A blurred image of a distant galaxy …

… lensed by four red galaxies.

(d)

figure 26-29

Gravitational Lensing (a) A massive object such as a galaxy can deflect light rays like a lens so that an observer sees more than one image of a more distant galaxy. (Compare with Figure 25-17, which shows the same effect on a much smaller scale.) (b), (c), (d) Three examples of gravitational lensing. In each case a single, distant blue galaxy is "lensed" by a closer red galaxy or galaxies. (b, c, d: Kavan Ratnatunga, Carnegie Mellon University, ESA, and NASA)

R **I** **V** U X G

results when the gravitational lens is not one but several massive galaxies.

Figure 26-30 shows a situation in which an entire cluster of galaxies acts as a gravitational lens. The image shows an ordinary-looking rich cluster of yellowish elliptical and spiral galaxies, but with a number of curious blue arcs. Reconstruction of the light paths through the cluster show that all these blue arcs are actually distorted images of a single galaxy that lies billions of light-years beyond the cluster.

By measuring the distortion of the images of such background galaxies, J. Anthony Tyson of Bell Laboratories and his colleagues have determined that dark matter, which constitutes about 90% of the cluster's mass, is distributed much like the visible matter in the cluster. In other words, the overall arrangement of visible galaxies seems to trace the location of dark matter.

Many proposals have been made to explain the nature of dark matter. One reasonable suggestion was that clusters might contain a large number of faint, red, low-mass (0.2 $M_\odot$ or less) stars. These faint stars could be located in extended halos surrounding individual galaxies or scattered throughout the spaces between the galaxies of a cluster. They would have escaped detection because their luminosity and hence apparent brightness would be very low. (The mass-luminosity relation for main-sequence stars, which we discussed in Section 19-9, tells us that low-mass stars are intrinsically very faint.) Searches for these stars around other galaxies as well as around the Milky Way have been carried out using the Hubble Space Telescope. None have yet been detected, so it is thought that faint stars are unlikely to constitute the majority of the dark matter in the universe.

As we described in Section 25-4, more exotic dark matter candidates include massive neutrinos, subatomic particles called WIMPs (weakly interacting massive particles), and MACHOs (massive compact halo objects, such as small black holes or brown dwarfs). To date, however, the true nature of dark matter remains unknown.

26-9 | Galaxies formed from the merger of smaller objects

How do galaxies form and how do they evolve? Astronomers can gain important clues about galactic evolution simply by looking deep into space. The more distant a galaxy is, the longer its light takes to reach us. As we examine galaxies that are at increasing distances from the Earth, we are actually looking further and further back in time. By looking into the past, we can see galaxies in their earliest stages.

The Hubble Space Telescope images in Figure 26-31 provide a glimpse of galaxy formation in the early universe. Figure 26-31a shows a number of galaxylike objects some 11 billion light-years (3400 Mpc) away and are thus seen as they were 11 billion years ago. These objects are smaller than even the smallest galaxies we see in the present-day universe and have unusual, irregular shapes (Figure 26-31b). Furthermore, these objects are scattered over an area only 600 kpc (2 million light-years) across—less than the distance between the Milky Way Galaxy and M31—making it quite probable that they would collide and merge with each other. These collisions would be aided by the dark matter associated with each subgalactic object, which increases the object's mass and, hence, the gravitational forces pulling the objects together. Such mergers would eventually give rise to a normal-sized galaxy.

Images such as those in Figure 26-31 lead astronomers to conclude that galaxies formed "from the bottom up"—that is, by the merger of smaller objects like those in Figure 26-31b to form full-size galaxies. (These same images appear to rule out an older idea that galaxies formed "from the top down"—that is, directly from immense, galaxy-sized clouds of material.) The blue color of the objects in Figure 26-31b indicates the presence of young stars. Observations indicate that the very first stars formed about 13.5 billion years ago, when the universe was only about 200 million years old. We will see evidence in Chapter 28 that the matter in the universe formed "clumps" even earlier than this. These clumps evolved into objects like those shown in Figure 26-31b, which in turn merged to form the population of galaxies that we see today.

Once a number of subgalactic units combine, they make an object called a *protogalaxy*. The rate at which stars form within a protogalaxy may determine whether this protogalaxy becomes a spiral or an elliptical. If stars form relatively slowly, the gas surrounding them has enough time to settle by collisions to form a

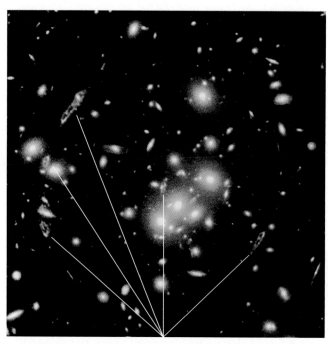

All of these blue arcs are images of the same distant galaxy.

Figure 26-30 R I **V** U X G
Gravitational Lensing by a Cluster of Galaxies
The blue arcs in this image of the rich cluster CL0024+1654 are distorted multiple images of a single more distant galaxy. These images are the result of gravitational lensing by the matter in CL0024+1654. The cluster is about 1600 Mpc (5 billion ly) from Earth; the blue galaxy is about twice as distant. The blue color of the remote galaxy suggests that it is very young and is actively forming stars. (W. N. Colley and E. Turner, Princeton University; J. A. Tyson, Bell Labs, Lucent Technologies; NASA)

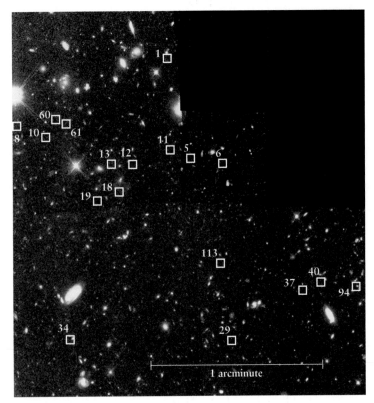

(a) A portion of the constellation Hercules

(b) Closeup images of the numbered objects in (a)

fi§ure 26-31 R I **V** U X G

The Building Blocks of Galaxies (a) In this Hubble Space Telescope image, the objects outlined by boxes are about 3400 Mpc (11 billion ly) from Earth and are only 600 to 900 pc (2000 to 3000 ly) across—larger than a star cluster but smaller than even dwarf elliptical galaxies like that shown in Figure 26-8. (b) If these objects were to coalesce, the result would be a full-sized galaxy such as we see in the nearby universe today. (Rogier Windhorst and Sam Pascarelle, Arizona State University; NASA)

flattened disk, much as happened on a much smaller scale in the solar nebula (see Section 8-4). Star formation continues because the disk contains an ample amount of hydrogen from which to make new stars. The result is a spiral or lenticular galaxy (Figure 26-32a). But if stars initially form in the protogalaxy at a rapid rate, virtually all of the available gas is used up to make stars before a disk can form. In this case what results is an elliptical galaxy (Figure 26-32b).

Figure 26-32c compares the stellar birthrate in the two types of galaxies. This graph helps us understand some of the differences between spiral and elliptical galaxies that we described in Section 26-3. Protogalaxies are thought to have been composed almost exclusively of hydrogen and helium gas, so the first stars were Population II stars with hardly any metals (that is, heavy elements). As stars die and form planetary nebulae or supernovae, they eject gases enriched in metals into the interstellar medium. In a spiral galaxy there is ongoing star formation in the disk, so these metals are incorporated into new generations of stars, making relatively metal-rich Population I stars like the Sun. By contrast, an elliptical galaxy has a single flurry of star formation when it is young, after which star formation ceases. Elliptical galaxies therefore contain only metal-poor Population II stars.

Figure 26-32c shows that both elliptical and spiral galaxies form stars most rapidly when they are young. This idea is borne out by the observation that very distant galaxies tend to be blue, which means that galaxies were bluer in the distant past than they are today. (Note the very blue colors of the distant, gravitationally lensed galaxies shown in Figure 26-29 and Figure 26-30, as well as those of the subgalactic objects shown in Figure 26-31b.) Spectro-

scopic studies of such galaxies in the 1980s by James Gunn and Alan Dressler demonstrated that most owe their blue color to vigorous star formation, often occurring in intense, episodic bursts. The hot, luminous, and short-lived O and B stars produced in these bursts of star formation give blue galaxies their characteristic color.

In addition to changes in galaxy colors, the character of the galactic population has also changed over the past several billion years. In nearby rich clusters, only about 5% of the galaxies are spirals. But observations of rich clusters at a redshift of $z = 0.4$—which corresponds to looking about 4 billion years into the past—show that about 30% of their galaxies were spirals.

Why were spiral galaxies more common in rich clusters in the distant past? Galactic collisions and mergers are probably responsible. During a collision, interstellar gas in the colliding galaxies is vigorously compressed, triggering a burst of star formation (see Figure 26-27). A succession of collisions produces a series of star-forming episodes that create numerous bright, hot O and B stars that become dispersed along arching spiral arms by the galaxy's rotation. Eventually, however, the gas is used up; star formation then ceases and the spiral arms become less visible. Furthermore, tidal forces tend to disrupt colliding galaxies, strewing their stars across intergalactic space until the galaxies are completely disrupted (see Figure 26-26).

A full description of galaxy formation and evolution must include the effects of dark matter. As we have seen, only about 10% of the mass of a galaxy—its stars, gas, and dust—emits electromagnetic radiation of any kind. We have no idea what the remaining 90% looks like or what it is made of. The dilemma of dark matter is one of the most challenging problems facing astronomers today.

1. Stars form gradually within a protogalaxy.

2. Gas not involved in star formation collapses to form a disk.

3. A spiral galaxy results.

(a) Formation of a spiral galaxy

1. Stars form rapidly within a protogalaxy.

2. Gas is quickly consumed to make stars.

3. A elliptical galaxy results.

(b) Formation of an elliptical galaxy

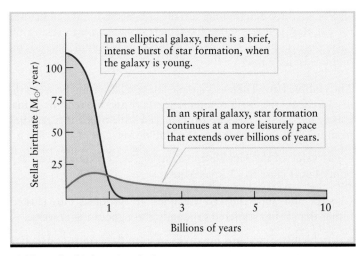

In an elliptical galaxy, there is a brief, intense burst of star formation, when the galaxy is young.

In an spiral galaxy, star formation continues at a more leisurely pace that extends over billions of years.

(c) The stellar birthrate in galaxies

figure 26-32

The Formation of Spiral and Elliptical Galaxies (a) If the initial star formation rate in a protogalaxy is low, it can evolve into a spiral galaxy with a disk. (b) If the initial star formation rate is rapid, no gas is left to form a disk. The result is an elliptical galaxy. (c) This graph shows how the rate of star birth (in solar masses per year) varies with age in spiral and elliptical galaxies.

Key Words

anisotropic, p. 584
barred spiral galaxy, p. 583
clusters (of galaxies), p. 592
dark-matter problem, p. 599
distance ladder, p. 588
dwarf elliptical galaxy, p. 584
elliptical galaxy, p. 583
fundamental plane, p. 587
galactic cannibalism, p. 598
giant elliptical galaxy, p. 584
gravitational lens, p. 600
groups (of galaxies), p. 592

Hubble classification, p. 580
Hubble constant, p. 590
Hubble flow, p. 589
Hubble law, p. 590
intracluster gas, p. 596
irregular cluster, p. 593
irregular galaxy, p. 585
isotropic, p. 584
lenticular galaxy, p. 584
Local Group, p. 592
maser, p. 588
poor cluster, p. 592

redshift, p. 589
regular cluster, p. 593
rich cluster, p. 592
spiral galaxy, p. 582
standard candle, p. 586
starburst galaxy, p. 596
supercluster, p. 594
Tully-Fisher relation, p. 587
tuning fork diagram, p. 584
void, p. 596

Key Ideas

The Hubble Classification: Galaxies can be grouped into four major categories: spirals, barred spirals, ellipticals, and irregulars.

• The disks of spiral and barred spiral galaxies are sites of active star formation.

• Elliptical galaxies are nearly devoid of interstellar gas and dust, and so star formation is severely inhibited.

• Lenticular galaxies are intermediate between spiral and elliptical galaxies.

• Irregular galaxies have ill-defined, asymmetrical shapes. They are often found associated with other galaxies.

Distance to Galaxies: Standard candles, such as Cepheid variables and the most luminous supergiants, globular clusters, H II regions, and supernovae in a galaxy, are used in estimating intergalactic distances.

• The Tully-Fisher relation, which correlates the width of the 21-cm line of hydrogen in a spiral galaxy with its luminosity, can also be used for determining distance. A method that can be used for elliptical galaxies is the fundamental plane, which relates the galaxy's size to its surface brightness distribution and to the motions of its stars.

The Hubble Law: There is a simple linear relationship between the distance from the Earth to a remote galaxy and the redshift of that galaxy (which is a measure of the speed with which it is receding from us). This relationship is the Hubble law, $v = H_0 d$.

• The value of the Hubble constant, H_0, is not known with certainty but is close to 71 km/s/Mpc.

Clusters and Superclusters: Galaxies are grouped into clusters rather than being scattered randomly throughout the universe.

• A rich cluster contains hundreds or even thousands of galaxies; a poor cluster, often called a group, may contain only a few dozen.

• A regular cluster has a nearly spherical shape with a central concentration of galaxies; in an irregular cluster, galaxies are distributed asymmetrically.

• Our Galaxy is a member of a poor, irregular cluster called the Local Group.

• Rich, regular clusters contain mostly elliptical and lenticular galaxies; irregular clusters contain spiral, barred spiral, and irregular galaxies along with ellipticals.

• Giant elliptical galaxies are often found near the centers of rich clusters.

Galactic Collisions and Mergers: When two galaxies collide, their stars pass each other, but their interstellar media collide violently, either stripping the gas and dust from the galaxies or triggering prolific star formation.

• The gravitational effects during a galactic collision can throw stars out of their galaxies into intergalactic space.

• Galactic mergers may occur; a large galaxy in a rich cluster may tend to grow steadily through galactic cannibalism, perhaps producing in the process a giant elliptical galaxy.

The Dark-Matter Problem: The luminous mass of a cluster of galaxies is not large enough to account for the observed motions of the galaxies; a large amount of unobserved mass must also be present. This situation is called the dark-matter problem.

• Hot intergalactic gases in rich clusters account for a small part of the unobserved mass. These gases are detected by their X-ray emission. The remaining unobserved mass is probably in the form of dark-matter halos that surround the galaxies in these clusters.

• Gravitational lensing of remote galaxies by a foreground cluster enables astronomers to glean information about the distribution of dark matter in the foreground cluster.

Formation and Evolution of Galaxies: Observations indicate that galaxies arose from mergers of several smaller gas clouds.

• Whether a protogalaxy evolves into a spiral galaxy or an elliptical galaxy depends on its initial rate of star formation.

Review Questions

1. Why did many nineteenth-century astronomers think that the "spiral nebulae" are part of the Milky Way?

2. What was the Shapley-Curtis "debate" all about? Was a winner declared at the end of the "debate"? Whose ideas turned out to be correct?

3. How did Edwin Hubble prove that the Andromeda "Nebula" is not a nebula within our Milky Way Galaxy?

4. An educational publication for children included the following statement: "The Sun is in fact the only star in our galaxy. All of the other stars in the sky are located in other galaxies." How would you correct this statement?

5. What is the Hubble classification scheme? Which category includes the largest galaxies? Which includes the smallest? Which category of galaxy is the most common?

6. Which is more likely to have a blue color, a spiral galaxy or an elliptical galaxy? Explain why.

7. Which types of galaxies are most likely to have new stars forming? Describe the observational evidence that supports your answer.

8. Explain why the apparent shape of an elliptical galaxy may be quite different from its real shape.

9. Are any galaxies besides our own visible with the naked eye from Earth? If so, which one(s)?

10. Why do astronomers suspect that the Hubble tuning fork diagram does not depict the evolutionary sequence of galaxies?

11. Why are Cepheid variable stars useful for finding the distances to galaxies? Are there any limitations on their use for this purpose?

12. What is the Tully-Fisher relation? How is it used for measuring distances? Can it be used for galaxies of all kinds? Why or why not?

13. What are masers? How can they be used to measure the distance to a galaxy?

14. What is the Hubble law? How can it be used to determine distances?

15. Why do you suppose it has been so difficult to determine the value of H_0?

16. Some galaxies in the Local Group exhibit blueshifted spectral lines. Why aren't these blueshifts violations of the Hubble law?

17. What are the differences between regular and irregular clusters?

18. What is the difference between a cluster and a supercluster? Are both clusters and superclusters held together by their gravity?

19. What measurements do astronomers make to construct three-dimensional maps of the positions of galaxies in space?

20. Describe what voids are and what they tell us about the large-scale structure of the universe.

21. Why is the intracluster gas in galaxy clusters at such high temperatures?

22. What are starburst galaxies? How are they related to collisions between galaxies?

23. Why do giant elliptical galaxies dominate rich clusters but not poor clusters?

24. What evidence is there for the existence of dark matter in clusters of galaxies?

25. What is gravitational lensing? Why don't we notice the gravitational lensing of light by ordinary objects on Earth?

26. What observations suggest that present-day galaxies formed from smaller assemblages of matter?

27. On what grounds do astronomers think that in the past, spiral galaxies were more numerous in rich clusters than they are today? What could account for this excess of spiral galaxies in the past?

Advanced Questions

Questions preceded by an asterisk (*) involve topics discussed in the Boxes.

Problem-solving tips and tools

Box 1-1 explains the small-angle formula, and Box 19-3 discusses the relationship among apparent magnitude, absolute magnitude, and distance. As Box 25-2 explains, a useful form of Kepler's third law is $M = rv^2/G$, where M is the mass within an orbit of radius r, v is the orbital speed, and G is the gravitational constant. Another form of Kepler's third law, particularly useful for two stars or two galaxies orbiting each other, is given in Section 19-9. The volume of a sphere of radius r is $4\pi r^3/3$. The mass of a hydrogen atom (^{1}H) is given in Appendix 7.

---

28. Hubble made his observations of Cepheids in M31 using the 100-inch (2.5-meter) telescope on Mount Wilson. Completed in 1917, this was the largest telescope in the world when Hubble carried out his observations in 1923. Why was it helpful to use such a large telescope?

29. The accompanying image shows the Small Magellanic Cloud (SMC), an irregular galaxy that orbits the Milky Way. The SMC is

(Anglo-Australian Observatory)   R I **V** U X G

63 kpc (200,000 ly) from Earth and 8 kpc (26,000 ly) across, and can be seen with the naked eye from southern latitudes. What features of this image indicate that there has been recent star formation in the SMC? Explain.

30. When the results from the Hipparcos mission were released, with new and improved measurements of the parallaxes of *nearby* stars within 500 pc, astronomers had to revise the distances to many *remote* galaxies millions of parsecs away. Explain why.

31. As Figure 21-17 shows, there are two types of Cepheid variables. Type I Cepheids are metal-rich stars of Population I, while Type II Cepheids are metal-poor stars of Population II. **(a)** Which type of Cepheid variables would you expect to be found in globular clusters? Which type would you expect to be found in the disk of a spiral galaxy? Explain your reasoning. **(b)** When Hubble discovered Cepheid variables in M31, the distinction between Type I and Type II Cepheids was not yet known. Hence, Hubble thought that the Cepheids seen in the disk of M31 were identical to those seen in globular clusters in our own Galaxy. As a result, his calculations of the distance to M31 were in error. Using Figure 21-17, explain whether Hubble's calculated distance was too small or too large.

*32. Astronomers often state the distance to a remote galaxy in terms of its distance modulus, which is the difference between the apparent magnitude m and the absolute magnitude M (see Box 19-3). **(a)** By measuring the brightness of supernova 1994I in the galaxy M51 (see Figure 26-2), the distance modulus for this galaxy was determined to be $m - M = 29.2$. Find the distance to M51 in megaparsecs (Mpc). **(b)** A separate distance determination, which involved measuring the brightnesses of planetary nebulae in M51, found $m - M = 29.6$. What is the distance to M51 that you calculate from this information? **(c)** What is the difference between your answers to parts (a) and (b)? Compare this difference with the 750-kpc distance from the Earth to M31, the Andromeda Galaxy. The difference between your answers illustrates the uncertainties involved in determining the distances to galaxies!

*33. Suppose you discover a Type Ia supernova in a distant galaxy. At maximum brilliance, the supernova reaches an apparent magnitude of +10. How far away is the galaxy? (*Hint:* See Box 26-1.)

34. The masers that orbit the center of the spiral galaxy M106 travel at an orbital speed of about 1000 km/s. Astronomers observed these masers at intervals of 4 months. **(a)** What distance does a single maser move during a 4-month period? Give your answer in kilometers and in AU. **(b)** During this period, a maser moving across the line of sight (like the maser shown in green in Figure 26-13) appeared to move through an angle of only 10^{-5} arcsec. Calculate the distance to the galaxy.

35. The average radial velocity of galaxies in the Hercules cluster pictured in Figure 26-19 is 10,800 km/s. **(a)** Using $H_0 = 71$ km/s/Mpc, find the distance to this cluster. Give your answer in megaparsecs and in light-years. **(b)** How would your answer to (a) differ if the Hubble constant had a smaller value? A larger value? Explain.

36. A certain galaxy is observed to be receding from the Sun at a rate of 7500 km/s. The distance to this galaxy is measured independently and found to be 1.4×10^8 pc. From these data, what is the value of the Hubble constant?

*37. In the spectrum of the galaxy NGC 4839, the K line of singly ionized calcium has a wavelength 403.2 nm. (**a**) What is the redshift of this galaxy? (*Hint:* See Box 26-2.) (**b**) Determine the distance to this galaxy using the Hubble law with H_0 = 71 km/s/Mpc.

*38. The galaxy RD1 has a redshift of z = 5.34. (**a**) Determine its recessional velocity v in km/s and as a fraction of the speed of light. (**b**) What recessional velocity would you have calculated if you had erroneously used the low-speed formula relating z and v? Would using this formula have been a small or large error? (**c**) According to the Hubble law, what is the distance from Earth to RD1? Use H_0 = 71 km/s/Mpc for the Hubble constant, and give your answer in both megaparsecs and light-years.

39. It is estimated that the Coma cluster (see Figure 26-16) contains about 10^{13} $M_\odot$ of intracluster gas. (**a**) Assuming that this gas is made of hydrogen atoms, calculate the total number of intracluster gas atoms in the Coma cluster. (**b**) The Coma cluster is roughly spherical in shape, with a radius of about 3 Mpc. Calculate the number of intracluster gas atoms per cubic centimeter in the Coma cluster. Assume that the gas fills the cluster uniformly. (**c**) Compare the intracluster gas in the Coma cluster with the gas in our atmosphere (3×10^{19} molecules per cubic centimeter, temperature 300 K); a typical gas cloud within our own Galaxy (a few hundred molecules per cubic centimeter, temperature 50 K or less); and the corona of the Sun (10^5 atoms per cubic centimeter, temperature 10^6 K).

*40. Two galaxies separated by 600 kpc are orbiting each other with a period of 40 billion years. What is the total mass of the two galaxies?

*41. Figure 26-28 shows the rotation curve of the Sa galaxy NGC 4378. Using data from that graph, calculate the orbital period of stars 20 kpc from the galaxy's center. How much mass lies within 20 kpc from the center of NGC 4378?

42. How might you determine what part of a galaxy's redshift is caused by the galaxy's orbital motion about the center of mass of its cluster?

43. According to Figure 26-32c, elliptical galaxies continue to form stars for about a billion years after they form. Give an argument why we might expect to find some Population I stars in an elliptical galaxy. (*Hint:* Table 21-1 gives the main-sequence lifetimes for stars of different masses.)

Discussion Questions

44. The Earth is composed principally of heavy elements, such as silicon, nickel, and iron. Would you be likely to find such planets orbiting stars in the disk of a spiral galaxy? In the nucleus of a spiral galaxy? In an elliptical galaxy? In an irregular galaxy? Explain your answers.

45. Discuss what observations you might make to determine whether or not the various Hubble types of galaxies represent some sort of evolutionary sequence.

46. Discuss the advantages and disadvantages of using the various standard candle distance indicators to obtain extragalactic distances.

47. How would you distinguish star images from unresolved images of remote galaxies on a CCD?

48. Describe what sorts of observations you might make to search for as-yet-undiscovered galaxies in our Local Group. How is it possible that such galaxies might still remain to be discovered? In what part of the sky would these galaxies be located? What sorts of observations might reveal these galaxies?

49. Explain why the dark matter in galaxy clusters could not be neutral hydrogen.

 Web/CD-ROM Questions

50. When galaxies pass close to one another, as should happen frequently in a rich cluster, tidal forces between the galaxies can strip away their outlying stars. The result should be a loosely dispersed sea of "intergalactic stars" populating the space between galaxies in a cluster. Search the World Wide Web for information about intergalactic stars. Have they been observed? Where are they found? What would our nighttime sky look like if our Sun were an intergalactic star?

 51. The Hubble Space Telescope (HST) has made extensive observations of very distant galaxies. Visit the HST Web site to learn about these investigations. How far back in time has HST been able to look? What sorts of early galaxies are observed? What is the current thinking about how galaxies formed and evolved?

 52. **The Formation of "The Mice."** Access and view the animation "The Formation of 'The Mice'" in Chapter 26 of the *Universe* Web site or CD-ROM. Based on this computer simulation, what will be the final fate of the two galaxies that make up NGC 4676 (see Figure 26-27)? Speculate on what might have happened if the two galaxies had collided at a much greater speed.

 53. **Radiation from a Rotating Galaxy.** Access and view the animation "Radiation from a Rotating Galaxy" in Chapter 26 of the *Universe* Web site or CD-ROM. Describe how the animation would have to be changed for a spiral galaxy of the same size but (**a**) greater mass and (**b**) smaller mass.

Observing Projects

 54. Using a telescope with an aperture of at least 30 cm (12 in.), observe as many of the spiral galaxies listed in the following table as you can. Use the *Starry Night Backyard*™ program to determine when they can best be viewed. Many of these galaxies are members of the Virgo cluster, which can best be seen from March through June. Because all galaxies are quite faint, be sure to schedule your observations for a moonless night. The best view is obtained when a galaxy is near the meridian. While at the eyepiece, make a sketch of what you see. Can you distinguish any spiral structure? After completing your observations, compare your sketches with photographs found in an online catalog of Messier objects (the "M" in the galaxy designations stands for Messier).

Spiral galaxy	Right ascension	Declination	Hubble type
M31 (NGC 224)	0^h 42.7^m	$+41°$ $16'$	Sb
M58 (NGC 4579)	12 37.7	+11 49	Sb
M61 (NGC 4303)	12 21.9	+4 28	Sc
M63 (NGC 5055)	13 15.8	+42 02	Sb
M64 (NGC 4826)	12 56.7	+21 41	Sb
M74 (NGC 628)	1 36.7	+15 47	Sc
M83 (NGC 5236)	13 37.0	−29 52	Sc
M88 (NGC 4501)	12 32.0	+14 25	Sb
M90 (NGC 4569)	12 36.8	+13 10	Sb
M91 (NGC 4548)	12 35.4	+14 30	SBb
M94 (NGC 4736)	12 50.9	+41 07	Sb
M98 (NGC 4192)	12 13.8	+14 54	Sb
M99 (NGC 4254)	12 18.8	+14 25	Sc
M100 (NGC 4321)	12 22.9	+15 49	Sc
M101 (NGC 5457)	14 03.2	+54 21	Sc
M104 (NGC 4594)	12 40.0	−11 37	Sa
M108 (NGC 3556)	11 11.5	+55 40	Sc

Note: The right ascensions and declinations are given for epoch 2000.

55. Using a telescope with an aperture of at least 30 cm (12 in.), observe as many of the following elliptical galaxies as you can. Six of these galaxies are in the Virgo cluster, which is conveniently located in the evening sky from March through June. Use the *Starry Night Backyard*™ program to help you determine when they can best be viewed. As in the previous exercise, be sure to schedule your observations for a moonless night, when the galaxies you wish to observe will be near the meridian. Do these elliptical galaxies differ in appearance from spiral galaxies?

Elliptical galaxy	Right ascension	Declination	Hubble type
M49 (NGC 4472)	12^h 29.8^m	$+8°$ $00'$	E4
M59 (NGC 4621)	12 42.0	+11 39	E3
M60 (NGC 4649)	12 43.7	+11 33	E1
M84 (NGC 4374)	12 25.1	+12 53	E1
M86 (NGC 4406)	12 26.2	+12 57	E3
M89 (NGC 4552)	12 35.7	+12 33	E0
M110 (NGC 205)	00 40.4	+41 41	E6

Note: The right ascensions and declinations are given for epoch 2000.

56. Using a telescope with an aperture of at least 30 cm (12 in.), observe as many of the following interacting galaxies as you can. As in the previous exercises, be sure to schedule your observations for a moonless night, when the galaxies you wish to observe will be near the meridian. While at the eyepiece, make a sketch of each galaxy. Can you distinguish hints of interplay among the galaxies? After completing your observations, compare your sketches with photographs found in an online catalog of Messier objects (the "M" in the galaxy designations stands for Messier).

Interacting galaxies	Right ascension	Declination
M51 (NGC 5194)	13^h 29.9^m	$+47°$ $12'$
NGC 5195	13 30.0	+47 16
M65 (NGC 3623)	11 18.9	+13 05
M66 (NGC 3627)	11 20.2	+12 59
NGC 3628	11 20.3	+13 36
M81 (NGC 3031)	9 55.6	+69 04
M82 (NGC 3034)	9 55.8	+69 41
M95 (NGC 3351)	10 44.0	+11 42
M96 (NGC 3368)	10 46.8	+11 49
M105 (NGC 3379)	10 47.8	+12 35

Note: The right ascensions and declinations are given for epoch 2000.

57. Use the *Starry Night Backyard*™ program to observe other galaxies. First display the entire celestial sphere (select **Guides > Atlas** in the **Go** menu). Use the **Find…** command in the **Edit** menu to center the field of view on each of the following galaxies in turn, and zoom in as needed until you can see the shape of the galaxy. In each case, use the image of the galaxy to determine its Hubble classification as best you can, and explain your reasoning: (i) M87; (ii) M101; (iii) M102; (iv) M104.

58. Use the *Deep Space Explorer*™ program to examine clusters of galaxies. First click on the "Home" button at the upper left of the *Deep Space Explorer*™ window. Then click on the "Highlight" tab on the left of the star field. A long list of galaxy clusters and other features will appear. Double click on **GA Virgo Cluster** under the heading **Supergalactic Eq Plane**. You will see a check mark next to the name, and the galaxies of the Virgo Cluster will appear in yellow. You can rotate the Virgo Cluster by putting the mouse cursor over the image, holding down the mouse button, and moving the mouse. (On a two-button mouse, hold down the left mouse button.) (**a**) Describe the general shape of the Virgo cluster. (**b**) As you rotate the Virgo cluster, you should notice other groupings of galaxies. Make a sketch of the screen and circle what you believe are other groups on your sketch. To see how astronomers have grouped these other galaxies, click on one of the other clusters (and clouds and extensions)

listed near Virgo in the "Highlight" list at the far left of the window. The cluster whose name you clicked will light up in yellow. Unclick this cluster and click another. Repeat until you have identified all of the clusters around Virgo. Sketch them on your drawing and compare with your original groupings. (c) Choose three of these clusters, study them by double clicking on each in the list, move space around, and describe their distributions of galaxies compared to Virgo. For example, what are their shapes and relative sizes compared to Virgo and to each other?

Collaborative Exercises

59. In the early twentieth century, there was considerable debate about the nature of spiral nebulae and their distance from us, but the debate was resolved by improvements in technology. As a group, list three issues that we, as a culture, did not understand in the past but understand today, and explain why we now have that understanding.

60. Even though there are billions of galaxies, there are not billions of different kinds. In fact, galaxies are classified according to their appearance. As a group, dig into your book bags and put all of the writing implements (pens, pencils, highlighters, and so on) you have in a central pile. Remember which ones are yours! Determine a classification scheme that sorts the writing implements into at least three to six piles. Write down the scheme and the number of items in each pile. Ask the group next to you to use your scheme and sort your materials. Correct any ambiguities before submitting your classification scheme.

61. Imagine your company, Astronomical Artistry, has been contracted by the local marching band to create a football half-time show about spiral galaxies. How exactly would you design the positions of the band members on the field to represent the different spiral galaxies of classes Sa, Sb, and Sc? Create two columns on your paper by drawing a line from top to bottom, drawing sketches in the left-hand column and writing a description of each sketch in the right-hand column. Also include what the band's opening formation and final formation should be.

27 Quasars, Active Galaxies, and Gamma-Ray Bursters

An ordinary star emits radiation primarily at ultraviolet, visible, and infrared wavelengths, in proportions that reflect the star's surface temperature. Ordinary galaxies, too, emit most strongly in these wavelength regions. But the object shown here, called 3C 273, is outrageously different: It emits strongly over an immense range of wavelengths from radio to X-ray. 3C 273 is also intensely luminous, with thousands of times the radiation output of an ordinary galaxy. And, as this X-ray image shows, some source of energy within 3C 273 causes it to eject a glowing, high-speed jet that extends outward for hundreds of thousands of light-years.

3C 273 is a *quasar*—one of many thousands of distant objects whose luminosity is far too great to come from starlight alone. Quasars are intimately related to *active galaxies,* whose tremendous luminosity can fluctuate substantially over a period of months, weeks, or even days.

What makes quasars and active galaxies so tremendously energetic? In this chapter, we will see evidence that a supermassive black hole lies at the center of each of these objects. As matter accretes onto the black hole, it releases fantastic amounts of energy and can produce powerful jets.

Quasars and active galaxies emit their intense radiation continuously. But from time to time astronomers detect flashes of high-energy gamma rays that for a brief time outshine even the most luminous quasar. For many years the nature of these *gamma-ray bursters* was a complete mystery to astronomers. As we will see, however, dramatic recent evidence suggests that these intense bursts are caused by stellar cataclysms taking place within distant galaxies.

A powerful X-ray–emitting jet blasts outward from quasar 3C 273. (NASA/CXC/SAO/H. Marshall et al.)
R I V U **X** G

Questions to Guide Your Reading

As you read the sections of this chapter, look for the answers to the following questions:

27-1 Why are quasars unusual? How did astronomers discover that they are extraordinarily distant and luminous?

27-2 What evidence showed a link between quasars and galaxies?

27-3 How are Seyfert galaxies and radio galaxies related to quasars?

27-4 How can material ejected from quasars appear to travel faster than light?

27-5 What could power the incredible energy output from active galaxies?

27-6 Why do many active galaxies emit ultrafast jets of material?

27-7 What are gamma-ray bursters? How did astronomers discover how far away they are?

27-1 | Quasars look like stars but have huge redshifts

SCIENCE IN PROCESS *Examining the spectra of quasars revealed that they are immensely distant*

Many of the most revolutionary discoveries in astronomy have been the result of advances in technology. So it was in 1610 when Galileo used a new device called the telescope to scan the sky, and in the process found evidence that shattered the ancient geocentric model of the universe (see Section 4-3). And so it was during the middle years of the twentieth century, when the new technique of radio astronomy revealed the existence of remote, dazzlingly luminous objects called quasi-stellar radio sources or quasars.

Grote Reber, a radio engineer and ham radio enthusiast, built the first true radio telescope in 1936 in his backyard in Illinois (see Section 6-6). By 1944 he had detected strong radio emissions from sources in the constellations Sagittarius, Cassiopeia, and Cygnus.

Two of these sources, named Sagittarius A and Cassiopeia A happen to lie in our own Galaxy; they are a supernova remnant (see Figure 22-23) and the center of the Galaxy (see Section 25-6). The nature of the third source, called Cygnus A, proved more elusive. The mystery only deepened in 1951, when Walter Baade and Rudolph Minkowski used the 200-inch (5-meter) optical telescope

on Palomar Mountain to discover a dim, strange-looking galaxy at the position of Cygnus A (see the inset in Figure 27-1).

When Baade and Minkowski photographed the spectrum of Cygnus A, they were surprised to find a number of bright *emission* lines. By contrast, a normal galaxy has *absorption* lines in its spectrum (see Section 26-3 and Figure 26-14). This absorption takes place in the atmospheres of the stars that make up the galaxy, as described in Section 19-5. In order for Cygnus A to have emission lines, something must be exciting and ionizing its atoms. Furthermore, the wavelengths of Cygnus A's emission lines are all shifted by 5.6% toward the red end of the spectrum. Astronomers use the letter z to denote redshift (see Section 26-5), and thus this object has $z = 0.056$, corresponding to a recessional velocity of 16,000 km/s.

If Cygnus A participates in the same Hubble flow as clusters of galaxies (described in Section 26-5), then this recessional velocity corresponds to a tremendous distance from the Earth—about 230 Mpc (750 million ly) if the Hubble constant equals 71 km/s/Mpc. Yet despite its tremendous distance from the Earth, radio waves from Cygnus A can be picked up by amateur astronomers with backyard equipment. This means that Cygnus A must be one of the most luminous radio sources in the sky. In fact, its radio luminosity is 10^7 times as great as that of an ordinary galaxy like the Milky Way. The object that creates the Cygnus A radio emission must be something quite extraordinary.

Cygnus A was not the only curious radio source to draw the attention of astronomers. In 1960, Allan Sandage used the 200-inch

(a) Radio image of Cygnus A

R I V U X G

figure 27-1

Cygnus A (3C 405) (a) This radio image from the Very Large Array shows that most of the emission from Cygnus A comes from luminous radio lobes located on either side of a peculiar galaxy. Each lobe extends about 70 kpc (230,000 ly) from the galaxy. (b) The galaxy at the heart of Cygnus A has a substantial redshift, so it must be extremely far from Earth (about 230 Mpc, or 750 million ly). To be so distant and yet be one of the brightest radio sources in the sky, Cygnus A must have an enormous energy output. (a: R. A. Perley, J. W. Dreher, and J. J. Cowan, NRAO; b: Palomar Observatory)

(b) Close-up of the central galaxy   R I **V** U X G

telescope to discover a "star" at the location of a radio source designated 3C 48 (Figure 27-2). (The "3C" refers to the *Third Cambridge Catalogue*, a compendium of radio sources.) Ordinary stars are not strong sources of radio emission, so 3C 48 must be something unusual. Like Cygnus A, its spectrum showed a series of emission lines, but astronomers were unable to identify the chemical elements that produced these lines.

Two years later, astronomers discovered a similar starlike optical counterpart to the radio source 3C 273. This "star" is even more unusual, with a luminous jet protruding from one side. (The image that opens this chapter shows this jet at X-ray wavelengths.) Like 3C 48, its visible spectrum contains a series of emission lines that no one could explain.

Although 3C 48 and 3C 273 were clearly oddballs, many astronomers thought they were just strange stars in our own Galaxy. A breakthrough occurred in 1963, when Maarten Schmidt at Caltech took another look at the spectrum of 3C 273. He realized that four of its brightest emission lines are positioned relative to one another in precisely the same way as four of the Balmer lines of hydrogen. However, these emission lines from 3C 273 were all shifted to much longer wavelengths than the usual wavelengths of the Balmer lines. Schmidt determined that 3C 273 has a redshift of $z = 0.158$, corresponding to a recessional velocity of 44,000 km/s (15% of the speed of light). No star could be moving this fast and remain within our Galaxy. Hence, Schmidt concluded that 3C 273 could not be a nearby star, but must lie outside the Milky Way.

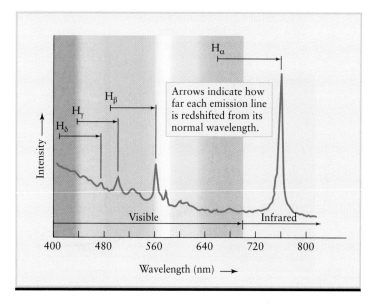

figure 27-3

The Spectrum of 3C 273 The visible-light and infrared spectrum of this quasar is dominated by four bright emission lines of hydrogen (see Section 5-8). The redshift is $z = 0.158$, so the wavelength of each line is 15.8% greater than for a sample of hydrogen on Earth. For example, the wavelength of H_β is shifted from 486 nm (blue-green) to $(1.158)(486 \text{ nm}) = 563$ nm (yellow).

According to the Hubble law, the recessional velocity of 3C 273 implies that it is about 620 Mpc (2 billion ly) away. To be detected at such distances, 3C 273 must be an extraordinarily powerful source of both visible light and radio radiation.

Figure 27-3 shows the visible and near-infrared spectrum of 3C 273. As we saw in Section 6-5 (review Figure 6-24), such a spectrum is a graph of intensity versus wavelength on which emission lines appear as peaks and absorption lines appear as valleys. Emission lines are caused by excited atoms, which emit radiation at specific wavelengths.

Upon learning how Schmidt deciphered the spectrum of 3C 273, two other Caltech astronomers, Jesse Greenstein and Thomas Matthews, found they could identify the spectral lines of 3C 48 as having suffered an even larger redshift of $z = 0.367$. That shift corresponds to a speed three-tenths that of light, which places 3C 48 twice as far away as 3C 273, or approximately 1300 Mpc (4 billion ly) from Earth.

Because of their strong radio emission and starlike appearance, 3C 48 and 3C 273 were dubbed *quasi-stellar radio sources*, a term soon shortened to **quasars**. After the first quasars were discovered by their radio emission, many similar, high-redshift, starlike objects were found that emit little or no radio radiation. These "radio-quiet" quasars were originally called *quasi-stellar objects*, or QSOs, to distinguish them from radio emitters. Today, however, the term "quasar" is often used to include both types. Only about 10% of quasars are "radio-loud" like 3C 48 and 3C 273.

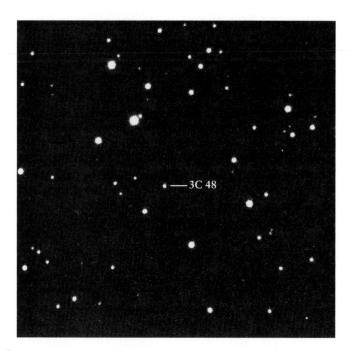

figure 27-2 R I **V** U X G

The Quasar 3C 48 This object in the constellation Triangulum was thought at first to be a nearby star that happened to emit radio waves. In fact, the redshift of 3C 48 is so great ($z = 0.367$) that, according to the Hubble law, it must be approximately 1300 Mpc (4 billion ly) away. (Alex G. Smith, Rosemary Hill Observatory, University of Florida)

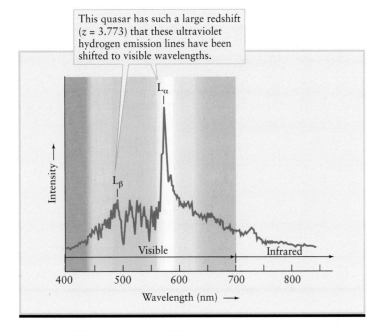

This quasar has such a large redshift ($z = 3.773$) that these ultraviolet hydrogen emission lines have been shifted to visible wavelengths.

figure 27-4

The Spectrum of a High-Redshift Quasar This quasar, known as PKS 2000-030, has a redshift $z = 3.773$. This is so large that the Lyman spectral lines L_α and L_β of hydrogen, which are normally at ultraviolet wavelengths, have been shifted into the visible part of the spectrum. The designation of the quasar refers to its position on the celestial sphere and to the Parkes Observatory in Australia, where it was first discovered (see Figure 6-25).

Thanks to sky surveys like those we discussed in Section 26-6, more than 100,000 quasars are now known. They all look rather like stars (see Figure 1-8), and all have large redshifts, ranging from 0.06 to at least 5.8 (Figure 27-4). Most quasars have redshifts of 0.3 or more, which implies that they are more than 1000 Mpc (3 billion light-years) from Earth. For example, the quasar PC 1247+3406 has a redshift of $z = 4.897$, which corresponds to a recessional velocity of more than 94% of the speed of light and a distance of roughly 3810 Mpc (12.4 billion ly) from Earth.

CAUTION A value of z greater than 1 does *not* mean that a quasar is receding from us faster than the speed of light. At high speeds, the relationship between redshift and recessional velocity must be modified by the special theory of relativity, as explained in Box 26-2. As the speed of a receding source approaches the speed of light, its redshift can become very much greater than 1. An infinite redshift ($z = \infty$) corresponds to a recessional velocity equal to the speed of light.

Light takes time to travel across space, so when we observe a very remote object, we are seeing it in the remote past. This means that for very remote objects, the relationship between redshift and distance from the Earth depends on how the universe has evolved over time. As we will see in Chapter 28, the Hubble law reveals that the universe is expanding. In other words, if you could watch the motions of widely separated clusters of galaxies over millions of years, you would see them gradually moving away from one another. Furthermore, in Chapter 28 we will see evidence that the universe has expanded at different rates at different times in the past. Table 27-1 relates the redshift to the recessional velocity and distance in a model consistent with our present understanding of the expansion of the universe.

CAUTION Since the universe is expanding, there is more than one way to state the distance between us and a distant quasar. When we say that quasar PC 1247+3406 is 12.4 billion light-years away, we mean that the light by which we see that quasar took 12.4 billion years to reach us. This elapsed time is called the *light travel time*. But the universe has continued to expand during that time. Hence, PC 1247+3406 is now much farther away from us, about 25.7 billion ly or 7880 Mpc. (This is called the *comoving radial distance* to the quasar, and is actually the correct distance d to use in the Hubble law $v = H_0 d$.) In this book, when we give the distance to a remote galaxy or quasar, we mean the light travel time multiplied by the speed of light—that is, the distance that light trav-

table 27-1	**Redshift and Distance**		WEB LINK 27.4
Redshift	**Recessional velocity**	**Distance**	
z	v/c	(Mpc)	(10^9 ly)
0	0	0	0
0.1	0.095	394	1.29
0.2	0.180	739	2.41
0.3	0.257	1040	3.39
0.4	0.324	1310	4.26
0.5	0.385	1540	5.02
0.75	0.508	2010	6.57
1	0.600	2370	7.73
1.5	0.724	2860	9.32
2	0.800	3170	10.3
3	0.882	3520	11.5
4	0.923	3710	12.1
5	0.946	3830	12.5
10	0.984	4040	13.2
Infinite	1	4190	13.7

This table assumes a Hubble constant $H_0 = 71$ km/s/Mpc, a matter density parameter $\Omega_m = 0.27$, and a dark energy density parameter $\Omega_\Lambda = 0.73$ (see Chapter 28). The distance in light-years is equal to the light travel time in years.

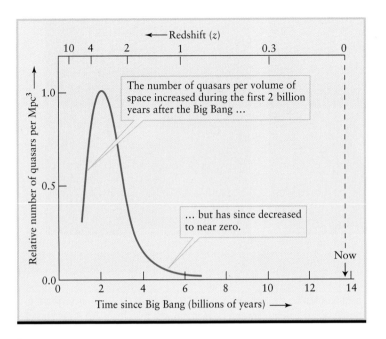

Figure 27-5

Quasars Are Extinct The greater the redshift of a quasar, the farther it is from Earth and the farther back in time we are seeing it. By observing the number of quasars found at different redshifts, astronomers can calculate how the density of quasars in the universe has changed over the history of the universe. The history of quasars is reminiscent of the history of the dinosaurs, which once populated the entire Earth but today are extinct. (Peter Shaver, European Southern Observatory)

eled to reach us from that galaxy or quasar. Then the distance to a remote object expressed in light-years tells you how many years into the past you are looking when you view that object. To avoid these issues about how best to define distances, many astronomers prefer simply to state the redshift z of a distant object; they then use the simple rule that the greater the redshift, the more distant the object (see Section 26-5).

Because there are no quasars with small redshifts, it follows that *there are no nearby quasars.* The nearest one is some 250 Mpc (800 million ly) from the Earth. Hence, the absence of nearby quasars means that there have been no quasars for 800 million years. Indeed, the number of quasars began to decline precipitously roughly 10 billion years ago (**Figure 27-5**). Quasars were a common feature of the universe in the distant past, but there are none in the present-day universe.

27-2 | Quasars are the ultraluminous centers of distant galaxies

If quasars are as distant as their redshifts show them to be, they must be extraordinarily luminous to be visible from Earth. What,

then, are these strange objects? Quasars cannot simply be very large and luminous galaxies, because their spectra are totally different: As we saw in Section 26-1, the visible-light and infrared spectrum of a galaxy is dominated by absorption lines from the galaxy's stars, while the corresponding spectrum of a quasar is dominated by emission lines. Quasars also emit much more of their light at ultraviolet wavelengths than do stars or galaxies. Hence, whatever the origin of a quasar's radiation, it is not simply starlight. While quasars are not galaxies, we will see compelling evidence that the two are intimately related: Quasars turn out to be ultraluminous objects located at the centers of remote galaxies.

A quasar's luminosity can be calculated from its apparent brightness and distance using the inverse-square law (see Section 19-2). For example, 3C 273 has a luminosity of about 10^{40} watts, which is equivalent to 2.5×10^{13} (25 trillion) Suns. Generally, quasar luminosities range from about 10^{38} watts up to nearly 10^{42} watts. For comparison, a typical large galaxy, like our own Milky Way, shines with a luminosity of 10^{37} watts, which equals 2.5×10^{10} (25 billion) Suns. Thus, a bright quasar can be many thousands of times more luminous than the entire Milky Way Galaxy.

 When quasars were first discovered, their energy output seemed so absurdly huge that a few astronomers began to question long-held ideas, such as the Hubble law. Traditionally, astronomers had used the Hubble law to calculate the distance to a galaxy from its redshift: the higher the redshift, the greater the distance. In the 1960s the American astronomer Halton C. Arp suggested that the Hubble law might not apply to quasars. If part of a quasar's redshift was caused by some yet-undiscovered phenomenon, then the quasar could be much closer to Earth than the Hubble law would have us believe. If so, a quasar's luminosity would not be so incredibly large.

In support of this theory, Arp and his collaborators drew attention to certain high-redshift quasars that seem to be located in or associated with low-redshift galaxies. These examples of "discordant redshifts" fueled a heated debate during the 1970s reminiscent of the Shapley-Curtis "debate" 50 years earlier (see Section 26-1). By the 1980s, however, the preponderance of evidence clearly favored the standard interpretation of redshifts according to the Hubble law. Most astronomers today regard Arp's discordant redshifts as simply a *projection effect,* whereby a distant quasar just happens to be in the same part of the sky as a nearby galaxy.

 An analogy to the projection effect sometimes happens in photography on Earth. If a photographer poses a person with a telephone pole in the background, the photo can give the misleading impression that the unlucky person has a telephone pole growing out of his head.

What largely ended the redshift debate of the 1970s were observations showing that quasars are associated with remote galaxies. Long-exposure images of quasars showed that they are often found in groups or clusters of galaxies, and that the redshift of each of these quasars is essentially the same as the redshifts of the galaxies that surround it. Thus, the quasar's redshift indicates its distance from Earth, just as do the redshifts of galaxies.

(a)   (b)   (c)

figure 27-6 R I **V** U X G

Quasars and Their "Host Galaxies" (a) Quasar PG 0052+251 is located at the center of an apparently normal spiral galaxy at redshift $z = 0.155$. Other quasars are found at the centers of ordinary-looking elliptical galaxies. (b) The galaxy that hosts quasar PG 1012+008 (redshift $z = 0.185$) is in the process of merging with a second luminous galaxy. The wispy material surrounding the quasar may have been pulled out of the galaxies by tidal forces (see Figures 26-26 and 26-27). The two merging galaxies are just 9500 pc (31,000 ly) apart. Another small galaxy to the left of the quasar may also be merging with the others. (c) A long tail of gas and dust extends upward in this image from quasar PG 0316-346 (redshift $z = 0.260$). This tail was presumably formed when the quasar's host galaxy collided with a second galaxy that lies beyond this image. (J. Bahcall, Institute for Advanced Study; M. Disney, University of Wales; NASA)

A second important link between quasars and galaxies was established in the 1980s. In that decade astronomers first had the technology to examine the faint "fuzz" seen around the images of some quasars. The spectrum of this fuzz shows stellar absorption lines, indicating that each of these quasars is embedded in a galaxy whose light is far fainter than that of the quasar itself. In each case, the absorption lines of the stars have the same redshift as the quasar's emission lines, further supporting the idea that quasars are at distances indicated by their redshifts and the Hubble law.

It is very difficult to observe the "host galaxy" in which a quasar is located because the quasar's light overwhelms light from the galaxy's stars. Nevertheless, painstaking observations have revealed some basic properties of these host galaxies (**Figure 27-6**). Relatively nearby radio-quiet quasars (with redshifts of less than 0.2, corresponding to distances of less than about 750 Mpc) tend to be located in spiral galaxies, whereas radio-loud quasars as well as more distant radio-quiet quasars tend to be located in ellipticals (Figure 27-6a). However, a large percentage of these host galaxies have distorted shapes or are otherwise peculiar. Many have nearby companion galaxies, suggesting a link between collisions or mergers and the quasar itself (Figures 27-6b and 27-6c). We will see in Section 27-6 that this link helps explain why quasars are found only at great distances from Earth—and hence only in the distant past of the universe.

While observations like those in Figure 27-6 show that quasars lie at the centers of galaxies, they do not really explain what quasars *are*. As we will see, important clues come from a class of galaxies that are intermediate in luminosity between quasars and normal galaxies.

27-3 | Seyferts and radio galaxies bridge the gap between normal galaxies and quasars

We have seen that some scientists in the 1970s preferred to challenge the Hubble law rather than accept the existence of highly luminous objects. One reason for their skepticism was the huge gap in energy output between normal galaxies and quasars. The gap was bridged, however, when astronomers realized that the centers of certain galaxies look like low-luminosity quasars.

The first "missing links" between quasars and ordinary galaxies were actually discovered before quasars themselves. In 1943, Carl Seyfert at the Mount Wilson Observatory made a systematic study of spiral galaxies with bright, compact nuclei that seem to show signs of intense and violent activity (**Figure 27-7**). Like quasars, the nuclei of these galaxies have strong emission lines in their spectra. These galaxies are now referred to as **Seyfert galaxies.**

A few percent of the most luminous spiral galaxies are Seyfert galaxies. These galaxies range in luminosity from about 10^{36} to 10^{38} watts, which makes the brightest Seyferts as luminous as faint quasars. Indeed, there is no sharp dividing line between the properties of Seyferts and those of quasars. Like radio-quiet quasars, Seyferts tend to have only weak radio emissions. And like quasars, some Seyferts are members of interacting pairs or exhibit the vestiges of mergers and collisions.

While Seyfert galaxies resemble dim, radio-quiet quasars, certain elliptical galaxies, called **radio galaxies** because of their strong radio emission, are like dim, radio-loud quasars. The energy output of radio galaxies covers approximately the same range as that of Seyferts. The first of these peculiar galaxies was discovered in 1918 by Heber D. Curtis. His short-exposure photograph of the giant elliptical galaxy M87 revealed a bright, starlike nucleus with a protruding jet. **Figure 27-8a** shows an overall view of the entire galaxy. The Hubble Space Telescope image in Figure 27-8b shows the jet extending outward from the nucleus of M87.

Most of the light from the central regions of M87 is **thermal radiation,** with a spectrum like that of a blackbody. This is radiation that is caused by the random thermal motion of the atoms and molecules that make up the emitting object. The spectrum of thermal radiation depends only on the object's temperature. For a given temperature, an object emits the maximum amount of thermal radiation at one particular frequency and emits less at frequencies above or below that value. There are also absorption lines in the spectrum of thermal radiation from M87's nucleus, which indicates that this radiation is due to a profusion of stars crowded around the galaxy's center.

By contrast, the light from M87's jet is **nonthermal radiation.** This radiation is *not* due to random thermal motion, and has a very different spectrum than does thermal radiation. In this case of M87's jet, the spectrum extends from radio to X-ray frequencies, which is a much broader range than thermal radiation.

The particular type of nonthermal radiation emitted from the jet is *synchrotron radiation* (**Figure 27-9**). As we saw in the discussion of the Crab Nebula in Section 23-4, synchrotron radiation is

figure 27-7 R I **V** U X G

A Seyfert Galaxy This spiral galaxy, called NGC 7742, is a Seyfert galaxy that lies some 22 Mpc (72 million ly) from Earth in the constellation Pegasus. The spectrum of this galaxy's intensely luminous nucleus shows emission lines of highly ionized atoms. (Hubble Heritage Team, AURA/STScI/NASA)

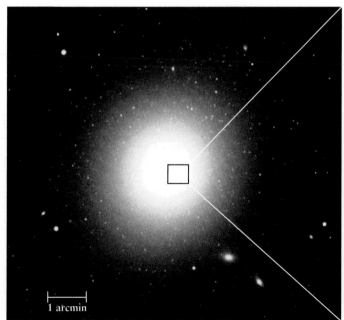

(a) The giant elliptical galaxy M87

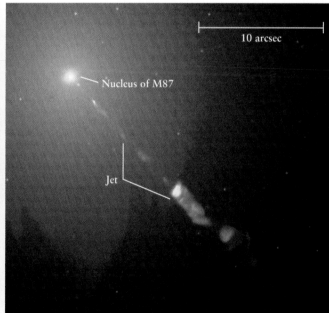

(b) A shorter exposure reveals M87's jet

figure 27-8 R I **V** U X G

The Radio Galaxy M87 (a) M87 measures about 100 kpc (300,000 ly) across and lies in the Virgo cluster about 15 Mpc (50 million ly) from Earth. The numerous fuzzy dots that surround the galaxy are globular clusters. (b) From the galaxy's tiny, bright nucleus—less than 2 pc (6.5 ly) in diameter—a jet extends outward some 1500 pc (5000 ly). (a: David Malin/ Anglo-Australian Observatory; b: NASA and The Hubble Heritage Team, STScI/AURA)

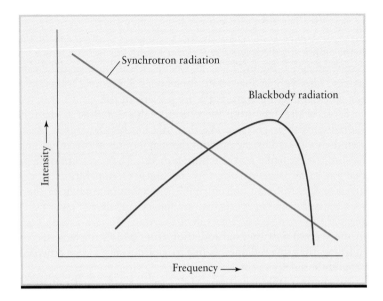

figure 27-9

Thermal and Nonthermal Spectra This schematic graph compares the spectra of synchrotron radiation and blackbody radiation. A blackbody spectrum, like the spectrum of a star or ordinary galaxy, always has a hump. By contrast, the nonthermal spectrum of synchrotron radiation shows a steady decline with frequency.

produced by relativistic electrons traveling in a strong magnetic field. (Recall that "relativistic" means "traveling near the speed of light.") As the electrons spiral around the magnetic field, they emit electromagnetic radiation. The presence of synchrotron radiation

coming from M87's jet indicates that relativistic particles are being ejected from the galaxy's nucleus and encountering a magnetic field.

One piece of evidence that the radiation from M87's jet is non-thermal is that some of it is **polarized radiation,** which means that the electric fields of the waves are oriented in a specific direction. By contrast, thermal radiation—like light from the Sun, stars, or light-bulbs—is *unpolarized* radiation. It consists of waves whose electric fields are oriented at random angles. (Most radio stations transmit signals that are polarized in a vertical plane. In order to detect these signals, the radio antennas on cars are oriented vertically rather than horizontally.)

During the 1960s and 1970s, astronomers found that many radio galaxies have a set of two **radio lobes,** one on each side of a parent galaxy. The radio lobes generally span a distance that is 5 to 10 times the size of the parent galaxy, although smaller and much larger examples are known. The parent galaxy—almost always a giant elliptical—sits midway between the radio lobes. Because of their overall shape, radio galaxies (and many similar quasars) are sometimes called **double radio sources.** Cygnus A, shown in Figure 27-1, is a fine example.

The spectrum and polarization of the radio-frequency emission from the lobes of a double radio source bear all the characteristics of synchrotron radiation. This suggests that radio galaxies should have jets of relativistic particles leading from the galaxy out to the radio lobes. A spectacular example of this is Centaurus A, one of the brightest radio sources in the southern sky (**Figure 27-10**). The peculiar parent galaxy of Centaurus A, shown in Figure 27-10*a*, has a broad dust lane studded with young, hot, massive stars. This galaxy, called NGC 5128, is thought to be an elliptical galaxy that has collided with a spiral galaxy. Radio waves pour from two lobes on opposite sides of the galaxy (Figure 27-10*b*). A second set of radio lobes farther from the galaxy (not shown in Figure 27-10)

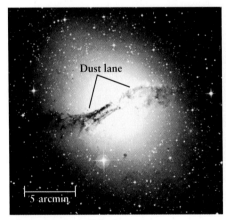

(a) Centaurus A: light from stars
R **V** U X G

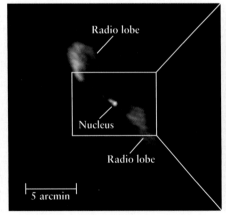

(b) Centaurus A: radio lobes
R I V U X G

(c) An X-ray-emitting jet emanates from the nucleus
R I V U **X** G

figure 27-10

WEB LINK 27.9

The Radio Galaxy Centaurus A (a) This elliptical galaxy, called NGC 5128, lies about 4 Mpc (13 million ly) from Earth at the location of the radio source Centaurus A. The dust lane is evidence of a collision with another galaxy. (b) At radio wavelengths we see synchrotron radiation from two radio lobes centered on the galaxy's nucleus. (c) A luminous jet extends from the nucleus directly toward one of the lobes. (a: Digital Sky Survey/U. K. Schmidt Image/STScI; b: NRAO/VLA/J. Condon et al.; c: NASA/SAO/R. Kraft et al.)

spans a volume 2 million light-years across. Figure 27-10*c* shows an X-ray–emitting jet emanating from the galaxy's nucleus. This jet, which is perpendicular to the galaxy's dust lane, is aimed toward one of the radio lobes.

The idea that jets of particles are found in double radio sources is reinforced by the existence of **head-tail sources.** These are radio galaxies with a concentrated source of radio emission (the "head") to which are attached two long radio-emitting streams (the "tails"). Figure 27-11 shows one such head-tail source. The "head" coincides with the position of the elliptical galaxy NGC 1265, which is a member of the Perseus cluster of galaxies. NGC 1265 is known to be moving through the cluster at 2500 km/s. Figure 27-11 shows that the two "tails" are swept backward, like smoke emerging from a fast-moving steam locomotive. This is just what we would expect if the "tails" are caused by jets emerging from the galaxy; as NGC 1265 moves through the intergalactic gas of the cluster, the jets feel a 2500-km/s wind blowing them back.

Like NGC 1265, many radio galaxies are found near the centers of rich clusters of galaxies and thus are probably subjected to collisions and mergers like the one that NGC 5128 has experienced (see Figure 27-10*a*).

Seyfert galaxies and radio galaxies share many properties with the more remote, more luminous quasars. A key difference is that we see many Seyferts and radio galaxies that are relatively close to us and hence in the recent past, whereas we only see quasars at great distances and hence in the remote past (Figure 27-12). Thus, quasar-like objects have not completely disappeared from the universe, but those that remain in the present-day universe—Seyferts and radio galaxies—are only a pale shadow of the intensely luminous quasars that populated the heavens when the universe was young.

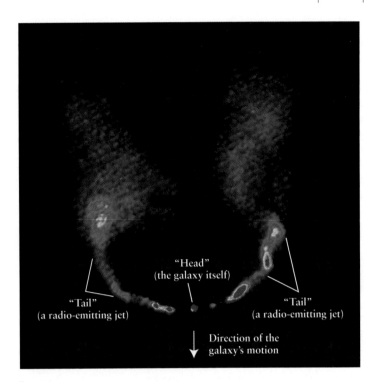

Figure 27-11 **R** I V U X G

The Head-Tail Source NGC 1265 The elliptical galaxy NGC 1265 would probably be an ordinary double radio source, except that it is moving at a high speed through the Perseus cluster of galaxies. Because of this motion, its two jets trail behind the galaxy, giving this radio source a distinctly windswept appearance. (NRAO)

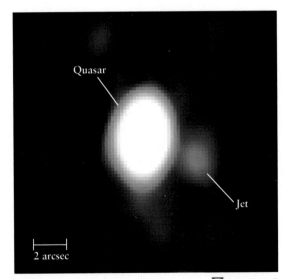

(a) A quasar jet at radio wavelengths... **R** I V U X G

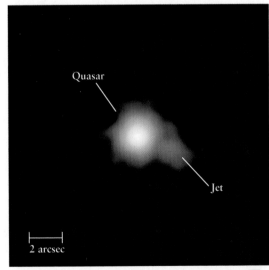

(b) ...and at X-ray wavelengths R I V U **X** G

WEB LINK 27.10

Figure 27-12

A Jet Emanating from a High-Redshift Quasar

Designated GB1508+5714, this quasar has a redshift $z = 4.3$, corresponding to a distance of 3700 Mpc (12 billion ly). We see this quasar as it was when the universe was only 10% of its present age. Quasars like this may have evolved into radio galaxies like those shown in Figures 27-8, 27-10, and 27-11. (a: NRAO/VLA/Brandeis/T. Cheung; b: NASA/CXC/SAO/ A. Siemiginowska et al.)

27-4 | Quasars, blazars, Seyferts, and radio galaxies are active galaxies

Inspired by the discovery of quasars, Seyferts, and luminous radio galaxies, astronomers during the 1960s and 1970s searched for clues to these powerful energy sources. One important clue turned out to be a new class of unusual objects called *blazars*. Like quasars, blazars are extraordinarily luminous objects that look like stars but prove to be the nuclei of distant galaxies. The difference is that unlike a quasar, the spectrum of a blazar is almost featureless, with hardly any absorption or emission lines at all!

The first of this new class of objects to be discovered was BL Lacertae, or BL Lac for short, in the constellation Lacerta (the Lizard). When discovered in 1929, it was mistaken for a variable star, largely because its brightness varies by a factor of 15 within only a few months. Careful examination, however, revealed some fuzz (visible in Figure 27-13) around its bright, starlike core. In the early 1970s, Joseph Miller at the University of California's Lick Observatory blocked out the light from the bright center of BL Lac and managed to obtain a spectrum of the fuzz. This spectrum, which contains stellar absorption lines, strongly resembles the spectrum of an elliptical galaxy. In other words, BL Lac is an elliptical galaxy with an intensely luminous center.

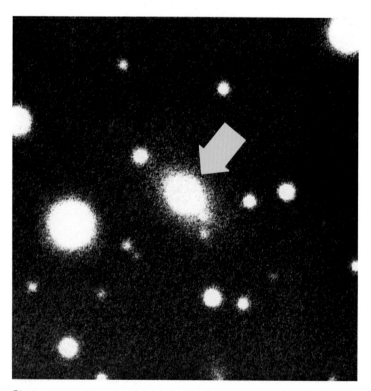

figure 27-13 R I **V** U X G

BL Lac This photograph shows fuzz around the blazar BL Lac. The redshift of this fuzz indicates that BL Lac is about 280 Mpc (900 million ly) from Earth. Blazars are giant elliptical galaxies with bright, starlike nuclei that have many quasarlike properties. (Courtesy of T. D. Kinman; Kitt Peak National Observatory)

Objects such as BL Lac are called **blazars.** They emit polarized light with a featureless, nonthermal spectrum typical of synchrotron radiation, as shown in Figure 27-9. (Careful observations have revealed emission and absorption lines in blazar spectra, but these are swamped by the intense synchrotron radiation.) Detailed radio observations show that blazars are probably double radio sources—but oriented so that we see their jets end-on. For example, high-resolution studies of blazars with the Very Large Array have revealed diffuse radio emission around a bright core in almost all cases. A faint radio halo around a bright radio core is exactly what you would expect if you were looking straight down the jet from a radio galaxy. As seen from this angle, the intense synchrotron radiation from the galaxy's bright nucleus is surrounded by weaker emission from its radio lobes.

WEB LINK 27.1 The idea that blazars are radio galaxies with their jets aimed toward the Earth was supported by the surprising discovery of movement that appeared to be faster than the speed of light. Such **superluminal motion** is also observed in some quasars, where very-long-baseline interferometry (described in Section 6-6) reveals a lumpy structure that changes with time. For example, Figure 27-14 shows four high-resolution images of the quasar 3C 273 spanning three years. During this interval, a "blob" moved away from the quasar at a rate of almost 0.001 arcsec per year. Taking into account the distance to the source, this rate of angular separation corresponds to a speed 10 times that of light!

Superluminal motion was puzzling when it was first discovered, because it seemed to violate one of the basic tenets of the special theory of relativity: Nothing can move faster than light. Astronomers soon realized, however, that superluminal motion can be explained as movement slower than light—once we take into account the angle at which we view the radio source. If a relativistic beam of material is aimed close to your line of sight, it can appear to be moving faster than the speed of light. Figure 27-15 shows an example. Quasars generally exhibit lower superluminal speeds (from c to $5c$) than blazars (from $5c$ to $10c$), probably because the relativistic jets from quasars are not aimed as close to our line of sight as are blazar jets.

Because of the many properties they share, quasars, blazars, Seyfert galaxies, and radio galaxies are now collectively called **active galaxies.** The activity of such a galaxy comes from an energy source at its center. Hence, astronomers say that these galaxies possess **active galactic nuclei,** or **AGNs.**

Table 27-2 summarizes the properties of different kinds of active galactic nuclei. As this table shows, one of the features that distinguishes different types of AGNs from each other is the width of the emission lines in their spectra. Figures 27-3 and 27-4 show the spectra of typical quasars in the visible and infrared parts of the spectrum. The "spikes" in these spectra are emission lines from hydrogen gas as well as other elements. The widths of these emission lines indicate that individual light-emitting gas clouds are moving within the quasar at very high speeds (10,000 km/s or so). Some of the clouds are moving toward us, causing the emitted light to have a shorter wavelength and higher frequency, while other clouds are moving away from us and emit light with longer wavelength and lower frequency. This explains why the emission

Quasar Blob   Quasar Blob   Quasar Blob   Quasar Blob

July 1977   March 1978   July 1979   July 1980

figure 27-14 **R** I V U X G

Superluminal Motion in 3C 273 These four images are false-color, high-resolution radio maps of the quasar 3C 273 (shown in the X-ray image that opens this chapter). They show a blob that seems to move away from the quasar at 10 times the speed of light. In fact, a beam of

relativistic particles from 3C 273 is aimed almost directly at the Earth, giving the illusion of faster-than-light motion. Each image is about 7 milliarcseconds (0.007 arcsec) across. (NRAO)

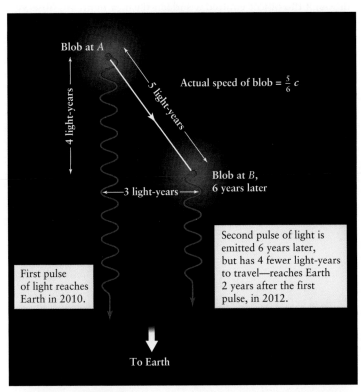

Blob at *A*

4 light-years

Actual speed of blob = $\frac{5}{6}$ *c*

5 light-years

Blob at *B*, 6 years later

3 light-years

First pulse of light reaches Earth in 2010.

Second pulse of light is emitted 6 years later, but has 4 fewer light-years to travel—reaches Earth 2 years after the first pulse, in 2012.

To Earth

(a) View from above

2010 → 2012

3 light-years

Apparent speed of blob = 1.5*c*

(b) View from Earth

lines of quasars are quite broad. By contrast, the emission lines of radio galaxies do not show the same kind of broadening. If there are fast-moving gas clouds in these galaxies, they are hidden from our sight. (In Section 27-6 we will see why this is.) In a similar way, astronomers distinguish between different subtypes of Seyfert galaxies (called *Seyfert 1* and *Seyfert 2*), depending on whether their dominant emission lines are broad or narrow.

One characteristic that is common to *all* types of active galactic nuclei is variability. For example, Figure 27-16 shows brightness fluctuations of the blazar 3C 279, determined by carefully examining old photographic plates on which it had been photographed inadvertently. Note the prominent outbursts that occurred around 1937 and 1943. During these outbursts, the luminosity of 3C 279 increased by a factor of at least 25. Other AGNs fluctuate more rapidly: Some quasars vary in brightness over a few weeks or months, and X-ray observations reveal that some blazars vary in brightness over time intervals as short as 3 hours.

figure 27-15

An Explanation of Superluminal Motion (a) If a blob of material ejected from a quasar moves at five-sixths of the speed of light, it covers the 5 ly from point *A* to point *B* in six years. In the case shown here, it moves 4 ly toward the Earth and 3 ly in a transverse direction. The light emitted by the blob at *A* reaches us in 2010. The light emitted by the blob at *B* reaches us in 2012. The light left the blob at *B* 6 years later than the light from *A* but had 4 fewer light-years to travel to reach us. (b) From Earth we can see only the blob's transverse motion across the sky, as in Figure 27-14. It appears that the blob has traveled 3 ly in just 2 years, so its apparent speed is $^3/_2$ of the speed of light, or 1.5*c*.

table 27-2 | Properties of Active Galactic Nuclei (AGNs)

Object	Found in which type of galaxy	Strength of radio emission	Type of emission lines in spectrum	Luminosity (watts)	(Milky Way Galaxy = 1)
Blazar	Elliptical	Strong	Weak (compared to synchrotron emission)	10^{38} to 10^{42}	10 to 10^5
Radio-loud quasar	Elliptical	Strong	Broad	10^{38} to 10^{42}	10 to 10^5
Radio galaxy	Elliptical	Strong	Narrow	10^{36} to 10^{38}	0.1 to 10
Radio-quiet quasar	Spiral or elliptical	Weak	Broad	10^{38} to 10^{42}	10 to 10^5
Seyfert 1	Spiral	Weak	Broad	10^{36} to 10^{38}	0.1 to 10
Seyfert 2	Spiral	Weak	Narrow	10^{36} to 10^{38}	0.1 to 10

These fluctuations in brightness allow astronomers to place strict limits on the maximum size of a light source. An object cannot vary in brightness faster than light can travel across that object. For example, an object that is one light-year in diameter cannot vary significantly in brightness over a period of less than one year.

To understand this limitation, imagine an object that measures 1 light-year across, as in Figure 27-17. Suppose the entire object emits a brief flash of light. Photons from that part of the object nearest the Earth arrive at our telescopes first. Photons from the middle of the object arrive at Earth 6 months later. Finally, light from the far side of the object arrives a year after the first photons.

Although the object emitted a sudden flash of light, we observe a gradual increase in brightness that lasts a full year. In other words, the flash is stretched out over an interval equal to the difference in the light travel time between the nearest and farthest observable regions of the object.

The rapid flickering exhibited by active galactic nuclei means that they emit their energy from a small volume, possibly less than one light-day across. In other words, a region no larger than our solar system can emit more energy per second than a thousand galaxies! Astrophysicists therefore face the challenge of explaining how so much energy can be produced in such a very small volume.

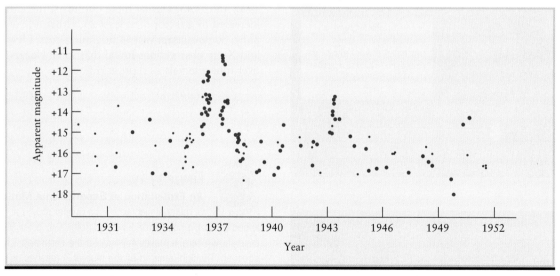

figure 27-16

Brightness Variations of an AGN This graph shows variations in apparent magnitude (a measure of apparent brightness) for the blazar 3C 279, located in the constellation Coma Berenices. The data were obtained by carefully examining old photographic plates at Harvard College Observatory. Note the large outburst in 1937 and the somewhat smaller one in 1943. (Adapted from L. Eachus and W. Liller)

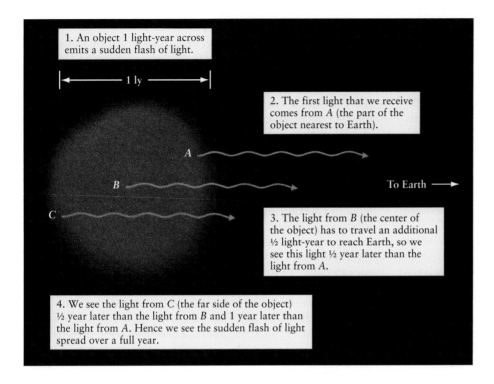

1. An object 1 light-year across emits a sudden flash of light.

2. The first light that we receive comes from *A* (the part of the object nearest to Earth).

3. The light from *B* (the center of the object) has to travel an additional ½ light-year to reach Earth, so we see this light ½ year later than the light from *A*.

4. We see the light from *C* (the far side of the object) ½ year later than the light from *B* and 1 year later than the light from *A*. Hence we see the sudden flash of light spread over a full year.

To Earth ⟶

figure 27-17

A Limit on the Speed of Variations in Brightness The rapidity with which the brightness of an object can vary significantly is limited by the time it takes light to travel across the object. If the object 1 light-year in size emits a sudden flash of light, the flash will be observed from Earth to last a full year. If the object is 2 light-years in size, brightness variations will last at least 2 years as seen from Earth, and so on.

27-5 | Supermassive black holes are the "central engines" that power active galactic nuclei

As long ago as 1968, the British astronomer Donald Lynden-Bell pointed out that a black hole lurking at the center of a galaxy could be the "central engine" that powers an active galactic nucleus. Lynden-Bell theorized that as gases fall onto a black hole, their gravitational energy would be converted into radiation. (As we saw in Section 24-3, a similar process produces radiation from black holes in close binary star systems.) To produce as much radiation as is seen from active galactic nuclei, the black hole would have to be very massive. But even a gigantic black hole would occupy a volume much smaller than our solar system—exactly what is needed to explain how active galactic nuclei can vary so rapidly in brightness.

How large a black hole would be needed to power an active galactic nucleus? You might think that what really matters is not the size of the black hole, but rather the amount of gas that falls onto it and releases energy. However, there is a natural limit to the luminosity that can be radiated by accretion onto a compact object like a black hole. This is called the **Eddington limit,** after the British astrophysicist Sir Arthur Eddington.

If the luminosity exceeds the Eddington limit, there is so much *radiation pressure*—the pressure produced by photons streaming outward from the infalling material—that the surrounding gas is pushed outward rather than falling inward onto the black hole. Without a source of gas to provide energy, the luminosity naturally decreases to below the Eddington limit, at which point gas can again fall inward. This limit allows us to calculate the minimum mass of an active galactic nucleus.

Numerically, the Eddington limit is:

The Eddington limit

$$L_{\text{Edd}} = 30,000 \left(\frac{M}{M_\odot} \right) L_\odot$$

L_{Edd} = maximum luminosity that can be radiated by accretion onto a compact object

M = mass of the compact object

$M_\odot$ = mass of the Sun

$L_\odot$ = luminosity of the Sun

The tremendous luminosity of an active galactic nucleus must be less than or equal to its Eddington limit, so this limit must be very high indeed. Hence, the mass of the black hole must also be quite large. For example, consider the quasar 3C 273 (shown in the image that opens this chapter), which has a luminosity of about $3 \times 10^{13} L_\odot$. To calculate the minimum mass of a black hole that could continue to attract gas to power the quasar, assume that the quasar's luminosity equals the Eddington limit. Inserting $L_{\text{Edd}} = 3 \times 10^{13} L_\odot$ into the above equation, we find that $M = 10^9 M_\odot$. Therefore, if a black hole is responsible for the energy output of 3C 273, its mass must be greater than a billion Suns.

Astronomers have indeed found evidence for such **supermassive black holes** at the centers of many nearby normal galaxies (see Section 24-4). As we saw in Section 25-6, at the center of our own Milky Way Galaxy lies what is almost certainly a black hole of about 2.6×10^6 solar masses—supermassive in comparison to a star but only about 1% the mass of the behemoth black hole at the center of 3C 273.

Theory suggests that unlike stellar-mass black holes, which require a supernova to produce them (see Figure 22-24), supermassive black holes can be produced without extreme pressures or densities. This may help to explain why they appear to be a feature of so many galaxies.

One galaxy that probably has a black hole at its center is the Andromeda Galaxy (M31), shown in the image that opens Chapter 26. M31 is only 750 kpc (2.5 million ly) from Earth, close enough that details in its core as small as 1 parsec across can be resolved under the best seeing conditions.

In the mid-1980s, astronomers made high-resolution spectroscopic observations of M31's core. By measuring the Doppler shifts of spectral lines at various locations in the core, we can determine the orbital speeds of the stars about the galaxy's nucleus.

Figure 27-18 plots the results for the innermost 80 arcseconds of M31. (At M31's distance, this corresponds to a linear distance of 290 pc or 950 ly.) Note that the rotation curve in the galaxy's nucleus does not follow the trend set in the outer core. Rather, there are sharp peaks—one on the approaching side of the galaxy and the other on the receding side—within 5 arcsec of the galaxy's center.

The most straightforward interpretation is that the peaks are caused by the orbital motions of stars around M31's center. Stars on one side of the galaxy's center are approaching us while stars on the other side are receding from us.

The high speeds of stars orbiting close to M31's center indicate the presence of a massive central object. Calculations using Newton's form of Kepler's third law (described in Section 4-7 and Box 4-4) show that there must be about 3×10^7 solar masses within 5 pc (16 ly) of the galaxy's center. That much matter confined to such a small volume strongly suggests the presence of a supermassive black hole. Observations of M31 with the Chandra X-ray Observatory are consistent with this picture.

By applying high-resolution spectroscopy to the cores of other nearby galaxies, astronomers have discovered a number of supermassive black holes like the one in M31. Unfortunately, this technique for identifying black holes is difficult to apply to quasars, which are very distant and have small angular sizes. The evidence for supermassive black holes in quasars is therefore circumstantial, yet compelling: No other known energy source could provide enough power to sustain a quasar's intense light output.

If most galaxies have supermassive black holes at their centers, and supermassive black holes are the "central engines" of active galactic nuclei, why aren't all galaxies active? Why are there no nearby quasars? Why do radio galaxies and some quasars have jets? As we will see in the next section, the answers to all these questions may be provided by a model of what happens in an active galactic nucleus.

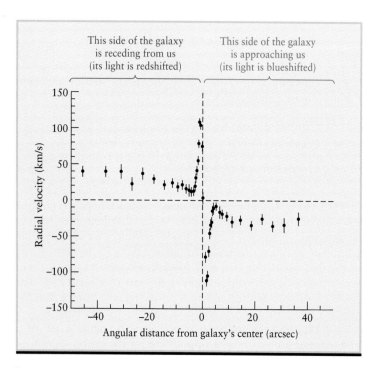

Figure 27-18

The Rotation Curve of the Core of M31 This graph plots radial velocity of matter in the core of M31 versus the angular distance from the galaxy's center. Note the sharp peaks, one blueshifted and one redshifted, within 5 arcsec of the galaxy's center. This indicates the presence of a compact, very massive object at the center of the galaxy. At the distance of M31, 1 arcsec corresponds to 3.6 pc (12 ly). (Adapted from J. Kormendy)

27-6 | Quasars, blazars, and radio galaxies may be the same kind of object seen from different angles

Accretion onto a supermassive black hole is the most likely explanation of the immense energy output of active galactic nuclei. The challenge to astrophysicists is to understand how that accretion takes place. A successful model of the accretion process must also explain other properties of active galactic nuclei, including their unusual spectra, variable light output, and energetic jets.

In the leading model of this process, at the heart of an active galaxy is a supermassive black hole surrounded by an **accretion disk,** an immense disk of matter captured by the hole's gravity and spiraling into it. We saw in Section 20-5 that accretion disks are found around stars in the process of formation. The accretion disks that astrophysicists envision around supermassive black holes are similar but far larger and far more dynamic.

Imagine a billion-solar-mass black hole sitting at the center of a galaxy, surrounded by a rotating accretion disk. According to Kepler's third law, the inner regions of this accretion disk would orbit the hole more rapidly than would the outer parts. Thus, the rapidly spinning inner regions would constantly rub against the slower moving gases in the outer regions. This friction, aided by magnetic forces within the disk, would cause the gases to lose energy and spiral inward toward the black hole.

As the gases move inward within the accretion disk, they are compressed and heated to very high temperatures. This causes the accretion disk to glow, thus producing the brilliant luminosity of an active galactic nucleus. (The temperature of material in the accretion disk reaches 100,000 K or more, which helps explain why many active galactic nuclei emit far more X-ray and gamma-ray radiation than do ordinary galaxies.) Any variations in the density of the gas

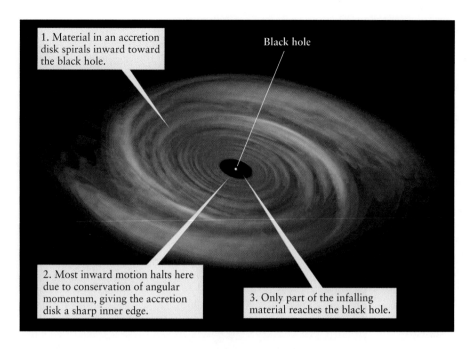

1. Material in an accretion disk spirals inward toward the black hole.

Black hole

2. Most inward motion halts here due to conservation of angular momentum, giving the accretion disk a sharp inner edge.

3. Only part of the infalling material reaches the black hole.

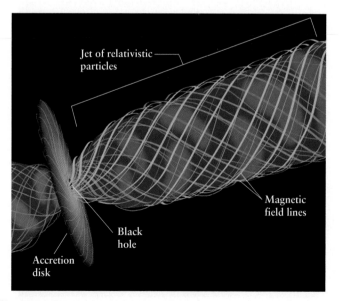

figure 27-19

The Inner Edge of an Accretion Disk This image from a computer simulation shows an accretion disk surrounding a black hole. As a consequence of the conservation of angular momentum, much of the material in the disk cannot fall into the black hole, but instead piles up at the disk's inner edge. (Courtesy of Michael Owen and John Blondin, North Carolina State University)

will cause the luminosity to fluctuate, giving rise to the brightness variations observed by astronomers (see Figure 27-16).

Supercomputer simulations can help us visualize the flow of matter in such an accretion disk. These simulations combine general relativity with equations describing gas flow. At first, matter accelerates to supersonic speeds as it spirals inward toward a black hole. But because the matter is rotating around the hole as well as moving toward it, this inward rush stops abruptly near the hole. The reason is the law of conservation of angular momentum (described in Section 8-4). Thanks to this law, rotating objects of all kinds tend to expand (like a spinning lump of dough that expands to make a pizza) rather than contract. The inward motion of gases stops where this tendency to expand outward balances the pull of the black hole's gravity. As a result, not all of the infalling gas reaches the black hole. Instead, part of the gas can become concentrated in high-speed orbits quite close to the hole. The sudden halt in the supersonic inflow creates a shock wave, which defines the inner edge of the accretion disk (Figure 27-19). Only a fraction of the material in the accretion disk can cross this inner edge and fall into the black hole.

Because of the constant inward crowding of hot gases, pressures climb rapidly in the inner accretion disk. These pressures relieve the congestion by expelling matter at extremely high speeds. This ejected material escapes moving at right angles to the plane of the accretion disk.

Magnetic forces play a crucial role in steering these fast-moving particles. These forces arise because the hot gases in the accretion disk are ionized, forming a plasma, and the motions of this plasma generate a magnetic field (see Section 18-9). As the plasma in the disk rotates around the black hole, it pulls the magnetic field along with it. But because the disk rotates faster in its inner regions than at its outer rim, the magnetic field becomes severely twisted. This twisted field forms two helix shapes, one on either side of the plane of the disk. Relativis-

tic particles flowing outward from the accretion disk tend to follow these magnetic field lines. The result is that the outflowing beams of particles are focused into two jets oriented perpendicular to the plane of the accretion disk, as **Figure 27-20** shows.

Jet of relativistic particles

Magnetic field lines

Black hole

Accretion disk

figure 27-20

Jets from a Supermassive Black Hole The rotation of the accretion disk surrounding a supermassive black hole twists the disk's magnetic field lines into a helix. The field then channels the flow of subatomic particles pouring out of the disk. Over a distance of about a light-year, this channeling changes a broad flow into a pair of tightly focused jets, one on each side of the disk. Figure 20-15 shows a similar process that takes place on a much smaller scale in protostars. (NASA and Ann Feild, Space Telescope Science Institute)

Figure 27-21 is an artist's conception of the accretion disk and jets as they might appear from a nearby planet.

Observations of active galaxies provide evidence in support of this model. One example is the radio galaxy NGC 4261, shown in Figure 24-13, whose radio lobes appear to emanate from an accretion disk around a supermassive black hole. While the accretion disk itself is too small to see, we can observe a dusty ring, or torus, about 250 pc (800 ly) in diameter orbiting the central black hole (Figure 27-22). The motions of this torus reveal the mass of the central black hole to be $1.2 \times 10^9 \, M_\odot$.

If there were no dusty torus around an accreting supermassive black hole, an observer could view such a black hole from any angle and see the intense radiation from the accretion disk. But the presence of such a torus seems to be a natural result of the accretion process. As a result, from certain angles the torus blocks the view of the innermost part of the active galactic nucleus. This idea offers a single explanation for several types of active galaxies. Many astronomers suspect that the main difference between blazars, quasars, and radio galaxies is only the angle at which the black hole "central engine" is viewed. This idea is called the *unified model* of active galactic nuclei.

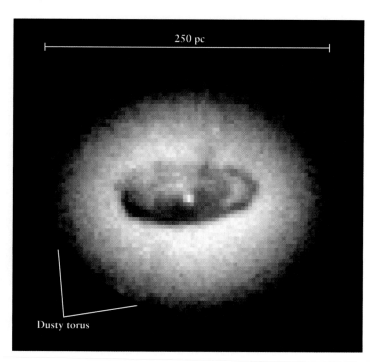

figure 27-22 R I **V** U X G

A Dusty Torus Around a Supermassive Black Hole This immense doughnut of dust and gas orbits the black hole at the center of the radio galaxy NGC 4261. The radio lobes of this galaxy appear to be the endpoints of jets emerging parallel to the plane of this torus (see Figure 24-13). (L. Ferrarese/Johns Hopkins University, NASA)

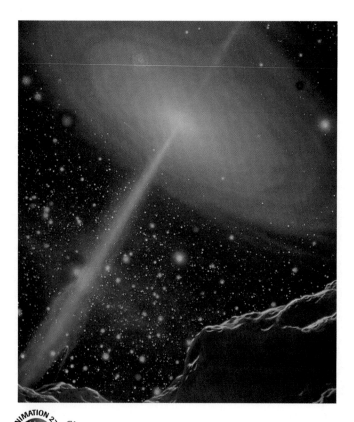

figure 27-21

At the Core of an Active Galaxy This artist's rendering shows a supermassive black hole surrounded by an accretion disk. In the inner regions of the accretion disk, matter crowding toward the hole is diverted outward along two oppositely directed jets. As the relativistic particles move in spiral paths within the jets, they emit synchrotron radiation. (*Astronomy*)

Figure 27-23 illustrates the unified model for a luminous active galactic nucleus with jets. If an observer looks straight down the axis of the jet, the observed radiation is dominated by synchrotron radiation from the jet. This has a continuous spectrum with no emission or absorption lines (see Figure 27-9). Hence, the observer sees a blazar, with a nearly featureless spectrum.

At a more oblique angle, the observer gets a clear view of the luminous accretion disk and the turbulent region around the black hole. Because gases move at many different velocities in this region, the observer sees spectral lines that have been broadened by the Doppler effect. The observer also sees intense thermal radiation from the accretion disk and synchrotron radiation from the jets. From this angle, what the observer sees is a radio-loud quasar.

If the observer looks nearly edge-on at the torus, the accretion disk will be completely hidden. Some of the light reaching the observer comes from hot gas flowing out of the accretion disk, and this light has an emission-line spectrum. But this gas is not moving rapidly either toward or away from the observer, so there is little Doppler shift and the emission lines are narrow. The synchrotron emission from the jets is still visible, and so our observer reports seeing a radio galaxy.

It may happen that no jets are present, which means that the active galactic nucleus will lack the powerful synchrotron radiation that particles in a jet produce. In this case, an observer viewing the

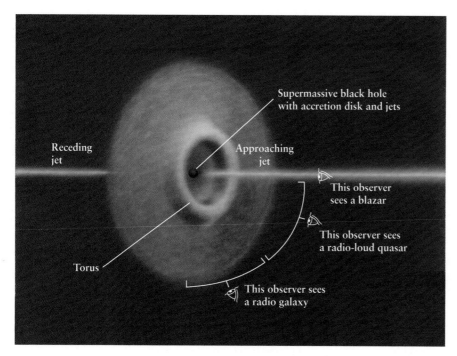

Supermassive black hole
with accretion disk and jets

Receding
jet

Approaching
jet

This observer
sees a blazar

This observer sees
a radio-loud quasar

Torus

This observer sees
a radio galaxy

Figure 27-23

A Unified Model of Active Galaxies Blazars, radio-loud quasars, and radio galaxies may be the same type of object—a supermassive black hole, its accretion disk, and its relativistic jets—viewed at different angles. (Compare Figure 24-12, which shows the similar environment of a stellar-mass black hole.)

"central engine" either face-on or at an oblique angle will see a radio-quiet quasar.

Unlike our imaginary observer, we cannot move the vast distances through space that would be needed to view a given active galaxy from different angles. Instead, we may see a given active galaxy as either a blazar, a quasar, or a radio galaxy, depending on how that galaxy's accretion disk and torus are oriented to our line of sight.

CAUTION! Note that we are really making use of *two* different but complementary models here. The unified model says that different types of active galactic nuclei are really different views of the same type of "central engine," while the accretion-disk model explains how the "central engine" works.

The accretion-disk idea helps to explain why there are no nearby quasars (see Figure 27-5). Over time, most of the available gas and dust surrounding a quasar's "central engine" is accreted onto the black hole. The "central engine" has less and less infalling matter to act as "fuel," and the quasar becomes much less active. The result is a relatively less luminous radio galaxy or Seyfert galaxy. (A Seyfert galaxy is essentially a low-luminosity, radio-quiet quasar. Whether it is a Seyfert 1 with predominantly broad emission lines or a Seyfert 2 with predominantly narrow lines may depend on whether we view its "central engine" more nearly face-on or edge-on.)

WEB LINK 27.14 Finally, the accretion-disk idea explains what happens when active galaxies collide or merge with other galaxies. Such collisions and mergers can transfer gas and dust from one galaxy to another, providing more "fuel" for a supermassive black hole. (The image in **Figure 27-24** shows a close-up view of a supermassive black hole about to take in more "fuel" from a

merger.) Thus, it is not surprising that many of the most luminous active galaxies are also those that have recently undergone collisions, such as the radio galaxy Centaurus A shown in Figure 27-10.

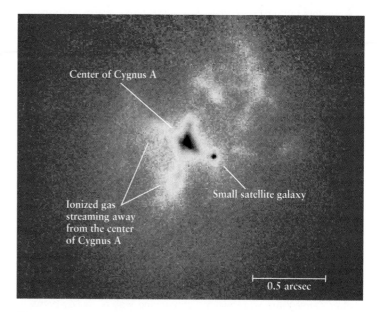

Center of Cygnus A

Small satellite galaxy

Ionized gas
streaming away
from the center
of Cygnus A

0.5 arcsec

Figure 27-24 R ■ V U X G

Providing Fresh "Fuel" for a Supermassive Black Hole A small satellite galaxy has fallen into the central regions of the radio galaxy Cygnus A (see Figure 27-1). Its material may eventually accrete onto the supermassive black hole at Cygnus A's very center. This extreme close-up was made using adaptive optics, described in Section 6-3. (Courtesy G. Canalizo, C. Max, D. Whysong, R. Antonucci, and S. E. Dahm)

Eventually an active galaxy will run out of "fuel" altogether, and its central supermassive black hole will become inactive. Perhaps galaxies such as our own Milky Way, which have supermassive black holes at their centers but are not presently active, once led much more dramatic lives as quasars or radio galaxies.

27-7 | Gamma-ray bursters produce amazingly intense flashes of radiation

 Astronomers in the 1960s used the large redshifts of quasars to show that these exotic objects are very distant. Astronomers in the 1990s used the same logic when they discovered large redshifts for **gamma-ray bursters**, mysterious objects that emit intense bursts of high-energy radiation. They concluded that gamma-ray bursters must, like quasars, be at tremendous distances from the Earth. Like quasars, gamma-ray bursters have been a major puzzle to astronomers, and only recently have we learned much about their origins.

Gamma-ray bursters were discovered in the late 1960s by the orbiting Vela satellites, whose detectors noticed flashes of gamma rays coming from random parts of the sky at random intervals. This discovery was an unexpected consequence of the Cold War. The Vela satellites were originally placed in orbit by the United States to look for high-energy photons coming from above-ground tests of nuclear weapons by the Soviet Union, tests that had been banned by treaty since 1963. (No such tests were ever detected.)

More than 3000 gamma-ray bursters have been discovered, and new ones are being found at a rate of about one per day. Gamma-ray bursters fall into two types. *Long-duration* gamma-ray bursters, which are more common, last from about 2 to about 1000 seconds before fading to invisibility. The less common *short-duration* gamma-ray bursters last from a few hundredths of a second to about 2 seconds, and tend to emit photons of shorter wavelength and hence higher energy. Unlike X-ray bursters (see Section 23-8), gamma-ray bursters of both types appear to emit only one burst in their entire history.

What are gamma-ray bursters, and how far away are they? These questions plagued astronomers for almost 30 years after the discovery of gamma-ray bursters. One clue is that gamma-ray bursters are seen with roughly equal probability in all parts of the sky, as **Figure 27-25** shows. This suggests that they are not in the disk of our Galaxy, because then most gamma-ray bursters would be in the plane of the Milky Way. One idea was that gamma-ray bursters are relatively close to us and lie in a spherical halo surrounding the Milky Way, rather like globular clusters (see Figure 25-7). Alternatively, gamma-ray bursters could be strewn across space like galaxies, with some of them billions of light-years away.

For many years there was no convincing way to decide between these competing models. The problem was that unlike the situation for quasars, astronomers were unable to find optical counterparts for gamma-ray bursters. With no spectral lines to examine, they could not tell whether gamma-ray bursters orbit our Galaxy (in which case their spectra should show little or no redshift) or are at galactic distances (in which case many of them should have substantial redshifts).

 Figure 27-25 R I V U X **G**
Gamma-Ray Bursters This map shows the locations of 1776 gamma-ray bursters detected by the Compton Gamma Ray Observatory (see Section 6-7). The entire celestial sphere is mapped onto an oval, with the Milky Way stretching horizontally across the center. Unlike X-ray bursters, which originate in the disk of the Milky Way Galaxy, gamma-ray bursters are seen in all parts of the sky. The colors in the order of the rainbow indicate the total amount of energy detected from each burst; bright bursts appear in red and weak bursts appear in violet. (Compare with Figure 6-34.)

This state of affairs changed dramatically in 1997, thanks to the Italian-Dutch BeppoSAX satellite. Unlike previous gamma-ray satellites, which could determine the position of a burst only to within a few degrees, BeppoSAX could pin down its position to within an arcminute. This made it far easier for astronomers using optical telescopes to search for a counterpart to the burster.

Just 21 hours after BeppoSAX observed a long-duration gamma-ray burster on February 28, 1997, astronomers led by Jan van Paradis of the University of Amsterdam located a visible-light "afterglow" of the burst. And when BeppoSAX detected another long-duration gamma-ray burster on May 8, Mark Metzger and his colleagues from Caltech were able to record the visible-light spectrum of its "afterglow." In the spectrum they found absorption lines of iron and magnesium, and with a substantial redshift: $z = 0.835$. The Hubble law immediately told the astronomers that the gamma-ray burster was about 2100 Mpc (7 billion ly) away. (The absorption lines could actually have been formed when light from the burster passed through a galaxy that happened to lie between the burster and the Earth. If so, the gamma-ray burster itself could have an even greater redshift and be even farther away.)

Since 1997, astronomers using large visible-light and infrared telescopes have detected the "afterglow" of many gamma-ray bursters. While the gamma-ray intensity of a typical long-duration burster fades away within 100 seconds or so, the intensity at visible-light and infrared wavelengths can persist for days (Figure 27-26). This enables astronomers to make detailed observations of the burster's spectrum and redshift. Gamma-ray bursters have now been seen with redshifts as great as $z = 4.5$, corresponding to a distance of 3800 Mpc (12 billion ly).

(a) December 1997

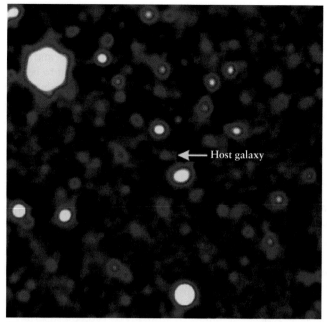

(b) February 1998

figure 27-26 R I **V** U X G

The Fading Light of a Gamma-Ray Burster (a) This false-color image shows the visible-light "afterglow" of gamma-ray burster GRB 971214, which was detected on December 14, 1997. This image was made two days later. This burster had a redshift $z = 3.418$, which implies that the burst of radiation was emitted about 12 billion years ago. (b) Three months later the "afterglow" had faded, revealing a faint galaxy at the position of the burster. (S. G. Djorgovski and S. R. Kulkarni, Caltech; the Caltech GRB Team; and W. M. Keck Observatory)

What could be the source of these intense bursts? An important clue has come from visible-light and infrared observations, which in a few cases have discerned a galaxy at the same location as a gamma-ray burster (see Figure 27-26). As **Figure 27-27** shows, the bursters do not appear at the centers of their host galaxies. Hence, they are unlikely to be a flare-up of an active galactic nucleus, but are likely to be the result of some cataclysmic event involving one of a galaxy's stars.

 Observations of a relatively close, long-duration gamma-ray burster have confirmed this idea. On March 29, 2003, the orbiting gamma-ray telescope HETE-2 (short for High Energy Transient Explorer) detected this burster—denoted GRB 030329 for the date it was observed—in the constellation Leo. (The burster was so intense that its high-energy photons temporarily ionized part of the Earth's atmosphere—the only event of its kind ever caused by an object beyond our own Galaxy.) Within 90 minutes a visible-light afterglow of GRB 030329 was found by astronomers in Australia and Japan, with spectral

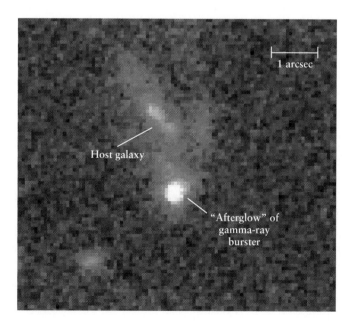

figure 27-27 R I **V** U X G

The Host Galaxy of a Gamma-Ray Burster This false-color image was made on February 8, 1999, 16 days after a gamma-ray burster was observed at this location. The host galaxy of the gamma-ray burster has a very blue color, indicating the presence of many recently formed stars. The gamma-ray burst may have been produced when one of the most massive of these stars became a supernova. (Andrew Fruchter, STScI; and NASA) lines that indicated a redshift $z = 0.1865$. This corresponds to a distance of 810 Mpc (2.65 billion ly)—close enough, and hence bright enough, that astronomers could measure a detailed spectrum of the afterglow. In short order, astronomers in the United States and Chile had obtained such a detailed spectrum, which turned out to be precisely that of a supernova. Further confirmation that the gamma-ray burster was associated with a supernova came from observations with the Hubble Space Telescope, which revealed an expanding shell of gas at the position of the gamma-ray burster. This shell is just what we would expect from a supernova explosion.

A long-duration gamma-ray burster cannot simply be an ordinary supernova, however. The supernovae that we discussed in Section 22-6 release about 10^{46} joules of energy, of which only about 0.03% is released as electromagnetic radiation. (The rest goes into neutrinos as well as into the energy of the debris expanding away from the supernova.) This radiation streams outward from the supernova in all directions equally, like light from the Sun. If we assume that a gamma-ray burster also emits light equally in all directions, the inverse-square law (see Section 19-2) tells us that the most energetic bursters would have to emit about 3×10^{47} joules of radiation in less than a minute. It would take 100,000 supernovae going off simultaneously to release this much radiation!

This dilemma can be resolved if we imagine a type of supernova that emits most of its radiation in narrow beams. If such a beam happened to be aimed toward Earth, we would detect a far more intense burst of radiation than we would from an ordinary supernova.

ANALOGY An ordinary flashlight is an example of beamed radiation. The lightbulb in a flashlight is very small and would produce only a weak light if you removed it from the flashlight housing. But the flashlight's curved mirror channels the bulb's light into a narrow beam. Thus, a flashlight produces an intense beam of light with only a small input of energy from the batteries.

ANIMATION 27.3 One theoretical model that would produce beamed radiation invokes a special type of supernova called a **collapsar** (also called a *hypernova*). These objects are thought to result from progenitor stars that are both very massive (more than about 30 $M_\odot$) and rotating rapidly. When thermonuclear reactions cease in such a star, the star's core can collapse and form a black hole before the outer layers have a chance to contract very much. The black hole formed in this way rotates very rapidly, quickly forming an accretion disk from the surrounding stellar material and ejecting some of that material in powerful jets. (Much the same mechanism acts in close binary star systems in which one member is a black hole, as we described in Section 24-3, and in active galactic nuclei, as we saw in Section 27-6.) The jets are so energetic that they reach and break through the surface of the star within 5 to 10 seconds after being formed. As they travel outward, the jets produce powerful shock waves that blow the star to pieces.

If one of the jets from a collapsar happens to be aimed toward Earth, we see an intense burst of gamma rays. These are produced as the relativistic particles in the jet slow down and convert their kinetic energy (energy of motion) into radiation. The burst is short-lived because the jets have only a brief existence: It takes the black hole only about 20 seconds to accrete the entire inner core of the star, after which there is no longer an accretion disk to produce the jets. We do not see a burster if the jets are aimed away from Earth. Hence, for each gamma-ray burster that we observe, there may be hundreds or thousands of collapsars that go undetected.

The collapsar model cannot explain short-duration gamma-ray bursters, whose lifetimes are far shorter than those of a collapsar's jets. Other models have been proposed, such as the energy released by the merger of two neutron stars.

WEB LINK 27.8 The newest tool in the search for answers about gamma-ray bursters is a NASA satellite called Swift, scheduled for launch in 2004. In addition to sensitive gamma-ray detectors to identify bursters, Swift will also carry telescopes that are sensitive to X rays, ultraviolet light, and visible light. Thus, instead of having to wait for ground-based telescopes to make follow-up observations of gamma-ray bursters, Swift will be able to make these observations itself as swiftly as possible (hence the spacecraft's name). Observations from Swift may help us better understand the nature of long-duration bursters, and provide important clues about the still enigmatic short-duration bursters.

Key Words

accretion disk, p. 622	nonthermal radiation, p. 615
active galactic nucleus (AGN), p. 618	polarized radiation, p. 616
	quasar, p. 611
active galaxy, p. 618	radio galaxy, p. 614
blazar, p. 618	radio lobes, p. 616
collapsar, p. 628	Seyfert galaxy, p. 614
double radio source, p. 616	superluminal motion, p. 618
Eddington limit, p. 621	supermassive black hole, p. 621
gamma-ray burster, p. 626	
head-tail source, p. 617	thermal radiation, p. 615

Key Ideas

Quasars: A quasar looks like a star but has a huge redshift. These redshifts show that quasars are several hundred megaparsecs or more from the Earth, according to the Hubble law.

• To be seen at such large distances, quasars must be very luminous, typically about 1000 times brighter than an ordinary galaxy.

• About 10% of all quasars are strong sources of radio emission and are therefore called "radio-loud"; the remaining 90% are "radio-quiet."

• Some of the energy emitted by quasars is synchrotron radiation produced by high-speed particles traveling in a strong magnetic field.

Seyfert Galaxies: Seyfert galaxies are spiral galaxies with bright nuclei that are strong sources of radiation. Seyfert galaxies seem to be nearby, low-luminosity, radio-quiet quasars.

Radio Galaxies: Radio galaxies are elliptical galaxies located midway between the lobes of a double radio source.

• Relativistic particles are ejected from the nucleus of a radio galaxy along two oppositely directed beams.

Blazars: Blazars are bright, starlike objects that can vary rapidly in their luminosity. They are probably radio galaxies or quasars seen end-on, with a jet of relativistic particles aimed toward the Earth.

Active Galaxies: Quasars, blazars, and Seyfert and radio galaxies are examples of active galaxies. The energy source at the center of an active galaxy is called an active galactic nucleus.

• Rapid fluctuations in the brightness of active galaxies indicate that the region that emits radiation is quite small.

Black Holes and Active Galactic Nuclei: The preponderance of evidence suggests that an active galactic nucleus consists of a supermassive black hole onto which matter accretes.

• As gases spiral in toward the supermassive black hole, some of the gas may be redirected to become two jets of high-speed particles that are aligned perpendicularly to the accretion disk.

• An observer sees a radio galaxy when the accretion disk is viewed nearly edge-on, so that its light is blocked by a surrounding torus. At a steeper angle, the observer sees a quasar. If one of the jets is aimed almost directly at the Earth, a blazar is observed.

Gamma-Ray Bursters: Short, intense bursts of gamma rays are observed at random times coming from random parts of the sky.

• By observing the afterglow of long-duration gamma-ray bursters, astronomers find that these objects have very large redshifts and appear to be located within distant galaxies. The bursts are correlated with supernovae, and may be due to an exotic type of supernova called a collapsar.

• The origin of short-duration gamma-ray bursters is unknown.

| Review Questions |

1. When quasi-stellar radio sources were first discovered and named, why were they called "quasi-stellar"?

2. How were quasars first discovered? How was it discovered that they are very distant objects?

3. Suppose you saw an object in the sky that you suspected might be a quasar. What sort of observations might you perform to test your suspicion?

4. Explain why astronomers cannot use any of the standard candles described in Section 26-4 to determine the distances to quasars.

5. How does the spectrum of a quasar differ from that of an ordinary galaxy? How do spectral lines help astronomers determine the distances to quasars?

6. If quasars lie at the centers of galaxies, why don't we see strong absorption lines from the galaxy's stars when we look at the spectrum of a quasar (like those shown in Figures 27-3 and 27-4)?

7. It was suggested in the 1960s that quasars might be compact objects ejected at high speeds from the centers of nearby ordinary galaxies. Explain why the absence of blueshifted quasars disproves this hypothesis.

8. Why were some astronomers skeptical that the redshifts of quasars gave a true indication of their distance?

9. How do astronomers know that quasars are located in galaxies? In what sorts of galaxies are they found?

10. What is a Seyfert galaxy? Why do astronomers think that Seyfert galaxies may be related to radio-quiet quasars?

11. What is a radio galaxy? What is a double radio source? Why do astronomers think these objects may be related to radio-loud quasars?

12. How would you distinguish between thermal and nonthermal radiation?

13. What is the difference between polarized and unpolarized radiation? How does the polarization of radiation from M87's jet show that the radiation is nonthermal radiation?

14. What are head-tail sources? How do they provide evidence that double radio sources include jets of fast-moving particles?

15. What is a blazar? What is unique about its spectrum? How is it related to other active galaxies?

16. Some blazars or quasars appear to be ejecting material at speeds faster than light. Is the material really moving that fast? If so, how is this possible? If not, why does the material appear to be traveling so fast?

17. What do astronomers learn from the widths of the spectral lines of quasars?

18. What do the brightness fluctuations of a particular active galaxy tell us about the size of the energy-emitting region within that galaxy?

19. How could a supermassive black hole, from which nothing—not even light—can escape, be responsible for the extraordinary luminosity of a quasar?

20. What is the Eddington limit? Explain how it can be used to set a limit on the mass of a supermassive black hole, and explain why this limit represents a minimum mass for the black hole.

21. Explain how the rotation curve of a galactic nucleus can help determine whether a supermassive black hole is present.

22. How does matter falling inward toward a central black hole find itself being ejected outward in a high-speed jet?

23. Explain how the unified model of active galaxies suggests that quasars, blazars, and radio galaxies are the same kind of object viewed from different angles.

24. Why do you suppose there are no quasars relatively near our Galaxy?

25. What is a gamma-ray burster? What is the evidence that gamma-ray bursters are not located in the disk of our Galaxy or in a halo surrounding our Galaxy?

26. Summarize the evidence that gamma-ray bursters result from a process involving a star in a distant galaxy.

27. What is a collapsar? How does the collapsar model account for the existence of long-duration gamma-ray bursters?

Advanced Questions

Questions preceded by an asterisk (*) involve topics discussed in the Boxes in Chapters 4, 24, and 26.

> Problem-solving tips and tools
>
> Table 27-1 gives the connection between redshift and distance, Box 26-2 discusses relativistic redshift, and Box 24-1 describes time dilation in the special theory of relativity. You may find it useful to recall from Section 1-7 that 1 light-year = 63,240 AU. A speed of 1 km/s is the same as 0.211 AU/yr. You can find Newton's form of Kepler's third law in Box 4-4 and the formula for the Schwarzschild radius in Box 24-2.

28. When we observe a quasar with redshift $z = 0.75$, how far into its past are we looking? If we could see that quasar as it really is right now (that is, if the light from the quasar could somehow reach us instantaneously), would it still look like a quasar? Explain why or why not.

*29. The quasar SDSS 1044–0125 has a redshift $z = 5.80$. At what speed does this quasar seem to be receding from us? Give your answer in km/s and as a fraction of the speed of light c.

*30. Suppose that an astronomer discovers a quasar with a redshift of 8.0. With what speed would this quasar seem to be receding from us? Give your answer in km/s and as a fraction of the speed of light.

31. In the quasar spectrum shown in Figure 27-4, there are many deep absorption lines to the left of the L_α emission line (that is, at shorter wavelengths). These lines, collectively called the *Lyman-alpha forest*, are due to remote gas clouds along our line of sight to the quasar. Hydrogen atoms in these clouds absorb L_α photons from the quasar. Explain why the L_α absorption lines due to these clouds are at shorter wavelengths than the L_α emission line from the quasar itself.

32. Explain how the existence of gravitational lenses involving quasars (review Section 26-8) constitutes evidence that quasars are located at the great distances inferred from the Hubble law.

33. The Seyfert galaxy NGC 1275 is actually two galaxies that are colliding. Images of NGC 1275 show a number of globular clusters with a distinctive blue color. Explain how this color shows that these clusters formed relatively recently, perhaps as a result of the collision.

34. Suppose the distance from point A to point B in Figure 27-15a is 26 light-years and the blob moves at 13/15 of the speed of light. As the blob moves from A to B, it moves 24 light-years toward the Earth and 10 light-years in a transverse direction. (a) How long does it take the blob to travel from A to B? (b) If the light from the blob at A reaches Earth in 2010, in what year does the light from B reach Earth? (c) As seen from Earth, at what speed does the blob appear to move across the sky?

*35. Suppose a blazar at $z = 1.00$ goes through a fluctuation in brightness that lasts one week (168 hours) as seen from Earth. (a) At what speed does the blazar seem to be moving away from us? (b) Using the idea of time dilation, determine how long this fluctuation lasted as measured by an astronomer within the blazar's host galaxy. (c) What is the maximum size (in AU) of the region from which this blazar emits energy?

36. (a) Calculate the maximum luminosity that could be generated by accretion onto a black hole of 3.7×10^6 solar masses. (This is the size of the black hole found at the center of the Milky Way, as described in Section 25-6.) Compare this to the total luminosity of the Milky Way, about 2.5×10^{10} $L_\odot$. (b) Speculate on what we might see if the center of our Galaxy became an active galactic nucleus with the luminosity you calculated in (a).

*37. Observations of a certain galaxy show that stars at a distance of 16 pc from the center of the galaxy orbit the center at a speed of 200 km/s. Use Newton's form of Kepler's third law to determine the mass of the central black hole.

*38. Calculate the Schwarzschild radius of a 10^9-solar-mass black hole. How does your answer compare with the size of our solar system (given by the diameter of Pluto's orbit)?

39. Figure 27-10 shows the double radio source Centaurus A. Is it possible that somewhere in the universe there is an alien astronomer who observes this same object as a blazar? Explain your answer with a drawing showing the relative positions of the Earth, the alien astronomer, and Centaurus A.

40. Contrast gamma-ray bursters with X-ray bursters (discussed in Section 23-8). From our models of what causes these energetic phenomena, explain why X-ray bursters emit repeated pulses but gamma-ray bursters apparently emit just once.

Discussion Questions

41. The accompanying image from the Very Large Array (VLA) shows the radio galaxy 3C 75 in the constellation Cetus. This galaxy has several radio-emitting jets. High-resolution optical photographs reveal that the galaxy has two nuclei, which are the

(NRAO) **R** I V U X G

two red spots near the center of the VLA image. Propose a scenario that might explain the appearance of 3C 75.

42. Some quasars show several sets of absorption lines whose redshifts are less than the redshifts of their emission lines. For example, the quasar PKS 0237-23 has five sets of absorption lines with redshifts in the range from 1.364 to 2.202, whereas the quasar's emission lines have a redshift of 2.223. Propose an explanation for these sets of absorption lines.

43. The Milky Way Galaxy is in the process of absorbing the satellite galaxy called the Canis Major Dwarf (see Section 26-6). Discuss whether this process could cause the Milky Way to someday become an active galaxy.

44. Figure 27-24 shows ionized gas streaming away from the "central engine" of the radio galaxy Cygnus A. Instead of spreading outward equally in all directions, the gas appears to be funneled into two oppositely directed cones. Discuss how this could be caused by a dusty torus surrounding a supermassive black hole at the center of Cygnus A.

Web/CD-ROM Questions

45. Search the World Wide Web for information about "microquasars." These are objects that are found *within* the Milky Way Galaxy. How are they detected? What are the similarities and differences between these objects and true quasars? Are they long-lasting or short-lived?

46. The *Lockman hole* is a region in the constellation Ursa Minor where the Milky Way's interstellar hydrogen is the thinnest. By observing in this part of the sky, astronomers get the clearest possible view of distant galaxies and quasars. Search the World Wide Web for information about observations of the Lockman hole. What has been learned through X-ray observations made by spacecraft such as ROSAT and XMM-Newton?

47. Search the World Wide Web for information about the supernova SN 1998bw. What was unusual about this supernova? What did SN 1998bw suggest about the origin of gamma-ray bursters? How have the conclusions drawn from SN 1998bw been reinforced by the gamma-ray burster observed on May 29, 2003 (designated GRB 030329) and its associated supernova, SN 2003dh?

 48. Relativistic Redshift. Access the Active Integrated Media Module "Relativistic Redshift" in Chapter 27 of the *Universe* Web site or CD-ROM. Use this to calculate the redshift and recessional velocity of a quasar in whose spectrum the H_α emission line of hydrogen (unshifted wavelength 656 nm) appears at a wavelength of (**a**) 937 nm and (**b**) 5000 nm.

Observing Projects

49. Use a telescope with an aperture of at least 20 cm (8 in.) to observe the Seyfert galaxy NGC 1068 (also known as M77). Located in the constellation Cetus (the Whale), this galaxy is most easily seen from September through January. The epoch 2000 coordinates are R.A. = 2h 2.7m and Decl. = –0° 01′. Sketch what you see. Is the galaxy's nucleus diffuse or starlike? How does this compare with other galaxies you have observed?

50. Use a telescope with an aperture of at least 20 cm (8 in.) to observe the two companions of the Andromeda Galaxy, M32 and M110. Both are small elliptical galaxies, but only M32 is suspected of harboring a supermassive black hole. They are located on opposite sides of the Andromeda Galaxy and are most easily seen from September through January. The epoch 2000 coordinates are:

Galaxy	Right ascension	Declination
M32 (NGC 221)	0^h 42.7^m	+40° 52′
M110 (NGC 205)	0 40.4	+41 41

Make a sketch of each galaxy. Can you see any obvious difference in the appearance of these galaxies? How does this difference correlate with what you know about these galaxies?

51. If you have access to a telescope with an aperture of at least 30 cm (12 in.), you should definitely observe the Sombrero Galaxy, M104, which is suspected to have a supermassive black hole at its center. It is located in Virgo and can most easily be seen from March through July. The epoch 2000 coordinates are R.A. = 12^h 40.0^m and Decl. = –11° 37′. Make a sketch of the galaxy. Can you see the dust lane? How does the nucleus of M104 compare with the centers of other galaxies you have observed?

52. If you have access to a telescope with an aperture of at least 30 cm (12 in.), observe M87 and compare it with the other two giant elliptical galaxies, M84 and M86, that dominate the central regions of the Virgo cluster. Use the *Starry Night Backyard*™ program to help you plan your observations. The epoch 2000 coordinates are:

Galaxy	Right ascension	Declination
M84 (NGC 4374)	12^h 25.1^m	+12° 53′
M86 (NGC 4406)	12 26.2	+12 57
M87 (NGC 4486)	12 30.8	+12 24

53. If you have access to a telescope with an aperture of at least 40 cm (16 in.), you might try to observe the brightest-appearing quasar, 3C 273, which has an apparent magnitude of nearly +13. It is located in Virgo at coordinates R.A. = 12^h 29^m 07^s and Decl. = +2° 03′ 07″.

 54. If the Milky Way had an active galactic nucleus, with an accretion disk around its central black hole, there might be a pair of relativistic jets emanating from its center. Use the *Starry Night Backyard*™ program to investigate how these jets might appear from Earth. In the Control Panel, set the date and time to June 15 of this year at 12:00:00 A.M. (midnight), when the center of the Milky Way is prominent in the sky. Use the **Find...** command in the **Edit** menu to center the field of view on the star TYC6840-25-1, whose position on the celestial sphere is less than 1° from the black hole at the center of the Milky Way. (To make the Milky Way itself more visible, click on the "View Options" tab on the left-hand side of the *Starry Night Backyard*™ window. Under the heading "Stars," make sure that the box labeled "Milky Way (Visible Spectrum)" is checked. Move the mouse over the title "Milky Way (Visible Spectrum)" so that the title changes to "Milky Way Options," and click on this title. Move the slider to "Brighter" and click the **OK** button. Then click again on the "View Options" tab so that the entire window is again devoted to a view of the sky.) From what you see, make a drawing showing how the night sky would appear on June 15 at 12:00:00 A.M. if our Galaxy had an active galactic nucleus. Label the Milky Way, the jets, the central black hole, and the accretion disk. Assume that the plane of the accretion disk is aligned with the plane of the Milky Way.

 55. Use the *Deep Space Explorer*™ program to examine the vicinity of the galaxy M87, shown in Figure 27-8. In the left-hand part of the window under the heading **Local Universe,** click on the triangle next to the word **Explore** and then click **Virgo Cluster.** (**a**) You can zoom in or out using the buttons at the upper left of the window (an upward-pointing triangle and a downward-pointing triangle). You can rotate the Virgo Cluster by putting the mouse cursor over the image, holding down the mouse button, and moving the mouse. (On a two-button mouse, hold down the left mouse button.) Use these controls to get a sense of the extent of the Virgo Cluster. The use the **Find...** command in the **Edit** menu to center the field of view on M87. Describe where M87 is located in the cluster. (**b**) Discuss how the position of M87 in the Virgo Cluster might relate to its being an active galaxy.

Collaborative Exercises

56. Make a labeled sketch clearly showing how the spectrum of 3C 273, shown in Figure 27-3, would be different if the object was moving toward us at the same velocity. Compare your sketch to that of another group and resolve any inconsistencies.

57. Consider the relationship between red shift and distance as shown in Table 27-1. If an object has a recessional velocity (v/c) of 0.902, how many light-years away is the object? Is this closer or farther than an object with a z of 3?

58. Two dramatic images of the radio galaxy M87 are shown in Figure 27-8. Invent an imaginary scenario that would be analogous to a local dance club where two pictures would show different aspects of the event and write a description of each. Be sure to include a description of how this is analogous to the images in Figure 27-8.

28 Cosmology: The Origin and Evolution of the Universe

So far in this book we have cataloged the contents of the universe. Our scope has ranged from submicroscopic objects, such as atomic nuclei, to superclusters of galaxies hundreds of millions of light-years across. In between, we have studied planets, moons, and stars.

But now we turn our focus beyond what we find in the universe to the nature of the universe itself—the subject of the science called *cosmology*. How large is the universe? What is its structure? How long has it existed, and how has it changed over time?

In this chapter we will see that the universe is expanding. This expansion began with an event at the beginning of time called the Big Bang. We will see direct evidence of the Big Bang in the form of microwave radiation from space. This radiation is the faint afterglow of a primordial fireball that filled all space shortly after the beginning of the universe.

Will the universe continue to expand forever, or will it eventually collapse back on itself? We will find that to predict the future of the universe, we must first understand what happened in the remote past. To this end, astronomers study luminous supernovae like the example shown in the accompanying image. These can be seen across billions of light-years, and so can tell us about conditions in the universe billions of years ago. We will see how recent results from such supernovae, as well as from studies of the Big Bang's afterglow, have revolutionized our understanding of cosmology and given us new insights into our place in the cosmos.

1995

2002

Supernova 2002dd
($z = 0.95$)

WEB LINK 28.1 Very distant supernovae—which we see as they were billions of years ago—help us understand the evolution of the universe. (NASA and J. P. Blakeslee/JHU) R I V U X G

Questions to Guide Your Reading

As you read the sections of this chapter, look for the answers to the following questions:

28-1 What does the darkness of the night sky tell us about the nature of the universe?

28-2 As the universe expands, what, if anything, is it expanding into?

28-3 Where did the Big Bang take place?

28-4 How do we know that the Big Bang was hot?

28-5 What was the universe like during its first 380,000 years?

28-6 What is "dark energy"? How does the curvature of the universe reveal its presence?

28-7 Has the universe always expanded at the same rate?

28-8 How reliable is our current understanding of the universe?

28-1 | The darkness of the night sky tells us about the nature of the universe

 Cosmology is the science concerned with the structure and evolution of the universe as a whole. One of the most profound and basic questions in cosmology may at first seem foolish: Why is the sky dark at night? This question, which haunted Johannes Kepler as long ago as 1610, was brought to public attention in the early 1800s by the German amateur astronomer Heinrich Olbers.

Olbers and his contemporaries pictured a universe of stars scattered more or less randomly throughout infinite space. Isaac Newton himself thought that no other model made sense. The gravitational forces between any *finite* number of stars, he argued, would in time cause them all to fall together, and the universe would soon be a compact blob.

Obviously, this has not happened. Newton concluded that we must be living amid a static, infinite expanse of stars. In this model, the universe is infinitely old, and it will exist forever without major changes in its structure. Olbers noticed, however, that a static, infinite universe presents a major puzzle.

If space goes on forever, with stars scattered throughout it, then any line of sight must eventually hit a star. In this case, no matter where you look in the night sky, you should ultimately see a star. The entire sky should be as bright as an average star, so, even at night, the sky should blaze like the surface of the Sun. **Olbers's paradox** is that the night sky is actually *dark* (Figure 28-1).

Olbers's paradox suggests that something is wrong with Newton's infinite, static universe. According to the classical, Newtonian picture of reality, space is like a gigantic flat sheet of inflexible, rectangular graph paper. (Space is actually three-dimensional, but it is easier to visualize just two of its three dimensions. In a similar way, an ordinary map represents the three-dimensional surface of the Earth, with its hills and valleys, as a flat, two-dimensional surface.)

This rigid, flat, Newtonian space stretches on and on, totally independent of stars or galaxies or anything else. The same is true of time in Newton's view of the universe; a Newtonian clock ticks steadily and monotonously forever, never slowing down or speeding up. Furthermore, Newtonian space and time are unrelated, in that a clock runs at the same rate no matter where in the universe it is located.

 Albert Einstein overturned this view of space and time. His special theory of relativity (recall Section 24-1 and Box 24-1) shows that measurements with clocks and rulers depend on the motion of the observer. What is more, Einstein's general theory of relativity (Section 24-2) tells us that gravity curves the fabric of space. As a result, the matter that occupies the universe influences the overall shape of space throughout the universe.

If we represent the universe as a sheet of graph paper, the sheet is not perfectly flat but has a dip wherever there is a concentration of mass, such as a person, a planet, or a star (see Figure 24-4). Because of gravitational effects, clocks run at different rates depending on whether they are close to or far from a massive object, as Figure 24-7a shows.

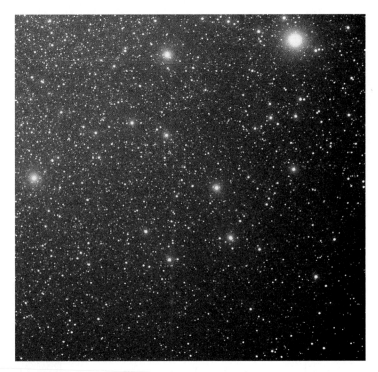

Figure 28-1   R I **V** U X G

The Dark Night Sky If the universe were infinitely old and filled uniformly with stars, the night sky would be ablaze with light. The observation that the night sky is dark, punctuated only by the light from isolated stars, tells us that the universe is quite different from this simple picture. (Okiro Fujii, *L'Astronomia*)

What does the general theory of relativity, with its many differences from the Newtonian picture, have to say about the structure of the universe as a whole? Einstein attacked this problem shortly after formulating his general theory in 1915. At that time, the prevailing view was that the universe was static, just as Newton had thought.

Einstein was therefore dismayed to find that his calculations could not produce a truly static universe. According to general relativity, the universe must be either expanding or contracting. In a desperate move to force his theory to predict a static universe, he added to the equations of general relativity a term called the **cosmological constant** (denoted by Λ, the capital Greek letter lambda). The cosmological constant was intended to represent a pressure that tends to make the universe expand as a whole. Einstein's idea was that this pressure would just exactly balance gravitational attraction, so that the universe would be static and not collapse.

 Einstein's cosmological constant is analogous to the pressure of gas inside a bicycle tire. This pressure exactly balances the inward force exerted by the stretched rubber of the tire itself, so the tire maintains the same size.

Unlike other aspects of Einstein's theories, the cosmological constant did not have a firm basis in physics. He just added it to make the general theory of relativity agree with the prejudice that the universe is static.

Because Einstein doubted his original equations, he missed an incredible opportunity: He could have postulated that we live in an expanding universe. Einstein has been quoted as saying in his later years that the cosmological constant was "the greatest blunder of my life." (In fact, the cosmological constant plays an important role in modern cosmology, although a very different one from what Einstein proposed. We will explore this in Section 28-7.)

Instead, the first hint that we live in an expanding universe came more than a decade later from the observations of Edwin Hubble. As we will see in Section 28-3, Hubble's discovery provides the resolution of Olbers's paradox.

28-2 | The universe is expanding

Hubble is usually credited with discovering that our universe is expanding. He found a simple linear relationship between the distances to remote galaxies and the redshifts of the spectral lines of those galaxies (review Section 26-5, especially Figures 26-14 and 26-15). This relationship, now called the *Hubble law,* states that the greater the distance to a galaxy, the greater is the galaxy's redshift. Thus, remote galaxies are moving away from us with speeds proportional to their distances. Specifically, the recessional velocity v of a galaxy is related to its distance d from the Earth by the equation

$$v = H_0 d$$

where H_0 is the Hubble constant. Because clusters of galaxies are getting farther and farther apart as time goes on, astronomers say that the universe is expanding.

What does it actually mean to say that the universe is expanding? According to general relativity, space itself is not rigid. The amount of space between widely separated locations changes over time. A good analogy is that of baking a chocolate chip cake, as in **Figure 28-2**. As the cake expands dur-

ing baking, the amount of space between the chocolate chips gets larger and larger. In the same way, as the universe expands, the amount of space between widely separated galaxies increases. The expansion of the universe *is* the expansion of space.

CAUTION It is important to realize that the expansion of the universe occurs primarily in the space that separates clusters of galaxies. Just as the chocolate chips in Figure 28-2 do not expand as the cake expands during baking, galaxies themselves do not expand. Einstein and others have established that an object that is held together by its own gravity, such as a galaxy, is always contained within a patch of nonexpanding space. A galaxy's gravitational field produces this nonexpanding region, which is indistinguishable from the rigid space described by Newton. Thus, the Earth and your body, for example, are not getting any bigger. Only the distance between widely separated galaxies increases with time.

The Hubble law is a direct proportionality—that is, a galaxy twice as far away is receding from us twice as fast. This is just what we would expect in an expanding universe. To see why this is so, imagine a grid of parallel lines (as on a piece of graph paper) crisscrossing the universe. **Figure 28-3a** shows a series of such gridlines 100 Mpc apart, along with five galaxies labeled A, B, C, D, and E that happen to lie where gridlines cross. As the universe expands in all directions, the gridlines and the attached galaxies spread apart. (This is just what would happen if the universe were a two-dimensional rubber sheet that was being pulled equally on all sides. Alternatively, you can imagine that Figure 28-3a depicts a very small portion of the chocolate chip cake in Figure 28-2, with galaxies taking the place of chocolate chips.)

Figure 28-3b shows the universe at a later time, when the gridlines are 50% farther apart (150 Mpc) and all the distances between galaxies are 50% greater than in Figure 28-3a. Imagine that A represents our galaxy, the Milky Way. The table accompanying Figure 28-3 shows how far each of the other galaxies have moved away from us during the expansion: Galaxies A and B and galaxies A and E were originally 100 Mpc apart, and have moved away from each other by an additional 50 Mpc; A and C, which were originally 200 Mpc apart, have increased their separation by an additional 100 Mpc; and the distance between A and D, originally

Figure 28-2

The Expanding Chocolate Chip Cake Analogy
The expanding universe can be compared to what happens inside a chocolate chip cake as the cake expands during baking. (The cake is floating weightlessly inside the oven of an orbiting spacecraft crewed by hungry astronauts.) All of the chocolate chips in the cake recede from one another as the cake expands, just as all the galaxies recede from one another as the universe expands.

100 Mpc

100 Mpc

(a) Five galaxies spaced 100 Mpc apart

150 Mpc

150 Mpc

(b) The expansion of the universe spreads the galaxies apart

	Original distance (Mpc)	Later distance (Mpc)	Change in distance (Mpc)
A–B	100	150	50
A–C	200	300	100
A–D	300	450	150
A–E	100	150	50

 Figure 28-3
The Expanding Universe and the Hubble Law
(a) Imagine five galaxies labeled A, B, C, D, and E. At the time shown here, adjacent galaxies are 100 Mpc apart. (b) As the universe expands, by some later time the spacing between adjacent galaxies has increased to 150 Mpc. The table shows that the greater the original distance between galaxies, the greater the amount that distance has increased. This agrees with the Hubble law.

300 Mpc, has increased by an additional 150 Mpc. In other words, the increase in distance between any pair of galaxies is in direct proportion to the original distance; if the original distance is twice as great, the increase in distance is also twice as great.

To see what this tells us about the recessional velocities of galaxies, remember that velocity is equal to the distance moved divided

by the elapsed time. (For example, if you traveled in a straight line for 360 kilometers in 4 hours, your velocity was (360 km)/(4 hours) = 90 kilometers per hour.) Because the distance that each galaxy moves away from A during the expansion is directly proportional to its original distance from A, it follows that the velocity v at which each galaxy moves away from A is also directly proportional to the original distance d. This is just the Hubble law, $v = H_0 d$.

CAUTION It may seem that if the universe is expanding, and if we see all the distant galaxies rushing away from us, then we must be in a special position at the very center of the universe. In fact, the expansion of the universe looks the same from the vantage point of *any* galaxy. For example, as seen from galaxy D in Figure 28-3, the initial distances to galaxies A, B, and C are 300 Mpc, 200 Mpc, and 100 Mpc, respectively. Between parts *a* and *b* of the figure, these distances increase by 150 Mpc, 100 Mpc, and 50 Mpc, respectively. So, as seen from D as well, the recessional velocity increases in direct proportion with the distance, and in the same proportion as seen from A. In other words, no matter which galaxy you call home, you will see all the other galaxies receding from you in accordance with the same Hubble law (and the same Hubble constant) that we observe from Earth.

WEB LINK 28.5 Figure 28-2 also shows that the expansion of the universe looks the same from one galaxy as from any other. An insect sitting on any one of the chocolate chips would see all the other chips moving away. If the cake were infinitely long, it would not actually have a center; as seen from any chocolate chip within such a cake, the cake would extend off to infinity to the left and to the right, and the expansion of the cake would appear to be centered on that chip. Likewise, because every point in the universe appears to be at the center of the expansion, it follows that our universe has no center at all. (Later in this chapter we will see evidence that the universe, like our imaginary cake, is indeed infinite.)

CAUTION "If the universe is expanding, what is it expanding into?" This commonly asked question arises only if we take our chocolate chip cake analogy too literally. In Figure 28-2, the cake (representing the universe) expands in three-dimensional space into the surrounding air. But the actual universe includes *all* space; there is nothing "beyond" it, because there is no "beyond." Asking "What lies beyond the universe?" is as meaningless as asking "Where on the Earth is north of the North Pole?"

The ongoing expansion of space explains why the light from remote galaxies is redshifted. Imagine a photon coming toward us from a distant galaxy. As the photon travels through space, the space is expanding, so the photon's wavelength becomes stretched. When the photon reaches our eyes, we see an increased wavelength: The photon has been redshifted. The longer the photon's journey, the more its wavelength will have been stretched. Thus, photons from distant galaxies have larger redshifts than those of photons from nearby galaxies, as expressed by the Hubble law.

A redshift caused by the expansion of the universe is properly called a **cosmological redshift**. It is *not* the same as a Doppler shift. Doppler shifts are caused by an object's *motion through space*,

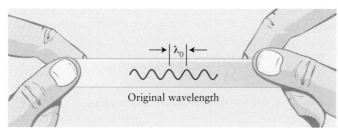

(a) A wave drawn on a rubber band ...

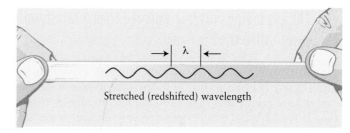

(b) ... increases in wavelength as the rubber band is stretched.

Figure 28-4

Cosmological Redshift A wave drawn on a rubber band stretches along with the rubber band. In an analogous way, a light wave traveling through an expanding universe "stretches," that is, its wavelength increases.

whereas a cosmological redshift is caused by the *expansion of space* (Figure 28-4).

We can calculate the factor by which the universe has expanded since some ancient time from the redshift of light emitted by objects at that time. As we saw in Section 26-5, redshift (z) is defined as

$$z = \frac{\lambda - \lambda_0}{\lambda_0}$$

where λ_0 is the unshifted wavelength of a photon and λ is the wavelength we observe. For example, λ_0 could be the wavelength of a particular emission line in the spectrum of light leaving a remote quasar. As the quasar's light travels through space, its wavelength is stretched by the expansion of the universe. Thus, at our telescopes we observe the spectral line to have a wavelength λ. The ratio λ/λ_0 is a measure of the amount of stretching. By rearranging terms in the preceding equation, we can solve for this ratio to obtain

$$\frac{\lambda}{\lambda_0} = 1 + z$$

For example, consider a quasar with a redshift $z = 3$. Since the time that light left that quasar, the universe has expanded by a factor of $1 + z = 1 + 3 = 4$. In other words, when the light left that quasar, representative distances between widely separated galaxies were only one-quarter as large as they are today. A representative volume of space, which is proportional to the cube of its dimensions, was only $(1/4)^3 = 1/64$ as large as it is today. Thus, the density of matter in such a volume was 64 times greater than it is today.

If you know the redshift z of a distant object such as a quasar, you can calculate that object's recessional velocity v (see Box 26-2 for how this is done). Then, using the Hubble law, you can determine the distance d to that object if you know the value of the Hubble constant H_0. This also tells you the **lookback time** of that object, that is, how far into the past you are looking when you see that object. For example, if the lookback time for a quasar is a billion years, that means the light from the quasar took a billion years to reach us and so we are seeing it as it was a billion years ago.

Distances and lookback times determined in this way are somewhat uncertain, because there is some uncertainty in the value of the Hubble constant. Furthermore, as we will see in Section 28-8, the universe has not always expanded at the same rate. This means that the value of the Hubble constant H_0 was different in the distant past. (In other words, the Hubble "constant" is not actually constant in time.)

To avoid dealing with these uncertainties and complications, astronomers commonly refer to times in the distant past in terms of redshift rather than years. For example, instead of asking, "How common were quasars 5 billion years ago?," an astronomer might ask, "How common were quasars at $z = 1.0$?" In this question, "at $z = 1.0$" is a shorthand way of saying "at the lookback time that corresponds to objects at $z = 1.0$." We will use this terminology later in this chapter. Remember that the greater the redshift, the greater the lookback time and hence the further back into time we are peering.

These ideas about the expanding universe demonstrate the central philosophy of cosmology. In cosmology, unlike other sciences, we cannot carry out controlled experiments or even make comparisons: There is only one observable universe. To make progress in cosmology, we must accept certain philosophical assumptions or abandon hope of understanding the nature of the universe. The Hubble law provides a classic example. It could be interpreted to mean that we are at the center of the universe. We reject this interpretation, however, because it violates a cosmological extension of Copernicus's belief that we do not occupy a special location in space.

When Einstein began applying his general theory of relativity to cosmology, he made a daring assumption: Over very large distances the universe is **homogeneous,** meaning that every region is the same as every other region, and **isotropic,** meaning that the universe looks the same in every direction. In other words, if you could stand back and look at a very large region of space, any one part of the universe would look basically the same as any other part, with the same kinds of galaxies distributed through space in the same way. The assumption that the universe is homogeneous and isotropic constitutes the **cosmological principle.** It gives precise meaning to the idea that we do not occupy a special location in space.

Models of the universe based on the cosmological principle have proven remarkably successful in describing the structure and evolution of the universe and in interpreting observational data. All the discussion about the universe in this chapter and the next assumes that the universe is homogeneous and isotropic on the largest scale.

28-3 | The expanding universe emerged from a cataclysmic event called the Big Bang

The Hubble flow shows that the universe has been expanding for billions of years. This means that in the past the matter in the universe must have been closer together and therefore denser than it is today. If we look far enough into the very distant past, there must have been a time when the density of matter was almost inconceivably high. This leads us to conclude that some sort of tremendous event caused ultradense matter to begin the expansion that continues to the present day. This event, called the **Big Bang,** marks the creation of the universe.

CAUTION It is not correct to think of the Big Bang as an explosion. When a bomb explodes, pieces of debris fly off *into space* from a central location. If you could trace all the pieces back to their origin, you could find out exactly where the bomb had been. This process is not possible with the universe, however, because the universe itself always has and always will consist of all space. As we have seen, the universe logically cannot have an edge (see the discussion of the expanding chocolate chip cake in Section 28-2).

How long ago did the Big Bang take place? To estimate an answer, imagine two galaxies that today are separated by a distance d and receding from each other with a velocity v. A movie of these galaxies would show them flying apart. If you run the movie backward, you would see the two galaxies approaching each other as time runs in reverse. We can calculate the time T_0 it will take for the galaxies to collide by using the equation

$$T_0 = \frac{d}{v}$$

This says that the time to travel a distance d at velocity v is equal to the ratio d/v. (As an example, to travel a distance of 360 km at a velocity of 90 km/h takes (360 km)/(90 km/h) = 4 hours.) If we use the Hubble law, $v = H_0 d$, to replace the velocity v in this equation, we get

$$T_0 = \frac{d}{H_0 d} = \frac{1}{H_0}$$

Note that the distance of separation, d, has canceled out and does not appear in the final expression. This means that T_0 is the same for *all* galaxies. This is the time in the past when all galaxies were crushed together, the time back to the Big Bang. In other words, the reciprocal of the Hubble constant H_0 gives us an estimate of the age of the universe, which is one reason why H_0 is such an important quantity in cosmology.

Observations suggest that H_0 = 71 km/s/Mpc to within a few percent, and this is the value we choose as our standard (see Section 26-5). Using this value, our estimate for the age of the universe is

$$T_0 = \frac{1}{71 \text{ km/s/Mpc}}$$

To convert this into units of time, we simply need to remember that 1 Mpc equals 3.09×10^{19} km and 1 year equals 3.156×10^7 s.

Using the technique we discussed in Box 1-3 for converting units, we get

$$T_0 = \frac{1}{71} \frac{\text{Mpc/s}}{\text{km}} \times \frac{3.09 \times 10^{19} \text{ km}}{1 \text{ Mpc}} \times \frac{1 \text{ year}}{3.156 \times 10^7 \text{ s}}$$

$$= 1.4 \times 10^{10} \text{ years} = 14 \text{ billion years}$$

By comparison, the age of our solar system is only 4.56 billion years, or about one-third the age of the universe. Thus, the formation of our home planet is a relatively recent event in the history of the cosmos.

The value of H_0 has an uncertainty of about 5%, so our simple estimate of the age of the universe is likewise uncertain by at least 5%. Furthermore, the formula $T_0 = 1/H_0$ is at best an approximation, because in deriving it we assumed that the universe expands at a constant rate. In Section 28-8 we will discuss how the expansion rate of the universe has changed over its history. When these factors are taken into consideration, we find that the age of the universe is 13.7 billion years, with an uncertainty of 0.2 billion years. This is remarkably close to our simple estimate.

 Whatever the true age of the universe, it must be at least as old as the oldest stars. The oldest stars that we can observe readily lie in the Milky Way's globular clusters (see Section 21-3 and Section 25-1). The most recent calculations based on the theory of stellar evolution indicate that these stars are 12.5 billion years old, with an uncertainty of about 10%. Encouragingly enough, this is less than the calculated age of the universe: The oldest stars in our universe are younger than the universe itself!

The Big Bang helps resolve Olbers's paradox. We know that the universe had a definite beginning, and thus its age is finite (as opposed to infinite). If the universe is 13.7 billion years old, then light from stars more than about 13.7 billion light-years away has just not had enough time to get here. As a result, we cannot see any objects that are more than about 13.7 billion light-years away. This is true even if the universe is infinite, with galaxies scattered throughout its limitless expanse.

You can think of the Earth as being at the center of an enormous sphere having a radius of 13.7 billion light-years (Figure 28-5). The surface of this sphere is called the **cosmic light horizon.** Our entire **observable universe** is located inside this sphere. We cannot see anything beyond the cosmic light horizon, because the time required for light to reach us from these incredibly remote distances is greater than the age of the universe. The size of this sphere increases as the universe ages. When the universe was 5 billion years old, for example, our observable universe was only 5 billion light-years in radius. Throughout the observable universe, galaxies are distributed sparsely enough that there are no stars along most of our lines of sight, which helps explain why the night sky is dark.

Besides the finite age of the universe, a second effect also contributes significantly to the darkness of the night sky—the redshift. According to the Hubble law, the greater the distance to a galaxy, the greater the redshift. When a photon is redshifted, its wavelength becomes longer, and its energy—which is inversely proportional to its wavelength (see Section 5-5)—decreases. Consequently, even

We cannot yet see this region of the universe.

Farthest we can see now

Present cosmic light horizon

Figure 28-5   R I **V** U X G

The Observable Universe The part of the universe that we can observe lies within a sphere centered on the Earth called the cosmic light horizon. Its radius is equal to the distance that light has traveled since the Big Bang, about 13.7 billion light-years. We cannot see objects beyond the cosmic light horizon, because their light has not had enough time to reach us. The galaxies that we can just barely make out with our most powerful telescopes lie inside the cosmic light horizon; we see them as they were less than a billion years after the Big Bang. (Inset: Robert Williams and the Hubble Deep Field Team, STScI; NASA)

though there are many galaxies far from the Earth, they have large redshifts and their light does not carry much energy. A galaxy nearly at the cosmic light horizon has a nearly infinite redshift, meaning that the light we receive from that galaxy carries practically no energy at all. This decrease in photon energy because of the expansion of the universe decreases the brilliance of remote galaxies, helping to make the night sky dark.

The concept of a Big Bang origin for the universe is a straightforward, logical consequence of having an expanding universe. If you can just imagine far enough back into the past, you can arrive at a time 13.7 billion years ago, when the density throughout the universe was infinite. As a result, throughout the universe space and time were completely jumbled up in a condition of infinite curvature similar to that at the singularity found at the center of a black hole (see Section 24-5). For this reason, a better name for the Big Bang is the **cosmic singularity**. Thanks to the infinite curvature, the usual laws of physics do not tell us exactly what happened at the moment of the Big Bang.

A very short time after the Big Bang, space and time began to behave in the way we think of them today. This short time interval, called the **Planck time** (t_P), is given by the following expression:

The Planck time

$$t_P = \sqrt{\frac{Gh}{c^5}} = 1.35 \times 10^{-43} \text{ s}$$

t_P = Planck time

G = universal constant of gravitation

h = Planck's constant

c = speed of light

We do not yet understand how space, time, and matter behaved in that brief but important interval from the beginning of the Big Bang to the Planck time, about

10^{-43} seconds later. Hence, the Planck time represents a limit to our knowledge of conditions at the very beginning of the universe.

28-4 | The microwave radiation that fills all space is evidence of a hot Big Bang

SCIENCE IN PROCESS *The afterglow of the Big Bang was first discovered by a happy coincidence*

One of the major advances in twentieth-century astronomy was the discovery of the origin of the heavy elements. We know today that essentially all the heavy elements are created by thermonuclear reactions at the centers of stars and in supernovae (see Chapter 22). The starting point of all these reactions is the fusion of hydrogen into helium, which we described in Section 18-1. But as astronomers began to understand the details of thermonuclear synthesis in the 1960s, they were faced with a dilemma: There is far more helium in the universe than could have been created by hydrogen fusion in stars.

For example, the Sun consists of about 74% hydrogen and 25% helium by mass, leaving only 1% for all the remaining heavier elements combined. This 1% can be understood as material produced inside earlier generations of massive stars that long ago cast these heavy elements out into space when they became supernovae. Some freshly made helium, produced by the thermonuclear fusion of hydrogen within the stars, certainly accompanied these heavy elements. But calculations showed that the amount of helium produced in this way was not nearly enough to account for one-quarter of the Sun's mass. Because it was thought that the universe originally contained only hydrogen—the simplest of all the chemical elements—the presence of so much helium posed a major dilemma.

Shortly after World War II, Ralph Alpher and Robert Hermann proposed that the universe immediately following the Big Bang must have been so incredibly hot that thermonuclear reactions occurred everywhere throughout space. Following up this idea in 1960, Princeton University physicists Robert Dicke and P. J. E. Peebles discovered that they could indeed account for today's high abundance of helium by assuming that the early universe had been at least as hot as the Sun's center, where helium is currently being produced. The hot early universe must therefore have been filled with many high-energy, short-wavelength photons. The properties of this radiation field depended on its temperature, as described by Planck's blackbody law (review Figure 5-10).

The universe has expanded so much since those ancient times that all those short-wavelength photons have had their wavelengths stretched by a tremendous factor. As a result, they have become low-energy, long-wavelength photons. The temperature of this cosmic radiation field is now only a few degrees above absolute zero. By Wien's law, radiation at such a low temperature should have its peak intensity at microwave wavelengths of approximately 1 millimeter. Hence, this radiation field, which fills all of space, is called the **cosmic microwave background** or **cosmic background radiation.** In the early 1960s, Dicke and his colleagues began designing an antenna to detect this microwave radiation.

Meanwhile, just a few miles from Princeton University, Arno Penzias and Robert Wilson of Bell Telephone Laboratories in New Jersey were working on a new microwave horn antenna designed to relay telephone calls to Earth-orbiting communications satellites (**Figure 28-6**). Penzias and Wilson were deeply puzzled when, no

matter where in the sky they pointed their antenna, they detected faint background noise. Thanks to a colleague, they happened to learn about the work of Dicke and Peebles and came to realize that they had discovered the cooled-down cosmic background radiation left over from the hot Big Bang.

You can actually detect cosmic background radiation using an ordinary television set. This radiation is responsible for about 1% of the random noise or "hash" that appears on the screen when you tune a television to a station that is off the air. Using far more sophisticated detectors than TV sets, scientists have made many measurements of the intensity of the background radiation at a variety of wavelengths. Unfortunately, the Earth's atmosphere is almost totally opaque to wavelengths between about 10 µm and 1 cm (see Figure 6-28), which is just the wavelength range in which the background radiation is most intense. As a result, scientists have had to place detectors either on high-altitude balloons (which can fly above the majority of the obscuring atmosphere) or, even better, on board orbiting spacecraft.

The first high-precision measurements of the cosmic microwave background came from the Cosmic Background Explorer (COBE, pronounced "coe-bee") satellite, which was placed in Earth orbit in 1989 (**Figure 28-7a**). Data from COBE's spectrometer, shown in Figure 28-7b, demonstrate that this ancient radiation has the spectrum of a blackbody with a temperature of 2.725 K.

An important feature of the microwave background is that its intensity is almost perfectly isotropic, that is, the same in all directions. In other words, we detect nearly the same background intensity from all parts of the sky. This is a striking confirmation of Einstein's assumption that the universe is isotropic (Section 28-2). However, extremely accurate measurements first made from high-flying airplanes, and later from high-altitude balloons and from COBE, reveal a very slight variation in temperature across the sky. The microwave background appears slightly warmer than average toward the constellation of Leo and slightly cooler than average in the opposite direction toward Aquarius. Between the warm spot in Leo and the cool spot in Aquarius, the background temperature declines smoothly across the sky. **Figure 28-8** is a map of the microwave sky showing this variation.

This apparent variation in temperature is caused by the Earth's motion through the cosmos. If we were at rest with respect to the microwave background, the radiation would be even more nearly isotropic. Because we are moving through this radiation field, however, we see a Doppler shift. Specifically, we see shorter-than-average wavelengths in the direction toward which we are moving, as drawn in **Figure 28-9**. A decrease in wavelength corresponds to an increase in photon energy and thus an increase in the temperature of the radiation. The slight temperature excess observed, about 0.00337 K, corresponds to a speed of 371 km/s. Conversely, we see longer-than-average wavelengths in that part of the sky from which we are receding. An increase in wavelength corresponds to a decline in photon energy and, hence, a decline in radiation temperature.

WEB LINK 28.7 **Figure 28-6**   R I 🅅 U X G

The Bell Labs Horn Antenna Using this microwave horn antenna, originally built for communications purposes, Arno Penzias and Robert Wilson detected a signal that seemed to come from all parts of the sky. After carefully removing all potential sources of electronic "noise" (including bird droppings inside the antenna) that could create a false signal, Penzias and Wilson realized that they were actually detecting radiation from space. This radiation is the afterglow of the Big Bang. (Bell Labs)

 WEB LINK 28.9 Our solar system is thus traveling away from Aquarius and toward Leo at a speed of 371 km/s. Taking into account the known velocity of the Sun around the center of our Galaxy, we find that the entire Local Group of galaxies,

(a) The COBE spacecraft

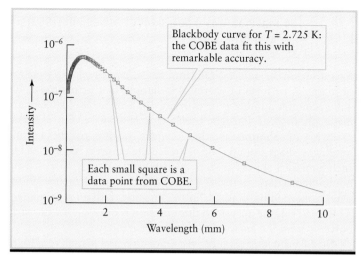

(b) The spectrum of the cosmic microwave background

figure 28-7
COBE and the Spectrum of the Cosmic Microwave Background (a) The Cosmic Background Explorer satellite (COBE), launched in 1989, measured the spectrum and angular distribution of the cosmic microwave background over a wavelength range from 1 μm to 1 cm. (b) A blackbody curve gives an excellent match to the COBE data. (a: Courtesy of J. Mather/NASA; b: Courtesy of E. Cheng/NASA COBE Science Team)

including our Milky Way Galaxy, is moving at about 620 km/s toward the Hydra-Centaurus supercluster. Observations show that thousands of other galaxies are being carried in this direction, as is the Hydra-Centaurus supercluster itself. This tremendous flow of matter is thought to be due to the gravitational pull of an enormous collection of visible galaxies and dark matter lying in that direction. This immense object, dubbed the *Great Attractor*, lies about 50 Mpc (150 million ly) from Earth (see Figure 26-20).

The existence of such concentrations of mass, as well as the existence of superclusters of galaxies (see Section 26-6), shows that the universe is rather "lumpy" on scales of 100 Mpc or smaller. It is only on larger scales that the universe is homogeneous and isotropic.

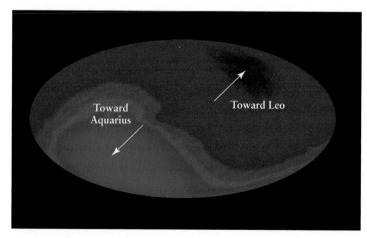

figure 28-8 ⓡ I V U X G

The Microwave Sky In this map of the entire sky made from COBE data, the plane of the Milky Way runs horizontally across the map, with the galactic center in the middle. Color indicates temperature—red is warm and blue is cool. The small temperature variation across the sky—only 0.0033 K above or below the average radiation temperature of 2.725 K—is caused by the Earth's motion through the microwave background. (NASA)

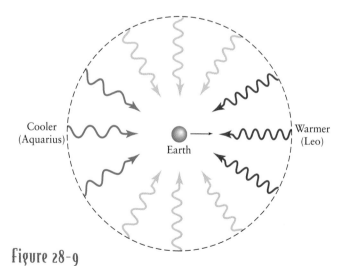

figure 28-9

Our Motion Through the Microwave Background Because of the Doppler effect, we detect shorter wavelengths in the microwave background and a higher temperature of radiation in that part of the sky toward which we are moving. This part of the sky is the area shown in red in Figure 28-8. In the opposite part of the sky, shown in blue in Figure 28-8, the microwave radiation has longer wavelengths and a cooler temperature.

28-5 | The universe was a hot, opaque plasma during its first 380,000 years

Everything in the universe falls into one of two categories—energy or matter. One form of energy is radiation, that is, photons. (We will encounter another type of energy in Section 28-7.) There are many photons of starlight traveling across space, but the vast majority of photons in the universe belong to the cosmic microwave background. The matter in the universe is contained in such luminous objects as stars, planets, and galaxies, as well as in nonluminous dark matter. A natural question to ask is this: Which plays a more important role in the universe, radiation or matter? As we will see, the answer to this question is different for the early universe from the answer for our universe today.

To make a comparison between radiation and matter, recall Einstein's famous equation $E = mc^2$ (see Section 18-1). We can think of the photon energy in the universe (E) as being equal to a quantity of mass (m) multiplied by the square of the speed of light (c). The amount of this equivalent mass in a volume V, divided by that volume, is the **mass density of radiation** (ρ_{rad}; say "rho sub rad").

We can combine $E = mc^2$ with the Stefan-Boltzmann law (see Section 5-4) to give the following formula:

Mass density of radiation

$$\rho_{rad} = \frac{4\sigma T^4}{c^3}$$

ρ_{rad} = mass density of radiation

T = temperature of radiation

σ = Stefan-Boltzmann constant = 5.67×10^{-8} W m^{-2} K^{-4}

c = speed of light = 3.00×10^8 m/s

For the present-day temperature of the cosmic background radiation, $T = 2.725$ K, this equation yields

$$\rho_{rad} = 4.6 \times 10^{-31} \text{ kg/m}^3$$

The **average density of matter** (ρ_m; say "rho sub em") in the universe is harder to determine. To find this density, we look at a large volume (V) of space, determine the total mass (M) of all the stars, galaxies, and dark matter in that volume, and divide the volume into the mass: $\rho_m = M/V$. (We emphasize that this is the *average* density of matter. It would be the actual density if all the matter in the universe were spread out uniformly rather than being clumped into galaxies and clusters of galaxies.) Determining how much mass is in a large volume of space is a challenging task. A major part of the challenge involves dark matter, which emits no electromagnetic radiation and can be detected only by its gravitational influence (see Section 25-4 and Section 26-8).

One method that appears to deal successfully with this challenge is to observe clusters of galaxies, within which most of the mass in the universe is concentrated. Rich clusters are surrounded by halos of hot, X-ray-emitting gas, typically at temperatures of 10^7 to 10^8 K (see Figure 26-23). Such a halo should be in hydrostatic equilibrium,

so that it neither expands nor contracts but remains the same size (see Section 18-2, especially Figure 18-3). The outward gas pressure associated with the halo's high temperature would then be balanced by the inward gravitational pull due to the total mass of the cluster. Thus, by measuring the temperature of the halo—which can be determined from the properties of the halo's X-ray emission—astronomers can infer the cluster's mass, including the contribution from dark matter.

From galaxy clusters and other measurements, the present-day average density of matter in the universe is thought to be

$$\rho_m = 2.6 \times 10^{-27} \text{ kg/m}^3$$

with an uncertainty of about 15%. The mass of a single hydrogen atom is 1.67×10^{-27} kg. Hence, if the mass of the universe were spread uniformly over space, there would be the equivalent of 1½ hydrogen atoms per cubic meter of space. By contrast, there are 5×10^{25} atoms in a cubic meter of the air you breathe! The very small value of ρ_m shows that our universe has very little matter in it.

Furthermore, by counting galaxies, astronomers determine that the average density of *luminous* matter (that is, the stars and gas within clusters of galaxies) is about 4.2×10^{-28} kg/m^3. This is only about 16% of the average density of matter of all forms. Thus, nonluminous dark matter is actually the predominant form of matter in our universe. The "ordinary" matter of which the stars, the planets, and ourselves are made is actually very exceptional!

Although the average density of matter in the universe is tiny by Earth standards, it is thousands of times larger than ρ_{rad}, the mass density of radiation. However, this was not always the case. Matter prevails over radiation today only because the energy now carried by microwave photons is so small. Nevertheless, the number of photons in the microwave background is astounding. From the physics of blackbody radiation it can be demonstrated that there are today 410 million (4.1×10^8) photons in every cubic meter of space. In other words, the photons in space outnumber atoms by roughly a billion (10^9) to one. In terms of total number of particles, the universe thus consists almost entirely of microwave photons. This radiation field no longer has much "clout," however, because its photons have been redshifted to long wavelengths and low energies after 13.7 billion years of being stretched by the expansion of the universe.

In contrast, think back toward the Big Bang. The universe becomes increasingly compressed, and thus the density of matter increases as we go back in time. The photons in the background radiation also become more crowded together as we go back in time. But, in addition, the photons become less redshifted and thus have shorter wavelengths and higher energy than they do today. Because of this added energy, the mass density of radiation (ρ_{rad}) increases more quickly as we go back in time than does the average density of matter (ρ_m). In fact, as Figure 28-10 shows, there was a time in the ancient past when ρ_{rad} equaled ρ_m. Before this time, ρ_{rad} was greater than ρ_m, and radiation thus held sway over matter. Astronomers call this state a **radiation-dominated universe.** After ρ_m became greater than ρ_{rad}, so that matter prevailed over radiation, our universe became a **matter-dominated universe.**

This transition from a radiation-dominated universe to a matter-dominated universe occurred about 2500 years after the Big Bang,

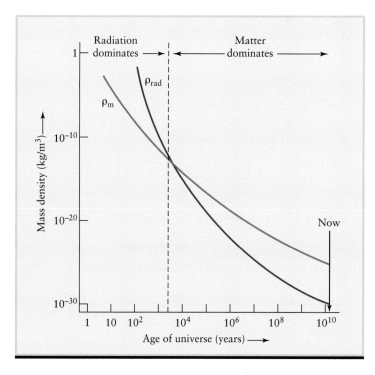

figure 28-10

The Evolution of Density For approximately 2500 years after the Big Bang, the mass density of radiation (ρ_{rad}, shown in red) exceeded the matter density (ρ_m, shown in blue), and the universe was radiation-dominated. Later, however, continued expansion of the universe caused ρ_{rad} to become less than ρ_m, at which point the universe became matter-dominated.

at a time that corresponds to a redshift of about $z = 25,000$. In other words, since that time the wavelengths of photons have been stretched by a factor of 25,000. Today, these microwave photons typically have wavelengths of about 1 mm. But when the universe was about 2500 years old, they had wavelengths of about 40 nm, in the ultraviolet part of the spectrum.

To calculate the temperature of the cosmic background radiation at the time of this transition from a radiation-dominated universe to a matter-dominated one, we use Wien's law (see Section 5-4). This law says that the wavelength of maximum emission (λ_{max}) of a blackbody is inversely proportional to its temperature (T): A *decrease* of λ_{max} by a factor of 2 corresponds to an *increase* of T by a factor of 2.

The present-day peak wavelength of the cosmic background radiation corresponds to a blackbody temperature of 2.725 K. Hence, a peak wavelength 25,000 times smaller corresponds to a temperature 25,000 times greater: $T = 25,000 \times 2.725$ K $= 68,000$ K. In other words, the temperature of the radiation background is in direct proportion to the factor $1 + z$ and has been declining over the ages, as Figure 28-11 shows.

As Figure 28-11 suggests, the nature of the universe changed again in a fundamental way 380,000 years after the Big Bang, when z was roughly 1100 and the temperature of the radiation background was about 1100×2.725 K $= 3000$ K. To see the significance

of this moment in cosmic history, recall that hydrogen is by far the most abundant element in the universe—hydrogen atoms outnumber helium atoms by about 12 to 1. A hydrogen atom consists of a single proton orbited by a single electron, and it takes relatively little energy to knock the proton and electron apart. In fact, ultraviolet radiation warmer than about 3000 K easily ionizes hydrogen. Thus, hydrogen atoms could not survive in the universe that existed before $z = 1100$. That is, in the first 380,000 years after the Big Bang, the background photons had energies great enough to prevent electrons and protons from binding to form hydrogen atoms (**Figure 28-12a**). Only since $z = 1100$ (that is, since $t = 380,000$ years) have the energies of these photons been low enough to permit hydrogen atoms to exist (Figure 28-12b).

The epoch when atoms first formed at $t = 380,000$ years is called the **era of recombination**. This refers to electrons "recombining" to form atoms. (The name is a bit misleading, because the electrons and protons had never before combined into atoms.)

Prior to $t = 380,000$ years, the universe was completely filled with a shimmering expanse of high-energy photons colliding vigorously with protons and electrons. This state of matter, called a

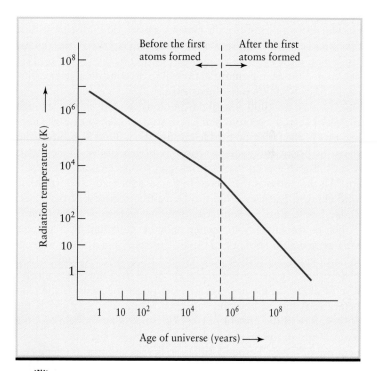

figure 28-11

The Evolution of Radiation Temperature As the universe expanded, the photons in the radiation background became increasingly redshifted and the temperature of the radiation fell. Approximately 380,000 years after the Big Bang, when the temperature fell below 3000 K, hydrogen atoms formed and the radiation field "decoupled" from the matter in the universe. After that point, the temperature of matter in the universe was not the same as the temperature of radiation. The time when the first atoms formed is called the era of recombination (see Figure 28-12).

(a) Before recombination:
- Temperatures were so high that electrons and protons could not combine to form hydrogen atoms.
- The universe was opaque: Photons underwent frequent collisions with electrons.
- Matter and radiation were at the same temperature.

(b) After recombination:
- Temperatures became low enough for hydrogen atoms to form.
- The universe became transparent: Collisions between photons and atoms became infrequent.
- Matter and radiation were no longer at the same temperature.

figure 28-12

The Era of Recombination (a) Before recombination, the energy of photons in the cosmic background was high enough to prevent protons and electrons from forming hydrogen atoms. (b) Some 380,000 years after the Big Bang, the energy of the cosmic background radiation became low enough that hydrogen atoms could survive.

plasma, is opaque, just like the glowing gases inside a discharge tube (like a neon advertising sign). The surface and interior of the Sun are also a hot, glowing, opaque plasma (see Section 18-9). P. J. E. Peebles coined the term **primordial fireball** to describe the universe during its first 380,000 years of existence.

After $t = 380,000$ years, the photons no longer had enough energy to keep the protons and electrons apart. As soon as the temperature of the radiation field fell below about 3000 K, protons and electrons began combining to form hydrogen atoms. These atoms do not absorb low-energy photons, and so space became transparent! All the photons that heretofore had been vigorously colliding with charged particles could now stream unimpeded across space. Today, these same photons constitute the microwave background.

Before recombination, matter and the radiation field had the same temperature, because photons, electrons, and protons were all in continuous interaction with one another. After recombination, photons and atoms hardly interacted at all, and thus the temperature of matter in the universe was no longer the same as the temperature of the background radiation. Thus, $T = 2.725$ K is the temperature of the present-day background radiation field, *not* the temperature of the matter in the universe. Note that while the temperature of the background radiation is very uniform, the temperature of matter in the universe is anything but: It ranges from hundreds of millions of kelvins in the interiors of giant stars to a few tens of kelvins in the interstellar medium.

 A good analogy is the behavior of a glass of cold water. If you hold the glass in your hand, the water will get warmer and your hand will get colder until both the water and your hand are at the same temperature. But if you set the glass down and do not touch it, so that the glass and your hand do not interact, their temperatures are decoupled: The water will stay cold and your hand will stay warm for much longer.

Because the universe was opaque prior to $t = 380,000$ years, we cannot see any further into the past than the era of recombination. In particular, we cannot see back to the era when the universe was radiation-dominated. The microwave background, whose photons have suffered a redshift of $z = 1100$, contains the most ancient photons we will ever be able to observe.

Careful analysis of COBE data showed that the cosmic background radiation is not completely isotropic. Even when the effects of the Earth's motion are accounted for, there remain variations in the temperature of the radiation field of about 100 μK (100 microkelvins, or 10^{-4} K) above or below the average 2.725 K temperature. These tiny temperature variations indicate that the matter and radiation in the universe were not totally uniform at the moment of recombination. Regions that were slightly denser than average were also slightly cooler than average; less dense regions were slightly warmer. When radiation decoupled from matter at the time of recombination, the radiation preserved a record of these variations in temperature and density.

Astronomers place great importance on studying temperature variations in the cosmic background radiation. The reason is that concentrations of mass in our present-day universe, such as superclusters of galaxies, are thought to have formed from the denser regions in the early universe. Within these immense concentrations formed the galaxies, stars, and planets. Thus, by studying these nonuniformities, we are really studying our origins.

Unfortunately, the detectors on board COBE had a relatively coarse angular resolution of 7° and thus could not give a detailed picture of these temperature variations. In 1998 two balloon-borne experiments, BOOMERANG and MAXIMA, carried new, high-resolution telescopes aloft to study the cosmic background radiation with unprecedented precision. The best all-sky coverage of the background radiation has

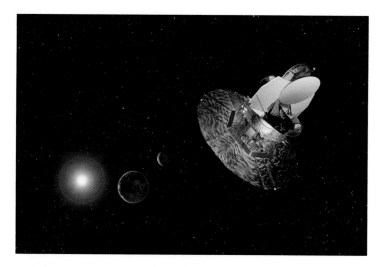

Figure 28-13

In Search of Ancient Photons This artist's impression shows the Wilkinson Microwave Anisotropy Probe (WMAP) en route to a location in space called L2, which lies about 1.5 million kilometers from Earth on the side opposite the Sun. At this position WMAP takes one year to orbit the Sun. The solar panels continually shade WMAP's detectors from sunlight, keeping them cold so that they can accurately measure the low-temperature photons ($T = 2.725$ K) of the cosmic background radiation. (NASA/WMAP Science Team)

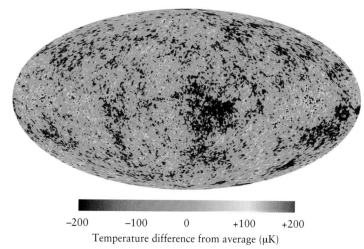

–200   –100   0   +100   +200
Temperature difference from average (μK)

 ## Figure 28-14   R I V U X G
Temperature Variations in the Cosmic Microwave Background This map from WMAP data shows small variations in the temperature of the cosmic background radiation across the entire sky. (The variations due to the Earth's motions through space, shown in Figure 28-8, have been factored out.) Lower-temperature regions (shown in blue) show where the early universe was slightly denser than average. Note that the variations in temperature are no more than 200 μK, or 2×10^{-4} K. (NASA/WMAP Science Team)

come from the state-of-the-art detectors on board the Wilkinson Microwave Anisotropy Probe (WMAP for short), a NASA spacecraft that was launched in 2001. Shown in **Figure 28-13**, the spacecraft is named for the late David Wilkinson of Princeton University, who was a pioneer in studies of the cosmic background radiation. **Figure 28-14** shows a map of the sky based on the first year of data taken by the WMAP detectors. This map shows us the state of the universe when it was less than 0.003% of its present age.

Temperature variations in the cosmic background radiation do more than show us the origins of large-scale structure in the universe. As we will see in the next two sections, they actually reveal the shape of the universe as a whole.

28-6 | The shape of the universe indicates its matter and energy content

 The geometry of space reveals that the universe is filled with dark energy

We have seen that by following the mass densities of radiation (ρ_{rad}) and of matter (ρ_m), we can learn about the evolution of the universe. But it is equally important to know the combined mass density of *all* forms of matter and energy. (In an analogous way, an accountant needs to know the overall financial status of a company, not just individual profits or losses.) Remarkably, we can do this by investigating the overall shape of the universe.

Einstein's general theory of relativity explains that gravity curves the fabric of space. Furthermore, the equivalence between matter and energy, expressed by Einstein's equation $E = mc^2$, tells that either matter or energy produces gravity. Thus, the matter and energy scattered across space should give the universe an overall curvature. The degree of curvature depends on the **combined average mass density** of all forms of matter and energy. This quantity, which we call ρ_0 (say "rho sub zero"), is the sum of the average mass densities of matter, radiation, and any other form of energy. Thus, by measuring the curvature of space, we should be able to determine the value of ρ_0 and, hence, learn about the content of the universe as a whole.

To see what astronomers mean by the curvature of the universe, imagine shining two powerful laser beams out into space so that they are perfectly parallel as they leave the Earth. Furthermore, suppose that nothing gets in the way of these two beams, so that we can follow them for billions of light-years as they travel across the universe and across the space whose curvature we wish to detect. There are only three possibilities:

1. We might find that our two beams of light remain perfectly parallel, even after traversing billions of light-years. In this case, space would not be curved: The universe would have **zero curvature**, and space would be **flat**.

2. Alternatively, we might find that our two beams of light gradually converge. In such a case, space would not be flat. Recall that lines of longitude on the Earth's surface are parallel at the equator but intersect at the poles. Thus, in this case the three-dimensional geometry

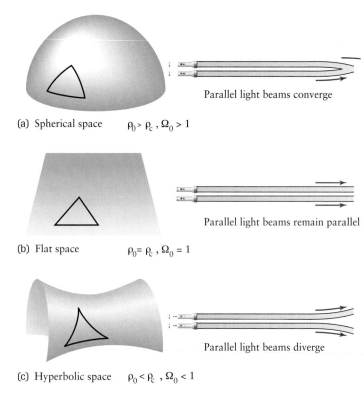

(a) Spherical space $\rho_0 > \rho_c$, $\Omega_0 > 1$

Parallel light beams converge

(b) Flat space $\rho_0 = \rho_c$, $\Omega_0 = 1$

Parallel light beams remain parallel

(c) Hyperbolic space $\rho_0 < \rho_c$, $\Omega_0 < 1$

Parallel light beams diverge

Figure 28-15

The Geometry of the Universe The curvature of the universe is either **(a)** positive, **(b)** zero, or **(c)** negative. The curvature depends on whether the combined mass density is greater than, equal to, or less than the critical density, or, equivalently, on whether the density parameter Ω_0 is greater than, equal to, or less than 1. In theory, the curvature could be determined by seeing whether two laser beams initially parallel to each other would converge, remain parallel, or spread apart.

of the universe would be analogous to the two-dimensional geometry of a spherical surface. We would then say that space is **spherical** and that the universe has **positive curvature.** Such a universe is also called **closed,** because if you travel in a straight line in any direction in such a universe, you will eventually return to your starting point.

3. Finally, we might find that the two initially parallel beams of light would gradually diverge, becoming farther and farther apart as they moved across the universe. In this case, the universe would still have to be curved, but in the opposite sense from the spherical model. We would then say that the universe has **negative curvature.** In the same way that a sphere is a positively curved two-dimensional surface, a saddle is a good example of a negatively curved two-dimensional surface. Parallel lines drawn on a sphere always converge, but parallel lines drawn on a saddle always diverge. Mathematicians say that saddle-shaped surfaces are hyperbolic. Thus, in a negatively curved universe, we would describe space as **hyperbolic.** Such a universe is also called **open** because if you were to travel in a straight line in any direction, you would never return to your starting point.

Figure 28-15 illustrates the three cases of positive curvature, zero curvature, and negative curvature. Real space is three-dimensional, but we have drawn the three cases as analogous, more easily visualized two-dimensional surfaces. Therefore, as you examine the drawings in Figure 28-15, remember that the real universe has one more dimension. For example, if the universe is in fact hyperbolic, then the geometry of space must be the (difficult to visualize) three-dimensional analog of the two-dimensional surface of a saddle.

Note that in accordance with the cosmological principle, none of these models of the universe has an "edge" or a "center." This is clearly the case for both the flat and hyperbolic universes, because they are infinite and extend forever in all directions. A spherical universe is finite, but it also lacks a center and an edge. You could walk forever around the surface of a sphere (like the surface of the Earth) without finding a center or an edge.

The curvature of the universe is determined by the value of the combined mass density ρ_0. If ρ_0 is greater than a certain value ρ_c (say "rho sub cee"), the universe has positive curvature and is closed. If ρ_0 is less than ρ_c, the universe has negative curvature and is open. In the special case that ρ_0 is exactly equal to ρ_c, the universe is flat. Because of its crucial role in determining the geometry of the universe, ρ_c is called the **critical density.** It is given by the expression

Critical density of the universe

$$\rho_c = \frac{3H_0{}^2}{8\pi G}$$

ρ_c = critical density of the universe

H_0 = Hubble constant

G = universal constant of gravitation

Using a Hubble constant $H_0 = 71$ km/s/Mpc, we get

$$\rho_c = 9.5 \times 10^{-27} \text{ kg/m}^3$$

A sample of hydrogen gas with this density would contain just 5.7 hydrogen atoms per cubic meter.

Many astronomers prefer to characterize the combined average mass density of the universe in terms of the **density parameter** Ω_0 (say "omega sub zero"). This is just the ratio of the combined average mass density to the critical density:

Density parameter

$$\Omega_0 = \frac{\rho_0}{\rho_c}$$

Ω_0 = density parameter

ρ_0 = combined average mass density

ρ_c = critical density

An open universe has a density parameter Ω_0 between 0 and 1, and a closed universe has Ω_0 greater than 1. In a flat universe, Ω_0 is equal to 1. Thus, we can use the value of Ω_0 as a measure of the curvature of the universe (Table 28-1).

table 28-1	The Geometry and Average Density of the Universe				
Geometry of space	Curvature of space	Type of universe	Combined average mass density (ρ_0)	Density parameter (Ω_0)	
Spherical	positive	closed	$\rho_0 > \rho_c$	$\Omega_0 > 1$	
Flat	zero	flat	$\rho_0 = \rho_c$	$\Omega_0 = 1$	
Hyperbolic	negative	open	$\rho_0 < \rho_c$	$\Omega_0 < 1$	

How can we determine the curvature of space across the universe? In theory, if you drew an enormous triangle whose sides were each a billion light-years long (see Figure 28-15), you could determine the curvature of space by measuring the three angles of the triangle. If their sum equaled 180°, space would be flat. If the sum was greater than 180°, space would be spherical. And if the sum of the three angles was less than 180°, space would be hyperbolic. Unfortunately, this direct method for measuring the curvature of space is not practical.

A way to determine the curvature of the universe that is both practical and precise is to see if light rays bend toward or away from each other, as shown in Figure 28-15. The greater the distance a pair of light rays has traveled, and, hence, the longer the time the light has been in flight, the more pronounced any such bending should be. Therefore, astronomers test for the presence of such bending by examining the oldest radiation in the universe: the cosmic microwave background.

If the cosmic microwave background were truly isotropic, so that equal amounts of radiation reached us from all directions in the sky, it would be impossible to tell whether individual light rays have been bent. However, as we saw in Section 28-5, there are localized "hot spots" in the cosmic microwave background due to density variations in the early universe. The apparent size of these hot spots depends on the curvature of the universe (Figure 28-16). If the universe is closed, the bending of light rays from a hot spot will make the spot appear larger (Figure 28-16a); if the universe is open, the light rays will bend the other way and the hot spots will appear smaller (Figure 28-16c). Only in a flat universe will the light rays travel along straight lines, so that the hot spots appear with their true size (Figure 28-16b).

By calculating what conditions were like in the primordial fireball, astrophysicists find that in a flat universe, the dominant "hot spots" in the cosmic background radiation should have an angular size of about 1°. (In Chapter 29 we will learn how this is deduced.) This is just what the BOOMERANG and MAXIMA experiments observed, and what the WMAP observations have confirmed (see Figure 28-14). Hence, the curvature of the universe must be very close to zero, and the universe must either be flat or very nearly so.

As Table 28-1 shows, once we know the curvature of the universe, we can determine the density parameter Ω_0 and hence the combined average mass density ρ_0. By analyzing the data shown in Figure 28-14, astrophysicists find that $\Omega_0 = 1.0$ with an uncertainty of about 2%. In other words, ρ_0 is within 2% of the critical density ρ_c.

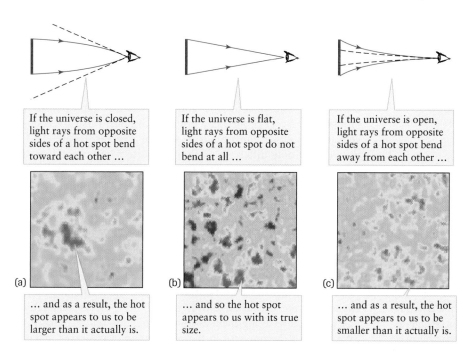

(a) If the universe is closed, light rays from opposite sides of a hot spot bend toward each other ... ... and as a result, the hot spot appears to us to be larger than it actually is.

(b) If the universe is flat, light rays from opposite sides of a hot spot do not bend at all ... ... and so the hot spot appears to us with its true size.

(c) If the universe is open, light rays from opposite sides of a hot spot bend away from each other ... ... and as a result, the hot spot appears to us to be smaller than it actually is.

figure 28-16   R I V U X G

The Cosmic Microwave Background and the Curvature of Space Temperature variations in the early universe appear as "hot spots" in the cosmic microwave background. The apparent size of these spots depends on the curvature of space. (The BOOMERANG Group, University of California, Santa Barbara)

The flatness of the universe poses a major dilemma. We saw in Section 28-5 that the average mass density of matter in the universe, ρ_m, is 2.6×10^{-27} kg/m³. This is only 0.27 of the critical density ρ_c. We can express this in terms of the **matter density parameter** Ω_m (say "omega sub em"), equal to the ratio of ρ_m to the critical density:

$$\Omega_m = \rho_m/\rho_0 = 0.27$$

If matter and radiation were all there is in the universe, the combined average mass density ρ_0 would be equal to ρ_m (plus a tiny contribution from radiation, which we can neglect because the average mass density of radiation is only about 1/5500 that of matter). Then the density parameter Ω_0 would be equal to Ω_m—that is, equal to 0.27—and the universe would be open. But the temperature variations in the cosmic microwave background clearly show that the universe is either flat or very nearly so. These variations also show that the density parameter Ω_0, which includes the effects of *all* kinds of matter and energy, is equal to 1.0. In other words, radiation and matter, including dark matter, together account for only 27% of the total density of the universe! The dilemma is this: What could account for the rest?

 The source of the missing density must be some form of energy that we cannot detect from its gravitational effects (the technique astronomers use to detect dark matter). It must also not emit detectable radiation of any kind. We refer to this mysterious energy as **dark energy.**

Just as we express the average density of matter and radiation by the matter density parameter Ω_m, we can express the average density of dark energy in terms of the **dark energy density parameter** Ω_Λ (say "omega sub lambda"). This is equal to the average mass density of dark energy, ρ_Λ, divided by the critical density ρ_c:

$$\Omega_\Lambda = \frac{\rho_\Lambda}{\rho_c}$$

We can determine the value of Ω_Λ by noting that the combined average mass density ρ_0 must be the sum of the average mass densities of matter, radiation, and dark energy. As we have seen, the contribution of radiation is so small that we can ignore it, so we have

$$\rho_0 = \rho_m + \rho_\Lambda$$

If we divide this through by the critical density ρ_c, we obtain

$$\Omega_0 = \Omega_m + \Omega_\Lambda$$

That is, the density parameter Ω_0 is the sum of the matter density parameter Ω_m and the dark energy density parameter Ω_Λ. Solving for Ω_Λ, we find

$$\Omega_\Lambda = \Omega_0 - \Omega_m$$

Since Ω_0 is close to 1.0 and Ω_m is 0.27, we conclude that Ω_Λ must be 1.0 − 0.27 = 0.73. Thus, whatever dark energy is, it accounts for 73% of the contents of the universe!

The concept of dark energy is actually due to Einstein. When he proposed the existence of a cosmological constant, he was suggesting that the universe is filled with a form of energy that by itself tends to make the universe expand (see Section 28-1). Unlike gravity, which tends to make objects attract, the energy associated with a cosmological constant would provide a form of "antigravity." Hence, it would not be detected in the same way as matter. (The subscript Λ in the symbol for the dark energy density parameter pays homage to the symbol that Einstein chose for the cosmological constant.)

If dark energy is in fact due to a cosmological constant, the value of this constant must be far larger than Einstein suggested. This is needed if we are to explain why Ω_Λ has a large value of 0.73. If Einstein felt he erred by introducing the idea of a cosmological constant, his error was giving it too small a value!

These ideas concerning dark energy are extraordinary, and extraordinary claims require extraordinary evidence to confirm them. As we will see in the next section, a crucial test is to examine how the rate of expansion of the universe has evolved over the eons.

28-7 | Observations of distant supernovae reveal that we live in an accelerating universe

We have seen that the universe is expanding. But does the rate of expansion stay the same? Because there is matter in the universe, and because gravity tends to pull the bits of matter in the universe toward one another, we would expect that the expansion should slow down with time. (In the same way, a cannonball shot upward from the surface of the Earth will slow down as it ascends because of the Earth's gravitational pull.) If there is a cosmological constant, however, its associated dark energy will exert an outward pressure that tends to accelerate the expansion.

To determine which of these effects is the more important, astronomers study the relationship between redshift and distance for extremely remote galaxies. We see these galaxies as they were billions of years ago. If the rate of expansion was the same in the distant past as it is now, the same Hubble law should apply to distant galaxies as to nearby ones. But if the rate of expansion has either increased or decreased, we will find important deviations from the Hubble law.

To see how astronomers approach this problem, first imagine two different parallel universes. Both Universe #1 and Universe #2 are expanding at a constant rate, so that there is a direct proportion between recessional velocity v and distance d as expressed by the Hubble law $v = H_0d$. Hence, a graph of distance versus recessional velocity for either universe is a straight line, as Figure 28-17a shows. The only difference is that Universe #1 is expanding at a slower rate than Universe #2. Hence, a galaxy at a certain distance from Earth in Universe #1 will have a slower recessional velocity than a galaxy at the same distance from Earth in Universe #2. As a result, the graph of distance versus recessional velocity for slowly expanding Universe #1 (shown in blue) has a steeper slope than the graph for rapidly expanding Universe #2 (shown in green). Keep this observation in mind: A slower expansion means a steeper slope on a graph of distance versus recessional velocity.

Now consider *our* universe and allow for the possibility that the expansion rate may change over time. If we observe very remote galaxies, we are seeing them as they were in the remote past. If the expansion of the universe in the remote past was slower or faster than it is now, the slope of the graph of distance versus recessional

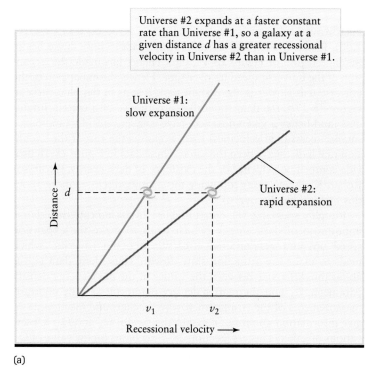

Universe #2 expands at a faster constant rate than Universe #1, so a galaxy at a given distance *d* has a greater recessional velocity in Universe #2 than in Universe #1.

Universe #1: slow expansion

Universe #2: rapid expansion

(a)

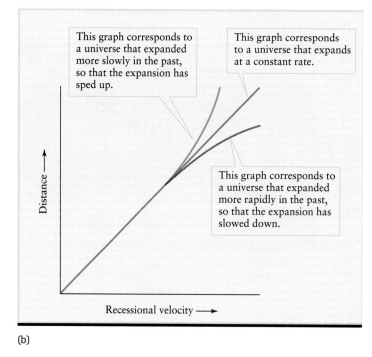

This graph corresponds to a universe that expanded more slowly in the past, so that the expansion has sped up.

This graph corresponds to a universe that expands at a constant rate.

This graph corresponds to a universe that expanded more rapidly in the past, so that the expansion has slowed down.

(b)

Figure 28-17

Varying Rates of Cosmic Expansion (a) Imagine two universes, #1 and #2. Each expands at its own constant rate. For a galaxy at a given distance, the recessional velocity will be greater in the more rapidly expanding universe. Hence, the graph of distance *d* versus recessional velocity *v* will have a shallower slope for the rapidly expanding universe and a steeper slope for the slowly expanding one. (b) If the rate of expansion of our universe was more rapid in the distant past, corresponding to remote distances, the graph of *d* versus *v* will have a shallower slope for large distances (green curve). If the expansion rate was slower in the distant past, the graph will have a steeper slope for large distances (blue curve).

velocity will be different for those remote galaxies. If the expansion was slower, then the slope will be steeper for distant galaxies (shown in blue in Figure 28-17b); if the expansion was faster, the slope will be shallower for distant galaxies (shown in green in Figure 28-17b). In either case, there will be a deviation from the straight-line Hubble law (shown in red in Figure 28-17b).

Figure 26-15 shows the observed relationship between distance and recessional velocity for galaxies. The data points appear to lie along a straight line, suggesting that the rate of cosmological expansion has not changed. (Figure 26-15 is actually a graph of recessional velocity versus distance, not the other way around. But a straight line on one kind of graph will be a straight line on the other, because in either case there is a direct proportion between the two quantities being graphed.) However, this graph was based on measurements of galaxies no farther than 400 Mpc (1.3 billion ly) from the Earth, which means we are looking only 1.3 billion years into the past. The straightness of the line in Figure 26-15 means only that the expansion of the universe has been relatively constant over the past 1.3 billion years—only 10% of the age of the universe, and a relatively brief interval on the cosmic scale.

Now suppose that you were to measure the redshifts and distances of galaxies *several* billion light-years from the Earth. The light from these galaxies has taken billions of years to arrive at your telescope, so your measurements will reveal how fast the universe was expanding billions of years ago. To do this, we need a technique that will allow us to find the distances to these very remote galaxies. We saw in Section 26-4 that one way to do this is to identify Type Ia supernovae in such galaxies. These supernovae are among the most luminous objects in the universe, and hence can be detected even at extremely large distances (see Figure 26-12). The maximum brightness of a supernova tells astronomers its distance through the inverse-square law for light, and the redshift of the supernova's spectrum tells them its recessional velocity. As an example, the image that opens this chapter shows a Type Ia supernova with a redshift $z = 0.95$, corresponding to a recessional velocity of 58% of the speed of light. It is at a distance of 2400 Mpc, or 8 billion ly. We see this supernova as it was when the universe was less than half of its present age.

In 1998, two large research groups—the Supernova Cosmology Project, led by Saul Perlmutter of Lawrence Berkeley National Laboratory, and the High-Z Supernova Search Team, led by Brian Schmidt of the Mount Stromlo and Siding Springs Observatories in Australia—reported their results from a survey of Type Ia supernovae in galaxies at redshifts of 0.2 or greater, corresponding to distances beyond 750 Mpc (2.4 billion ly). **Figure 28-18** shows some of their data, along with more recent observations, on a graph of apparent magnitude versus redshift. A greater apparent magnitude corresponds to a dimmer supernova (see Section 19-3), which means that the supernova

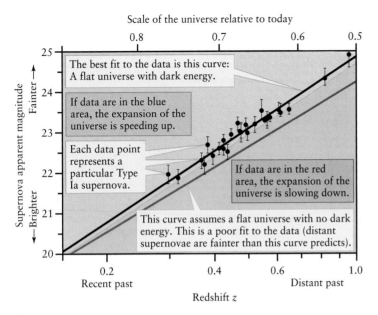

Scale of the universe relative to today

figure 28-18

The Hubble Diagram for Distant Supernovae This graph shows apparent magnitude versus redshift for supernovae in distant galaxies. The greater the apparent magnitude, the dimmer the supernova and the greater the distance to it and its host galaxy. If the expansion of the universe is speeding up, the data will lie in the blue area; if it is slowing down, the data will lie in the red area. The data show that the expansion is in fact speeding up. (The Supernova Cosmology Project/R. A. Knop et al.)

is more distant. A greater redshift implies a greater recessional velocity. Hence, this graph is basically the same as those in Figure 28-17.

To interpret these results, we need guidance from **relativistic cosmology**. This is a theoretical description of the universe and its expansion, based on Einstein's general theory of relativity and developed in the 1920s by Alexander Friedmann in Russia, Georges Lemaître in France, and Willem de Sitter in the Netherlands. Given values of the mass density parameter Ω_m and the dark energy density parameter Ω_Λ, cosmologists can use the equations of relativistic cosmology to predict how the expansion rate of the universe should change over time. Such predictions can be expressed as curves on a graph of distance versus redshift such as Figure 28-18.

The lower, gray curve in Figure 28-18 shows what would be expected in a flat universe with $\Omega_0 = 1.00$ but with no dark energy, so that $\Omega_\Lambda = 0$ and $\Omega_m = \Omega_0$ (that is, the density parameter is due to matter and radiation alone). In this model, and in fact in any model whose curve lies in the red area in Figure 28-18, the absence of dark energy means that gravitational attraction between galaxies would cause the expansion of the universe to slow down with time. Hence, the expansion rate would have been greater in the past (compare with the green curve in Figure 28-17b).

In fact, the data points in Figure 28-18 are almost all in the *blue* region of the graph, and agree very well with the curve shown in black. This curve also assumes a flat universe, but with an amount of dark energy consistent with the results from the cosmic

microwave background ($\Omega_m = 0.23$, $\Omega_\Lambda = 0.73$, $\Omega_0 = \Omega_m + \Omega_\Lambda = 1.00$). In this model, and indeed in any model whose curve lies in the blue region of Figure 28-18, dark energy has made the expansion of the universe speed up over time. Hence, the expansion of the universe was slower in the distant past. Just as in the blue curve in Figure 28-17b, the data show that supernovae of a certain brightness (and hence a given distance) have smaller redshifts (and hence smaller recessional velocities) than would be the case if the expansion rate had always been the same. These data in Figure 28-18 provide compelling evidence of the existence of dark energy. (Professor Robert Kirshner's essay following this chapter discusses more about this remarkable discovery.)

Roughly speaking, the data in Figure 28-18 indicate the relative importance of dark energy (which tends to make the expansion speed up) and gravitational attraction between galaxies (which tends to make the attraction slow down). Thus, these data tell us about the *difference* between the values of the dark energy density parameter Ω_Λ and the matter density parameter Ω_m. By contrast, measurements of the cosmic microwave background (Section 28-6) give information about Ω_0, equal to the *sum* of Ω_Λ and Ω_m. Observations of galaxy clusters (Section 28-5) set limits on the value of Ω_m by itself. By combining these three very different kinds of observations as shown in **Figure 28-19**, we can set more stringent limits on both Ω_Λ and Ω_m.

Taken together and combined with other observations, all these data suggest the following values.

$$\Omega_m = 0.27 \pm 0.04$$

$$\Omega_\Lambda = 0.73 \pm 0.04$$

$$\Omega_0 = \Omega_m + \Omega_\Lambda = 1.02 \pm 0.02$$

In each case, the number after the ± sign is the uncertainty in the value.

This collection of numbers points to a radically different model of the universe from what was suspected just a few years ago. In the 1980s there was no compelling evidence for an accelerating expansion of the universe, so it was widely assumed that $\Omega_\Lambda = 0$. Evidence from distant galaxies suggested a flat universe, so it was presumed that $\Omega_m = 1$. Figure 28-19 shows that modern data rule out this model.

The model we are left with is one in which the universe is suffused with a curious dark energy due to a cosmological constant. Unlike matter or radiation, whose average densities decrease as the universe expands and thins out, the average density of this dark energy remains constant throughout the history of the universe (**Figure 28-20**). The dark energy was relatively unimportant over most of the early history of the universe. Today, however, the density of dark energy is greater than that of matter (Ω_Λ is greater than Ω_m). In other words, we live in a **dark-energy-dominated universe.**

As Figure 28-20 shows, the dominance of dark energy is a relatively recent development in the history of the universe. Prior to about 5 billion years ago, the density of matter should have been greater than that of dark energy. Hence, we would expect that up until about 5 billion years ago, the expansion of the universe

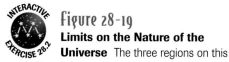

figure 28-19

Limits on the Nature of the Universe The three regions on this graph show values of the mass density parameter Ω_m and the dark energy density parameter Ω_Λ that are consistent with various types of observations. Galaxy cluster measurements (in brown) set limits on Ω_m. Observations of the cosmic microwave background (in green) set limits on the sum of Ω_m and Ω_Λ: A larger value of Ω_m (to the right in the graph) implies a smaller value of Ω_Λ (downward in the graph) to keep the sum the same, which is why this band slopes downward. Observations of Type Ia supernovae (in blue) set limits on the difference between Ω_m and Ω_Λ; this band slopes upward since a larger value of Ω_m implies a larger value of Ω_Λ to keep the difference the same. The best agreement to all these observations is where all three regions overlap (the red dot). (The Supernova Cosmology Project/R. A. Knop et al.)

should have been slowing down rather than speeding up. Recently, astronomers have found evidence of this picture by using the Hubble Space Telescope to observe extremely distant Type Ia supernovae with redshifts z close to 1. (The image that opens this chapter shows one of these supernovae.) When astronomers compare the distance to these supernovae (determined from their brightness) to their redshifts, they find that the redshifts are *greater* than would be the case if the expansion of the universe had always been at the same rate or had always been speeding up (see Figure 28-17). This is just what would be expected if the expansion was slowing down in the very early universe. After about 5 billion years ago, the effects of dark energy became dominant and the expansion began to speed up.

figure 28-20

The Evolution of Density, Revisited The average mass density of matter, ρ_m, and the average mass density of radiation, ρ_{rad}, both decrease as the universe expands and becomes more tenuous. But if the dark energy is due to a cosmological constant, its average mass density ρ_Λ remains constant. In this model, our universe became dominated by dark energy about 5 billion years ago.

If you see a red light up ahead while driving, you would probably apply the brakes to make the car slow down. But if the light then turns green before your car comes to a stop, and the road ahead is clear, you would step on the gas to make the car speed up again. The expansion of the universe has had a similar history. The mutual gravitational attraction of all the matter in the universe means that "the brakes were on" for about the first 9 billion years after the Big Bang, so that the expansion slowed down. But for about the past 5 billion years, dark energy has "had its foot on the gas," and the expansion has been speeding up.

If dark energy truly is a cosmological constant, the density of dark energy will continue to remain constant, as shown in Figure 28-20. Due to the effects of this dark energy, the universe will keep on expanding forever, and the rate of expansion will continue to accelerate. Eventually, some 30 billion years from now, the universe will have expanded so much that only a thousand or so of the nearest galaxies will still be visible. The billions of other galaxies that we can observe today will have moved so far away from us that their light will have faded to invisibility. Furthermore, they will be moving away from us so rapidly that what light we do receive from them will have been redshifted out of the visible range.

There may be other explanations for dark energy besides a cosmological constant, however. Several physicists have proposed a type of dark energy whose density decreases slowly as the universe expands. Depending on how the density of dark energy evolves over time, the universe could continue to expand or could eventually recollapse on itself. Future observations, including space-based measurements of both the cosmic background radiation and of Type Ia supernovae, should help resolve the nature of the mysterious dark energy.

28-8 | Primordial sound waves help reveal the character of the universe

We have seen how studying the "hot spots" in the cosmic background radiation reveals that we live in a flat universe. In fact, temperature variations reveal more: They give us a window on conditions in the early universe, and actually help us pin down the values of other important quantities such as the Hubble constant and the density of matter in the universe. The key to extracting this additional information from the cosmic background radiation is recognizing that the hot and cold spots in a map such as Figure 28-14 are actually sound waves.

Sound waves can travel in fluids of all kinds. Sound waves in air are used in human speech and hearing, while whales communicate underwater using high-frequency underwater clicks and whistles. If you could take a snapshot of a sound wave, you would see that at any moment there are some regions, called **compressions**, where the fluid is squeezed together, and other regions, called **rarefactions**, where the fluid is thinned out or rarefied. (When a sound wave enters a human ear, the air next to the eardrum is alternately compressed and rarefied, which makes the eardrum flex back and forth. The ear detects this flexing and translates it into an electrical signal that is sent to the brain.) Even a seemingly quiet fluid, such as the still air inside a room, has sound waves in it that are triggered by random motions of the fluid. Such random sounds in still air are too faint to be detected by your ear. This is a good thing; it would be very annoying to hear a continuous background noise from the air!

 There would also have been sound waves in the early universe, which was filled with a fluid composed primarily of photons, electrons, and protons, with a density more than 10^9 times greater than that of our present-day universe. Just as water molecules in a glass of water collide with each other, photons and particles collided frequently with each other in this primordial fluid, triggering random sound waves with compressions and rarefactions.

Because there was more mass in a compression than in a rarefaction, photons emerging from a compression experienced a greater gravitational redshift than did photons emerging from a rarefaction (see Section 24-2). As a result, while the light we see from either a compression or a rarefaction has a blackbody spectrum, the light from a compression is shifted to slightly longer wavelengths. We saw in Sections 5-3 and 5-4 that a blackbody spectrum dominated by longer wavelengths corresponds to a lower blackbody temperature (see Figure 5-10). Hence, we see compressions as the cold spots in Figure 28-14, and we see rarefractions as the hot spots. The overall pattern of cold and hot spots is thus a record of the sound waves that were present just as the universe became transparent, some 380,000 years after the Big Bang.

The nature of a sound wave depends on the material through which it passes. For example, sound waves travel faster in helium than they do in air (because helium is less dense) and faster still in water (which, while denser than air, is much more resistant to compression). So by studying the primordial sound waves recorded in Figure 28-14, we can learn about the properties of the fluid that made up the early universe. These properties include the average densities of matter and dark energy in the fluid, as well as the value of the Hubble constant (which helps determine how rapidly the fluid was expanding and thinning out as the universe expanded). We can also determine the age of the universe at the time that the cosmic background radiation was emitted, since this determines the maximum size to which a hot spot (rarefaction) or cold spot (compression) could have grown since the Big Bang.

Figure 28-21 shows an important way in which astronomers systematize their data about hot and cold spots in the cosmic background radiation. This graph shows the number of observed hot or cold spots of different angular sizes, with larger spots on the left and smaller spots on the right. The presence of peaks in the graph shows that spots of certain sizes are more common than others. Different cosmological models predict different shapes for the curve shown in Figure 28-21. Astronomers determine the best model by seeing which one gives a curve that best fits the data points. Table 28-2 summarizes the results of a model that yields the particular curve shown in Figure 28-21.

Even more information can be obtained from the *polarization* of light in cosmic microwave background. Ordinary light from a lightbulb or from the Sun is *unpolarized*, which means that the electric fields of the light waves are oriented in random directions. But when light collides with and bounces off an object, it tends to

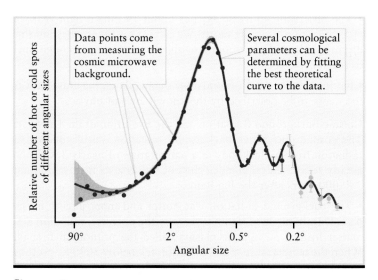

Figure 28-21

Sound Waves in the Early Universe Observations of the cosmic background radiation show that hot and cold spots of certain angular sizes are more common than others. A model that describes these observations helps to constrain the values of important cosmological parameters. Most of the data shown here is from the Wilkinson Microwave Anisotropy Probe; the data for the smallest angles (at the right of the graph) come from the CBI detector in the Chilean Andes and the ACBAR detector at the South Pole. (NASA/WMAP Science Team)

become *polarized,* with its electric fields oriented in a specific direction. (As an example, the electric fields in sunlight reflected from the ground are mostly oriented horizontally. Polarizing sunglasses work by screening out electric fields of this orientation only, which helps eliminate distracting reflections.) In a similar way, the cosmic background radiation acquires a polarization when it scatters from material in a large hot spot. The amount of polarization turns out to be a very sensitive probe of conditions in the early universe, and, hence, of the nature of the universe itself.

In 2002 astronomers using the DASI microwave telescope at the South Pole reported the first detection of polarized light in the cosmic background radiation. Such measurements are very challenging to do. But the rewards for cosmology are very great, which is why researchers are devoting great effort to further studies of the polarization of the background radiation. (Professor John Ruhl's essay following Chapter 29 describes more about polarization measurements.)

Our understanding of the universe as a whole has increased tremendously over the past several years. We have found compelling evidence that dark energy exists, and that it is the dominant form of energy in the universe. Studies of supernovae, galaxy clusters, and the cosmic background radiation have provided us with so much high-quality data that we can now express the key parameters of the universe (Table 28-2) with very high accuracy. When we look back to the situation in the 1980s, when the value of the Hubble constant was uncertain by at least 50%, it is no exaggeration to say that we have entered an age of *precision* cosmology.

table 28-2 | Some Key Properties of the Universe

Quantity	Significance	Value*
Hubble constant, H_0	Present-day expansion rate of the universe	$71\,^{+4}_{-3}$ km/s/Mpc
Density parameter, Ω_0	Combined mass density of all forms of matter *and* energy in the universe, divided by the critical density	1.02 ± 0.02
Matter density parameter, Ω_m	Combined mass density of all forms of matter in the universe, divided by the critical density	0.27 ± 0.04
Density parameter for ordinary matter, Ω_b	Mass density of ordinary atomic matter in the universe, divided by the critical density	0.044 ± 0.004
Dark energy density parameter, Ω_Λ	Mass density of dark energy in the universe, divided by the critical density	0.73 ± 0.04
Age of the universe, T_0	Elapsed time from the Big Bang to the present day	$(1.37 \pm 0.02) \times 10^{10}$ years
Age of the universe at the time of recombination	Elapsed time from the Big Bang to when the universe became transparent, releasing the cosmic background radiation	$(3.79\,^{+0.08}_{-0.07}) \times 10^5$ years
Redshift z at the time of recombination	Since the cosmic background radiation was released, the universe has expanded by a factor $1 + z$	1089 ± 1

*Values are from the first year of WMAP data. (NASA/WMAP Science Team)

Yet many questions remain unanswered. What is the nature of dark matter? What actually is dark energy? Can these mysterious entities be detected and studied in the laboratory? These and other questions will continue to occupy cosmologists for many years to come.

Key Words

average density of matter, p. 642

Big Bang, p. 638

closed universe, p. 646

combined average mass density, p. 645

compression, p. 652

cosmic background radiation, p. 640

cosmic microwave background, p. 640

cosmic light horizon, p. 638

cosmic singularity, p. 639

cosmological constant, p. 634

cosmological principle, p. 637

cosmological redshift, p. 636

cosmology, p. 634

critical density, p. 646

dark energy, p. 648

dark energy density parameter, p. 648

dark-energy-dominated universe, p. 650

density parameter, p. 646

era of recombination, p. 643

flat space, p. 645

homogeneous, p. 637

hyperbolic space, p. 646

isotropic, p. 637

lookback time, p. 637

mass density of radiation, p. 642

matter density parameter, p. 648

matter-dominated universe, p. 642

negative curvature, p. 646

observable universe, p. 638

Olbers's paradox, p. 634

open universe, p. 646

Planck time, p. 639

plasma, p. 644

positive curvature, p. 646

primordial fireball, p. 644

radiation-dominated universe, p. 642

rarefaction, p. 652

relativistic cosmology, p. 650

spherical space, p. 646

zero curvature, p. 645

Key Ideas

The Expansion of the Universe: The Hubble law describes the continuing expansion of space. The redshifts that we see from distant galaxies are caused by this expansion, not by the motions of galaxies through space.

• The redshift of a distant galaxy is a measure of the scale of the universe at the time the galaxy emitted its light.

• It is meaningless to speak of an edge or center to the universe or of what lies beyond the universe.

The Cosmological Principle: Cosmological theories are based on the idea that on large scales, the universe looks the same at all locations and in every direction.

The Big Bang: The universe began as an infinitely dense cosmic singularity which began its expansion in the event called the Big Bang, which can be described as the beginning of time.

• The observable universe extends about 14 billion light-years in every direction from the Earth. We cannot see objects beyond this distance because light from these objects has not had enough time to reach us.

• During the first 10^{-43} second after the Big Bang, the universe was too dense to be described by the known laws of physics.

Cosmic Background Radiation and the Evolution of the Universe: The cosmic microwave background radiation, corresponding to radiation from a blackbody at a temperature of nearly 3 K, is the greatly redshifted remnant of the hot universe as it existed about 380,000 years after the Big Bang.

• The background radiation was hotter and more intense in the past. During the first 380,000 years of the universe, radiation and matter formed an opaque plasma called the primordial fireball. When the temperature of the radiation fell below 3000 K, protons and electrons could combine to form hydrogen atoms and the universe became transparent.

• The abundance of helium in the universe is explained by the high temperatures in its early history.

The Geometry of the Universe: The curvature of the universe as a whole depends on how the combined average mass density ρ_0 compares to a critical density ρ_c.

• If ρ_0 is greater than ρ_c, the density parameter Ω_0 has a value greater than 1, the universe is closed, and space is spherical (with positive curvature).

• If ρ_0 is less than ρ_c, the density parameter Ω_0 has a value less than 1, the universe is open, and space is hyperbolic (with negative curvature).

• If ρ_0 is equal to ρ_c, the density parameter Ω_0 is equal to 1 and space is flat (with zero curvature).

Cosmological Parameters and Dark Energy: Observations of temperature variations in the cosmic microwave background indicate that the universe is flat or nearly so, with a combined average mass density equal to the critical density. Observations of galaxy clusters suggest that the average density of matter in the universe is about 0.27 of the critical density. The remaining contribution to the average density is called dark energy.

• Measurements of Type Ia supernovae in distant galaxies show that the expansion of the universe is speeding up. This may be due to the presence of dark energy in the form of a cosmological constant, which provides a pressure that pushes the universe outward.

Cosmological Parameters and Primordial Sound Waves: Temperature variations in the cosmic background radiation are a record of sound waves in the early universe. Studying the character of these sound waves, and the polarization of the background radiation that they produce, helps constrain models of the universe.

Review Questions

1. Why did Isaac Newton conclude that the universe was static? Was he correct?

2. What is Olbers's paradox? How can it be resolved?

3. What is a cosmological constant? Why did Einstein introduce it into cosmology?

4. What does it mean when astronomers say that we live in an expanding universe? What is actually expanding?

5. Would it be correct to say that due to the expansion of the universe, the Earth is larger today than it was 4.56 billion years ago? Why or why not?

6. Using a diagram, explain why the expansion of the universe as seen from a distant galaxy would look the same as seen from our Galaxy.

7. How does modern cosmology preclude the possibility of either a center or an edge to the universe?

8. Explain the difference between a Doppler shift and a cosmological redshift.

9. Explain how redshift can be used as a measure of lookback time. In what ways is it superior to time measured in years?

10. By what factor has the universe expanded since $z = 1$? Explain your reasoning.

11. What does it mean to say that the universe is homogeneous? That it is isotropic?

12. What is the cosmological principle? How is it justified?

13. How was the Big Bang different from an ordinary explosion? Where in the universe did it occur?

14. Some people refer to the Hubble constant as "the Hubble variable." In what sense is this justified?

15. What is meant by "the observable universe"?

16. How did the abundance of helium in the universe suggest the existence of the cosmic background radiation?

17. Can you see the cosmic background radiation with the naked eye? With a visible-light telescope? Explain why or why not.

18. If the universe continues to expand forever, what will eventually become of the cosmic background radiation?

19. How can astronomers measure the average mass density of the universe?

20. What does it mean to say that the universe was once radiation-dominated? What happened when the universe changed from being radiation-dominated to being matter-dominated? When did this happen?

21. What was the era of recombination? What significant events occurred in the universe during this era? Was the universe matter-dominated or radiation-dominated during this era?

22. Describe two different ways in which the cosmic microwave background is not isotropic.

23. What is meant by the critical density of the universe? Why is this quantity important to cosmologists?

24. Explain why the universe does not have a center.

25. Describe how astronomers use the cosmic background radiation to determine the geometry of the universe.

26. Explain why it is important to measure how the expansion rate of the universe has changed over time. How is this rate measured?

27. What is dark energy? Describe two ways that we can infer its presence.

28. What does it mean to say that the universe is dark-energy-dominated? What happened when the universe changed from being matter-dominated to being dark-energy-dominated?

29. How can we detect the presence of sound waves in the early universe? What do these sound waves tell us?

Advanced Questions

> **Problem-solving tips and tools**
>
> We discussed Wien's law in Section 5-4. You may find it useful to recall that 1 parsec equals 3.26 light-years, that 1 Mpc equals 3.09×10^{19} km, and that a year contains 3.16×10^7 seconds.

30. (a) For what value of the redshift z were representative distances between galaxies only 20% as large as they are now? (b) Compared to representative distances between galaxies in the present-day universe, how large were such distances at $z = 8$? Compared to the density of matter in the present-day universe, what was the density of matter at $z = 8$? (c) If dark energy is in the form of a cosmological constant, how does its present-day density compare to the density of dark energy at $z = 2$? At $z = 5$? Explain your answers.

31. The galaxy RD1 has a redshift $z = 5.34$. The light from this galaxy includes the Lyman-alpha (L_α) spectral line of hydrogen, with an unshifted wavelength of 121.6 nm. Calculate the wavelength at which we detect the Lyman-alpha photons from RD1. In what part of the electromagnetic spectrum does this wavelength lie?

32. Estimate the age of the universe for a Hubble constant of (a) 50 km/s/Mpc, (b) 75 km/s/Mpc, and (c) 100 km/s/Mpc. On the basis of your answers, explain how the ages of globular clusters could be used to place a limit on the maximum value of the Hubble constant.

33. Some so-called "creation scientists" claim that the universe came into being about 6000 years ago. Find the value of the Hubble constant for such a universe. Is this a reasonable value for H_0? Explain.

34. The quasar HS 1946+7658 has a redshift $z = 3.02$. At the time when the light we see from HS 1946+7658 left the quasar, how many times more dense was the matter in the universe than it is today?

35. Use Wien's law (Section 5-4) to calculate the wavelength at which the cosmic microwave background ($T = 2.725$ K) is most intense.

36. If the mass density of radiation in the universe were 625 times larger than it is now, what would the background temperature be?

37. Suppose that the present-day temperature of the cosmic background radiation were somehow increased by a factor of 100, from 2.725 K to 272.5 K. (**a**) Calculate ρ_{rad} in this situation. (**b**) If the average density of matter (ρ_m) remained unchanged, would it be more accurate to describe our universe as matter-dominated or radiation-dominated? Explain.

38. Calculate the mass density of radiation (ρ_{rad}) in each of the following situations, and explain whether each situation is matter-dominated or radiation-dominated: (**a**) the photosphere of the Sun ($T = 5800$ K, $\rho_m = 3 \times 10^{-4}$ kg/m³); (**b**) the center of the Sun ($T = 1.55 \times 10^7$ K, $\rho_m = 1.6 \times 10^5$ kg/m³); (**c**) the solar corona ($T = 2 \times 10^6$ K, $\rho_m = 5 \times 10^{-13}$ kg/m³).

39. What would be the critical density of matter in the universe (ρ_c) if the value of the Hubble constant were (**a**) 50 km/s/Mpc? (**b**) 100 km/s/Mpc?

40. Consider the galaxy RD1 (see Advanced Question 31), which has $z = 5.34$. (**a**) Suppose that in the present-day universe, two clusters of galaxies are 500 Mpc apart. At the time that the light was emitted from RD1 to produce an image tonight, how far apart were those two clusters? (**b**) What was the average density of matter (ρ_m) at that time? Assume that in today's universe, $\rho_m = 2.6 \times 10^{-27}$ kg/m³. (**c**) What were the temperatures of the cosmic background radiation and the mass density of radiation (ρ_{rad}) at that time? (**d**) At this time in the remote past, was the universe matter-dominated, radiation-dominated, or dark-energy-dominated? Explain.

41. Whether the expansion of the universe is speeding up or slowing down can be expressed in term of a quantity called the *deceleration parameter,* denoted by q_0. The expansion is slowing down if q_0 is positive and speeding up if q_0 is negative; if $q_0 = 0$, the expansion proceeds at a constant rate. If we assume that the dark energy is due to a cosmological constant, the deceleration parameter can be calculated using the formula

$$q_0 = \frac{1}{2}\Omega_0 - \frac{3}{2}\Omega_\Lambda$$

(Recall that the density parameter Ω_0 is equal to $\Omega_m + \Omega_\Lambda$.) (**a**) Show that if there is no cosmological constant, the expansion of the universe must slow down. (**b**) Using the values of Ω_m and Ω_Λ given in Table 28-2, find the value of the deceleration parameter for our universe. Based on this, is the expansion of the universe speeding up or slowing down? (**c**) Imagine a universe that has the same value of Ω_m as our universe but in which the expansion of the universe is neither speeding up nor slowing down. What would be the value of Ω_Λ in such a universe? Which would be dominant in such a universe, matter or dark energy? Explain.

42. In general, the deceleration parameter (see Advanced Question 41) is not constant but varies with time. For a flat universe, the deceleration parameter at a redshift z is given by the formula

$$q_z = \frac{1}{2} - \frac{3}{2}\left[\frac{\Omega_\Lambda}{\Omega_\Lambda + (1 - \Omega_\Lambda)(1 + z)^3}\right]$$

where Ω_Λ is the dark energy density parameter. Using the value of Ω_Λ given in Table 28-2, find the value of q_z for (**a**) $z = 0.5$ and (**b**) $z = 1.0$. (**c**) Explain how your results show that the expansion of the universe was actually decelerating at $z = 1.0$, but changed from deceleration to acceleration between $z = 1.0$ and $z = 0.5$.

43. The dark energy density parameter Ω_Λ is related to the value of the cosmological constant Λ by the formula

$$\Omega_\Lambda = \frac{\Lambda c^2}{3H_0^2}$$

where $c = 3.00 \times 10^8$ m/s is the speed of light. Determine the value of Λ if Ω_Λ and H_0 have the values given in Table 28-2. (*Hint:* You will need to convert units to eliminate kilometers and megaparsecs.)

Discussion Questions

44. Suppose we were living in a radiation-dominated universe. Discuss how such a universe would be different from what we now observe.

45. How can astronomers be certain that the cosmic microwave background fills the entire cosmos, not just the vicinity of the Earth?

46. Do you think there can be "other universes," regions of space and time that are not connected to our universe? Should astronomers be concerned with such possibilities? Why or why not?

Web/CD-ROM Questions

47. Before the discovery of the cosmic microwave background, it seemed possible that we might be living in a "steady-state universe" with overall properties that do not change with time. The steady-state model, like the Big Bang model, assumes an expanding universe, but does not assume a "creation event." Instead, matter is assumed to be created continuously everywhere in space to ensure that the average density of the universe remains constant. Search the World Wide Web for information about the "steady-state" theory. Explain why the existence of the cosmic microwave background was a fatal blow to the steady-state theory.

48. Search the World Wide Web for information on a European Space Agency mission called Planck. In what ways will Planck be an improvement over the WMAP mission? What do scientists hope to learn from Planck? When is it scheduled to be launched?

49. Temperatures in the Early Universe. Access the Active Integrated Media Module "Blackbody Curves" in Chapter 28 of the *Universe* Web site or CD-ROM. (**a**) Use the module to determine by trial and error the temperature at which a blackbody spectrum has its peak at a wavelength of 1 μm. (**b**) At the time when the temperature of the cosmic background radiation was equal to the value you found in (a), was the universe matter-dominated or radiation-dominated? Explain your answer.

Observing Projects

50. Use the *Starry Night Backyard*™ program to determine how the solar system moves through the cosmic microwave background. First display the entire celestial sphere (select **Guides > Atlas** in the **Go** menu). Use the **Find...** command in the **Edit** menu to center the field of view on the constellation Leo. (To make the constellation itself visible, click on the "View Options" tab at the far left of the main window. Under "Constellations," click on "Boundaries" and "Labels." Then click again on the "View Options" tab so that the view of the sky again fills the screen.) (**a**) Draw a sketch showing the Sun, the plane in which the Earth orbits the Sun, and the direction in which the solar system moves through the cosmic microwave background. (**b**) Use the time controls in the Control Panel to step through the months of the year. In which month is the Sun placed most nearly in front of the Earth as the solar system travels through the cosmic background radiation?

51. Use the *Deep Space Explorer*™ program to examine clusters of galaxies. First click on the "Home" button at the upper left of the *Deep Space Explorer*™ window. Start by clicking on **Home.** Now, click and hold the "zoom out" (upward-pointing triangle) near the "Spaceship" button until the image on the screen stops with a "Big Bang" essay in the upper-left corner of the screen. Keep in mind that the Milky Way is located in the center of the sphere you see. Does that mean that we are at the center of the universe? Why or why not? Read the essay. What does the blue haze inside the outermost spherical shell physically repre-

sent? What is and is not occurring in the region between the outermost shell and the orange shell inside it? Now "zoom in" until the "CMB" essay is visible in the upper left corner. Read the essay and state what the orange-yellow represents. Now "zoom in" until the "young galaxies" essay is visible. Read it and write down what the wavy blue lines represent. Move in until the orange-yellow blobs are visible. What do these represent?

Collaborative Exercises

52. As a group, create a four- to six-panel cartoon strip showing a discussion between two individuals describing why the sky is dark at night.

53. Imagine your firm, Creative Cosmologists Coalition, has been hired to create a three-panel, folded brochure describing the principal observations that astronomers use to infer the existence of a Big Bang. Create this brochure on an 8½ × 11 piece of paper. Be sure each member of your group supervises the development of a different portion of the brochure and that the small print acknowledges who in your group was primarily responsible for which portion.

54. The three potential geometries of the universe are shown in Figure 28-15. To demonstrate this, ask one member of your group to hold a piece of paper in one of the positions while another member draws two parallel lines that never change in one geometry, eventually cross in another geometry, and eventually diverge in another.

ROBERT P. KIRSHNER

The Extravagant Universe

Robert Kirshner is Clowes Professor of Science at the Harvard-Smithsonian Center for Astrophysics in Cambridge, Massachusetts. Born in Sudbury, Massachusetts, in 1949, he obtained his A.B. degree from Harvard College in 1970 and his Ph.D from the California Institute of Technology in 1975. Kirshner was elected to the National Academy of Sciences in 1998 and elected president of the American Astronomical Society in 2003. At Harvard, he teaches a large course for students who are not science concentrators, entitled "Matter in the Universe." He is also master of Quincy House, one of Harvard's undergraduate residences. In 2003, Princeton University Press published Kirshner's popular-level book *The Extravagant Universe: Exploding Stars, Dark Energy, and the Accelerating Cosmos.*

Exploding stars halfway across the observable universe reveal a surprising fact. Judging the distances to distant supernovae from their apparent brightness, the rate of cosmic expansion has been speeding up in the last 5 billion years. While gravitation acts to slow cosmic expansion, these observations require something else to make the universe accelerate. We call this the "dark energy," though, in truth, we do not know what it is. Perhaps it is the modern version of Einstein's notorious cosmological constant.

This result was a big surprise to people working on the problem. Early in 1998, I wrote a e-mail to the members of our High-Z Supernova Team, saying, "In your heart you know this is wrong, though your head tells you [that] you don't care and you're just reporting the observations." One reason for our reluctance was a contrary result. The Supernova Cosmology Project, based at the Lawrence Berkeley Lab, had published a paper in the *Astrophysical Journal* in July 1997 claiming that distant supernovae showed the universe was slowing down. Our methods weren't that different, but our results were pointing the other way.

Another reason to be wary was Einstein's bad experience with this idea, invented to make a static universe. Did we think we were smarter than Einstein? Einstein never liked the cosmological constant, as he wrote, "I am unable to believe that such an ugly thing should be realized in nature." It became a kind of theoretical poison ivy, touched only by the unwary for about 65 years. But the data were leading us to reconsider. As my fellow team member Adam Riess wrote to the rest of the team, "Approach these results not with your heart or your head, but with your eyes. We are observers, after all!"

Technology is a lot better now than it was a century ago, when Henrietta Leavitt made her studies of Cepheid variables (see Section 25-1). While she painstakingly examined photographic plates under a microscope to find variable stars, we use fast computers to scan digital images of the sky and pick out the objects that change. We've developed a pipeline system that pops out the supernova candidates about an hour after we take the images. This is important because supernovae are like fish—after about 3 days, they begin to lose their freshness. If you want to see the peak of the light curve prompt action is essential.

In 1998, we saw the signature of cosmic acceleration in supernovae at redshift 0.5: light that had been en route for about 5 billion years. Now we're using the Hubble Space Telescope to search for even fainter and more distant supernovae at redshift 1.5, roughly 9 billion light-years away, when we expect deceleration from dense dark matter. Technically, a change in velocity is called acceleration, and a change in acceleration is called (to the unending amusement of physics and engineering students) "jerk." A preliminary report on the highest-redshift supernovae ran in the *New York Times* recently. Over a not-too-flattering picture of Adam Riess, a headline writer wrote, "A 'cosmic jerk' that reversed the universe."

We would like to understand better the nature of the dark energy. Is it really Einstein's cosmological constant, retrieved from the dumpster of history, smoothed out and made new again? Or is it some more general "quintessence" whose energy density changes over time? Better measurements will show whether the dark energy comes from a source that is constant, or one that changes subtly as the universe expands. Either way will be very interesting.

Faint light from distant stellar catastrophes traces the history of cosmic expansion. It is not what we expected to see. The universe contains more parts than the simplest universe we could imagine: atoms that glow, atoms that don't, neutrinos with mass, and another dark matter particle with more mass, something that made the universe expand exponentially in the era of inflation, and something more that is now making the universe accelerate. Perhaps some day in the future all of this will seem essential, but at the moment, it seems we live in a recklessly extravagant universe, with extra parts whose function we do not yet fathom.

When people ask what science is for, the answer is often framed in terms of economic development, or national security, or improved life span. Nobody is against being rich, or safe, or healthy, but there is something deeper at work when we study how the universe began and where we are going.

29 Exploring the Early Universe

In Chapter 28 we traced the history of our expanding universe back almost to its very birth in the Big Bang. Unfortunately, the early universe was so opaque that we cannot observe it directly. Can we nonetheless reconstruct some of the events of those first 380,000 years? The answer is yes, because many aspects of today's universe are relics of the earliest events in the cosmos.

Immediately after the Big Bang, the universe was so dense and hot that particles were constantly colliding at nearly the speed of light. These violent collisions shattered particles down to the most fundamental building blocks of matter. Physicists today have learned to break particles apart in much the same way using enormous particle accelerators.

The accompanying image shows the result of a head-on collision between two gold nuclei, both of which were traveling at nearly the speed of light. When the nuclei collided (at the center of the image), much of the energy of their motion was transformed into a host of new particles, which made the tracks shown. Experiments on the creation of matter from energy with particle accelerators may therefore hold important clues to the formation of our universe.

And so, cosmology, the study of the universe on the largest scale, is now intimately tied to *elementary particle physics,* the study of matter on the smallest scale. Advances in cosmology also benefit elementary particle physics. Not even the most powerful accelerators can match the energies that prevailed during the Big Bang. Because we cannot duplicate these energies

ourselves, physicists cannot easily test many of their predictions. Instead, they look to the Big Bang as a laboratory in which we can explore the most deeply hidden secrets of nature.

 Simulating the very early universe: A high-energy collision between two gold nuclei produces a shower of subatomic particles. (Brookhaven National Laboratory)

Questions to Guide Your Reading

As you read the sections of this chapter, look for the answers to the following questions:

29-1 Has the universe always expanded as it does today, or might it have suddenly "inflated"?

29-2 How did the fundamental forces of nature and the properties of empty space change during the first second after the Big Bang?

29-3 What is antimatter? How can it be created, and how is it destroyed?

29-4 Why is antimatter so rare today?

29-5 What materials in today's universe are remnants of nuclear reactions in the hot early universe?

29-6 How did the first galaxies form?

29-7 Are scientists close to developing an all-encompassing "theory of everything"?

29-1 | The newborn universe may have undergone a brief period of vigorous expansion

With the discovery of the cosmic microwave background, astronomers had direct evidence that the universe began with a hot Big Bang (see Section 28-4). But the properties of the microwave background pose a substantial challenge to the standard theory of the expansion of the universe. The difficulty is that the microwave background is incredibly uniform, or *isotropic*, across the sky. If we subtract the effects of our own motion through the microwave background (see Figure 28-8), we find that the temperature of the microwave background is the same in all parts of the sky to an accuracy of 1 part in 10,000.

To appreciate why this poses a problem, think about two opposite parts of the sky, labeled A and B in Figure 29-1. Both of these points lie on a sphere centered on the Earth, called our **cosmic light horizon**. We can see light coming from any object on or inside our cosmic light horizon, but we cannot yet see objects beyond this horizon. Even traveling at the speed of light—the ultimate speed limit— no information of any kind from objects beyond our cosmic light horizon has had time to reach us over the entire 13.7-billion-year history of the universe. Thus, the cosmic light horizon defines the limits of our observable universe. (As time passes, our cosmic light horizon expands, so eventually we will receive light from objects that are beyond the present-day horizon.)

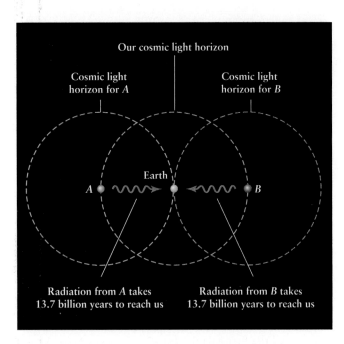

figure 29-1

The Isotropy Problem Regions A and B, both of which lie on our cosmic light horizon, are so far apart that they seem never to have been in communication over the lifetime of the universe. Yet the cosmic background radiation from A and B, and from all other parts of the sky, shows that these disconnected regions have almost exactly the same temperature. The dilemma of why this should be is called the isotropy problem.

Points A and B lie just on our cosmic light horizon, so when we look at these points we are seeing as far back into the past as possible—that is, the light we see from these points is the cosmic background radiation. In order for the radiation that reaches us from A and B to be nearly the same, the material of the early universe at A must have had the same temperature as at B. But for two objects to be at the same temperature, they should have been in contact and able to exchange heat with each other. (A hot cup of coffee and a cold spoonful of cream reach the same temperature only after you pour the cream into the coffee.)

The problem is that the widely separated regions at A and B have absolutely no connection with each other. As Figure 29-1 shows, point A lies outside the cosmic light horizon for point B, and point B lies outside the cosmic light horizon for point A. (Put another way, point A lies outside the observable universe of point B and vice versa.) Hence, no information has had time to travel between points A and B over the entire history of the universe. How is it possible, then, that these unrelated parts of the universe have almost exactly the same temperature? This dilemma is called the **isotropy problem** or the **horizon problem.**

The density of our universe presents us with a second enigma. Recall that the geometry of our universe depends on the density parameter Ω_0, which is the ratio of the combined average mass density in the universe (ρ_0) to the critical density (ρ_c) (see Section 28-6). Observations of temperature variations in the cosmic microwave background indicate that Ω_0 is very close to 1, which corresponds to a flat universe.

For the density parameter Ω_0 to be close to 1 today, it must have been *extremely* close to 1 during the Big Bang. In other words, the density of the early universe was almost exactly equal to the critical density. (The density was much higher than it is today, but the value of the critical density was also much higher. In a flat universe, the average mass density and the critical density decrease together as the universe expands, so that they remain equal at all times.)

The equations for an expanding universe show that any deviation from exact equality would have mushroomed within a fraction of a second. Had the average mass density been slightly less than ρ_c, the universe would have expanded so rapidly that matter could never have clumped together to form galaxies. Without galaxies, there would be no stars or planets, and humans would never have evolved. If, on the other hand, the density had been slightly greater than ρ_c, the universe would soon have become tightly packed with matter. Had this been the case, the gravitational attraction of this matter would long ago have collapsed the entire cosmos in a reversed Big Bang or "Big Crunch," and again humans would never have evolved.

In other words, immediately after the Big Bang the fate of the universe hung in the balance. The tiniest deviation from the precise equality $\rho_0 = \rho_c$ would have rapidly propelled the universe away from the special case of $\Omega_0 = 1$. Had there been such a deviation, we would not be here to contemplate the nature of the universe.

Happily, our universe is one in which galaxies, stars, planets, and humans do exist, the Big Crunch has not taken place, and Ω_0 is approximately 1. This tells us that the density of the universe immediately after the Big Bang must have been equal to the critical density to an incredibly high order of precision. In order for space to be

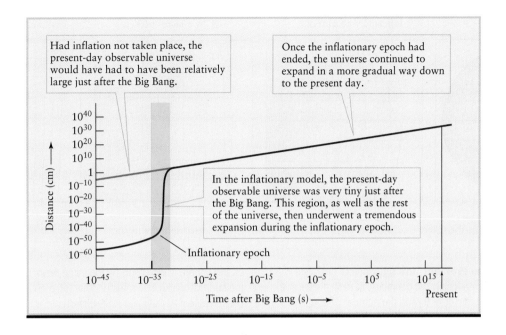

figure 29-2

The Observable Universe With and Without Inflation According to the inflationary model (shown in red), the universe expanded by a factor of about 10^{50} shortly after the Big Bang. This growth in the size of the present-day observable universe—that portion of the universe that lies within our present cosmic light horizon—occurred during a very brief interval, as indicated by the vertical shaded area on the graph. The blue line shows the projected size of the present-day observable universe soon after the Big Bang if inflation had not taken place. (Adapted from A. Guth)

as flat as it is today, the value of ρ_0 right after the Big Bang must have been equal to ρ_c to more than 50 decimal places!

What could have happened during the first few moments of the universe to ensure that $\rho_0 = \rho_c$ to such an astounding degree of accuracy? Because $\rho_0 = \rho_c$ means that space is flat, this enigma is called the **flatness problem.**

 In the early 1980s, a remarkable solution was proposed to both the isotropy problem and the flatness problem. Working independently, Alexei Starobinsky at the L. D. Landau Institute of Theoretical Physics in Moscow and Alan Guth at Stanford University suggested that the universe might have experienced a brief period of **inflation,** or extremely rapid expansion, shortly after the Planck time. (As we saw in Section 28-3, before the Planck time the normal laws of physics do not properly describe the behavior of space, time, and matter.) During this **inflationary epoch,** the universe expanded outward in all directions by a factor of about 10^{50}. This epoch of dramatic expansion may have lasted only about 10^{-32} s (Figure 29-2).

Inflation accounts for the isotropy of the microwave background. During the inflationary epoch, much of the material that was originally near our location was moved out to tremendous distances. Over the past 13.7 billion years, our cosmic light horizon has expanded so that we can see radiation from these distant regions. Hence, when we examine microwaves from opposite parts of the sky, we are seeing radiation from parts of the universe that were originally in intimate contact with one another. This common origin is why all parts of the sky have almost exactly the same temperature.

An inflationary epoch can also account for the flatness of the universe. To see why, think about a small portion of the Earth's surface, such as your backyard. For all practical purposes, it is impossible to detect the Earth's curvature over such a small area, and your backyard looks flat. Similarly, the observable universe is such a tiny fraction of the entire inflated universe that any overall curvature in it is virtually undetectable (Figure 29-3). Like your backyard, the segment of space we can observe looks flat.

It is important to note that the concept of inflation does not violate Einstein's dictum that nothing can travel faster than the speed of light. Remember that the expansion of the universe is the expansion of space itself, not the motion of objects through space. During the inflationary epoch, the distances between particles increased enormously, but this was entirely the result of a sudden vigorous *expansion of space*. Particles did not move through space; space itself inflated.

As we will see later in this chapter, the inflationary model helps us to understand a number of other aspects of the universe. This is an important reason why many astronomers have confidence that inflation really did take place. However, other cosmological models have been proposed to explain the isotropy and flatness problems. It turns out that a stringent test on such models, as well as on different types of inflationary models, is what they predict about the polarization of the cosmic background radiation. (We discussed this polarization in Section 28-8.) Professor John Ruhl discusses efforts to measure this polarization in the essay "New Horizons in the Cosmic Microwave

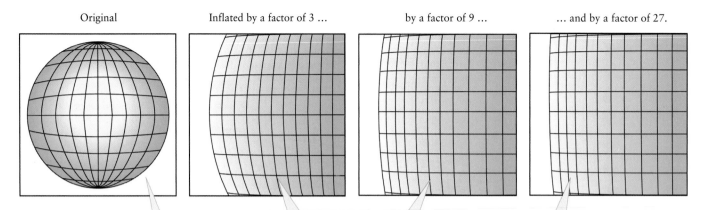

Original   Inflated by a factor of 3 ...   by a factor of 9 ...   ... and by a factor of 27.

As the sphere is inflated, its curvature eventually becomes undetectable and its surface appears flat.

Figure 29-3

Inflation Solves the Flatness Problem This sequence of drawings shows how inflation can produce a locally flat geometry. In each successive frame, the sphere is inflated by a factor of 3 (and the number of grid lines on the sphere is increased by the same factor). Note how the curvature of the surface quickly becomes undetectable on the scale of the illustration. (Adapted from A. Guth and P. Steinhardt)

Background" that follows this chapter. Such measurements should help us decide whether the inflationary model is the best description of the very early universe.

29-2 | Inflation was one of several profound changes that occurred in the very early universe

SCIENCE IN PROCESS *Experimental and theoretical research into the basic forces of nature helps us understand the evolution of the cosmos*

If the universe went through an episode of extreme inflation, what could have triggered it? Our understanding is that inflation was one of a sequence of remarkable events during the first 10^{-12} second after the Big Bang. In each of these events there was a fundamental transformation of the basic physical properties of the universe. To understand what happened during that brief moment of time, when the universe was a hot, dense sea of fast-moving particles and energetic photons colliding with each other, we must first understand how particles interact at very high energies.

Just *four* physical forces—gravity, electromagnetism, and the strong and weak forces—explain the interactions of everything in the universe. Of these forces, gravity is the most familiar. It is a long-range force that dominates the universe over astronomical distances. The electromagnetic force is also a long-range force, but it is intrinsically much stronger than the gravitational force. For example, the electromagnetic force between an electron and a proton is about 10^{39} times stronger than the gravitational force between those two particles. That is why the electromagnetic force, not the gravitational force, is responsible for holding electrons in orbit about the nuclei in atoms.

We do not generally observe longer-range effects of the electromagnetic force, because there is usually a negative electric charge for every positive charge and a south magnetic pole for every north magnetic pole. Thus, over great volumes of space the effects of electromagnetism effectively cancel out. No similar canceling occurs with gravity because there is no equivalent "negative mass." This explains why the force that holds the Moon in orbit about the Earth is gravitational, not electromagnetic.

The **strong force** holds protons and neutrons together inside the nuclei of atoms. It is said to be a *short-range force:* Its influence extends only over distances less than about 10^{-15} m, about the diameter of a proton. Without the strong force, nuclei would disintegrate because of the electromagnetic repulsion of the positively charged protons. In fact, the strong force overpowers the electromagnetic forces inside nuclei.

The **weak force,** which also is a short-range force, is at work in certain kinds of radioactive decay. An example is the transformation of a neutron (n) into a proton (p), in which an electron (e^-) is released along with a nearly massless particle called an antineutrino ($\bar{\nu}$):

$$n \rightarrow p + e^- + \bar{\nu}$$

Numerous experiments in nuclear physics demonstrate that protons and neutrons are composed of more basic particles called **quarks,** the most common varieties being "up" (u) quarks and "down" (d) quarks. A proton is composed of two up quarks and one down quark, a neutron of two down quarks and one up quark.

In the 1970s the concept of quarks led to a breakthrough in our understanding of the strong and weak forces. The strong force holds quarks together, while the weak force is at work whenever a quark changes from one variety to another. For example, when a neutron decays into a proton, one of the neutron's down quarks changes into an up quark. Thus, the weak force is responsible for transformations such as

$$d \rightarrow u + e^- + \bar{\nu}$$

 In the 1940s, physicists Richard P. Feynman and Julian S. Schwinger (working independently in the United States) and Sin-Itiro Tomonaga (in Japan) succeeded in developing a basic description of what we mean by force. Focusing their attention on the electromagnetic force, they tried to describe exactly what happens when two charged particles interact. According to their theory, now called **quantum electrodynamics,** charged particles interact by exchanging photons that cannot be observed directly, because they exist for extremely short time intervals. (We will explore such *virtual* photons in Section 29-3.)

Quantum electrodynamics has proved the most successful theory in modern physics. It describes with remarkable accuracy many details of the electromagnetic interaction between charged particles. Inspired by these successes, physicists have tried to develop similar theories for the other three forces. In these theories, the weak force occurs when particles exchange **intermediate vector bosons,** the gravitational force occurs when particles exchange **gravitons,** and quarks stick together by exchanging **gluons.** Table 29-1 summarizes these features of the four physical forces.

 In the 1970s, physicists made important progress in understanding the weak force. Steven Weinberg, Sheldon Glashow, and Abdus Salam proposed a theory with three types of intermediate vector bosons, which are exchanged in various manifestations of the weak force. These three particles were actually discovered in experiments in the 1980s, providing strong support for the theory.

A startling prediction of the Weinberg-Glashow-Salam theory is that the weak force and the electromagnetic force should be identical to each other for particles with energies greater than 100 GeV. (One GeV equals 10^9, or 1 billion, electron volts; see Section 5-5.) In other words, if particles are slammed together with a total energy greater than 100 GeV, then electromagnetic interactions become indistinguishable from weak interactions. We say that above 100 GeV the electromagnetic force and the weak force are "unified" into a single **electroweak force.**

This unification occurs because the three types of intermediate vector bosons behave just like photons above 100 GeV. At such high energies, these three particles actually lose their mass, and the weak force becomes a long-range force with the same intrinsic strength as electromagnetism. Physicists describe this similarity by saying that "symmetry is restored" above 100 GeV.

In the world around us, however, the typical energies with which particles interact are very much lower, on the order of 1 eV or less. Below 100 GeV, intermediate vector bosons behave like massive particles, but photons are always massless. Because intermediate vector bosons and photons are not similar at low energies, we say that "symmetry is broken" below 100 GeV, which is why the electromagnetic and the weak forces behave so differently in the world around us. In the language of physics, a **spontaneous symmetry breaking** occurs.

In the 1970s, Sheldon Glashow, Howard Georgi, Jogesh Pati, and Abdus Salam proposed **grand unified theories** (or GUTs), which predict that the strong force becomes unified with the weak and electromagnetic forces at energies above 10^{14} GeV. In other words, if particles were to collide at energies greater than 10^{14} GeV, the strong, weak, and electromagnetic interactions would all be long-range forces and would be indistinguishable from each other.

Many physicists suspect that all four forces may be unified at energies greater than 10^{19} GeV (Figure 29-4). That is, if particles were to collide at these colossal energies, there would be no difference between the gravitational, electromagnetic, and nuclear forces. However, no one has yet succeeded in working out the details of such a **supergrand unified theory,** which is sometimes called a **theory of everything** (or **TOE**).

Physicists use particle accelerators to examine the unification of the weak and electromagnetic forces by slamming particles together at energies around 100 GeV. There is probably no hope, however, of ever constructing machines capable of making particles collide with energies approaching 10^{19} GeV. It may thus be impossible to test directly the grand and supergrand unified theories in a laboratory. However, the universe immediately after the Big Bang was so hot and its particles were moving so fast that they did indeed collide with energies on the order of 10^{19} GeV. The evolution of the universe during its earliest moments has thus become a laboratory for testing some of the most elegant, sophisticated theories in physics.

Figure 29-4 shows how the various forces changed during the first fraction of a second after the Big Bang. Before the Planck time (from $t = 0$ s to $t = 10^{-43}$ s), particles collided with energies greater than 10^{19} GeV, and all four forces were unified. Because we do not yet have a TOE that properly describes the behavior of gravity, we

table 29-1	The Four Forces				
Force	Relative strength	Particles exchanged	Particles on which the force can act	Range	Example
Strong	1	gluons	quarks	10^{-15} m	holding protons, neutrons, and nuclei together
Electromagnetic	$\frac{1}{137}$	photons	charged particles	infinite	holding atoms together
Weak	10^{-4}	intermediate vector bosons	quarks, electrons, neutrinos	10^{-16} m	radioactive decay
Gravitational	6×10^{-39}	gravitons	everything	infinite	holding the solar system together

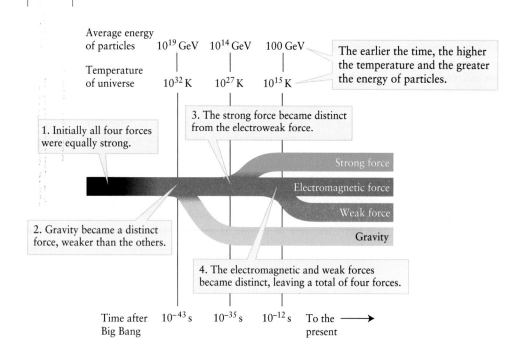

Average energy
of particles 10^{19} GeV 10^{14} GeV 100 GeV

Temperature
of universe 10^{32} K 10^{27} K 10^{15} K

The earlier the time, the higher the temperature and the greater the energy of particles.

1. Initially all four forces were equally strong.

3. The strong force became distinct from the electroweak force.

Strong force

Electromagnetic force

Weak force

Gravity

2. Gravity became a distinct force, weaker than the others.

4. The electromagnetic and weak forces became distinct, leaving a total of four forces.

Time after 10^{-43} s 10^{-35} s 10^{-12} s To the →
Big Bang present

Figure 29-4
Unification of the Four Forces
The strength of the four physical forces depends on the energy of the particles that interact. As shown in this schematic diagram, the higher the energy, the more the forces resemble each other. Also included here are the temperature of the universe and the time after the Big Bang when the strengths of the forces are thought to have been equal.

remain ignorant of what was going on during the first 10^{-43} second of the universe's existence. We know, however, that by the end of the Planck time, the expansion and cooling of the universe had caused the energy of particles to fall to 10^{19} GeV. At energies below this, gravity is thought not to be unified with the other three forces.

In the language of physics, at $t = 10^{-43}$ s there was a spontaneous symmetry breaking in which gravity was "frozen out" of the otherwise unified hot soup that filled all space. In such a "soup," the typical energy of a particle (E) is related to temperature (T) by $E = kT$, where k is the Boltzmann constant (roughly 10^{-4} eV/K, or 10^{-13} GeV/K). Thus, the temperature of the universe was 10^{32} K when gravity emerged as a separate force.

As the universe expanded, its temperature decreased and the energy of particles decreased as well. (We discussed this property of

gases in Box 21-1.) At $t = 10^{-35}$ s, the energy of particles in the universe had fallen to 10^{14} GeV, equivalent to a temperature of 10^{27} K. At energies and temperatures below these, the strong force is no longer unified with the electromagnetic and weak forces. Thus, at $t = 10^{-35}$ s, there was a second spontaneous symmetry breaking, at which time the strong force "froze out."

The inflationary epoch is thought to have begun at this point. Physicists hypothesize that before the strong force decoupled from the electroweak force, the universe was in an unstable state called a **false vacuum**. In this state, physicists hypothesize that the energy associated with a quantity called the *inflaton field* had a nonzero value (Figure 29-5a). (Just as the space around a magnet is permeated by a magnetic field, like that shown in Figure 7-11a, the entire universe is thought to be permeated by the inflation field.) This state

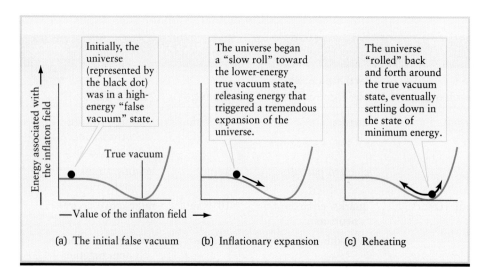

Initially, the universe (represented by the black dot) was in a high-energy "false vacuum" state.

The universe began a "slow roll" toward the lower-energy true vacuum state, releasing energy that triggered a tremendous expansion of the universe.

The universe "rolled" back and forth around the true vacuum state, eventually settling down in the state of minimum energy.

Energy associated with the inflaton field

True vacuum

Value of the inflaton field →

(a) The initial false vacuum   (b) Inflationary expansion   (c) Reheating

Figure 29-5

Inflation: Transitioning from the False Vacuum to the True Vacuum (a) The energy of the vacuum is thought to be determined by the value of a quantity called the inflaton field. As shown by the red curve in this diagram, this energy is at a minimum for a certain value of this field. The state of lowest energy is called the true vacuum. (b) Inflation is associated with the universe making a transition from its initial false-vacuum state to a true-vacuum state. This expansion took place over a period of about 10^{-32} s, during which the universe cooled to a temperature of about 3 K. (c) As the universe "settled in" to the true vacuum state, energy was released that reheated the universe to a temperature of 10^{27} K.

was unstable in the same sense as a ball perched atop a cone with a pointed top: The ball will stay there if left undisturbed, but will roll downhill if even slightly disturbed. In an analogous way, it is thought that at the time that the strong and electroweak forces decoupled, the universe "rolled downhill" to the *true vacuum*, a state of lower energy. The universe's transition to the true vacuum released energy that caused it to expand tremendously in a brief interval of time (Figure 29-4b). By the time the inflationary epoch had ended, about 10^{-32} s after the Big Bang, the universe had increased in scale by a factor of roughly 10^{50}.

The rapid expansion of the universe also gave rise to a rapid cooling. At the end of the inflationary epoch, the temperature of the universe may have dropped to about 3 K, about the same as the temperature that the cosmic background radiation has today. But as the universe finally settled in to the vacuum-energy state, an additional amount of energy was released that went into *reheating* the universe to a temperature of 10^{27} K — about the same as it had before inflation began (Figure 29-5c). Thus inflation caused the universe to expand tremendously while having no net effect on its temperature.

After the end of the inflationary epoch at $t = 10^{-32}$ s, the universe continued to expand and cool at a more sedate rate. At $t = 10^{-12}$ s, the temperature of the universe had dropped to 10^{15} K, the energy of the particles had fallen to 100 GeV, and there was a final spontaneous symmetry breaking and "freeze-out" that separated the electromagnetic force from the weak force. From that moment on, all four forces have interacted with particles essentially as they do today.

29-3 | During inflation, all the mass and energy in the universe burst forth from the vacuum of space

Inflation was a brief but stupendous expansion of the universe soon after the beginning of time. Physicists now realize that inflation helps explain where all the matter and radiation in the universe came from. To see how a violent expansion of space could create particles, we must first understand what quantum mechanics tells us about space.

 Quantum mechanics is the branch of physics that explains the behavior of nature on the atomic scale and smaller. For example, quantum mechanics tells us how to calculate the structure of atoms and the interactions between atomic nuclei. **Elementary particle physics** is the branch of quantum mechanics that deals with individual subatomic particles and their interactions, including the strong, weak, and electromagnetic forces that we discussed in Section 29-2.

The submicroscopic world of quantum mechanics is significantly different from the ordinary world around us. In the ordinary world we have no trouble knowing where things are. You know where your house is; you know where your car is; you know where this book is. In the subatomic world of electrons and nuclei, however, you can no longer speak with this same confidence. A certain amount of fuzziness, or uncertainty, enters into the description of reality at the incredibly small dimensions of the quantum world.

To appreciate the reasons for this uncertainty, imagine trying to measure the position of a single electron. To find out where it is located, you must observe it. And to observe it, you must shine a light on it. However, the electron is so tiny and has such a small mass that the photons in your beam of light possess enough energy to give the electron a mighty kick. As soon as a photon strikes the electron, the electron recoils in some unpredictable direction. Consequently, no matter how carefully you try to measure the precise location of an electron, you necessarily introduce some uncertainty into the speed of that electron.

 These ideas are at the heart of the **Heisenberg uncertainty principle**, first formulated in 1927 by the German physicist Werner Heisenberg, one of the founders of quantum mechanics. This principle states that there is a reciprocal uncertainty between position and momentum (equal to the mass of a particle times its velocity). The more precisely you try to measure the position of a particle, the more unsure you become of how the particle is moving. Conversely, the more accurately you determine the momentum of a particle, the less sure you are of its location. These restrictions are not a result of errors in making measurements; they are fundamental limitations imposed by the nature of the universe.

There is an analogous uncertainty involving energy and time. You cannot know the energy of a system with infinite precision at every moment in time. Over short time intervals, there can be great uncertainty about the amounts of energy in the subatomic world. Specifically, let ΔE be the smallest possible uncertainty in energy measured over a short interval of time Δt. (Astronomers and physicists often use the capital Greek letter delta, Δ, as a prefix to denote a small quantity or a small change in a quantity.)

Heisenberg uncertainty principle for energy and time

$$\Delta E \times \Delta t = \frac{h}{2\pi}$$

ΔE = uncertainty in energy

Δt = time interval over which energy is measured

h = Planck's constant = 6.625×10^{-34} J s

This says that the shorter the time interval Δt, the greater the energy uncertainty ΔE must be in order to ensure that the product of ΔE and Δt is equal to $h/2\pi$.

We might look upon the Heisenberg uncertainty principle as merely an unfortunate limitation on our ability to know everything with infinite precision. But, in fact, this principle provides startling insights into the nature of the universe.

We have seen that one of the important conclusions of Einstein's special theory of relativity is the equivalence of mass and energy: $E = mc^2$ (see Section 18-1). There is nothing uncertain about the speed of light (c), which is an absolute constant. Therefore, any uncertainty in the energy of a physical system can be attributed to an uncertainty Δm in the mass. Thus,

$$\Delta E = \Delta m \times c^2$$

Combining this expression with the previous equation, we obtain

Heisenberg uncertainty principle for mass and time

$$\Delta m \times \Delta t = \frac{h}{2\pi c^2}$$

Δm = uncertainty in mass

Δt = time interval over which mass is measured

h = Planck's constant = 6.625×10^{-34} J s

c = speed of light = 3.00×10^8 m/s

This result is astonishing. It means that over a very brief interval Δt of time, we cannot be sure how much matter there is in a particular location, even in "empty space." During this brief moment, matter can spontaneously appear and then disappear. The greater the amount of matter (Δm) that appears spontaneously, the shorter the time interval (Δt) it can exist before disappearing into nothingness. This bizarre state of affairs is a natural consequence of quantum mechanics.

No particle can appear spontaneously by itself, however. For each particle created, so is a second, almost identical **antiparticle.** In other words, equal amounts of matter and **antimatter** come into existence and then disappear.

Despite its exotic name, there is actually nothing terribly mysterious about antimatter. A particle and an antiparticle are identical in almost every respect; their main distinction is that they carry opposite electric charges. For example, an ordinary electron (e^-) carries a negative charge; the corresponding antiparticle has the same mass but a positive charge, which is why it is called a **positron** (e^+). Because particles and antiparticles come and go in pairs, the total electric charge in the universe remains constant. Particles that have no electric charge can also have corresponding antiparticles. An example is the neutrino (ν); we met its antiparticle, the antineutrino ($\bar{\nu}$), in Section 28-2. The antineutrino is also electrically neutral, but differs from the neutrino in having opposite values of other, more subtle physical properties.

A spontaneously created particle-antiparticle pair lasts for only an incredibly brief time. For example, consider an electron and a positron, each with a mass 9.11×10^{-31} kg. If we rewrite the Heisenberg uncertainty principle for mass and time to solve for Δt and then substitute the combined mass of $2 \times 9.11 \times 10^{-31}$ kg, we find that a spontaneously created electron-positron pair can last for a time

$$\Delta t = \frac{1}{\Delta m}\frac{h}{2\pi c^2} = \frac{1}{2 \times 9.11 \times 10^{-31}} \times \frac{6.625 \times 10^{-34}}{2\pi(3.00 \times 10^8)^2}$$

$$= 6.44 \times 10^{-22}\,\text{s}$$

In other words, an electron and a positron can spontaneously appear and then disappear without violating any laws of physics—but they can remain in existence for no longer than 6.44×10^{-22} s. The more massive the particle, the shorter the time interval it can exist. For example, the proton has about 2000 times more mass

than the electron. Pairs of protons and their antiparticles, called **antiprotons,** can appear and disappear spontaneously, but they exist for only $1/2000$ as long as pairs of electrons and positrons do.

Spontaneous creation can and does happen absolutely anywhere and at any time, not just under the unusual conditions of the early universe. (It is happening right now in the space between this book and your eyes.) Quantum mechanics tells us that if a process is not strictly forbidden, then it must occur. Pairs of every conceivable particle and antiparticle are constantly being created and destroyed at every location across the universe. However, we have no way of observing these pairs directly without violating the uncertainty principle. For this reason, they are called **virtual pairs.** They do not "really" exist in the same sense as ordinary particles; they "virtually" exist. The particles that are exchanged in the four fundamental forces (see Section 29-2) are also virtual particles.

Although virtual pairs of particles and antiparticles cannot be observed directly, their effects have nonetheless been detected. Imagine, for example, an electron in orbit about the nucleus of an atom, such as a hydrogen atom. Ideally, the electron should follow its orbit in a smooth, unhampered fashion. However, because of the constant brief appearance and disappearance of pairs of particles and antiparticles, minuscule electric fields exist for extremely short intervals of time. These tiny, fleeting electric fields cause the electron to jiggle slightly in its orbit. This jiggling produces slight changes in the energies of different electron orbits in the hydrogen atom, which manifest themselves as a minuscule shift in the wavelengths of the hydrogen spectral lines. (We discussed the connection between the energies of electron orbits in hydrogen and the hydrogen spectrum in Section 5-8.)

This shift was first detected in 1947 by the American physicists Willis Lamb and R. C. Retherford and today is known as the **Lamb shift.** The Lamb shift provides powerful support for the idea that every point in space, all across the universe, is seething with virtual pairs of particles and antiparticles. In this sense, "empty space" is actually not empty at all. **Figure 29-6** sketches the constant appearance and disappearance of virtual particles and antiparticles.

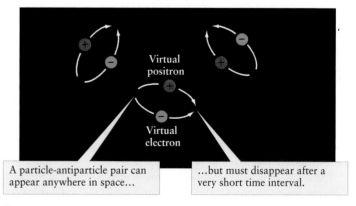

A particle-antiparticle pair can appear anywhere in space...

...but must disappear after a very short time interval.

Virtual positron

Virtual electron

figure 29-6

Virtual Pairs Pairs of particles and antiparticles can appear and then disappear anywhere in space provided that each pair exists only for a very short time interval, as dictated by the uncertainty principle. In this sketch, electrons are shown in green and positrons are shown in red.

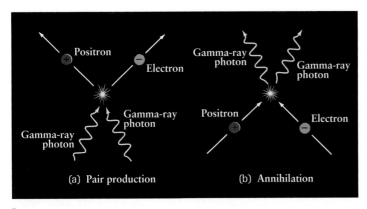

figure 29-7

Pair Production and Annihilation (a) Pairs of virtual particles can be converted into real particles by high-energy gamma-ray photons. In this illustration, an electron (shown in green) and a positron (in red) are produced. This process can take place only if the combined energy of the two photons is no less than Mc^2, where M is the total mass of the electron and positron. (b) Conversely, a particle and an antiparticle can annihilate each other and be transformed into energy in the form of gamma rays.

In some circumstances, virtual pairs can become *real* pairs of particles and antiparticles, a phenomenon called **pair production.** It has been known for years that highly energetic gamma rays (photons) can convert their energy into pairs of particles and antiparticles. Quite simply, the gamma ray disappears upon colliding with a second photon, and a particle and an antiparticle appear in its place. These particles and antiparticles come from nature's ample supply of virtual pairs. The gamma rays provide a virtual pair with so much energy that the virtual particles can appear as real particles in the real world.

Pair production is routinely observed in high-energy nuclear experiments (see the image that opens this chapter). Indeed, it is one of the ways in which physicists manufacture exotic species of particles and antiparticles. The only requirement is that nature's balance sheet be satisfied. To create a particle and an antiparticle having a total mass M, the incoming gamma-ray photons must possess an amount of energy E that is greater than or equal to Mc^2. If the photons carry too little energy (less than Mc^2), pair production will not occur. Likewise, the more energetic the photons, the more massive the particles and antiparticles that can be manufactured. Figure 29-7a shows this process of pair production. Figure 29-7b shows the inverse process of **annihilation,** in which a particle and antiparticle collide with each other and are converted into high-energy gamma rays.

Around 1980 physicists began applying these ideas to their thinking about the creation of the universe. During the inflationary epoch, space was expanding with explosive vigor. As we have seen, however, all space is seething with virtual pairs of particles and antiparticles. Normally, a particle and an antiparticle have no trouble getting back together in a time interval (Δt) short enough to be in compliance with the uncertainty principle. During inflation, however, the universe expanded so fast that particles were rapidly separated from their corresponding antiparticles. Deprived of the opportunity to recombine, these virtual particles had to become *real* particles in the real world. In this way, the universe was flooded with particles and antiparticles created by the violent expansion of space.

29-4 | As the early universe expanded and cooled, most of the matter and antimatter annihilated each other

As soon as the flood of matter and antimatter appeared in the universe, collisions between particles and antiparticles began to produce numerous high-energy gamma rays. As these gamma rays collided, they promptly turned back into the particles and antiparticles from which they came. As a result, the rate of pair production soon equaled the rate of annihilation. For example, for every electron and positron that annihilated each other to create gamma rays (Figure 29-7b), two gamma rays collided elsewhere to produce an electron and a positron (Figure 29-7a). In other words, annihilation and pair production reactions proceeded with equal vigor, and as many particles and antiparticles were being created as were being destroyed.

As the universe continued to expand, all the gamma-ray photons became increasingly redshifted. As a result, the temperature of the radiation field fell. Due to their frequent interaction, radiation and particles of all kinds were in **thermal equilibrium:** All particle species, including photons, were at the same temperature. Hence, as the radiation temperature decreased, the temperature of particles of different types decreased as well.

The first change in the population of particles and antiparticles occurred at $t = 10^{-6}$ s, when the temperature was 10^{13} K and particles were colliding with energies of roughly 1 GeV. Prior to this moment, particles collided so violently that individual protons and neutrons could not exist, being constantly fragmented into quarks. After this time, appropriately called the period of **quark confinement,** quarks were finally able to stick together and became confined within individual protons and neutrons.

 As the universe continued to expand, temperatures eventually became so low that the gamma rays no longer had enough energy to create particular kinds of particles and antiparticles. We say that the temperature dropped below the particular particle's **threshold temperature.** Collisions between these types of particles and antiparticles continued to add photons to the cosmic-radiation background, but collisions between photons could no longer replenish the supply of particles and antiparticles.

At the same time that quark confinement became possible so that protons and neutrons appeared, the universe also became cooler than the 10^{13}-K threshold temperatures of both protons and neutrons. No new protons or neutrons were formed by pair production, but the annihilation of protons by antiprotons and of neutrons by antineutrons continued vigorously everywhere throughout space. This wholesale annihilation dramatically lowered the matter content (particles and antiparticles) of the universe, while simultaneously increasing the radiation (photon) content.

A little later, when the universe was about 1 second old, its temperature fell below 6×10^9 K, the threshold temperature for electrons and positrons. A similar annihilation of pairs of electrons and positrons further decreased the matter content of the universe while raising its radiation content. This radiation field, which fills all space, is the *primordial fireball* discussed in Section 28-5. This fireball, which dominated the universe for the next 380,000 years, therefore derived much of its energy from the annihilation of particles and antiparticles during the first second after the Big Bang.

Now we have a dilemma. If there had been perfect symmetry between particles and antiparticles, then for every proton there should have been an antiproton. For every electron, there should likewise have been a positron. Consequently, by the time the universe was 1 second old, every particle would have been annihilated by an antiparticle, leaving no matter at all in the universe.

Obviously, this did not happen. The planets, stars, and galaxies we see in the sky are made of matter, not antimatter. (If there were substantial amounts of antimatter in the universe, it would eventually collide with ordinary matter. We would then see copious amounts of gamma rays being emitted from the entire sky. While we do observe gamma-ray photons from various locations in the universe, they are neither numerous enough nor of the right energy to indicate the presence of much antimatter.) Thus, there must have been an excess of matter over antimatter immediately after the Big Bang, so that the particles outnumbered the antiparticles.

We can estimate the extent of this asymmetry between matter and antimatter. As noted in Section 28-5, there are roughly 10^9 photons today in the microwave background for each proton and neutron in the universe. Thus, for every 10^9 antiprotons, there must have been 10^9 plus one ordinary protons, leaving one surviving proton after annihilation. Similarly, for every 10^9 positrons, there must have been 10^9 plus one ordinary electrons. Theories of elementary particles and their interactions do indeed predict a slight preference for matter over antimatter of just this sort.

29-5 | A background of neutrinos and most of the helium in the universe are relics of the primordial fireball

The early universe must have been populated with vast numbers of neutrinos (ν) and their antiparticles, the antineutrinos ($\bar{\nu}$). These particles have a very small mass, so their threshold temperature is quite low. These particles take part in the nuclear reactions that transform neutrons into protons and vice versa. For example, a neutron can decay into a proton by emitting an electron and an antineutrino:

$$n \rightarrow p + e^- + \bar{\nu}$$

This radioactive decay happens quickly (its half-life is about 10.5 minutes), which is why we do not find free neutrons floating around in the universe today. In the first 2 seconds after the Big Bang, however, neutrons were also created by collisions between protons and electrons:

$$p + e^- \rightarrow n + \nu$$

This reaction kept the number of neutrons approximately equal to the number of protons. This balance was maintained only as long as collisions between protons and electrons were frequent. But the number of electrons decreased precipitously as the temperature fell below 6×10^9 K and electrons and positrons annihilated each other (Section 29-4). By the time the universe was about 2 seconds old, no new neutrons were being formed, the natural tendency for neutrons to decay into protons took over, and the number of neutrons began to decline.

Before many of the neutrons could decay into protons, they began to combine with protons to form nuclei. Nuclei of helium, the first element more massive than hydrogen, consist of either two protons and two neutrons (^{4}He) or two protons and a single neutron (^{3}He). It is exceedingly improbable that two protons and one or two neutrons should all simultaneously collide with one another to form a helium nucleus. Instead, helium nuclei are built in a series of steps. The first step is to have a single proton and a single neutron combine to form deuterium (^{2}H), sometimes called "heavy hydrogen." A photon is emitted in this process, so we write this reaction as

$$p + n \rightarrow {}^2H + \gamma$$

Forming deuterium does not immediately lead to the formation of helium, however. The problem is that deuterium nuclei are easily destroyed, because a proton and a neutron do not stick together very well. Indeed, in the early universe, high-energy gamma rays easily broke deuterium nuclei back down into independent protons and neutrons. As a result, the synthesis of helium could not get beyond the first step. This block to the creation of helium is called the **deuterium bottleneck.**

When the universe was about 3 minutes old, the background radiation had cooled enough that its photons no longer had enough energy to break up the deuterium. By this time, most of the neutrons had decayed into protons, and protons outnumbered neutrons by about six to one. Because deuterium nuclei could now survive, the remaining neutrons combined with protons and rapidly produced helium. (Figure 18-2 depicts a similar sequence of reactions that produce helium from hydrogen in the core of the present-day Sun.)

The result was what we find in the universe today—about one helium atom for every ten hydrogen atoms. In addition to helium, nuclei of lithium (Li, which has three protons) and beryllium (Be, which has four) were also produced in small numbers. The process of building up nuclei such as deuterium and helium from protons and neutrons is called **nucleosynthesis** (**Figure 29-8**).

Because nuclei have positive electric charges, bringing them together to form more massive nuclei requires that they overcome their mutual electric repulsion. They are unable to do so if they are moving too slowly, which will be the case if the temperature is too low. As a result, by about 15 minutes after the Big Bang the universe was no longer hot enough for nucleosynthesis to take place. Only the four lightest elements (hydrogen, helium, lithium, and beryllium) were present in appreciable numbers. The heavier elements would be formed only much later, once stars had formed and nuclear reactions within those stars could manufacture carbon, nitrogen, oxygen, and all the other elements.

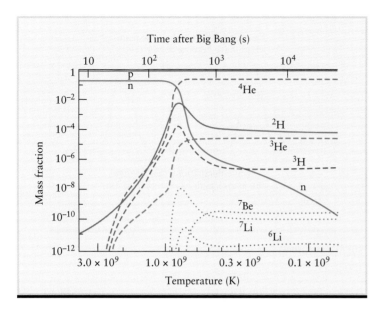

figure 29-8

Nucleosynthesis in the Early Universe This graph shows how nuclei were produced between 10 seconds and 10 hours after the Big Bang. The vertical axis shows the fraction of the total mass that was in each type of particle or nucleus (p = proton, n = neutron, ^{2}H = deuterium, ^{3}He and ^{4}He = helium, ^{6}Li and ^{7}Li = lithium, ^{7}Be = beryllium). Very few nuclei were formed before the universe was 10 seconds old, thanks to the phenomenon of the deuterium bottleneck, which occurred at times earlier than those shown here. By about 10^3 seconds (roughly 15 minutes) after the Big Bang, the temperature had dropped below 4×10^8 K, no further nucleosynthesis was possible, and the relative amounts of different nuclei stabilized. The number of free neutrons declined rapidly as these particles decayed into protons, electrons, and antineutrinos. (Adapted from R. V. Wagoner)

While nuclei were being formed in the early universe, what happened to all those primordial neutrinos and antineutrinos that had interacted so vigorously with the protons and neutrons before the universe was 2 seconds old? The answer is that by $t = 2$ s, matter was sufficiently spread out so that the universe became transparent to neutrinos and antineutrinos. From that time on, neutrinos and antineutrinos could travel across the universe unimpeded. (Recall from Section 18-4 that the Earth itself is virtually transparent to neutrinos from the Sun.)

The neutrinos and antineutrinos that were liberated at $t = 2$ s should now fill the universe much as the cosmic microwave background does. Indeed, these ancient neutrinos and antineutrinos may be about as populous today as the photons in the microwave background (of which there are 4.1×10^8 per cubic meter). The neutrino-antineutrino background should be slightly cooler than the photon background, which received extra energy from electron-positron annihilations. Physicists estimate that the current temperature of the neutrino-antineutrino background is about 2 K, as opposed to 2.725 K for the microwave background. Unfortunately, because neutrinos and antineutrinos are so difficult to detect,

we do not yet have direct evidence of the neutrino-antineutrino background.

29-6 | Galaxies formed from density fluctuations in the early universe

The distribution of matter in the universe today is quite lumpy. Stars are grouped together in galaxies, galaxies into clusters, and clusters into superclusters that stretch across 50 Mpc (150 million ly) or more (see Section 26-6). Furthermore, galaxies seem to be concentrated along enormous sheets, which in turn surround voids measuring 30 to 120 Mpc (100 million to 400 million ly) across. These features, which characterize the large-scale structure of the universe, are shown in Figures 26-21 and 26-22.

Although there is a lot of lumpiness in the universe today, the early universe must have been exceedingly smooth. To see why, think back to the era of recombination that occurred 380,000 years after the Big Bang (see Section 28-5). Before recombination, high-energy photons were constantly and vigorously colliding with charged particles throughout all space. After recombination, the universe became transparent, and these photons stopped interacting with the matter in the universe. Astronomers say that matter "decoupled" from radiation during the era of recombination. Because the cosmic microwave background is extremely isotropic, we can conclude that the matter with which these photons once collided so frequently must also have been spread smoothly across space.

The distribution of matter during the early universe could not have been *perfectly* uniform, however. If it had been, it would still have to be absolutely uniform today; there would now be only a few atoms per cubic meter of space, with no stars and no galaxies. Consequently, there must have been slight lumpiness, or **density fluctuations**, in the distribution of matter in the early universe. These are thought to have originated in the very early universe, even before the inflationary epoch. Infinitesimally small quantum fluctuations in density, which are allowed by the Heisenberg uncertainty principle (see Section 29-3), were stretched during inflation to appreciable size. Through the action of gravity, these fluctuations eventually grew to become the galaxies and clusters of galaxies that we see today throughout the universe. As we saw in Section 28-5 and Section 28-8, the pattern of density fluctuations became imprinted on the cosmic background radiation during the era of recombination. Figure 28-14 shows a map of these fluctuations obtained from the WMAP microwave background spacecraft.

Our understanding of how gravity can amplify density fluctuations dates back to 1902, when the British physicist James Jeans solved a problem first proposed by Isaac Newton. Suppose that you have a gas with only very tiny fluctuations in density, as shown in Figure 29-9a. These regions of higher density will then gravitationally attract nearby material and thus gain mass. As this happens, however, the pressure of the gas inside these regions will also increase, which can make these regions expand and disperse. The problem that Jeans attacked was this: Under what conditions does gravity overwhelm gas pressure so that a permanent object can form?

Jeans proved that an object will grow from a density fluctuation provided that the fluctuation extends over a distance that exceeds the so-called **Jeans length** (L_J):

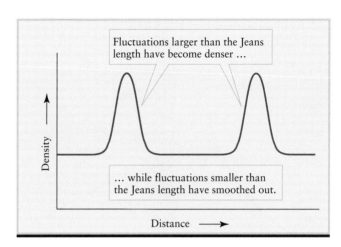

(a) At an early time

(b) At a later time

figure 29-9

The Growth of Density Fluctuations (a) This conceptual illustration shows small density fluctuations in the distribution of matter shortly after the era of recombination. (b) If the size of a fluctuation is greater than the Jeans length (L_J), it becomes gravitationally unstable and can grow in amplitude.

Jeans length for density fluctuations

$$L_J = \sqrt{\frac{\pi k T}{m G \rho_m}}$$

L_J = Jeans length

k = Boltzmann constant = 1.38×10^{-23} J/K

T = temperature of the gas (in kelvins)

m = mass of a single particle in the gas (in kilograms)

G = universal constant of gravitation = 6.67×10^{-11} N • m^2 /kg^2

ρ_m = average density of matter in the gas

Density fluctuations that extend across a distance larger than the Jeans length tend to grow, while fluctuations smaller than L_J tend to disappear (Figure 29-9b).

We can apply the Jeans formula to the conditions that prevailed during the era of recombination, when T = 3000 K and ρ_m = 10^{-18} kg/m^3. Taking m to be the mass of the hydrogen atom (m = 1.67×10^{-27} kg), we find that L_J = 100 light-years, the diameter of a typical globular cluster (Figure 29-10). Furthermore, the mass contained in a cube whose sides are 1 Jeans length in size (equal to the product of the density ρ_m and the volume of the cube L_J^3) is about 5×10^5 M$_\odot$, equal to the mass of a typical globular cluster.

As we saw in Section 21-3, globular clusters contain the most ancient stars we can find in the sky. For these reasons, Robert Dicke and P. J. E. Peebles at Princeton University proposed that globular clusters were among the first objects to form after matter decoupled from radiation. Objects the size of globular clusters may have merged to form still larger collections of matter. Over time, such mergers may have led to the population of galaxies we see today (see the discussion in Section 26-9, especially Figure 26-31).

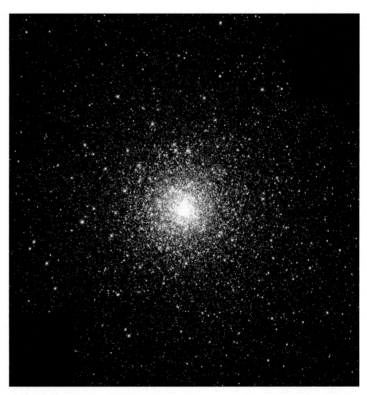

figure 29-10   R I **V** U X G

A Globular Cluster A typical globular cluster contains 10^5 to 10^6 stars, each with an average mass of about 1 M$_\odot$, so the total mass of a typical cluster is 10^5–10^6 M$_\odot$. Cluster diameters range from about 6 to 120 pc (20 to 400 ly). Because these masses and diameters are comparable to the Jeans length (L_J) during the era of recombination, astronomers suspect that globular clusters were among the first objects to form in the universe. (Hubble Heritage Team/AURA/STScI/NASA)

If large-scale structure did indeed develop from density fluctuations, how did this development take place? One issue that complicates this matter is the presence of dark energy, which acts to accelerate the expansion of the universe (see Section 28-6). This accelerated expansion pulls clumps of material away from each other and makes it more difficult for them to coalesce into larger structures. Another complication is that about 85% of the mass in the universe is in the form of dark matter, whose nature is not known (see Section 26-8). Researchers have hypothesized different types of dark matter in the hope of explaining the large-scale structure that we see. Neutrinos are an example of **hot dark matter,** so named because it consists of lightweight particles traveling at high speeds. **Cold dark matter,** on the other hand, consists of massive particles traveling at slow speeds. Examples include WIMPs (which we discussed in Section 25-4), as well as other exotic, speculative particles.

 Scientists use supercomputer simulations to see how different types of dark matter would influence the development of large-scale structure. Figure 29-11 shows the results of such a simulation for a flat universe with dark energy and *cold* dark matter. The simulation follows the motions of 2 million particles of cold dark matter in a box that expands as the universe expands. The box at the lower right of the figure, representing the present time (redshift $z = 0$) is 43 Mpc (160 million ly) a side. At earlier times, the box represents a volume whose side is smaller by a factor $1/(1 + z)$. For example, each side of the box for $z = 0.99$ is actually 1/1.99 as long as the box for $z = 0$.

The simulation begins 120 million years after the Big Bang with an almost perfectly uniform distribution of particles, mimicking the tiny density fluctuations that must have been present just after inflation. A supercomputer then calculates how these particles move, based on Newton's laws in an expanding universe. As time goes on, the fluctuations grow into small, bright clumps whose sizes and masses are similar to those of galaxies. A large filament also forms, spanning the entire box from left to right. The simulation shows that no additional structures formed after the universe was about 6 billion years old, corresponding to redshift $z = 1$. The explanation is that after this time, the accelerating expansion of the universe becomes more important than gravitational attraction. The final frame of the simulation strongly

$z = 27.36$ Universe 120 million years old $z = 9.83$ Universe 490 million years old $z = 4.97$ Universe 1.2 billion years old

$z = 2.97$ Universe 2.2 billion years old $z = 0.99$ Universe 6.0 billion years old $z = 0.00$ Universe 13.7 billion years old

 Figure 29-11
A Cold Dark Matter Simulation with Dark Energy These six views show the evolution of dark matter particles in a large, box-shaped volume of space. The box actually expands with time to follow the expansion of the universe; in this figure, the boxes have been rescaled so they all appear at the same size. Small fluctuations in density are put into the simulation at the beginning (at upper left); these evolve over time to form structures that resemble those actually observed in our present-day universe ($z = 0.00$, shown at the lower right). (Simulations performed at the National Center for Supercomputer Applications by Andrey Kravtsov/University of Chicago and Anatoly Klypin/NMSU)

resembles actual maps of galaxies in our present-day universe (see Figure 26-22).

Simulations similar to those in Figure 29-11 have also been carried out using *hot* dark matter in the form of neutrinos. A massless neutrino would always travel at the speed of light, just as a photon does. However, experiments show that neutrinos do have a small mass. (This nonzero mass is what allows one type of neutrino to transform into another. As we saw in Section 18-4, such transformations provided the explanation to the long-standing solar neutrino problem.) Hence neutrinos travel slower than light, and slow down as the universe expands and cools. Slow-moving neutrinos would accumulate over time within density fluctuations, and the gravitational pull of these neutrinos on surrounding matter could eventually lead to the formation of clusters of galaxies.

A primary difference between simulations based on cold and hot dark matter is the way in which galaxies form. In calculations based on cold dark matter, the formation of galaxies takes place from the "bottom up." In these simulations, the densest gas undergoes collapse early in the history of the universe and stars begin to form. The regions of star formation stream along the filaments (Figure 29-12). When they meet at the intersections between filaments, they merge and group together into galaxies, then clusters of galaxies, then superclusters. But in calculations based on hot dark matter, galaxies form from the "top down." Huge supercluster-sized sheets of matter form first and then fragment into galaxies. Observations of remote galaxies show that galaxies actually formed from the "bottom up" scenario. One piece of evidence for this is the image in Figure 29-13*a*, which shows a handful of "galaxy building blocks" at $z = 3.04$ (when the universe was 2.2 billion years old). These "building blocks," which have not yet coalesced into

Figure 29-12

"Bottom-Up" Galaxy Formation: Simulation This image is taken from a cold dark matter simulation like that shown in Figure 29-11. A portion of the universe is shown at a time 2.2 billion years after the Big Bang, corresponding to redshift $z = 3.04$. The colors indicate the density of gas: Yellow is highest, red is medium, and blue is the lowest density. Over time, the gas tends to pile up at points where filaments intersect, forming galaxies and clusters of galaxies. (T. Theuns, MPA Garching/ESO)

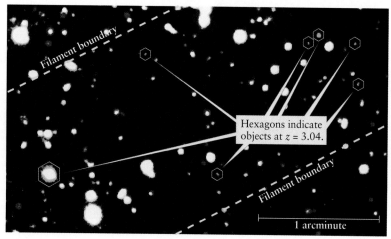

(a) High-redshift objects that lie within a filament

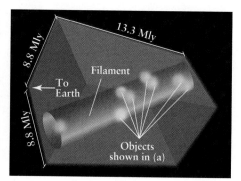

(b) Illustration of the filament

Figure 29-13   R I **V** U X G

"Bottom-Up" Galaxy Formation: Observation (a) The hexagons in this image from the Very Large Telescope show the positions of a number of sub-galaxy-sized objects at a redshift $z = 3.04$, the same as in the simulation shown in Figure 29-12. Excited hydrogen atoms in these objects emit ultraviolet photons, which are redshifted to visible wavelengths. This gives these objects a characteristic green color. (The object at lower left actually lies in front of a much brighter quasar.) (b) The objects in (a) all lie within an immense filament. The purple box shows the volume of space studied in this observing program. The dimensions are given in millions of light-years (Mly). (European Southern Observatory)

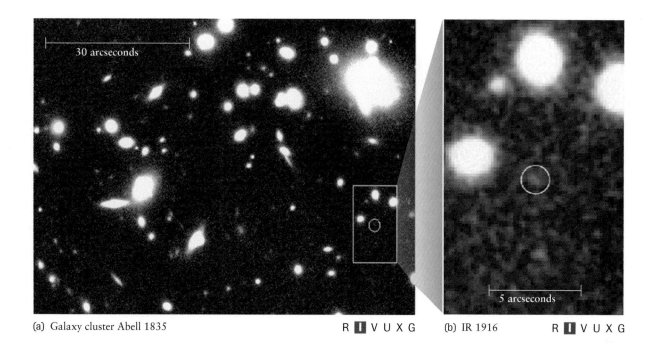

(a) Galaxy cluster Abell 1835   R **I** V U X G   (b) IR 1916   R **I** V U X G

IR 1916 cannot be seen at visible wavelengths...

...but is observable at longer wavelengths.

$\lambda = 540$ nm $\lambda = 1260$ nm $\lambda = 1650$ nm $\lambda = 2160$ nm

(c) Determining the redshift of IR 1916   R **I** **V** U X G

Figure 29-14   R I **V** U X G

A Galaxy "Building Block" at z = 10 (a) This galaxy cluster in Virgo acts as a gravitational lens. (b) IR 1916, one of the objects lensed by this galaxy cluster, lies within the circle in this close-up, false-color infrared image. (c) To determine the redshift of IR 1916, astronomers observed it using different filters. They conclude that hydrogen emission from IR 1916 is redshifted from 122 nm (ultraviolet) to about 1650 nm (infrared), indicating a redshift of about $z = 10$. IR 1916 appears to be a galaxy building block rather than a complete galaxy; it is estimated to be only about 1 kpc (3000 ly) across, less than one-tenth the size of our Milky Way. (European Southern Observatory)

galaxies, lie within a long filament (see Figure 29-13b) that resembles those shown in the simulation of Figure 29-12. (Figure 26-31 is another view of "galaxy building blocks" of this kind.) These observations strongly suggest that the dominant form of dark matter is cold, not hot.

Recent observations indicate that the first galaxy building blocks formed very early in the history of the universe. Figure 29-14a is an infrared image of a galaxy cluster named Abell 1835 that lies some 920 Mpc (3 billion ly) from Earth. This cluster acts as a grav-itational lens that allows us to see even more distant objects that lie beyond it (see Section 26-8, especially Figures 26-29 and 26-30). Figure 29-14b shows one such distant object named IR 1916, which appears brightest at infrared wavelengths around 1650 nm (Figure 29-14c). The light from IR 1916 is thought to be Lyman-α (L_α) radiation produced when an electron in a hydrogen atom drops from the $n = 2$ orbit to the $n = 1$ orbit (see Section 5-8, especially Figure 5-22). To stretch the wavelength of Lyman-α photons to about 1650 nm from the 122-nm wavelength at which they are

In our flat universe dominated by dark energy, matter coalesced along filaments separated by large voids.

(a) A flat universe with dark energy:
$\Omega_m = 0.3$, $\Omega_\Lambda = 0.7$

Similar structures would appear in an open, matter-dominated universe with the same matter density as our own.

(b) A open universe without dark energy:
$\Omega_m = 0.3$, $\Omega_\Lambda = 0$

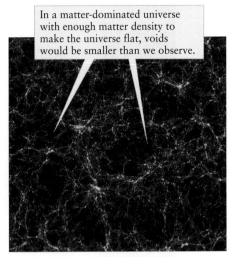

In a matter-dominated universe with enough matter density to make the universe flat, voids would be smaller than we observe.

(c) A flat universe without dark energy:
$\Omega_m = 1.0$, $\Omega_\Lambda = 0$

figure 29-15

Using Simulations to Constrain the Matter Density of the Universe Cold dark matter simulations like those in Figures 29-11 and 29-12 help astronomers determine the value of the matter density parameter Ω_m. These three simulations show a portion of the universe at $z = 0$. (a) A simulation with $\Omega_m = 0.3$ and $\Omega_\Lambda = 0.7$, close to the values for our universe, gives a good match to the observed distribution of filaments and voids. (b) Nearly as good a match is obtained if we keep $\Omega_m = 0.3$, but eliminate dark energy so that $\Omega_\Lambda = 0$. (c) If we use a larger value of Ω_m, the distribution of matter in the simulation is a poor match to our universe. (Simulation by the Virgo Supercomputing Center using computers based at the Computer Center of the Max-Planck-Institute in Garching and at the Edinburgh Parallel Computer Centre)

emitted requires a redshift of about $z = 10$. This redshift means that we are seeing IR 1916 as it was a mere 480 million years after the Big Bang, when the universe was less than 4% of its present age.

How might the universe have evolved if it had contained different amounts of cold dark matter and dark energy? Figure 29-15 shows some simulations designed to explore these possibilities. If the density of matter in the universe is kept constant, simulations predict approximately the same structure for different values of the dark energy density parameter Ω_Λ defined in Section 28-6 (see Figure 29-15a and Figure 29-15b). But if too large a matter density is used in the simulation, the voids between galaxies are smaller than what we actually observe in our universe (Figure 29-15c). Hence, observations of galaxy clustering coupled with supercomputer simulations of galaxy formation help determine the matter density of our universe. (We made use of this idea in Section 28-7. The brown band in Figure 28-19 shows the constraints on cosmological parameters from these observations and simulations of galaxies.)

The best match to the observed distribution of galaxy clusters and to the cosmic background radiation data (see Figure 28-21) is a model like that shown in Figure 29-11, with dark energy and cold dark matter in the proportions listed in Table 28-2.

Figure 29-16 summarizes the past history of our universe down to the present day. As we discussed in Section 28-7, the *future* of our universe is less certain, and depends on the detailed character of dark energy. More detailed data about galaxy clusters and the cosmic background radiation will be needed to pin down the future evolution of our universe.

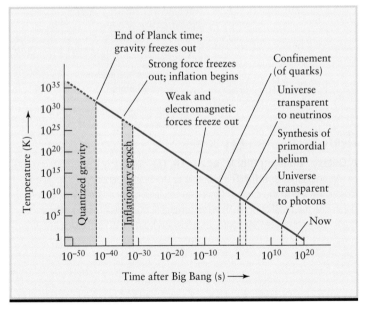

figure 29-16

The History of the Universe As the universe cooled, the four forces "froze out" of their unified state as a result of spontaneous symmetry breaking. Neutrons and protons froze out of the hot "quark soup" during the quark confinement stage that occurred 10^{-6} second after the Big Bang. The inflationary epoch lasted from 10^{-35} to 10^{-32} s after the Big Bang. The universe became transparent to light 380,000 years after the Big Bang.

29-7 | Theories that attempt to unify the physical forces predict that the universe may have 11 dimensions

While we have a growing understanding of the early universe, there remains a veil obscuring the first 10^{-43} s after the Big Bang. These very *first* moments in the history of the universe, whose duration was the Planck time, determined everything that would come after. To understand this brief interval, we need to construct a quantum-mechanical theory that unifies gravity with the other physical forces of nature and that reconciles quantum mechanics with gravity. While this remains an unfinished task, remarkable progress has been made in recent years. One major breakthrough is that physicists have had to abandon the idea that there are only three dimensions of space.

In his special and general theories of relativity, Einstein combined time with the three dimensions of ordinary space, resulting in a four-dimensional combination called *spacetime* (see Section 24-1). In 1919, the Polish physicist Theodor Kaluza proposed the existence of a *fifth* dimension. Kaluza hoped to describe both gravity and electromagnetism in terms of the curvature of five-dimensional space-time, just as Einstein had explained gravity by itself in terms of the curvature of four-dimensional spacetime (see Section 24-2).

A particle always follows the straightest possible path in the four space dimensions of Kaluza's theory. But in the three dimensions of ordinary space, the path appears curved. Hence, it appears to us that the particle has been deflected by gravitational and electromagnetic forces. Kaluza's hypothetical fifth dimension exists at every point in ordinary space but is curled up so tightly, like a very tiny loop, that it is not directly observable.

In 1926, the Swedish physicist Oskar Klein attempted to make Kaluza's five-dimensional theory compatible with quantum mechan-

ics. While he was not successful, Klein discovered that particles of different masses could be identified with different vibrations of the tiny loop of Kaluza's fifth dimension. Today, any quantum-mechanical theory that uses more than four dimensions to provide a unified description of the forces of nature is called a **Kaluza-Klein theory**.

When Kaluza and Klein developed their theories, gravity and electromagnetism were the only known forces of nature. Today we know of four physical forces, which suggests that modern Kaluza-Klein theories should have even more than five dimensions. Edward Witten at Princeton University has argued that a geometric theory for describing all four forces would work best with 11 dimensions, ten of space and one of time. At every point in ordinary space and at every moment of time, the seven "extra" dimensions must be rolled up, like the loops of Kaluza and Klein, into compact structures far too tiny for us to detect (Figure 29-17).

While we do not yet have a definite 11-dimensional theory, physicists know something of how it would work. Particles would travel along the straightest possible paths in spacetime, but their paths in ordinary three-dimensional space would appear curved. From our perspective, we interpret these curved paths as the result of the four forces acting on the particles.

 Theoretical physicists have shown that if there are indeed 11 dimensions, there must exist particles so massive that they have not yet been discovered. These speculative particles may be the dark matter that pervades the universe and holds together clusters of galaxies. Even more bizarre, the new theories no longer regard fundamental particles, such as electrons and quarks, as tiny points of mass. Instead, these particles may actually be multidimensional membranes, wrapped so tightly around the extra dimensions of space that they appear to us as points.

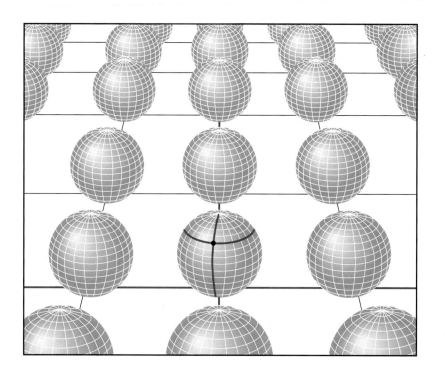

figure 29-17

Hidden Dimensions of Space Hidden dimensions of space might exist provided they are curled up so tightly that we cannot observe them. This drawing shows how an ordinary two-dimensional plane might contain two additional dimensions. At every point on the plane, there is a very tiny sphere so small that it cannot be seen. To pinpoint a particular location, you need to give not only a position on the plane but also a position on the sphere that is tangent to the plane at that point, as indicated by the red lines. (Adapted from D. Freedman and P. Van Nieuwenhuizen)

Andrew Strominger at Harvard University has shown that some of these membranes may fold on themselves in such a way that not even light can escape from them. In other words, these membranes may be black holes! This theory of membranes in 11 dimensions, or **M-theory,** may well explain the most exotic aspects of quantum mechanics, cosmology, and gravity.

One of the many speculative ideas inspired by M-theory is the *cyclic model* (known in an earlier version as the *ekpyrotic model,* from the Greek word for "conflagration"). In this model, our universe is one of two four-dimensional membranes (three space dimensions and one time dimension) that move with respect to each other along a fifth, "hidden" dimension. When the two membranes collide, a Big Bang results. The physics of the collision is such that each membrane automatically has the critical density, so that our universe is automatically flat. This provides a solution to the flatness problem that does not require inflation, so there is no inflationary epoch in the cyclic model. The cyclic model offers a natural explanation for dark matter: It is matter in the other membrane, the particles of which we cannot see but can detect through their gravitation. Dark energy in the cyclic model is a manifestation of the field that controls the interaction between the two membranes. The model is called *cyclic* because the membranes move apart and then back together in a rhythmic way. We would see a universe that expands for a time after the Big Bang, then eventually collapses back

on itself in a Big Crunch. This is followed by another Big Bang, and the cycle repeats.

The cyclic model and the inflationary model are both *scientific* theories, because they both make predictions about things that can be observed. In particular, the two models make different predictions about the redshifts of very distant galaxies and about the polarization of the cosmic background radiation. Improved measurements of both of these will help us understand which of these models—each exotic in its own right—is a better description of our universe.

There is still much work to be done to understand the fundamental nature of the universe. At present, M-theory is only the outline of a "theory of everything." We do not yet know how to describe a quark or an electron in terms of the higher dimensions of spacetime, nor can we yet explain the interactions among such particles. A full description of the first 10^{-43} s after the Big Bang is still far off. However, theoretical physicists around the world are devoting their efforts toward improving our understanding.

We will first understand
How simple the universe is
When we realize
How strange it is.
        —Anonymous

Key Words

Key Ideas

Cosmic Inflation: A brief period of rapid expansion, called inflation, is thought to have occurred immediately after the Big Bang. During a tiny fraction of a second, the universe expanded to a size many times larger than it would have reached through its normal expansion rate.

• Inflation explains why the universe is nearly flat and the 2.725-K microwave background is almost perfectly isotropic.

The Four Forces and Their Unification: Four basic forces—gravity, electromagnetism, the strong force, and the weak force—explain all the interactions observed in the universe.

• Grand unified theories (GUTs) are attempts to explain three of the forces in terms of a single consistent set of physical laws. A supergrand unified theory would explain all four forces.

• GUTs suggest that all four physical forces were equivalent just after the Big Bang. However, because we have no satisfactory supergrand unified theory, we can as yet say nothing about the nature of the universe during this period before the Planck time ($t = 10^{-43}$ s after the Big Bang).

• At the Planck time, gravity froze out to become a distinctive force in a spontaneous symmetry breaking. During a second spontaneous symmetry breaking, the strong nuclear force became a distinct force; this transition triggered the rapid inflation of the universe. A final spontaneous symmetry breaking separated the electromagnetic force from the weak nuclear force; from that moment on, the universe behaved as it does today.

Particles and Antiparticles: Heisenberg's uncertainty principle states that the amount of uncertainty in the mass of a subatomic particle increases as it is observed for shorter and shorter time periods.

• Because of the uncertainty principle, particle-antiparticle pairs can spontaneously form and disappear within a fraction of a second. These pairs, whose presence can be detected only indirectly, are called virtual pairs.

• A virtual pair can become a real particle-antiparticle pair when high-energy photons collide. In this process, called pair production, the photons disappear, and their energy is replaced by the mass of the particle-antiparticle pair. In the process of annihilation, a colliding particle-antiparticle pair disappears and high-energy photons appear.

The Origin of Matter: Just after the inflationary epoch, the universe was filled with particles and antiparticles formed by pair production and with numerous high-energy photons formed by annihilation. A state of thermal equilibrium existed in this hot plasma.

• As the universe expanded, its temperature decreased. When the temperature fell below the threshold temperature required to produce each kind of particle, annihilation of that kind of particle began to dominate over production.

• Matter is much more prevalent than antimatter in the present-day universe. This is because particles and antiparticles were not created in exactly equal numbers just after the Planck time.

Nucleosynthesis: Helium could not have been produced until the cosmological redshift eliminated most of the high-energy photons. These photons created a deuterium bottleneck by breaking down deuterons before they could combine to form helium.

Density Fluctuations and the Origin of Galaxies: Galaxies are generally located on the surfaces of roughly spherical voids. Astronomers use supercomputers to simulate how this large-scale structure of the universe arose from primordial density fluctuations. Models based on dark energy and cold dark matter give good agreement with details of the large-scale structure.

The Frontier of Knowledge: The search for a theory that unifies gravity with the other physical forces suggests that the universe actually has 11 dimensions (ten of space and one of time), seven of which are folded on themselves so that we cannot see them. The idea of higher dimensions has motivated alternative cosmological models.

Review Questions

1. What is the horizon problem? What is the flatness problem? How can these problems be resolved by the idea of inflation?

2. The inflationary epoch lasted a mere 10^{-32} second. Why, then, is it worthy of so much attention by scientists?

3. In what ways is inflation similar to the present-day expansion of the universe? In what ways is it different?

4. Describe an example of each of the four basic types of interactions in the physical universe. Do you think it possible that a fifth force might be discovered someday? Explain your answer.

5. If gravity is intrinsically so weak compared to the strong force, why do we say that gravity rather than the strong force keeps the planets in orbit around the Sun?

6. What is the Heisenberg uncertainty principle? How does it lead to the idea that all space is filled with virtual particle-antiparticle pairs?

7. What is the difference between an electron and a positron?

8. Is it possible for a single hydrogen atom, with a positively charged proton and a negatively charged electron, to be created as a virtual pair? Why or why not?

9. Which can exist for a longer time, a virtual electron-positron pair or a virtual proton-antiproton pair? Explain your reasoning.

10. Explain why antimatter was present in copious amounts in the early universe but is very rare today.

11. What is meant by the threshold temperature of a particle?

12. Explain the connection between the fact that humans exist and the imbalance between matter and antimatter in the early universe.

13. Explain the connection between particles and antiparticles in the early universe and the cosmic microwave background that we observe today.

14. What is the deuterium bottleneck? Why was it important during the formation of nuclei in the early universe?

15. Why were only the four lightest chemical elements produced in the early universe?

16. The first stars in the universe are thought to have appeared some 200 million (2×10^8) years after the Big Bang. Once these stars formed, thermonuclear fusion reactions began in their interiors. Explain why these were the first fusion reactions to occur since the universe was 15 minutes old.

17. Why is it reasonable to suppose that all space is filled with a neutrino background analogous to the cosmic microwave background?

18. Why do astronomers suspect that globular clusters were among the first objects to form in the history of the universe? Why not something larger and more massive?

19. Describe the large-scale structure of the universe as revealed by the distribution of clusters and superclusters of galaxies.

20. What is the difference between hot and cold dark matter? How do astronomers decide which was more important in the formation of large-scale structures such as clusters of galaxies?

21. How did the presence of dark energy help to "turn off" the process of structure formation in the universe?

22. Describe the observational evidence for (**a**) the Big Bang, (**b**) the inflationary epoch, (**c**) the confinement of quarks, and (**d**) the era of recombination.

Advanced Questions

> **Problem-solving tips and tools**
>
> The mass of a proton is 1.67×10^{-27} kg, the mass of an electron is 9.11×10^{-31} kg, and the mass of the Sun is 1.99×10^{30} kg. As explained in Section 28-6, the critical density ρ_c is equal to $3H_0^2/8\pi G$; that is, ρ_c is proportional to the square of the Hubble constant H_0. If $H_0 = 71$ km/s/Mpc, the critical density is equal to about 9.5×10^{-27} kg/m^3. It is also useful to know that 1 m^3 = 10^6 cm^3, that 1 light-year = 9.46×10^{15} m, and that 1 GeV = 10^3 MeV = 10^9 eV.

23. An electron has a lifetime of 1.0×10^{-8} s in a given energy state before it makes a transition to a lower state. What is the uncertainty in the energy of the photon emitted in this process?

24. How many times stronger than the weak force is the electromagnetic force? How many times stronger than the electromagnetic force is the strong force? Use this information to suggest one reason why the electromagnetic and weak forces can become unified at a lower energy than do the electroweak and strong forces.

25. How long can a proton-antiproton pair exist without violating the principle of the conservation of mass?

26. The mass of the intermediate vector boson W$^+$ (and of its antiparticle, the W$^-$) is 85.6 times the mass of the proton. The weak nuclear force involves the exchange of the W$^+$ and the W$^-$. (**a**) Find the rest energy of the W$^+$. Give your answer in GeV. (**b**) Find the threshold temperature for the W$^+$ and W$^-$. (**c**) From Figure 29-4, how long after the Big Bang did W$^+$ and W$^-$ particles begin to disappear from the universe? Explain.

27. Using the physical conditions present in the universe during the era of recombination ($T = 3000$ K and $\rho_m = 10^{-18}$ kg/m^3), show by calculation that the Jeans length for the universe at that time was about 100 ly and that the total mass contained in a sphere with this diameter was about 4×10^5 M$_\odot$.

28. (**a**) If the Hubble constant is 71 km/s/Mpc, the critical density ρ_c is 9.5×10^{-27} kg/m^3. The average density of dark matter is known to be 0.23 times the critical density. Suppose that massive neutrinos constitute this dark matter, and the average density of neutrinos throughout space is 100 neutrinos per cubic centimeter. (In fact, the density of neutrinos is far less than this.) Under these assumptions, what must be the mass of the neutrino? Give your answers in kilograms and as a fraction of the mass of the electron. (**b**) Why do astronomers think that massive neutrinos are *not* the dominant type of dark matter in the universe?

29. A typical dark nebula (see Figure 20-3) has a temperature of 30 K and a density of about 10^{-12} kg/m^3. (**a**) Calculate the Jeans length for such a dark nebula, assuming that the nebula is mostly composed of hydrogen. Express your answer in meters and in light-years. (**b**) A typical dark nebula is several light-years across. Is it likely that density fluctuations within such nebulae will grow with time? (**c**) Explain how your answer to (**b**) relates to the idea that protostars form within dark nebulae (see Section 20-3).

30. At the time labeled $z = 4.97$ in Figure 29-11, how large was the length of each side of the box used in the simulation compared to its size in the present day ($z = 0$)? How much greater was the density at $z = 4.97$ than the present-day density?

Discussion Questions

31. If you hold an iron rod next to a strong magnet, the rod will become magnetized; one end will be a north pole and the other a south pole. But if you heat the iron rod to 1043 K (770°C = 1418°F) or higher, it will lose its magnetization and there will be no preferred magnetic direction in the rod. This demagnetization is an example of *restoring* a spontaneously broken symmetry. Explain why.

32. Some GUTs predict that the proton is unstable, although with a half-life far longer than the present age of the universe. What would it be like to live at a time when protons were decaying in large numbers?

 ## Web/CD-ROM Questions

33. Search the World Wide Web for information about the top quark. What kind of particle is it? How does it compare with the up and down quarks found in protons and neutrons? Why did physicists work so hard to try to find it?

34. Search the World Wide Web for information about primordial deuterium (that is, deuterium that was formed in the very early universe). Why are astronomers interested in knowing how abundant primordial deuterium is in the universe? What techniques do they use to detect it?

35. Search the World Wide Web for information about the South Pole Telescope. What is the purpose of this telescope? Why is it to be sited at the South Pole? How will it help us understand the early universe?

Observing Projects

 36. Use the *Starry Night Backyard*™ program to observe globular clusters. First display the entire celestial sphere (select **Guides** > **Atlas** in the **Go** menu). Use the **Find...** command in the **Edit** menu to examine the following globular clusters. (**a**) In each case, find the approximate angular diameter of the cluster: (i) M3; (ii) M12; (iii) M13. (**b**) Speculate on how these clusters would appear if you could see them at the same distance at the time of recombination, before the first stars formed.

37. Use the *Deep Space Explorer*™ program to examine the distribution of galaxies in our local universe. In the left-hand part of the window under the heading **Universe,** click on the triangle next to the word **Explore** and then click on **Tully Collection.** You will see a box showing the positions of the 28,000 galaxies nearest the Milky Way. (The Milky Way is at the center of the box.) You can zoom in or out using the buttons at the upper left of the window (an upward-pointing triangle and a downward-pointing triangle). You can rotate the box by putting the mouse cursor over the image, holding down the mouse button, and moving the mouse. (On a two-button mouse, hold down the left mouse button.) Compare the box to the simulated present-day universe shown at the lower right of Figure 28-10. What are the similarities? What are the differences?

Collaborative Exercises

38. The four fundamental forces of nature are the strong force, the weak force, the gravitational force, and the electromagnetic force. List four things at your school that rely on one of these fundamental forces, and explain how each thing is dependent on one of the fundamental forces.

39. Consider the following hypothetical scenario adapted from a daytime, cable television talk show. Chris states that Pat borrowed Chris's telescope without permission. Tyler purchased balloons and a new telescope eyepiece without telling Chris. Sean borrowed star maps from the library, with the library's permission, but without telling Pat. Eventually, when the four met on Sunday evening, Chris was crying and speechless. Can you create a "grand unified theory" that explains this entire situation?

40. A history of the universe is shown in Figure 29-16. Create a similar history of your class, starting with estimated outside temperature on the vertical axis and number of days since the beginning of the academic term on the horizontal axis. Include dates for major exams and assignments up through today. In different color ink, show your predictions for temperatures, days, and events from today until the end of the course.

JOHN RUHL

New Horizons in the Cosmic Microwave Background

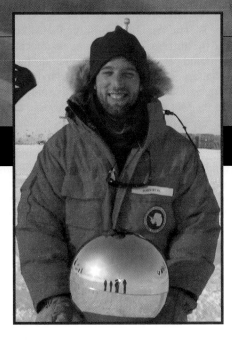

Professor **John Ruhl** has been studying the cosmic microwave background for over a decade, using a variety of ground-based and balloon-borne instruments. His work has taken him to Antarctica eight times, including many seasons at the "high and dry" South Pole, one of the world's best sites for millimeter-wave astronomy. Professor Ruhl picked up the astrophysics bug as an undergraduate physics major at the University of Michigan, and did his graduate work in cosmology at Princeton University. Since 2002 he has been a professor of physics and astronomy at Case Western Reserve University.

The WMAP image of the cosmic microwave background radiation (CMB) in Figure 28-14 gives us a beautiful vision of the universe when it was only a few hundred thousand years old. This map was not—in any sense of the phrase—made overnight. We knew for more than a decade that measurements like these would be a wonderful tool for cosmology. Many independent groups toiled, and many years passed, before cutting-edge technology and hard-earned experience made these measurements possible. The result, using data from WMAP and many other ground-based and balloon-borne instruments with higher angular resolution, has helped pin down parameters of the standard cosmological model such as the density parameter Ω_0, the dark energy density parameter Ω_Λ, and the Hubble constant H_0 to a few percent. This is truly precision cosmology—which not long ago would have been an oxymoron!

While the temperature of the CMB (2.725 K) has been well measured, a new frontier awaits. We expect, and in fact two groups of researchers have now found, that the CMB photons are slightly polarized. The level of polarization is extremely small (a few percent of the temperature fluctuations) and very challenging to measure, but it provides us with a new way of probing the early universe. More precise measurements of that polarization will be able to tell us a few very interesting things about the early universe.

First, the prediction of polarization assumes that our current model of the very young universe is correct—specifically, that we understand the nature of the lumps that eventually collapsed into galaxies and clusters of galaxies, and that we understand the process of the decoupling of photons and matter that occurred when the universe was a few hundred thousand years old. We are always looking for ways to test this model, and the polarization provides a very strong one! Second, polarization information can be combined with the measured temperature anisotropies to improve our measurements of cosmological parameters. We recently flew an instrument around Antarctica to make more precise measurements of the CMB polarization—we are analyzing the data even now—and other groups are ramping up their efforts as well.

There is another, very long-shot—and long-term—goal of polarization measurements that, if successful, would lead to an incredible discovery. The simplest models of inflation predict that gravitational waves (see Section 24-2) were produced as inflation ends. These waves can impress their unique "fingerprints" in the pattern of CMB polarization. If inflation occurred at a sufficiently high energy scale (near 10^{16} GeV), this gravity wave "fingerprint" may be strong enough for us to detect. But if inflation happened at a much lower energy scale, there would be too few gravitational waves for us to see the signature.

This measurement is very challenging, and not for the faint of heart. But if that signature can be detected, it would give us a fantastic probe of inflation!

30 The Search for Extraterrestrial Life

One of the most compelling questions in science is also one of the simplest: Are we alone? That is, does life exist beyond the Earth? As yet, we have no definitive answer to this question. None of our spacecraft has found life elsewhere in the solar system, and radio telescopes have yet to detect signals of intelligent origin coming from space. Reports of aliens visiting our planet and abducting humans make compelling science fiction, but none of these reports has ever been verified.

Yet there are reasons to suspect that life might indeed exist beyond the Earth. One is that biologists find living organisms in some of the most "unearthly" environments on our planet. An example (shown here) is at the bottom of the Gulf of Mexico, where the crushing pressure and low temperature make methane—normally a gas—form into solid, yellowish mounds. Amazingly, these mounds teem with colonies of pink, eyeless, alien-looking worms the size of your thumb. If life can flourish here, might it not also flourish in the equally hostile conditions found on other worlds?

In this chapter we will look for places in our solar system where life may once have originated, and where it may exist today. We will see how scientists estimate the chances of finding life beyond our solar system, and how they search for signals from other intelligent species. And we will learn how a new generation of telescopes may make it possible to detect the presence of even single-celled organisms on worlds many light-years away.

 An "alien" life-form on Earth: pink, eyeless worms in an underwater mound of yellow solid methane. (Dr. Charles Fisher, Eberly College of Science, Pennsylvania State University) R I **V** U X G

Questions to Guide Your Reading

As you read the sections of this chapter, look for the answers to the following questions:

30-1 What role could comets and meteorites have played in the origin of life on Earth?

30-2 Have spacecraft found any evidence for life elsewhere in our solar system?

30-3 Do meteorites from Mars give conclusive proof that life originated there?

30-4 How likely is it that other civilizations exist in our Galaxy?

30-5 How do astronomers search for evidence of civilizations on planets orbiting other stars?

30-6 Will it ever be possible to see Earthlike planets orbiting other stars?

30-1 | The chemical building blocks of life are found throughout space

Suppose you were the first visitor to a new and alien planet. How would you recognize which of the strange objects around you were living, and which were inanimate? Questions such as these are central to **astrobiology** (also called **exobiology**), the study of life in the universe. Most astrobiologists suspect that if we find living organisms on other worlds, they will be "life as we know it"—that is, their biochemistry will be based on the unique properties of the carbon atom, as is the case for all terrestrial life.

Why carbon? The reason is that carbon has the most versatile chemistry of any element. Carbon atoms can form chemical bonds to create especially long and complex molecules. These carbon-based compounds, called **organic molecules,** include all the molecules of which living organisms are made.

Organic molecules can be linked together to form elaborate structures, such as chains, lattices, and fibers. Some of these structures are capable of complex, self-regulating chemical reactions. Furthermore, the primary constituents of organic molecules—carbon, hydrogen, nitrogen, oxygen, sulfur, and phosphorus—are among the most abundant elements in the universe. The versatility and abundance of carbon suggest that extraterrestrial life is also likely to be based on organic chemistry.

If life is based on organic molecules, then these molecules must initially be present on a planet in order for life to arise from nonliving matter. We now understand that many organic molecules originate from nonbiological processes in interstellar space. One such molecule is carbon monoxide (CO), which is made when a carbon atom and an oxygen atom collide and bond together. Carbon monoxide is found in abundance within giant interstellar clouds that lie along the spiral arms of our Milky Way Galaxy as well as other galaxies (see Figure 1-7). Carbon atoms have also combined with other elements to produce an impressive array of interstellar organic molecules, including ethyl alcohol (CH_3CH_2OH), formaldehyde (H_2CO), methyl cyanoacetylene (CH_3C_3N), and acetaldehyde (CH_3CHO). Radio astronomers have detected these by looking for the telltale microwave emission lines of carbon-based chemicals in interstellar clouds.

The planets of our solar system formed out of interstellar material (see Section 8-5), and some of the organic molecules in that material must have ended up on the planets' surfaces. Evidence for this comes from meteorites called **carbonaceous chondrites,** like the one shown in Figure 30-1. These are ancient meteorites that date from the formation of the solar system and that are often found to contain a variety of carbon-based molecules. The spectra of comets (see Section 7-5)—which are also among the oldest objects in the solar system—show that they, too, contain an assortment of organic compounds.

Comets and meteoroids were much more numerous in the early solar system than they are today, and they were correspondingly more likely to collide with a planet. These collisions would have seeded the planets with organic compounds from the very beginning of our solar system's history. Similar processes are thought to take place in other planetary systems, which are thought to form in

Figure 30-1 R I **V** U X G

A Carbonaceous Chondrite Carbonaceous chondrites are primitive meteorites that date back to the very beginning of the solar system. Chemical analyses of newly fallen specimens disclose that they are rich in organic molecules, many of which are the chemical building blocks of life. This sample is a piece of the Allende meteorite, a large carbonaceous chondrite that fell in Mexico in 1969. (From the collection of Ronald A. Oriti)

basically the same way as did our own (see the image that opens Chapter 8).

Comets and meteorites would not have been the only sources of organic material on the young planets of our solar system. In 1952, the American chemists Stanley Miller and Harold Urey demonstrated that under conditions that are thought to have prevailed on the primitive Earth, simple chemicals can combine to form the chemical building blocks of life. In a closed container, they prepared a sample of "atmosphere": a mixture of hydrogen (H_2), ammonia (NH_3), methane (CH_4), and water vapor (H_2O), the most common molecules in the solar system. Miller and Urey then exposed this mixture of gases to an electric arc (to simulate atmospheric lightning) for a week. At the end of this period, the inside of the container had become coated with a reddish-brown substance rich in amino acids and other compounds essential to life.

Since Miller and Urey's original experiment, most scientists have come to the conclusion that Earth's primordial atmosphere was composed of carbon dioxide (CO_2), nitrogen (N_2), and water vapor outgassed from volcanoes, along with some hydrogen. Modern versions of the Miller-Urey experiment (Figure 30-2) using these common gases have also succeeded in synthesizing a wide variety of organic compounds. The combination of comets and meteorites falling from space and chemical synthesis in the atmosphere could have made the chemical building blocks of life available in substantial quantities on the young Earth.

It is important to emphasize that scientists have *not* created life in a test tube. While organic molecules may have been available on the ancient Earth, biologists have yet to figure

Figure 30-2

An Updated Miller-Urey Experiment Modern versions of this classic experiment prove that numerous organic compounds important to life can be synthesized from gases that were present in the Earth's primordial atmosphere. This experiment supports the hypothesis that life on the Earth arose as a result of ordinary chemical reactions.

out how these molecules gathered themselves into cells and developed systems for self-replication. Nevertheless, because so many chemical components of life are so easily synthesized under conditions that simulate the primordial Earth, it seems reasonable to suppose that life could have originated as the result of chemical processes. Furthermore, because the molecules that combine to form these compounds are rather common, it seems equally reasonable that life could have originated in the same way on other planets.

Organic building blocks are commonplace throughout the universe, but this does not guarantee that life is equally commonplace. If a planet's environment is hostile, life may never get started or may

quickly be extinguished. But we now have evidence that Jupiter-sized planets orbit other stars (see Section 8-6 and Geoff Marcy's essay "Alien Planets" following Chapter 8) and that additional planetary systems are forming around young stars (see Section 8-4, especially Figure 8-5). It seems probable that there are Earthlike planets orbiting other stars, and that conditions on some of these worlds may be suitable for life as we know it.

30-2 | Europa and Mars have the potential for life to have evolved

If life evolved on Earth from nonliving organic molecules, might the same process have taken place elsewhere in our solar system? Scientists have carefully scrutinized the planets and satellites in an attempt to answer this question, and most of the answers have been disappointing.

One major problem is that liquid water is essential for the survival of life as we know it. The water need not be pleasant by human standards—terrestrial organisms have been found in water that is boiling hot, fiercely acidic, or ice cold (see the image that opens this chapter)—but it must be liquid. In order for water on a planet's surface to remain liquid, the temperature cannot be too hot or too cold. Furthermore, there must be a relatively thick atmosphere to provide enough pressure to keep liquid water from evaporating. Of all the worlds of the present-day solar system, only Earth has the right conditions for water to remain liquid on its surface.

However, there is now compelling evidence that Europa, one of the large satellites of Jupiter (see Table 7-2), has an ocean of water *beneath* its icy surface. As it orbits Jupiter, Europa is caught in a tug-of-war between Jupiter's gravitational influence and those of the other large satellites. This flexes the interior of Europa, and this flexing generates enough heat to keep subsurface water from freezing. Chunks of ice on the surface can float around on this underground ocean, rearranging themselves into a pattern that reveals the liquid water beneath.

No one knows whether life exists in Europa's ocean. But interest in this exotic little world is great, and scientists have proposed several missions to explore Europa in more detail.

The next best possibility for the existence of life is Mars. The present-day Martian atmosphere is so thin that water can exist only as ice or as a vapor. However, images made from Martian orbit show dried-up streambeds, flash flood channels, and sediment deposits. These features are evidence that the Martian atmosphere was once thicker and that water once coursed over the planet's surface. Could life have evolved on Mars during its "wet" period? If so, could life—even in the form of microorganisms—have survived as the Martian atmosphere thinned and the surface water either froze or evaporated?

In 1976, two spacecraft landed in different parts of Mars in search of answers to these questions. *Viking Lander 1* and *Viking Lander 2* each carried a scoop at the end of a mechanical arm to retrieve surface samples (**Figure 30-3**). These samples were deposited into a compact on-board biological laboratory that carried out three different tests for Martian microorganisms.

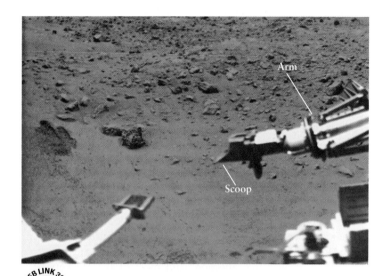

Figure 30-3 R I **V** U X G

Digging in the Martian Surface This view from the *Viking Lander 1* spacecraft shows the mechanical arm with its small scoop against the backdrop of the Martian terrain. The scoop was able to dig about 30 cm (12 in.) beneath the surface. (NASA)

1. The *gas-exchange experiment* was designed to detect any processes that might be broadly considered as respiration. A surface sample was placed in a sealed container along with a controlled amount of gas and nutrients. The gases in the container were then monitored to see if their chemical composition changed.

2. The *labeled-release experiment* was designed to detect metabolic processes. A sample was moistened with nutrients containing radioactive carbon atoms. If any organisms in the sample consumed the nutrients, their waste products should include gases containing the telltale radioactive carbon.

3. The *pyrolytic-release experiment* was designed to detect photosynthesis, the biological process by which terrestrial plants use solar energy to help synthesize organic compounds from carbon dioxide. In the *Viking* experiments, a surface sample was placed in a container along with radioactive carbon dioxide and exposed to artificial sunlight. If plantlike photosynthesis occurred, microorganisms in the sample would take in some of the radioactive carbon from the gas.

The first data returned from these experiments caused great excitement, for in almost every case, rapid and extensive changes were detected inside the sealed containers. Further analysis of the data, however, led to the conclusion that these changes were due solely to nonbiological chemical processes. It appears that the Martian surface is rich in unstable chemicals that react with water to release oxygen gas. Because the present-day surface of Mars is bone-dry, these chemicals had nothing to react with until they were placed inside the moist interior of the *Viking Lander* laboratory.

At best, the results from the *Viking Lander* biological experiments were inconclusive. Perhaps life never existed on Mars at all.

Or perhaps it did originate there, but failed to survive the thinning of the Martian atmosphere, the unstable chemistry of the planet's surface, and exposure to ultraviolet radiation from the Sun. (Unlike Earth, Mars has no ozone layer to block ultraviolet rays.) Another possibility is that Martian microorganisms have survived only in certain locations that the *Viking Lander*s did not sample, such as isolated spots on the surface or deep beneath the ground. And yet another option is that there is life on Mars, but the experimental apparatus on board the *Viking Lander* spacecraft was not sophisticated enough to detect it.

An entirely different set of biological experiments were designed for the British spacecraft *Beagle 2*, which landed on Mars in December 2003. (The spacecraft's name commemorated HMS *Beagle*, the survey ship from which Charles Darwin made many of the observations that led to the theory of evolution.)

Unlike the *Viking Lander* experiments, the apparatus on board *Beagle 2* was designed to test for the presence of either living or dead microorganisms. To do this, the spacecraft was to bore into the interiors of rocks to gather pristine, undisturbed samples, then heat these samples in the presence of oxygen gas. All carbon compounds decompose and form carbon dioxide (CO_2) when treated in this way, but biologically important molecules signal their presence by decomposing at a lower temperature. As a further test for the chemicals of life, *Beagle 2* was to check to see how many of the CO_2 molecules contain the isotope ^{12}C (which appears preferentially in biological molecules) and how many contain ^{13}C (which does not). (See Box 5-5 for a description of isotopes.) A positive result to these experiments would indicate that Martian rocks contain microorganisms that either survive to the present day or that died out at some point in the past.

Another *Beagle 2* experiment was to search for traces of methane in the Martian atmosphere. Microorganisms on Earth can gain energy by converting carbon dioxide to methane, and presumably Martian microorganisms could do the same. Left to itself, methane rapidly decomposes in the Martian atmosphere. Hence, if any methane is found on Mars, it must necessarily have been freshly formed—which would strongly suggest that life exists on Mars today.

Unfortunately, scientists on Earth were unable to establish contact with *Beagle 2* after landing, and so no results were returned from these experiments. But scientists plan to send a replacement lander to Mars, using a set of experiments like those on board *Beagle 2*, as early as 2007.

While the promise of *Beagle 2* will have to be fulfilled by a future mission, scientists searching for evidence of Martian life have been encouraged by other spacecraft that have provided new evidence of water on Mars. The European Space Agency's *Mars Express* spacecraft, which went into orbit around Mars in December 2003, used its infrared cameras to examine the ice cap at the Martian south pole (Figure 30-4). These cameras allowed scientists to see through the ice cap's surface layer of frozen carbon dioxide and reveal an underlying layer with the characteristic spectrum of water ice. (Figure 7-4 shows a similar spectrum obtained from Europa.) In January 2004, NASA successfully landed two robotic rovers named *Spirit* and *Opportunity* at two very different sites on opposite sides of Mars

Craters

South polar
ice cap

The surface layer
is frozen carbon
dioxide ("dry ice").

Beneath the surface
is a layer of frozen
water.

(a) Visible-light image   (b) Infrared image of   (c) Infrared image of
carbon dioxide ice   water ice

R I **V** U X G   R **I** V U X G   R **I** V U X G

WEB LINK 30.7 Figure 30-4
Water at the Martian South Pole (a) This visible-light
image shows the south polar cap of Mars, but does not indicate its chemical
composition. But by using a camera tuned to different wavelengths of
infrared light, the *Mars Express* spacecraft was able to identify the distinctive
reflections of (b) an upper layer of carbon dioxide ice and (c) a lower layer
of water ice. Other observations have shown that there is also water ice at
the Martian north pole. (ESA-OMEGA)

(a) *Spirit* landing site in Gusev Crater

(b) *Opportunity* landing site in Meridiani Planum

WEB LINK 30.8 Figure 30-5   R I **V** U X G
Two Martian Landing Sites Both of the NASA rovers
that reached Mars in 2004 landed at locations that may once have been
covered in water. (a) *Spirit* landed on a rocky plain that may be an ancient
lake bed (compare the similar terrain at the *Viking Lander 1* landing site in
Figure 30-3). (b) The dark terrain at the *Opportunity* landing site is unlike
any other yet seen on Mars. Minerals found at this site are of a sort that
forms in watery environments. (NASA/JPL/Cornell)

(Figure 30-5). Measurements made by *Opportunity* confirm that
some of the very dark surface material at its landing site contains an
iron-rich mineral called gray hematite (Figure 30-5b). On Earth,
deposits of gray hematite are commonly found at the bottoms of
lakes or mineral hot springs. The presence of gray hematite at the
Opportunity site reinforces the argument that Mars once had liquid
water on its surface, and helps hold open the possibility that living
organisms could have evolved on Mars.

In 1976, while the *Viking Lander*s were carrying out their biological experiments on the Martian surface, the companion *Viking
Orbiter* spacecraft photographed some surface features that at first

glance seemed to have been crafted by *intelligent* life on Mars. The
Viking Orbiter 1 image in Figure 30-6a shows what appears to be
a humanlike face, perhaps the product of an advanced and artistic
civilization. However, when the more advanced *Mars Global Surveyor* spacecraft viewed the surface in 1998 using a superior camera (Figure 30-6b), it found no evidence for facial features.

Scientists are universally convinced that the "face" and other
apparent patterns in the *Viking Orbiter* images were created by
shadows on wind-blown hills. Microscopic life may once have
existed on Mars, and may yet exist today, but there is no evidence
that the red planet has ever been the home of intelligent beings.

(a)

(b)

(c)

WEB LINK 30.9

Figure 30-6 R I **V** U X G

A "Face" on Mars? (a) This 1976 image from *Viking Orbiter 1* shows a Martian surface feature that resembles a human face. Some suggested that this feature might have been made by intelligent beings. (b) This 1998 *Mars Global Surveyor* (MGS) image, made under different lighting conditions with a far superior camera, reveals the "face" to be just an eroded hill. (c) This MGS image shows features of natural origin within a 215-km (134-mi) wide crater on Mars. Can you see this "face"? (a: NSSDC/NASA and Dr. Michael H. Carr; b, c: Malin Space Science Systems/NASA)

30-3 | Meteorites from Mars have been scrutinized for life-forms

While spacecraft can carry biological experiments to other worlds such as Mars, many astrobiologists look forward to the day when a spacecraft will return Martian samples to laboratories on Earth. Until that day arrives, we have the next best thing: A dozen meteorites that appear to have formed on Mars have been found at a variety of locations on the Earth.

These meteorites are called **SNC meteorites** after the names given to the first three examples found (Shergotty, Nakhla, and Chassigny). What identifies SNC meteorites as having come from Mars is the chemical composition of trace amounts of gas trapped within them. This composition is very different from that of the Earth's atmosphere, but is a nearly perfect match to the composition of the Martian atmosphere found by the *Viking Landers*.

How could a rock have traveled from Mars to Earth? When a large piece of space debris collides with a planet's surface and forms an impact crater, most of the material thrown upward by the impact falls back onto the planet's surface. But some extraordinarily powerful impacts produce large craters—on Mars, roughly 100 km in diameter or larger. These tremendous impacts eject some rocks with such speed that they escape the planet's gravitational attraction and fly off into space.

There are numerous large craters on Mars, so a good number of Martian rocks have probably been blasted into space over the planet's history. These ejected rocks then go into elliptical orbits around the Sun. A few such rocks will have orbits that put them on a collision course with the Earth, and these are the ones that scientists find as SNC meteorites.

Using the radioactive age-dating technique (see Section 8-3), scientists find that most SNC meteorites are between 200 million and 1.3 billion years old, much younger than the 4.56-billion-year age of the solar system. But one SNC meteorite, denoted by the serial number ALH 84001 and found in Antarctica in 1984, was discovered in 1993 to be 4.5 billion years old (Figure 30-7a). Thus, ALH 84001 is a truly ancient piece of Mars. Analysis of ALH 84001 suggests that it was fractured by an impact between 3.8 and 4.0 billion years ago, was ejected from Mars by another impact 16 million years ago, and landed in Antarctica a mere 13,000 years ago.

ALH 84001 is the only known specimen of a rock that was on Mars during the era when liquid water existed on the planet's surface. Scientists have therefore investigated its chemical composition carefully, in the hope that this rock may contain clues to the amount of water that once flowed on the Martian surface. One such clue is the presence of rounded grains of minerals called carbonates, which can form only in the presence of water.

In 1996, David McKay and Everett Gibson of the NASA Johnson Space Center, along with several collaborators, reported the results of a two-year study of the carbonate grains in ALH 84001. They made three remarkable findings. First, in and around the carbonate grains were large numbers of elongated, tubelike structures resembling fossilized microorganisms (Figure 30-7b). Second, the carbonate grains contain very pure crystals of iron sulfide and magnetite. These two compounds are rarely found together (especially in the presence of carbonates) but can be produced by certain types of bacteria. Indeed, about one-fourth of the magnetite crystals

(a)

(b)

Figure 30-7 R I V U X G

A Meteorite from Mars (a) This 1.9-kg meteorite, known as ALH 84001, formed on Mars some 4.5 billion years ago. About 16 million years ago a massive impact blasted it into space, where it drifted in orbit around the Sun until landing in Antarctica 13,000 years ago. The small cube at lower right is 1 cm (0.4 in.) across. (b) This electron microscope image, magnified some 100,000 times, shows tubular structures about 100 nanometers (10^{-7} m) in length found within the Martian meteorite ALH 84001. One controversial interpretation is that these are the fossils of microorganisms that lived on Mars billions of years ago. (a: NASA Johnson Space Center; b: *Science*, NASA)

found in ALH 84001 are of a type that on Earth are formed only by bacteria. Third, the carbonates contain organic molecules—just the sort, in fact, that result from the decay of microorganisms.

McKay and Gibson concluded that the structures seen in Figure 30-7*b* are fossilized remains of microorganisms. If so, these organisms lived and died on Mars billions of years ago, during the era when liquid water was abundant.

Are McKay and Gibson's conclusions correct? Their claims of ancient life on Mars are extraordinary, and they require extraordinary proof. With only one rock like ALH 84001 known to science, however, such proof is hard to come by, and many scientists are skeptical. They argue that the structures found in ALH 84001 could have been formed in other ways that do not require the existence of Martian microorganisms. Future spacecraft may help resolve the controversy by examining rocks on the Martian surface.

For now, the existence of microscopic life on Mars in the distant past remains an open question. What is without question, however, is that ALH 84001 has fired the imagination of scientists and the public and generated new excitement about the search for extraterrestrial life.

30-4 | The Drake equation helps scientists estimate how many civilizations may inhabit our Galaxy

We have seen that only a few locations in our solar system may have been suitable for the origin of life. But what about other planetary systems? The development of life on the Earth seems to suggest that extraterrestrial life, including intelligent species, might evolve on terrestrial planets around other stars, given sufficient time and hospitable conditions. How can we learn whether such worlds exist, given the tremendous distances that separate us from them? This is the great challenge facing the **search for extraterrestrial intelligence,** or **SETI.**

A tenet of modern folklore is the belief that alien civilizations do exist, and that their spacecraft have visited Earth. Indeed, surveys show that between one-third and one-half of all Americans believe in unidentified flying objects (UFOs). A somewhat smaller percentage believes that aliens have landed on Earth. But, in fact, there is *no* scientifically verifiable evidence of alien visitations. As an example, many UFO proponents believe that the U.S. government is hiding evidence of an alien spacecraft that crashed near Roswell, New Mexico, in 1947. However, the bits of "spacecraft wreckage" found near Roswell turn out to be nothing more than remnants of an unmanned research balloon. To find real evidence of the presence or absence of intelligent civilizations on worlds orbiting other stars, we must look elsewhere.

With our present technology, sending even a small unmanned spacecraft to another star requires a flight time of tens of thousands of years. Speculative design studies have been made for unmanned probes that could reach other stars within a century or less, but these are prohibitively expensive. Instead, many astronomers hope to learn about extraterrestrial civilizations by detecting radio transmissions from them. Radio waves are a logical choice for interstellar communication because they can travel immense distances without being significantly degraded by the interstellar medium, the thin gas and dust found between the stars (see Section 8-1).

Over the past several decades, astronomers have proposed various ways to search for alien radio transmissions, and several searches have been undertaken. In 1960, Frank Drake first used a radio telescope at the National Radio Astronomy Observatory in West Virginia to listen to two Sunlike stars, Tau Ceti and Epsilon Eridani, without success. More than 60 more extensive SETI searches have taken place since then, using radio telescopes around the world. Occasionally, a search has detected an unusual or powerful signal. But none has ever repeated, as a signal of intelligent origin might be expected to do. To date, we have no confirmed evidence of radio transmissions from another world.

 Should we be discouraged by this failure to make contact? What are the chances that a radio astronomer might someday detect radio signals from an extraterrestrial civilization? The first person to tackle this issue was Frank Drake, who proposed that the number of technologically advanced civilizations in the Galaxy could be estimated by a simple equation. This is now called the **Drake equation:**

Drake equation

$$N = R_* \, f_p \, n_e \, f_l \, f_i \, f_c \, L$$

N = number of technologically advanced civilizations in the Galaxy whose messages we might be able to detect

R_* = the rate at which solar-type stars form in the Galaxy

f_p = the fraction of stars that have planets

n_e = the number of planets per solar system that are Earthlike (that is, suitable for life)

f_l = the fraction of those Earthlike planets on which life actually arises

f_i = the fraction of those life-forms that evolve into intelligent species

f_c = the fraction of those species that develop adequate technology and then choose to send messages out into space

L = the lifetime of a technologically advanced civilization

The Drake equation is enlightening because it expresses the number of extraterrestrial civilizations in a simple series of terms. We can estimate some of these terms from what we know about stars and stellar evolution. For example, the first two factors, R_* and f_p, can be determined by observation. In estimating R_*, we should probably exclude stars with masses greater than about 1.5 times that of the Sun. These more massive stars use up the hydrogen in their cores in 3 billion (3×10^9) years or less. On Earth, by contrast, human intelligence developed only within the last million years or so, some 4.56 billion years after the formation of the solar system. If that is typical of the time needed to evolve higher life-forms, then a star of 1.5 solar masses or more probably fades away or explodes

into a supernova before creatures as intelligent as we can evolve on any of that star's planets.

Although stars less massive than the Sun have much longer lifetimes, they, too, seem unsuited for life because they are so dim. Only planets very near a low-mass star would be sufficiently warm for life as we know it, and a planet that close is subject to strong tidal forces from its star. We saw in Section 4-8 how the Earth's tidal forces keep the Moon locked in synchronous rotation, with one face continually facing the Earth. In the same way, a planet that orbits too close to its star would have one hemisphere that always faced the star, while the other hemisphere would be in perpetual, frigid darkness.

This leaves us with stars not too different from the Sun. (Like Goldilocks sampling the three bears' porridge, we must have a star that is not too hot and not too cold, but just right.) Based on statistical studies of star formation in the Milky Way, some astronomers estimate that roughly one of these Sunlike stars forms in the Galaxy each year, thus setting R_* at 1 per year.

As we saw in Sections 8-4 and 8-5, the planets in our solar system formed as a natural consequence of the birth of the Sun. We have also seen evidence suggesting that planetary formation may be commonplace around single stars (see Figure 8-5). Many astronomers suspect that most Sunlike stars probably have planets, and so they give f_p a value of 1.

Unfortunately, the rest of the terms in the Drake equation are very uncertain. Let's play with some hypothetical values. The chances that a planetary system has an Earthlike world suitable for life are not known. Were we to consider our own solar system as representative, we could put n_e at 1. Let's be more conservative, however, and suppose that one in ten solar-type stars is orbited by a habitable planet, making $n_e = 0.1$. From what we know about the evolution of life on the Earth, we might assume that, given appropriate conditions, the development of life is a certainty, which would make $f_l = 1$. This is an area of intense interest to astrobiologists.

For the sake of argument, we might also assume that evolution might naturally lead to the development of intelligence (a conjecture that is hotly debated) and also make $f_i = 1$. It's anyone's guess as to whether these intelligent extraterrestrial beings would attempt communication with other civilizations in the Galaxy, but were we to assume they would, f_c would be put at 1 also.

The last variable, L, involving the longevity of a civilization, is the most uncertain of all. Looking at our own example, we see a planet whose atmosphere and oceans are increasingly polluted by creatures that possess nuclear weapons. If we are typical, perhaps L is as short as 100 years. Putting all these numbers together, we arrive at

$$N = 1/\text{year} \times 1 \times 0.1 \times 1 \times 1 \times 1 \times 100 \text{ years} = 10$$

In other words, out of the hundreds of billions of stars in the Galaxy, we would estimate that there are only ten technologically advanced civilizations from which we might receive communications.

A wide range of values has been proposed for the terms in the Drake equation, and these various guesses produce vastly different estimates of N. Some scientists argue that there is exactly one advanced civilization in the Galaxy and that we are it. Others speculate that there may be hundreds or thousands of planets inhabited by intelligent creatures. If we wish to know whether our Galaxy is devoid of other intelligence, teeming with civilizations, or something in between, we must keep searching the skies.

30-5 | Radio searches for alien civilizations are under way

Even if only a few alien civilizations are scattered across the Galaxy, we have the technology to detect radio transmissions from them. But if other civilizations are trying to communicate with us using radio waves, what frequency are they using? This is an important question, because if we fail to tune our radio telescopes to the right frequency, we might never know whether the aliens are out there.

A reasonable choice would be a frequency that is fairly free of interference from extraneous sources. SETI pioneer Bernard Oliver was the first to draw attention to a range of relatively noise-free frequencies in the neighborhood of the microwave emission lines of hydrogen (H) and hydroxide (OH) (**Figure 30-8**). This region of the microwave spectrum is called the **water hole**, because H and OH together make H_2O, or water.

In 1989, NASA began work on the High Resolution Microwave Survey (HRMS), an ambitious project to scan the entire sky at frequencies spanning the water hole from 10^3 to 10^4 MHz. HRMS would have observed more than 800 nearby solar-type stars over a narrower frequency range in the hope of detecting signals that were either pulsed (like Morse code) or continuous (like the carrier wave for a TV or radio broadcast). The sophisticated signal-processing technology of HRMS would have been able to sift through tens of millions of individual frequency channels simultaneously. It would even have been able to detect the minute Doppler shifts in a signal coming from an alien planet as that planet spun on its axis and moved around its star.

Sadly, just one year after HRMS began operation in 1992, the U.S. Congress imposed a mandate requiring that NASA no longer support HRMS or any other radio searches for extraterrestrial intelligence. This decision, which was made on budgetary grounds, saved a few million dollars—an entirely negligible amount compared to the total NASA budget. Ironically, the senator who spearheaded this move was from the state of Nevada, where tax dollars have been spent to signpost a remote desert road as "The Extraterrestrial Highway."

 Even though NASA funding is no longer available, several teams of scientists remain actively involved in SETI programs. Funding for these projects has come from nongovernmental organizations such as the Planetary Society and from private individuals. Since 1995 the SETI Institute in California has been carrying out Project Phoenix, the direct successor to HRMS. When complete, this project will have surveyed a thousand Sunlike stars within 200 light-years at millions of radio frequencies. At Harvard University, BETA (the *Billion-channel ExtraTerrestrial Assay*) is scanning the sky at even more individual frequencies within the water hole. Other multifrequency searches are being carried out under the auspices of the University of Western Sydney in Australia and the University of California.

 A major challenge facing SETI is the tremendous amount of computer time needed to analyze the mountains of data returned by radio searches. To this end, scientists at the University of California, Berkeley, have recruited nearly 5 million personal computer users to participate in a project called SETI@home. Each user receives actual data from a detector called SERENDIP IV (*Search for Extraterrestrial Radio Emissions from Nearby, Developed, Intelligent Populations*) and a data analysis program that also acts as a screensaver. When the computer's screensaver is on, the program runs, the data are

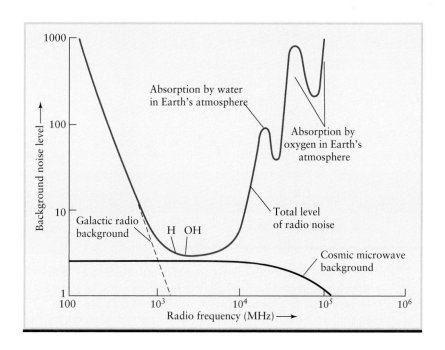

Figure 30-8

The Water Hole This graph shows the background noise level from the sky at various radio and microwave frequencies. The so-called water hole is a range of radio frequencies from about 10^3 to 10^4 megahertz (MHz) in which there is little noise and little absorption by the Earth's atmosphere. Some scientists suggest that this noise-free region would be well suited for interstellar communication. Within the water hole itself, the principal source of noise is the afterglow of the Big Bang, called the cosmic microwave background. To put this graph in perspective, a frequency of 100 MHz corresponds to "100" on a FM radio, and 10^3 MHz is a frequency used for various types of radar. (Adapted from C. Sagan and F. Drake) analyzed, and the results are reported via the Internet to the researchers at Berkeley. The program then downloads new data to be analyzed. As of early 2004, SETI@home users had provided as much computer time as a single computer working full-time for 1,800,000 years!

All current SETI projects make use of existing radio telescopes and must share telescope time with astronomical researchers. But by 2005, the SETI Institute plans to put into operation a radio telescope that will be dedicated solely to the search for intelligent signals. This telescope, called the Allen Telescope Array, will actually be hundreds of relatively small and inexpensive radio dishes working together. Perhaps this new array will be the first to detect a signal from a distant civilization. (In the essay that follows this chapter, SETI Institute senior astronomer Seth Shostak further discusses the search for extraterrestrial intelligence.)

30-6 | Infrared telescopes in space will soon begin searching for Earthlike planets

Although no longer involved in SETI, NASA is planning a major effort to search for Earthlike planets suitable for the evolution of an advanced civilization. Such a search poses a major challenge. Astronomers have discovered many Jupiter-sized planets by detecting the "wobble" that these planets produce in their parent star (see Section 8-6). But a planet the size of the Earth would exert only a weak gravitational force on its parent star, and the resulting "wobble" is too small for us to detect with present technology. Earth-sized planets are also too dim to be seen in visible light against the glare of their parent star.

An alternative technique will be used by an orbiting telescope called Kepler, currently targeted for launch in 2007. If a star is orbited by a planet whose orbital plane is oriented edge-on to our line of sight, once per orbit the planet will pass in front of the star in an event called a *transit*. This causes a temporary dimming of the light we see from that star. As we discussed in Section 8-6, astronomers have used Earth-based telescopes to observe dimming of this kind from Jupiter-sized planets transiting their parent stars. To detect the much slighter dimming caused by the transit of a small, Earth-sized planet, Kepler will use specialized detectors. Furthermore, by observing from orbit using a telescope with a wide field of view, Kepler will be able to continuously monitor thousands of stars at once.

If Kepler detects stars that dim slightly as expected, astronomers will have to make sure that the dimming really is due to a transiting planet and not some other cause. (One key test will be whether the dimming repeats with a definite period, as would be expected for a planet in a periodic orbit.) Once this is done, astronomers will be able to determine the transiting planets' sizes (which determines how much dimming takes place), as well as the sizes of their orbits (which can be calculated from Kepler's third law by using the orbital period of the planet, which is the same as the time interval between successive transits). Given the distance from the planet to its parent star and the star's luminosity, astronomers will even be able to estimate the planet's average temperature.

The results from Kepler may help astronomers select stars to study in more detail using Terrestrial Planet Finder, a more advanced orbiting telescope targeted for launch in 2012 or later. While the technologies to be used in the mission have not yet been chosen, one proposed form of Terrestrial Planet Finder would search for Earthlike planets by detecting their infrared radiation. The rationale is that stars like the Sun emit much less infrared radiation than visible light, while planets are relatively strong emitters of infrared. Hence, observing in the infrared makes it less difficult (although still technically challenging) to detect planets orbiting a star.

Terrestrial Planet Finder will search for planets around some 200 stars within 45 light-years of the Sun. It will also analyze the infrared spectra of any planets that it finds, in the hope of seeing the characteristic absorption of atmospheric gases such as ozone, carbon dioxide, and water vapor (**Figure 30-9**). The relative amounts

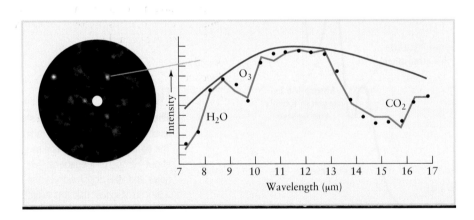

Figure 30-9

The Spectrum of a Simulated Planet The image on the left is a simulation of what the Terrestrial Planet Finder infrared telescope might see once it is launched. The white dot at the center is a nearby Sunlike star, and the smaller dots around it are planets orbiting the star. On the right is the simulated infrared spectrum of one of the planets, showing broad absorption lines of water vapor (H_2O), ozone (O_3), and carbon dioxide (CO_2). While all these molecules can be created by nonbiological processes, the presence of life will change the relative amounts of each molecule in the planet's atmosphere. Thus, the infrared spectrum of such planets will make it possible to identify worlds on which life may have evolved. (Jet Propulsion Laboratory) of these gases, as determined from a planet's spectrum, can reveal whether life is present on that planet.

 Terrestrial Planet Finder will need to achieve enough resolution to detect individual planets. One proposed mission design makes use of interferometry. We discussed this technique for improving the resolution of telescopes in Section 6-6. By combining the light from four widely spaced 3- to 4-m dishes, Terrestrial Planet Finder will make the sharpest infrared images of any telescope in history. The European Space Agency has proposed a similar planet-finding interferometry mission called Darwin.

A more speculative project using interferometry is Planet Imager. If funded, this will be an infrared telescope with sufficient resolution that some detail would be visible in the image of an extrasolar planet. One concept for such a mission would consist of five Terrestrial Planet Finder–type telescopes flying in a geometrical formation some 6000 km across (equal to the radius of the Earth). All five telescopes would collect light from the same extrasolar planet, then reflect it onto a single 8-m mirror. The combined light would go to detectors on board a sixth spacecraft. The technology to build Planet Imager does not yet exist, but may become available within a few decades.

Sometime in the next few decades, missions such as Terrestrial Planet Finder and Planet Imager may answer the question "Are there worlds like Earth orbiting other stars?" If the answer is yes, radio searches for intelligent signals will gain even more impetus.

The potential rewards from such searches are great. Detecting a message from an alien civilization could dramatically change the course of our own civilization, through the sharing of scientific information with another species or an awakening of social or humanistic enlightenment. In only a few years our technology, industry, and social structure might advance the equivalent of centuries into the future. Such changes would touch every person on the Earth. Mindful of these profound implications, scientists push ahead with the search for extraterrestrial intelligence.

Key Words

astrobiology, p. 682

carbonaceous chondrite, p. 682

Drake equation, p. 688

exobiology, p. 682

organic molecules, p. 682

search for extraterrestrial intelligence (SETI), p. 687

SNC meteorite, p. 686

water hole, p. 689

Key Ideas

Organic Molecules in the Universe: All life on Earth, and presumably on other worlds, depends on organic (carbon-based) molecules. These molecules occur naturally throughout interstellar space.

• The organic molecules needed for life to originate were probably brought to the young Earth by comets or meteorites. Another likely source for organic molecules is chemical reactions in the Earth's primitive atmosphere. Similar processes may occur on other worlds.

Life in the Solar System: Besides Earth, only two worlds in our solar system—the planet Mars and Jupiter's satellite Europa—may have had the right conditions for the origin of life.

• Mars once had liquid water on its surface, though it has none today. Life may have originated on Mars during the liquid water era.

• The *Viking Lander* spacecraft searched for microorganisms on the Martian surface, but found no conclusive sign of their presence. The unsuccessful *Beagle 2* mission to Mars was to carry out a different set of biological experiments on samples taken from the interiors of rocks; these experiments may be attempted again on a future mission.

• An ancient Martian rock that came to Earth as a meteorite shows circumstantial evidence that microorganisms once existed on Mars. Additional rock samples are needed to provide corroboration.

• Europa appears to have extensive liquid water beneath its icy surface. Future missions may search for the presence of life there.

Radio Searches for Extraterrestrial Intelligence: Astronomers have carried out a number of searches for radio signals from other stars. No signs of intelligent life have yet been detected, but searches are continuing and using increasingly sophisticated techniques.

• The Drake equation is a tool for estimating the number of intelligent, communicative civilizations in our Galaxy.

Telescope Searches for Earthlike Planets: A new generation of orbiting telescopes may be able to detect terrestrial planets around nearby stars. If such planets are found, their infrared spectra may reveal the presence or absence of life.

Review Questions

1. Why are extreme life-forms on Earth, such as those shown in the photograph that opens this chapter, of interest to astrobiologists?

2. What is meant by "life as we know it"? Why do astrobiologists suspect that extraterrestrial life is likely to be of this form?

3. How have astronomers discovered organic molecules in interstellar space? Does this discovery mean that life of some sort exists in the space between the stars?

4. Mercury, Venus, and the Moon are all considered unlikely places to find life. Suggest why this should be.

5. Summarize the differences in philosophy between the biological experiments on board the *Viking Landers* and those on board *Beagle 2*.

6. What arguments can you give against the idea that the "face" on Mars (Figure 30-6) is of intelligent origin? What arguments can you give in favor of this idea?

7. Suppose someone brought you a rock that he claimed was a Martian meteorite. What scientific tests would you recommend be done to test this claim?

8. Why are most searches for extraterrestrial intelligence made using radio telescopes? Why are most of these carried out at frequencies between 10^3 MHz and 10^4 MHz?

9. Explain why infrared telescopes like those proposed for Terrestrial Planet Finder need to be placed in space.

Advanced Questions

> **Problem-solving tips and tools**
>
> The small-angle formula, discussed in Box 1-1, will be useful. Section 5-2 gives the relationship between wavelength and frequency, while Section 5-9 and Box 5-6 discuss the Doppler effect. Section 6-3 gives the relationship between the angular resolution of a telescope, the telescope diameter, and the wavelength used. You will find useful data about the planets in Appendix 1.

10. In 1802, when it seemed likely to many scholars that there was life on Mars, the German mathematician Karl Friedrich Gauss proposed that we signal the Martian inhabitants by drawing huge geometric patterns in the snows of Siberia. His plan was never carried out. (**a**) Suppose patterns had been drawn that were 1000 km across. What minimum diameter would the objective of a Martian telescope need to have to be able to resolve these patterns? Assume that the observations are made at a wavelength of 550 nm, and assume that Earth and Mars are at their minimum separation. (**b**) Ideally, the patterns used would be ones that could not be mistaken for natural formations. They should also indicate that they were created by an advanced civilization. What sort of patterns would you have chosen?

11. Assume that all the terms in the Drake equation have the values given in the text, except for N and L. (**a**) If there are 1000 civilizations in the Galaxy today, what must be the average lifetime of a technological civilization? (**b**) What if there are a million such civilizations?

12. (**a**) Of the visually brightest stars in the sky listed in Appendix 5, which might be candidates for having Earthlike planets on which intelligent civilizations have evolved? Explain your selection criteria. (**b**) Repeat part (a) for the nearest stars, listed in Appendix 4.

13. It has been suggested that extraterrestrial civilizations would choose to communicate at a wavelength of 21 cm. Hydrogen atoms in interstellar space naturally emit at this wavelength, so astronomers studying the distribution of hydrogen around the Galaxy would already have their radio telescopes tuned to receive extraterrestrial signals. (**a**) Calculate the frequency of this radiation in megahertz. Is this inside or outside the water hole? (**b**) Discuss the merits of this suggestion.

14. Imagine that a civilization in another planetary system is sending a radio signal toward Earth. As our planet moves in its orbit around the Sun, the wavelength of the signal we receive will change due to the Doppler effect. This gives SETI scientists a way to distinguish stray signals of terrestrial origin (which will not show this kind of wavelength change) from interstellar signals. (**a**) Use the data in Appendix 1 to calculate the speed of the Earth in its orbit. For simplicity, assume the orbit is circular. (**b**) If the alien civilization is transmitting at a frequency of 3000 MHz,

what wavelength (in meters) would we receive if the Earth were moving neither toward nor away from their planet? (**c**) The maximum Doppler shift occurs if the Earth's orbital motion takes it directly toward or directly away from the alien planet. How large is that maximum wavelength shift? Express your answer both in meters and as a percentage of the unshifted wavelength you found in (b). (**d**) Discuss why it is important that SETI radio receivers be able to measure frequency and wavelength to very high precision.

15. Astronomers have proposed using interferometry to make an extremely high-resolution telescope. This proposal involves placing a number of infrared telescopes in space, separating them by thousands of kilometers, and combining the light from the individual telescopes. One design of this kind has an effective diameter of 6000 km and uses infrared radiation with a wavelength of 10 µm. If it is used to observe an Earthlike planet orbiting the star Epsilon Eridani, 3.22 parsecs (10.5 light-years) from Earth, what is the size of the smallest detail that this system will be able to resolve on the face of that planet? Give your answer in kilometers.

Discussion Questions

16. Suppose someone told you that the *Viking Landers* failed to detect life on Mars simply because the tests were designed to detect terrestrial life-forms, not Martian life-forms. How would you respond?

17. Science-fiction television shows and movies often depict aliens as looking very much like humans. Discuss the likelihood that intelligent creatures from another world would have (**a**) a biochemistry similar to our own, (**b**) two legs and two arms, and (**c**) about the same dimensions as a human.

18. The late, great science-fiction editor John W. Campbell exhorted his authors to write stories about organisms that think as well as humans, but not *like* humans. Discuss the possibility that an intelligent being from another world might be so alien in its thought processes that we could not communicate with it.

19. If a planet always kept the same face toward its star, just as the Moon always keeps the same face toward Earth, most of the planet's surface would be uninhabitable. Discuss why.

20. How do you think our society would respond to the discovery of intelligent messages coming from a civilization on a planet orbiting another star? Explain your reasoning.

21. What do you think will set the limit on the lifetime of our technological civilization? Explain your reasoning.

22. The first of all Earth spacecraft to venture into interstellar space were *Pioneer 10* and *Pioneer 11*, which were launched in 1972 and 1973, respectively. Their missions took them past Jupiter and Saturn and eventually beyond the solar system. Both spacecraft carry a metal plaque with artwork (reproduced on page 693) that shows where the spacecraft is from and what sort of creatures designed it. If an alien civilization were someday to find one of these spacecraft, which of the features on the plaque do you think would be easily understandable to them? Explain.

Hydrogen emitting radiation at a wavelength of 21 cm

Silhouette of spacecraft

Binary equivalent of the decimal number 8

Position of the Sun relative to 14 reference stars (pulsars) and the center of the Galaxy

Planets of the solar system and their relative distances from the Sun (shown in binary code)

Web/CD-ROM Questions

23. Any living creatures in the subsurface ocean of Europa would have to survive without sunlight. Instead, they might obtain energy from Europa's inner heat. Search the World Wide Web for information about "black smokers," which are associated with high-temperature vents at the bottom of Earth's oceans. What kind of life is found around black smokers? How do these lifeforms differ from the more familiar organisms found in the upper levels of the ocean?

24. Search the World Wide Web for information about the *Mars Express* orbiter and the *Spirit* and *Opportunity* rovers. What discoveries have these missions made about water on Mars? Have they found any evidence that liquid water has existed on Mars in the recent past? Describe the evidence, if any.

25. Like other popular media, the World Wide Web is full of claims of the existence of "extraterrestrial intelligence"—namely, UFO sightings and alien abductions. (**a**) Choose a Web site of this kind and analyze its content using the idea of *Occam's razor*, the principle that if there is more than one viable explanation for a phenomenon, one should choose the simplest explanation that fits all the observed facts. (**b**) Read what a skeptical Web site has to say about UFO sightings. A good example is the Web site of the Committee for the Scientific Investigation of Claims of the Paranormal, or CSICOP. After considering what you have read on both sides of the UFO debate, discuss your opinions about whether aliens really have landed on Earth.

26. The Drake Equation. Access the Active Integrated Media Module "The Drake Equation" in Chapter 30 of the *Universe* Web site or CD-ROM. (**a**) For each of the terms in the Drake equation, choose a value that seems reasonable to you. How did you choose these values? Using the module, what

do you find for the number of civilizations in our Galaxy? From your calculation, are civilizations common or uncommon in our Galaxy? (**b**) Using the Module, choose a set of values that give $N = 10^6$ (a million civilizations). What values did you use? Which of these seem reasonable to you, and why?

Observing Projects

27. Use the *Starry Night Backyard*™ program to view the Earth as it might be seen by a visiting spacecraft. First select **Viewing Location...** in the **View** menu and set the viewing location to your city or town in the list of cities provided. (Alternatively, click on the "Map" tab and use the mouse to click on your approximate position on the world map.) Then click the **Set Location** button. In the Control Panel at the top of the window, set the local time to 12:00:00 P.M. (noon). To see the Earth from space, use the elevation buttons (the ones that look like arrows pointing up and down) in the Control Panel to raise yourself off the surface until you can see the entire Earth. (You may want to use the scrollbars on the right side and bottom of the window so that you are looking directly at the Earth.) (**a**) Describe any features you see that suggest life could exist on Earth. Explain your reasoning. (**b**) Using the controls at the right-hand end of the Control Panel, zoom in to show more detail around your city or town. The amount of detail is comparable to the view from a spacecraft a few million kilometers away. Can you see any evidence that life does exist on Earth? (**c**) From a distance of a few million kilometers, are there any measurements that a spacecraft could carry out to prove that life exists on Earth? Explain your reasoning.

28. Use the *Deep Space Explorer*™ program to examine the planet Mars. In the left-hand part of the window under the heading **Solar System,** click on the triangle next to the word **Explore** and then click on **Mars.** You can zoom in or out using the buttons at the upper left of the window (an upward-pointing triangle and a downward-pointing triangle). You can rotate Mars by putting the mouse cursor over the image, holding down the mouse button, and moving the mouse. (On a two-button mouse, hold down the left mouse button.) Rotate Mars and zoom in and out to familiarize yourself with the different surface features. Based on what you observe, where on the Martian surface would you choose to land a spacecraft to search for the presence of life? Explain how you made your choice.

Collaborative Exercise

29. Imagine that astronomers had discovered intelligent life in a nearby star system and your group is submitting a proposal for who on Earth should speak for the planet and what 50-word message should be conveyed. Prepare a maximum one-page proposal that states (**a**) who should speak for Earth and why; (**b**) what this person should say in 50 words; and (**c**) why this message is the most important compared to other things that could be said. Only serious responses receive full credit.

SETH SHOSTAK

Searching for Extraterrestrial Life

Dr. Seth Shostak is Senior Astronomer at the SETI Institute in California, and participates in the world's most comprehensive search for extraterrestrial intelligence. He has a physics degree from Princeton University and a Ph.D. in astronomy from the California Institute of Technology. Dr. Shostak studied galaxies at observatories and universities worldwide before joining the SETI Institute in 1990. In addition to his research, he keeps busy promoting science to the general public with lectures, writings, and a weekly radio show.

For more than four decades, a small number of researchers have been training their telescopes on the skies, looking for signals that would prove that someone out there is at least as smart as we are. It's a project known as SETI, the Search for Extraterrestrial Intelligence, and it's something I've been doing for more than a dozen years.

Of course, looking for aliens with aptitude is only a very small part of the hunt for extraterrestrial life in general. Our spacecraft continue to scout the rusty surface of Mars, hoping to find conclusive evidence that this planet once had (or maybe still has) some simple biology. We've also mounted robotic expeditions to a few of the Jovian moons, where cryptic life might possibly eke out a living. This continuing effort to uncover living things nearby is quite exciting for me. Even though such life would probably be less impressive than bathtub mold, the discovery of even a bit of biology on another world of our solar system would strongly suggest that cosmic life is as common as crabgrass.

If life springs up in many places, the chances for intelligent life also improve. But of course it's possible that even if biology is widespread, intelligence is not. Perhaps the ability to think is a rare evolutionary development. Perhaps most intelligent beings, once they've mastered technology, quickly self-destruct. Despite such uncertainties, it's difficult for me to imagine (and frankly, it's impossible to prove) that *Homo sapiens* is the smartest species in the Galaxy, let alone the universe.

In fact, I'm quite upbeat about the chances that we may soon find a signal that will certify that we have cosmic company. My optimism derives from the continuing improvement of our instruments. Today's SETI scientists are building new telescopes and greatly refining the techniques of signal detection. It's true that, so far, we've carefully examined only a few thousand star systems for either radio or light transmissions. But that number will increase to a few million in the coming two decades as we extend our search to a distance of approximately 1000 light-years.

If we find an extraterrestrial transmission, what would happen? Some people worry that a signal received from another world might be perceived as dangerous, and expect that the news would be covered up. My work in SETI has convinced me that this isn't true. There's no policy of secrecy, and no mechanism for enforcing it. Within a few days of finding an interesting signal, astronomers around the world (and their friends and relatives) would all know about it, and so would you. Every telescope on Earth would be pointed in the direction of the broadcast.

A few facts about the signal would quickly emerge. On the basis of the slow change in frequency, we could deduce some parameters of the aliens' world—for example, the length of their day and year. We could speedily identify their sun, allowing us to estimate the temperature of their home planet.

But what about the message? No self-respecting extraterrestrial would transmit an empty signal; it will surely contain information. If they are deliberately beaming their broadcasts our way—if they are trying to get in touch—then they may fashion the content in a way that just about anyone, including us, can understand. They could, as an example, send picture dictionaries followed by the contents of their libraries.

This could be enormously instructive because it would be coming from a civilization considerably beyond our own. No society we detect will be less advanced than we are (after all, they're able to transmit across interstellar distances). And it's unlikely we'll ever pick up a signal unless many technological civilizations survive for long periods of time. Consequently, the chances are high that the discovered aliens will be thousands of years beyond us or more. Their message, if we can fathom it, would be a revelation.

It's also possible that we won't be able to decode the detected signal, either because it wasn't intended for us or because we're simply not clever enough to figure it out. Consider how difficult it would be for Marconi, one of radio's inventors, to decipher the digitized transmissions from a modern telecommunications satellite. We may well discover that the aliens are there, but their message remains a puzzle. Even so, it would be important to know that we're not the only thinking inhabitants of the Galaxy. Indeed, it's hard to imagine a more profoundly exciting discovery.

Appendices

Appendix 1 — The Planets: Orbital Data

Planet	Semimajor axis (10⁶ km)	Semimajor axis (AU)	Sidereal period (years)	Sidereal period (days)	Synodic period (days)	Average orbital speed (km/s)	Orbital eccentricity	Inclination of orbit to ecliptic (°)
Mercury	57.9	0.387	0.241	87.969	115.88	47.9	0.206	7.00
Venus	108.2	0.723	0.615	224.70	583.92	35.0	0.007	3.39
Earth	149.6	1.000	1.000	365.256	—	29.79	0.017	0.00
Mars	227.9	1.524	1.88	686.98	779.94	24.1	0.093	1.85
Jupiter	778.3	5.203	11.86		398.9	13.1	0.048	1.30
Saturn	1429	9.554	29.46		378.1	9.64	0.053	2.48
Uranus	2871	19.194	84.10		369.7	6.83	0.043	0.77
Neptune	4498	30.066	164.86		367.5	5.5	0.010	1.77
Pluto	5915	39.537	248.60		366.7	4.7	0.250	17.15

Appendix 2 — The Planets: Physical Data

Planet	Equatorial diameter (km)	Equatorial diameter (Earth = 1)	Mass (kg)	Mass (Earth = 1)	Average density (kg/m³)	Rotation period* (solar days)	Inclination of equator to orbit (°)	Surface gravity (Earth = 1)	Albedo	Escape speed (km/s)
Mercury	4880	0.383	3.302×10^{23}	0.0553	5430	58.646	0.5	0.38	0.12	4.3
Venus	12,104	0.949	4.868×10^{24}	0.815	5243	243.01^R	177.4	0.91	0.59	10.4
Earth	12,756	1.000	5.974×10^{24}	1.000	5515	0.997	23.45	1.00	0.39	11.2
Mars	6794	0.533	6.418×10^{23}	0.107	3934	1.026	25.19	0.38	0.15	5.0
Jupiter	142,984	11.209	1.899×10^{27}	317.8	1326	0.414	3.12	2.36	0.44	60.2
Saturn	120,536	9.449	5.685×10^{26}	95.16	687	0.444	26.73	1.1	0.47	35.5
Uranus	51,118	4.007	8.682×10^{25}	14.53	1318	0.718^R	97.86	0.92	0.56	21.3
Neptune	49,528	3.883	1.024×10^{26}	17.15	1638	0.671	29.56	1.1	0.51	23.5
Pluto	2300	0.180	1.3×10^{22}	0.0021	2000	6.387^R	122.52	0.07	0.5	1.2

* For Jupiter, Saturn, Uranus, and Neptune, the internal rotation period is given. A superscript R means that the rotation is retrograde (opposite the planet's orbital motion).

Appendix 3 | Satellites of the Planets

Planet	Satellite	Date of discovery	Average distance from center of planet (km)	Orbital (sidereal) period (days)*	Orbital eccentricity	Size of satellite (km)**	Mass (kg)
EARTH	Moon	—	384,400	27.322	0.0549	3476	7.349×10^{22}
MARS	Phobos	1877	9378	0.319	0.01	$28 \times 23 \times 20$	1.1×10^{16}
	Deimos	1877	23,460	1.263	0.00	$16 \times 12 \times 10$	1.8×10^{15}
JUPITER	Metis	1979	128,000	0.295	0.0012	44	1×10^{17}
	Adrastea	1979	129,000	0.298	0.0018	$24 \times 20 \times 16$	1.9×10^{16}
	Amalthea	1892	181,400	0.498	0.0031	$270 \times 200 \times 155$	7.2×10^{18}
	Thebe	1979	221,900	0.675	0.0177	98	8×10^{17}
	Io	1610	421,600	1.769	0.0041	3642	8.932×10^{22}
	Europa	1610	670,900	3.551	0.0094	3120	4.791×10^{22}
	Ganymede	1610	1,070,000	7.155	0.0011	5268	1.482×10^{23}
	Callisto	1610	1,883,000	16.689	0.0074	4800	1.077×10^{23}
	Themisto	2000	7,284,000	130.02	0.2426	9	(?)
	Leda	1974	11,165,000	240.92	0.1636	18	6×10^{15}
	Himalia	1904	11,461,000	250.56	0.1623	184	1×10^{19}
	Lysithea	1938	11,717,000	259.2	0.1124	38	8×10^{16}
	Elara	1905	11,741,000	259.64	0.2174	78	8×10^{17}
	S/2000 J11	2000	12,555,000	287.0	0.2480	4	(?)
	S/2003 J20	2003	16,989,000	456.1	0.4297	3	(?)
	S/2003 J12	2003	17,582,000	489.5^R	0.5095	1	(?)
	Euporie	2001	19,304,000	550.74^R	0.1432	2	(?)
	S/2003 J3	2003	20,221,000	583.88^R	0.197	2	(?)
	S/2003 J18	2003	20,514,000	596.59^R	0.0221	2	(?)
	Orthosie	2001	20,720,000	622.56^R	0.2808	2	(?)
	Euanthe	2001	20,797,000	620.49^R	0.2321	3	(?)
	Harpalyke	2000	20,858,000	623.31^R	0.2268	4	(?)
	Praxidike	2000	20,907,000	625.38^R	0.2308	7	(?)
	Thyone	2001	20,939,000	627.21^R	0.2286	4	(?)
	S/2003 J16	2003	20,957,000	616.36^R	0.2246	2	(?)
	Iocaste	2000	21,061,000	631.6^R	0.216	5	(?)
	S/2003 J21	2003	21,069,000	620.04^R	0.2273	2	(?)
	Hermippe	2001	21,131,000	633.9^R	0.2096	4	(?)
	S/2003 J22	2003	21,162,000	628.09^R	0.2206	2	(?)
	S/2003 J6	2003	21,263,000	634.77^R	0.1558	4	(?)
	Ananke	1951	21,276,000	629.77^R	0.2435	28	4×10^{16}
	S/2003 J15	2003	22,627,000	689.77^R	0.191	2	(?)
	Eurydome	2001	22,865,000	717.33^R	0.2759	3	(?)
	S/2002 J1	2002	22,931,000	723.9^R	0.2588	3	(?)

Appendix 3 | Satellites of the Planets

Planet	Satellite	Date of discovery	Average distance from center of planet (km)	Orbital (sidereal) period (days)*	Orbital eccentricity	Size of satellite (km)**	Mass (kg)
JUPITER (continued)	S/2003 J17	2003	22,992,000	714.47^R	0.2378	2	(?)
	Pasithee	2001	23,004,000	719.44^R	0.2675	2	(?)
	S/2003 J10	2003	23,041,000	716.25^R	0.4295	2	(?)
	Chaldene	2000	23,100,000	723.7^R	0.2519	4	(?)
	Isonoe	2000	23,155,000	726.25^R	0.2471	4	(?)
	Erinome	2000	23,196,000	728.51^R	0.2665	3	(?)
	Kale	2001	23,217,000	729.47^R	0.2599	2	(?)
	Aitne	2001	23,229,000	730.18^R	0.2643	3	(?)
	Taygete	2000	23,280,000	732.41^R	0.2525	5	(?)
	S/2003 J9	2003	23,384,000	733.29^R	0.2632	1	(?)
	Carme	1938	23,404,000	734.17^R	0.2533	46	1×10^{17}
	Sponde	2001	23,487,000	748.34^R	0.3121	2	(?)
	Magaclite	2000	23,493,000	752.88^R	0.4197	6	(?)
	S/2003 J5	2003	23,495,000	738.73^R	0.2478	4	(?)
	S/2003 J19	2003	23,533,000	740.42^R	0.2556	2	(?)
	S/2003 J23	2003	23,563,000	732.44^R	0.2714	2	(?)
	Kalyke	2000	23,566,000	742.03^R	0.2465	5	(?)
	Pasiphae	1908	23,624,000	743.63^R	0.4090	58	(?)
	S/2003 J1	2003	23,661,000	746.39^R	0.2721	4	(?)
	S/2003 J4	2003	23,930,000	755.24^R	0.3618	2	(?)
	Sinope	1914	23,939,000	758.9^R	0.2495	38	8×10^{16}
	S/2003 J8	2003	23,947,000	739.6^R	0.3276	3	(?)
	S/2003 J13	2003	23,951,000	751.94^R	0.4116	2	(?)
	S/2003 J7	2003	23,981,000	761.5^R	0.4322	4	(?)
	S/2003 J14	2003	24,011,000	779.18^R	0.3351	2	(?)
	S/2003 J11	2003	24,043,000	764.73^R	0.2640	2	(?)
	Autonoe	2001	24,046,000	760.95^R	0.3168	4	(?)
	Callirrhoe	1999	24,103,000	758.77^R	0.2828	7	(?)
	S/2003 J2	2003	29,541,000	979.99^R	0.2255	2	(?)
SATURN	Pan	1981	133,600	0.575	0	20	(?)
	Atlas	1980	137,700	0.602	0	32	(?)
	Prometheus	1980	139,400	0.613	0.0023	140 × 100 × 80	3×10^{17}
	Pandora	1980	141,700	0.629	0.0044	110 × 90 × 70	2×10^{17}
	Epimetheus	1980	151,400	0.694	0.0205	140 × 120 × 100	6×10^{17}
	Janus	1980	151,500	0.695	0.0073	220 × 200 × 160	2×10^{18}
	Mimas	1789	185,600	0.942	0.0206	397	3.8×10^{19}
	Enceladus	1789	238,100	1.37	0.0001	499	7.3×10^{19}

Appendix 3 | Satellites of the Planets

Planet	Satellite	Date of discovery	Average distance from center of planet (km)	Orbital (sidereal) period (days)*	Orbital eccentricity	Size of satellite (km)**	Mass (kg)
SATURN (continued)	Tethys	1684	294,700	1.888	0.0001	1060	6.2×10^{20}
	Telesto	1980	294,700	1.888	0.0010	$24 \times 22 \times 22$	(?)
	Calypso	1980	294,700	1.888	0.0005	$34 \times 28 \times 26$	(?)
	Dione	1684	377,400	2.737	0.0002	1118	1.1×10^{21}
	Helene	1980	377,400	2.737	0.0001	$36 \times 32 \times 30$	(?)
	Rhea	1672	527,100	4.518	0.0009	1528	2.3×10^{21}
	Titan	1655	1,221,900	15.95	0.0288	5150	1.34×10^{23}
	Hyperion	1848	1,464,100	21.28	0.0175	$410 \times 260 \times 220$	2×10^{19}
	Iapetus	1671	3,560,800	79.33	0.0284	1436	1.6×10^{21}
	Kiviuq	2000	11,365,000	449.22	0.3336	16	(?)
	Ijiraq	2000	11,442,000	451.47	0.3215	12	(?)
	Phoebe	1898	12,944,300	548.21[R]	0.1644	120	4×10^{18}
	Paaliaq	2000	15,198,000	686.94	0.3632	22	(?)
	Skadi	2000	15,641,000	728.18[R]	0.2690	8	(?)
	Albiorix	2000	16,394,000	783.47	0.4791	32	(?)
	Erriapo	2000	17,604,000	871.25	0.4740	10	(?)
	Siarnaq	2000	18,195,000	895.55	0.2962	40	(?)
	Tarvos	2000	18,239,000	926.13	0.5365	15	(?)
	S/2003 S1	2003	18,719,000	956.19[R]	0.3522	7	(?)
	Mundilfari	2000	18,722,000	951.56[R]	0.2078	7	(?)
	Suttung	2000	19,465,000	1016.51[R]	0.1140	7	(?)
	Thrym	2000	20,219,000	1091.76[R]	0.4852	7	(?)
	Ymir	2000	23,130,000	1315.33[R]	0.3339	14	(?)
URANUS	Cordelia	1986	49,800	0.335	0.0003	40	(?)
	Ophelia	1986	53,800	0.376	0.0099	42	(?)
	Bianca	1986	59,200	0.435	0.0009	51	(?)
	Cressida	1986	61,800	0.464	0.0004	80	(?)
	Desdemona	1986	62,700	0.474	0.0001	64	(?)
	Juliet	1986	64,400	0.493	0.0007	93	(?)
	Portia	1986	66,100	0.513	0.0001	135	(?)
	Rosalind	1986	69,900	0.558	0.0001	72	(?)
	Belinda	1986	75,300	0.624	0.0001	80	(?)
	Puck	1985	86,000	0.762	0.0001	162	(?)
	Miranda	1948	129,900	1.413	0.0013	471	6.6×10^{19}
	Ariel	1851	190,900	2.52	0.0012	1158	1.4×10^{21}
	Umbriel	1851	266,000	4.144	0.0039	1169	1.2×10^{21}

Appendix 3 | Satellites of the Planets

Planet	Satellite	Date of discovery	Average distance from center of planet (km)	Orbital (sidereal) period (days)*	Orbital eccentricity	Size of satellite (km)**	Mass (kg)
URANUS (continued)	Titania	1787	436,300	8.706	0.0011	1578	3.5×10^{21}
	Oberon	1787	583,500	13.46	0.0014	1522	3.0×10^{21}
	S/2001 U3	2001	4,276,000	266.56^R	0.1459	12	(?)
	Caliban	1997	7,231,000	579.73^R	0.1587	98	(?)
	Stephano	1999	8,004,000	677.36^R	0.2292	20	(?)
	Trinculo	2001	8,504,000	749.24^R	0.2200	10	(?)
	Sycorax	1997	12,179,000	1288.3^R	0.5224	190	(?)
	S/2003 U3	2003	14,345,000	1687.01^R	0.6608	11	(?)
	Prospero	1999	16,256,000	1978.29^R	0.4448	30	(?)
	Setebos	1999	17,418,000	2225.21^R	0.5914	30	(?)
	S/2001 U2	2001	20,901,000	2887.21^R	0.3682	12	(?)
NEPTUNE	Naiad	1989	48,200	0.294	0.0003	58	(?)
	Thalassa	1989	50,100	0.311	0.0002	80	(?)
	Despina	1989	52,500	0.335	0.0001	148	(?)
	Galatea	1989	62,000	0.429	0.0001	158	(?)
	Larissa	1989	73,500	0.555	0.0014	192	(?)
	Proteus	1989	117,600	1.122	0.0004	416	(?)
	Triton	1846	354,800	5.877^R	0.0000	2706	2.15×10^{22}
	Nereid	1949	5,513,400	360.14	0.7512	340	(?)
	S/2002 N1	2002	15,686,000	1874.83^R	0.5720	60	(?)
	S/2002 N2	2002	22,452,000	2918.94	0.2973	38	(?)
	S/2002 N3	2002	22,580,000	2982.27	0.4789	38	(?)
	S/2002 N4	2002	46,570,000	8863.08^R	0.5273	60	(?)
	S/2003 N1	2003	46,738,000	9136.11^R	0.4503	38	(?)
PLUTO	Charon	1978	19,410	6.387	0.0002	1190	1.9×10^{21}

This table was compiled from data provided by the Jet Propulsion Laboratory.

A superscript R means that the satellite orbits in a retrograde direction (opposite to the planet's rotation).

**The size of a spherical satellite is equal to its diameter.*

Appendix 4 | The Nearest Stars

Name*	Parallax (arcsec)	Distance (parsecs)	Distance (light-years)	Spectral type	Proper motion (arcsec/yr)	Apparent visual magnitude	Absolute visual magnitude	Mass (Sun = 1)
Proxima Centauri	0.772	1.30	4.22	M5.5 V	3.853	+11.09	+15.53	0.107
Alpha Centauri A	0.747	1.34	4.36	G2 V	3.710	−0.01	+4.36	1.144
Alpha Centauri B	0.747	1.34	4.36	K0 V	3.724	+1.34	+5.71	0.916
Barnard's Star	0.547	1.83	5.96	M4.0 V	10.358	+9.53	+13.22	0.166
Wolf 359	0.419	2.39	7.78	M6.0 V	4.696	+13.44	+16.55	0.092
Lalande 21185	0.393	2.54	8.29	M2.0 V	4.802	+7.47	+10.44	0.464
Sirius A	0.380	2.63	8.58	A1 V	1.339	−1.43	+1.47	1.991
Sirius B	0.380	2.63	8.58	white dwarf	1.339	+8.44	+11.34	0.500
UV Ceti	0.374	2.68	8.73	M5.5 V	3.368	+12.54	+15.40	0.109
BL Ceti	0.374	2.68	8.73	M6.0 V	3.368	+12.99	+15.85	0.102
Ross 154	0.337	2.97	9.68	M3.5 V	0.666	+10.43	+13.07	0.171
Ross 248	0.316	3.16	10.32	M5.5 V	1.617	+12.29	+14.79	0.121
Epsilon Eridani	0.310	3.23	10.52	K2 V	0.977	+3.73	+6.19	0.850
Lacaille 9352	0.304	3.29	10.74	M1.5 V	6.896	+7.34	+9.75	0.529
Ross 128	0.299	3.35	10.92	M4.0 V	1.361	+11.13	+13.51	0.156
EZ Aquarii A	0.290	3.45	11.27	M5.0 V	3.254	+13.33	+15.64	0.105
EZ Aquarii B	0.290	3.45	11.27	—	3.254	+13.27	+15.58	0.106
EZ Aquarii C	0.290	3.45	11.27	—	3.254	+14.03	+16.34	0.095
Procyon A	0.286	3.50	11.40	F5 IV-V	1.259	+0.38	+2.66	1.569
Procyon B	0.286	3.50	11.40	White Dwarf	1.259	+10.70	+12.98	0.500
61 Cygni A	0.286	3.50	11.40	K5.0 V	5.281	+5.21	+7.49	0.703
61 Cygni B	0.286	3.50	11.40	K7.0 V	5.172	+6.03	+8.31	0.630
GJ725 A	0.283	3.53	11.53	M3.0 V	2.238	+8.90	+11.16	0.351
GJ725 B	0.283	3.53	11.53	M3.5 V	2.313	+9.69	+11.95	0.259
GX Andromedae	0.281	3.56	11.62	M1.5 V	2.918	+8.08	+10.32	0.486
GQ Andromedae	0.281	3.56	11.62	M3.5 V	2.918	+11.06	+13.30	0.163
Epsilon Indi A	0.276	3.63	11.82	K5 V	4.704	+4.69	+6.89	0.766
Epsilon Indi B	0.276	3.63	11.82	T1.0	4.823			0.044
Epsilon Indi C	0.276	3.63	11.82	T6.0	4.823			0.028
DX Cancri	0.276	3.63	11.83	M6.5 V	1.290	+14.78	+16.98	0.087
Tau Ceti	0.274	3.64	11.89	G8 V	1.922	+3.49	+5.68	0.921
RECONS 1	0.272	3.68	11.99	M5.5 V	0.814	+13.03	+15.21	0.113
YZ Ceti	0.269	3.72	12.13	M4.5 V	1.372	+12.02	+14.17	0.136
Luyten's Star	0.264	3.79	12.37	M3.5 V	3.738	+9.86	+11.97	0.257
Kapteyn's Star	0.255	3.92	12.78	M1.5 V	8.670	+8.84	+10.87	0.393
AX Microscopium	0.253	3.95	12.87	M0.0 V	3.455	+6.67	+8.69	0.600

This table, compiled from data reported by the Research Consortium on Nearby Stars, lists all known stars within 4.00 parsecs (13.05 light-years).
* Stars that are components of multiple star systems are labeled A, B, and C.

Appendix 5 | The Visually Brightest Stars

Name	Designation	Distance (parsecs)	Distance (light-years)	Spectral type	Radial velocity (km/s)*	Proper motion (arcsec/year)	Apparent visual magnitude	Apparent visual brightness (Sirius = 1)**	Absolute visual magnitude
Sirius A	α CMa A	2.63	8.58	A1 V	–7.6	1.34	–1.43	1.000	+1.46
Canopus	α Car	95.9	313	F0 II	+20.5	0.03	–0.72	0.520	–5.63
Arcturus	α Boo	11.3	36.7	K1.5 III	–5.2	2.28	–0.04	0.278	–0.30
Alpha Centauri A	α Cen A	1.34	4.36	G2 V	–25	3.71	–0.01	0.270	+4.36
Vega	α Lyr	7.76	25.3	A0 V	–13.9	0.35	+0.03	0.261	+0.58
Capella	α Aur	12.9	42.2	G5 III	+30.2	0.43	+0.08	0.249	–0.48
Rigel	α Ori A	237	773	B8 Ia	+20.7	0.002	+0.12	0.240	–6.75
Procyon	α CMi A	3.50	11.4	F5 IV–V	–3.2	1.26	+0.34	0.196	+2.62
Achernar	α Eri	44.1	144	B3 V	+16	0.10	+0.50	0.169	–2.72
Betelgeuse	α Ori	131	427	M1 Iab	+21	0.03	+0.58	0.157	–5.01
Hadar	β Cen	161	525	B1 III	+5.9	0.04	+0.60	0.154	–5.43
Altair	α Aql	51.4	168	A7 V	–26.1	0.66	+0.77	0.132	–2.79
Aldebaran	α Tau A	20.0	65.1	K5 III	+54.3	0.20	+0.85	0.122	–0.65
Spica	α Vir	80.4	262	B1 III-IV	+1	0.05	+1.04	0.103	–3.49
Antares	α Sco A	185	604	M1.5 Iab	–3.4	0.03	+1.09	0.098	–5.25
Pollux	β Gem	10.3	33.7	K0 IIIb	+3.3	0.63	+1.15	0.093	+1.08
Fomalhaut	α PsA	7.69	25.1	A3 V	+6.5	0.37	+1.16	0.092	+1.73
Deneb	α Cyg	990	3230	A2 Ia	–4.5	0.002	+1.25	0.085	–8.73
Mimosa	β Cru	108	353	B0.5 IV	+15.6	0.05	+1.297	0.081	–3.87
Regulus	α Leo A	23.8	77.5	B7 V	+5.9	0.25	+1.35	0.077	–0.53

Data in this table was compiled from the SIMBAD database operated at the Centre de Données Astronomiques de Strasbourg, France.

* A positive radial velocity means the star is receding; a negative radial velocity means the star is approaching.

** This is the ratio of the star's apparent brightness to that of Sirius, the brightest star in the night sky.

Note: Acrux, or α Cru (the brightest star in Crux, the Southern Cross) appears to the naked eye as a star of apparent magnitude +0.87, the same as Aldebaran. However, it does not appear in this table because Acrux is actually a binary star system. The blue-white component stars of this binary system have apparent magnitudes of +1.4 and +1.9, and so they are dimmer than any of the stars listed here.

Appendix 6	Some Important Astronomical Quantities

Astronomical unit	$1\text{ AU} = 1.4960 \times 10^{11}$ m
	$= 1.4960 \times 10^{8}$ km
Light-year	$1\text{ ly} = 9.4605 \times 10^{15}$ m
	$= 63{,}240$ AU
Parsec	$1\text{ pc} = 3.2616$ ly
	$= 3.0857 \times 10^{16}$ m
	$= 206{,}265$ AU
Year	$1\text{ y} = 365.2564$ days
	$= 3.156 \times 10^{7}$ s
Solar mass	$1\text{ M}_{\odot} = 1.989 \times 10^{30}$ kg
Solar radius	$1\text{ R}_{\odot} = 6.9599 \times 10^{8}$ m
Solar luminosity	$1\text{ L}_{\odot} = 3.90 \times 10^{26}$ W

Appendix 8	Some Useful Mathematics

Area of a rectangle of sides a and b	$A = ab$
Volume of a rectangular solid of sides a, b, and c	$V = abc$
Hypotenuse of a right triangle whose other sides are a and b	$c = \sqrt{a^2 + b^2}$
Circumference of a circle of radius r	$C = 2\pi r$
Area of a circle of radius r	$A = \pi r^2$
Surface area of a sphere of radius r	$A = 4\pi r^2$
Volume of a sphere of radius r	$V = 4\pi r^3/3$
Value of π	$\pi = 3.1415926536$

Appendix 7	Some Important Physical Constants

Speed of light	$c = 2.9979 \times 10^{8}$ m/s
Gravitational constant	$G = 6.673 \times 10^{-11}$ N m^2/kg^2
Planck's constant	$h = 6.6261 \times 10^{-34}$ J s
	$= 4.1357 \times 10^{-15}$ eV s
Boltzmann constant	$k = 1.3807 \times 10^{-23}$ J/K
	$= 8.6173 \times 10^{-5}$ eV/K
Stefan-Boltzmann constant	$\sigma = 5.6704 \times 10^{-8}$ W m^{-2} K^{-4}
Mass of electron	$m_{\mathrm{e}} = 9.1094 \times 10^{-31}$ kg
Mass of proton	$m_{\mathrm{p}} = 1.6726 \times 10^{-27}$ kg
Mass of neutron	$m_{\mathrm{n}} = 1.6749 \times 10^{-27}$ kg
Mass of hydrogen atom	$m_{\mathrm{H}} = 1.6735 \times 10^{-27}$ kg
Rydberg constant	$R = 1.0973 \times 10^{7}$ m^{-1}
Electron volt	$1\text{ eV} = 1.6022 \times 10^{-19}$ J

Glossary

You can find more information about each term in the indicated chapter or chapters. See the Key Words section in each chapter for the specific page on which the meaning of a given term is described.

A ring One of three prominent rings encircling Saturn. (Chapter 14)

absolute magnitude The apparent magnitude that a star would have if it were at a distance of 10 parsecs. (Chapter 19)

absolute zero A temperature of –273°C (or 0 K), at which all molecular motion stops; the lowest possible temperature. (Chapter 5)

absorption line spectrum Dark lines superimposed on a continuous spectrum. (Chapter 5)

acceleration The rate at which an object's velocity changes due to a change in speed, a change in direction, or both. (Chapter 4)

accretion The gradual accumulation of matter in one location, typically due to the action of gravity. (Chapter 8, Chapter 20)

accretion disk A disk of gas orbiting a star or black hole. (Chapter 27)

active galactic nucleus (AGN) The center of an active galaxy. (Chapter 27)

active galaxy A galaxy that is emitting exceptionally large amounts of energy; a Seyfert galaxy, radio galaxy, or quasar. (Chapter 27)

active optics A technique for improving a telescopic image by altering the telescope's optics to compensate for variations in air temperature or flexing of the telescope mount. (Chapter 6)

adaptive optics A technique for improving a telescopic image by altering the telescope's optics in a way that compensates for distortion caused by the Earth's atmosphere. (Chapter 6)

aerosol Tiny droplets of liquid dispersed in a gas. (Chapter 15)

AGB star See *asymptotic giant branch star.*

AGN See *active galactic nucleus.*

albedo The fraction of sunlight that a planet, asteroid, or satellite reflects. (Chapter 9)

alpha particle The nucleus of a helium atom, consisting of two protons and two neutrons. (Chapter 21)

amino acids The chemical building blocks of proteins. (Chapter 17)

andesite Low-density, silicon-rich volcanic rocks found in the young northern lowlands of Mars. (Chapter 13)

angle The opening between two lines that meet at a point. (Chapter 1)

angular diameter The angle subtended by the diameter of an object. (Chapter 1)

angular distance The angle between two points in the sky. (Chapter 1)

angular measure The size of an angle, usually expressed in degrees, arcminutes, and arcseconds. (Chapter 1)

angular momentum See *conservation of angular momentum.*

angular resolution The angular size of the smallest feature that can be distinguished with a telescope. (Chapter 6)

angular size See *angular diameter.*

anisotropic Possessing unequal properties in different directions; not isotropic. (Chapter 26)

annihilation The process by which the masses of a particle and antiparticle are converted into energy. (Chapter 29)

annular eclipse An eclipse of the Sun in which the Moon is too distant to cover the Sun completely, so that a ring of sunlight is seen around the Moon at mid-eclipse. (Chapter 3)

anorthosite Rock commonly found in ancient, cratered highlands on the Moon. (Chapter 10)

Antarctic Circle A circle of latitude 23½° north of the Earth's south pole. (Chapter 2)

antimatter Matter consisting of antiparticles such as antiprotons, antineutrons, and positrons. (Chapter 29)

antiparticle A particle with the same mass as an ordinary particle but with other properties, such as electric charge, reversed. (Chapter 29)

antiproton A particle with the same mass as a proton but with a negative electric charge. (Chapter 29)

aphelion The point in its orbit where a planet is farthest from the Sun. (Chapter 4)

apogee The point in its orbit where a satellite or the Moon is farthest from the Earth. (Chapter 3)

apparent brightness The flux of a star's light arriving at the Earth. (Chapter 19)

apparent magnitude A measure of the brightness of light from a star or other object as measured from Earth. (Chapter 19)

apparent solar day The interval between two successive transits of the Sun's center across the local meridian. (Chapter 2)

apparent solar time Time reckoned by the position of the Sun in the sky. (Chapter 2)

arcminute One-sixtieth (1/60) of a degree, designated by the symbol ′. (Chapter 1)

arcsecond One-sixtieth (1/60) of an arcminute or 1/3600 of a degree, designated by the symbol ″. (Chapter 1)

Arctic Circle A circle of latitude 23½° south of the Earth's north pole. (Chapter 2)

asteroid One of tens of thousands of small, rocky, planetlike objects in orbit about the Sun. Also called minor planets. (Chapter 7, Chapter 17)

asteroid belt A region between the orbits of Mars and Jupiter that encompasses the orbits of many asteroids. (Chapter 7, Chapter 17)

asthenosphere A warm, plastic layer of the mantle beneath the lithosphere of the Earth. (Chapter 9)

astrobiology The study of life in the universe. (Chapter 30)

astrometric method A technique for detecting extrasolar planets by looking for stars that "wobble" periodically. (Chapter 8)

astronomical unit (AU) The semimajor axis of the Earth's orbit; the average distance between the Earth and the Sun. (Chapter 1)

asymptotic giant branch A region on the Hertzsprung-Russell diagram occupied by stars in their second and final red-giant phase. (Chapter 22)

asymptotic giant branch star (AGB star) A red giant star that has ascended the giant branch for the second and final time. (Chapter 22)

atmosphere (atm) A unit of atmospheric pressure. (Chapter 9)

atmospheric pressure The force per unit area exerted by a planet's atmosphere. (Chapter 9)

atom The smallest particle of an element that has the properties characterizing that element. (Chapter 5)

atomic number The number of protons in the nucleus of an atom of a particular element. (Chapter 5, Chapter 8)

AU See *astronomical unit*.

aurora (plural **aurorae**) Light radiated by atoms and ions in the Earth's upper atmosphere, mostly in the polar regions. (Chapter 9)

aurora australis Aurorae seen from southern latitudes; the southern lights. (Chapter 9)

aurora borealis Aurorae seen from northern latitudes; the northern lights. (Chapter 9)

autumnal equinox The intersection of the ecliptic and the celestial equator where the Sun crosses the equator from north to south. Also used to refer to the date on which the Sun passes through this intersection. (Chapter 2)

average density The mass of an object divided by its volume. (Chapter 7)

average density of matter In cosmology, the average density of all the material substance in the universe. (Chapter 28)

B ring One of three prominent rings encircling Saturn. (Chapter 14)

Balmer line An emission or absorption line in the spectrum of hydrogen caused by an electron transition between the second and higher energy levels. (Chapter 5)

Balmer series The entire pattern of Balmer lines. (Chapter 5)

Barnard object One of a class of dark nebulae discovered by E. E. Barnard. (Chapter 20)

barred spiral galaxy A spiral galaxy in which the spiral arms begin from the ends of a "bar" running through the nucleus rather than from the nucleus itself. (Chapter 26)

baseline In interferometry, the distance between two telescopes whose signals are combined to give a higher-resolution image. (Chapter 6)

belt A dark band in Jupiter's atmosphere. (Chapter 14)

Big Bang An explosion of all space that took place roughly 13.7 billion years ago and that marks the beginning of the universe. (Chapter 1, Chapter 28)

binary star (also **binary star system** or **binary**) Two stars orbiting about each other. (Chapter 19)

biosphere The layer of soil, water, and air surrounding the Earth in which living organisms thrive. (Chapter 9)

bipolar outflow Oppositely directed jets of gas expelled from a young star. (Chapter 20)

black hole An object whose gravity is so strong that the escape speed exceeds the speed of light. (Chapter 1, Chapter 24)

black hole evaporation The process by which black holes emit particles. (Chapter 24)

blackbody A hypothetical perfect radiator that absorbs and re-emits all radiation falling upon it. (Chapter 5)

blackbody curve The intensity of radiation emitted by a blackbody plotted as a function of wavelength or frequency. (Chapter 5)

blackbody radiation The radiation emitted by a perfect blackbody. (Chapter 5)

blazar A type of active galaxy whose nucleus has a featureless spectrum without prominent spectral lines. (Chapter 27)

blueshift A decrease in the wavelength of photons emitted by an approaching source of light. (Chapter 5)

Bohr orbits In the model of the atom described by Niels Bohr, the only orbits in which electrons are allowed to move about the nucleus. (Chapter 5)

Bok globule A small, roundish, dark nebula. (Chapter 20)

bright terrain (Ganymede) Young, reflective, relatively crater-free terrain on the surface of Ganymede. (Chapter 15)

brightness See *apparent brightness*.

brown dwarf A starlike object that is not massive enough to sustain hydrogen fusion in its core. (Chapter 8, Chapter 19)

brown ovals Elongated, brownish features usually seen in Jupiter's northern hemisphere. (Chapter 14)

C ring One of three prominent rings encircling Saturn. (Chapter 14)

capture theory The hypothesis that the Moon was gravitationally captured by the Earth. (Chapter 10)

carbon fusion The thermonuclear fusion of carbon to produce heavier nuclei. (Chapter 22)

carbon star A peculiar red giant star whose spectrum shows strong absorption lines of carbon and carbon compounds. (Chapter 22)

carbonaceous chondrite A type of meteorite that has a high abundance of carbon and volatile compounds. (Chapter 17, Chapter 30)

Cassegrain focus An optical arrangement in a reflecting telescope in which light rays are reflected by a secondary mirror to a focus behind the primary mirror. (Chapter 6)

Cassini division An apparent gap between Saturn's A and B rings. (Chapter 14)

CCD See *charge-coupled device*.

celestial equator A great circle on the celestial sphere 90° from the celestial poles. (Chapter 2)

celestial poles The points about which the celestial sphere appears to rotate. See also *north celestial pole* and *south celestial pole*. (Chapter 2)

celestial sphere An imaginary sphere of very large radius centered on an observer; the apparent sphere of the sky. (Chapter 2)

center of mass The point between a star and a planet, or between two stars, around which both objects orbit. (Chapter 8, Chapter 10, Chapter 19)

central bulge A spherical distribution of stars around the nucleus of a spiral galaxy. (Chapter 25)

Cepheid variable A type of yellow, supergiant, pulsating star. (Chapter 21)

Cerenkov radiation The radiation emitted by a particle traveling through a substance at a speed greater than the speed of light in that substance. (Chapter 22)

Chandrasekhar limit The maximum mass of a white dwarf. (Chapter 22)

charge-coupled device (CCD) A type of solid-state device designed to detect photons. (Chapter 6)

chemical composition A description of which chemical substances make up a given object. (Chapter 7)

chemical differentiation The process by which the heavier elements in a planet sink toward its center while lighter elements rise toward its surface. (Chapter 8)

chemical element See *element*.

chondrule A glassy, roughly spherical blob found within meteorites. (Chapter 8)

chromatic aberration An optical defect whereby different colors of light passing through a lens are focused at different locations. (Chapter 6)

chromosphere A layer in the atmosphere of the Sun between the photosphere and the corona. (Chapter 18)

circumpolar A term describing a star that neither rises nor sets but appears to rotate around one of the celestial poles. (Chapter 2)

circumstellar accretion disk An accretion disk that surrounds a protostar. (Chapter 20)

close binary A binary star system in which the stars are separated by a distance roughly comparable to their diameters. (Chapter 21)

closed universe A universe with positive curvature, so that its geometry is analogous to that of the surface of a sphere. (Chapter 28)

cluster of galaxies A collection of galaxies containing a few to several thousand member galaxies. (Chapter 26)

cluster of stars A group of stars that formed together and that have remained together due to their mutual gravitational attraction. (Chapter 20)

CNO cycle A series of nuclear reactions in which carbon is used as a catalyst to transform hydrogen into helium. (Chapter 18)

cocoon nebula The nebulosity surrounding a protostar. (Chapter 20)

co-creation theory The hypothesis that the Earth and the Moon formed at the same time from the same material. (Chapter 10)

cold dark matter Slowly moving, weakly interacting particles presumed to constitute the bulk of matter in the universe. (Chapter 29)

collapsar A proposed type of supernova in which a black hole forms at the center of a dying star before the outer layers of the star have time to collapse. (Chapter 27)

collisional ejection theory The hypothesis that the Moon formed from material ejected from the Earth by the impact of a large asteroid. (Chapter 10)

color-magnitude diagram A plot of the apparent magnitudes (that is, apparent brightnesses) of stars in a cluster versus their color indices (a measure of their surface temperatures). (Chapter 21)

color ratio The ratio of the apparent brightness of a star measured in one spectral region to its brightness measured in a different region. (Chapter 19)

coma (of a comet) The diffuse gaseous component of the head of a comet. (Chapter 17)

coma (optical) The distortion of off-axis images formed by a parabolic mirror. (Chapter 6)

combined average mass density The average density in the observable universe of all forms of matter and energy, measured in units of mass per volume. (Chapter 28)

comet A small body of ice and dust in orbit about the Sun. While passing near the Sun, a comet's vaporized ices give rise to a coma and tail. (Chapter 7, Chapter 17)

condensation temperature The temperature at which a particular substance in a low-pressure gas condenses into a solid. (Chapter 8)

conduction The transfer of heat by directly passing energy from atom to atom. (Chapter 18)

compression A region in a sound wave where the medium carrying the wave is compressed. (Chapter 28)

conic section The curve of intersection between a circular cone and a plane; this curve can be a circle, ellipse, parabola, or hyperbola. (Chapter 4)

conjunction The geometric arrangement of a planet in the same part of the sky as the Sun, so that the planet is at an elongation of 0°. (Chapter 4)

conservation of angular momentum A law of physics stating that in an isolated system, the total amount of angular momentum—a measure of the amount of rotation—remains constant. (Chapter 8)

constellation A configuration of stars in the same region of the sky. (Chapter 2)

contact binary A binary star system in which both members fill their Roche lobes. (Chapter 21)

continuous spectrum A spectrum of light over a range of wavelengths without any spectral lines. (Chapter 5)

convection The transfer of energy by moving currents of fluid or gas containing that energy. (Chapter 9, Chapter 18)

convection cell A circulating loop of gas or liquid that transports heat from a warm region to a cool region. (Chapter 9)

convection current The pattern of motion in a gas or liquid in which convection is taking place. (Chapter 9)

convective zone The region in a star where convection is the dominant means of energy transport. (Chapter 18)

core (of the Earth) The iron-rich inner region of the Earth's interior. (Chapter 9)

core accretion model The hypothesis that each of the Jovian planets formed by accretion of gas onto a rocky core. (Chapter 8)

core helium fusion The thermonuclear fusion of helium at the center of a star. (Chapter 21, Chapter 22)

core hydrogen fusion The thermonuclear fusion of hydrogen at the center of a star. (Chapter 21)

corona (of the Sun) The Sun's outer atmosphere, which has a high temperature and a low density. (Chapter 18)

coronal hole A region in the Sun's corona that is deficient in hot gases. (Chapter 18)

coronal mass ejection An event in which billions of tons of gas from the Sun's corona is suddenly blasted into space at high speed. (Chapter 9, Chapter 18)

cosmic background radiation See *cosmic microwave background*.

cosmic censorship, law of See *law of cosmic censorship*.

cosmic light horizon An imaginary sphere, centered on the Earth, whose radius equals the distance light has traveled since the Big Bang. (Chapter 28, Chapter 29)

cosmic microwave background An isotropic radiation field with a blackbody temperature of about 2.725 K that permeates the entire universe. (Chapter 28)

cosmic rays Fast-moving subatomic particles that enter our solar system from interstellar space. (Chapter 13)

cosmic singularity The Big Bang. (Chapter 28)

cosmological constant (Λ) In the equations of general relativity, a quantity signifying a pressure throughout all space that helps the universe expand; a type of dark energy. (Chapter 28)

cosmological principle The assumption that the universe is homogeneous and isotropic on the largest scale. (Chapter 28)

cosmological redshift A redshift that is caused by the expansion of the universe. (Chapter 28)

cosmology The study of the structure and evolution of the universe. (Chapter 28)

coudé focus An optical arrangement with a reflecting telescope. A series of mirrors is used to direct light to a remote focus away from the moving parts of the telescope. (Chapter 6)

crater See *impact crater*.

critical density The value of the combined average mass density throughout the universe for which space would be flat. (Chapter 28)

crust (of a planet) The surface layer of a terrestrial planet. (Chapter 9)

crustal dichotomy (Mars) The contrast between the young northern lowlands and older southern highlands on Mars. (Chapter 13)

crystal A material in which atoms are arranged in orderly rows. (Chapter 9)

current sheet A broad, flat region in Jupiter's magnetosphere that contains an abundance of charged particles. (Chapter 14)

D ring One of several faint rings encircling Saturn. (Chapter 14)

dark energy A form of energy that appears to pervade the universe and causes the expansion of the universe to accelerate, but has no discernible gravitational effect. (Chapter 28)

dark energy density parameter (Ω_Λ) The ratio of the average mass density of dark energy in the universe to the critical density. (Chapter 28)

dark-energy-dominated universe A universe in which the mass density of dark energy exceeds both the average density of matter and the mass density of radiation. (Chapter 28)

dark matter Nonluminous matter that is the dominant form of matter in galaxies and throughout the universe. (Chapter 25)

dark-matter problem The enigma that most of the matter in the universe seems not to emit radiation of any kind and is detectable by its gravity alone. (Chapter 26)

dark nebula A cloud of interstellar gas and dust that obscures the light of more distant stars. (Chapter 20)

dark terrain (Ganymede) Older, heavily cratered, dark-colored terrain on the surface of Ganymede. (Chapter 15)

decametric radiation Radiation from Jupiter whose wavelength is about 10 meters. (Chapter 14)

decimetric radiation Radiation from Jupiter whose wavelength is about a tenth of a meter. (Chapter 14)

declination Angular distance of a celestial object north or south of the celestial equator. (Chapter 2)

deferent A stationary circle in the Ptolemaic system along which another circle (an epicycle) moves, carrying a planet, the Sun, or the Moon. (Chapter 4)

degeneracy The phenomenon, due to quantum mechanical effects, whereby the pressure exerted by a gas does not depend on its temperature. (Chapter 21)

degenerate electron (neutron) pressure The pressure exerted by degenerate electrons (neutrons). (Chapter 21, Chapter 23)

degree A basic unit of angular measure, designated by the symbol °. (Chapter 1)

degree Celsius A basic unit of temperature, designated by the symbol °C and used on a scale where water freezes at 0° and boils at 100°. (Chapter 5)

degree Fahrenheit A basic unit of temperature, designated by the symbol °F and used on a scale where water freezes at 32° and boils at 212°. (Chapter 5)

density See *average density*.

density fluctuation A variation of density from one place to another. (Chapter 29)

density parameter (Ω_0) The ratio of the average density of matter and energy in the universe to the critical density. (Chapter 28)

density wave In a spiral galaxy, a localized region in which matter piles up as it orbits the center of the galaxy; this is a proposed explanation of spiral structure. (Chapter 25)

detached binary A binary star system in which neither star fills its Roche lobe. (Chapter 21)

deuterium bottleneck The situation during the first 3 minutes after the Big Bang when deuterium (an isotope of hydrogen whose nucleus contains one proton and one neutron) inhibited the formation of heavier elements. (Chapter 29)

differential rotation The rotation of a nonrigid object in which parts adjacent to each other at a given time do not always stay close together. (Chapter 14, Chapter 18)

differentiated asteroid An asteroid in which chemical differentiation has taken place, so that denser material is toward the asteroid's center. (Chapter 17)

differentiation See *chemical differentiation*.

diffraction The spreading out of light passing through an aperture or opening in an opaque object. (Chapter 6)

diffraction grating An optical device, consisting of thousands of closely spaced lines etched in glass or metal, that disperses light into a spectrum. (Chapter 6)

direct motion The apparent eastward movement of a planet seen against the background stars. (Chapter 4)

disk (of a galaxy) The disk-shaped distribution of Population I stars that dominates the appearance of a spiral galaxy. (Chapter 25)

disk instability model The hypothesis that gases in the solar nebula coalesced rapidly to form the Jovian planets. (Chapter 8)

distance ladder The sequence of techniques used to determine the distances to very remote galaxies. (Chapter 26)

distance modulus The difference between the apparent and absolute magnitudes of an object. (Chapter 19)

diurnal motion Any apparent motion in the sky that repeats on a daily basis, such as the rising and setting of stars. (Chapter 2)

Doppler effect The apparent change in wavelength of radiation due to relative motion between the source and the observer along the line of sight. (Chapter 5)

double radio source An extragalactic radio source characterized by two large regions of radio emission, typically located on either side of an active galaxy. (Chapter 27)

double star A pair of stars located at nearly the same position in the night sky. Some, but not all, double stars are binary stars. (Chapter 19)

Drake equation An equation used to estimate the number of intelligent civilizations in the Galaxy with whom we might communicate. (Chapter 30)

dredge-up The transporting of matter from deep within a star to its surface by convection. (Chapter 22)

dust devil Whirlwinds found in dry or desert areas on both Earth and Mars. (Chapter 13)

dust grain A microscopic bit of solid matter found in interplanetary or interstellar space. (Chapter 20)

dust tail The tail of a comet that is composed primarily of dust grains. (Chapter 17)

dwarf elliptical galaxy A low-mass elliptical galaxy that contains only a few million stars. (Chapter 26)

dynamo The mechanism whereby electric currents within an astronomical body generate a magnetic field. (Chapter 7)

E ring A very broad, faint ring encircling Saturn. (Chapter 14)

earthquake A sudden vibratory motion of the Earth's surface. (Chapter 9)

eccentricity A number between 0 and 1 that describes the shape of an ellipse. (Chapter 4)

eclipse The cutting off of part or all the light from one celestial object by another. (Chapter 3)

eclipse path The track of the tip of the Moon's shadow along the Earth's surface during a total or annular solar eclipse. (Chapter 3)

eclipse year The interval between successive passages of the Sun through the same node of the Moon's orbit. (Chapter 3)

eclipsing binary A binary star system in which, as seen from Earth, the stars periodically pass in front of each other. (Chapter 19)

ecliptic The apparent annual path of the Sun on the celestial sphere. (Chapter 2)

Eddington limit A constraint on the rate at which a black hole can accrete matter. (Chapter 27)

electromagnetic radiation Radiation consisting of oscillating electric and magnetic fields. Examples include gamma rays, X rays, visible light, ultraviolet and infrared radiation, radio waves, and microwaves. (Chapter 5)

electromagnetic spectrum The entire array of electromagnetic radiation. (Chapter 5)

electromagnetism Electric and magnetic phenomena, including electromagnetic radiation. (Chapter 5)

electron A subatomic particle with a negative charge and a small mass, usually found in orbits about the nuclei of atoms. (Chapter 5)

electron volt (eV) The energy acquired by an electron accelerated through an electric potential of one volt. (Chapter 5)

electroweak force A unification of the electromagnetic and weak forces that occurs at high energies. (Chapter 29)

element A chemical that cannot be broken down into more basic chemicals. (Chapter 5)

elementary particle physics The branch of quantum mechanics that deals with individual subatomic particles and their interactions. (Chapter 29)

ellipse A conic section obtained by cutting completely through a circular cone with a plane. (Chapter 4)

elliptical galaxy A galaxy with an elliptical shape and no conspicuous interstellar material. (Chapter 26)

elongation The angular distance between a planet and the Sun as viewed from Earth. (Chapter 4)

emission line spectrum A spectrum that contains bright emission lines. (Chapter 5)

emission nebula A glowing gaseous nebula whose spectrum has bright emission lines. (Chapter 20)

Encke gap A narrow gap in Saturn's A ring. (Chapter 14)

energy flux The rate of energy flow, usually measured in joules per square meter per second. (Chapter 5)

energy level In an atom, a particular amount of energy possessed by an atom above the atom's least energetic state. (Chapter 5)

energy-level diagram A diagram showing the arrangement of an atom's energy levels. (Chapter 5)

epicenter The location on the Earth's surface directly over the focus of an earthquake. (Chapter 9)

epicycle A moving circle in the Ptolemaic system about which a planet revolves. (Chapter 4)

epoch The date used to define the coordinate system for objects on the sky. (Chapter 2)

equal areas, law of See *Kepler's second law.*

equilibrium resurfacing hypothesis (Venus) A proposed explanation for the young age of Venus's surface. It hypothesizes that volcanic eruptions on Venus are continually covering up craters at about the same rate that craters are formed by impact. (Chapter 12)

equinox One of the intersections of the ecliptic and the celestial equator. Also used to refer to the date on which the Sun passes through such an intersection. (Chapter 2) See also *autumnal equinox* and *vernal equinox.*

equivalence principle In the general theory of relativity, the principle that in a small volume of space, the downward pull of gravity can be accurately and completely duplicated by an upward acceleration of the observer. (Chapter 24)

era of recombination The moment, approximately 380,000 years after the Big Bang, when the universe had cooled sufficiently to permit the formation of hydrogen atoms. (Chapter 28)

ergoregion The region of space immediately outside the event horizon of a rotating black hole where it is impossible to remain at rest. (Chapter 24)

escape speed The speed needed by an object (such as a spaceship) to leave a second object (such as a planet or star) permanently and to escape into interplanetary space. (Chapter 7)

eV See *electron volt.*

event horizon The location around a black hole where the escape speed equals the speed of light; the surface of a black hole. (Chapter 24)

evolutionary track The path on an H-R diagram followed by a star as it evolves. (Chapter 20)

excited state A state of an atom, ion, or molecule with a higher energy than the ground state. (Chapter 5)

exobiology The biology of life on other worlds. (Chapter 30)

exponent A number placed above and after another number to denote the power to which the latter is to be raised, as n in 10^n. (Chapter 1)

extinction (interstellar) See *interstellar extinction.*

extrasolar planet A planet orbiting a star other the Sun. (Chapter 8)

eyepiece lens A magnifying lens used to view the image produced at the focus of a telescope. (Chapter 6)

F ring A thin, faint ring encircling Saturn just beyond the A ring. (Chapter 14)

false color In astronomical images, color used to denote different values of intensity, temperature, or other quantities. (Chapter 6)

false vacuum Empty space with an abnormally high energy density that is thought to have filled the universe immediately after the Big Bang. (Chapter 29)

far-infrared The part of the infrared spectrum most different in wavelength from visible light. (Chapter 25)

far side (of the Moon) The side of the Moon that faces perpetually away from the Earth. (Chapter 10)

favorable opposition An opposition of Mars that affords good Earth-based views of the planet. (Chapter 13)

filament A portion of the Sun's chromosphere that arches to high altitudes. (Chapter 18)

first quarter moon The phase of the Moon that occurs when the Moon is 90° east of the Sun. (Chapter 3)

fission theory The hypothesis that the Moon was pulled out of a rapidly rotating proto-Earth. (Chapter 10)

flare See *solar flare.*

flat space Space that is not curved; space with zero curvature. (Chapter 28)

flatness problem The dilemma posed by the fact that the combined average mass density of the universe is very nearly equal to the critical density. (Chapter 29)

flocculent spiral galaxy A spiral galaxy with fuzzy, poorly defined spiral arms. (Chapter 25)

fluorescence A process in which high-energy ultraviolet photons are absorbed and the absorbed energy is radiated as lower-energy photons of visible light. (Chapter 20)

focal length The distance from a lens or mirror to the point where converging light rays meet. (Chapter 6)

focal plane The plane in which a lens or mirror forms an image of a distant object. (Chapter 6)

focal point The point at which a lens or mirror forms an image of a distant point of light. (Chapter 6)

focus (of an ellipse) (plural **foci**) One of two points inside an ellipse such that the combined distance from the two foci to any point on the ellipse is a constant. (Chapter 4)

focus (of a lens or mirror) The point to which light rays converge after passing through a lens or being reflected from a mirror. (Chapter 6)

force A push or pull that acts on an object. (Chapter 4)

frequency The number of crests or troughs of a wave that cross a given point per unit time. Also, the number of vibrations per unit time. (Chapter 5)

full moon A phase of the Moon during which its full daylight hemisphere can be seen from Earth. (Chapter 3)

fundamental plane A relationship among the size of an elliptical galaxy, the average motions of its stars, and how the galaxy's brightness appears distributed over its surface. (Chapter 26)

fusion crust The coating on a stony meteorite caused by the heating of the meteorite as it descended through the Earth's atmosphere. (Chapter 17)

G ring A thin, faint ring encircling Saturn. (Chapter 14)

galactic cannibalism A collision between two galaxies of unequal mass and size in which the smaller galaxy seems to be absorbed by the larger galaxy. (Chapter 26)

galactic cluster See *open cluster.*

galactic nucleus The center of a galaxy. (Chapter 25)

galaxy A large assemblage of stars, nebulae, and interstellar gas and dust. (Chapter 1, Chapter 25)

Galilean satellites The four large moons of Jupiter. (Chapter 15)

gamma-ray bursters Objects found in all parts of the sky that emit a one-time intense burst of high-energy radiation. (Chapter 27)

gamma rays The most energetic form of electromagnetic radiation. (Chapter 5)

general theory of relativity A description of gravity formulated by Albert Einstein. It explains that gravity affects the geometry of space and the flow of time. (Chapter 24)

geocentric model An Earth-centered theory of the universe. (Chapter 4)

giant A star whose diameter is typically 10 to 100 times that of the Sun and whose luminosity is roughly that of 100 Suns. (Chapter 19)

giant elliptical galaxy A large, massive, elliptical galaxy containing many billions of stars. (Chapter 26)

giant molecular cloud A large cloud of interstellar gas and dust in which temperatures are low enough and densities high enough for atoms to form into molecules. (Chapter 20)

glitch A sudden speedup in the period of a pulsar. (Chapter 23)

global catastrophe hypothesis (Venus) A proposed explanation for the young age of Venus's surface. It hypothesizes that the entire planet was resurfaced over a short period with fresh lava, covering up any older craters. (Chapter 12)

global warming The upward trend of the Earth's average temperature caused by increased amounts of greenhouse gases in the atmosphere. (Chapter 9)

globular cluster A large spherical cluster of stars, typically found in the outlying regions of a galaxy. (Chapter 21, Chapter 25)

gluon A particle that is exchanged between quarks. (Chapter 29)

grand-design spiral galaxy A galaxy with well-defined spiral arms. (Chapter 25)

grand unified theory (GUT) A theory that describes and explains the four physical forces. (Chapter 29)

granulation The rice grain–like structure found in the solar photosphere. (Chapter 18)

granule A convective cell in the solar photosphere. (Chapter 18)

grating See *diffraction grating*.

gravitational force See *gravity*.

gravitational lens A massive object that deflects light rays from a remote source, forming an image much as an ordinary lens does. (Chapter 26)

gravitational radiation (gravitational waves) Oscillations of space produced by changes in the distribution of matter. (Chapter 24)

gravitational redshift The increase in the wavelength of a photon as it climbs upward in a gravitational field. (Chapter 24)

graviton The particle that is responsible for the gravitational force. (Chapter 29)

gravity The force with which all matter attracts all other matter. (Chapter 4)

Great Red Spot A prominent high-pressure system in Jupiter's southern hemisphere. (Chapter 14)

greatest eastern elongation The configuration of an inferior planet at its greatest angular distance east of the Sun. (Chapter 4, Chapter 11)

greatest western elongation The configuration of an inferior planet at its greatest angular distance west of the Sun. (Chapter 4, Chapter 11)

greenhouse effect The trapping of infrared radiation near a planet's surface by the planet's atmosphere. (Chapter 9)

greenhouse gas A substance whose presence in a planet's atmosphere enhances the greenhouse effect. (Chapter 9)

ground state The state of an atom, ion, or molecule with the least possible energy. (Chapter 5)

group (of galaxies) A poor cluster of galaxies. (Chapter 26)

GUT See *grand unified theory*.

H I Neutral, unionized hydrogen. (Chapter 25)

H II region A region of ionized hydrogen in interstellar space. (Chapter 20)

half-life The time required for one-half of a quantity of a radioactive substance to decay. (Chapter 8)

halo (of a galaxy) A spherical distribution of globular clusters and Population II stars that surround a spiral galaxy. (Chapter 25)

head-tail source A radio source consisting of a bright "head" and a long, dimmer "tail." (Chapter 27)

Heisenberg uncertainty principle A principle of quantum mechanics that places limits on the precision of simultaneous measurements. (Chapter 24, Chapter 29)

heliocentric model A Sun-centered theory of the universe. (Chapter 4)

helioseismology The study of the vibrations of the Sun as a whole. (Chapter 18)

helium fusion The thermonuclear fusion of helium to form carbon and oxygen. (Chapter 21)

helium flash The nearly explosive beginning of helium fusion in the dense core of a red giant star. (Chapter 21)

helium shell flash A brief thermal runaway that occurs in the helium-fusing shell of a red supergiant. (Chapter 22)

Herbig-Haro object A small, luminous nebula associated with the end point of a jet emanating from a young star. (Chapter 20)

Hertzsprung-Russell (H-R) diagram A plot of the luminosity (or absolute magnitude) of stars against their surface temperature (or spectral type). (Chapter 19)

high-velocity star A star traveling at an exceptionally high speed relative to the Sun. (Chapter 25)

highlands (on Mars) See *southern highlands*.

highlands (on the Moon) See *lunar highlands*.

Hirayama family A group of asteroids that have nearly identical orbits about the Sun. (Chapter 17)

homogeneous Having the same property in one region as in every other region. (Chapter 28)

horizon problem See *isotropy problem*.

horizontal branch A group of stars on the color-magnitude diagram of a typical globular cluster that lie near the main sequence and having roughly constant luminosity. (Chapter 22)

horizontal-branch star A low-mass, post–helium-flash star on the horizontal branch. (Chapter 21)

hot dark matter Dark matter consisting of low-mass particles moving at high speeds. (Chapter 29)

hot spot (on Jupiter) An unusually warm and cloud-free part of Jupiter's atmosphere. (Chapter 14)

hot-spot volcanism Volcanic activity that occurs over a hot region buried deep within a planet. (Chapter 12)

H-R diagram See *Hertzsprung-Russell diagram*.

Hubble classification A method of classifying galaxies as spirals, barred spirals, ellipticals, or irregulars according to their appearance. (Chapter 26)

Hubble constant (H_0) In the Hubble law, the constant of proportionality between the recessional velocities of remote galaxies and their distances. (Chapter 26)

Hubble flow The recessional motions of remote galaxies caused by the expansion of the universe. (Chapter 26)

Hubble law The empirical relationship stating that the redshifts of remote galaxies are directly proportional to their distances from Earth. (Chapter 26)

hydrocarbon Any one of a variety of chemical compounds composed of hydrogen and carbon. (Chapter 15)

hydrogen envelope A huge, tenuous sphere of gas surrounding the head of a comet. (Chapter 17)

hydrogen fusion The thermonuclear conversion of hydrogen into helium. (Chapter 18)

hydrostatic equilibrium A balance between the weight of a layer in a star and the pressure that supports it. (Chapter 18)

hyperbola A conic section formed by cutting a circular cone with a plane at an angle steeper than the sides of the cone. (Chapter 4)

hyperbolic space Space with negative curvature. (Chapter 28)

hypothesis An idea or collection of ideas that seems to explain a specified phenomenon; a conjecture. (Chapter 1)

ice rafts (Europa) Segments of Europa's icy crust that have been moved by tectonic disturbances. (Chapter 15)

ices Solid materials with low condensation temperatures, including ices of water, methane, and ammonia. (Chapter 7)

ideal gas A gas in which the pressure is directly proportional to both the density and the temperature of the gas; an idealization of a real gas. (Chapter 21)

igneous rock A rock that formed from the solidification of molten lava or magma. (Chapter 9)

imaging The process of recording the image made by a telescope of a distant object. (Chapter 6)

impact breccia A type of rock formed from other rocks that were broken apart, mixed, and fused together by a series of meteoritic impacts. (Chapter 10)

impact crater A circular depression on a planet or satellite caused by the impact of a meteoroid. (Chapter 7, Chapter 10)

inertia, law of See *Newton's first law.*

inferior conjunction The configuration when an inferior planet is between the Sun and Earth. (Chapter 4)

inferior planet A planet that is closer to the Sun than the Earth is. (Chapter 4)

inflation A sudden expansion of space. (Chapter 29)

inflationary epoch A brief period shortly after the Big Bang during which the scale of the universe increased very rapidly. (Chapter 29)

infrared radiation Electromagnetic radiation of wavelength longer than visible light but shorter than radio waves. (Chapter 5)

inner core (of the Earth) The solid innermost portion of the Earth's iron-rich core. (Chapter 9)

inner Lagrangian point The point between the two stars comprising a binary system where their Roche lobes touch; the point across which mass transfer can occur. (Chapter 21)

instability strip A region of the H-R diagram occupied by pulsating stars. (Chapter 21)

interferometry A technique of combining the observations of two or more telescopes to produce images better than one telescope alone could make. (Chapter 6)

intermediate-period comet A comet with an orbital period between 20 and 200 years. (Chapter 17)

intermediate vector boson The particle that is responsible for the weak force. (Chapter 29)

internal rotation period The period with which the core of a Jovian planet rotates. (Chapter 14)

interstellar extinction The dimming of starlight as it passes through the interstellar medium. (Chapter 20, Chapter 25)

interstellar medium Gas and dust in interstellar space. (Chapter 8, Chapter 20)

interstellar reddening The reddening of starlight passing through the interstellar medium as a result of blue light being scattered more than red. (Chapter 20)

intracluster gas Gas found between the galaxies that make up a cluster of galaxies. (Chapter 26)

inverse-square law The statement that the apparent brightness of a light source varies inversely with the square of the distance from the source. (Chapter 19)

Io torus A doughnut-shaped ring of gas circling Jupiter at the distance of Io's orbit. (Chapter 15)

ion tail The relatively straight tail of a comet produced by the solar wind acting on ions. (Chapter 17)

ionization The process by which a neutral atom becomes an electrically charged ion through the loss or gain of electrons. (Chapter 5)

iron meteorite A meteorite composed primarily of iron. (Chapter 17)

irregular cluster (of galaxies) A sprawling collection of galaxies whose overall distribution in space does not exhibit any noticeable spherical symmetry. (Chapter 26)

irregular galaxy An asymmetrical galaxy having neither spiral arms nor an elliptical shape. (Chapter 26)

isotope Any of several forms for the same chemical element whose nuclei all have the same number of protons but different numbers of neutrons. (Chapter 5)

isotropic Having the same property in all directions. (Chapter 26, Chapter 28)

isotropy problem The dilemma posed by the fact that the cosmic microwave background is isotropic; also called the horizon problem. (Chapter 29)

Jeans length The smallest scale over which a density fluctuation in a medium will contract to form a gravitationally bound object. (Chapter 29)

jet An extended line of fast-moving gas ejected from the vicinity of a star or a black hole. (Chapter 8)

joule (J) A unit of energy. (Chapter 5)

Jovian planet Low-density planets composed primarily of hydrogen and helium, including Jupiter, Saturn, Uranus, and Neptune. (Chapter 7)

Jupiter-family comet A comet with an orbital period of less than 20 years. (Chapter 17)

Kaluza-Klein theory A theory that attempts to describe the four physical forces in terms of the geometry of space-time with more than four dimensions. (Chapter 29)

kelvin (K) A unit of temperature on the Kelvin temperature scale, equivalent to a degree Celsius. (Chapter 5)

Kelvin-Helmholtz contraction The contraction of a gaseous body, such as a star or nebula, during which gravitational energy is transformed into thermal energy. (Chapter 8)

Kepler's first law The statement that each planet moves around the Sun in an elliptical orbit with the Sun at one focus of the ellipse. (Chapter 4)

Kepler's second law The statement that a planet sweeps out equal areas in equal times as it orbits the Sun; also called the law of equal areas. (Chapter 4)

Kepler's third law A relationship between the period of an orbiting object and the semimajor axis of its elliptical orbit. (Chapter 4)

kiloparsec (kpc) One thousand parsecs; about 3260 light-years. (Chapter 1)

kinetic energy The energy possessed by an object because of its motion. (Chapter 7)

Kirchhoff's laws Three statements about circumstances that produce absorption lines, emission lines, and continuous spectra. (Chapter 5)

Kirkwood gaps Gaps in the spacing of asteroid orbits, discovered by Daniel Kirkwood. (Chapter 17)

Kuiper belt A region that extends from around the orbit of Pluto to about 500 AU from the Sun where many icy objects orbit the Sun. (Chapter 7, Chapter 16, Chapter 17)

Kuiper belt object A member of the Kuiper belt. (Chapter 7)

Lamb shift A tiny shift in the spectral lines of hydrogen caused by the presence of virtual particles. (Chapter 29)

lava Molten rock flowing on the surface of a planet. (Chapter 9)

law of cosmic censorship The hypothesis that all singularities must be surrounded by an event horizon. (Chapter 24)

law of equal areas See *Kepler's second law.*

law of inertia See *Newton's first law.*

law of universal gravitation A formula deduced by Isaac Newton that expresses the strength of the force of gravity that two masses exert on each other. (Chapter 4)

laws of physics A set of physical principles with which we can understand natural phenomena and the nature of the universe. (Chapter 1)

length contraction In the special theory of relativity, the shrinking of an object's length along its direction of motion. (Chapter 24)

lenticular galaxy A galaxy with a central bulge and a disk but no spiral structure; an S0 galaxy. (Chapter 26)

libration An apparent rocking of the Moon whereby an Earth-based observer can, over time, see slightly more than one-half the Moon's surface. (Chapter 10)

light curve A graph that displays how the brightness of a star or other astronomical object varies over time. (Chapter 19)

light-gathering power A measure of the amount of radiation brought to a focus by a telescope. (Chapter 6)

light pollution Light from cities and towns that degrades telescope images. (Chapter 6)

light scattering The process by which light bounces off particles in its path. (Chapter 5, Chapter 14)

light-year (ly) The distance light travels in a vacuum in one year. (Chapter 1)

limb darkening The phenomenon whereby the Sun looks darker near its apparent edge, or limb, than near the center of its disk. (Chapter 18)

line of nodes The line where the plane of the Earth's orbit intersects the plane of the Moon's orbit. (Chapter 3)

liquid metallic hydrogen Hydrogen compressed to such a density that it behaves like a liquid metal. (Chapter 7, Chapter 14)

lithosphere The solid, upper layer of the Earth; essentially the Earth's crust. (Chapter 9)

Local Bubble A large cavity in the interstellar medium within which the Sun and nearby stars are located. (Chapter 25)

Local Group The cluster of galaxies of which our Galaxy is a member. (Chapter 26)

local meridian See *meridian.*

long-period comet A comet that takes hundreds of thousands of years or more to complete one orbit of the Sun. (Chapter 17)

long-period variable A variable star with a period longer than about 100 days. (Chapter 21)

lookback time How far into the past we are looking when we see a particular object. (Chapter 28)

Lorentz transformations Equations that relate the measurements of different observers who are moving relative to each other at high speeds. (Chapter 24)

lower meridian The half of the meridian that lies below the horizon. (Chapter 2)

lowlands (on Mars) See *northern lowlands.*

luminosity The rate at which electromagnetic radiation is emitted from a star or other object. (Chapter 5, Chapter 18, Chapter 19)

luminosity class A classification of a star of a given spectral type according to its luminosity. (Chapter 19)

luminosity function The numbers of stars of differing brightness per cubic parsec. (Chapter 19)

lunar eclipse An eclipse of the Moon by the Earth; a passage of the Moon through the Earth's shadow. (Chapter 3)

lunar highlands Ancient, high-elevation, heavily cratered terrain on the Moon. (Chapter 10)

lunar month See *synodic month.*

lunar phase The appearance of the illuminated area of the Moon as seen from Earth. (Chapter 3)

Lyman series A series of spectral lines of hydrogen produced by electron transitions to and from the lowest energy state of the hydrogen atom. (Chapter 5)

M-theory A theory that attempts to describe fundamental particles as 11-dimensional membranes. (Chapter 29)

MACHO See *massive compact halo object.*

magma Molten rock beneath a planet's surface. (Chapter 9)

magnetic axis A line connecting the north and south magnetic poles of a planet or star possessing a magnetic field. (Chapter 16)

magnetic-dynamo model A theory that explains the solar cycle as a result of the Sun's differential rotation acting on the Sun's magnetic field. (Chapter 18)

magnetic reconnection An event where two oppositely directed magnetic fields approach and cancel, thus releasing energy. (Chapter 18)

magnetic resonance imaging (MRI) A technique for viewing the interior of the human body that makes use of how protons respond to magnetic fields. (Chapter 25)

magnetogram An image of the Sun that shows regions of different magnetic polarity. (Chapter 18)

magnetometer A device for measuring magnetic fields. (Chapter 7)

magnetopause That region of a planet's magnetosphere where the magnetic field counterbalances the pressure from the solar wind. (Chapter 9)

magnetosphere The region around a planet occupied by its magnetic field. (Chapter 9)

magnification The factor by which the apparent angular size of an object is increased when viewed through a telescope. (Chapter 6)

magnifying power See *magnification*.

magnitude scale A system for denoting the brightnesses of astronomical objects. (Chapter 19)

main sequence A grouping of stars on the Hertzsprung-Russell diagram extending diagonally across the graph from hot, luminous stars to cool, dim stars. (Chapter 19)

main-sequence lifetime The total time that a star spends fusing hydrogen in its core, and hence the total time that it will spend as a main-sequence star. (Chapter 21)

main-sequence star A star whose luminosity and surface temperature place it on the main sequence on an H-R diagram; a star that derives its energy from core hydrogen fusion. (Chapter 19)

major axis (of an ellipse) The longest diameter of an ellipse. (Chapter 4)

mantle (of a planet) That portion of a terrestrial planet located between its crust and core. (Chapter 9)

mare (plural maria) Latin for "sea"; a large, relatively crater-free plain on the Moon. (Chapter 10)

mare basalt A type of lunar rock commonly found in the mare basins. (Chapter 10)

maser An interstellar cloud in which water molecules emit intensely at microwave wavelengths. (Chapter 26)

mass A measure of the total amount of material in an object. (Chapter 4)

mass density of radiation The energy possessed by a radiation field per unit volume divided by the square of the speed of light. (Chapter 28)

mass loss A process by which a star gently loses matter. (Chapter 21)

mass-luminosity relation A relationship between the masses and luminosities of main-sequence stars. (Chapter 19)

mass-radius relation A relationship between the masses and radii of white dwarf stars. (Chapter 22)

mass transfer The flow of gases from one star in a binary system to the other. (Chapter 21)

massive compact halo object (MACHO) A dim star or low-mass black hole that may comprise part of the unseen dark matter. (Chapter 25)

matter density parameter (Ω_m) The ratio of the average density of matter in the universe to the critical density. (Chapter 28)

matter-dominated universe A universe in which the average density of matter exceeds both the mass density of radiation and the mass density of dark energy. (Chapter 28)

mean solar day The interval between successive meridian passages of the mean Sun; the average length of a solar day. (Chapter 2)

mean Sun A fictitious object that moves eastward at a constant speed along the celestial equator, completing one circuit of the sky with respect to the vernal equinox in one tropical year. (Chapter 2)

medium (plural media) A material through which light travels. (Chapter 6)

medium, interstellar: See *interstellar medium*.

megaparsec (Mpc) One million parsecs. (Chapter 1)

melting point The temperature at which a substance changes from solid to liquid. (Chapter 9)

meridian (or local meridian) The great circle on the celestial sphere that passes through an observer's zenith and the north and south celestial poles. (Chapter 2)

meridian transit The crossing of the meridian by any astronomical object. (Chapter 2)

mesosphere A layer in the Earth's atmosphere above the stratosphere. (Chapter 9)

metal In reference to stars and galaxies, any elements other than hydrogen and helium; in reference to planets, a material such as iron, silver, and aluminum that is a good conductor of electricity and of heat. (Chapter 19)

metal-poor star A star that, compared to the Sun, is deficient in elements heavier than helium; also called a Population II star. (Chapter 21)

metal-rich star A star whose abundance of heavy elements is roughly comparable to that of the Sun; also called a Population I star. (Chapter 21)

metamorphic rock A rock whose properties and appearance have been transformed by the action of pressure and heat beneath the Earth's surface. (Chapter 9)

meteor The luminous phenomenon seen when a meteoroid enters the Earth's atmosphere; a "shooting star." (Chapter 17)

meteor shower Many meteors that seem to radiate from a common point in the sky. (Chapter 17)

meteorite A fragment of a meteoroid that has survived passage through the Earth's atmosphere. (Chapter 8, Chapter 17)

meteoritic swarm A collection of meteoroids moving together along an orbit about the Sun. (Chapter 17)

meteoroid A small rock in interplanetary space. (Chapter 7)

microlensing A phenomenon in which a compact object such as a MACHO acts as a gravitational lens, focusing the light from a distant star. (Chapter 25)

microwaves Short-wavelength radio waves. (Chapter 5)

mid-mass black hole A black hole with a mass of hundreds of Suns. (Chapter 24)

Milky Way Galaxy Our Galaxy; the band of faint stars seen from the Earth in the plane of our Galaxy's disk. (Chapter 25)

millisecond pulsar A pulsar with a period of roughly 1 to 10 milliseconds. (Chapter 23)

mineral A naturally occurring solid composed of a single element or chemical combination of elements, often in the form of crystals. (Chapter 9)

minor planet See *asteroid*.

minute of arc See *arcminute*.

model A hypothesis that has withstood experimental or observational tests; or, the results of a theoretical calculation that gives the values of temperature, pressure, density, and so forth throughout the interior of an object such as a planet or star. (Chapter 1)

molecule A combination of two or more atoms. (Chapter 7) moonquake Sudden, vibratory motion of the Moon's surface. (Chapter 10)

MRI See *magnetic resonance imaging.*

nanometer (nm) One billionth of a meter: $1 \text{ nm} = 10^{-9}$ meter $= 10^{-6}$ millimeter $= 10^{-3}$ μm. (Chapter 5)

neap tide An ocean tide that occurs when the Moon is near first-quarter or third-quarter phase. (Chapter 4)

near-Earth object (NEO) An asteroid whose orbit lies wholly or partly within the orbit of Mars. (Chapter 17)

near-infrared The part of the infrared spectrum closest in wavelength to visible light. (Chapter 25)

nebula A cloud of interstellar gas and dust. (Chapter 1, Chapter 20)

nebular hypothesis The idea that the Sun and the rest of the solar system formed from a cloud of interstellar material. (Chapter 8)

nebulosity See *nebula.*

negative curvature The curvature of a surface or space in which parallel lines diverge and the sum of the angles of a triangle is less than 180°. (Chapter 28)

negative hydrogen ion A hydrogen atom that has acquired a second electron. (Chapter 18)

NEO See *near-Earth object.*

neon fusion The thermonuclear fusion of neon to produce heavier nuclei. (Chapter 22)

neutrino A subatomic particle with no electric charge and very little mass, yet one that is important in many nuclear reactions. (Chapter 18)

neutron A subatomic particle with no electric charge and with a mass nearly equal to that of the proton. (Chapter 5)

neutron capture The buildup of neutrons inside nuclei. (Chapter 22)

neutron star A very compact, dense star composed almost entirely of neutrons. (Chapter 23)

new moon The phase of the Moon when the dark hemisphere of the Moon faces the Earth. (Chapter 3)

Newtonian mechanics The branch of physics based on Newton's laws of motion. (Chapter 1, Chapter 4)

Newtonian reflector A reflecting telescope that uses a small mirror to deflect the image to one side of the telescope tube. (Chapter 6)

Newton's first law of motion The statement that a body remains at rest, or moves in a straight line at a constant speed, unless acted upon by a net outside force; the law of inertia. (Chapter 4)

Newton's form of Kepler's third law A relationship between the period of two objects orbiting each other, the semimajor axis of their orbit, and the masses of the objects. (Chapter 4)

Newton's second law of motion A relationship between the acceleration of an object, the object's mass, and the net outside force acting on the mass. (Chapter 4)

Newton's third law of motion The statement that whenever one body exerts a force on a second body, the second body exerts an equal and opposite force on the first body. (Chapter 4)

no-hair theorem A statement of the simplicity of black holes. (Chapter 24)

noble gas An element whose atoms do not combine into molecules. (Chapter 14)

node See *line of nodes.*

nonthermal radiation Radiation other than that emitted by a heated body. (Chapter 14, Chapter 27)

north celestial pole The point directly above the Earth's north pole where the Earth's axis of rotation, if extended, would intersect the celestial sphere. (Chapter 2)

northern lights See *aurora borealis.*

northern lowlands (on Mars) Relatively young and crater-free terrain in the Martian northern hemisphere. (Chapter 13)

nova (plural novae) A star that experiences a sudden outburst of radiant energy, temporarily increasing its luminosity roughly a thousandfold. (Chapter 23)

nuclear density The density of matter in an atomic nucleus; about $4 \times 10^{17} \text{ kg/m}^3$. (Chapter 22)

nucleosynthesis The process of building up nuclei such as deuterium and helium from protons and neutrons. (Chapter 29)

nucleus (of an atom) The massive part of an atom, composed of protons and neutrons, about which electrons revolve. (Chapter 5)

nucleus (of a comet) A collection of ices and dust that constitute the solid part of a comet. (Chapter 17)

nucleus (of a galaxy) See *galactic nucleus.*

OB association A grouping of hot, young, massive stars, predominantly of spectral types O and B. (Chapter 20)

OBAFGKM The temperature sequence (from hot to cold) of spectral classes. (Chapter 19)

objective lens The principal lens of a refracting telescope. (Chapter 6)

objective mirror The principal mirror of a reflecting telescope. (Chapter 6)

oblate Flattened at the poles. (Chapter 14)

oblateness A measure of how much a flattened sphere (or spheroid) differs from a perfect sphere. (Chapter 14)

observable universe That portion of the universe inside our cosmic light horizon. (Chapter 28)

Occam's razor The notion that a straightforward explanation of a phenomenon is more likely to be correct than a convoluted one. (Chapter 4)

occultation The eclipsing of an astronomical object by the Moon or a planet. (Chapter 15, Chapter 16)

oceanic rift A crack in the ocean floor that exudes lava. (Chapter 9)

Olbers's paradox The dilemma associated with the fact that the night sky is dark. (Chapter 28)

1-to-1 spin-orbit coupling See *synchronous rotation.*

Oort cloud A presumed accumulation of comets and cometary material surrounding the Sun at distances of roughly 50,000 AU. (Chapter 8, Chapter 17)

open cluster A loose association of young stars in the disk of our Galaxy; a galactic cluster. (Chapter 20)

open universe A universe with negative curvature, so that its geometry is analogous to a saddle-shaped hyperbolic surface. (Chapter 28)

opposition The configuration of a planet when it is at an elongation of 180° and thus appears opposite the Sun in the sky. (Chapter 4)

optical double star Two stars that lie along nearly the same line of sight but are actually at very different distances from us. (Chapter 19)

optical telescope A telescope designed to detect visible light. (Chapter 6)

optical window The range of visible wavelengths to which the Earth's atmosphere is transparent. (Chapter 6)

organic molecules Molecules containing carbon, some of which are the molecules of which living organisms are made. (Chapter 30)

outer core (of the Earth) The outer, molten portion of the Earth's iron-rich core. (Chapter 9)

outgassing The release of gases into a planet's atmosphere by volcanic activity. (Chapter 9)

overcontact binary A close binary system in which the two stars share a common atmosphere. (Chapter 21)

oxygen fusion The thermonuclear fusion of oxygen to produce heavier nuclei. (Chapter 22)

ozone A type of oxygen whose molecules contain three oxygen atoms. (Chapter 9)

ozone hole A region of the Earth's atmosphere over Antarctica where the concentration of ozone is abnormally low. (Chapter 9)

ozone layer A layer in the Earth's upper atmosphere where the concentration of ozone is high enough to prevent much ultraviolet light from reaching the surface. (Chapter 9)

P wave One of three kinds of seismic waves produced by an earthquake; a primary wave. (Chapter 9)

pair production The creation of a particle and its antiparticle from energy. (Chapter 23, Chapter 29)

parabola A conic section formed by cutting a circular cone at an angle parallel to one of the sides of the cone. (Chapter 4)

parallax The apparent displacement of an object due to the motion of the observer. (Chapter 4, Chapter 19)

parsec (pc) A unit of distance; 3.26 light-years. (Chapter 1, Chapter 19)

partial lunar eclipse A lunar eclipse in which the Moon does not appear completely covered. (Chapter 3)

partial solar eclipse A solar eclipse in which the Sun does not appear completely covered. (Chapter 3)

Paschen series A series of spectral lines of hydrogen produced by electron transitions between the third and higher energy levels. (Chapter 5)

Pauli exclusion principle A principle of quantum mechanics stating that no two electrons can have the same position and momentum. (Chapter 21)

penumbra (of a shadow) (plural **penumbrae**) The portion of a shadow in which only part of the light source is covered by an opaque body. (Chapter 3)

penumbral eclipse A lunar eclipse in which the Moon passes only through the Earth's penumbra. (Chapter 3)

perigee The point in its orbit where a satellite or the Moon is nearest the Earth. (Chapter 3)

perihelion The point in its orbit where a planet or comet is nearest the Sun. (Chapter 4)

period (of a planet) The interval of time between successive geometric arrangements of a planet and an astronomical object, such as the Sun. (Chapter 4)

period-luminosity relation A relationship between the period and average density of a pulsating star. (Chapter 21)

periodic table A listing of the chemical elements according to their properties, invented by Dmitri Mendeleev. (Chapter 5)

permafrost Frozen soil that lies just underneath the surface. (Chapter 13)

photodisintegration The breakup of nuclei by high-energy gamma rays. (Chapter 22)

photoelectric effect The phenomenon whereby certain metals emit electrons when exposed to short-wavelength light. (Chapter 5)

photometry The measurement of light intensities. (Chapter 6, Chapter 19)

photon A discrete unit of electromagnetic energy. (Chapter 5)

photosphere The region in the solar atmosphere from which most of the visible light escapes into space. (Chapter 18)

photosynthesis A biochemical process in which solar energy is converted into chemical energy, carbon dioxide and water are absorbed, and oxygen is released. (Chapter 9)

physics, laws of See *laws of physics.*

pixel A picture element. (Chapter 6)

plage A bright region in the solar atmosphere as observed in the monochromatic light of a spectral line. (Chapter 18)

Planck time A fundamental interval of time defined by basic physical constants. (Chapter 28)

Planck's law The relationship between the energy of a photon and its wavelength or frequency; $E = hc/\lambda = h\nu$. (Chapter 5)

plane of the ecliptic The plane in which the Earth orbits the Sun. (Chapter 3)

planetary nebula A luminous shell of gas ejected from an old, low-mass star. (Chapter 22)

planetesimal One of many small bodies of primordial dust and ice that combined to form the planets. (Chapter 8)

plasma A hot ionized gas. (Chapter 14, Chapter 18, Chapter 28)

plastic The attribute of being nearly solid yet able to flow. (Chapter 9)

plate A large section of the Earth's lithosphere that moves as a single unit. (Chapter 9)

plate tectonics The motions of large segments (plates) of the Earth's surface over the underlying mantle. (Chapter 9)

polarized radiation Electromagnetic waves whose electric fields are aligned in a particular direction. (Chapter 27)

polymer A long molecule consisting of many smaller molecules joined together. (Chapter 15)

poor cluster (of galaxies) A cluster of galaxies with very few members; a group of galaxies. (Chapter 26)

Population I star A star whose spectrum exhibits spectral lines of many elements heavier than helium; a metal-rich star. (Chapter 21)

Population II star A star whose spectrum exhibits comparatively few spectral lines of elements heavier than helium; a metal-poor star. (Chapter 21)

positional astronomy The study of the apparent positions of the planets and stars and how those positions change. (Chapter 2)

positive curvature The curvature of a surface or space in which parallel lines converge and the sum of the angles of a triangle is greater than 180°. (Chapter 28)

positron An electron with a positive rather than negative electric charge; the antiparticle of the electron. (Chapter 18, Chapter 29)

power of ten The exponent n in 10^n. (Chapter 1)

powers-of-ten notation A shorthand method of writing numbers, involving 10 followed by an exponent. (Chapter 1)

precession (of the Earth) A slow, conical motion of the Earth's axis of rotation caused by the gravitational pull of the Moon and Sun on the Earth's equatorial bulge. (Chapter 2)

precession of the equinoxes The slow westward motion of the equinoxes along the ecliptic due to precession of the Earth. (Chapter 2)

primary mirror See *objective mirror.*

prime focus The point in a telescope where the objective focuses light. (Chapter 6)

primitive asteroid See *undifferentiated asteroid.*

primordial black hole A type of black hole that may have formed in the very early universe. (Chapter 24)

primordial fireball The extremely hot gas that filled the universe immediately following the Big Bang. (Chapter 28)

principle of equivalence See *equivalence principle.*

progenitor star A star that later explodes into a supernova. (Chapter 22)

prograde orbit An orbit of a satellite around a planet that is in the same direction as the rotation of the planet. (Chapter 15)

prograde rotation A situation in which an object (such as a planet) rotates in the same direction that it orbits around another object (such as the Sun). (Chapter 12)

prominence Flamelike protrusions seen near the limb of the Sun and extending into the solar corona. (Chapter 18)

proper distance See *proper length.*

proper length A length measured by a ruler at rest with respect to an observer. (Chapter 24)

proper motion The angular rate of change in the location of a star on the celestial sphere, usually expressed in arcseconds per year. (Chapter 19)

proper time A time interval measured with a clock at rest with respect to an observer. (Chapter 24)

proplyd See *protoplanetary disk.*

proton A heavy, positively charged subatomic particle that is one of two principal constituents of atomic nuclei. (Chapter 5)

proton-proton chain A sequence of thermonuclear reactions by which hydrogen nuclei are built up into helium nuclei. (Chapter 18)

protoplanet A Moon-sized object formed by the coalescence of planetesimals. (Chapter 8)

protoplanetary disk (proplyd) A disk of material encircling a protostar or a newborn star. (Chapter 8, Chapter 20)

protostar A star in its earliest stages of formation. (Chapter 20)

protosun The part of the solar nebula that eventually developed into the Sun. (Chapter 8)

Ptolemaic system The definitive version of the geocentric cosmogony of ancient Greece. (Chapter 4)

pulsar A pulsating radio source thought to be associated with a rapidly rotating neutron star. (Chapter 1, Chapter 23)

pulsating variable star A star that pulsates in size and luminosity. (Chapter 21)

pulsating X-ray source An object that emits pulses of X rays at regular intervals of a few seconds. (Chapter 23)

quantum electrodynamics A theory that describes details of how charged particles interact by exchanging photons. (Chapter 29)

quantum mechanics The branch of physics dealing with the structure and behavior of atoms and their constituents as well as their interaction with light. (Chapter 5, Chapter 29)

quark One of several particles thought to be the internal constituents of certain heavy subatomic particles such as protons and neutrons. (Chapter 29)

quark confinement The permanent bonding of quarks to form particles like protons and neutrons. (Chapter 29)

quasar A very luminous object with a very large redshift and a starlike appearance. (Chapter 1, Chapter 27)

radial velocity That portion of an object's velocity parallel to the line of sight. (Chapter 5, Chapter 19)

radial velocity curve A plot showing the variation of radial velocity with time for a binary star or variable star. (Chapter 19)

radial velocity method A technique used to detect extrasolar planets by observing Doppler shifts in the spectrum of the planet's star. (Chapter 8)

radiant (of a meteor shower) The point in the sky from which meteors of a particular shower seem to originate. (Chapter 17)

radiation darkening The darkening of methane ice by electron impacts. (Chapter 16)

radiation-dominated universe A universe in which the mass density of radiation exceeds both the average density of matter and the mass density of dark energy. (Chapter 28)

radiation pressure Pressure exerted on an object by radiation falling on the object. (Chapter 17)

radiative diffusion The random migration of photons from a star's center toward its surface. (Chapter 18)

radiative zone A region within a star where radiative diffusion is the dominant mode of energy transport. (Chapter 18)

radio galaxy A galaxy that emits an unusually large amount of radio waves. (Chapter 27)

radio lobe A region near an active galaxy from which significant radio radiation emanates. (Chapter 27)

radio telescope A telescope designed to detect radio waves. (Chapter 6)

radio waves The longest-wavelength electromagnetic radiation. (Chapter 5)

radio window The range of radio wavelengths to which the Earth's atmosphere is transparent. (Chapter 6)

radioactive age-dating A technique for determining the age of a rock sample by measuring the radioactive elements and their decay products in the sample. (Chapter 8)

radioactive decay The process whereby certain atomic nuclei spontaneously transform into other nuclei. (Chapter 8)

rarefaction A region in a sound wave where the medium carrying the wave is spread out or rarefied. (Chapter 28)

recombination The process in which an electron combines with a positively charged ion. (Chapter 20)

red giant A large, cool star of high luminosity. (Chapter 19, Chapter 21)

red-giant branch The region of an H-R diagram occupied by stars in their first red-giant phase. (Chapter 22)

reddening (interstellar) See *interstellar reddening*.

redshift The shifting to longer wavelengths of the light from remote galaxies and quasars; the Doppler shift of light from a receding source. (Chapter 5, Chapter 26)

reflecting telescope A telescope in which the principal optical component is a concave mirror. (Chapter 6)

reflection The return of light rays by a surface. (Chapter 6)

reflection nebula A comparatively dense cloud of dust in interstellar space that is illuminated by a star. (Chapter 20)

reflector A reflecting telescope. (Chapter 6)

refracting telescope A telescope in which the principal optical component is a lens. (Chapter 6)

refraction The bending of light rays when they pass from one transparent medium to another. (Chapter 6)

refractor A refracting telescope. (Chapter 6)

refractory element An element with high melting and boiling points. (Chapter 10)

regolith The layer of rock fragments covering the surface of the Moon. (Chapter 10)

regular cluster (of galaxies) A spherical cluster of galaxies. (Chapter 26)

relativistic cosmology A cosmology based on the general theory of relativity. (Chapter 28)

residual polar cap An ice-covered polar region on Mars that does not completely evaporate during the Martian summer. (Chapter 13)

respiration A biological process that produces energy by consuming oxygen and releasing carbon dioxide. (Chapter 9)

retrograde motion The apparent westward motion of a planet with respect to background stars. (Chapter 4)

retrograde orbit An orbit of a satellite around a planet that is in the direction opposite to which the planet rotates. (Chapter 15)

retrograde rotation A situation in which an object (such as a planet) rotates in the direction opposite to which it orbits around another object (such as the Sun). (Chapter 12)

rich cluster (of galaxies) A cluster of galaxies containing many members. (Chapter 26)

rift valley A feature created when a planet's crust breaks apart along a line. (Chapter 13)

right ascension A coordinate for measuring the east-west positions of objects on the celestial sphere. (Chapter 2)

ring particles Small particles that constitute a planetary ring. (Chapter 14)

ringlet One of many narrow bands of particles of which Saturn's ring system is composed. (Chapter 14)

Roche limit The smallest distance from a planet or other object at which a second object can be held together by purely gravitational forces. (Chapter 14)

Roche lobe A teardrop-shaped volume surrounding a star in a binary inside which gases are gravitationally bound to that star. (Chapter 21)

rock A mineral or combination of minerals. (Chapter 9)

rotation curve A plot of the orbital speeds of stars and nebulae in a galaxy versus distance from the center of the galaxy. (Chapter 25)

RR Lyrae variable A type of pulsating star with a period of less than one day. (Chapter 21)

runaway greenhouse effect A greenhouse effect in which the temperature continues to increase. (Chapter 12)

runaway icehouse effect A situation in which a decrease in atmospheric temperature causes a further decrease in temperature. (Chapter 13)

S wave One of three kinds of seismic waves produced by an earthquake; a secondary wave. (Chapter 9)

Sagittarius A* A powerful radio source at the center of our Galaxy. (Chapter 25)

saros A particular cycle of similar eclipses that recur about every 18 years. (Chapter 3)

scarp A line of cliffs formed by the faulting or fracturing of a planet's surface. (Chapter 11)

scattering of light See *light scattering*.

Schwarzschild radius (R_{Sch}) The distance from the singularity to the event horizon in a nonrotating black hole. (Chapter 24)

scientific method The basic procedure used by scientists to investigate phenomena. (Chapter 1)

seafloor spreading The separation of plates under the ocean due to lava emerging in an oceanic rift. (Chapter 9)

search for extraterrestrial intelligence (SETI) The scientific search for evidence of intelligent life on other planets. (Chapter 30)

second of arc See *arcsecond*.

sedimentary rock A rock that is formed from material deposited on land by rain or winds, or on the ocean floor. (Chapter 9)

seeing disk The angular diameter of a star's image. (Chapter 6)

seismic wave A vibration traveling through a terrestrial planet, usually associated with earthquake-like phenomena. (Chapter 9)

seismograph A device used to record and measure seismic waves, such as those produced by earthquakes. (Chapter 9)

self-propagating star formation The process by which the formation of stars in one location in a galaxy stimulates the formation of stars in a neighboring location. (Chapter 25)

semidetached binary A binary star system in which one star fills its Roche lobe. (Chapter 21)

semimajor axis One-half of the major axis of an ellipse. (Chapter 4)

SETI See *search for extraterrestrial intelligence*.

Seyfert galaxy A spiral galaxy with a bright nucleus whose spectrum exhibits emission lines. (Chapter 27)

shell helium fusion The thermonuclear fusion of helium in a shell surrounding a star's core. (Chapter 22)

shell hydrogen fusion The thermonuclear fusion of hydrogen in a shell surrounding a star's core. (Chapter 21)

shepherd satellite A satellite whose gravity restricts the motions of particles in a planetary ring, preventing them from dispersing. (Chapter 14)

shield volcano A volcano with long, gently sloping sides. (Chapter 12)

shock wave An abrupt, localized region of compressed gas caused by an object traveling through the gas at a speed greater than the speed of sound. (Chapter 9)

SI units The International System of Units, based on the meter (m), the second (s), and the kilogram (kg). (Chapter 1)

sidereal clock A clock that measures sidereal time. (Chapter 2)

sidereal day The interval between successive meridian passages of the vernal equinox. (Chapter 2)

sidereal month The period of the Moon's revolution about the Earth with respect to the stars. (Chapter 3)

sidereal period The orbital period of one object about another as measured with respect to the stars. (Chapter 4)

sidereal time Time reckoned by the location of the vernal equinox. (Chapter 2)

sidereal year The orbital period of the Earth about the Sun with respect to the stars. (Chapter 2)

silicon fusion The thermonuclear fusion of silicon to produce heavier elements, especially iron. (Chapter 22)

singularity A place of infinite space-time curvature; the center of a black hole. (Chapter 24)

small-angle formula A relationship between the angular and linear sizes of a distant object. (Chapter 1)

SNC meteorite A meteorite that came to Earth from Mars. (Chapter 30)

solar constant The average amount of energy received from the Sun per square meter per second, measured just above the Earth's atmosphere. (Chapter 5)

solar corona Hot, faintly glowing gases seen around the Sun during a total solar eclipse; the uppermost regions of the solar atmosphere. (Chapter 3)

solar cycle See *22-year solar cycle.*

solar eclipse An eclipse of the Sun by the Moon; a passage of the Earth through the Moon's shadow. (Chapter 3)

solar flare A sudden, temporary outburst of light from an extended region of the solar surface. (Chapter 18)

solar nebula The cloud of gas and dust from which the Sun and solar system formed. (Chapter 8)

solar neutrino A neutrino emitted from the core of the Sun. (Chapter 18)

solar neutrino problem The discrepancy between the predicted and observed numbers of solar neutrinos. (Chapter 18)

solar system The Sun, planets and their satellites, asteroids, comets, and related objects that orbit the Sun. (Chapter 1)

solar transit The passage of an object in front of the Sun. (Chapter 11)

solar wind An outward flow of particles (mostly electrons and protons) from the Sun. (Chapter 8, Chapter 18)

south celestial pole The point directly above the Earth's south pole where the Earth's axis of rotation, if extended, would intersect the celestial sphere. (Chapter 2)

southern highlands (on Mars) Older, cratered terrain in the Martian southern hemisphere. (Chapter 13)

southern lights See *aurora australis.*

space velocity How fast and in what direction a star moves through space. (Chapter 19)

spacetime A four-dimensional combination of time and the three dimensions of space. (Chapter 24)

special theory of relativity A description of mechanics and electromagnetic theory formulated by Albert Einstein, which explains that measurements of distance, time, and mass are affected by the observer's motion. (Chapter 24)

spectral analysis The identification of chemical substances from the patterns of lines in their spectra. (Chapter 5)

spectral class A classification of stars according to the appearance of their spectra. (Chapter 19)

spectral line In a spectrum, an absorption or emission feature that is at a particular wavelength. (Chapter 5)

spectral type A subdivision of a spectral class. (Chapter 19)

spectrograph An instrument for photographing a spectrum. (Chapter 6)

spectroscopic binary A binary star system whose binary nature is deduced from the periodic Doppler shifting of lines in its spectrum. (Chapter 19)

spectroscopic parallax The distance to a star derived by comparing its apparent brightness to a luminosity inferred from the star's spectrum. (Chapter 19)

spectroscopy The study of spectra and spectral lines. (Chapter 5, Chapter 6, Chapter 7)

spectrum (plural spectra) The result of dispersing a beam of electromagnetic radiation so that components with different wavelengths are separated in space. (Chapter 5)

spectrum binary A binary star whose binary nature is deduced from the presence of two sets of incongruous spectral lines. (Chapter 19)

speed Distance traveled divided by the time elapsed to cover that distance. (Chapter 4)

spherical aberration The distortion of an image formed by a telescope due to differing focal lengths of the optical system. (Chapter 6)

spherical space Space with positive curvature. (Chapter 28)

spicule A narrow jet of rising gas in the solar chromosphere. (Chapter 18)

spin The intrinsic angular momentum possessed by certain particles. (Chapter 25)

spin-flip transition A transition in the ground state of the hydrogen atom, which occurs when the orientation of the electron's spin changes. (Chapter 25)

spin-orbit coupling See *1-to-1 spin-orbit coupling* and *3-to-2 spin-orbit coupling.*

spiral arms Lanes of interstellar gas, dust, and young stars that wind outward in a plane from the central regions of a galaxy. (Chapter 25)

spiral galaxy A flattened, rotating galaxy with pinwheel-like spiral arms winding outward from the galaxy's nucleus. (Chapter 26)

spontaneous symmetry breaking A process by which certain symmetries in the mathematics of particle physics are suddenly altered to produce new particles and forces. (Chapter 29)

spring tide An ocean tide that occurs at new moon and full moon phases. (Chapter 4)

stable Lagrange points Locations along Jupiter's orbit where the combined gravitational effects of the Sun and Jupiter cause asteroids to collect. (Chapter 17)

standard candle An astronomical object of known intrinsic brightness that can be used to determine the distances to other galaxies. (Chapter 26)

starburst galaxy A galaxy that is experiencing an exceptionally high rate of star formation. (Chapter 26)

star cluster See *cluster of stars.*

stationary absorption line An absorption line in the spectrum of a binary star that does not show the same Doppler shift as other lines, indicating that it originates in the interstellar medium. (Chapter 20)

Stefan-Boltzmann law A relationship between the temperature of a blackbody and the rate at which it radiates energy. (Chapter 5)

stellar association A loose grouping of young stars. (Chapter 20)

stellar evolution The changes in size, luminosity, temperature, and so forth that occur as a star ages. (Chapter 20)

stellar-mass black hole A black hole with a mass comparable to that of a star. (Chapter 24)

stellar parallax The apparent displacement of a star due to the Earth's motion around the Sun. (Chapter 19)

stony iron meteorite A meteorite composed of both stone and iron. (Chapter 17)

stony meteorite A meteorite composed of stone. (Chapter 17)

stratosphere A layer in the Earth's atmosphere directly above the troposphere. (Chapter 9)

strong force The force that binds protons and neutrons together in nuclei. (Chapter 29)

subduction zone A location where colliding tectonic plates cause the Earth's crust to be pulled down into the mantle. (Chapter 9)

subtend To extend over an angle. (Chapter 1)

summer solstice The point on the ecliptic where the Sun is farthest north of the celestial equator. Also used to refer to the date on which the Sun passes through this point. (Chapter 2)

sunspot A temporary cool region in the solar photosphere. (Chapter 18)

sunspot cycle The semiregular 11-year period with which the number of sunspots fluctuates. (Chapter 18)

sunspot maximum/minimum That time during the sunspot cycle when the number of sunspots is highest/lowest. (Chapter 18)

supercluster A collection of clusters of galaxies. (Chapter 26)

superconductivity The phenomenon whereby a flowing electric current does not experience any electrical resistance. (Chapter 23)

superfluidity The phenomenon whereby a fluid flows without experiencing any viscosity. (Chapter 23)

supergiant A very large, extremely luminous star of luminosity class I. (Chapter 19, Chapter 22)

supergrand unified theory A complete description of all forces and particles, as well as the structure of space and time; a "theory of everything" (TOE). (Chapter 29)

supergranule A large convective feature in the solar atmosphere, usually outlined by spicules. (Chapter 18)

superior conjunction The configuration of a planet being behind the Sun as viewed from the Earth. (Chapter 4)

superior planet A planet that is more distant from the Sun than the Earth is. (Chapter 4)

superluminal motion Motion that appears to involve speeds greater than the speed of light. (Chapter 27)

supermassive black hole A black hole with a mass of a million or more Suns. (Chapter 24, Chapter 27)

supernova (plural **supernovae**) A stellar outburst during which a star suddenly increases its brightness roughly a millionfold. (Chapter 1, Chapter 17, Chapter 22)

supernova remnant The gases ejected by a supernova. (Chapter 20, Chapter 22)

supersonic Faster than the speed of sound. (Chapter 20)

surface wave A type of seismic wave that travels only over the Earth's surface. (Chapter 9)

synchronous rotation The rotation of a body with a period equal to its orbital period; also called 1-to-1 spin-orbit coupling. (Chapter 3, Chapter 10)

synchrotron radiation A type of nonthermal radiation emitted by charged particles moving through a magnetic field. (Chapter 14, Chapter 23)

synodic month The period of revolution of the Moon with respect to the Sun; the length of one cycle of lunar phases. Also called the lunar month. (Chapter 3)

synodic period The interval between successive occurrences of the same configuration of a planet. (Chapter 4)

T Tauri stars Young variable stars associated with interstellar matter that show erratic changes in luminosity. (Chapter 20)

T Tauri wind A flow of particles away from a T Tauri star. (Chapter 8)

tail (of a comet) Gas and dust particles from a comet's nucleus that have been swept away from the comet's head by the radiation pressure of sunlight and the solar wind. (Chapter 17)

tangential velocity That portion of an object's velocity perpendicular to the line of sight. (Chapter 19)

temperature See *degree Celsius, degree Fahrenheit,* and *kelvin.*

terminator The line dividing day and night on the surface of the Moon or a planet; the line of sunset or sunrise. (Chapter 10)

terrae Cratered lunar highlands. (Chapter 10)

terrestrial planet High-density worlds with solid surfaces, including Mercury, Venus, Earth, and Mars. (Chapter 7)

theory A hypothesis that has withstood experimental or observational tests. (Chapter 1)

theory of everything See *supergrand unified theory.*

thermal equilibrium A balance between the input and outflow of heat in a system. (Chapter 18, Chapter 29)

thermal pulse A brief burst in energy output from the helium-fusing shell of an aging low-mass star. (Chapter 22)

thermal radiation The radiation naturally emitted by any object that is not at absolute zero. Blackbody radiation is an idealized case of thermal radiation. (Chapter 14, Chapter 27)

thermonuclear fusion The combining of nuclei under conditions of high temperature in a process that releases substantial energy. (Chapter 18)

thermosphere A region in the Earth's atmosphere between the mesosphere and the exosphere. (Chapter 9)

third quarter moon The phase of the Moon that occurs when the Moon is 90° west of the Sun. (Chapter 3)

3-to-2 spin-orbit coupling The rotation of Mercury, which makes three complete rotations on its axis for every two complete orbits around the Sun. (Chapter 11)

threshold temperature The temperature above which photons spontaneously produce particles and antiparticles of a particular type. (Chapter 29)

tidal force A gravitational force whose strength and/or direction varies over a body and thus tends to deform the body. (Chapter 4, Chapter 14)

tidal heating The heating of the interior of a satellite by continually varying tidal stresses. (Chapter 15)

time dilation The slowing of time due to relativistic motion. (Chapter 24)

time zone A region on the Earth where, by agreement, all clocks have the same time. (Chapter 2)

TOE See *supergrand unified theory*.

total lunar eclipse A lunar eclipse during which the Moon is completely immersed in the Earth's umbra. (Chapter 3)

total solar eclipse A solar eclipse during which the Sun is completely hidden by the Moon. (Chapter 3)

totality (lunar eclipse) The period during a total lunar eclipse when the Moon is entirely within the Earth's umbra. (Chapter 3)

totality (solar eclipse) The period during a total solar eclipse when the disk of the Sun is completely hidden. (Chapter 3)

transit An event in which an astronomical body moves in front of another. See also *meridian transit* and *solar transit*. (Chapter 8)

transit method A method for detecting extrasolar planets that come between us and their parent star, dimming the star's light. (Chapter 8)

triple alpha process A sequence of two thermonuclear reactions in which three helium nuclei combine to form one carbon nucleus. (Chapter 21)

Trojan asteroid One of several asteroids that share Jupiter's orbit about the Sun. (Chapter 17)

Tropic of Cancer A circle of latitude 23½° north of the Earth's equator. (Chapter 2)

Tropic of Capricorn A circle of latitude 23½° south of the Earth's equator. (Chapter 2)

tropical year The period of revolution of the Earth about the Sun with respect to the vernal equinox. (Chapter 2)

troposphere The lowest level in the Earth's atmosphere. (Chapter 9)

Tully-Fisher relation A correlation between the width of the 21-cm line of a spiral galaxy and the total luminosity of that galaxy. (Chapter 26)

tuning fork diagram A diagram that summarizes Edwin Hubble's classification scheme for spiral, barred spiral, and elliptical galaxies. (Chapter 26)

turnoff point The point on an H-R diagram where the stars in a cluster are leaving the main sequence. (Chapter 21)

21-cm radio emission Radio radiation emitted by neutral hydrogen atoms in interstellar space. (Chapter 25)

22-year solar cycle The semiregular 22-year interval between successive appearances of sunspots at the same latitude and with the same magnetic polarity. (Chapter 18)

Type I Cepheid A metal-rich Cepheid variable star. (Chapter 21)

Type I supernova A supernova whose spectrum lacks hydrogen lines. Type Ia supernovae are further classified as Type Ia, Ib, or Ic. (Chapter 22)

Type Ia supernova A supernova whose spectrum lacks hydrogen lines but has a strong absorption line of ionized silicon. (Chapter 22)

Type Ib supernova A supernova whose spectrum lacks hydrogen and silicon lines but has a strong helium absorption line. (Chapter 22)

Type Ic supernova A supernova whose spectrum is almost devoid of emission or absorption lines. (Chapter 22)

Type II Cepheid A metal-poor Cepheid variable star. (Chapter 21)

Type II supernova A supernova with hydrogen emission lines in its spectrum, caused by the explosion of a massive star. (Chapter 22)

UBV photometry A system for determining the surface temperature of a star by measuring the star's brightness in the ultraviolet (U), blue (B), and visible (V) spectral regions. (Chapter 19)

ultramafic lava A type of lava enriched in magnesium and iron. These give the lava a higher melting temperature. (Chapter 15)

ultraviolet radiation Electromagnetic radiation of wavelengths shorter than those of visible light but longer than those of X rays. (Chapter 5)

umbra (of a shadow) (plural **umbrae**) The central, completely dark portion of a shadow. (Chapter 3)

undifferentiated asteroid An asteroid within which chemical differentiation did not occur. (Chapter 17)

universal constant of gravitation (G) The constant of proportionality in Newton's law of gravitation. (Chapter 4)

upper meridian The half of the meridian that lies above the horizon. (Chapter 2)

Van Allen belts Two doughnut-shaped regions around the Earth where many charged particles (protons and electrons) are trapped by the Earth's magnetic field. (Chapter 9)

velocity The speed and direction of an object's motion. (Chapter 4)

vernal equinox The point on the ecliptic where the Sun crosses the celestial equator from south to north. Also used to refer to the date on which the Sun passes through this intersection. (Chapter 2)

very-long-baseline interferometry (VLBI) A method of connecting widely separated radio telescopes to make very high-resolution observations. (Chapter 6)

virtual pair A particle and antiparticle that exist for such a brief interval that they cannot be observed. (Chapter 24, Chapter 29)

visible light Electromagnetic radiation detectable by the human eye. (Chapter 5)

visual binary A binary star in which the two components can be resolved through a telescope. (Chapter 19)

VLBI See *very-long-baseline interferometry.*

void A large volume of space, typically 30 to 120 Mpc (100 to 400 million light-years) in diameter, that contains very few galaxies. (Chapter 26)

volatile element An element with low melting and boiling points. (Chapter 10)

waning crescent moon The phase of the Moon that occurs between third quarter and new moon. (Chapter 3)

waning gibbous moon The phase of the Moon that occurs between full moon and third quarter. (Chapter 3)

water hole A range of frequencies in the microwave spectrum suitable for interstellar radio communication. (Chapter 30)

watt A unit of power, equal to one joule of energy per second. (Chapter 5)

wavelength The distance between two successive wave crests. (Chapter 5)

wavelength of maximum emission The wavelength at which a heated object emits the greatest intensity of radiation. (Chapter 5)

waxing crescent moon The phase of the Moon that occurs between new moon and first quarter. (Chapter 3)

waxing gibbous moon The phase of the Moon that occurs between first quarter and full moon. (Chapter 3)

weak force The short-range force that is responsible for transforming certain particles into other particles, such as the decay of a neutron into a proton. (Chapter 29)

weakly interacting massive particle (WIMP) A hypothetical massive particle that may comprise part of the unseen dark matter. (Chapter 25)

weight The force with which gravity acts on a body. (Chapter 4)

white dwarf A low-mass star that has exhausted all its thermonuclear fuel and contracted to a size roughly equal to the size of the Earth. (Chapter 19, Chapter 22)

white ovals Round, whitish feature usually seen in Jupiter's southern hemisphere. (Chapter 14)

Widmanstätten patterns Crystalline structure seen in certain types of meteorites. (Chapter 17)

Wien's law A relationship between the temperature of a blackbody and the wavelength at which it emits the greatest intensity of radiation. (Chapter 5)

WIMP See *weakly interacting massive particle.*

winding dilemma The problem of why the spiral arms of a galaxy like the Milky Way do not disappear. (Chapter 25)

winter solstice The point on the ecliptic where the Sun reaches its greatest distance south of the celestial equator. Also used to refer to the date on which the Sun passes through this point. (Chapter 2)

wormhole A conjectured connection between different regions of spacetime made possible by the gravitational effects of a black hole. (Chapter 24)

X-ray burster A nonperiodic X-ray source that emits powerful bursts of X rays. (Chapter 1, Chapter 23)

X rays Electromagnetic radiation whose wavelength is between that of ultraviolet light and gamma rays. (Chapter 5)

ZAMS See *zero-age main sequence.*

Zeeman effect A splitting or broadening of spectral lines due to a magnetic field. (Chapter 18)

zenith The point on the celestial sphere directly overhead an observer. (Chapter 2)

zero-age main sequence The main sequence of young stars that have just begun to burn hydrogen at their cores. (Chapter 21)

zero-age main-sequence star A newly formed star that has just arrived on the main sequence. (Chapter 21)

zero curvature The curvature of a surface or space in which parallel lines remain parallel and the sum of the angles of a triangle is exactly 180°. (Chapter 28)

zodiac A band of 12 constellations around the sky centered on the ecliptic. (Chapter 2)

zonal winds The pattern of alternating eastward and westward winds found in the atmospheres of Jupiter and Saturn. (Chapter 14)

zone A light-colored band in Jupiter's atmosphere. (Chapter 14)

Answers to Selected Questions

How to Get the Most from UNIVERSE

1. Paragraphs labeled by the Caution! icon 2. Paragraphs labeled by the Analogy icon 3. Tools of the Astronomer's Trade and The Heavens on the Earth 4. On the CD-ROM that accompanies this book, or on the UNIVERSE Web site 5. *Starry Night Backyard*™ is a planetarium program, and *Deep Space Explorer*™ is a program that allows you to examine astronomical objects from different perspectives; both are on the CD-ROM that accompanies this book; they come with the book at no additional cost 6. It shows whether the image was made with Radio waves, Infrared radiation, Visible light, Ultraviolet light, X rays, or Gamma rays 7. On the CD-ROM that accompanies this book, or on the UNIVERSE Web site 8. (a) page A-8 (b) page A-1 (c) page A-5 (d) page A-6 9. (a) page 391 (b) page 120 (c) page 83 (d) page 71 10. 656.9 nm 11. At the back of the book, following the Index

 Only mathematical answers are given in the following, not answers that require interpretation or discussion. Your instructor will expect you to show the steps required to reach each mathematical answer.

Chapter 1

20. 8.5×10^3 km 21. 2.8×10^7 Suns 22. About 3×10^{36} times larger 23. 8.94×10^{56} hydrogen atoms 24. (a) 1.581×10^{-5} ly (b) 4.848×10^{-6} pc 25. 4.99×10^2 s 26. 4.3×10^9 km 27. (a) 1.08×10^{14} km (b) 11.4 years 28. 4.4×10^{17} s 30. (a) 1.5 m (b) 89 m (c) 5.4×10^3 m 31. 6.9 m 32. 3.4×10^3 km 33. 3.7 km 34. 0.387 arcmin

Chapter 2

24. Around 8:02 P.M. 29. (a) About 9 hours 35. 26½° 40. October 25, 1917 43. 50° 44. 3:50 A.M. local time 46. (a) 6:00 P.M. (b) September 21

Chapter 3

27. (a) 0.91 hour (b) 11° 28. 49 arcsec 40. (a) December 14, 2020; southern Pacific and South America (b) January 5, 2057

Chapter 4

16. Semimajor axis = 0.25 AU, period = 0.125 year 17. Average = 25 AU, farthest = 50 AU 20. 6 newtons, 4 m/s² 23. 1/9 as strong 30. 87.97 days 35. (a) 4 AU (b) 8 years 36. (a) 16.0 AU (b) 0.5 AU 37. (a) 1.26 AU (b) 258 days 38. Earth exerts a force of 1.98×10^{20} newtons on the Moon 39. ¼ as much as on Earth 40. Forces are approximately the same, Earth's acceleration is 100 times greater 41. 62 newtons, 0.13 42. (a) 24 hours (b) 43,200 km 43. 0.5 year 44. 119 minutes 46. (a) 5 years (b) 2.92 AU 47. (a) 3.43×10^{-5} newton (b) 3.21×10^{-5} newton (c) 2.2×10^{-6} newton

Chapter 5

1. 7.5 times 2. 500 s 6. 0.340 m 7. 2.61×10^{14} Hz 13. 135 nm 14. 3400 K 22. 9400 nm 23. 9980°F 24. About 10 μm 25. 2.9 nm 26. 3.9×10^{26} W 27. 190 times more 28. (a) 4890 nm = 4.89 μm (b) 540 times more 29. (a) 5.75×10^8 W/m² (b) 10,000 K 31. 2.43×10^{-3} nm 32. (a) 1005 nm 33. (a) 371.2 nm 33. Possible transitions: energy = 1 eV, λ = 1240 nm; energy = 2 eV, λ = 620 nm; energy = 3 eV, λ = 414 nm 37. 656.9 nm 38. Coming toward us at 13.0 km/s 39. 8.6×10^4 km/s

Chapter 6

30. ¹/₂₅ = 0.04 31. (a) Subaru has 12 times the light-gathering power 32. (a) 222× (b) 100× (c) 36× (d) 0.75 arcsec 34. 300 km (Hubble Space Telescope, Jupiter's moons); 110 km (human eye, our Moon) 35. (a) 34 ly (b) 37 km 37. (a) 0.18 m (b) 9.0×10^{-4} arcsec 38. (a) 5.39×10^{-4} m (c) 0.56 m

Chapter 7

21. Mass = 6.4×10^{23} kg, average density = 3900 kg/m³ 23. (a) 1.0×10^{13} kg (b) 1.2 m/s 24. (a) 3.3×10^{21} J (b) Equivalent to 3.9×10^7 Hiroshima-type weapons 25. (a) 2.02 km/s (b) 19.4 km/s 26. 1.2 km/s 27. (a) 618 km/s 29. 1.76 years 31. (a) 1000 years (b) 17 days 33. In 100 years, probability is 8.3×10^{-8} (one chance in 12 million); in 10^6 years, probability is 8.3×10^{-4} (one chance in 1200)

Chapter 8

22. 0.40 kg after 1.3 billion years; 0.20 kg after 2.6 billion years; 0.10 kg after 3.9 billion years 23. 2.6 billion years 25. (a) About 180 AU 26. 2.9×10^7 AU = 140 parsecs = 460 light-years 27. (a) About 600 AU (b) About 5×10^{40} cubic meters (c) About 10^{55} atoms (d) About 3×10^{14} atoms per cubic meter 28. 860 years 30. 2.2×10^{30} kg = 1.1 times the mass of the Sun 31. (a) 12 m/s (b) 1.3×10^{-3} arcsec (c) 9.0×10^{-5} arcsec

Chapter 9

23. (a) 6.8×10^{16} W (b) 1.07×10^{17} W (c) 209 watts per square meter (d) 246 K = −27°C 24. 32 K = −248°C 25. 4 km 26. Core: 17%; mantle: 82%; crust: 1% 27. 0.020 (2% of the total mass) 28. (b) About 15,000 kg/m³ 30. 2.2×10^8 years ago 35. 2.22 eV

Chapter 10

21. (a) 4671 km (b) 1707 km below the surface (c) 449 km 26. 130 newtons on the Moon, 780 newtons on Earth 27. (a) 5.78×10^{-5} newton (b) 4.15×10^{-5} newton (c) 1.39 33. 2.56 seconds 34. 5.5×10^5 km 36. (b) 10^6 times greater

Chapter 11

23. 13.0 arcsec **25.** 730 km **27.** For $T = 430°C$, $\lambda_{max} = 4.1$ μm
29. (a) 3.03 m/s (b) 1.26 nm **30.** 300 newtons on Mercury, 130 newtons on the Moon, 780 newtons on Earth **32.** (a) About 30 **33.** 0.615 AU

Chapter 12

29. 0.23 nm **30.** For $T = 460°C$, $\lambda_{max} = 4.0$ μm **37.** 920 m
39. 0.5 arcminute

Chapter 13

33. Earth completes 2.10 orbits, Mars completes 1.12 orbits
36. (a) 0.16 AU **37.** (a) 43 km (b) 4.3 km **38.** 2.0 cm **42.** (a) 3770 km
(b) 370 km **44.** Difference between round-trip times is 2.0×10^{-4} s
55. Phobos: about 16 arcminutes; Deimos: about 2.7 arcminutes
56. (a) Radius = 20,400 km, altitude = 17,000 km

Chapter 14

39. 12.7 km/s **42.** 8.5×10^{53} hydrogen atoms, 7.1×10^{52} helium atoms
43. Roughly 600 km/h **44.** 127 K **47.** 59.5 km/s **48.** 8300 newtons
49. 7.1×10^4 kg/m^3 **52.** (a) 14.4 hours for inner edge of A ring,
7.9 hours for inner edge of B ring

Chapter 15

37. 760 km **38.** 0.046 arcsec **42.** 2.8×10^{11} (280 billion) years
43. About 4×10^{-11} (one part in 3×10^{10}) **44.** About 8.1 minutes
47. Escape speed = 2.6 km/s; mass of molecule = 2.0×10^{-26} kg, corresponding to a molecular weight of 12 **50.** 0.09 arcsec
52. (a) 65.7 hours (b) 27.6 arcsec

Chapter 16

25. Sun-Uranus force = 1.39×10^{21} newtons, Neptune-Uranus force = 2.24×10^{17} newtons; Neptune reduces the sunward gravitational pull on Uranus by 1.61×10^{-4}, or 0.0161% **27.** (a) 2900 kg/m^3
28. (a) 8.4 hours **32.** (a) 1.13×10^{-3} as bright as on Earth
(b) 4.10×10^{-4} as bright as on Earth (c) 2.77 times as bright **33.** About 7200 km **35.** 0.95 arcsec **38.** Without the boost, a one-way trip would take 30.5 years

Chapter 17

29. 2.82 AU **31.** 30 km **33.** (a) 3.6×10^{12} kg (b) 0.83 m/s **35.** About 4% **36.** 3.6×10^7 km = 0.24 AU **37.** (a) Period = 350 years, lifetime = 3.5×10^4 years (b) Period = 1.1×10^4 years, lifetime = 1.1×10^6 years (c) Period = 3.5×10^5 years, lifetime = 3.5×10^7 years (d) Period = 1.1×10^7 years, lifetime = 1.1×10^9 years **40.** (a) 10^{15} kg (b) 10^{-16} kg/m^3

Chapter 18

27. (a) 1.8×10^{-9} J (b) 9.0×10^{16} J (c) 5.4×10^{41} J **28.** (a) 4.6×10^{-36} s
(b) 2.3×10^{-10} s (c) 1.4×10^{15} s = 4.4×10^7 years **29.** 0.048 (4.8%) of the Sun's mass will be converted from hydrogen to helium; chemical composition of the Sun (by mass) will be 69% hydrogen, 30% helium
30. (a) 8.8×10^{29} kg of hydrogen consumed, 6.2×10^{26} kg lost
32. (a) 1.64×10^{-13} J (b) 2.43×10^{-3} nm **33.** 1.4×10^{13} kg/s
37. 98,600 kg/m^3 **38.** 1.9×10^{-7} nm **40.** (a) 1700 nm **42.** For the photosphere, 500 nm; for the chromosphere, 58 nm; for the corona, 1.9 nm **45.** For the umbra, 670 nm; for the penumbra, 580 nm

46. (a) (Flux from patch of penumbra)/(flux from patch of photosphere) = 0.55 (b) (Flux from patch of penumbra)/(flux from patch of umbra) = 1.8

Chapter 19

31. Average Sun-Pluto distance = 1.92×10^{-4} pc; Proxima Centauri is 6780 times farther away **32.** (a) 9.7 pc (b) 0.10 arcsec **33.** 6.54 pc
34. (a) 161 km/s (b) 294 km/s **35.** Distance = 105 pc; tangential velocity would have to be about 5200 km/s **36.** 110 km/s **37.** (a) +59.4 km/s
(c) 486.23 nm **39.** The Sun is 2.71×10^{-3} as bright on Uranus as it is on Neptune **40.** Star B is 16 times more luminous than star A **41.** Star D appears 25 times brighter than star C **42.** The farther star is 81 times more luminous than the closer star **43.** 37.0 AU **44.** 6.8 L$_\odot$ **45.** 0.38 pc = 7.9×10^4 AU **46.** (a) +13.8 (b) (Luminosity of HIP 72509)/(luminosity of Sun) = 2.4×10^{-4} **47.** +17 **48.** 6300 pc **49.** (a) Brightest star has $M = -2.37$; dimmest star has $M = +4.32$ (b) $M = +0.79$ **52.** (b) $m_B - m_V$ = -0.23 (Bellatrix), +0.68 (Sun), +1.86 (Betelgeuse) **53.** 99 R$_\odot$
56. Star Q is 4 times more luminous than star P **57.** The radius of star X is 17 times larger than the radius of star Y **58.** Radius increases by a factor of 2, luminosity increases by a factor of 64 **60.** $T = 10,000$ K, $R = 3.8$ R$_\odot$ **61.** $L = 35$ L$_\odot$, so distance = 14 pc **62.** (b) 0.26 R$_\odot$ = 1.8×10^5 km **63.** (a) 5.0 pc (b) 22.5 AU (c) 1.5 M$_\odot$ **64.** (a) 40 M$_\odot$
(b) $M_1 = 32$ M$_\odot$, $M_2 = 8$ M$_\odot$ **66.** About 2500 pc, about 125 times greater volume

Chapter 20

29. 0.34% **30.** 3.4×10^4 atoms per cm^3 **33.** 1100 R$_\odot$ = 7.4×10^8 km = 4.9 AU **35.** 250 AU = 3.7×10^{10} km **37.** 1.3×10^{31} m^3 **38.** 27 arcmin **39.** 2.2×10^3 km/s, or 7.5×10^{-3} (0.75%) of the speed of light

Chapter 21

28. 657,000 km **29.** (a) 618 km/s (b) 61.8 km/s **30.** (a) 12.0 km/s
(b) 9.3 km/s **31.** 2.3×10^{29} kg, or 0.15 (15%) of the original mass of hydrogen **34.** (a) 4.9×10^7 years (b) 3.8×10^{11} years **36.** About 1900 K (1600°C, or 2900°F) **37.** 3.4×10^7 years **42.** About 650 pc **43.** About 370 pc **47.** (a) 1.65×10^6 km

Chapter 22

31. 0.11 R$_\odot$ **33.** 1.2 pc = 2.5×10^5 AU **34.** We see the nebula about 5900 to 8200 years after the central star shed its outer layers
36. (a) 97 nm **37.** (a) 1.8×10^9 kg/m^3 (b) 6.5×10^3 km/s **38.** About 4×10^6 kg **40.** (a) 1.7×10^{30} kg = 0.84 M$_\odot$ (b) 1.1×10^{12} newtons
(c) 1.5×10^8 m/s, or 0.5 (50%) of the speed of light **41.** (a) 6.4×10^5 m/s
42. 3.2×10^9 km = 22 AU **43.** (Maximum luminosity of SN 1993J)/(maximum luminosity of SN 1987A) = 4.5 **45.** (a) 7×10^{-7} b$_\odot$
(b) It would be about 700 times brighter than Venus **46.** 1.3×10^8 pc = 130 Mpc **47.** 1100 km/s, or 3.8×10^{-3} (0.38%) of the speed of light

Chapter 23

28. About 5500 B.C. **30.** (a) About 700 A.D. **31.** (a) Radius = 4.6 ly, diameter = 9.2 ly **32.** (b) About 40 pc **33.** (b) Maximum correction = 10^{-4} of the pulsar period **34.** (a) 1250 years **35.** (a) Density of matter in a neutron = 4×10^{17} kg/m^3 **39.** (a) 0.097 nm (b) Luminosity of neutron star = 5.8×10^{31} W = 1.5×10^5 L$_\odot$ **40.** (b) 48 nm (c) 14 km

Chapter 24

29. 0.6 of the speed of light **30.** 8.3×10^{-8} s **31.** 0.8 of the speed of light **32.** (a) 25 years (b) 20 light-years as measured by an Earth observer, 12 light-years as measured by the astronaut **34.** 2.8 $M_\odot$ **35.** 5.7×10^8 years **36.** (a) 2.01×10^6 km **37.** 0.32 year **38.** 0.84 m **39.** (a) R_{Sch} = 8.9 mm, density = 2.0×10^{30} kg/m^3 (b) R_{Sch} = 3.0 km, density = 1.8×10^{19} kg/m^3 (c) R_{Sch} = 3.5×10^9 km = 24 AU, density = 13 kg/m^3 **40.** 7.4×10^{30} kg **41.** 2.9×10^{17} kg/m^3 **42.** 2.7×10^{38} kg = 1.4×10^8 $M_\odot$

Chapter 25

28. (a) 1.2×10^{12} cubic parsecs (b) 1.1×10^8 cubic parsecs (c) Probability = 9.6×10^{-5}; we can expect to see a supernova within 300 pc once every 350,000 years **29.** (a) 8.3×10^9 AU (b) 4.0×10^4 pc **31.** 9.5×10^{-25} J; it takes 3.2×10^5 such photons to equal the energy of one H_α photon **32.** (b) 5700 pc **33.** 21 times **35.** A 10% error in radius results in a 10% error in mass; a 10% error in velocity results in a 20% error in mass **36.** (a) 3.1×10^8 years (b) 7.4×10^{11} $M_\odot$ **37.** 2.7×10^{11} $M_\odot$ **41.** (a) 1.1×10^7 km = 0.073 AU (b) 9.1×10^{-6} arcsec (c) 251 arcsec **42.** (a) 8 kpc **43.** (a) 2.5×10^3 AU for S0-1, 1.0×10^3 AU for S0-2 (b) 0.31 arcsec for S0-2, 0.13 arcsec for S0-2 **44.** (a) 6.3 years (b) 3.7×10^6 $M_\odot$

Chapter 26

32. (a) 6.9 Mpc (b) 8.3 Mpc (c) 1.4 Mpc **33.** 9.5 Mpc **34.** (a) 1.1×10^{10} km = 70 AU (b) 7.0 Mpc **35.** (a) 152 Mpc = 4.96×10^8 ly **36.** 54 km/s/Mpc **37.** (a) z = 0.0252 (b) 106 Mpc **38.** (a) 2.85×10^5 km/s, or 0.951 (95.1%) of the speed of light (b) 1.60×10^6 km/s (c) 4020 Mpc = 1.31×10^{10} ly **39.** (a) 1.2×10^{70} atoms (b) 3.6×10^{-6} atoms per cm^3 **40.** 1.2×10^{12} $M_\odot$ **41.** Period = 4.4×10^8 years; mass = 3.6×10^{11} $M_\odot$

Chapter 27

29. 2.87×10^5 km/s, or 0.958 (95.8%) of the speed of light **30.** 2.93×10^5 km/s, or 0.976 (97.6%) of the speed of light **34.** (a) 30 years (b) 2006 (c) 5/3 of the speed of light **35.** (a) 1.80×10^5 km/s, or 0.600 (60.0%) of the speed of light (b) 134 hours (c) 970 AU **36.** (a) 1.2×10^{11} $L_\odot$ **37.** 1.5×10^8 $M_\odot$ **38.** 2.9×10^9 km = 20 AU

Chapter 28

30. (a) z = 4 (b) Distances were $^1/_9$ as great and the density of matter was 729 times greater **31.** 771 nm **32.** (a) 20 billion years (b) 13 billion years (c) 9.7 billion years **33.** 1.6×10^8 km/s/Mpc **34.** 65.0 times denser **35.** 1.06×10^{-3} m = 1.06 mm **36.** 13.6 K **37.** (a) 4.6×10^{-23} kg/m^3 **38.** (a) 9.5×10^{-18} kg/m^3 (b) 4.8×10^{-4} kg/m^3 (c) 1.3×10^{-7} kg/m^3 **39.** (a) 4.7×10^{-27} kg/m^3 (b) 1.9×10^{-26} kg/m^3 **40.** (a) 79 Mpc (b) 6.6×10^{-25} kg/m^3 (c) Temperature = 17.3 K, ρ_{rad} = 7.4×10^{-28} kg/m^3 **41.** (b) q_0 = −0.595 (c) Ω_Λ = 0.135 **42.** (a) −0.17 (b) +0.12 **43.** 1.3×10^{-52} m^{-2}

Chapter 29

23. 1.1×10^{-26} J **25.** 3.5×10^{-25} s **26.** (a) 80.5 GeV (b) 9.34×10^{14} K **28.** (a) 2.2×10^{-35} kg = 2.4×10^{-5} times the mass of the electron **29.** (a) 1.1×10^{14} m = 0.011 ly **30.** Length was 0.168 of its present-day value; density was 213 times the present-day value

Chapter 30

10. (a) 3.7 cm **11.** (a) 10,000 years (b) 10 million years **13.** (a) 1430 MHz **14.** (a) 29.8 km/s (b) 0.10 m (c) 9.9×10^{-6} m, or 9.9×10^{-3} % of the unshifted wavelength **15.** 200 km

Index

Page numbers in *italics* indicate figures.

Star Charts

The following set of star charts, one for each month of the year, are from *Griffith Observer* magazine. They are useful in the northern hemisphere only. For a set of star charts suitable for use in the southern hemisphere, see the *Universe* CD-ROM or Web site.

To use these charts, first select the chart that best corresponds to the date and time of your observations. Hold the chart vertically and turn it so that the direction you are facing shows at the bottom.

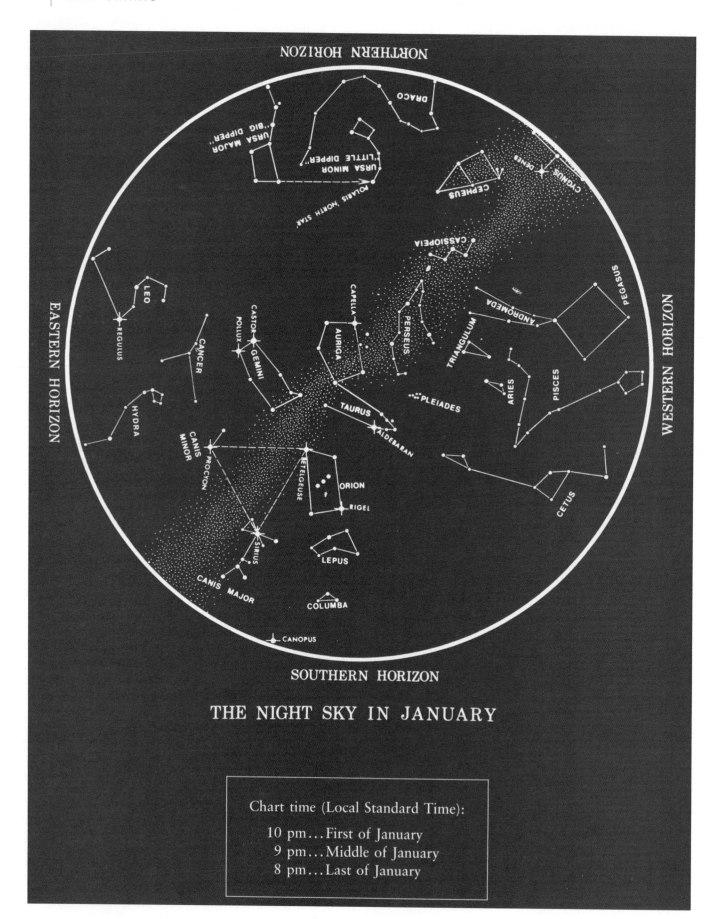

THE NIGHT SKY IN JANUARY

Chart time (Local Standard Time):

10 pm...First of January
9 pm...Middle of January
8 pm...Last of January

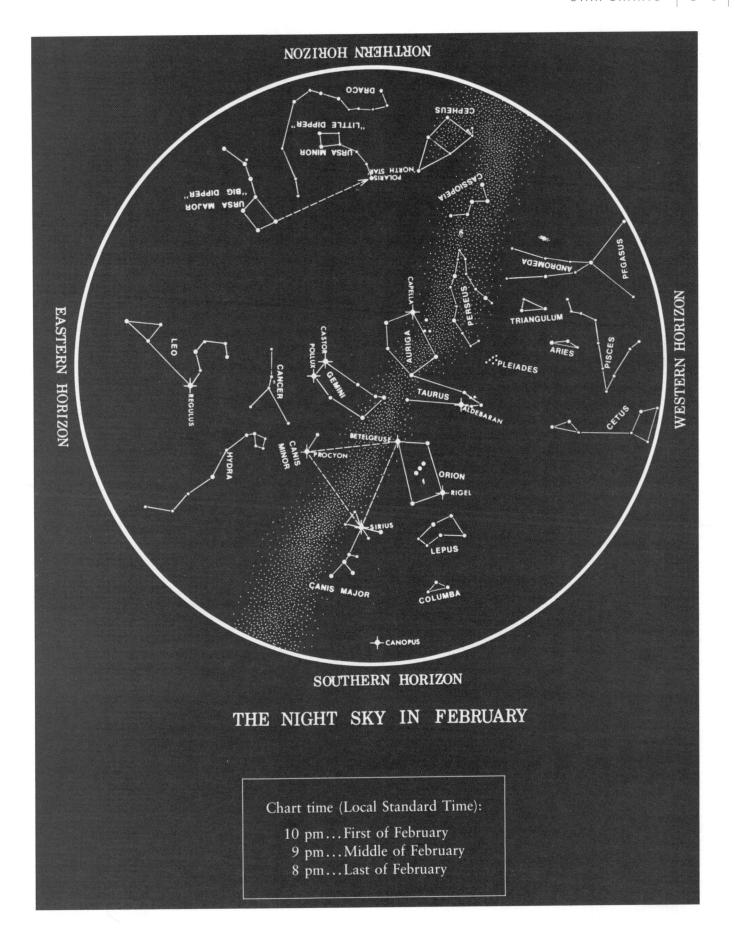

THE NIGHT SKY IN FEBRUARY

Chart time (Local Standard Time):

10 pm...First of February
9 pm...Middle of February
8 pm...Last of February

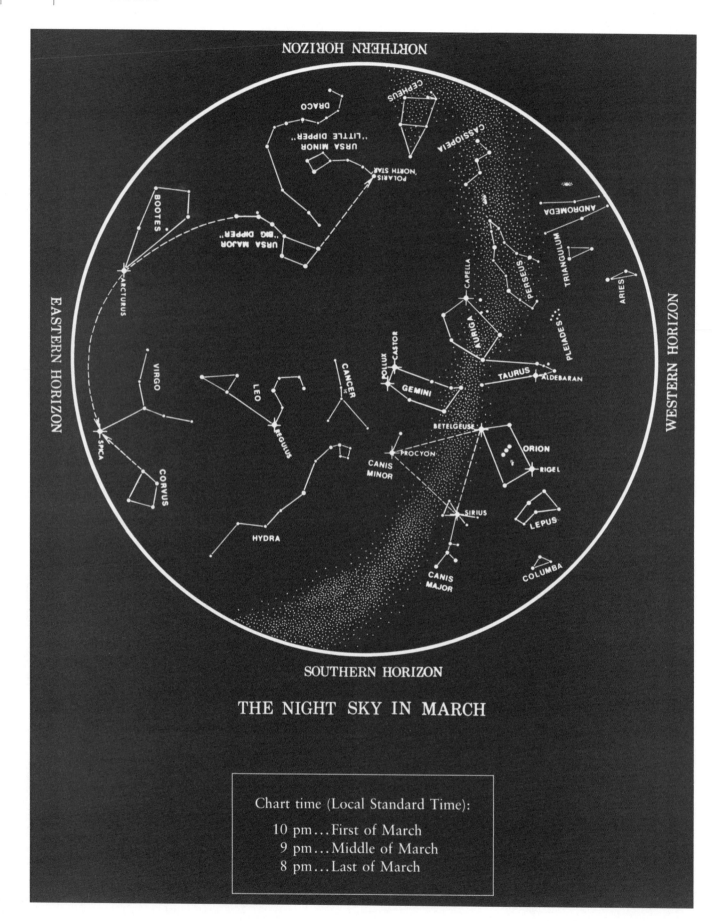

THE NIGHT SKY IN MARCH

Chart time (Local Standard Time):

10 pm...First of March
9 pm...Middle of March
8 pm...Last of March

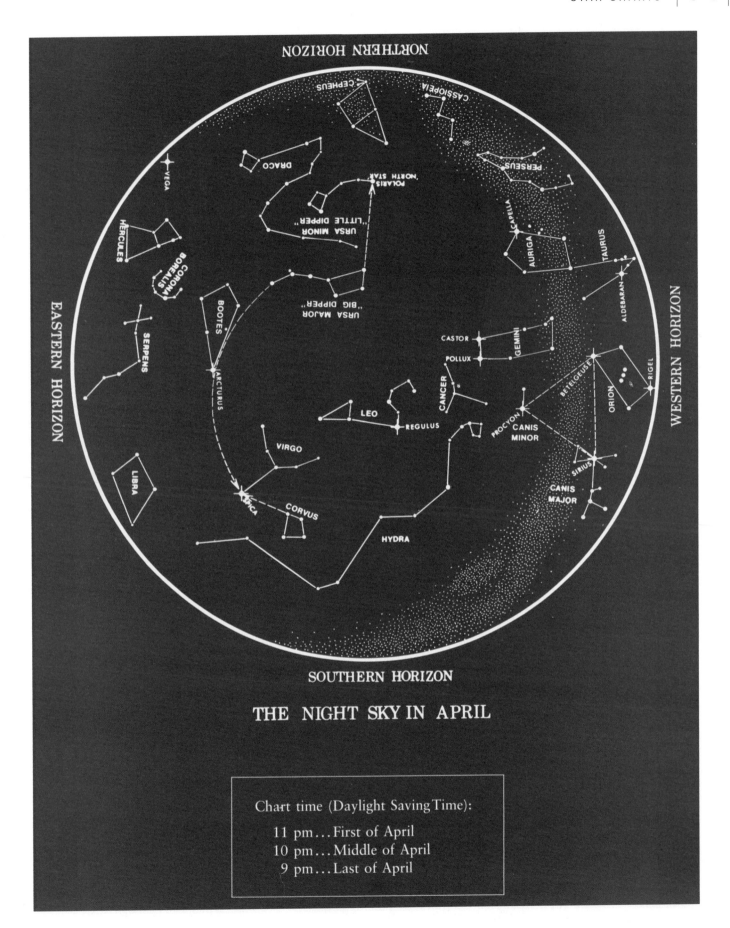

THE NIGHT SKY IN APRIL

Chart time (Daylight Saving Time):

11 pm ... First of April

10 pm ... Middle of April

9 pm ... Last of April

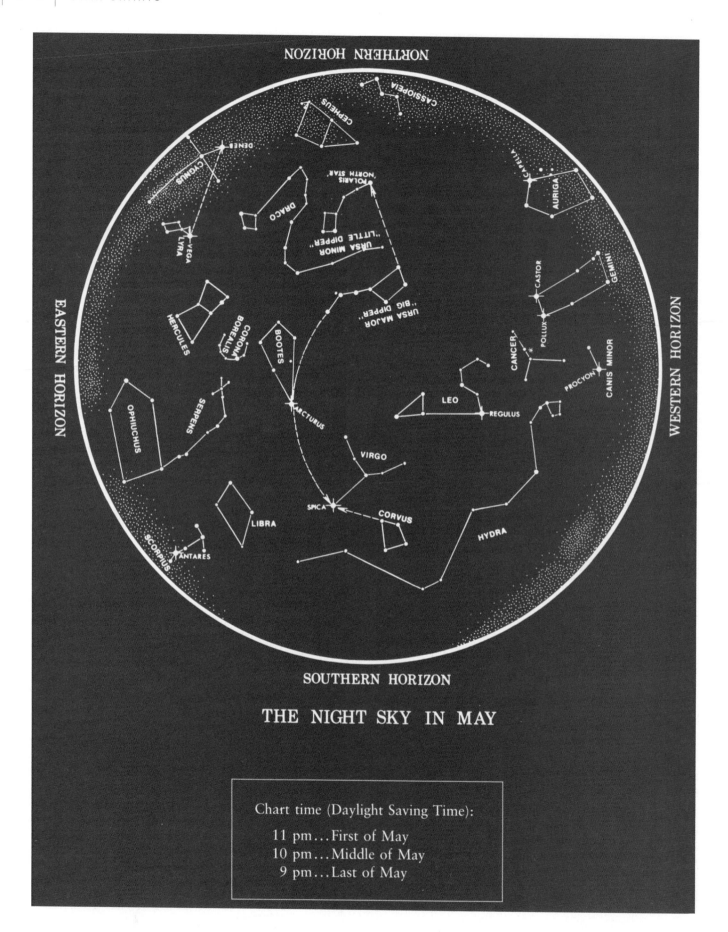

THE NIGHT SKY IN MAY

Chart time (Daylight Saving Time):

11 pm...First of May
10 pm...Middle of May
9 pm...Last of May

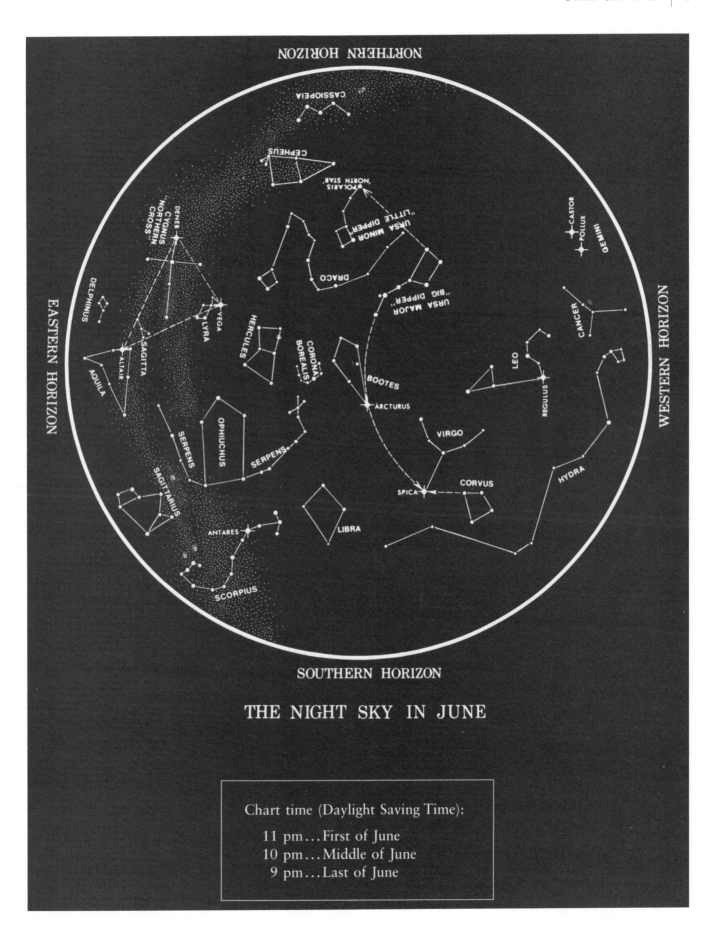

THE NIGHT SKY IN JUNE

Chart time (Daylight Saving Time):

11 pm...First of June
10 pm...Middle of June
9 pm...Last of June

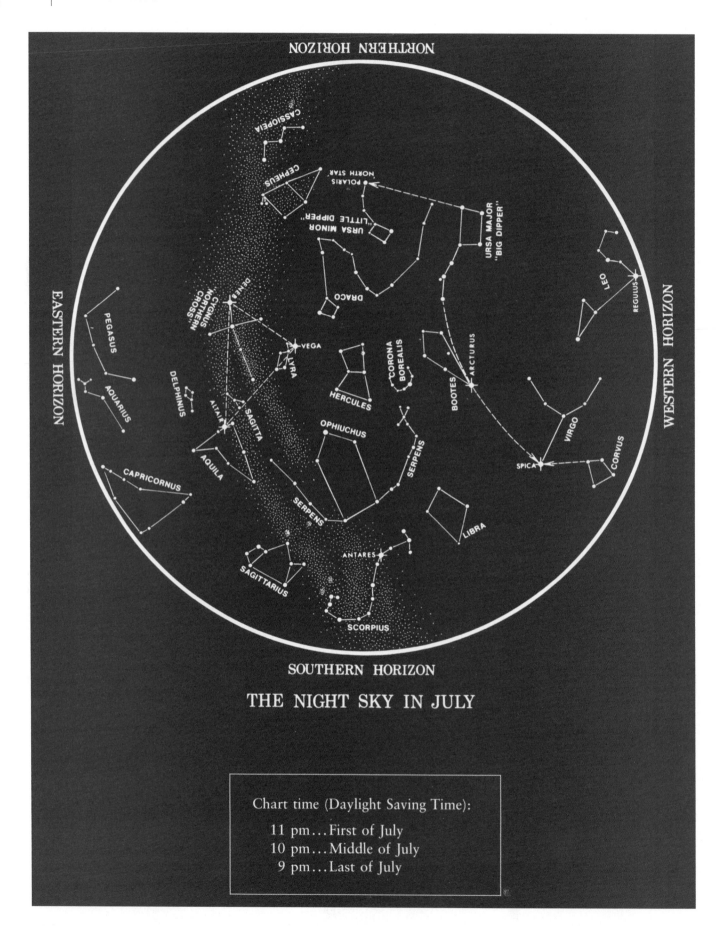

THE NIGHT SKY IN JULY

Chart time (Daylight Saving Time):

11 pm...First of July

10 pm...Middle of July

9 pm...Last of July

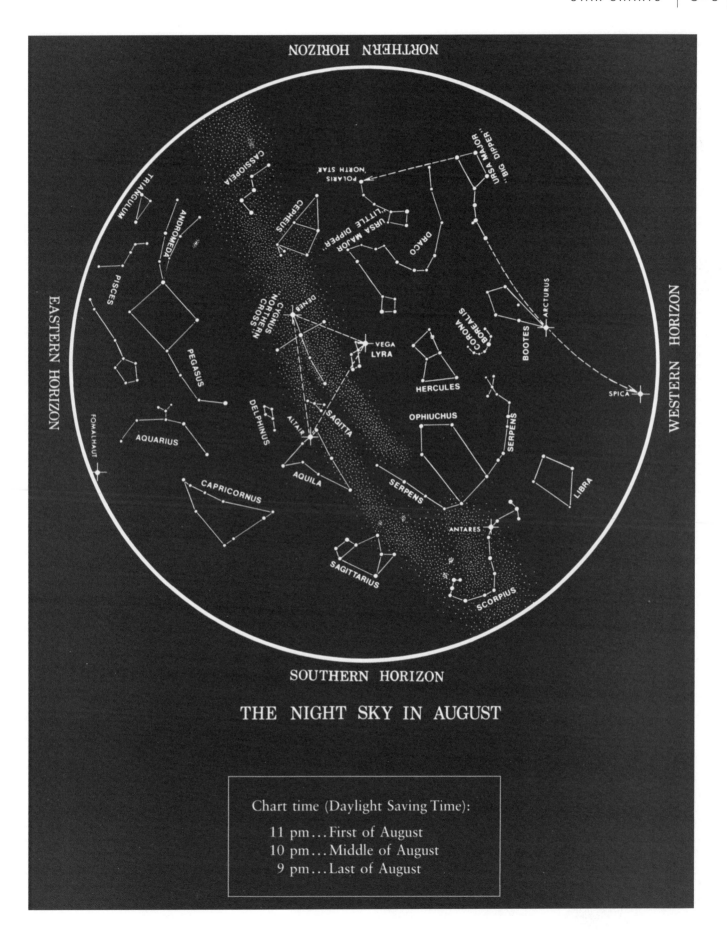

THE NIGHT SKY IN AUGUST

Chart time (Daylight Saving Time):

11 pm...First of August
10 pm...Middle of August
9 pm...Last of August

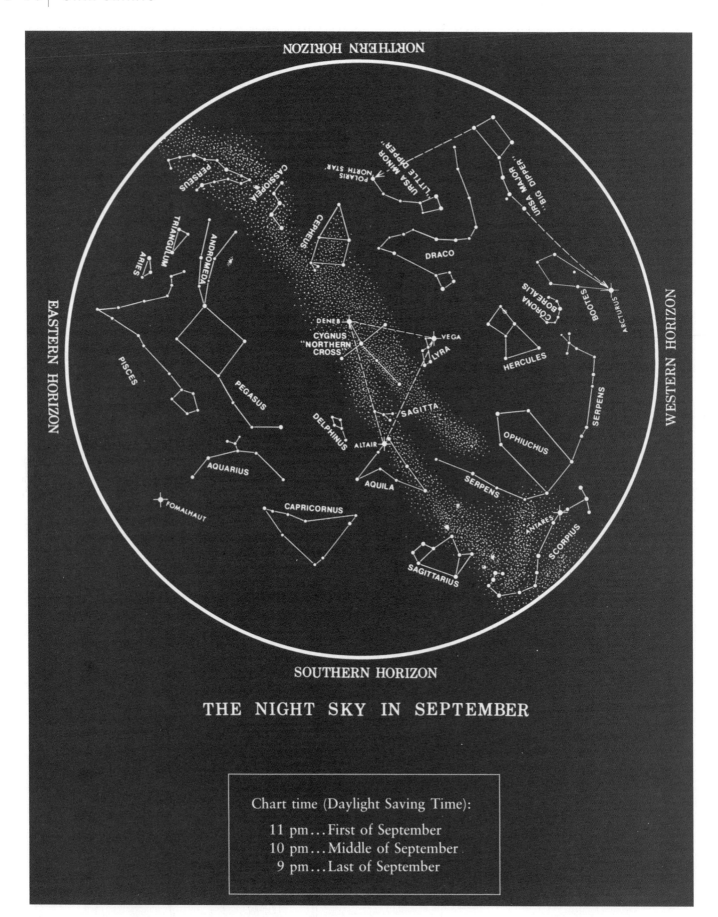

THE NIGHT SKY IN SEPTEMBER

Chart time (Daylight Saving Time):

11 pm...First of September
10 pm...Middle of September
9 pm...Last of September

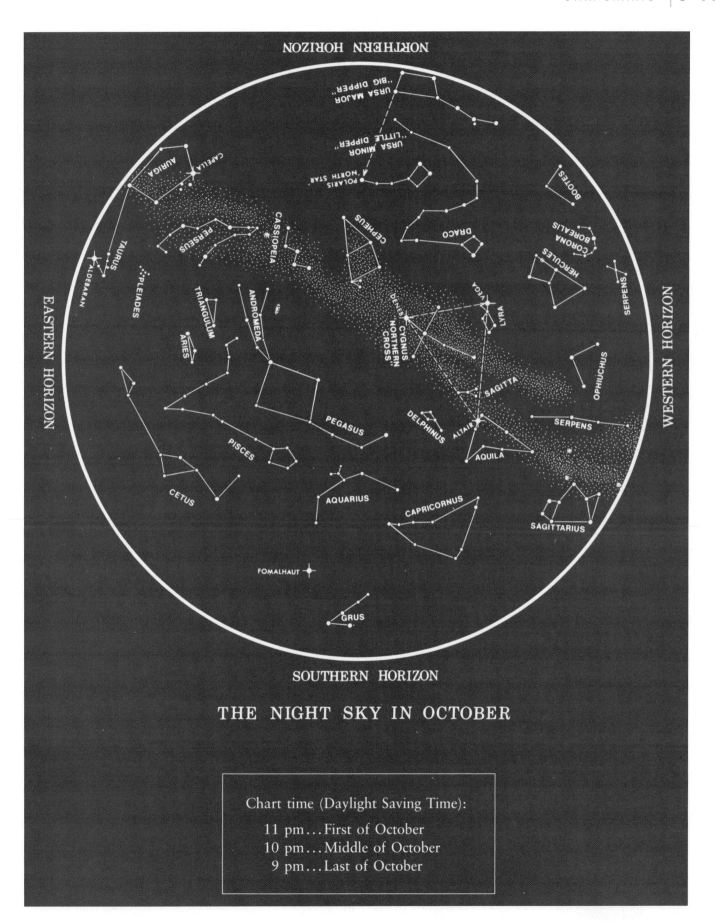

THE NIGHT SKY IN OCTOBER

Chart time (Daylight Saving Time):

11 pm...First of October
10 pm...Middle of October
9 pm...Last of October

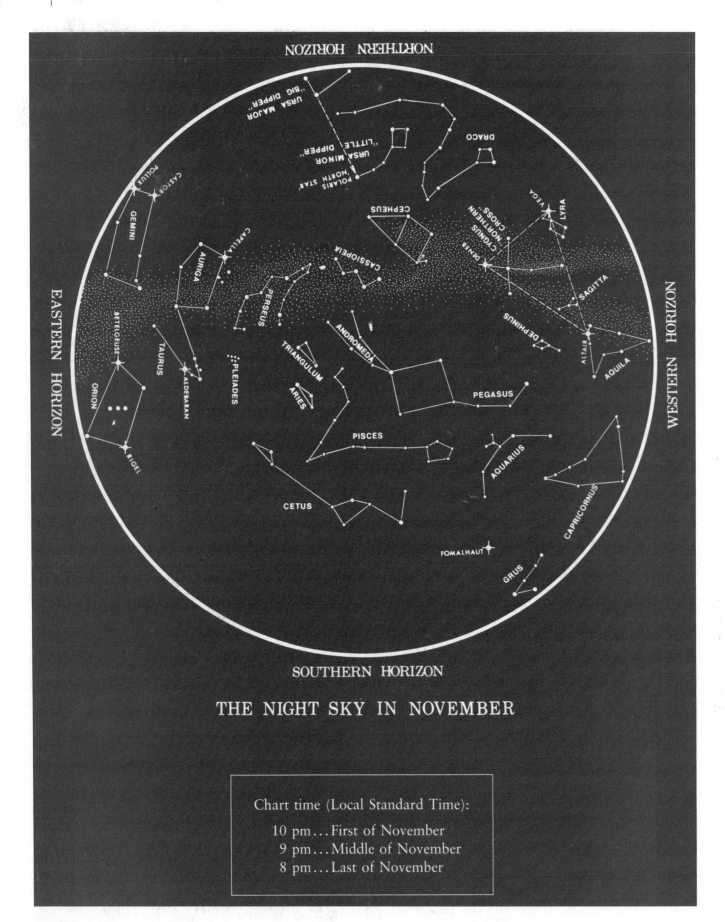

THE NIGHT SKY IN NOVEMBER

Chart time (Local Standard Time):

10 pm...First of November
9 pm...Middle of November
8 pm...Last of November

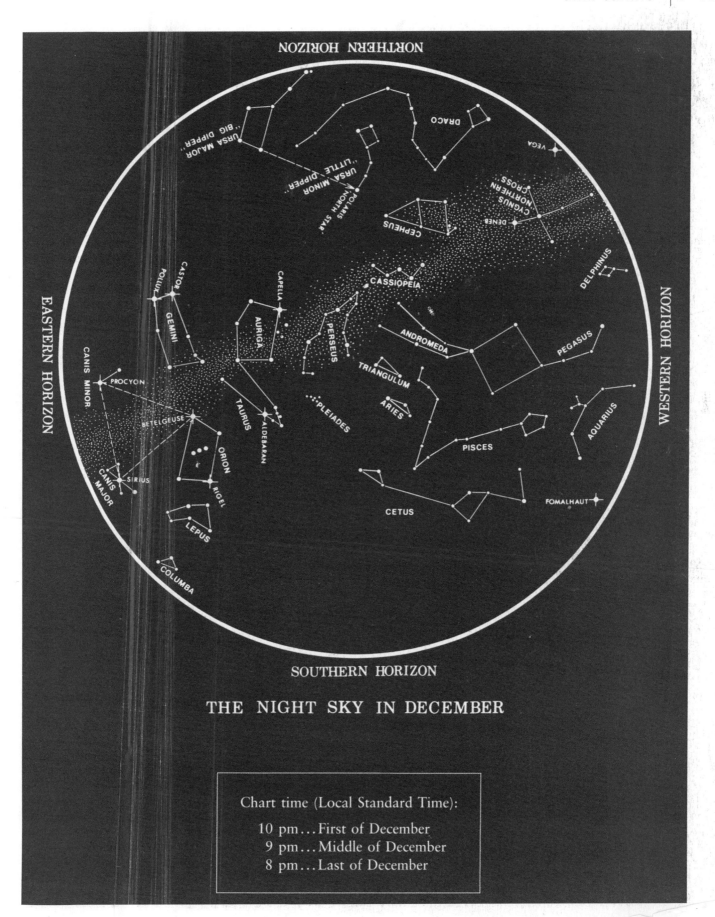

THE NIGHT SKY IN DECEMBER

Chart time (Local Standard Time):

10 pm...First of December
9 pm...Middle of December
8 pm...Last of December